Advances in Global Change Research

Volume 69

More information about this series at http://www.springer.com/series/5588

Vincenzo Levizzani • Christopher Kidd
Dalia B. Kirschbaum • Christian D. Kummerow
Kenji Nakamura • F. Joseph Turk

Editors

Satellite Precipitation Measurement

Volume 2

 Springer

Editors
Vincenzo Levizzani
CNR-ISAC
Bologna, Italy

Dalia B. Kirschbaum
Code 617
NASA Goddard Space Flight Center
Greenbelt, MD, USA

Kenji Nakamura
Department of Economics on
Sustainability
Dokkyo University
Saitama, Japan

Christopher Kidd
Earth System Science Interdisciplinary Center
University of Maryland and NASA Goddard
Space Flight Center
Greenbelt, MD, USA

Christian D. Kummerow
Department of Atmospheric Science
Colorado State University
Fort Collins, CO, USA

F. Joseph Turk
Jet Propulsion Laboratory
California Institute of Technology
Pasadena, CA, USA

ISSN 1574-0919 ISSN 2215-1621 (electronic)
Advances in Global Change Research
ISBN 978-3-030-35797-9 ISBN 978-3-030-35798-6 (eBook)
https://doi.org/10.1007/978-3-030-35798-6

Sognatore è un uomo con i piedi fortemente appoggiati sulle nuvole*

Ennio Flaiano (1910–1972)

**Dreamer is a man with his feet firmly resting on the clouds*

In memory of
Arthur Y. Hou (1947–2013)

Preface

This book is published 13 years after the book *Measuring Precipitation from Space: EURAINSAT and the Future* (V. Levizzani, P. Bauer, and F. J. Turk, Eds., Springer, ISBN 978-1-4020-5835-6), but it is not a revised edition of the previous. It is a new book that aims to construct a quasi-complete picture of the science and applications of satellite-derived precipitation measurements at the present time.

The book comes out at the end of a very exciting era of precipitation measurements from space. The Tropical Rainfall Measuring Mission (TRMM), launched in November 1997, ended its long life in space in April 2015 providing an unprecedented 17-year-long dataset of tropical precipitation and lightning. The Global Precipitation Measurement (GPM) mission, launched in February 2014, is now in space as TRMM's natural successor with a more global perspective that extends precipitation radar observations to the Arctic and Antarctic circles. At the same time, the CloudSat mission, launched in April 2006, is in its 13th year in space and focuses on cloud structure, which is essential for improving precipitation retrievals. These are just a few examples of precipitation-oriented missions that continuously provide data from geostationary and low Earth orbits in a truly cooperative effort worldwide. This effort involves many agencies and a broad range of countries who collaborate in a genuine way to observe global precipitation.

It is by realizing the significance of this historical moment and the need to think about what is important for the future that the community joined in the effort of writing a book with the goal of serving the precipitation community itself, the scholars, the students, the stakeholders, the end users, and all the readers interested in knowing the progress of satellite precipitation studies. The most recent achievements in precipitation monitoring from space drive us into the future of measuring not only heavy rainfall but less intense rainfall, snowfall, and even hailfall. Such a scientific framework would not have even been conceivable 13 years ago and is only possible thanks to the relentless effort of the worldwide space and precipitation communities.

Naturally, we realize that at the time of the printing of this book, the field will already have made advances and thus part of the material may already be a bit outdated. However, in this era of rapidly evolving technological developments, sensors that take years to design, build, and launch are already considered old. This is particularly true nowadays when the progress in approaching new scientific challenges is particularly fast.

Since 2007, science has made substantial progresses toward transforming satellite rainfall "estimates" into accurate "measurements" and producing operational rainfall products readily available for a wide field of applications ranging from climate research and numerical weather prediction to hydrology, agriculture, health, civil protection, and much more. Satellite-derived precipitation products are now being considered as a valuable tool for a number of applications that benefit society and save lives. This is perhaps the most important achievement of all.

This book represents a significant effort, and each author has provided high-quality material in the topics of current and future mission contributions, observations of precipitation using the suite of precipitation satellites, retrieval techniques, validation, and applications. The result is a book that not only photographs the state of the art of the discipline but also projects it into the future.

Bologna, Italy   Vincenzo Levizzani
Greenbelt, MD, USA   Christopher Kidd
Greenbelt, MD, USA   Dalia B. Kirschbaum
Fort Collins, CO, USA   Christian D. Kummerow
Saitama, Japan   Kenji Nakamura
Pasadena, CA, USA   F. Joseph Turk
9 March 2020

Acknowledgments

The first acknowledgment goes to Springer Nature for asking us to start this project and for being very patient with us for the considerable amount of time it took to put the material together.

All the colleagues who spent their precious time contributing their ideas and results deserve special gratitude. They are all very busy scientists, and this is why their contribution is particularly valuable. We deem the book to be a first-hand image of the achievements of the whole community at this time while also providing an important glimpse into future developments.

Then we feel that we need to thank the readers who have already made the previous 2007 Springer book a success, thus de facto making it possible to start writing the new one. We hope you will get from this new book even more inspiration than you got from its predecessor. While some concepts and details will surely become outdated as time goes by, it is our hope that the material contained herein is sufficiently broad that it will always serve as a springboard to understand and put into context the latest research and findings.

It would be almost impossible to thank all the people and organizations behind this effort. You realize this simple truth by looking at the list of contributors and seeing the very long list of institutes, research organizations, university departments, and operational agencies that allowed their members to spend a substantial amount of time writing and correcting the chapters of the book. We thank, in particular, our home institutions that were very supportive in understanding the importance of our work for the community: CNR, Colorado State University, Dokkyo University, JPL-Caltech, NASA, and University of Maryland.

It is very important to remember all the colleagues who are no longer with us and who worked very hard until the last minute providing an essential contribution. This book is dedicated to the memory of a friend of all of us, Arthur Y. Hou (1947–2013). Arthur was not only the US Project Scientist of the Global Precipitation Measurement (GPM) mission, he was a man of a truly global vision who now is in place with the GPM constellation. More than that, he made great efforts to establish an international science cooperation through his gentle and unique way of approaching each one of us. Other colleagues left us in recent times, and we want to honor them as well: David (Dave) H. Staelin (1938–2011), David I. F. Grimes (1951–2011), and James (Jim) A. Weinman (1930–2012). They all left us much too soon, and we miss them, but their work is here to testify to their essential contribution to the advancement of science and to meet the needs of mankind.

Two major international organizations gave us the opportunity to work together with a global strategy for the future: the International Precipitation Working Group (IPWG) and the World Meteorological Organization (WMO).

The senior editor (Vincenzo Levizzani) would like to recognize the ceaseless work of his coeditors in effectively putting together the material of their respective sections: F. Joseph (Joe) Turk for Section 1, Christian (Chris) D. Kummerow for Sections 2 and 3, Christopher (Chris) Kidd for Section 4, Kenji Nakamura for Section 5, and Dalia B. Kirschbaum for Section 6. Their commitment and competence largely influenced the quality level of this book.

Finally, our families are part of the project through their understanding and their moral and practical support. Without them, the writing of this book would have never even started.

Bologna, Italy   Vincenzo Levizzani
Greenbelt, MD, USA   Christopher Kidd
Greenbelt, MD, USA   Dalia B. Kirschbaum
Fort Collins, CO, USA   Christian D. Kummerow
Saitama, Japan   Kenji Nakamura
Pasadena, CA, USA   F. Joseph Turk
9 March 2020

Contents of Volume 2

Part IV Validation

25 **The IPWG Satellite Precipitation Validation Effort** 453
Christopher Kidd, Shoichi Shige, Daniel Vila, Elena Tarnavsky,
Munehisa K. Yamamoto, Viviana Maggioni, and Bathobile Maseko

26 **The GPM Ground Validation Program** 471
Walter A. Petersen, Pierre-Emmanuel Kirstetter, Jianxin Wang,
David B. Wolff, and Ali Tokay

27 **The GPM DPR Validation Program** 503
Riko Oki, Toshio Iguchi, and Kenji Nakamura

28 **Error and Uncertainty Characterization** 515
Christian Massari and Viviana Maggioni

29 **Multiscale Evaluation of Satellite Precipitation Products:
Effective Resolution of IMERG** 533
Clément Guilloteau and Efi Foufoula-Georgiou

30 **Remote Sensing of Orographic Precipitation** 559
Ana P. Barros and Malarvizhi Arulraj

31 **Integrated Multi-satellite Evaluation for the Global Precipitation
Measurement: Impact of Precipitation Types on Spaceborne
Precipitation Estimation** 583
Pierre-Emmanuel Kirstetter, Walter A. Petersen,
Christian D. Kummerow, and David B. Wolff

32 **Hydrologic Validation and Flood Analysis** 609
Witold F. Krajewski, Felipe Quintero, Mohamed El Saadani,
and Radoslaw Goska

33 **Global-Scale Evaluation of 22 Precipitation Datasets Using
    Gauge Observations and Hydrological Modeling** 625
    Hylke E. Beck, Noemi Vergopolan, Ming Pan, Vincenzo Levizzani,
    Albert I. J. M. van Dijk, Graham P. Weedon, Luca Brocca,
    Florian Pappenberger, George J. Huffman, and Eric F. Wood

34 **OceanRAIN – The Global Ocean Surface-Reference Dataset
    for Characterization, Validation and Evaluation of the Water
    Cycle** 655
    Christian Klepp, Paul A. Kucera, Jörg Burdanowitz, and Alain Protat

Part V Observed Characteristics of Precipitation

35 **GPCP and the Global Characteristics of Precipitation** 677
    Robert F. Adler, Guojun Gu, George J. Huffman,
    Mathew R. P. Sapiano, and Jian-Jian Wang

36 **Global Snowfall Detection and Measurement** 699
    Mark S. Kulie, Lisa Milani, Norman B. Wood,
    and Tristan S. L'Ecuyer

37 **Snowfall Detection by Spaceborne Radars** 717
    Atsushi Hamada, Toshio Iguchi, and Yukari N. Takayabu

38 **On the Duration and Life Cycle of Precipitation Systems
    in the Tropics** 729
    Rémy Roca, Dominique Bouniol, and Thomas Fiolleau

39 **Observational Characteristics of Warm-Type Heavy Rainfall** 745
    Byung-Ju Sohn, Geun-Hyeok Ryu, and Hwan-Jin Song

40 **Satellite Precipitation Measurement and Extreme Rainfall** 761
    Olivier P. Prat and Brian R. Nelson

41 **Rainfall Trends in East Africa from an Ensemble of IR-Based
    Satellite Products** 791
    Elsa Cattani, Andrés Merino, and Vincenzo Levizzani

42 **Heavy Precipitation Systems in the Mediterranean Area:
    The Role of GPM** 819
    Giulia Panegrossi, Anna Cinzia Marra, Paolo Sanò, Luca Baldini,
    Daniele Casella, and Federico Porcù

43 **Dryland Precipitation Climatology from Satellite Observations** 843
    Efrat Morin, Francesco Marra, and Moshe Armon

44 **Hailfall Detection** 861
    Ralph R. Ferraro, Daniel Cecil, and Sante Laviola

45 **Improving High-Latitude and Cold Region Precipitation
    Analysis** 881
    Ali Behrangi

46 **Latent Heating Retrievals from Satellite Observations** 897
Yukari N. Takayabu and Wei-Kuo Tao

Part VI Applications

47 **Operational Applications of Global Precipitation Measurement
Observations** 919
Anita LeRoy, Emily Berndt, Andrew Molthan, Bradley Zavodsky,
Matthew Smith, Frank LaFontaine, Kevin McGrath, and Kevin Fuell

48 **Assimilation of Precipitation Observations from Space into
Numerical Weather Prediction (NWP)** 941
Sid-Ahmed Boukabara, Erin Jones, Alan Geer, Masahiro Kazumori,
Kevin Garrett, and Eric Maddy

49 **Precipitation Ensemble Data Assimilation in NWP Models** 983
Takemasa Miyoshi, Shunji Kotsuki, Koji Terasaki, Shigenori Otsuka,
Guo-Yuan Lien, Hisashi Yashiro, Hirofumi Tomita, Masaki Satoh,
and Eugenia Kalnay

50 **PERSIANN-CDR for Hydrology and Hydro-climatic
Applications** 993
Phu Nguyen, Hamed Ashouri, Mohammed Ombadi, Negin Hayatbini,
Kuo-Lin Hsu, and Soroosh Sorooshian

51 **Soil Moisture and Precipitation: The SM2RAIN Algorithm
for Rainfall Retrieval from Satellite Soil Moisture** 1013
Luca Ciabatta, Stefania Camici, Christian Massari, Paolo Filippucci,
Sebastian Hahn, Wolfgang Wagner, and Luca Brocca

52 **Drought Risk Management Using Satellite-Based
Rainfall Estimates** 1029
Elena Tarnavsky and Rogerio Bonifacio

53 **Two Decades of Urban Hydroclimatological Studies
Have Yielded Discovery and Societal Benefits** 1055
J. Marshall Shepherd, Steven J. Burian, Menglin Jin, Chuntao Liu,
and Bradford Johnson

54 **Validation of Climate Models** 1073
Francisco J. Tapiador

55 **Extreme Precipitation in the Himalayan Landslide Hotspot** 1087
Thomas Stanley, Dalia B. Kirschbaum, Salvatore Pascale,
and Sarah Kapnick

56 **The Value of Satellite Rainfall Estimates in Agriculture
and Food Security** 1113
Tufa Dinku

57   Using Satellite Estimates of Precipitation for Fire
     Danger Rating 1131
     Robert D. Field

58   Variability of Satellite Sea Surface Salinity Under Rainfall 1155
     Alexandre Supply, Jacqueline Boutin, Gilles Reverdin,
     Jean-Luc Vergely, and Hugo Bellenger

Contents of Volume 1

Part I Status of Observations and Satellite Programs

1 **The Global Precipitation Measurement (GPM) Mission** 3
Christopher Kidd, Yukari N. Takayabu, Gail M. Skofronick-Jackson,
George J. Huffman, Scott A. Braun, Takuji Kubota,
and F. Joseph Turk

2 **Status of the CloudSat Mission** 25
Matthew D. Lebsock, Tristan S. L'Ecuyer, Norman B. Wood,
John M. Haynes, and Mark A. Smalley

3 **The Megha-Tropiques Mission After Seven Years in Space** 45
Rémy Roca, Michel Dejus, Philippe Chambon, Sophie Cloché,
and Michel Capderou

4 **Microwave Sensors, Imagers and Sounders** 63
Kazumasa Aonashi and Ralph R. Ferraro

5 **Microwave and Sub-mm Wave Sensors:**
A European Perspective 83
Christophe Accadia, Vinia Mattioli, Paolo Colucci, Peter Schlüssel,
Salvatore D'Addio, Ulf Klein, Tobias Wehr, and Craig Donlon

6 **Plans for Future Missions** 99
Christian D. Kummerow, Simone Tanelli, Nobuhiro Takahashi,
Kinji Furukawa, Marian Klein, and Vincenzo Levizzani

Part II Retrieval Techniques, Algorithms and Sensors

7 **Introduction to Passive Microwave Retrieval Methods** 123
Christian D. Kummerow

**8   The Goddard Profiling (GPROF) Precipitation Retrieval
        Algorithm** 141
       David L. Randel, Christian D. Kummerow, and Sarah Ringerud

**9   Precipitation Estimation from the Microwave Integrated
        Retrieval System (MiRS)** 153
       Christopher Grassotti, Shuyan Liu, Quanhua Liu, Sid-Ahmed
       Boukabara, Kevin Garrett, Flavio Iturbide-Sanchez,
       and Ryan Honeyager

10   Introduction to Radar Rain Retrieval Methods 169
       Toshio Iguchi and Ziad S. Haddad

**11   Dual-Frequency Precipitation Radar (DPR) on the Global
        Precipitation Measurement (GPM) Mission's Core Observatory** ... 183
       Toshio Iguchi

12   DPR Dual-Frequency Precipitation Classification 193
       V. Chandrasekar and Minda Le

13   Triple-Frequency Radar Retrievals 211
       Alessandro Battaglia, Simone Tanelli, Frederic Tridon,
       Stefan Kneifel, Jussi Leinonen, and Pavlos Kollias

**14   Precipitation Retrievals from Satellite Combined Radar
        and Radiometer Observations** 231
       Mircea Grecu and William S. Olson

15   Scattering of Hydrometeors 249
       Stefan Kneifel, Jussi Leinonen, Jani Tyynelä, Davide Ori,
       and Alessandro Battaglia

16   Radar Snowfall Measurement 277
       Guosheng Liu

**17   A 1DVAR-Based Snowfall Rate Algorithm for Passive
        Microwave Radiometers** 297
       Huan Meng, Cezar Kongoli, and Ralph R. Ferraro

18   X-Band Synthetic Aperture Radar Methods 315
       Saverio Mori, Frank S. Marzano, and Nazzareno Pierdicca

Part III   Merged Precipitation Products

**19   Integrated Multi-satellite Retrievals for the Global
        Precipitation Measurement (GPM) Mission (IMERG)** 343
       George J. Huffman, David T. Bolvin, Dan Braithwaite, Kuo-Lin Hsu,
       Robert J. Joyce, Christopher Kidd, Eric J. Nelkin,
       Soroosh Sorooshian, Erich F. Stocker, Jackson Tan,
       David B. Wolff, and Pingping Xie

**20 Global Satellite Mapping of Precipitation (GSMaP) Products
in the GPM Era** 355
Takuji Kubota, Kazumasa Aonashi, Tomoo Ushio, Shoichi Shige,
Yukari N. Takayabu, Misako Kachi, Yoriko Arai, Tomoko Tashima,
Takeshi Masaki, Nozomi Kawamoto, Tomoaki Mega,
Munehisa K. Yamamoto, Atsushi Hamada, Moeka Yamaji,
Guosheng Liu, and Riko Oki

**21 Improving PERSIANN-CCS Using Passive Microwave
Rainfall Estimation** 375
Kuo-Lin Hsu, Negar Karbalee, and Dan Braithwaite

22 TAMSAT 393
Ross Maidment, Emily Black, Helen Greatrex, and Matthew Young

**23 Algorithm and Data Improvements for Version 2.1
of the Climate Hazards Center's InfraRed Precipitation
with Stations Data Set** 409
Chris Funk, Pete Peterson, Martin Landsfeld, Frank Davenport,
Andreas Becker, Udo Schneider, Diego Pedreros, Amy McNally,
Kristi Arsenault, Laura Harrison, and Shraddhanand Shukla

**24 Merging the Infrared Fleet and the Microwave Constellation
for Tropical Hydrometeorology (TAPEER) and Global
Climate Monitoring (GIRAFE) Applications** 429
Rémy Roca, Adrien Guérou, Rômulo A. Jucá Oliveira,
Philippe Chambon, Marielle Gosset, Sophie Cloché,
and Marc Schröder

List of Figures

Fig. 25.1   Distribution of past and current IPWG validation regions and
             their surface reference data sets 457
Fig. 25.2   Example of IPWG validation over South Africa for gauge data
             vs the WRF model output on 14 July 2018 458
Fig. 25.3   Example of the IPWG validation of the CMORPH product over
             the United States, utilising both surface gauge and radar data.
             The comparison of the gauge and radar provides a useful
             measure of the retrieval ability of the surface reference data
             sets 459
Fig. 25.4   Time series of spatial correlation over Japan between the Japan
             Meteorological Agency (JMA) gauge-calibrated radar data and
             various satellite estimates and numerical model short-range
             prediction over the period 1 January 2011 to 31 December 2014.
             The time series is displayed as an 11-day running mean to
             improve clarity 461
Fig. 25.5   Temporal correlation over Japan between the JMA gauge-
             calibrated radar data and various satellite estimates and
             numerical model short-range prediction for the December–
             February period during 2011–2014 462
Fig. 25.6   Example of longer-term analysis of precipitation products over
             Northern Brazil of the IPWG South America validation site. The
             upper right plot shows the daily gauge-observed precipitation
             (blue) together with the bias of the GSMaP and TMPA products
             for 27 December 2014 through 18 August 2016. The lower left
             plot shows the range of the RMSE for each satellite precipitation
             product, while the lower right plot shows the bias for the same
             products 463
Fig. 25.7   Example of instantaneous, full retrieval resolution validation of
             the GMI precipitation product (GPROF Level 2 version 5) over
             the European IPWG validation region for 26 January 2016 at
             0834 UTC 466

Fig. 26.1   Translation of high-quality precipitation measurements to
             satellite footprint and swath measurement scales for direct and
             physical validation 473
Fig. 26.2   VN verification of DPR calibration adjustment. The DPR
             Ku-band radar calibration was adjusted by JAXA from Version
             4 (V4; left) to Version 5 (V5; right). The V5 increase of DPR
             Ku-Band radar reflectivity (Z_e) by ~1.2 dB (y-axis) is clearly
             evident in relative frequency histograms (shaded, percent) when
             plotted against the Ku-adjusted VN Z_e (Ground Radar, GR:
             X-axis) 480
Fig. 26.3   Same as Fig. 26.1, but D_m (mm). Displayed are the V4 (left) and
             V5 (right) DPR D_m relative to GV. Adjusting the DPR
             calibration in V5 resulted in a slight, but perceptible positive
             bias shift in the DPR D_m relative to GV. NASA L1SRs for
             D_m (± 0.5 mm) indicated by dashed lines 480
Fig. 26.4   DPR Measured (left) and PIA-corrected (right) V5 Ku-Band
             radar reflectivity (y-axis) in convective precipitation plotted
             against VN radar reflectivity (x-axis) for layers of precipitation
             below the height of the 0 °C level. Similar results are attained
             for stratiform precipitation (not shown) 481
Fig. 26.5   Coincident and DPR footprint-matched VN-estimated SWER
             using the Marquette, Michigan WSR-88D radar (KMQT) and
             Pluvio network in the DPR swath, 15 Apr. 2018. The DPR
             single frequency estimate (left) is compared against the DP radar
             (center) and PQPE median estimators (right). The MQT Pluvio
             network is located west through south of KMQT and within 15
             km range. Range rings are illustrated at 50 and 100 km from
             KMQT 482
Fig. 26.6   Scatter density plots for winter 2017/18 DPR normal swath
             (y-axis) SWER plotted against VN-estimated SWER (x-axis) for
             the PQPE 25th% (left), dual-pol (center), and 50th% (right)
             SWER estimators. It is clear that the DPR SWER estimate best
             matches the VN dual-polarimetric SWER 482
Fig. 26.7   Evolution of GPM versions. Joint histograms of GV-MRMS
             reference rain rate (x-axis) for all rain types plotted against
             matched DPR 2AKu-algorithm IVOV rain rate estimates
             (y-axis) V04 (left), V05 (center) and V06 (right) 483
Fig. 26.8   Same as Fig. 26.7, but matched-swath dual-frequency combined
             radar-radiometer algorithm 483
Fig. 26.9   Same as Fig. 26.8, but V5 combined algorithm for stratiform
             (left) and convective (right) precipitation. The KuPR algorithm
             strongly resembles this plot as well 484

Fig. 26.10   KuPR (top left) and GMI-GPROF (top right) 50 km gridded
              mean rate rates (mm h^{-1}) plotted against their matched
              sample of GV-MRMS (indicated as "Q3") rain rates in
              bottom panels   485
Fig. 26.11   GPM V6 combined radar-radiometer algorithm matched-swath
              rain rates (x-axis) vs. normalized error (%; y-axis) at 50 km grid
              scales over the CONUS L1SR region. Solid line is relative bias,
              dashed line is NMAE. Green shading represents L1SR
              requirement (Skofronick-Jackson et al. 2018 for V5 result)   486
Fig. 26.12   Similar to Fig. 26.11 (i.e., V6) but for instrument footprint scales
              in the Kwajalein (K-pol; left), and PAIH (right) oceanic
              domains. The red line represents bias (%) and the blue line
              indicates normalized mean absolute error (NMAE; %) at EFOV
              scales. The solid black line indicates the RMSE (%) for EFOVs
              scaled to 50 km using Steiner et al. (2003)   487
Fig. 26.13   March 2014 to October 2017 V5b IMERG vs. RAD-RAR
              statistics for Early (IMERG_E), Late (IMERG_L), Final without
              gauge adjustment (IMERG_F_U), and Final with gauge
              adjustment (IMERG_F_G). Top panels: Pearson correlation
              coefficient (Correlation), Bias (%), NMAE, and NRMSE.
              Bottom: contingency scores for hits (green), misses (light blue),
              false alarms (purple), and correct negatives (black). POD, FAR
              and HSS are also indicated   492
Fig. 26.14   Difference in daily rainfall between RAD-RAR and IMERG-E,
              L, F-U and F-C, products   492
Fig. 26.15   Error components of IMERG-F V5b Final for March 2014–
              October 2017 broken down by total bias (E), hit bias (H), biases
              due to missing precipitation (M), and false precipitation (F)   493
Fig. 26.16   CONUS-wide comparisons of V04A (left) and V5b (right)
              IMERG-L 30 min rain rates to GV-MRMS for the year
              of 2015   494
Fig. 26.17   Same as Fig. 26.14, but CONUS IMERG-L compared to
              GV-MRMS June 2014–August 2017   494
Fig. 26.18   Hurricane Florence total rain accumulation (mm) for 10–16
              September 2018. Totals for the GV-MRMS, IMERG-L, and the
              difference in accumulation between the two (MRMS-IMERG)
              are displayed in the left, center and right panels, respectively.
              The location of the maximum accumulations for GV-MRMS
              and IMERG are indicated by the dark circle   495
Fig. 26.19   Same as Fig. 26.6 and Fig. 26.7, but 30 min rainrates for domain
              of Fig. 26.16 for IMERG "Precipcal" (left), HQPrecip
              (microwave; center), and IR (right) during Hurricane
              Florence   495

Fig. 27.1   Schematic illustration of the measured radar reflectivity profiles
             by the dual Ka-band radar system. Z_{m1} and Z_{m2} are measured
             radar reflectivity factors and r is the range 507
Fig. 27.2   Scatter plots of k and Z_e for snow events in 2012. Each figure is
             with different surface air temperature. Black dots are 1 min data,
             and dark lines and crosses are fitting curves and averaged values.
             (Adapted from Nishikawa et al. 2015) 508
Fig. 27.3   The area of comparison over Japan. 0.5° longitude/latitude
             boxes were set and the entire area was divided into six climatic
             sub areas considering the total amount of precipitation and near
             surface air temperature 509
Fig. 27.4   The differences between DPR and AMeDAS rain gauge in six
             sub areas. Despite slight differences from region to region, the
             results meet the success criterion that the difference is within
             ±10% in all cases 511
Fig. 27.5   The area of comparison over the US. The area includes total of
             871 1° latitude/longitude boxes 511
Fig. 27.6   The difference between DPR product and MRMS/NMQ by 24
             1°latitude bins. The results meet the success criterion of
             difference within ±10% in 22 bins among 24 bins 512
Fig. 27.7   The rain/snow distribution by JMA's visual reports and DPR
             data for the case of 18 January 2016 (0900 JST) over north-east
             part of Japan. The agreement is quite good 513

Fig. 28.1   Map showing the distance to nearest GPCC gauge, typical of all
             regular and reliable gauge measurements; blank areas in the
             figure are beyond 100 km from the nearest gauge. (Adapted
             from Kidd et al. 2017) 522
Fig. 28.2   Global correlation of the 3B42RT (**a**), CMORPH (**b**),
             SM2RAIN (**c**) and ERA-Interim (**d**) products obtained by TCA
             using Triplet A (ERA-Interim-SM2RAIN-3B42RT) for
             3B42RT, ERA-Interim and SM2RAIN and Triplet B
             (ERA-Interim-SM2RAIN-CMORPH) for CMORPH. (Adapted
             from Massari et al. 2017) 523

Fig. 29.1   MRMS radar quality index. The quality index considers the
             distance to the closest radar and the beam blockage by the
             relief (see Zhang et al. 2016 for the definition of the quality
             index) 536

Fig. 29.2   (**a**) MRMS hourly precipitation over South-Eastern US, on 30
             Nov. 2016 from 0400 to 0500 UTC. (**b**) Corresponding IMERG
             hourly precipitation. (**c**) IMERG error relative to MRMS.
             (**d**) Wavelet energy spectra of MRMS (thick black line with
             circles), IMERG (thick blue line with triangles) and of the error
             IMERG-MRMS (dashed red line with crosses). The energy
             spectra are normalised by the total energy (sum of squared
             values) of the MRMS field   540
Fig. 29.3   Seasonal cumulative precipitation from Nov. 2016 to Apr.
             2017 (cold season) and May – Oct. 2017 (warm season) for
             MRMS and IMERG. The fields are spatially smoothed through
             sliding window averaging with a 640 km wide window
             (corresponding to the smoothing function associated with the
             Haar wavelet)   541
Fig. 29.4   Average wavelet energy spectra of MRMS (black line with
             circles), and IMERG (blue line with triangles) hourly
             precipitation fields and of the error IMERG-MRMS (dashed red
             line with crosses), computed over CONUS Nov. 2016 – Apr.
             2017 (cold season) and May – Oct. 2017 (warm season). The
             spectra are computed at each time step and summed over the
             analysed period. The energy spectra are normalised by the total
             energy (sum of squared values) of the MRMS fields. Only the
             pixels with a MRMS radar quality index >0.9 are retained for the
             computation of the spectra   542
Fig. 29.5   Ratio of the spectral energy of IMERG hourly precipitation
             fields over the spectral energy of MRMS hourly precipitation
             fields during the cold and warm seasons. The spectral energies
             are computed at each time step and summed. A ratio close to one
             at all scales indicates the agreement of the IMERG and MRMS
             wavelet energy spectra   544
Fig. 29.6   Ratio of the spectral energy the IMERG hourly error (i.e.,
             IMERG-MRMS) over the spectral energy of MRMS hourly
             precipitation during the cold and warm seasons. The spectral
             energies are computed at each time step and summed. A low
             spectral energy of the error indicates the agreement between
             IMERG and MRMS spatial patterns (gradients) at the
             corresponding scale   546
Fig. 29.7   Linear correlation between the wavelet coefficients of the
             IMERG and MRMS hourly precipitation fields. A high
             correlation of the wavelet coefficients indicates the agreement
             between IMERG and MRMS spatial patterns (gradients) at the
             corresponding scale. Contrary the spectral energy of the error,
             the correlation is not sensitive to potential biases in the
             magnitude of the wavelet coefficients   548

Fig. 29.8   Effective resolution of IMERG hourly precipitation evaluated
             against MRMS during the cold and warm seasons. The criterion
             (29.3) is used to define the effective resolution 550
Fig. 29.9   Same as Fig. 29.4 for the cold season, but computed only over
             the South-Eastern region where the effective resolution is found
             finer than 160 km 551
Fig. 29.10   Effective resolution of IMERG daily precipitation evaluated
             against MRMS during the cold and warm seasons. The criterion
             (29.3) is used to define the effective resolution 552
Fig. 29.11   Representation of the spatial convolution filters used to perform
             the two-dimensional Haar wavelet decomposition. A and B and
             C are the wavelet functions ψ^H, ψ^V and ψ^D used to compute the
             horizontal (HL), vertical (LH) and diagonal (HH) wavelet
             coefficients. D is the smoothing function ϕ^{2D} used to compute
             the smoothing (LL) coefficients. The four functions are
             orthogonal, meaning that the scalar product of one with any of
             the other three equals zero 555
Fig. 29.12   Illustration of the discrete wavelet decomposition process of a
             two-dimensional field (using the Haar wavelet) 555

Fig. 30.1   Histogram of daily rainfall observed along an altitudinal
             raingauge transect on the eastern slopes of the Peruvian Andes
             (Barros 2013). Stations with incomplete records were removed.
             A, B, C refer to different pixels of TRMM 3B42 and GPM
             IMERG products. The red (tropical montane forest) and orange
             (cloud forest) lines represent the rainfall envelope for the two
             main ecosystems 561
Fig. 30.2   Conceptual representation of orographic precipitation
             mechanisms in the Southern Appalachian Mountains. (Adapted
             and modified from Wilson and Barros 2017) 562
Fig. 30.3   Example of ground-clutter effects on radar measurements of
             low-level rainfall. Left: Vertical structure of MRR reflectivity
             above ground level (AGL) with GPM overpass marked by the
             pink dashed line in right panel. Right: Reflectivity cross-section
             over the Southern Appalachian Mountains. Black dashed lines
             indicate the boundaries of near-nadir scan (angle <7 deg). Pink
             line marks the position of the MRR where rainfall was missed by
             the GPM DPR algorithm. The white band between the terrain
             (dark black line) and the GPM reflectivity lower reflectivity
             measurements (red band) is due to the terrain mask used in the
             algorithm 565

Fig. 30.4 Dependence of detection errors to the satellite orientation. Detection errors varying with (**a**) satellite geometry and (**b**) local viewing angle. Near-surface precipitation estimates of Level 2, version 5A products of GPM Ku-PR compared with ground-based rain-gauges (RG) that lie within 2.5 km of the center of DPR pixels. Time period considered for analysis: March 2014–May 2017. Note- YY is when both GPM and RG detects precipitation. FA is false alarm when GPM detects precipitation while RG did not detect any precipitation. MD is missed detection where RG detects rain and GPM misses the detection 566

Fig. 30.5 Location of IPHEx GV (left) rain-gauges and (right) Parsivel disdrometers in the SAM. Triangles denote ridge (elevation >900 m) and circles denote valley locations 567

Fig. 30.6 Spatial distributions of AQUA MODIS LLCF (CTH <5 MSL, confident cloudy only; $0.05° \times 0.05°$) during daytime (Left Column) and nighttime (Right Column) overpasses in summer (June–July-August; top row) and winter (December–January-February; bottom row). (Adapted from Duan and Barros 2017) 568

Fig. 30.7 Fingerprinting SFI on DSD metrics. Left: diurnal cycle of Dm (ratio of fourth moment to the third moment of the diameter) at P4 (western ridge) and P6 (foothills). Right: diurnal cycle of rain and fog occurrences at P4. Note that Dm (P6) always $\geq$ Dm (P4). *FCC:* Fog and Cap Clouds. *LLC:* Layered Low Level Clouds 569

Fig. 30.8 Left: diurnal cycle of GPM revisit overpasses over the region shown in Fig. 30.5. Right: Spatial distribution of rainfall detection. The blue tones in the inner region valleys and over the complex terrain at low elevations along the eastern ridges of the SAM are indicative of low frequency bias 570

Fig. 30.9 DSD nonstationarity on altitudinal gradients: (**a**) disdrometer (continuous lines) and GPM Ku-PR estimates (dashed lines) on the western slopes of the SAM; (**b**) and (**c**) DSD statistics on the eastern slopes of the SAM 570

Fig. 30.10 (**a**) Seasonal and (**b**) diurnal distribution of GPM Ku-PR detection errors in the Southern Appalachian Mountains for ascending and descending modes of the GPM 570

Fig. 30.11 (**a**) Detection and (**b**) estimation errors varying with rain-gauge rain-rate observations for ascending and descending modes of the GPM Ku-PR. $Bias = 10\log_{10}\left(\dfrac{\sum R_{GPM,i}}{\sum R_{RG,i}}\right)$ 571

Fig. 30.12   Average GPM Ku-PR Level 2A DSD parameters in the SAM for
              4.5 years since GPM launch. Top row: spatial distribution of (a)
              Dm and (b) Nw. Bottom row: vertical structure along the
              transect at 83.5°W (black line in b) of (c) Nw and (d) Dm. The
              pink circles and triangles mark respectively valley and ridge
              disdrometer locations   572
Fig. 30.13   Phase-space maps of (Dm, Nw) from GPM Level 2A products
              since launch and from GV Parsivel disdrometer P18 during
              IPHEx   573
Fig. 30.14   Left: sensitivity analysis of surface rainfall accumulation to the
              drop size distribution (DSD) spectra of fog in the simulation of
              RoF SFI at P6 in the western foothills of the SAM. RG–dashed
              black lines are local raingauge observations. After Duan and
              Barros (2017). Right: simulated microphysical trajectory for
              stratiform rainfall with and without RoF SFI at P6 corresponding
              to Fog#4 DSD and fog depth of 400 m. Circles describe
              microphysical state at 1-min intervals from start to end of the
              event as per the legend   574
Fig. 30.15   Composites of GPM Level 2A DSD parameters and reflectivity
              along a cross-section designed to capture the hot-spots of
              precipitation in Central Himalayas (84.4°E, red circle) during
              the monsoon. Orographic precipitation on the upwind slopes is
              highlighted by the boxes. The map with TRMM Precipitation
              Features (PFs) on the left panel is adapted from Barros et al.
              (2004)   574
Fig. 30.16   Left panels: diurnal cycle of the spatial distribution of the
              climatology of Probability of Detection (PoD) of IMERG
              precipitation (~11 km spatial resolution) compared with STAGE
              IV (~4 km spatial resolution) (http://www.emc.ncep.noaa.gov/
              mmb/ylin/pcpanl/stage4/, last accessed 4 Dec. 2018). Time
              period of analysis- March 2014 to February 2017. Right panels:
              diurnal cycle of Parsivel disdrometer DSD measurements at
              locations marked in the left panel   575
Fig. 30.17   Illustrative sketching of typical DSD evolution in the lower
              troposphere under when different Seeder-Feeder
              Interactions (SFI) scenarios. ABGL Above Ground Level,
              D Diameter, N(D) number concentration   577
Fig. 30.18   Contour frequency by altitude synthesis of reflectivity
              profiles derived from Weather Research Forecast (WRF)
              model V3.8 simulations using the standard continental aerosol
              in the model (left) and using aerosol properties measured in
              Central Nepal. Details of the simulation can be found in Barros
              et al. (2018)   577

Fig. 31.1   Research framework and overview flowchart to bridge from the
             Global Precipitation Measurement mission core satellite to the
             combined gridded IMERG product using ground radar-based
             Multi-Radar/Multi-Sensor QPE. An example of Multi-Radar/
             Multi-Sensor instantaneous precipitation rates at 0725 UTC on
             11 April 2011 is shown   585

Fig. 31.2   Comparison framework across Level-2 and Level-3 satellite
             QPE to bridge from the GPM core satellite to IMERG   594

Fig. 31.3   Reference (a) and DPR (b) rainfall rate distributions
             (mm h^{-1}) as functions of the CPI (%). The thick black line
             represents the median (50% quantile), the dark grey-shaded
             region represents the area between the 25 and 75% quantiles, the
             light grey-shaded region represents the area between the 10 and
             90% quantiles   595

Fig. 31.4   DPR residual distributions (mm h^{-1}) as functions of the CPI
             (%). The thick black line represents the median (50% quantile),
             the dark grey-shaded region represents the area between the 25
             and 75% quantiles, the light grey-shaded region represents the
             area between the 10 and 90% quantiles   596

Fig. 31.5   Conditional bias of spaceborne radars relative to the reference as
             a function of the CPI (%) for the GPM-DPR/Ku (black),
             GPM-DPR/Ka (blue), GPM-DPR/Ka-Ku (red) and TRMM-PR
             (grey)   597

Fig. 31.6   Reference and GPROF-GMI cumulative distribution of
             convective contribution   598

Fig. 31.7   GPROF-GMI residual distributions (mm h^{-1}) as functions of the
             CPI (%). The thick black line represents the median
             (50% quantile), the dark grey-shaded region represents the area
             between the 25% and 75% quantiles, the light grey-shaded
             region represents the area between the 10% and 90%
             quantiles   599

Fig. 31.8   GPROF-GMI (left) systematic part –conditional median - and
             (right) random part –interquantile 10%–90%- of error   599

Fig. 31.9   GPROF-GMI (left) correlation and (right) relative bias as a
             function of convective contribution difference between GPROF-
             GMI and the reference   600

Fig. 31.10   IMERG residual distributions (mm h^{-1}) as functions of the CPI
             (%). The thick black line represents the median (50% quantile),
             the dark grey-shaded region represents the area between the 25
             and 75% quantiles, the light grey-shaded region represents the
             area between the 10 and 90% quantiles   601

Fig. 31.11   IMERG (left) systematic part – conditional median – and (right)
             random part – interquantile 10–90% – of error for the PMW
             component (red) and IR component (blue)   602

Fig. 32.1   Correlation coefficient between hydrologic model prediction
             and streamflow observations at 130 gauges in Iowa. The
             IMERG forced model results are on the left, the MRMS-forced
             model output is on the right 612
Fig. 32.2   Northeast Iowa, the site of the 2013 Iowa Flood Studies field
             campaign. The outlined river basins include Turkey River,
             Cedar River and Iowa River 614
Fig. 32.3   The width function for the Cedar River basin with the outlet at
             Cedar Rapids, Iowa. In constructing the plot, we assumed a
             constant stream velocity of 0.75 m s^{-1}. It is clear that the streams
             that are located 200–400 km from the outlet, contribute most to
             the discharge 614
Fig. 32.4   The panels on the left correspond to the gridded representation
             of rainfall for radar-based (top) and satellite-based (middle)
             products and the difference (bottom). On the right, we show the
             same rainfall but calculated for the areas upstream from a given
             link of the river network for several watersheds in northeast
             Iowa 616
Fig. 32.5   Panel (**a**) shows network integrated rainfall difference between a
             reference and a product of interest. Panel (**b**) shows the
             differences in the peak flow for all streams in the network based
             on a simple linear model; panel (**c**) shows the same but for a
             nonlinear model 617
Fig. 32.6   Network-integrated rainfall for the reference product (MRMS)
             and a satellite product (IMERG). The third panel show the
             relative differences between the two products 619
Fig. 32.7   Standard deviation (in mm) of the rainfall product discrepancies
             calculated in Euclidean space (black), and over the river network
             for flow-connected segments (red) and flow-unconnected
             segments. The inset illustrates the flow-connected and
             unconnected cases 620

Fig. 33.1   For a selection of the evaluated uncorrected P datasets, temporal
             correlations between 3-day mean gauge- and dataset-based P
             time series (R_{3day}). Each data point represents a gauge. See Beck
             et al. (2017c) for global maps of the other performance
             metrics 636
Fig. 33.2   Calibration NSE scores obtained using P time series from (**a**)
             CHIRPS V2.0, (**b**) CMORPH-CRT V1.0, (**c**) CPC Unified, (**d**)
             MSWEP V1.2, (**e**) MSWEP V2.0, (**f**) PERSIANN-CDR V1R1,
             (**g**) TMPA 3B42 V7, and (**h**) WFDEI-CRU. Each data point
             represents a catchment centroid. Only the eight best performing
             P datasets are shown 641

Fig. 33.3   For each catchment, the P dataset with the highest calibration
             NSE. Each data point represents a catchment centroid. Only the
             seven best performing P datasets (excluding MSWEP V1.2 due
             to its similarity to V2.0) are considered. Note that CHIRPS
             V2.0, CMORPH-CRT V1.0, PERSIANN-CDR V1R1, and
             TMPA 3B42 V7 do not provide data beyond 50, 60, 60, and 50°
             latitude, respectively   642

Fig. 34.1   OceanRAIN shipboard data ingest and post-processing flow
             chart. (Adapted from Klepp et al. 2018)   660
Fig. 34.2   OceanRAIN data distribution for all eight ships (**a**), seasons
             (**b**), years (**c**) and precipitation occurrence for type
             (rain, snow, mixed) and true-zeros (**d**). (Adapted from
             Klepp et al. 2018)   662
Fig. 34.3   Latitudinal fraction of the precipitation phase with red for rain,
             blue for snow and green for mixed phase (**a**). The scatter
             diagram shows individual precipitation minutes for phases
             against latitude as a function of precipitation rate (mm h^{-1}).
             Panels (**b**) shows the joint histogram of the precipitation rate in
             mm h^{-1} and the orange line indicates bin-wise mean for 2°
             latitude bands. Frequency of occurrence in % of all cases is
             shown with colors. Panel (**c**) shows the Rayleigh reflectivity log
             (Z) and log (R) distribution of rainfall for all 696,740 events
             (black), 139,557 stratiform events (blue) and 15,823 convective
             events (red). Panel (**d**) denotes the mean number concentration
             PSDs (thick marks) and their latitudinal variability (thin marks)
             for all rainfall (red), snowfall (blue) and mixed-phase
             precipitation (green). (Adapted from Klepp et al. 2018)   665
Fig. 34.4   Maps of average precipitation-rate difference of HOAPS (R_H)
             subtracted by adjusted OceanRAIN (R_O**) in mm h^{-1} per 2° ×
             2° grid-box for (**a**) hits, (**b**) misses, (**c**) false detections and d) all
             cases. Gray boxes have no data available. (**e**) Map of occurrence
             of precipitation from 24,990 HOAPS–OceanRAIN collocations
             from available long-term RVs in OceanRAIN from 06/2010–12/
             2015 (cyan dots), misses ($R_H = 0$ and $R_O^{**} > 0$; purple) and hits
             ($R_H > 0$ and $R_O^{**} > 0$; orange) using ±20 km and ± 30 min as
             collocation boundaries. (Adapted from Burdanowitz et al.
             2018)   667
Fig. 34.5   Location of matched OceanRAIN-IMERG pairs for the period
             20 March 2014–28 February 2017. The "x" symbol indicates
             rain events, and "+" symbol indicates snow eents. Ship tracks of
             OceanRAIN database. R/V Investigator (red), R/V Merian
             (magenta), R/V Meteor (blue), R/V Polarstern (cyan), R/V
             Revelle (green), R/V Sonne II (yellow), R/V World (black)   669

Fig. 34.6   Scatterplot of matched IMERG and OceanRAIN pairs. The
             colors indicate the ship that observed the precipitation.
             The "+" symbol indicates rain pairs and "×" symbol
             indicates snow pairs 670
Fig. 34.7   Plot of cross-correlation of rainfall for neighboring IMERG
             grids 670
Fig. 34.8   Scatter plot of matched pairs for convective and stratiform
             rainfall 671
Fig. 35.1   GPCP climatological mean precipitation (mm day^{-1}) during
             1979–2017 680
Fig. 35.2   Latitudinal profiles of mean zonal precipitation (1979–2017) for
             land, ocean and combined 680
Fig. 35.3   Seasonal mean GPCP precipitation (mm day^{-1}). (a) December–
             January-February (DJF), (b) March–April-May (MAM), (c)
             June–July-August (JJA), and (d) September–October-
             November (SON) 681
Fig. 35.4   Time series (January 1979–December 2017) of global mean
             (land+ocean) precipitation (black line), and corresponding
             ENSO (red) and volcanic effects (blue) 684
Fig. 35.5   Composite precipitation anomaly changes (%) between ENSO
             warm (defined as Niño 3.4 ≥ 0.34) and cold phases (Niño 3.4 ≤
             −0.45). The number of months for either phase is 155 (one third
             of months) during 1979–2017 686
Fig. 35.6   Long-term trends in (a) GISS surface temperature
             (ts; 1979–2017), (b) SSMI/SSMIS columnar water vapor
             (CWV; 1988–2014), and (c) GPCP precipitation (1979–2017).
             (d) Long-term trend in precipitation without the effect of PDO
             and AMO 687
Fig. 35.7   Zonal mean precipitation trend during 1979–2017 from GPCP
             (black curve). Also shown is the zonal mean profile of
             climatological mean precipitation (red curve; scaled by 50 with
             resulting units of mm day^{-1}). Green and blue curves are the
             trend curves for CMIP and AMIP climate model ensembles for
             similar periods 689
Fig. 35.8   Long-term trends (mm day^{-1} per decade) in (a) GPCP
             precipitation (1979–2017), (b) GPCP precipitation with no PDO
             effect, (c) AMIP precipitation (1979–2008), and (d) CMIP5
             historical full forcing precipitation (1979–2012) 690

Fig. 35.9   Annual anomalies of (**a**, **b**, **c**) precipitation percentiles and
(**d**) mean precipitation determined by GPCP monthly rain-rates
between $30°$N and $30°$S. Also shown are their corresponding
linear trends (mm day^{-1} per decade), and those followed
by "$*$" are statistically significant at the 5% level. For selected
percentiles and the mean, the trends are also shown in
%/decade 691

Fig. 35.10   Annual anomalies of the 95th, 30th precipitation percentiles and
mean values between $30°$N and $30°$S for GPCP (black), AMIP
(green), and Hist-Full (red) monthly rain rates. Also shown are
their corresponding linear trends (mm day^{-1} per decade), and
those followed by "$*$" are statistically significant 692

Fig. 36.1   CloudSat observations per $0.1°$ latitude/longitude grid box for
the 2006–2016 time period (left panel). Zonally-averaged
($1°$ latitude bins) number of CloudSat observations for the
2006–2016 (black), July 2006–April 2011 (dark gray), and
December 2011–December 2016 (light gray) time periods
(right panel) 705

Fig. 36.2   Zonally-averaged conditional snow (black), rain (dark gray) and
mixed precipitation (dot-dash) occurrence fractions from the
2C-PRECIP-COLUMN product. Unconditional snowfall
occurrence fractions (light solid gray) are also shown.
All results shown are averaged over $1°$ latitudinal belts
for the 2006–2016 period 706

Fig. 36.3   Conditional (top) and unconditional (bottom) snow occurrence
fractions ($1°$ latitude/longitude grid boxes) from the
2C-PRECIP-COLUMN product for the 2006–2016 period 707

Fig. 36.4   Near-surface bin CPR normalized reflectivity distributions for
nimbostratus (left) and cumuliform (right) snowfall events.
Distributions are partitioned between Northern (NH) and
Southern Hemisphere (SH) land and oceanic surfaces 708

Fig. 36.5   Left panel: unconditional snowfall occurrence fraction with
(gray) and without (black) ground clutter quality control applied.
Right panel: mean annual snowfall rates with (solid black) and
without (gray dash-dot) ground clutter quality control applied
for all snowfall events over the 2006–2016 and 2011–2016
(dark gray dash-dot) periods. Cumuliform snowfall events with
quality control applied for the 2006–2016 (gray dash) and mean
fractional snowfall rate uncertainty (light gray solid) also
shown 709

Fig. 36.6   2C-SNOW-PROFILE mean annual snowfall rate in $1°$ grid
boxes for the 2006–2016 period 711

Fig. 36.7 Monthly snowfall rate [mm d^{-1}] (top panels) and unconditional snowfall fraction (bottom panels) for the Northern (left) and Southern (right) Hemispheres. Median (line), 25–75 percentile range (gray shaded), and minimum/maximum values extending up to 1.5 times the 25–75 interquartile range (thin lines) are shown. Outlier data points are indicated by circles. The Southern Hemisphere panels are restricted to the 2006–2011 time period, while the Northern Hemisphere panels use the entire 2006–2016 CloudSat dataset 712

Fig. 37.1 Geographical distribution of the percentage of snow in the total precipitation. The grid box is 0.25 ° × 0.25 °. (Adapted of Adhikari et al. 2018) 719

Fig. 37.2 Three-dimensional snapshot of GPM DPR observations near Costa Rica, 1113 UTC 1 Jul 2014, by setting different minimum detectable reflectivities: (a) 12 and (b) 18 dBZ. (c) Along-track cross section of reflectivity at the 28th angle bin [indicated by a white arrow in (a)]. The thin black line shows the topography contour. (d) Same as in (c), but at the third angle bin. (Adapted from Hamada and Takayabu 2016) 720

Fig. 37.3 (a) GPM DPR observation of rainfall rate on 17 March 2014 (orbit #000272). Circled areas A, B, and C represent snow, stratiform rain, and convective rain, respectively. (b) Averaged reflectivity profiles as well as dual-frequency ratio profile for snow. (c) Same as (b) for stratiform rain. (d) Same as (b) for convective rain. (Adapted from Le et al. 2017) 723

Fig. 37.4 Frequencies at which flagHeavyIcePrecip was set with different conditions over 2° × 2° boxes in 2015. (a) Percentage of occurrences in which flagHeavyIcePrecip is set by "condition A [DFR$_m$ > 7 dB, Z$_m$(Ku) > 27 dBZ above −10°C level]". (b) Percentage of occurrences in which the maximum of Z$_m$ of KuPR above the −10 °C height exceeds 40 dBZ. (c) Percentage of occurrences in which flagHeavyIcePrecip is set by condition A and the surface air temperature is below 1 °C. Percentages are calculated relative to the total number of the measured pixels that include no rain pixels. (Adapted from Iguchi et al. 2018) 725

Fig. 38.1 (a) Climatology of occurrence of MCS (b) Climatology of precipitation. (Adapted from Huang et al. 2018) 732

Fig. 38.2 Duration statistics for the 2012–2014 summer seasons. (a) Duration PDF, (b) duration CDF, and (c) duration CDF weighted by the MCS cold cloudiness. Maximum extent (d) CDF and (e) CDF weighted by the MCS cold cloudiness. Shown are the entire tropical belt (solid), continent (short dashed), and ocean (long dashed). (Adapted from Roca et al. 2017) 734

Fig. 38.3   (top) CDF of the tropical rainfall as a function of duration, and
             (bottom) as a function of duration. Solid lines are for all the
             tropics, long dashed for land and short dashed for ocean
             conditions. Black refers to the TAPEER 1.0 product (Chambon
             et al. 2013), blue to the 3B42v7 (Huffman et al. 2007) and red to
             the GPCP 1DD v1.2 (Huffman et al. 2001). (Adapted from Roca
             et al. 2014)   735
Fig. 38.4   Four examples of convective cold shield time evolution.
             (Adapted from Roca et al. 2017)   736
Fig. 38.5   Evolution of the mean profile of reflectivity (first row) and of the
             reflectivity interquartile range (second row) for (a) continental
             West Africa (AF) and (b) Atlantic Ocean (ATL). Solid upper
             line with squares and dashed line are the mean cloud-top height
             obtained from radar and radar–lidar, respectively. The lower line
             with squares is the mean cloud base. (Adapted from Bouniol
             et al. 2016)   738
Fig. 38.6   Evolution of (a) latent heating profile in K h^{-1}, (b) shortwave
             radiative heating profiles at 0130 LT in K day^{-1} and (c)
             longwave radiative heating profiles in K day^{-1} for the different
             MCS sub-regions over the life cycle of a continental MCS. Each
             panel in (a) corresponds to a life step with convective heating in
             solid line and stratiform heating in dashed line. In (b) and (c)
             each panel corresponds to an MCS sub-region (convective,
             stratiform and cirriform), the color level showing the life stages.
             The black line shows the clear sky radiative heating profile   739

Fig. 39.1   Scatterplots of PR rain rate and TMI polarization corrected
             temperature at 85 GHz (PCT85) over Oklahoma (US-OK, left)
             and the Korean Peninsula (right)   747
Fig. 39.2   Probability distribution functions (PDFs) of (a, b) TMI
             polarization corrected temperature at 85GHz (PCT85) and
             (c, d) PR storm height classified by rain rate intensity for
             Oklahoma (US-OK, blue line) and Korea (red line)   749
Fig. 39.3   The Contoured Frequency by Altitude Diagrams (CFADs) of the
             PR reflectivity classified by rain rate intensity for (a, b)
             Oklahoma (US-OK) and (c, d) Korea. Percentages refer to the
             occurrence of each rain class   750
Fig. 39.4   Vertical structure (a, b) and occurrence frequency distribution
             (c, d) of the cold-type and warm-type heavy rain classified by
             K-means clustering analysis for the period June–August
             1998–2014. The percentages indicate the total occurrence of
             each type. The area displayed in the top left of the upper figures
             represents the mean heavy rain area at a $5° \times 5°$ grid over the
             entire domain. The 17-year summer mean geopotential height
             (gpm, solid lines) and moisture flux (m s^{-1}, arrows) at 850 hPa
             are shown in the bottom-right figure   752

Fig. 39.5   (a) Diurnal variation of cold-type (blue line) and warm-type (red
              line) heavy rain (>10 mm h^{-1}) occurrence. (b) Diurnal
              variations of the warm-type over the oceanic area north of 30°N
              (solid line) and south of 30°N (dashed line) 753
Fig. 39.6   Spatial distributions of 17-year summer mean (a, b) convective
              available potential energy (CAPE), (c, d) total precipitable water
              (TPW), and (e, f) vertically-integrated water vapor flux
              convergence (ConQ) for cold-type and warm-type heavy rain
              events. Regions showing occurrence frequency lower than 20%
              are excluded from the analysis 754
Fig. 39.7   The summer (June–August) mean equivalent potential
              temperature (color, K) and relative humidity (dotted line, %)
              profiles for (a) cold-type and (b) warm-type heavy rain events
              over the Korean peninsula. (Adapted from Song and Sohn
              2015) 755
Fig. 39.8   Temporal evolution of (a, b) radar reflectivity (Z), (c, d)
              raindrop mean-mass diameter (D_R), and (e, f) raindrop number
              concentration (N_R) profiles for the cold-type and warm-type
              heavy rain. (Adapted from Song et al. 2017) 757
Fig. 40.1   Average precipitation (first column) and maximum daily
              precipitation (second column) for PERSIANN (a, b), CMORPH
              (c, d), and GPCP (e, f). (Adapted from Prat et al. 2017) 765
Fig. 40.2   Cumulative distribution function of daily rainfall derived from
              three satellite products GPCP, CMORPH, and PERSIANN-
              CDR for the period 1998–2013. (Adapted from Prat et al.
              2017) 766
Fig. 40.3   Spatial distribution of daily rainfall above the 90th (first column)
              and 99th (second column) percentiles for PERSIANN (a, b),
              CMORPH (c, d), and GPCP (e, f). Values of the percentiles of
              each satellite product are defined according to the distribution
              presented in Fig. 40.2. (Adapted from Prat et al. 2017) 767
Fig. 40.4   (a) Number of rainy days (R > 0 mm day^{-1}), (b) Wet Millimeter
              Days (R > 17.8 mm day^{-1}: WMMD), (c) Rainy Days with
              accumulation greater than 2 in day^{-1} (R > 50.8 mm day^{-1}:
              EPD2), and (d) Rainy Days with accumulation greater than 4 in
              day^{-1} (R > 101.6 mm day^{-1}: EPD4) for GHCN-D (first
              column), TMPA 3B42 (second column), and TMPA 3B42RT
              (third column). (Adapted from Prat and Nelson 2015) 772
Fig. 40.5   (a) Average annual tropical cyclone rainfall, and (b) annual
              tropical cyclone rainfall contribution. (c) Average annual
              tropical cyclone number of rainy days with rainfall R > 0, and
              (d) tropical cyclone contribution to daily rainfall R > 0. (e) and
              (f) same as (c) and (d) for EPD2. (g) and (h) same as (c) and (d)
              for EPD4. (Adapted from Prat and Nelson 2016) 774

Fig. 40.6   (**a–c**) Distribution of daily rainfall for events observed simultaneously by USCRN stations and satellites. (**a**) PERSIANN-CDR, (**b**) CMORPH, and (**c**) GPCP. (**d–f**) Same as (**a–c**) but for the cumulative distribution function of USCRN stations and (**d**) PERSIANN-CDR, (**e**) CMORPH, and (**f**) GPCP for the period 2007–2015. (Adapted from Prat et al. 2017)   781

Fig. 40.7   Distribution of annual maximum daily rainfall for events observed simultaneously by USCRN stations and satellites. (**a**) PERSIANN-CDR, (**b**) CMORPH, and (**c**) GPCP for the period 2007–2015. (Adapted from Prat et al. 2017)   782

Fig. 40.8   Value of the 99th percentile daily rainfall retrieved from in-situ data (USCRN), and satellite products (PERSIANN-CDR, CMORPH, GPCP) for the period 2007–2015   783

Fig. 40.9   Accumulated rainfall (mm) derived from PERSIANN-CDR, TRMM TMPA, and GPM IMERG during Hurricane Matthew (25 Sept.-10 Oct., 2016). (Figure courtesy of Andrew Shannon)   784

Fig. 41.1   Maps of the seasonal climatology of PRCPTOT: JF (**a**); MAM (**b**); JJAS (c); OND (**d**). For each satellite product the seasonal PRCPTOT climatology was firstly computed, followed by the average over the three products (ensemble mean). The black line in (**a**) identifies two EA regions (Western East Africa – WEA and Eastern East Africa – EEA) characterized by different rainfall annual cycles, monomodal with a single wet season – WEA, and bimodal with two distinct wet seasons – EEA, respectively. (Adapted from Cattani et al. 2016)   798

Fig. 41.2   MAM rainfall index time series for the three satellite products, ARC2 (long dashed), TARCAT3 (solid), and CHIRPS (short dashed): PRCPTOT (**a**); R1 (**b**); CDD (**c**). Time series were averaged over EEA   800

Fig. 41.3   Same as in Fig. 41.2 but for OND   801

Fig. 41.4   Same as in Fig. 41.2 but for JJAS. Time series were averaged over WEA   802

Fig. 41.5   Percentage of the 1000 randomly generated time series providing significant trend at confidence level $\geq$95% (PVAL%) for PRCPTOT (**a**), CWD (**b**), SDII (**c**), and R20 (**d**)   805

Fig. 41.6   Mean values of the Sen slope estimator for PRCPTOT (**a**), CWD (**b**), SDII (**c**), and R20 (**d**)   806

Fig. 41.7   Same as in Fig. 41.6 but for the standard deviation associated with the distribution of the 1000 Sen slope values computed for each grid cell   807

Fig. 41.8   Maps of PVAL% (**a**), Mean Sen slope estimator (**b**), and Sen slope SD (**c**) relative to the JF R1 index   808

Fig. 41.9   MAM PRCPTOT mean anomalies time series over WEA (**a**),
              and EEA (**b**). Anomalies are computed from the mean MAM
              PRCPTOT time series of the two regions. 808
Fig. 41.10   PVAL% values relative to MAM R1(**a**), CWD (**b**), and SDII
              (**c**) 808
Fig. 41.11   MAM mean values of the Sen slope estimator for R1(**a**), CWD
              (**b**), and SDII (**c**) 809
Fig. 41.12   MAM Sen slope SD values for R1(**a**), CWD (**b**),
              and SDII (**c**) 809
Fig. 41.13   JJAS PVAL% values for R1(**a**), PRCPTOT (**b**),
              and CDD (**c**). 810
Fig. 41.14   JJAS mean values of the Sen slope estimator for R1(**a**),
              PRCPTOT (**b**), and CDD (**c**) 810
Fig. 41.15   JJAS Sen slope SD values for R1(**a**), PRCPTOT (**b**),
              and CDD (**c**) 811
Fig. 41.16   OND PVAL% values for PRCPTOT (**a**), SDII (**b**),
              and R20 (**c**) 811
Fig. 41.17   OND mean values of the Sen slope estimator for PRCPTOT (**a**),
              SDII (**b**), and R20 (**c**) 812
Fig. 41.18   OND Sen slope SD values for PRCPTOT (**a**), SDII (**b**),
              and R20 (**c**) 812

Fig. 42.1   Livorno flash-flood: cumulated precipitation measured in 24 h
              by the Italian raingauge network between 1200 UTC 9 Sept. and
              1200 UTC 10 Sept., 2017 823
Fig. 42.2   Livorno flash-flood: GPM-CO overpass at 0117 UTC 10 Sept.
              GMI TBs at 89 GHz (V-pol) (first panel on the left) and 166
              GHz (V-pol) (second panel), DPC radar SRI product at 0120
              UTC (third panel), and PNPR-GMI rainfall rate estimate (last
              panel on the right) over the area covered by the radar. The black-
              box corresponds to the whole area shown in Fig. 42.1 824
Fig. 42.3   24-h cumulated precipitation from 1200 UTC 9 Sept. to 1200
              UTC 10 Sept. 2017 (on a regular 0.1° × 0.1° grid): raingauges
              and radar (top panels), PMW products (middle panels), and
              MW/IR products (see text for details) 826
Fig. 42.4   Temporal evolution of the mean instantaneous precipitation (top
              panel) and mean cumulated precipitation (bottom panel),
              computed over the Livorno area (black box in Fig. 42.3)
              (see text for details) 827
Fig. 42.5   Naples hailstorm on 5 Sept. 2015 0847 UTC: TBs at 166 GHz
              (V-pol) (left panel) and at 19 GHz (V-pol) (right panel) for the
              GMI orbit 8630. Ku- and Ka-band DPR swaths are evidenced
              in green and magenta, respectively. The position of the
              DPR-Ku scan 4941 (cross-section shown in Fig. 42.7) is shown
              in black 829

Fig. 42.6   Left panel: Map of DPR Ku-band median corrected reflectivity
             factor (Z_c) and LINET strokes (IC and CG) registered during a 4
             min time interval around 0847 UTC with the position of
             DPR-Ku scan 4941 of Fig. 42.7 shown as a black line. Right
             panel: heavy ice precipitation flag available in the JAXA DPR
             V05 products determined with different criteria 830
Fig. 42.7   DPR across-track scan 4941 of Ku measured reflectivity (Z_m)
             (DPR-NS V05). GMI TBs measured at four window channels
             (V-pol) and 183.31 GHz channels, corrected for parallax effects,
             are superimposed. On top the DPR Ku-band ray number along
             the cross-section is indicated. See Figs. 42.5 and 42.6 as
             reference for the position of the section 830
Fig. 42.8   Surface precipitation rate estimated by the GPROF V05
             algorithm for four Numa overpasses by AMSR2 (middle panels)
             and GMI (left and right panels) 833
Fig. 42.9   (a) GPM-CO overpass on 16 Nov. 1350 UTC. Left panel: GMI
             TB at 37 GHz. Middle panel: GMI TB at 89 GHz (Ku-band
             swath is delimited by black lines). Right panel: Cloud Top
             Height (km), with LINET strokes, registered in 1 h around the
             time of the overpass, shown by black crosses and the position of
             Ray 38 (Fig. 42.10) shown in pink. (b) GPM-CO overpass on 18
             Nov. 0359 UTC. Left panel: GMI TB at 37 GHz. Right panel:
             GMI TB at 89 GHz 834
Fig. 42.10   Cross section along Ray 38 of DPR Ku-band Z_c (left panel) and
             DPR Ku particle size distribution D_m (right panel) 835
Fig. 43.1   (a) Aridity index map based on the UNEP (1992) definition.
             (b) Magnification of the northern Sahara with (blue) the
             southernmost line of winter precipitation (i.e., with maximal
             monthly rainfall occurring in winter) and (red) the northernmost
             line of summer precipitation. Aridity index: Climatic Research
             Unit of University of East Anglia (New et al. 2002); Pressure:
             NCEP Reanalysis 1948–2018 [NOAA/OAR/ESRL PSD,
             Boulder, Colorado, USA, https://www.esrl.noaa.gov/psd/, last
             accessed 3 Apr. 2019 (Kalnay et al. 1996)]; Precipitation:
             Tropical Rainfall Measuring Mission [TRMM (TMPA/3B43)
             Rainfall Estimate L3 1 month 0.25° × 0.25° V7, NASA
             Goddard Earth Sciences Data and Information Services Center,
             Accessed 06/2018 https://doi.org/10.5067/TRMM/TMPA/
             MONTH/7] 846
Fig. 43.2   (a) An example of a convective rain-cell derived from a C-Band
             weather radar and a fitted ellipse (Belachsen et al. 2017). (b)
             Spatial and temporal autocorrelation of convective rainfall fields
             in an arid region of the eastern Mediterranean as observed from
             60 m, 1 min resolution X-Band weather radar estimates.
             (Adapted from Marra and Morin 2018) 848

Fig. 43.3   (**a**) Average relation between rain intensity and return period for
             arid, semi-arid and Mediterranean climates as observed from
             rain gauges and a C-Band weather radar in the eastern
             Mediterranean (adapted from Marra and Morin 2015). (**b**)
             Spatial correlation coefficient between quantiles derived from
             C-Band weather radar and satellite-based estimates (high-
             resolution CMORPH) over arid, semi-arid and Mediterranean
             climates in the eastern Mediterranean. (Adapted from Marra et
             al. 2017) 849
Fig. 43.4   (**a**) Mean annual precipitation computed for 19 hydrological
             years (1 Sept. 1998–31 Aug. 2017). Colors are mapped in log10
             scale to better emphasize dryland regions. Solid line labeled
             squares delimit regions used for autocorrelation analysis (see
             Fig. 43.6). Precipitation estimation artifacts (dashed line
             squares) are spotted near the shoreline of the Atacama Desert (**b**)
             and the Nile and the Red Sea (**c**). Precipitation data are based on
             TMPA Precipitation L3 1 day – $0.25° \times 0.25°$ V7, Edited by
             Andrey Savtchenko, Goddard Earth Sciences Data and
             Information Services Center, Accessed 06/2018 https://doi.org/
             10.5067/TRMM/TMPA/DAY/7 851
Fig. 43.5   Box plots of mean annual precipitation (**a**), coefficient of
             variation (CV) of annual precipitation (**b**), and, mean number of
             rainy days (> 0.1 mm day^{-1}) per year, comparing five
             climatological classes: hyper-arid, arid, semi-arid,
             dry-subhumid and non-dryland. Each box shows statistics of all
             pixels of a given climatological classification. Outliers are not
             shown 852
Fig. 43.6   Spatial autocorrelation curves, using an exponential function, of
             daily precipitation over $10° \times 10°$ windows (shown in Fig.
             43.4a with matching labels) for three climatic categories: hyper-
             arid/arid, semi-arid/dry-subhumid, and, non-dryland. Two
             curves for each climatic class 853
Fig. 43.7   Same as Fig. 43.5 but for the shape parameters of: a) the Weibull
             distribution for all the non-zero daily precipitation, and b) the
             GEV distribution for annual daily maxima 853

Fig. 44.1   Hail storm schematic illustrating the combined effect of hail
             location, size distribution and vertical extent which impact the
             satellite measurement. Courtesy of http://agatelady.blogspot.
             com/2013/08/all-about-hail.html (last accessed 26 Oct. 2018) .. 862
Fig. 44.2   Percentage of TMI brightness temperature local minima
             associated with large (at least 2 cm diameter) hail reports in the
             US. (Adapted from Cecil 2009) 864
Fig. 44.3   Extinction coefficient per 1 g m^{-3} concentration of spherical
             hailstones as a function of diameter, at selected GMI
             frequencies. (Adapted from Mroz et al. 2017) 864

Fig. 44.4   Hailstorm frequency of occurrence estimated from AMSR-E at
              36-GHz. (Adapted from Cecil and Blankenship 2012) 865
Fig. 44.5   GPM cases from Mar 2014 – Feb 2017 satisfying the large hail
              thresholds from Ni et al. (2017). Cases with near-surface
              temperature below 10 °C are screened out to avoid snow and ice
              contamination. (Adapted from Ni et al. 2017) 866
Fig. 44.6   Median profiles of maximum Ku-band radar reflectivity from
              TRMM cases that satisfy the 230 K threshold at 37 GHz. Red
              lines are for tropical regions (Maritime Continent, tropical South
              America, and west-central Africa). Blue lines are for subtropical
              regions (southeastern US, South Africa and northern
              Argentina) 867
Fig. 44.7   Fraction of GPM radar profiles with mean mixed-phase Ku-band
              reflectivity >40.42 dBZ, Apr 2014 – Mar 2016. (Adapted from
              Mroz et al. 2017) 868
Fig. 44.8   MetOp-A MHS, 20 February 2010 1220 UTC. Top: signal
              perturbation (K) with respect to clear sky conditions induced by
              scattering hydrometeors on the MHS brightness temperatures in
              the channels at (a) 184, (b) 186 and (c) 190 GHz. Bottom:
              diagrams corresponding to the computed percent variation of the
              signal at the above frequencies. Values lower than 10% are
              generally associated with water vapor absorption while the
              spikes identify the scattering due to precipitating clouds. Values
              close to zero correspond to low water vapor absorption in the
              upper, middle and lower atmosphere at (d) 184, (e) 186 and (f)
              190 GHz, respectively 870
Fig. 44.9   20 February 2010 1220 UTC. The MWCC product (left)
              identifies as shallow convection (CO1) the main core of
              precipitation as retrieved by the 183-WSL (right). Stratiform
              clouds (ST1 and ST2) surround the precipitation with low rain
              intensity to no-rain 871
Fig. 44.10   4 September 2015 1253 UTC. Brightness temperatures during
              the hailstorm over Sardinia and Central Italy. From (a) to (d), the
              MHS-N19 measurements at 89, 157, 186 and 190 GHz,
              respectively. The black circle indicates the hailstorm; the light
              green arrow marks the deep convection that evolved in a strong
              hailstorm during the successive hours. The dashed rectangle in
              (b) groups the longitudes where the main cores of the storm
              formed 873

Fig. 46.6   Simulated and (LFM- and GPM-) retrieved LH along the GPM
             orbit shown in Fig. 46.5 905
Fig. 46.7   Three-dimensional SLH latent heating structure derived from
             the GPM DPR observation case on 16 Apr. 2016, the same case
             as in Fig. 46.5. Northward is to the right and southward is to the
             left. Top and bottom panels show different cross sections of the
             cyclone. Reddish colors show the heating and bluish colors
             show the cooling 905
Fig. 46.8   Geographic locations of the twelve field campaigns that provide
             data to drive and evaluate CRM simulations. These campaigns
             are: ARM-SGP campaigns conducted in summer 1997, spring
             2000 and summer 2002; GATE (1974); KWAJEX (1999);
             TOGA-COARE (1992 and 1993); TWP-ICE (2006); SCSMEX/
             NESA and SESA (1998); AMMA (2006; African Monsoon
             Multidisciplinary Analysis experiment); AMIE/DYNAMO
             (2011; ARM Madden-Julian Oscillation Experiment/Dynamics
             of the Madden-Julian Oscillation). (Adapted from Zeng et al.
             2009) 908
Fig. 46.9   Vertical cross sections of latent heating/cooling rates and air
             temperature (contour lines) along a SW-NE direction over
             eastern CONUS, (a) directly from the NU-WRF simulation at
             1800 UTC 16 March 2014, (b) retrieved using a look-up table
             (LUT) based on the NU-WRF simulation results at this time
             only, and (c) retrieved using a LUT based on the NU-WRF
             simulation results from every 10 min for the whole 24 h,
             simulation period, except for the first 6 h. (d–f) Vertical cross
             sections along a W-E direction over eastern CONUS 912

Fig. 47.1   SPoRT research-to-operations/operations-to-research paradigm.
             (Adapted from Jedlovec 2013) 924
Fig. 47.2   GPM Level 2 swath rain rates were typically estimated by
             forecasters to be the same as radar, model estimated rain, other
             satellite products, and gauge data for the events they evaluated.
             Number of events in which comparisons were performed is
             shown on the y-axis. (Adapted from LeRoy et al. 2017b) 927
Fig. 47.3   IMERG 6 h precipitation accumulation (inches) 1200–1800
             UTC 18 August 2015. IMERG imagery gave forecasters insight
             into the heaviest rain rates outside of radar coverage. IMERG
             was used by forecasters to analyze an event that resulted in a
             mudslide in the Juneau WFO's area of responsibility on 18
             August 2015 (annotations by Aaron Jacobs, NWS WFO AJK) . 929
Fig. 47.4   Daily precipitation (inches) from Early IMERG for precipitation
             ending 1200 UTC 13 July 2015. Gallup, NM, is labeled with a
             white star, and the ABQ radar is labeled with a red star 931

Fig. 47.5   (a) IMERG 1-h precipitation accumulation (inches) 1100 UTC 9
             September 2016, and (b) swath rain rate estimates (in h^{-1}) 1041
             UTC 9 September 2016 offshore of southeast Alaska helped a
             forecaster issue a flood advisory by showing rain rate intensity
             produced by the system prior to landfall 932
Fig. 47.6   (a) HQPrecip precipitation (inches) and (b) IMERG-Early
             precipitation (inches) for 0800 UTC on 07 February 2019 over
             Alaska 934

Fig. 48.1   Measuring the environmental state (including precipitation)
             from space. The signal measured at the sensor level is a
             composite signal that includes the effect of absorption and
             scattering from the surface, the dry atmosphere, cloud, and
             precipitation. The degree of sensitivity of the signal to these
             parameters depends heavily on the operating frequencies and
             other sensor characteristics such as viewing angle, polarization,
             etc. See text for more details 943
Fig. 48.2   Effect of rain and ice/graupel amounts and particle effective
             radius on the microwave spectrum from 3 to 330 GHz using a
             US Tropical Atmosphere. The top panels show the actual
             simulations of measurements for different scaling factors of
             hydrometeors amounts and two specific particle sizes (500 and
             1500 μm), while the bottom panels show the departures of those
             simulations from clear-sky (hydrometeor-free) simulations. In
             the top panels, the black line represents the microwave spectrum
             in the absence of cloud or precipitation. HyMS represents a
             hypothetical hyperspectral microwave sensor with
             measurements ranging from 3 to 330 GHz with a 0.1 GHz
             spectral resolution. See text for more details 947
Fig. 48.3   Sensitivity of the precipitation-impacted measurement at 19
             GHz, defined as the difference between clear-sky simulation and
             precipitation-based simulation, as a function of hydrometeors
             and multiple surrounding environment parameters. The
             hydrometeors and surrounding environment parameters include
             (clockwise, from top left) surface emissivity, total precipitable
             water (TPW), effective particle radius of liquid rain, effective
             particle radius of frozen graupel, amount of graupel (GWP), and
             cloud fraction. The x-axes on the plots represent parameters of
             hydrometeors and their surrounding environment, and the
             y-axes represent scaling factors for rain and graupel
             hydrometeor amounts (with 0 being no hydrometeors used in
             simulation and 2 being double the original amounts of liquid and
             frozen precipitation used). Contours represent signal amounts .. 950
Fig. 48.4   Same as in Fig. 48.3, but for a 37 GHz simulation 951

Fig. 48.5   Overview of the different driving factors impacting how the
             precipitation observations are handled in a physical retrieval
             and/or DA context. These driving factors include observations
             characteristics, modeling capabilities, and signal representation
             issues. These driving factors determine what approach and
             techniques are adopted for the assimilation and physical
             inversion of precipitation observations (what methodology to
             use, what to include in the state vector, etc.)   952
Fig. 48.6   Evolution of simulated brightness temperature as a function of
             hydrometeor amount (RWP – top left, and ice-graupel water
             amount – top right) highlighting the non-linearity (but
             incrementally locally-linear) of the RT. The different colors/
             symbols represent different channels with precipitation-sensitive
             frequencies in the AMSU/MHS sensor pair. Middle plots show
             the spatial evolution (left) and temporal evolution (right) of real
             NOAA-18 MHS data at different frequencies from 8 September
             2011. A cross section at 28°N is highlighted (in red) in the left
             bottom map showing the field of 190 GHz brightness
             temperature. The temporal evolution is computed by averaging
             the area highlighted by a red circle on the map, using several
             passes of NOAA-18 over multiple days. The non-linearities due
             to spatial and temporal evolutions of space observations are
             highly non-linear and their linearization (incrementally) would
             be challenging due to abrupt variations with local first degree
             derivative discontinuities   961
Fig. 48.7   Simulated brightness temperature differences from 1–330 GHz,
             between clear and cloudy/precipitating cases with varying
             assumed cloud and rain effective radii   963
Fig. 48.8   FSOI in ECMWF operational forecasts, for different categories
             of observations, for February 2018. Microwave WV, cloud, and
             precipitation observations (the "Microwave WV" category)
             account for 21% of short-range forecast impact, with "moist"
             assimilation having slightly more impact than "dry" assimilation
             ("Microwave T")   968
Fig. 48.9   Impact of all-sky AMSR2 radiances in the ECMWF operational
             system during July and August 2016 (see Geer et al. 2017), in
             terms of FSOI (forecast impact) and number of observations, all
             given as a percentage of the total (FSOI or number of
             observations) for AMSR2. Scenes have been characterized
             according to whether they are clear-sky, cloudy or precipitating.
             Although the number of precipitating scenes is relatively small,
             they have large impact on the 24 h forecast quality   968

Fig. 48.10   Comparisons of tropical cyclone prediction (central pressure) for
             Hurricane Jimena in 2015. (a) Comparison of TEST2 (all-sky
             assimilation plus outer-loop iteration) and CNTL. (b)
             Comparison of TEST1 (outer-loop iteration) and CNTL. Red
             lines denote the predicted central pressure of TEST and blue
             lines denote those of CNTL. Black solid line is NOAA best track
             data. The period is from 25 August to 13 September 2015   971
Fig. 48.11   Comparisons of analyzed TPW fields for Hurricane Jimena in
             2015. The analysis is for 26 August 2015 1200 UTC. (a)
             analyzed TPW of TEST2, (b) TEST1, and (c) CNTL. The units
             are mm. The contour lines denote analyzed sea surface pressure
             in hPa. Small black circles indicate locations of assimilated
             microwave radiance data. (d) TPW difference of TEST2 and
             CNTL, (e) TPW difference TEST1 and CNTL. The units are
             mm. (f) surface wind vector difference of TEST2 and CNTL, (g)
             surface wind vector difference TEST1 and CNTL. The units are
             m s^{-1}   971
Fig. 48.12   Comparisons of 6 h accumulated precipitation forecasting from
             0600 UTC 29 August 2015 initial time (tropical cyclone
             developing stage). (a) TEST2, (b) TEST1, (c) CNTL, and (d)
             GSMaP. The units are mm. Contours denote sea surface
             pressure in hPa   972
Fig. 48.13   Example Analysis Cycle at 2300 UTC 7 August 2006. (a)
             Background-G5NR shows large displacements (dipoles) in
             TPW field, (b) GSI analysis-G5NR reduces magnitude of
             dipoles slightly where SNPP ATMS data exists (red trapezoid),
             (c) Environmental Data Fusion (EDF) analysis through
             MIIDAPS-based background adjustment removes most of
             dipole feature and reduces TPW differences where SNPP ATMS
             data exists   975
Fig. 48.14   Highlighting the impact of using a 1D-Var system as
             preprocessing to DA. One of the main features is the generation
             (when absent) and displacement correction of hydrometeors in
             the analysis (shown is the case of Hurricane Matthew over the
             US Floridian coast, from October 2016). Top panels show the
             evolution of the convergence metric (from iteration#0:
             background, to iteration 4 to the last iteration 7 in the 1D-Var
             preprocessor). The second row shows the evolution of the cloud
             field within the iterative process, again starting with the
             background to the final cloud field retrieved. The third and
             fourth rows show the evolution of temperature field at 500 hPa
             and the vertical cross-section of temperature respectively.
             Instead of the temperature itself, the departure from the reference
             (ECMWF analyses) is shown. Both spatial and vertical
             displacements are corrected. See text for more details   976

Fig. 49.1   Time-series of the global-mean root mean square differences
             (RMSD) for temperature (K) relative to the ERA-Interim. (**a**)
             LL-LETKF, (**b**) ICO-LETKF, and (**c**) the difference between
             LL-LETKF and ICOLETKF. The horizontal and vertical axes
             represent the date in 2011 and pressure level (hPa), respectively.
             Negative values of panel (**c**) corresponds to ICO-LETKF's
             advantage. (Adapted from Terasaki et al. 2015) 985
Fig. 49.2   The probability density function (PDF; **a** and **b**) and cumulative
             density function (CDF; **c** and **d**) of the original precipitation and
             the transformed precipitation based on the 10-year (2001–2010)
             model (red) and observation (green) climatologies at a grid point
             in the extratropics (39.08°N, 76.98°W; near Maryland). All
             plots correspond to the 11–20 Jan period. The procedure of the
             Gaussian transformation is indicated by the arrows, i.e., (**a**) to (**c**)
             to (**d**) to (**b**). The open circles correspond to zero precipitation
             probability and the solid circles correspond to the half value
             (median) of zero precipitation probability. (Adapted from Lien
             et al. 2016a) 986
Fig. 49.3   Innovation statistics 0000–1800 UTC 1 November 2014 for (**a**)
             original precipitation data (mm/6 h) and (**b**) precipitation after
             applying the GT (standard deviation). Bold lines show the
             normal distributions computed by the mean and standard
             deviation of the innovation samples. (Adapted from Kotsuki
             et al. 2017a) 987
Fig. 49.4   Time series of RMSDs for zonal wind (m s^{-1}) at 500 hPa
             relative to the ERA Interim reanalysis. (black lines) control
             experiment without assimilating GSMaP and (red lines) test
             experiment with assimilation of GSMaP with the GT. Bold and
             thin lines denote the analyses and the 5-day forecasts from the
             eight initial times, respectively. The abscissa shows month/date
             in 2014. (Adapted from Kotsuki et al. 2017a) 987
Fig. 49.5   Global precipitation patterns (mm/6 h) at the first data
             assimilation step (0000 UTC 1 November 2014). (**a**) original
             NICAM forecast, (**b**) original GSMaP, (**c**) analysis without
             applying GT, (**d**) GT'ed. NICAM forecast, (**e**) GT'ed. GSMaP
             and (**f**) analysis with applying the GT. Panels (**g–i**) demonstrate
             impacts of the inverse GT. Panels (**g**) and (**i**) are obtained by
             applying observation-CDF-based inverse GT to panels (**d**) and
             (**f**). Panel (**h**) is obtained by applying model-CDF-based inverse
             GT to panel (**h**). Gray and black colors show missing value and
             negative precipitation value, respectively. (Modified from
             Kotsuki et al. 2017a) 988

Fig. 49.6   Global precipitation forecasts (mm/6 h) at 0000 UTC 16 June
            2014. (**a**) control experiment with the default model setting, (**b**)
            test experiment with the model parameter estimation, and (right)
            GSMaP_Gauge observation data, respectively. Overproduced
            precipitation over ocean in the control experiment is
            successfully mitigated by the model parameter estimation.
            (Adapted and modified from Kotsuki et al. 2018)   988

Fig. 50.1   CHRS RainSphere interface: (1) Navigation Bar, (2) Map
            Layers, (3) Rain Information, (4) Rain Layers, (5) Rain
            Comparison, (6) Rain Statistics, (7) Reference Map, (8) Search
            Location, (9) Map Canvas (URL: http://rainsphere.eng.uci.edu/,
            last accessed 24 Nov. 2018). (Nguyen et al. 2017a, 2018)   996

Fig. 50.2   Downloadable Rain Query Report including: Rain Linear Trend,
            Rain Average, Mann-Kendall Test. [http://rainsphere.eng.uci.
            edu/, last accessed 24 Nov. 2018]   997

Fig. 50.3   Rainfall (mm day^{-1}) over land during Hurricane Katrina on 29
            August 2005 from: (**a**) PERSIANN-CDR, (**b**) Stage IV Radar (Lin
            and Mitchell 2005), and (**c**) TMPA v7 (Huffman et al. 2007). Black
            and gray pixels show radar blockages and zero precipitation,
            respectively. (Adapted from Ashouri et al. 2015)   998

Fig. 50.4   Annual average count of days where rainfall $\geq$10 mm (left
            column) and rainfall $\geq$20 mm (right column) for CPC (top), and
            PERSIANN-CDR (bottom) for 1983–2011   999

Fig. 50.5   Scatter plots of the annual average count of days where rainfall
            $\geq$10 mm (left column) and rainfall $\geq$20 mm (right) for
            PERSIANN-CDR against CPC. Correlation coefficient, RMSE,
            and Bias are shown on the plots   1000

Fig. 50.6   The 99th and 95th percentile indices of extreme daily
            precipitation from the EA data set (first column), and
            PERSIANN-CDR (second column). The spatial correlation
            distribution and the scatterplots of the indices from the EA and
            PERSIANN-CDR data sets are shown in the third and fourth
            columns, respectively. The stippled areas in the third column
            show the significant correlation coefficient at the 95% level
            (Miao et al. 2015)   1000

Fig. 50.7   (**a**) The 26 continent-climate zone groups, (**b**) Correlation (left)
            and relative RMSE (right) for precipitation indices in Warm
            Temperate continent-climate zone (CCZ) groups (statistical
            insignificance at 0.05 in hatched boxes). (Adapted from Nguyen
            et al. 2017b)   1002

Fig. 50.8   Precipitation comparison plots SLOA4 basins between
            PERSIANN, PERSIANN-CDR, and TMPA against Stage IV
            gauge-adjusted radar data for 2003–2010. (Adapted from
            Ashouri et al. 2016a, b)   1003

Fig. 50.9   Simulated and observed streamflow hydrographs and respective
             scatterplots at the outlet of SAVOY basin using (from top to
             bottom) Stage IV, TMPA, PERSIANN, and PERSIANN-CDR
             precipitation products. The solid black line shows the USGS
             observations. (Adapted from Ashouri et al. 2016a, b) 1004
Fig. 50.10   Boxplots of satellite-based IDF relative error for durations of
             (1, 2 and 3) days and return periods of 25, 50 and 100 years.
             (Adapted from Ombadi et al. 2018) 1006

Fig. 51.1   Bottom-up vs. top-down perspective for rainfall retrieval from
             remote sensing assuming no error in the satellite measurements
             and in the retrieval algorithms. Due to the satellite overpass
             during low rainfall intensities, the "top down" method may fail
             in estimating the accumulated rainfall whereas the "bottom up"
             approach accurately reproduces the observations even with a
             lower number of overpasses 1015
Fig. 51.2   Comparison in terms of Pearson's correlation, R, and root mean
             square error, RMSE, between SM2RAIN-ASCAT and TMPA-
             RT satellite-based rainfall products as compared with GPCC
             gauge-based dataset used as benchmark. The analysis is carried
             out at 1-day and 1° temporal spatial resolution. The maps clearly
             show the differences in the accuracy of the two products, with
             SM2RAIN-ASCAT performing well in the eastern US, Brazil,
             the Sahel, south-eastern Asia and Australia (green colors) 1018
Fig. 51.3   Boxplot of KGE for the six investigated rainfall products and for
             the 600 basins. For each box, the red line represents the median
             values and the blue box represents the 25th and 75th percentile,
             the black dotted whiskers extend to the most extreme data points
             and cross symbols represent outliers. The numbers in the top
             boxes indicate the mean value for each rainfall product. The
             integration of SM2RAIN-ASCAT (SM2R) with TMPA-RT
             (TMPA) and CMORPH (CMOR) provides the best performance
             close to the ones obtained with high-quality raingauge
             observations (EOBS) 1021
Fig. 51.4   Spatial distribution of KGE performance for TMPA-RT and
             SM2RAIN-ASCAT rainfall products over 600 basins in Europe.
             Red colours mean high KGE values and, hence, better
             performances. Overall, SM2RAIN-ASCAT is performing better
             over 77% of basins with TMPA-RT showing better results over
             some basins in Italy (mainly close to the western Alps), in
             central France and in South-Eastern Europe (Balkans) 1021

Fig. 51.5   Simulation of 3-year discharge at four basins across Europe by
             using SM2RAIN-ASCAT (red line), TMPA-RT (blue line) and
             SM2RAIN-ASCAT+TMPA-RT (black line) as input rainfall.
             The comparison with in situ discharge (green area) clearly
             underlined the benefit of integrating top-down and bottom-up
             approaches for flood simulation with a better reproduction of
             both high and low flows 1022

Fig. 52.1   Approaches to drought risk management using satellite-based
             rainfall estimates (*EWS* Early Warning Systems, *WII* Weather
             Index-based Insurance). Complexity refers to both the
             complexity of the modelling approach and its operational
             application over large areas at high spatial and temporal
             timescales 1035

Fig. 52.2   Example of rainfall anomalies from the World Food Programme
             Vulnerability Assessment and Mapping (VAM) unit's Seasonal
             Explorer for the Matam region in Senegal 1037

Fig. 52.3   Example of rainfall anomalies from the World Food Programme
             Vulnerability Assessment and Mapping (VAM) unit's Seasonal
             Explorer for part of the Limpopo River Basin. Note: std. is
             standard deviation 1041

Fig. 52.4   Example of the (**a**) Water Requirement Satisfaction Index
             (WRSI) for a selected season across Tanzania and (**b**) WRSI
             time series for the location depicted in (**a**) (after research in
             Tarnavsky et al. 2018). WRSI was calculated with CHIRPS as
             the rainfall input dataset 1043

Fig. 53.1   Precipitation anomalies stated in terms of a concentration factor
             analysis of Hand and Shepherd (2009). The concentration factor
             relates spatial variations in rainfall amount and intensity to
             frequency of occurrence of a certain wind direction. The
             climatological "downwind" region of Oklahoma City is East-
             Northeast 1058

Fig. 53.2   MODIS aerosol optical depth for July 2005. The two boxes
             represent urban land and oceanic study regions 1059

Fig. 53.3   Ratio of precipitation types over the period 1996–2016 (Nov–
             Apr) in the Washington, Baltimore, Philadelphia, and New York
             Urban Climate Archipelago. (Adapted from Johnson and
             Shepherd 2018) 1061

Fig. 53.4   Probability (%) of liquid and frozen precipitation 2015
             (January–February) in the DC to NYC Urban Climate
             Archipelago (UCA) via the GPM IMERG product 1062

Fig. 53.5   Pre-urban and Post Urban rainfall fields (mm) in the Houston
             area as represented in a GIS 1063

Fig. 53.6   An analysis of runoff values (acre-ft) from a global hydrological
              model (GHM) and stream gauges. The various groups represent
              areas that were overestimated or underestimated. Most error was
              associated with flood control mechanisms upstream. The use of
              TRMM data products as initialization significantly mitigates
              these errors 1063
Fig. 53.7   Estimated rainfall rate (mm) for the September 2013 Colorado
              flood event over the period 8–17 September using the TRMM
              multi-satellite precipitation analysis 1064
Fig. 53.8   Mean precipitation per person (ton/year) in $0.1° \times 0.1°$ boxes
              over a 3-year period 1066
Fig. 53.9   Seasonal estimates of P-E minus consumption over a 4-year
              period 1066

Fig. 54.1   Mean annual precipitation as depicted in five observational
              databases (CRU, GPCC, CPC, CMAP and GPCP) compared
              with the ensemble average of several Regional Climate Models
              (RCMs: first and second rows), and a few individual RCMs
              (third and fourth rows). The ensemble average compares best
              with the observations than most of the models but differences
              are apparent between the observations themselves. [*CRU*
              Climate Research Unit, *GPCC* Global Precipitation Climatology
              Centre, *CMAP* CPC Merged Analysis of Precipitation, *GPCP*
              Global Precipitation Climatology Project.] (Adapted from
              Tapiador 2010) 1076
Fig. 54.2   Seasonal precipitation climatologies (1961–2000, columns)
              derived from ten RCMs nested on reanalysis (ENSEMBLES
              project data, 25 km grid) compared with ten RCMs forced with
              GCMs (PRUDENCE project, 50 km grid), and three
              observational databases at 50 km resolution: CRU rain gauge
              data, GPCC rain gauge data, and CPC (Climate Prediction
              Center) rain gauge and satellite-merged data. Units are mm
              year^{-1}. (Adapted from Tapiador et al. 2011) 1077

Fig. 55.1   The Global Landslide Catalog reveals a band of terrain with a
              high number of landslides (black dots indicate landslides from
              2007–2016). The Himalayan Mountains and Foothills
              experience intense monsoon precipitation and occasional
              seismicity, while comprising some of the Earth's steepest
              terrain. We focused our analysis on this region by identifying the
              sites with the greatest elevation difference from neighboring
              pixels at the half-degree scale (black box). Unless otherwise
              noted, results are presented for this region 1092

Fig. 55.2   Mean annual values of extreme precipitation indices: (**a**) CWD,
               (**b**) R10mm, (**c**) R20mm, (**d**) R95pTOT, (**e**) R99pTOT, (**f**)
               Rx1day, (**g**) Rx5day for GFDL FLOR (left) and TMPA (right).
               The data products show similar values and geographic
               distributions for all indices except CWD 1095
Fig. 55.3   Trends in extreme precipitation indices from 2000–2016: (**a**)
               CWD, (**b**) R10mm, (**c**) R20mm, (**d**) R95pTOT, (**e**) R99pTOT,
               (**f**) Rx1day, (**g**) Rx5day for FLOR (left) and TMPA (right) 1096
Fig. 55.4   Mean annual Rx1day (x-axis) from FLOR is not highly
               correlated with landslide activity from the GLC (y-axis).
               However, the first 3 years of the GLC (2007–2009) may have
               been underreported. If so, the relationship between extreme
               precipitation and landslide frequency might be strongly
               positive 1097
Fig. 55.5   The monthly pattern of landslide activity mirrors the extreme
               precipitation indices Rx1day (light colors) and Rx5day (dark
               colors). Landslide activity is represented by the total number of
               events recorded within the study area for each month from 2007
               to 2016 in the GLC (orange). Daily precipitation was obtained
               from FLOR (green) and TMPA (blue) 1098
Fig. 55.6   The monthly means for Rx1day and Rx5day derived from
               FLOR are highly correlated. Precipitation is typically most
               intense during the monsoon, but February 2013 and March 2007
               are exceptions 1099
Fig. 55.7   The study area was divided into roughly equal halves at the 80th
               meridian, which left 516 (560) landslides in the western
               (eastern) section. The monthly distribution of landslides is
               similar, but the western portion of the study area experiences a
               secondary peak of landslide activity in March. Both FLOR and
               TMPA show a lesser peak in extreme precipitation during
               February, especially west of 80° 1099
Fig. 55.8   Distribution of the GLC by month and year. June, July, August,
               and September dominate the record, but some years show little
               landslide activity in one or more of these months 1100

Fig. 55.9   FLOR monthly mean values of Rx1day are weakly correlated to the monthly number of landslides reported in the GLC. Some months that combine extreme precipitation with relatively few landslides (March 2007, July 2007, and February 2013) might be explained by underreporting during the first year of the GLC (March 2007 and July 2007) or the dominance of frozen precipitation (March 2007 and February 2013), which has less of an influence on landslide hazard than rainfall. It is harder to explain the outliers for which more than 60 landslides were reported but mean values of Rx1day were less than 40 mm (September 2010, July 2011, and July 2013). It is possible that pre-conditioning of the soil in previous months that had above average rainfall may have contributed to the clustering of landslides during the following months. The nine months with more than 30 reported landslides are all associated with the summer monsoon, as well as above-average daily precipitation intensity 1101

Fig. 55.10   Map of Rx1day monthly estimates (mm) from FLOR and reported landslides across the study region for (**a**) September 2010 (63 landslides), (**b**) July 2011 (62 landslides), (**c**) July 2013 (65 landslides), (**d**) March 2007 (5 landslides), (**e**) July 2007 (11 landslides), and (**f**) February 2013 (0 landslides) 1102

Fig. 55.11   Landslides appear to be slightly more common in grid cells with higher mean values of Rx1day (TMPA shown) and other ETCCDI indices, but the relationship does not appear to be very strong 1104

Fig. 55.12   The mean value of R10mm (TMPA) (top) exhibits a smooth spatial distribution due to its emphasis on less extreme precipitation events. In contrast, R99pTOT (TMPA) (bottom) exhibits a noisy spatial pattern, which suggests that the record is too short to contain the most extreme events at every location . 1106

Fig. 56.1   Average (2001–2010) number of stations per 100 km × 100 km grid box used by GPCC gridded rainfall product. (Data source: GPCC) 1118

Fig. 56.2   Time series of average number of stations used in the GPCC full-data product over Africa (15°W to 45°E, and 30°S to 30°N). (Data source: GPCC, https://www.dwd.de/EN/ourservices/gpcc/gpcc.html, last accessed 28 Nov. 2018) 1119

Fig. 56.3   Average number of weather stations (mostly raingauges) reporting each year for Uganda 1119

Fig. 56.4   Average number of weather stations reporting each year over Rwanda 1120

Fig. 56.5   Station observations (top left) of rainfall from the operational
            network in Ethiopia are combined with the TAMSAT satellite
            rainfall estimates (top right) to produce a spatially complete and
            more accurate merged product (bottom) 1122
Fig. 56.6   Climate Analysis and Application map room for Ethiopia,
            consisting of Climate and Application maprooms. (http://www.
            ethiometmaprooms.gov.et:8082/maproom/) 1123
Fig. 56.7   An example of PICSA graph showing historical onset dates
            interactively generated using the online ENACTS Maproom ... 1124

Fig. 57.1   Fire danger rating sign for public display at the Tinanggea
            Manggala Agni Local Fire Center, South Konawe Regency,
            South East Sulawesi Province, Indonesia. Fire danger is set daily
            using the Fine Fuel Moisture Code of the Indonesian Fire
            Danger Rating System calculated from on-site weather data
            (Credit: Israr Albar, Indonesian Ministry of Environment and
            Forestry) 1132
Fig. 57.2   Structure of the Fire Weather Index System. 1200 pm local time
            surface temperature, relative humidity, wind speed and 24-hour
            precipitation are used to calculate three moisture codes and three
            fire behavior indices. (Adapted from de Groot et al. 2007) 1134
Fig. 57.3   Fire Weather Index calculated from rain gauges (left) and
            IMERG-F (right) for 15 (top) and 18 July 2017 (bottom). Grey
            squares indicate grid cells with reporting rain gauges 1137
Fig. 57.4   May–September 2017 time series of daily MODIS fire activity
            and FWI from rain gauges and IMERG-F for central BC (left)
            and south-central BC (right), corresponding to the regions
            identified by the grey boxes in Fig. 57.3 1138
Fig. 57.5   Weekly Fire Weather Index (FWI) and MODIS active fires over
            areas in Chile affected by catastrophic wildfires during January,
            2017. FWI categories are taken from those of the Global
            Wildfire Information System (GWIS, http://gwis.jrc.ec.europa.
            eu/about-gwis/technical-background/fire-danger-forecast/, last
            accessed 28 Nov. 2018) with the addition of a 'Catastrophic'
            danger class for FWI > 80 1140
Fig. 57.6   Weekly Fire Weather Index (FWI) and MODIS active fire
            counts from 1 August 12,016 to 31 May 2017 over the Maule
            and region of Chile 1141
Fig. 57.7   Fire Weather Index (FWI) on 23 July 2018 over southern Greece
            computed from rain gauges (left) and IMERG precipitation
            (right) 1143
Fig. 57.8   Daily FWI and MODIS active fire totals over the Athens region
            shown in Fig. 57.7 calculated using IMERG-L and rain gauge
            precipitation estimates 1143

Fig. 57.9   Elevation map of western Indonesia, showing the three
                 low-lying regions in Jambi province (1), South Sumatra (2), and
                 Central Kalimantan (3), with the highest fire activity (Field et al.
                 2016) and CO emissions (Huijnen et al. 2016) during the 2015
                 fire season 1145
Fig. 57.10   Daily Fine Fuel Moisture Code (FFMC) computed using
                 MERRA2 (left) and IMERG-F (right) precipitation for 1 May
                 2015 to 31 March 2018 over the regions in Fig. 57.9: (1) Jambi,
                 (2) South Sumatra, and (3) Central Kalimantan. The vertical
                 grey line shows the estimated change point, and the dashed blue
                 line the fitted change-point model 1147

Fig. 58.1   Cumulative Distribution Function (CDF) of rain rates
                 considered during the study (between 50°S and 50°N) obtained
                 with three different algorithms: UMORA (solid line), GPROF
                 (dashed line) and IMERG (dash-dotted line). Only RR higher
                 than 1 mm h^{-1} are considered because RR distribution between
                 UMORA, GPROF and IMERG between 0 and 1 mm h^{-1}
                 strongly differ due to the difficulty of identifying very low rain
                 rates and expected corresponding freshening are within the error
                 of satellite salinities 1161
Fig. 58.2   Percentage of measurements retained during 2016 after filtering
                 based on monthly values of $\overline{\Delta S}$ and σ_S. Black lines delimit study
                 areas 1162
Fig. 58.3   Study case, 20 January 2016, (a) SMOS salinity anomaly (b)
                 SMAP salinity anomaly (black line are RR isolines in mm h^{-1}
                 from IMERG collocated at less than 15 min with SMOS) 1163
Fig. 58.4   (a) Idealized Rain event (b) Auto-correlation function of RR for
                 idealized rain event (c) Idealized salinity freshening due to the
                 idealized rain event without influence of rain history (line) and
                 with hypothetical influence of rain history (dashed-line). (d)
                 Cross correlation function between RR and salinity freshening
                 without (line) and with (dashed line) influence of time history.
                 In b) and d) negative time lags concern rain before ΔS 1164

Fig. 58.5   Relationship between ΔS and RR (**a**) for SMOS and RR_{UMORA}
             (**b**) for SMAP and RR_{UMORA} (**c**) for SMOS and RR_{GPROF} (**d**) for
             SMAP and RR_{GPROF} (**e**) for SMOS and RR_{IMERG} (**f**) for SMAP
             and RR_{IMERG} after filtering (black points are average per class of
             RR and error bar standard deviation per class of RR. Grey lines
             are linear regression fit. The color scale corresponds to the log10
             of the number of occurrences). Pearson correlation coefficient
             are computed for RR higher than 1 mm h^{-1} because RR
             distribution between UMORA, GPROF and IMERG between
             0 and 1 mm h^{-1} strongly differ due to the difficulty of
             identifying very low rain rates. The magnitude of the error at the
             origin derived from linear regression is 10E-4 for all cases 1166
Fig. 58.6   For each study area (**a, b, c, d, e**): (triangle) Temporal cross-
             correlation between ΔS and RR (individual triangles are linked
             with a dashed line) (dots) Estimated cross-correlation between
             ΔS and RR inferred from RR auto-correlation and correlation
             between ΔS and instantaneous RR (individual dots are linked
             with a line). Grey colored areas show the 2σ confidence interval
             for the ΔS versus RR cross-correlation. Negative time lags
             concern rain before considered freshening. (**f**) RR auto-
             correlations for study areas 1168
Fig. 58.7   For NTPa study area and SMAP: (**a**) all WS (**b**) WS under
             5.5 m s^{-1} (**c**) WS above 8 m s^{-1}. For the three plots: (triangle)
             Temporal cross-correlation between ΔS and RR (individual
             triangles are linked with a dashed line) (dots) Estimated cross-
             correlation between ΔS and RR inferred from RR auto-
             correlation and correlation between ΔS and instantaneous RR
             (individual dots are linked with a line). Grey colored areas show
             the 2σ confidence interval for the ΔS versus RR cross-
             correlation. Negative time lags concern rain before considered
             freshening 1169
Fig. 58.8   Relationship between ΔS and WS$_{SMAP}$ per class of RR: (**a**) with
             ΔS_{SMOS} and RR_{UMORA} (**b**) with ΔS_{SMOS} and RR_{GPROF} (**c**) with
             ΔS_{SMOS} and RR_{IMERG} (**d**) with ΔS_{SMAP} and RR_{UMORA} (**e**) with
             ΔS_{SMAP} and RR_{GPROF} (**f**) with ΔS_{SMAP} and RR_{IMERG}. Points
             correspond to ΔS and RR average for a given class of RR. Points
             are colored with a color corresponding to RR class 1170

List of Tables

Table 25.1   Comparison of typical ground validation programs and the IPWG validation effort 456

Table 26.1   NASA GV primary US data products routinely produced from GPM launch to present 475

Table 26.2   Routinely provided or extended international datasets shared with and processed by NASA GV, including collaborative instrument deployments 477

Table 26.3   NASA physical-validation field deployment efforts 489

Table 27.1   Averaged accumulated rain amount for AMeDAS, Ku product (250 km swath data) and dual-frequency DPR product (125 km swath data) for 2 years from April 2014 to March 2016 510

Table 27.2   Averaged accumulated rain amount for MRMS/NMQ, Ku product and dual-frequency DPR product from June 2014 to May 2015 512

Table 28.1   Contingency table commonly used for characterizing detection errors of SPPs 520

Table 28.2   Summary of the main validation studies of SPPs 526

Table 33.1   Overview of the 22 (quasi-)global (sub-)daily gridded P datasets evaluated in this study 627

Table 33.2   Median values of the performance metrics for the uncorrected P datasets based on daily P observations from 76,086 gauges around the globe 633

Table 33.3   Median calibration NSE scores for the gauge-corrected P datasets obtained using HBV 640

Table 35.1   Mean oceanic precipitation estimates (mm day^{-1}) between 35°N and 35°S from various products 682

Table 35.2   Global oceanic precipitation estimates (mm day^{-1}) 683

Table 40.1   Physical principles of satellite precipitation retrieval methods
              along with advantages and known caveats along with a list of
              corresponding products 763
Table 40.2   Long term evaluation of gridded (level III) satellite precipitation
              products (SPPs) in relationship with extreme precipitation 769
Table 40.3   Examples of recent studies using gridded (level III) satellite
              precipitation products (SPPs) for applications in relationship
              with extreme precipitation 777

Table 41.1   List of the rainfall indices exploited in the analysis 795

Table 42.1   Analysis of Precipitation Features (PFs) found in 49 months of
              global observations (03/2014–03/2018, http://atmos.tamucc.
              edu/trmm/data/gpm, last accessed 27 Oct. 2018) based on
              minimum TB and PCT values 831

Table 43.1   Spatial statistical properties of daily precipitation 851

Table 46.1   Evolution of the SLH algorithm with a brief summary of
              references 901
Table 46.2   LUTs constructed from 8 extratropical cyclone cases simulated
              using the JMA LFM 904
Table 46.3   Evolution of the Goddard CSH algorithm with key
              improvements and references 909
Table 46.4   Key parameters used for the new CSH cold season LH look-up
              table. The look-up table is built from NU-WRF simulated LH
              profiles associated with six synoptic storm events 911

Table 50.1   Evaluation results of the two regional frequency approaches:
              Index Flood Method (IFM) and Regional Rainfall Frequency
              Analysis using Satellite Precipitation (RRFA-S) 1007

Table 52.1   Advantages and disadvantages of gauge-based rainfall
              observations 1032
Table 52.2   Advantages and disadvantages of satellite-based rainfall
              estimates 1033
Table 52.3   Selected evaluation and inter-comparison studies for satellite-
              based rainfall estimates (SREs) 1034

Table 54.1   Basic, extended and full set of requirements of data and climate
              models involved in validation 1078
Table 54.2   A checklist of known issues that must be considered in the
              interdisciplinary field of validation of precipitation outputs from
              climate models 1080

Table 55.1   Data sources 1090
Table 55.2   Extreme precipitation indices selected for relevance to
              landslides 1093

Table 57.1   Different versions of the Global Fire Weather Database 1136

Table 57.2   Coefficient of determination (r^2) for fitted linear change-point
             regression models (Field et al. 2016) between Fine Fuel
             Moisture Code (FFMC) and MODIS active fire totals over three
             regions in Indonesia. The FFMC was computed using four
             different daily precipitation estimates: MERRA2, MERRA2
             with rain gauge correction (MERRA2-C), IMERG Late with
             climatological rain gauge correction (IMERG-L) and IMERG
             Final with enhanced rain gauge correction (IMERG-F).
             Estimates were computed for FFMC averaged, and active fires
             totaled, over 1 month, 7 days, 2 days, and 1 day. For each
             region, precipitation estimate and averaging period, the highest
             r^2 is shown in bold 1146

Table 58.1   Dataset used in this study 1159
Table 58.2   Statistics at global scale of ΔS versus RR_{UMORA} relationship for
             different filtering (Slope is obtained with a linear regression (Y

             $= \Delta S$, $X = RR_{UMORA}$) and Root Mean Squared Error $RMSD =$

             $\sqrt{(\Delta S - \Delta S_e)^2}) = \sqrt{(\Delta S - 0.16\,RR_{UMORA})^2}$ with ΔS_e
             estimated considering a relationship between ΔS
             and RR_{UMORA}) 1165
Table 58.3 $\mathbf{a}$ and $\mathbf{b}$ coefficient for Eq. (58.2) considering RR_{UMORA},
             RR_{GPROF} and RR_{IMERG} 1170

Contributors

Christophe Accadia European Organization for the Exploitation of Meteorological Satellites (EUMETSAT), Darmstadt, Germany

Robert F. Adler Earth System Science Interdisciplinary Center (ESSIC), University of Maryland, College Park, MD, USA

Kazumasa Aonashi Meteorological Research Institute (MRI), Japan Meteorological Agency (JMA), Tsukuba, Japan

Yoriko Arai Remote Sensing Technology Center of Japan (RESTEC), Tokyo, Japan

Moshe Armon The Fredy & Nadine Herrmann Institute of Earth Sciences, Hebrew University of Jerusalem, Jerusalem, Israel

Kristi Arsenault SAIC, Inc., McLean, VA, and National Aeronautics and Space Administration (NASA), Goddard Space Flight Center (GSFC), Greenbelt, MD, USA

Malarvizhi Arulraj Pratt School of Engineering, Civil & Environmental Engineering, Duke University, Durham, NC, USA

Hamed Ashouri Department of Civil and Environmental Engineering, Center for Hydrometeorology and Remote Sensing (CHRS), University of California, Irvine, CA, USA

Luca Baldini Institute of Atmospheric Sciences and Climate (ISAC), National Research Council (CNR), Roma, Italy

Ana P. Barros Pratt School of Engineering, Civil & Environmental Engineering, Duke University, Durham, NC, USA

Alessandro Battaglia Department of Physics and Astronomy, University of Leicester, Leicester, UK

Hylke E. Beck Department of Civil and Environmental Engineering, Princeton University, Princeton, NJ, USA

Andreas Becker Global Precipitation Climatology Center (GPCC), Deutscher Wetterdienst (DWD), Offenbach, Germany

Ali Behrangi Department of Hydrology and Atmospheric Sciences, University of Arizona, Tucson, AZ, USA

Hugo Bellenger Laboratoire de Météorologie Dynamique/IPSL, CNRS, Sorbonne Université, École Normale Supérieure, École Polytechnique, Paris, France
Japan Agency for Marine-Earth Science and Technology (JAMSTEC), Yokosuka, Japan

Emily Berndt NASA, Marshall Space Flight Center (MSFC), Huntsville, AL, USA

Emily Black Department of Meteorology, University of Reading, Reading, UK

David T. Bolvin Science Systems and Applications, Inc., Lanham, MD, and NASA/GSFC, Greenbelt, MD, USA

Rogerio Bonifacio World Food Programme, Vulnerability Assessment and Mapping Unit, Roma, Italy

Sid-Ahmed Boukabara NOAA/NESDIS/STAR, College Park, MD, USA

Dominique Bouniol Météo France, Centre National de Recherches Météorologiques (CNRM), Groupe de Modélisation et d'Assimilation pour la Prévision (GMAP), OBS, Toulouse, France

Jacqueline Boutin Sorbonne Université, CNRS, Institut de Recherche pour le Développement (IRD), Muséum National d'Histoire Naturelle (MNHN), Laboratoire d'Océanographie et du Climat, Expérimentations et Approches Numériques (LOCEAN), Paris, France

Dan Braithwaite Department of Civil and Environmental Engineering/CHRS, University of California, Irvine, CA, USA

Scott A. Braun NASA/GSFC, Greenbelt, MD, USA

Luca Brocca Research Institute for Geo-Hydrological Protection (IRPI), National Research Council (CNR), Perugia, Italy

Jörg Burdanowitz Institute for Meteorology, University of Hamburg, Hamburg, Germany

Steven J. Burian Department of Civil and Environmental Engineering, University of Utah, Salt Lake City, UT, USA

Stefania Camici CNR/IRPI, Perugia, Italy

Michel Capderou CNRS/LMD, Palaiseau, France

Daniele Casella CNR/ISAC, Roma, Italy

Elsa Cattani CNR/ISAC, Bologna, Italy

Daniel Cecil NASA/MSFC, Huntsville, AL, USA

Philippe Chambon Météo France, CNRM/GMAP/OBS, Toulouse, France

Venkatachalam Chandrasekar Department of Electrical and Computer Engineering, Colorado State University, Ft. Collins, CO, USA

Luca Ciabatta CNR/IRPI, Perugia, Italy

Sophie Cloché CNRS/IPSL, Palaiseau, France

Paolo Colucci EUMETSAT, Darmstadt, Germany

Salvatore D'Addio European Space Agency (ESA), European Space Research and Technology Centre (ESTEC), Noordwijk, The Netherlands

Frank Davenport Climate Hazards Group (CHG), University of California, Santa Barbara, CA, USA

Michel Dejus Centre National d'Études Spatiales (CNES), Toulouse, France

Tufa Dinku International Research Institute for Climate and Society (IRI), The Earth Institute at Columbia University, Palisades, NY, USA

Craig Donlon ESA/ESTEC, Noordwijk, The Netherlands

Mohamed El Saadani Department of Civil Engineering, University of Louisiana Lafayette, Lafayette, LA, USA

Ralph R. Ferraro NOAA/NESDIS/STAR, College Park, MD, USA

Robert D. Field Department of Applied Physics and Applied Mathematics, Columbia University, and NASA Goddard Institute for Space Studies, New York, NY, USA

Paolo Filippucci CNR/IRPI, Perugia, Italy

Thomas Fiolleau CNRS, Laboratoire d'Études en Géophysique et Océanographie Spatiales (LEGOS), Toulouse, France

Efi Foufoula-Georgiou Department of Civil and Environmental Engineering and Department of Earth Science, University of California, Irvine, CA, USA

Kevin Fuell Earth System Science Center (ESSC), University of Alabama in Huntsville, Huntsville, AL, USA

Chris Funk United States Geological Survey (USGS), Earth Resources Observation and Science (EROS) Center, Sioux Falls, SD, and CHG, University of California, Santa Barbara, CA, USA

Phu Nguyen Department of Civil and Environmental Engineering/CHRS, University of California, Irvine, CA, USA

Riko Oki Earth Observation Research Center (EORC)/Japan Aerospace Exploration Agency (JAXA), Ibaraki, Japan

William S. Olson University of Maryland Baltimore County, Baltimore, MD, and NASA/GSFC, Greenbelt, MD, USA

Mohammed Ombadi Department of Civil and Environmental Engineering/CHRS, University of California, Irvine, CA, USA

Davide Ori Institute for Geophysics and Meteorology, University of Cologne, Cologne, Germany

Shigenori Otsuka RIKEN Center for Computational Science, Kobe, Japan

Ming Pan Department of Civil and Environmental Engineering, Princeton University, Princeton, NJ, USA

Giulia Panegrossi CNR/ISAC, Roma, Italy

Florian Pappenberger ECMWF, Shinfield Park, Reading, UK

Salvatore Pascale Department of Earth System Science, Stanford University, Stanford, CA, USA

Diego Pedreros USGS/EROS, Sioux Falls, SD, USA

Walter A. Petersen NASA/MSFC, Huntsville, AL, USA

Pete Peterson CHG, University of California, Santa Barbara, CA, USA

Nazzareno Pierdicca DIET, Sapienza University of Roma, Roma, Italy

Federico Porcù Department of Physics and Astronomy, University of Bologna, Bologna, Italy

Olivier P. Prat Cooperative Institute for Satellite Earth System Studies (CISESS), North Carolina State University, Asheville, NC, USA

Alain Protat Bureau of Meteorology (BoM), Melbourne, VIC, Australia

Felipe Quintero IIHR-Hydroscience & Engineering, University of Iowa, Iowa City, IA, USA

David L. Randel Department of Atmospheric Science, Colorado State University, Ft. Collins, CO, USA

Gilles Reverdin Sorbonne Université, CNRS/IRD/MNHN/LOCEAN, Paris, France

Sarah Ringerud ESSIC, University of Maryland, College Park, MD, and NASA/GSFC, Greenbelt, MD, USA

Rémy Roca CNRS/LEGOS, Toulouse, France

Geun-Hyeok Ryu National Meteorological Satellite Center, Korea Meteorological Administration, Seoul, South Korea

Paolo Sanò CNR/ISAC, Roma, Italy

Mathew R. P. Sapiano Sapiano Statistical Services, Atlanta, GA, USA

Masaki Satoh Atmosphere and Ocean Research Institute (AORI), The University of Tokyo, Chiba, Japan

Peter Schlüssel EUMETSAT, Darmstadt, Germany

Udo Schneider GPCC/DWD, Offenbach, Germany

Marc Schröder DWD, Offenbach, Germany

J. Marshall Shepherd Department of Geography, University of Georgia, Athens, GA, USA

Shoichi Shige Division of Earth and Planetary Sciences, Graduate School of Science, Kyoto University, Kyoto, Japan

Shraddhanand Shukla CHG, University of California, Santa Barbara, CA, USA

Gail M. Skofronick-Jackson NASA, Headquarters, Science Mission Directorate, Washington, DC, USA

Mark A. Smalley JPL/Caltech, Pasadena, CA, USA

Matthew Smith Information Technology and Systems Center, University of Alabama in Huntsville, Huntsville, AL, USA

Byung-Ju Sohn School of Earth and Environmental Sciences, Seoul National University, Seoul, South Korea

Hwan-Jin Song National Institute of Meteorological Sciences, Korea Meteorological Administration, Seoul, South Korea

Soroosh Sorooshian Department of Civil and Environmental Engineering/CHRS, University of California, Irvine, CA, USA

Thomas Stanley Universities Space Research Association (USRA), Columbia, MD, and NASA/GSFC, Greenbelt, MD, USA

Erich F. Stocker NASA/GSFC, Greenbelt, MD, USA

Alexandre Supply Sorbonne Université, CNRS/IRD/MNHN/LOCEAN, Paris, France

Nobuhiro Takahashi Institute for Space-Earth Environmental Research, Nagoya University, Nagoya, Japan

Yukari N. Takayabu AORI/The University of Tokyo, Chiba, Japan

Jackson Tan USRA, Columbia, MD, and NASA/GSFC, Greenbelt, MD, USA

Simone Tanelli JPL/Caltech, Pasadena, CA, USA

Wei-Kuo Tao NASA/GSFC, Greenbelt, MD, USA

Francisco J. Tapiador University of Castilla-La Mancha, Toledo, Spain

Elena Tarnavsky Department of Meteorology, University of Reading, Reading, UK

Tomoko Tashima EORC/JAXA, Ibaraki, Japan

Koji Terasaki RIKEN Center for Computational Science, Kobe, Japan

Ali Tokay University of Maryland Baltimore County, Baltimore, MD, and NASA/GSFC, Greenbelt, MD, USA

Hirofumi Tomita RIKEN Center for Computational Science, Kobe, Japan

Frederic Tridon Earth Observation Science, Department of Physics and Astronomy, University of Leicester, Leicester, UK

F. Joseph Turk JPL/Caltech, Pasadena, CA, USA

Jani Tyynelä Finnish Meteorological Institute (FMI), Helsinki, Finland

Tomoo Ushio Tokyo Metropolitan University, Tokyo, Japan

Albert I. J. M. van Dijk Fenner School of Environment & Society, The Australian National University, Canberra, Australia

Jean-Luc Vergely ACRI-st, Guyancourt, France

Noemi Vergopolan Department of Civil and Environmental Engineering, Princeton University, Princeton, NJ, USA

Daniel Vila Instituto Nacional de Pesquisas Espaciais (IPE), Centro de Previsão de Tempo e Estudos Climaticos (CPTEC), Cachoeira Paulista, Brazil

Wolfgang Wagner Department of Geodesy and Geoinformation, Research Group Remote Sensing, TU Wien, Vienna, Austria

Jian-Jian Wang ESSIC, University of Maryland, College Park, MD, USA

Jianxin Wang Science Systems Applications International and NASA/GSFC, Greenbelt, MD, USA

Graham P. Weedon Joint Centre for Hydro-Meteorological Research, Met Office, Wallingford, UK

Tobias Wehr ESA/ESTEC, Noordwijk, The Netherlands

David B. Wolff NASA/GSFC, Wallops Flight Facility, Wallops Island, VA, USA

Eric F. Wood Department of Civil and Environmental Engineering, Princeton University, Princeton, NJ, USA

Norman B. Wood Space Science and Engineering Center (SSEC), University of Wisconsin-Madison, Madison, WI, USA

Pingping Xie NOAA/NWS/CPC, College Park, MD, USA

Moeka Yamaji EORC/JAXA, Ibaraki, Japan

Munehisa K. Yamamoto Division of Earth and Planetary Sciences, Graduate School of Science, Kyoto University, Kyoto, Japan

Hisashi Yashiro Satellite Observation Center, National Institute for Environmental Studies, Tsukuba, Japan

Matthew Young Department of Meteorology, University of Reading, Reading, UK

Bradley Zavodsky NASA/MSFC, Huntsville, AL, USA

Acronyms

ABI	Advanced Baseline Imager (GOES)
ACE	Aerosol-Clouds-Ecosystem Mission (NASA)
AD	Analog-to-Digital converter
ADDA	Amsterdam DDA
AET	Actual Evapotranspiration
AGL	Above Ground Level
AHPS	Advanced Hydrologic Prediction Service (NWS)
AI	Aridity Index
AIP	Algorithm Intercomparison Programme
Air-MSPI	Airborne Multi-angle SpectroPolarimetric Imager (NASA)
AIRS	Atmospheric Infrared Sounder (NASA)
AKDT	Alaska Daylight Time
ALEXI	Atmosphere-Land Exchange Inverse
AMeDAS	Automated Meteorological Data Acquisition System (JMA)
AMIE/DYNAMO	ARM Madden-Julian Oscillation Investigation Experiment/ Dynamics of the Madden-Julian Oscillation
AMIP	Atmospheric Model Intercomparison Project (WCRP)
AMMA	African Monsoon Multidisciplinary Analysis experiment
AMO	Atlantic Meridional Oscillation
AMPR	Advanced Microwave Precipitation Radiometer (NASA)
AMS	Annual Maximum Series (of rainfall)
AMSL	Above Mean Sea Level
AMSR	Advanced Microwave Scanning Radiometer (JAXA)
AMSR-E	AMSR-EOS (NASA)
AMSU	Advanced Microwave Sounding Unit (NOAA and EUMETSAT)
AMW	Active Microwave
APHRODITE	Asian Precipitation—Highly Resolved Observational Data Integration Towards Evaluation of Water Resources (Japan)

APR-3	Airborne Third Generation Precipitation Radar (NASA)
APSIM	Agricultural Production Systems sIMulator
AR	Atmospheric River
ARC (1)	Africa Rainfall Climatology (NOAA)
ARC (2)	Active Radar Calibrator
ARM	Atmospheric Radiation Measurement (DoE)
ARM-SGP	ARM Southern Great Plains (DoE)
ARMAR	Airborne Rain-Mapping Radar (NASA and JPL)
ASCAT	Advanced SCATterometer (ESA)
ASCII	American Standard Code for Information Interchange
ASI	Italian Space Agency
ASL	Above Sea Level (a.s.l.)
ASTRAIA	Analyse Stereoscopique par Radar Aeroporte (CNRS)
ATBD	Algorithm Theoretical Basis Document
ATMS	Advanced Technology Microwave Sounder (NASA/NOAA)
AWARE	ARM West Antarctic Radiation Experiment
AWIPS	Advanced Weather Interactive Processing System (NWS and UNIDATA)
BAECC	Biogenic Aerosols-Effects on Clouds and Climate Experiment (ARM)
BB	Bright Band
BC	British Columbia
BCS	Bias Correction Scheme
BMKG	Badan Meteorologi, Klimatologi, dan Geofisika (Indonesia)
BoM	Bureau of Meteorology (Australia)
BRAIN	Bayesian Rain Algorithm Including Neural Networks (Megha-Tropiques)
BSA	Backscatter Alignment
BUFR	Binary Universal Form for the Representation of Meteorological Data
BUI	Buildup Index (FWI)
CAPE	Convective Available Potential Energy
CAPRICORN	Clouds, Aerosols, Precipitation, Radiation, and Atmospheric Composition over the Southern Ocean
CARE	Centre for Atmospheric Research Experiments (Environment Canada)
CATDS	Centre Aval de Traitement des Données SMOS
CC	Correlation Coefficient
CCD	Cold Cloud Duration
CCDF	Complementary Cumulative Distribution Function
CCI	Climate Change Initiative (ESA)
CCl	Commission for Climatology (WMO)
CCP (1)	Clouds, Convection, and Precipitation
CCP (2)	Cloud and Precipitation Process Mission

CCZ	Continent-Climate Zone
CDD	Consecutive Dry Days Index (ETCCDI)
CDF	Cumulative Density Function
CDR	Climate Data Record
CDRD	Cloud Dynamics and Radiation Database
CEMADEN	Centro Nacional de Monitoramento e Alertas de Desastres Naturais (Brazil)
CEOS	Committee on Earth Observation Satellites
CERES	Clouds and the Earth's Radiant Energy System (NASA)
CESM	Community Earth System Model
CFAD	Contoured Frequency by Altitude Diagram
CG	Cloud-to-Ground Lightning
CGMS	Coordination Group for Meteorological Satellites
CHC	Climate Hazards Center (University of California, Santa Barbara)
CHIRPS	Climate Hazards center InfraRed Precipitation with Stations
CHPcli	Climate Hazards Group's Precipitation Climatology
CIMR	Copernicus Imaging Microwave Radiometer Mission (EU)
CIndO	Central Equatorial Indian Ocean index
CIRA	Cooperative Institute for Research in the Atmosphere (CSU)
CLC	Corine Land Cover
CLIVAR	Climate Variability and Predictability (WMO)
CLW	Cloud Liquid Water
CLWC	Cloud Liquid Water Content
CMA	China Meteorological Administration
CMAP	CPC Merged Analysis of Precipitation
CMIP	Coupled Model Intercomparison Project (WCRP)
CMIP-5	CMIP Phase 5
CMORPH	CPC MORPHing algorithm
CMORPH-CRT	CMORPH Bias Corrected
CMORPH-KF	CMORPH-Kalman Filter
CM-SAF	Climate Monitoring-SAF (EUMETSAT)
CNES	Centre National D'Études Spatiales (France)
CNRM/GAME	Centre National de Recherches Météorologiques—Groupe d'études de l'Atmosphère Météorologique (Météo France)
CNR	Consiglio Nazionale delle Ricerche (Italy)
CNR-IRPI	CNR-Istituto di Ricerca per la Protezione Idrogeologica
CNR-ISAC	CNR-Istituto di Scienze dell'Atmosfera e del Clima
CNRS	Centre National de la Recherche Scientifique (France)
CNTL	Control Run
COADS	Comprehensive Ocean Atmosphere Data Set
ConQ	Moisture Flux Convergence
CONUS	Conterminous US
CORRA	Combined Radar-Radiometer Product (GPM)

CoSMIR	Conical Scanning Millimeter-wave Imaging Radiometer (NASA)
CPC	Climate Prediction Center (NOAA)
CPI	Convective Percent Index
CPL	Cloud Physics Lidar (NASA)
CPR	Cloud Profiling Radar (NASA)
CRS	Cloud Remote Sensing Radar (NASA)
CrIS	Cross-Track Infrared Sounder (NASA)
CRM	Cloud-Resolving Model
CRS	Cloud Radar System (NASA)
CRTM	Community Radiative Transfer Model
CRU	Climate Research Unit (Univ. of East Anglia)
CSA	Climate Service for Agriculture (Rwanda)
CSH	Convective and Stratiform Heating
CSI	Critical Success Index
CSK	COSMO-SkyMed (ASI)
CSP	Climate Services Partnership
CSU	Colorado State University
CSPP	Community Satellite Processing Package
CT	Cloud Thickness
CTH	Cloud Top Height
CWD	Consecutive Wet Days (ETCCDI)
CWV	Columnar Water Vapor
CYGNSS	Cyclone Global Navigation Satellite System (NASA)
DA	Data Assimilation
DAR	Differential Absorption Radar
DB	Dark Band
DBNet	Direct Broadcast Network
DC	Drought Code (FWI)
DD	Downward Decreasing
DDA	Discrete Dipole Approximation
DDSCAT	Discrete Dipole Scattering
DEM	Digital Elevation Model
DFR	Dual-Frequency Ratio
DI	Downward Increasing
DJF	December-January-February
DKRZ	Deutsches Klimarechenzentrum (Germany)
DLR	Deutschen Zentrums für Luft- und Raumfahrt (Germany)
DMC	Duff Moisture Code (FWI)
DMIP2	Distributed Hydrologic Model Intercomparison Project–Phase 2 (NWS)
DMSP	Defense Meteorological Satellite Program (US Navy)
DNN	Deep Neural Network
DoE	Department of Energy

HRPP	High-Resolution Precipitation Product
HRWS	High-Resolution Wide Swath
H-SAF	Support to Operational Hydrology and Water Management (EUMETSAT)
HSS	Heidke Skill Score
HyMeX	Hydrological Cycle in Mediterranean Experiment
IC	Intra-cloud Lightning
ICDC	Integrated Climate Data Center (University of Hamburg)
ICE-POP	International Collaborative Experiment for the PyeongChang Olympics and Paralympics Experiment 2018
ICHARM	International Centre for Water Hazard and Risk Management
ICI	Ice Cloud Imager (EUMETSAT)
ICO-LETKF	Icosahedral LETKF
IDF	Intensity-Duration-Frequency Curve
IDW	Inverse Distance Weighting
IFAS	Integrated Flood Analysis System
IFM	Index Flood Method
IMERG	Integrated Multi-satellitE Retrievals for GPM
IMERG_E	IMERG Early Run (near real time with a latency of 6 h)
IMERG_F	IMERG Final Run (gauged-adjusted with a latency of 4 months)
IMERG_L	IMERG Late Run (reprocessed near real time with a latency of 18 h)
IOD	Indian Ocean Dipole
IPHEx	Integrated Precipitation and Hydrology Experiment (GPM)
IPS	Institut Pierre Simon Laplace (CNRS)
IPWG	International Precipitation Working Group (CGMS)
IR	Infrared
IRI	International Research Institute for Climate and Society (Columbia University)
IRP	Infrared Precipitation Estimate
ISI	Initial Spread Index (FWI)
ISRO	Indian Space Research Organisation
ISS	International Space Station
ITCZ	Intertropical Convergence Zone
IWP	Ice Water Path
I&Q	In-Phase and Quadrature Signal
JAXA	Japan Aerospace Exploration Agency
JCOMM	Joint Technical Commission for Oceanography and Marine Meteorology (WMO)
JCSDA	Joint Center for Satellite Data Assimilation (NOAA)
JERD	JPSS NESDIS ESPC Requirements Document
JF	January-February
JJA	June-July-August

JJAS	June-July-August-September
JMA	Japan Meteorological Agency
JPL	Jet Propulsion Laboratory
JPSS	Joint Polar Satellite System
JRA55	Japanese 55-year Reanalysis
JRC	Joint Research Centre (EC)
KGE	Kling-Gupta Efficiency
KMA	Korea Meteorological Administration
KWAJEX	Kwajalein Experiment
LACA&D	Latin American Climate Assessment & Dataset
LDM	Local Data Manager (UNIDATA)
LDR	Linear Depolarization Ratio
LEGOS	Laboratoire d'Études en Géophysique et Océanographie Spatiales
LEO	Low Earth Orbit
LETKF	Local Ensemble Transform Kalman Filter (RIKEN)
LFM	Local Forecast Model
LH	Latent Heating
LHASA	Landslide Hazard Assessment for Situational Awareness (NASA)
LIA	Local Incidence Angle
LLCF	Low-Level Clouds and Fog
LL-LETKF	Latitude-Longitude LETKF
LMD	Laboratoire de Météorologie Dynamique (CNRS)
LMODEL	Lagrangian Model (UC Irvine and University of Hull)
LPVEx	Light Precipitation Validation Experiment (GPM)
LR	Logistic Regression
LST	Land Surface Temperature
LUT	Lookup Table
LWP	Liquid Water Path
LZA	Local Zenith Angle
L1SR	Level 1 Science Requirements (GPM)
MADRAS	Microwave Analysis and Detection of Rain and Atmospheric Structures (Megha-Tropiques)
MAE	Mean Absolute Error
MAFF	Ministry of Agriculture, Forestry and Fisheries (Japan)
MAM	March-April-May
MARSOP	Monitoring Agricultural ResourceS Operational (JRC)
MBE	Mean Bias Error
MCS	Mesoscale Convective System
MCTA	Merged CloudSat, TRMM, and AMSR product
MC3E	Mid-latitude Continental Convective Clouds Experiment
ME	Mean Error
Medicane	Mediterranean hurricane

DO-Op	Daylight Only Operations (CloudSat)
DoW	Doppler on Wheels (Center for Severe Weather Research)
DP	Dual Polarimetric Radar
DPC (1)	Data Processing Center (CloudSat)
DPC (2)	Department of Civil Protection of Italy
DPCA	Displaced Phase Center Antenna
DPR	Dual-frequency Precipitation Radar (GPM)
DRC	Democratic Republic of the Congo
DryMOD	Dryland hydrological MODel (University of Reading)
DSD	Drop Size Distribution
DSI	Drought Severity Index
DSSAT	Decision Support System for Agrotechnology Transfer
DWR	Dual-Wavelength Ratio
DYNAMO	Dynamics of the MJO experiment
D3R	Dual-Frequency Dual-Polarized Doppler Radar (NASA)
EA	East Africa
EAF	East Africa
EarthCARE	Earth Clouds, Aerosols, and Radiation Explorer (ESA-JAXA)
EASE	Equal-Area Scalable Earth
EBCM	Extended Boundary Condition Method
EC	European Commission
ECCC	Environment and Climate Change Canada
ECDI	Enhanced Combined Drought Index
ECMWF	European Centre for Medium-Range Weather Forecasts
EDF	Environmental Data Fusion
EDOP	ER-2 Doppler Radar
EDR	Environmental Data Record
EEA	Eastern East Africa
EIA	Earth Incident Angle
ELDORA	Electra Doppler Radar (NCAR)
EM	Electromagnetic
EMA	Effective Medium Approximations
eMAs	extended MODIS Airborne Simulator (NASA)
EMSR	Electronically Scanning Microwave Radiometer (NOAA)
ENACTS	Enhancing National Climate Services
ENSO	El Niño Southern Oscillation
EOF	Empirical Orthogonal Function
EOS	Earth Observing System (NASA)
EPD2	Extreme Precipitation Day > 2 in day^{-1}
EPD4	Extreme Precipitation Day > 4 in day^{-1}
EPS	EUMETSAT Polar System
EPSAT-SG	Estimation of Precipitation by Satellite Second Generation (CNRS-LMD)
EPS-SG	EPS Second Generation

ERA	ECMWF Reanalysis
ESA	European Space Agency
ESA-CCI	ESA Climate Change Initiative
ESMR	Electronically Scanned Microwave Radiometer (NOAA)
ESPC	Environmental Satellite Processing Center (NESDIS)
ESSIC	Earth System Science Interdisciplinary Center
ET	Evapotranspiration
ETCCDI	Expert Team on Climate Change Detection and Indices
EU	European Union
EUMETCast	EUMETSAT's Multicast Distribution System
EUMETSAT	European Organization for the Exploitation of Meteorological Satellites
EVI-3	Earth Venture Instrument-3 program (NASA)
EVT	Extreme Value Theory
EWFN	Energy-Water-Food Nexus
EWS	Early Warning System
EXRAD	ER-2 X-band Radar (NASA)
FAO	Food and Agriculture Organization (UN)
FAR	False Alarm Rate
FAS	Foreign Agricultural Service (USDA)
FB	Frequency Bias
FCDR	Fundamental Climate Data Record
FD	Frost Days
FDRS	Fire Danger Rating Systems
FEWS NET	Famine Early Warning Systems Network
FFMC	Fine Fuel Moisture Code (FWI)
FL	Freezing Level
FLDAS	FEWS NET Land Data Assimilation System
FLOR	Forecast-oriented Low Ocean Resolution model (GFDL)
FMI	Finnish Meteorological Institute
FNMOC	Fleet Numerical Meteorology and Oceanography Center (US Navy)
FoV	Field of View
FP	Forward Processing (GMAO)
FRMSE	Fractional RMSE
FSOI	Forecast Sensitivity Observation Impact
FWI	Fire Weather Index
GAGES	Geospatial Attributes of Gages for Evaluating Streamflow (USGS)
GANAL	Global Analysis (JMA)
GATE	Global Atmospheric Research Program Atlantic Tropical Experiment
GC	Ground Clutter
GCEM	Goddard Cumulus Ensemble Model (NASA)

GCM	General Circulation Model
GCOM	Global Change Observation Mission (JAXA)
GCOM-W	GCOM-Water
GCOS	Global Climate Observing System (WMO)
GCPEx	GPM Cold Season Precipitation Experiment
GDAP	GEWEX Data and Assessment Panel
GDAS/GFS	Global Data Assimilation System/Global Forecast System (NOAA)
GEC	Geocoded Ellipsoid Corrected
GEO (1)	Geostationary orbit
GEO (2)	Group on Earth Observations
GEOGLAM	GEO Global Agricultural Monitoring
GEOS	Goddard Earth Observing System (NASA)
GEOS FP	Global Earth Observing System Forward Processing
GEOSS	Group on Earth Observation System of Systems
GeoSTAR	Geostationary Synthetic Thinned Aperture Radiometer (NASA)
GES DISC	Goddard Earth Sciences Data and Information Services Center
GEV	Generalized Extreme Value Distribution
GEWEX	Global Energy and Water Exchanges (WCRP)
GFCS	Global Framework for Climate Services (WMO)
GFDL	Geophysical Fluid Dynamics Laboratory (NOAA and Princeton University)
GFS	Global Forecast System (NCEP)
GFWED	Global Fire Weather Database (NASA GISS)
GHA	Greater Horn of Africa
GHCN-D	Global Historical Climatology Network-Daily (NOAA)
GHM	Global Hydrological Model
GHRC	Global Hydrology Resource Center (NASA)
GIRAFE	Global Interpolated RAinFall Estimation
GIS	Geographic Information System
GISS	Goddard Institute for Space Studies (NASA)
GLC	Global Landslide Catalog (NASA)
GMa	Gulf of Mexico area
GMAO	Global Modeling and Assimilation Office (NASA)
GMI	GPM Microwave Imager
GMM	Generalized Multiparticle Mie-solution
GoAMAZON	Green Ocean Amazon Experiment
GOES	Geostationary Operational Environmental Satellite (NOAA)
GOSAT-3	Greenhouse Gases Observing Satellite (JAXA)
GOTM	General Ocean Turbulence Model
GPC	Global Precipitation Climatology Center (DWD)
GPCP	Global Precipitation Climatology Project (GEWEX)
GPI	GOES Precipitation Index

GPM	Global Precipitation Measurement mission (NASA and JAXA)
GPM-CO	GPM Core Observatory
GPP	Gross Primary Production
GPROF	Goddard Profiling Algorithm (NASA)
GPS	Global Positioning System
GRACE	Gravity Recovery and Climate Experiment (NASA and DLR)
GRACE-FO	GRACE Follow-On
GRDC	Global Runoff Data Centre
Grid-Sat	Gridded Satellite Data (NOAA)
GRIP	Genesis and Rapid Intensification Processes (NASA)
GRISO	Rainfall Generator of Spatial Interpolation from Observation
GS	Grain Size
GSFC	Goddard Space Flight Center (NASA)
GSI	Gridpoint Statistical Interpolation
GSM	Global Spectral Model
GSMaP	Global Satellite Mapping of Precipitation (Japan)
GSMaP-MVK	GSMaP Motion Vector Kalman
GSMaP-NRT	GSMaP Near Real Time
GSOD	Global Surface Summary of the Day (NOAA)
GT	Gaussian Transformation
GTC	Geocoded Terrain Corrected
GTS	Global Telecommunication System
GV	Ground Validation
GW	Global Warming
GWIS	Global Wildfire Information System
GWP	Graupel Water Path
G5NR	GEOS-5 Nature Run (NASA)
HADS	Hydrometeorological Automated Data System (NWS)
HAMSR	High-Altitude MMIC Sounding Radiometer (JPL)
HDF	Hierarchical Data Format
HEMT	High Electron Mobility Transistor
HEPEX	Hydrological Ensemble Prediction Experiment
HID	Hydrometeor Identification algorithms
HIRS	High-Resolution Infrared Radiation Sounder (NOAA)
HISA	Hurricane Intensity and Structure Algorithm
HIWRAP	High-Altitude Imaging Wind and Rain Airborne Profiler (NASA)
HLM	Hillslope-Link Model
HMA	High Mountain Asia
HOAPS	Hamburg Ocean Atmosphere Parameters and Fluxes from Satellite Data (University of Hamburg)
HQPrecip	High-Quality Precipitation (GPM)
HR	Hit Rate

MERRA	Modern-Era Retrospective analysis for Research and Applications (NASA)
MGD	Multilook Ground Detected
MGDSST	Merged Satellite and In Situ Data Global Daily SST (JMA)
MHEMT	Metamorphic HEMT
MHOPrEx	Monsoon Himalaya Orographic Precipitation Experiment
MHS	Microwave Humidity Sounder (EUMETSAT)
MicroMAS-2	Micro-sized Microwave Atmospheric Satellite-2
MIIDAPS	Multi-instrument Inversion and Data Assimilation Preprocessing System (NOAA)
MIR	Middle IR
MIRA	Microwave/Infrared Rainfall Algorithm
MIRAS	Microwave Imaging Radiometer using Aperture Synthesis (SMOS)
MiRS	Microwave Integrated Retrieval System (NOAA)
MISDc	Modello Idrologico Semi-Distribuito in continuo (CNR-IRPI)
MLP	Melting Level Precipitation
MLS	Microwave Limb Sounder (EOS)
MMIC	Millimeter-Wave Monolithic Integrated Circuits
MODIS	Moderate Resolution Imaging Spectroradiometer (NASA)
MPE	Multisensor Precipitation Estimator
MP-M	Max-Planck-Institut für Meteorologie (Germany)
MRR	Micro Rain Radar
MREA	Modified Regression Empirical Algorithm
MRMS	Multi-Radar/Multisensor Precipitation Data
MREA	Modified Regression Empirical Algorithm
MRMS	Multi-Radar Multisensor system (NOAA)
MS	Multiple Scattering
MSG	Meteosat Second Generation (EUMETSAT)
MSG-CPP	MSG Cloud Physical Properties (EUMETSAT)
MSLP	Mean Sea-Level Pressure
MSPPS	Microwave Surface and Precipitation Products System (NOAA)
MSU	Microwave Sounding Unit (NOAA)
MSWEP	Multisource Weighted-Ensemble Precipitation
MTSAT	Multifunctional Transport Satellites (JMA)
MW	Microwave
MWCC	Microwave Cloud Classification
MWCOMB	Combined MW Rainfall Retrieval (CPC)
MWHS-2	Microwave Humidity Sounder-2 (CMA)
MWI	Microwave Imager (EUMETSAT)
MWR (1)	Microwave Radiometer
MWR (2)	Moving Window Regression

MWS	Microwave Sounder
NAa	North Atlantic Area
NAMMA	NASA African Monsoon Multidisciplinary Analyses
NAS	National Academies of Sciences (US)
NASA	National Aeronautics and Space Administration
NCA	North and Central America
NCAR	National Center for Atmospheric Research
NCE	National Centers for Environmental Information (NOAA)
NCEI	National Centers for Environmental Information
NCEP	National Centers for Environmental Prediction (NOAA)
NCEP-CFSR	Climate Forecast System Reanalysis
NCL	NCAR Command Language
NDVI	Normalized Difference Vegetation Index
NESDIS	National Environmental Satellite, Data, and Information Service (NOAA)
netCDF	Network Common Data Form
NEWS	NASA Energy and Water Cycle Study
NEXRAD	Next-Generation Weather Doppler Radar (NOAA/NWS)
NH	Northern Hemisphere
NHM	National Hydrometeorological Service
NICAM	Nonhydrostatic Icosahedral Atmospheric Model (RIKEN)
NIR	Near IR
NM	National Meteorology Agency
NMAE	Normalized Mean Absolute Error
NMHS	National Meteorological and Hydrological Service
NMQ	National Mosaic Quantitative Precipitation Estimation (NOAA/NSSL)
NU-WRF	NASA-Unified WRF
NPP	Net Primary Production
NRCS	Normalized Radar Cross Section
NOAA	National Oceanic and Atmospheric Administration
NOP	Numerical Ocean Prediction
NPOL	NASA S-Band Dual Polarimetric
NRL	Naval Research Laboratory (US Navy)
NRMSE	Normalized Root Mean Square Error
NRT	Near Real Time
NSE	Nash-Sutcliffe Efficiency
NSMC	National Satellite Meteorological Center (CMA)
NSSL	National Severe Storms Laboratory (NOAA)
NTPa	North Tropical Pacific Area
NUBF	Nonuniform Beam Filling
NU-WRF	NASA-Unified Weather Research and Forecasting
NWP	Numerical Weather Prediction
NWS	National Weather Service (NOAA)

OBCT	On-Board Calibration Target
OCE	Oceania
OceanRain	Ocean Rainfall And Ice-Phase Precipitation Measurement Network
OE	Optimal Estimation
OI	Optimal Interpolation
OLYMPEX	Olympic Mountains Experiment
OM	Observatoire Midi-Pyrénées
OND	October-November-December
OSCAR	Observing Systems Capability Analysis and Review (WMO)
OSPO	Office of Satellite and Product Operations (NOAA)
OSSE	Observing Systems Simulation Experiment
OZA	Observation Zenith Angle
PAW	Percentage Available Water
PCT	Polarization Corrected Temperature
PDF (1)	Probability Density Function
PDF (2)	Particle Distribution Function
PDO	Pacific Decadal Oscillation
PDSI	Palmer Drought Severity Index
PERHPP	Program for the Evaluation of High-Resolution Precipitation Products
PERSIANN	Precipitation Estimation from Remotely Sensed Information using Artificial Neural Networks (UC Irvine)
PERSIANN-CCS	PERSIANN-Cloud Classification System
PERSIANN-CDR	PERSIANN-Climate Data Records
PERSIANN-MSA	PERSIANN-Multispectral Analysis
PF	Precipitation Feature
PHEMT	Pseudomorphic HEMT
PHIVOLCS	Philippine Institute of Volcanology and Seismology
PIA	Path-Integrated Attenuation
PICSA	Participatory Integrated Climate Services
PIP	Precipitation Imaging Package
PMA	Probability Matching Algorithm
PMI	Polarimetric Microwave Imager (CMA)
PMM (1)	Precipitation Measurement Mission (NASA)
PMM (2)	Probability Matching Method
PMP	Probable Maximum Precipitation
PMW	Passive Microwave
PNPR	Passive Microwave Neural Network Precipitation Retrieval (CNR-ISAC)
POD	Probability of Detection
POES	Polar Operational Environmental Satellites (NOAA)
POFD	Probability of False Detection
PoP	Probability of Precipitation

POS	Probability of Snowfall
PPI	Plan Position Indicator
PPP	Precipitation Per Person
PPS	Precipitation Processing System (IMERG)
PQPE	Probabilistic Quantitative Precipitation Estimation
PR	Precipitation Radar (TRMM)
PRCPTOT	Total Rainfall Amount (ETCCDI)
PRISM	Parameter-Elevation Regressions on Independent Slopes Model (Oregon State University)
PRPS	Precipitation Retrieval and Profiling Scheme
PSD	Particle Size Distribution
PSS	Practical Salinity Scale
PTH	Precipitation Top Height
PUSH	Precipitation Uncertainties for Satellite Hydrology
PV	Physical Validation (GPM)
PW	Precipitable Water
PWC	Precipitable Water Content
QA	Quality Assurance
QC	Quality Control
QCLCD	Quality Controlled Local Climatological Data
QI	Quality Index
QPE	Quantitative Precipitation Estimation
RADAP	Radar Data Processor (NOAA)
RADAR	Radio Detection and Ranging
RADEX	Radar Definition Experiment (OLYMPEX)
RCM	Regional Climate Model
RCS	Radar Cross Section
REA	Regressive Empirical Algorithm
REFAME	Rain Estimation Using Forward Adjusted-Advection of Microwave Estimates (UC Irvine)
RFC	River Forecast Center (NWS)
RFE	Rainfall Estimate (FAO)
RFI	Radio Frequency Interference
RGA	Rayleigh-Gans Approximation
RGB	Red Green Blue
RH	Relative Humidity
RHI	Range Height Indicator
RICO	Rain in Cumulus over the Ocean
RIM	Rain Impact Model (salinity)
RMS	Root Mean Square
RMSD	RMS Deviation
RMSE	RMS Error
RoF	Rain on Fog
RoFCC	RoF and Cap Clouds

ROI	Region of Interest
RoLLC	Rain on Low-Level Clouds
ROSA	Radio Occultation Sensor for Atmosphere
RQI	Radar Quality Index
RR	Rain Rate
RRFA-S	Regional Rainfall Frequency Analysis using Satellite Precipitation
RSS	Remote Sensing Systems Inc.
RTH	Rain Top Height
RTM	Radiative Transfer Model
RTTOV	Radiative Transfer for TOVS
RV (or (R/V)	Research Vessel
RWH	Rainwater Harvesting
RW	Rainwater Path
R2O/O2	Research-to-Operations/Operations-to-Research paradigm (SPoRT)
SAa	South Atlantic Area
SAF	Satellite Application Facility (EUMETSAT)
SAM (1)	System for Atmospheric Modeling
SAM (2)	Southern Appalachian Mountains
SAPHIR	Sondeur Atmosphérique du Profile d'Humidité Intertropicale par Radiométrie (Megha-Tropiques)
SAR	Synthetic Aperture Radar
SARRA-H	Systéme d'Analyse Régional des Risques Agroclimatiques-Habillé
SBA	Split-Based Approach
SCA	Scatterometer (ESA)
SCaMPR	Self-Calibrating Multivariate Precipitation Retrieval (NESDIS)
ScaRaB	Scanner for Radiation Budget (Megha-Tropiques)
scPDSI	Self-Calibrated PDSI
SCS	Single-Look Complex Slant Products
SCSMEX	South China Sea Monsoon Experiment
SCSMEX/NESA	SCSMEX Northern Enhanced Sounding Array
SCSMEX/SESA	SCSMEX Southern Enhanced Sounding Array
SD (1)	Snowfall Detection
SD (2)	Standard Deviation
SDCI	Scaled Drought Condition Index
SDG	Sustainable Development Goal
SDII	Simple Daily Intensity Index (ETCCDI)
SEA	Southeast and East Asia
SEAK	Southeast Alaska
SEM	Semiempirical Model

SEVIRI	Spinning Enhanced Visible and Infrared Imager (EUMETSAT)
SFI	Seeder-Feeder Interactions
SFR	Snowfall Rate (NOAA)
SH	Southern Hemisphere
SI	Snow Index
SIA	Sea Ice Age
SIC	Sea Ice Concentration
SID	Sea Ice Drift
SIDOC	SAR Images Dark Object Classifier
SIT	Sea Ice Thickness
SLH	Spectral Latent Heating
SM	Soil Moisture
SMA	Soil Moisture Anomaly
SMAP	Soil Moisture Active Passive (NASA)
SMMR	Scanning Multichannel Microwave Radiometer (NOAA)
SMOS	Soil Moisture and Ocean Salinity (ESA)
SM2RAIN	Soil Moisture to Rain Algorithm (CNR-IRPI)
S-NPP	Suomi National Polar-Orbiting Partnership (NASA/NOAA)
SNR	Signal-to-Noise Ratio
SON	September-October-November
SOS	Start of Season
SPA	Specific Power Attenuation
SPCZ	South Pacific Convergence Zone
SPEI	Standardized Precipitation Evapotranspiration Index
SPI	Standardized Precipitation Index
SPICE	Solid Precipitation Intercomparison Experiment (WMO)
SPoRT	Short-term Prediction Research and Transition Center (NASA)
SPP (1)	Satellite Precipitation Product
SPP (2)	Seasonal Performance Probability
SPS	Special Weather Statement (NWS)
SPURS	Salinity Processes in the Upper Ocean Regional Study
SR	Success Ratio
SRE	Satellite-Based Rainfall Estimate
SREM2D	Two-Dimensional Satellite Rainfall Error Model
SRI	Surface Rainfall Intensity
SRT	Surface Reference Technique
SSA (1)	Space Situational Awareness
SSA (2)	Sub-Saharan Africa
SSN	Spatial Stream Network
SSM/I	Special Sensor Microwave/Imager (DMSP)
SSMIS	Special Sensor Microwave Imager Sounder (DMSP)
SSM/T	Special Sensor Microwave Temperature (DMSP)
SSP	Surface Salinity Profiler

SSRGA	Self-Similar Rayleigh-Gans Approximation
SSS	Sea Surface Salinity
SST	Sea Surface Temperature
SSU	Stratospheric Sounding Unit (NOAA)
SSW	Sea Surface Wind
STAR	Center for Satellite Applications and Research (NOAA-NESDIS)
STDV	Standard Deviation
STH	Storm-Top Height
STIa	South Tropical Indian Area
STPa	South Tropical Pacific Area
SWA	South and West Asia
SWB	Soil Water Balance
SWE	Snow Water Equivalent
SWER	Snow Water Equivalent Rate
SYNOP	Surface Synoptic Observation (WMO)
TAMSAT	Tropical Applications of Meteorology Using Satellite and Ground-Based Observations
TAPEER	Tropical Amount of Precipitation with an Estimate of Errors
TARCAT	TAMSAT African Rainfall Climatology and Time Series
TB	Brightness Temperature
TC	Tropical Cyclone
TCA	Triple Collocation Analysis
TCC	TRMM Composite Climatology
TC4	Tropical Composition, Cloud and Climate Coupling Experiment
TDTS	Time-Dependent Two-Stream Method
TEMPEST	Temporal Experiment for Storms and Tropical Systems
TEMPEST-D	TEMPEST Technology Demonstration
TIR	Thermal Infrared
TLC	Tropical-Like Cyclone
TMD	Thai Meteorological Department (Thailand)
TMI	TRMM Microwave Imager
TMPA	TRMM Multisatellite Precipitation Analysis (NASA)
TOOCAN	Tracking Of Organized Convection Algorithm through a 3-D segmentatioN
TOGA-COARE	Tropical Ocean Global Atmosphere Coupled Ocean-Atmosphere Response Experiment
TOVS	TIROS Operational Vertical Sounder
TPW	Total Precipitable Water
TQV	Total Precipitable Water Vapor
TRMM	Tropical Rainfall Measuring Mission
TROPICS	Time-Resolved Observations of Precipitation structure and storm Intensity with a Constellation of Smallsats (NASA)

TSX	TerraSAR-X (DLR)
TVA	Tennessee Valley Authority
TWP-ICE	Tropical Warm Pool – International Cloud Experiment
TWSA	Terrestrial Water Storage Anomaly
T2M	Two-Meter Temperature
UCA	Urban Climate Archipelago
UCLM	Universidad de Castilla-La Mancha
UH	University of Helsinki
UHI	Urban Heat Island
UMORA	Unified Microwave Ocean Retrieval Algorithm
UN	United Nations
UNEP	United Nations Environment Programme (UN)
UNESCO	United Nations Educational, Scientific and Cultural Organization (UN)
US	United States of America
USCRN	US Climate Reference Network
USDA	US Department of Agriculture
USDM	US Drought Monitor
USGS	US Geological Survey
UTC	Universal Time Coordinated
UWSI	Urban Water Stress Index per Individual
VAM	Vulnerability Assessment and Mapping (WFP)
VarBC	Variable Bias Correction
VI	Vegetation Index
VMI	Vertical Maximum Intensity
VN	Validation Network (GPM)
WATCH	Water and Global Change (EU)
WCOM	Water Cycle Observation Mission (CMA)
WCRP	World Climate Research Programme (WMO)
WDCC	World Data Center for Climate (DKRZ)
WDM6	WRF Double-Moment 6-Class
WEA	Western East Africa
WFF	Wallops Flight Facility (NASA)
WFO	Weather Forecast Office (NWS)
WFP	World Food Programme
WFDEI-CRU	WATCH Forcing Data ERA Interim-CRU
WII	Weather Index-Based Insurance
WMMD	Wet Millimeter Days
WMO	World Meteorological Organization (UN)
WR	Weather Radar
WRF	Weather Research and Forecasting Model
WRF-ARW	WRF-Advanced Research WRF
WRSI	Water Requirements Satisfaction Index
WS	Wind Speed

WSR-57	Weather Surveillance Radar 1957 (NWS)
WV	Water Vapor
XCAL	Intersatellite Calibration Working Group (GPM)
ZAR	Zones À Risque model
2DVD	2D Video Disdrometer

Part IV
Validation

Chapter 25
The IPWG Satellite Precipitation Validation Effort

Christopher Kidd, Shoichi Shige, Daniel Vila, Elena Tarnavsky, Munehisa K. Yamamoto, Viviana Maggioni, and Bathobile Maseko

Abstract The estimation of precipitation (rainfall and snowfall) across the Earth's surface is important for both science and user applications, ranging from understanding and improving our knowledge of the global energy and water cycle, to water resources and hydrological modelling, and to societal applications such as water availability and monitoring of waterborne diseases (see Kirschbaum DB, Huffman GJ, Adler RF, Braun S, Garrett K, Jones E, McNally A, Skofronick-Jackson G, Stocker E, Wu H, Zaitchik BF, Bull Am Meteorol Soc 98:1169–1194, 2017). The global mapping of precipitation through conventional means is essentially limited to land areas due to the reliance upon rain (and snow) gauges and/or radar (see Kidd C, Becker A, Huffman GJ, Muller CL, Joe P, Skofronick-Jackson G, Kirschbaum DB, Bull Am Meteorol Soc 98:69–78, 2017a). For truly global precipitation mapping satellite observations must be used. A range of techniques, algorithms and schemes have been developed to exploit these satellite observations and generate quantitative precipitation products, many with (quasi-) global

C. Kidd (✉)
Earth System Science Interdisciplinary Center, University of Maryland, College Park, MD, USA

NASA, Goddard Space Flight Center, Greenbelt, MD, USA
e-mail: chris.kidd@nasa.gov

S. Shige · M. K. Yamamoto
Division of Earth and Planetary Sciences, Graduate School of Science, Kyoto University, Kyoto, Japan

D. Vila
INPE/CPTEC, Cachoeira Paulista, Brazil

E. Tarnavsky
Department of Meteorology, University of Reading, Reading, UK

V. Maggioni
Sid and Reva Dewberry Department of Civil, Environmental, and Infrastructure Engineering, George Mason University, Fairfax, VA, USA

B. Maseko
South African Weather Service, Pretoria, South Africa

© Springer Nature Switzerland AG 2020
V. Levizzani et al. (eds.), *Satellite Precipitation Measurement*, Advances in Global Change Research 69, https://doi.org/10.1007/978-3-030-35798-6_1

coverage. Alongside these techniques, there is a need for the inter-comparison, verification, and validation of such products in order to quantify their accuracy and performance (and consistency) for both developers and users. The International Precipitation Working Group (IPWG) has supported a long-term effort to inter-compare and validate precipitation products through the exploitation of large-scale regional surface reference data sets. Here, we present the current and future validation efforts of the IPWG together with examples of satellite-surface inter-comparisons.

Keywords Precipitation · Rainfall · Snowfall · Validation · Inter-comparison · IPWG · Microwave · Infrared · Raingauge · Radar

25.1 Introduction

The generation of quantitative precipitation estimates (QPE) on a global scale is only possible through the exploitation of satellite observations or through numerical prediction (NWP) models. Satellite-based sensors are able to provide regular and frequent observations of the Earth and its atmosphere using a range of different sensors: for precipitation these include visible, infrared (IR), and passive and active microwave (MW) sensors. While simple techniques may be used to relate these observations to surface precipitation, more complex, physically-based techniques are now commonly used to exploit not only the multi-spectral capabilities of the sensors, but also multi-sensor/multi-satellite techniques which combine all available information to improve the precipitation estimates. The multi-sensor/multi-satellite approach is crucial in establishing precipitation estimates over the range of time scales and spatial resolutions required by the different user communities.

The IPWG has been established to foster the development of better precipitation measurements and improve their utilization, to improve scientific understanding of precipitation, and to develop international partnerships (Turk and Bauer 2006; Kidd et al. 2010). The IPWG builds upon the expertise of scientists currently involved in precipitation measurements from satellites with an emphasis on the development of precipitation products (Levizzani et al. 2018). A major activity of the IPWG is the verification, validation and inter-comparison of precipitation products to enable product developers and users to continually assess the performance of the available products. Consequently, a concerted effort has been made to devise an ongoing validation program, exploiting available surface reference data sets and satellite precipitation products in order to better inform product developers and the user community.

A number of key inter-comparison projects have been undertaken in the past to assess the ability of satellite observations and their associated products to provide quantitative estimates of precipitation. These have concentrated on regional and global assessment (as opposed to the limited-period field-based approach of microphysical validation studies). A number of past inter-comparisons, outlined below, have provided the basis of the ongoing validation approach of the IPWG:

(i) The Algorithm Inter-comparison Programme (AIP), established by the Global Precipitation Climatology Project (GPCP), assessed the (then) current satellite rainfall products and understand the difference between them in order to improve the retrieval algorithms (Ebert et al. 1996). Three inter-comparison projects were carried out focusing upon small regions for which both satellite and surface reference data was available:

   AIP-1: Centred over Japan (23°–46°N, 120°–147°E) for June–August 1989, using the Radar-AMeDAS as surface reference data (see Arkin and Xie 1994, Barrett and Bellerby 1992),

   AIP-2: Located over Western Europe (43°–55°N, 11°W–9°E) for February–April 1991, using the UKMO nimrod radar product as surface reference data (see Allam et al. 1993), and

   AIP-3: Located over the Tropical Pacific (4°S–1°N, 153°–158°E) during the Tropical Ocean Global Atmosphere Coupled Ocean-Atmosphere Response Experiment (TOGA-COARE) from November 1992–February 1993. Surface measurements were derived from two shipborne weather radars (see Ebert 1996).

(ii) Precipitation Inter-comparison Projects (PIP), were established through the NASA WetNet project to evaluate the current algorithms and techniques:

   PIP-1: Global, 0.5° resolution products for August–November 1987 (see Barrett et al. 1994);

   PIP-2: Full resolution, instantaneous products for selected cases between July 1987 and February 1993 (see Smith et al. 1998), and;

   PIP-3: Global, 0.25° resolution, all of 1992, but with inter-annual comparisons for 1991/2/3 for January and July. Gauge data forms the basis of surface reference data sets, with high-density regional gauges networks, while over the oceans the Pacific Atoll data set (Morrissey and Greene 1991) was utilised (see Adler et al. 2001). The frequency of the occurrence of precipitation was compared against the Comprehensive Ocean Atmosphere Data Set (COADS, Woodruff et al. 1987).

Since 2002 the IPWG has identified and established a number of validation regions based primarily upon the availability of surface reference data sets. Unlike many validation programmes (e.g., Skofronick-Jackson et al. 2015; Houze et al. 2017), the IPWG does not require access to the validation data sets: the developers of the various precipitation data products make these available, usually in near real time, to the validators who carry out the analysis and generate the results. In this way surface data sets that would not normally be publicly accessible can be fully utilised, exploiting regional large-scale surface reference data sets available in the form of national or international radar and gauge data sets. The data products are typically analysed at the 0.25° × 0.25°, daily scale, although more recently an effort has been

Table 25.1 Comparison of typical ground validation programs and the IPWG validation effort

Criteria	Ground validation programs	IPWG validation
Purpose of validation	Physical understanding of precipitation processes and systems	Assessment of precipitation products for developers and user communities
Type of validation	Priority on physical, but also statistical metrics	Focus on descriptive indicators and statistical metrics
Sources of validation data	Specifically collected for, or by project investigators	Contributed by validation groups from established sources
Source of precipitation products	Specific products provided by developers	Developers provide products directly to validation groups
Types of validation data	Diverse, multi-tier specialist instrumentation (gauge, radar, aircraft), in specific locations	Gauges and/or radars, usually part of a national network over specific regions of interest
Types of precipitation products	Full-resolution/instantaneous, single-sensor or model products	Time/area averaged model or blended satellite products

made to provide inter-comparisons at the full, instantaneous resolution of the products (see Kidd et al. 2018).

The IPWG validation effort can be seen as complementary to that of the specially targeted ground validation (GV) campaigns that seek a better understanding of precipitation at the microphysical scale (Skofronick-Jackson et al. 2015; Houze et al. 2017; and Chap. 26). Table 25.1 summarises the characteristics of the GV approaches and the IPWG validation approach. While the GV approach provides a detailed, but generally short-lived validation effort, the IPWG seeks to provide an ongoing validation effort using a consistent set of statistics and skill scores.

The IPWG validation regions (Fig. 25.1) are operated on a best-effort basis with only a few regions receiving funding to provide continual analysis of the data products. Regions that have been recently added include South Africa, which exploits a good surface rain gauge network, and is also aimed to developing their fledgling weather radar work and modelling capabilities over that region. Work is presently ongoing to help develop a validation region over the Indian subcontinent through collaboration with the India Meteorological Department.

Beyond the mechanical validation of precipitation products, the IPWG validation effort also informs the developers and user community about the inter-comparison and validation process itself, not least since comparisons of satellite-derived precipitation estimates with surface reference data sets is not an exact science. In particular, the nature of the different observing systems must be taken into consideration when selecting data sets, in the data processing and analysis and in the final interpretation of the results. The IPWG inter-comparisons rely upon large-scale regional surface reference data sets at moderate spatial (e.g., $0.25° \times 0.25°$) and temporal (e.g., daily) resolutions.

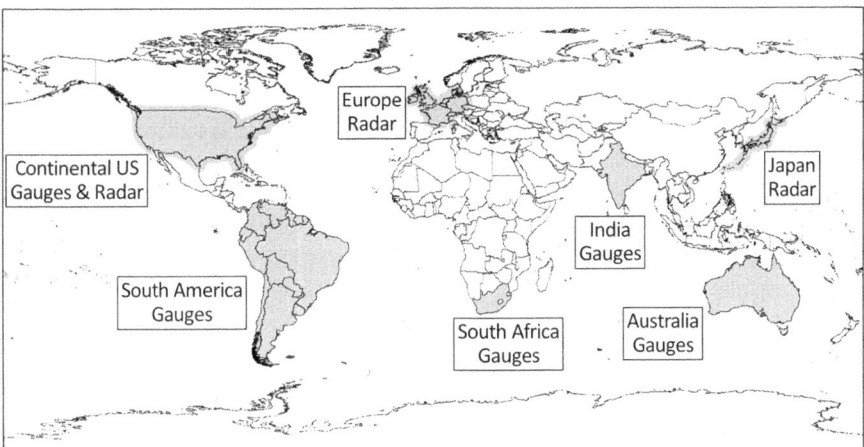

Fig. 25.1 Distribution of past and current IPWG validation regions and their surface reference data sets

Two main surface reference data sources used in the IPWG comparisons are:

Radar data, which provides instantaneous spatial measurements of precipitation usually at the lowest beam elevation. These instantaneous measurements need aggregating into daily totals, and also usually require averaging to the coarser 0.25° resolution of the IPWG comparisons. While weather radar undoubtedly provided valuable information at the instantaneous scale, artefacts are notable when aggregated spatially and temporally, such as range effects often missing low level precipitation (see Kidd et al. 2012), beam blockage, and anomalous propagation errors.

Gauge data, which is usually a time-integrated point measurement, the integration time being from a few minutes to hourly, daily or longer depending on the type of gauge and accessibility. While gauges provide a physical measure of precipitation at a 'point' (typically c.100 mm diameter), they try to represent the precipitation of the local area. However, given the temporal and spatially variability of precipitation, the distance over which their measurements are representative is somewhat limited: it is therefore good practice to only consider regions with networks of dense gauges that when gridded contribute multiple gauges to values within the grid boxes. It is worth noting that this leaves many 0.25° × 0.25° grid boxes with few or no surface-based observations, restricting the number of locations available for robust validation.

For the IPWG inter-comparisons radar data sets, as used over Japan, Europe, and the United States, are provided as gridded products, while gauge data sets, as used over United States, South America, South Africa, India and Australia, are generated from gauge reports with suitable thresholds set for the number of contributing gauges. Critically, these inter-comparisons rely upon the local expertise of the providers of the surface reference data sets to carry out the analyses and generate the results.

25.2 Current Validation Work: Data and Methodological Approach

The current ongoing IPWG validation work provides visual and statistical analysis of satellite/model-based vs surface data sets at a daily 0.25° × 0.25° resolution in near-real time. These results are available through web sites set up by the relevant validation region, which are accessible through the main IPWG cal/val web page (http://ipwg.isac.cnr.it/calval.html, last accessed 19 Oct. 2018). Each site provides online results as images for each product vs validation analysis, an example for South Africa is shown in Fig. 25.2 and another for the validation of CMORPH over the US in Fig. 25.3. While the layout of the results images varies between the different sites, the information available remains broadly similar.

The information contained in each results image generally includes:

- *Visual display* of the surface validation data and of the satellite-/model-based precipitation product, co-registered with map outline and latitude/longitude graticules. Inclusion of such imagery is deemed essential to place the statistics in the context of particular meteorological events;
- *Scatterplot* of satellite/surface data to provide an overview of the distribution and highlight any outliers;
- *Cumulative distribution plot* of the accumulation and occurrence of both satellite-/model-based product and surface reference data, aimed at identifying the ability of the precipitation product to faithfully represent the characteristics of the surface validation data (Europe, South Africa and India regions);

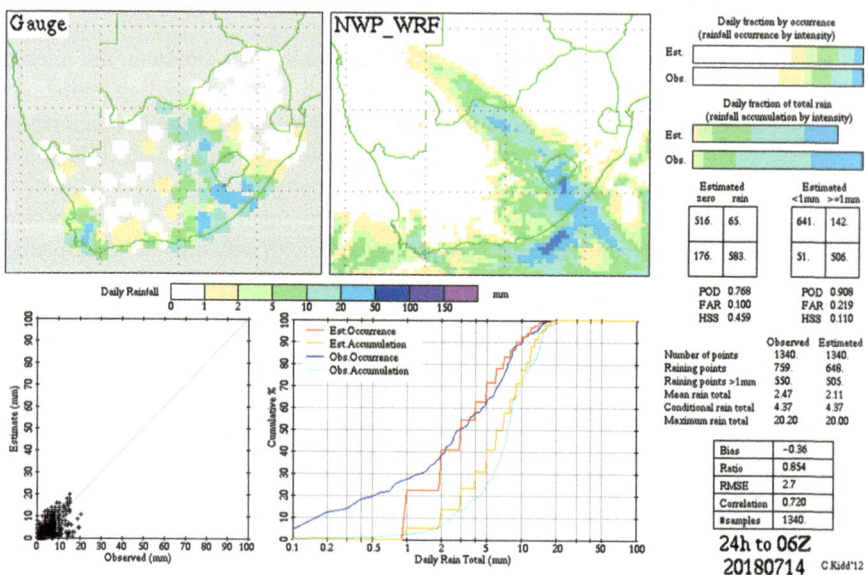

Fig. 25.2 Example of IPWG validation over South Africa for gauge data vs the WRF model output on 14 July 2018

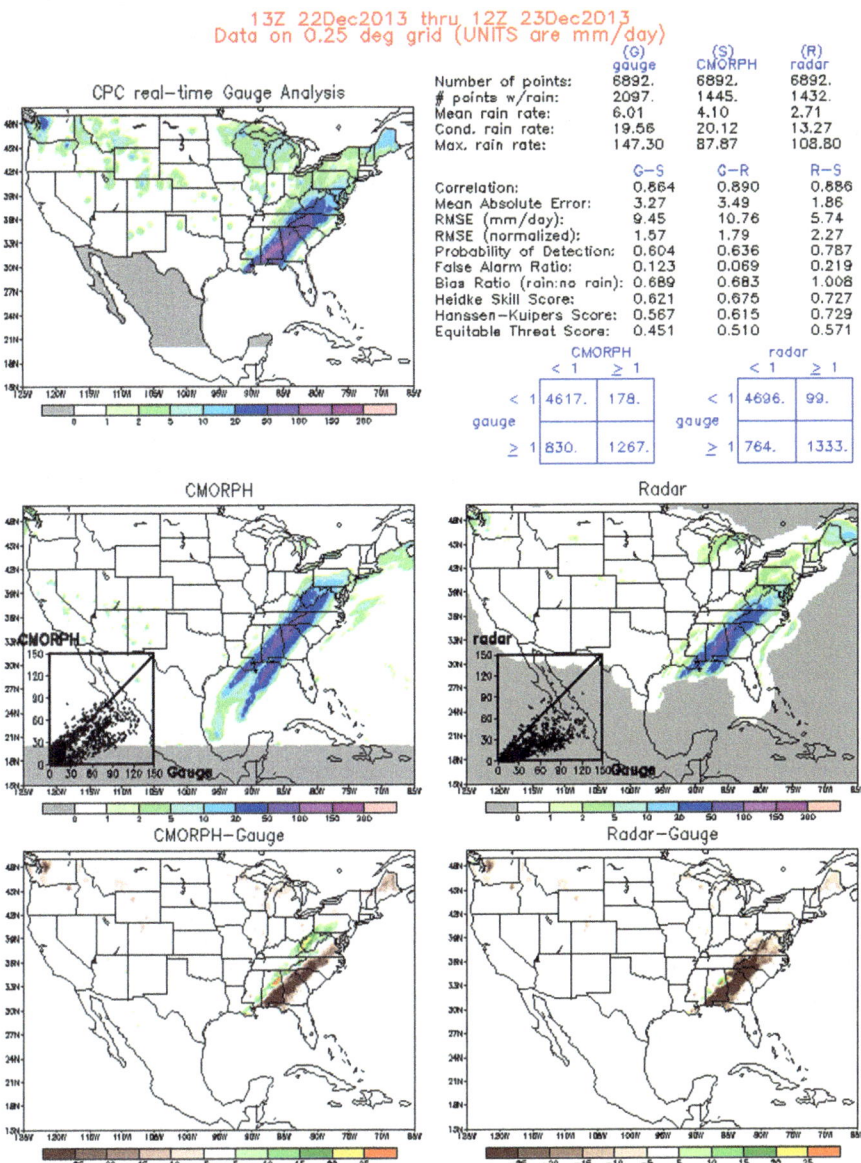

Fig. 25.3 Example of the IPWG validation of the CMORPH product over the United States, utilising both surface gauge and radar data. The comparison of the gauge and radar provides a useful measure of the retrieval ability of the surface reference data sets

- *Bar plots* of fractions of precipitation occurrence and accumulation by intensity to help identify differences between the product and validation data;
- *Categorical statistics* of rain/no-rain for both 0 and 1 mm day^{-1} thresholded precipitation, with scores of Probability of Detection (POD), False Alarm Ratio (FAR) and Heidke Skill Score (HSS);
- *Descriptive statistics* for the observed (validation) and estimate (product) precipitation, for number of values used in the comparison, number of raining values, number of raining values greater than 1 mm day^{-1}, conditional mean rain rate (i.e., when raining), conditional rain total (i.e., when raining), and maximum rain total (mm day^{-1});
- *Statistical scores* of bias, ratio (product/validation), RMSE, correlation and number of samples.

The IPWG inter-comparisons have generally followed a strategy of keeping any analysis clear and simple, not least to ensure that the results of the statistical comparison are understandable and pertinent to the user community. In addition, while the usefulness of some statistical scores are undoubtedly limited, particularly for assessing precipitation due to the skewness of the precipitation intensities, the limitations of these statistics are widely known and can therefore be taken into account in the interpretation of the results. More advanced comparison techniques, such as entity-based (contiguous rain area) verification (see Ebert and Gallus 2009), have been considered but have yet to be fully implemented in routine IPWG inter-comparison activities. In addition, some of the inter-comparison web sites also provide longer-term analysis of the results, including monthly and seasonal performances: two examples are included in the next section.

25.3 Examples of IPWG Validation

In this section a number of validation summaries are presented based upon the IPWG validation studies for the Japan and Brazil regions, providing longer-term comparisons and assessments of the different precipitation products against the available surface reference data.

25.3.1 Regional Analysis Over Japan Region

In 2003, the IPWG Japan validation site started producing standard, daily validation metrics for several satellite precipitation products in post-real time, with reference to a gauge-calibrated radar data-set provided by the Japan Meteorological Agency (JMA, Makihara et al. 1996; Makihara 2007), termed "Radar-AMeDAS" data. Based on the validation for the year 2004, Kubota et al. (2009) showed poor performance of satellite precipitation products over mountainous regions and coastlines, leading to the improvements of microwave radiometer (MWR) retrievals in the Global Satellite Mapping of Precipitation (GSMaP) over mountainous regions (Shige et al. 2013, 2014; Taniguchi et al. 2013; Yamamoto and Shige 2015; Yamamoto et al. 2017) and over coastlines (Mega and Shige 2016).

Near-real time validation for the satellite products started tentatively in 2008 (officially in 2009). In 2010, several operational numerical weather prediction (NWP) models were included. Here, a summary of the results during 2011–2014 when most of model forecasts were archived are given. This study focuses on two near real-time satellite products (GSMaP_NRT and 3B42RT), and four model forecasts (JMA, BoM, ECMWF, and METEOFR). One important caveat is that datasets that were available at the time of the comparisons, therefore the results do not necessarily reflect recent improvement of the datasets. For example, the improvements of MWR retrievals over mountainous regions (Shige et al. 2013, 2014; Taniguchi et al. 2013; Yamamoto and Shige 2015; Yamamoto et al. 2017) and over coastlines (Mega and Shige 2016) in GSMaP_NRT are not reflected. Validation results for the latest GSMaP products are described in Chap. 20 of volume 1.

Figure 25.4 shows time series of spatial correlation. Note that there was no snowfall estimation in the GSMaP products at the time of the comparison, but the threshold surface temperature above 4 °C for non-snowfall regions adopted by Kubota et al. (2009) is not used here. Satellite products generally exhibit a seasonal cycle in correlation with better statics during the warm season and poorer statistics during the cool season. Interestingly, seasonal fluctuations are much larger than those from the IPWG European validation site (see Figure 10a of Ebert et al. 2007) and the IPWG US validation site (see Figure 3 of Sapiano et al. 2010). This reflects various precipitation regimes over Japan, including "Baiu" frontal rainfall, tropical convective rainfall, and heavy snow associated with shallow convective clouds. The better performance of model forecasts during the cool season than the warm season is also consistent with that from other IPWG validation sites. However, the seasonal

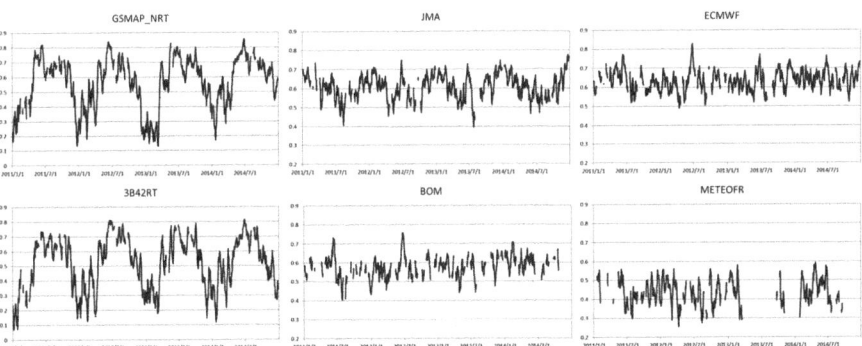

Fig. 25.4 Time series of spatial correlation over Japan between the Japan Meteorological Agency (JMA) gauge-calibrated radar data and various satellite estimates and numerical model short-range prediction over the period 1 January 2011 to 31 December 2014. The time series is displayed as an 11-day running mean to improve clarity

fluctuations of the model forecasts are much smaller than those the satellite products. This behaviour is in contrast to that of the models over the US (see Figure 3 of Sapiano et al. 2010). The difference may be explained by the dominant type of heavy rainfall (Sohn et al. 2013 and Chap. 39) over East Asia (warm-type convection) and over the North American continent (cold-type convection). In particular, the models show relatively good performance in June when "Baiu" frontal rainfall occurs, indicating that the models are quite skillful in predicting the large-scale stratiform rainfall.

Validation of snowfall was beyond the scope of Kubota et al. (2009) and remains challenging. Figure 25.5 presents the spatial distribution of temporal correlation for the cool season. Striking features include large differences in performance of the satellite algorithms between the Pacific coast and the Japan sea coast. The latter is well-known for heavy snowfall due to "lake-effect" over the Japan Sea where cold air outbreaks from the Eurasian continent receive a great amount heat and moisture. Heavy snow associated with shallow cumuliform clouds (Murakami et al. 1994) is not detected by the satellite algorithms. Correlations for the BoM and ECMWF models are small over the Japan sea coast, but the models, in general, show good performance over Japan in winter except for METEOFR. The results here suggest that an important focus for future work will be the improvement of the satellite

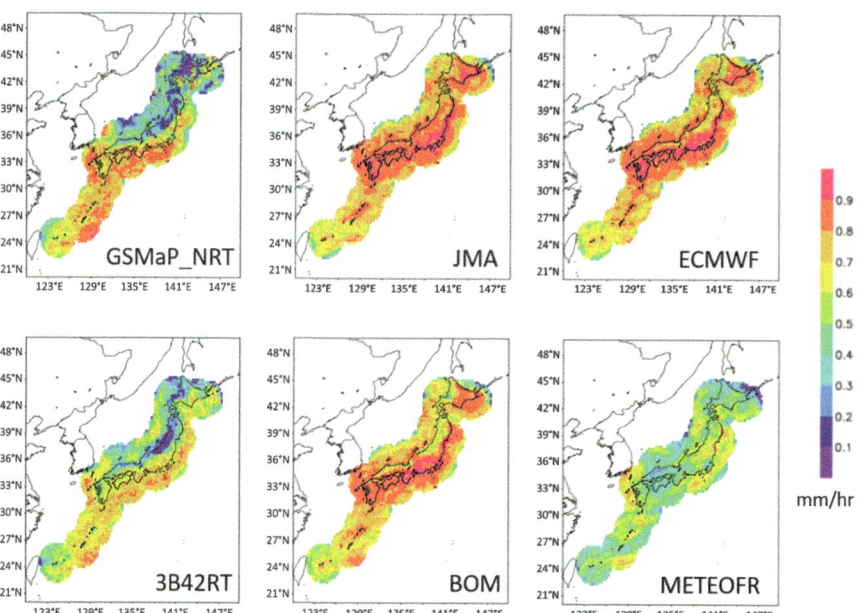

Fig. 25.5 Temporal correlation over Japan between the JMA gauge-calibrated radar data and various satellite estimates and numerical model short-range prediction for the December–February period during 2011–2014

algorithms for heavy snow associated with shallow cumuliform clouds, which occurs pervasively around the world (Kulie et al. 2016).

25.3.2   Seasonal Studies Over South America

The IPWG Brazil validation site exploits rain gauge data from across South America. While access to the daily gauge data in near-real time is often difficult, analysis of the data sets is later possible, which is particularly useful for assessing product performance at the monthly or seasonal scale.

In the example below (Fig. 25.6) a number of satellite products have been evaluated over the period from 1 Jan. 2015 through to 30 Nov. 2016. The algorithms evaluated here include Global Satellite Mapping of Precipitation (GSMaP) Near Real Time (-NRT) and Motion Vector Kalman (-MVK) products (see Shige et al. 2013, 2014; Yamamoto et al. 2017), TRMM Merged Precipitation Analysis (TMPA, Huffman et al. 2007), Hydro-estimator (Scofield and Kuligowski 2003), and Precipitation Estimation from Remotely Sensed Information using Artificial Neural Networks (PERSIANN, Hong et al. 2004). Daily gauge data (1200–1200 UTC), obtained from national and regional networks over Brazil and from GTS data form

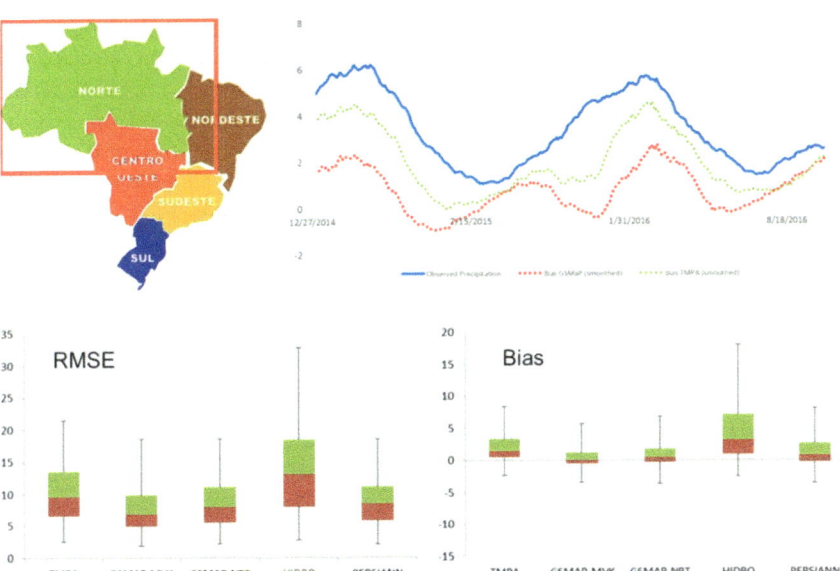

Fig. 25.6 Example of longer-term analysis of precipitation products over Northern Brazil of the IPWG South America validation site. The upper right plot shows the daily gauge-observed precipitation (blue) together with the bias of the GSMaP and TMPA products for 27 December 2014 through 18 August 2016. The lower left plot shows the range of the RMSE for each satellite precipitation product, while the lower right plot shows the bias for the same products

other South American countries, has been mapped to 0.25° resolution. As with other IPWG validation sites categorical statistics of POD, FAR, HSS, and bias score are calculated, together with quantitative statistics including the bias, RMSE, and correlation.

The results show seasonal variations in the performance of the different satellite products, together with the overall statistical range of the products for the period of the study. Such information is useful not only to the user community, but also to the product developers to help further improve and develop their algorithms. Indeed, product developers are mindful of feedback from the user communities as to the suitability of their products to specific applications.

25.3.3   Examples of IPWG-Related Studies

A number of studies have benefitted from the IPWG validation strategy and support. One of the goals of the IPWG has been the validation of products at resolutions that are commensurate with those needed by the user community, and as such the IPWG instigated the Program for the Evaluation of High Resolution Precipitation Products (PERHPP). At the time high resolution precipitation products were to deemed to have resolutions of 3-hourly, 0.25° × 0.25°, typified by the 3B42RT and similar products. Kidd et al. (2012) compared six satellite precipitation products, together with the ECMWF operational forecast model over Western Europe, using radar as the surface reference data set. Despite notable radar range effects present in the analysis, the results showed the seasonal variation in the performance of the satellite products, producing poorer results in the winter/cold season, due to frozen/shallow precipitation. Conversely, the model product produced poorer results in the summer months due to poorer identification of convective features. The ability of the model to adequately represent the diurnal cycle of precipitation was also highlighted by Kidd et al. (2013) who compared the 3B42RT, ECMWF model and TRMM Precipitation Radar: the satellite products were generally in agreement, while the model had difficulty in reproducing the complexity of the diurnal cycle in certain regions of the world.

As part of the Journal of Hydrometeorology IPWG-7 special collection, Maggioni et al. (2016) reviewed the main satellite precipitation validation efforts during the Tropical Rainfall Measuring Mission (TRMM) era. They presented major results from previous work and generalized the consistent findings for seven high-resolution satellite liquid precipitation products (TMPA 3B42RT research and real-time products, CMORPH, GSMaP near-real-time and the Motion Vector Kalman filter products, PERSIANN, and PERSIANN-CCS) over different regions of the world (Africa, North America, South America, Asia, Europe, Australia, and oceans). The performance of these satellite precipitation products was shown to be influenced by topography, seasonality, and climatology, particularly in terms of detection probability and bias. Mountainous regions are usually characterized by poor rain detection and magnitude-dependent mean errors, whereas semiarid areas are affected by low probability of detection. The largest biases are commonly observed during winter months, which are associated with lighter rainfalls, snow, and mixed-phase precipitation.

25.4   Future Validation Efforts

As noted above, the inter-comparison and validation of precipitation products is one of the IPWG core activities, and one that evolves over time as the scope of the IPWG develops, responding to ongoing improvements and as a result of the developer and user communities.

The IPWG has been quite successful in expanding the number of validation sites to include a range of meteorological and climatological regions, including the most recent additions of South Africa and India. The IPWG remains open to include more regions through the involvement of more nations in IPWG activities, and actively encourages participation from under-represented countries (e.g., Asia, Africa) through contacts with the relevant National Meteorological Agencies (NMAs) – and activity that has been pursued through the IPWG's validation working group agenda.

In addition to the 'standard' IPWG validation assessments at the $0.25° \times 0.25°$ daily scale, efforts are now underway to provide near real time evaluation and assessment of precipitation products at their full resolution, instantaneous scales. Initial comparisons have been set up for the European region (see http://ipwg.isac. cnr.it/calval.html, last accessed 19 Oct. 2018) where the GPROF products from the Global Precipitation Mission (GPM) constellation have been compared with surface radar data (see Fig. 25.7). As well as the usual IPWG results images, users may also download (copy/paste) daily/monthly tables of the statistical performance of the precipitation products to further analyse the results. The results of this fine-scale analysis appear in Kidd et al. (2018) and, together with analysis of the H-SAF precipitation products in Kidd et al. (2017b). It is anticipated that over time these comparisons will be expanded to include other finer spatial- and/or temporal-resolution products.

Development of future validation strategies include:

- *Inclusion of errors and uncertainties* in the validation process, including how to interpret and assess the different measures of uncertainty associated with satellite precipitation products. In addition, how errors and uncertainties within Level 2 (retrieval resolution/instantaneous scales) should transition through to Level 3 (gridded daily/monthly) products.
- *Validation of snowfall*, through establishing validation methodologies for snowfall through collaboration with the existing rainfall validation community.
- *Expansion of validation to rainfall over the open ocean* where possible, such as use of ocean buoy data; and
- *Large-scale validation using established data sets* such as the Global Precipitation Climatology Centre (GPCC) gauge data, at daily/monthly and $1°$ scales.

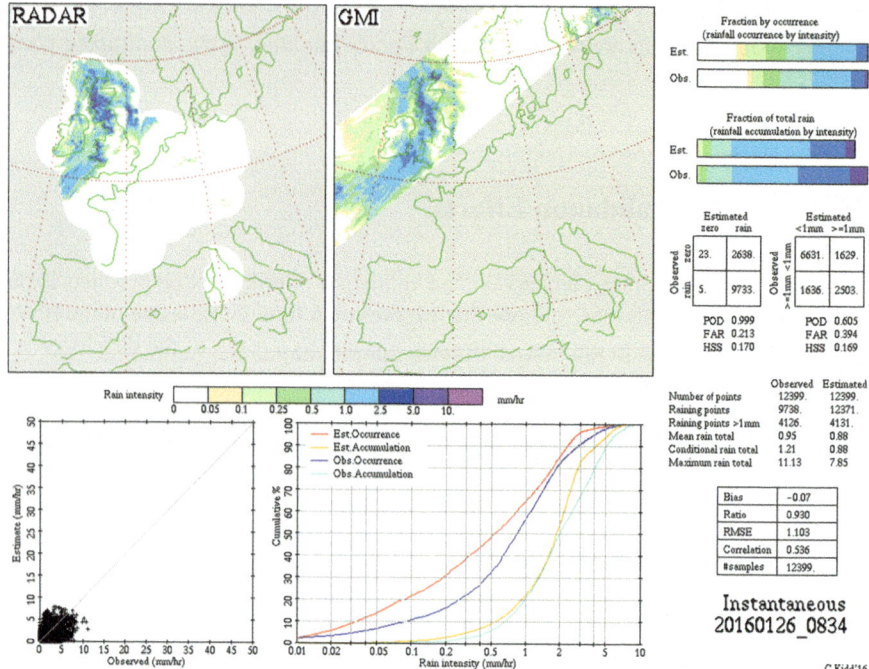

Fig. 25.7 Example of instantaneous, full retrieval resolution validation of the GMI precipitation product (GPROF Level 2 version 5) over the European IPWG validation region for 26 January 2016 at 0834 UTC

25.5 Conclusions

The IPWG provides a forum for both precipitation product developers and their user community. A core element of the IPWG is the ongoing validation of the precipitation products that allow the product developers and the user community access to assessments of products in near real time. Although these comparisons are generally at coarser temporal and spatial scales than targeted field validation campaigns, the longevity and coverage of the IPWG assessments provide a complementary approach. Further development of the IPWG validation effort has started with the analysis of products at the instantaneous, full resolution scales, which provides a better indication of the performance of the retrieval techniques, particularly across different precipitation systems and regimes (e.g., consistency in measures of the ratio and correlation). The IPWG validation is also looking to also provide longer-term global comparisons through the use of established data sets such as the GPCC gauge product. It is worth noting however that the IPWG validation work remains largely unfunded, and depends heavily on levering off existing validation efforts and through more widespread satellite precipitation estimation programs in general. In addition, it is not expected that the validation requirements of the diverse user communities for different applications would necessarily be met by the current IPWG validation programme. Some application-specific validation examples are discussed in other chapters of this volume (e.g., see Chap. 52), and the integration of application-relevant metrics could be a theme for future development of the IPWG validation effort.

The validation science working group of the IPWG has identified a number of key areas that need attention over the coming years, including the inclusion of errors and uncertainties within precipitation products and how to assess these within the current validation framework; the validation of snowfall products, and how to usefully assess multi-sensor precipitation products that include surface data sets.

The validation sites remain an important part of the IPWG activities for product developers who are keen to check the performance of their techniques against other algorithms and products, and for the users of such products who seek to establish which are most appropriate for their particular application(s) (e.g., Maggioni and Massari 2018).

References

Adler, R. F., Kidd, C., Petty, G., Morrissey, M., & Goodman, H. M. (2001). Inter- comparison of global precipitation products: The third Precipitation Intercomparison Project (PIP-3). *Bulletin of the American Meteorological Society, 82*, 1377–1396. https://doi.org/10.1175/1520-0477 (2001)082<1377:IOGPPT>2.3.CO;2.

Allam, R. E., Holpin, G., Jackson, P., & Liberti, G. L. (1993). Second Algorithm Intercomparison Project (AIP-2), U.K. and Northwest Europe, February–April 1991, Pre-Workshop Report. 439 pp. Available from Satellite Image Applications Group, U.K. Meteorological Office, Bracknell, Berkshire, RG12 2SZ, United Kingdom.

Arkin, P. A., & Xie, P. (1994). The global precipitation climatology project: First algorithm intercomparison project. *Bulletin of the American Meteorological Society, 75*, 401–419. https://doi.org/10.1175/1520-0477(1994)075<0401:TGPCPF>2.0.CO;2.

Barrett, E. C., & Bellerby, T. J. (1992). The application of satellite infrared and passive microwave rainfall estimation techniques to Japan – results from the 1st GPCP Algorithm Intercomparison Project. *Meteor Magazine, 121*, 34–46, ISSN:0026-1149. Available at https://digital.nmla. metoffice.gov.uk/digitalFile_ec4ad38c-2f61-4d24-b71b-ca885d78c32a/. Last accessed 29 Nov 2018.

Barrett, E. C., Dodge, J., Goodman, M., Janowiak, J., Kidd, C., & Smith, E. A. (1994). The first WetNet precipitation intercomparison project. *Remote Sensing Reviews, 11*, 49–60. https://doi. org/10.1080/02757259409532258.

Ebert, E. E. (1996). *Results of the 3rd Algorithm Intercomparison Project (AIP-3) of the Global Precipitation Climatology Project (GPCP). Revision 1.* Bureau of Meteorology Research Centre, 199 pp. Available at https://catalogue.nla.gov.au/Record/1350958. Last accessed 19 Oct 2018.

Ebert, E. E., & Gallus, W. A. (2009). Toward better understanding of the contiguous rain area (CRA) method for spatial forecast verification. *Weather and Forecasting, 24*, 1401–1415. https://doi.org/10.1175/2009WAF2222252.1.

Ebert, E. E., Manton, M. J., Arkin, P. A., Allam, R. J., Holpin, G. E., & Gruber, A. (1996). Results from the GPCP algorithm intercomparison programme. *Bulletin of the American Meteorological Society, 77*, 2875–2887. https://doi.org/10.1175/1520-0477(1996)077<2875:RFTGAI>2.0. CO;2.

Ebert, E. E., Janowiak, J. E., & Kidd, C. (2007). Comparison of near real time precipitation estimates from satellite observations and numerical models. *Bulletin of the American Meteorological Society, 88*, 47–64. https://doi.org/10.1175/BAMS-88-1-47.

Hong, Y., Hsu, K.-L., Sorooshian, S., & Gao, X. (2004). Precipitation estimation from remotely sensed imagery using an artificial neural network cloud classification system. *Journal of Applied Meteorology, 43*, 1834–1853. https://doi.org/10.1175/JAM2173.1.

Houze, R. A., Jr., McMurdie, L. A., Petersen, W. A., Schwaller, M. R., Baccus, W., Lundquist, J. D., Mass, C. F., Nijssen, B., Rutledge, S. A., Hudak, D. R., Tanelli, S., Mace, G. G., Poellot, M. R., Lettenmaier, D. P., Zagrodnik, J. P., Rowe, A. K., DeHart, J. C., Madaus, L. E., Barnes, H. C., & Chandrasekar, V. (2017). The Olympic Mountains Experiment (OLYMPEX). *Bulletin of the American Meteorological Society, 98*, 2167–2188. https://doi.org/10.1175/BAMS-D-16-0182.1.

Huffman, G. J., Bolvin, D. T., Nelkin, E. J., Wolff, D. B., Adler, R. F., Gu, G., Hong, Y., Bowman, K. P., & Stocker, E. F. (2007). The TRMM Multisatellite Precipitation Analysis (TMPA): Quasi-global, multiyear, combined-sensor precipitation estimates at fine scales. *Journal of Hydrometeorology, 8*, 38–55. https://doi.org/10.1175/JHM560.1.

Kidd, C., Ferraro, R. R., & Levizzani, V. (2010). The Fourth International Precipitation Working Group workshop. *Bulletin of the American Meteorological Society, 91*(8), 1095–1099. https://doi.org/10.1175/2009BAMS2871.1.

Kidd, C., Bauer, P., Turk, F. J., Huffman, G. J., Joyce, R., Hsu, K.-L., & Braithwaite, D. (2012). Inter-comparison of high-resolution precipitation products over Northwest Europe. *Journal of Hydrometeorology, 13*, 67–83. https://doi.org/10.1175/JHM-D-11-042.

Kidd, C., Dawkins, E., & Huffman, G. J. (2013). Comparison of precipitation derived from the ECMWF operational forecast model and satellite precipitation datasets. *Journal of Hydrometeorology, 14*, 1463–1482. https://doi.org/10.1175/JHM-D-12-0182.1.

Kidd, C., Becker, A., Huffman, G. J., Muller, C. L., Joe, P., Skofronick-Jackson, G., & Kirschbaum, D. B. (2017a). So, how much of the Earth's surface is covered by rain gauges? *Bulletin of the American Meteorological Society, 98*, 69–78. https://doi.org/10.1175/BAMS-D-14-00283.1.

Kidd, C., Panegrossi, G., Sanò, P., Ringerud, S., Casella, D., & Stocker, E. (2017b). Inter-comparison of precipitation products over Western Europe from the EUMETSAT H-SAF and NASA PPS. In Proceedings of EUMESAT 2017 Conference, Rome, Italy. [Available at https://www.eumetsat.int/website/home/News/ConferencesandEvents/DAT_3212307.html, last accessed 24 July 2018].

Kidd, C., Tan, J., Kirstetter, P.-E., & Petersen, W. A. (2018). Validation of the version 05 level 2 precipitation products from the GPM Core Observatory and constellation satellite sensors. *Quarterly Journal of the Royal Meteorological Society, 144*(S1), 313–328. https://doi.org/10.1002/qj.3175.

Kirschbaum, D. B., Huffman, G. J., Adler, R. F., Braun, S., Garrett, K., Jones, E., McNally, A., Skofronick-Jackson, G., Stocker, E., Wu, H., & Zaitchik, B. F. (2017). NASA's remotely-sensed precipitation: A reservoir for applications users. *Bulletin of the American Meteorological Society, 98*, 1169–1194. https://doi.org/10.1175/BAMS-D-15-00296.1.

Kubota, T., Ushio, T., Shige, S., Kida, S., Kachi, M., & Okamoto, K. (2009). Verification of high resolution satellite-based rainfall estimates around Japan using gauge-calibrated ground radar dataset. *Journal of the Meteorological Society of Japan, 87A*, 203–222. https://doi.org/10.2151/jmsj.87A.203.

Kulie, M. S., Milani, L., Wood, N. B., Tushaus, S. A., Bennartz, R., & L'Ecuyer, T. S. (2016). A shallow cumuliform snowfall census using spaceborne radar. *Journal of Hydrometeorology, 17*, 1261–1279. https://doi.org/10.1175/JHM-D-15-0123.1.

Levizzani, V., Kidd, C., Aonashi, K., Bennartz, R., Ferraro, R. R., Huffman, G. J., Roca, R., Turk, F. J., & Wang, N.-Y. (2018). The activities of the International Precipitation Working Group. *Quarterly Journal of the Royal Meteorological Society, 144*(S1), 3–15. https://doi.org/10.1002/qj.3214.

Maggioni, V., & Massari, C. (2018). On the performance of satellite precipitation products in riverine flood modeling: A review. *Journal of Hydrology, 558*, 214–224. https://doi.org/10.1016/j.jhydrol.2018.01.039.

Maggioni, V., Meyers, P. C., & Robinson, M. D. (2016). A review of merged high-resolution satellite precipitation product accuracy during the Tropical Rainfall Measuring Mission (TRMM) era. *Journal of Hydrometeorolgy, 17*, 1101–1117. https://doi.org/10.1175/JHM-D-15-0190.1.

Makihara, Y. (2007). Steps towards decreasing heavy rain disasters by short-range precipitation and land-slide forecast using weather radar accompanied by improvement of meteorological operational activities (in Japanese). *Tenki, 54*, 21–33.

Makihara, Y., Uekiyo, N., Tabata, A., & Abe, Y. (1996). Accuracy of radar-AMeDAS precipitation. *IEICE Transactions* on *Communications, 79*, 751–762. [Available at https://search.ieice.org/bin/summary.php?id=e79-b_6_751, last accessed 29 Nov 2018].

Mega, T., & Shige, S. (2016). Improvements of rain/no-rain classification methods for microwave radiometer over coasts by dynamic surface-type classification. *Journal of Atmospheric and Oceanic Technology, 33*, 1257–1270. https://doi.org/10.1175/JTECH-D-15-0127.1.

Morrissey, M. L., & Greene, J. S. (1991). *The Pacific Atoll Raingage data set.* Joint Institute for Marine and Atmospheric Research, University of Hawaii at Manoa, Contrib. No. 91-242, 445 pp.

Murakami, M., Matsuo, T., Mizuno, H., & Yamada, Y. (1994). Mesoscale and microscale structures of snow clouds over the Sea of Japan Part I: Evolution of microphysical structures in short-lived convective snow clouds. *Journal of the Meteorological Society of Japan, 72*, 671–694. https://doi.org/10.2151/jmsj1965.72.5_671.

Sapiano, M. R. P., Janowiak, J. E., Shi, W., Higgins, R. W., & Silva, V. B. (2010). Regional evaluation through independent precipitation measurements: USA. In M. Gebremichael & F. Hossain (Eds.), *Satellite rainfall applications for surface hydrology* (pp. 169–191). New York: Springer. https://doi.org/10.1007/978-90-481-2915-7_10.

Scofield, R. A., & Kuligowski, R. J. (2003). Status and outlook of operational satellite precipitation algorithms for extreme-precipitation events. *Monthly Weather Review, 18*, 1037–1051. https://doi.org/10.1175/1520-0434(2003)018<1037:SAOOOS>2.0.CO;2.

Shige, S., Kida, S., Ashiwake, H., Kubota, T., & Aonashi, K. (2013). Improvement of TMI rain retrievals in mountainous areas. *Journal of Applied Meteorology and Climatology, 52*, 242–254. https://doi.org/10.1175/JAMC-D-12-074.1.

Shige, S., Yamamoto, M. K., & Taniguchi, A. (2014). Improvement of TMI rain retrieval over the Indian subcontinent. In V. Lakshmi (Ed.), *Remote sensing of the terrestrial water cycle* (Geophysical monograph series) (Vol. 206, pp. 27–42). Washington, DC: American Geophysical Union. https://doi.org/10.1002/9781118872086.ch2.

Skofronick-Jackson, G., Hudak, D., Petersen, W. A., Nesbitt, S. W., Chandrasekar, V., Durden, S., Gleicher, K. J., Huang, G.-J., Joe, P., Kollias, P., Reed, K. A., Schwaller, M. R., Stewart, R., Tanelli, S., Tokay, A., Wang, J. R., & Wolde, M. (2015). Global Precipitation Measurement Cold Season Precipitation Experiment (GCPEx): For measurement sake let it snow. *Bulletin of the American Meteorological Society, 96*, 1719–1741. https://doi.org/10.1175/BAMS-D-13-00262.1.

Smith, E. A., Lamm, J. E., Adler, R., Alishouse, J., Aonashi, K., Barrett, E., Bauer, P., Berg, W., Chang, A., Ferraro, R., Ferriday, J., Goodman, S., Grody, N., Kidd, C., Kniveton, D., Kummerow, C., Liu, G., Marzano, F., Mugnai, A., Olson, W., Petty, G., Shibata, A., Spencer, R., Wentz, F., Wilheit, T., & Zipser, E. (1998). Results of WetNet PIP-2 project. *Journal of the Atmospheric Sciences, 55*, 1483–1536. https://doi.org/10.1175/1520-0469(1998)055<1483:ROWPP>2.0.CO;2.

Sohn, B. J., Ryu, G.-H., Song, H.-J., & Ou, M.-L. (2013). Characteristic features of warm-type rain producing heavy rainfall over the Korean peninsula inferred from TRMM measurements. *Monthly Weather Review, 141*, 3873–3888. https://doi.org/10.1175/MWR-D-13-00075.1.

Taniguchi, A., Shige, S., Yamamoto, M. K., Mega, T., Kida, S., Kubota, T., Kachi, M., Ushio, T., & Aonashi, K. (2013). Improvement of high-resolution satellite rainfall product for typhoon Morakot (2009) over Taiwan. *Journal of Hydrometeorology, 14*, 1859–1871. https://doi.org/10.1175/JHM-D-13-047.1.

Turk, F. J., & Bauer, P. (2006). The International Precipitation Working Group and its role in the improvement of quantitative precipitation measurements. *Bulletin of the American Meteorological Society, 87*, 643–647. https://doi.org/10.1175/BAMS-87-5-643.

Woodruff, S. D., Slutz, R. J., Jenne, R. L., & Steurer, P. M. (1987). A comprehensive ocean–atmosphere data set. *Bulletin of the American Meteorological Society, 68*, 1239–1250. https://doi.org/10.1175/1520-0477(1987)068,1239:ACOADS.2.0.CO;2.

Yamamoto, M. K., & Shige, S. (2015). Implementation of an orographic/nonorographic rainfall classification scheme in the GSMaP algorithm for microwave radiometers. *Atmospheric Research, 163*, 36–47. https://doi.org/10.1016/j.atmosres.2014.07.024.

Yamamoto, M. K., Shige, S., Yu, C.-K., & Cheng, L.-W. (2017). Further improvement of the heavy orographic rainfall retrievals in the GSMaP algorithm for microwave radiometers. *Journal of Applied Meteorology and Climatology, 56*, 2607–2619. https://doi.org/10.1175/JAMC-D-16-0332.1.

Chapter 26
The GPM Ground Validation Program

Walter A. Petersen, Pierre-Emmanuel Kirstetter, Jianxin Wang,
David B. Wolff, and Ali Tokay

Abstract We present a detailed overview of the structure and activities associated with the NASA-led ground-validation component of the NASA-JAXA Global Precipitation Measurement (GPM) mission. The overarching philosophy and approaches for NASA's GV program are presented with primary focus placed on aspects of direct validation and a summary of physical validation campaigns and results. We describe a spectrum of key instruments, methods, field campaigns and data products developed and used by NASA's GV team to verify GPM level-2 precipitation products in rain and snow. We describe the tools and analysis framework used to confirm that NASA's Level-1 science requirements for GPM are met by the GPM Core Observatory. Examples of routine validation activities related to verification of Integrated Multi-satellitE Retrievals for GPM (IMERG) products for two different regions of the globe (Korea and the US) are provided, and a brief analysis related to IMERG performance in the extreme rainfall event associated with Hurricane Florence is discussed.

Keywords Precipitation · Rainfall · Snowfall · Validation · GPM · Field campaign · Radar · Drop size distribution · Cloud microphysics · Path integrated attenuation · Convective · Stratiform · Snow water equivalent rate

W. A. Petersen (✉)
NASA, Marshall Space Flight Center, Huntsville, AL, USA
e-mail: walt.petersen@nasa.gov

P.-E. Kirstetter
School of Meteorology and School of Civil Engineering and Environmental Sciences and Advanced Radar Research Center, University of Oklahoma and NOAA/National Severe Storms Laboratory, Norman, OK, USA

J. Wang
Science Systems Applications International and NASA/GSFC, Greenbelt, MD, USA

D. B. Wolff
NASA/GSFC, Wallops Flight Facility, Wallops Island, VA, USA

A. Tokay
University of Maryland Baltimore County and NASA, GSFC, Greenbelt, MD, USA

© Springer Nature Switzerland AG 2020
V. Levizzani et al. (eds.), *Satellite Precipitation Measurement*, Advances in Global Change Research 69, https://doi.org/10.1007/978-3-030-35798-6_2

26.1 Overview

The overarching philosophy for NASA's implementation of GPM ground validation (GV) revolves around three highly complementary approaches (Hou et al. 2014; Skofronick-Jackson et al. 2017). These approaches include "direct", "physical" and "integrated" validation. Direct validation uses routinely available instrument networks and data products to assess convergence between and sources of uncertainty in GV ground- and GPM space-based precipitation estimates provided by the Dual-Frequency Precipitation Radar (DPR) and GPM Microwave Imager (GMI). Because versions of GPM products evolve as the mission progresses and statistical validation of orbital data should generally improve with the duration of sampling that occurs as the mission progresses, direct GV datasets are generally comprised of sustained data collections over regional to continental scales. As such, direct GV relies on contributions of high quality, calibrated ground observations from both operational and research instruments such as rain gauge and radar networks, regional and continental scale precipitation and hydrological products, and related activities on regional to continental scales.

Physical validation operates within the larger envelope of direct validation, but is focused on the execution of field campaigns designed to study targeted precipitation processes for the testing and improvement of retrieval algorithms. This is accomplished by use of combined airborne and ground-based field measurements of cloud microphysical and precipitation properties through the atmospheric column. Common instruments used include airborne in situ cloud microphysical probes, high altitude airborne and ground-based multi-frequency and/or polarimetric radar, and airborne multi-frequency microwave radiometers spanning microwave frequencies sampled by instruments in the GPM satellite constellation. Field campaign observations are often further used for testing of coupled atmospheric simulations designed for active and passive microwave retrieval algorithm testing (e.g., Matsui et al. 2013). Finally, integrated validation is focused on the assessment and utility of GPM products and their uncertainties in hydrometeorological and related applications. In a sense, integrated validation represents a bridge between GPM measurements, validation, and the hydrologic applications community.

Because many GPM applications and related "utilities" ultimately focus on quality of the products in a "direct" sense, in addition to outlining the NASA GPM GV program as a whole, this chapter will place a primary focus on relating GV datasets to direct validation activities with a brief review of results related to physical validation campaigns.

26.1.1   GV Measurement Synergy

NASA GV instrument components provide complementary precipitation measurements at a range of scales (Fig. 26.1) in time and space. From a direct GV

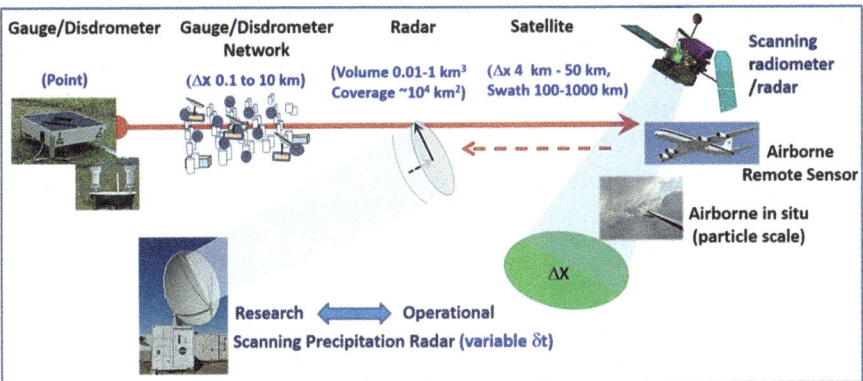

Fig. 26.1 Translation of high-quality precipitation measurements to satellite footprint and swath measurement scales for direct and physical validation

perspective, the requirement to validate myriad orbit-level GPM precipitation products (rain detection and intensity, drop size distribution, detection of snow etc.) at nearly instantaneous timescales for instrument fields of view (IFOV) of ~5 km or larger, requires the combined use of volume scanning multi-parameter radar and gauge instrumentation. From the perspective of physical GV (precipitation processes, algorithm physics) and integrated GV (hydrologic studies) sub-IFOV spatial scales are required.

Operating within and between the aforementioned scales and to accommodate the GV requirements, individual or spatially distributed rain gauge and disdrometer point measurements reference scanning dual polarimetric (DP) radar observations. In turn, DP radars translate the gauge and disdrometer measurements to instantaneously-viewed and spatially-distributed footprint, volume, and swath scales sampled by GPM satellite instruments. High quality national DP radar networks such as the WSR-88D radar network operated by the National Weather Service in the US, provide a continuous, reliable, and regional to continental scale "heartbeat" of statistical sampling, while research-grade DP radars provide the capability to flexibly target 3-D precipitation structure from sub-footprint to regional scales at high space-time resolution. In turn, these ground observations support airborne in situ and remote sensing measurements collected during physical validation campaigns- a collective bridging of ground, atmospheric column and space-based measurements of the precipitation process.

26.2 Validation Instruments, Data, and Examples

Here we describe primary GV datasets with examples of direct GPM level-2 (individual orbit) DPR and GMI validation including GPM Core Observatory (hereafter, "GPM CO") "Level-1" science requirements (L1SR; Skofronick-Jackson et al. 2017). L1SRs can be summarized as follows: (1) quantify rainrates over the range [0.22–110] mm h^{-1} ([0.2–60] mm h^{-1}) and at effective resolution of 5 km (15 km) for DPR (GMI); (2) detect falling snow at the respective instrument effective resolutions for both DPR and GMI; (3) estimate rain rates at intensities of 1 and 10 mm h^{-1} within bias and uncertainty levels that do not exceed 50% and 25%, respectively, at a resolution of 50 km for the GPM CO; and (4) estimate the mass-weighted mean diameter (D_m) of the drop size distribution (DSD) to within ± 0.5 mm of GV measurements for DPR.

26.2.1 Primary Datasets

US and international GV datasets collected and routinely *processed* by the NASA GV team are summarized in Tables 26.1 and 26.2, respectively (see also Skofronick-Jackson et al. 2018).

These data are currently collected, or have been collected in an extended mode (1 year or longer) for targeted continental and oceanic locations covering a variety of regimes, and from modes of near continuous continental scale monitoring to short-term periodic regional campaigns. Note that the international datasets shown in Table 26.2 represent only those data that have been or are routinely processed by NASA GPM GV; they do not represent the full complement of instrument, analysis, and product datasets collected by numerous international NASA Precipitation Measurement Mission (PMM) and GPM collaborating partners contributing to GPM GV around the globe (Hou et al. 2014; Skofronick-Jackson et al. 2017). Many of these datasets can be accessed on the GPM GV website found at https://gpm-gv.gsfc.nasa.gov/ (last accessed 18 Nov. 2018).

We highlight for discussion three particular activities and associated datasets.

(i) **Validation Network (VN):** Following Chandrasekar et al. 2008, a critical component of the direct validation strategy involves use of dual-polarimetric (DP) radar products (Fig. 26.1). More than 75 US VN operational WSR-88D (land and ocean) and contributing international DP radars are used for statistical comparisons of VN equivalent radar reflectivity (Z_e), DP moments, and DP-derived precipitation parameters such as rain rate, the DSD (mass-weighted mean diameter, D_m, normalized intercept parameter, N_w), and hydrometeor types. VN databases are built by carefully geolocating and matching coincident VN and GPM DPR radar pulse volumes along all DPR rays (Schwaller and Morris 2011). While the VN can be applied to virtually any DP radar and GPM overpass, the majority of VN radars used are located in the continental US and were selected to minimize obvious radar quality issues associated with excessive beam blockage and clutter. As such, the majority of the VN radars used are National Weather Service WSR88-D DP radars located in the central to eastern continental US. Oceanic locations include DP radars in Kwajalein, Middleton Island, Hawaii, Guam, and Puerto Rico.

Table 26.1 NASA GV primary US data products routinely produced from GPM launch to present

Product	Instrument	Data product description
U.S. Continental Scale/Multi-Regime Direct Validation		
GV Multi-Radar Multi-Sensor (MRMS) rain intensity and accumulation products	WSR-88D radar and collective U.S. rain gauge networks	GV-processed CONUS NOAA MRMS (Zhang et al. 2016) radar/rain gauge bias-adjusted precipitation rates and types. Resolution of 2-minute and $0.01° \times 0.01°$ for CONUS-coincident GPM orbits launch to present, and continuous 30 min accumulations (cf. Chap. 31)
Polarimetric Radar Validation Network (VN)	WSR-88D operational and Research grade (NPOL, KPOL) radars with GPM Satellite	For 70+ sites; CONUS/Offshore (including Kwajalein Atoll; KPOL) quality controlled DP radar and 3-D volumes of GPM Core satellite-coincident geo and volume-matched reflectivity, DP variables, rain-rates, hydrometeor types, rain drop size distributions, and match statistics within 100 km range of ground radars.
NASA Wallops Super Site (WFF) Datasets		Supersite/instrument base established at WFF and the surrounding region. Coordinated, sustained collection of GPM satellite and GV overpass datasets.
Scanning S-band polarimetric radar volumetric precipitation characteristics	NASA NPOL S-Band Radar	GPM orbit-coincident/weather event PPIs, RHIs, vertically pointing volume scans (DP moments, hydrometeor types, rain rates, drop size distributions). $1°$ beam $\times$ 125 m range gates. (Gerlach and Petersen 2011; Wolff et al. 2015)
Scanning dual-frequency (Ka/Ku) polarimetric radar volumetric precipitation characteristics	NASA Dual-frequency, Dual-pol. Doppler Radar (D3R)	GPM orbit-coincident/weather event PPIs, RHIs, vertically pointing volume scans and cross-sections (Ku/Ka-Band I&Q data, polarimetric moments, reflectivity, dual-frequency ratio, hydrometeor types, rain rates, drop size distributions). Radar ray $1° \times 150$ m range gates (Vega et al. 2014).

(continued)

Table 26.1 (continued)

Product	Instrument	Data product description
MRR-2/Pro precipitation profiles	K-Band Micro Rain Radar II/Pro	Continuous vertically pointing K-band radar; 2° beam × 30–60 m resolution for 30 gates; reflectivity, velocity spectra, DSD, precipitation rate;
Disdrometer network hydro-meteor size distributions, types, rain rate	2D-Video and Parsivel Disdrometers network	Disdrometer network continuous hydrometeor size distribution, types, shapes, and precipitation rates. 1–minute resolution, multi--instrument networks.
Precipitation imaging, rain and snow size distributions	Precipitation Imaging Package (PIP)	Continuous hydrometeor images, size distributions, fall-speeds, bulk density and water equivalent precipitation rate; deployments at WFF, Canada, Finland, Antarctica, Republic of Korea.
Dense rain gauge network rain intensity/accumulation	Single/dual/triple platform tipping bucket rain gauges	Continuous, quality-controlled, 1-minute average and accumulated rainfall, time of tip. 25 dual-platform gauges in network diameter of ~6 km, and numerous platforms in a broader 0.5° regional grid.
Snow, rain water accumulation rates	Pluvio 200, 400 weighing gauge network(s)	Continuous weighing-gauge frozen and liquid precipitation accumulations. Temporal scale 1–10 min accumulations. Includes WFF, 12-gauge Marquette, Michigan network, international deployments.

Instruments at the Wallops Flight Facility (WFF) Supersite are often deployed for extended periods in targeted regimes or to augment external instrument networks of opportunity

Specific to processing (Pippitt et al. 2015), VN DP radar data first undergo quality control, and estimation of rain and hydrometeor properties. DP clutter removal, multiple radar rain rate estimators, DSD and hydrometeor identification algorithms are all applied using approaches well established in the literature. For example, precipitation estimates in VN radar volumes are accomplished using hybrid DP precipitation estimation approaches based various combinations of reflectivity (Z), differential phase (KDP), and differential reflectivity (ZDR) as summarized in Bringi et al. (2004), Cifelli et al. (2011), Chen et al. (2017), and Bukovčić et al. (2018). DSD retrievals derive from a heritage of radar modeling approaches of 2D Video Disdrometer (2DVD) data (e.g., Brandes et al. 2004; Thurai et al. 2012). The NASA GV team has collected 2DVD DSD datasets in a broad sample of precipitation

Table 26.2 Routinely provided or extended international datasets shared with and processed by NASA GV, including collaborative instrument deployments

Country	Product/instrument description
Austria	WegenerNet dense rain gauge network for GPM level-2 and IMERG footprint-scale validation (cf. Kirchengast et al. 2014)
Brazil	CEMADEN; routine/operational S-band dual-polarimetric radar volumes (9 radars) VN-processed; national rain gauge network
Canada	ECCC-deployed NASA PIP, Pluvio gauges and routine data collection at multi-instrument sites for snowfall studies
Finland	FMI/U. Helsinki C-Band polarimetric radar and supporting snow water equivalent data from NASA PIP, Pluvio, hot-plate instruments deployed at Hytiaala site for continuous collection of snow products (cf. von Lerber et al. 2018).
Republic of Korea	KMA hourly gauge-adjusted radar estimates (RAD-RAR) of rain accumulation for IMERG validation (see Suk et al. 2013)
Netherlands	KNMI C-band dual-pol radar data for VN processing
Spain-UCLM	UCLM-Toledo extended deployment of TREx dual-pol radar at NASA WFF site for coordinated multi-frequency radar scanning

WegenerNet University of Graz, Austria; *CEMADEN* Brazilian Center for Natural Disaster Monitoring and Early Warnings, *ECCC* Environment Climate Change Canada, *FMI* Finnish Meteorological Institute, *KMA* Korean Meteorological Administration, *KNMI* Royal Netherlands Meteorological Institute, *UCLM* University of Castilla-La Mancha

regimes associated with locations of numerous physical-validation field efforts (Sect. 26.5) and extended multi-year observation periods at both Wallops Flight Facility (WFF) in Virginia and the Southeastern US in Huntsville, Alabama. The 2DVD-diagnosed N_w and D_m are used to model polarimetric radar moments of Z and ZDR and create subsequent empirical functions relating the DSD observables to simulated ZDR and Z (e.g., Tokay et al. 2020). The derived equations for D_m [f (ZDR)] and N_w [f(Z,ZDR)] are then used with VN polarimetric data to generate volume fields of the DSD for ray and range gate comparisons to DPR observations. DP fuzzy hydrometeor identification algorithms (HID) are applied within the VN processing stream (e.g., Dolan and Rutledge 2009 and references therein). After VN radar parameter retrievals are completed, GPM overpass and VN polarimetric data for precipitation events occurring within 100 km of a given radar are volume-matched and compared at geometric intersections of the GPM DPR and ground radar scans (Bolen and Chandrasekar 2003). To date, more than 45,000 volumes of coincident VN DP and GPM DPR radar data and derived products have been processed and matched since GPM launch. VN matched data files and intermediate DP radar and GPM data files are all archived.

(ii) ***GV-Multi-Radar Multi-Sensor (MRMS) products***: In parallel with the VN, NOAA/University of Oklahoma MRMS (cf., Zhang et al. 2016) radar-based estimates of precipitation are post-processed to provide a GPM GV dataset for continental scale statistical validation of GPM rain and falling snow water equivalent rates (SWER) over the US (130°–60°W, 20°–55°N). It comprises precipitation rate and type (liquid, frozen, convective, stratiform) together with radar data quality and gauge-bias ratio metrics at 0.01° spatial resolution. The resultant product, termed Level-2 (L2) GV-MRMS, is suited to "instantaneous" orbit-level validation of GPM precipitation estimates for all CONUS GPM overpasses (Kirstetter et al. 2014). A 30-min "level-3 (L3)" precipitation accumulation dataset (also including dominant precipitation type, radar data quality etc.) is also produced for validating products such as the GPM Integrated Merged GPM satellite rainfall product (IMERG; Tan et al. 2016, 2017a; O and Kirstetter 2017; Huffman et al. 2018).

The creation of a "best" GV-MRMS reference critically depends on additional adjustment and filtering of the datasets. For the GV-MRMS rain rate dataset, MRMS hourly gauge-bias adjustments are applied to 2-minute radar-estimated rain rates using a spatially-variable multiplicative bias field (e.g., Amitai et al. 2012; Kirstetter et al. 2012). A conservative filtering is applied in instances when the radar and gauge have significant quantitative disagreement (i.e., radar–rain gauge hourly ratios outside of the range 0.1–10) and by using a radar quality index (RQI). Only GV-MRMS rain pixels associated with RQI values of 1.0 are used to validate GPM rain rates. Snow water equivalent rates (SWER) (cf. Zhang et al. 2016) are processed at the same temporal/spatial scales as rain but with no bias correction or RQI-based filter applied, instead, additional beam height filters are used (cf. Sect. 26.2.4). Use of GV-MRMS data is constrained to a large region of CONUS where beam heights are ≤2 km (1.5 km) for rain (snow) for evaluating L1SR near surface precipitation rate criteria.

While the filtering approaches mentioned may not eliminate all errors in the GV-MRMS reference, they standardize the reference product across regions of the US. The evaluation of MRMS rain rates at fine scale (0.01° and 5 km) by Kirstetter et al. (2015a, b) demonstrated that biases are significantly mitigated using these qualitative and quantitative filtering procedures. Additional evaluation of the GV-MRMS using spatially-dense NASA WFF Pocomoke and Nassawadox rain gauge networks (Table 26.1) demonstrated GV-MRMS biases of 10%–15% or less and random errors of 35–40% for hourly rain rate estimates at the nominal footprint of the DPR. For L1SR verification we assume that the MRMS bias relative to that observed with the dense GV gauge networks will remain at or below ~15% at the 50 km scale, and that random errors will substantially decrease with spatial averaging (e.g., Steiner et al. 2003). For GPM L1SRs related to detection of snow, initial evaluations of MRMS rain/snow delineation are favorable, though cases of weak horizontal temperature gradients can be challenging (Chen et al. 2016). The high spatial resolution of GV-MRMS allows matching the resolution of any L2 and L3 GPM precipitation and estimate area-mean precipitation rates along with sub-IFOV precipitation occurrence, variability and types for direct validation of GPM DPR, GPROF, and IMERG products (Kirstetter et al. 2012, 2014, 2015a, b; Tan et al. 2016, 2017a; Kidd et al. 2018).

(iii) ***NASA WFF GV Precipitation Research Facility Products and Support Activities***: GPM GV maintains a world-class instrumentation network based at the NASA Wallops Flight Facility (WFF) on the mid-Atlantic Eastern Shore of Virginia. Instruments operated as part of this "supersite" include NASA's S-band DP radar (NPOL; Gerlach and Petersen 2011; Wolff et al. 2015), the Ku-Ka band Dual-frequency Dual-Polarimetric Doppler Radar (D3R, Vega et al. 2014), Micro Rain Radars (MRR-2, MRR-Pro), Precipitation Imaging Package(s) (PIP; e.g., Newman et al. 2009; von Lerber et al. 2018), Pluvio-2 weighing gauges, and networks of multi-tipping bucket rain gauge platforms and disdrometers (2D Video, Parsivel).

NPOL radar data are quality controlled and used for GPM direct statistical and physical process studies by science team members. D3R datasets are stored in raw in-phase and quadrature (I&Q) signals form and further processed on a case-by-case basis by team members at Colorado State University. D3R data products include attenuation corrected Ku/Ka-band reflectivity, derived quantities such as rain rates, hydrometeor types and DSD. When not deployed for field campaigns, the instruments are operated within the WFF network (or other domestic partner sites) to collect data during GPM overpasses and other weather-related targets of opportunity. Subsets of these instruments are also routinely operated with international GV partners (Table 26.2). WFF rain gauge network datasets are quality controlled and processed to create time-of-tip and one-minute splined rain rate products (e.g., Wang et al. 2008). The products are used to estimate area-mean rain rate at footprint scales and compile intra-footprint variability statistics, as an independent means of verifying GV-MRMS rain rate estimates (e.g., Marks et al. 2017), and direct validation of GPM DPR, GPROF, and IMERG products (Tan et al. 2016, 2017b; Kidd et al. 2018). Disdrometer network data (2D Video, Parsivel) are used for verifying GV radar calibration, developing DSD retrieval algorithms for GV polarimetric radars in the course of evaluating GPM DSD science requirements, and for studies of global to footprint scale DSD variability (Williams et al. 2014; Liao et al. 2014; Gatlin et al. 2015; Bringi et al. 2015; Thurai et al. 2017; Tokay et al. 2017).

For snow, PIP data are used to quantify the snow particle size distribution, bulk density, and to verify GV SWER products (e.g., Huang et al. 2015; Kneifel et al. 2015; von Lerber et al. 2018). Liao et al. (2016) used WFF PIP data to demonstrate GPM DPR dual-frequency ratio application to retrievals of bulk snow water contents and equivalent rates independent of derived particle size distribution parameters.

26.2.2 Example Applications of VN Datasets

VN datasets serve as a check on satellite and ground-based radar calibration, algorithm performance, and derived parameter stability between GPM product versions. Figure 26.2 illustrates one example wherethe VN detects the DPR reflectivity calibration change that occurred with release of DPR Version 5 (V5) products. In Fig. 26.2, VN and DPR sample volumes are compared for stratiform precipitation observed above the melting layer. Interestingly, the DPR reflectivity shift also impacted retrieved parameters such as D_m (Fig. 26.3), evident as a slight

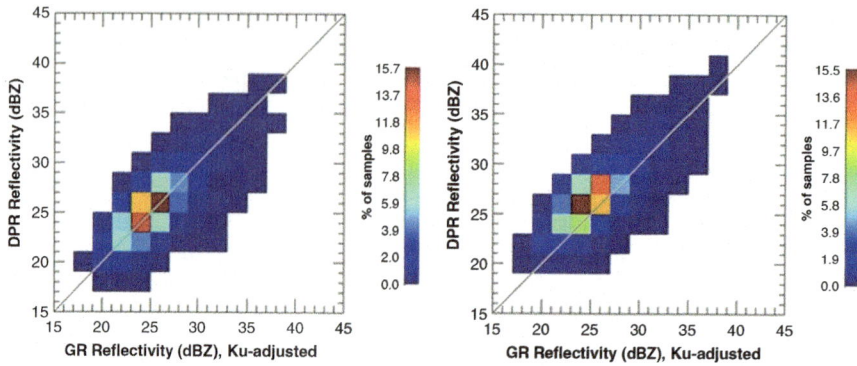

Fig. 26.2 VN verification of DPR calibration adjustment. The DPR Ku-band radar calibration was adjusted by JAXA from Version 4 (V4; left) to Version 5 (V5; right). The V5 increase of DPR Ku-Band radar reflectivity (Z_e) by ~1.2 dB (y-axis) is clearly evident in relative frequency histograms (shaded, percent) when plotted against the Ku-adjusted VN Z_e (Ground Radar, GR: X-axis)

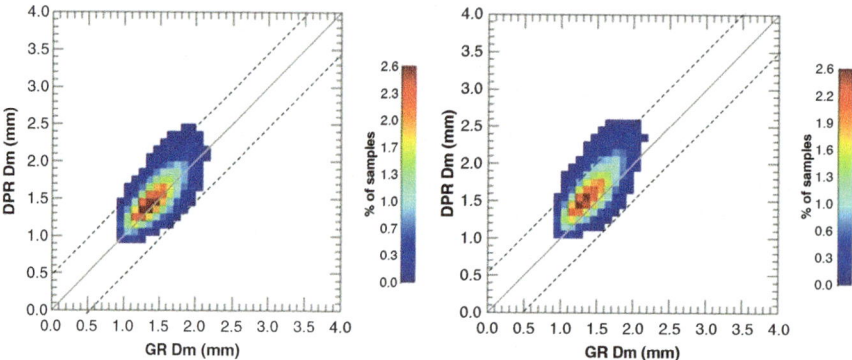

Fig. 26.3 Same as Fig. 26.1, but D_m (mm). Displayed are the V4 (left) and V5 (right) DPR D_m relative to GV. Adjusting the DPR calibration in V5 resulted in a slight, but perceptible positive bias shift in the DPR D_m relative to GV. NASA L1SRs for D_m (± 0.5 mm) indicated by dashed lines

(~0.2 mm) shift in the bias between V4 and V5 D_m relative to VN. Figure 26.3 also demonstrates that the GPM retrieval of D_m satisfies L1SRs.

Another example of the application of the VN includes direct verification of DPR algorithm corrections of Path Integrated Attenuation (PIA) as a function of precipitation type (e.g., convective/stratiform). Indeed, PIA-corrected reflectivity as a function of precipitation type is fundamental to DPR precipitation estimation and DSD retrieval (e.g., Seto and Iguchi 2015). The correction of DPR reflectivity and subsequent retrieval of precipitation rates can be especially challenging for convective precipitation at the frequencies used by the DPR. However, the VN data provide a robust verification of GPM PIA algorithm ability to correct convective reflectivity profiles (e.g., Fig. 26.4).

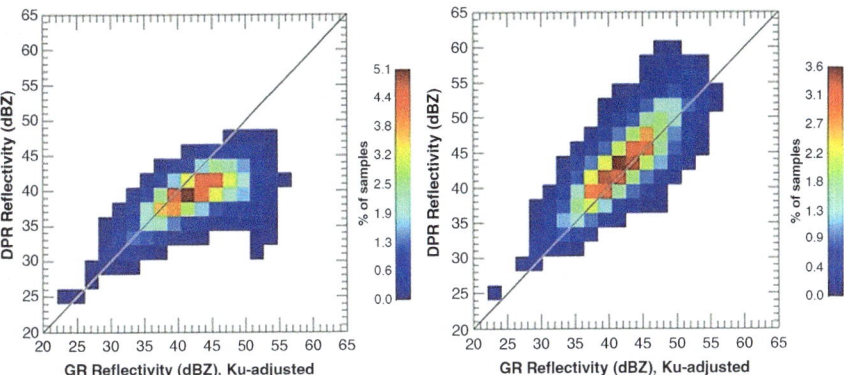

Fig. 26.4 DPR Measured (left) and PIA-corrected (right) V5 Ku-Band radar reflectivity (y-axis) in convective precipitation plotted against VN radar reflectivity (x-axis) for layers of precipitation below the height of the 0 °C level. Similar results are attained for stratiform precipitation (not shown)

More recently we have been modifying the VN for direct validation of GPM SWER. The approach recognizes the intrinsic difficulties in measuring an instantaneous SWER (even at the ground) at IFOV scales. The methodology relies on multiple radar-based estimators, seeking only first-order agreement between the DPR and the VN estimates. We leverage VN DP radar hydrometeor HID fields to identify the occurrence of snow and then estimate SWER using polarimetric estimators (e.g., capturing physical variability in the SWER; Bukovčić et al. 2018), or reflectivity-SWER (Z-S) equations based on probabilistic quantitative precipitation estimation (PQPE; e.g., Kirstetter et al. 2015a, b). The VN estimation "ensemble" is thus constructed from a single DP estimator, and three PQPE estimators (representing the 25th, 50th, and 75th percentile of Z-S SWER relationships). Verification of VN SWERs (Fig. 26.5) is being tested against 11 Pluvio snowgauges deployed in ~ 15 km footprint located over and to the southwest of the Marquette, Michigan WSR-88D radar.

Preliminary results (Fig. 26.5) indicate that the DPR estimates generally fall within the bounds of the VN estimators (all within a factor of two on SWER). Pluvio network reference data (20-min mean and median SWER) for the 15 km footprint indicate a mean SWER of 2.4 mm h^{-1}, with an individual gauge range of 1.4–3.3 mm h^{-1}. Of the VN radar estimates, the PQPE median (50th percentile) estimator was the highest, with the DP radar estimator providing the best match to the Pluvio estimates. The DPR estimate was slightly lower than the VN DP estimator. When upscaled to the CONUS-wide VN network for the full winter of 2018 (Fig. 26.6), a similar trend is found. That is, the DPR markedly underestimates the 50th percentile PQPE SWER but more closely resembles the DP SWER and even the 25th% PQPE SWER. Further verification of the VN estimators using the Pluvio network and other select measurement sites will enable more quantitative estimation of the uncertainty in VN SWER estimates resulting in improved application to GPM validation efforts.

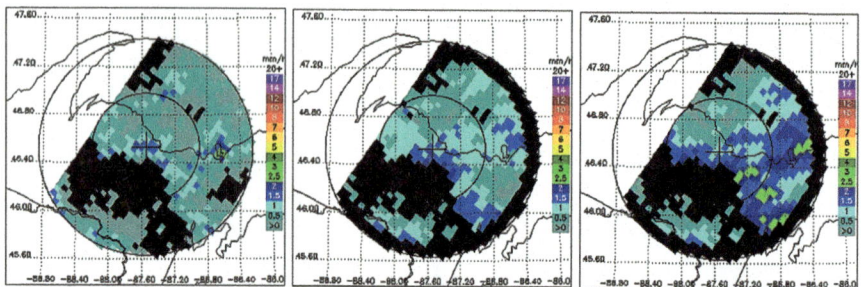

Fig. 26.5 Coincident and DPR footprint-matched VN-estimated SWER using the Marquette, Michigan WSR-88D radar (KMQT) and Pluvio network in the DPR swath, 15 Apr. 2018. The DPR single frequency estimate (left) is compared against the DP radar (center) and PQPE median estimators (right). The MQT Pluvio network is located west through south of KMQT and within 15 km range. Range rings are illustrated at 50 and 100 km from KMQT

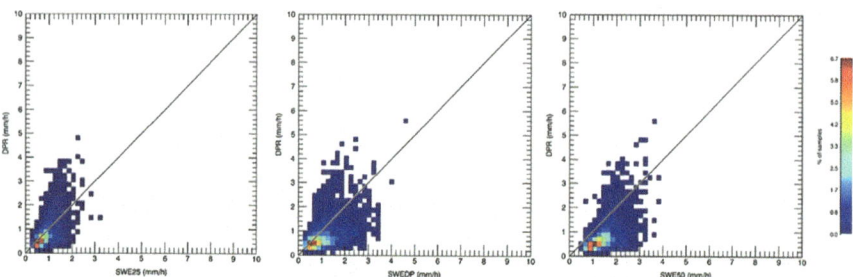

Fig. 26.6 Scatter density plots for winter 2017/18 DPR normal swath (y-axis) SWER plotted against VN-estimated SWER (x-axis) for the PQPE 25th% (left), dual-pol (center), and 50th% (right) SWER estimators. It is clear that the DPR SWER estimate best matches the VN dual-polarimetric SWER

26.2.3   Validation Using GV-MRMS

In parallel with VN datasets, high resolution GV-MRMS data serve as the primary contiguous continental scale reference for verifying GPM near surface rain rate estimation and snowfall detection. The data provide an efficient and statistically robust means to evaluate IFOV behavior of DPR estimates of rain rate for individual algorithm versions, precipitation regimes and types, and to monitor evolution of algorithm versions as the mission progresses. Relative to evaluating the overall quality and evolution of the DPR retrievals using the GV-MRM, consider Figs. 26.7 and 26.8. Both figures provide scatter density plots for different radar-based retrieval algorithms. For example, data shown in the single frequency KuPR algorithm scatter plots of Fig. 26.7 suggest reasonable but somewhat similar comparisons between KuPR and the GV-MRMS across V04 to V06. The net relative bias behavior in the Ku-PR estimate relative to GV-MRMS for all rain types evolves from a value of +2% in V04 to −10% in V05, with correlations of 0.53–0.51, respectively. Alternatively, in Fig. 26.8, GV-MRMS products indicate that GPM

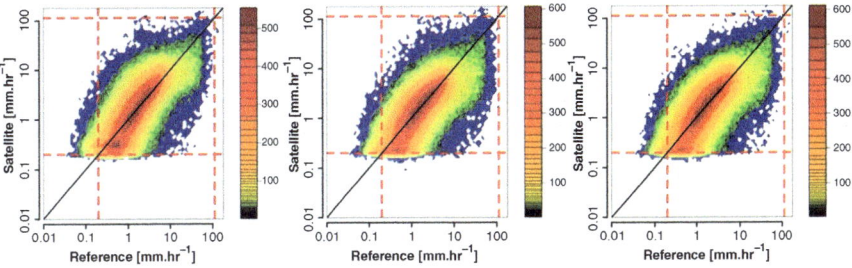

Fig. 26.7 Evolution of GPM versions. Joint histograms of GV-MRMS reference rain rate (x-axis) for all rain types plotted against matched DPR 2AKu-algorithm IVOV rain rate estimates (y-axis) V04 (left), V05 (center) and V06 (right)

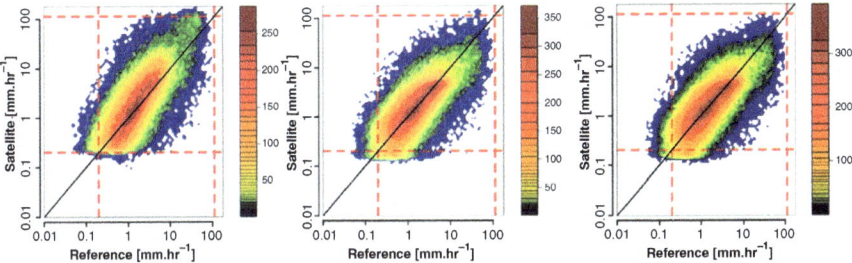

Fig. 26.8 Same as Fig. 26.7, but matched-swath dual-frequency combined radar-radiometer algorithm

rain rates estimated using the combined dual-frequency capability of the DPR radar and the GMI radiometer (e.g., the combined-radar radiometer algorithm; Grecu et al. 2016) more clearly improve with versions. A much better match to the GV-MRMS is evident in V05 and V06 relative to V04, with a relatively large V04 sample bias of +61% that markedly decreases to −3.6% in V05, and is +5.8% V06. Correlations were similar between versions at 0.59, 0.55 and 0.56 for V04, V05 and V06, respectively.

Importantly, the data points in Figs. 26.7 and 26.8 are dominated by the occurrence of stratiform precipitation (e.g., rates typically below 10 mm h^{-1}). It is therefore interesting to examine the same data points partitioned as a function of their convective (C) or stratiform (S) rain rate classification (e.g., Fig. 26.9). Figure 26.9 displays the C and S precipitation comparisons for the combined algorithm, but the same behavior is also observed in the KuPR product. Also, recall from the VN example shown in Fig. 26.4 that the retrieval algorithm does a reasonable job of correcting the radar reflectivity in convection. In Fig. 26.9, the stratiform rain rate scatter is reasonably well behaved, resembling that of Fig. 26.8. However, the convective rain rates of the GPM combined algorithm exhibit more error relative to the GV-MRMS, most notably at higher (>5 mm h^{-1}) and lower (<1 mm h^{-1}) rain rates. These departures from the GV-MRMS in convective rain contribute to the trends in Figs. 26.7 and 26.8, but the underlying behavior is masked.

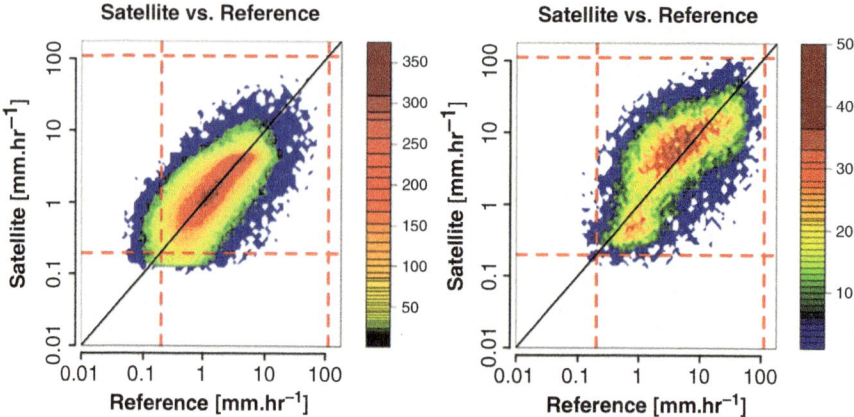

Fig. 26.9 Same as Fig. 26.8, but V5 combined algorithm for stratiform (left) and convective (right) precipitation. The KuPR algorithm strongly resembles this plot as well

The behavior of the DPR-derived rain rates is important to track as they serve as an a priori database for radiometer retrievals in the Goddard Profiling (GPROF) algorithm (Kummerow et al. 2015). Consider Fig. 26.10 which displays gridded comparisons of the KuPR and GMI-GPROF rain estimates over the continental US to the GV-MRMS product. The KuPR product compares well to the GV-MRMS, but is most notably high-biased (on average 25–30%) in the southern and central mid-section of the US. In tandem, the GPROF product also displays a high bias, though more pronounced, in the mid-section of the US. Conversely, over the eastern seaboard and neighboring Appalachian region, the GPROF is slightly lower than the GV-MRMS.

While GPROF precipitation estimates over land do rely on DPR-retrieved precipitation profiles, the attribution of error in GPROF rain rate is complicated. This is because the algorithm is intrinsically related to underlying ice-scattering relationships to rainfall, land surface emission characteristics, forecast model analysis errors and biases in relation to environment state, and even potential GV-MRMS matchup errors associated with parallax assumptions (e.g., Guilloteau et al. 2018). As is evident in Fig. 26.10 (contrast behavior in the west to that of the central US) the aforementioned rain and surface characteristics shift with geography (see also Kidd et al. 2018), and these are also likely tied to precipitation type/regime. On the other hand, understanding something about the underlying state of a given environment that supports a given rainfall regime can be used to improve the algorithm (e.g., Petkovic and Kummerow 2017).

26.2.4 Application to Verification of GPM L1SRs

As discussed previously, NASA's GPM L1SRs (this section; Skofronick-Jackson et al. 2017) define specific measurement range and error standards for retrieved

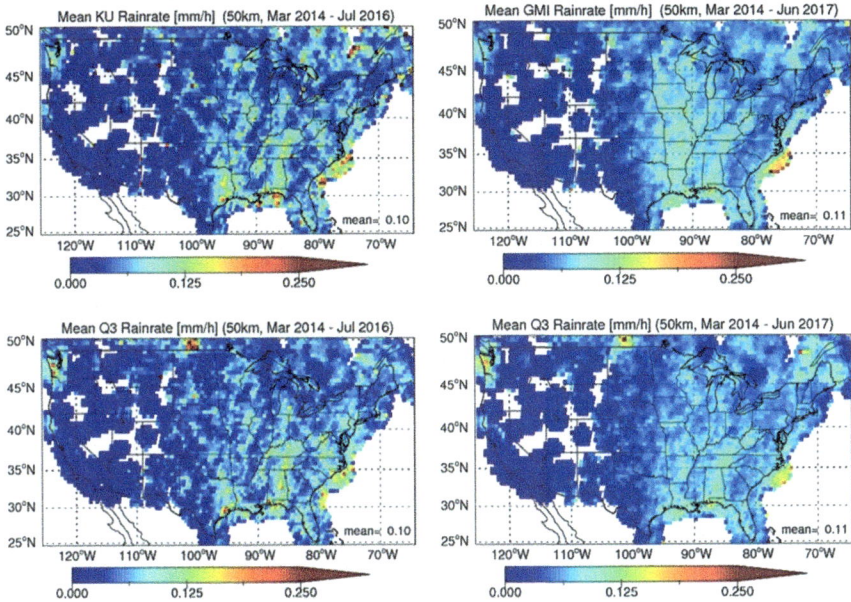

Fig. 26.10 KuPR (top left) and GMI-GPROF (top right) 50 km gridded mean rate rates (mm h^{-1}) plotted against their matched sample of GV-MRMS (indicated as "Q3") rain rates in bottom panels

precipitation parameters such as rain rate and raindrop size distribution, and somewhat more binary requirements for the detection falling snow (as opposed to estimation of water equivalent rate). Figures 26.3, 26.5, 26.6, 26.7 and 26.8, while used to illustrate GV monitoring of the evolution of GPM product versions or evaluation of SWER, also provide robust examples of L1SR verification of GPM L2 products. As noted previously, Fig. 26.3 illustrates satisfaction of the GPM L1SR pertaining to D_m estimation using VN GV data (Tokay et al. 2020). GV-MRMS data in Figs. 26.6, 26.7 and 26.8 confirm L1SR rain rate intensity ranges sampled at IFOV scales (e.g., 5 km for the DPR, 15 km GMI- not shown) and consistency between GPM product versions.

Figure 26.10 provides one example of GV-MRMS product use for examining the bias and uncertainty errors between GV and GPM over CONUS for L1SR grid scales (50 km). Note that these maps are continuously updated as GV-MRMS datasets and GPM CO level-2 data are produced and downloaded and as GPM versions change. For the formal verification process for L1SRs, we conservatively selected a large sub-region of the central and southern Plains of the US where GV-MRMS beam heights and rain gauge density requirements were optimal. A comparison for this region is exhibited in Fig. 26.11 for the GPM V6 Combined Dual-Frequency Radar-Radiometer algorithm. As illustrated in Skofronick-Jackson et al. (2018) for V5, V6 also easily meets L1SRs over the CONUS region sampled.

As in Figs. 26.9 and 26.11 shows a low bias (25%) for rain rates on the convective (>5–10 mm h^{-1}) and light rain ends (<0.6 mm h^{-1}) of the rain rate spectrum.

Fig. 26.11 GPM V6 combined radar-radiometer algorithm matched-swath rain rates (x-axis) vs. normalized error (%; y-axis) at 50 km grid scales over the CONUS L1SR region. Solid line is relative bias, dashed line is NMAE. Green shading represents L1SR requirement (Skofronick-Jackson et al. 2018 for V5 result)

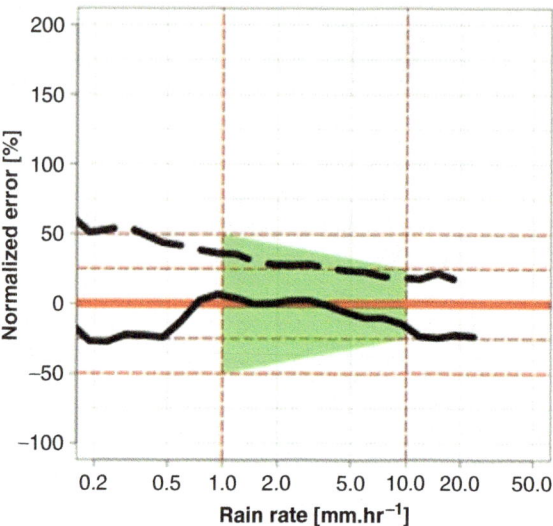

Though under investigation at this time, candidate explanations for the low biases include a combination of impacts ranging from approaches to constrain the DSD in the algorithms (e.g., Grecu et al. 2016), non-uniform beam filling and/or possible multiple scattering impacts on correction of path integrated attenuation, related DSD biases observed in convective rain (e.g., Petersen et al. 2018), or incorrect assumptions about the functional form of the DSD applied at the light rain rate end of the spectrum (e.g., Thurai et al. 2017).

In addition to comparisons reported for the CONUS (over land), we have also tested L1SRs over the ocean (e.g., Fig. 26.12). Here we selected two oceanic dual-polarimetric radar sites representing distinctly different regime types; i.e., tropical and mid/high-latitude climates. The radars and sites included the Kwajalein Atoll dual-pol (K-pol) radar operated by the US Army (with a strong tradition of GV use during the NASA TRMM era; e.g., Marks and Wolff, 2011), and the WSR-88D radar located on Middleton Island, Alaska (PAIH). Both radars were selected for their relatively open view of the ocean at ranges of 100 km or greater. Internal polarimetric consistency checks (Marks and Wolff, 2011), calibration adjustments (Wolff et al. 2015), and VN matched-volume comparisons of radar reflectivity between K-Pol, and GPM were used intermittently to monitor, verify and if necessary correct the reflectivity calibration on both platforms.

The quality-controlled data were subsequently interpolated to a 1-km Cartesian grid, with gridded rain rate files containing rain estimates based on the aforementioned DP radar retrieval methods (Cifelli et al. 2011, used for Fig. 26.12). In contrast to Fig. 26.12 and use of GV-MRMS over CONUS, because the number of 50×50 km^2 grid boxes in a single 100 km radar–range domain is fairly limited. Hence, to demonstrate verification of L1SRs we use matched GV radar – GPM IFOVs. For each GPM overpass of a given site, the GV radar volume scan occurring

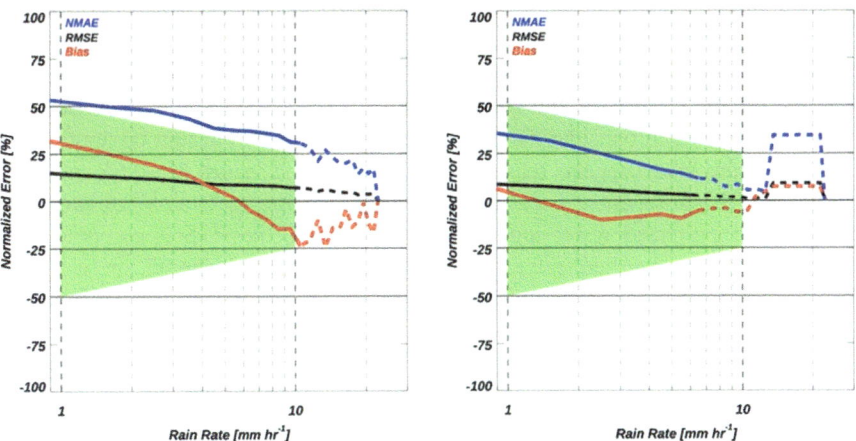

Fig. 26.12 Similar to Fig. 26.11 (i.e., V6) but for instrument footprint scales in the Kwajalein (K-pol; left), and PAIH (right) oceanic domains. The red line represents bias (%) and the blue line indicates normalized mean absolute error (NMAE; %) at EFOV scales. The solid black line indicates the RMSE (%) for EFOVs scaled to 50 km using Steiner et al. (2003)

closest in time to each overpass was identified and 1-km gridded rain rates at the 1 km height level located in each IFOV pixel were averaged. For the collection of matched pixels, we then calculated the bias, root mean square error (RMSE), mean absolute error, and NMAE. To evaluate L1SR criteria we assumed that the bias computed at the IFOV scale does not increase at 50 km grid scales, that the L1SR uncertainties could be computed using RMSE computed for IFOV scales and scale-adjusted via Steiner et al. (2003).

An example is shown in Fig. 26.12 for the combined radar-radiometer algorithm matched-swath (MS) product. The quality of the combined radar-radiometer algorithm meets L1SRs for estimation of the rain rate in both of the contrasting oceanic rain regimes. However, there are differences in behavior between bias and uncertainty trends of the tropical and mid-latitude regimes of Fig. 26.12, and similarly between the oceanic regimes and land regime sampled in Fig. 26.11. Considered with Fig. 26.10, these differences emphasize the non-global nature of rain rate retrieval errors in relation to the convergence of ground and space-based estimates of rainfall (Kidd et al. 2018).

In addition to rain rate and DSD, L1SRs also require that GPM CO instruments demonstrate "detection" of snow at GPM CO instrument IFOVs. Demonstration that GPM meets this L1SR has been accomplished in several ways. For example, we can directly compare ground-identified occurrences of snowfall coincident with satellite IFOVs (e.g., Figs. 26.5 and 26.6; see also von Lerber et al. 2018). We can also use larger datasets such as the GV-MRMS to statistically determine the delineation of snow from rain when precipitation is detected by the GV-MRMS and the CO. Additionally, we can attempt to identify a lower GV-MRMS snow water equivalent rate (SWER) threshold defining the satellite detection capabilities.

For the latter two tests the Heidke Skill Score (HSS) has proven useful for managing the tradeoff between increasing the percentage of correct detections (snow versus rain for delineation, positive versus zero snow rate for detection) and minimizing misses. An HSS value of one (zero) indicates perfect (no) contingency skill. For GPM L1SRs, when considering the delineation of snow from rain computation of the maximum HSS suggests reasonable performance at values of ~0.65 and ~0.47 for the DPR and GMI, respectively. For evaluating the "detection" of snow as a function of estimated water-equivalent snowfall rate, the HSS is computed as a function of the detected reference water-equivalent rate (based on a Z-S of $Z = 75 \times S^2$, random error suspected to be large). Maximizing the HSS enables us to identify the approximate lower threshold SWER for snow detection and to evaluate this model. Using this approach, the maximum HSS = 0.65 for the CO and occurs at a water equivalent rate of 0.53 mm h^{-1} for the DPR MS product. The maximum HSS for the GPROF GMI product is 0.34 at a rate of ~ 0.6 mm h^{-1}. Both DPR and GMI GPROF HSS appear to exhibit some very limited skill down to an SWER of ~0.3 mm hr.$^{-1}$. Interestingly, the HSS-based SWER threshold values of 0.5–0.6 mm h^{-1} for the GMI are similar to those computed in previous theoretical work conducted by Skofronick-Jackson et al. (2013). You et al. (2016) presented probability of detection results for the GMI that are consistent with our results-demonstrating GMI radiometer detection capability maximized by use of the higher frequency 166 GHz channel. Though numerous ways exist to approach the problem, GPM satisfies its L1SR for snow detection (Skofronick-Jackson et al., 2017, 2018).

26.3   Physical Validation Activities

Physical validation (PV) datasets originate from a plethora of pre and post-launch GPM GV field campaigns (Table 26.3). Data to validate retrieval algorithm assumptions and methodologies have been collected for a range of precipitation regimes and types. In the broadest sense PV data necessarily include most, if not all of the ground-based direct-statistical datasets discussed above (including processing), for both rain and snow regimes, with the important distinction that almost all PV datasets (except WFF, the Iowa Flood Studies campaign) also include high-altitude airborne (ER-2, DC-8) remote sensing data, airborne microphysical measurements, and supporting sounding profiles of atmospheric thermodynamic state (e.g., Skofronick-Jackson et al. 2015; Jensen et al. 2016; Houze et al. 2017). Airborne remote sensing datasets collected in PV campaigns serve as "proxy GPM satellite measurements" and depending on the situation, were coordinated with GPM over-passes when possible. The PV datasets generally include coincident downward looking multi-frequency radar collections at all or several of W, Ka, Ku, and X-band frequencies, and passive microwave radiometer data covering the 10–183 GHz frequency ranges. In situ airborne microphysical data were collected in coordination with remote sensing and ground-based instruments and used

Table 26.3 NASA physical-validation field deployment efforts

Field campaign (Partners)	Description
LPVEX 2010 (NASA, FMI, UH)	Light Precipitation Validation Experiment. High latitude cold rain over ocean and continental land surfaces. U. Wyoming King Air microphysics aircraft also carrying U. Wyoming W-band radar. FMI/UH C-band polarimetric radars, MRRs, radiometers, rawinsonde, snow video imager, disdrometer and rain gauge network.
MC3E 2011 NASA, DOE	Mid-latitude Continental Convective Clouds Experiment. Warm-season mid-latitude convective and stratiform precipitation. NASA ER-2 high altitude aircraft carrying HIWRAP Ka/Ku band radar, AMPR 10–85 GHz radiometer, UND Citation aircraft with microphysics suite, multi-frequency polarimetric radar network including NASA NPOL and DOE X/C-band radars, MRRs, dense rain gauge and disdrometer networks, high temporal resolution rawinsonde.
GCPEX 2012 (NASA, ECCC)	GPM Cold Season Precipitation Experiment. Mid-latitude synoptic and lake-effect snow. NASA DC-8 aircraft carrying 50–183 GHz CoSMIR radiometer, APR-2 Ka/Ku band radar. UND Citation microphysics aircraft, NRC C580 microphysics aircraft and W-band radar, C-band dual-pol radar, NASA D3R radar, MRR, Pluvio snow gauge and PIP network, rawinsonde.
IFloodS 2013 (NASA, U. Iowa)	Iowa Floods Studies. Warm-season mid-latitude mesoscale precipitation events and hydrologic validation. NASA NPOL and D3R radars, U. Iowa X-band radars, extensive gauge and disdrometer networks, soil moisture network.
IPHEx 2014 (NASA, Duke U.)	Integrated Precipitation and Hydrology Experiment. Warm-season orographic precipitation and hydrologic validation, coastal oceanic precipitation. NASA ER-2 carrying EXRAD, HIWRAP and CRS radars (X/Ka/W band), AMPR (10–85 GHz) and CoSMIR (50–183 GHz) radiometer suite, NASA NPOL and D3R radars, MRRs, ACHIEVE W-band radar, extensive disdrometer and rain gauge networks.
OLYMPEX 2015/16 (NASA, U. Washington)	Olympic Mountains Experiment. Cold season orographic and oceanic rainfall and snow; integrated hydrologic validation. NASA ER-2 carrying EXRAD, HIWRAP and CRS radars (X/Ku/Ka/W bands), AMPR (10–85 GHz) radiometer, AirMSPI polarimeter, eMAS VIS/IR imager, CPL backscatter lidar; NASA DC-8 carrying APR-3 (Ku/Ka/W-bands) radar, CoSMIR radiometer (50–183 GHz), MASC radiometer (118, 183 GHz) and dropsondes. NASA NPOL and D3R radars, NSF DOW radar, ECCC X-band radar, MRRs, PIPs, extensive rain gauge and disdrometer network, high temporal resolution rawinsondes.

Partners indicated in the first column [*FMI* Finnish Meteorological Institute, *UH* University of Helsinki, *DOE* US Department of Energy, *ECCC* Environment Climate Change Canada]. *HIWRAP* High Altitude Imaging Wind and Rain Airborne Profiler, *EXRAD* ER-2 X-band Radar, *CRS* Cloud Remote Sensing radar, *CoSMIR* Conical Scanning Millimeter Imaging Radiometer, *AMPR* Advanced Microwave Precipitation Radiometer, *DoW* Doppler on Wheels, *CPL* Cloud Physics Lidar, *AirMSPI* Airborne Multi-angle Spectro Polarimetric Imager, *eMAS* extended MODIS Airborne Simulator. (Also see Skofronick-Jackson et al. (2018, their Table 4))

standard suites of microphysical probes that span measurement scales of aerosol (0.1 μm) to large hydrometeors (2 cm).

Airborne and ground-based PV data have supported a variety of algorithm applications related to the physics and spatial variability of rain DSDs (e.g., Williams et al. 2014; Liao et al. 2014; Bringi et al. 2015; Gatlin et al. 2015; Thurai et al. 2017; Zagrodnick et al. 2018; Tokay et al. 2017, 2020) to include regime variability (Dolan et al. 2018), radar multiple-scattering and detection of the multiple scattering "knee" at DPR frequencies in strong convection (Heymsfield et al. 2014; Battaglia et al. 2014, 2016), ice hydrometeor profiles and radiometer response (Leppert and Cecil 2015), and ice and snow scattering at multiple radar frequencies (Molthan and Petersen 2011; Olson et al. 2016; Kneifel et al. 2015; Chase et al. 2018; Leinonen et al. 2018). PV datasets have also supported development of new GV methods for radar calibration (Wolff et al. 2015; Louf et al. 2018), creation of multi-parameter rain rate reference products (Seo et al. 2018), and multi-dataset fusion and analysis (e.g., Wingo et al. 2018). New analyses relative to snowfall estimation include reference estimation of bulk snow density and snow water equivalent rates (e.g., Huang et al. 2015; von Lerber et al. 2018) and the potential limitations of snow water equivalent rate estimation using passive microwave (Skofronick-Jackson et al. 2013). Finally, the field data have supported tests of cloud resolving model physics (e.g., Shi et al. 2010; Tao et al. 2013, 2014, 2016; Lang et al. 2014; Iguchi et al. 2014; Colle et al. 2017), and column microphysical impacts on satellite remote sensing simulators (Matsui et al. 2013).

Data from GPM PV campaigns are archived at the NASA Global Hydrology Resource Center (GHRC, https://ghrc.nsstc.nasa.gov/home/field-campaigns).

In addition to the larger campaigns mentioned in Table 26.3, the NASA GV effort has deployed instrumentation in several international-led campaigns, most recently the International Collaborative Experiment–PyeongChang Olympics and Paralympics Experiment 2018 (ICE-POP 2018) led by the Korean Meteorological Administration to study heavy orographic snow.

26.4 Validation of the GPM IMERG Product

The NASA GV team conducts routine verification of IMERG Early (E), Late (L) and Final (F) products. Accordingly, herein we provide a few examples of analysis types routinely conducted for two geographic regions (Korea and CONUS) and for one extreme event, Hurricane Florence. Because IMERG is such a popular product, many other examples of IMERG validation and analysis are available in the broader community for different areas of the world (e.g., Tang et al. 2015; Tan et al. 2017a; Dezfuli et al. 2017; O et al. 2017; Rios Gaona et al. 2017; O and Kirstetter 2017).

IMERG can be briefly described as a global gridded precipitation product that unifies microwave measurements from the GPM *network* of satellites with geostationary VIS/IR sensors sampling at higher temporal and spatial resolution (Huffman et al. 2018). IMERG covers latitudes of ±60° with high resolution 0.1° spatial and

30 min temporal sampling. To accommodate various user requirements for data latency and accuracy, the IMERG is generated in the form of near-real-time (IMERG-E, and IMERG-L; approximately 4 and 12-hour latency, respectively) and post-real-time research data that also incorporates a rain gauge-bias correction (IMERG-F). IMERG has undergone several episodes of development with many continuing improvements; here we examine Version-5b (V5b). Multi-scale validation of IMERG products for many versions and regions is essential to improving the product and its applications.

GV datasets including the GV-MRMS L3 products for CONUS, and KMA Real-time Adjusted Radar-AWS Rain rate (RAD-RAR) hourly rain accumulations (Table 26.2) are used to conduct comparisons with IMERG. Due to differences in spatial and temporal resolution among products, the validation is carried out using an approach that matches and temporally resamples specific GV and IMERG data to common grids (nominally 0.1°, and either 30 min for GV-MRMS, or 1 h for RAD-RAR); missing data are excluded. Standard statistical metrics such as Correlation Coefficient (CC), Relative Bias (Bias), Normalized Mean Absolute Error (NMAE), Normalized Root Mean Square Error (NRMSE) are used to quantitatively compare the performance of IMERG products. Comparisons between IMERG and the ground reference are conducted for individual extreme events (e.g., hurricanes), regional to continental spatial, and diurnal to multi-annual temporal scales. Common approaches using probability density and cumulative density functions (PDF, CDF, respectively) for rain occurrence and volume, scatter density plots, time series of daily rainfall, and instantaneous hourly rainfall are routinely produced. To evaluate precipitation detection capability, three widely applied categorical statistical metrics are also employed in the analyses: Probability of Detection (POD), False Alarm Ratio (FAR) and Heidke Skill Score (HSS). Contributions to errors are also evaluated as a function of the estimator type (e.g., IR or passive microwave), and fraction of contribution to the relative bias is examined by partitioning error into hit, miss and false alarm biases. A full collection of IMERG analyses with updates as new data are received can be viewed at the NASA GPM-GV website https://gpm-gv.gsfc.nasa.gov/ (last accessed 29 Nov. 2018).

26.4.1 Examples of IMERG Validation Over South Korea

Validation of IMERG using Korean RAD-RAR data is performed over the latitude/longitude box region enclosed by [125°–129.5°E, 34°–38°N]. Results herein focus on a commonly available period from March 2014 to October 2017. All data are resampled to 0.1°/1-h resolutions. The NASA GV team routinely updates results at: https://wallops-prf.gsfc.nasa.gov/KoreanQPE.V05/index.html (last accessed 18 Nov. 2018).

Categorical statistics (Fig. 26.13) suggest that IMERG products compare reasonably well with the RAD-RAR over the region of South Korea as a whole. IMERG-F appears to be in the best agreement with the RAD-RAR in terms of correlation and

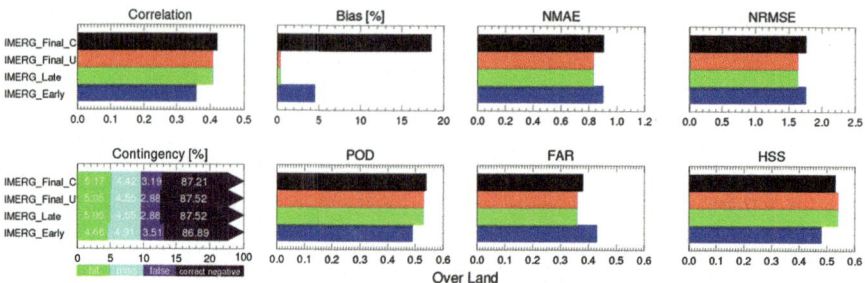

Fig. 26.13 March 2014 to October 2017 V5b IMERG vs. RAD-RAR statistics for Early (IMERG_E), Late (IMERG_L), Final without gauge adjustment (IMERG_F_U), and Final with gauge adjustment (IMERG_F_G). Top panels: Pearson correlation coefficient (Correlation), Bias (%), NMAE, and NRMSE. Bottom: contingency scores for hits (green), misses (light blue), false alarms (purple), and correct negatives (black). POD, FAR and HSS are also indicated

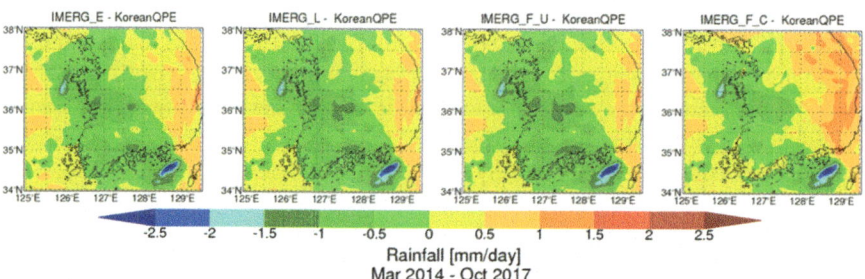

Fig. 26.14 Difference in daily rainfall between RAD-RAR and IMERG-E, L, F-U and F-C, products

POD. The expected improvement in skill between the IMERG-E and -L products is also evident, likely due to the inclusion of more microwave data and forward/backward morphing in IMERG-L. Note that IMERG-E and IMERG-L outperform the gauge bias-corrected IMERG-F, at least in terms of relative bias and uncertainty. Indeed, Fig. 26.14 suggests that monthly gauge-adjustments in the IMERG-F enhance disagreement with the RAD-RAR, increasing overestimation of precipitation rates- especially along the orography and coastline of the eastern Korean Peninsula. This is further demonstrated in Fig. 26.15 when the bias errors are decomposed into hit, miss, and false precipitation errors. Figure 26.15 suggests that the IMERG hit-bias is the primary driver of the bias error pattern for the period examined.

Categorical statistics (Fig. 26.13) suggest that IMERG products compare reasonably well with the RAD-RAR over the region of South Korea as a whole. IMERG-F appears to be in the best agreement with the RAD-RAR in terms of correlation and POD. The expected improvement in skill between the IMERG-E and L products is also evident, likely due to the inclusion of more microwave data and forward/backward morphing in IMERG-L. It is interesting to note that IMERG-E and

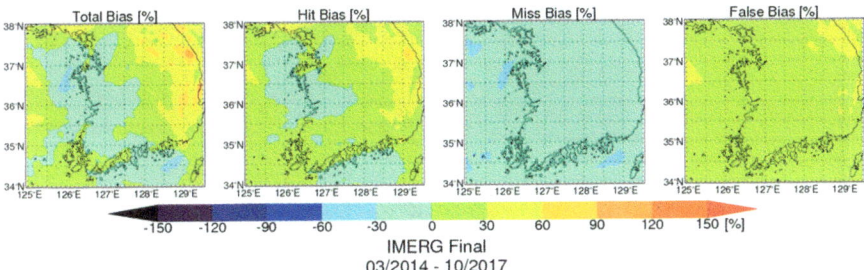

Fig. 26.15 Error components of IMERG-F V5b Final for March 2014–October 2017 broken down by total bias (E), hit bias (H), biases due to missing precipitation (M), and false precipitation (F)

IMERG- L outperform the gauge bias-corrected IMERG-F, at least in terms of relative bias and uncertainty. Indeed, Fig. 26.14 suggests that monthly gauge-adjustments in the IMERG-F enhance disagreement with the RAD-RAR, increasing overestimation of precipitation rates- especially along the orography and coastline of the eastern Korean Peninsula. This is further demonstrated in Fig. 26.15 when the bias errors are decomposed into hit, miss, and false precipitation errors. Figure 26.15 suggests that the IMERG hit-bias is the primary driver of the bias error pattern for the period examined.

26.4.2   Selected Examples of IMERG Validation Over CONUS

For verification studies of IMERG over CONUS we use half-hourly gauge-adjusted GV-MRMS rain accumulations at 0.01° resolution (e.g., Tan et al. 2017a, O and Kirstetter 2017). The comparisons are conducted at various spatial and temporal scales with respect to different precipitation intensities, and filtered with GV-MRMS radar quality index (RQI) thresholds as needed. As a first example, consider the scatter-density plots of Fig. 26.16. These plots demonstrate both the general performance of IMERG vs. GV-MRMS, but also enable tracking of performance between versions. Figure 26.16 demonstrates the improved performance of IMERG-L in V5b relative to V4 and GV-MRMS. The discretization of IMERG values in V4 is associated with the presence of underlying rain rate modes that existed in V4 of the passive microwave GPROF estimates. This issue was fixed in GPROF V5. Figure 26.16 suggests IMERG V5 exhibits a slightly high bias for rain rate values greater than ~1 mm h^{-1} when considered at CONUS scale. Figure 26.17 suggests that the bias indicated in Fig. 26.16 may be due to enhanced false-alarms over the central US combined with hit-bias reflecting IMERG sensitivity to underlying passive microwave estimates observed over the central portions of the US (e.g., Fig. 26.10).

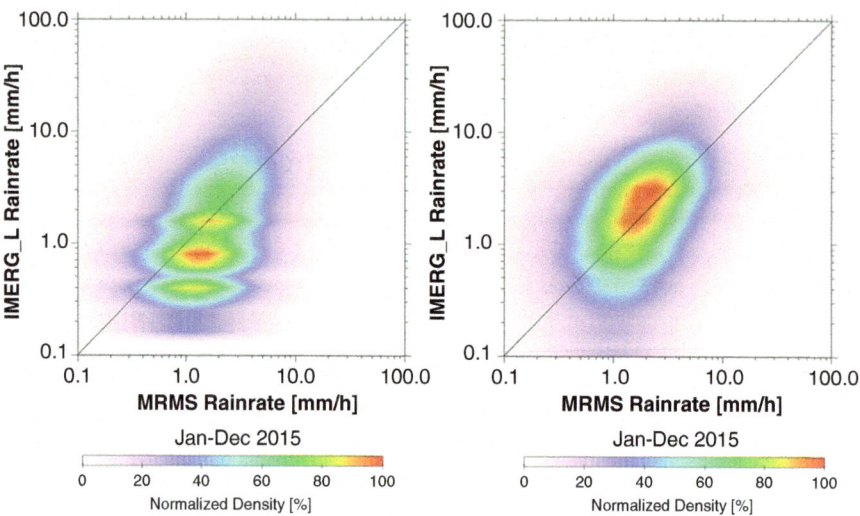

Fig. 26.16 CONUS-wide comparisons of V04A (left) and V5b (right) IMERG-L 30 min rain rates to GV-MRMS for the year of 2015

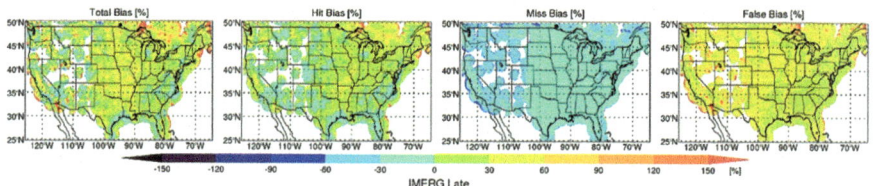

Fig. 26.17 Same as Fig. 26.14, but CONUS IMERG-L compared to GV-MRMS June 2014–August 2017

Of great interest to the applications community is the relative performance of IMERG in extreme rain events. Accordingly, we provide an example of GV-MRMS use for evaluating the performance of IMERG in estimating rain accumulations associated with the land-fall of Hurricane Florence. Florence made landfall near Wrightsville Beach, North Carolina (NC) on 14 Sept. 2018 as a Category-1 hurricane on the Saffir-Simpson scale. However, Florence stalled along the Carolina coastline just prior to and during its landfall, resulting in several days of excessive rain that occurred over a persistent storm surge- the combined effect being extensive and record-setting flooding.

Event total rainfall measured at one gauge location near Elizabethtown in southeastern North Carolina exceeded 900 mm, and more than 850 mm was measured at one gauge along the NC central coastline. An example of the GV-MRMS and IMERG-L product total rain accumulation for a seven-day period bracketing the period of heaviest rain over the Carolinas (10–17 September) is shown in Fig. 26.18 together with a map of the difference in accumulation between GV-MRMS and

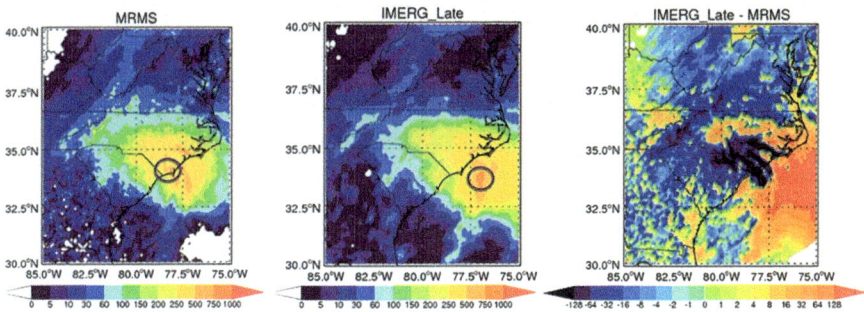

Fig. 26.18 Hurricane Florence total rain accumulation (mm) for 10–16 September 2018. Totals for the GV-MRMS, IMERG-L, and the difference in accumulation between the two (MRMS-IMERG) are displayed in the left, center and right panels, respectively. The location of the maximum accumulations for GV-MRMS and IMERG are indicated by the dark circle

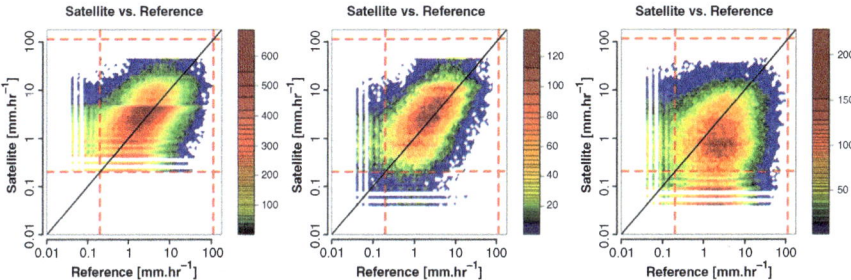

Fig. 26.19 Same as Fig. 26.6 and Fig. 26.7, but 30 min rainrates for domain of Fig. 26.16 for IMERG "Precipcal" (left), HQPrecip (microwave; center), and IR (right) during Hurricane Florence

IMERG. The GV-MRMS maximum accumulations are located in approximately the right locations relative to reported rain gauge maxima. The GV-MRMS maximum was 890 mm, close to that of gauge reports. In contrast, the IMERG-L maximum accumulation was located offshore at a value of 862 mm (~ 120 mm larger than the IMERG-E estimate, not shown). It is clear from the difference field in Fig. 26.18 that the IMERG-L bias structure was affected by the bands of the hurricane and associated training precipitation features along those bands. IMERG estimated much lower rainfall than the GV MRMS in bands located over southern NC and larger amounts in a band located over northern NC and offshore (some of the apparent bias 150 km or more offshore may be the result of GV-MRMS radar estimates overshooting precipitation at distant ranges). Another interesting bias feature in the difference plot of Fig. 26.18 is the prominent band of low bias long the foothills of the Appalachian Mountains. Here IMERG appears to have underestimated an orographically-forced component of the rainfall that was detected by GV-MRMS radars and gauges.

Figure 26.19 focuses on the contribution of two primary rain estimators to the IMERG calibrated precipitation estimate for Florence ("Precipcal" variable in the data files). Here there is a tendency for the microwave estimates to be slightly positively biased, but even more clear is the tendency for IR estimates to be very low biased. The observed trend of low IR-bias for Florence is also consistent with that noted by our team for other hurricanes such as Harvey, Irma and Michael (not shown). Further decomposition of error components by estimator will likely illuminate shortcomings and potential pathways to improve IMERG.

26.5   Summary and Moving Forward

The NASA GV team has successfully constructed high-quality standardized tools such as the GV-MRMS and VN for accomplishing fundamental direct validation, and forged international collaborations for analysis of precipitation datasets in a variety of regimes. The team developed multi-parameter radar and supporting gauge and disdrometer instrument infrastructure and a base to operate instruments in a supersite. During the pre and post-launch GPM phases GV planned and executed numerous field campaigns deploying supporting instruments on the ground and in the air. These physical validation datasets provided data enabling confirmation of basic approaches to DSD retrieval, scattering impacts on higher radar and radiometer frequencies, and basic variability of rain and snow properties to include applications in cloud modeling. In the post-launch phase, the GV team has been able to demonstrate conformance of GPM core observatory measurements to L1SRs, and provided data and analysis for numerous other applications ranging from multiple scattering and hail detection, drop size distribution issues in convection, and falling snow measurement. Moving forward, GPM GV will continue to improve reference measurements of SWER for use in radiometer and radar algorithm retrieval testing and verification. GV will work to improve measurements and representation in algorithms of light rain, conduct more refined studies and resolution of the impacts of non-uniform beam filling on multiple physical aspects of precipitation retrieval-especially in convection. Continuing emphasis will be placed on investigating orographic precipitation structure, variability, and environmental controls, especially as they pertain to the coupling of remote sensing and algorithm retrieval methodologies. A renewed emphasis will be placed on validation of IMERG products, especially in the context of integrated hydrologic validation.

Acknowledgements We are grateful to all our US and international GPM GV partners for their outstanding collaboration in GPM GV research. We specifically acknowledge the Korean Meteorological Administration for provision of RAD-RAR datasets enabling the IMERG validation and analysis discussed in this chapter. The NASA GPM and PMM Programs, specifically, Dr. Ramesh Kakar (retired), Dr. Gail Skofronick-Jackson, and Dr. Scott Braun, are acknowledged for their support of this research.

References

Amitai, E., Petersen, W. A., Llort, X., & Vasilof, S. (2012). Multi-platform comparisons of rain intensity for extreme precipitation events. *IEEE Transactions on Geoscience and Remote Sensing, 50*, 675–686. https://doi.org/10.1109/TGRS.2011.2162737.

Battaglia, A., Tanelli, S., Heymsfield, G. M., & Tian, L. (2014). The dual wavelength ratio knee: A signature of multiple scattering in airborne Ku–Ka observations. *Journal of Applied Meteorology and Climatology, 53*, 1790–1808. https://doi.org/10.1175/JAMC-D-13-0341.1.

Battaglia, A., Mroz, K., Lang, T., Tridon, F., Tanelli, S., Tian, L., & Heymsfield, G. M. (2016). Using a multiwavelength suite of microwave instruments to investigate the microphysical structure of deep convective cores. *Journal of Geophysical Research, 121*, 9356–9381. https://doi.org/10.1002/2016JD025269.

Bolen, S., & Chandrasekar, V. (2003). Methodology for aligning and comparing spaceborne radar and ground-based radar observations. *Journal of Atmospheric and Oceanic Technology, 20*, 647–659. https://doi.org/10.1175/1520-0426(2003)20<647:MFAACS>2.0.CO;2.

Brandes, E. A., Zhang, G., & Vivekanandan, J. (2004). Drop size distribution retrieval with polarimetric radar: Model and application. *Journal of Applied Meteorology, 43*, 461–475. https://doi.org/10.1175/1520-0450(2004)043<0461:DSDRWP>2.0.CO;2.

Bringi, V. N., Tang, T., & Chandrasekar, V. (2004). Evaluation of a new polarimetrically based Z–R relation. *Journal of Atmospheric and Oceanic Technology, 21*, 612–623. https://doi.org/10.1175/1520-0426(2004)021<0612:EOANPB>2.0.CO;2.

Bringi, V. N., Thurai, M., Tolstoy, L., & Petersen, W. A. (2015). Estimation of spatial correlation of rain drop size distribution parameters and rain rate using NASA's S-band polarimetric radar and 2D-video disdrometer network: Two case studies from MC3E. *Journal of Hydrometeorology, 16*, 1207–1221. https://doi.org/10.1175/JHM-D-14-0204.1.

Bukovčić, P., Ryzhkov, A., Zrnić, D., & Zhang, G. (2018). Polarimetric radar relations for quantification of snow based on disdrometer data. *Journal of Applied Meteorology and Climatology, 57*, 103–121. https://doi.org/10.1175/JAMC-D-17-0090.1.

Chandrasekar, V., Hou, A. Y., Smith, E. A., Bringi, V. N., Rutledge, S. A., Gorgucci, E., Petersen, W. A., & Skofronick-Jackson, G. (2008). Potential role of dual-polarization radars in the validation of satellite precipitation measurements: Rationale and opportunities. *Bulletin of the American Meteorological Society, 89*, 1127–1145. https://doi.org/10.1175/2008BAMS2177.1.

Chase, R., Finlon, J. A., Borque, P., McFarquhar, G. M., Nesbitt, S. W., Tanelli, S., Sy, O. O., Durden, S. L., & Poellot, M. R. (2018). Evaluation of triple-frequency radar retrieval of snowfall properties using coincident airborne in-situ observations during OLYMPEX. *Geophysical Research Letters, 45*, 5752–5760. https://doi.org/10.1029/2018GL077997.

Chen, S., Hong, Y., Kulie, M., Behrangi, A., Stepanian, P. M., Cao, Q., You, Y., Zhang, J., Hu, J., & Zhang, X. (2016). Comparison of snowfall estimates from the NASA CloudSat Cloud Profiling Radar and NOAA/NSSL Multi-Radar Multi-Sensor System. *Journal of Hydrology, 541*, 862–872. https://doi.org/10.1016/j.jhydrol.2016.07.047.

Chen, H., Chandrasekar, V., & Bechini, R. (2017). An improved dual-polarization radar rainfall algorithm (DROPS2.0): Application in NASA IFloodS field campaign. *Journal of Hydrometeorology, 18*, 917–937. https://doi.org/10.1175/JHM-D-16-0124.1.

Cifelli, R. C., Chandrasekar, V., Lim, S., Kennedy, P. C., Wang, Y., & Rutledge, S. A. (2011). A new dual-polarization radar rainfall algorithm: Application in Colorado precipitation events. *Journal of Atmospheric and Oceanic Technology, 28*, 352–364. https://doi.org/10.1175/2010JTECHA1488.1.

Colle, B. A., Naeger, A. R., & Molthan, A. (2017). Structure and evolution of a warm frontal precipitation band during the GPM Cold Season Precipitation Experiment (GCPEx). *Monthly Weather Review, 145*, 473–493. https://doi.org/10.1175/MWR-D-16-0072.1.

Dezfuli, A. K., Ichoku, C. M., Huffman, G. J., Mohr, K. I., Selker, J. S., van de Giesen, N., Hochreutener, R., & Annor, F. O. (2017). Validation of IMERG precipitation in Africa. *Journal of Hydrometeorology, 18*, 2817–2825. https://doi.org/10.1175/JHM-D-17-0139.1.

Dolan, B., & Rutledge, S. A. (2009). A theory-based hydrometeor identification algorithm for X-band polarimetric radars. *Journal of Atmospheric and Oceanic Technology, 26*, 2071–2088. https://doi.org/10.1175/2009JTECHA1208.1.

Dolan, B., Fuchs, B., Rutledge, S. A., Barnes, E. A., & Thompson, E. J. (2018). Primary modes of global drop size distributions. *Journal of the Atmospheric Sciences, 75*, 1453–1476. https://doi.org/10.1175/JAS-D-17-0242.1.

Gatlin, P., Bringi, V. N., Thurai, M., Petersen, W. A., Wolff, D. B., Tokay, A., Carey, L. D., & Wingo, M. (2015). Searching for large raindrops: A global summary of two-dimensional video disdrometer observations. *Journal of Applied Meteorology and Climatology, 54*, 1059–1069. https://doi.org/10.1175/JAMC-D-14-0089.1.

Gerlach, J., & Petersen, W. A. (2011). NPOL: The NASA transportable S-band dual-polarimetric Radar. Antenna system upgrades, performance and deployment during MC3E. In Preparation of 35th Conference Radar Meteorology, AMS, 25–30 September, Pittsburgh, PA, USA. Available at https://ams.confex.com/ams/35Radar/webprogram/Paper191918.html. Last accessed 29 Nov 2018.

Grecu, M., Olson, W. S., Munchak, S. J., Ringerud, S., Liao, L., Haddad, Z. S., Kelley, B. L., & McLaughlin, S. F. (2016). The GPM combined algorithm. *Journal of Atmospheric and Oceanic Technology, 33*, 2225–2245. https://doi.org/10.1175/JTECH-D-16-0019.1.

Guilloteau, C., Foufoula-Georgiou, E., Kummerow, C., & Petković, V. (2018). Resolving surface rain from GMI high-frequency channels: Limits imposed by the three-dimensional structure of precipitation. *Journal of Atmospheric and Oceanic Technology, 35*. https://doi.org/10.1175/JTECH-D-18-0011.1.

Heymsfield, G. M., Tian, L., Li, L., McLinden, M., & Cervantes, J. I. (2014). Airborne radar observations of severe hailstorms: Implications for future spaceborne radar. *Journal of Applied Meteorology and Climatology, 52*, 1851–1867. https://doi.org/10.1175/JAMC-D-12-0144.1.

Hou, A. Y., Kakar, R. K., Neeck, S., Azarbarzin, A. A., Kummerow, C. D., Kojima, M., Oki, R., Nakamura, K., & Iguchi, T. (2014). The global precipitation measurement mission. *Bulletin of the American Meteorological Society, 95*, 701–722. https://doi.org/10.1175/BAMS-D-13-00164.1.

Houze, R. A., McMurdie, L., Petersen, W. A., Schwaller, M. R., Baccus, W., Lundquist, J. D., Mass, C. F., Nijssen, B., Rutledge, S. A., Hudak, D. R., Tanelli, S., Mace, G. G., Poellot, M. R., Lettenmaier, D. P., Zagrodnik, J. P., Rowe, A. K., DeHart, J. C., Madaus, L. E., Barnes, H. C., & Chandrasekar, V. (2017). The Olympic Mountains Experiment (OLYMPEX). *Bulletin of the American Meteorological Society, 98*, 2167–2188. https://doi.org/10.1175/BAMS-D-16-0182.1.

Huang, G., Bringi, V. N., Moisseev, D., Petersen, W. A., Bliven, L., & Hudak, D. (2015). Use of 2D-video disdrometer to derive mean density-size and Z_e-SR relations: Four snow cases from the light precipitation validation experiment. *Atmospheric Research, 153*, 34–48. https://doi.org/10.1016/j.atmosres.2014.07.013.

Huffman, G. J., Bolvin, D. T., Braithwaite, D., Hsu, K., Joyce, R., Kidd, C., Nelkin, E., Sorooshian, S., Tan, J., & Xie, P. (2018). NASA Global Precipitation Measurement (GPM) Integrated Multi-Satellite Retrievals for GPM (IMERG) algorithm theoretical basis Doc. version 5.2. NASA GSFC, 30 pp. Available at https://pmm.nasa.gov/sites/default/files/document_files/IMERG_ATBD_V5.2.pdf. Last accessed 18 Nov 2018.

Iguchi, T., Matsui, T., Tao, W.-K., Khain, A. P., Phillips, V. T. J., Kidd, C., L'Ecuyer, T., Braun, S. A., & Hou, A. Y. (2014). WRF–SBM simulations of melting-layer structure in mixed-phase: Precipitation events observed during LPVEx. *Journal of Applied Meteorology and Climatology, 53*, 2710–2731. https://doi.org/10.1175/JAMC-D-13-0334.1.

Jensen, M., Petersen, W. A., Bansemer, A., Bharadwaj, N., Carey, L. D., Cecil, D. J., Collis, S. M., Del Genio, A. D., Dolan, B., Gerlach, J., Giangrande, S. E., Heymsfield, A., Heymsfield, G., Kollias, P., Lang, T. J., Nesbitt, S. W., Neumann, A., Poellot, M., Rutledge, S. A., Schwaller, M., Tokay, A., Williams, C. R., Wolff, D. B., Xie, S., & Zipser, E. J. (2016). The Midlatitude continental convective clouds experiment (MC3E). *Bulletin of the American Meteorological Society, 97*, 1667–1686. https://doi.org/10.1175/BAMS-D-14-00228.1.

Kidd, C., Tan, J., Kirstetter, P., & Petersen, W. A. (2018). Validation of the version 05 precipitation products from the GPM Core Observatory and constellation satellite sensors. *Quarterly Journal of the Royal Meteorological Society, 144*(S1), 313–328. https://doi.org/10.1002/qj.3175.

Kirchengast, G., Kabas, T., Leuprecht, A., Bichler, C., & Truhetz, H. (2014). WegenerNet – A pioneering high-resolution network for monitoring weather and climate. *Bulletin of the American Meteorological Society, 95*, 227–242. https://doi.org/10.1175/BAMS-D-11-00161.1.

Kirstetter, P.-E., Hong, Y., Gourley, J. J., Chen, S., Flamig, Z., Zhang, J., Schwaller, M., Petersen, W. A., & Amitai, E. (2012). Toward a framework for systematic error modeling of spaceborne precipitation radar with NOAA/NSSL ground radar-based National Mosaic QPE. *Journal of Hydrometeorology, 13*, 1285–1300. https://doi.org/10.1175/JHM-D-11-0139.1.

Kirstetter, P.-E., Hong, Y., Gourley, J. J., Cao, Q., Schwaller, M., & Petersen, W. A. (2014). A research framework to bridge from the global precipitation measurement mission core satellite to the constellation sensors using ground radar-based National Mosaic QPE. In L. Venkataraman (Ed.), *Remote sensing of the terrestrial water cycle* (AGU books geophysical monograph series, Chapman monograph on remote sensing). Hoboken: Wiley. ISBN: 1118872037.

Kirstetter, P.-E., Gourley, J. J., Hong, Y., Zhang, J., Moazamigoodarzi, S., Langston, C., & Arthur, A. (2015a). Probabilistic precipitation rate estimates with ground-based radar networks. *Water Resources Research, 51*, 1422–1442. https://doi.org/10.1002/2014WR015672.

Kirstetter, P.-E., Hong, Y., Gourley, J. J., Schwaller, M., Petersen, W. A., & Cao, Q. (2015b). Impact of sub-pixel rainfall variability on spaceborne precipitation estimation: Evaluating the TRMM 2A25 product. *Quarterly Journal of the Royal Meteorological Society, 141*, 953–966. https://doi.org/10.1002/qj.2416.

Kneifel, S., von Lerber, A., Tiira, J., Moisseev, D., Kollias, P., & Leinonen, J. (2015). Observed relations between snowfall microphysics and triple-frequency radar measurements: Triple frequency signatures of snowfall. *Journal of Geophysical Research, 120*, 6034–6055. https://doi.org/10.1002/2015JD023156.

Kummerow, C. D., Randel, D. L., Kulie, M., Wang, N.-Y., Ferraro, R., Munchak, S. J., & Petkovic, V. (2015). The evolution of the Goddard profiling algorithm to a fully parametric scheme. *Journal of Atmospheric and Oceanic Technology, 32*, 2265–2280. https://doi.org/10.1175/JTECH-D-15-0039.1.

Lang, S., Tao, W.-K., Chern, J., Wu, D., & Li, X. (2014). Benefits of a 4th ice class in the simulated radar reflectivities of convective systems using a bulk microphysics scheme. *Journal of the Atmospheric Sciences, 71*, 3583–3612. https://doi.org/10.1175/JAS-D-13-0330.1.

Leinonen, J., Lebsock, M. D., Tanelli, S., Sy, O. O., Dolan, B., Chase, R. J., Finlon, J. A., von Lerber, A., & Moisseev, D. (2018). Retrieval of snowflake microphysical properties from multi-frequency radar observations. *Atmospheric Measurement Techniques, 11*, 5471–5488. https://doi.org/10.5194/amt-2018-73.

Leppert, K. D., & Cecil, D. J. (2015). Signatures of hydrometeor species from airborne passive microwave data for frequencies 10–183 GHz. *Journal of Applied Meteorology and Climatology, 54*, 1313–1334. https://doi.org/10.1175/JAMC-D-14-0145.1.

Liao, L., Meneghini, R., & Tokay, A. (2014). Uncertainties of GPM DPR rain estimates caused by DSD parameterizations. *Journal of Applied Meteorology and Climatology, 53*, 2524–2537. https://doi.org/10.1175/JAMC-D-14-0003.1.

Liao, L., Meneghini, R., Tokay, A., & Bliven, L. F. (2016). Retrieval of snow properties from Ku- and Ka-band dual-frequency radar. *Journal of Applied Meteorology and Climatology, 55*, 1845–1858. https://doi.org/10.1175/JAMC-D-15-0355.1.

Louf, V., Protat, A., Warren, R. A., Collis, S. M., Wolff, D. B., Raunyiar, S., Jakob, C., & Petersen, W. A. (2018). An integrated approach to weather radar calibration and monitoring using ground clutter and satellite comparisons. *Journal of Atmospheric and Oceanic Technology, 36*, 17–39. https://doi.org/10.1175/JTECH-D-18-0007.1.

Marks, D., & Wolff, D. B. (2011). Quality control and calibration of the dual-polarization radar at Kwajalein, RMI. *Journal of Atmospheric and Oceanic Technology, 28*, 181–196. https://doi.org/10.1175/2010JTECHA1462.1.

Marks, D., Wolff, D. B., Pabla, C. S., Petersen, W. A., Kirstetter, P. -E., Tokay, A., Pippitt, J. L., & Wang, J. (2017). Evaluation of S-Band radar rain rate retrieval algorithms and precipitation variability over a dense rain gauge network. In *Preparation of 38th Conference Radar Meteorology*, AMS, 27 Aug–1 Sept, Chicago, IL. Available at https://ams.confex.com/ams/38RADAR/webprogram/Paper320743.html. Last accessed 29 Nov 2018.

Matsui, T., Iguchi, T., Li, X., Han, M., Tao, W.-K., Petersen, W. A., L'Ecuyer, T., Meneghini, R., Olson, W., Kummerow, C. D., Hou, A. Y., Schwaller, M. R., Stocker, E. F., & Kwiatkowski, J. (2013). GPM satellite simulator over ground validation sites. *Bulletin of the American Meteorological Society, 94*, 1653–1660. https://doi.org/10.1175/BAMS-D-12-00160.1.

Molthan, A., & Petersen, W. A. (2011). Incorporating ice crystal scattering databases in the simulation of millimeter wavelength radar reflectivity. *Journal of Atmospheric and Oceanic Technology, 26*, 2257–2269. https://doi.org/10.1175/2010JTECHA1511.1.

Newman, A., Kucera, P., & Bliven, L. (2009). Presenting the Snowflake Video Imager (SVI). *Journal of Atmospheric and Oceanic Technology, 26*, 167–179. https://doi.org/10.1175/2008JTECHA1148.1.

O, S., & Kirstetter, P. E. (2017). Evaluation of diurnal variation of GPM IMERG-derived summer precipitation over the contiguous US using MRMS data. *Quarterly Journal of the Royal Meteorological Society, 144*(S1), 270–281. https://doi.org/10.1002/qj.3218.

O, S., Foelsche, U., Kirchengast, G., Fuchsberger, J., Tan, J., & Petersen, W. A. (2017). Evaluation of GPM IMERG early, late, and final rainfall estimates using WegenerNet gauge data in southeastern Austria. *Hydrology and Earth System Sciences, 21*, 6559–6572. https://doi.org/10.5194/hess-21-6559-2017.

Olson, W. S., Tian, L., Grecu, M., Kuo, K.-S., Johnson, B. T., Heymsfield, A. J., Bansemer, A., Heymsfield, G. M., Wang, J. R., & Meneghini, R. (2016). The microwave radiative properties of falling snow derived from nonspherical ice particle models. Part II: Initial testing using radar, radiometer and in situ observations. *Journal of Applied Meteorology and Climatology, 55*, 709–722. https://doi.org/10.1175/JAMC-D-15-0131.1.

Petersen, W. A., Gatlin, P. N., Wolff, D. B., Tokay, A., & Grecu, M. (2018). A radar-based evaluation of GPM retrievals of the rain drop size distribution. In 10th European Conference on Radar in Meteorology and Hydrology (ERAD 2018), 1–6 July, Ede-Wageningen, The Netherlands, https://doi.org/10.18174/454537.

Petkovic, V., & Kummerow, C. D. (2017). Understanding the sources of satellite passive microwave rainfall retrieval systematic errors over land. *Journal of Applied Meteorology and Climatology, 56*, 597–614. https://doi.org/10.1175/JAMC-D-16-0174.1.

Pippitt, J., Wolff, D. B., Petersen, W. A., & Marks, D. (2015). Data and operational processing for NASA's GPM ground validation program. In 37th Conference on radar meteorology, Norman, OK, AMS, [Available at https://ams.confex.com/ams/37RADAR/webprogram/Paper275627.html, last accessed 18 Nov 2018].

Rios Gaona, M. F., Overeem, A., Brasjen, A. M., Meirink, J. F., Leijnse, H., & Uijlenhoet, R. (2017). Evaluation of rainfall products derived from satellites and microwave links for The Netherlands. *IEEE Transactions on Geoscience and Remote Sensing, 55*, 6849–6859. https://doi.org/10.1109/TGRS.2017.2735439.

Schwaller, M. R., & Morris, K. R. (2011). A ground validation network for the global precipitation measurement mission. *Journal of Atmospheric and Oceanic Technology, 28*, 301–319. https://doi.org/10.1175/2010JTECHA1403.1.

Seo, B.-C., Krajewski, W. F., Quintero, F., El Saadani, M., Goska, R., Cunha, L. K., Dolan, B., Wolff, D. B., Smith, J. A., Rutledge, S. A., & Petersen, W. A. (2018). Comprehensive evaluation of the IFloodS radar-rainfall products for hydrologic applications. *Journal of Hydrometeorology, 19*, 1793–1813. https://doi.org/10.1175/JHM-D-18-0080.1.

Seto, S., & Iguchi, T. (2015). Intercomparison of attenuation correction methods for the GPM dual-frequency precipitation radar. *Journal of Atmospheric and Oceanic Technology, 32*, 915–926. https://doi.org/10.1175/JTECH-D-14-00065.1.

Shi, R. W.-K., Tao, T., Matsui, R., Cifelli, A. Y., Hou, S., Lang, A., Tokay, N.-Y., Wang, C., Peters-Lidard, G., Skofronick-Jackson, S. R., & Petersen, W. A. (2010). WRF simulations of the 20-22 January 2007 snow events over Eastern Canada: Comparison with in-situ and satellite observations. *Journal of Applied Meteorology and Climatology, 49*, 2246–2266. https://doi.org/10.1175/2010JAMC2282.1.

Skofronick-Jackson, G., Johnson, B. T., & Munchak, S. J. (2013). Detection thresholds of falling snow from satellite-borne active and passive sensors. *IEEE Transactions on Geoscience and Remote Sensing, 51*, 4177–4189. https://doi.org/10.1109/TGRS.2012.2227763.

Skofronick-Jackson, G., Hudak, D., Petersen, W. A., Nesbitt, S. W., Chandrasekar, V., Durden, S., Gleicher, K. J., Huang, G.-J., Joe, P., Kollias, P., Reed, K. A., Schwaller, M. R., Stewart, R., Tanelli, S., Tokay, A., Wang, J. R., & Wolde, M. (2015). Global Precipitation Measurement Cold Season Precipitation Experiment (GCPEx): For measurement sake let it snow. *Bulletin of the American Meteorological Society, 96*, 1719–1741. https://doi.org/10.1175/BAMS-D-13-00262.1.

Skofronick-Jackson, G., Hudak, D., Petersen, W. A., Nesbitt, S. W., Chandrasekar, V., Durden, S., Gleicher, K. J., Huang, G.-J., Joe, P., Kollias, P., Reed, K. A., Schwaller, M. R., Stewart, R., Tanelli, S., Tokay, A., Wang, J. R., & Wolde, M. (2017). The Global Precipitation Measurement (GPM) mission for science and society. *Bulletin of the American Meteorological Society, 98*, 1675–1699. https://doi.org/10.1175/BAMS-D-15-00306.1.

Skofronick-Jackson, G., Kirschbaum, D., Petersen, W. A., Huffman, G. J., Kidd, C., Stocker, E. F., & Kakar, R. (2018). GPM scientific achievements and societal contributions: Reviewing three years of advanced rain and snow measurements. *Quarterly Journal of the Royal Meteorological Society, 144*(S1), 27–48. https://doi.org/10.1002/qj.3313.

Steiner, M., Bell, T. L., Zhang, Y., & Wood, E. F. (2003). Comparison of two methods for estimating the sampling-related uncertainty of satellite rainfall averages based on large radar dataset. *Journal of Climate, 16*, 3759–3778. https://doi.org/10.1175/1520-0442(2003)016,3759:COTMFE.2.0.CO;2.

Suk, M.-K., Chang, K.-H., Cha, J.-W., & Kim, K.-E. (2013). Operational real-time adjustment of radar rainfall estimation over the South Korea region. *Journal of the Meteorological Society of Japan, 91*, 545–554. https://doi.org/10.2151/jmsj.2013-409.

Tan, B.-Z., Petersen, W. A., & Tokay, A. (2016). A novel approach to identify sources of errors in IMERG for GPM ground validation. *Journal of Hydrometeorology, 17*, 2477–2491. https://doi.org/10.1175/JHM-D-16-0079.1.

Tan, B.-Z., Petersen, W. A., Kirstetter, P.-E., & Tian, Y. (2017a). Performance of IMERG as a function of spatiotemporal scale. *Journal of Hydrometeorology, 18*, 307–319. https://doi.org/10.1175/JHM-D-16-0174.1.

Tan, J., Petersen, W. A., Kirchengast, G., Goodrich, D. C., & Wolff, D. B. (2017b). Evaluation of global precipitation measurement rainfall estimates against three dense gauge networks. *Journal of Hydrometeorology, 19*, 517–532. https://doi.org/10.1175/JHM-D-17-0174.1.

Tang, G., Zeng, Z., Long, D., & Guo, X. (2015). Statistical and hydrological comparisons between TRMM and GPM Level-3 products over a midlatitude basin: Is day-1 IMERG a good successor for TMPA 3B42V7? *Journal of Hydrometeorology, 17*, 121–137. https://doi.org/10.1175/JHM-D-15-0059.1.

Tao, W.-K., Wu, D., Matsui, T., Peters-Lidard, C., Lang, S., Hou, A. Y., Reinecker, M., & Petersen, W. A. (2013). The diurnal variation of precipitation during MC3E: A modeling study. *Journal of Geophysical Research, 118*, 7199–7218. https://doi.org/10.1002/jgrd.50410/asset/jgrd50410.

Tao, W.-K., Lang, S., Zeng, X., Li, X., Matsui, T., Mohr, K., Posselt, D., Chern, J., Peters-Lidard, C., Norris, P. M., Kang, I.-S., Choi, I., Hou, A. Y., Lau, K.-M., & Yang, Y.-M. (2014). The Goddard Cumulus Ensemble model (GCE): Improvements and applications for studying precipitation processes. *Atmospheric Research, 143*, 392–424. https://doi.org/10.1016/j.atmosres.2014.03.005.

Tao, W.-K., Wu, D., Lang, S., Chern, J., Peters-Lidard, C., Fridlind, A., & Matsui, T. (2016). High-resolution NU-WRF simulations of a deep convective-precipitation system during MC3E: Further improvements and comparisons between Goddard microphysics schemes and observations. *Journal of Geophysical Research: Atmospheres, 121*, 1278–1305. https://doi.org/10.1002/2015JD023986.

Thurai, M., Bringi, V. N., Carey, L. D., Gatlin, P., Schultz, E., & Petersen, W. A. (2012). Estimating the accuracy of polarimetric radar-based retrievals of drop size distribution parameters and rain rate: An application of error variance separation using radar-derived spatial correlations. *Journal of Hydrometeorology, 13*, 1066–1079. https://doi.org/10.1175/JHM-D-11-070.1.

Thurai, M., Gatlin, P. N., Bringi, V. N., Petersen, W. A., Kennedy, P., Notaros, B., & Carey, L. D. (2017). Towards completing the rain drop size spectrum: Case studies involving 2D-video disdrometer, droplet spectrometer, and polarimetric radar measurements. *Journal of Applied Meteorology, 56*, 877–896. https://doi.org/10.1175/JAMC-D-16-0304.1.

Tokay, A., D'Adderio, L., Porcù, F., Wolff, D., & Petersen, W. A. (2017). A field study of footprint-scale variability of raindrop size distribution. *Journal of Hydrometeorology, 16*, 1855–1868. https://doi.org/10.1175/JHM-D-15-0159.1.

Tokay, A., D'Aderio, L. P., Wolff, D., & Petersen, W. (2020). Development and evaluation of the raindrop size distribution parameters for the NASA Global Precipitation Measurement Mission Ground Validation Program. *Journal of Atmospheric and Oceanic Technology, 37*, 115-138. https://doi.org/10.1175/JTECH-D-18-0071.1.

Vega, M., Chandrasekar, V., Carswell, J., Beauchamp, R. M., Schwaller, M. R., & Nguyen, C. M. (2014). Salient features of the dual-frequency, dual-polarized, Doppler radar for remote sensing of precipitation. *Radio Science, 49*, 1087–1105. https://doi.org/10.1002/2014RS005529.

von Lerber, A., Moisseev, D., Marks, D., Petersen, W. A., Harri, A., & Chandrasekar, V. (2018). Validation of satellite-based snowfall products by using a combination of weather radar and surface observations. *Journal of Applied Meteorology and Climatology, 57*, 797–820. https://doi.org/10.1175/JAMC-D-17-0176.1.

Wang, J., Fisher, B. L., & Wolff, D. B. (2008). Estimating rain rates from tipping-bucket rain gauge measurements. *Journal of Atmospheric and Oceanic Technology, 25*, 43–56. https://doi.org/10.1175/2007JTECHA895.1.

Williams, C. R., Bringi, V. N., Carey, L., Gatlin, P., Haddad, Z. S., Munchak, S. J., Petersen, W. A., Meneghini, R., Nesbitt, S. W., Tanelli, S., Tokay, A., Thurai, M., Wilson, A., & Wolff, D. B. (2014). Describing the shape of raindrop size distributions using uncorrelated raindrop mass spectrum parameters. *Journal of Applied Meteorology and Climatology, 53*, 1282–1296. https://doi.org/10.1175/JAMC-D-13-076.1.

Wingo, S.-M., Petersen, W. A., Pabla, C. S., Gatlin, P. N., Marks, D. A., & Wolff, D. B. (2018). The system for integrating multi-platform data to build the atmospheric column (SIMBA) precipitation observation fusion framework. *Journal of Atmospheric and Oceanic Technology, 35*, 1353–1374. https://doi.org/10.1175/JTECH-D-17-0187.1.

Wolff, D. B., Marks, D., & Petersen, W. A. (2015). General application of the Relative Calibration Adjustment (RCA) technique for monitoring and correcting radar reflectivity calibration. *Journal of Atmospheric and Oceanic Technology, 32*, 496–506. https://doi.org/10.1175/JTECH-D-13-00185.1.

You, Y., Wang, N.-Y., Ferraro, R. R., & Rudlosky, S. (2016). Quantifying the snowfall detection performance of the Global Precipitation Measurement (GPM) Microwave Imager channels over land. *Journal of Hydrometeorology, 17*, 1101–1117. https://doi.org/10.1175/JHM-D-16-0190.1.

Zagrodnick, J. P., McMurdie, L. A., & Houze, R. A., Jr. (2018). Stratiform precipitation processes in cyclones passing over a coastal mountain range. *Journal of the Atmospheric Sciences, 75*, 983–1004. https://doi.org/10.1175/JAS-D-17-0168.1.

Zhang, J., Howard, K., Langston, C., Kaney, B., Qi, Y., Tang, L., Grams, H., Wang, Y., Cocks, S., Martinaitis, S., Arthur, A., Cooper, K., Brogden, J., & Kitzmiller, D. (2016). Multi-Radar Multi-Sensor (MRMS) quantitative precipitation estimation: Initial operating capabilities. *Bulletin of the American Meteorological Society, 97*, 621–638. https://doi.org/10.1175/BAMS-D-14-00174.1.

Chapter 27
The GPM DPR Validation Program

Riko Oki, Toshio Iguchi, and Kenji Nakamura

Abstract The validation of the DPR onboard the GPM Mission's core observatory was carried out in three ways; calibration of the DPR, verification of the appropriateness of the attenuation and scattering parameters in the Ka band, and through comparisons of the precipitation estimates from the DPR with corresponding ground-based measurements. The DPR was proved to have been stably operated since the launch of the GPM core observatory, and both the Ku and Ka radars were calibrated within ± 1 dB which is the specification of the DPR. A ground-based dual Ka-band radar system was developed for obtaining the attenuation and scattering characteristics of snow particles in the Ka band. Observations with this system show that the range of parameters in the k-Ze relationship adopted by the DPR standard algorithm is reasonable. Comparisons of the DPR rain estimates with the AMeDAS rain gauge in Japan and MRMS/NMQ data over the continental US show that the differences between the DPR and ground measurements were mostly within 10%. Also, the results of classification on rain/snow included in the DPR Level 2 data were verified by comparison with visual observation at ground sites. Overall, the rain rates and parameters estimated by the DPR were found to be reasonable.

Keywords Precipitation · Rainfall · Snowfall · Snow · Calibration · Validation · GPM · DPR · JAXA · AMeDAS · MRMS · Radar · Reflectivity · Raingauge

R. Oki (✉)
Earth Observation Research Center (EORC)/Japan Aerospace Exploration Agency (JAXA), Ibaraki, Japan
e-mail: oki.riko@jaxa.jp

T. Iguchi
National Institute of Information and Communications Technology (NICT), Koganei, Japan

K. Nakamura
Dokkyo University, Saitama, Japan

© Springer Nature Switzerland AG 2020
V. Levizzani et al. (eds.), *Satellite Precipitation Measurement*, Advances in Global Change Research 69, https://doi.org/10.1007/978-3-030-35798-6_3

27.1   Introduction

This chapter describes calibration and validation of the Dual-Frequency Precipitation Radar (DPR) onboard the Global Precipitation Measurement (GPM) Mission's core observatory. The first section is concerned with calibration of the DPR. The second section outlines the ground-based Ka-radar experiment to characterize the attenuation of the Ka-band radio waves by melting snow particles for the GPM/DPR Ka radar which is the first spaceborne Ka precipitation radar. The third section shows the results of comparisons between the precipitation rates estimated by DPR and rain gauges in Japan and the US. A result of case study comparing a rain/snow assessment by DPR with ground observation is also described.

27.2   Calibration

Absolute calibration of the DPR is essential in the validation process because the accuracy of rain retrieval heavily depends upon it. In fact, the rain retrieval algorithm assumes that the accurate apparent radar reflectivity factor that includes the attenuation effect is available. Calibration of radar is to determine the conversion factor that relates the radar received signal to the radar reflectivity factor.

As explained in Chap. 10 of volume 1, the reflectivity of a target is calculated from the received power, the range to the target, and the radar constant that consists of radar system parameters, physical constants and the transmitting power. Specifically, the measured radar reflectivity factor $Z_m(r)$ at range r is given by

$$Z_m(r) = P_r(2r/c) \times \frac{r^2}{C_r} \tag{27.1}$$

where $P_r(t)$ is the received power at time t after the pulse transmission, and c is the speed of light. The radar constant C_r is defined by

$$C_r = \frac{\pi^3 |K_w|^2}{\lambda^2 2^{10} \ln 2} P_{t0} G_0^2 \theta_{w1} \theta_{w2} c \tau \tag{27.2}$$

where P_{t0} is the transmitting power, G_0 the antenna gain, θ_{w1} and θ_{w2} the beam widths in two orthogonal directions, τ the effective pulse width in time, λ is the wavelength, and $|K_w|^2$ the dielectric factor of water. To calculate Z_m, it is necessary to know P_r, r, and C_r that include several parameters. Of these parameters, c is a fixed constant, λ is a known constant, r can be accurately determined from the delay time of the echo, and $|K_w|^2$ is treated as a constant in the retrieval algorithm. The DPR calibrations are carried out to determine P_r, P_{t0}, G_0, θ_{w1}, θ_{w2}, and τ.

The received signal of the DPR at each sample gate for each transmitted pulse is log-detected and converted to an integer by an analog-to-digital (AD) converter.

About one hundred samples at each gate are averaged and sent to the ground station. The received signal for each beam of the DPR consists of these integers or count values. There are two kinds of calibration for the DPR; internal and external calibrations. The internal calibration feeds signals identical to the transmitting pulses into the receiver channel at variety of signal levels through attenuators. The internal calibration data are used to determine the slope of the conversion line that relates the count values of the received signal to the power. The external calibration has two modes; transmitting and receiving modes. The transmitting mode of the external calibration transmits calibrated signals from ground to the GPM satellite and compares the corresponding count values recorded with the DPR, thus determining the intercept of the conversion line so that the received power at the DPR antenna can be calculated from the count value. In other words, the internal calibration is for the calibration excluding the antenna and the radar receiver front end whereas the external calibration is for the absolute calibration of the whole system. The internal calibration is carried out about once a day on orbit in order to monitor the change of radar performance. The external calibration is carried out once in every few months.

To convert the received power into a radar reflectivity factor, several parameters used in C_r must be known. They are determined by the external calibration. Of these parameters, the range can be known very accurately from the delay of echoes or the satellite altitude and attitude data. The equivalent pulse width after the receiver's bandpass filter is determined from the data taken on ground before launch and from the shape of sea surface echo at nadir incidence. The equivalent beam widths were determined by measuring the beam pattern by fitting the Taylor beam pattern to the measured distribution of received power for various transmitting directions and then by converting them to the equivalent Gaussian beam widths. The gains of the antenna are determined by measuring the DPR's transmitted signal at ground, and by comparing the count values recoded on the DPR for the known power of signal transmitted from the ground. Note that the external calibration actually determines the products $P_t G_t$ and P_r/G_r instead of P_t, P_r, G_t and G_r separately. This is not an issue because they appear in the product form in the radar equation.

About two years of calibration data were used to determine the parameter values and the calibration factors for both KuPR and KaPR. These data are stable from the launch and their variations are within the specification of ± 1 dB. However, the calibration results after the launch are somewhat different from the corresponding values measured on ground before launch. In fact, the received powers of KuPR and KaPR were changed by +0.13 dB and + 1.13 dB, and their transmitting powers were changed by -0.29 dB and $- 0.95$ dB, respectively, according to the calibration results after launch. These changes correspond to the change in Z_m by +1.3 dB and + 1.2 dB for KuPR and KaPR, respectively, and are implemented in the processing version 05 of the DPR algorithm.

Similar calibration was carried out with the Precipitation Radar on the TRMM satellite. After applying these calibration results to both the PR and the DPR, data from the PR and the KuPR agree well with each other. For example, sea surface cross sections derived from these instruments agree well with the recent cross section results derived by the Ku-band altimeter on the Jason satellite.

27.3 Ground Ka-Radar Experiment

27.3.1 Measurements

The DPR is a 13.6/35.5 GHz radar system. The 35.5 GHz (Ka-band) radar is a new addition to the spaceborne precipitation radar. For the full utilization of the Ka-band radar, scattering characteristics of radio waves at the Ka band by precipitation particles must be well understood. The scattering cross sections for liquid particles can be simply calculated, because the shape is approximately spherical and the permittivity is essentially the same as pure water. The GPM's core observatory covers almost the whole globe with the DPR designed to observe not only liquid precipitation but also solid precipitation. The scattering characteristics of solid particles are complicated, since their shape and density vary significantly. When snow particles are partially melted, the scattering characteristics become very complicated. Although many model results have been reported, it is not clear which model can be assumed in the retrieval algorithm. Measurements of scattering and attenuation characteristics of real precipitation are desirable.

Attenuation of radio waves in the Ka band is significant, and conventional Ka-band radars measure only the combined characteristics of scattering and attenuation. In order to measure scattering and attenuation characteristics of precipitation particles separately, a dual Ka-band radar system was developed by JAXA (Nakamura et al. 2011). This system consists of two identically designed Ka-band radars that are separately deployed at a certain distance and facing each other. When a precipitation system comes between the two radars, they observe it from opposite directions. From the two radar signatures, specific attenuation (k) and equivalent radar reflectivity factor (Z_e) can be estimated. This idea was successfully applied to an observation with air-borne and ground-based 94 GHz cloud radars (Li et al. 2001). The dual Ka-band radar system was used for snow observations and showed scattering characteristics of snow particles (Nishikawa et al. 2015).

The method to estimate the specific attenuation k and the equivalent radar reflectivity factor Z_e has been reported in Nishikawa et al. (2015) and Nakamura et al. (2018). Here, the estimation method is briefly described.

When two radars observe a precipitation system, two measured radar signatures in dBZ, $Z_{m1}(r)$ and $Z_{m2}(r)$, show different profiles because of the attenuation as illustrated in Fig. 27.1. Here, r is the range from Radar 1, and "measured" means the equivalent radar reflectivity factor without attenuation correction.

The specific attenuation k between points at r and r + Δ can be estimated as

$$k = \frac{Z_{m1}(r) - Z_{m2}(r) - Z_{m1}(r+\Delta) + Z_{m2}(r+\Delta)}{2\Delta} \qquad (27.3)$$

After obtaining the specific attenuation, equivalent radar reflectivity factor can be obtained by correcting the attenuation provided that the radar is well calibrated. The calibration was performed by a comparison of the measured radar reflectivity factors with those estimated from the drop size distribution data obtained simultaneously by a disdrometer located near the radar in rain cases.

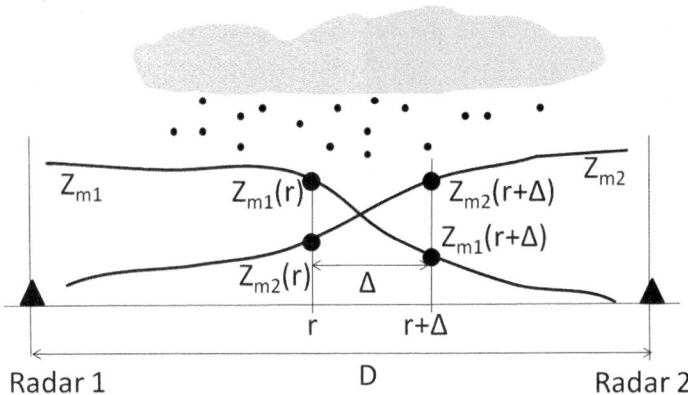

Fig. 27.1 Schematic illustration of the measured radar reflectivity profiles by the dual Ka-band radar system. Z_{m1} and Z_{m2} are measured radar reflectivity factors and r is the range

27.3.2 Results

Here, results of snow observations in Nagaoka City (about 37.4°N, 138.8°E), Japan in 2012 are described according to Nishikawa et al. (2015). The observation area experiences significant snowfall in winter. It is known that k and Z_e are both small for dry snow, but they show large variations in transition to rain. Surface air temperature is a good index for stratifying the data. One of the good methods to show the scattering and attenuation characteristics is to produce k-Z_e plots. The k–Z_e plots of all snow events are divided by the surface air temperature and then categorized by snow type. The k–Z_e plots of all snow events are shown for different temperature ranges with 1 °C increments in Fig. 27.2. When the surface temperature was below 0 °C, both k and Z_e were low, and the k–Z_e plots showed no relation. This case was dominated by dry snow and/or graupel. When the surface temperature was between 0 and 1 °C, k increased rapidly with Z_e. This suggests that dry snow started to melt in this temperature range. As the surface air temperature increased from 1 to 3 °C, both k and Z_e increased and the k–Z_e plots featured greater data scatter and larger k values. However, the mean k–Z_e relationship itself changed only slightly in the temperature range. These cases were dominated by wet snow. The temperature ranges are reasonable because model and direct observation studies have shown that the mean melted fraction rapidly approaches a value of 1 as the temperature rises from 0 to 1 °C (e.g., Brandes et al. 2008). When the surface air temperature exceeded 3 °C, the k–Z_e plot was less scattered and approached that of rain. This suggests that wet snow was almost completely melted and became rain or sleet, which is at least partly verified by the surface meteorological data. Thus, k and Z_e show clear differences according to the phase of the precipitation particles.

In the DPR rain retrieval algorithms, k and Z_e characteristics are incorporated. The results from the field measurements can provide one of the bases of the algorithms and can also be used to validate the algorithms.

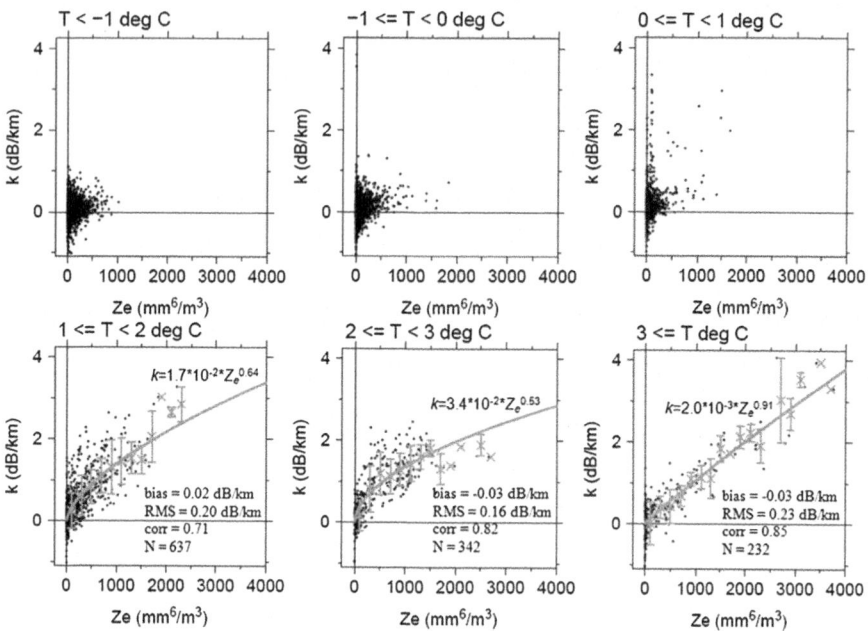

Fig. 27.2 Scatter plots of k and Z_e for snow events in 2012. Each figure is with different surface air temperature. Black dots are 1 min data, and dark lines and crosses are fitting curves and averaged values. (Adapted from Nishikawa et al. 2015)

27.4 Comparisons of DPR Products with Ground Observations

Precipitation rates estimated by GPM/DPR were compared with the AMeDAS (Automated Meteorological Data Acquisition System) rain gauge network data over Japan and with MRMS/NMQ (Multi-Radar Multi-Sensor/National Mosaic and Multi-Sensor Quantitative Precipitation Estimation) rain data over the United States which was based on ground radar measurements calibrated by rain gauges. JAXA had set up the success criteria for judging the success of the GPM/DPR project. The minimum success criterion was that the difference between DPR estimated rain rates and the AMeDAS rain rates was within approximately ±10% when their annual rain amounts were compared over Japan. The full success criterion was that the difference between DPR estimated rain rates and ground rain observations became within ±10% when their long-term averaged rain rates were compared in areas other than Japan. MRMS/NMQ data was used for the full success judgment.

27.4.1   GPM/DPR Ground Validation Comparing with Rain Gauge Data Over Japan

The AMeDAS data include rain gauge data provided by the Japan Meteorological Agency (JMA). There are about 1300 observation points over Japan with a mean distance between adjacent observation points of about 17 km. We used 10 min accumulations. The AMeDAS rain gauge data used here were for 2 years from April 2014 to March 2016. The GPM/DPR standard products consist of three kinds of rain estimates, that is, the Ku product, Ka product and DPR product. The version of the DPR product used here is ITE113 which is essentially identical to the version 5 standard products. The lowest layer observed by the DPR is about from 0.5 to 2.0 km above the surface and the rain rate retrieved at this height is designated as "near surface rain". The product also contains "estimated surface rain" which is the rain rate estimated at the actual surface level. Here, the estimated surface rain is used for comparisons with ground measurements.

To calculate the annual rain amount from both DPR and AMeDAS, 200 of the 0.5° latitude/longitude grid boxes were selected from all over Japan as shown in Fig. 27.3. DPR data were accumulated in each box whenever the DPR overpassed the box. For the comparison, AMeDAS rain gauge data taken in the first 10-min time interval after the time of the DPR over-pass were used. This is because the fall-time of rain from the "near surface" altitude level to ground takes a few minutes. All AMeDAS rain gauge data in each box were averaged. The average number of AMeDAS points in each box is about 6. The difference (*Diff*) between DPR and AMeDAS data is defined as follows:

Fig. 27.3 The area of comparison over Japan. 0.5° longitude/latitude boxes were set and the entire area was divided into six climatic sub areas considering the total amount of precipitation and near surface air temperature

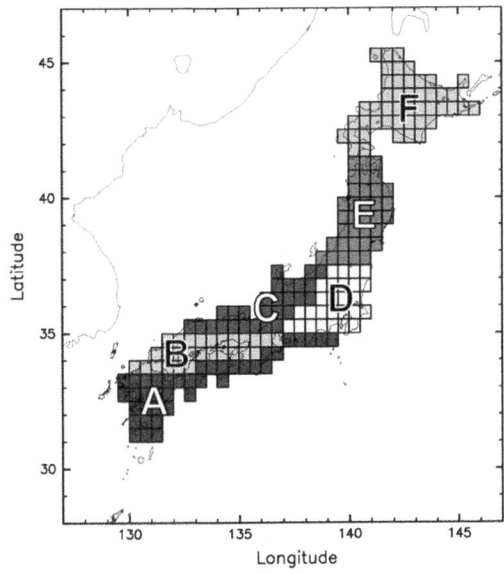

Table 27.1 Averaged accumulated rain amount for AMeDAS, Ku product (250 km swath data) and dual-frequency DPR product (125 km swath data) for 2 years from April 2014 to March 2016

	Accumulated rain amount	Difference (%)
AMeDAS	1843 mm/year	–
Ku product	1549 mm/year	−16.0%
DPR product	1624 mm/year	−11.9%

Accumulation was performed for the DPR and AMeDAS data when the DPR passed over

$$Diff = \frac{DPR\ 12\ months\ rain - AMeDAS\ 12\ months\ rain}{AMeDAS\ 12\ months\ rain} \qquad (27.4)$$

Table 27.1 shows the results of the comparison of annual rain amounts from AMeDAS, single-frequency Ku product (250 km swath), and dual-frequency DPR product (125 km swath) for 2 years from April 2014 to March 2016. Hereafter, "DPR product" means dual-frequency product which is generated by using both Ku and Ka radar echoes within the 125-km inner swath.

The difference between DPR and AMeDAS data over Japan is −16.0% for Ku product and − 11.9% for DPR product. Although both radar products are lower than the AMeDAS data, the DPR estimates are better than the Ku only estimates. In order to investigate further details, we set six sub areas based on the temperature and annual rain amount as shown in Fig. 27.3. In order to remove the snow cases which seemed to cause underestimation, we filtered only when the surface air temperature was above 6 °C. By filtering out low temperature cases, 10 to 30% of data were removed in northern parts of Japan. The results are shown in Fig. 27.4. Despite slight differences from region to region, the results satisfy the success criterion of the difference within ±10% in all sub-regions.

27.4.2 GPM/DPR Ground Validation with MRMS/NMQ Data Over the US

Precipitation rates estimated by GPM/DPR were also compared with MRMS/NMQ rain data over the continental US (Zhang et al. 2016). MRMS/NMQ data were provided every 2 min by 0.01° longitude/latitude. The data period of MRMS/NMQ data used here is 1 year from June 2014 to May 2015.

A total of 871 1° longitude/latitude boxes were set all over the continental US as shown in Fig. 27.5. The rain data by both DPR and MRMS/NMQ were accumulated when the DPR passed over the box. The differences between DPR and MRMS/NMQ were calculated in each box. The results are shown in Table 27.2.

As seen in the case of Japan, the DPR estimates show tendency of underestimation for Ku product and DPR product. But here again, the DPR product shows better result than Ku product.

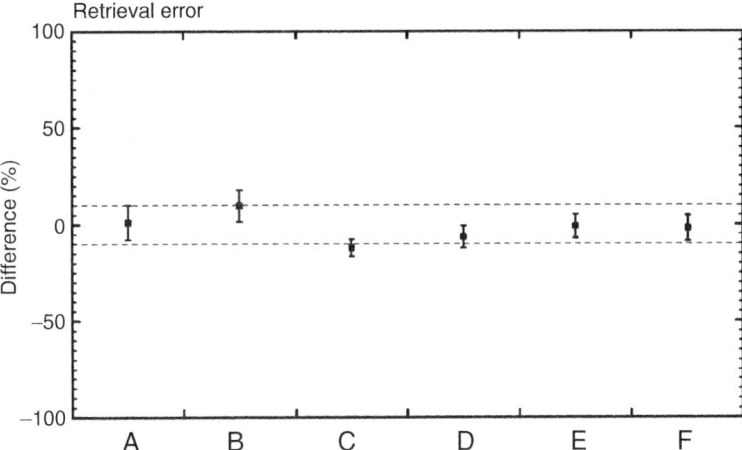

Fig. 27.4 The differences between DPR and AMeDAS rain gauge in six sub areas. Despite slight differences from region to region, the results meet the success criterion that the difference is within ±10% in all cases

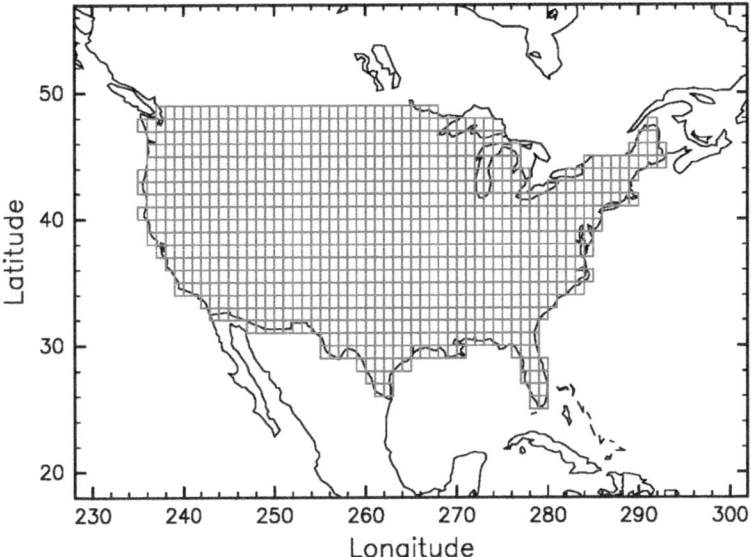

Fig. 27.5 The area of comparison over the US. The area includes total of 871 1° latitude/longitude boxes

To see the differences by regions, we divided the whole area into 24 1° latitude bins (see Fig. 27.6).

Table 27.2 Averaged accumulated rain amount for MRMS/NMQ, Ku product and dual-frequency DPR product from June 2014 to May 2015

	Accumulated rain amount	Difference (%)
MRMS/NMQ	784 mm/year	–
Ku product	658 mm/year	−16.1%
DPR product	744 mm/year	−4.8%

Accumulation was performed for the DPR and MRMS/NMQ data when the DPR passed over

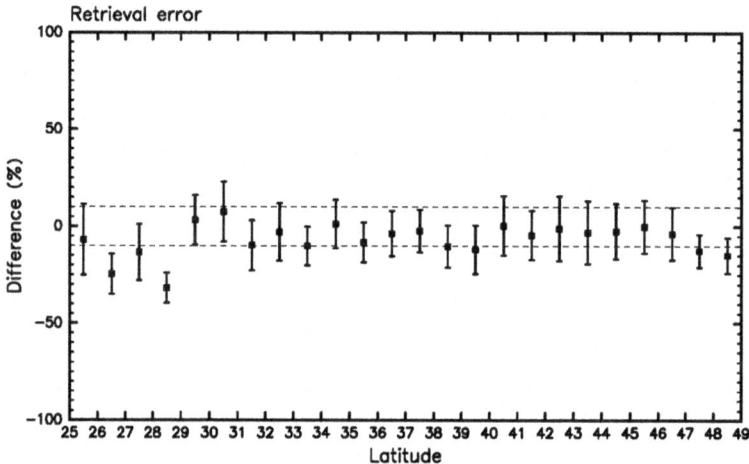

Fig. 27.6 The difference between DPR product and MRMS/NMQ by 24 1°latitude bins. The results meet the success criterion of difference within ±10% in 22 bins among 24 bins

27.4.3 Surface Snow Flag in the Latest DPR Product

The DPR level 2 product in the version 05 includes a new flag called "flagSurfaceSnowfall" which indicates whether the precipitation is rain or snow at surface (Chap. 11 of volume 1). Its detection is based on dual frequency radar signatures in addition to the surface air temperature. We checked whether the rain/ snow discrimination is functioning correctly by comparing the flags with ground observations in Japan.

Some JMA's observatories give visual reports of rain/sleet/snow discrimination. We compared such visual reports with DPR estimated rain/snow flag status. Figure 27.7 shows rain/snow status given by JMA and DPR for the case of 18 January 2016 (0900 JST) over north-east part of Japan. The discrimination flag by DPR is given only in the narrow swath area where the Ku and Ka matched beam observation is realized. The figure shows quite a good agreement. The JMA rain/ sleet/snow reports are available only a few times a day. Although the opportunities

Fig. 27.7 The rain/snow distribution by JMA's visual reports and DPR data for the case of 18 January 2016 (0900 JST) over north-east part of Japan. The agreement is quite good

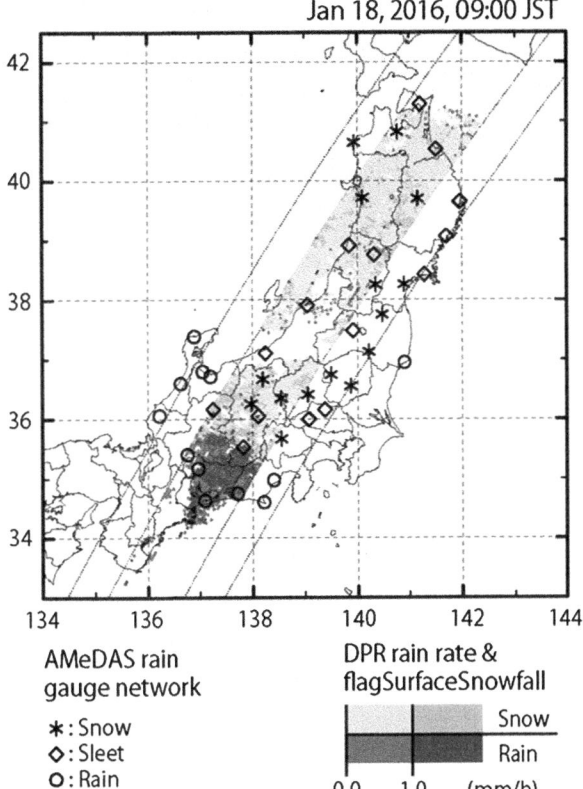

for such comparisons is very limited considering the frequency of simultaneous DPR observation and snow events, validation of the rain/snow discrimination should be increased in the future.

27.5 Summary

Calibration and validation of the DPR onboard the GPM Mission's core observatory have been carried out. As the result of the calibration, it was shown that the DPR has been stably operated since the launch of the GPM core observatory, and both the Ku and Ka radars were calibrated within ± 1 dB which is the specification of the DPR. The Precipitation Radar (PR) on the TRMM satellite was also calibrated to be consistent with the DPR.

The results of the ground experiments with the dual Ka-band radar system show that the range of parameters in the k-Ze relation adopted by the DPR standard algorithm is reasonable.

Since the verification of snowfall is generally difficult, the first priority was given to the validation of rainfall. As a result, compared with the AMeDAS rain gauge in Japan, it was found that the difference between the DPR and the AMeDAS rain gauge was within 10%. The result using MRMS/NMQ over the U.S. also showed that the difference between the DPR and MRMS/NMQ was also within 10%. The estimated rain intensity by DPR was reasonable.

The results of classification on rain/snow included in the DPR Level 2 data were verified by comparison with visual observation at ground sites and it was confirmed that the DPR product output reasonable information.

Acknowledgments The authors are grateful to Mr. Takeshi Masaki, Ms. Yuki Kaneko and Dr. Nozomi Kawamoto for helping with DPR data analyses.

References

Brandes, E. A., Ikeda, K., Thompson, G., & Schönhuber, M. (2008). Aggregate terminal velocity/ temperature relations. *Journal of Applied Meteorology and Climatology, 47*, 2729–2736. https://doi.org/10.1175/2008JAMC1869.1.

Li, I., Sekelsky, S. M., Reising, S. C., Swift, C. T., Durden, S. L., Sadowy, G. A., Dinardo, S. J., Li, F. K., Huffman, A., Stephens, G. L., Babb, D. M., & Rosenberger, H. W. (2001). Retrieval of atmospheric attenuation using combined ground-based and airborne 95-GHz cloud radar measurements. *Journal of Atmospheric and Oceanic Technology, 18*, 1345–1353. https://doi.org/10.1175/1520-0426(2001)018<1345:ROAAUC>2.0.CO;2.

Nakamura, K., Shimizu, S., Nakagawa, K., & Hanado H. (2011). Dual Ka-band radar experiment for the DPR algorithm development. In *Proceedings of IEEE international geoscience and remote sensing symposium*, Vancouver, Canada, pp 1554–1557, https://doi.org/10.1109/IGARSS.2011.6049366.

Nakamura, K., Kaneko, Y., Nakagawa, K., Hanado, H., & Nishikawa, M. (2018). Measurement method for specific attenuation in the melting layer using a Dual Ka-band radar system. *IEEE Transactions on Geoscience and Remote Sensing, 56*(6), 3511–3519. https://doi.org/10.1109/TGRS.201801291.

Nishikawa, M., Nakamura, K., Fujiyoshi, Y., Nakagawa, K., Hanado, H., Minda, H., Nakai, S., Kumakura, T., & Oki, R. (2015). Radar attenuation and reflectivity measurements of snow with Dual Ka-band radar. *IEEE Transactions on Geoscience and Remote Sensing, 54*(2), 714–722. https://doi.org/10.1109/TGRS.2015.2464099.

Zhang, J., Howard, K., Langston, C., Kaney, B., Qi, Y., Tang, L., Grams, H., Wang, Y., Cocks, S., Martinaitis, S., Arthur, A., Cooper, K., Brogden, J., & Kitzmiller, D. (2016). Multi-Radar Multi-Sensor (MRMS) quantitative precipitation estimation: Initial operating capabilities. *Bulletin of the American Meteorological Society, 97*, 621–638. https://doi.org/10.1175/BAMS-D-14-00174.1.

Chapter 28
Error and Uncertainty Characterization

Christian Massari and Viviana Maggioni

Abstract Quantifying errors and uncertainties associated with satellite precipitation products (SPPs) is fundamental to guarantee their correct use in several applications, including hydrological predictions, climate studies, and water resource management. Numerous factors affect the accuracy and precision of these products, including the sensor frequencies and channels, the type of precipitation, the heterogeneity of precipitation within the sensor footprint, as well as the choice of algorithm that transfers the sensor retrieval information to a precipitation rate. This chapter analyses these sources and summarizes the most common methods to estimate, quantify, and model errors and uncertainties associated with SPPs.

Keywords Precipitation · Rainfall · Snowfall · Validation · Error characterization · Metrics · Uncertainty · Snow · Ice · Orographic precipitation · HEPEX · Raingauges · Radar · Ground truth · OceanRAIN

28.1 Uncertainty Sources of Satellite Precipitation Products

Satellite precipitation products (SPPs) are affected by both errors and uncertainties. Errors are commonly defined as the difference between the satellite product and a reference considered to be the "truth". SPP errors are characterized by a systematic component – a shift of the mean SPP from the reference – and a random component – which varies in an unpredictable way. Uncertainty on the other hand represents the range of values within which the true value lies with some level of confidence.

C. Massari (✉)
National Research Council, Research Institute for Geo-Hydrological Protection (CNR-IRPI),
Perugia, Italy
e-mail: christian.massari@irpi.cnr.it

V. Maggioni
Sid and Reva Dewberry Department of Civil, Environmental, and Infrastructure Engineering,
George Mason University, Fairfax, VA, USA
e-mail: vmaggion@gmu.edu

© Springer Nature Switzerland AG 2020   515
V. Levizzani et al. (eds.), *Satellite Precipitation Measurement*, Advances in Global
Change Research 69, https://doi.org/10.1007/978-3-030-35798-6_4

Although error and uncertainty are two very distinct concepts, their characterization is not an easy task with the result that in the literature, the two terms are often used interchangeably. Although this is not formally correct, in this chapter, the two terms are indistinctly used when speaking about general concepts, while their formal scientific meaning will be used whenever this becomes necessary.

The most common sources of uncertainty in SPPs include: (i) the sampling uncertainty due to the limited satellite overpasses at a given location within a day; (ii) parameter calibration errors, (iii) retrieval algorithm formulations; and (iv) errors in the a priori databases (Kummerow et al. 2006; Stephens and Kummerow 2007). The spectral and design characteristics of each remote sensing technique (sensor, frequency, channel) along with the specific design of the mission determine both what kind of information about precipitation is retrieved and the associated type of uncertainties.

Geostationary infrared (IR) sensors only detect cloud-top characteristics, whose relationship with rainfall rate is physically indirect, drawing a large source of uncertainty in the precipitation estimates (Bellerby and Sun 2005). Conversely, high-frequency passive microwave (PMW) channels provide information related to ice content, and low-frequency PMW ones provide information on the cloud water content. PMW-based precipitation estimates are physically based and relatively accurate over oceans. However, over land, they often rely on empirical relationships between brightness temperature and rain rates due to the complexity of land surface emissivity.

The sampling error is mainly determined by the satellite orbit and swath width. In this respect, both the number and the temporal distribution of the different samples during the day are particularly important resulting in increasing sampling errors with increasing sampling intervals (Nijssen and Lettenmaier 2003; Ciabatta et al. 2017) and diurnal-cycle bias effects (Gebremichael and Krajewski 2004).

Many recent attempts to estimate precipitation from satellites merge observations from different sensors at different frequencies and channels, affecting retrieval errors associated with each individual observation and introduce several types of error sources, which non-linearly interact with each other. As a result, quantifying the overall uncertainty in satellite precipitation products is a challenging problem that requires an understanding of the contribution of several factors. To further compli-cate matters, the uncertainties associated with precipitation estimates are region, season, and rain rate dependent, which adds additional uncertainty to the final products (Stephens and Kummerow 2007; Dinku and Anagnostou 2005; Kummerow et al. 2006).

For example, both precipitation type and intensity play an important role in determining the sampling error magnitude with 1) isolated convective systems being more sensitive to the spatial and temporal sampling characteristics of the sensor than precipitation from a stratiform systems (see Figure 1 in Behrangi and Wen 2017); and 2) a reduction of the error for higher rainfall rates, larger domain sizes, and longer time integration (Huffman 1997; Hossain et al. 2004).

Several past studies showed the dependence of satellite precipitation uncertainty on region, rain rate, storm type, season, resolution, rain rate, storm type, and background surface (e.g., Tian and Peters-Lidard 2010; Tang and Hossain 2012; Gottschalck et al. 2005; Tian et al. 2007; Ebert et al. 2007; Oliveira et al. 2016). In this respect, Ebert et al. (2007) compared the performance of different SPPs and numerical weather models in Australia, US, and Northwestern Europe finding that SPPs perform better than models for convective storms (summer) and from the tropics to mid latitudes. In these cases, retrieval uncertainty is the primary error source, mainly caused by the IR inaccuracy in estimating stratiform precipitation (Bytheway and Kummerow 2013). Ebert et al. (2007) also demonstrated that seasonality and surface type significantly affect the magnitude of errors in precipitation rate and accumulation with a relatively larger underestimation for higher rain rates during warm months and misidentification of light rain above cold land surfaces (i.e., snow and ice at the ground) that scatter radiation similarly to a precipitating cloud (Villarini and Krajewski 2007; AghaKouchak et al. 2011; Stampoulis et al. 2013). Wind and temperature may also introduce inaccuracy both in rainfall intensity and distributions. For example, Ebert et al. (2007) found a large overestimation of rainfall in the semi-arid mountain regions of western coastal Mexico during the summer monsoon that was explained by an effect of rainfall evaporation before it reaches the ground. Similar results were obtained by Moazami et al. (2013) in the semi-arid region of Iran where raindrops tend to evaporate before they reach the surface with a consequent high false alarm rate. Uncertainty reduces when precipitation is accumulated over larger spatial and temporal scales (Steiner et al. 2003). Tian and Peters-Lidard (2010) showed uncertainties as large as 100% for rainfall accumulations of 1 mm day^{-1} and as little as 30% for accumulations larger than 30 mm day^{-1}.

Because of snow and ice cover, orography-enhanced precipitation, and large weather and climate variability, high latitudes and complex terrain add a challenge in estimating precipitation (Maggioni et al. 2016a). However, these regions are the ones typically under-monitored (due to the lack of in-situ instrumentation) and prone to extreme precipitation triggered events, like flash floods and landslides, whose consequences can be devastating. IR-based products are more prone to errors in mountainous environments due to the presence of warm orographic rain. IR algorithms use cloud-top temperature thresholds that are too cold for orographic clouds, thus leading to significant underestimation (Hirpa et al. 2010; Dinku et al. 2007). Nevertheless, PMW products also suffer from two major challenges. The first is the warm orographic rain, which may not produce enough ice aloft. As over-land PMW rain retrievals are based mainly on scattering by ice aloft, this may lead to an underestimation of rain at the surface. The second challenge comes from very cold surfaces and ice over mountain-tops, which could be confused with precipitation.

In summary, satellite precipitation estimates are characterized by a number of uncertainty and error sources, whose evaluation is complicated by the natural temporal and spatial variability of precipitation. In addition, SPPs are affected by both systematic and random errors, whose knowledge is paramount for developing bias reduction techniques to improve precipitation retrieval algorithms (Maggioni et al. 2016b). The next section discusses state-of-the-art techniques to assess the quality and limitations of SPPs. This can give useful information to help algorithm developers to improve their products, and help users to understand the accuracy and limitations of those products.

28.2 Methods for Assessing Satellite Precipitation Products

The various sources of uncertainty summarized in the previous section cause satellite-based precipitation estimates to be erroneous, i.e., different from the "true" precipitation value. Quantifying the "distance" between the satellite estimate and the true precipitation means finding the error (often denoted as accuracy) associated with the precipitation product. The assessment of this error (or validation) is fundamental to guarantee SPPs' correct use in applications like hydrological predictions, climate studies, and water resource management (Serrat-Capdevila et al. 2014; Maggioni and Massari 2018). However, we do not have knowledge of the "true" precipitation field, therefore the SPPs' validation is commonly carried out using an independent reference or benchmark, such as observations from rain gauges and/or ground radars, assuming that the latters are characterized by a much lower error than the SPP's. In this context, the Hydrologic Ensemble Prediction EXperiment (HEPEX, Schaake et al. 2007) is an international project established by the hydrological and meteorological communities and has recently proposed a joint (GEWEX-GDAP/CGMS-IPWG) precipitation assessment that will start in 2018 and produce a completed assessment of SPPs by 2021.

28.2.1 The Benchmark

The main advantage of rain gauges is their ability to directly provide a direct cumulative estimate, which is typically unrepresentative of the areal and instantaneous precipitation observed by satellites. On the other hand, ground radars provide a "snapshot" type of measurements with spatial resolution more similar to satellites. However, radars also provide an indirect rainfall estimate and are prone to errors and significant biases.

Rain gauge and radar data are typically combined with the aim of providing gauge-corrected radar estimates (e.g., Todini 2001). On top of inherent characteristics of gauge and radar observations, uncertainty is added to the validation process by the need to map and aggregate these observations onto a common grid. This may result in reduced spatial detail and reduced maximum rain rates (i.e., smoothing of extreme events). It is also important to note that validation results depend on the spatial scale at which validation is performed, with coarser grids generally producing better results.

The choice of one type of rainfall source to be used as benchmark depends on, other than the data availability, the type of products to be validated and the specific objective of the validation study. For instance, for instantaneous and high spatial resolution estimates, gauge-corrected radar estimates are generally preferable to gauge observations. On the other hand, for larger regions and timescales (6 h to daily), rain-gauge analyses or combined gauge/radar analyses are more accurate and should be preferred to raw gauge or radar observations (Ebert et al. 2007).

The presence of errors in the benchmark dataset increases the apparent error of the satellite estimates and thus must be considered when validating SPPs. A common assumption is that as long as the observational error is random and is much smaller than the SPP error, then the reference can be reliably used to intercompare estimates from different products (Huffman et al. 2007). However, this assumption is not always met. For instance, Tian et al. (2007) showed that the bias in the North-Western US changes from positive to negative if either the gauge only or the radar + gauge dataset is considered. Anagnostou et al. (2010) demonstrated the need to benchmark reference data sources prior to their quantitative use in validating remote sensing retrievals. Ali et al. (2005) used different gauge networks for performing an empirical evaluation of (1) the error of the reference network (often used in validation of global products as the ground truth) and (2) the covariance between the errors in the product and the errors in the reference. They showed that the "true" error of evaluated global products might be significantly lower than when these two terms are neglected.

Several studies in the past have tried to assess the error in gauge-based datasets and their efficiency as benchmark to evaluate the performance of SPPs (Xie and Arkin 1995; Villarini et al. 2008; Roca et al. 2010). Xie and Arkin (1995) concluded that the biases in gauge observations are small when compared to biases in satellite estimates. Villarini et al. (2008) showed that temporal sampling uncertainties increase with the sampling interval according to a scaling law. They also showed that spatial sampling uncertainties tend to decrease with increasing accumulation time, with no strong dependence on the gauge location within the pixel or the gauge elevation.

Herold et al. (2015) compared different non-zero land precipitation products showing that a substantial disparity exists in a simple index of daily precipitation intensity within and between observational and reanalysis products. The most salient result of this study is that differences in precipitation intensity among observational datasets are as large as those among reanalysis products. This is relevant since the gauge reference can be easily obtained in the US, Europe, Australia, and China thanks to the relatively high rain gauge density, but numerous regions of the world still lack a sufficiently dense coverage, which is one of the main obstacles of a proper evaluation of SPPs (Kidd et al. 2017; Massari et al. 2017).

Direct measurements of precipitation over oceans are even more uncommon than over land and are limited to gauge data either mounted on buoys or onboard ships. These observations are affected by deficiencies due to high wind speeds and snowfall. One effort is the Ocean Rainfall and Ice-Phase Precipitation Measurement Network (OceanRAIN), which has been sampling precipitation from optical disdrometers carried by various research vessels since 2010 (Klepp et al. 2018). These disdrometers are exclusively developed for ship usage under all weather conditions and represents a great improvement for observing oceanic precipitation.

28.2.2 Verification Metrics

This section briefly introduces common scores to quantify errors and uncertainty associated with SPPs. For more detail, the reader should refer to Jolliffe and Stephenson (2012) and Ebert (2007). Continuous verification metrics measure the accuracy of a continuous variable, such as rain amount or intensity, whereas categorical verification metrics measure the correspondence between the estimated and observed occurrence of events through a contingency table (Table 28.1). Both types of scores provide a measure of the precipitation error/uncertainty in a conceptually different manner therefore any meaningful validation should assess both. Additionally, diagnostic verification methods, scale decomposition methods, and entity-based methods are used for product validation. For a brief introduction of these methods the reader is referred to Ebert (2007).

Among the continuous verification metrics, the most common scores are:

- Bias: The difference between the SPP mean and the mean of the reference observations. Also known as overall bias, systematic bias, or unconditional bias. Perfect score is zero and it has the units of precipitation.
- Bias ratio: The ratio between the SPP mean and the mean of the reference observations. Perfect score is 1.
- Correlation Coefficient: A measure of the linear association between SPP and reference independent of the mean and variance of the marginal distributions. Pearson Correlation Coefficient and Spearman Rank Correlation are the most widely used ones. Perfect score is 1.
- Mean Absolute Error (MAE): The average of the absolute differences between SPP and reference. It is considered a more robust measure than Mean Square Error (MSE) that is sensitive to large outlier errors. The perfect score is 0 and it has the units of precipitation.
- Mean Error (ME): The average difference between SPP and reference. The perfect score is zero and it has the units of precipitation.
- Root Mean Square Error (RMSE): The square root of the average of the squared differences between SPP and reference. It puts a greater influence on large errors than smaller errors, which may be good if large errors are especially undesirable, but may also encourage conservative precipitation estimation. The perfect score is zero and it has the units of precipitation.

A common way to simultaneously display several continuous statistics at once is the Taylor diagram (Taylor 2001), which summarizes correlation coefficient, standard deviation, and the root mean-square error on a single plot.

Table 28.1 Contingency table commonly used for characterizing detection errors of SPPs

Reference		Satellite	
		RSat $\geq$ th	RSat < th
	RRef $\geq$ th	H	M
	RRef < th	F	Z

Categorical verification metrics are often based on the use of a contingency table (also known as confusion matrix; Table 28.1) that displays the (multivariate) frequency distribution of two variables (reference and SPP). H represents hit cases, when both SPP and reference are greater than or equal to the rain/no-rain threshold (*th*); F represents false alarms, when SPP is greater than or equal to *th*, but the reference is less than *th*; M represents missed events, when the reference is greater than or equal to *th*, but SPP is less than *th*; and Z represents correct no-rain detection, when both SPP and reference are less than *th*. N is the sample size, i.e., the total number of observed events and N = H + M + F + Z.

Categorical statistics are defined based on the contingency table to quantify detection errors associated with SPPs. Some common categorical statistics are listed below:

- Hit rate: $HR = (H + Z)/N$; it credits correct detection (whether rain or no rain) cases equally. This metric is not recommended for extreme event analysis, where the number of correct zeroes is high, resulting in a high score, even if the probability of correctly detecting an extreme event (e.g., tornado) is low. The perfect H value is 1.
- Critical success index or threat score: $CSI = TS = H/(H + M + F)$; it is a better metric than HR when events are rare, since it removes the effect of Z. The perfect value of CSI/TS is 1.
- Probability of detection: $POD = H/(H + M)$; it measures the likelihood of SPP to detect an event when it in fact occurs.
- Success Ratio: $SR = H/(H + F)$; it measures the likelihood of SPP being correct, when detecting rain. The perfect SR value is 1.
- False alarm ratio: $FAR = F/(F + H)$; it measures the likelihood that a precipitation event does not occur when SPP estimates rain. The perfect FAR value is zero.
- Frequency Bias: $FB = (H + F)/(H + M)$; it measures the total number of events estimated by SPP divided by the total number of events observed by the reference. The perfect FB value is 1.
- Heidke Skill Score: $HSS = 2 \times (HZ-FM)/[(H + M)(M + Z) + (H + F)(F + Z)]$; it measures the fractional improvement of the SPP over the reference. HSS ranges between $-\infty$ and 1. Negative values indicate that the chance estimate is better, 0 means no skill, and a perfect SPP would obtains an HSS of 1.

A common technique to display three categorical statistics on the same plot is the performance diagram introduced by Roebber (2009), which concisely presents probability of detection, success ratio, critical success index, and the hit bias.

SPP detection capabilities were shown to depend on the retrieval algorithm, topography, and seasonality. For instance, TMPA 3B42 was shown to be superior to the other SPPs in terms of false alarm rates, but to be affected by low probability of detection (even less than 50%) in several regions of the world, including the contiguous US, the South American Equatorial region, and Australia (Maggioni et al. 2016a, b). In regions characterized by complex terrain and sharp precipitation gradients, rain detection is weak, mainly because of the poor ability of the IR sensors to distinguish between raining and non-raining clouds. Seasonality and climatology also affect the performance of SPPs, especially in terms of probability of detection: semi-arid areas are characterized by weak rain detection, but better performances were observed in tropical and equatorial regions. Winters, which are commonly characterized by lighter rain events (harder to detect from satellites), show low detection ability. In semi-arid climates (like northern Mexico), high false alarm rates in SPPs are observed because of raindrop evaporation happening before reaching the surface.

28.2.3   Triple Collocation Analysis

The ability to properly validate SPPs depends on the existence of a reliable benchmark dataset. However, in regions like Africa, Asia, and South America the density of rain gauges is very poor (Fig. 28.1) and common validation methods are often infeasible.

To overcome this issue, some studies have tried to use alternative techniques for validating SPPs in absence of ground-based observations. One of this technique is known as Triple Collocation Analysis (TCA). The first applications of TCA concerned geophysical variables such as ocean wind speed, wave height (Stoffelen 1998), and soil moisture (Dorigo et al. 2010; Crow and van den Berg 2010), while only recently has the TCA been applied to uncertainties in rainfall estimates. TCA is based on the following assumptions. Given three estimates of the same variable

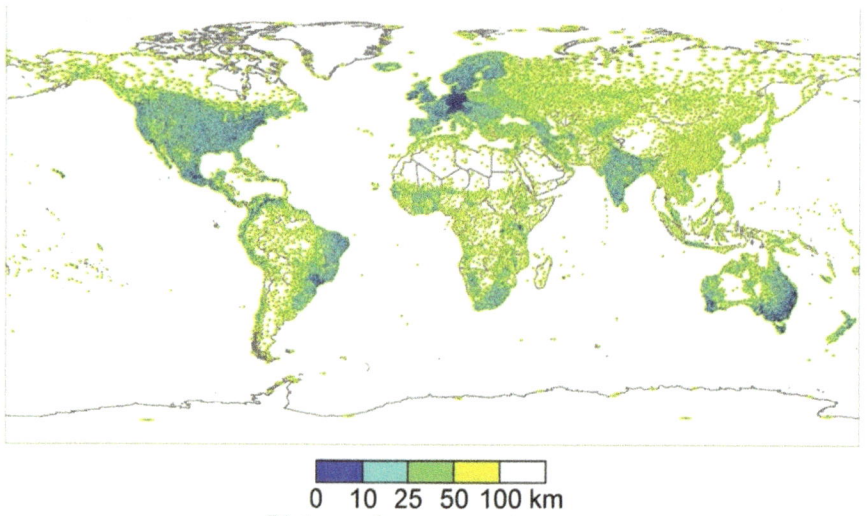

0   10 25 50 100 km
Distance from nearest gauge

Fig. 28.1 Map showing the distance to nearest GPCC gauge, typical of all regular and reliable gauge measurements; blank areas in the figure are beyond 100 km from the nearest gauge. (Adapted from Kidd et al. 2017)

characterized by (i) stationarity of the statistics, (ii) linearity between the three estimates (vs. the same target) across all timescales, and (iii) existence of uncorrelated error among the three estimates, TCA is able to provide error and correlation of each of the three datasets. For example, Roebeling et al. (2012) determined the spatial and temporal uncertainty characteristics of three precipitation datasets over Europe (a visible/near-infrared data set, a weather radar data set and gridded rain gauge products). The authors ensured an error Gaussian distribution by averaging the datasets over a sufficiently long period (10 days) and re-gridding them to a sufficiently coarse spatial resolution (about 25 km). Alemohammad et al. (2015), by assuming a multiplicative error model of precipitation, applied TCA to 14-day cumulated rainfall estimates derived from satellite, gauges, radars, and models in order to retrieve the RMSE and the correlation of each data set across the US. Applying TCA to state-of-the-art SPPs (with more than two of them in the triplet) can be problematic given the highly overlapping set of common sensors used in the algorithms (that inherently introduces mutually independent errors within the datasets). To overcome this issue, Massari et al. (2017) applied TCA to the rainfall accumulation estimates derived from a reanalysis product, a TRMM-era satellite-based rainfall product and an independent satellite-based rainfall product obtained via inversion of the Advanced SCATterometer satellite soil moisture (i.e., SM2RAIN, Brocca et al. 2014). On this basis, TCA was applied on a global scale for the first time to provide global correlations of the aforementioned products without using a ground-based benchmark data set (Fig. 28.2).

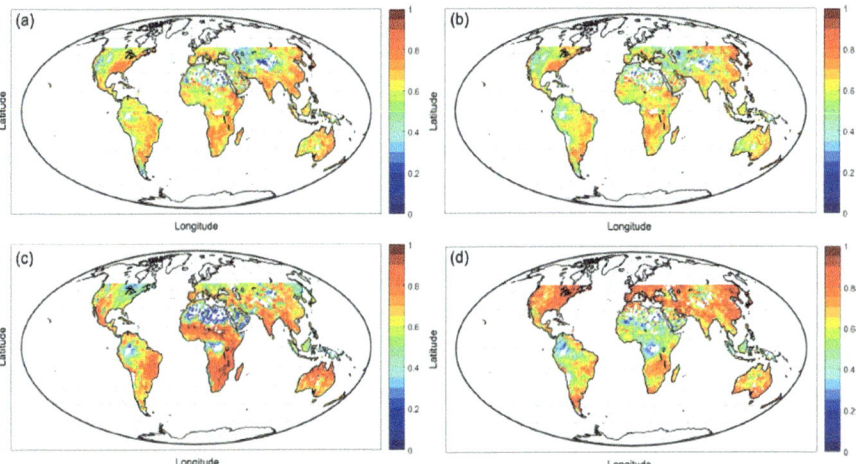

Fig. 28.2 Global correlation of the 3B42RT (**a**), CMORPH (**b**), SM2RAIN (**c**) and ERA-Interim (**d**) products obtained by TCA using Triplet A (ERA-Interim-SM2RAIN-3B42RT) for 3B42RT, ERA-Interim and SM2RAIN and Triplet B (ERA-Interim-SM2RAIN-CMORPH) for CMORPH. (Adapted from Massari et al. 2017)

The analysis using daily rainfall accumulation products conveys the relatively high performance of SPPs in eastern North and South America, southern Africa, southern and eastern Asia, eastern Australia, and southern Europe, as well as complementary performances between the reanalysis product and SM2RAIN, with the first performing reasonably well in the Northern Hemisphere and the second providing good performance in the Southern Hemisphere. Although the TCA is a valuable technique for the assessment of SPPs within sparse-data regions, it is not exhaustive in the sense that it does not provide information on the bias of the products and only delivers a relative evaluation of the products. In addition, it requires three products with mutually uncorrelated errors, which is often not feasible within SPPs.

28.3   Error and Uncertainty Models

Accurate modeling of errors and uncertainties associated with SPPs is indispensable for the proper use of those products in hydrological modeling, water resources management, and climate studies. Some methods rely on a reference to estimate the SPP errors/uncertainties, whereas other methods focus only on the uncertainty component and do not require a benchmark. For instance, Tian and Peters-Lidard (2010) followed a method proposed by Adler et al. (2009) to assess SPP uncertainties using the spread of coincidental and co-located estimates from an ensemble of six different SPPs. Although this method presents a global view of the SPP uncertainty characteristics and their regional and seasonal variations, this is only a relative analysis and some SPPs are not completely independent. This section describes stochastic methods that model systematic and random errors associated with SPPs.

Error and uncertainty models are fundamental to gain knowledge of SPP accuracy and precision in regions and during periods in which in-situ measurements are not available. Two types of error models – additive and multiplicative – are commonly used to assess errors and uncertainties in precipitation measurements. Tian et al. (2013) investigated the suitability of both models for estimating errors in daily SPPs and concluded that the multiplicative error model is a much better choice according to three criteria:

1. ability in separating the systematic and random components of the error;
2. applicability to the large range of variability in daily precipitation; and
3. predictive skills.

Products like the Global Precipitation Climatology Project (GPCP) analysis and the Tropical Rainfall Measuring Mission (TRMM) Multi-satellite Precipitation Analysis (TMPA) provide an error estimate at every grid box, based on the method proposed by Huffman et al. (1997). However, these error estimates depend on the samples being functionally independent, which does not hold true when considering the 3-hourly resolution. Other studies have investigated the SPP sampling error and adopted the power law model to quantify the standard deviation of the sampling error (e.g., Bell et al. 1990; Gebremichael and Krajewski 2004; Steiner et al. 2003; Hong et al. 2006; Gebregiorgis and Hossain 2013). The main assumption here is that precipitation errors follow a lognormal distribution, which can be unrealistic, especially at high rain rates (Gebremichael and Krajewski 2005).

Hossain and Anagnostou (2004) proposed a different approach, namely the two-dimensional satellite rainfall error model (SREM2D), which generates an ensemble for a given rain field. SREM2D is based on a stochastic space–time formulation and models the joint probability of successful delineation of rainy and non-rainy areas. SREM2D has been demonstrated to be adequate to correct SPPs for their applications in streamflow simulations and debris flow predictions (Maggioni et al. 2013; Falck et al. 2015; Nikolopoulos et al. 2017). Another attempt to estimate SPP errors/uncertainties at fine temporal and spatial scales (3-hourly/25 km) is the non- parametric model introduced by Gebremichael et al. (2011). This method is based on SPP conditional density functions at each grid box that are calibrated using a ground reference (i.e., gauge-adjusted ground radar observations).

The Precipitation Uncertainties for Satellite Hydrology (PUSH) framework was introduced by Maggioni et al. (2014) as an alternative method to model errors associated with SPPs. PUSH considers the SPP error as a combination of a combination of random and systematic components and models missed precipitation cases, false alarms, and hit biases. PUSH was later modified by Oliveira et al. (2018) to account for factors like seasonality and surface type (i.e., land and water).

Ad-hoc solutions have been proposed to separate the random and systematic component of the error (AghaKouchak et al. 2012; Maggioni et al. 2016a, b). For example, AghaKouchak et al. (2012) proposed a method to separate the random and systematic components of the total mean squared error. They showed that the systematic error spatial distribution had similar patterns for all precipitation products analyzed, whereas the random part increased for sub-daily scales with respect to daily accumulation. They also found a dependence of the systematic component on the season being higher during winter (while the random component reduced). Following the work of AghaKouchak et al. (2012), Maggioni et al. (2016a, b) quantified the systematic and the random component of the RMSE metric of different SPPs over the contiguous United States. They found a strong dependence of both systematic and random error on satellite-derived rain rates – with larger error components at larger rain rates. The findings of these studies highlight that bias removal methods should take into account the error spatio-temporal characteristics, as well as the proportionality of error to the magnitude of rain rate.

28.4   Summary of the Performance of the Main Satellite Precipitation Products

Table 28.2 adapted from Maggioni et al. (2016b) summarizes the performance of the main SPPs around the world. Overall, SPPs have the enormous advantage of providing an estimate anytime/anywhere, even in impervious regions of the world

Table 28.2 Summary of the main validation studies of SPPs

Region	Error magnitude	Detection
Africa	Superiority of 3B42-V6 across Ethiopia, Somalia, Kenya, Uganda, and Sahel; In the Nile Basin, better performance of CMORPH over 3B42-V6; General underestimation of heavy precipitation for 3B42-V6, GSMaP-MVK, CMORPH, and PERSIANN over Eastern Africa; Overestimation of PERSIANN during the rainy season and over tropical zones; Overall better performance over tropical/equatorial regions than over semi-arid and in mountainous regions.	Large underestimation of GSMaP-MVK in the number of rainy days per year; Poor detection over complex terrain and in semi-arid areas; Considerable overestimation of the number of rainy days over tropical zones for all products except for GSMaP-MKV in the Volta and Zambezi basins.
North and Central America	Better agreement of SPPs with ground-radar products than with rain gauges in central US; Across CONUS, season-dependent bias for PERSIANN and CMORPH; Lowest bias shown by 3B42-V6; Elevation-dependent biases for CMORPH and PERSIANN-CCS with an underestimation of light precipitation.	3B42: low POD and low FAR; CMORPH: high POD and largest FAR; PERSIANN: high FAR and better POD than 3B42, but not as good as CMORPH; No dependence of FAR on seasonality, but POD is higher during the convective warm season.
South America	Overestimation of PERSIANN and 3B42-V6 in the equatorial region; Overestimation of daily 3B42 across Colombia, but good performance over the Peruvian Andes, the Amazon, and Brazil; Overestimation of CMORPH in Brazil, but better performance over in the Amazon and in Colombia; Location-dependent bias for PERSIANN, with a significant overestimation in Colombia and an underestimation in Brazil; Good performance of GSMaP-MVK in terms of mean errors and correlations in Colombia; 3B42-V7 outperforms 3B42-V6 over the Andean-Amazon River Basin.	Weak detection of PERSIANN and 3B42-V6.
Asia	China: positive bias in 3B42RT during the cold season, positive bias in CMORPH during the warm season, 3B42-V7 better than V6 and RT; India: during the summer monsoon, 3B42-V7 has small underestimation, high correlation, and low RMSE. Underestimation of 3B42-V6 in western India but high correlations in heavy rainfall regions; PERSIANN well reproduces seasonal and daily variability;	3B42 research product underestimates precipitation occurrence across Northwest China.

(continued)

Table 28.2 (continued)

Region	Error magnitude	Detection
	Tibet and Himalaya: high correlations and low RMSE for 3B42 and CMORPH, over(under)-estimation of light(heavy) rain for PERSIANN; Japan: GSMaP-MVK and CMORPH have the highest correlations with gauge data; Korea: overestimation of 3B42 of light rain, negative bias in CMORPH and PERSIANN; Cambodia and Indonesia: poor relationship between gauges and GSMaP-NRT; Middle East: 3B42-V7 outperforms CMORPH. Region-dependent bias in PERSIANN.	
Europe	Overall overestimation of 3B42-V6; CMORPH exhibits a large and spatially consistent underestimation, with season-dependent biases; SPPs with no ground information underestimate precipitation above 35°N; Magnitude-dependent mean error of 3B42, CMORPH, and PERSIANN; Complex terrain: underestimation of 3B42-V6 and CMORPH, good performance of PERSIANN over Sicily and the Eastern Alps.	No single SPP was found ideal for detecting heavy precipitation over the Eastern Italian Alps; Over the western mountainous Mediterranean region, CMORPH captures convective events better than PERSIANN.
Australia	3B42-V6 outperforms PERSIANN and CMORPH, whereas PERSIANN presents the worst RMSEs; Daily 3B42-RT (V7), CMORPH, and PERSIANN show increasing mean biases and error variances with increasing rain rates; 3B42-RT, CMORPH, PERSIANN are fairly unbiased in the tropics and have a positive bias in the mid-latitudes; 3B42-V7 shows good agreement with ground data during intense tropical cyclones; Large bias errors are associated with 3B42-RT, CMORPH, PERSIANN during the winter season at the daily/1o scale.	PERSIANN presents the worst FARs, PODs; Better detection at larger rain rates, and larger false alarms at small rain rates and during dry seasons; Poor detection for 3B42-RT for rainfall smaller than 3 mm/day and POD less than 50%; CMORPH and PERSIANN have worse detection capability and larger false alarms than 3B42-V7.
Oceans	Pacific: general underestimation of SPPs, particularly over the eastern region, with a rain rate dependent error in 3B42-RT and V6; Indian Ocean: 3B42-V7 underestimates heavy and light precipitation and shows the highest correlation with buoy data	Similar detection for 3B42, CMORPH, and PERSIANN over the East Pacific; 3B42 has lower POD and FARs in the West Pacific than CMORPH and PERSIANN; Indian Ocean: 3B42-V7 misses intermittent warm precipitation events.

(continued)

Table 28.2 (continued)

Region	Error magnitude	Detection
	during the monsoon season; For 3B42, RMSEs are minimum in the Indian and maximum in the Atlantic; 3B42-V7 improving RMSE and bias with respect to V6 during intense precipitation events over the three tropical oceans.	

Adapted from Maggioni et al. (2016b)

(e.g., oceans, forests, and deserts). However, SPPs are indirect estimates and, as such, are affected by errors and uncertainties that depends on several factors, including seasonality, topography, sensor type, among others. Numerous studies in the past attempted to validate SPPs in different regions and under different conditions. During the TRMM-era, the TMPA 3B42 product was shown to be superior to the other SPPs in terms of false alarm rates, correlations, and mean errors, but low probability of detection (< 50%) across several regions of the globe (e.g., CONUS, South America, Australia).

In complex terrain regions, characterized by low rain detection and magnitude-dependent mean errors, there is no consensus on which SPP performs best. Both climatology and seasonality were shown to play a role in SPP errors/uncertainties. Weak rain detection is common in semiarid areas, but tropical and equatorial regions show good SPP performance. Warm seasons, normally characterized by convective systems, show higher correlations and lower biases, while winter light rain events are more difficult to detect.

References

Adler, R. F., Wang, J.-J., Gu, G., & Huffman, G. J. (2009). A ten-year tropical rainfall climatology based on a composite of TRMM products. *Journal of the Meteorological Society of Japan, 87*, 281–293. https://doi.org/10.2151/jmsj.87A.281.

AghaKouchak, A., Behrangi, A., Sorooshian, S., Hsu, K., & Amitai, E. (2011). Evaluation of satellite-retrieved extreme precipitation rates across the Central United States. *Journal of Geophysical Research: Atmospheres, 116*(D2). https://doi.org/10.1029/2010JD014741.

AghaKouchak, A., Mehran, A., Norouzi, H., & Behrangi, A. (2012). Systematic and random error components in satellite precipitation data sets. *Geophysical Research Letters, 39*, L09406. https://doi.org/10.1029/2012GL051592.

Alemohammad, S. H., McColl, K. A., Konings, A. G., Entekhabi, D., & Stoffelen, A. (2015). Characterization of precipitation product errors across the United States using multiplicative triple collocation. *Hydrology and Earth System Sciences, 19*, 3489–3503. https://doi.org/10.5194/hess-19-3489-2015.

Ali, A., Amani, A., Diedhiou, A., & Lebel, T. (2005). Rainfall estimation in the Sahel. Part II: Evaluation of rain gauge networks in the CILSS countries and objective intercomparison of rainfall products. *Journal of Applied Meteorology, 44*, 1707–1722. https://doi.org/10.1175/JAM2305.1.

Anagnostou, E. N., Maggioni, V., Nikolopoulos, E. I., Meskele, T., Hossain, F., & Papadopoulos, A. (2010). Benchmarking high resolution global satellite rainfall products to radar and rain-gauge rainfall estimates. *IEEE Transactions on Geoscience and Remote Sensing, 48*, 1667–1683. https://doi.org/10.1109/TGRS.2009.2034736.

Behrangi, A., & Wen, Y. (2017). On the spatial and temporal sampling errors of remotely sensed precipitation products. *Remote Sensing, 9*, 1127. https://doi.org/10.3390/rs9111127.

Bell, T. L., Abdullah, A., Martin, R. L., & North, G. R. (1990). Sampling errors for satellite-derived tropical rainfall: Monte Carlo study using a space-time stochastic model. *Journal of Geophysical Research, 95*, 2195–2205. https://doi.org/10.1029/JD095iD03p02195.

Bellerby, T., & Sun, J. (2005). Probabilistic and ensemble repre- sentations of the uncertainty in an IR/microwave satellite precipitation product. *Journal of Hydrometeorology, 6*, 1032–1044. https://doi.org/10.1175/JHM454.1.

Brocca, L., Ciabatta, L., Massari, C., Moramarco, T., Hahn, S., Hasenauer, S., Kidd, R., Dorigo, W., Wagner, W., & Levizzani, V. (2014). Soil as a natural rain gauge: Estimating global rainfall from satellite soil moisture data. *Journal of Geophysical Research, 119*, 5128–5141. https://doi.org/10.1002/2014JD021489.

Bytheway, J. L., & Kummerow, C. D. (2013). Inferring the uncertainty of satellite precipitation estimates in data-sparse regions over land. *Journal of Geophysical Research, 118*, 9524–9533. https://doi.org/10.1002/jgrd.50607.

Ciabatta, L., Marra, A. C., Panegrossi, G., Casella, D., Sanò, P., Dietrich, S., Massari, C., & Brocca, L. (2017). Daily precipitation estimation through different microwave sensors: Verification study over Italy. *Journal of Hydrology, 545*, 436–450. https://doi.org/10.1016/j.jhydrol.2016.12.057.

Crow, W. T., & van den Berg, M. J. (2010). An improved approach for estimating observation and model error parameters in soil moisture data assimilation. *Water Resources Research, 46*, W12519. https://doi.org/10.1029/2010WR009402.

Dinku, T., & Anagnostou, E. N. (2005). Regional differences in overland rainfall estimation from PR-calibrated TMI algorithm. *Journal of Applied Meteorology, 44*, 189–205. https://doi.org/10.1175/JAM2186.1.

Dinku, T., Ceccato, P., Grover-Kopec, E., Lemma, M., Connor, S. J., & Ropelewski, C. F. (2007). Validation of satellite rainfall products over East Africa's complex topography. *International Journal of Remote Sensing, 28*, 1503–1526. https://doi.org/10.1080/01431160600954688.

Dorigo, W. A., Scipal, K., Parinussa, R. M., Liu, Y. Y., Wagner, W., de Jeu, R. A. M., & Naeimi, V. (2010). Error characterisation of global active and passive microwave soil moisture datasets. *Hydrology and Earth System Sciences, 14*, 2605–2616. https://doi.org/10.5194/hess-14-2605-2010.

Ebert, E. E. (2007). Methods for verifying satellite precipitation estimates. In V. Levizzani, P. Bauer, & F. J. Turk (Eds.), *Measuring precipitation from space* (Advances global change research) (Vol. 28, pp. 345–356). Dordrecht: Springer. ISBN: 978-1-4020-5835-6.

Ebert, E. E., Janowiak, J. E., & Kidd, C. (2007). Comparison of near-real-time precipitation estimates from satellite observa- tions and numerical models. *Bulletin of the American Meteorological Society, 88*, 47–64. https://doi.org/10.1175/BAMS-88-1-47.

Falck, A. S., Maggioni, V., Tomasella, J., Vila, D. A., & Diniz, F. L. R. (2015). Propagation of satellite precipitation uncertainties through a distributed hydrologic model: A case study in the Tocantins-Araguaia basin in Brazil. *Journal of Hydrology, 527*, 943–957. https://doi.org/10.1016/j.jhydrol.2015.05.042.

Gebregiorgis, A. S., & Hossain, F. (2013). Understanding the dependence of satellite rainfall uncertainty on topography and climate for hydrologic model simulation. *IEEE Transactions on Geoscience and Remote Sensing, 51*, 704–718. https://doi.org/10.1109/TGRS.2012.2196282.

Gebremichael, M., & Krajewski, W. F. (2004). Characterization of the temporal sampling error in space-time-averaged rainfall estimates from satellites. *Journal of Geophysical Research, 109*, D11110. https://doi.org/10.1029/2004JD004509.

Gebremichael, M., & Krajewski, W. F. (2005). Modeling distribution of temporal sampling errors in area-time-averaged rainfall estimates. *Atmospheric Research, 73*, 243–259. https://doi.org/10.1016/j.atmosres.2004.11.004.

Gebremichael, M., Liao, G.-Y., & Yan, J. (2011). Nonparametric error model for a high resolution satellite rainfall product. *Water Resources Research, 47*, W07504. https://doi.org/10.1029/2010WR009667.

Gottschalck, J., Meng, J., Rodell, M., & Houser, P. (2005). Analysis of multiple precipitation products and preliminary assessment of their impact on global land data assimilation system land surface states. *Journal of Hydrometeorology, 6*, 573–598. https://doi.org/10.1175/JHM437.1.

Herold, N., Alexander, L. V., Donat, M. G., Contractor, S., & Becker, A. (2015). How much does it rain over land? *Geophysical Research Letters, 43*, 341–348. https://doi.org/10.1002/2015GL066615.

Hirpa, F. A., Gebremichael, M., & Hopson, T. (2010). Evaluation of high-resolution satellite precipitation products over very complex terrain in Ethiopia. *Journal of Applied Meteorology and Climatology, 49*, 1044–1051. https://doi.org/10.1175/2009JAMC2298.1.

Hong, Y., Hsu, K.-L., Moradkhani, H., & Sorooshian, S. (2006). Uncertainty quantification of satellite precipitation estimation and Monte Carlo assessment of the error propagation into hydrologic response. *Water Resources Research, 42*, W08421. https://doi.org/10.1029/2005WR004398.

Hossain, F., & Anagnostou, E. N. (2004). Assessment of current passive-microwave- and infrared-based satellite rainfall remote sensing for flood prediction. *Journal of Geophysical Research, 109*, D07102. https://doi.org/10.1029/2003JD003986.

Hossain, F., Anagnostou, E. N., Dinku, T., & Borga, M. (2004). Hydrological model sensitivity to parameter and radar rainfall estimation uncertainty. *Hydrological Processes, 18*, 3277–3291. https://doi.org/10.1002/hyp.5659.

Huffman, G. J. (1997). Estimates of root-mean-square random error for finite samples of estimated precipitation. *Journal of Applied Meteorology, 36*, 1191–1201. https://doi.org/10.1175/1520-0450(1997)036,1191:EORMSR.2.0.CO;2.

Huffman, G. J., Adler, R. F., Arkin, P., Chang, A., Ferraro, R., Gruber, A., Janowiak, J., McNab, A., Rudolf, B., & Schneider, U. (1997). The global precipitation climatology project (GPCP) combined precipitation dataset. *Bulletin of the American Meteorological Society, 78*(1), 5–20. https://doi.org/10.1175/1520-0477(1997)078<0005:TGPCPG>2.0.CO;2.

Huffman, and Coauthors. (2007). The TRMM multisatellite Precipita- tion analysis (TMPA): Quasi-global, multiyear, combined- sensor precipitation estimates at fine scales. *Journal of Hydrometeorology, 8*, 38–55. https://doi.org/10.1175/JHM560.1.

Jolliffe, I. T., & Stephenson, D. B. (2012). *Forecast verification: A practitioner's guide in atmospheric science* (2nd ed.). Somerset. 274 pp: Wiley. https://doi.org/10.1002/9781119960003.

Kidd, C., Becker, A., Huffman, G. J., Muller, C. L., Joe, P., Skofronick-Jackson, G., & Kirschbaum, D. B. (2017). So, how much of the earth's surface is covered by rain gauges? *Bulletin of the American Meteorological Society, 98*, 69–78. https://doi.org/10.1175/BAMS-D-14-00283.1.

Klepp, C., Michel, S., Protat, A., Burdanowitz, J., Albern, N., Kähnert, M., Dahl, A., Louf, V., Bakan, S., & Buehler, S. A. (2018). OceanRAIN, a new in-situ shipboard global ocean surface-reference dataset of all water cycle components. *Scientific Data, 5*, 180122. https://doi.org/10.1038/sdata.2018.122.

Kummerow, C. D., Berg, W., Thomas-Stahle, J., & Masunaga, H. (2006). Quantifying global uncertainties in a simple microwave rainfall algorithm. *Journal of Atmospheric and Oceanic Technology, 23*, 23–37. https://doi.org/10.1175/JTECH1827.1.

Maggioni, V., & Massari, C. (2018). On the performance of satellite precipitation products in riverine flood modeling: A review. *Journal of Hydrology, 558,* 214–224. https://doi.org/10.1016/J.JHYDROL.2018.01.039.

Maggioni, V., Vergara, H. J., Anagnostou, E. N., Gourley, J. J., Hong, Y., & Stampoulis, D. (2013). Investigating the applicability of error correction ensembles of satellite rainfall products in river flow simulations. *Journal of Hydrometeorology, 14,* 1194–1211. https://doi.org/10.1175/JHM-D-12-074.1.

Maggioni, V., Sapiano, M. R. P., Adler, R. F., Tian, Y., & Huffman, G. J. (2014). An error model for uncertainty quantification in high-time-resolution precipitation products. *Journal of Hydrometeorology, 15,* 1274–1292. https://doi.org/10.1175/JHM-D-13-0112.1.

Maggioni, V., Meyers, P. C., & Robinson, M. D. (2016a). A review of merged high-resolution satellite precipitation product accuracy during the tropical rainfall measuring mission (TRMM) era. *Journal of Hydrometeorology, 17,* 1101–1117. https://doi.org/10.1175/JHM-D-15-0190.1.

Maggioni, V., Sapiano, M. R. P., & Adler, R. F. (2016b). Estimating uncertainties in high-resolution satellite precipitation products: Systematic or random error? *Journal of Hydrometeorology, 17,* 1119–1129. https://doi.org/10.1175/JHM-D-15-0094.1.

Massari, C., Crow, W., & Brocca, L. (2017). An assessment of the performance of global rainfall estimates without ground-based observations. *Hydrology and Earth System Sciences, 21,* 4347–4361. https://doi.org/10.5194/hess-21-4347-2017.

Moazami, S., Golian, S., Kavianpour, M. R., & Hong, Y. (2013). Comparison of PERSIANN and V7 TRMM multi-satellite precipitation analysis (TMPA) products with rain gauge data over Iran. *International Journal of Remote Sensing, 34,* 8156–8171. https://doi.org/10.1080/01431161.2013.833360.

Nijssen, B., & Lettenmaier, D. P. (2003). Effect of precipitation sampling error on simulated hydrological fluxes and states: Anticipating the global precipitation measurement satellites. *Journal of Geophysical Research, 109,* D02103. https://doi.org/10.1029/2003JD003497.

Nikolopoulos, E. I., Destro, E., Maggioni, V., Marra, F., & Borga, M. (2017). Satellite rainfall estimates for debris flow prediction: An evaluation based on rainfall accumulation–duration thresholds. *Journal of Hydrometeorology, 18,* 2207–2214. https://doi.org/10.1175/JHM-D-17-0052.1.

Oliveira, R., Maggioni, V., Vila, D., & Morales, C. (2016). Characteristics and diurnal cycle of GPM rainfall estimates over the Central Amazon region. *Remote Sensing, 8,* 544. https://doi.org/10.3390/rs8070544.

Oliveira, R., Maggioni, V., Vila, D., & Porcacchia, L. (2018). Using satellite error modeling to improve GPM-level 3 rainfall estimates over the Central Amazon region. *Remote Sensing, 10,* 336. https://doi.org/10.3390/rs10020336.

Roca, R., Chambon, P., Jobard, I., Kirstetter, P.-E., Gosset, M., & Bergès, J. C. (2010). Comparing satellite and surface rainfall products over West Africa at meteorologically relevant scales during the AMMA campaign using error estimates. *Journal of Applied Meteorology and Climatology, 49,* 715–731. https://doi.org/10.1175/2009JAMC2318.1.

Roebber, P. J. (2009). Visualizing multiple measures of forecast quality. *Weather and Forecasting, 24,* 601–608. https://doi.org/10.1175/2008WAF2222159.1.

Roebeling, R. A., Wolters, E. L. A., Meirink, J. F., & Leijnse, H. (2012). Triple collocation of summer precipitation retrievals from SEVIRI over Europe with gridded rain gauge and weather radar data. *Journal of Hydrometeorology, 13,* 1552–1566. https://doi.org/10.1175/JHM-D-11-089.1.

Schaake, J. C., Hamill, T. M., Buizza, R., & Clark, M. (2007). HEPEX: The hydrological ensemble prediction experiment. *Bulletin of the American Meteorological Society, 88,* 1541–1548. https://doi.org/10.1175/BAMS-88-10-1541.

Serrat-Capdevila, A., Valdes, J. B., & Stakhiv, E. Z. (2014). Water management applications for satellite precipitation products: Synthesis and recommendations. *Journal of the American Water Resources Association, 50,* 509–525. https://doi.org/10.1111/jawr.12140.

Stampoulis, D., Anagnostou, E. N., & Nikolopoulos, E. I. (2013). Assessment of high-resolution satellite-based rainfall estimates over the Mediterranean during heavy precipitation events. *Journal of Hydrometeorology, 14*, 1500–1514. https://doi.org/10.1175/JHM-D-12-0167.1.

Steiner, M., Bell, T. L., Zhang, Y., & Wood, E. F. (2003). Comparison of two methods for estimating the sampling-related uncertainty of satellite rainfall averages based on a large radar dataset. *Journal of Climate, 16*(22), 3759–3778. https://doi.org/10.1175/1520-0442(2003) 016<3759:COTMFE>2.0.CO;2.

Stephens, G. L., & Kummerow, C. D. (2007). The remote sensing of clouds and precipitation from space: A review. *Journal of the Atmospheric Sciences, 64*, 3742–3765. https://doi.org/10.1175/ 2006JAS2375.1.

Stoffelen, A. (1998). Toward the true near-surface wind speed: Error modeling and calibration using triple collocation. *Journal of Geophysical Research, 103*, 7755–7766. https://doi.org/10. 1029/97JC03180.

Tang, L., & Hossain, F. (2012). Investigating the similarity of satellite rainfall error metrics as a function of Köppen climate classification. *Atmospheric Research, 104–105*, 182–192. https:// doi.org/10.1016/j.atmosres.2011.10.006.

Taylor, K. E. (2001). Summarizing multiple aspects of model performance in a single diagram. *Journal of Geophysical Research, 106*, 7183–7192. https://doi.org/10.1029/2000JD900719.

Tian, Y., & Peters-Lidard, C. D. (2010). A global map of uncertainties in satellite-based precipitation measurements. *Geophysical Research Letters, 37*, L24407. https://doi.org/10.1029/ 2010GL046008.

Tian, Y., Peters-Lidard, C. D., Choudhury, B. J., & Garcia, M. (2007). Multitemporal analysis of TRMM-based satellite precipitation products for land data assimilation applications. *Journal of Hydrometeorology, 8*, 1165–1183. https://doi.org/10.1175/2007JHM859.1.

Tian, Y., Huffman, G. J., Adler, R. F., Tang, L., Sapiano, M., Maggioni, V., & Wu, H. (2013). Modeling errors in daily precipitation measurements: Additive or multiplicative? *Geophysical Research Letters, 40*, 2060–2065. https://doi.org/10.1002/grl.50320.

Todini, E. (2001). A Bayesian technique for conditioning radar precipitation estimates to rain-gauge measurements. *Hydrology and Earth System Sciences, 5*, 187–199. https://doi.org/10.5194/ hess-5-187-2001.

Villarini, G., & Krajewski, W. F. (2007). Evaluation of the research version TMPA three-hourly three-hourly $0.25° × 0.25°$ rainfall estimates over Oklahoma. *Geophysical Research Letters, 34*, L05402. https://doi.org/10.1029/2006GL029147.

Villarini, G., Mandapaka, P. V., Krajewski, W. F., & Moore, R. J. (2008). Rainfall and sampling uncertainties: A rain gauge perspective. *Journal of Geophysical Research, 113*, W07504. https://doi.org/10.1029/2007JD009214.

Xie, P., & Arkin, P. A. (1995). An intercomparison of gauge observations and satellite estimates of monthly precipitation. *Journal of Applied Meteorology, 34*, 1143–1160. https://doi.org/10. 1175/1520-0450(1995)034<1143:AIOGOA>2.0.CO;2.

Chapter 29
Multiscale Evaluation of Satellite Precipitation Products: Effective Resolution of IMERG

Clément Guilloteau and Efi Foufoula-Georgiou

Abstract Satellite precipitation products are essential for global analysis of water cycle dynamics as well as for regional analyses in regions where no ground observations are available. For any climatic or hydrologic application, it is important to know down to which scale a gridded satellite precipitation product can accurately resolve the spatial patterns of precipitation. This scale, which we call "effective resolution", is a complex combination of the instrument resolution (especially so for multisensor products such as IMERG), the multi-sensor retrieval or merging algorithm, and the type of the precipitating system, and it can differ substantially from the grid size of the satellite product. Here, we use a wavelet-based framework to quantitatively define the effective resolution of the IMERG multi-satellite product by comparison with the MRMS ground radar product at the hourly time scale over the continental United States. Our findings show that the effective resolution varies across geographical areas, seasons and types of precipitation and provide insight for the use of those products in hydrologic applications and for algorithmic improvements.

Keywords Precipitation · Rainfall · Evaluation · Validation · Error modeling · GPM · Raingauges · Nominal resolution · Effective resolution · Spectral analysis · Wavelets

C. Guilloteau (✉)
Department of Civil and Environmental Engineering, University of California, Irvine, CA, USA
e-mail: cguillot@uci.edu

E. Foufoula-Georgiou
Department of Civil and Environmental Engineering, University of California, Irvine, CA, USA

Department of Earth Science, University of California, Irvine, CA, USA

V. Levizzani et al. (eds.), *Satellite Precipitation Measurement*, Advances in Global Change Research 69, https://doi.org/10.1007/978-3-030-35798-6_5

29.1   Introduction

IMERG (Integrated Multi-Satellite Retrievals for GPM, Huffman et al. 2015) is the operational quasi-global multi-satellite quantitative precipitation estimation product (QPE) of the NASA Global Precipitation Measurement (GPM) program. IMERG V05 is provided on a spatial grid covering the globe from latitude 60°N to latitude 60°S with a 0.1° latitude and longitude increment, and a temporal sampling of 30 min. It strives to resolve the sub-mesoscale patterns of precipitation, which typically requires a resolution of a few km or a few dozen km. Obviously, a fine grid increment is necessary for such purpose. But in fact, although the grid increment is often referred to as the product's "resolution" (or "nominal resolution"), there is no guarantee that the precipitation patterns are actually resolved down to the finest scale allowed by the grid. The IMERG estimates rely on the measurements of a dozen different instruments in the microwave and infrared domains. The spatial resolution of these instruments varies from 70 km (SSM/I at 19 GHz) to 2 km (Himawari-8 infrared channels). Given this, it is a-priori uncertain which scales can be resolved by the IMERG algorithm. Moreover, the performance of IMERG at various scales is expected to vary because of the irregular temporal sampling from the different instruments, the evolving configuration of the GPM constellation and the climatic spatial and temporal variability. One must also consider the potential filtering effects associated with the retrieval procedure of each instrument and with the merging procedure.

The question of the "effective resolution" of numerical weather prediction models has been raised several times in the literature (e.g., Pielke 1991; Grasso 2000) and is typically assessed by comparison with observations in the Fourier spectral domain (Skamarock 2004; Frehlich and Sharman 2008; Wong and Skamarock 2016), or in a wavelet spectral domain (Bousquet et al. 2006; Vasić et al. 2007). For numerical models, the assessment of the effective resolution consists of verifying if the output fields show the right spectral energy at all frequencies or scales. The effective resolution is quantified as the finest scale (or highest frequency) for which the power spectrum of the modelled field matches the power spectrum of the observed field. However, the agreement of the power spectra does not guarantee the agreement of the two fields in terms of the spatial location of the observed patterns. In this paper, the effective resolution of IMERG is assessed by comparison with the MRMS gauge-adjusted radar QPE over the Contiguous United States (CONUS) in the wavelet scale domain, which in contrast to the Fourier analysis provides a localised assessment. Moreover, in addition to the wavelet power spectra of the two fields, the spectrum of their difference and the correlation of the local wavelet coefficients of the two fields at various scales are analysed and used in defining the effective resolution.

29.2 Data

29.2.1 IMERG

IMERG is the quasi-global time-continuous precipitation product of the NASA GPM program. To allow time-continuous coverage IMERG combines the observations of a dozen passive microwave and infrared sensors. The algorithm implements the CMORPH Kalman filter approach (Joyce and Xie 2011) to dynamically merge the various instantaneous passive microwave and infrared precipitation fields. It is therefore labelled as a level-three product. The PERSIANN-CCS (Hong et al. 2004) and GPROF (Kummerow et al. 2015) algorithms are used to respectively retrieve the instantaneous infrared and microwave precipitation fields (level two products) on which IMERG relies. The product's grid resolution is 0.1° and the temporal sampling is 30 min. In this study, the Final IMERG product (v5), which includes model-based corrections and gauge adjustment is evaluated at the hourly scale.

29.2.2 MRMS Gauge-Adjusted Radar QPE

The NOAA Multi-Radar Multi-Sensor system (Zhang et al. 2016) integrates the data from 176 radars and more than 7000 automatic rain gauges over CONUS and Southern Canada to generate a suite of precipitation estimation products. The product used in this study as a reference for the evaluation of IMERG is the 1-h MRMS gauge-adjusted radar QPE. The MRMS gauge-adjusted QPE has a native grid increment of 0.01°. It is aggregated here at 0.1° to be compared with IMERG. In spite of the large number of radars composing the MRMS network, some areas remain poorly covered. Figure 29.1 shows the MRMS radar quality index, which is a function of the distance to the closest radar and also takes into account the beam blockage from the relief (Zhang et al. 2016). In this study, only the pixels (at 0.1°) for which the quality index is higher than 0.5 are retained. For the computation of some specific scores a higher threshold is retained (e.g., Fig. 29.4).

29.3 Method: Spectral Analysis in the Wavelet Domain

29.3.1 Rationale

Many studies have demonstrated that the performance of satellite estimation products strongly depends on the spatial and temporal scales at which they are evaluated (Hossain and Huffman 2008; Turk et al. 2009; Sohn et al. 2010; Scheel et al. 2011). When evaluated at spatio-temporal scales approaching their full nominal resolution, finely-gridded products may show mediocre performances, to the point that the

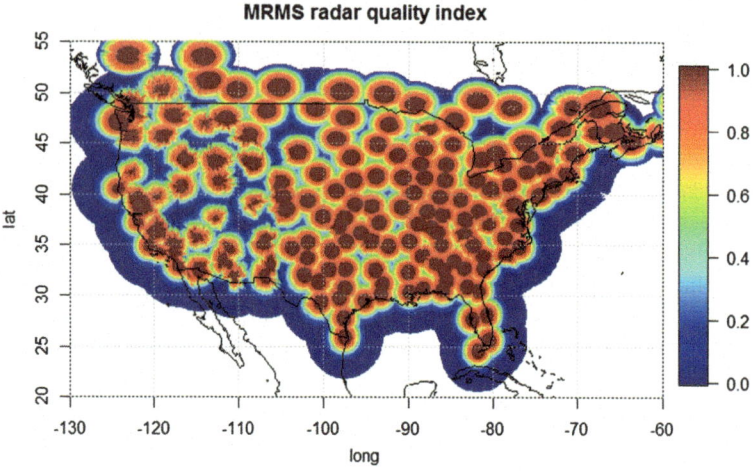

Fig. 29.1 MRMS radar quality index. The quality index considers the distance to the closest radar and the beam blockage by the relief (see Zhang et al. 2016 for the definition of the quality index)

variance of the retrieval error may be in the same order of magnitude as the statistical variance of the reference precipitation (Shen et al. 2010; Haile et al. 2013; Rios Gaona et al. 2016). Under the hypotheses that the retrieval error is purely random, has a zero expected value and is independent from one pixel to another (i.e., is spatially and temporally independent), its variance naturally decreases through spatial and temporal averaging with a N^{-1} decrease rate, where N is the number of averaged individual estimates (i.e., number of averaged pixels). However, these hypotheses, are generally not verified, particularly the hypothesis of spatial and temporal independence of the error. With correlated errors, the decrease rate of the error variance is significantly slower (von Storch 1999; Hossain and Anagnostou 2006). This makes necessary to evaluate precipitation products at multiple scales.

The validation of satellite products is typically performed by comparison with a trusted reference dataset. The most straightforward way to perform the multiscale evaluation is to coarsen the compared fields by aggregation at multiple scales and to perform a complete analysis at each scale (Turk et al. 2009; Sohn et al. 2010; Scheel et al. 2011). However, this analysis is highly redundant: all the information contained in the coarse-resolution fields is necessarily present in the fine-resolution fields too. This can make the interpretation of the multiresolution scores ambiguous. For example, if the retrieval performance appears to be identical at all scales there may be two different explanations: (1) The product performs identically well at capturing fine-scale and coarse-scale variations. (2) The fine-scale (high-frequency) variations are negligible compared to the coarse-scale (low-frequency) variations in both the evaluated product and the reference dataset; in that case the high-resolution fields and low-resolution fields are in fact very similar (the high-resolution fields being simply "oversampled" versions of the low-resolution ones) and the retrieval of the fine-scale variations is not really evaluated.

The spectral analysis in the Fourier domain or wavelet domain allows us to decompose the fields into several frequency bands or several scales and to analyse them independently. While the multi-resolution analysis by successive coarsening can be seen as equivalent to analysing the fields through a suite of low-pass spatial filters, the spectral analysis can be seen as equivalent to analysing the fields through a suite of band-pass filters. The essential question we want to answer through the spectral analysis is: "what supplementary information does the fine-resolution fields contain relatively to the coarse resolution fields?". We chose here to perform the spectral analysis in the wavelet scale domain, using the two-dimensional discrete orthogonal Haar wavelet decomposition. This choice is guided by the fact that: (1) The wavelet coefficients are interpretable as local differences or as local gradients. In particular, the Haar wavelet coefficients are simply computed as the difference between the spatially averaged values of the analysed variable (here precipitation rate) in two adjacent rectangular areas. (2) Unlike the Fourier coefficients, the wavelet coefficients are spatially localised, allowing to study the spatial variations of the spectral properties of the fields. This in particular overcomes the issue of spatial and temporal non-stationarity of the precipitation fields. (3) With the discrete orthogonal wavelet decomposition, the wavelet coefficients are spatially uncorrelated and uncorrelated across scales. Because of this, the wavelet power spectrum, which is simply the statistical variance of the wavelet coefficients as a function of the scale, is unambiguous. Weniger et al. (2017) provides a review on the use of wavelet transforms for spatial verification.

The discrete orthogonal wavelet decomposition is performed as an iterative process over a finite number of levels corresponding to dyadically increasing spatial scales. At the first level, the original field is decomposed into one coarse-scale component (smoothing coefficients) obtained by convolution of the field with a low-pass spatial filter, and three fine-scale components (wavelet coefficients) obtained by convolution of the field with high-pass directional filters (see Appendix). The fine-scale components are retained and the coarse-scale component is further decomposed at the second level, etc. At the end, since the decomposition has a finite number of levels, one residual coarse-scale component must be retained along with the wavelet coefficients at all scales to allow the reconstruction of the original field. The low-pass filter used for the Haar decomposition is a simple rectangular averaging filter (see Appendix A). Consequently, analysing the Haar smoothing coefficients would be strictly equivalent to a multi-resolution analysis by successive aggregations. Instead, we analyse the high-pass wavelet coefficients at each scale. In fact, the Haar wavelet coefficients at scale λ contain exactly the information that is lost when the field is coarsened by averaging from the resolution λ to the resolution $2 \times \lambda$. Because there is no redundancy of information across scales in the wavelet coefficients space, the spatial variations at each scale can be evaluated independently. We can therefore actually estimate the added value of a high-resolution field relatively to the same field at a coarser resolution in terms of information content.

In this article, the "information content" is quantified through the statistical variance or energy (i.e., sum of the squared values) of the original fields and of the wavelet coefficients. In information theory, the "information content" is classically quantified through the Shannon entropy. One could analyse the Shannon entropy and mutual information of the wavelet coefficients (Starck et al. 1998; Labat 2005; Labat et al. 2005). However, we chose here to rather use the variance and covariance metrics because we also analyse the error field and its wavelet decomposition. The Shannon entropy depends of the probability of observing a given value in the field but ignores the numerical value itself. Therefore, the Shannon entropy of the error field would ignore the amplitude of the errors (Petty 2018).

29.3.2 Implementation

The IMERG and MRMS hourly precipitation fields at 0.1° (about 10 km) are decomposed through a two-dimensional discrete orthogonal decomposition with the Haar wavelet. The decomposition is performed at six levels, with resulting dyadic scales being 10, 20, 40, 80, 160 and 320 km. The residual low-pass component of this decomposition is the original field smoothed at the 640 km spatial resolution. We analyse the wavelet coefficients of IMERG and MRMS at each scale in terms of their variance (or energy) and covariance as well as the variance of their difference. The orthogonal wavelet decomposition is energy conservative, i.e. the sum of the energy of the wavelet coefficients at all dyadic scales plus the energy of the residual low-pass component equals the energy of the original field. Because the wavelet decomposition is a linear operation, the wavelet coefficients coming from the decomposition of the difference of the two fields are equal to the difference of the wavelet coefficients of the two fields. Consequently, the squared difference between the two fields equals the sum of the spectral energy of the difference of their wavelet coefficients at all scales plus the energy of the difference of their low-pass components.

We noted previously that, in two dimensions, three series of directional wavelet coefficients are produced at each scale. The first two series of coefficients encode the variation of the field along the vertical (North-South) and horizontal (East-West) direction. The third series of coefficients, sometimes referred to as the "diagonal" coefficients encodes the coupling between the horizontal and vertical variations (see appendix). One can analyse the three series independently, which allows characterising anisotropic features in the fields (Kumar and Foufoula-Georgiou 1993; Kumar 1995; Perica and Foufoula-Georgiou 1996). Here, the three series of coefficients are not differenced, and the spectral energy at each scale is computed as the sum of the energy of the three series of coefficients. The discrete wavelet energy spectrum (or power spectrum) $S(y, \lambda)$ of the field y shows the energy (or variance) of the wavelet coefficients as a function of the scale λ.

Besides comparing the energy spectra of MRSM and IMERG and analysing the spectrum of their difference, we also analyse the linear correlation and covariance between the wavelet coefficients of the two precipitation fields. We note that at each scale λ, the correlation of the wavelet coefficients is related to the spectral energies:

$$Wcor(y_1, y_2, \lambda) = \frac{CS(y_1, y_2, \lambda)}{\sqrt{S(y_1, \lambda)\, S(y_2, \lambda)}} \qquad (29.1)$$

and

$$2 \times CS(y_1, y_2, \lambda) = S(y_1, \lambda) + S(y_2, \lambda) - S(y_2 - y_1, \lambda) \qquad (29.2)$$

where $Wcor(y_1, y_2, \lambda)$ is the linear correlation between the wavelet coefficients of the y_1 and y_2 fields as a function of the scale λ; and $CS(y_1, y_2, \lambda)$ is the cross-spectrum of y_1 and y_2, i.e. co-spectral energy (sum of the products of the wavelet coefficients) of y_1 and y_2 as a function of λ.. The first relation is true because the expected value of the wavelet coefficients at any scale is equal to zero.

Our analysis aims at determining at which scales the spatial variations of precipitation are actually resolved by the IMERG product. The following criterion is chosen to assess the effective resolution. The scale λ is considered resolved if:

$$S(y_{retr} - y_{ref}, \lambda) < 0.5 \times S(y_{ref}, \lambda) \qquad (29.3)$$

i.e. if the spectral energy of the error is less than half the spectral energy of the reference field. Here, y_{retr} denotes the evaluated retrieved precipitation field (IMERG in our case) and y_{ref} the reference precipitation field (MRMS in our case). The effective resolution is the finest resolved scale.

29.3.3   Illustrative Case Study

Figure 29.2 shows the 1-h cumulative IMERG and MRMS precipitation fields over the South-Eastern part of CONUS on 30 Nov. 2016 from 0400 to 0500 UTC, along with the error field (IMERG-MRMS). The Haar wavelet power spectrum of the three fields is shown on the last panel. One can see that the IMERG precipitation field is smooth compared to the MRMS field, specifically, it does not reproduce the fine-scale structures and sharp transitions observed in the MRMS field. Small very active cells, showing precipitation rates between 35 and 50 mm h^{-1} are observed in the MRMS field but not in the IMERG field. Consequently, while the two fields have about the same average hourly precipitation amount (0.60 mm for MRMS and 0.57 mm for IMERG), for the MRMS field, precipitation is concentrated in a smaller area (in the MRMS field, 60% of the precipitation is concentrated in a 50,000 km^2 area, while in the IMERG field, 60% of the precipitation is concentrated in an 80,000 km^2 area). These characteristics are reflected on the wavelet power spectrum: the IMERG field shows a deficit of energy (i.e., spatial variability) at scales finer than 160 km. The wavelet coefficients are also poorly correlated at fine scales. For example, at scales of 10 and 20 km the linear correlations between IMERG and MRMS wavelet coefficients are 0.15 and 0.36, respectively. One direct consequence

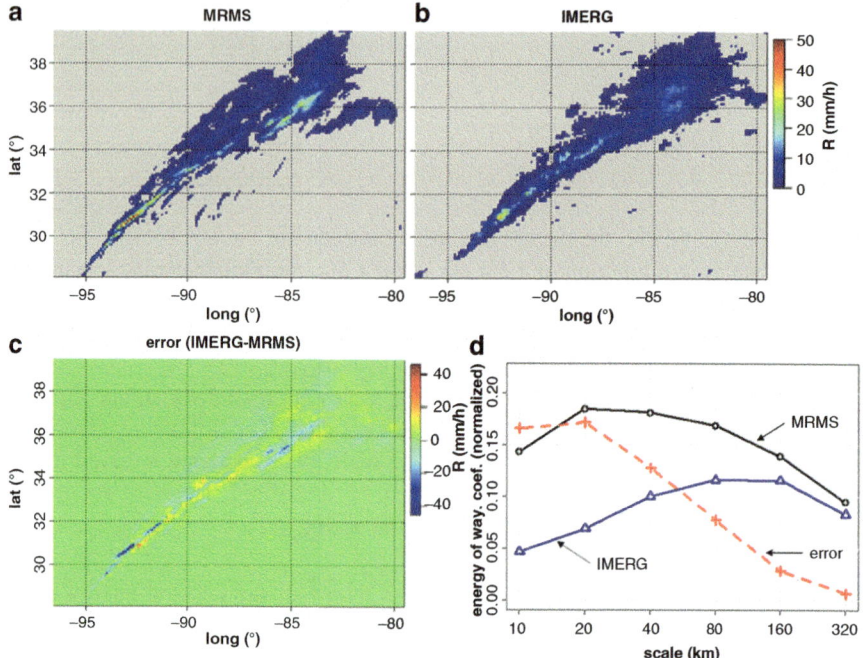

Fig. 29.2 (**a**) MRMS hourly precipitation over South-Eastern US, on 30 Nov. 2016 from 0400 to 0500 UTC. (**b**) Corresponding IMERG hourly precipitation. (**c**) IMERG error relative to MRMS. (**d**) Wavelet energy spectra of MRMS (thick black line with circles), IMERG (thick blue line with triangles) and of the error IMERG-MRMS (dashed red line with crosses). The energy spectra are normalised by the total energy (sum of squared values) of the MRMS field

of this is the high spectral energy of the error at these scales. The spectrum of the error reveals that 94% of the squared error is explained by the misrepresentation of the spatial variations at scales finer than 160 km (and 58% only for the scales finer than 40 km). Applying criterion (29.3), the effective resolution is found to be close to 80 km for this case study.

29.4   Results

The approach described previously is extended over CONUS to analyse the performance of IMERG over a complete year (November 2016 to October 2017). The cold season (November–April) and warm season (May–October) are separated for the analysis. Figure 29.3 shows the cumulative precipitation of IMERG and MRMS for both periods. The cumulative precipitation fields are smoothed using a 640 km sliding window averaging to preserve only the coarse-scale patterns. One can see that IMERG and MRMS show similar patterns, but also that IMERG notably

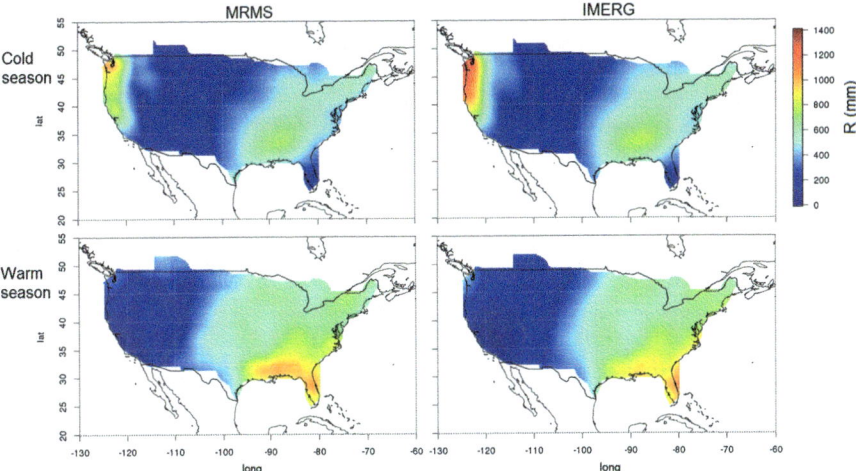

Fig. 29.3 Seasonal cumulative precipitation from Nov. 2016 to Apr. 2017 (cold season) and May – Oct. 2017 (warm season) for MRMS and IMERG. The fields are spatially smoothed through sliding window averaging with a 640 km wide window (corresponding to the smoothing function associated with the Haar wavelet)

(relatively) overestimates the cumulative precipitation amount in the North-West region along the Pacific Coastline during the cold season. For each period, the wavelet power spectra of the hourly IMERG and MRMS fields and of the error field are cumulated at every time step. Figure 29.4 shows the cumulative spectra. One can see that, the power spectra of IMERG and MRMS are relatively close even if IMERG slightly overestimates the amplitude of the spatial variations at the 80–160 km scales for the cold season, and underestimates the amplitude of the spatial variations at scales finer than 80 km during the warm season. The MRMS cold season and warm season spectra show notable differences. The spectral energy of MRMS is maximal around 40~80 km during the cold season and around 20~40 km during the warm season. These scales correspond to the typical dimension of the most prominent structures observed in the MRMS hourly precipitation fields during the two periods. For IMERG, the maximum of energy is found consistently around 80 km for both seasons: the IMERG spectrum shows less seasonal variations than the MRMS spectrum.

The mean squared difference between IMERG and MRMS fields is 1.16 times the mean squared value (power) of MRMS during the cold season and 0.83 times the mean squared value of MRMS during the warm season. These numbers may appear very high; however, Fig. 29.4 shows that the ratio of the spectral energy of the error over the spectral energy of the MRMS reference varies considerably across scales. During the warm season, the spectral energy of the error increases with finer scales. During the cold season, the spectral energy of the error peaks at the 40 km scale. At the 10 and 20 km scales, the spectral energy of the error is systematically higher than the spectral energy of MRMS, meaning that erasing the fine-scale variation in

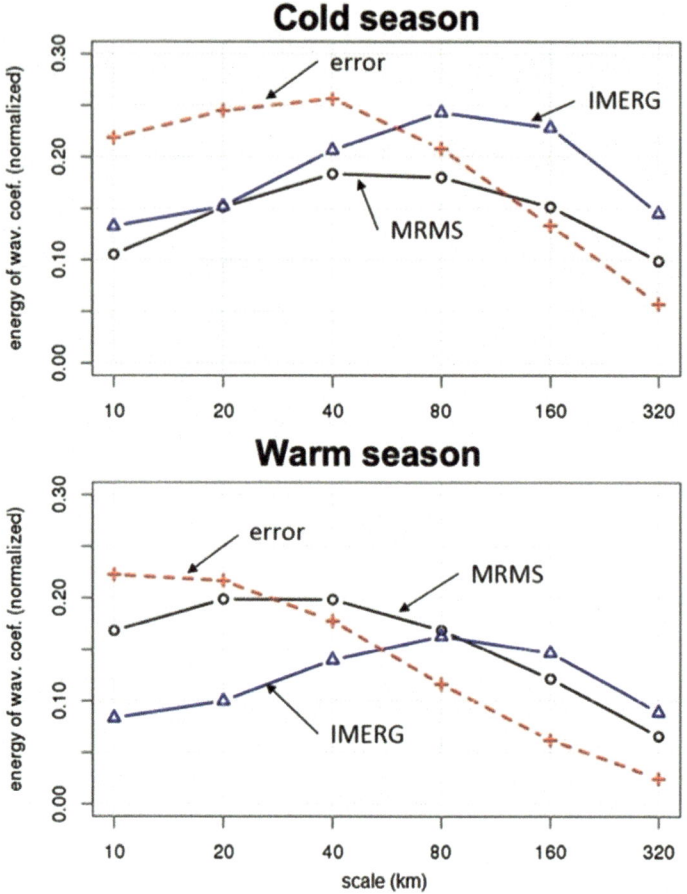

Fig. 29.4 Average wavelet energy spectra of MRMS (black line with circles), and IMERG (blue line with triangles) hourly precipitation fields and of the error IMERG-MRMS (dashed red line with crosses), computed over CONUS Nov. 2016 – Apr. 2017 (cold season) and May – Oct. 2017 (warm season). The spectra are computed at each time step and summed over the analysed period. The energy spectra are normalised by the total energy (sum of squared values) of the MRMS fields. Only the pixels with a MRMS radar quality index >0.9 are retained for the computation of the spectra

IMERG (e.g., using a sliding window averaging to smooth the field) would reduce the mean squared difference between IMERG and the 10 km resolution MRMS fields. The scales 10 and 20 km contribute together 40% of the total squared error during the cold season and 53% during the warm season. The scales 40 and 80 km contribute together 40% of the total squared error during the cold season and 35% during the warm season. Applying the criterion (29.3), we find an effective resolution around 160 km over CONUS during the warm season. During the cold season, the spectral energy of the error is relatively high at all scales, and not even the 320 km scale is resolved according to criterion (29.3). However, the spectra in Fig. 29.4 are averaged over various climatic zones, including mountainous areas, and over 6-month periods during which various types of precipitation systems occur (including snow storms and hail which are known to be challenging in terms of satellite retrieval). A local precipitation feature can significantly affect the CONUS-averaged spectra. For example, the overestimation of precipitation in the North-West is essentially the reason why IMERG shows more spectral energy than MRMS at all scales during the cold season. The energy spectra, because they are quadratic measures are strongly influenced by extreme values in the fields (same is true for the mean squared error and the linear correlation), and consequently a few extreme precipitation events may have a strong influence on the computed statistics. A more detailed analysis is therefore necessary.

The spatial localisation of the wavelet coefficients allows to study the spatial variations of the spectral properties of the fields, and to perform a detailed regionalised analysis over CONUS. Figure 29.5 shows as a map the local ratio of IMERG spectral energy over MRMS spectral energy for all dyadic scales between 10 and 320 km. Overestimation by IMERG of the spectral energy at scales finer than 320 km is observed in particular in the North East during the cold season. During the warm season, IMERG overestimates the spatial variability at all scales along the Pacific Coast. The ratio between IMERG and MRMS spectral energy is close to 1 at all scales in the East during the warm season and in the South East during the cold season, demonstrating the agreement between MRMS and IMERG energy spectra in these regions. Strong underestimation by IMERG of the spatial variability at scales finer than 80 km is observed around −100°W and 30°N (between Texas and New Mexico) during the cold season, over the South-West (excluding coastal areas) during the warm season and locally along the Canada-US border during both seasons. For the 80 km scale and finer scales, a dependence on the distance to the closest ground radar can be observed: the apparent underestimation of the spatial variability by IMERG increases when the distance to the radar decreases. This is explained by the fact that the radar can better capture the fine-scale variability at shorter ranges. For Fig. 29.4 only the pixels with a quality index higher than 0.9 (i.e., pixels close to a ground radar) have been retained to mitigate this effect.

Figure 29.6 shows the regionalised ratio between the spectral energy of the error and the spectral energy of the MRMS reference. For the scales 40 km and coarser, a ratio lower than 1 is found in most areas (except locally, along coastal areas and along the US-Canada border) during the warn season. This is also true in the South-East during the cold season. In contrast, the spectral energy of the error is larger than the spectral energy of the reference at all scales in the North and over the Rocky Mountains for the cold season. This is likely related to the presence of frozen precipitation during winter in these regions, which is challenging for both the radar reference and the satellite retrieval. During the warm season, the highest relative error is found at all scales along the Pacific coastline. However, one must note that the Pacific coastline (and in particular its southern portion) is exposed to an extremely low amount of precipitation during the warm season (Fig. 29.3); therefore, even if the relative errors are large their absolute amplitude is very low. Figure 29.7 shows the local linear correlation coefficients between IMERG and MRMS wavelets coefficients at all scales. The patterns mostly follow the patterns shown in Fig. 29.6,

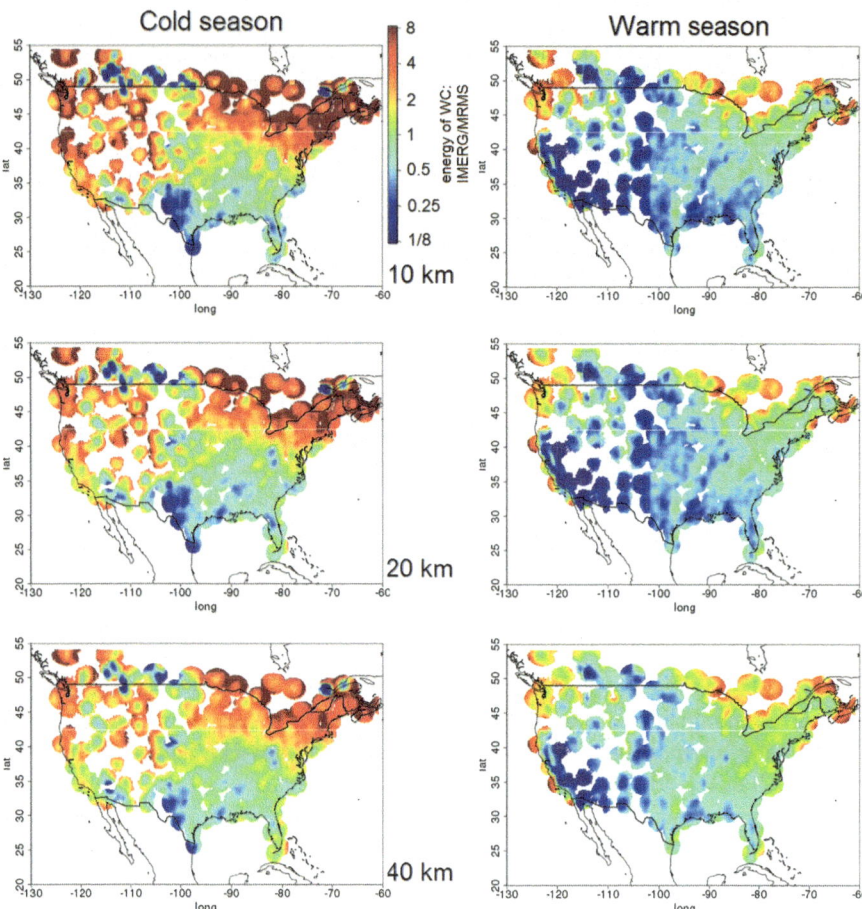

Fig. 29.5 Ratio of the spectral energy of IMERG hourly precipitation fields over the spectral energy of MRMS hourly precipitation fields during the cold and warm seasons. The spectral energies are computed at each time step and summed. A ratio close to one at all scales indicates the agreement of the IMERG and MRMS wavelet energy spectra

high correlation of the wavelet coefficients corresponding to low spectral energy of the error. Nevertheless, one will note that, particularly during the cold season, the spectral energy of the error at scales 320 and 160 km along the north part of the Pacific coastline is high in spite of the relatively good correlation of the wavelet coefficients at these scales. This shows that, in this region, the error mostly comes from a systematic bias: the amplitude of the spatial variations of the precipitation fields is systematically overestimated, which comes directly from the overestimation of the cumulative precipitation in winter in this region. Figure 29.8 shows the local effective resolution of IMERG, estimated using criterion (29.3). During the cold season, the effective resolution is generally found between 80 and 320 km in the

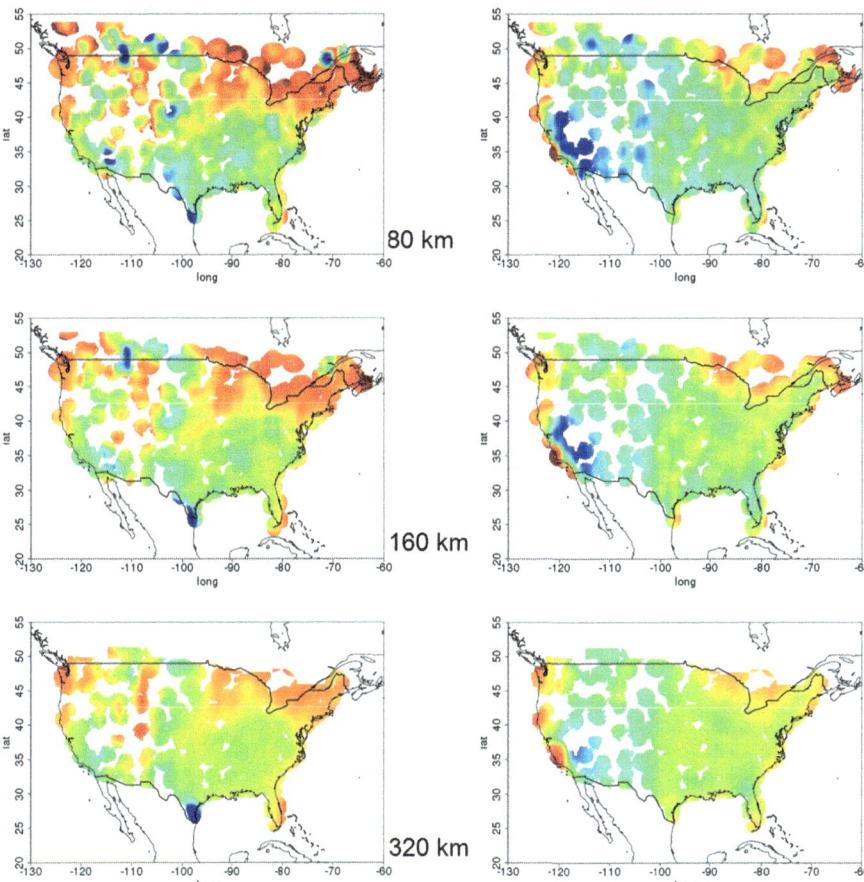

Fig. 29.5 (continued)

South-East and in the southern part of California, and locally between 40 and 80 km. In contrast, an effective resolution coarser than 320 km is found in the northern regions and over the mountains. During the warm season, the effective resolution is found finer than 160 km over most of CONUS, between 160 and 320 km in North-West and over the mountains, and coarser than 320 km in the northernmost regions. Coastal areas appear to be challenging, with an effective resolution generally coarser than 320 km during both seasons, except for the Northern coastline of the Gulf of Mexico where the performance of IMERG is remarkably good even at the 40~80 km scales.

Figure 29.9 shows the averaged spectra of IMERG, MRSM and of the error during the cold season computed only over the South-Eastern region where the effective resolution is finer than 160 km. The cold-season spectra in the South-East are similar to the all-CONUS warm season spectra. During the warm season and also

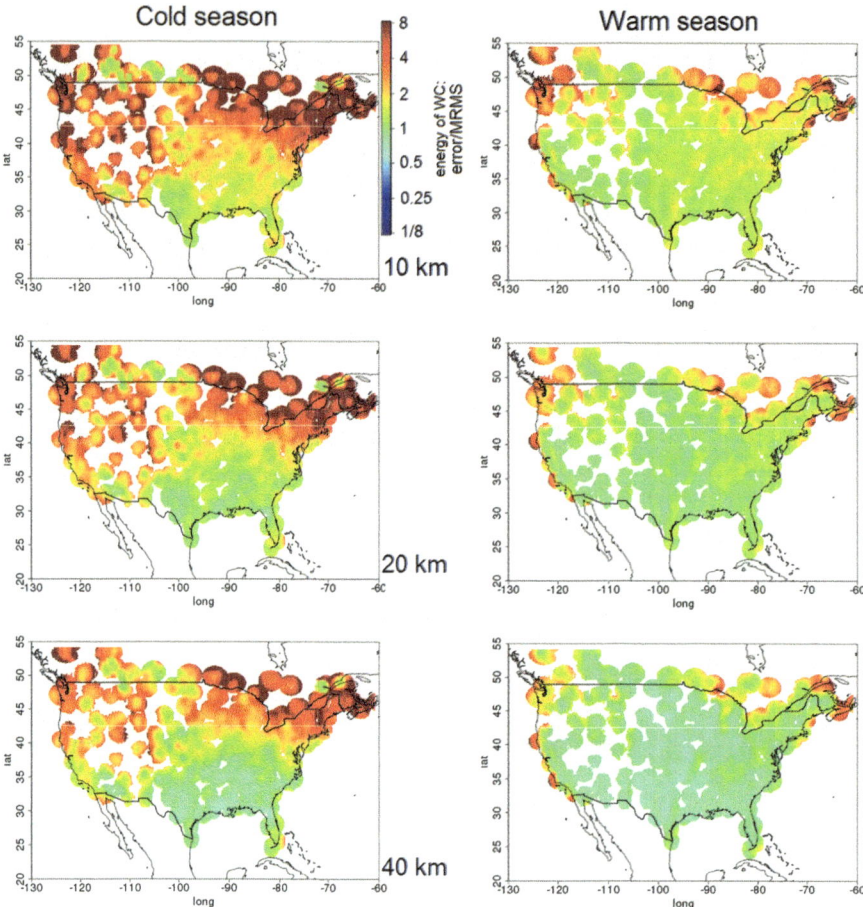

Fig. 29.6 Ratio of the spectral energy the IMERG hourly error (i.e., IMERG-MRMS) over the spectral energy of MRMS hourly precipitation during the cold and warm seasons. The spectral energies are computed at each time step and summed. A low spectral energy of the error indicates the agreement between IMERG and MRMS spatial patterns (gradients) at the corresponding scale

during the cold season in the South-West IMERG slightly underestimates the spatial variability of precipitation at scales finer than 80 km but the IMERG spectrum still shows substantial spectral energy at scales 10–40 km. This shows that IMERG hourly precipitation fields are not dramatically smoother than the MRMS hourly precipitation fields. Consequently, the high value of the spectral energy of the error at scales finer than 80 km is not caused by the absence of information in the IMERG fields at these scales but is rather due to the fact that this information does not match the information in the MRMS fields at the same scales. Both IMERG and MRMS fields show substantial fine-scale patterns but these patterns are different: they may be spatially shifted, oriented in different directions or completely independent.

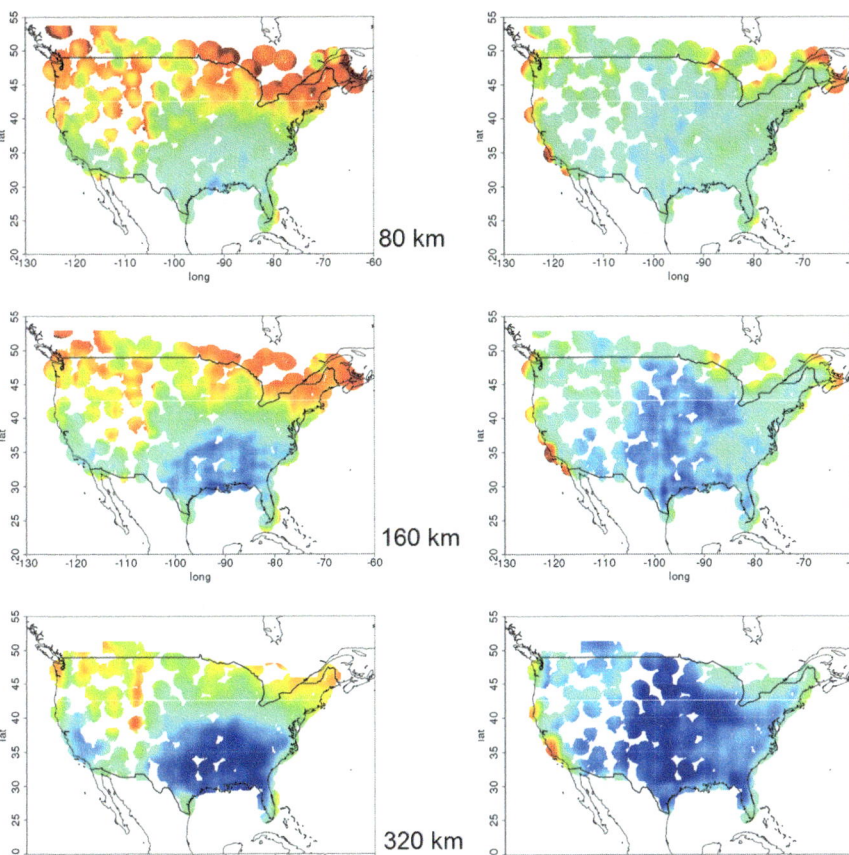

Fig. 29.6 (continued)

29.5 Conclusions

The performed spectral comparison of IMERG and MRMS hourly cumulative precipitation reveals that over the continental United States IMERG can resolve scales down to 40~80 km. However, in certain areas, and particularly where frozen precipitation is present, the retrieval of spatial variations of precipitation at scales between 80 and 320 km can be challenging for IMERG. The spatial variability a scales 640 km and larger scales was not evaluated against the radar because only a few geographical areas with such large dimensions are continuously covered by the MRMS radar network.

The wavelet energy spectra of IMERG and MRMS show generally little difference. IMERG shows a maximum of energy at a coarser scale than MRMS (around 80 km for IMERG and around 20 or 40 km for MRMS depending on the season). For regions and seasons dominated by frozen precipitation, IMERG appears to slightly

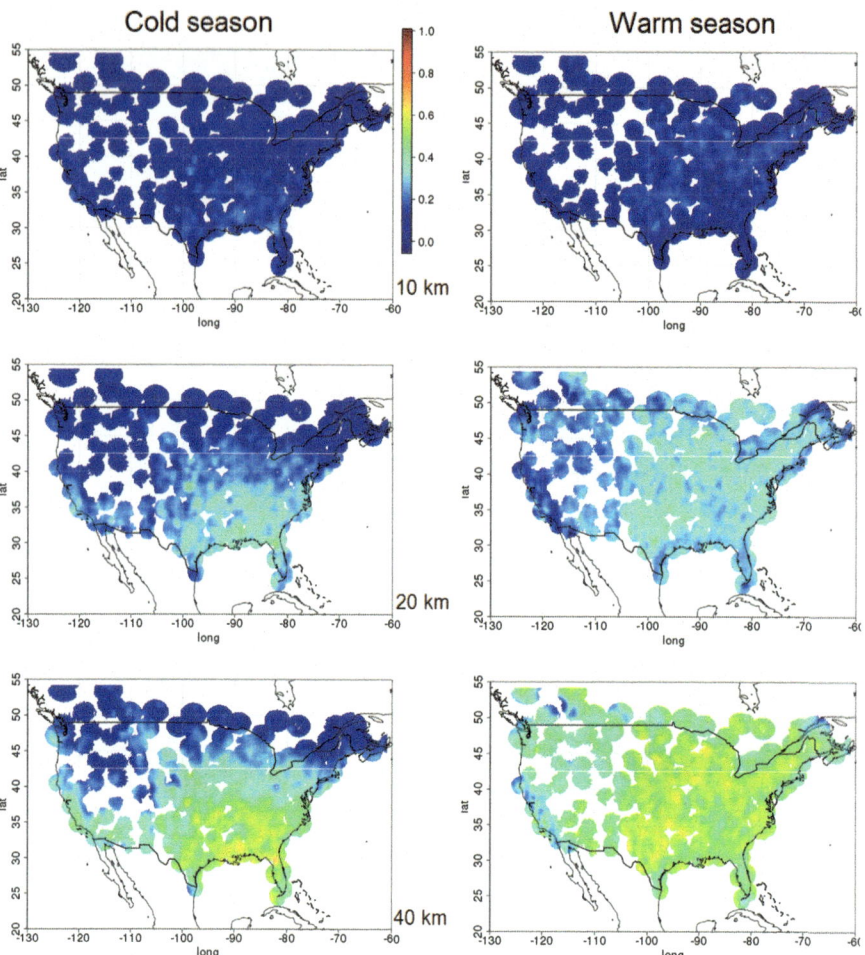

Fig. 29.7 Linear correlation between the wavelet coefficients of the IMERG and MRMS hourly precipitation fields. A high correlation of the wavelet coefficients indicates the agreement between IMERG and MRMS spatial patterns (gradients) at the corresponding scale. Contrary the spectral energy of the error, the correlation is not sensitive to potential biases in the magnitude of the wavelet coefficients

overestimate the amplitude of the spatial variation of the precipitation rates at scales between 80 and 320 km. When liquid precipitation is dominant, IMERG tends to slightly underestimate the spatial variability of precipitation at scales finer than 80 km, producing marginally smoother fields than MRSMS. The error field systematically shows high spectral energy at scales finer than 80 km, sometimes higher than the spectral energy of the radar reference. This is caused by the fact that the fine-scale variations of the IMERG fields do not match the variations of MRMS as revealed by

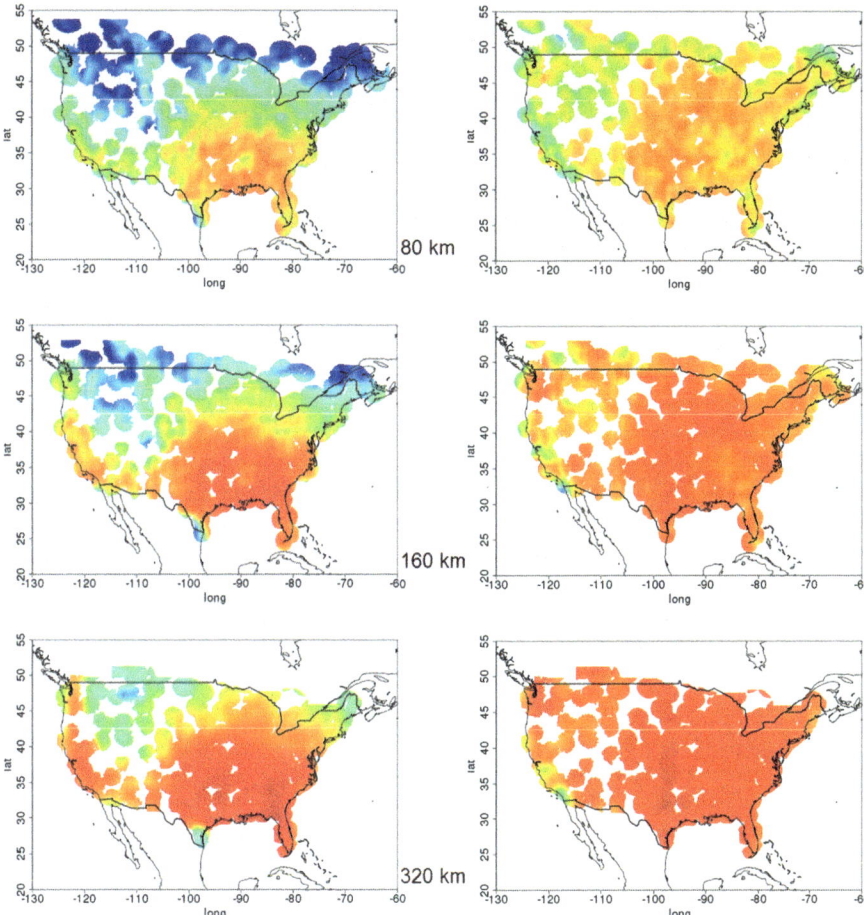

Fig. 29.7 (continued)

the low correlation of the wavelet coefficients. During the warm season, the spectral energy of the error is more concentrated in the fine scales than during the cold season: scales coarser than 40 km account for only 25% of the total squared error during the warm season (against 40% during the cold season).

We note that, while the gradients of IMERG and MRMS at scales finer than 80 km do not generally match, they are still statistically similar (in terms of variance of the wavelet coefficients). Considering that the IMERG merging algorithm relies on Kalman filtering, one could have expected IMERG retrievals to be dramatically smoother than the radar reference at fine scales. Moreover, a similar analysis performed on the GPROF passive microwave instantaneous retrievals on which IMERG partially relies revealed that GPROF retrievals are significantly smoother than radar observations (Guilloteau et al. 2017). In contrast, only a marginal deficit of spectral energy is observed at scales 10 km–40 km for IMERG compared to

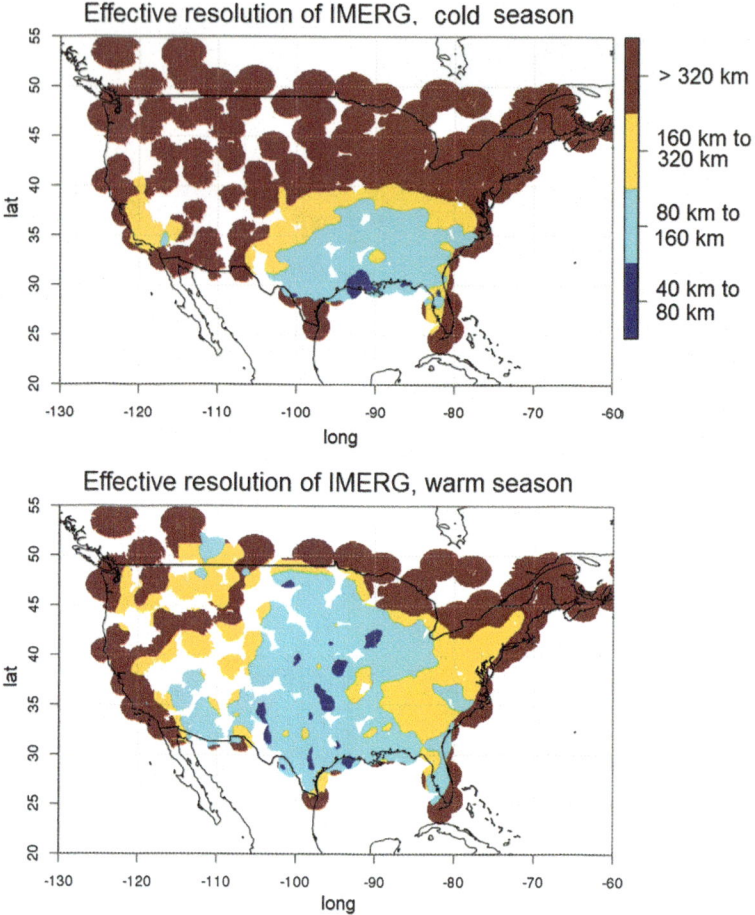

Fig. 29.8 Effective resolution of IMERG hourly precipitation evaluated against MRMS during the cold and warm seasons. The criterion (29.3) is used to define the effective resolution

MRMS. This may be partially due to the sampling variance introduced by the limited temporal frequency of the satellite observations. Indeed, while each MRMS hourly precipitation field is computed as the average of 30 instantaneous precipitation fields (one observation every 2 min), IMERG hourly precipitation is derived from only a few instantaneous satellite observations (2 infrared observations and no more than one or two microwave observations in 1 h).

In terms of mean squared error, it is more penalising to retrieve variations uncorrelated with the reference than to retrieve no variation at all at a given scale. This can be related to the concept of "double penalty" (Rossa et al. 2008; Mittermaier 2014), which states that mislocating a feature is more penalising than not detecting it in terms of mean squared error. Consequently, a filtering operation

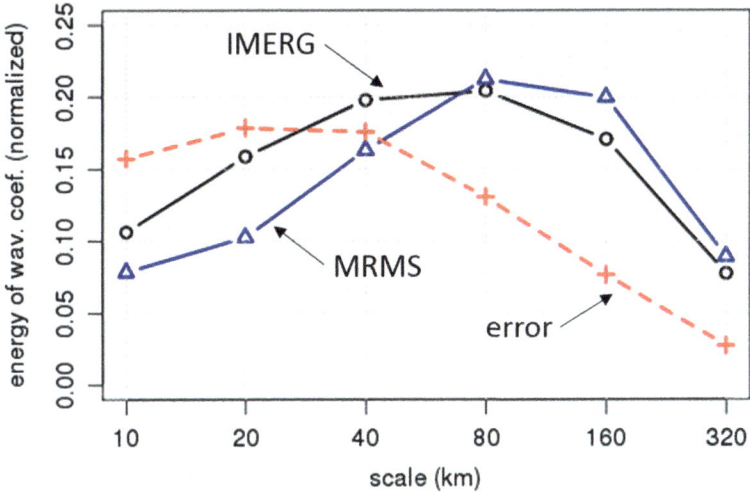

Fig. 29.9 Same as Fig. 29.4 for the cold season, but computed only over the South-Eastern region where the effective resolution is found finer than 160 km

reducing the fine-scale variability of IMERG would reduce the mean squared error (relatively to the MRMS reference) (Turner et al. 2004). However, this would produce unrealistically smooth fields. The suitability of such a filtering procedure depends of the targeted application; some applications such as, for example, rainfall-runoff modelling over a large basin, may tolerate mislocated fine-scale features better than others. Moreover, the smoothing of the fields would necessarily reduce the statistical variance and erase local extremes, which may have a considerable effect on the computed rainfall-runoff values for example (Harris et al. 2001; Smith et al. 2004; Nikolopoulos et al. 2010; Vergara et al. 2014).

The analysis of the spatial patterns is performed here at the hourly temporal scale. The poor ability of IMERG to resolve scales finer than 80 km at the hourly time scale does not mean that this product cannot resolve fine-scale patterns at longer time scales or resolve the fine-scale climatology. Indeed, the effective spatial resolution is expected to vary with the desired temporal resolution. Sampling-related noise in particular is expected to decrease at coarse time scales (Nijssen and Lettenmaier 2004; Gebremichael and Krajewski 2004). Figure 29.10 shows the effective resolution of IMERG evaluated against MRMS at the daily scale; the effective resolution is generally found to be improved by a factor of about two compared to the hourly scale. A comprehensive assessment would require performing the analysis at more temporal scales. Temporal averaging generally tends to reduce the fine-scale spatial variability, giving rise to smoother fields. However, some regions may still show strong climatic gradients at scales of a few km or a few dozen km (Hirose and Okada 2018). We note that, as well as for the spatial variations, a (one-dimensional) wavelet spectral analysis can be performed to assess the ability of a product to capture the

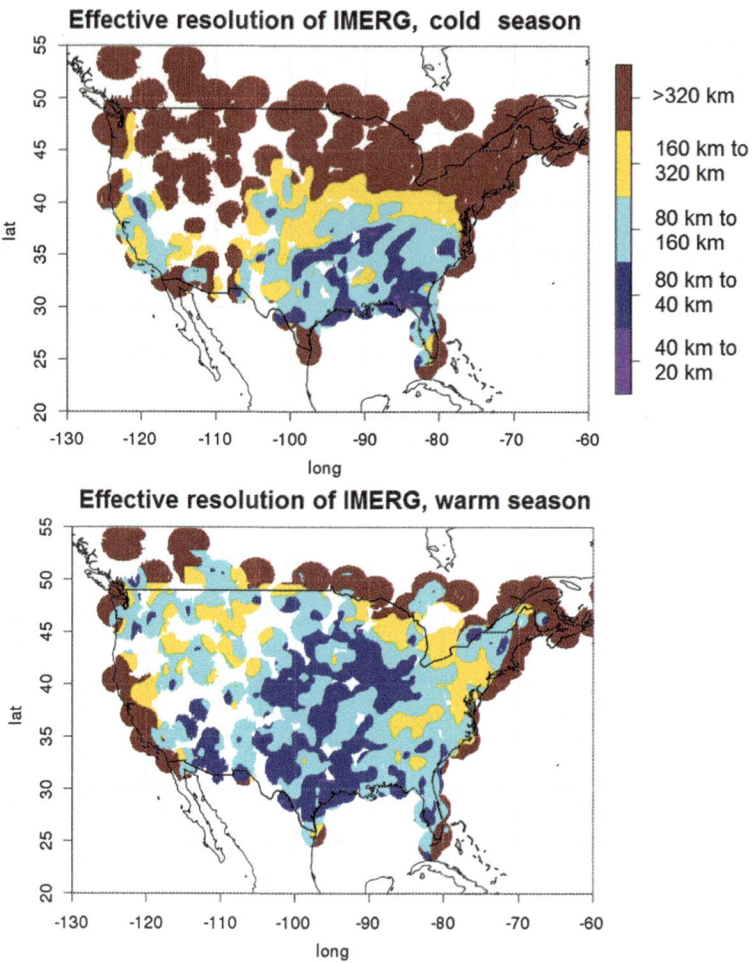

Fig. 29.10 Effective resolution of IMERG daily precipitation evaluated against MRMS during the cold and warm seasons. The criterion (29.3) is used to define the effective resolution

temporal variations of precipitation at a given location (Whitcher et al. 2000; De Jongh et al. 2006). One can also study the coupling between the temporal and spatial variations by performing successively the spatial and temporal wavelet decompositions (Guilloteau et al. 2016).

Acknowledgments This work was supported by the NASA Global Precipitation Measurement Program under Grants 80NSSC19K0684 and NNX16AO56G, and by NSF grants DMS-1839336 and ECCS-1839441. The wavelet decomposition procedure was implemented in R distributed under GNU General Public License, using the Waveslim 1.7.5 package (Whitcher 2015).

Appendix: Two-Dimensional Discrete Orthogonal Decomposition with the Haar Wavelet

Wavelets Functions in One Dimension and N Dimensions

In one-dimensional or multi-dimensional spaces, the wavelet transform is obtained through the convolution of the analysed signal with specific analysing functions called wavelets. Wavelets are locally oscillating functions; to be admitted as a wavelet a given function must meet several requirements such as having a zero mean and being square integrable (Kumar and Foufoula-Georgiou 1997; Mallat and Peyré 2008).

The function obtained by the dilation and/or translation of a wavelet is also a wavelet; multiple "daughter wavelets" can then be generated from a "mother wavelet", allowing multiscale analyses. In one dimension:

$$\psi_{a,b}(x) = a^{-\frac{1}{2}} \psi\left(\frac{x-b}{a}\right) \tag{29.4}$$

where $\psi_{a,\ b}$ is the daughter wavelet, ψ is the mother wavelet, $a \in \mathbb{R}$ is the dilation coefficient (or scale factor) and $b \in \mathbb{R}$ is the translation coefficient.

In N dimensions:

$$\psi_{a,\vec{B}}\left(\vec{X}\right) = a^{-\frac{1}{2N}} \psi\left(\frac{\vec{X} - \vec{B}}{a}\right) \tag{29.5}$$

where $\vec{B}$ is the translation vector belonging to $\mathbb{R}^N$.

The wavelet transform of the analysed function $y\left(\vec{X}\right)$ is obtained by computing the inner products with the wavelets:

$$T\left(a,\vec{B}\right) = \left\langle y, \psi_{a,\vec{B}} \right\rangle = \int_{\mathbb{R}^N} y\left(\vec{X}\right) \psi_{a,\vec{B}}^{*}\left(\vec{X}\right) d\vec{X} \tag{29.6}$$

where T is the wavelet transform of y, * denotes the complex conjugate operator and $\langle\ \rangle$ denotes the canonical inner product of $L_2(\mathbb{R}^N)$.

The Haar Discrete Orthogonal Wavelets in One and Two Dimensions

In one dimension, the Haar mother wavelet is defined as:

$$\psi(x) = \begin{cases} 1 & if \ 0 \leq x < 0.5 \\ -1 & if \ 0.5 \leq x < 1 \\ 0 \ otherwise \end{cases} \tag{29.7}$$

The Haar wavelet is associated to the smoothing function (or scaling function) ϕ:

$$\phi(x) = \begin{cases} 1 & if \ 0 \leq x < 1 \\ 0 \ otherwise \end{cases} \tag{29.8}$$

ψ and ϕ are orthogonal functions as their inner product equals zero. As for the wavelet function, multiple smoothing functions $\phi_{a, b}$ can be generated by dilatation and translation of ϕ (Eq. 29.4). While the wavelet ψ is a high-pass (H) convolution filter, the smoothing function ϕ is a low-pass (L) convolution filter.

By discretising the scaling and translation coefficients a and b, such as $a \in \{2^i, i \in \mathbb{Z}\}$ and $b \in \{k \times a, k \in \mathbb{Z}\}$ the ensemble of the Haar daughter wavelets $\{\psi_{a, b}\}$ forms an orthogonal basis of $L_2(\mathbb{R})$.

In two dimensions, the Haar wavelet and scaling functions are defined as follows:

$$\phi^{2D}(x_1, x_2) = \phi(x_1)\phi(x_2)$$
$$\psi^H(x_1, x_2) = \psi(x_1), \phi(x_2)$$
$$\psi^V(x_1, x_2) = \phi(x_1), \psi(x_2)$$
$$\psi^D(x_1, x_2) = \psi(x_1), \psi(x_2)$$

ψ^H, ψ^V and ψ^D are the horizontal (HL), vertical (LH) and diagonal (HH) two-dimensional Haar wavelets. ϕ^{2D} is the two-dimensional Haar smoothing (LL) function. The four functions are graphically represented in Fig. 29.11.

The ensemble of the two-dimensional Haar wavelets $\left\{\psi^H_{a,\vec{B}}, \psi^V_{a,\vec{B}}, \psi^D_{a,\vec{B}}\right\}$ with $a \in \{2^i, i \in \mathbb{Z}\}$ and $\vec{B} \in \left\{\vec{k} \times a, \vec{k} \in \mathbb{Z}2\right\}$ forms an orthogonal basis of $L_2(\mathbb{R}^2)$. The decomposition of a two-dimensional precipitation field in this basis is illustrated in Fig. 29.12.

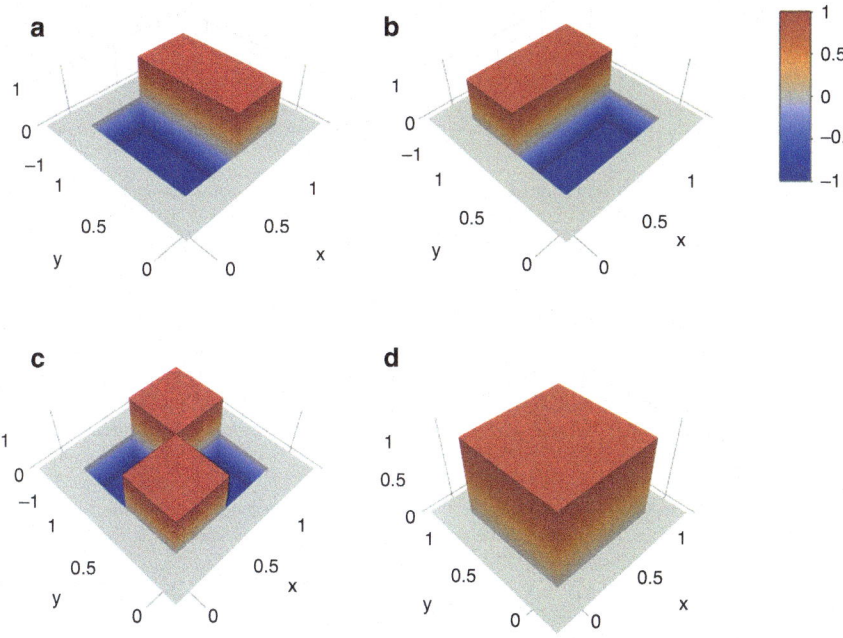

Fig. 29.11 Representation of the spatial convolution filters used to perform the two-dimensional Haar wavelet decomposition. A and B and C are the wavelet functions ψ^H, ψ^V and ψ^D used to compute the horizontal (HL), vertical (LH) and diagonal (HH) wavelet coefficients. D is the smoothing function ϕ^{2D} used to compute the smoothing (LL) coefficients. The four functions are orthogonal, meaning that the scalar product of one with any of the other three equals zero

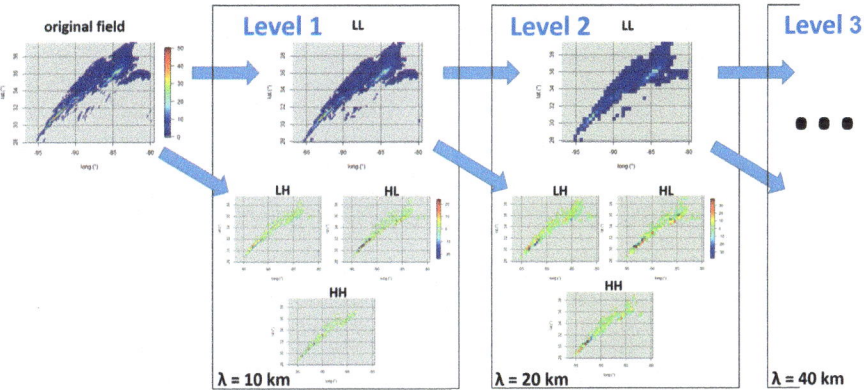

Fig. 29.12 Illustration of the discrete wavelet decomposition process of a two-dimensional field (using the Haar wavelet)

References

Bousquet, O., Lin, C. A., & Zawadzki, I. (2006). Analysis of scale dependence of quantitative precipitation forecast verification: A case-study over the Mackenzie river basin. *Quarterly Journal of the Royal Meteorological Society, 132*, 2107–2125. https://doi.org/10.1256/qj.05. 154.

De Jongh, I. L., Verhoest, N. E., & De Troch, F. P. (2006). Analysis of a 105-year time series of precipitation observed at Uccle, Belgium. *International Journal of Climatology, 26*, 2023–2039. https://doi.org/10.1002/joc.1352.

Frehlich, R., & Sharman, R. (2008). The use of structure functions and spectra from numerical model output to determine effective model resolution. *Monthly Weather Review, 136*, 1537–1553. https://doi.org/10.1175/2007MWR2250.1.

Gebremichael, M., & Krajewski, W. F. (2004). Characterization of the temporal sampling error in space-time-averaged rainfall estimates from satellites. *Journal of Geophysical Research, 109* (D11), D11110. https://doi.org/10.1029/2004JD004509.

Grasso, L. D. (2000). The differentiation between grid spacing and resolution and their application to numerical modeling. *Bulletin of the American Meteorological Society, 81*(3), 579. https://doi.org/10.1175/1520-0477(2000)081<0579:CAA>2.3.CO;2.

Guilloteau, C., Roca, R., & Gosset, M. (2016). A multiscale evaluation of the detection capabilities of high-resolution satellite precipitation products in West Africa. *Journal of Hydrometeorology, 17*(7), 2041–2059. https://doi.org/10.1175/JHM-D-15-0148.1.

Guilloteau, C., Foufoula-Georgiou, E., & Kummerow, C. D. (2017). Global multiscale evaluation of satellite passive microwave retrieval of precipitation during the TRMM and GPM eras: Effective resolution and regional diagnostics for future algorithm development. *Journal of Hydrometeorology, 18*(11), 3051–3070. https://doi.org/10.1175/JHM-D-17-0087.1.

Haile, A. T., Habib, E., & Rientjes, T. (2013). Evaluation of the Climate Prediction Center (CPC) morphing technique (CMORPH) rainfall product on hourly time scales over the source of the Blue Nile River. *Hydrological Processes, 27*(12), 1829–1839. https://doi.org/10.1002/hyp. 9330.

Harris, D., Foufoula-Georgiou, E., Droegemeier, K. K., & Levit, J. J. (2001). Multiscale statistical properties of a high-resolution precipitation forecast. *Journal of Hydrometeorology, 2*(4), 406–418. https://doi.org/10.1175/1525-7541(2001)002<0406:MSPOAH>2.0.CO;2.

Hirose, M., & Okada, K. (2018). A 0.01 degree resolving TRMM PR precipitation climatology. *Journal of Applied Meteorology and Climatology, 57*, 1645–1661. https://doi.org/10.1175/JAMC-D-17-0280.1.

Hong, Y., Hsu, K.-L., Sorooshian, S., & Gao, X. (2004). Precipitation estimation from remotely sensed imagery using an artificial neural network cloud classification system. *Journal of Applied Meteorology and Climatology, 43*(12), 1834–1853. https://doi.org/10.1175/JAM2173.1.

Hossain, F., & Anagnostou, E. N. (2006). A two-dimensional satellite rainfall error model. *IEEE Transactions on Geoscience and Remote Sensing, 44*(6), 1511–1522. https://doi.org/10.1109/TGRS.2005.863866.

Hossain, F., & Huffman, G. J. (2008). Investigating error metrics for satellite rainfall data at hydrologically relevant scales. *Journal of Hydrometeorology, 9*(3), 563–575. https://doi.org/10.1175/2007JHM925.1.

Huffman, G. J., Bolvin, D. T., Braithwaite, D., Hsu, K., Joyce, R., Kidd, C., Nelkin, E. J., & Xie, P. (2015). *NASA global precipitation measurement (GPM) integrated multi-satellitE retrievals for GPM (IMERG)*. Algorithm Theoretical Basis Doc., v4.5, 26 pp. Available at https://pmm. nasa.gov/sites/default/files/document_files/IMERG_ATBD_V4.5.pdf. Last accessed 27 Oct 2018.

Joyce, R. J., & Xie, P. (2011). Kalman filter–based CMORPH. *Journal of Hydrometeorology, 12* (6), 1547–1563. https://doi.org/10.1175/JHM-D-11-022.1.

Kumar, P. (1995). A wavelet based methodology for scale-space anisotropic analysis. *Geophysical Research Letters, 22*(20), 2777–2780. https://doi.org/10.1029/95GL02934.

Kumar, P., & Foufoula-Georgiou, E. (1993). A multicomponent decomposition of spatial rainfall fields: 1. Segregation of large-and small-scale features using wavelet transforms. *Water Resources Research, 29*(8), 2515–2532. https://doi.org/10.1175/JHM-D-11-022.1.

Kumar, P., & Foufoula-Georgiou, E. (1997). Wavelet analysis for geophysical applications. *Reviews of Geophysics, 35*(4), 385–412. https://doi.org/10.1029/97RG00427.

Kummerow, C. D., Randel, D. L., Kulie, M., Wang, N.-Y., Ferraro, R. R., Munchak, S. J., & Petkovic, V. (2015). The evolution of the Goddard profiling algorithm to a fully parametric scheme. *Journal of Atmospheric and Oceanic Technology, 32*, 2265–2280. https://doi.org/10.1175/JTECH-D-15-0039.1.

Labat, D. (2005). Recent advances in wavelet analyses: Part 1. A review of concepts. *Journal of Hydrology, 314*(1–4), 275–288. https://doi.org/10.1016/j.jhydrol.2005.04.003.

Labat, D., Ronchail, J., & Guyot, J. L. (2005). Recent advances in wavelet analyses: Part 2—Amazon, Parana, Orinoco and Congo discharges time scale variability. *Journal of Hydrology, 314*(1–4), 289–311. https://doi.org/10.1016/j.jhydrol.2005.04.004.

Mallat, S., & Peyré, G. (2008). Time meets frequency. In *A wavelet tour of signal processing: The sparse way* (pp. 89–150). Boston: Academic. ISBN: 0123743702.

Mittermaier, M. P. (2014). A strategy for verifying near-convection-resolving model forecasts at observing sites. *Weather and Forecasting, 29*(2), 185–204. https://doi.org/10.1175/WAF-D-12-00075.1.

Nijssen, B., & Lettenmaier, D. P. (2004). Effect of precipitation sampling error on simulated hydrological fluxes and states: Anticipating the global precipitation measurement satellites. *Journal of Geophysical Research, 109*(D2), D02103. https://doi.org/10.1029/2003JD003497.

Nikolopoulos, E. I., Anagnostou, E. N., Hossain, F., Gebremichael, M., & Borga, M. (2010). Understanding the scale relationships of uncertainty propagation of satellite rainfall through a distributed hydrologic model. *Journal of Hydrometeorology, 11*(2), 520–532. https://doi.org/10.1175/2009JHM1169.1.

Perica, S., & Foufoula-Georgiou, E. (1996). Model for multiscale disaggregation of spatial rainfall based on coupling meteorological and scaling descriptions. *Journal of Geophysical Research, 101*(D21), 26347–26361. https://doi.org/10.1029/96JD01870.

Petty, G. W. (2018). On some shortcomings of Shannon entropy as a measure of information content in indirect measurements of continuous variables. *Journal of Atmospheric and Oceanic Technology, 35*(5), 1011–1021. https://doi.org/10.1175/JTECH-D-17-0056.1.

Pielke, R. A. (1991). A recommended specific definition of "resolution". *Bulletin of the American Meteorological Society, 72*(12), 1914–1914. https://doi.org/10.1175/1520-0477-72.12.1914.

Rios Gaona, M. F., Overeem, A., Leijnse, H., & Uijlenhoet, R. (2016). First-year evaluation of GPM rainfall over the Netherlands: IMERG day 1 final run (V03D). *Journal of Hydrometeorology, 17*, 2799–2814. https://doi.org/10.1175/JHM-D-16-0087.1.

Rossa, A., Nurmi, P., & Ebert, E. E. (2008). Overview of methods for the verification of quantitative precipitation forecasts. In S. C. Michaelides (Ed.), *Precipitation: Advances in measurement, estimation and prediction* (pp. 419–452). Berlin/Heidelberg: Springer. ISBN: 978-3-540-77655-0.

Scheel, M. L. M., Rohrer, M., Huggel, C., Santos Villar, D., Silvestre, E., & Huffman, G. J. (2011). Evaluation of TRMM multi-satellite precipitation analysis (TMPA) performance in the Central Andes region and its dependency on spatial and temporal resolution. *Hydrology and Earth System Sciences, 15*(8), 2649–2663. https://doi.org/10.5194/hess-15-2649-2011.

Shen, Y., Xiong, A., Wang, Y., & Xie, P. (2010). Performance of high-resolution satellite precipitation products over China. *Journal of Geophysical Research, 115*(D2), D02114. https://doi.org/10.1029/2009JD012097.

Skamarock, W. C. (2004). Evaluating mesoscale NWP models using kinetic energy spectra. *Monthly Weather Review, 132*(12), 3019–3032. https://doi.org/10.1175/MWR2830.1.

Smith, M. B., Koren, V. I., Zhang, Z., Reed, S. M., Pan, J. J., & Moreda, F. (2004). Runoff response to spatial variability in precipitation: An analysis of observed data. *Journal of Hydrology, 298*(1–4), 267–286. https://doi.org/10.1016/j.jhydrol.2004.03.039.

Sohn, B. J., Han, H. J., & Seo, E. K. (2010). Validation of satellite-based high-resolution rainfall products over the Korean peninsula using data from a dense rain gauge network. *Journal of Applied Meteorology and Climatology, 49*(4), 701–714. https://doi.org/10.1175/2009JAMC2266.1.

Starck, J. L., Murtagh, F., & Gastaud, R. (1998). A new entropy measure based on the wavelet transform and noise modeling. *IEEE Transactions Circuits and Systems-II: Analog and Digital Signal Processing, 45*(8), 1118–1124. https://doi.org/10.1109/82.718822.

Turk, F. J., Sohn, B. J., Oh, H. J., Ebert, E. E., Levizzani, V., & Smith, E. A. (2009). Validating a rapid-update satellite precipitation analysis across telescoping space and time scales. *Meteorology and Atmospheric Physics, 105*(1–2), 99–108. https://doi.org/10.1007/s00703-009-0037-4.

Turner, B. J., Zawadzki, I., & Germann, U. (2004). Predictability of precipitation from continental radar images. Part III: Operational nowcasting implementation (MAPLE). *Journal of Applied Meteorology, 43*(2), 231–248. https://doi.org/10.1175/1520-0450(2004)043<0231:POPFCR>2.0.CO;2.

Vasić, S., Lin, C. A., Zawadzki, I., Bousquet, O., & Chaumont, D. (2007). Evaluation of precipitation from numerical weather prediction models and satellites using values retrieved from radars. *Monthly Weather Review, 135*(11), 3750–3766. https://doi.org/10.1175/2007MWR1955.1.

Vergara, H., Hong, Y., Gourley, J. J., Anagnostou, E. N., Maggioni, V., Stampoulis, D., & Kirstetter, P.-E. (2014). Effects of resolution of satellite-based rainfall estimates on hydrologic modeling skill at different scales. *Journal of Hydrometeorology, 15*(2), 593–613. https://doi.org/10.1175/JHM-D-12-0113.1.

von Storch, H. (1999). Misuses of statistical analysis in climate research. In H. von Storch & A. Navarra (Eds.), *Analysis of climate variability* (pp. 11–26). Berlin/Heidelberg: Springer. https://doi.org/10.1007/978-3-662-03167-4_2.

Weniger, M., Kapp, F., & Friederichs, P. (2017). Spatial verification using wavelet transforms: A review. *Quarterly Journal of the Royal Meteorological Society, 143*(702), 120–136. https://doi.org/10.1002/qj.2881.

Whitcher, B. (2015). Waveslim: Basic wavelet routines for one-,two- and three-dimensional signal processing, version 1.7.5. R package. [Available at https://cran.r-project.org/web/packages/waveslim/index.html, last accessed 30 Nov 2018].

Whitcher, B., Guttorp, P., & Percival, D. B. (2000). Wavelet analysis of covariance with application to atmospheric time series. *Journal of Geophysical Research, 105*(D11), 14941–14962. https://doi.org/10.1029/2000JD900110.

Wong, M., & Skamarock, W. C. (2016). Spectral characteristics of convective-scale precipitation observations and forecasts. *Monthly Weather Review, 144*(11), 4183–4196. https://doi.org/10.1175/MWR-D-16-0183.1.

Zhang, J., Howard, K., Langston, C., Kaney, B., Qi, Y., Tang, L., Grams, H., Wang, Y., Cocks, S., Martinaitis, S., Arthur, A., Cooper, K., Brogden, J., & Kitzmiller, D. (2016). Multi-radar multi-sensor (MRMS) quantitative precipitation estimation: Initial operating capabilities. *Bulletin of the American Meteorological Society, 97*, 621–638. https://doi.org/10.1175/BAMS-D-14-00174.1.

Chapter 30
Remote Sensing of Orographic Precipitation

Ana P. Barros and Malarvizhi Arulraj

Abstract Quantitative precipitation estimation (QPE) in mountainous regions remains a challenging task owing to its high spatiotemporal variability. Satellite-based radar observations at high resolution have the best potential to capture the spatial patterns of precipitation, but there is high uncertainty in the interpretation of low-level measurements due to ground clutter effects, observing geometry, and sub-grid scale vertical and horizontal heterogeneity of precipitation systems that result from interactions among orographic clouds and propagating storm systems. In the high elevation tropics and in middle mountains everywhere, the landscape is often immersed in multi-layered cloud systems that modify precipitation significantly at low levels in a complex manner depending on time of day and location very different from the classical understanding of orographic precipitation enhancement with elevation, and are not easily parameterized or corrected for in QPE algorithms. Here, a review of challenges to remote sensing of orographic precipitation with a focus on the physical-basis of rainfall estimation errors is presented using radar measurements and precipitation products from the Tropical Rainfall Measurement Mission (TRMM) and Global Precipitation Measurement (GPM) satellites, ground-validation (GV) data in the Andes, the Himalayas, and the Southern Appalachian Mountains, and model simulations. Emphasis is placed on spatial and temporal variability of rainfall and associated cloud systems with a focus on water cycle research and hydrological applications in the tropics and in the mid-latitudes.

Keywords Precipitation · Rainfall · Snowfall · Validation · Orographic precipitation · Mountains · Natural hazards · Landslide · Flash flood · Advection · Seeder-feeder interaction · Cloud microphysics · Snow melting · Rain radar · Reflectivity

A. P. Barros (✉) · M. Arulraj
Pratt School of Engineering, Civil & Environmental Engineering, Duke University, Durham, NC, USA
e-mail: barros@duke.edu

© Springer Nature Switzerland AG 2020
V. Levizzani et al. (eds.), *Satellite Precipitation Measurement*, Advances in Global Change Research 69, https://doi.org/10.1007/978-3-030-35798-6_6

30.1 Introduction

Orographic precipitation is the primary source of freshwater for more than half of the world's population, it provides more than 40% of all water used for irrigation in adjacent lowlands going up to 90% in semi-arid regions impacting up to 70% of the world's population, and it is a major source of renewable energy through hydropower (Garrido and Dinar 2009; Nellemann et al. 2009; Tapiador et al. 2011). Further, because of the preponderance of rain-fed agriculture in regions of complex terrain such as the Andes and the Himalayas, inter-annual food security of mountain populations is tightly linked to precipitation, and thus water resilience of downstream low-lying landscapes in the face of global change (FAO 2014, 2016; Falkenmark and Rockström 2010). Heavy rainfall triggers flashfloods and landslides that can destroy crops and erode high-quality arable soils, causing roads and infrastructure to fail, thereby disrupting transportation tied to a myriad of socioeconomic impacts. In the US, orographic precipitation (rain and snow) is the key water resource providing 70% of all water resources in the West, it is the key source of runoff for the dams operated by Tennessee Valley Authority (TVA), and essential to regional scale groundwater recharge east of Appalachian Divide (e.g., Barros et al. 2017).

Mountain landscapes can be described as altitudinal scaffolds of topographically delineated collectors of precipitation (i.e. watersheds) interlinked through a system of converging channel connectors (the river network). Because of erosional processes, the very spatial structure of mountain landscapes reflects the co-evolution of geological mountain building processes and regional climate, in particular precipitation patterns (e.g., Lowman and Barros 2014; Barros et al. 2006). Depending on the latitude, season, and topography, and environmental conditions, orographic precipitation can be classified as stratiform, convective, or stratiform with embedded convection (Houze 2012). Whereas light stratiform rainfall and fog in mountains environments are essential for regional freshwater sustainability (Bruijnzeel et al. 2011; Barros 2013), especially during drought, heavy precipitation associated with myriad convective systems from thunderstorms to tropical storms is not only a key source of freshwater but it is also the main driver of natural hazards, flashfloods, landslides and other natural hazards. The value of orographic precipitation measurement is strongly tied therefore to its hydrological utility at the desired (application dependent) spatial and temporal scales: for example, flood (days) or flashflood (<6 h) and landslides (<1 h) forecasting versus water budget studies (inter-annual to decadal, e.g., Lowman and Barros 2016; Lowman et al. 2018) with or without (sub-seasonal to seasonal) groundwater recharge and transboundary exchanges (e.g., Tao and Barros 2009, 2013, 2014; Tao et al. 2016).

Whereas a general understanding of the classical mechanisms of orographic precipitation enhancement by which topography modifies the advection of moist air masses, modulates cloud development and impacts precipitation intensity and accumulation patterns is well established (e.g., Barros and Lettenmaier 1994; Barros 2013), the goal in orographic precipitation measurement and prediction is to quantify precipitation everywhere at any time (when, how fast, how long, and where in the landscape) given the high spatial and temporal variability of nonlinear land–atmosphere interactions that dynamically redistribute precipitation from one watershed to another (e.g., Barros 2013). Because of remoteness and access, mountainous regions remain among the least observed regions of the planet, and even where observations are available long-term science-grade observations are rare due to the difficulties of maintaining instruments and collecting data (Barros and Lettenmaier 1994; Viviroli et al. 2011; Barros 2013).

Transformative advancements in precipitation science and precipitation measurement have been possible under the auspices of the TRMM and GPM missions over the last 20 years (Skofronick-Jackson et al. 2017). This is well illustrated in Fig. 30.1 that shows remarkable improvement in the precipitation estimation from TRMM 3B42 (gray, Huffman et al. 2007) to IMERG (blue, Huffman et al. 2018) along the rain-gauge transect maintained by Duke University in the Eastern Andes (Barros 2013), and which cannot be explained simply by improved spatial resolution in IMERG compared to 3B42. The TRMM Precipitation Radar (PR) made possible unprecedented and systematic monitoring of precipitation in Middle and Low Mountains probing into the inner regions of complex terrain to map the vertical reflectivity profiles of orographic precipitation systems, including highly localized extreme events (e.g. Barros et al. 2000; Barros et al. 2004; Eghdami and Barros 2019).

The measurement of orographic precipitation through ground-based networks is a challenging task due to the heterogeneity of the terrain requiring a highly dense observational network in order to be representative of the resulting spatial variability in precipitation patterns that is difficult to achieve due to remoteness of and difficult access to high elevations sites. Ground-based point observations of the space-time structure of rainfall in the Himalayas (Barros et al. 2000) and in Southern

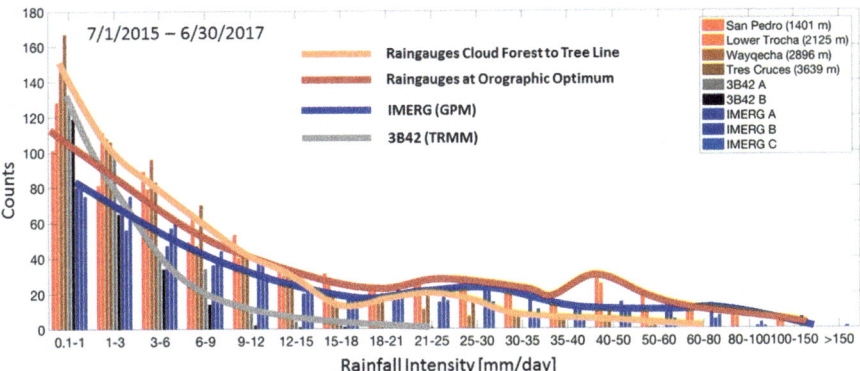

Fig. 30.1 Histogram of daily rainfall observed along an altitudinal raingauge transect on the eastern slopes of the Peruvian Andes (Barros 2013). Stations with incomplete records were removed. A, B, C refer to different pixels of TRMM 3B42 and GPM IMERG products. The red (tropical montane forest) and orange (cloud forest) lines represent the rainfall envelope for the two main ecosystems

Appalachian Mountains (SAM) show strong spatial gradients with elevation and landform, as well as large temporal variability at diurnal, monthly, seasonal and inter-annual scales (Wilson and Barros 2014, 2015). In the SAM, light rainfall contributes up to 50% of the total annual precipitation. In particular, Wilson and Barros (2014) identified low-level precipitation systems with high-frequency light rainfall between 1100 and 1600 LT (i.e., mid-day) in the inner regions of the SAM in all seasons.

High heterogeneity between the valley and the ridge precipitation in the SAM is attributed to the spatial and temporal variation in the precipitation microphysics (Prat and Barros 2010a; Wilson and Barros 2014, 2017; Duan and Barros 2017). Interestingly, as opposed to the classical orographic enhancement of precipitation with elevation on upwind slopes (e.g., Barros and Lettenmaier 1994), there are many instances in the SAM where mountain valleys receive more precipitation than ridges, up to one order of magnitude, due to Seeder-Feeder Interactions (SFI; Wilson and Barros 2014) among rainfall and multilayer cloud systems resulting in the reverse-orographic enhancement effect identified by Wilson and Barros (2015). SFI refers to the process where by the raindrops produced by the high-level clouds (seeders) interact with the low-level clouds and fog (feeders) to enhance coalescence and accelerate raindrop growth, and thus surface precipitation at low-levels. The mid-afternoon SFI is due to the interaction between stratiform and low-level orographic clouds and advection fog, and nighttime and early morning SFI is governed by interactions among passing stratiform systems, convective clouds and radiation fog (Wilson and Barros 2014). A synthesis of these processes is presented in Fig. 30.2.

Besides the spatial and temporal variability linked to landform and weather, aerosol-cloud-precipitation interactions can be another important source of variability. Specifically, changes in CCN activation (Shrestha et al. 2010, 2012) behavior on

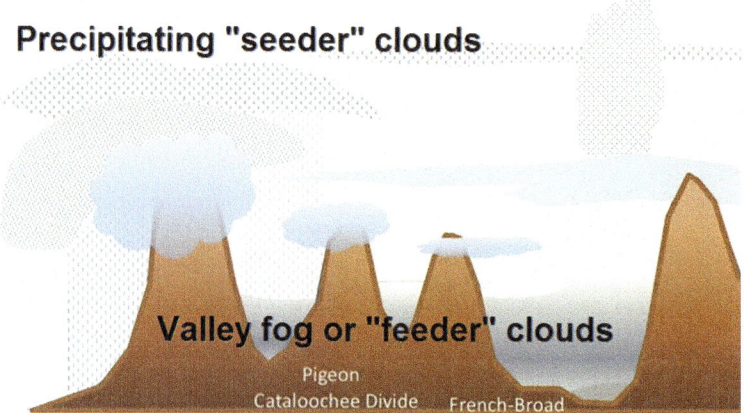

Fig. 30.2 Conceptual representation of orographic precipitation mechanisms in the Southern Appalachian Mountains. (Adapted and modified from Wilson and Barros 2017)

orographic precipitation processes (Barros et al. 2018; Duan and Barros 2019) can result in significant spatial shifts in cumulative rainfall distributions impacting flashflood forecasting, water resources management, and hydropower production.

30.2 Orographic Precipitation Measurement

Satellite remote sensing and the recent introduction of multi-frequency radar-based precipitation products have the potential to improve the estimation of orographic precipitation. Radars are active sensors that transmit and receive monochromatic radiation. The transmitted radiation undergoes scattering and absorption as it interacts with hydrometeors in the atmosphere. Notionally, radar retrieval consists of estimating the size distribution N(D) (spectrum of number of particles N as a function of equivalent diameter D) of the hydrometeors from backscatter measurements which can be used subsequently to derive bulk rainfall properties such as rainfall volume and intensity. Because of complex microphysics, N(D) can change significantly in the vertical and in time in the atmosphere (see Testik and Barros 2007), and thus a key challenge in radar measurements is to infer the profiles of rainfall from temporal measurements of backscatter as the transmitted signal propagates in the atmosphere. Typically, single scattering, a first order process whereby the incident photons are scattered only once, is assumed in operational radar algorithms (Marzano et al. 2003). However, the incident photons can undergo multiple-scattering based on the incoming wavelength of the photons, horizontal and vertical distribution of hydrometeors, and the optical properties of the hydrometeors. Multiple scattering (MS) becomes significant at higher frequencies, larger particle sizes, larger antenna footprints, and with asymmetry in the scattering phase function that describes volume scattering (Battaglia et al. 2005, 2015). Satellite-based precipitation radars operate at microwave frequencies and capture the 3D structure of precipitation storms within the radar beam. Reflected and backscattered electromagnetic signals from the vertical distribution of hydrometeors are used to estimate the vertical profiles of precipitation rate. The reflectivity factor Z at height z is related to the diameter (assuming spherical shapes) and concentration of atmospheric hydrometeors at the same height (Bringi and Chandrasekar 2001):

$$Z_z = \int D^6 N(D) dD \qquad (30.1)$$

Therefore, rainfall rates are related to Z via a power law in retrieval algorithms, with different strategies used for stratiform and convective rainfall. The accuracy of satellite-based precipitation estimates in mountainous areas is affected by multiple sources of error including bright-band (BB) and dark-band (DB) effects, attenuation and non-uniform beam-filling (NUBF) artifacts further enhanced by the complex spatial variability of rainfall systems, complex multiple-scattering processes (MS), and ground-clutter (GC) artifacts that result in significant underestimation of low-level precipitation systems.

The melting of ice and snow at the 0-degree isotherm in the atmosphere leads to the formation of a mixed phase layer of stratiform precipitation. Changes in size due to coalescence of liquid and ice hydrometeors and or partially melted particles, and changes in the dielectric constant due to phase changes lead to an increase in the observed reflectivity factor, the BB effect (Smyth and Illingworth 1998). The BB is absent in convective precipitation as strong updrafts and warm temperatures keep the ice particles aloft, whereas in the case of large frozen hydrometeors like graupel and hail, fall velocities are very high and may only melt near the ground, or not melt at all. The absence of a BB is used to distinguish convective from stratiform precipitation (e.g., Awaka et al. 1997). Underestimation and False Alarm (FA) errors in the case of stratiform precipitation are linked to a decrease in the gradient of reflectivity profiles near surface (Duan et al. 2015). W-Band radars (95 GHz) observe a "dark-band" effect near the 0-degree isotherm, where there is a sudden increase in the reflectivity profiles at the melting layer without being followed by a decrease in the reflectivity values (i.e., no BB), because the average attenuation in the melting layer is comparable with the attenuation in the rain layer underneath (Kollias and Albrecht 2005; hereafter KA05). The reflectivity factor Z is directly proportional to the sixth moment of the drop diameter (e.g., Eq. 30.1) in the Rayleigh [wavelength is larger than hydrometeor size] scattering regime and exhibits an oscillating behavior for drop-sizes larger than 1-mm in Mie [wavelength is smaller than hydrometeor size] scattering regime (Kollias et al. 2002). For lower frequency radar, the assumption of Rayleigh scattering behavior holds well for large drop-sizes and the reflectivity shows sensitivity in the melting layer. KA05 noted that the total backscattering coefficient does not show variability for drop diameters greater than 1 mm in the W-band (~ 3.2mm wavelength) explaining the BB absence. KA05 also points out though that in the case of very small drop sizes (less than 0.8 mm), the DB vanishes aloft and a BB signature is present at W-band.

Non-Uniform Beam Filling (NUBF) refers to the integration of the radar signal over a volume that is not completely filled with hydrometeors due to the heterogeneity in the vertical and horizontal structure of precipitation systems within the radar beam, including clear-sky and non-rainy clouds. Thus, the magnitude of the radar backscatter signal is an underestimation of the reflectivity if only the rainy regions are considered. As the distance between the radar and the storm increases, the heterogeneity increases, and thus there can be significant underestimation in the precipitation retrievals generally (Durden et al. 1998), and low-level and small-scale local orographic precipitation systems in particular (Duan et al. 2015). MS effects are dominant in mesoscale convective precipitation systems leading to the overestimation of precipitation rate (Marzano et al. 2003; Bouniol et al. 2008). Monte-Carlo simulations suggest that the CloudSat (satellite-based W-Band radar) retrievals of stratiform precipitation with intensities greater than 5 mm h^{-1} suffer from 30% to 50% overestimation (Matrosov et al. 2008). Previous studies found that the TRMM-PR precipitation estimates overestimate deep convective systems in the SAM (Prat and Barros 2010b; Duan et al. 2015).

Specific details on the operational radar retrieval algorithm, and how it approaches the different sources of error are provided by Iguchi et al. (2017) specifically for GPM Dual Polarization Radar (DPR) measurements. GC effects are illustrated in Fig. 30.3. The high reflectivity measurements (red band) in the

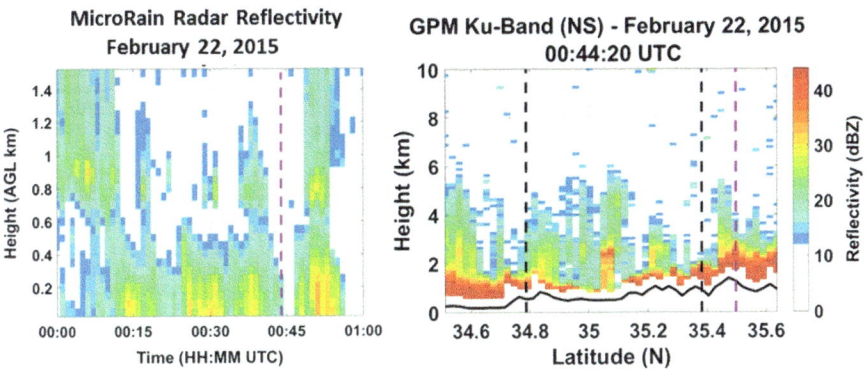

Fig. 30.3 Example of ground-clutter effects on radar measurements of low-level rainfall. Left: Vertical structure of MRR reflectivity above ground level (AGL) with GPM overpass marked by the pink dashed line in right panel. Right: Reflectivity cross-section over the Southern Appalachian Mountains. Black dashed lines indicate the boundaries of near-nadir scan (angle <7 deg). Pink line marks the position of the MRR where rainfall was missed by the GPM DPR algorithm. The white band between the terrain (dark black line) and the GPM reflectivity lower reflectivity measurements (red band) is due to the terrain mask used in the algorithm

right panel are contaminated by ground-clutter and cannot be used to estimate rainfall. This results in frequently eliminating reflectivity measurements in the lower 2 km (low level rainfall) and even higher for large viewing angles unless additional constraints are introduced by using multiple sensors for example (Arulraj and Barros 2017; Arulraj and Barros, 2019). The low-level vertical structure of reflectivity from the ground-based Micro Rain Radar (MRR) in the left panel is indicative of missed rainfall when ground-clutter corrections are applied to the satellite-based measurements.

Note the intermittency of the MRR fine reflectivity structure that illustrates the question of temporal representativeness of satellite measurements (instantaneous) and spatial representativeness of ground-based measurements (point scale) that must be reconciled in the evaluation of precipitation products (Prat and Barros 2010b; Duan et al. 2015).

Attempts to address GC by "filling" the reflectivity profiles below the height at which no GC effects are detected often do not work either because of underestimation of SFI and cloud layering but also because they increase the False Alarm Rate (FAR $= \frac{FA}{YY+FA}$), even if they increase the Probability of Detection (POD $= \frac{YY}{YY+MD}$). Further, it is important to note that rainfall detection and measurement sensitivity depend strongly on the geometry of the measurement proper as shown in Fig. 30.4 as a function of satellite orbit (descending and ascending overpasses) and viewing angle.

This also illustrates the potential for retrieval ambiguity as complex precipitation processes appear different depending on measurement geometry, which explains why simple calibration and, or optimization of retrieval algorithms without a physical basis does not result in QPE improvements (e.g., compare Prat and Barros 2010b; Duan et al. 2015). Indeed, inspection of Fig. 30.1 shows that the IMERG product severely underestimates the frequency of light rainfall days from the cloud

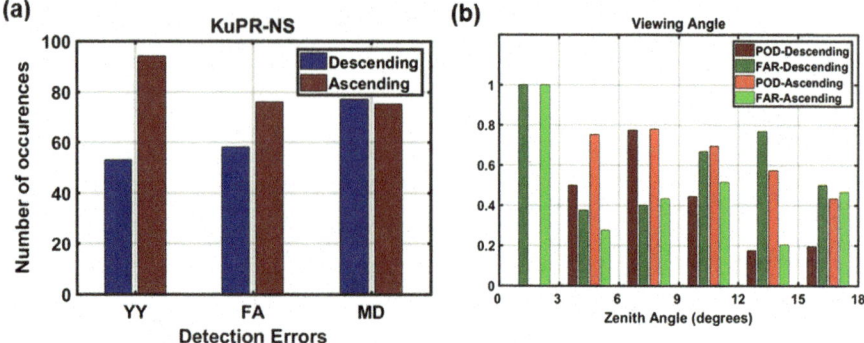

Fig. 30.4 Dependence of detection errors to the satellite orientation. Detection errors varying with (**a**) satellite geometry and (**b**) local viewing angle. Near-surface precipitation estimates of Level 2, version 5A products of GPM Ku-PR compared with ground-based rain-gauges (RG) that lie within 2.5 km of the center of DPR pixels. Time period considered for analysis: March 2014–May 2017. Note- YY is when both GPM and RG detects precipitation. FA is false alarm when GPM detects precipitation while RG did not detect any precipitation. MD is missed detection where RG detects rain and GPM misses the detection

forest up to tree line corresponding to about 50% of annual rainfall at high elevations the Andes, and it underestimates moderate rainfall at the orographic optimum at lower elevation during the monsoon, which combined with missed detection of light rainfall amounts to 2.5 m of precipitation that is approximately 40% of annual rainfall in the rainforest altitudinal band. Deep convection forms at the foothills of the Andes and in the Altiplano and occasionally at nighttime and early morning at the confluence of river networks (Giovanettone and Barros 2009). Most moderate rainfall events on the Andean orographic envelope (as in the Himalayas) are stratiform with embedded shallow convection (<5 km AMSL), and thus contamination of radar measurements in the lower 2 km on steep altitudinal gradients and complex terrain geometry is one major source of error. Whereas only GC is specific to mountainous regions, the heterogeneous vertical organization of hydrometeors depending on storm systems and topography, and the organization of precipitation processes on the terrain are unique to mountainous regions (e.g., Fig. 30.2).

30.3 Ground-Validation

A ground-validation (GV) strategy relying on intense ground-based data collection of precipitation accumulations and precipitation structure during limited-duration field campaigns such as the Integrated Precipitation and Hydrology Experiment (IPHEx; Barros et al. 2014) and the Monsoon Himalaya Orographic Precipitation Experiment (MHOPrEx; Barros and Lang 2003) in the Central Himalayas including the deployment of long-term science-grade observing systems (e.g., Barros et al. 2000; Prat and Barros 2010a) enables scientific data analysis and discovery to inform detailed evaluation of precipitation retrievals and GV synthesis in the past.

From synthesis, a geography of nonstationary retrieval errors emerges exhibiting robust spatial modes and diurnal and seasonal cycles tied to physical processes that may vary from one region to another, and from one location to another within the same region (e.g., Barros et al. 2004; Prat and Barros 2010b; Duan et al. 2015). The GV framework for error analysis and attribution therefore sets the stage for process studies, including models and observations, to elucidate the physical-basis of error toward ultimately improving quantitative precipitation estimation (QPE).

Next, we rely on IPHEx data over the SAM (Fig. 30.5) to examine GPM DPR measurements and precipitation products aiming at elucidating the physical under-pinnings of retrieval errors.

In the Southern Appalachians (SAM), seeder-feeder interactions (SFI) between rainfall and low-level clouds and fog (LLCF) modify the vertical structure of rainfall enhancing drop coalescence efficiency that results in increasing the number of larger raindrops and the number of drops overall (Wilson and Barros 2014). Detailed studies within sub-regions of the SAM to characterize SFI using spatially distributed observations to monitor different microclimates (Fig. 30.2) indicate that there is

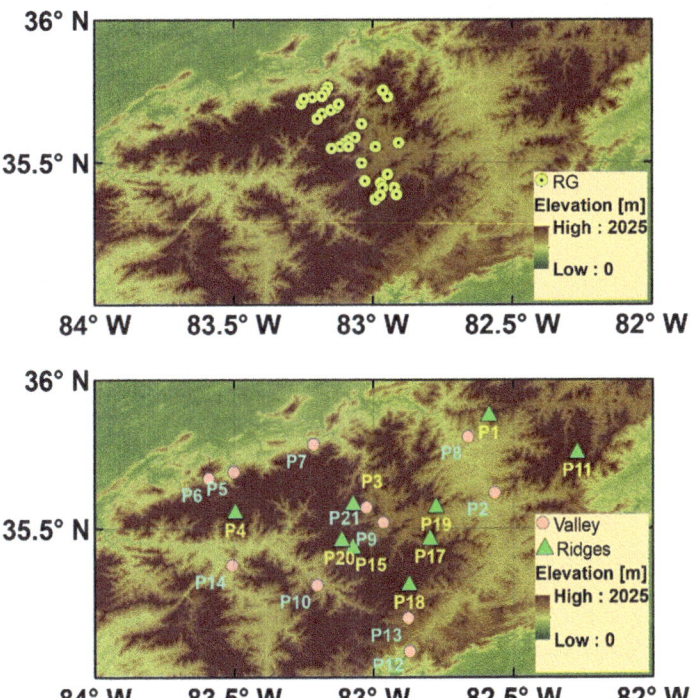

Fig. 30.5 Location of IPHEx GV (left) rain-gauges and (right) Parsivel disdrometers in the SAM. Triangles denote ridge (elevation >900 m) and circles denote valley locations

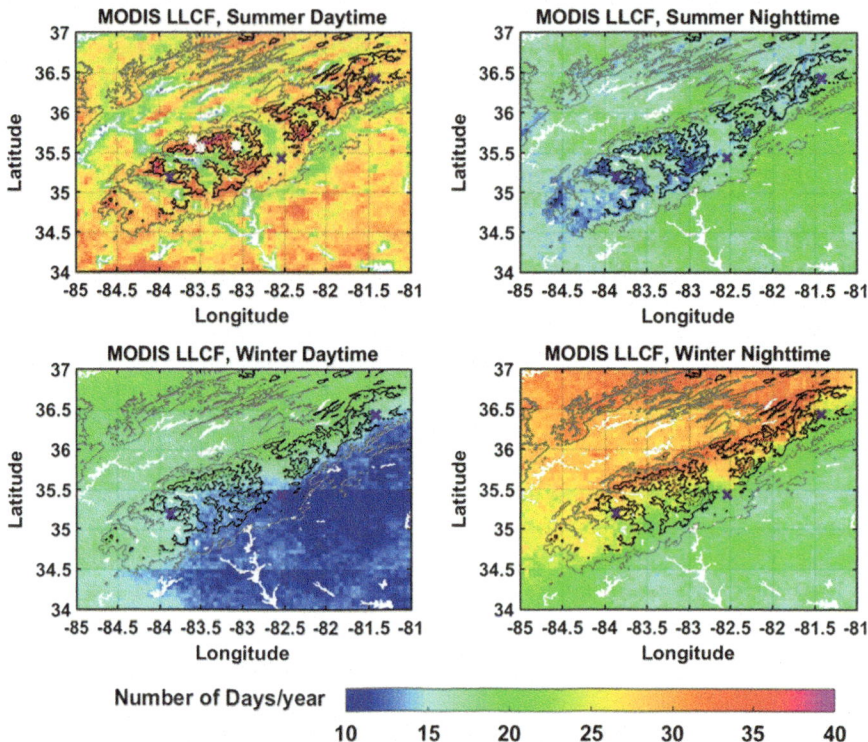

Fig. 30.6 Spatial distributions of AQUA MODIS LLCF (CTH <5 MSL, confident cloudy only; 0.05° × 0.05°) during daytime (Left Column) and nighttime (Right Column) overpasses in summer (June–July-August; top row) and winter (December–January-February; bottom row). (Adapted from Duan and Barros 2017)

substantial heterogeneity in the spatial and temporal organization of LLCF, which in turn leads to variability in rainfall microphysics from one location to another and, even at the same location, depending on time of day and season as illustrated by the regional climatology of LLCF in Fig. 30.6 (Wilson and Barros 2017; Duan and Barros 2017).

Figure 30.7 summarizes analysis of rainfall microphysics from two disdrometer locations, P4 and P6 (see Fig. 30.5 for location). From mid-morning to midnight, the number of drops is approximately the same in the ridge and in the foothills (not shown), but Dm is larger by 20% at P6 due to a deeper atmospheric column for SFI among low level layered clouds (Rain on LLC).

The value of Nw, the normalized DSD intercept, is the same in the ridge and in the valley from midmorning to the evening hours, but there is a 3–5 dB increase at P4 between 0300 and 0700 AM (purple). Between 0300 and 0700 AM, fog forms on the ridge resulting in a significant increase in Nw at P4 (Rain on Fog and Cap Clouds; Rain on FCC), but the fog layer is not deep enough to impact drop dynamics and significantly change Dm.

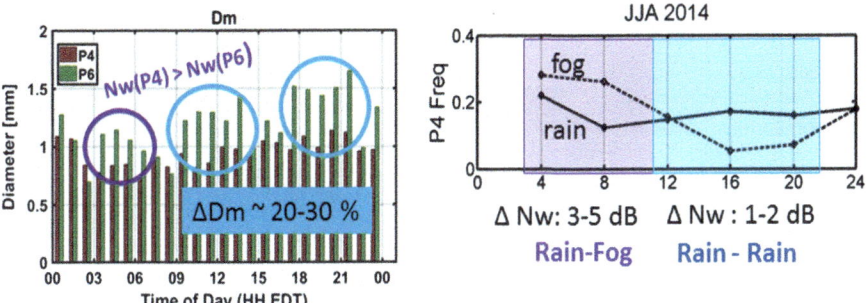

Fig. 30.7 Fingerprinting SFI on DSD metrics. Left: diurnal cycle of Dm (ratio of fourth moment to the third moment of the diameter) at P4 (western ridge) and P6 (foothills). Right: diurnal cycle of rain and fog occurrences at P4. Note that Dm (P6) always ≥ Dm (P4). *FCC:* Fog and Cap Clouds. *LLC*: Layered Low Level Clouds

30.4   Physical-Basis of Retrieval Errors

In addition to measurement geometry (Fig. 30.4), overpass fequency is also an important source of variability in satellite-based remote sensing of precipitation as demonstrated in Fig. 30.8, which reconciles the diurnal cycle of revisit time frequency over the IPHEx region globally (left panel) and its spatial distribution for times when it is raining (right panel).

Note the low frequency revisit bias in the morning (left panel) and the inner mountain region (right panel), where SFI are prominent and strongly impact the diurnal cycle of rainfall.

The underestimation of rainfall along the western slopes of the SAM results from underestimating the number of small and intermediate sized raindrops (D < 2 mm) at low elevations as shown in Fig. 30.9(a) for locations P5 and P6 (see Fig. 30.5 for locations). Note the much higher Dm for May and June compared to the remainder summer months, consistent with the regional climatology of mesoscale convective systems that produce heavy rain with larger size drops in spring. Consequently, the seasonal and diurnal cycle of detection errors reflects the relative contribution of SFI enhancement versus classical orographic enhancement and spatial modulation of stratiform versus convective precipitation processes (Figs. 30.10 and 30.11): high FARs in the winter (DJF) and in the morning (high elevation) and in the summer (JJA) at mid-day in contrast with the improved detection for convective processes generally; overestimation of light rainfall rates due to NUBF and overcorrection of GC effects, and underestimation of convective rainfall due to lack of sensitivity for high rainfall rates.

The DSD of liquid hydrometeors in the GPM Level 2A products is a gamma distribution function with specified shape factor $\mu = 3.0$ as follows:

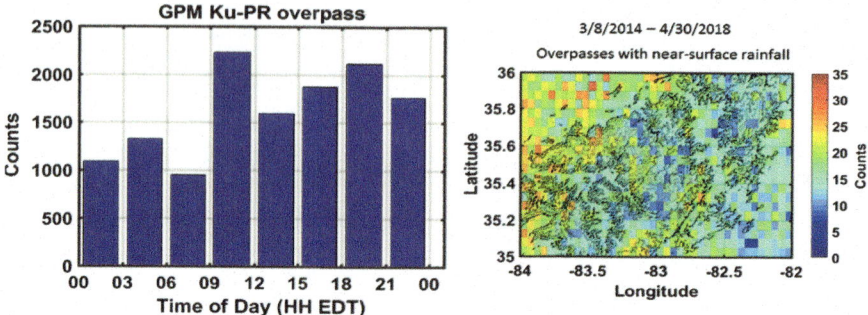

Fig. 30.8 Left: diurnal cycle of GPM revisit overpasses over the region shown in Fig. 30.5. Right: Spatial distribution of rainfall detection. The blue tones in the inner region valleys and over the complex terrain at low elevations along the eastern ridges of the SAM are indicative of low frequency bias

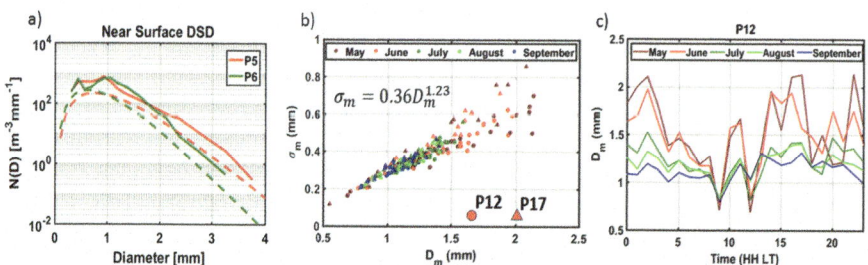

Fig. 30.9 DSD nonstationarity on altitudinal gradients: (**a**) disdrometer (continuous lines) and GPM Ku-PR estimates (dashed lines) on the western slopes of the SAM; (**b**) and (**c**) DSD statistics on the eastern slopes of the SAM

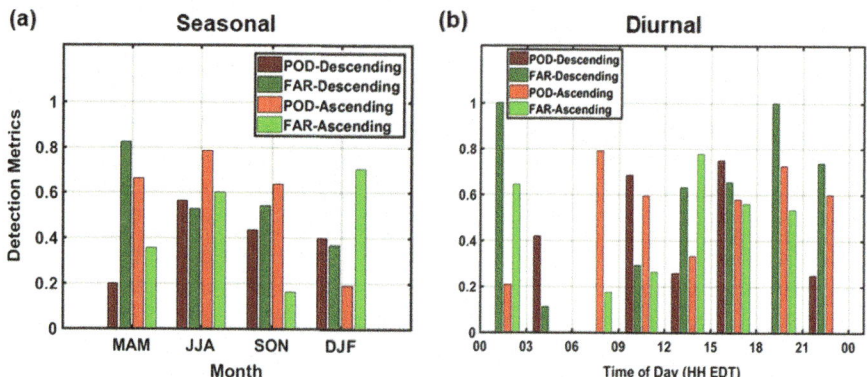

Fig. 30.10 (**a**) Seasonal and (**b**) diurnal distribution of GPM Ku-PR detection errors in the Southern Appalachian Mountains for ascending and descending modes of the GPM

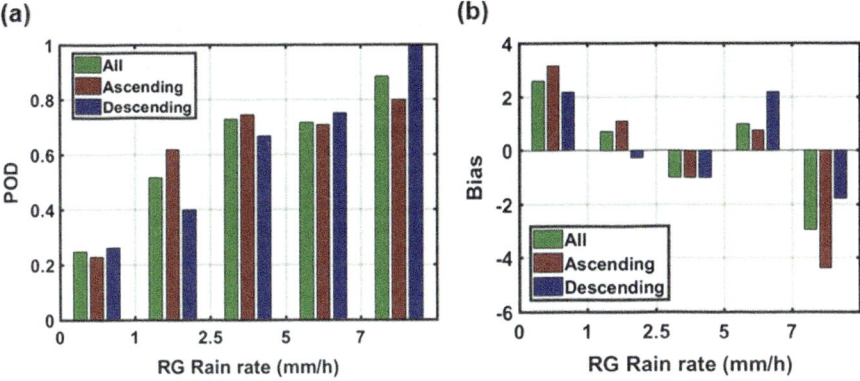

Fig. 30.11 (a) Detection and (b) estimation errors varying with rain-gauge rain-rate observations for ascending and descending modes of the GPM Ku-PR. $Bias = 10\log_{10}\left(\dfrac{\sum R_{GPM,i}}{\sum R_{RG,i}}\right)$

$$N(D) = N_w f(\mu)\left(\frac{D}{D_m}\right)^\mu \exp\left(-\frac{(4+\mu)D}{D_m}\right) \tag{30.2}$$

where $f(\mu) = \frac{6(4+\mu)^{(\mu+4)}}{4^4\Gamma(\mu+4)}$ and Γ the gamma function.

The mass-weighted diameter D_m is defined as the ratio of the fourth moment to the third moment of diameter with the drop diameter varying between D_{min} and D_{max} with interval dD:

$$D_m = \frac{\sum\limits_{D_{min}}^{D_{max}} N(D)D^4 dD}{\sum\limits_{D_{min}}^{D_{max}} N(D)D^3 dD} \tag{30.3}$$

The normalized number concentration or intercept factor, N_w (m^{-3} mm^{-1}) is calculated from the liquid water content q (g m^{-3}) [where $q = \frac{\pi}{6}\rho_w \sum\limits_{D_{min}}^{D_{max}} N(D)D^3 dD$, ρ_w is the density of water (10^{-3} g mm^{-3})] and Dm:

$$N_w = \frac{4^4}{\pi\rho_w}\left(\frac{q}{D_m^4}\right) \tag{30.4}$$

One approach to elucidate the physical basis of retrieval error is to investigate whether the conceptual microphysics model underlying radar retrieval is representative of observed microphysics. Figure 30.12 shows the spatial structure of the average D_m and N_w retrieved since GPM launch. One salient feature in the time-

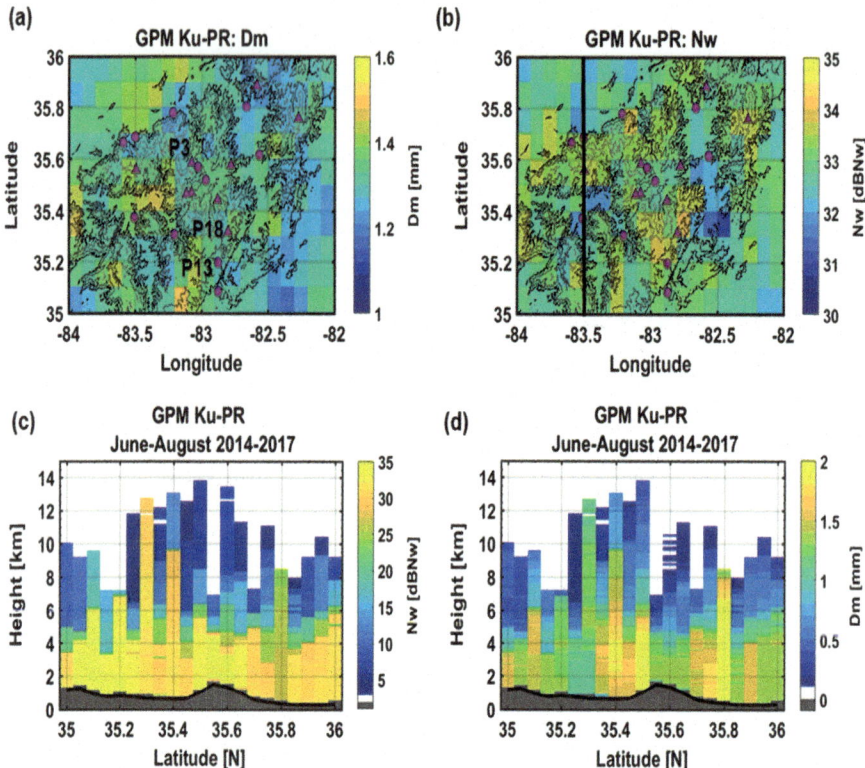

Fig. 30.12 Average GPM Ku-PR Level 2A DSD parameters in the SAM for 4.5 years since GPM launch. Top row: spatial distribution of (**a**) Dm and (**b**) Nw. Bottom row: vertical structure along the transect at 83.5°W (black line in b) of (**c**) Nw and (**d**) Dm. The pink circles and triangles mark respectively valley and ridge disdrometer locations

average spatial distribution of KU-PR Dm and Nw[1] is that they vary in opposite direction within their respective ranges. This results in higher Nw values over the ridges than in the valleys, and larger Dm values in the foothills on the eastern and western slopes of the SAM as well as at the confluences of wide-open tributaries in the inner region (Little Tennessee and the French-Broad). The spatial distribution is consistent with disdrometer observations at some, but not all times of day on the western sector, and not everywhere due to widespread mid-day and afternoon LLCF in the inner region (not shown; Duan and Barros 2017; Wilson and Barros 2014). Further, higher (lower) values of Nw (Dm) in the sub-regions of low frequency bias in Fig. 30.8 suggest that this is not rooted in physics but rather an artifact due to the lack of adequate physical constraints to infer the gamma distribution parameters with

---

[1]Level 2A data with spatial resolution of 5 Km (~0.05°) were spatially averaged to 10 km in the two leftmost panels in the top row.

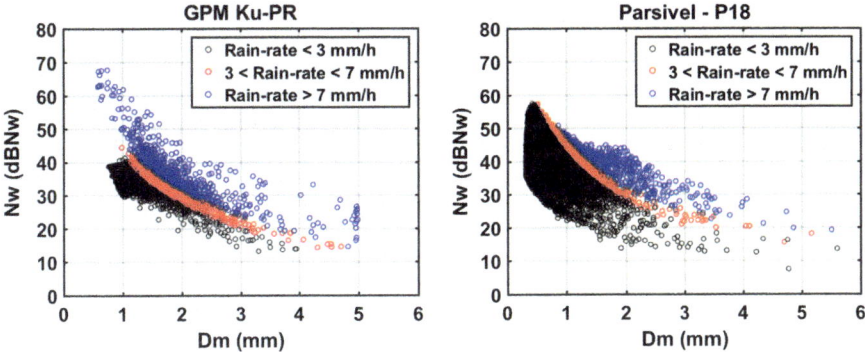

Fig. 30.13 Phase-space maps of (Dm, Nw) from GPM Level 2A products since launch and from GV Parsivel disdrometer P18 during IPHEx

fixed $\mu = 3.0$ as in the Level 2A products. However, retrievals are based not on mean statistical values but rather on physically-based local covariation of N_w and D_m depending on the hydrometeorological environment.

Figure 30.13 contrasts the Ku-PR estimates (top right panel) with the disdrometer observations at P18 (see Fig. 30.5 for location) which further supports this hypothesis. The DSD parameters at P18 are calculated from the disdrometer DSD distributions for each time using Eqs. (30.3) and (30.4). The Ku-PR parameters exhibit steep trade-offs in (Dm, Nw) space over a relatively narrow region governed by the inverse relationship in Eq. (30.2) also apparent in the spatial distributions in Fig. 30.12 (see also Prat and Barros 2009). This behavior is not consistent with the disdrometer at either low (black) or moderate to high rainfall intensity (blue), and only to a much lesser degree for intermediate rain rates (red). The closed brown contour line delineates a region of (Dm, Nw) space filled by the multiplicity of microphysical states (variable μ) observed within the disdrometer measurement capabilities at high temporal resolution [1 min], in line with the near-instantaneous duration of the satellite overpass. In addition, increases in disdrometer Dm values do not imply substantial increases or decreases in Nw indicating convective rainfall (see P12 in Fig. 30.9) or SFI on layered LLC as previously discussed at P6 (Fig. 30.7).

Figure 30.14 shows simulation results of Rain on Fog simulations at P6 (see Fig. 30.7) using an existing model (Prat and Barros 2007; Prat et al. 2012) that describes stochastic raindrop dynamics in the atmospheric column modified to simulate SFI through layered low level clouds, orographic cap clouds and, or fog (Wilson and Barros 2014; Duan and Barros 2017; Arulraj and Barros 2019). The left panel shows the sensitivity of simulated rainfall intensity to the DSD spectra of fog, and thus the specific fog microphysical regime matters in order to capture the observed rainfall at the collocated rain-gauge. Detailed description of simulations is provided in Duan and Barros (2017). The right panel shows that the trajectory of an individual precipitation event occupies a large area of phase-space from start to end of the Rain on Fog event. Thus, the space-filling behavior exhibited by the disdrometer data in Fig. 30.13 reflects changes in rainfall intensity that emerge as the microphysics change during a storm. This nonlinear behavior translates into changes

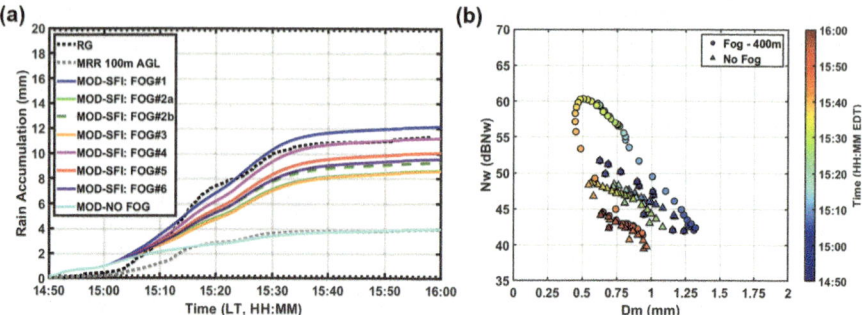

Fig. 30.14 Left: sensitivity analysis of surface rainfall accumulation to the drop size distribution (DSD) spectra of fog in the simulation of RoF SFI at P6 in the western foothills of the SAM. RG–dashed black lines are local raingauge observations. After Duan and Barros (2017). Right: simulated microphysical trajectory for stratiform rainfall with and without RoF SFI at P6 corresponding to Fog#4 DSD and fog depth of 400 m. Circles describe microphysical state at 1-min intervals from start to end of the event as per the legend

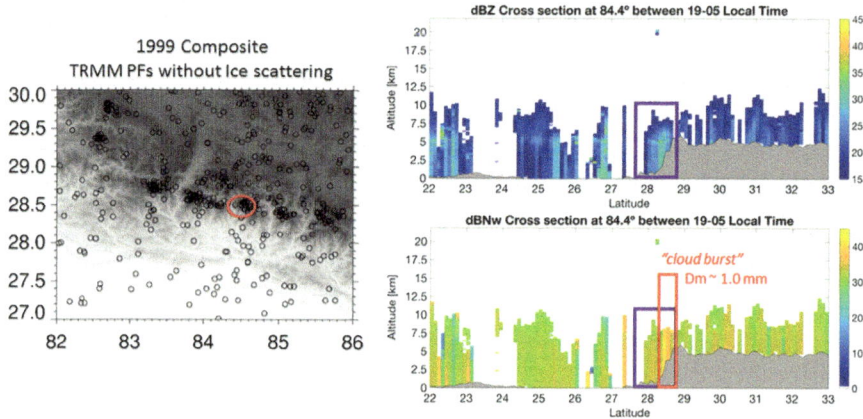

Fig. 30.15 Composites of GPM Level 2A DSD parameters and reflectivity along a cross-section designed to capture the hot-spots of precipitation in Central Himalayas (84.4°E, red circle) during the monsoon. Orographic precipitation on the upwind slopes is highlighted by the boxes. The map with TRMM Precipitation Features (PFs) on the left panel is adapted from Barros et al. (2004)

in the shape parameter μ that characterize the evolution raindrop population collision dynamics (coalescence versus breakup) at difference stages of the event (Testik and Barros 2007; Prat and Barros 2010a; Testik et al. 2011; Prat et al. 2012).

GPM Level 2A composites of DSD parameters in the Central Himalayas (Barros et al. 2000; Barros and Lang 2003; Shrestha et al. 2010) are shown in Fig. 30.15 along a cross-section that includes one of the hot-spots of precipitation in Central Nepal (red circle, regionally referred to as cloud bursts) identified by Barros et al. (2004). Note how the location where GPM's Nw is the highest (inside red rectangle on the right) corresponds to the orographic optimum in Central Nepal. The highest

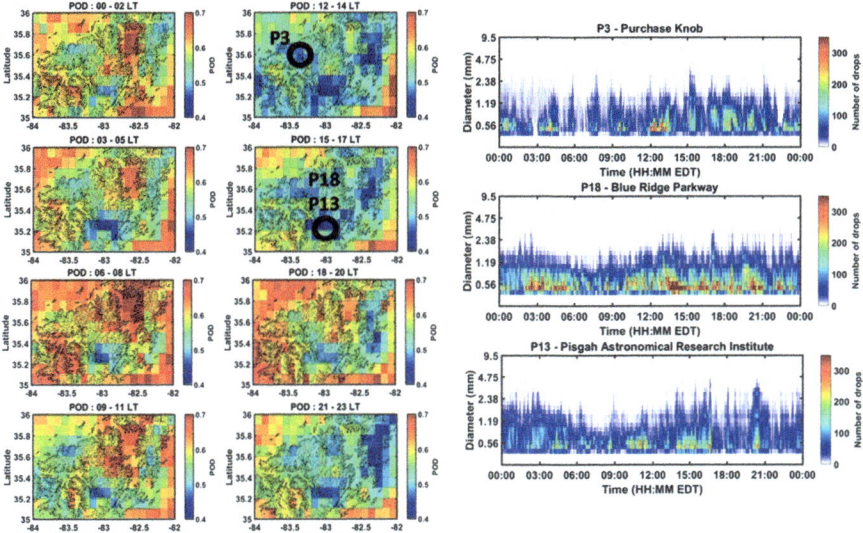

Fig. 30.16 Left panels: diurnal cycle of the spatial distribution of the climatology of Probability of Detection (PoD) of IMERG precipitation (~11 km spatial resolution) compared with STAGE IV (~4 km spatial resolution) (http://www.emc.ncep.noaa.gov/mmb/ylin/pcpanl/stage4/, last accessed 4 Dec. 2018). Time period of analysis- March 2014 to February 2017. Right panels: diurnal cycle of Parsivel disdrometer DSD measurements at locations marked in the left panel

rainfall accumulations are linked to shallow embedded convection (red circle) and warm microphysics are dominant. Whereas rainfall intensities and accumulations are at least twice as high in the Central Himalayas compared to the SAM, the GPM Dm maximum is close to 1 mm compared to 2 mm in the SAM, and the Nw is larger by approximately 10 dBNw. This illustrates a significant unintended consequence of artificially imposing a fixed microphysical mode of interaction ($\mu = 3$) in the present GPM retrieval algorithm that translates into severe underestimation of rainfall rates in the region.

The space-time time organization of errors by topography and microphysical regime propagates from GPM Level 2A to higher level products such as IMERG QPE (Huffman et al. 2018). This is demonstrated in Fig. 30.16 for the SAM by comparing IMERG to the Stage IV combined radar-rain-gauge product (March 2014–February 2017). The spatial maps show the diurnal cycle climatology of POD. Note the concurrence between the spatial patterns in Figs. 30.8 and 30.16 and the temporal patterns in Fig. 30.4. The time-series graphs on the right show the DSD climatologies at three Parsivel disdrometer locations (P3, P18, and P13) marked in the spatial maps. The lowest skill of the IMERG product is at mid-day (1200–1700 LT) in the inner ridge and eastern slopes of the SAM, which are times during which there is significant rainfall with small Dm (<1 mm) at very high concentrations and large spatial variability. This outcome while illustrative of the challenges of orographic QPE, clearly makes the case for the importance of capturing the underlying physical processes in retrieval which cannot be statistically corrected.

30.5 Summary

The grand challenge in remote sensing of orographic precipitation, and indeed precipitation in regions of complex terrain generally, is to measure rainfall extremes, both light and heavy rainfall rates, at the relevant hydrometeorological spatial and temporal scales with the skill required to understand the water cycle and to address multisector societal needs ranging from water supply to food security, energy production, and water hazard prevention and safety.

Beginning with the first precipitation products derived from passive microwave satellite observations nearly 50 years ago (e.g., Rao et al. 1977; Wilheit 1986) to the latest passive-active measurement architecture embodied by GPM (Skofronick-Jackson et al. 2017), remarkable progress has been achieved in attaining global coverage of precipitation measurement (e.g., Adler et al. 2001) with increasingly higher spatial and temporal resolution in the TRMM (Huffman et al. 2007) and GPM eras (Huffman et al. 2018). Although improvements in skill have followed improvements in resolution, errors remain large in mountainous regions due to radar limitations such as ground-clutter, non-uniform beam filling, and parallax errors as discussed in Sects. 30.2, 30.3, and 30.4, and due to the lack of ground-based observations to correct quantitative precipitation estimates. Overall, missed detection and underestimation errors dominate, which can be classified into three different categories: light rainfall (instantaneous rain rates <1 mm h^{-1}), shallow precipitation, and layered precipitation. Light rainfall errors are the most difficult to overcome as they are tied to measurement sensitivity. Shallow precipitation errors can be tackled by improving detection through the use of multi-frequency multi-platform observing systems taking advantage of parallax for example (e.g., Arulraj and Barros 2017). Layered precipitation systems are associated with "bottom heavy" profiles of hydrometeors in the lower troposphere that result from microphysical interactions (e.g. SFI) between cloud layers and precipitation, and errors can be addressed by physically-based retrieval if the presence of layered clouds can be unambiguously identified (Arulraj and Barros 2019).

Systematic statistical correction of layered precipitation errors is frustrated both by nonlinearity as illustrated in Fig. 30.17 for various modes of seeder-feeder interactions, and geographical nonstationarity linked to aerosol type and abundance that have a critical role on the formation of frozen hydrometeors, specifically hail and graupel, and thus play an important role in the vertical structure of orographic clouds and precipitation systems as shown in Fig. 30.18 (Barros et al. 2018; Duan et al. 2019).

A path to address shallow and layered precipitation errors is the integration of remote sensing measurements and models either via data-assimilation of radar reflectivities and brightness temperatures and, or via physically-based retrieval as discussed in Sect. 30.4. The need for integration of measurements and process models is at the core of the vision for remote sensing in Earth Sciences articulated by the recent Decadal Survey (NAS 2018) that aims to inspire the development of new observations from space in the next decade.

Fig. 30.17 Illustrative sketching of typical DSD evolution in the lower troposphere under when different Seeder-Feeder Interactions (SFI) scenarios. *ABGL* Above Ground Level, *D* Diameter, *N(D)* number concentration

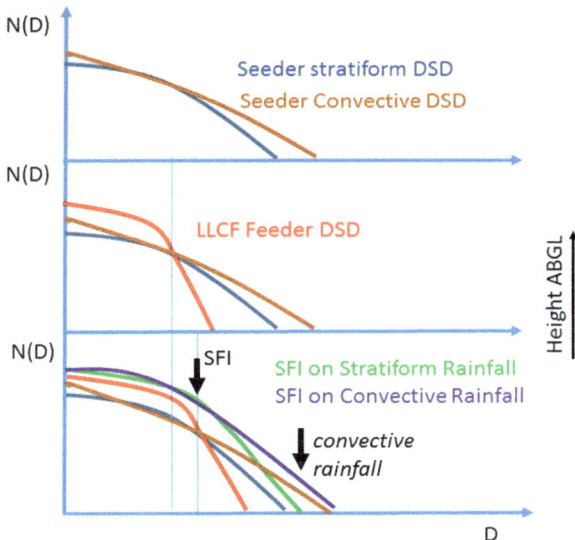

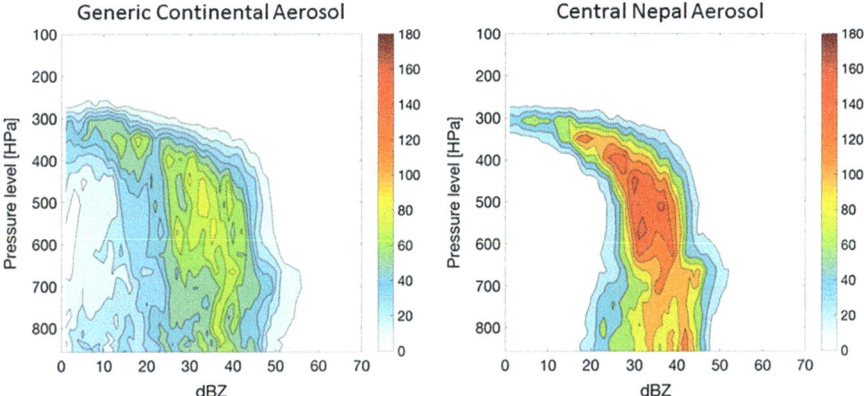

Fig. 30.18 Contour frequency by altitude synthesis of reflectivity profiles derived from Weather Research Forecast (WRF) model V3.8 simulations using the standard continental aerosol in the model (left) and using aerosol properties measured in Central Nepal. Details of the simulation can be found in Barros et al. (2018)

References

Adler, R. F., Kidd, C., Petty, G., Morrissey, M., & Goodman, H. M. (2001). Intercomparison of global precipitation products: The third precipitation intercomparison project (PIP-3). *Bulletin of the American Meteorological Society, 82*, 1377–1396. https://doi.org/10.1175/1520-0477 (2001)082<1377:IOGPPT>2.3.CO;2.

Arulraj, M., & Barros, A. P. (2017). Shallow precipitation detection and classification using multifrequency radar observations and model simulations. *Journal of Atmospheric and Oceanic Technology, 34*, 1963–1983. https://doi.org/10.1175/JTECH-D-17-0060.1.

Arulraj, M., & Barros, A. P. (2019). Improving quantitative precipitation estimates in mountainous regions by modeling low level seeder-feeder interactions constrained by GPM DPR measurements. *Remote Sensing of the Environment, 231*, 111213. https://doi.org/10.1016/j.rse.2019.111213.

Awaka, J., Iguchi, T., Kumagai, H., & Okamoto, K. (1997). Rain type classification algorithm for TRMM precipitation radar. In: *Proceedings of IEEE IGARSS*, Singapore, pp. 317–319, https://doi.org/10.1109/IGARSS.1997.608993.

Barros, A. P. (2013). Orographic precipitation, freshwater resources, and climate vulnerabilities in mountainous regions. In R. A. Pielke Sr. (Ed.), *Climate vulnerability: Understanding and addressing threats to essential resources* (pp. 57–78). Dordrecht: Academic/Elsevier. ISBN:9780123847034.

Barros, A. P., & Lang, T. J. (2003). Monitoring the monsoon in the Himalayas: Observations in Central Nepal, June 2001. *Monthly Weather Review, 131*, 1408–1427. https://doi.org/10.1175/1520-0493(2003)131<1408:MTMITH>2.0.CO;2.

Barros, A. P., & Lettenmaier, D. P. (1994). Dynamic modeling of orographically induced precipitation. *Reviews of Geophysics, 32*(3), 265–284. https://doi.org/10.1029/94RG00625.

Barros, A. P., Joshi, M., Putkonen, J., & Burbank, D. W. (2000). A study of the 1999 monsoon rainfall in a mountainous region in Central Nepal using TRMM products and raingauge observations. *Geophysical Research Letters, 27*, 3683–3686. https://doi.org/10.1029/2000GL011827.

Barros, A. P., Kim, G., Williams, E., & Nesbitt, S. W. (2004). Probing orographic controls in the Himalayas during the monsoon using satellite imagery. *Natural Hazards and Earth System Sciences, 4*, 29–51. https://doi.org/10.5194/nhess-4-29-2004.

Barros, A. P., Chiao, S., Lang, T. J., Burbank, D. W., & Putkonen, J. (2006). From weather to climate—Seasonal and interannual variability of storms and implications for erosion processes in the Himalaya. In S. D. Willett, N. Hovius, M. T. Brandon, & D. Fisher (Eds.), *Tectonics, climate, and landscape evolution* (Geological Society of America, special paper) (Vol. 398, pp. 17–38). Boulder: Geological Society of America. https://doi.org/10.1130/2006.2398(02).

Barros A. P., Petersen, W. A., Schwaller, M., Cifelli, R., Mahoney, K., Peters-Lidard, C., Shepherd, J. M., Nesbitt, S., Wolff, D., Heymsfield, G., & Starr, D. (2014). *NASA GPM-ground validation: Integrated precipitation and hydrology experiment 2014 science plan*. EPL/Duke University, 64 pp. Available at https://doi.org/10.7924/G8CC0XMR, Last accessed 4 Dec 2018.

Barros, A. P., Hodes, J., & Arulraj, M. (2017). Decadal climate variability and the spatial organization of deep hydrological drought. *Environmental Research Letters, 12*. https://doi.org/10.1088/1748-9326/aa81de.

Barros, A. P., Shrestha, P., Chavez, S., & Duan, Y. (2018). Modeling aerosol-cloud-precipitation interactions in mountainous regions – Challenges in the representation of indirect microphysical effects with impacts at sub-regional scales. In *Rainfall - Extremes, Distribution and Properties*, IntechBookOpen (Pub.), Abbott, J. (Ed.), Chapter 5, 22pp, ISBN: 978-1-78984-735-2. https://doi.org/10.5772/intechopen.80025.

Battaglia, A., Ajewole, M. O., & Simmer, C. (2005). Multiple scattering effects due to hydrometeors on precipitation radar systems. *Geophysical Research Letters, 32*, L19801. https://doi.org/10.1029/2005GL023810.

Battaglia, A., Tanelli, S., Mroz, K., & Tridon, F. (2015). Multiple scattering in observations of the GPM dual-frequency precipitation radar: Evidence and impact on retrievals. *Journal of Geophysical Research, 120*, 4090–4101. https://doi.org/10.1002/2014JD022866.

Bouniol, D., Protat, A., Plana-Fattori, A., Giraud, M., Vinson, J., & Grand, N. (2008). Comparison of airborne and spaceborne 95-GHz radar reflectivities and evaluation of multiple scattering effects in spaceborne measurements. *Journal of Atmospheric and Oceanic Technology, 25*, 1983–1995. https://doi.org/10.1175/2008JTECHA1011.1.

Bringi, V., & Chandrasekar, V. (2001). *Polarimetric Doppler weather radar: Principles and applications* (636 pp). Cambridge: Cambridge University Press. https://doi.org/10.1017/CBO9780511541094.

Bruijnzeel, L. A., Mulligan, M., & Scatena, F. N. (2011). Hydrometeorology of tropical montane cloud forests: Emerging patterns. *Hydrological Processes, 25*, 465–498. https://doi.org/10.1002/hyp.7974.

Duan, Y., & Barros, A. P. (2017). Understanding how low-level clouds and fog modify the diurnal cycle of orographic precipitation using in situ and satellite observations. *Remote Sensing, 9*, 920. https://doi.org/10.3390/rs9090920.

Duan, Y., Petters,M.D., & Barros, A.P. (2019). Understanding aerosol-cloud interactions through modelling the development of orographic cumulus congestus during IPHEx. *Atmospheric Chemistry and Physics*, 19, 1–25. https://doi.org/10.5194/acp-19-1-2019.

Duan, Y., Wilson, A. M., & Barros, A. P. (2015). Scoping a field experiment: Error diagnostics of TRMM precipitation radar estimates in complex terrain as a basis for IPHEx2014. *Hydrology and Earth System Sciences, 19*, 1501–1520. https://doi.org/10.5194/hess-19-1501-2015.

Durden, S. L., Haddad, Z. S., Kitiyakara, A., & Li, F. K. (1998). Effects of nonuniform beam filling on rainfall retrieval for the TRMM precipitation radar. *Journal of Atmospheric and Oceanic Technology, 15*, 635–646. https://doi.org/10.1175/1520-0426(1998)015<0635:EONBFO>2.0.CO;2.

Eghdami, M., & Barros, A. P. (2019). Extreme orographic rainfall tied to Cold Air Intrusions in the eastern Andes. *Frontiers in Earth Sciences, 7*, 101. https://doi.org/10.3389/fenvs.2019.00101.

Falkenmark, M., & Rockström, J. (2010). Building water resilience in the face of global change: From a blue-only to a green-blue water approach to land-water management. *Journal of Water Resources Planning and Management, 136*, 606–610. https://doi.org/10.1061/(ASCE)WR.1943-5452.0000118.

FAO. (2014). *The state of food and agriculture. Innovation in family farming* (139 pp). Rome: FAO. ISBN: ISBN 978-92-5-108536-3. Available at http://www.fao.org/3/a-i4040e.pdf. Last accessed 14 Dec 2018.

FAO. (2016). *Climate change and food security: Risks and responses* (100 pp). Rome: FAO. ISBN: 978-92-5-108998-9. Available at http://www.fao.org/3/a-i5188e.pdf. Last accessed 14 Nov 2018.

Garrido, A., & Dinar, A. (2009). Overcoming the constraints for a more integrated and adaptive water management. In: Garrido, A., and Dinar, A. (eds.) Managing Water Resources in a Time of Global Change: Mountains, Valleys and Flood Plains (pp. 288). New York, NY: Taylor & Francis, (Pub). https://doi.org/10.4324/9780203884386.

Giovannettone, J. P., & Barros, A. P. (2009). Probing regional landform controls of cloudiness and precipitation in the Central Andes using satellite data. *Journal of Hydrometeorolgy, 10*, 167–182. https://doi.org/10.1175/2008JHM973.1.

Houze, R. A. (2012). Orographic effects of precipitating clouds. *Reviews of Geophysics, 50*, RG1001. https://doi.org/10.1029/2011RG000365.

Huffman, G. J., Bolvin, D. T., Nelkin, E. J., Wolff, D. B., Adler, R. F., Gu, G., Hong, Y., Bowman, K. P., & Stocker, E. F. (2007). The TRMM multisatellite precipitation analysis (TMPA): Quasi-global, multiyear, combined-sensor precipitation estimates at fine scales. *Journal of Hydrometeorology, 8*, 38–55. https://doi.org/10.1175/JHM560.1.

Huffman, G. J., Bolvin, D. T., & Nelkin, E. J. (2018). Integrated Multi-satellitE Retrievals for GPM (IMERG) technical documentation. NASA/GSFC, 31 pp. Available at https://pmm.nasa.gov/sites/default/files/document_files/IMERG_ATBD_V5.2_0.pdf. Last accessed 4 Dec 2018.

Iguchi, T., Seto, S., Meneghini, R., Yoshida, N., Awaka, J., Le, M., Chandrasekar, V., & Kubota, T. (2017). *GPM/DPR level-2 algorithm theoretical basis document, JAXA-NASA* (Technical Report), 81 pp. Available online at https://pmm.nasa.gov/resources/documents/gpmdpr-level-2-algorithm-theoretical-basis-document-atbd. Last accessed 9 Oct 2018.

Kollias, P., & Albrecht, B. A. (2005). Why the melting layer radar reflectivity is not bright at 94 GHz. *Geophysical Research Letters, 32*(24), L24818. https://doi.org/10.1029/2005GL024074.

Kollias, P., Albrecht, B. A., & Marks, F. (2002). Why Mie? Accurate observations of vertical air velocities and raindrops using a cloud radar. *Bulletin of the American Meteorological Society, 83*, 1471–1484. https://doi.org/10.1175/BAMS-83-10-1471.

Lowman, L. E. L., & Barros, A. P. (2014). Investigating links between climate and orogeny in the Central Andes: Coupling erosion and precipitation using a physical- statistical model. *Journal of Geophysical Research, 119*, 1322–1353. https://doi.org/10.1002/2013JF002940.

Lowman, L. E. L., & Barros, A. P. (2016). Interplay of drought and tropical cyclone activity in SE US gross primary productivity. *Journal of Geophysical Research, 120*, 1540–1567. https://doi.org/10.1002/2015JG003279.

Lowman, L. E. L., Wei, T. M., & Barros, A. P. (2018). Rainfall variability, wetland persistence, and water-carbon cycle coupling in the Upper Zambezi River Basin in Southern Africa. *Remote Sensing, 10*(5), 692. https://doi.org/10.3390/rs10050692.

Marzano, F. S., Roberti, L., Di Michele, S., Mugnai, A., & Tassa, A. (2003). Modeling of apparent radar reflectivity due to convective clouds at attenuating wavelengths. *Radio Science, 38*(1), 1002. https://doi.org/10.1029/2002RS002613.

Matrosov, S. Y., Battaglia, A., & Rodriguez, P. (2008). Effects of multiple scattering on attenuation-based retrievals of stratiform rainfall from CloudSat. *Journal of Atmospheric and Oceanic Technology, 25*, 2199–2208. https://doi.org/10.1175/2008JTECHA1095.1.

NAS. (2018). *Thriving on our changing planet: A decadal strategy for Earth observation from space*. Washington, DC: The National Academies Press. Available at https://www.nap.edu/catalog/24938/thriving-on-our-changing-planet-a-decadal-strategy-for-earth. Last accessed 14 Dec 2018.

Nellemann, C., MacDevette, M., Manders, T., Eickhout, B., Svihus, B., Prins, A., & Kaltenborn, B. (Eds.). (2009). *The environmental food crisis. The environment's role in averting future food crises. A UNEP rapid response assessment*. Arendal: UNDP. Available at http://www.grida.no/publications/154. Last accessed 14 Dec 2018.

Prat, O. P., & Barros, A. P. (2007). A robust solution for the stochastic collection-breakup equation in warm rain. *Journal of Applied Meteorology, 46*, 1480–1497. https://doi.org/10.1175/JAM2544.1.

Prat, O. P., & Barros, A. P. (2009). Exploring the transient behavior of Z-R relationships – Implications for radar rainfall estimation. *Journal of Applied Meteorology and Climatology, 48*, 2127–2143. https://doi.org/10.1175/2009JAMC2165.1.

Prat, O. P., & Barros, A. P. (2010a). Ground observations to characterize the spatial gradients and vertical structure of orographic precipitation – Experiments in the inner region of the Great Smoky Mountains. *Journal of Hydrology, 391*, 141–196. https://doi.org/10.1016/j.jhydrol.2010.07.013.

Prat, O. P., & Barros, A. P. (2010b). Assessing satellite-based precipitation estimates in the Southern Appalachian Mountains using rain gauges and TRMM PR. *Advances in Geosciences, 25*, 143–153. https://doi.org/10.5194/adgeo-25-143-2010.

Prat, O. P., Barros, A. P., & Testik, F. (2012). On the influence of raindrop collision outcomes on equilibrium size distributions. *Journal of the Atmospheric Sciences, 69*, 1534–1546. https://doi.org/10.1175/JAS-D-11-0192.1.

Rao, M. S., & Theon, J. S. (1977). New features of global climatology revealed by satellite-derived oceanic rainfall maps. *Bulletin of the American Meteorological Society, 58*, 1285–1288. https://doi.org/10.1175/1520-0477(1977)058<1285:NFOGCR>2.0.CO;2.

Shrestha, P., Barros, A. P., & Khlystov, A. (2010). Chemical composition and aerosol size distribution of the middle mountain range in the Nepal Himalayas during the 2009 pre-monsoon season. *Atmospheric Chemistry and Physics, 10*, 11605–11621. https://doi.org/10.5194/acp-10-11605-2010.

Shrestha, P. A., Barros, P., & Khlystov, A. (2012). CCN estimates from bulk hygroscopic growth factors of ambient aerosols during the pre-monsoon season over Central Nepal. *Atmospheric Environment, 67*, 120–129. https://doi.org/10.1016/j.atmosenv.2012.10.042.

Skofronick-Jackson, G., Petersen, W. A., Berg, W., Kidd, C., Stocker, E. F., Kirschbaum, D. B., Kakar, R., Braun, S. A., Huffman, G. J., Iguchi, T., Kirstetter, P.-E., Kummerow, C. D., Meneghini, R., Oki, R., Olson, W. S., Takayabu, Y. N., Furukawa, K., & Wilheit, T. T. (2017). The Global Precipitation Measurement (GPM) mission for science and society. *Bulletin of the American Meteorological Society, 98*, 1679–1695. https://doi.org/10.1175/BAMS-D-15-00306.1.

Smyth, T. J., & Illingworth, A. J. (1998). Correction for attenuation of radar reflectivity using polarization data. *Quarterly Journal of the Royal Meteorological Society, 124*, 2393–2415. https://doi.org/10.1002/qj.49712455111.

Tao, K., & Barros, A. P. (2009). Fractal downscaling of satellite precipitation products for hydrometeorological applications. *Journal of Atmospheric and Oceanic Technology, 27*, 409–427. https://doi.org/10.1175/2009JTECHA1219.1.

Tao, J., & Barros, A. P. (2013). Prospects for flash flood forecasting in mountainous regions – An investigation of tropical storm fay in the southern Appalachians. *Journal of Hydrology, 506*, 69–89. https://doi.org/10.1016/j.jhydrol.2013.02.052.

Tao, J., & Barros, A. P. (2014). Coupled prediction of flood response and debris flows initiation during warm and cold season events in the Southern Appalachians, USA. *Hydrology and Earth System Sciences, 18*, 1–14. https://doi.org/10.5194/hess-18-1-2014.

Tao, J., Wu, D., Gourley, J. J., Zhang, S. Q., Crow, W., Peters-Lidard, C., & Barros, A. P. (2016). Operational hydrological forecasting during the IPHEx-IOP campaign – Meet the challenge. *Journal of Hydrology, 541*, 434–456. https://doi.org/10.1016/j.jhydrol.2016.02.019.

Tapiador, F. J., Hou, A. Y., Castro, M., Checa, R., Cuartero, F., & Barros, A. P. (2011). Precipitation estimates for hydroelectricity. *Energy and Environmental Science, 4*, 4435–4448. https://doi.org/10.1039/C1EE01745D.

Testik, F. Y., & Barros, A. P. (2007). Toward elucidating the microstructure of warm rainfall: A survey. *Reviews of Geophysics, 45*, RG2003. https://doi.org/10.1029/2005RG000182.

Testik, F. Y., Barros, A. P., & Bliven, L. F. (2011). Toward a physical characterization of raindrop collision outcome regimes. *Journal of the Atmospheric Sciences, 68*, 1097–1113. https://doi.org/10.1175/2010JAS3706.1.

Viviroli, D., Archer, D. R., Buytaert, W., Fowler, H. J., Greenwood, G. B., Hamlet, A. F., Huang, Y., Koboltschnig, G., Litaor, M. I., López-Moreno, J. I., Lorentz, S., Schädler, B., Schreier, H., Schwaiger, K., Vuille, M., & Woods, R. (2011). Climate change and mountain water resources: Overview and recommendations for research, management and policy. *Hydrology and Earth System Sciences, 15*, 471–504. https://doi.org/10.5194/hess-15-471-2011.

Wilheit, T. T. (1986). Some comments on passive microwave measurement of rain. *Bulletin of the American Meteorological Society, 67*, 1126–1132. https://doi.org/10.1175/1520-0477(1986)067<1226:SCOPMM>2.0.CO;2.

Wilson, A. M., & Barros, A. P. (2014). An investigation of warm rainfall microphysics in the Southern Appalachians: Orographic enhancement via low-level seeder-feeder interactions. *Journal of the Atmospheric Sciences, 71*, 1783–1805. https://doi.org/10.1175/JAS-D-13-0228.1.

Wilson, A. M., & Barros, A. P. (2015). Landform controls on low level moisture convergence and the diurnal cycle of warm season orographic rainfall in the Southern Appalachians. *Journal of Hydrology, 531*, 475–493. https://doi.org/10.1016/j.jhydrol.2015.10.068.

Wilson, A. M., & Barros, A. P. (2017). Orographic land-atmosphere interactions and the diurnal cycle of low level clouds and fog. *Journal of Hydrometeorology, 18*, 1513–1533. https://doi.org/10.1175/JHM-D-16-0186.1.

Chapter 31
Integrated Multi-satellite Evaluation for the Global Precipitation Measurement: Impact of Precipitation Types on Spaceborne Precipitation Estimation

Pierre-Emmanuel Kirstetter, Walter A. Petersen, Christian D. Kummerow, and David B. Wolff

Abstract An integrated multi-sensor assessment is proposed as a novel approach to advance satellite precipitation validation in order to provide users and algorithm developers with an assessment adequately coping with the varying performances of merged satellite precipitation estimates. Gridded precipitation rates retrieved from space sensors with quasi-global coverage feed numerous applications ranging from water budget studies to forecasting natural hazards caused by extreme events. Characterizing the error structure of satellite precipitation products is recognized as a major issue for the usefulness of these estimates. The Global Precipitation Measurement (GPM) mission aims at unifying precipitation measurements from a constellation of low-earth orbiting (LEO) sensors with various capabilities to detect, classify and quantify precipitation. They are used in combination with geostationary observations to provide gridded precipitation accumulations. The GPM Core Observatory satellite serves as a calibration reference for consistent precipitation retrieval algorithms across the constellation. The propagation of QPE uncertainty from LEO active/passive microwave (PMW) precipitation estimates to gridded QPE is addressed in this study, by focusing on the impact of precipitation typology on QPE from the Level-2 GPM Core Observatory Dual-frequency Precipitation Radar (DPR) to the Microwave Imager (GMI) to Level-3 IMERG precipitation over the

P.-E. Kirstetter (✉)
School of Meteorology and School of Civil Engineering and Environmental Sciences and Advanced Radar Research Center, University of Oklahoma and NOAA/National Severe Storms Laboratory, Norman, OK, USA
e-mail: pierre.kirstetter@noaa.gov

W. A. Petersen
NASA, Marshall Space Flight Center, Huntsville, AL, USA

C. D. Kummerow
Department of Atmospheric Science, Colorado State University, Ft. Collins, CO, USA

D. B. Wolff
NASA/GSFC, Wallops Flight Facility, Wallops Island, VA, USA

© Springer Nature Switzerland AG 2020   583
V. Levizzani et al. (eds.), *Satellite Precipitation Measurement*, Advances in Global Change Research 69, https://doi.org/10.1007/978-3-030-35798-6_7

Conterminous U.S. A high-resolution surface precipitation used as a consistent reference across scales is derived from the ground radar-based Multi-Radar/Multi-Sensor. While the error structure of the DPR, GMI and subsequent IMERG is complex because of the interaction of various error factors, systematic biases related to precipitation typology are consistently quantified across products. These biases display similar features across Level-2 and Level-3, highlighting the need to better resolve precipitation typology from space and the room for improvement in global-scale precipitation estimates. The integrated analysis and framework proposed herein applies more generally to precipitation estimates from sensors and error sources affecting low-earth orbiting satellites and derived gridded products.

Keywords Precipitation · Rainfall · Evaluation · Validation · GPM · Radar · DPR · IPWG · IMERG · GPROF · Microwave · MRMS · Particle size distribution · Non-uniform beam filling · Convective · Stratiform · Convective percent index

31.1   Introduction

Precipitation is key to the Earth hydrologic and energy fluxes through its occurrence, type, and quantity. A host of scientific questions impacting society are concerned with the distribution of precipitation characteristics, from extreme events such as droughts and hurricanes to how the availability of fresh water evolves under a non-stationary climate. Precipitation physical processes take place over a range of spatial and temporal scales and drive its highly variable intermittency, intensity, areal extent, and duration. This variability poses challenges for observations specifically from remote sensing for hydrologic, meteorological, and climatic applications. Their quasi-global coverage promotes the use of satellite-based quantitative precipitation estimates (QPE) for such purposes. However, converting satellite measurements into QPE poses challenges, as it depends on the spatial heterogeneity of the precipitation fields, the indirect nature of the measurement, the sensor resolution and sensitivity, and the retrieval algorithm. Hence satellite-based precipitation estimates suffer from poorly characterized and quantified sources of uncertainty, which currently limit their assimilation into hydrologic and atmospheric models (Bauer et al. 2011; Stephens and Kummerow 2007; Weng et al. 2007). To improve the satellite estimates and maximize their usability, their uncertainty must be evaluated in terms of precipitation characteristics including intermittency, distribution of types (e.g., stratiform, convective, snow, hail) and rates, as highlighted by the International Precipitation Working Group (IPWG; see http://ipwg.isac.cnr.it/, last accessed 30 Nov. 2018; Levizzani et al. 2018) (Turk et al. 2008; Yang et al. 2006; Zeweldi and Gebremichael 2009; Sapiano and Arkin 2009; Kummerow et al. 2006; Wolff and Fisher 2009; Grimes and Diop 2003; Lebel et al. 2009). Comprehensive characterization of the satellite precipitation error structure relies on ground-validation research to ensure proper accuracy of spaceborne QPE missions (Petersen and Schwaller 2008) like the former Tropical Rainfall Measurement Mission (TRMM; http://trmm.gsfc.nasa.gov, last accessed 11 Dec. 2018) and the current

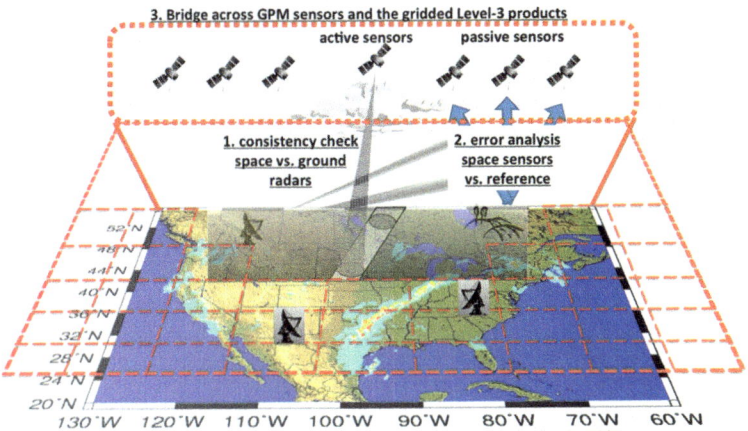

Fig. 31.1 Research framework and overview flowchart to bridge from the Global Precipitation Measurement mission core satellite to the combined gridded IMERG product using ground radar-based Multi-Radar/Multi-Sensor QPE. An example of Multi-Radar/Multi-Sensor instantaneous precipitation rates at 0725 UTC on 11 April 2011 is shown

Global Precipitation Measurement Mission (GPM; http://gpm.gsfc.nasa.gov, last accessed 30 Nov. 2018).

Current high-resolution and gridded satellite-based QPE have been developed by combining active and passive microwave (PMW) and infrared (IR) sensors into multi-sensor precipitation retrievals. PMW-based estimates from Low Earth Orbiting (LEO) satellites provide higher accuracy but limited spatial and temporal resolution. They are used to calibrate IR-based estimates from geosynchronous Earth-orbiting (GEO) satellite platforms with comparatively higher spatio-temporal resolution and lower latency (Joyce et al. 2004; Huffman et al. 2007; Ushio et al. 2006; Ebert 2007). As of today, our understanding on how uncertainties originate from LEO platforms and propagate through such combined products is critically lacking, and there is a need for comparison across product levels (Fig. 31.1). While single satellite precipitation products are typically evaluated independently in the literature, a consistent end-to-end error quantification tracking the uncertainty from Level-2 through Level-3 products is needed to fully understand the errors across scales and assess the room for improvement in global precipitation estimation.

During the last couple of decades, individual satellite-based precipitation products have been evaluated over various regions (e.g., Carr et al. 2015; Derin et al. 2016; Derin and Yilmaz 2014; Dinku et al. 2010; Golian et al. 2015; Grecu and Anagnostou 2001), surfaces (e.g., Bhuiyan et al. 2017), precipitation regimes (e.g., Kirstetter et al. 2012, 2013a, b, 2014, 2015b; Smalley et al. 2017; Kidd et al. 2018; Khan et al. 2018; O and Kirstetter 2018), seasons (e.g., Gebregiorgis et al. 2017, 2018; Tang et al. 2014), scales (e.g., Tan et al. 2017), precipitation intensities (e.g., Kirstetter et al. 2012; Panegrossi et al. 2016; Wolff and Fisher 2009), and for applications in flood prediction (e.g., Hossain and Anagnostou 2004; Vergara et al. 2014) because their effective use necessitates the characterization of their uncertainties at fine spatio-temporal scales (e.g., Anagnostou 2004, Huffman et al. 2007, Joyce et al. 2004, Munchak and Skofronick-Jackson 2013, Tian et al. 2007, Sorooshian et al. 2011, Kirstetter et al. 2018). However, the extensive body of literature on satellite precipitation validation generally provides limited insight in their error characteristics for several reasons. Common assessment practices typically use a surface-based precipitation reference and bulk comparison metrics (e.g., correlation, bias) to assess performances over a given spatial and temporal domain. First, an objective assessment requires an independent precipitation reference often derived from ground-based sensors, usually gauges. Independence may not be verified when gauges already incorporated in the satellite product are re-used to assess its accuracy. Second, metrics such as correlation, bias, contingency statistics are often applied without necessarily checking the relevance of such criteria. For example, the definition of bias may be ambiguous as it can be defined as an additive satellite QPE-to-reference difference or a multiplicative difference (ratio), sometimes based on conditional (positive) precipitation values. The linear correlation is generally insufficient to describe the non-linear and heteroscedastic dependence structure between the satellite QPE and the reference. Third, the satellite product is often assumed to be consistent and display homogeneous performances over the spatial and temporal domain of comparison. Bulk metrics (correlation, bias, contingency, etc.) are computed over samples actually gathering a variety of precipitation characteristics (intermittency, typology, rates) for which the satellite algorithm (or combination of algorithms for merged products) is likely to behave differently. More generally the comparison is always performed with precipitation estimates ambiguously derived from the satellite sensor observation through the retrieval algorithm and associated assumptions. Individual PMW/IR retrievals are underconstrained by nature and sensitive to unobserved atmospheric parameters (Stephens and Kummerow 2007). The combined products inherit the varying PMW/IR performances and create additional uncertainties with temporal/spatial resampling. Hence bulk error metrics depict averaged space/time properties while the errors tend to be non-stationary and sensitive to parameters not accounted for in the assessment formulation. Fourth, the representativeness of any overall satellite QPE assessment or error model is confined to the time and space domain over which it is performed. It tends to be specific to the satellite sensor (e.g., resolution), the retrieval algorithm (and associated version), the space-time-scale and the accuracy of the reference, with limited applicability for other precipitation regimes, regions, products, etc. As a result, the actual benefit of these analyses to satellite precipitation users and developers is limited. Integrated multi-sensor assessment is necessary to track the origin of uncertainties and their propagation through various Level-2 active, passive then Level-3 merged precipitation estimates. More informative assessment and information to algorithm improvement can be gained by stratifying (conditioning) the assessment according to relevant factors driving the state of the satellite estimation error. Hence targeting the most significant factors is essential to characterize uncertainties in satellite QPE and lead to a generalization of their assessment.

Generic issues have been identified for both LEO active and passive and subsequent gridded satellite-based precipitation retrieval algorithms and motivate ongoing and future research such as variability of the precipitation inside the resolution volumes (Non-Uniform Beam Filling effects - NUBF), precipitation typology and phase. In this chapter we focus on the impact of precipitation typology. While sensors measurements from ground-based radars, spaceborne radar and PWM and IR sensors differ in terms of frequency, polarization state, beam geometry, and incidence angle, they are physically consistent through the identification of hydrometeors and estimation of particle size distribution (PSD). The precipitation retrieval errors caused by the variability in the cloud vertical structure and the observed signal (e.g., reflectivity for radars, brightness temperatures for PMW and IR sensors)-to-precipitation relationship are related due to their dependence on the underlying precipitation microphysics. Precipitation type is a fundamental characteristic that drives the interpretation of the observed signal for precipitation estimation through the particle size distribution (PSD), hydrometeor properties and their evolution through precipitation microphysical processes. The satellite algorithms' capabilities to classify precipitation systems (e.g., Grams et al. 2016) therefore present great potential for the generalization of the uncertainty characteristics, specifically at coarser scales. Within the Global Precipitation Measurement (GPM) mission (https://www.nasa.gov/mission_pages/GPM/main/, last accessed 30 Nov. 2018; Hou et al. 2014, Skofronick-Jackson et al. 2017), precipitation type is currently used only to constrain the space-borne radar precipitation retrievals. Since subsequent passive sensor retrievals and combined precipitation products do not include this fundamental characteristic, there is an interest in documenting how the absence of constraints on precipitation type impacts these precipitation products.

Targeting significant uncertainty factors for a quantitative and detailed characterization necessitate working at the primary satellite QPE scale for LEO (Level 2) through to gridded precipitation products (Level-3). This task is often impaired by the difficulty of obtaining a consistent reference precipitation commensurate with the various scales of such products. To the best of our knowledge, no satellite assessment has been designed at the fine space–time resolutions of these primary satellite QPE scales. We propose an original framework to tackle these challenges. The problem is addressed by comparing the satellite QPE overall accuracy with respect to an external, independent reference precipitation product adapted to each type of space-based sensors and product. In order to match the resolution of various precipitation estimates, the primary resolution of the reference needs to finer than any satellite footprint or resolution. The reference has also to perform better than the space sensors regarding precipitation detection to ensure proper evaluation, and a correct reference precipitation classification (type) is required to target the physical factors contributing to erroneous satellite precipitation retrievals. A high-resolution surface rainfall product is used as a consistent reference across scales for robust comparison and evaluation over the CONUS (Fig. 31.1). It is derived from the Multi-Radar/Multi-Sensor (MRMS) developed at the NOAA National Severe Storms Laboratory and the University of Oklahoma from the NEXRAD ground-based radar network observations. Joint precipitation observations with NEXRAD and from space sensors provide unique opportunities for comparison of QPE as estimated from various sensors. Measurements from NEXRAD and GPM are physically consistent through the identification of hydrometeors and estimation of particle size distribution. We build on previous research performed on the ground-based MRMS data at unprecedented high resolution and accuracy and the high-resolution spaceborne precipitation measuring techniques to analyze the impact and propagation of uncertainty related to precipitation typology from the GPM Dual-frequency Precipitation Radar (DPR) to passive sensors such as GPM Microwave Imager (GMI) to the gridded Integrated Multi-Satellite Retrievals for GPM (IMERG).

The aim of this study is to propose a novel approach for satellite QPE assessment by addressing the propagation of uncertainties related to precipitation type from LEO platforms through combined precipitation products. End-to-end integrated error quantification tracking the uncertainty from Level-2 through Level-3 precipitation products is particularly relevant for understanding the origin and impact of uncertainty. It addresses the critical problem that precipitation algorithm developers and users need more than just an overall assessment to adequately cope with the varying performances of precipitation products. To the best of our knowledge this approach offers, for the first time, uncertainty assessment across scales and products. It advances practices in the evaluation of remote sensing precipitation estimates by consistently analyzing precipitation estimation across levels. The focus on precipitation types is particularly relevant in the context of the current GPM mission and assesses the potential benefit of introducing precipitation type constrains to the PMW Level-2 QPE and subsequent Level-3 precipitation. Working at the primary satellite QPE scale benefits from the explicit integration of factors directly impacting the uncertainty. This framework is tested over the conterminous United States (CONUS) covered by the NEXRAD radars from which MRMS is derived (Fig. 31.1). This study uses a period of 2 years (from 2014 to 2016) of satellite and ground-based precipitation observations to obtain representative samples characterizing precipitation for various conditions of climatology, occurrence, type and rate.

DPR, GMI, IMERG data and steps required to refine the MRMS ground-based precipitation to arrive at a consistent reference precipitation across scales used for comparisons are presented in Sect. 31.2. Section 31.3 assesses the impact of precipitation typology on DPR, GMI and IMERG precipitation retrievals. The chapter is closed with concluding remarks in Sect. 31.4.

31.2 Spaceborne and Ground-Based Precipitation Datasets

Following the Tropical Rainfall Measuring Mission (TRMM), the Global Precipitation Measurement mission led by NASA and JAXA aims to provide a comprehensive description of precipitation at the global scale (Hou et al. 2014). It consists of a core satellite in non-sun-synchronous orbit to serve as a physics observatory to gain insights into precipitation systems and as a calibration reference to unify precipitation estimates from a constellation of research and operational satellites involving passive microwave (PMW) sensors. The core satellite carries the Ku/Ka-band Dual-

Frequency Precipitation Radar (DPR) measuring reflectivity profiles and the GPM microwave radiometer (GMI) measuring brightness temperatures (T_b). The complementary Ku- and Ka-band measurements provide estimates of the shape and size of hydrometeors and higher sensitivity for detection of light rain and snow than the single frequency radar capabilities of the TRMM Precipitation Radar (PR; Ku-band). The GMI extends the range of frequencies on the TRMM Microwave Imager to provide brightness temperatures covering the range of frequencies on GPM constellation radiometers. Precipitation estimates from the constellation sensors enable the creation of a quasi-global-scale and gridded merged precipitation product, the Integrated Multi-Satellite Retrievals for GPM (IMERG).

For each precipitation product (DPR, GMI, or IMERG), the satellite rainfall estimate $R(A, T)$ is compared with a reference rainfall $R_{ref}(A, T)$ over a spatial domain A (satellite footprint or pixel for DPR and GMI or IMERG, respectively) over the time period T (snapshot or 30-min timescale for DPR and GMI or IMERG, respectively). The reference rainfall $R_{ref}(A, T)$ is a proxy of the true (and unknown) area-averaged rainfall rate over the same area A and time period T. While we do not know the truth at the ground we need to correctly assess the reference's uncertainties for a reliable quantitative comparison of precipitation products.

31.2.1 Dual-Frequency Phased Array Radar

The DPR version 5 product used in this work is described in Iguchi et al. (2009, 2017). It provides 3-D reflectivity and 2-D precipitation rate fields at ground. The product classifies snowfall and rain into stratiform and convective (Awaka et al. 2007). The DPR algorithm relies on a hybrid attenuation correction method that combines the surface reference technique and Hitschfeld-Bordan method (Iguchi et al. 2000; Meneghini et al. 2000, 2004; Takahashi et al. 2006). It uses models to describe the hydrometeor particle size distributions (PSD) depending on the precipitation type, which are adjusted to match the observed dual-frequency Path Integrated Attenuation. The DPR is a well-calibrated and very stable radar. The scan geometry and sampling rate of the DPR lead to footprints spaced approximately 5 km across- and along-track, over 245-km and 120-km wide swaths at Ku-band and Ka-band respectively, centered within the 885-km-wide GMI swath. DPR observations provide a more direct measurement of the precipitation rates than GMI. The minimum theoretical detectable precipitation rate by the DPR is fixed by its sensitivity and is better than 18 dBZ or 0.5 mm h^{-1}, at Ku-band and around 12 dBZ at Ka-band or about 0.2 mm h^{-1}. Through the combined DPR and GMI-based retrievals, the DPR may be regarded as a "calibrator" of the PMW precipitation estimates, while the passive sensors, already a component of several polar-orbiting observatories (e.g., SSMI, SSMI-S, AMSR-2), provide more extensive sampling of precipitation events over the globe. Uncertainty sources affecting the DPR-based precipitation estimates propagate into PMW estimates.

DPR observations provide a more direct measurement of the precipitation rates than PMW sensors. However, similar to the TRMM-PR, primary errors in rainfall retrievals are mainly attributed to attenuation correction of the radar signal and conversion from reflectivity-to-precipitation intensity. Both involve incorrect physical assumptions related to snowfall and convective versus stratiform rainfall classification and assumed particle size distribution. Contamination by surface backscatter and NUBF are other challenges in correctly interpreting the radar signal into precipitation (Wolff and Fisher 2008; Iguchi et al. 2009). Retrieval of precipitation rate requires knowledge of precipitation type, on which the PSD depends (Battan 1973; Awaka et al. 2007), and which has profound impacts on the accuracy of the quantitative retrievals. It drives the vertical model of microphysics used to correct for the attenuation of the DPR signal, to estimate the vertical profile of reflectivity, and the rainfall rate at ground (Iguchi et al. 2009). Satellite precipitation classification relies partly on subjective analysis based on interpretation of reflectivity spatial variability. While the DPR algorithm classifies rain into three categories: convective, stratiform and others, there is actually a mixture of processes and types within the footprint (Kirstetter et al. 2015b). Classification capabilities and their impact have not been evaluated extensively (Porcacchia et al. 2019) .

Our dataset covers 30 months (June 2014–October 2016) of satellite overpasses over the CONUS. The variable precipRateESurface (estimated surface rain) from the matched scan GPM-DPR/Ka-Ku (hereafter DPR) was extracted from the DPR files as the DPR surface QPE. The use of gridded MRMS data for reference provided a large sample size totaling 1,012,151 non-zero DPR-reference pairs including 798,155 pixels classified as stratiform by the DPR and 196,801 convective pixels. Along with the GPM-DPR/Ka-Ku, other products used in this study are the GPM-DPR/Ku and GPM-DPR/Ka. A TRMM-PR dataset is also used for comparison, covering the period from March to October 2011 (Kirstetter et al. 2015b).

31.2.2   GPM Microwave Imager

The GMI measures brightness temperatures T_b at 13 microwave frequencies ranging from 10 GHz to 183 GHz, which are used by the Goddard Profiling Algorithm (GPROF) to estimate surface rain rates (Kummerow 2017). The GPROF uses a Bayesian approach and an a priori database populated with combined DPR and GMI-based precipitation. Note the DPR precipitation types are inputs for the combined DPR and GMI-based precipitation retrievals hence impact the GPROF a priori database.

Sources of systematic errors in GPROF are related to the a priori database and ancillary information used to subset the a priori database. Over land the hydrometeor information is entangled with highly variable surface characteristics in T_b. The main information used for the retrieval comes from the scattering channels because of the radiometrically warm land surface and variable surface emissivity. As the T_b at these frequencies are mostly sensitive to the scattering processes in the higher regions of the cloud (Wilheit et al. 2003), the available information for the retrieval is not directly related to surface precipitation (the correlation between ice aloft and surface rainfall is variable). Hence, there are particular needs for a detailed assessment of GPROF performances especially regarding the quantitative retrieval (Gopalan et al. 2010). Currently GPROF does not condition precipitation retrievals by precipitation types (convective/stratiform), although recent works indicate that atmospheric sta-bility and precipitating system structure impact its performance (Petković and Kummerow 2017; Petković et al. 2018; Henderson et al. 2017).

The GPM constellation includes a number of satellites with GMI-like radiometers or microwave sounding instruments, including the DMSP F16, F17, F18 and F19 (US; imager), GCOM-W1 (JAXA; imager), Megha–Tropiques (CNES/ ISRO; sounder), MetOp A and B (EUMETSAT; sounder) and NOAA 18, 19, 20 and NPP (NOAA; sounder) satellites. The GPM core satellite sensors provide self-consistent radiometric observations across the constellation. The DPR-calibrated GMI is, in turn, used as the calibrator for other passive microwave sensors as GPROF is consistently applied on the GPM constellation PMW sensors that collec-tively enable the creation of quasi-global-scale combined precipitation products. Thus, DPR and GMI have fundamental impacts on satellite-based precipitation estimates from other low Earth-orbiting passive microwave measurements and a number of satellite-based, high-resolution precipitation products.

The GPROF-GMI V05 product is used here. The datasets for GMI covers the same period as for DPR (from June 2014 to October 2016) with a sample size totaling 3,782,453 non-zero GPROF-GMI-reference pairs.

31.2.3   *Integrated Multi-satellitE Retrievals for GPM*

The radiometer-based precipitation estimates are incorporated into the merged precipitation product IMERG by creating a uniformly gridded precipitation product at the global scale. IMERG blends complementary satellite-based precipitation estimates, i.e. from IR and PMW sensors (Huffman et al. 2018). IR data have a more indirect relationship with surface precipitation occurrence and rates than PMW observations (Kirstetter et al. 2018). IMERG take advantage of the complementary more accurate GPROF retrievals from all constellation sensors and more frequent IR-based Precipitation Estimation from the Remotely Sensed Information using Artificial Neural Networks-Cloud Classification System algorithm (PERSIANN-CCS; Hong et al. 2004) to produce near-real time estimates at a high spatial and temporal resolution (0.1°, 0.5 h) with quasi-global coverage. To increase their coverage PMW estimates are morphed with the Climate Prediction Center (CPC) Morphing-Kalman Filter cloud motion from infrared imagery (Joyce et al. 2004) before being merged with PERSIANN-CCS precipitation. IMERG has three sepa-rate products called Early with 4 h latency used e.g. for flood and landslide monitoring, Late with 12 h latency for drought monitoring and agricultural applica-tions, and Final with 2 months latency for research applications. The IMERG Early and Late products are adjusted to climatological sets of coefficients and IMERG final product to gauge observations.

Whereas IMERG provides an approach to high-resolution precipitation estimation, uncertainty increases with resolution due to the combined effect of higher variability of precipitation and more indirect information provided by higher-resolution IR relative to PMW remote sensing. While fine-resolution quantitative precipitation estimates (QPEs) are needed for many applications, they come at the expense of degraded accuracy. Satellite estimates currently display limits in accurately capturing extreme precipitation events often related to convective precipitation, which impedes short-term warning systems based on satellite precipitation.

IMERG version V05 Late is employed in this study. The datasets for IMERG covers the same period (from June 2014 to October 2016) with a sample size totaling 11,796,935 non-zero IMERG-reference pairs. The information about the data used in this study is archived in the following fields in the IMERG files: *precipCal* for the merged products, *HQprecipitation* for the MW component, and *IRprecipitation* for the IR component (Huffman et al. 2018).

31.2.4 Ground-Based Reference Precipitation

Over the US, Kirstetter et al. (2012, 2014) set up a framework to compare the GPM constellation sensors using the ground-based Multi-Radar/Multi-Sensor (MRMS) system (Zhang et al. 2016). It provides an independent reference for space-based precipitation products regarding resolution, accuracy and sample size, and a consistent database in time and space up to 55°N covering various geographical (plains, mountains) and meteorological (subtropical to midlatitudes) conditions for robust comparison (Fig. 31.1). To perform end-to-end uncertainty characterization the ground reference derived from MRMS is designed to bridge the scales of multiples precipitation measurements from local gauges to regional ground-based radar coverage to satellite DPR and GMI field of view (FOV) and swaths to the resolution of IMERG gridded products. It combines the fine spatial and temporal depiction of precipitation variability by ground-based radars with the quantitative accuracy of gauges and match this information at the scale of the satellite precipitation estimates.

The NOAA/NSSL and University of Oklahoma (OU) MRMS system (https://mrms.nssl.noaa.gov/, last accessed 30 Nov. 2018) incorporates data from all NEXRAD S-band polarimetric radars and automated rain gauge networks in the CONUS (Zhang et al. 2016). The NEXRAD radars' sensitivity (about −5 dBZ at a distance of 50 km) allows provides high performance for rainfall detection at least at confined ranges from the radar (e.g., <75 km) compared to satellite sensors. At S-band the radar signal is relatively unattenuated. Dual-polarization improves the radar data quality and enables targeting specific microphysical situations where the ground-radar estimates are the most reliable (Ryzhkov et al. 2005, 2014;

Chandrasekar et al. 2008). The MRMS system generates and high-resolution 3-D reflectivity mosaic grids and a suite of severe weather and QPE products at a 0.01° horizontal resolution and 2-min update cycle: rain types, precipitation phase and rate, freezing level height, etc. The geographical extension and time period covered by MRMS compiles a broad sample of precipitation situations (stratiform, convective, orographically enhanced precipitation, etc.) from a stable and uniform observation system (Kirstetter et al. 2012, 2014). At the hourly time step, MRMS adjusts radar estimates with automated rain gauge networks using a spatially variable bias multiplicative factor. A radar quality index (RQI) is produced to represent the radar QPE uncertainty associated with reflectivity changes with height and near the melting layer (Zhang et al. 2011).

One should note that it is not possible to "validate" satellite precipitation estimates in a strict sense because independent precipitation estimates with no uncertainty do not exist (Kirstetter et al. 2015a). Yet trustworthy values of the MRMS rainfall estimates within the satellite pixel are needed to evaluate the satellite estimates. A reference matched to satellite sampling properties is derived for Level-2 and Level-3 precipitation. Post-processing complements procedures already in place within the MRMS system to further refines and standardize the reference dataset. Several quality control steps are applied in order to use ground radar data only where confidence in skill is very high and artifacts (beam blockage due to mountain, range effects, etc.) are essentially minimized and deemed negligible. Blending techniques build on the locally reliable rain gauge measurement and the space-time resolution of the radar (Kirstetter et al. 2012, 2013a). A conservative approach is followed by (i) filtering out instances when the radar and gauge have significant quantitative disagreement (i.e., radar-rain gauge ratios outside of the range [0.1–10]) and (ii) by retaining only the best measurement conditions (i.e., no beam blockage and radar beam below the melting layer) using the RQI product as described in Kirstetter et al. (2012, 2013a). These data quality controls standardize the reference product and filter out the less trustworthy MRMS estimates, e.g. radar measurements at far range and in the Rockies. These improvements may not screen out all possible errors in ground-based radar estimates. Kirstetter et al. (2012) showed the increased consistency between TRMM-PR and the MRMS-based reference following sequential MRMS data quality control steps including bias correction using rain gauges and filtering using the RQI product. This finding highlights the importance of matching the scales and refining the accuracy of the reference dataset as much as possible before reaching meaningful conclusions about the satellite sensor's accuracy.

Because of the highly variable precipitation processes, the resolution of the ground reference should match the satellite products resolution to reduce noise when comparing satellite retrievals to $R_{ref}(A, T)$. The spatial resolution of MRMS is finer than any satellite sensor, allowing the resolution of the reference to be specifically adapted to each product using spatial and temporal sampling techniques (Kirstetter et al. 2012). To determine the reference rainfall $R_{ref}(A, T)$ over the sensor pixel A and time period T, a block-MRMS precipitation pixel matches each sensor/product pixel. For Level-2 products the MRMS products closest in time to the GPM core satellite local overpass schedule time are used. The time difference between the MRMS data and satellite data (max 1 min) could add random noise (but not bias) in the comparison, although other factors like the difference in resolution may have more significant impacts. For Level-3 products the MRMS products are aggregated to match the grid spacing at $0.1°$ and the 30-min temporal accumulation. All of the MRMS pixels (rainy and non-rainy) found within a footprint sensor or pixels are located to compute unconditional mean rain rates for the MRMS at the sensor pixel scale. When more than 25% of MRMS pixels have missing values, the data pair is discarded from the comparison. Matched DPR/GMI/IMERG and $R_{ref}(A, T)$ estimates only exist at locations where both the satellite and ground radars have taken actual observations. The satellite products remain untouched, hence preserving their characteristics: the total rainfall amount, the total rainy area, the convective/stratiform contribution and the PDF shapes, and may therefore be compared to the reference at once. An extended description of the reference precipitation regarding types of rainfall within the satellite sensor's FOV or pixel grid is assessed through a Convective Percent Index (CPI) quantifying the volume contribution of convective rainfall to $R_{ref}(A, T)$. Note that the MRMS convective and stratiform classification is indirect since no updraft speed information is currently available. The CPI is expressed in percent between 0% (purely stratiform rainfall within the FOV or pixel) to 100% (purely convective rainfall). CPI values between 0% and 100% indicate mixed precipitation types. Given its spatial extension and native resolution, the MRMS products provide large samples of matched ground-satellite comparison pairs (Kirstetter et al. 2012, 2013a, b, 2014, 2015a, b, 2018). This allows judicious selections of data pairs to monitor the comparison quality. All significant rain fields observed coincidentally by DPR and GMI overpasses and IMERG and the NEXRAD radar network from June 2014 to October 2016 are collected.

Given the highly variable precipitation processes, a consistent reference derived from MRMS is instrumental to bridge in-depth characterization across Level 2 constellation sensors from the GPM core to the constellation to IMERG (Level 3), directly assessing the influence of precipitation types (see Fig. 31.2).

Fig. 31.2 Comparison framework across Level-2 and Level-3 satellite QPE to bridge from the GPM core satellite to IMERG

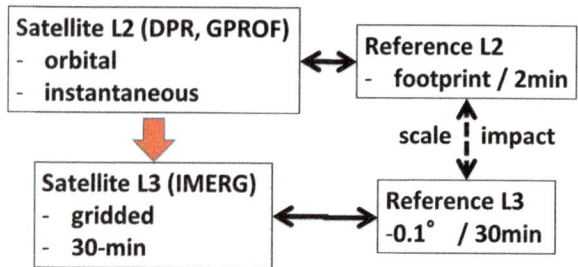

31.3 Impact of Precipitation Typology on Satellite-Based Active, Passive and Merged Precipitation Estimation

Precipitation error characterization is conditioned on precipitation types by taking full advantage of the reference insights into precipitation across scales (from sub-satellite Level-2 FOV to IMERG grid). The state dependency of the error is diagnosed by stratifying the error according to the CPI across sensors and products.

31.3.1 DPR QPE and Precipitation Typology

The DPR algorithm uses different reflectivity-to-precipitation rate relationships in the convective/stratiform profiling components. The CPI continuous classification provides a fine assessment of the DPR categorical classification and its impact on rate retrievals. Figure 31.3 shows the reference and DPR rainfall rate distributions as functions of the convective contribution CPI. All coincident and collocated DPR values are considered and sorted according to the reference sample. Figure 31.3 shows a shift toward higher rainfall rates as CPI increases, as we would expect. Despite these consistencies, we note rain rate distributions indicating higher rainfall rates for the reference compared to those of DPR (i.e., Fig. 31.3a compared to Fig. 31.3b). The dynamic ranges of rain rate distribution are greater for the reference than for the DPR. Such differences, which will undoubtedly result in some bias, could be related to the DPR reflectivity-to-precipitation relationships. The difference

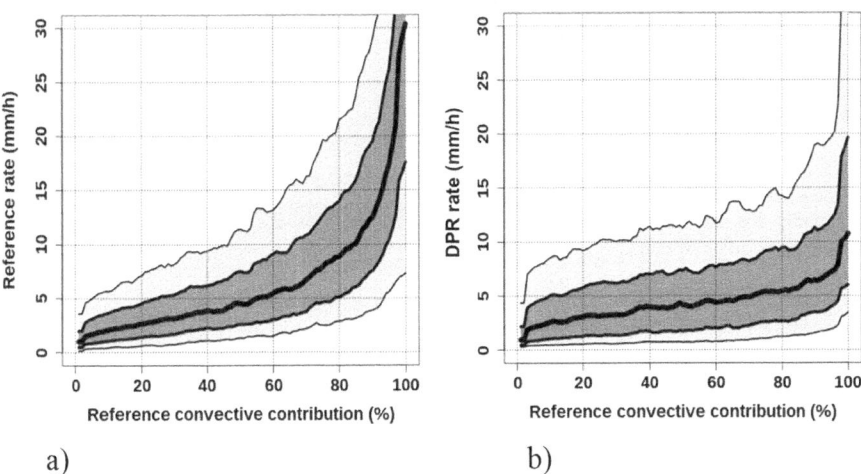

a)   b)

Fig. 31.3 Reference (a) and DPR (b) rainfall rate distributions (mm h^{-1}) as functions of the CPI (%). The thick black line represents the median (50% quantile), the dark grey-shaded region represents the area between the 25 and 75% quantiles, the light grey-shaded region represents the area between the 10 and 90% quantiles

is larger for convective contributions >80%, e.g. for CPI = 90% the conditional reference median rate is 15 mm h^{-1} while the DPR median rate is 7 mm h^{-1}. It is apparent that the DPR profiling algorithm lacks sufficient dynamics to deal with extreme rainfall amounts.

The precipitation type impacts the discrepancies of the DPR relative to the reference. In order to provide some insight into the influence of precipitation typing on the DPR error, the departures of DPR estimates from the reference values are analyzed as functions of the convective contribution on a point-to-point basis. The residuals are defined as the difference between the reference rainfall (R_{ref}) and the satellite estimates (R): $\epsilon = (R - R_{ref})$. Only pairs for which R_{ref} and R are both non-zero are considered in the calculations, so as to remove any discrepancies related to detectability. Figure 31.4 shows the residuals as a function of the convective contribution CPI. All coincident and co-located PR values are considered and sorted according to the reference sample.

The conditional PDFs of residuals ϵ present a high conditional shift from the 0 line and a large conditional spread. The spread increases with CPI, indicating larger uncertainties in quantifying the reference convective rainfall. The panel shows also a tendency to overestimate rain rates for low CPI values (CPI < 30%, the conditional median of residuals is positive), a shift toward lower rainfall rates as CPI increases, and rain rates underestimation for CPI values higher than 50% (negative median of residuals). As an example, the DPR model overestimates at CPI = 20% (mainly stratiform) rain rates with an occurrence of 55% and underestimates at CPI = 100% (convective) rain rates with an occurrence of 80%.

Because of the asymmetric density of residuals and for a better representativeness and intercomparison across products, we consider the conditional mean relative bias of the residuals ($MRE = 100\frac{\overline{R_{ref}-R}}{\overline{R_{ref}}}$ in percent) to compare the systematic error

Fig. 31.4 DPR residual distributions (mm h^{-1}) as functions of the CPI (%). The thick black line represents the median (50% quantile), the dark grey-shaded region represents the area between the 25 and 75% quantiles, the light grey-shaded region represents the area between the 10 and 90% quantiles

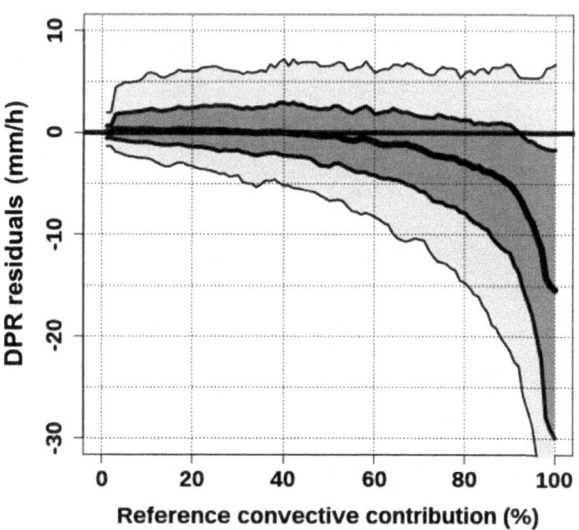

Fig. 31.5 Conditional bias
of spaceborne radars relative
to the reference as a function
of the CPI (%) for the
GPM-DPR/Ku (black),
GPM-DPR/Ka (blue),
GPM-DPR/Ka-Ku (red) and
TRMM-PR (grey)

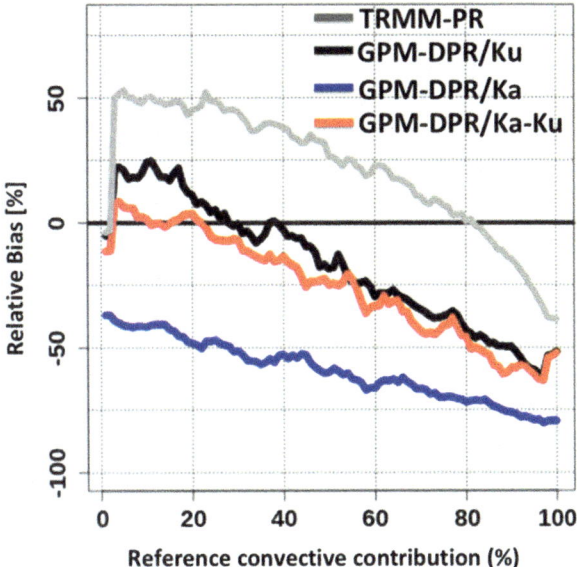

components. Figure 31.5 shows the conditional biases as a function of the CPI for
GPM-DPR/Ku, GPM-DPR/Ka, GPM-DPR/Ka-Ku and TRMM-PR. The conditional
biases are distinct according to the satellite-based radar product but also shows
similar features. Except for GPM-DPR/Ka, all biases are slightly negative and within
10% for CPI = 0% (stratiform precipitation). These biases display an abrupt shift
toward higher values (e.g., +50% for TRMM-PR, +23% for GPM-DPR/Ku
and + 11% for GPM-DPR/Ka-Ku) for CPI shifting from 0% to 3%. This might be
due to the satellite profiling algorithms interpreting this mixed microphysics within
the FOV with the stratiform parameterization. For CPI < 20% the GPM-DPR/Ka-Ku
bias covers a narrower range (from −10 to +11%) than TRMM-PR and GPM-DPR/
Ku. It probably illustrates the benefits of using two frequencies to interpret the
vertical structure of PSD and hydrometeor properties and estimate surface precipitation rates. All biases decrease with CPI and reach negative values for CPI > 80%.
It is apparent that all profiling algorithms lack sufficient dynamics to deal with
extreme rainfall rates. In addition to the reflectivity-to-precipitation rate relationships, the consistent underestimation at high rainfall rates associated with high CPI
values could also be related to the attenuation correction algorithm, or even total loss
of the radar signal, specifically for the GPM-DPR/Ka as the signal at this frequency
is more attenuated than at Ku-band. The TRMM-PR bias presents a shift towards
higher values compared to GPM-DPR and present relatively limited biases for high
CPI values (biases within 50% for CPI > 70%), which is consistent with the design
and application of the TRMM-PR profiling algorithm on intertropical precipitation.
Note that the PR estimates surface rain rates over the southern US up to a latitude of
37°N, which is dominated by deep convection during the warm season. The
GPM-DPR algorithms display relatively limited biases for light CPI values (biases
within 25% for CPI < 50%) consistent with their application to mid-latitude

precipitation. GPM-DPR/Ka shows systematic underestimation (biases ranging from -77% to -35%) possibly related to the strong attenuation of the signal at Ka-band.

To summarize, the categorical classification of precipitation types in the space-borne radar algorithms does not handle mixed types and microphysics within the FOV and appear to lack of sufficient dynamics to deal with extreme rainfall amounts.

31.3.2 GMI QPE and Precipitation Typology

Currently GPROF does not condition precipitation retrievals by precipitation types, although recent works indicate that atmospheric stability and precipitating system structure impact its performance (Petković and Kummerow 2017; Petković et al. 2018, 2019; Henderson et al. 2017). However, GPROF-GMI quantifies the convective contribution in precipitation rate retrievals (CPI_{GPROF}). CPI provides a reliable assessment of the output convective contribution CPI_{GPROF} inside a passive sensor FOV. We focus here on cases when both GPROF-GMI and the reference precipitation are greater than zero over the CONUS dataset, so that precipitation detection is not a factor.

Figure 31.6 shows the cumulative distributions of the reference and GPROF-GMI convective contributions. The reference stratiform precipitation (CPI = 0%) dominates at 81% the reference precipitation typology in terms of the fraction of type occurrence, while the fraction of convective/stratiform mix (0% < CPI < 80%) is 15%, and the fraction of primarily convective precipitation (CPI > 80%) is only 4%. The GPROF-GMI cumulative distribution is significantly different as stratiform precipitation (CPI = 0%) is not frequently retrieved (< 1%), the fraction of convective/stratiform mix is significantly higher than the reference at 96%, and the fraction

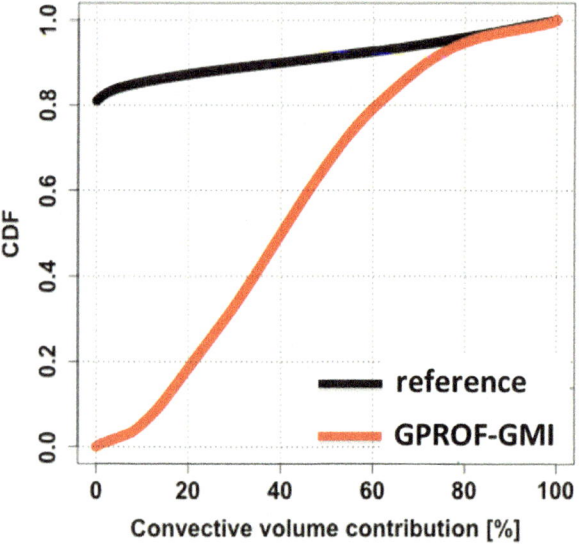

Fig. 31.6 Reference and GPROF-GMI cumulative distribution of convective contribution

Fig. 31.7 GPROF-GMI residual distributions (mm h^{-1}) as functions of the CPI (%). The thick black line represents the median (50% quantile), the dark grey-shaded region represents the area between the 25% and 75% quantiles, the light grey-shaded region represents the area between the 10% and 90% quantiles

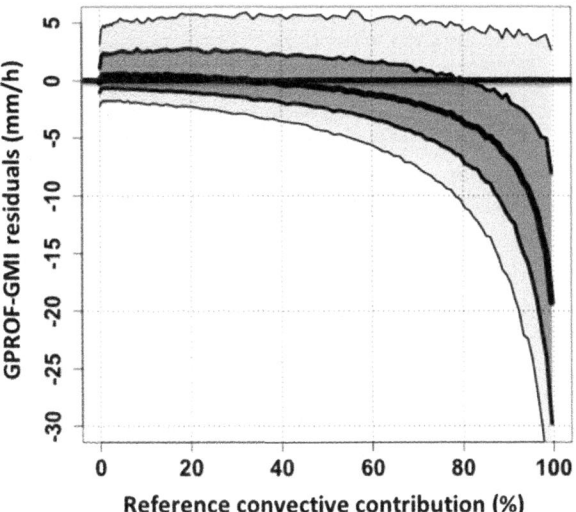

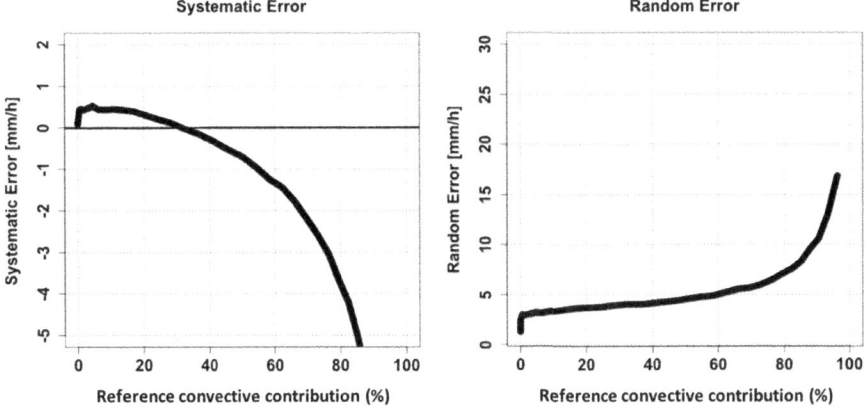

Fig. 31.8 GPROF-GMI (left) systematic part –conditional median - and (right) random part – interquantile 10%–90%- of error

of primarily convective precipitation (CPI > 80%) is 4%. While the Bayesian methodology in the GPROF-GMI yields estimates that are well constrained as designed from the a-priori database, this approach apparently doesn't accommodate precipitation types.

A conditional error analysis is applied to target biases over different precipitation type conditions. Figure 31.7 illustrates GPROF-GMI residuals as a function of the convective contribution CPI. All coincident and collocated GPROF-GMI values are considered and sorted according to the reference sample.

The systematic error is extracted as the conditional median of residuals for a better representativeness given the asymmetric density of residuals. The random error is assessed with the interquantile (90% - 10%) value. Both systematic and random errors are displayed in Fig. 31.8. Similar to the DPR (Fig. 31.4), the conditional PDFs of residuals ϵ present a high conditional shift from the 0 line and a large conditional spread. The spread increases with CPI, indicating larger uncertainties in quantifying the reference convective rainfall. It is confirmed with the random error increasing with CPI (Fig. 31.8). As for DPR for stratiform situations (CPI $= 0\%$), the systematic error is low, and the random error minimal. Systematic error displays an abrupt shift toward higher values for CPI from 0% to 3%. There is a systematic overestimation of rain rates for low CPI values (CPI $< 25\%$, the conditional median of residuals is positive in Fig. 31.8), a shift toward lower rainfall rates as CPI increases, and rain rates underestimation for CPI values higher than 50% (negative median of residuals, see Fig. 31.8). As an example, at CPI $= 20\%$ (mainly stratiform), the GPROF-GMI retrieval overestimates rain rates with an occurrence of 55%, and at CPI $= 100\%$ (convective) it underestimates rain rates with an occurrence of 85%. The similarities between the GPROF-GMI and DPR residuals as functions of CPI (Fig. 31.4 and Fig. 31.8) may be caused by the propagation of precipitation type related errors from the DPR profiling algorithm to the GPROF-GMI retrieval database. In particular DPR interprets mixed microphysics within the FOV with a specific parameterization type (e.g., stratiform) and lacks sufficient dynamics to deal with extreme rainfall amounts.

To investigate the potential improvement in QPE if GPROF-GMI correctly estimates the convective contribution, the CPI difference $\epsilon_{CPI} = CPI_{GPROF} - CPI$ is computed. A positive (negative) ϵ_{CPI} indicates that the GPROF-GMI retrieval is more convective (stratiform) than the reference. Figure 31.9 shows the GPROF surface precipitation mean relative bias and correlation as functions of ϵ_{CPI}.

The precipitation rate correlation between GPROF-GMI and the reference displays a strong dependence with ϵ_{CPI}, taking on lower values when GPROF-GMI and the reference precipitation types are significantly different. The correlation is above 0.65 when ϵ_{CPI} ranges from -1% to $+11\%$ and maximizes at 0.71 when ϵ_{CPI} is $+8\%$. The mean relative bias of GPROF-GMI increases with ϵ_{CPI} and shows

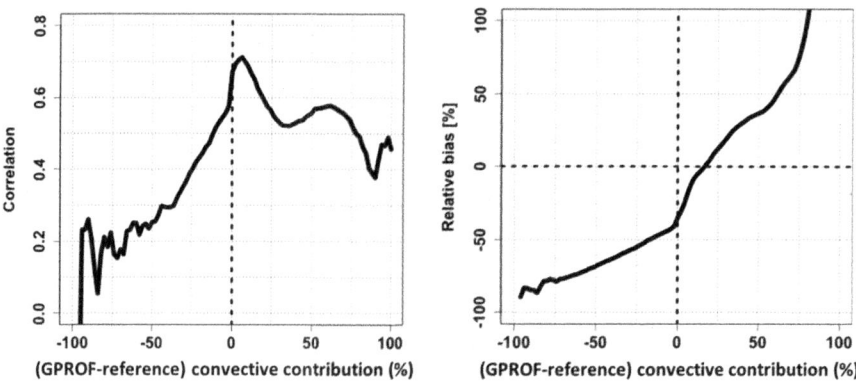

Fig. 31.9 GPROF-GMI (left) correlation and (right) relative bias as a function of convective contribution difference between GPROF-GMI and the reference

underestimation (overestimation) when GPROF is more stratiform (convective) than the reference, as one might expect.

Considering that GPROF-GMI retrieves much less stratiform precipitation than the reference (Fig. 31.6), it can partly explain the positive precipitation rate bias for reference CPI < 30% (Fig. 31.8). The bias is 0 for the $\epsilon_{CPI} = +8\%$, which is also where the correlation maximizes. Improving both the bias and correlation is noteworthy. Ciach et al. (2000) show that post-processing optimization of a precipitation product relative to a reference can be done by improving the bias (systematic error) or the mean square error (random error), but not both. Matching the precipitation types output from GPROF-GMI shows improvements in both bias and correlation, which demonstrate the potential benefit in conditioning precipitation rate retrievals by precipitation types in the algorithm.

31.3.3 IMERG and Precipitation Typology

To investigate the impact of precipitation types on IMERG, Fig. 31.10 displays the residuals of IMERG estimates relative to the reference as functions of the convective contribution on a pixel-to-pixel basis. The residuals are consistently defined as the difference between the IMERG estimates (R) the reference rainfall (R_{ref}): $\epsilon = (R - R_{ref})$ at the resolution [0.1°, 30-min]. Only pairs for which R_{ref} and R are both nonzero are considered in the calculations, so as to remove any discrepancies related to detectability. All coincident and collocated IMERG values are considered and sorted according to the reference sample.

Consistent with DPR and GPROF-GMI, IMERG conditional PDFs of residuals ϵ present a high conditional shift from the 0 line and a large conditional spread. The

Fig. 31.10 IMERG residual distributions (mm h^{-1}) as functions of the CPI (%). The thick black line represents the median (50% quantile), the dark grey-shaded region represents the area between the 25 and 75% quantiles, the light grey-shaded region represents the area between the 10 and 90% quantiles

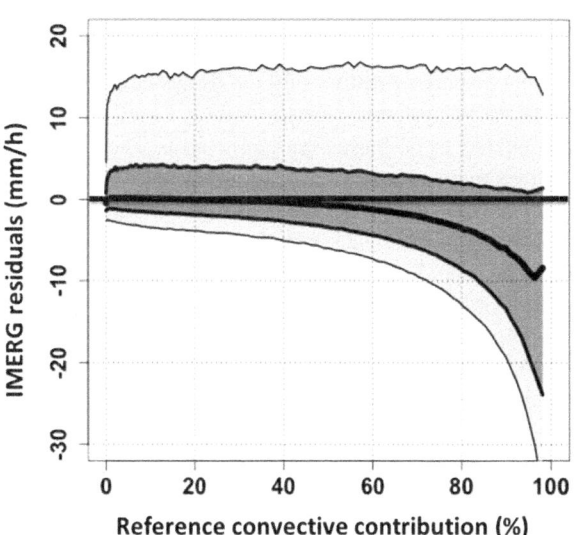

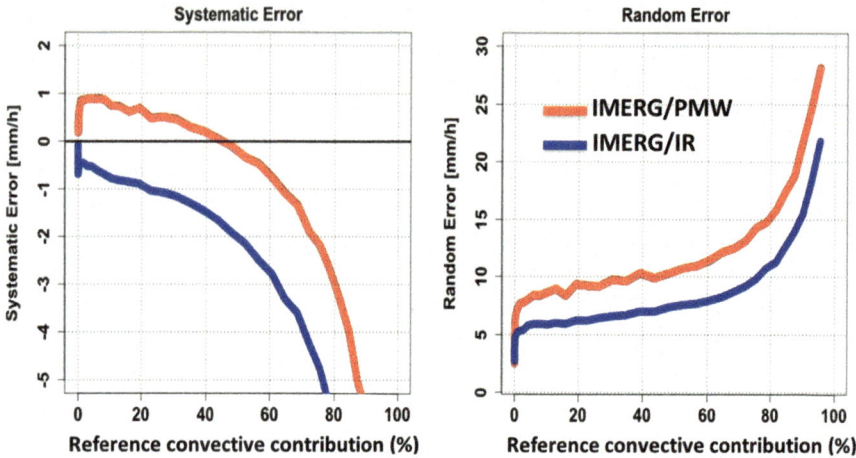

Fig. 31.11 IMERG (left) systematic part – conditional median – and (right) random part – interquantile 10–90% – of error for the PMW component (red) and IR component (blue)

spread increases with CPI, indicating larger uncertainties in quantifying the reference convective rainfall at the IMERG scale. The panel shows a shift toward lower rainfall rates as CPI increases, and rain rates underestimation for CPI > 50% (negative median of residuals). It illustrates the propagation of the GPROF biases with respect to precipitation types into IMERG.

Further insight is provided in Fig. 31.11 showing the systematic error (median of residuals) and random error (interquantile 90–10%) of the PMW and IR components of IMERG. The conditional biases are distinct according to the IMERG components. Both PMW and IR systematic and random errors are minimized for stratiform situations (CPI = 0%). Over the range of CPI values, systematic and random PMW errors show similar behavior to GROF-GMI (Fig. 31.8), especially for CPI values from 0% to a few percent. Over the same CPI range, while PMW systematic errors display a positive shift the IR systematic errors display the opposite behavior. The PMW systematic biases present a shift towards higher values (+1.5 mm h^{-1}) compared to IR biases for CPI < 50%. They are both decreasing functions of the convective contribution CPI, with the PMW component overestimating at CPI < 45% while the IR component underestimates over the whole CPI range. Both components lack sufficient dynamics to deal with extreme rainfall amounts. The random discrepancies increase consistently with CPI for both components, and it is worth noting that the PMW random error is greater (~+5 mm h^{-1}) than the IR component.

This analysis confirms that satellite estimates are currently limited in capturing extreme precipitation events often related to convective precipitation.

31.4 Conclusion

An integrated multi-sensor assessment is proposed as a novel approach to advance satellite QPE validation, in order to provide precipitation algorithm developers and users with more than just an overall assessment and adequately cope with the varying performances of satellite precipitation estimates. End-to-end integrated error quantification tracking the origin and propagation of uncertainty from Level-2 active, passive then through Level-3 precipitation products is particularly relevant for understanding the origin and impact of uncertainty. Precipitation typology is addressed in this context as a relevant factor driving the state of the GPM satellite estimation error and leading to a generalization of the assessment.

Such assessment is performed at the primary satellite QPE scale across products for a quantitative and detailed characterization. It is tested over a multi-year and multi-scale data sample of TRMM-PR, DPR, GMI and IMERG satellite precipitation products and a high-quality precipitation reference derived from MRMS. The reference convective contribution was adapted at multiple resolutions to perform precipitation error characterization conditioned on precipitation types, by taking full advantage of the reference insights into precipitation across scales (from sub-satellite Level-2 FOV to IMERG grid). The state dependency of the error is diagnosed by stratifying the error according to the CPI across sensors and products.

For DPR (and TRMM-PR), incorrect physical assumptions related to convective versus stratiform rainfall classification are confirmed to be a primary error impacting the conversion from reflectivity-to-rainfall intensity. The non-uniformity of precipitation types within the FOV is a driving error contributor. Both DPR stratiform and convective profiling algorithms seem to lack sufficient dynamics to deal with extreme rainfall amounts. For stratiform FOV-filling conditions DPR algorithms underestimate precipitation rates relative to the reference, overestimates mixed stratiform dominated conditions and underestimates for convective situations. Uncertainty in estimating precipitation rates increases in case of convective precipitation. Results from the error analysis presented here provide insights into significant characteristics of DPR rainfall retrieval errors that need to be considered to improve the retrieval and when such data are used in applications.

Based on passive sensors observations and conditioned on DPR outputs, GPROF-GMI retrievals display similar error features as DPR with respect to precipitation types. It probably inherits the lack of sufficient dynamics to deal with extreme rainfall amounts observed with the active profiling algorithm. This may be amplified by the GPROF Bayesian approach used to retrieve precipitation in GPROF-GMI. Room for improvement is shown to reside in conditioning the GPROF retrievals by precipitation types, and this can be considered for future versions of the algorithm. GPROF retrievals systematic and random errors with respect to precipitation types propagate into the IMERG PMW component.

References

Anagnostou, E. N. (2004). Overview of overland satellite rainfall estimation for hydrometeorolognical applications. *Surveys in Geophysics, 25*, 511–537. https://doi.org/10.1007/s10712-004-5724-6.

Awaka, J., Iguchi, T., & Okamoto, K. (2007). Rain type classification algorithm. In V. Levizzani, P. Bauer, & F. J. Turk (Eds.), *Measuring precipitation from space* (pp. 213–224). Dordrecht: Springer. ISBN: 978-1-4020-5834-9.

Battan, L. J. (1973). *Radar observations of the atmosphere*. Chicago: University of Chicago Press. 324 pp, ISBN: 978-1878907271.

Bauer, P., Auligné, T., Bell, W., Geer, A., Guidard, V., Heilliette, S., Kazumori, M., Kim, M.-J., Liu, E. H.-C., McNally, A. P., Macpherson, B., Okamoto, K., Renshaw, R., & Riishøjgaard, L. P. (2011). Satellite cloud and precipitation assimilation at operational NWP centres. *Quarterly Journal of the Royal Meteorological Society, 137*, 1934–1951. https://doi.org/10.1002/qj.905.

Bhuiyan, M. A. E., Anagnostou, E. N., & Kirstetter, P.-E. (2017). A non-parametric statistical technique for modeling overland TMI (2A12) rainfall retrieval error. *IEEE Geoscience and Remote Sensing Letters, 14*(11), 1898–1902. https://doi.org/10.1109/LGRS.2017.2728658.

Carr, N., Kirstetter, P.-E., Hong, Y., Gourley, J. J., Schwaller, M., Petersen, W. A., Wang, N.-Y., Ferraro, R. R., & Xue, X. (2015). The influence of surface and precipitation characteristics on TRMM microwave imager rainfall retrieval uncertainty. *Journal of Hydrometeorology, 16*, 1596–1614. https://doi.org/10.1175/JHMD-14-0194.1.

Chandrasekar, V., Hou, A. Y., Smith, E. A., Bringi, V. N., Rutledge, S. A., Gorgucci, E., Petersen, W. A., & Skofronick Jackson, G. (2008). Potential role of dual-polarization radar in the validation of satellite precipitation measurements: Rationale and opportunities. *Bulletin of the American Meteorological Society, 89*, 1127–1145. https://doi.org/10.1175/2008BAMS2177.1.

Derin, Y., & Yilmaz, K. K. (2014). Evaluation of multiple satellite-based precipitation products over complex topography. *Journal of Hydrometeorology, 15*, 1498–1516. https://doi.org/10.1175/JHM-D-13-0191.1.

Derin, Y., Anagnostou, E. N., Berne, A., Borga, M., Boudevillain, B., Buytaert, W., Chang, C., Delrieu, G., Hong, Y., Hsu, Y. C., Lavado-Casimiro, W., Manz, B., Moges, S., Nikolopoulos, E. I., Sahlu, D., Salerno, F., Rodríguez-Sánchez, J., Vergara, H. J., & Yilmaz, K. K. (2016). Multiregional satellite precipitation products evaluation over complex terrain. *Journal of Hydrometeorology, 17*, 1817–1836. https://doi.org/10.1175/JHM-D-15-0197.1.

Dinku, T., Connor, S. J., & Ceccato, P. (2010). Comparison of CMORPH and TRMM-3B42 over mountainous regions of Africa and South America. In M. Gebremichael & F. Hossain (Eds.), *Satellite rainfall applications for surface hydrology* (pp. 193–204). Dordrecht: Springer. ISBN: 978-90-481-2915-7.

Ebert, E. E. (2007). Methods for verifying satellite precipitation estimates. In V. Levizzani, P. Bauer, & F. J. Turk (Eds.), *Measuring precipitation from space* (pp. 345–356). Dordrecht: Springer. ISBN: 978-1-4020-5834-9.

Gebregiorgis, A., Kirstetter, P.-E., Hong, Y., Carr, N., Gourley, J. J., & Zheng, Y. (2017). Understanding multi-sensor satellite precipitation error structure in level-3 TRMM products. *Journal of Hydrometeorology, 18*, 285–306. https://doi.org/10.1175/JHM-D-15-0207.1.

Gebregiorgis, A., Kirstetter, P.-E., Hong, Y., Gourley, J. J., Huffman, G. J., Petersen, W. A., Xue, X., & Schwaller, M. (2018). To what extent is the Day-1 IMERG satellite rainfall estimate improved as compared to TMPA-RT? *Journal of Geophysical Research, 123*, 1694–1707. https://doi.org/10.1002/2017JD027606.

Golian, S., Moazami, S., Kirstetter, P.-E., & Hong, Y. (2015). Evaluating the performance of merged multi-satellite precipitation products over a complex terrain. *Water Resources Management, 29*(13), 4885–4901. https://doi.org/10.1007/s11269-015-1096-6.

Gopalan, K., Wang, N.-Y., Ferraro, R. R., & Liu, C. (2010). Status of the TRMM 2A12 land precipitation algorithm. *Journal of Atmospheric and Oceanic Technology, 27*, 1343–1354. https://doi.org/10.1175/2010JTECHA1454.1.

Grams, H. M., Kirstetter, P.-E., & Gourley, J. J. (2016). Naïve Bayesian precipitation type retrieval from satellite using a cloud-top and ground-radar matched climatology. *Journal of Hydrometeorology, 17*, 2649–2665. https://doi.org/10.1175/JHM-D-16-0058.1.

Grecu, M., & Anagnostou, E. N. (2001). Overland precipitation estimation from TRMM passive microwave observations. *Journal of Applied Meteorology, 40*, 1367–1380. https://doi.org/10.1175/1520-0450(2001)040<1367:OPEFTP>2.0.CO;2.

Grimes, D. I. F., & Diop, M. (2003). Satellite-based rainfall estimation for river flow forecasting in Africa. I: Rainfall estimates and hydrological forecasts. *Hydrological Sciences Journal, 48*, 567–584. https://doi.org/10.1623/hysj.48.4.567.51410.

Henderson, D. S., Kummerow, C. D., Marks, D. A., & Berg, W. (2017). A regime-based evaluation of TRMM oceanic precipitation biases. *Journal of Atmospheric and Oceanic Technology, 34*, 2613–2635. https://doi.org/10.1175/JTECH-D-16-0244.1.

Hong, Y., Hsu, K.-L., Sorooshian, S., & Gao, X. (2004). Precipitation estimation from remotely sensed imagery using an artificial neural network cloud classification system. *Journal of Applied Meteorology, 43*, 1834–1853. https://doi.org/10.1175/JAM2173.1.

Hossain, F., & Anagnostou, E. N. (2004). Assessment of current passive-microwave and infrared satellite rainfall remote sensing for flood prediction. *Journal of Geophysical Research, 109*, D07102. https://doi.org/10.1029/2003JD003986.

Hou, A. Y., Kakar, R. K., Neeck, S. A., Azarbarzin, A., Kummerow, C. D., Kojima, M., Oki, R., Nakamura, K., & Iguchi, T. (2014). The global precipitation measurement Mission. *Bulletin of the American Meteorological Society, 95*, 701–722. https://doi.org/10.1175/BAMS-D-13-00164.1.

Huffman, G. J., Bolvin, D. T., Nelkin, E. J., Wolff, D. B., Adler, R. F., Gu, G., Hong, Y., Bowman, K. P., & Stocker, E. F. (2007). The TRMM multisatellite precipitation analysis (TMPA): Quasi-global, multiyear, combined-sensor precipitation estimates at fine scales. *Journal of Hydrometeorology, 8*, 38–55. https://doi.org/10.1175/JHM560.1.

Huffman, G. J., Bolvin, D. T., Braithwaite, D., Hsu, K.-L., Joyce, R., Kidd, C., Nelkin, E. J., Sorooshian, S., Tan, J., & Xie, P. (2018). Algorithm theoretical basis document (ATBD) Version 5.2 for NASA global precipitation measurement (GPM) Integrated Multi-satellitE Retrievals for GPM (IMERG). *NASA*, 34 pp. Available at https://docserver.gesdisc.eosdis.nasa.gov/public/project/GPM/IMERG_ATBD_V5.pdf. Last accessed 28 Nov. 2018.

Iguchi, T., Kozu, T., Kwiatkowski, J., Meneghini, R., Awaka, J., & Okamoto, K. (2009). Uncertainties in the rain profiling algorithm for the TRMM precipitation radar. *Journal of the Meteorological Society of Japan, 87A*, 1–30. https://doi.org/10.2151/jmsj.87A.1.

Iguchi, T., Kozu, T., Meneghini, R., Awaka, J., & Okamoto, K. (2000). Rain-profiling algorithm for the TRMM precipitation radar. *Journal of Applied Meteorology, 39*, 2038–2052. https://doi.org/10.1175/1520-0450(2001)040<2038:RPAFTT>2.0.CO;2.

Iguchi, T., Seto, S., Meneghini, R., Yoshida, N., Awaka, J., Le, M., Chandrasekar, V., & Kubota, T. (2017). GPM/DPR Level-2 algorithm theoretical basis document. *JAXA-NASA Tech. Rep.*, 81 pp. Available online at https://pmm.nasa.gov/resources/documents/gpmdpr-level-2-algorithm-theoretical-basis-document-atbd. Last accessed 9 Oct. 2018.

Joyce, R., Janowiak, J., Arkin, P. A., & Xie, P. (2004). CMORPH: A method that produces global precipitation estimates from passive microwave and infrared data at high spatial and temporal resolution. *Journal of Hydrometeorology, 5*, 487–503. https://doi.org/10.1175/1525-7541(2004)005<0487:CAMTPG>2.0.CO;2.

Khan, S., Maggioni, V., & Kirstetter, P.-E. (2018). Investigating the potential of using satellite-based precipitation radars as reference for evaluating multisatellite merged products. *Journal of Geophysical Research, 123*, 8646–8660. https://doi.org/10.1029/2018JD028584.

Kidd, C., Tan, J., Kirstetter, P.-E., & Petersen, W. A. (2018). Validation of the version 05 level 2 precipitation products from the GPM Core Observatory and constellation satellite sensors. *Quarterly Journal of the Royal Meteorological Society, 144*(S1), 313–328. https://doi.org/10.1002/qj.3175.

Kirstetter, P.-E., Hong, Y., Gourley, J. J., Cao, Q., Schwaller, M., & Petersen, W. A. (2014). A research framework to bridge from the global precipitation measurement mission core satellite to the constellation sensors using ground radar-based National Mosaic QPE. In

L. Venkataraman (Ed.), *Remote sensing of the terrestrial water cycle* (AGU geophysical monograph series, Chapman monograph on remote sensing). Hoboken: Wiley. ISBN:1118872037.

Kirstetter, P.-E., Hong, Y., Gourley, J. J., Chen, S., Flamig, Z., Zhang, J., Howard, K., & Petersen, W. A. (2012). Toward a framework for systematic error modeling of spaceborne precipitation radar with NOAA/NSSL ground radar-based National Mosaic QPE. *Journal of Hydrometeorology, 13*(4), 1285–1300. https://doi.org/10.1175/JHM-D-11-0139.1.

Kirstetter, P.-E., Gourley, J. J., Hong, Y., Zhang, J., Moazamigoodarzi, S., Langston, C., & Arthur, A. (2015a). Probabilistic precipitation rate estimates with ground-based radar networks. *Water Resources Research, 51*, 1422–1442. https://doi.org/10.1002/2014WR015672.

Kirstetter, P.-E., Hong, Y., Gourley, J. J., Schwaller, M., Petersen, W. A., & Cao, Q. (2015b). Impact of sub-pixel rainfall variability on spaceborne precipitation estimation: Evaluating the TRMM 2A25 product. *Quarterly Journal of the Royal Meteorological Society, 141*, 953–966. https://doi.org/10.1002/qj.2416.

Kirstetter, P.-E., Hong, Y., Gourley, J. J., Schwaller, M., Petersen, W. A., & Zhang, J. (2013a). Comparison of TRMM 2A25 products version 6 and version 7 with NOAA/NSSL ground radar-based National Mosaic QPE. *Journal of Hydrometeorology, 14*(2), 661–669. https://doi.org/10.1175/JHM-D-12-030.1.

Kirstetter, P.-E., Viltard, N., & Gosset, M. (2013b). An error model for instantaneous satellite rainfall estimates: Evaluation of BRAIN-TMI over West Africa. *Quarterly Journal of the Royal Meteorological Society, 139*, 894–911. https://doi.org/10.1002/qj.1964.

Kirstetter, P.-E., Karbalaee, N., Hsu, K.-L., & Hong, Y. (2018). Probabilistic precipitation rate estimates with space-based infrared sensors. *Quarterly Journal of the Royal Meteorological Society, 144*(S1), 191–205. https://doi.org/10.1002/qj.3243.

Kummerow, C. D. (2017). *GPROF2017 version 1. Algorithm theoretical basis document*, 63 pp. Available at https://pmm.nasa.gov/sites/default/files/document_files/ATBD_GPM_GPROF_June1_2017.pdf. Last accessed 30 Nov. 2018.

Kummerow, C. D., Berg, W., Thomas-Stahle, J., & Masunaga, H. (2006). Quantifying global uncertainties in a simple microwave rainfall algorithm. *Journal of Atmospheric and Oceanic Technology, 23*, 23–37. https://doi.org/10.1175/JTECH1827.1.

Lebel, T., Cappelaere, C., Galle, S., Hanan, N., Kergoat, L., Levis, S., Vieux, B., Descroix, L., Gosset, M., & Mougin, E. (2009). AMMA-CATCH studies in the Sahelian region of West-Africa: An overview. *Journal of Hydrology, 375*, 3–13. https://doi.org/10.1016/j.jhydrol.2009.03.020.

Levizzani, V., Kidd, C., Aonashi, K., Bennartz, R., Ferraro, R. R., Huffman, G. J., Roca, R., Turk, F. J., & Wang, N.-Y. (2018). The activities of the International Precipitation Working Group. *Quarterly Journal of the Royal Meteorological Society, 144*(S1), 3–15. https://doi.org/10.1002/qj.3214.

Meneghini, R., Iguchi, T., Kozu, T., Liao, L., Okamoto, K., Jones, J. A., & Kwiatkowski, J. R. (2000). Use of the surface reference technique for path attenuation estimates from the TRMM precipitation radar. *Journal of Atmospheric and Oceanic Technology, 40*, 2053–2070. https://doi.org/10.1175/1520-0450(2001)040<2053:UOTSRT>2.0.CO;2.

Meneghini, R., Jones, J. A., Iguchi, T., Okamoto, K., & Kwiatkowski, J. R. (2004). A hybrid surface reference technique and its application to the TRMM precipitation radar. *Journal of Atmospheric and Oceanic Technology, 21*, 1645–1658. https://doi.org/10.1175/JTECH1664.1.

Munchak, S. J., & Skofronick-Jackson, G. (2013). Evaluation of precipitation detection overvarious surfaces from passive microwave imagers and sounders. *Atmospheric Research*, 131, 81–94. doi:https://doi.org/10.1016/j.atmosres.2012.10.011.

O, S., & Kirstetter, P.-E. (2018). Evaluation of diurnal variation of GPM IMERG derived summer precipitation over the contiguous US using MRMS data. *Quarterly Journal of the Royal Meteorological Society, 144*(S1), 270–281. https://doi.org/10.1002/qj.3218.

Panegrossi, G., Casella, D., Dietrich, S., Marra, A. C., Sanò, P., Mugnai, A., Baldini, L., Roberto, N., Adirosi, E., Cremonini, R., Bechini, R., Vulpiani, G., Petracca, M., & Porcù, F. (2016). Use of the GPM constellation for monitoring heavy precipitation events over the Mediterranean region. *IEEE JSTAR, 9*, 2733–2753. https://doi.org/10.1109/JSTARS.2016.2520660.

Petersen, W. A., & Schwaller, M. R. (2008). NASA GPM ground validation science implementation plan. In *NASA Report*, 41 pp. Available at https://pmm.nasa.gov/resources/documents/gpm-ground-validation-science-implementation-plan. Last accessed 30 Nov. 2018.

Petković, V., & Kummerow, C. D. (2017). Understanding the sources of satellite passive microwave rainfall retrieval systematic errors over land. *Journal of Applied Meteorology and Climatology, 56*, 597–614. https://doi.org/10.1175/JAMC-D-16-0174.1.

Petković, V., Kummerow, C. D., Randel, D. L., Pierce, J. R., & Kodros, J. K. (2018). Improving the quality of heavy precipitation estimates from satellite passive microwave rainfall retrievals. *Journal of Hydrometeorology, 19*, 69–85. https://doi.org/10.1175/JHM-D-17-0069.1.

Petković, V., Orescanin, M., Kirstetter, P., Kummerow, C., & Ferraro, R. (2019). Enhancing PMW Satellite Precipitation Estimation: Detecting convective class. *Journal of Atmospheric and Oceanic Technology, 36*, 2349–2363. https://doi.org/10.1175/JTECH-D-19-0008.1.

Porcacchia, L., Kirstetter, P.-E., Maggioni, V., & Tanelli, S. (2019). Investigating the GPM Dual-frequency Precipitation Radar signatures of low-level precipitation enhancement. *Quarterly Journal of the Royal Meteorological Society, 145*, 3161–3174. https://doi.org/10.1002/qj.3611.

Ryzhkov, A., Diederich, M., Zhang, P., & Simmer, C. (2014). Potential utilization of specific attenuation for rainfall estimation, mitigation of partial beam blockage, and radar networking. *Journal of Atmospheric and Oceanic Technology, 31*, 599–619. https://doi.org/10.1175/JTECH-D-13-00038.1.

Ryzhkov, A. V., Schuur, T. J., Burguess, D. W., Heinselman, P. L., Giangrande, S. E., & Zrnić, D. S. (2005). The joint polarization experiment: Polarimetric rainfall measurements and hydrometeor classification. *Bulletin of the American Meteorological Society, 86*, 809–824. https://doi.org/10.1175/BAMS-86-6-809.

Sapiano, M. R. P., & Arkin, P. A. (2009). An intercomparison and validation of high-resolution satellite precipitation estimates with 3-hourly gauge data. *Journal of Hydrometeorology, 10*, 149–166. https://doi.org/10.1175/2008JHM1052.1.

Skofronick-Jackson, G., Petersen, W. A., Berg, W., Kidd, C., Stocker, E. F., Kirschbaum, D. B., Kakar, R., Braun, S. A., Huffman, G. J., Iguchi, T., Kirstetter, P.-E., Kummerow, C. D., Meneghini, R., Oki, R., Olson, W. S., Takayabu, Y. N., Furukawa, K., & Wilheit, T. T. (2017). The global precipitation measurement (GPM) mission for science and society. *Bulletin of the American Meteorological Society, 98*, 1679–1695. https://doi.org/10.1175/BAMS-D-15-00306.1.

Smalley, M., Kirstetter, P.-E., & L'Ecuyer, T. S. (2017). How frequent is precipitation over the contiguous United States? Perspectives from ground-based and spaceborne radars. *Journal of Hydrometeorology, 18*, 1657–1672. https://doi.org/10.1175/JHM-D-16-0242.1.

Sorooshian, S., AghaKouchak, A., Arkin, P. A., Eylander, J., Foufoula-Georgiou, E., Harmon, R., Hendrickx, J. M., Imam, B., Kuligowski, R., Skahill, B., & Skofronick-Jackson, G. (2011). Advanced concepts on remote sensing of precipitation at multiple scales. *Bulletin of the American Meteorological Society, 92*, 1353–1357. https://doi.org/10.1175/2011BAMS3158.1.

Stephens, G. L., & Kummerow, C. D. (2007). The remote sensing of clouds and precipitation from space: A review. *Journal of the Atmospheric Sciences, 64*, 3742–3765. https://doi.org/10.1175/2006JAS2375.1.

Takahashi, N., Hanado, H., & Iguchi, T. (2006). Estimation of path-integrated attenuation and its nonuniformity from TRMM/PR range profile data. *IEEE Transactions on Geoscience and Remote Sensing, 44*, 3276–3283. https://doi.org/10.1109/TGRS.2006.876295.

Tan, J., Petersen, W. A., Kirstetter, P.-E., & Tian, Y. (2017). Performance of IMERG as a function of spatiotemporal scale. *Journal of Hydrometeorology, 18*, 307–319. https://doi.org/10.1175/JHM-D-16-0174.1.

Tang, L., Tian, Y., & Lin, X. (2014). Validation of precipitation retrievals over land from satellite based passive microwave sensors. *Journal of Geophysical Research, 119*, 4546–4567. https://doi.org/10.1002/2013JD020933.

Tian, Y., Peters-Lidard, C. D., Chaudhury, B. J., & Garcia, M. (2007). Multitemporal analysis of TRMM-based satellite precipitation products for land data assimilation applications. *Journal of Hydrometeorology, 8,* 1165–1183. https://doi.org/10.1175/2007JHM859.1.

Turk, F. J., Arkin, P. A., Ebert, E. E., & Sapiano, M. R. P. (2008). Evaluating high-resolution precipitation products. *Bulletin of the American Meteorological Society, 89,* 1911–1916. https://doi.org/10.1175/2008BAMS2359.1.

Ushio, T., Okamoto, K., Kubota, T., Hashizume, H., Shige, S., Noda, S., Iida, Y., Aonashi, K., Inoue, T., Oki, R., Kachi, M., Takahashi, N., & Iguchi, T. (2006). A combined microwave and infrared radiometer approach for a high-resolution global precipitation mapping in the GSMAP project Japan. In *3rd IPWG Workshop on Precipitation Measurements,* Melbourne, Australia. Available at http://ipwg.isac.cnr.it/meetings/melbourne-2006/pres/Ushio.ppt. Last accessed 30 Nov. 2018.

Vergara, H., Hong, Y., Gourley, J. J., Anagnostou, E. N., Maggioni, V., Stampoulis, D., & Kirstetter, P.-E. (2014). Effects of resolution of satellite-based rainfall estimates on hydrologic modeling skill at different scales. *Journal of Hydrometeorology, 15,* 593–613. https://doi.org/10.1175/JHM-D-12-0113.1.

Weng, F., Zhu, T., & Yan, B. (2007). Satellite data assimilation in numerical weather prediction models. Part II: Uses of rain affected radiances from microwave observations for hurricane vortex analysis. *Journal of the Atmospheric Sciences, 64,* 3910–3925. https://doi.org/10.1175/2006JAS2051.1.

Wilheit, T. T., Kummerow, C. D., & Ferraro, R. R. (2003). Rainfall algorithms for AMSR-E. *IEEE Transactions on Geoscience and Remote Sensing, 41,* 204–214. https://doi.org/10.1109/TGRS.2002.808312.

Wolff, D. B., & Fisher, B. L. (2008). Comparisons of instantaneous TRMM ground validation and satellite rain-rate estimates at different spatial scales. *Journal of Applied Meteorology and Climatology, 47,* 2215–2237. https://doi.org/10.1175/2008JAMC1875.1.

Wolff, D. B., & Fisher, B. L. (2009). Assessing the relative performance of microwave-based satellite rain-rate retrievals using TRMM ground validation data. *Journal of Applied Meteorology and Climatology, 48,* 1069–1099. https://doi.org/10.1175/2008JAMC2127.1.

Yang, S., Olson, W. S., Wang, J. J., Bell, T. L., Smith, E. A., & Kummerow, C. D. (2006). Precipitation and latent heating distributions from satellite passive microwave radiometry. Part II: Evaluation of estimates using independent data. *Journal of Applied Meteorology and Climatology, 45,* 721–739. https://doi.org/10.1175/JAM2370.1.

Zeweldi, D. A., & Gebremichael, M. (2009). Sub-daily scale validation of satellite-based high-resolution rainfall products. *Atmospheric Research, 92,* 427–433. https://doi.org/10.1016/j.atmosres.2009.01.001.

Zhang, J., Qi, Y., Howard, K., Langston, C., & Kaney, B. (2011). Radar quality index (RQI) – A combined measure of beam blockage and VPR effects in a national network. In *Proceedings of an international symposium weather radar and hydrology,* IAHS Publications, 351. Available at https://www.nssl.noaa.gov/projects/q2/tutorial/images/mosaic/WRaH_Proceedings_Zhang-et-al_v3.pdf. Last accessed 30 Nov. 2018.

Zhang, J., Howard, K., Langston, C., Kaney, B., Qi, Y., Tang, L., Grams, H., Wang, Y., Cocks, S., Martinaitis, S., Arthur, A., Cooper, K., Brogden, J., & Kitzmiller, D. (2016). Multi-Radar Multi-Sensor (MRMS) quantitative precipitation estimation: Initial operating capabilities. *Bulletin of the American Meteorological Society, 97,* 621–638. https://doi.org/10.1175/BAMS-D-14-00174.1.

Chapter 32
Hydrologic Validation and Flood Analysis

Witold F. Krajewski, Felipe Quintero, Mohamed El Saadani, and Radoslaw Goska

Abstract Hydrologic validation centers on examining the utility of satellite rainfall products for applications that concern terrestrial waters. In this chapter the authors focus on streamflow and flood forecasting. As basin response to rainfall is a complex interaction of runoff generation and water transport in the drainage network, satellite rainfall products should be evaluated in that context. This basin-centric view of the validation problem provides new performance metrics that can guide satellite product developers. The authors outline a spatio-temporal framework for studies of satellite-rainfall products. The framework involves a ground-based rainfall reference product available at the basin scale as well as a hydrologic model that can faithfully mimic the basin response to rainfall. As rainfall generated runoff propagates through river network so do the product uncertainties. The authors discuss a statistical analysis of the dependency scale of the river "transported" discrepancies. The results indicate that errors propagate through much longer distances than what a traditional geostatistical analysis would reveal.

Keywords Precipitation · Rainfall · Validation · Floods · Hydrology · Hydrological model · Rainfall-runoff model · Streamflow forecasting · Basin scale · Uncertainty · Raingauges · Radar · Hydrograph · Drainage basin · GPM · Hillslope link model

32.1 Introduction

The ability to predict floods anywhere around the globe is a main attraction of using space-based precipitation mapping. For this ability to be firmly established, we need a better understanding of the issues of the interplay of space-time scales involved in

W. F. Krajewski (✉) · F. Quintero · R. Goska
IIHR-Hydroscience & Engineering, University of Iowa, Iowa City, IA, USA
e-mail: witold-krajewski@uiowa.edu

M. El Saadani
University of Louisiana Lafayette, Lafayette, LA, USA

© Springer Nature Switzerland AG 2020   609
V. Levizzani et al. (eds.), *Satellite Precipitation Measurement*, Advances in Global
Change Research 69, https://doi.org/10.1007/978-3-030-35798-6_8

flood genesis and uncertainties of hydrologic models and relevant observations. Here, we outline a framework that allows us to gain insights into the features of rainfall estimates and their consequences for skillful flood prediction.

Traditionally the validation of rainfall products, i.e. maps with certain space-time resolution, has been approached through comparisons with reference rainfall data sets. The most established reference remains ground-based rain gauge networks but radar networks offer an attractive alternative due to their extensive coverage in a growing number of regions. Methodological issues involved in validation using rain gauge data focused on the problem of "representativeness" of point observations and space-time variability of rainfall (Ciach et al. 2007; Ciach and Krajewski 1999a, b; Gebremichael and Krajewski 2004; Habib et al. 2001; Krajewski 2007; Seo and Krajewski 2011; Villarini and Krajewski 2010). The final product of such validation is a statistical characterization of the rainfall product's uncertainty structure (e.g., Aghakouchak et al. 2011, 2012; Ashouri et al. 2016; Krajewski et al. 2000). The simplest characterization is in terms of overall bias and error variance. More complete descriptions of the products uncertainty would include space-time dependence, taking advantage of radar-rainfall data, but these are still lacking. The ultimate goal of statistical validation is a multidimensional probability distribution across space-time scales but this seems to be years away.

In this chapter, we focus on hydrologic applications and specifically on streamflow forecasting as a fundamental element of flood prediction systems. As the research community has speculated for some time, skillful predictions at small scales is unlikely due to coarse resolution in time and space and intrinsic inaccuracy of satellite products. This aspect is changing quickly with the increasing number of satellites equipped with sensors capable of observing precipitating cloud systems that result in higher temporal and spatial resolution precipitation products. The key scientific problem is that of characterizing uncertainties across spatial scales and the corresponding skill of streamflow prediction. Addressing this problem requires an adequate framework. Here, we outline such a framework and discuss its various aspects. Our goal is not to precisely answer the scale vs. skill question as a long-term concerted effort of a larger community is necessary to accomplish that. We illustrate the framework with examples of hydrologic validation studies conducted in the past few years over watersheds of Iowa, a region frequently affected by flooding. Outside of the coastal areas of the United States, Iowa suffers most flood-related economic losses.

The potential for wide-spread use of satellite-based rainfall in hydrologic applications and in particular in flood forecasting, has stimulated vigorous research by several groups around the world (e.g., Behrangi et al. 2011; Demaria et al. 2014; Gebremichael and Hossain 2010; Gourley et al. 2011; Harris and Hossain 2008; Maggioni et al. 2011, 2013; Nikolopoulos et al. 2010, 2013; Thiemig et al. 2013; Vergara et al. 2014; Wu et al. 2012, 2014). Despite numerous studies, a clear view of the spatio-temporal perspective on the flood forecasting problem and the associated questions linked to basin scale and uncertainties in rainfall estimates and hydrologic models is still deficient.

32.2   Space-Time Validation Framework

32.2.1   Point Based Evaluation

A traditional and straightforward approach for assessing the utility of satellite-based precipitation products is to drive a hydrologic rainfall-runoff model with a product and evaluate streamflow forecasting skill at an outlet of a basin of interest. As the basin size increases, the time from the rainfall event to the time of the peak of the basin response varies from hours to weeks or even months and the contribution of the rainfall information becomes less and less relevant to the skill of the forecast (see Zalenski et al. 2017). Insights gained by evaluating hydrographs at a basin outlet or at a limited number of stream gauges inside of the basin are insufficient and can lead to erroneous conclusions regarding the ability of satellite-based rainfall products to represent hydrologic response over a drainage basin. If two different rainfall products are used, the one leading to a better simulation of the streamflow observations is considered superior.

If such analysis is performed over a significant number of stream gauges and over an extended period of time, one may get a meaningful assessment of the predictive skill. Below, we show such an assessment in which the exact same hydrologic model was used for streamflow prediction over Iowa at some 130 stream gauges operated by the United States Geological Survey (USGS). We used two different rainfall inputs to force the model and compare the results.

The satellite rainfall product used is the Integrated Multi-satellitE Retrievals for GPM (IMERG), a high resolution global coverage precipitation product (Huffman et al. 2015). Precipitation estimates for IMERG are obtained through combing and merging precipitation estimates from satellite microwave estimates, infrared (IR) satellite estimates, and rain gauge precipitation estimates. The spatial and temporal resolutions of IMERG are 0.1° and 30 min, respectively. The most reliable data for IMERG are those obtained from the GPM core observatory and the Passive Microwave (PMW) LEO satellite sensors. The gaps between the observations from these satellites are filled by using geo-IR-based feature motion advection of PMW estimates and microwave calibrated IR estimates; where the GPM core observatory serves as a calibration and evaluation tool for all other satellite products (IR and PMW) used in IMERG. The IMERG Final Run product undergoes bias correction using rain gauge rainfall estimates provided by the Global Precipitation Climatology Centre (GPCC) (Liu 2016).

Our benchmark product (ground reference) in this example is the Multi-Radar/ Multi-Sensor (MRMS) Quantitative Precipitation Estimate (QPE), now being operationally produced by the National Center for Environmental Prediction (NCEP). The ground based radar product is bias adjusted using hourly quality controlled rain gauge data provided by the Hydro-meteorological Automated Data System HADS, (Zhang et al. 2016). The product integrates observations from 180 operational radars (seven over Iowa) and has a high spatial resolution of 0.01° and hourly temporal resolution for the gauge-corrected version.

In the examples in this chapter we use the distributed hydrologic model called the Hillslope Link Model (HLM). This is the same model that the Iowa Flood Center uses for real-time streamflow prediction over the state of Iowa (Krajewski et al. 2017). Rainfall conversion to runoff takes place on the hillslopes where rainfall is partitioned into surface and subsurface runoff and infiltration. Evapotranspiration removes water from the hillslopes. Excess runoff water flows to the nearest channel (a link in the river network) where it is routed downstream using a velocity model (Ghimire et al. 2018; ElSaadani et al. 2018a, b). Mathematically, the HLM has the form of a large system of nonlinear ordinary differential equations, some used to describe the storage and flux processes on the hillslope and the rest describing water transport via the river network. Small et al. (2013) developed an efficient algorithm for solving the system of HLM equations. For the state of Iowa, the model requires solving two million equations.

We performed state wide hydrologic model runs for the 2016 wet season using the HLM. We allowed a proper initialization period for both model runs and calculated all our statistics for the period between June through September, 2016. We compared the models' outputs to the observed stream flows from 130 available USGS stream gauges unaffected by reservoir operation from across the state of Iowa.

In Fig. 32.1 we show the correlation coefficient between the modeled and observed discharges calculated with time interval of 15 min. In general, the MRMS product performed better than IMERG as a model input. However, the spatial distribution of the performance of both model runs is similar. Both performed much better in the eastern side of the state than they did in the western part of the state. This spatial distribution of the model performance seems to imply that the hydrologic model has difficulty dealing with the topology of the landscape, and the extensive agriculture on the Missouri River floodplains. There is also a downgraded performance of one of the radars that covers that part of the state (partial beam blockage). Still, keep in mind that this is the "best" version of IMERG as it is rain gauge corrected. The early version, appropriate for real-time application is likely to perform not as well.

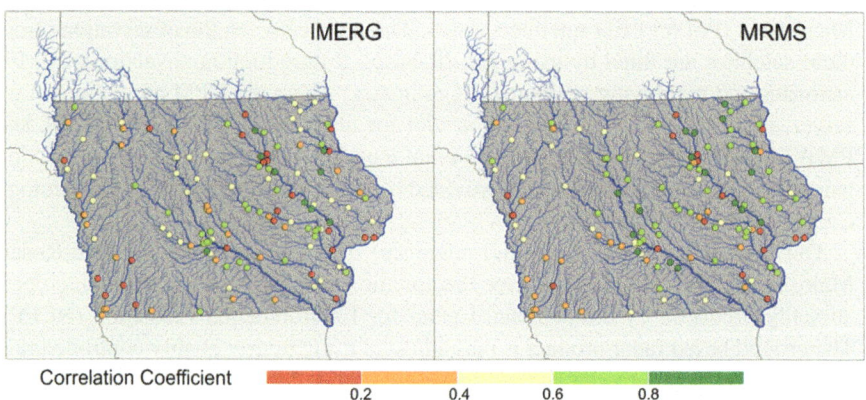

Fig. 32.1 Correlation coefficient between hydrologic model prediction and streamflow observations at 130 gauges in Iowa. The IMERG forced model results are on the left, the MRMS-forced model output is on the right

32.2.2 River Network Based Multiscale Valuation

Quintero et al. (2016) argue that a point-hydrograph based approach is overly simplistic and may even be misleading. One reason for the potential misinterpretation of the results is the fact that a river network acts as a low-pass filter. An overestimation by a given product in one branch of the network can be compensated by an underestimation in another with the result of averaging the errors while runoff is transported downstream. This suggests that more complete insights can be gained by considering simulated streamflow over the entire river network for a given basin.

As in many other approaches to validation of remote sensing products, including the example discussed above, the analyses we propose require a reference standard. We need a high-quality, high-resolution ground-based rainfall product as well as a hydrologic model that can faithfully capture key aspects of hydrologic transformation of rainfall into runoff and discharge. Clearly, this validation framework cannot be implemented anywhere; it requires the existence of a serious observational and modeling infrastructure. The observational infrastructure exists in some highly developed regions of the world and any eventual shortcomings that exist can be supplemented through periodic ground-based campaigns (Barros et al. 2014; Hou et al. 2014). The modeling infrastructure is easier to achieve but it has its own requirements as well. The first is that the hydrologic rainfall-runoff model should not be calibrated. The components that address various processes can be calibrated as long as the rainfall products, both the reference and the one being evaluated, are not involved in the calibration. Without going into a detailed discussion of the pros and cons of model calibration (e.g., Quintero et al. 2016), in our context we do not want the model to "compensate" for the uncertainties of the input. While the model needs to have skill in mimicking the key processes involved in flood genesis, a validation application is different from an operational application where absolute accuracy is the most important.

Below, we discuss an example of the above spatio-dynamic framework. Consider a river basin with its river network. The streams and rivers drain the basin to its outlet. In Fig. 32.2 we show several river basins in north-eastern Iowa in the central United States. The largest basin is that of Cedar River draining to Cedar Rapids, Iowa. The basin has a drainage area of approximately 17,000 km^2 and a water travel time of up to a week. The topology of the river network is important in propagating runoff as streamflow. If the hydrologic model used represents water transport via the river network, as it should, any modeling uncertainties are also propagated through the network. The topology is well characterized by the width function, i.e. a distribution of the channel links that comprise the river network with respect to the outlet. The width function can be interpreted as a distribution of distances from various parts of the basin or, assuming a constant velocity of water, as a distribution of travel times from across the basin. From Fig. 32.3 that shows the width function for the Cedar River basin, we see that most of the runoff will travel between 4–5 days and relatively little water passing through the outlet originates from rainfall fallen very close or very far from Cedar Rapids.

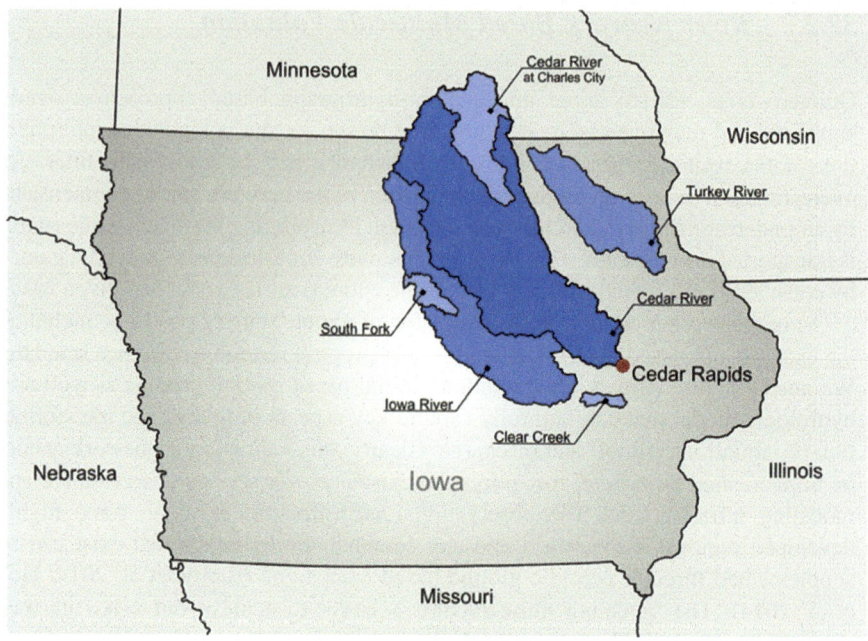

Fig. 32.2 Northeast Iowa, the site of the 2013 Iowa Flood Studies field campaign. The outlined river basins include Turkey River, Cedar River and Iowa River

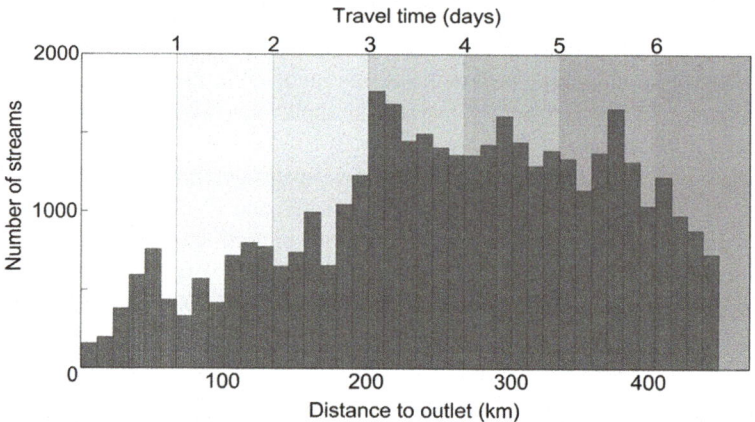

Fig. 32.3 The width function for the Cedar River basin with the outlet at Cedar Rapids, Iowa. In constructing the plot, we assumed a constant stream velocity of 0.75 m s^{-1}. It is clear that the streams that are located 200–400 km from the outlet, contribute most to the discharge

Consider now the insight gained by looking at rainfall from the river network perspective. First let us show "standard" gridded rainfall products within the basin boundaries. The products are the Multi Radar Multi Sensor radar-rainfall adjusted with the local rain gauge data (Zhang et al. 2016) and a multi satellite TMPA product (Huffman et al. 2015).

In Fig. 32.4 we display rainfall accumulations during the period of a 10-day period from 24 May to 2 June 2013 for the MRMS rainfall and the TMPA rainfall. Since the MRMS is available at considerably higher spatial resolution, we have aggregated it in space to match the resolution of the TMPA as necessary for the analyses and the comparisons. We also show the normalized differences (in %) between the two rainfall estimates. Large discrepancies are evident between the two products but little can be said about their hydrologic effect, on streamflow in particular.

Now contrast this with a river-network based presentation of the very same rainfall. In Fig. 32.4 in the right-hand side column of panels we show river network segments color-coded according to the mean area rainfall accumulated upstream from a given link. We use the same color scale in all corresponding panels. The color codes are for the links of the drainage network with the lowest order of three (due to limited size and resolution of the plot). Particularly interesting is the panel showing relative differences. We see that larger rivers are more likely to be "white" implying that relative differences decrease with the increasing scale. In the gridded product analysis (the difference field), there is no basis to make the same statement. Therefore, a clear insight from simply plotting rainfall-to-rainfall but considering the topology of the river is that random errors that affect small basins may average out for the large basins.

In the above example, the rainfall was a 10-day accumulation of substantial quantity that generated significant flooding throughout the basin. In principle, rainfall analysis as the one we demonstrated above could be done for any temporal accumulation scale. However, if the scale is too short, or if we display rainfall intensity, the hydrologic interpretation might not be very meaningful. This is because the multiplicity of basins spanning a wide range of spatial scales, ranging over five orders of magnitude, from ~0.1 to 10,000 km^2, respond over a wide range of temporal scales. From the flooding perspective, the most important aspects of this response are peak flows and discharge volumes. While observations of these quantities are not available over the basins representing the range of scales we discussed, their estimates can be obtained using a hydrologic rainfall-runoff model.

Using a hydrologic rainfall-runoff model provides additional understanding of the error propagation across spatial scales. In the example that follows we use various versions of the modeling system based on the HLM concept we introduced earlier. Components of the model can have different levels of complexity, for example, one can represent runoff simply as a portion of the rainfall. Similarly, water velocity could be modelled as constant or as a non-linear function of discharge. All these cases can be expressed in the form of ordinary differential equations. For the Cedar River basin of interest here, the HLM system comprises some 100,000 equations.

Integrating the model in time, provides a hydrograph (streamflow as function of time) at each and every link in the river network. Among the many variables that can be derived from the solution are peak flows. They correspond to the highest extent of the potential flooding in case the discharge exceeds the capacity of the respective channel.

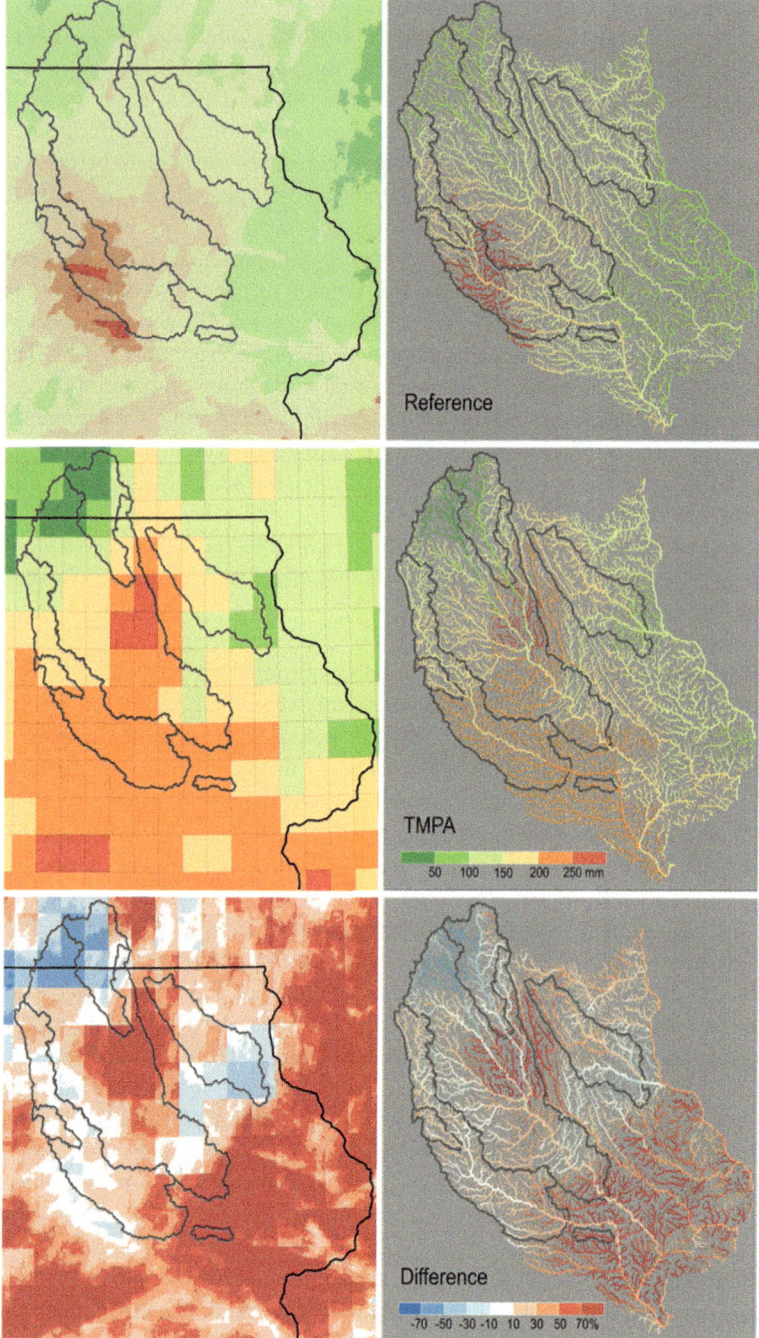

Fig. 32.4 The panels on the left correspond to the gridded representation of rainfall for radar-based (top) and satellite-based (middle) products and the difference (bottom). On the right, we show the same rainfall but calculated for the areas upstream from a given link of the river network for several watersheds in northeast Iowa

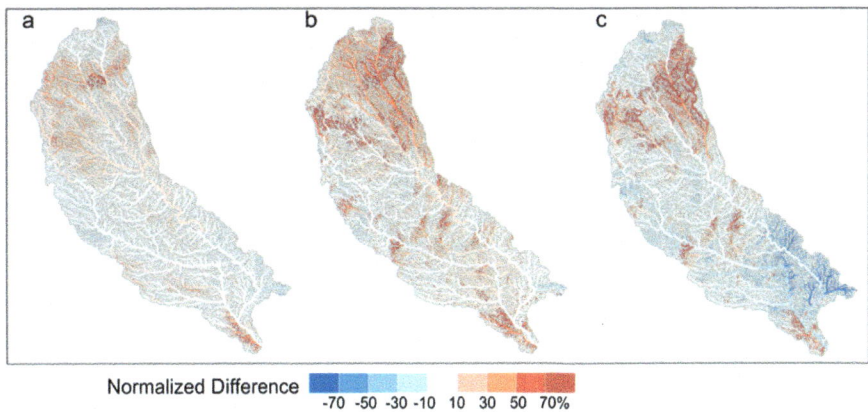

Fig. 32.5 Panel (**a**) shows network integrated rainfall difference between a reference and a product of interest. Panel (**b**) shows the differences in the peak flow for all streams in the network based on a simple linear model; panel (**c**) shows the same but for a nonlinear model

In Fig. 32.5 we show three panels with color-coded drainage network for the Cedar River basin in Eastern Iowa. In the left panel (a), we show normalized differences (in %) between the rainfall product of interest (TMPA rainfall) and the reference product (Stage IV) calculated at all basins above a given link. The rainfall is accumulated over a 10-day window (May–June 2013).

Again, it is clear that while the small-scale stream segments (links) are colored, the large rivers remain white indicating that the random differences of the respective upstream rainfall are small. The middle panel (b) shows the normalized difference between the peak flows calculated at all segments using a linear version of the HLM driven by the product of interest and the reference product.

In the linear version, we assume a runoff coefficient and a constant water velocity. In this case, we can see the errors propagating away from the regions where the rainfall differences occur due to the transport of runoff in the river network. Since the hydrologic model is rather simple, the interpretation of the results is clear. We see that, in general, as the river becomes larger, the small scale (tributaries) differences do not have much effect. In the right panel (c), we show the results of using a non-linear model. The model now has "memory" in that runoff coefficient being a function of the past rainfall thus soil moisture exerts influence on the flood peak. We can see this in the south-east part of the basin where a previous storm conditioned the soil moisture so that it responds differently than the linear model indicates. The non-linear model also has a non-linear velocity formulation, reflecting increasing flow velocities as channel volume and depth increase (Ghimire et al. 2018).

These three views allow us to better understand the impact of rainfall variability and its propagation through hydrologic systems. Regarding the dynamic aspects of the hydrologic transformation, one should examine two basic characteristics:

(1) time-aggregated discharge over various time scales, and (2) instantaneous peak flows, as these correspond to the maximum societal impact of floods. Peak flows and their statistical properties are the quantities used in engineering design and for flood mitigation strategies.

It should be clear by now why hydrologic validation of satellite-based rainfall products should not be based on a simple point-wise hydrograph. What we observe at a single station may happened there by chance, either due to a canceling of errors in different branches of the river network, or by an amplification due to a "traffic jam" in the river network where overestimated flows meet.

Tracking the rainfall products as they propagate through a river network requires a specialized implementation and visualization system. If we want to follow details of the hydrographs everywhere in the network, including at the smallest scales, we need to solve the model equation frequently and accurately, which is computationally expensive. To visualize the output, we need a mapping system of river network and other points of interest as well as zoom and time control. The static images we presented herein serve well as examples but do not do the justice regarding the insights that one may get from using a fully dynamic and distributed system.

While useful, the visualization of the network-based rainfall and model-based hydrologic quantities is not sufficient to statistically quantify uncertainties associated with the satellite-based products. In particular, the question about the space and time scale dependence as measured by the correlation function or the variogram, requires a specialized approach not based on Euclidean geometry. As we demonstrate in the next chapter, Euclidean geometry-based characterization of dependence as a function of separation distance is misleading in the context of river processes.

32.2.3 River Network Based Error Dependence

River network-based characterization of the satellite error dependence provides a hydrologically meaningful measure of product performance. An example of this approach is introduced in ElSaadani et al. (2018a, b). This approach is similar to stochastic interpolation known as kriging but not over Euclidean space but rather on a fractal space given by the river network. Two points in a river basin may be close to each other in Euclidean geometry but far from each other if the path connecting them is along a tree-like network. In hydrologic applications uncertainties in rainfall products are tied to river network structure. The appropriate description of the spatial dependence of the errors should take the topology of the network into account. Early results given by ElSaadani et al. (2018a, b) indicate that the error variance as a function of distance is reduced if considered along the flow path but the correlation distance far exceeds that taken in Euclidean space.

The method we use is described by Ver Hoef and Peterson (2010) and called the Spatial Stream Network (SSN). SSN accounts for the nested nature of the stream network by using stream distances and site connectivity information. The stream distances are separated into two categories: (1) flow-connected distances between sites, i.e. discharge flows directly from the upstream site location to the downstream site location; and (2) flow-unconnected distances between sites that are only connected through a common downstream junction in the stream network, i.e. discharge cannot flow from one site to another but rather from each site to the nearest connecting downstream junction.

The key tool that needs to be estimated is the Torgegram (Zimmerman and Ver Hoef 2017), an equivalent to the semi-variogram in classical geostatistical methods. It requires the quantities to be characterized organized on relevant river network. The complete semi-variogram produced from this analysis consists of two distinct semi-variograms called the Torgegram when combined for flow-connected and flow-unconnected categories. If the spatial dependence of the variable of interest is not impacted by network connectivity, the flow-connected and flow-unconnected portions of the Torgegram should be similar; otherwise, they could be quite different.

In addition, Ver Hoef and Peterson (2010) developed theoretical spatial covariance models valid for stream networks. For a detailed discussion of modeling of the satellite rainfall dependency along river network interested readers are referred to ElSaadani et al. (2018a, b). Here we limit discussion to an example of an empirical Torgegram of satellite errors.

Consider the extreme rainfall event that happened in September 2016 in north central Iowa. In Fig. 32.6 we show the rainfall accumulations during the flood event period (14–27 September 2016) obtained from MRMS (a), IMERG (b). We process the rainfall data in the same way as we discussed above in Sect. 32.2.2. The data organized on the river network represents discrepancies between the IMERG product and a high-quality ground-based reference. For the sake of illustration, if we ignore the uncertainties associated with the reference, we can call these discrepancies as errors.

To reduce the computational burden of calculating the Tergogram, we sampled the errors at 2000 points (links) of the network. With such a large sample, we can

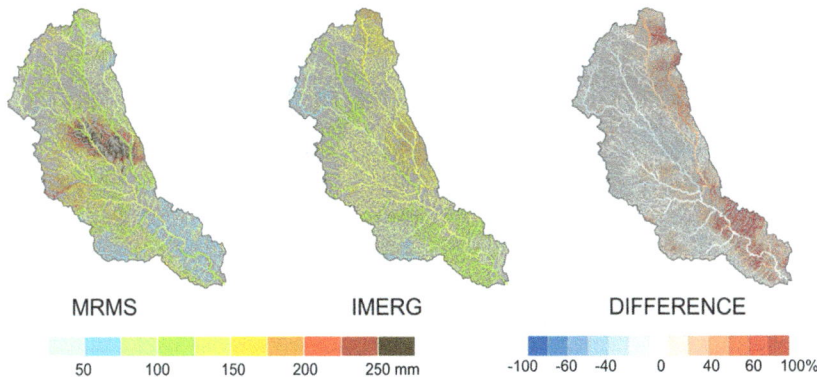

Fig. 32.6 Network-integrated rainfall for the reference product (MRMS) and a satellite product (IMERG). The third panel show the relative differences between the two products

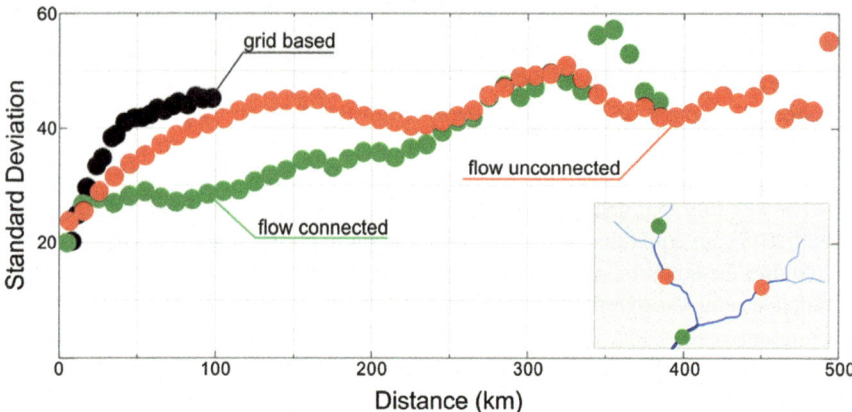

Fig. 32.7 Standard deviation (in mm) of the rainfall product discrepancies calculated in Euclidean space (black), and over the river network for flow-connected segments (red) and flow-unconnected segments. The inset illustrates the flow-connected and unconnected cases

reliably calculate the semi-variogram of the flow-connected, flow-unconnected, and Euclidean distance-based cases. We have many, often over 1000 pairs for each distance class. Prior to the calculations, we have removed the linear trend from the data to avoid biasing the final product. We show the Tergogram with the classical semivariogram in Fig. 32.7. For easier interpretation of the results, we also plot the square root of the variogram values.

It is clear that the standard characterization of the satellite errors severely underestimates the spatial dependence. The Euclidean variogram suggest that after about 50–70 km the errors are uncorrelated while network-based analysis, with the flow-connected view extends this distance considerably. The above analyses and discussion imply that river basins are efficient in effectively filtering out random errors but they also propagate the systematic errors over long distance.

32.3   Conclusions

The presented validation framework allows deeper insights into the space-time dynamics of satellite error propagation. While the main objective of product validation is the determination of the degree to which the product is valid, i.e. relevant to the true but unknown quantity of interest, in our case the interest is in a derived quantity, i.e. streamflow and flooding. As rainfall is the agent of flooding, it is important to understand its estimation (retrieval) uncertainty in the context of the application, in this case a streamflow forecasting system that uses a rainfall-runoff model.

The framework provides an avenue to develop performance metrics for rainfall product development. We have clearly demonstrated that basins and river networks draining them are efficient filters of random errors. We also demonstrated that errors persist in space and time. Therefore, it seems that for applications for larger basins, minimizing error variance is not the best strategy. Instead, minimizing biases should lead to better results. For small basins, both error variance and the biases are important. The tradeoff between them should be application specific.

The framework allows addressing the central question about the basin scale at which the satellite-based rainfall products have sufficient skill for making decisions regarding flood mitigation. However, the framework is not sufficient to definitely answer the question. This is because risk tolerance and resiliency of various communities is different and what works in one place may not work in another. Also, the nature of rainfall regime as well as the local terrestrial processes differ from place to place and impacts the results.

An implementation of the framework requires a proper informatics and visualization infrastructure. Analysts should have ready access to time-varying results for streams at multiple scales. Developing such an infrastructure is not easy as the variability scale for rainfall and streamflow is quite different. Also, the infrastructure should allow previewing of long periods, not just specific events. This is because events at large scale are aggregates of the events that happened faster and more frequently at the small scales and because the initial conditions affect spatial scales differently.

The framework we proposed herein is not a panacea for the hydrologic validation of satellite precipitation problem. It merely complements other existing approaches and provides additional insights. The framework is particularly relevant to questions of scale as the spatial scaling properties of river networks are well studied.

References

Aghakouchak, A., Behrangi, A., Sorooshian, S., Hsu, K.-L., & Amitai, E. (2011). Evaluation of satellite-retrieved extreme precipitation rates across the Central United States. *Journal of Geophysical Research, 116*, D02115. https://doi.org/10.1029/2010JD014741.

Aghakouchak, A., Mehran, A., Norouzi, H., & Behrangi, A. (2012). Systematic and random error components in satellite precipitation data sets. *Geophysical Research Letters, 39*, L09406. https://doi.org/10.1029/2012GL051592.

Ashouri, H., Nguyen, P., Thorstensen, A., Hsu, K.-L., Sorooshian, S., & Braithwaite, D. (2016). Assessing the efficacy of high-resolution satellite-based PERSIANN-CDR precipitation product in simulating streamflow. *Journal of Hydrometeorology, 17*, 2061–2076. https://doi.org/10.1175/JHM-D-15-0192.1.

Barros, A., Petersen, W. A., Schwaller, M., Cifelli, R., Mahoney, K., Peters-Lidard, C., Shepherd, J. M., Nesbitt, S., Wolff, D., Heymsfield, G., & Starr, D. (2014). *NASA GPM-ground validation: Integrated precipitation and hydrology experiment 2014 science plan*. Available at https://dukespace.lib.duke.edu/dspace/handle/10161/8991?show=full. Last accessed 1 Dec 2018.

Behrangi, A., Khakbaz, B., Jaw, T. C., AghaKouchak, A., Hsu, K.-L., & Sorooshian, S. (2011). Hydrologic evaluation of satellite precipitation products over a mid-size basin. *Journal of Hydrology, 397*, 225–237. https://doi.org/10.1016/j.jhydrol.2010.11.043.

Ciach, G. J., & Krajewski, W. F. (1999a). Radar–rain gauge comparisons under observational uncertainties. *Journal of Applied Meteorology, 38*, 1519–1525. https://doi.org/10.1175/1520-0450(1999)038<1519:RRGCUO>2.0.CO;2.

Ciach, G. J., & Krajewski, W. F. (1999b). On the estimation of radar rainfall error variance. *Advances in Water Resources, 22*, 585–595. https://doi.org/10.1016/S0309-1708(98)00043-8.

Ciach, G. J., Krajewski, W. F., & Villarini, G. (2007). Product-error-driven uncertainty model for probabilistic quantitative precipitation estimation with NEXRAD data. *Journal of Hydrometeorology, 8*, 1325–1347. https://doi.org/10.1175/2007JHM814.1.

Demaria, E. M. C., Nijssen, B., Valdés, J. B., Rodriguez, D. A., & Su, F. (2014). Satellite precipitation in southeastern South America: How do sampling errors impact high flow simulations? *International Journal of River Basin Management, 12*, 1–13. https://doi.org/10.1080/15715124.2013.865637.

ElSaadani, M., Krajewski, W. F., Goska, R., & Smith, M. (2018a). An investigation of errors in the NFIE-hydro frameworks' stream discharge prediction due to channel routing. *Journal of the American Water Resources Association, 54*(3), 742–751. https://doi.org/10.1111/1752-1688.12627.

ElSaadani, M., Krajewski, W. F., & Zimmerman, D. L. (2018b). River network based characterization of errors in remotely sensed rainfall products in hydrological applications. *Remote Sensing Letters, 9*, 743–752. https://doi.org/10.1080/2150704X.2018.1475768.

Gebremichael, M., & Hossain, F. (2010). *Satellite rainfall applications for surface hydrology*. Dordrecht: Springer. 327 pp, ISBN:978-90-481-2914-0.

Gebremichael, M., & Krajewski, W. F. (2004). Assessment of the statistical characterization of small-scale rainfall variability from radar: Analysis of TRMM ground validation datasets. *Journal of Applied Meteorology, 43*, 1180–1199. https://doi.org/10.1175/1520-0450(2004)043<1180:AOTSCO>2.0.CO;2.

Ghimire, G., Krajewski, W. F., & Mantilla, R. (2018). A power law model for river flow velocity in Iowa basins. *Journal of the American Water Resources Association, 54*, 1055–1067. https://doi.org/10.1111/1752-1688.12665.

Gourley, J. J., Hong, Y., Flamig, Z. L., Wang, J., Vergara, H., & Anagnostou, E. N. (2011). Hydrologic evaluation of rainfall estimates from radar, satellite, gauge, and combinations on Ft. Cobb Basin, Oklahoma. *Journal of Hydrometeorology, 12*, 973–988. https://doi.org/10.1175/2011JHM1287.1.

Habib, E., Krajewski, W. F., & Kruger, A. (2001). Sampling errors of tipping-bucket rain gauge measurements. *Journal of Hydrologic Engineering, 6*, 159–166. https://doi.org/10.1061/(ASCE)1084-0699(2001)6:2(159).

Harris, A., & Hossain, F. (2008). Investigating the optimal configuration of conceptual hydrologic models for satellite-rainfall-based flood prediction. *IEEE Geoscience and Remote Sensing Letters, 5*, 532–536. https://doi.org/10.1109/LGRS.2008.922551.

Hou, A. Y., Kakar, R. K., Neeck, S., Azarbarzin, A. A., Kummerow, C. D., Kojima, M., Oki, R., Nakamura, K., & Iguchi, T. (2014). The Global Precipitation Measurement mission. *Bulletin of the American Meteorological Society, 95*, 701–722. https://doi.org/10.1175/BAMS-D-13-00164.1.

Huffman, G. J., Bolvin, D. T., Braithwaite, D., Hsu, K., Joyce, R., Kidd, C., Nelkin, E. J., & Xie, P. (2015). *NASA global precipitation measurement (GPM) integrated multi-satellitE retrievals for GPM (IMERG)*. Algorithm Theoretical Basis Doc., v4.5, 26 pp. Available at https://pmm.nasa.gov/sites/default/files/document_files/IMERG_ATBD_V4.5.pdf. Last accessed 27 Oct 2018.

Krajewski, W. F. (2007). Ground networks: Are we doing the right thing? In V. Levizzani, P. Bauer, & F. J. Turk (Eds.), *Measuring precipitation from space* (Advances global change research) (Vol. 28, pp. 403–417). Dordrecht: Springer. ISBN:978-1-4020-5835-6.

Krajewski, W. F., Ciach, G. J., McCollum, J. R., & Bacotiu, C. (2000). Initial validation of the global precipitation climatology project monthly rainfall over the United States. *Journal of Applied Meteorology, 39*, 1071–1086. https://doi.org/10.1175/1520-0450(2000)039<1071:IVOTGP>2.0.CO;2.

Krajewski, W. F., Ceynar, D., Demir, I., Goska, R., Kruger, A., Langel, C., Mantilla, R., Niemeier, J., Quintero, F., Seo, B. C., Small, S. J., Weber, L. J., & Young, N. C. (2017). Real-time flood forecasting and information system for the state of Iowa. *Bulletin of the American Meteorological Society, 98*, 539–554. https://doi.org/10.1175/BAMS-D-15-00243.1.

Liu, Z. (2016). Comparison of integrated multisatellite retrievals for GPM (IMERG) and TRMM multisatellite precipitation analysis (TMPA) monthly precipitation products: Initial results. *Journal of Hydrometeorology, 17*, 777–790. https://doi.org/10.1175/JHM-D-15-0068.1.

Maggioni, V., Reichle, R. H., & Anagnostou, E. N. (2011). The effect of satellite rainfall error modeling on soil moisture prediction uncertainty. *Journal of Hydrometeorology, 12*, 413–428. https://doi.org/10.1175/2011JHM1355.1.

Maggioni, V., Vergara, H. J., Anagnostou, E. N., Gourley, J. J., Hong, Y., & Stampoulis, D. (2013). Investigating the applicability of error correction ensembles of satellite rainfall products in river flow simulations. *Journal of Hydrometeorology, 14*, 1194–1211. https://doi.org/10.1175/JHM-D-12-074.1.

Nikolopoulos, E. I., Anagnostou, E. N., Hossain, F., Gebremichael, M., & Borga, M. (2010). Understanding the scale relationships of uncertainty propagation of satellite rainfall through a distributed hydrologic model. *Journal of Hydrometeorology, 11*, 520–532. https://doi.org/10.1175/2009JHM1169.1.

Nikolopoulos, E. I., Anagnostou, E. N., & Borga, M. (2013). Using high-resolution satellite rainfall products to simulate a major flash flood event in northern Italy. *Journal of Hydrometeorology, 14*, 171–185. https://doi.org/10.1175/JHM-D-12-09.1.

Quintero, F., Krajewski, W. F., Mantilla, R., Small, S., & Seo, B.-C. (2016). A spatial–dynamical framework for evaluation of satellite rainfall products for flood prediction. *Journal of Hydrometeorology, 17*, 2137–2154. https://doi.org/10.1175/JHM-D-15-0195.1.

Seo, B.-C., & Krajewski, W. F. (2011). Investigation of the scale-dependent variability of radar-rainfall and rain gauge error covariance. *Advances in Water Resources, 34*, 152–163. https://doi.org/10.1016/j.advwatres.2010.10.006.

Small, S. J., Jay, L. O., Mantilla, R., Curtu, R., Cunha, L. K., Fonley, M., & Krajewski, W. F. (2013). An asynchronous solver for systems of ODEs linked by a directed tree structure. *Advances in Water Resources, 53*, 23–32. https://doi.org/10.1016/j.advwatres.2012.10.011.

Thiemig, V., Rojas, R., Zambrano-Bigiarini, M., & De Roo, A. (2013). Hydrological evaluation of satellite-based rainfall estimates over the Volta and Baro-Akobo basin. *Journal of Hydrology, 499*, 324–338. https://doi.org/10.1016/j.jhydrol.2013.07.012.

Ver Hoef, J. M., & Peterson, E. E. (2010). A moving average approach for spatial statistical models of stream networks. *Journal of the American Statistical Association, 105*, 6–18. https://doi.org/10.1198/jasa.2009.ap08248.

Vergara, H., Hong, Y., Gourley, J. J., Anagnostou, E. N., Maggioni, V., Stampoulis, D., & Kirstetter, P.-E. (2014). Effects of resolution of satellite-based rainfall estimates on hydrologic modeling skill at different scales. *Journal of Hydrometeorology, 15*, 593–613. https://doi.org/10.1175/JHM-D-12-0113.1.

Villarini, G., & Krajewski, W. F. (2010). Review of the different sources of uncertainty in single polarization radar-based estimates of rainfall. *Surveys in Geophysics, 31*, 107–129. https://doi.org/10.1007/s10712-009-9079-x.

Wu, H., Adler, R. F., Hong, Y., Tian, Y., & Policelli, F. (2012). Evaluation of global flood detection using satellite-based rainfall and a hydrologic model. *Journal of Hydrometeorology, 13*, 1268–1284. https://doi.org/10.1175/JHM-D-11-087.1.

Wu, H., Adler, R. F., Tian, Y., Huffman, G. J., Li, H., & Wang, J. (2014). Real-time global flood estimation using satellite-based precipitation and a coupled land surface and routing model. *Water Resources Research, 50*, 2693–2717. https://doi.org/10.1002/2013WR014710.

Zalenski, G., Krajewski, W. F., Quintero, F., Restrepo, P., & Buan, S. (2017). Analysis of national weather service stage forecast errors. *Weather and Forecasting, 32*, 1441–1465. https://doi.org/10.1175/WAF-D-16-0219.1.

Zhang, J., Howard, K., Langston, C., Kaney, B., Qi, Y., Tang, L., Grams, H., Wang, Y., Cocks, S., Martinaitis, S., Arthur, A., Cooper, K., Brogden, J., & Kitzmiller, D. (2016). Multi-Radar Multi-Sensor (MRMS) quantitative precipitation estimation: Initial operating capabilities. *Bulletin of the American Meteorological Society, 97,* 621–638. https://doi.org/10.1175/BAMS-D-14-00174.1.

Zimmerman, D. L., & Ver Hoef, J. M. (2017). The Torgegram for fluvial variography: Characterizing spatial dependence on stream networks. *Journal of Computational and Graphical Statistics, 26,* 253–264. https://doi.org/10.1080/10618600.2016.1247006.

Chapter 33
Global-Scale Evaluation of 22 Precipitation Datasets Using Gauge Observations and Hydrological Modeling

Hylke E. Beck, Noemi Vergopolan, Ming Pan, Vincenzo Levizzani, Albert I. J. M. van Dijk, Graham P. Weedon, Luca Brocca, Florian Pappenberger, George J. Huffman, and Eric F. Wood

Abstract We undertook a comprehensive evaluation of 22 gridded (quasi-)global (sub-)daily precipitation (P) datasets for the period 2000–2016. Thirteen non-gauge-corrected P datasets were evaluated using daily P gauge observations from 76,086 gauges worldwide. Another nine gauge-corrected datasets were evaluated using hydrological modeling, by calibrating the conceptual model HBV against streamflow records for each of 9053 small to medium-sized (<50,000 km^2) catchments worldwide, and comparing the resulting performance. Marked differences in spatio-temporal patterns and accuracy were found among the datasets. Among the uncorrected P datasets, the satellite- and reanalysis-based MSWEP-ng V1.2 and V2.0 datasets generally showed the best temporal correlations with the gauge observations, followed by the reanalyses (ERA-Interim, JRA-55, and NCEP-

H. E. Beck (✉) · N. Vergopolan · M. Pan · E. F. Wood
Department of Civil and Environmental Engineering, Princeton University, Princeton, NJ, USA
e-mail: hylke.beck@gmail.com

V. Levizzani
National Research Council of Italy, Institute of Atmospheric Sciences and Climate (CNR-ISAC), Bologna, Italy

A. I. J. M. van Dijk
Fenner School of Environment & Society, The Australian National University, Canberra, Australia

G. P. Weedon
Met Office, Joint Centre for Hydro-Meteorological Research, Wallingford, UK

L. Brocca
Research Institute for Geo-Hydrological Protection (IRPI), National Research Council (CNR), Perugia, Italy

F. Pappenberger
European Centre for Medium-range Weather Forecasts, Reading, UK

G. J. Huffman
NASA, Goddard Space Flight Center, Greenbelt, MD, USA

© Springer Nature Switzerland AG 2020
V. Levizzani et al. (eds.), *Satellite Precipitation Measurement*, Advances in Global Change Research 69, https://doi.org/10.1007/978-3-030-35798-6_9

CFSR) and the satellite- and reanalysis-based CHIRP V2.0 dataset, the estimates based primarily on passive microwave remote sensing of rainfall (CMORPH V1.0, GSMaP V5/6, and TMPA 3B42RT V7) or near-surface soil moisture (SM2RAIN-ASCAT), and finally, estimates based primarily on thermal infrared imagery (GridSat V1.0, PERSIANN, and PERSIANN-CCS). Two of the three reanalyses (ERA-Interim and JRA-55) unexpectedly obtained lower trend errors than the satellite datasets. Among the corrected P datasets, the ones directly incorporating daily gauge data (CPC Unified and MSWEP V1.2 and V2.0) generally provided the best calibration scores, although the good performance of the fully gauge-based CPC Unified is unlikely to translate to sparsely or ungauged regions. Next best results were obtained with P estimates directly incorporating temporally coarser gauge data (CHIRPS V2.0, GPCP-1DD V1.2, TMPA 3B42 V7, and WFDEI-CRU), which in turn outperformed the one indirectly incorporating gauge data through another multi-source dataset (PERSIANN-CDR V1R1). Our results highlight large differences in estimation accuracy, and hence, the importance of P dataset selection in both research and operational applications. The good performance of MSWEP emphasizes that careful data merging can exploit the complementary strengths of gauge-, satellite- and reanalysis-based P estimates.

Keywords Precipitation · Rainfall · Evaluation · Validation · Raingauges · Streamflow · Satellite products · Reanalysis · Radar

33.1   Introduction

Precipitation (P) is arguably the most important driver of the hydrological cycle, but also one of the most challenging to estimate (Daly et al. 2008; Michaelides et al. 2009; Kidd and Levizzani 2011; Tapiador et al. 2012). Over recent decades, several gridded P datasets have been developed that are suitable for large-scale hydrological applications (for overviews, see Table 33.1, Beck et al. 2017b, http://ipwg.isac.cnr.it, and http://reanalyses.org, last accessed 20 Oct. 2018). The datasets differ in terms of design objective (temporal homogeneity, instantaneous accuracy, or both), data sources (radar, gauge, satellite, analysis, or reanalysis, or combinations thereof), spatial resolution (from 0.05° to 2.5°), spatial coverage (from continental to fully global), published temporal resolution (from 30 min to monthly), temporal span (from ~1 to 115 years), and latency (from ~3 h to several years).

   A plethora of studies addressed the important task of evaluating these P datasets to understand their respective advantages and limitations (see reviews by Gebremichael 2010, Maggioni et al. 2016). Most studies assessed accuracy using independent gauge observations (e.g., Hirpa et al. 2010; Buarque et al. 2011; Bumke et al. 2016; Alijanian et al. 2017) or gauge-adjusted radar fields (e.g., AghaKouchak et al. 2011; Islam et al. 2012), while others merely compared their spatio-temporal patterns (e.g., Kidd et al. 2013). Still others quantified the performance of different P datasets using hydrological modeling, by comparing simulated and observed

Table 33.1 Overview of the 22 (quasi-)global (sub-)daily gridded P datasets evaluated in this study

Short name	Full name and details	Data source(s)	Spatial resolution	Spatial coverage	Temporal resolution	Temporal coverage	References
Non-gauge-corrected datasets							
CHIRP V2.0	Climate Hazards group Infrared Precipitation (CHIRP) V2.0 (http://chg.ucsb.edu/data/chirps/)	S, R	0.05°	Land, <50°	Daily	1981–NRT	Funk et al. (2015a)
CMORPH V1.0	CPC MORPHing technique (CMORPH) V1 (www.cpc.ncep.noaa.gov)	S	0.07°	<60°	30 min	1998–NRT	Joyce et al. (2004)
ERA-interim	European Centre for Medium-range Weather Forecasts ReAnalysis Interim (ERA-interim; https://www.ecmwf.int/en/research/climate-reanalysis/era-interim)	R	~0.75°	Global	3 hourly	1979–2017	Dee et al. (2011)
GSMaP V5/6	Global Satellite Mapping of Precipitation (GSMaP) Moving Vector with Kalman (MVK) standard V5 and V6 (http://sharaku.eorc.jaxa.jp/GSMaP/)	S	0.1°	<60°	Hourly	2000–NRT	Ushio et al. (2009)
GridSat V1.0	P derived from the Gridded Satellite (GridSat) B1 thermal infrared archive v02r01 (Knapp et al. 2011; https://www.ncdc.noaa.gov/gridsat/)	S	0.1°	<50°	3 hourly	1983–2016	Beck et al. (2019b)
JRA-55	Japanese 55-year ReAnalysis (JRA-55; \urljra.kishou.go.jp/JRA-55)	R	~0.56°	Global	3 hourly	1959–NRT	Kobayashi et al. (2015)
MSWEP-ng V1.2	Multi-Source Weighted-Ensemble Precipitation (MSWEP) no-gauge (ng) V1.2 (www.gloh2o.org)	S, R	0.25°	Global	3 hourly	1979–2015	Beck et al. (2017b)
MSWEP-ng V2.0	Multi-Source Weighted-Ensemble Precipitation (MSWEP) no-gauge (ng) V2.0 (www.gloh2o.org)	S, R	0.1°	Global	3 hourly	1979–NRT	Beck et al. (2019b)
NCEP-CFSR	National Centers for Environmental Prediction (NCEP) Climate Forecast System Reanalysis (CFSR; http://cfs.ncep.noaa.gov/cfsr/)	R	~0.31°	Global	Hourly	1979–2010	Saha et al. (2010)
PERSIANN	Precipitation Estimation from Remotely Sensed Information using Artificial Neural Networks (PERSIANN; http://chrs.web.uci.edu)	S	0.25°	<60°	Hourly	2000–NRT	Sorooshian et al. (2000)
PERSIANN-CCS	Precipitation Estimation from Remotely Sensed Information using Artificial Neural Networks (PERSIANN) Cloud Classification System (CCS: http://chrs.web.uci.edu)	S	0.04°	<60°	Hourly	2003–NRT	Hong et al. (2004)

(continued)

Table 33.1 (continued)

Short name	Full name and details	Data source(s)	Spatial resolution	Spatial coverage	Temporal resolution	Temporal coverage	References
SM2RAIN-ASCAT	P inferred from Advanced Scatterometer (ASCAT) satellite near-surface soil moisture (http://hydrology.irpi.cnr.it)	S	0.5°	Land	Daily	2007–2015	Brocca et al. (2014)
TMPA 3B42RT V7	TRMM Multi-satellite Precipitation Analysis (TMPA) 3B42RT V7 (https://mirador.gsfc.nasa.gov)	S	0.25°	<50°	3 hourly	2000–NRT	Huffman et al. (2007)
Gauge-corrected datasets							
CHIRPS V2.0	Climate hazards group infrared precipitation with stations (CHIRPS) V2.0 (http://chg.ucsb.edu/data/chirps/)	G, S, R	0.05°	Land, <50°	Daily	1981–NRT	Funk et al. (2015a)
CMORPH-CRT V1.0	CPC MORPHing technique (CMORPH) bias corrected (CRT) V1.0 (www.cpc.ncep.noaa.gov)	G, S	0.07°	<60°	30 min	1998–2015	Xie et al. (2017)
CPC unified	Climate Prediction Center (CPC) unified V1.0 and RT (https://www.esrl.noaa.gov/psd/data/gridded/)	G	0.5°	Land	Daily	1979–NRT	Chen et al. (2008)
GPCP-1DD V1.2	Global Precipitation Climatology Project (GPCP) 1-Degree Daily (1DD) combination V1.2 (https://precip.gsfc.nasa.gov)	G, S	1°	Global	Daily	1996–2015	Huffman et al. (2001)
MSWEP V1.2	Multi-Source Weighted-Ensemble Precipitation (MSWEP) V1.2 (www.gloh2o.org)	G, S, R	0.25°	Global	3 hourly	1979–2015	Beck et al. (2017b)
MSWEP V2.0	Multi-Source Weighted-Ensemble Precipitation (MSWEP) V2.0 (www.gloh2o.org)	G, S, R	0.1°	Global	3 hourly	1979–NRT	Beck et al. (2019b)
PERSIANN-CDR V1R1	Precipitation Estimation from Remotely Sensed Information using Artificial Neural Networks (PERSIANN) Climate Data Record (CDR) V1R1 (http://chrs.web.uci.edu)	G, S	0.25°	<60°	6 hourly	1983–2016	Ashouri et al. (2015)
TMPA 3B42 V7	TRMM Multi-satellite Precipitation Analysis (TMPA) 3B42 V7 (https://mirador.gsfc.nasa.gov/)	G, S	0.25°	<50°	3 hourly	2000–2017	Huffman et al. (2007)
WFDEI-CRU	WATCH Forcing Data ERA-Interim (WFDEI; www.eu-watch.org)	G, R	0.5°	Global	3 hourly	1979–2015	Weedon et al. (2014)

Abbreviations in the data source(s) column defined as follows: *G* gauge, *S* satellite, *R* reanalysis. The acronym NRT in the temporal coverage column stands for near real time. In the spatial coverage column, "global" indicates fully global coverage including ocean areas, while "land" indicates that the coverage is limited to the terrestrial surface

values of river discharge (Q; e.g., Collischonn et al. 2008, Behrangi et al. 2011, Bitew et al. 2012, Falck et al. 2015) or soil moisture (e.g., Pan et al. 2010; Albergel et al. 2013; Martens et al. 2017). More recently, Massari et al. (2017) assessed the performance of different P datasets using triple collocation. Marked differences in spatio-temporal P patterns and accuracy have been found among the datasets, even among those employing the same data sources. This highlights the critical importance of dataset choice for research and operational applications alike.

Previous evaluation studies used a wide variety of evaluation approaches and performance metrics (Ebert 2007; Gebremichael 2010; Loew et al. 2017). However, many studies considered only a single P dataset (e.g., Scheel et al. 2011; Nair and Indu 2017) or disregarded (re)analysis-based P datasets (e.g., Moazami et al. 2013; Mei et al. 2014; Zambrano-Bigiarini et al. 2017), despite their demonstrated superior performance in cold climates (Ebert et al. 2007; Beck et al. 2017b; Massari et al. 2017). In addition, some studies re-used gauge observations already incorporated in some of the P datasets to determine their accuracy (e.g., Chen et al. 2013; Ashouri et al. 2016; Zambrano-Bigiarini et al. 2017), precluding independent validation. Furthermore, to our knowledge, so far no study has accounted for differences in the exact UTC boundary of the 24 h accumulation period of daily gauge reports when evaluating P datasets, potentially confounding the results. Moreover, studies employing hydrological modeling generally used Q observations from a small number of catchments (e.g., Bitew et al. 2012, Tang et al. 2016; both used only one) and did not attempt to recalibrate the hydrological model for each P dataset individually (e.g., Su et al. 2008; Li et al. 2013), leading to combined rainfall and model uncertainty that is not easily interpreted. Finally, many have a regional (subcontinental) focus (Maggioni et al. 2016), and therefore it is not clear to what extent the results can be generalized.

Nevertheless, there have also been several (quasi-)global P dataset evaluation studies that produced general insights (e.g., Adler et al. 2001; Fekete et al. 2004; Voisin et al. 2008; Bosilovich et al. 2008; Tian and Peters-Lidard 2010; Lorenz and Kunstmann 2012; Yong et al. 2015; Herold et al. 2015; Gehne et al. 2016; Massari et al. 2017). These studies revealed that satellites (reanalyses) exhibit superior performance at low (high) latitudes dominated by intense, localized convective (persistent, large-scale stratiform) P systems. However, none of these studies took advantage of the vast number of P gauge observations contained in the freely available GHCN-D (Menne et al. 2012) and GSOD (https://data.noaa.gov, last accessed 20 Oct. 2018) databases. Among the only two studies employing hydrological modeling, Fekete et al. (2004) performed monthly simulations and did not compare the results against observed Q, while Voisin et al. (2008) used monthly observed Q data from only nine very large catchments (>290,000 km^2). Moreover, several promising recently released or revised P datasets, such as CHIRPS V2.0, MSWEP V2.0, and PERSIANN-CDR V1R1 (see Table 33.1), have not been thoroughly evaluated yet at a (quasi-)global scale.

Our objective was to undertake the most comprehensive global-scale P dataset evaluation to date. We evaluated 13 non-gauge-corrected P datasets using daily P gauge observations from 76,086 gauges worldwide. Nine more gauge-corrected

P datasets were evaluated using hydrological modeling for 9053 catchments ($<$50,000 km^2) worldwide, by calibrating a hydrological model. The expectation is that such a large number of P datasets and observations should lead to more generally valid conclusions and allow us to explicitly compare the performance among climate types and regions (Andréassian et al. 2007; Gupta et al. 2014).

33.2   Data and Methods

33.2.1   P Datasets

Table 33.1 presents the 22 gridded P datasets included in the evaluation. The datasets were classified as either uncorrected, meaning that their temporal dynamics depend entirely on satellite and/or reanalysis data, or gauge-corrected, meaning that their temporal dynamics depend at least partly on gauge data (hence precluding an independent evaluation using P gauge observations). We included seven datasets based exclusively on satellite data (CMORPH V1.0, GSMaP, GridSat V1.0, PERSIANN, PERSIANN-CCS, SM2RAIN-ASCAT, and TMPA 3B42RT V7), three based exclusively on reanalysis data (ERA-Interim, JRA-55, and NCEP-CFSR), and three incorporating both satellite and reanalysis data (CHIRP V2.0, and MSWEP-ng V1.2 and V2.0). Among the gauge-corrected datasets, four combined gauge and satellite data (CMORPH-CRT, GPCP-1DD V1.2, PERSIANN-CDR V1R1, and TMPA 3B42 V7), one combined gauge and reanalysis data (WFDEI-CRU), while three combined gauge, satellite, and reanalysis data (CHIRPS V2.0, and MSWEP V1.2 and V2.0). We also included a fully gauge-based dataset (CPC Unified). For clarity and reproducibility, we report dataset version numbers throughout the study for the datasets for which this information was available. We only included datasets with a temporal span of $>$8 years.

33.2.2   Performance Evaluation Using Gauge Observations

The performance of the 13 uncorrected P datasets (see Table 33.1) was evaluated using daily gauge observations from across the globe. Our collection of gauge observations was compiled from the Global Historical Climatology Network-Daily (GHCN-D) database (Menne et al. 2012), the Global Surface Summary of the Day (GSOD) database (https://data.noaa.gov), the Latin American Climate Assessment & Dataset (LACA&D) database (http://lacad.ciifen.org, last accessed 20 Oct. 2018), the Chile Climate Data Library (http://www.climatedatalibrary.cl, last accessed 20 Oct. 2018), and national databases for Mexico, Brazil, Peru, and Iran. To discard erroneous observations, each gauge record was subjected to several quality checks as described in Beck (2017). Only gauges with $>$365 days of valid data (not necessarily consecutive) during 2000–2016 were retained. To minimize temporal mismatches in gauge and gridded P time series, we used the gauge reporting times from Beck (2017) to shift the records of gauges with reporting times $>+12$ h UTC backward by 1 day, and the records of gauges with reporting times <-12 h UTC forward by 1 day. In total 76,086 gauges had sufficient quality-controlled data for the evaluation.

We considered the following five performance metrics to evaluate the P datasets in terms of temporal dynamics: (i) Pearson linear correlation coefficient (R) calculated for 3-day means (R_{3day}); (ii) R calculated for monthly means ($R_{monthly}$); (iii) R calculated for 6-month Standardized Precipitation Index values (R_{SPI-6}; Hayes et al. 1999); (iv) mean absolute error (MAE; mm month^{-1}) for monthly means; and (v) the trend error (the difference between gauge- and dataset-based linear regression slopes calculated from annual anomalies; % year^{-1}). We opted for MAE instead of the more widely used root mean square error (RMSE) because the errors are unlikely to follow a normal distribution (Chai and Draxler 2014; Willmott et al. 2017). We used 3-day rather than daily means for R_{3day} to minimize the impact of any residual mismatches in the UTC boundary of the 24 h accumulation period between the gauges and datasets. The R_{3day} metric was only calculated if ≥ 60 3-day contemporaneous gauge and dataset values were available, while the $R_{monthly}$, R_{SPI-6}, and MAE were only calculated if ≥ 12 monthly contemporaneous gauge and dataset values were available.

To evaluate P datasets in terms of long-term mean climate indices, we considered the following four metrics: (i) long-term relative bias, defined as $[s - o]/[s + o]$, where s and o represent the dataset- and gauge-based long-term means, respectively; (ii) annual number of dry days error (using a 0.5 mm d^{-1} threshold to identify dry days, as in Akinremi et al. 1999, Haylock et al. 2008, and Driouech et al. 2009); and (iii) 99th and 99.9th percentile daily P error (mm d^{-1}). Bias and trend error metrics were calculated if >5 years of daily simultaneous gauge and dataset values were available.

33.2.3 Performance Evaluation Using Hydrological Modeling

The performance of the nine gauge-corrected P datasets (Table 33.1) was evaluated using hydrological modeling for 9053 catchments. Our collection of Q observations was compiled from the same three sources as Beck et al. (2015), viz. (i) the US Geological Survey (USGS) Geospatial Attributes of Gages for Evaluating Streamflow (GAGES)-II database (Falcone et al. 2010); (ii) the Global Runoff Data Centre (GRDC; http://www.bafg.de/GRDC/, last accessed 20 Oct. 2018); and (iii) the Australian Peel et al. (2000) database. We only used catchments $<50,000$ km^2 because applying a daily lumped hydrological model in very large catchments would result in spatial averaging of the forcings over very large areas, confounding the daily runoff generation and water balance calculations. In addition, catchments were required to have a Q record length >365 days (not necessarily consecutive) during 2000–2012 (the common temporal coverage of the P datasets), resulting in 9053 catchments that were suitable for the evaluation (5th, 50th, and 95th catchment-size percentiles equal to 9633, and 18,468 km^2, respectively).

For each catchment, the HBV conceptual hydrological model (Bergström 1992; Seibert and Vis 2012) was calibrated in a lumped fashion against Q observations using daily P time series from each of the datasets to force the model. The model was selected because of its agility, computational efficiency, and widespread successful application (e.g., Te Linde et al. 2008; Deelstra et al. 2010; Plesca et al. 2012; Beck et al. 2013, 2017a; Valéry et al. 2014; Vetter et al. 2015). For the calibration, we employed the $(\mu + \lambda)$ evolutionary algorithm (Ashlock 2010; Fortin et al. 2012) with the population size (μ) set to 20, the recombination pool size (λ) set to 40, and the number of generations set to 12 (amounting to 480 model runs per catchment per P dataset and approximately 40 million model runs in total). See Beck et al. (2016, 2017b) for more details on the hydrological model, calibration algorithm, model parameter ranges, Q observations, potential evaporation (Ep) forcing, and air temperature (Ta) forcing. We recognize that using data from different sources may bias results as the water balances are unlikely to be closed.

As objective function we used the Nash and Sutcliffe (1970) efficiency (NSE) computed between 3-day mean simulated and observed Q time series. We used the NSE, despite the criticism it has received (e.g., Schaefli and Gupta 2007; Jain and Sudheer 2008; Criss and Winston 2008; Gupta et al. 2009) for three reasons. First, the NSE is highly sensitive to peak flows (Krause et al. 2005), which is desirable for this study given that peak flows are primarily driven by the P forcing, whereas low flows are primarily driven by the hydrological model structure and parameters. Second, besides peak flows, NSE is also sensitive to the long-term bias (Gupta et al. 2009), another important feature of the hydrograph primarily influenced by the precipitation forcing. Third, most hydrologists and meteorologists are familiar with the NSE (Moriasi et al. 2007), facilitating the interpretation of the obtained values. We used 3-day rather than daily mean Q time series for the NSE calculation to reduce the impact of temporal mismatches in simulated and observed Q peaks. A higher calibration NSE generally implies that the P dataset in question is more consistent with the Q observations and Ep and Ta estimates and thus that the P dataset is more accurate.

33.3 Results and Discussion

33.3.1 Performance in Terms of Temporal Dynamics

The temporal dynamics of the 13 uncorrected P datasets were evaluated using daily P observations from 76,086 gauges around the globe. Table 33.2 presents summary statistics separately for the gauges located at latitudes <40° for all datasets, and for the gauges located at latitudes ≥40° only for the datasets covering the entire

Table 33.2 Median values of the performance metrics for the uncorrected P datasets based on daily P observations from 76,086 gauges around the globe

	CHIRP V2.0	CMORPH V1.0	ERA-Interim	GridSat V1.0	GSMaP V5/6	JRA-55	MSWEP-ng V1.2	MSWEP-ng V2.0	NCEP-CFSR	PERS.	PERS.-CCS	SM2R-ASCAT	TMPA-3B42RT V7
Gauges located at latitudes <40° (n = 51,271)													
R_{3day} (−)	0.55	0.53	0.59	0.44	0.54	0.56	0.67	0.64	0.57	0.47	0.42	0.52	0.52
$R_{monthly}$ (−)	0.74	0.69	0.75	0.6	0.69	0.75	0.82	0.81	0.75	0.62	0.59	0.68	0.69
R_{SPI-6} (−)	0.71	0.65	0.74	0.6	0.67	0.72	0.81	0.8	0.72	0.58	0.56	0.68	0.66
MAE (mm month^{-1})	30.54	37.81	31.41	43.79	36.1	32.87	26.96	27.99	32.32	42.53	45.51	36.67	37.46
Trend error (% year^{-1})	1.87	2.23	1.97	2.34	3.34	1.91	1.61	1.53	3.56	2.68	2.46	3.39	2.14
Bias (−)	0.06	0.14	0.11	0.07	0.13	0.11	0.06	0.06	0.1	0.17	0.17	0.14	0.11
Annual dry days error (days)	73.85	15.77	47.49	21.55	20.9	43.22	65.06	10.46	37.95	27.65	28.49	112.36	17.63
99th percentile error (mm d^{-1})	13.02	7.27	13.73	4.71	7.54	8.71	11.01	4.59	7.37	9.69	8.97	26	6.18
99.9th percentile error (mm d^{-1})	34.65	17.21	27.82	15.87	18.54	24.66	29.3	14.9	16.09	21.64	20.24	63.38	15.83
Gauges located at latitudes ≥ 40° (n = 24,815)													
R_{3day} (−)	−	−	0.68	−	−	0.67	0.74	0.72	0.66	−	−	−	−
$R_{monthly}$ (−)	−	−	0.78	−	−	0.79	0.84	0.83	0.73	−	−	−	−
R_{SPI-6} (−)	−	−	0.77	−	−	0.78	0.82	0.82	0.73	−	−	−	−
MAE (mm month^{-1})	−	−	21.56	−	−	24.17	19.25	19.7	26.6	−	−	−	−
Trend error (% year^{-1})	−	−	1.41	−	−	1.35	1.27	1.2	2.2	−	−	−	−

(continued)

Table 33.2 (continued)

	CHIRP V2.0	CMORPH V1.0	ERA-Interim	GridSat V1.0	GSMaP V5/6	JRA-55	MSWEP-ng V1.2	MSWEP-ng V2.0	NCEP-CFSR	PERS.	PERS.-CCS	SM2R.-ASCAT	TMPA-3B42RT V7
Bias (−)	–	–	0.09	–	–	0.1	0.05	0.05	0.11	–	–	–	–
Annual dry days error (days)	–	–	45.85	–	–	41.93	58.14	7.79	55.79	–	–	–	–
99th percentile error (mm d^{-1})	–	–	6.26	–	–	3.8	6.1	3.06	3.59	–	–	–	–
99.9th percentile error (mm d^{-1})	–	–	15.95	–	–	12.52	16.8	9.22	9.83	–	–	–	–

Statistics were not shown for the satellite-based P datasets for the group of gauges located at latitudes $\geq 40°$. For all performance metrics, with the exception of R_{3day}, $R_{monthly}$, and RSPI−6, a lower value represents better performance

terrestrial surface (i.e., MSWEP-ng V1.2 and V2.0, and the reanalyses). In terms of temporal correlations (R_{3day}, $R_{monthly}$, and R_{SPI-6}), the satellite- and reanalysis-based MSWEP-ng datasets performed overall slightly better overall than the reanalyses (ERA-Interim, JRA-55, and NCEP-CFSR) and the satellite- and reanalysis-based CHIRP V2.0 dataset, which in turn performed slightly better than the satellite datasets based primarily on passive microwave retrievals (CMORPH V1.0, GSMaP V5/6, and TMPA 3B42RT V7) and near-surface soil moisture (SM2RAIN-ASCAT), which in turn performed slightly better than the satellite datasets based primarily on thermal infrared imagery (GridSat V1.0, PERSIANN and PERSIANN-CCS). The high correlations obtained using both versions of MSWEP-ng underscore the effectiveness of merging multiple satellite and reanalysis datasets (Beck et al. 2017b). Indeed, Ciabatta et al. (2017) found the soil moisture-based rainfall dataset SM2RAIN-CCI to exhibit considerably better 5-day correlations with MSWEP V1.2 than with the comprehensive gauge-based GPCC dataset (Schneider et al. 2014), even though the latter was used to train the SM2RAIN algorithm. In agreement with our results, Stillman et al. (2016) found reanalyses to outperform infrared- and passive microwave-based satellite datasets in Arizona. SM2RAIN-ASCAT was found to perform similarly to TMPA 3B42RT V7, in agreement with Brocca et al. (2014), suggesting that soil moisture-based approaches provide a promising additional source of rainfall estimates.

The better performance of the microwave-based datasets compared to infrared-based ones is in line with previous evaluations (e.g., Hirpa et al. 2010; Peña Arancibia et al. 2013; Cattani et al. 2016) and attributed to the indirect relationship between cloud-top infrared brightness temperatures and surface rainfall (Stephens and Kummerow 2007). Contrary to expectation, PERSIANN-CCS attained lower median correlations than both GridSat V1.0 and PERSIANN, despite using a more sophisticated algorithm and higher spatial resolution (Hong et al. 2004). This indicates that a higher spatial resolution does not necessarily lead to more skillful estimates, and that there may be limited additional value to be gained from extracting cloud-patch characteristics. GridSat V1.0 P estimates have been derived by a cumulative distribution function (CDF) matching the entire period of infrared data to a reference P distribution. Better results might be obtained by CDF matching on a monthly or seasonal climatological basis, to account for intra-annual variability in the infrared–P relationship.

Figure 33.1 presents global R_{3day} maps for a selection of eight P datasets, permitting a geographical interpretation of the results. All datasets performed relatively poorly ($R_{3day} < 0.5$) in arid and tropical regions, due to the often highly localized and short-lived nature of the convective rainfall that dominates (Cecil et al. 2014). Sub-cloud evaporation of falling rain potentially constitutes an additional confounding factor in arid regions (Dinku et al. 2016). Africa showed the lowest R_{3day} values overall, probably due to the high prevalence of convective rain events over most of the continent (Cecil et al. 2014). Conversely, all datasets performed relatively well ($R_{3day} \geq 0.5$) in moist mid-latitude regions with mild winters (e.g., the southeastern US, eastern South America, and eastern China). In accordance with several previous global evaluations (e.g., Barrett et al. 1994; Xie and Arkin 1997; Adler et al. 2001; Ebert et al. 2007; Massari et al. 2017), the reanalyses exhibited

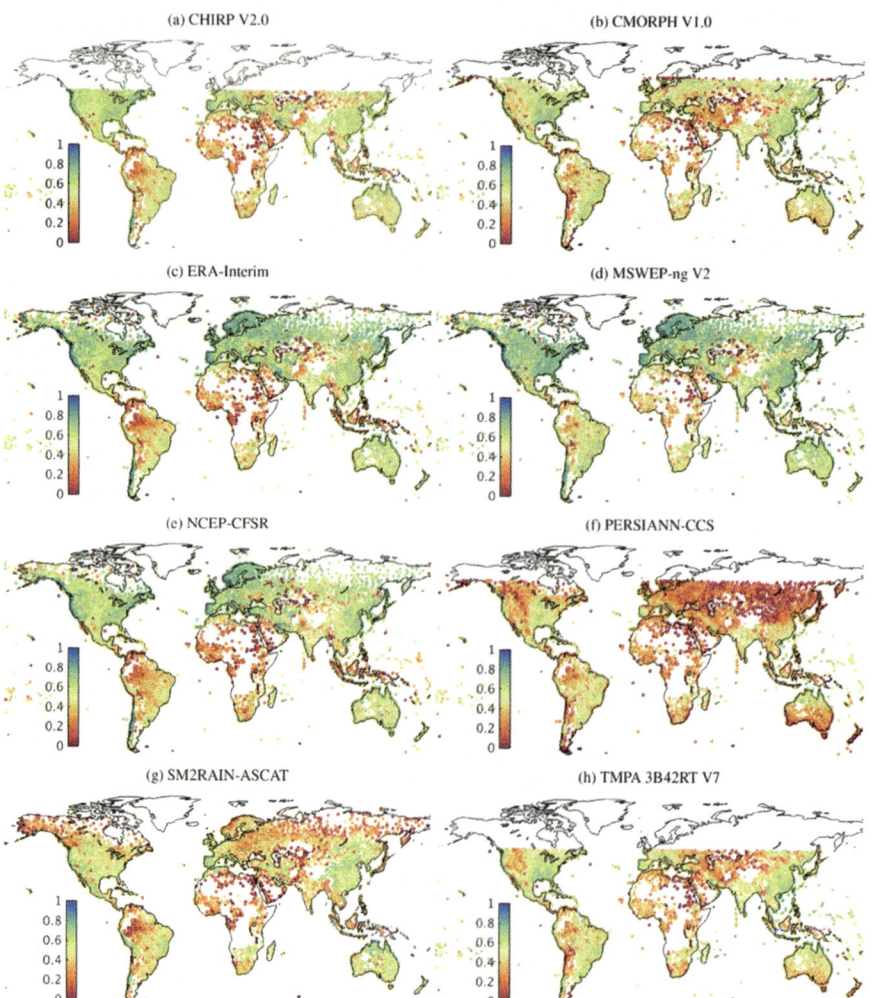

Fig. 33.1 For a selection of the evaluated uncorrected P datasets, temporal correlations between 3-day mean gauge- and dataset-based P time series (R_{3day}). Each data point represents a gauge. See Beck et al. (2017c) for global maps of the other performance metrics

lower skill levels than the microwave- and infrared-based satellite datasets in the tropics, whereas the opposite is true for colder regions (latitudes >40°). The comparatively high skill of the reanalyses in colder regions reflects the ability of atmospheric models to simulate synoptic-scale weather systems (Haiden et al. 2012; Zhu et al. 2014). The comparatively low skill of the reanalyses in the tropics is attributable to deficiencies in the sub-grid convection parameterization schemes (Arakawa 2004), as well as issues in the land surface parameterization and unrealistic strengthening and northward displacement of the monsoon cycle (Di Giuseppe et al. 2013). Multi-scale modeling frameworks incorporating high-resolution

(<4 km), convection-permitting models, which negate the need for sub-grid convection parameterization schemes, provide a promising way forward in this regard (Prein et al. 2015; Clark et al. 2016).

MSWEP V2.0 obtained lower mean annual P trend errors than the other P datasets (Table 33.2). Two of the three reanalyses (ERA-Interim and JRA-55) provided more reliable trends than the satellite datasets, contrary to the common assumption that reanalyses tend to contain temporal discontinuities due to changes in the assimilated observations (Bengtsson et al. 2004; Lorenz and Kunstmann 2012; Kang and Ahn 2015). However, our evaluation covers a relatively short period (2000–2016) during which the assimilated observations did not change considerably (Saha et al. 2010; Dee et al. 2011; Kobayashi et al. 2015). Among the satellite datasets, SM2RAIN-ASCAT provided the least accurate P trends, probably due to the use of two ASCAT sensors after 2013 (on-board MetOp-A and MetOp-B) which artificially increased rainfall amounts obtained using SM2RAIN (separate calibrations for 2007–2012 and 2013–2015 are necessary but yet to be performed). Among the reanalyses, NCEP-CFSR performed worst. Following previous authors (Saha et al. 2010; Wang et al. 2013), we speculate that this may be attributable to the six parallel-run streams of analysis covering different periods, which have been combined to generate the final dataset. The relatively small mean annual P trend errors obtained for the different datasets (ranging from 1.53 to 3.56% year^{-1}) provide some confidence in the ability to infer significant trends from the various datasets. However, trends for variables measured over shorter temporal scales (e.g., annual maxima or percentiles) are likely to be subject to much greater uncertainty. We expect the dataset performance ranking to be similar for the period prior to the year 2000; however, additional studies are necessary to confirm this.

33.3.2 Performance in Terms of Climate Indices

The performance of the 13 uncorrected P datasets in terms of several long-term climate indices is summarized in Table 33.2, listing summary statistics for P gauges at latitudes <40° and ≥40° (for the five datasets covering the entire terrestrial surface), respectively. In terms of bias, the reanalyses performed better overall than the satellite datasets (Table 33.2). Although CHIRP V2.0, GridSat V1.0, and MSWEP-ng V1.2 and V2.0 obtained the best bias scores, these datasets use the gauge-based CHPclim (Funk et al. 2015a) or WorldClim (Fick and Hijmans 2017) datasets to determine their long-term mean. The spread in the range of bias scores among the datasets was generally greatest over topographically complex regions (notably the Rockies, Andes, and Hindu Kush), and in arid regions (notably the Sahara and the Arabian and Gobi deserts), demonstrating the particular difficulty of estimating P in these regions (Fekete et al. 2004; Hirpa et al. 2010; Xu et al. 2017; Kim et al. 2017). All fully global datasets exhibited positive biases at high northern latitudes, probably because the P gauge data used for evaluation were not corrected for wind-induced under-catch (Groisman and Legates 1994; Rasmussen et al. 2012; Kauffeldt et al. 2013).

In terms of the annual number of dry days, the datasets exhibited a particularly large spread in performance, with MSWEP-ng V2.0 outperforming the other datasets by a substantial margin (Table 33.2). The dramatic improvement in MSWEP-ng V2.0 compared to V1.2 is mainly attributable to the CDF corrections introduced in V2, which eliminate the drizzle caused by averaging multiple data sources (Beck et al. 2019a). The infrared- and microwave-based satellite datasets also performed reasonably well, although the P frequency was generally overestimated at low and mid latitudes and underestimated at high latitudes, reflecting the difficulty of detecting P signals at high latitudes (Ferraro et al. 1998; Ebert et al. 2007; Kidd and Levizzani 2011; Kidd et al. 2012; Laviola et al. 2013). Conversely, the reanalyses consistently underestimated the number of dry days across the globe, due to the presence of spurious drizzle caused by deficiencies in the representation and/or parameterization of the physical processes governing P generation (Zolina et al. 2004; Lopez 2007; Sun et al. 2006; Skok et al. 2015). SM2RAIN-ASCAT also consistently underestimated the number of dry days due to the presence of spurious drizzle, in this case due to the relatively noisy soil moisture retrievals (Crow et al. 2011; Brocca et al. 2014) and the use of the already fairly wet ERA-Interim dataset for the algorithm calibration. CHIRP V2.0 also exhibited too few dry days, which is attributed to the use of linear regression equations to estimate 5-day mean P from infrared-based cold-cloud duration values (Funk et al. 2015b). Forcing a hydrological model with P data overestimating the frequency of low-intensity rainfall events is likely to result in overestimated evaporation and underestimated runoff, particularly in regions with high soil or canopy water storage capacities.

The 99th and 99.9th percentile daily P errors measure the error in the magnitude of storms with return periods of 100 days and 2.7 years, respectively (Table 33.2). MSWEP-ng V2.0 performed best in this respect, whereas CHIRP V2.0, the reanalyses, MSWEP-ng V1.2, and particularly SM2RAIN-ASCAT consistently underestimated the 99th and 99.9th percentile storm magnitudes. However, some degree of underestimation would be expected, given the spatial-scale mismatch between gauge observations and grid-cell averages (e.g., Maraun 2013), particularly for P datasets with a coarse spatial resolution (see Table 33.1). Nevertheless, for the reanalyses the underestimation is probably primarily attributable to the aforementioned model uncertainties. For MSWEP-ng V1.2, it is due to the attenuating effect of merging multiple data sources (Beck et al. 2017b). For SM2RAIN-ASCAT, the strong underestimation of storm magnitudes may at least partly be due to signal loss induced by soil saturation (Brocca et al. 2014). Among the microwave- and infrared-based satellite datasets, PERSIANN-CCS showed the greatest spatial variability in storm magnitude bias. The generally strong differences in spatial performance patterns among datasets highlight the difficulty of generalizing the findings of regional (sub-continental) evaluation studies.

33.3.3 Performance Evaluation Using Hydrological Modeling

The performance of the nine gauge-corrected P datasets (see Table 33.1) was evaluated using hydrological modeling for 9053 catchments around the globe. Table 33.3 presents median calibration NSE scores obtained using the different P datasets for different climate zones. The overall performance ranking of the datasets from best to worst (% of catchments in which the dataset performed best between parentheses) is MSWEP V2.0 (45.5%), MSWEP V1.2 (21.5%), CPC Unified (15.9%), WFDEI-CRU (5.0%), TMPA 3B42 V7 (3.3%), CMORPH-CRT V1.0 (2.6%), CHIRPS V2.0 (2.5%), PERSIANN-CDR V1R1 (2.1%), and GPCP-1DD V1.2 (1.6%). Thus, the datasets directly incorporating daily gauge data (CPC Unified, and MSWEP V1.2 and V2.0) overall outperformed the ones directly incorporating 5-day (CHIRPS V2.0) or monthly (GPCP-1DD V1.2, TMPA 3B42 V7, and WFDEI-CRU) gauge data, which in turn outperformed PERSIANN-CDR V1R1. Rather than using gauge observations directly for corrections, PERSIANN-CDR V1R1 is adjusted to match the satellite- and gauge-based GPCP dataset (monthly temporal and 2.5° spatial resolution). Note that some of the datasets, such as CHIRPS V2.0 and PERSIANN-CDR V1R1, have not been specifically designed to provide the best instantaneous accuracy, but rather to achieve the most temporally homogeneous record possible. Furthermore, the good performance of the exclusively gauge-based CPC Unified is unlikely to generalize to regions with sparse rain gauge networks.

Figure 33.2 presents global maps with calibration NSE values obtained for a selection of the best performing P datasets, while Fig. 33.3 shows which of these P datasets obtained the highest calibration NSE for each catchment. All P datasets provided low calibration NSE scores (<0.3) over the US Great Plains, consistent with several previous studies using different hydrological models and forcing datasets (e.g., Newman et al. 2015; Bock et al. 2016; Essou et al. 2016). It reflects the spatio-temporally highly intermittent rainfall regime combined with a strongly nonlinear rainfall–runoff response (Pilgrim et al. 1988). Low calibration scores were also found in northern Alaska, presumably due to P underestimation (Kauffeldt et al. 2013); in Namibia and Zambia, probably partly due to the importance of convective rainfall and partly due to the Q data quality (Li et al. 2013); and in Hawaii, which we suspect are due to flow overestimations caused by (i) erroneous rating curves, as visual inspection of the records revealed the presence of drift errors, and (ii) submarine groundwater discharge (Garrison et al. 2003), which is not explicitly accounted for by HBV. In North America, Europe, Japan, Australia, New Zealand, and southern and western Brazil, MSWEP V2.0 generally exhibited the best performance, whereas in Central America, and in central and eastern Brazil, CHIRPS V2.0 tended to perform best. No obvious best estimate could be identified for Africa, emphasizing the challenge of hydrological modeling in Africa (Sylla et al. 2013; Beck et al. 2017a). In summary, there are some P datasets that consistently outperform others regionally, but there is not one that performs best everywhere (Barrett et al. 1994).

Table 33.3 Median calibration NSE scores for the gauge-corrected P datasets obtained using HBV

Köppen-Geiger climate zone	Number of catchments	CHIRPS V2.0	CMORPH-CRT V1.0	CPC Unified	GPCP-1DD V1.2	MSWEP V1.2	MSWEP V2.0	PERSIANN-CDR V1R1	TMPA 3B42RT V7	WFDEI-CRU
All	8220	0.45	0.17	0.54	0.27	0.58	0.62	0.31	0.41	0.35
Tropical (A)	289	0.4	0.31	0.25	0.22	0.43	0.53	0.26	0.31	0.13
Dry (B)	384	0.17	0.12	0.23	0.12	0.25	0.26	0.12	0.18	0.17
Temperate (C)	3491	0.48	0.44	0.59	0.27	0.6	0.67	0.3	0.45	0.3
Cold (D)	4041	0.44	−0.05	0.53	0.28	0.58	0.61	0.33	0.39	0.42
Polar (E)	14	0.17	−2.62	−0.14	0.23	0.52	0.42	0.19	0.17	0.32

Only the catchments with calibration NSE values for all P datasets are considered. Thus, catchments at latitudes >50° have been excluded. The results are grouped according to the five broadest Köppen–Geiger climate categories, commonly referred to using the letters A–E

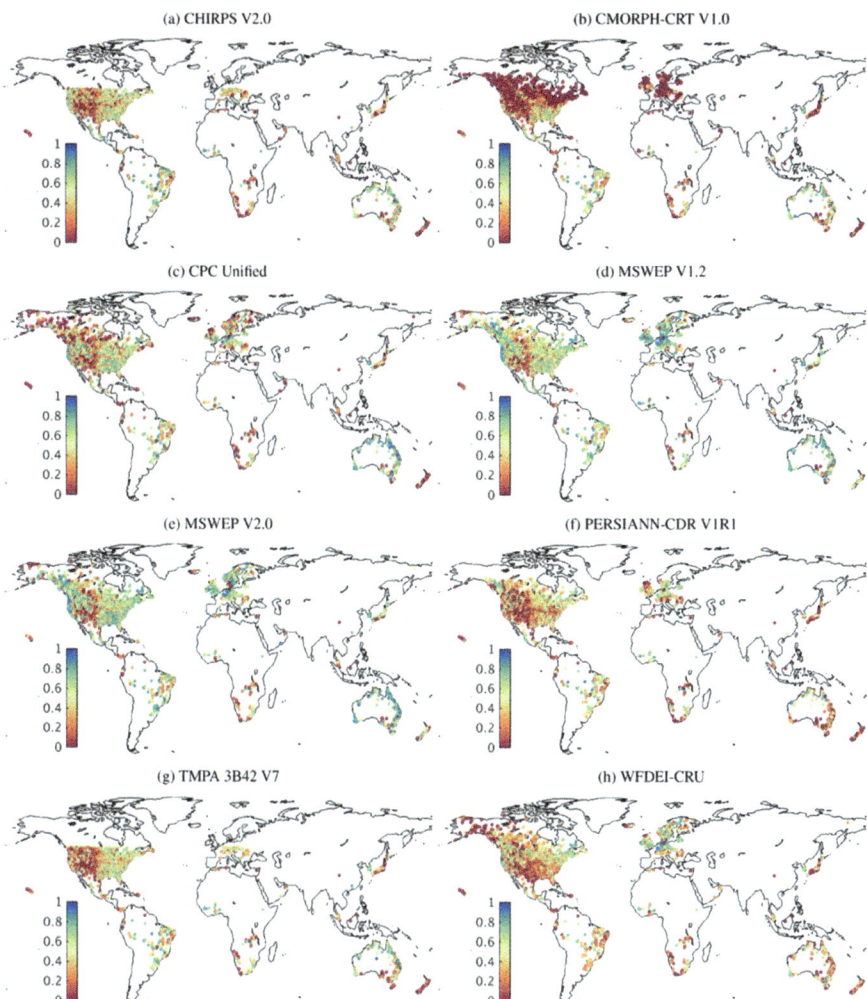

Fig. 33.2 Calibration NSE scores obtained using P time series from (**a**) CHIRPS V2.0, (**b**) CMORPH-CRT V1.0, (**c**) CPC Unified, (**d**) MSWEP V1.2, (**e**) MSWEP V2.0, (**f**) PERSIANN-CDR V1R1, (**g**) TMPA 3B42 V7, and (**h**) WFDEI-CRU. Each data point represents a catchment centroid. Only the eight best performing P datasets are shown

The good performance obtained for CPC Unified, CHIRPS V2.0, and MSWEP V1.2 and V2.0 underscores the importance of using sub-monthly gauge observations to improve Q simulations. Few P datasets currently incorporate sub-monthly gauge data, possibly because of the better global-scale availability of monthly gauge data, the lack of reliable information on the 24 h accumulation time for the large majority of gauges across the globe, and the difficulty of applying daily rather than monthly

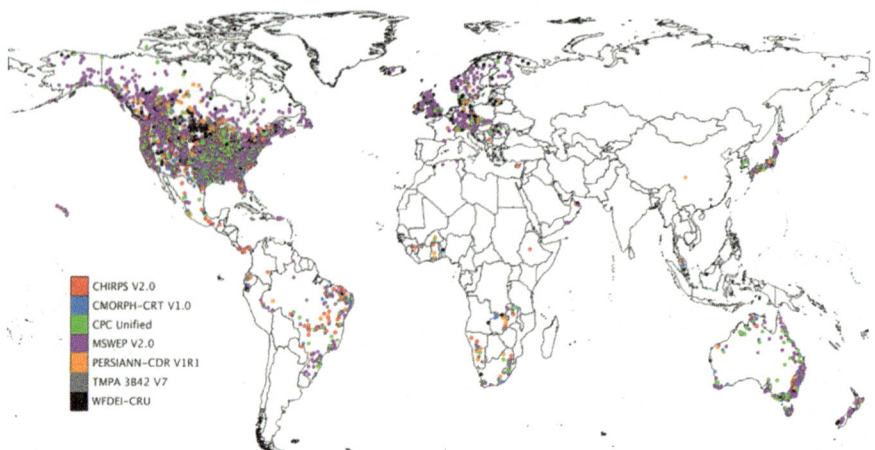

Fig. 33.3 For each catchment, the P dataset with the highest calibration NSE. Each data point represents a catchment centroid. Only the seven best performing P datasets (excluding MSWEP V1.2 due to its similarity to V2.0) are considered. Note that CHIRPS V2.0, CMORPH-CRT V1.0, PERSIANN-CDR V1R1, and TMPA 3B42 V7 do not provide data beyond 50, 60, 60, and 50° latitude, respectively

gauge corrections (Vila et al. 2009). However, a wealth of daily gauge data is currently freely available (Menne et al. 2012; Funk et al. 2015b), and sub-daily satellite and reanalysis P estimates provide an efficient and consistent means to infer the most probable UTC boundary of the 24 h accumulation period for any gauge with observations during the satellite era (1979–present; Beck et al. 2019a).

Most previous studies using hydrological modeling to evaluate the accuracy of P datasets had a regional or sub-continental focus, used Q observations from a relatively small number of catchments, considered only a few P datasets, did not consider reanalysis-based P datasets, or did not re-calibrate the hydrological model for each P dataset (e.g., Voisin et al. 2008; Su et al. 2008; Bitew et al. 2012; Tang et al. 2016). Here, we used 9053 catchments covering all climate zones and latitudes, considered a diverse range of P datasets, and re-calibrated the model for each P dataset, to maximize the generalizability of our findings. Nevertheless, our catchments are predominantly located in regions with dense P gauge networks (i.e., the conterminous US, Europe, and parts of Australia). Therefore, our results may not unequivocally generalize to regions with sparse P gauge networks. Use of another calibration objective function, hydrological model, or Ta or Ep forcing may lead to slightly different results, although we consider it unlikely to change the overall performance ranking of the P datasets. Finally, a poor score for a particular P dataset may also simply reflect a systematic bias that could be easily corrected.

33.4 Conclusions

This study may represent the most comprehensive global-scale P dataset evaluation to date. We evaluated 13 uncorrected P datasets using P observations from 76,086 gauges, and 9 gauge-corrected ones using hydrological modeling for 9053 catchments (<50,000 km^2). Our results can be summarized as follows:

1. Among the non-gauge-corrected P datasets, MSWEP-ng V1.2 and V2.0, based on optimal merging of multiple satellite and reanalysis P datasets, provided the best temporal correlations overall. They were followed, in order, by reanalyses, estimates based on microwave remote sensing of rainfall and near-surface soil moisture, and estimates based on thermal IR remote sensing. MSWEP-ng V2.0 obtained considerably lower mean annual P trend errors than the other datasets. Contrary to expectations, two of the three reanalyses (ERA-Interim and JRA-55) provided, on average, more reliable mean annual P trends than the satellite datasets.

2. Among the uncorrected P datasets, CHIRP V2.0 and MSWEP-ng V1.2 and V2.0 yielded the most accurate long-term P means, primarily due to the use of high-resolution gauge-based climatic datasets to determine their long-term mean. The reanalyses also provided reasonably accurate long-term means. The uncertainty in long-term means among the datasets was generally greatest in topographically complex and arid regions. In terms of the annual number of dry days, MSWEP-ng V2.0 exhibited markedly better performance than the other datasets, due to the use of CDF corrections after data merging. The satellite datasets also performed quite well in this respect, while CHIRP V2.0, the reanalyses, MSWEP-ng V1.2, and the soil moisture remote sensing-based SM2RAIN-ASCAT consistently underestimated the number of dry days. The satellite-based datasets generally exhibited difficulties in detecting P signals at high latitudes.

3. Among the gauge-corrected P datasets, the datasets directly incorporating daily gauge data (CPC Unified and the MSWEP versions) outperformed those directly incorporating temporally coarser gauge data. These in turn outperformed the datasets that only indirectly incorporated gauge data. This highlights the benefit of explicit and careful incorporation of daily gauge data. The good performance of the fully gauge-based CPC Unified is unlikely to generalize to sparse or ungauged regions. In general, the performance was best in temperate regions, due to the presence of dense monitoring networks, and worst in arid regions, due to the convective rainfall and the highly non-linear rainfall–runoff response.

So, which P dataset should one use? While this depends on the region under consideration and the specific user needs or application, in most cases MSWEP V2.0 appears to be a good choice: it has a long temporal record (1979–2016), a fully global coverage (including ocean areas), a comparatively high temporal (3-hourly) and spatial (0.1°) resolution, daily gauge corrections, and, as demonstrated in the current study, comparatively good performance for all performance metrics for all climate types. However, for tropical regions, CHIRPS V2.0 also presents a viable choice, if a daily temporal resolution suffices, and if the peak magnitude underestimation and spurious drizzle are less critical. In regions with dense rain gauge networks, CPC Unified also offers good performance. For some regions, notably Africa, it remains difficult to provide reliable recommendations due to the limited availability and quality of rain gauge and Q data, highlighting the critical importance of maintaining and expanding data collection efforts.

Acknowledgements We gratefully acknowledge the P dataset developers for producing and making available their datasets. The Water Center for Arid and Semi-Arid Zones in Latin America and the Caribbean (CAZALAC) and the Centro de Ciencia del Clima y la Resiliencia (CR)2 (FONDAP 15110009) are thanked for sharing the Mexican and Chilean gauge data, respectively. We also acknowledge the gauge data providers in the Latin American Climate Assessment & Dataset (LACA&D) project: IDEAM (Colombia), INAMEH (Venezuela), INAMHI (Ecuador), SENAMHI (Peru), SENAMHI (Bolivia), and DMC (Chile). We further wish to thank Ali Alijanian, Koen Verbist, and Piyush Jain for providing additional gauge data. The Global Runoff Data Centre (GRDC) and the United States Geological Survey (USGS) are gratefully acknowledged for providing the majority of the observed Q data. We thank Mauricio Zambrano Bigiarini, Pete Peterson, Hamed Ashouri, Tomoo Ushio, Louise Slater, and three anonymous reviewers for thoughtful comments and suggestions which helped improve the quality of the paper. Graham P. Weedon was supported by the Joint DECC and Defra Integrated Climate Program – DECC/Defra (GA01101). Vincenzo Levizzani wishes to acknowledge funding from the European Union Seventh Framework Programme (FP7/2007–2013) under grant agreement no. 603608, "Global Earth Observation for integrated water resource assessment": earth2Observe, and from the "Progetto di Interesse NextData" of the Italian Ministry of Education, University, and Research (MIUR). The work was supported through IPA support for the first author from the US Army Corps of Engineers' International Center for Integrated Water Resources Management (ICIWaRM), under the auspices of UNESCO.

References

Adler, R. F., Kidd, C., Petty, G., Morrissey, M., & Goodman, H. M. (2001). Intercomparison of global precipitation products: The third precipitation intercomparison project (PIP-3). *Bulletin of the American Meteorological Society, 82*, 1377–1396. https://doi.org/10.1175/1520-0477 (2001)082<1377:IOGPPT>2.3.CO;2.

AghaKouchak, A., Behrangi, A., Sorooshian, S., Hsu, K., & Amitai, E. (2011). Evaluation of satellite-retrieved extreme precipitation rates across the Central United States. *Journal of Geophysical Research, 116*, D02115. https://doi.org/10.1029/2010JD014741.

Akinremi, O. O., McGinn, S. M., & Cutforth, H. W. (1999). Precipitation trends on the Canadian prairies. *Journal of Climate, 12*, 2996–3003. https://doi.org/10.1175/1520-0442(1999) 012<2996:PTOTCP>2.0.CO;2.

Albergel, C., Dorigo, W., Reichle, R. H., Balsamo, G., de Rosnay, P., Muñoz Sabater, J. M., Isaksen, L., de Jeu, R. A. M., & Wagner, W. (2013). Skill and global trend analysis of soil moisture from reanalyses and microwave remote sensing. *Journal of Hydrometeorology, 14*, 1259–1277. https://doi.org/10.1175/JHM-D-12-0161.1.

Alijanian, M., Rakhshandehroo, G. R., Mishra, A. K., & Dehghani, M. (2017). Evaluation of satellite rainfall climatology using CMORPH, PERSIANN-CDR, PERSIANN, TRMM, MSWEP over Iran. *International Journal of Climatology, 37*, 4896–4914. https://doi.org/10. 1002/joc.5131.

Andréassian, V., Lerat, J., Loumagne, C., Mathevet, T., Michel, C., Oudin, L., & Perrin, C. (2007). What is really undermining hydrologic science today? *Hydrological Procedure, 21*, 2819–2822. https://doi.org/10.1002/hyp.6854.

Arakawa, A. (2004). The cumulus parameterization problem: Past, present, and future. *Journal of Climate, 17*, 2493–2525. https://doi.org/10.1175/1520-0442(2004)017<2493:RATCPP>2.0. CO;2.

Ashlock, D. (2010). *Evolutionary computation for modeling and optimization.* New York: Springer. 471 pp, ISBN: 978-0-387-31909-4.

Ashouri, H., Hsu, K., Sorooshian, S., Braithwaite, D. K., Knapp, K. R., Cecil, L. D., Nelson, B. R., & Pratt, O. P. (2015). PERSIANN-CDR: Daily precipitation climate data record from multisatellite observations for hydrological and climate studies. *Bulletin of the American Meteorological Society, 96*, 69–83. https://doi.org/10.1175/BAMS-D-13-00068.1.

Ashouri, H., Nguyen, P., Thorstensen, A., Hsu, K., Sorooshian, S., & Braithwaite, D. (2016). Assessing the efficacy of high-resolution satellite-based PERSIANN-CDR precipitation product in simulating streamflow. *Journal of Hydrometeorology, 17*, 2061–2076. https://doi.org/10. 1175/JHM-D-15-0192.1.

Barrett, E. C., Adler, R. F., Arpe, K., Bauer, P., Berg, W., Chang, A., Ferraro, R., Ferriday, J., Goodman, S., Hong, Y., Janowiak, J., Kidd, C., Kniveton, D., Morrissey, M., Olson, W., Petty, G., Rudolf, B., Shibata, A., Smith, E. A., & Spencer, R. (1994). The first WetNet precipitation intercomparison project (PIP-1): Interpretation of results. *Remote Sensing Reviews, 11*, 303–373. https://doi.org/10.1080/02757259409532268.

Beck, H. E. (2017). MSWEP Version 2 documentation, Tech. Rep., Princeton University, www. gloh2o.org. Last accessed 6 Mar 2020.

Beck, H. E., van Dijk, A. I. J. M., & de Roo, A. (2015). Global maps of streamflow characteristics based on observations from several thousand catchments. *Journal of Hydrometeorology, 16*, 1478–1501. https://doi.org/10.1175/JHM-D-14-0155.1.

Beck, H. E., Bruijnzeel, L. A., van Dijk, A. I. J. M., McVicar, T. R., Scatena, F. N., & Schellekens, J. (2013). The impact of forest regeneration on streamflow in 12 mesoscale humid tropical catchments. *Hydrology and Earth System Sciences, 17*, 2613–2635. https://doi.org/10.5194/ hess-17-2613-2013.

Beck, H. E., van Dijk, A. I. J. M., de Roo, A., Miralles, D. G., McVicar, T. R., Schellekens, J., & Bruijnzeel, L. A. (2016). Global-scale regionalization of hydrologic model parameters. *Water Resources Research, 52*, 3599–3622. https://doi.org/10.1002/2015WR018247.

Beck, H. E., van Dijk, A. I. J. M., Levizzani, V., Schellekens, J., Miralles, D. G., Martens, B., & de Roo, A. (2017a). MSWEP: 3-hourly 0.25° global gridded precipitation (1979–2015) by merging gauge, satellite, and reanalysis data. *Hydrology and Earth System Sciences, 21*, 589–615. https://doi.org/10.5194/hess-21-589-2017.

Beck, H. E., van Dijk, A. I. J. M., de Roo, A., Dutra, E., Fink, G., Orth, R., & Schellekens, J. (2017b). Global evaluation of runoff from 10 state-of-the-art hydrological models. *Hydrology and Earth System Sciences, 21*, 2881–2903. https://doi.org/10.5194/hess-21-2881-2017.

Beck, H. E., Vergopolan, N., Pan, M., Levizzani, V., van Dijk, A. I. J. M., Weedon, G. P., Brocca, L., Pappenberger, F., Huffman, G. J., & Wood, E. F. (2017c). Global-scale evaluation of 22 precipitation datasets using gauge observations and hydrological modeling. *Hydrology and Earth System Sciences, 21*, 6201–6217. https://doi.org/10.5194/hess-21-6201-2017.

Beck, H. E., Wood, E. F., Pan, M., Fisher, C. K., Miralles, D. M., van Dijk, A. I. J. M., McVicar, T. R., & Adler, R. F. (2019a). MSWEP 2 global 3-hourly 0.1° precipitation: Methodology and quantitative assessment. *Bulletin of the American Meteorological Society, 100*, 473–500. https:// doi.org/10.1175/BAMS-D-17-0138.1.

Beck, H. E., Pan, M., Roy, T., Weedon, G. P., Pappenberger, F., van Dijk, A. I. J. M., Huffman, G. J., Adler, R. F., & Wood, E. F. (2019b). Daily evaluation of 26 precipitation datasets using Stage-IV gauge-radar data for the CONUS. *Hydrology and Earth System Sciences, 23*, 207–224. https://doi.org/10.5194/hess-23-207-2019.

Behrangi, A., Khakbaz, B., Jaw, T. C., AghaKouchak, A., Hsu, K.-L., & Sorooshian, S. (2011). Hydrologic evaluation of satellite precipitation products over a mid-size basin. *Journal of Hydrology, 397*, 225–237. https://doi.org/10.1016/j.jhydrol.2010.11.043.

Bengtsson, L., Hagemann, S., & Hodges, K. I. (2004). Can climate trends be calculated from reanalysis data? *Journal of Geophysical Research, 109*, D11111. https://doi.org/10.1029/2004JD004536.

Bergström, S. (1992). The HBV model – Its structure and applications. *SMHI Reports RH 4*, Swedish Meteorological and Hydrological Institute (SMHI), Norrköping, Sweden. Available at https: //www.smhi.se/en/publications/the-hbv-model-its-structure-and-applications-1.83591. Last accessed 8 Nov 2018.

Bitew, M. M., Gebremichael, M., Ghebremichael, L. T., & Bayissa, Y. A. (2012). Evaluation of high-resolution satellite rainfall products through streamflow simulation in a hydrological modeling of a small mountainous watershed in Ethiopia. *Journal of Hydrometeorology, 13*, 338–350. https://doi.org/10.1175/2011JHM1292.1.

Bock, A. R., Hay, L. E., McCabe, G. J., Markstrom, S. L., & Atkinson, R. D. (2016). Parameter regionalization of a monthly water balance model for the conterminous United States. *Hydrology and Earth System Sciences, 20*, 2861–2876. https://doi.org/10.5194/hess-20-2861-2016.

Bosilovich, M. G., Chen, J., Robertson, F. R., & Adler, R. F. (2008). Evaluation of global precipitation in reanalyses. *Journal of Applied Meteorology and Climatology, 47*, 2279–2299. https://doi.org/10.1175/2008JAMC1921.1.

Brocca, L., Ciabatta, L., Massari, C., Moramarco, T., Hahn, S., Hasenauer, S., Kidd, R., Dorigo, W., Wagner, W., & Levizzani, V. (2014). Soil as a natural rain gauge: estimating global rainfall satellite soil moisture data. *Journal of Geophysical Research, 119*, 5128–5141. https://doi.org/10.1002/2014JD021489.

Buarque, D. C., de Paiva, R. C. D., Clarke, R. T., & Mendes, C. A. B. (2011). A comparison of Amazon rainfall characteristics derived from TRMM, CMORPH and the Brazilian national rain gauge network. *Journal of Geophysical Research, 116*, D19105. https://doi.org/10.1029/2011JD016060.

Bumke, K., König-Langlo, G., Kinzel, J., & Schröder, M. (2016). HOAPS and ERA-interim precipitation over the sea: Validation against shipboard in situ measurements. *Atmospheric Measurement Techniques, 9*, 2409–2423. https://doi.org/10.5194/amt-9-2409-2016.

Cattani, E., Merino, A., & Levizzani, V. (2016). Evaluation of monthly satellite-derived precipitation products over East Africa. *Journal of Hydrometeorology, 17*, 2555–2573. https://doi.org/10.1175/JHM-D-0042.1.

Cecil, D. J., Buechler, D. E., & Blakeslee, R. J. (2014). Gridded lightning climatology from TRMM-LIS and OTD: Dataset description. *Atmospheric Research, 135*, 404–414. https://doi.org/10.1016/j.atmosres.2012.06.028.

Chai, T., & Draxler, R. R. (2014). Root mean square error (RMSE) or mean absolute error (MAE)? – Arguments against avoiding RMSE in the literature. *Geoscientific Model Development, 7*, 1247–1250. https://doi.org/10.5194/gmd-7-1247-2014.

Chen, M., Shi, W., Xie, P., Silva, V. B. S., Kousky, V. E., Higgins, R. W., & Janowiak, J. E. (2008). Assessing objective techniques for gauge-based analyses of global daily precipitation. *Journal of Geophysical Research, 113*, D04110. https://doi.org/10.1029/2007JD009132.

Chen, S., Hong, Y., Gourley, J. J., Huffman, G. J., Tian, Y., Cao, Q., Yong, B., Kirstetter, P.-E., Hu, J., Hardy, J., Li, Z., Khan, S. I., & Xue, X. (2013). Evaluation of the successive V6 and V7 TRMM multisatellite precipitation analysis over the continental United States. *Water Resources Research, 49*, 8174–8186. https://doi.org/10.1002/2012WR012795.

Ciabatta, L., Massari, C., Brocca, L., Gruber, A., Reimer, C., Hahn, S., Paulik, C., Dorigo, W., Kidd, R., & Wagner, W. (2017). SM2RAIN-CCI: A new global long-term rainfall data set derived from ESA CCI soil moisture. *Earth System Science Data, 10*, 267–280. https://doi.org/10.5194/essd-10-267-2018.

Clark, P., Roberts, N., Lean, H., Ballard, S. P., & Charlton-Perez, C. (2016). Convection-permitting models: A step-change in rainfall forecasting. *Meteorological Applications, 23*, 165–181. https://doi.org/10.1002/met.1538.

Collischonn, B., Collischonn, W., & Tucci, C. E. M. (2008). Daily hydrological modeling in the Amazon basin using TRMM rainfall estimates. *Journal of Hydrology, 360*, 207–216. https://doi. org/10.1016/j.jhydrol.2008.07.032.

Criss, R. E., & Winston, W. E. (2008). Do Nash values have value? Discussion and alternate proposals. *Hydrological Procedure, 22*, 2723–2725. https://doi.org/10.1002/hyp.7072.

Crow, W. T., van den Berg, M. J., Huffman, G. J., & Pellarin, T. (2011). Correcting rainfall using satellite-based surface soil moisture retrievals: The soil moisture analysis rainfall tool (SMART). *Water Resources Research, 47*, W08521. https://doi.org/10.1029/2011WR010576.

Daly, C., Halbleib, M., Smith, J. I., Gibson, W. P., Doggett, M. K., Taylor, G. H., Curtis, J., & Pasteris, P. P. (2008). Physiographically sensitive mapping of climatological temperature and precipitation across the conterminous United States. *International Journal of Climatology, 28*, 2031–2064. https://doi.org/10.1002/joc.1688.

Dee, D. P., Uppala, S. M., Simmons, A. J., Berrisford, P., Poli, P., Kobayashi, S., Andrae, U., Balmaseda, M. A., Balsamo, G., Bauer, P., Bechtold, P., Beljaars, A. C. M., van de Berg, L., Bidot, J., Bormann, N., Delsol, C., Dragani, R., Fuentes, M., Geer, A. J., Haimberger, L., Healy, S. B., Hersbach, H., Hólm, E. V., Isaksen, L., Kallberg, P., Köhler, M., Matricardi, M., McNally, A. P., Monge-Sanz, B. M., Morcrette, J.-J., Park, B.-K., Peubey, C., de Rosnay, P., Tavolato, C., Thépaut, J.-N., & Vitart, F. (2011). The ERA-interim reanalysis: Configuration and performance of the data assimilation system. *Quarterly Journal of the Royal Meteorological Society, 137*, 553–597. https://doi.org/10.1002/qj.828.

Deelstra, J., Farkas, C., Engebretsen, A., Kværnø, S., Beldring, S., Olszewska, A., & Nesheim, L. (2010). Can we simulate runoff from agriculture dominated watersheds? Comparison of the DrainMod, SWAT, HBV, COUP and INCA models applied for the Skuterud catchment. *Bioforsk FOKUS, 5*, 119–128.

Di Giuseppe, F., Molteni, F., & Dutra, E. (2013). Real-time correction of ERA-interim monthly rainfall. *Geophysical Research Letters, 40*, 3750–3755. https://doi.org/10.1002/grl.50670.

Dinku, T., Ceccato, P., & Connor, S. J. (2016). Challenges of satellite rainfall estimation over mountainous and arid parts of East Africa. *International Journal of Remote Sensing, 32*, 5965–5979. https://doi.org/10.1080/01431161.2010.499381.

Driouech, F., Déqué, M., & Mokssit, A. (2009). Numerical simulation of the probability distribution function of precipitation over Morocco. *Climate Dynamics, 32*, 1055–1063. https://doi.org/10.1007/s00382-008-0430-6.

Ebert, E. E. (2007). Methods for verifying satellite precipitation estimates. In Measuring Precipitation from Space, V. Levizzani, P. Bauer, & F. J. Turk (Eds.), *Advances Global Change Research* (Vol. 28, pp. 345–356). Dordrecht: Springer. ISBN: 978-1-4020-5835-6.

Ebert, E. E., Janowiak, J. E., & Kidd, C. (2007). Comparison of near-real-time precipitation estimates from satellite observations and numerical models. *Bulletin of the American Meteorological Society, 88*, 47–64. https://doi.org/10.1175/BAMS-88-1-47.

Essou, G. R. C., Arsenault, R., & Brissette, F. P. (2016). Comparison of climate datasets for lumped hydrological modeling over the continental United States. *Journal of Hydrology, 537*, 334–345. https://doi.org/10.1016/j.jhydrol.2016.03.063.

Falck, A. S., Maggioni, V., Tomasella, J., Vila, D. A., & Diniz, F. L. R. (2015). Propagation of satellite precipitation uncertainties through a distributed hydrologic model: A case study in the Tocantins-Araguaia basin in Brazil. *Journal of Hydrology, 527*, 943–957. https://doi.org/10.1016/j.jhydrol.2015.05.042.

Falcone, J. A., Carlisle, D. M., Wolock, D. M., & Meador, M. R. (2010). GAGES: A stream gage database for evaluating natural and altered flow conditions in the conterminous United States. *Ecology, 91*, 621. https://doi.org/10.1890/09-0889.1.

Fekete, B. M., Vörösmarty, C. J., Roads, J. O., & Willmott, C. J. (2004). Uncertainties in precipitation and their impacts on runoff estimates. *Journal of Climate, 17*, 294–304. https://doi.org/10.1175/1520-0442(2004)017<0294:UIPATI>2.0.CO;2.

Ferraro, R. R., Smith, E. A., Berg, W., & Huffman, G. J. (1998). A screening methodology for passive microwave precipitation retrieval algorithms. *Journal of the Atmospheric Sciences, 55*, 1583–1600. https://doi.org/10.1175/1520-0469(1998)055<1583:ASMFPM>2.0.CO;2.

Fick, S. E., & Hijmans, R. J. (2017). WorldClim 2: New 1-km spatial resolution climate surfaces for global land areas. *International Journal of Climatology, 37,* 4302–4315. https://doi.org/10. 1002/joc.5086.

Fortin, F., De Rainville, F., Gardner, M., Parizeau, M., & Gagné, C. (2012). DEAP: Evolutionary algorithms made easy. *Journal of Machine Learning Research, 13,* 2171–2175.

Funk, C., Verdin, A., Michaelsen, J., Peterson, P., Pedreros, D., & Husak, G. (2015a). A global satellite-assisted precipitation climatology. *Earth System Science Data, 7,* 275–287. https://doi. org/10.5194/essd-7-275-2015.

Funk, C., Peterson, P., Landsfeld, M., Pedreros, D., Verdin, J., Shukla, S., Husak, G., Rowland, J., Harrison, L., Hoell, A., & Michaelsen, J. (2015b). The climate hazards infrared precipitation with stations – A new environment record for monitoring extremes. *Scientific Data, 2,* 150066. https://doi.org/10.1038/sdata.2015.66.

Garrison, G. H., Glenn, C. R., & McMurtry, G. M. (2003). Measurement of submarine groundwater discharge in Kahana Bay, Oahu, Hawaii. *Limnology and Oceanography, 48,* 920–928. https:// doi.org/10.4319/lo.2003.48.2.0920.

Gebremichael, M. (2010). Framework for satellite rainfall product evaluation. In F. Y. Testik & M. Gebremichael (Eds.), *Rainfall: State of the science* (Geophys. Monogr. Series) (Vol. 191, pp. 265–275). Washington, DC: American Geophysical Union. https://doi.org/10.1029/2010GM000974.

Gehne, M., Hamill, T. M., Kiladis, G. N., & Trenberth, K. E. (2016). Comparison of global precipitation estimates across a range of temporal and spatial scales. *Journal of Climate, 29,* 7773–7795. https://doi.org/10.1175/JCLI-D-15-0618.1.

Groisman, P. Y., & Legates, D. R. (1994). The accuracy of United States precipitation data. *Bulletin of the American Meteorological Society, 72,* 215–227. https://doi.org/10.1175/1520-0477 (1994)075<0215:TAOUSP>2.0.CO;2.

Gupta, H. V., Kling, H., Yilmaz, K. K., & Martinez, G. F. (2009). Decomposition of the mean squared error and NSE performance criteria: Implications for improving hydrological modelling. *Journal of Hydrology, 377,* 80–91. https://doi.org/10.1016/j.jhydrol.2009.08.003.

Gupta, H. V., Perrin, C., Blöschl, G., Montanari, A., Kumar, R., Clark, M., & Andréassian, V. (2014). Large-sample hydrology: A need to balance depth with breadth. *Hydrology and Earth System Sciences, 18,* 463–477. https://doi.org/10.5194/hess-18-463-2014.

Haiden, T., Rodwell, M. J., Richardson, D. S., Okagaki, A., Robinson, T., & Hewson, T. (2012). Intercomparison of global model precipitation forecast skill in 2010/11 using the SEEPS score. *Monthly Weather Review, 140,* 2720–2733. https://doi.org/10.1175/MWR-D-11-00301.1.

Hayes, M. J., Svoboda, M. D., Wilhite, D. A., & Vanyarkho, O. V. (1999). Monitoring the 1996 drought using the standardized precipitation index. *Bulletin of the American Meteorological Society, 80,* 429–438. https://doi.org/10.1175/1520-0477(1999)080<0429:MTDUTS>2.0. CO;2.

Haylock, M. R., Hofstra, N., Klein Tank, A. M. G., Klok, E. J., Jones, P., & New, M. (2008). A European daily high-resolution gridded data set of surface temperature and precipitation for 1950–2006. *Journal of Geophysical Research, 113,* D20119. https://doi.org/10.1029/2008JD010201.

Herold, N., Alexander, L. V., Donat, M. G., Contractor, S., & Becker, A. (2015). How much does it rain over land? *Geophysical Research Letters, 43,* 341–348. https://doi.org/10.1002/2015GL066615.

Hirpa, F. A., Gebremichael, M., & Hopson, T. (2010). Evaluation of high-resolution satellite precipitation products over very complex terrain in Ethiopia. *Journal of Applied Meteorology and Climatology, 49,* 1044–1051. https://doi.org/10.1175/2009JAMC2298.1.

Hong, Y., Hsu, K.-L., Sorooshian, S., & Gao, X. (2004). Precipitation estimation from remotely sensed imagery using an artificial neural network cloud classification system. *Journal of Applied Meteorology, 43,* 1834–1853. https://doi.org/10.1175/JAM2173.1.

Huffman, G. J., Adler, R. F., Morrissey, M. M., Bolvin, D. T., Curtis, S., Joyce, R., McGavock, B., & Susskind, J. (2001). Global precipitation at one-degree daily resolution from multi-satellite observations. *Journal of Hydrometeorology, 2*, 36–50. https://doi.org/10.1175/1525-7541 (2001)002<0036:GPAODD>2.0.CO;2.

Huffman, G. J., Bolvin, D. T., Nelkin, E. J., Wolff, D. B., Adler, R. F., Gu, G., Hong, Y., Bowman, K. P., & Stocker, E. F. (2007). The TRMM multisatellite precipitation analysis (TMPA): Quasi-global, multiyear, combined-sensor precipitation estimates at fine scales. *Journal of Hydrometeorology, 8*, 38–55. https://doi.org/10.1175/JHM560.1.

Islam, T., Rico-Ramirez, M. A., Han, D., Srivastava, P. K., & Ishak, A. M. (2012). Performance evaluation of the TRMM precipitation estimation using ground-based radars from the GPM validation network. *Journal of Atmospheric and Solar – Terrestrial Physics, 77*, 194–208. https://doi.org/10.1016/j.jastp.2012.01.001.

Jain, S. K., & Sudheer, K. P. (2008). Fitting of hydrologic models: A close look at the Nash–Sutcliffe index. *Journal of Hydrologic Engineering, 13*, 981–986. https://doi.org/10.1061/(ASCE)1084-0699(2008)13:10(981).

Joyce, R. J., Janowiak, J. E., Arkin, P. A., & Xi, P. (2004). CMORPH: A method that produces global precipitation estimates from passive microwave and infrared data at high spatial and temporal resolution. *Journal of Hydrometeorology, 5*, 487–503. https://doi.org/10.1175/1525-7541(2004)005<0487:CAMTPG>2.0.CO;2.

Kang, S., & Ahn, J.-B. (2015). Global energy and water balances in the latest reanalyses. *Asia-Pacific Journal of Atmospheric Sciences, 51*, 293–302. https://doi.org/10.1007/s13143-015-0079-0.

Kauffeldt, A., Halldin, S., Rodhe, A., Xu, C.-Y., & Westerberg, I. K. (2013). Disinformative data in large-scale hydrological modelling. *Hydrology and Earth System Sciences, 17*, 2845–2857. https://doi.org/10.5194/hess-17-2845-2013.

Kidd, C., & Levizzani, V. (2011). Status of satellite precipitation retrievals. *Hydrology and Earth System Sciences, 15*, 1109–1116. https://doi.org/10.5194/hess-15-1109-2011.

Kidd, C., Dawkins, E., & Huffman, G. J. (2013). Comparison of precipitation derived from the ECMWF operational forecast model and satellite precipitation datasets. *Journal of Hydrometeorology, 14*, 1463–1482. https://doi.org/10.1175/JHM-D-12-0182.1.

Kidd, C., Bauer, P., Turk, F. J., Huffman, G. J., Joyce, R., Hsu, K.-L., & Braithwaite, D. (2012). Intercomparison of high-resolution precipitation products over Northwest Europe. *Journal of Hydrometeorology, 13*, 67–83. https://doi.org/10.1175/JHM-D-11-042.1.

Kim, K., Park, J., Baik, J., & Choi, M. (2017). Evaluation of topographical and seasonal feature using GPM IMERG and TRMM 3B42 over far-East Asia. *Atmospheric Research, 187*, 95–105. https://doi.org/10.1016/j.atmosres.2016.12.007.

Knapp, K. R., Ansari, S., Bain, C. L., Bourassa, M. A., Dickinson, M. J., Funk, C., Helms, C. N., Hennon, C. C., Holmes, C. D., Huffman, G. J., Kossin, J. P., Lee, H.-T., Loew, A., & Magnusdottir, G. (2011). Globally gridded satellite observations for climate studies. *Bulletin of the American Meteorological Society, 92*, 893–907. https://doi.org/10.1175/2011BAMS3039.1.

Kobayashi, S., Ota, Y., Harada, Y., Ebita, A., Moriya, M., Onoda, H., Onogi, K., Kamahori, H., Kobayashi, C., Endo, H., Miyaoka, K., & Takahashi, K. (2015). The JRA-55 reanalysis: General specifications and basic characteristics. *Journal of the Meteorological Society of Japan, 93*, 5–48. https://doi.org/10.2151/jmsj.2015-001.

Krause, P., Boyle, D. P., & Bäse, F. (2005). Comparison of different efficiency criteria for hydrological model assessment. *Advances in Geosciences, 5*, 89–97. https://doi.org/10.5194/adgeo-5-89-2005.

Laviola, S., Levizzani, V., Cattani, E., & Kidd, C. (2013). The 183-WSL fast rainrate retrieval algorithm. Part II: Validation using ground radar measurements. *Atmospheric Research, 134*, 77–86. https://doi.org/10.1016/j.atmosres.2013.07.013.

Li, L., Ngongondo, C. S., Xu, C.-Y., & Gong, L. (2013). Comparison of the global TRMM and WFD precipitation datasets in driving a large-scale hydrological model in southern Africa. *Hydrology Research, 44*, 770–788. https://doi.org/10.2166/nh.2012.175.

Loew, A., Bell, W., Brocca, L., Bulgin, C. E., Burdanowitz, J., Calbet, X., Donner, R. V., Ghent, D., Gruber, A., Kaminski, T., Kinzel, J., Klepp, C., Lambert, J.-C., Schaepman-Strub, H., & Schröder, M. (2017). Validation practices for satellite based earth observation data across communities. *Reviews of Geophysics, 55*, 779–817. https://doi.org/10.1002/2017RG000562.

Lopez, P. (2007). Cloud and precipitation parameterizations in modeling and variational data assimilation: A review. *Journal of the Atmospheric Sciences, 64*, 3766–3784. https://doi.org/10.1175/2006JAS2030.1.

Lorenz, C., & Kunstmann, H. (2012). The hydrological cycle in three state-of-the-art reanalyses: Intercomparison and performance analysis. *Journal of Hydrometeorology, 13*, 1397–1420. https://doi.org/10.1175/JHM-D-11-088.1.

Maggioni, V., Meyers, P. C., & Robinson, M. D. (2016). A review of merged high resolution satellite precipitation product accuracy during the Tropical Rainfall Measuring Mission (TRMM)-era. *Journal of Hydrometeorology, 17*, 1101–1117. https://doi.org/10.1175/JHM-D-15-0190.1.

Maraun, D. (2013). Bias correction, quantile mapping, and downscaling: Revisiting the inflation issue. *Journal of Climate, 26*, 2137–2143. https://doi.org/10.1175/JCLI-D-12-00821.1.

Martens, B., Miralles, D. G., Lievens, H., van der Schalie, R., de Jeu, R. A. M., Fernández-Prieto, D., Beck, H. E., Dorigo, W. A., & Verhoest, N. E. C. (2017). GLEAM v3: satellite-based land evaporation and root-zone soil moisture. *Geoscientific Model Development, 10*, 1903–1925. https://doi.org/10.5194/gmd-10-1903-2017.

Massari, C., Crow, W., & Brocca, L. (2017). An assessment of the performance of global rainfall estimates without ground-based observations. *Hydrology and Earth System Sciences, 21*, 4347–4361. https://doi.org/10.5194/hess-21-4347-2017.

Mei, Y., Anagnostou, E. N., Nikolopoulos, E. I., & Borga, M. (2014). Error analysis of satellite precipitation products in mountainous basins. *Journal of Hydrometeorology, 15*, 1778–1793. https://doi.org/10.1175/JHM-D-13-0194.1.

Menne, M. J., Durre, I., Vose, R. S., Gleason, B. E., & Houston, T. G. (2012). An overview of the global historical climatology network-daily database. *Journal of Atmospheric and Oceanic Technology, 29*, 897–910. https://doi.org/10.1175/JTECH-D-11-00103.1.

Michaelides, S., Levizzani, V., Anagnostou, E. N., Bauer, P., Kasparis, T., & Lane, J. E. (2009). Precipitation: Measurement, remote sensing, climatology and modeling. *Atmospheric Research, 94*, 512–533. https://doi.org/10.1016/j.atmosres.2009.08.017.

Moazami, S., Golian, S., Kavianpour, M. R., & Hong, Y. (2013). Comparison of PERSIANN and V7 TRMM multi-satellite precipitation analysis (TMPA) products with rain gauge data over Iran. *International Journal of Remote Sensing, 34*, 8156–8171. https://doi.org/10.1080/01431161.2013.833360.

Moriasi, D. N., Arnold, J. G., van Liew, M. W., Bingner, R. L., Harmel, R. D., & Veith, T. L. (2007). Model evaluation guidelines for systematic quantification of accuracy in watershed simulations. *Transactions of the ASABE, 50*, 885–900. https://doi.org/10.13031/2013.23153.

Nair, A. S., & Indu, J. (2017). Performance assessment of Multi-Source Weighted-Ensemble Precipitation (MSWEP) product over India. *Climate, 5*, 2. https://doi.org/10.3390/cli5010002.

Nash, J. E., & Sutcliffe, J. V. (1970). River flow forecasting through conceptual models. Part I – A discussion of principles. *Journal of Hydrology, 10*, 282–290. https://doi.org/10.1016/0022-1694(70)90255-6.

Newman, A. J., Clark, M. P., Sampson, K., Wood, A., Hay, L. E., Bock, A., Viger, R. J., Blodgett, D., Brekke, L., Arnold, J. R., Hopson, T., & Duan, Q. (2015). Development of a large-sample watershed-scale hydrometeorological data set for the contiguous USA: Data set characteristics and assessment of regional variability in hydrologic model performance. *Hydrology and Earth System Sciences, 19*, 209–223. https://doi.org/10.5194/hess-19-209-2015.

Pan, M., Li, H., & Wood, E. F. (2010). Assessing the skill of satellite-based precipitation estimates in hydrologic applications. *Water Resources Research, 46*, W09535. https://doi.org/10.1029/2009WR008290.

Peel, M. C., Chiew, F. H. S., Western, A. W., & McMahon, T. A. (2000). Extension of unimpaired monthly streamflow data and regionalisation of parameter values to estimate streamflow in ungauged catchments. *Report prepared for the Australian National Land and Water Resources Audit,* Centre for Environmental Applied Hydrology, University of Melbourne, Australia.

Peña Arancibia, J. L., van Dijk, A. I. J. M., Renzullo, L. J., & Mulligan, M. (2013). Evaluation of precipitation estimation accuracy in reanalyses, satellite products, and an ensemble method for regions in Australia and south and East Asia. *Journal of Hydrometeorology, 14,* 1323–1333. https://doi.org/10.1175/JHM-D-12-0132.1.

Pilgrim, D. H., Chapman, T. G., & Doran, D. G. (1988). Problems of rainfall-runoff modelling in arid and semiarid regions. *Hydrological Sciences Journal, 33,* 379–400. https://doi.org/10. 1080/02626668809491261.

Plesca, I., Timbe, E., Exbrayat, J. F., Windhorst, D., Kraft, P., Crespo, P., Vachéa, K. B., Frede, H. G., & Breuer, L. (2012). Model intercomparison to explore catchment functioning: Results from a remote montane tropical rainforest. *Ecological Modelling, 239,* 3–13. https://doi.org/10. 1016/j.ecolmodel.2011.05.005.

Prein, A. F., Langhans, W., Fosser, G., Ferrone, A., Ban, N., Goergen, K., Keller, M., Tölle, M., Gutjahr, O., Feser, F., Brisson, E., Kollet, S., Schmidli, J., van Lipzig, N. P. M., & Leung, R. (2015). A review on regional convection-permitting climate modeling: Demonstrations, prospects, and challenges. *Reviews of Geophysics, 53,* 323–361. https://doi.org/10.1002/ 2014RG000475.

Rasmussen, R. M., Baker, B., Kochendorfer, J., Meyers, T., Landolt, S., Fischer, A. P., Black, J., Thériault, J. M., Kucera, P., Gochis, D., Smith, C., Nitu, R., Hall, M., Ikeda, K., & Gutmann, E. (2012). How well are we measuring snow: The NOAA/FAA/NCAR winter precipitation test bed. *Bulletin of the American Meteorological Society, 93,* 811–829. https://doi.org/10.1175/ BAMS-D-11-00052.1.

Saha, S., Moorthi, S., Pan, H.-L., Wu, X., Wang, J., Nadiga, S., Tripp, P., Kistler, R., Woollen, J., Behringer, D., Liu, H., Stokes, D., Grumbine, R., Gayno, G., Wang, J., Hou, Y.-T., Chuang, H.-Y., Juang, H.-M. H., Sela, J., Iredell, M., Treadon, R., Kleist, D., Van Delst, P., Keyser, D., Derber, J., Ek, M., Meng, J., Wei, H., Yang, R., Lord, S., van den Dool, H., Kumar, A., Wang, W., Long, C., Chelliah, M., Xue, Y., Huang, B., Schemm, J.-K., Ebisuzaki, W., Lin, R., Xie, P., Chen, M., Zhou, S., Higgins, W., Zou, C.-Z., Liu, Q., Chen, Y., Han, Y., Cucurull, L., Reynolds, R. W., Rutledge, G., & Goldberg, M. (2010). The NCEP climate forecast system reanalysis. *Bulletin of the American Meteorological Society, 91,* 1015–1057. https://doi.org/10. 1175/2010BAMS3001.1.

Schaefli, B., & Gupta, H. V. (2007). Do Nash values have value? *Hydrological Procedure, 21,* 2075–2080. https://doi.org/10.1002/hyp.6825.

Scheel, M. L. M., Rohrer, M., Huggel, C., Santos Villar, D., Silvestre, E., & Huffman, G. J. (2011). Evaluation of TRMM multi-satellite precipitation analysis (TMPA) performance in the Central Andes region and its dependency on spatial and temporal resolution. *Hydrology and Earth System Sciences, 15,* 2649–2663. https://doi.org/10.5194/hess-15-2649-2011.

Schneider, U., Becker, A., Finger, P., Meyer-Christoffer, A., Ziese, M., & Rudolf, B. (2014). GPCC's new land surface precipitation climatology based on quality-controlled in situ data and its role in quantifying the global water cycle. *Theoretical and Applied Climatology, 115,* 15–40. https://doi.org/10.1007/s00704-013-0860-x.

Seibert, J., & Vis, M. J. P. (2012). Teaching hydrological modeling with a user-friendly catchment-runoff-model software package. *Hydrology and Earth System Sciences, 16,* 3315–3325. https:// doi.org/10.5194/hess-16-3315-2012.

Skok, G., Žagar, N., Honzak, L., Žabkar, R., Rakovec, J., & Ceglar, A. (2015). Precipitation intercomparison of a set of satellite- and raingauge-derived datasets, ERA interim reanalysis, and a single WRF regional climate simulation over Europe and the North Atlantic. *Theoretical and Applied Climatology, 123,* 217–232. https://doi.org/10.1007/s00704-014-1350-5.

Sorooshian, S., Hsu, K.-L., Gao, X., Gupta, H. V., Imam, B., & Braithwaite, D. (2000). Evaluation of PERSIANN system satellite-based estimates of tropical rainfall. *Bulletin of the American*

Meteorological Society, 81, 2035–2046. https://doi.org/10.1175/1520-0477(2000)081<2035: EOPSSE>2.3.CO;2.

Stephens, G. L., & Kummerow, C. D. (2007). The remote sensing of clouds and precipitation from space: A review. *Journal of the Atmospheric Sciences, 64*, 3742–3765. https://doi.org/10.1175/2006JAS2375.1.

Stillman, S., Zeng, X., & Bosilovich, M. G. (2016). Evaluation of 22 precipitation and 23 soil moisture products over a semiarid area in southeastern Arizona. *Journal of Hydrometeorology, 17*, 211–230. https://doi.org/10.1175/JHM-D-15-0007.1.

Su, F., Hong, Y., & Lettenmaier, D. P. (2008). Evaluation of TRMM multisatellite precipitation analysis (TMPA) and its utility in hydrologic prediction in the La Plata Basin. *Journal of Hydrometeorology, 9*, 622–640. https://doi.org/10.1175/2007JHM944.1.

Sun, Y., Solomon, S., Dai, A., & Portmann, R. W. (2006). How often does it rain? *Journal of Climate, 19*, 916–934. https://doi.org/10.1175/JCLI3672.1.

Sylla, M. B., Giorgi, F., Coppola, E., & Mariotti, L. (2013). Uncertainties in daily rainfall over Africa: Assessment of gridded observation products and evaluation of a regional climate model simulation. *International Journal of Climatology, 33*, 1805–1817. https://doi.org/10.1002/joc.3551.

Tang, G., Zeng, Z., Long, D., Guo, X., Yong, B., Zhang, W., & Hong, Y. (2016). Statistical and hydrological comparisons between TRMM and GPM Level-3 products over a midlatitude basin: Is day-1 IMERG a good successor for TMPA 3B42V7? *Journal of Hydrometeorology, 17*, 121–137. https://doi.org/10.1175/JHM-D-15-0059.1.

Tapiador, F. J., Turk, F. J., Petersen, W. A., Hou, A. Y., García-Ortega, E., Machado, L. A. T., Angelis, C. F., Salio, P., Kidd, C., Huffman, G. J., & de Castro, M. (2012). Global precipitation measurement: Methods, datasets and applications. *Atmospheric Research, 104–105*, 70–97. https://doi.org/10.1016/j.atmosres.2011.10.021.

te Linde, A. H., Aerts, J. C. J. H., Hurkmans, R. T. W. L., & Eberle, M. (2008). Comparing model performance of two rainfall-runoff models in the Rhine basin using different atmospheric forcing data sets. *Hydrology and Earth System Sciences, 12*, 943–957. https://doi.org/10.5194/hess-12-943-2008.

Tian, Y., & Peters-Lidard, C. D. (2010). A global map of uncertainties in satellite-based precipitation measurements. *Geophysical Research Letters, 37*, L24407. https://doi.org/10.1029/2010GL046008.

Ushio, T., Kubota, T., Shige, S., Okamoto, K., Aonashi, K., Inoue, T., Takahashi, N., Iguchi, T., Kachi, M., Oki, R., Morimoto, T., & Kawasaki, Z. (2009). A Kalman filter approach to the global satellite mapping of precipitation (GSMaP) from combined passive microwave and infrared radiometric data. *Journal of the Meteorological Society of Japan, 87A*, 137–151. https://doi.org/10.2151/jmsj.87A.137.

Valéry, A., Andréassian, V., & Perrin, C. (2014). "As simple as possible but not simpler": What is useful in a temperature-based snow-accounting routine? Part 1 – Comparison of six snow accounting routines on 380 catchments. *Journal of Hydrology, 517*, 1166–1175. https://doi.org/10.1016/j.jhydrol.2014.04.059.

Vetter, T., Huang, S., Aich, V., Yang, T., Wang, X., Krysanova, V., & Hattermann, F. (2015). Multi-model climate impact assessment and intercomparison for three large-scale river basins on three continents. *Earth System Dynamics, 6*, 17–43. https://doi.org/10.5194/esd-6-17-2015.

Vila, D. A., de Goncalves, L. G. G., Toll, D. L., & Rozante, J. R. (2009). Statistical evaluation of combined daily gauge observations and rainfall satellite estimates over continental South America. *Journal of Hydrometeorology, 10*, 533–543. https://doi.org/10.1175/2008JHM1048.1.

Voisin, N., Wood, A. W., & Lettenmaier, D. P. (2008). Evaluation of precipitation products for global hydrological prediction. *Journal of Hydrometeorology, 9*, 388–407. https://doi.org/10.1175/2007JHM938.1.

Wang, W., Xie, P., Yoo, S.-H., Xue, Y., Kumar, A., & Wu, X. (2013). An assessment of the surface climate in the NCEP climate forecast system reanalysis. *Climate Dynamics, 37*, 1601–1620. https://doi.org/10.1007/s00382-010-0935-7.

Weedon, G. P., Balsamo, G., Bellouin, N., Gomes, S., Best, M. J., & Viterbo, P. (2014). The WFDEI meteorological forcing data set: WATCH forcing data methodology applied to ERA-interim reanalysis data. *Water Resources Research, 50*, 7505–7514. https://doi.org/10.1002/2014WR015638.

Willmott, C. J., Robeson, S. M., & Matsuura, K. (2017). Climate and other models may be more accurate than reported. *Eos, 98*. https://doi.org/10.1029/2017EO074939.

Xie, P., & Arkin, P. A. (1997). Global precipitation: A 17-year monthly analysis based on gauge observations, satellite estimates, and numerical model outputs. *Bulletin of the American Meteorological Society, 78*, 2539–2558. https://doi.org/10.1175/1520-0477(1997)078<2539:GPAYMA>2.0.CO;2.

Xie, P., Joyce, R., Wu, S., Yoo, S.-H., Yarosh, Y., Sun, F., & Lin, R. (2017). Reprocessed, bias-corrected CMORPH global high-resolution precipitation estimates from 1998. *Journal of Hydrometeorology, 18*, 1617–1641. https://doi.org/10.1175/JHM-D-16-0168.1.

Xu, R., Tian, F., Yang, L., Hu, H., Lu, H., & Hou, A. (2017). Ground validation of GPM IMERG and TRMM 3B42V7 rainfall products over southern Tibetan plateau based on a high-density rain gauge network. *Journal of Geophysical Research, 122*, 910–924. https://doi.org/10.1002/2016JD025418.

Yong, B., Liu, D., Gourley, J. J., Tian, Y., Huffman, G. J., Ren, L., & Hong, Y. (2015). Global view of real-time TRMM multisatellite precipitation analysis: Implications for its successor global precipitation measurement mission. *Bulletin of the American Meteorological Society, 96*, 283–296. https://doi.org/10.1175/BAMS-D-14-00017.1.

Zambrano-Bigiarini, M., Nauditt, A., Birkel, C., Verbist, K., & Ribbe, L. (2017). Temporal and spatial evaluation of satellite-based rainfall estimates across the complex topographical and climatic gradients of Chile. *Hydrology and Earth System Sciences, 21*, 1295–1320. https://doi.org/10.5194/hess-21-1295-2017.

Zhu, H., Wheeler, M. C., Sobel, A. H., & Hudson, D. (2014). Seamless precipitation prediction skill in the tropics and extratropics from a global model. *Monthly Weather Review, 142*, 1556–1569. https://doi.org/10.1175/MWR-D-13-00222.1.

Zolina, O., Kapala, A., Simmer, C., & Gulev, S. K. (2004). Analysis of extreme precipitation over Europe from different reanalyses: A comparative assessment. *Global and Planetary Change, 44*, 129–161. https://doi.org/10.1016/j.gloplacha.2004.06.009.

Chapter 34
OceanRAIN – The Global Ocean Surface-Reference Dataset for Characterization, Validation and Evaluation of the Water Cycle

Christian Klepp, Paul A. Kucera, Jörg Burdanowitz, and Alain Protat

Abstract OceanRAIN—the Ocean Rainfall And Ice-phase precipitation measurement Network—is the first comprehensive in-situ surface-reference dataset comprising all water cycle parameters to meet the requirements of the international scientific teams for high-quality precipitation data over the global oceans. OceanRAIN is aimed at satellite retrieval calibration and product validation of the GPM (Global Precipitation Measurement) era, to improve the representation of precipitation and air-sea interactions in re-analyses and models, and to enhance our understanding of water cycle processes over the global oceans. For this purpose, OceanRAIN version 1.0 provides in-situ data of precipitation, evaporation and the resulting freshwater flux at 1-min resolution from June 2010 to April 2017. More than 6.83 million minutes with 75 parameters from 8 ships cover all routinely measured atmospheric and oceanographic state variables along with those required to derive the turbulent heat fluxes. The precipitation parameter is based on measurements of the optical disdrometer ODM470 specifically designed for all-weather shipboard operations. The occurrence, intensity and accumulation of rain, snow and mixed-phase precipitation are derived from particle size distributions. Additionally, all relevant microphysical precipitation and radar-related parameters are provided.

Keywords Precipitation · Rainfall · Snowfall · GPM · TRMM · OceanRAIN · HOAPS · Ocean · Water cycle · Precipitation microphysics · Freshwater flux · Particle size distribution · SeaFlux · OceanObs · Disdrometer · Validation

C. Klepp (✉)
Max Planck Institute for Meteorology, Hamburg, Germany
e-mail: christian.klepp@mpimet.mpg.de

P. A. Kucera
University Corporation for Atmospheric Research (UCAR), Boulder, CO, USA

J. Burdanowitz
Institute of Meteorology, University of Hamburg, Hamburg, Germany

A. Protat
Bureau of Meteorology (BoM), Melbourne, VIC, Australia

© Springer Nature Switzerland AG 2020   655
V. Levizzani et al. (eds.), *Satellite Precipitation Measurement*, Advances in Global
Change Research 69, https://doi.org/10.1007/978-3-030-35798-6_10

34.1 Introduction

Significant improvements in the quality and uncertainty estimation of satellite remote sensing precipitation retrievals and products over the global oceans require novel, high-quality surface-reference validation data. Combining satellite remote sensing and in-situ monitoring of the global ocean water cycle is crucial for a successful understanding of the climate system. The net gain (precipitation) or loss (evaporation) of water through the ocean surface is described by the freshwater flux, linking the global water cycle to the energy budget through latent heat exchange. The freshwater flux couples the ocean to the atmosphere and drives oceanic and atmospheric circulations with impacts on sea surface temperature and salinity. Since the oceans cover about 71% of the Earth's surface, the precipitation parameter is probably the most essential climate variable to be characterized (Bojinski et al. 2014; Kidd and Huffman 2011) and accounts for an estimated 78% of global precipitation. However, several factors complicate its measurement and thus explain its large uncertainties, such as its intermittent nature, inhomogeneous spatial distribution, intensity variations and phase changes as well as technical detection limits for very light precipitation and solid precipitation. Accurate measurements of precipitation processes and associated statistical properties such as phase-dependent frequency of precipitation, instantaneous intensity and accumulation are of crucial importance for hydrological applications, numerical weather prediction and climate change studies (Trenberth et al. 2003; Stephens et al. 2010).

Recent active and passive satellite remote sensing sensors and their products provide unprecedented spatio-temporal coverage of rainfall. Satellite missions, such as TRMM (Tropical Rainfall Measurement Mission; Kummerow et al. 1998), CloudSat (Ellis et al. 2009) and Global Precipitation Measurement (GPM; Hou et al. 2014), aided by their spaceborne precipitation radars TRMM-PR, CloudSat-CPR and GPM-DPR boosted the emergence of precipitation products (Kidd and Huffman 2011). New products, such as the Integrated Multi-satellitE Retrieval for GPM (IMERG) retrievals additionally discriminate between rain and snow and feature increased sensitivity to light rain processes.

However, all of these products inevitably exhibit sensitivity issues as well as considerably large differences between them, especially for very light and heavy rainfall as well as high-latitude frozen precipitation (Andersson et al. 2011; Béranger et al. 2006; Petty 1997; Stephens et al. 2012). Furthermore, model intercomparisons reveal that, although accumulations at different temporal scales are generally reasonably accurate, general circulation models tend to overestimate the frequency and underestimate the intensity of precipitation, with different behavior of the models in different regions and latitudes (Stephens et al. 2010).

Prior to the recent release of the OceanRAIN-1.0 dataset (Klepp et al. 2018) the lack of global ocean surface-reference precipitation data remained critical and substantially hindered the required validation efforts. This was mainly due to the unavailability of a capable instrumentation, e.g. disdrometers for shipboard operation and the large effort required establishing such a network (Klepp et al. 2010). Common gauge-type devices onboard ships and buoy networks are generally not well suited for this task because of large wind-induced rainfall undercatch resulting from superimposed ship's speed, surface wind speed and turbulent flow distortion around the ship superstructure (Weller et al. 2008). Adding to the problem, snow and mixed-phase precipitation is either blown over or clog the gauge orifice resulting in erroneous measurements.

In 2008, this motivated the initiation of the OceanRAIN project to tackle the challenge of developing, operating and maintaining a long-term monitoring network of automated disdrometer systems, designed for shipboard operation under all-weather conditions, onboard globally operating research vessels to close this fundamental observation gap. By 2017, OceanRAIN is the first in-situ global ocean shipboard dataset comprising 75 meteorological and oceanographic parameters including consistently derived along-track precipitation, evaporation, the resulting freshwater flux as well as surface turbulent fluxes. The precipitation parameters include rain, snow and mixed-phase precipitation frequency of occurrence, intensity and accumulation, all derived through PSDs based on automated ODM470 optical disdrometers that were specifically designed to meet all-weather shipboard requirements. All relevant microphysical precipitation properties and radar-related reflectivity parameters at different radar operating frequencies are additionally provided. These datasets are collected during the ongoing long-term installations and special campaigns onboard eight research vessels from June 2010 to April 2017 covering all latitudes, oceanic basins and seasons and comprise more than 6.83 million minutes of which 696,740 min contain precipitation. OceanRAIN-1.0 data (Klepp et al. 2017a, b, c) is publicly available via the project website https://www.oceanrain.org/ and the World Data Center for Climate (WDCC).

OceanRAIN is primarily designed to

- provide global ocean precipitation characterization to broaden our understanding of precipitation and its microphysical properties,
- to validate satellite data retrievals and precipitation products and for assessing and reducing their biases and uncertainties with special focus on the cold-season high-latitudes and light precipitation,
- to provide the full suite of water cycle parameters, including the along-track freshwater flux in a consistent framework,
- to supply the international science teams of IPWG, GPM-GV, SeaFlux and OceanObs as well as the Global Climate Observing System (GCOS) with surface-reference validation data, and
- to evaluate air-sea flux processes in re-analysis and general circulation model data.

34.2   The OceanRAIN Optical Disdrometer ODM470

In-situ precipitation particle size distribution (PSD) data for satellite retrieval calibration can only be provided by disdrometers (Klepp 2015). However, most of the
existing disdrometers are not designed for all-weather shipboard operation and thus
do not meet the requirements for operation in strong turbulence, frequently varying
wind directions and sea state. Consequently, the backbone of the OceanRAIN
project is the optical disdrometer ODM470, initially developed by GEOMAR in
Kiel, Germany, because it was specifically designed for all-weather shipboard
operation. Within OceanRAIN it was further developed into a fully automated
measurement system (Klepp 2015). An infrared light emitting diode homogeneously
illuminates the cylindrical measuring volume. Hydrometeors passing this volume
cause light extinction proportional to their cross-sectional area. The reduction in the
light detected is stored as an activation voltage. During the integration time of 60 s
all hydrometeors are counted and sorted into size bins ranging from 0.39 to 22 mm in
diameter to obtain the PSD. A cup anemometer additionally measures relative wind
speed and a precipitation detector activates the disdrometer during precipitation
events only. This reduces the risk of artificial signals caused by vibration of the
instrument due to sea state, ice-breaking activity, gusty winds or the ship's engine.
The OceanRAIN instruments are installed in the highest parts of the ship's mast and
in front of the ship's funnel, to minimize the exposure to sea spray, wave water
and soot.

   Disdrometer calibration is performed in a twofold procedure that first comprises a
lab hardware calibration of the optical axis and the reference voltage using steel ball
bearings. This is followed by an outdoor rainfall calibration at low wind speeds using
a reference rain gauge for accumulation comparison. Both instruments typically
differ in the order of 2% rain accumulation. The calibration drift after shipboard
operation is negligible in most cases because the reference voltage is continuously
checked and adjusted during the cruises as necessary.

   The main advantages of the ODM470 system over other existing disdrometers
can be specified in five points,

- the measurement volume of the disdrometer has a cylindrical shape causing
  precipitation particles to be independent of their incidence angle,
- the measurement volume is always kept perpendicular to the local wind direction.
  This is achieved by a wind vane that pivots the instrument around a vertical axis.
  Consequently, local up- and downdrafts as well as turbulence induced by the ship
  superstructure cause a minimal impact on the measurement of the PSD,
- the high-resolution optical unit allows discriminating hydrometeors into 128 size
  bins with a logarithmically increased resolution towards smaller size bins,
- an automated precipitation phase detection algorithm allows to process PSDs
  using either the rainfall or snowfall algorithm, and
- the instrument is fully automated, robust and requires minimal maintenance
  during operation. Thus, it is ideally suited for long-term monitoring of precipitation over the global oceans in all-weather conditions.

The ODM470 operated reliably during snowfall at -30 °C near Antarctica, torrential rainfall in the tropics with a record value of 367 mm h^{-1} and during gale force wind conditions with severe sea states in mid-latitude storms. A detailed description of the instrumentation and its measurement principles is provided in Klepp (2015) and Klepp et al. (2018).

34.3   Data Ingest and Data Set Construction

Research vessels (RVs) with long-term OceanRAIN installations offer a perfect measurement platform because they operate in all climate-relevant areas and seasons over the global remote oceans including the high-latitudes and do not routinely circumvent high-impact weather. This largely avoids the so-called fair-weather bias that occurs when merchant ships or cruise liners circumvent high-impact weather along their main shipping routes. For these reasons the OceanRAIN data set comprises the entire spectrum of weather events including extreme values in remote locations. The dataset is complemented by RVs providing short-term campaign data with special emphasis on water cycle analysis, e.g. the tropical Pacific SPURS (Salinity Processes in the Upper Ocean Regional Study) campaign onboard RV Roger Revelle, and the CAPRICORN (Clouds, Aerosols, Precipitation Radiation and atmospherIc Composition Over the southeRN ocean) cruises onboard RV Investigator with a focus on GPM and CloudSat satellite validation using collocated ship underpasses.

The data acquisition and automated data processing chain to derive all water cycle and precipitation parameters is visualized in the flow chart of Fig. 34.1. Four time-synchronized data streams are recorded separately onboard the ships and are ingested into the OceanRAIN database. These are the temporally continuous ship-board navigational data (NAV), the automated underway surface atmospheric and oceanographic data (MET), three-hourly manual synoptic weather observations (SYN) including precipitation type and intensity observations as well as the temporally discontinuous OceanRAIN ODM470 precipitation raw data (ODM). The raw ODM470 PSD data contains the number of particles per bin size with an integration time of 1 min, stored into daily files that are designed empty if no precipitation occurred. While the ODM470 records temporally discontinuous precipitation data, the OceanRAIN post-processing also includes the true-zero precipitation information. The no-precipitation signal carries important information for the analysis of precipitation frequency of occurrence, false alarm statistics as well as retrieval behavior analysis in satellite-derived precipitation products. Additionally, metadata protocols on special weather and instrument conditions (META) are provided to identify reasons for instrument problems, outages or unusual spikes in the data. All individual data streams are into a navigated 1-min resolution match-up database. A cascading automated and visual quality control procedure flags, accepts or rejects data to ensure that all measured as well as derived parameters rely on data that

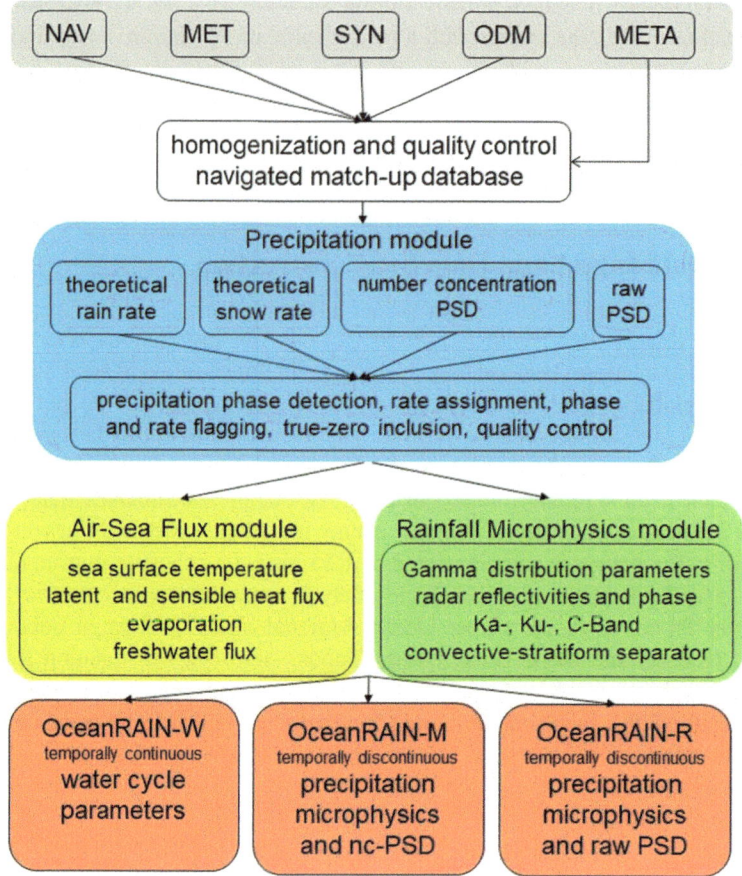

Fig. 34.1 OceanRAIN shipboard data ingest and post-processing flow chart. (Adapted from Klepp et al. 2018)

successfully passed the quality control test. All quality control changes applied to the precipitation data can be traced throughout the data set.

As visualized in the flow chart of Fig. 34.1, this quality-controlled navigated match-up database is applied to three modules to derive the precipitation, rainfall microphysics and air-sea flux parameters. In the following they are summarized briefly while all algorithms and methods are described in detail in Klepp (2015) and Klepp et al. (2018).

For each minute with ODM470 precipitation occurrence the PSD density in m^{-3} is calculated using the product of the measurement volume, the integration time and the geometrical sum of the relative wind speed and the parameterized terminal fall velocity. Additionally, normalization by the non-constant bin width leads to the number concentration PSD (m^{-3} mm^{-1}). The 1-min resolution number concentration PSD (nc-PSD) is stored in the OceanRAIN-M files while the raw particle count

PSD is stored in the OceanRAIN-R files (raw PSD). Both files are temporally discontinuous and contain precipitation minutes only.

The calculation of the rainfall and snowfall rate in mm h^{-1} requires the mass (liquid water equivalent) of the particles and the terminal fall velocity. For rainfall with near-spherical drop shapes and constant drop density, the parameterizations are well known. For the complex problem of snowfall, a common lump graupel parameterization was developed, relating the measured cross-sectional area to the maximum dimension of the particle. Therein, the product of the terminal fall velocity and the mass as a function of the cross-sectional area remains in the same order of magnitude for a variety of different snow crystals. Because lump graupel is nearly spherical in shape it circumvents the need for a transfer function between the cross-sectional area and the maximum dimension of the particles.

Each precipitation minute is ingested into the rainfall and snowfall algorithms to derive the rainfall and snowfall intensity in mm h^{-1}. The precipitation-phase assignment into rain, snow or mixed-phase precipitation including a precipitation phase probability is calculated using an automated OceanRAIN precipitation phase distinction algorithm (Burdanowitz et al. 2016) using air temperature, relative humidity and the 99th percentile of the PSD particle diameter. Because GPM and CloudSat products provide model-estimated stratification of rain and snow, we decided to introduce mixed-phase precipitation to OceanRAIN so that the dataset can be stratified accordingly.

A precipitation flag (flag1) assigns the precipitation phase into rainfall (0), snowfall (1) and mixed-phase (2). Because the data in the OceanRAIN-M and OceanRAIN-R files are temporally discontinuous (precipitation-only events) there can be no true-zero value. However, the OceanRAIN-W files are temporally continuous and therefore contain the true–zero information (value 3). This is especially important for very light precipitation because two sources for 0.00 mm h^{-1} precipitation rates exist that need to be discriminated. A precipitation minute can either be zero because no precipitation occurred or, precipitation occurred but resulted numerically in zero accumulation (below the detection threshold of 0.01 mm h^{-1}). The data is further stratified into precipitation events and intensities (flag 2) to discriminate between spurious (electronic artefacts due to vibration of the instrument), possible (less than 20 particles and less than 5 bins occupied; precipitation rates insignificant or zero) and real precipitation.

The output of the precipitation module comprises precipitation occurrence including true-zero minutes, intensity, accumulation, precipitation phase and probability for rain, snow and mixed-phase, 99th percentile of particle diameter, theoretical rain and snow intensities, flag1, flag2, number of particles and number of bins as well as raw PSDs and number concentration PSDs.

The rainfall microphysics module provides normalized gamma distributions of rainfall using the intercept of the distribution, the median volume diameter, and the shape parameter. The PSDs are fitted using the normalized gamma formulation after Testud et al. (2001) and Bringi et al. (2002) when at least 10 size bins are filled with data. Once The 1-min resolution PSD is further classified as convective or stratiform rain using the physically-based convective-stratiform classification proposed by

Thurai et al. (2010). The intercept parameter of the standard gamma distribution is additionally provided. Because the three parameters of the normalized gamma distribution are not statistically independent, this caused issues in the TRMM and GPM satellite retrievals of rainfall rate. Therefore, the mass-weighted mean diameter and the standard deviation of the mass spectrum are also included in the OceanRAIN files.

Additionally, the radar reflectivity Z, differential reflectivity ZDR, and specific differential phase KDP at different frequencies (Rayleigh at 3 GHz, C-band at 5.6 GHz, Ku-band at 13.6 GHz and Ka-band at 35 GHz) are calculated using the pyTmatrix tool after Leinonen (2014). Applications for such data comprise satellite radar retrievals of rainfall rate, GPM satellite radar validation, or high-resolution model evaluation.

The air-sea flux module adds the turbulent heat fluxes and the evaporation to obtain the freshwater flux. The evaporation is derived from along-track ship data using a bulk formulation according to Fairall et al. (1996). This requires the shipboard air temperature, relative humidity, water temperature and absolute wind speed. From these values the sea surface temperature (SST) and latent as well as sensible heat fluxes are parameterized using the COARE bulk flux algorithm after Fairall et al. (2003). The evaporation in mm h^{-1} is derived from the turbulent latent heat flux. Finally, the difference between the evaporation E and the precipitation P yields the ocean surface freshwater flux E-P in mm h^{-1} into the atmosphere.

Figure 34.2 provides an overview of the data recorded and processed for OceanRAIN version 1.0. The geographical distribution of all eight RVs (a) is color-coded for seasons (b), years (c) and precipitation phase occurrence including true-zero precipitation information (d). Figure 34.2c allows for an easy

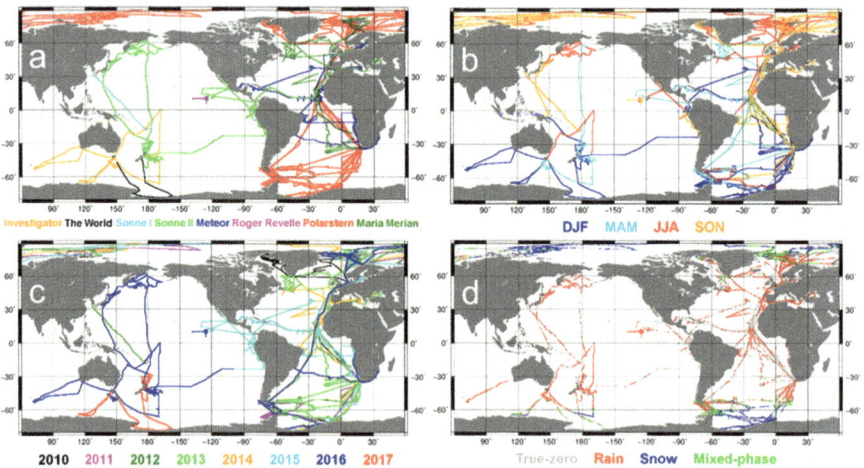

Fig. 34.2 OceanRAIN data distribution for all eight ships (**a**), seasons (**b**), years (**c**) and precipitation occurrence for type (rain, snow, mixed) and true-zeros (**d**). (Adapted from Klepp et al. 2018)

discrimination into the data volume of the pre and post GPM period before and after 2014. The dataset comprises a total of more than 6.83 million minutes of data with 696,740 precipitation minutes, of which 414,807 min contain rain, 232,358 snow and 49,575 mixed-phase precipitation including 4,699,282 true-zero minutes for all seasons over the global oceans including the cold-season high-latitude Arctic Ocean and the Southern Oceans.

34.4   OceanRAIN Data Sets

OceanRAIN provides users with three dataset versions for each of the eight ships in the fleet from June 2010 to April 2017 so that users can choose which dataset best meets their research needs.

The OceanRAIN-W (Klepp et al. 2017a) files contain the 1-min resolution water cycle components of evaporation, precipitation and the freshwater flux along with all meteorological and oceanographic state variables required to derive these fluxes. The dataset is continuous in time and contains 73 parameters and more than 6.83 million minutes of data. Typical applications comprise of process studies and statistical analysis as well as satellite validation and re-analysis or model evaluation. The OceanRAIN along-track point data can serve as the surface reference and can be collocated with satellite or model data to analyse and improve their error characteristics. Therefore, it is important to highlight that the RVs sampled on the global oceans during all seasons including the cold-season Southern Oceans.

OceanRAIN-M (Klepp et al. 2017b) and OceanRAIN-R (Klepp et al. 2017c) focus on minutes containing precipitation and are therefore discontinuous in time. Both datasets comprise 37 precipitation-relevant parameters plus the 128 size bin number concentration PSDs (OceanRAIN-M) and raw number count PSDs (OceanRAIN-R) for 696,740 min in total with rain, snow or mixed-phase precipitation. The precipitation-related parameters are identical in the three versions of the dataset. Applications for these datasets especially comprise of satellite retrieval performance evaluation for liquid and solid precipitation. For this purpose, OceanRAIN-M and OceanRAIN-R supply the user with a convective versus stratiform precipitation classification and contain the main PSD characteristics and the radar reflectivities at important frequencies for radar rainfall studies. This is of special importance for users looking at TRMM, CloudSat and GPM product and retrieval validation since these satellite missions carry spaceborne radars.

The OceanRAIN version 1.0 datasets are freely available as netCDF (Network Common Data Form) and ASCII (American Standard Code for Information Interchange) files through the World Data Center for Climate (WDCC), the Integrated Climate Data Center (ICDC) at the University of Hamburg via http://icdc.cen.uni-hamburg.de/1/daten/atmosphere/oceanrain/ (last accessed 22 Oct. 2018) and through the OceanRAIN project website http://www.oceanrain.org/ (last accessed 22 Oct. 2018). A ReadMe file is provided at each of the data repositories to assist the user on ingesting and using the data.

34.5 Applications and Validation

34.5.1 Precipitation Characterization

The overall precipitation occurrence in the OceanRAIN-1.0 database across all latitudes and ocean basins varies significantly depending on the threshold used to define a precipitation minute. Using all minutes containing precipitation the absolute occurrence is 14.8% with 8.8% rainfall, 4.9% snowfall and 1.1% mixed-phase precipitation. This includes minutes with insignificant precipitation amounts that correspond to rates less than 0.01 mm h^{-1} that are numerically set to 0.00 mm h^{-1}. Defining precipitation with a lower threshold of 0.01 mm h^{-1} the occurrence drops to 9.8% (6.9% rain, 2.6% snow and 0.3% mixed). Using a threshold of 0.1 mm h^{-1}, the typical threshold of what gauge-type sensors are capable to measure, the occurrence drops to 6.2% (4.9% rain, 1.2% snow and 0.1% mixed). This consolidates the importance of sensors to be able to detect very light precipitation and demonstrates how predominant very light precipitation is (Klepp et al. 2018). Moreover, the light precipitation fraction increases from the tropics (56.8%) towards the polar latitudes (88.1%). However, the very light precipitation has little effect on the precipitation accumulation. Rainfall contributes 92.1% to the total accumulated precipitation (7.4% snow and 0.5% mixed). The occurrence of intense rainfall beyond 5 mm h^{-1} shows a maximum of 10.8% in the inner tropics and accounts for an accumulation of 76.0% (Klepp et al. 2018).

Figure 34.3 depicts the precipitation characteristics of the OceanRAIN-1.0 dataset. The latitudinal precipitation fraction (panel a) shows the occurrence of mixed-phase precipitation poleward of 40°N and 45°S and snowfall poleward of 44°N and 50°S. Solid precipitation is the predominant phase beyond 80°N and 53°S. The disparity of the latitudes on both hemispheres is due to the boreal summer sampling of the RVs in the Arctic while the RVs sampled all seasons in the Southern Oceans. The joint histogram of the precipitation intensity in panel (b) shows that the subtropics are characterized by significantly fewer rainfall events. The mid-latitude rainfall begins at approx. 35° latitude on both hemispheres. The rainfall intensities in the mid-latitudes are comparable with those measured in the tropics. First, this indicates the importance of mid-latitude cyclones that contribute a major fraction of the global rainfall with high intensities at high wind speeds. Second, this may indicate an undersampling of the intertropical convergence zone rainfall in the OceanRAIN dataset. Comprehensive sampling of the inner tropical convection cells requires more ships and more years of data collection. The next version of the OceanRAIN dataset will largely improve this sampling. This is also supported by the bin-wise mean for 2° latitude bands. The zonal mean curve shows a strong latitudinal fluctuation and therefore is strongly precipitation event driven. The climatology requirements are only met where the sampling is high, as indicated by the darker colors of the frequency of occurrence in percent. In turn, the zonal mean does not meet the climatology value wherever the sampling is low (e.g., in the inner tropics).

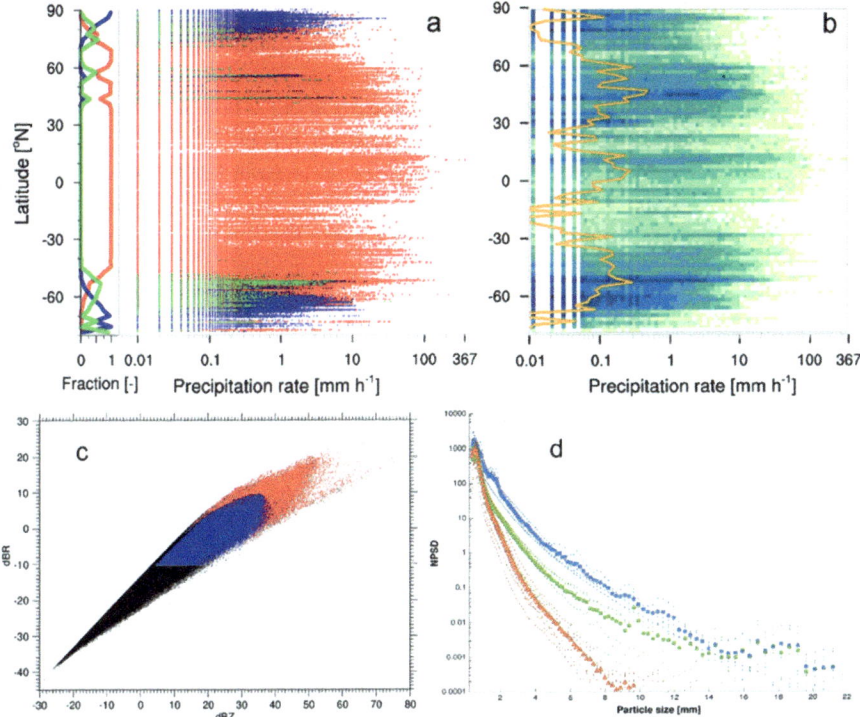

Fig. 34.3 Latitudinal fraction of the precipitation phase with red for rain, blue for snow and green for mixed phase (**a**). The scatter diagram shows individual precipitation minutes for phases against latitude as a function of precipitation rate (mm h^{-1}). Panels (**b**) shows the joint histogram of the precipitation rate in mm h^{-1} and the orange line indicates bin-wise mean for 2° latitude bands. Frequency of occurrence in % of all cases is shown with colors. Panel (**c**) shows the Rayleigh reflectivity log(Z) and log (R) distribution of rainfall for all 696,740 events (black), 139,557 stratiform events (blue) and 15,823 convective events (red). Panel (**d**) denotes the mean number concentration PSDs (thick marks) and their latitudinal variability (thin marks) for all rainfall (red), snowfall (blue) and mixed-phase precipitation (green). (Adapted from Klepp et al. 2018)

The precipitation microphysics of the OceanRAIN-M database are also summarized in Fig. 34.3. Panel (c) shows the log(Z) and log(R) values in dBZ and dBR for all 696,740 rainfall events, as well as for subsets of stratiform (139,557 min in blue) and convective rainfall (15,823 min in red). The black part of the distributions holds the number of 1-min spectra with less than 10 bin-sizes occupied in the PSD. The distribution reaches from one-drop spectra at 0.39 mm (−40 dBR, −35 dBZ) to extreme precipitation events at 20 dBR and 55 dBZ. From all enclosed points or any subset of the dataset it is possible to derive Z (Rayleigh reflectivity) – R (rainfall rate) relationships to constrain, validate or improve radar-derived and modelled rainfall rate estimates. Panel (d) depicts the mean number concentration PSDs (bold marks) and their latitudinal variability (small marks) for all 414,807 min of rainfall (red), 232,358 snowfall (blue) and 49,575 mixed-phase precipitation (green). The rainfall PSD peaks at a diameter of 0.39 mm with a number concentration of about 1000 m^{-3} mm^{-1} and steeply decreases down to about 0.003 m^{-3} mm^{-1} for the largest drops of about 6.5 mm diameter. Single drops, however, can reach a diameter of about 9.7 mm and were measured during convective rainfall in the tropics. The snowfall curve peaks at 1100 m^{-3} mm^{-1} and descends more gradually towards largest measured snowflakes of 22 mm in diameter. The mixed-phase precipitation curve resembles that of rainfall for smaller particle diameters while for larger diameters it converges towards the snowfall PSD. The number concentration at 0.39 mm is higher for frozen particles than for rain drops because rain drops tend to evaporate faster than frozen particles.

34.5.2 The Point-to-Area Representativeness

Particularly for precipitation, the validation of satellite-derived as well as the evaluation of model-based areal precipitation estimates often uses pointwise surface precipitation data that leads to a number of challenges to be tackled. Loew et al. (2017) describes these challenges in a comprehensive overview and provide ways how to address them properly.

Resolution differences in space and time mainly limit the representativeness of point data compared to areal satellite data. This long-standing point-to-area (p2a) problem influences both—if and how much precipitation is observed by the point measurement within the area of a satellite pixel. How much precipitation the point data estimates can be statistically adjusted to what a satellite sensor would see. For OceanRAIN, Burdanowitz et al. (2017) derived statistical adjustments for along-track ship data matched to areal satellite pixels by conducting a synthetic study with weather radar data from the Rain In Cumulus over the Ocean (RICO) campaign on Barbuda. This idealized study reduces the representativeness error and has been applied in a validation study of Hamburg Ocean Atmosphere Parameters and fluxes from Satellite (HOAPS; Andersson et al. 2010) data. In this application, Burdanowitz et al. (2018) revealed from geostationary Meteosat Second Generation Cloud Physical Properties (MSG-CPP) data that the point-to-area effect can stronger influence the OceanRAIN–HOAPS difference than the precipitation regime as classified using precipitation area and intensity from MSG-CPP.

Statistical adjustment for OceanRAIN are particularly crucial for rather coarsely resolved satellite estimates of 0.5° and above, marking an important step towards a more representative precipitation validation of satellite data over the ocean (Fig. 34.4).

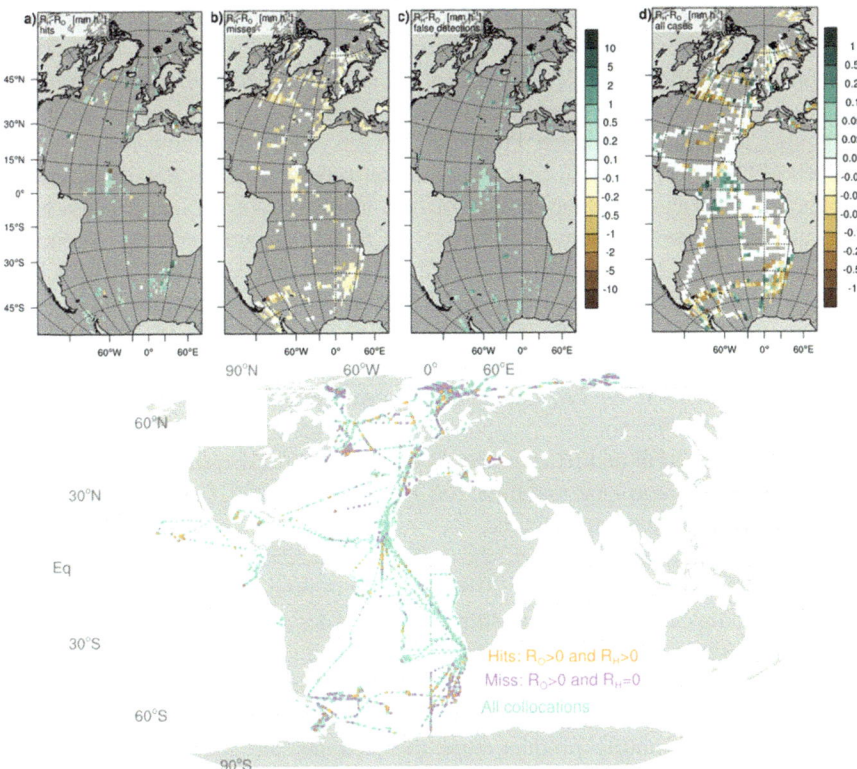

Fig. 34.4 Maps of average precipitation-rate difference of HOAPS (R_H) subtracted by adjusted OceanRAIN (R_O**) in mm h^{-1} per $2° \times 2°$ grid-box for (**a**) hits, (**b**) misses, (**c**) false detections and d) all cases. Gray boxes have no data available. (**e**) Map of occurrence of precipitation from 24,990 HOAPS–OceanRAIN collocations from available long-term RVs in OceanRAIN from 06/2010–12/2015 (cyan dots), misses ($R_H = 0$ and R_O** > 0; purple) and hits ($R_H > 0$ and R_O** > 0; orange) using ± 20 km and ± 30 min as collocation boundaries. (Adapted from Burdanowitz et al. 2018)

34.5.3   *IMERG Validation*

OceanRAIN provides a comprehensive database of high-quality in-situ global oceanic phase-dependent precipitation observations through observed PSDs. OceanRAIN also provides estimates of evaporation and freshwater flux data for water cycle analysis. These data are used for evaluating NASA's Integrated Multi-satellitE Retrievals for GPM (IMERG) satellite precipitation products over the ocean.

The OceanRAIN Water Cycle dataset was used to evaluate the IMERG precipitation estimates using OceanRAIN data from seven ships. The ships that overlapped the IMERG dataset include Investigator, Merian, Meteor, Polarstern, Revelle, Sonne II, and World. IMERG is the unified US algorithm that provides a multi-satellite precipitation product for the US GPM team. The precipitation estimates from the various precipitation-relevant satellite passive microwave (PMW) sensors comprising the GPM constellation are computed using the 2017 version of the Goddard Profiling Algorithm (GPROF 2017), gridded, intercalibrated to the GPM Combined Instrument product, and combined into half-hourly $0.1° \times 0.1°$ fields. These are provided to both the Climate Prediction Center (CPC) Morphing-Kalman Filter (CMORPH-KF) Lagrangian time interpolation scheme and the Precipitation Estimation from Remotely Sensed Information using Artificial Neural Networks – Cloud Classification System (PERSIANN-CCS) re-calibration scheme. In parallel, CPC assembles the zenith-angle-corrected, intercalibrated "even-odd" geo-IR fields and forward them to PPS for use in the CMORPH-KF Lagrangian time interpolation scheme and the PERSIANN-CCS computation routines. The PERSIANN-CCS estimates are computed (supported by an asynchronous re-calibration cycle) and sent to the CMORPH-KF Lagrangian time interpolation scheme. The CMORPH-KF Lagrangian time interpolation (supported by an asynchronous KF weights updating cycle) uses the PMW and IR estimates to create half-hourly estimates. The IMERG system is run twice in near-real time. See Huffman et al. (2018) for more details on the IMERG algorithm and datesets.

The IMERG "Final-Run" (e.g., the calibrated research quality data) precipitation product record starts on 20 March 2014. The current version (V05B) was used which provides complete precipitation estimates limited to $\pm60°$ north and south of the equator in latitude. At higher latitudes (polar regions), precipitation estimated are not morphed and only providing in microwave swaths for non-snowy/non-icy surfaces (Huffman et al. 2018). The matched OceanRAIN and IMERG V05B Final Run $0.1° \times 0.1°$ half-hourly product were matched for the coincident data period: 20 March 2014–28 February 2017. The datasets were matched using two main criteria. OceanRAIN had to have at least 10 min continuous precipitation at the 1-min resolution. Different time thresholds of (15, 30, and 60 min) were also computed to examine dependency of results on size of precipitation events. However, for this initial study, only the 10-min continuous threshold cases were examined. OceanRAIN was also stratified by precipitation rate threshold. The OceanRAIN observations had to have rainfall rates of greater than 1 mm h^{-1} for the 10-min period. The goal was to find significant precipitation events that should be able to be observed by IMERG. The satellite product could possibly not observe precipitation below these temporal and intensity thresholds. Mixed phase precipitation was not examined at this time.

A total of 5590 cases were matched globally for the 10-min, 1 mm h^{-1} threshold. Of those cases, a total of 2954 rainfall cases were observed while the total number of snow cases identified was 2636. A global map of all the cases are shown in Fig. 34.5. Most of the snow cases were observed above $\pm60°$ north and south latitudes. Most of the snow cases are above this latitude. Therefore, about 3500 cases used in the analysis with only a few 100 snow cases below $\pm60°$ latitudes.

The following discussion provides an example of the initial intercomparison between IMERG and OceanRAIN. Figure 34.6 shows the overall results and reveal

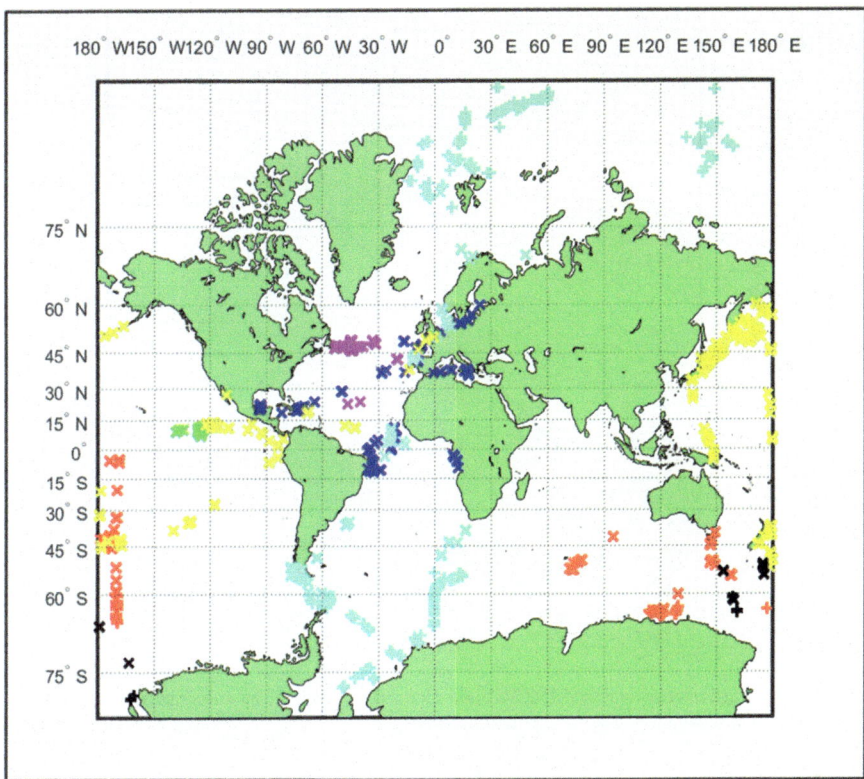

Fig. 34.5 Location of matched OceanRAIN-IMERG pairs for the period 20 March 2014–
28 February 2017. The "x" symbol indicates rain events, and "+" symbol indicates snow eents.
Ship tracks of OceanRAIN database. R/V Investigator (red), R/V Merian (magenta), R/V Meteor
(blue), R/V Polarstern (cyan), R/V Revelle (green), R/V Sonne II (yellow), R/V World (black)

a large scatter in the OceanRAIN-IMERG grid comparison. The results show that
IMERG misses the small, intense tropical rain systems, which are the values along
the bottom x-axis. There is an IMERG mismatch for many of cases which are
observed along the y-axis. The following analysis is focused on trying to identify
the source of the errors. Figure 34.7 shows the cross correlation of IMERG precipitation. Precipitation in the neighboring 11×11 IMERG grid boxes were compared
to OceanRAIN. Results indicate a NW-SE signal in IMERG precipitation. It
is possible that morphing technique needs adjustment in NW-SE directions.
Figure 34.8 shows the comparison of IMERG and OceanRAIN stratified by convective and stratiform precipitation. IMERG underestimates the precipitation of
convective events while IMERG overestimates precipitation for stratiform events.

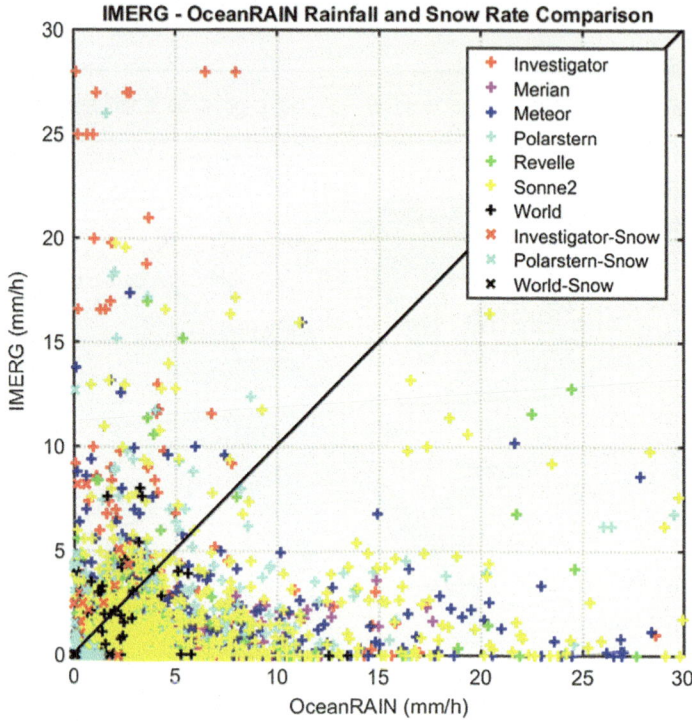

Fig. 34.6 Scatterplot of matched IMERG and OceanRAIN pairs. The colors indicate the ship that observed the precipitation. The "+" symbol indicates rain pairs and "×" symbol indicates snow pairs

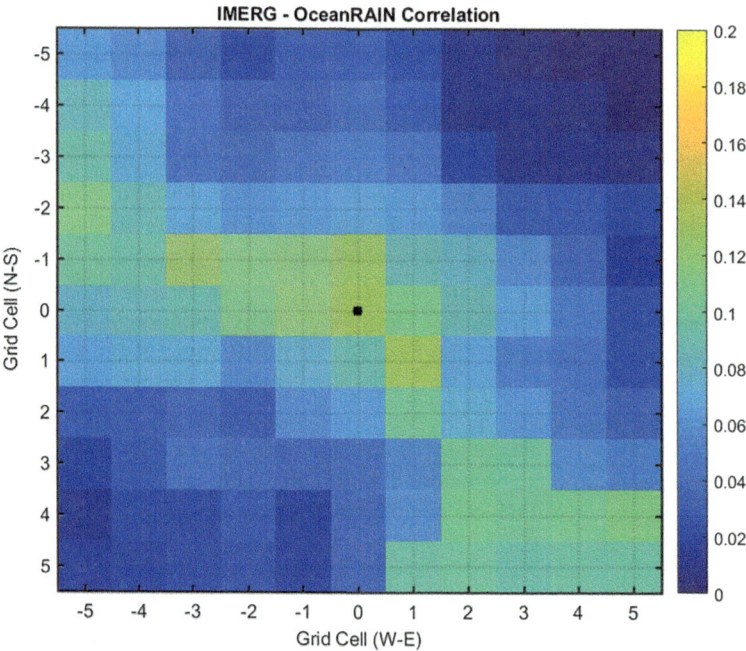

Fig. 34.7 Plot of cross-correlation of rainfall for neighboring IMERG grids

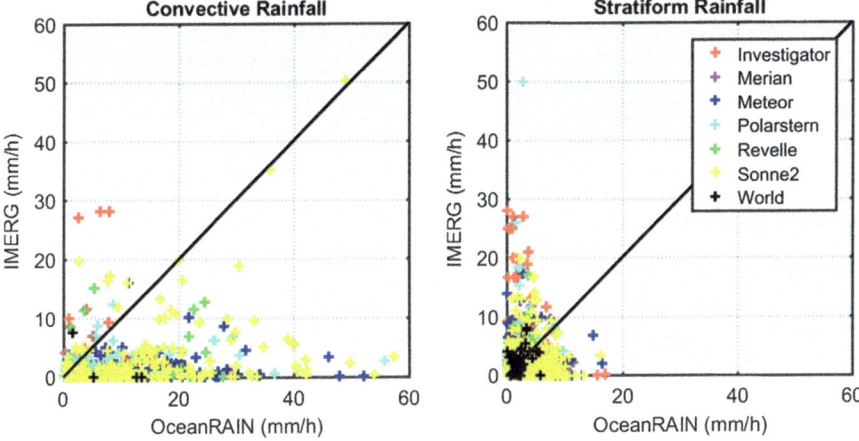

Fig. 34.8 Scatter plot of matched pairs for convective and stratiform rainfall

34.6   Conclusions and Future Outlook

OceanRAIN has been successful in developing and producing a novel in-situ global ocean surface-reference dataset for all water cycle parameters. With special emphasis on the precipitation parameter, OceanRAIN provides rain, snow and mixed-phase precipitation occurrence, intensity and accumulation including microphysical properties derived from optical disdrometer PSD data. The OceanRAIN-1.0 dataset has been collected from June 2010 to April 2017 onboard 8 ships covering all oceanic basins, latitudes and seasons. The data is freely disseminated via the project website www.oceanrain.org with application foci on satellite retrieval calibration and product calibration of the GPM era as well as re-analysis and model evaluation. However, OceanRAIN also serves to improve our understanding about oceanic precipitation and water cycle characterization.

Alongside the presented point-to-area analysis and IMERG validation work, OceanRAIN is involved in the following ongoing projects:

- Within the CAPRICORN Southern Ocean cruises onboard the Australian RV Investigator collocated CloudSat and GPM measurements are investigated using the shipboard OceanRAIN as well as a vertically pointing precipitation radar and a horizontally scanning weather radar. One of the research goals is to analyze how different Southern Ocean PSD properties are from that elsewhere over the global oceans. Such results may help to explain why rainfall retrievals south of 40°S diverge.

- In September 2013, there was a historical flood in the Colorado front range. An ODM470 was installed at the NCAR Marshall field site for an instrument intercomparison study and was operational during the flood event. These data provided a comprehensive observation dataset to evaluate the variability of DSD's. These data are being used to improve the radar rainfall estimates for future radar algorithm implementation

Acknowledgements The sustained funding by Initiative Pro Klima (Mabanaft, Mabanaft Deutschland, Petronord and OIL! Tankstellen GmbH & Co. KG) enabled the development, operation and research on OceanRAIN. We greatly appreciate this support and gratefully thank Tanja Thiele, Gerhard Grambow, Volker Tiedemann, Ulrich Freudental and Jan Falke. The project was hosted and co-funded by CliSAP/CEN, Universität Hamburg, and the Max Planck Institute for Meteorology, Hamburg, Germany.

References

Andersson, A., Fennig, K., Klepp, C., Bakan, S., Graßl, H., & Schulz, J. (2010). The Hamburg Ocean atmosphere parameters and fluxes from satellite data – HOAPS-3. *Earth System Science Data, 2*, 215–234. https://doi.org/10.5194/essd-2-215-2010.

Andersson, A., Klepp, C., Fennig, K., Bakan, S., Graßl, H., & Schulz, J. (2011). Evaluation of HOAPS-3 ocean surface freshwater flux components. *Journal of Applied Meteorology and Climatology, 50*, 379–398. https://doi.org/10.1175/2010JAMC2341.1.

Béranger, K., Barnier, B., Gulev, S., & Crepon, M. (2006). Comparing 20 years of precipitation estimates from different sources over the world ocean. *Ocean Dynamics, 56*, 104–138. https://doi.org/10.1007/s10236-006-0065-2.

Bojinski, S., Verstraete, M., Peterson, T. C., Richter, C., Simmons, A., & Zemp, M. (2014). The concept of essential climate variables in support of climate research, applications, and policy. *Bulletin of the American Meteorological Society, 95*, 1431–1443. https://doi.org/10.1175/BAMS-D-13-00047.1.

Bringi, V. N., Huang, G.-J., Chandrasekar, V., & Gorgucci, E. (2002). A methodology for estimating the parameters of a gamma raindrop size distribution model from polarimetric radar data: Application to a squall-line event from the TRMM/Brazil campaign. *Journal of Atmospheric and Oceanic Technology, 19*, 633–645. https://doi.org/10.1175/1520-0426(2002)019<0633:AMFETP>2.0.CO;2.

Burdanowitz, J., Klepp, C., & Bakan, S. (2016). An automatic precipitation-phase distinction algorithm for optical disdrometer data over the global ocean. *Atmospheric Measurement Techniques, 9*, 1637–1652. https://doi.org/10.5194/amt-9-1637-2016.

Burdanowitz, J., Klepp, C., Bakan, S., & Buehler, S. A. (2017). Simulation of ship-track versus satellite-sensor differences in oceanic precipitation using an island-based radar. *Remote Sensing, 9*, 593. https://doi.org/10.3390/rs9060593.

Burdanowitz, J., Klepp, C., Bakan, S., & Buehler, S. A. (2018). Towards an along-track validation of HOAPS precipitation using OceanRAIN optical disdrometer data over the Atlantic Ocean. *Quarterly Journal of the Royal Meteorological Society, 144*(S1), 235–254. https://doi.org/10.1002/qj.3248.

Ellis, T. D., L'Ecuyer, T. S., Haynes, J. M., & Stephens, G. L. (2009). How often does it rain over the global oceans? The perspective from CloudSat. *Geophysical Research Letters, 36*, L03815. https://doi.org/10.1029/2008GL036728.

Fairall, C. W., Bradley, E. F., Edson, J., Rogers, D., & Young, G. (1996). Bulk parameterization of air-sea fluxes for tropical ocean-global atmosphere coupled-ocean atmosphere response experiment. *Journal of Geophysical Research, 101*(C2), 3747–3764. https://doi.org/10.1029/95JC0320.

Fairall, C. W., Bradley, E. F., Hare, J. E., Grachev, A., & Edson, J. (2003). Bulk parameterization of air-sea fluxes: Updates and verification for the COARE algorithm. *Journal of Climate, 16*, 571–591. https://doi.org/10.1175/1520-0442(2003)016<0571:BPOASF>2.0.CO;2.

Hou, A. Y., Kakar, R. K., Neeck, S., Azarbarzin, A. A., Kummerow, C. D., Kojima, M., Oki, R., Nakamura, K., & Iguchi, T. (2014). The global precipitation measurement mission. *Bulletin of the American Meteorological Society, 95*, 701–722. https://doi.org/10.1175/BAMS-D-13-00164.1.

Huffman, G. J., Bolvin, D. T., Braithwaite, D., Hsu, K.-L., Joyce, R., Kidd, C., Nelkin, E. J., Sorooshian, S., Tan, J., & Xie, P. (2018). *Integrated multi-satellitE retrievals for GPM (IMERG), algorithm theoretical basis document*, 31 pp. [Available at https://pmm.nasa.gov/sites/default/files/document_files/IMERG_ATBD_V5.2_0.pdf, last accessed 22 Oct 2018].

Kidd, C., & Huffman, G. J. (2011). Global precipitation measurement. *Meteorological Applications, 18*, 334–353. https://doi.org/10.1002/met.284.

Klepp, C. (2015). The oceanic shipboard precipitation measurement network for surface validation – OceanRAIN. *Atmospheric Research, 163*, 74–90. https://doi.org/10.1016/j.atmosres.2014.12.014.

Klepp, C., Bumke, K., Bakan, S., & Bauer, P. (2010). Ground validation of oceanic snowfall detection in satellite climatologies during LOFZY. *Tellus A, 62*, 469–480. https://doi.org/10.1111/j.1600-0870.2010.00459.x.

Klepp, C., Michel, S., Protat, A., Burdanowitz, J., Albern, N., Louf, V., Bakan, S., Dahl, A., & Thiele, T. (2017a). *Ocean rainfall and ice-phase precipitation measurement network – OceanRAIN-W*. World Data Center for Climate (WDCC) at DKRZ. https://doi.org/10.1594/WDCC/OceanRAIN-W

Klepp, C., Michel, S., Protat, A., Burdanowitz, J., Albern, N., Louf, V., Bakan, S., Dahl, A., & Thiele, T. (2017b). *Ocean rainfall and ice-phase precipitation measurement network – OceanRAIN-M*. World Data Center for Climate (WDCC) at DKRZ. https://doi.org/10.1594/WDCC/OceanRAIN-M

Klepp, C., Michel, S., Protat, A., Burdanowitz, J., Albern, N., Louf, V., Bakan, S., Dahl, A., & Thiele, T. (2017c). *Ocean rainfall and ice-phase precipitation measurement network – OceanRAIN-R*. World Data Center for Climate (WDCC) at DKRZ. https://doi.org/10.1594/WDCC/OceanRAIN-R

Klepp, C., Michel, S., Protat, A., Burdanowitz, J., Albern, N., Dahl, A., Kähnert, M., Louf, V., Bakan, S., & Buehler, S. A. (2018). OceanRAIN, a new in-situ shipboard global ocean surface-reference dataset of all water cycle components. *Scientific Data, 5*, 180122. https://doi.org/10.1038/sdata.2018.122.

Kummerow, C., Barnes, W., Kozu, T., Shiue, J., & Simpson, J. (1998). The tropical rainfall measuring mission (TRMM) sensor package. *Journal of Atmospheric and Oceanic Technology, 15*, 809–817. https://doi.org/10.1175/1520-0426(1998)015<0809:TTRMMT>2.0.CO;2.

Leinonen, J. (2014). High-level interface to T-matrix scattering calculations: Architecture, capabilities and limitations. *Optics Express, 22*, 1655. https://doi.org/10.1364/OE.22.001655.

Loew, A., Bell, W., Brocca, L., Bulgin, C. E., Burdanowitz, J., Calbet, X., Donner, R. V., Ghent, D., Gruber, A., Kaminski, T., Kinzel, J., Klepp, C., Lambert, J.-C., Schaepman-Strub, G., Schröder, M., & Verhoelst, T. (2017). Validation practices for satellite-based earth observation data across communities. *Reviews of Geophysics, 55*, 779–817. https://doi.org/10.1002/2017RG000562.

Petty, G. W. (1997). An inter-comparison of oceanic precipitation frequencies from 10 special sensor microwave/imager rain rate algorithms and shipboard present weather reports. *Journal of Geophysical Research, 102*, 1757–1777. https://doi.org/10.1029/96JD03000.

Stephens, G. L., L'Ecuyer, T., Forbes, R., Gettelmen, A., Golaz, J.-C., Bodas-Salcedo, A., Suzuki, K., Gabriel, P., & Haynes, J. (2010). Dreary state of precipitation in global models. *Journal of Geophysical Research, 115*, D24211. https://doi.org/10.1029/2010JD014532.

Stephens, G. L., Li, J., Wild, M., Clayson, C. A., Loeb, N., Kato, S., L'Ecuyer, T., Stackhouse, P. W., Jr., Lebsock, M., & Andrews, T. (2012). An update on earth's energy balance in light of the latest global observations. *Nature Geoscience, 5*, 691–696. https://doi.org/10.1038/ngeo1580.

Testud, J., Oury, S., Black, R. A., Amayenc, P., & Dou, X. K. (2001). The concept of "normalized" distribution to describe raindrop spectra: A tool for cloud physics and cloud remote sensing. *Journal of Applied Meteorology, 40*, 1118–1140. https://doi.org/10.1175/1520-0450(2001)040,1118:TCONDT.2.0.CO;2.

Thurai, M., Bringi, V. N., & May, P. T. (2010). CPOL radar-derived drop size distribution statistics of stratiform and convective rain for two regimes in Darwin, Australia. *Journal of Atmospheric and Oceanic Technology, 27*, 932–942. https://doi.org/10.1175/2010JTECHA1349.1.

Trenberth, K. E., Dai, A., Rasmussen, R. M., & Parsons, D. B. (2003). The changing character of precipitation. *Bulletin of the American Meteorological Society, 84*, 1205–1217. https://doi.org/10.1175/BAMS-84-9-1205.

Weller, R. A., Bradley, E. F., Edson, J. B., Fairall, C. W., Brooks, I., Yelland, M. J., & Pascal, R. W. (2008). Sensors for physical fluxes at the sea surface: Energy, heat, water, salt. *Ocean Science, 4*, 247–263. https://doi.org/10.5194/os-4-247-2008.

Part V
Observed Characteristics of Precipitation

Chapter 35
GPCP and the Global Characteristics of Precipitation

Robert F. Adler, Guojun Gu, George J. Huffman, Mathew R. P. Sapiano, and Jian-Jian Wang

Abstract Global precipitation means and variations over the last 40 years are described using information from the Global Precipitation Climatology Project (GPCP). The GPCP international program is outlined along with the satellite and conventional data and techniques used. Global climatological values and patterns are examined and are shown to compare generally well to recent measurements from new satellite missions and with results from water balance studies. Global precipitation totals are tied to ENSO events, with small global increases during El Niños, and noticeable global decreases associated with major volcanic eruptions. A very slight significant trend is noted in the global precipitation mean value during this period of planetary warming, with a rate of ~1%/K. However, there is a pattern of larger positive and negative trends across the planet with increases over tropical oceans and decreases over some middle latitude regions. Significant trends in intensity at the monthly time scale are noted with the GPCP analyses, with larger rainfall magnitudes increasing, moderate rainfall values decreasing and dry areas expanding. These results indicate the value of the GPCP data set, but also call for continuing revision and improvement as the record lengthens, and as new, more sophisticated observations are integrated into this type of analysis system.

Keywords Precipitation · Global · Microwave · Infrared · GPCP · WCRP · GPCC · ENSO · Climate data record · ITCZ · SPCZ · Seasonal scale · Monthly scale · Trends · Ocean · Land · GPM · TRMM · PR · TMI · CloudSat · AMSR · TRMM composite climatology · Volcanic effects · SST · AMIP · CMIP · PDO · AMO · Water cycle

R. F. Adler (✉) · G. Gu · J.-J. Wang
Earth System Science Interdisciplinary Center (ESSIC), University of Maryland,
College Park, MD, USA
e-mail: radler@umd.edu

G. J. Huffman
NASA, Goddard Space Flight Center, Greenbelt, MD, USA

M. R. P. Sapiano
Sapiano Statistical Services, Atlanta, GA, USA

© Springer Nature Switzerland AG 2020   677
V. Levizzani et al. (eds.), *Satellite Precipitation Measurement*, Advances in Global
Change Research 69, https://doi.org/10.1007/978-3-030-35798-6_11

35.1 Introduction

Over many land areas, especially populated areas, early raingauge observations were used to construct climatologies (see Park et al. 2017 for an historical review). Over oceans precipitation information was originally limited to islands and ships, and before the advent of satellites, quasi-global climatologies were constructed from gauges over land and estimates of precipitation estimated from shipboard weather observations (e.g., Jaeger 1983). With satellite-based precipitation estimates becoming available in the latter part of the twentieth century, the Global Precipitation Climatology Project (GPCP) was formed in 1986 as an international effort under the World Climate Research Program (WCRP) to attempt to provide global (land plus ocean) and regional climatologies of precipitation and information about large-scale inter-annual variations (WCRP 1986; Arkin and Xie 1994). An early focus of the project was to work with various satellite agencies to archive and distribute relevant polar and geosynchronous satellite data (from the US, Europe and Japan) for use in applying precipitation algorithms at production centers at NASA and NOAA to calculate regional or quasi-global estimates of satellite-based mean monthly precipitation. Collection and analysis of global raingauge data were also a focus and led to the initiation of the Global Precipitation Climatology Center (GPCC) of the Deutscher Wetterdienst (DWD), whose successful activities in this area continue today. The combination of the satellite-based estimates over land and ocean (primarily from polar-orbiting microwave and geosynchronous infrared imagers) and from the land-based gauge information was accomplished by a hierarchical merging technique to produce the initial GPCP Monthly analysis (Huffman et al. 1997). Improvements to that original version have been made at intervals over the past years (Adler et al. 2003, 2018; Huffman et al. 2009), with the current Version 2.3 being available since 2017. The GPCP Monthly product provides a consistent, globally complete, analysis of precipitation from an integration of these various data sets. It also is used as a constraint on GPCP analyses at finer time resolutions (Xie et al. 2003; Huffman et al. 2001). The GPCP V2.3 is a NOAA Climate Data Record (CDR) and a preliminary monthly estimate is available within 10 days of the end of a month as an Interim CDR. The data is available at http://gpcp.umd.edu or https://www.ncdc.noaa.gov/cdr/atmospheric/precipitation-gpcp-monthly (last accessed 24 Oct. 2018). The GPCP Monthly product is widely used in the scientific community and is the basis for this discussion of the climatology and large-scale variations of global precipitation during the satellite era (1979–2017).

35.2 GPCP Monthly Analysis Technique

Archived on a global 2.5° × 2.5° grid, the current V2.3 Monthly analysis covers the period beginning in 1979 and is produced by merging a variety of data sources, including passive microwave-based rainfall retrievals from the Special Sensor

Microwave/Imager (SSM/I) and the Special Sensor Microwave Imager Sounder (SSMIS), infrared (IR) rainfall estimates from geostationary and polar-orbiting satellites, and surface rain gauges. The passive microwave data dominates over tropical and mid-latitude oceans and is also used over land. For example, over most of the period the ocean estimates (up to 40–50° latitude) are driven by passive microwave estimates using SSM/I and more recently SSMIS data, with IR data being used to increase sampling and therefore smoothness and to help incorporate the diurnal cycle. At higher latitudes, over both ocean and land, empirical precipitation estimates from temperature-moisture sounders are utilized. Over land, at all latitudes where available, the gauge information (adjusted for undercatch) is utilized, and dominates where gauge spacing is dense. Although many of the input data sets are now available on finer spatial scales, the 2.5° × 2.5° analysis resolution allows for easier incorporation of historical data sets and is still commensurate with a monthly time-scale. However, development of a next-generation of the GPCP analysis is underway and will be at a finer spatial scale (~1°). The combination procedures are designed to harness the strengths of individual inputs, specifically in terms of bias reduction (Adler et al. 2003; Huffman et al. 2009). The GPCP Version 2.3 has recently been released and corrects some small inter-calibration and analysis errors that increase ocean precipitation somewhat compared to the previous Version 2.2, especially in middle latitudes during the last decade of analysis. The new version also includes the Version 7 Full Data monthly gauge analysis from GPCC (Schneider et al. 2014). More details on Version 2.3 can be found in Adler et al. (2017, 2018).

35.3 Global Precipitation Climatology

The long-term mean map of annual mean precipitation (Fig. 35.1) for the satellite era (1979–2017) displays the usual dramatic features, including the Inter-Tropical Convergence Zone (ITCZ) across the Pacific Ocean and over the Atlantic and Indian Oceans and the land areas of Africa, the Maritime continent and South America. The South Pacific Convergence Zone (SPCZ) extending from the Equatorial regions southeastward across the South Pacific Ocean is also very evident, along with a similar feature in the South Atlantic, extending from South America. Figure 35.2 shows the latitudinal profiles of the zonal mean precipitation climatology for land, ocean and total areas. The sub-tropic minima along the latitudes of 20° are located with prominent deserts of those regions and dry areas of the eastern regions of the oceans at that latitude. In mid-latitudes of the northern hemisphere rainfall maxima along and off the east coasts of the continents are noted extending northeastward, although in the mid-latitudes of the southern hemisphere, a weaker, but continuous precipitation maximum is observed circling the hemisphere (Fig. 35.1). These mid-latitude regional features aggregate to mid-latitude secondary peaks (Fig. 35.2) in the zonal mean plots. The peak over oceans at 60°S is not

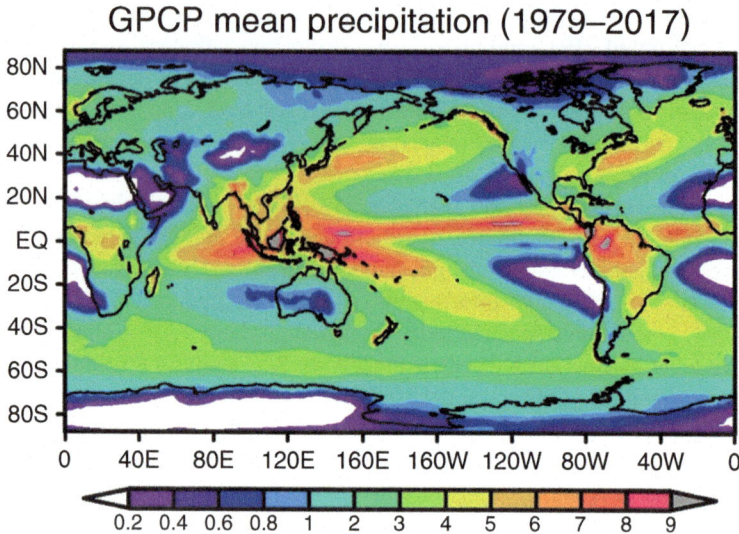

Fig. 35.1 GPCP climatological mean precipitation (mm day^{-1}) during 1979–2017

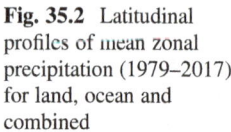

Fig. 35.2 Latitudinal profiles of mean zonal precipitation (1979–2017) for land, ocean and combined

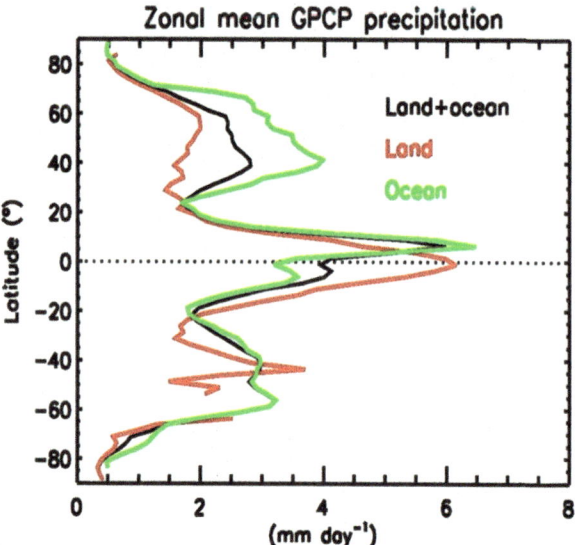

confirmed by CloudSat information (Behrangi et al. 2014) and is likely an artifact in the GPCP analysis related to the transition going poleward from passive microwave to IR sounder input information.

There are significant variations on the seasonal scale (Fig. 35.3). Over both the ocean and land, there is the seasonal thermal inertia process operating, with a larger amplitude of latitudinal variation over land. The areas of sub-tropical minima and

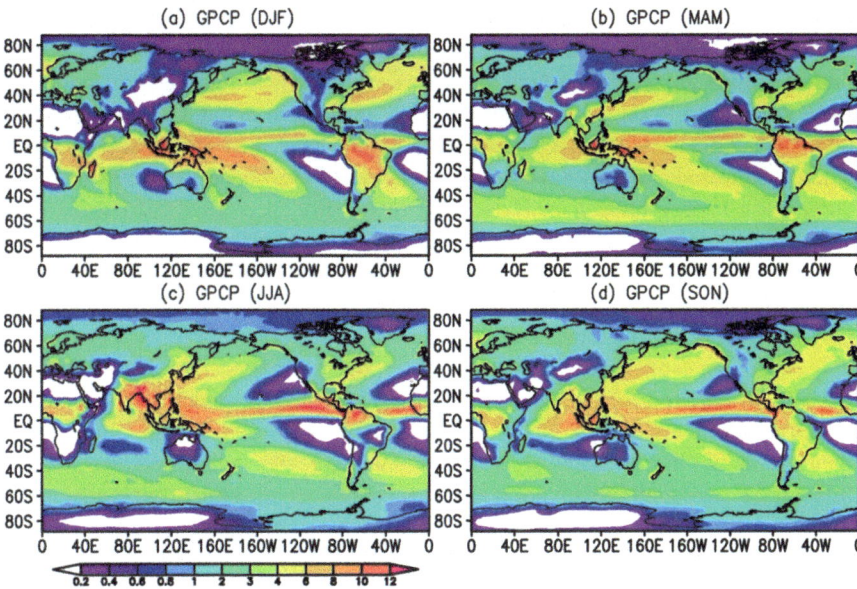

Fig. 35.3 Seasonal mean GPCP precipitation (mm day^{-1}). (**a**) December–January-February (DJF), (**b**) March–April-May (MAM), (**c**) June–July-August (JJA), and (**d**) September–October-November (SON)

mid-latitude maxima also evidence seasonal variation in their magnitudes and positions. This is especially evident over the monsoon areas from Australia to southeast Asia, but also over Africa and South America and, with smaller latitudinal shifts, over the tropical oceans.

Integrating over the entire period of the GPCP record and pole-to-pole the grand total estimate for global precipitation is 2.69 mm day^{-1} with an estimated error of approximately $\pm 7\%$ based on the variation among estimates from a number of other products (see Adler et al. 2012). The ocean and land totals are 2.89 and 2.24 mm day^{-1}, respectively. Factoring in the difference in area between ocean and land over the globe, it can easily be seen that more than 2.5 times as much precipitation by volume falls over ocean as does over land.

Because determination of ocean precipitation is very largely dependent on satellite-based estimates, it is critical to determine our confidence in those estimates to understand the magnitude of our planet's water cycle. Researchers have used a variety of data sets, including satellite data sets (and models) to estimate the components of the water cycle (precipitation, evaporation, transport, etc.) to try and balance, or "close" the water cycle on a global or large regional (e.g., continental or ocean basin) scale. For example, Rodell et al. (2015) examined the mean global water cycle for the first decade of the twenty-first century by using various satellite-based and conventional data sets, with reanalysis used to fill certain gaps. Initially they used the observed magnitudes of the components, but the study then used an objective method to adjust the various component magnitudes to achieve closure, based on the magnitude of the estimated errors for each of the different data sets. The earlier V2.2 GPCP regional and global means were used in this exercise and the budget closure procedure required that the mean global precipitation over that period from GPCP had to be increased by 5% over ocean and about 1% over land. The GPCP V2.3 mean value over ocean is slightly larger than the V2.2 value, so a smaller adjustment would have been necessary if it had been used instead. In the companion energy budget analysis paper to the Rodell et al. (2015) water cycle study, L'Ecuyer et al. (2015), find an energy balance using the same GPCP values as Rodell et al. (2015). These GPCP adjustments (up to 5% over ocean) fall within the estimated bias errors for GPCP (Adler et al. 2012) and give some confidence that the GPCP global-scale magnitudes are close to the actual values, or at least fit comfortably with the estimates of the other water cycle components. The very small adjustment required for balance over land in these studies reflects the value of the gauge data (Schneider et al. 2014) and the careful analyses thereof in GPCP, before their blending with the satellite data. It should be remembered, however, that the final GPCP over-land precipitation analyses contain an adjustment for gauge undercatch and are therefore slightly larger than straight gauge-based estimates.

Recently there have been new mission entrants into precipitation estimation from space. These include the Tropical Rainfall Measuring Mission (TRMM) (e.g., Kummerow et al. 2000) and CloudSat (Stephens et al. 2002) mission. TRMM carried both a Precipitation Radar (PR) at 14 GHz and the TRMM Microwave Imager (TMI). CloudSat has a higher (94 GHz) frequency radar that is valuable for detecting light precipitation, more dominant at higher latitudes. Results from the Global Precipitation Measurement (GPM) mission launched in 2014 (Hou et al. 2014; Skofronick-Jackson et al. 2017) are not included here as retrieval algorithms are still maturing and mean estimates are still being evaluated, but will shortly become very important in terms of climatological values.

So, how do the estimates from GPCP compare in terms of very large-scale means to these new estimates involving TRMM and CloudSat? First, we will focus on the tropics, and on the oceans. The tropics are important because a large fraction of global precipitation falls there, and oceans are critical, because of the lack of simple, but relatively accurate raingauge observations. For the ocean area between 35°N and 35°S the GPCP numbers compare quite closely with those of TRMM (Version 7; see Table 35.1). The TRMM program has three different algorithm products, one based on the passive microwave instrument, the TMI; one based on the radar, the PR, and one based on a combination of the two instruments. The three estimates have been combined into a TRMM Composite Climatology (TCC) (Adler et al. 2009; Wang

Table 35.1 Mean oceanic precipitation estimates (mm day^{-1}) between 35°N and 35°S from various products

PR TRMM precipitation radar	TMI TRMM microwave imager	TCC TRMM composite climatology	GPCP (1998–2014)	PR + CloudSat
2.9	2.8	2.9	2.99	3.0

TRMM (Version 7) and GPCP values are for 1998–2014; The CloudSat value is for 2007–2009 and is from Behrangi et al. (2014)

et al. 2014), where the local spread among the estimates provides a measure of confidence in the result. As indicated by the spread of estimates in some locations (and the maps in the referenced articles), there can be significant differences between these three estimates in certain regions, but when averaged over the large area of the tropical oceans, they are within a few percent of each other. A separate study (Behrangi et al. 2014) used the TRMM PR data and the CloudSat radar and obtained an independent estimate from that of the TCC and the separate TRMM instrument estimates (also in Table 35.1). Again, that TRMM/CloudSat radar estimate for the tropics is very close to the GPCP number.

Outside the tropics the GPCP values are dependent on a mixture of passive microwave-based estimates and empirical relations between cloud and moisture information from satellite sounders and from gauge information, with the influence of the microwave–based estimates decreasing poleward from $40°$. Behrangi et al. (2014) also made an estimate from $60°N–60°S$ over ocean using the TRMM and CloudSat radars in the tropics and a combination of passive microwave data from the Advanced Microwave Scanning Radiometer (AMSR) on the AQUA satellite and the CloudSat radar in middle and high latitudes. Table 35.2 indicates that this new estimate of total ocean precipitation is about the same as GPCP (V2.3) for the same years, but slightly higher by about 3% compared to the earlier V2.2 number. This 3% difference is within the error bars (Adler et al. 2012) of GPCP, but is also of the same sign (and rough magnitude) of the adjustment needed for water cycle closure (Rodell et al. 2015). Thus, for the same time period the new V2.3 numbers seem to compare very well with the Behrangi et al. (2014) satellite estimates and the Rodell et al. (2015) water balance values.

Both the tropical and total ocean precipitation estimates from GPCP are therefore confirmed (within about $\pm5\%$) by the newer, more sophisticated estimates using TRMM and CloudSat. It should be pointed out that these very large area totals mask some significant regional differences that need closer attention (Behrangi et al. 2014, 2015).

However, these results do not indicate an end to the discussion of mean ocean rainfall; there is still more to be done. The GPM mission has been launched in 2014 and will provide new, more sophisticated data to continue to refine our knowledge of the absolute magnitude of global precipitation once the GPM precipitation products mature and are connected (in the tropics) to TRMM-based estimates to provide a 20+ year radar/passive microwave data set for careful analysis. Work is underway to incorporate these new data sets into analyses such as GPCP and these newer radar and passive microwave observations are likely to become a foundation of the long-term analyses.

Table 35.2 Global oceanic precipitation estimates (mm day^{-1})

	GPCP (1998–2014)	PR/AMSR/CloudSat
$60°N–60°S$	3.04	3.13

The PR/AMSR/CloudSat value is from Behrangi et al. (2014)

35.4   Variations in Global Mean Precipitation (1979–2017)

Precipitation is highly variable, especially when estimated over small areas and/or short time periods. But, how does it vary on a very large scale, for example, for the whole planet? And since we know the planet's surface temperature has been increasing during the satellite era, how is that affecting global precipitation. Figure 35.4 shows anomalies of global total GPCP precipitation (ocean plus land) in the black curves with a 3-month running mean. The global precipitation shows a small trend (about 0.01 mm day^{-1}/decade). With a significant surface temperature trend (0.16 K/decade) during the period, results show an increase of global precipitation of 1.3%/K. Although this precipitation trend number is close to some estimates based on climate models (e.g., Allen and Ingram 2002; Held and Soden 2006; Sun et al. 2007), it is highly sensitive to the length of record used and other factors (see Gu and Adler 2013). The global precipitation curve in Fig. 35.4 shows significant inter-annual variations. The variation, or σ, is about 0.03 mm day^{-1}, roughly 1% of the mean precipitation. Some of the variance probably comes from noise in the original data and the analysis techniques; however, because we are averaging over the entire planet using strong efforts at maintaining homogeneity of the records, most of this variation is likely real and associated with variations in storage in the ocean and land.

A portion of these global-scale, inter-annual variations are related to El Niño-Southern Oscillation (ENSO) variations and the effect of volcanoes. The planetary-scale precipitation signals related to these effects (the red and blue lines in Fig. 35.4) have been estimated using the Niño 3.4 Index as an indicator of ENSO and a stratospheric aerosol index as an indicator of volcano impacts along with a linear regression approach (with relevant time-lags) (Gu et al. 2007; Gu and Adler 2011). El Niños and La Niñas are accompanied by global increases and decreases, respectively, in surface temperature. Higher surface temperature leads to greater evaporation (especially over ocean) and greater instability; therefore, the inter-annual

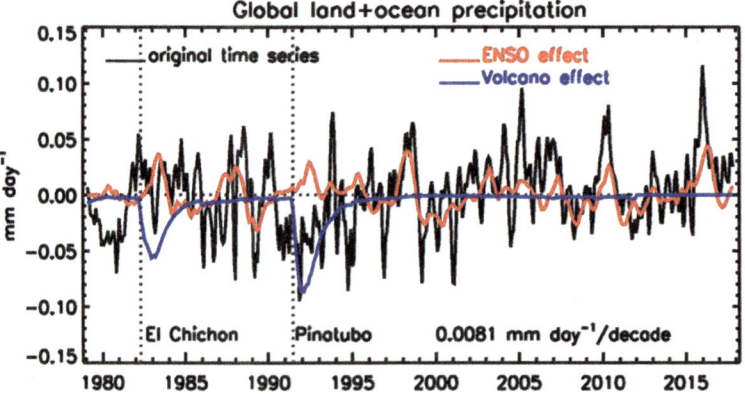

Fig. 35.4 Time series (January 1979–December 2017) of global mean (land+ocean) precipitation (black line), and corresponding ENSO (red) and volcanic effects (blue)

variations in surface temperature, even if averaged across the globe, tend to be related to variations of the same sign in global precipitation. For ENSO one can see a global increase in precipitation (Fig. 35.4) for El Niños (1983, 1988, 1992, 1998, 2010, 2016) and negative global anomalies during La Niñas (e.g., 1999–2001). The maximum amplitude of the ENSO-related precipitation signal is +0.04 mm day^{-1} (1.5%) during the 1983 El Niño, and is made up of a larger volume increase over oceans and a smaller decrease over land areas, resulting in the net increase of precipitation. For the two major volcanic eruptions during the period, in 1982 (El Chichón) and in 1991 (Pinatubo), there is a sharp drop in surface temperature due to decreased solar radiation at the surface at the time of, and immediately after, the eruptions, and then a gradual return to normal mean surface temperature over a two to three-year period as the volcanic aerosols filtered back to Earth. The temperature anomaly magnitude for Pinatubo (the larger event) is −0.35 °C, a larger magnitude than the +0.15 °C estimated global impact of the 1998 El Niño. For the two volcano events, amplitudes of precipitation change of −0.06 and − 0.09 mm day^{-1} (3%) are estimated, again larger than that for ENSO-related changes. For volcanoes the decrease in precipitation occurs over both ocean and land.

The signs and magnitude of the precipitation variations are directly correlated with the amplitude of the surface temperature anomalies (Gu and Adler 2013). There is also a lag (precipitation lagging temperature) of ~6 months for these global signals (Gu and Adler 2011). For ENSO, a signal of ~9%/°C is calculated for the global precipitation/surface temperature variation, a little higher, but close to, the Clausius-Clapeyron relation of 7%/C. For the volcano impact the combined value is ~8%/°C. Thus, although the 35-year trend signal for mean precipitation and surface temperature indicates ~+ 1.5%/°C, these global variations at the shorter, inter-annual time scale indicate processes closer to having the Clausius-Clapeyron relation.

35.5 Patterns of Precipitation Variation and Trends

In the last section one could see that even when precipitation was averaged across the entire planet, we could pick out variations on inter-annual and inter-decadal time-scales and that phenomena such as ENSO and volcanoes could be detected on a planetary scale. We also noted that while surface temperature has increased significantly during the satellite era, the calculated trend in mean precipitation from the GPCP data set is only slightly positive. In this section we look at regional changes, including inter-annual and inter-decadal changes, and patterns of trends during this period.

Patterns of precipitation anomalies, focusing on ENSO, have been the subject of numerous studies (e.g., Dai and Wigley 2000; Held and Soden 2006; Xie et al. 2010; Curtis and Adler 2003; Gu and Adler 2013). Figure 35.5 shows a summary map of ENSO anomalies using GPCP. It is constructed by splitting the Niño 3.4 index based on SST anomalies in the Central Pacific into three categories: an El Niño is defined

GPCP (Nino 3.4)

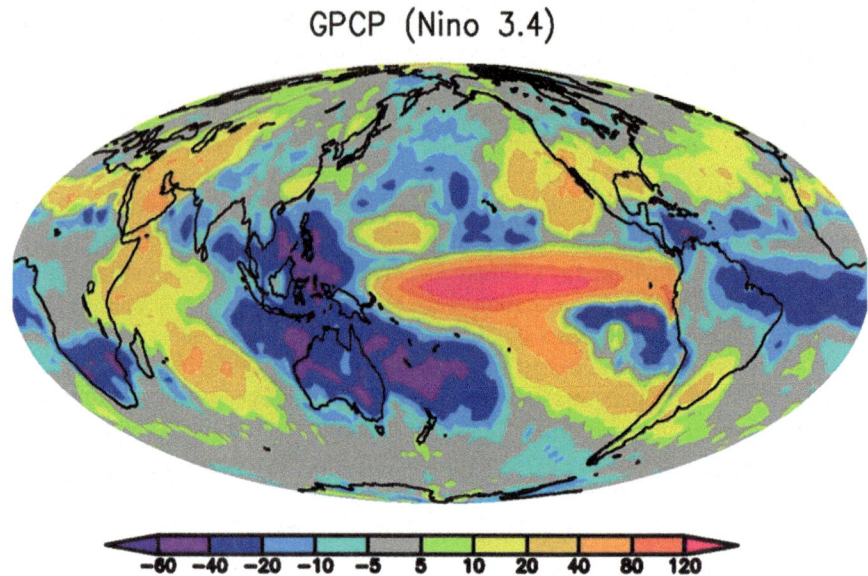

Fig. 35.5 Composite precipitation anomaly changes (%) between ENSO warm (defined as Niño 3.4 ≥ 0.34) and cold phases (Niño 3.4 ≤ −0.45). The number of months for either phase is 155 (one third of months) during 1979–2017

when the index is in the highest one-third months of the index and La Niña is defined when the index is in the lowest one-third of the distribution. The middle third is defined as neutral. Figure 35.5 is the result of taking the mean anomalies of the El Niño months and subtracting the mean anomalies of the La Niña months. Since the La Niña patterns tend to be a mirror image of El Niño pattern, they tend to reinforce each other. The magnitudes are presented in terms of percentages of the climatological means at the location.

The result in Fig. 35.5 shows a dramatic, alternating pattern of large positive and negative anomalies along the Equator, with the largest (over 5 mm day^{-1}, 120%) positive peak in the central Pacific, associated with the typical maximum (with El Niño) surface temperature anomaly. The largest negative, but weaker, anomaly is over the Maritime Continent to the west. The positive anomaly stretches eastward from its core right up to the coast of South America, where there is a quick shift to a deficit of rainfall over Amazonia. There is also a weaker positive anomaly over the western Indian Ocean, but these Equatorial features decrease in strength as they approach the opposite longitude of the primary maximum at about 180° longitude. Moving off the Equator into the Southern Hemisphere there are strong extensions of the Equatorial feature-oriented northwest to southeast. The negative anomaly over the Maritime Continent can be seen extending across a long swath of the South Pacific Ocean to the area between Tierra del Fuego and Antarctica at 60°S. From the central Pacific positive anomaly one can faintly trace a feature across southern South America, into the South Atlantic and below the tip of Africa. The positive anomaly over East Africa and the Equatorial Indian Ocean also extends southeastward. The Amazon feature extends across the Equatorial Atlantic into Africa.

In the Northern Hemisphere the features are more fragmented, probably due to the impact of the larger continents and associated mountain ranges, especially in Asia. However, the extensive positive feature across the sub-tropical eastern Pacific into the southern US and beyond, all the way to western Europe, seems to be linked to the original, largest positive anomaly on the Equator at the Dateline. This map and the features therein are composites of a number both El Niños and La Niñas. Each individual event is different, but the fact that the composites have features that cover vast areas at their core and extend across oceans and continents indicates the impact of ENSO on the geographic distribution of precipitation.

Variations of regional precipitation due to ENSO are significant, but since they are associated with an inter-annual phenomenon, these variations do not significantly affect long-term regional trends. As well, although the global mean precipitation (and even that averaged over land and ocean separately) has a near zero trend, that does not mean that regional trends are zero, and they are not. Figure 35.6 shows the regional trends of surface temperature (Fig. 35.6a), total column water vapor (Fig. 35.6b, over the ocean) and precipitation (Fig. 35.6c) during the satellite era. Immediately obvious is that the surface temperature trend pattern is not uniform (e.g., Hansen et al. 1999). While the global trend number is +0.1 °C/decade for the period in question, there is significant spatial variability across the globe, varying

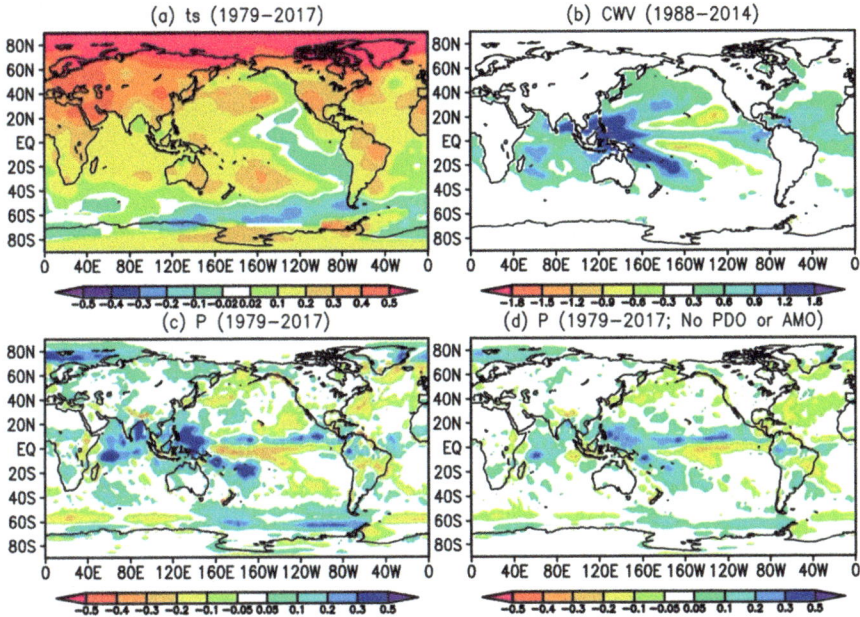

Fig. 35.6 Long-term trends in (**a**) GISS surface temperature (ts; 1979–2017), (**b**) SSMI/SSMIS columnar water vapor (CWV; 1988–2014), and (**c**) GPCP precipitation (1979–2017). (**d**) Long-term trend in precipitation without the effect of PDO and AMO

from about +0.5 °C/decade in the north polar regions to about −0.2 °C/decade in the eastern Pacific and across the Southern Ocean at high latitudes. The oceanic water vapor trend pattern shows a pattern similar to the SST change pattern, with the largest water vapor increases linked to the maxima in increases in surface temperature in the tropical western Pacific, Indian and Atlantic Oceans. A significant difference is the narrow water vapor increase along the ITCZ across the Pacific, with areas of decreases north and south of the ITCZ. These water vapor trends over the ocean are related to increased evaporation associated with increased surface temperature, but also to moisture transport, including transport into the ITCZ.

The precipitation trend pattern (Fig. 35.6c) is noisier than either the surface temperature or water vapor pattern, but resembles the water vapor pattern over the tropical oceans. For example, it shows a very narrow feature of increase along the ITCZ in the Pacific, the western Pacific area of increase, along with the increase in the SPCZ, and increases in the tropical Indian and Atlantic Oceans. Decreases are noted on either side of the Pacific ITCZ, again nearly co-located with the water vapor change. Positive-change extensions toward mid-latitudes are seen emanating from the western Pacific maximum aligned approximately with the water vapor and SST positive change areas. There is also an area of positive change at 60°S in the southern Pacific. Over land there is positive/negative pattern of change over South America going from north to south, a weak pattern over Africa and a general area of increase over land along 60°N, especially evident over Asia. Across North America there is weak negative change across the southwest US and Mexico and extending to the East Coast. Weak increases are indicated from Alaska, across Canada and the Great Lakes to the east.

Figure 35.6c is arguably the best, globally complete, observation-based estimate of precipitation trend over this long a period (36 years). While short-term processes up through inter-annual time scale presumably average out, the relatively short period of the satellite era could allow effects of inter-decadal processes, such as the Pacific Decadal Oscillation (PDO) and the Atlantic Meridional Oscillation (AMO), to be evident. Recent work (Gu and Adler 2013; Gu et al. 2016) has sought to separate the signal associated with these inter-decadal processes and isolate the global warming (GW) signal as much as possible using a statistical approach and indices of the PDO and AMO. The updated results of this exercise (Fig. 35.6d) results in a retention of the overall pattern seen in the raw trend map (Fig. 35.6c), but expands the area of the increase in the ITCZ across the Pacific, shows weaker negative trends north and south of the ITCZ, and tends to decrease the magnitude of a number of the other trend features. Comparison of the precipitation trend map in Fig. 35.6 with the climatological map in Fig. 35.1 tends to provide support to the idea of wet areas (e.g., ITCZ) getting wetter and dry areas (e.g., subtropical zones) getting drier, at least in the tropics.

Figure 35.7 shows a summary of the zonal-mean trends in precipitation during the period in question, along with the mean precipitation. The observed precipitation trend (black curve) shows a sharp peak just north of the Equator and a negative feature just south of that point. The positive trend peak is co-located with the maximum mean precipitation (red curve) along the ITCZ. In the Northern

Fig. 35.7 Zonal mean precipitation trend during 1979–2017 from GPCP (black curve). Also shown is the zonal mean profile of climatological mean precipitation (red curve; scaled by 50 with resulting units of mm day^{-1}). Green and blue curves are the trend curves for CMIP and AMIP climate model ensembles for similar periods

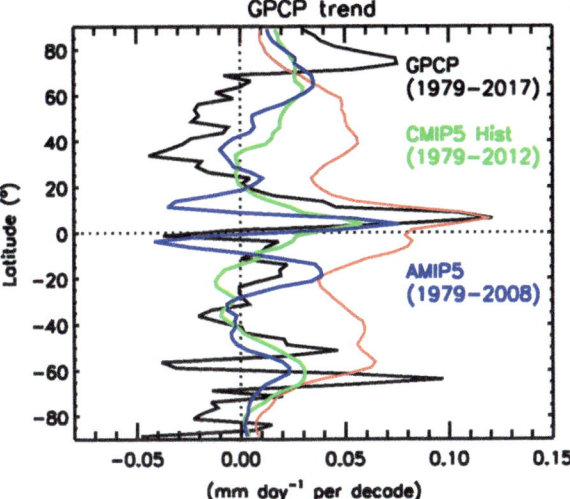

Hemisphere the trend shows a relatively broad negative region in mid-latitudes and then a secondary peak at higher latitudes. In the Southern Hemisphere the trend is very variable, but there is a sharp negative feature between 0 and 10°S, which together with the positive trend at the ITCZ peak and the zone of positive trend up to 20°N indicates a shift of the ITCZ further into the northern hemisphere, where warming is accented (see Fig. 35.6a).

Since the satellite period is concurrent with a period of significant global warming, one key question is whether the observed pattern is largely the result of that planetary warming? Having such an observed pattern in which we have confidence would be useful for understanding the recent past, for assuming a similar future pattern of change and for comparison and validation with climate models of global warming covering the past and projecting future precipitation patterns. Comparisons with climate models covering roughly the same period can be seen in Fig. 35.8, where the GPCP trend maps (repeated from Fig. 35.6) are compared with ensemble results from the AMIP (Atmospheric Model Intercomparison Project, SST-driven) and CMIP (Coupled Model Intercomparison Project, coupled historical runs). The model trend maps are similar to the observations, but with distinct differences. The AMIP results, driven by observed SST patterns are similar to the observations (Fig. 35.8a), but with much larger negative features in the central Pacific, with a smaller area of positive trend there. This difference may be due to the AMIP runs overemphasizing the effect of the PDO during this period. The CMIP trend results (Fig. 35.8d) also show a similar general pattern, but with much smaller magnitudes (note the scale difference for CMIP). A case can be made that the trend pattern in Fig. 35.8b (GPCP with PDO effect removed) matches better with that derived from climate models for the same period using GW forcing, i.e., the CMIP runs (Gu et al. 2016).

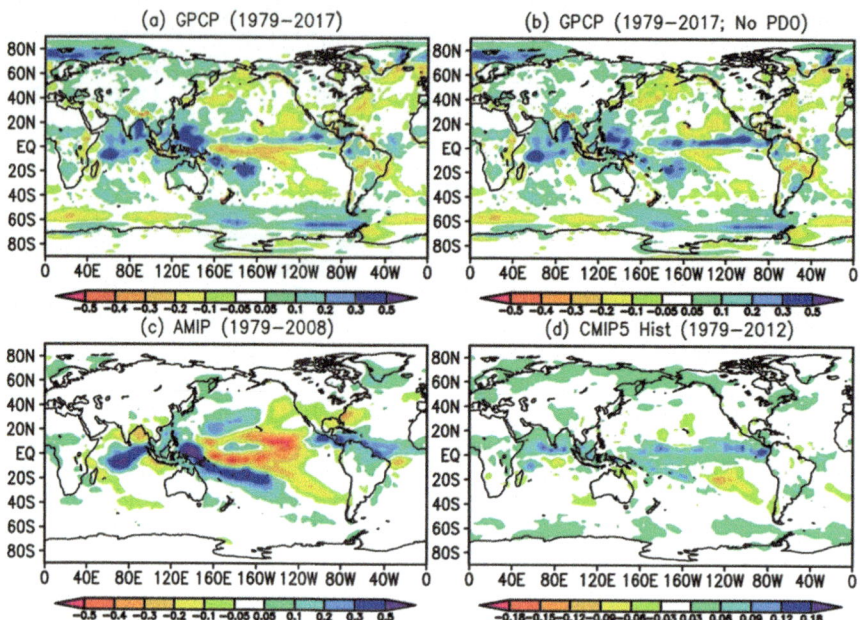

Fig. 35.8 Long-term trends (mm day^{-1} per decade) in (**a**) GPCP precipitation (1979–2017), (**b**) GPCP precipitation with no PDO effect, (**c**) AMIP precipitation (1979–2008), and (**d**) CMIP5 historical full forcing precipitation (1979–2012)

Zonal averages of the climate model trends (in Fig. 35.7) eliminate some of the differences in the longitudinal direction and provide better overall agreement between the models and observations than the trend maps. The GPCP observed trend shows the increase in the ITCZ and then a falloff toward the sub-tropics, with a continued negative trend in the NH mid-latitudes. The models agree well in terms of the location of the ITCZ trend peak, but with a significantly smaller magnitude, and generally agree up to about 30°N and into the SH. In northern mid-latitudes, however, the models and observations disagree even in sign, calling for closer examination of both the models and the observations. For the future, the improvement of climate models in handling precipitation changes and trends, along with the lengthening of the satellite record should allow for better understanding of regional trends in mean precipitation in the both the past and future through the comparison of climate models and observations such as GPCP.

35.6 Trends in Precipitation Intensity at the Monthly Scale

As seen in the last sections the global trend in mean precipitation is small and slightly positive, with a stronger overall positive trend in the tropics (see Fig. 35.8), but with strong regional variations (Fig. 35.6c). It is also interesting to examine if and how

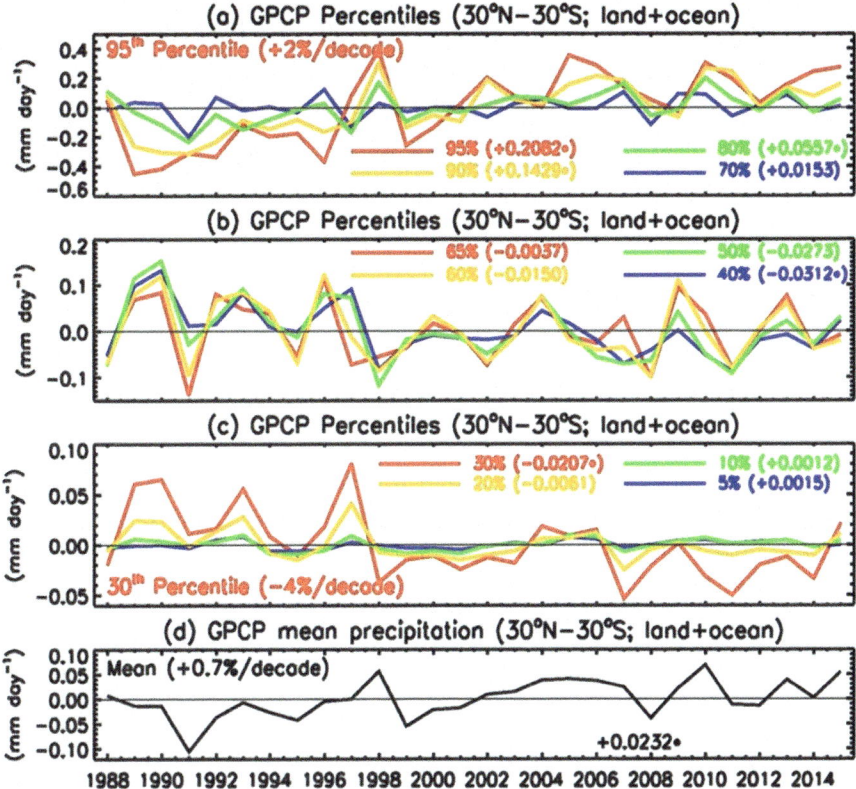

Fig. 35.9 Annual anomalies of (**a, b, c**) precipitation percentiles and (**d**) mean precipitation determined by GPCP monthly rain-rates between 30˚N and 30˚S. Also shown are their corresponding linear trends (mm day^{-1} per decade), and those followed by "∗" are statistically significant at the 5% level. For selected percentiles and the mean, the trends are also shown in %/decade

precipitation intensity is changing comparing to mean precipitation during the satellite period and how the observations (GPCP) match up with climate model estimates. Figure 35.9 (adapted from Gu and Adler 2018) shows anomalies in various percentiles over the tropical (30°N–30°S) region (land and ocean combined) during the 1988–2015 period derived from the monthly GPCP monthly analysis. Also shown is the variation of the tropical mean precipitation during the period. The post-1997 period is focused on here because this is the period when microwave data are the key satellite measurement driving the GPCP tropical estimates. The trend in the tropical mean rainfall is significantly positive (at the 5% level) and is about 0.7%/decade (Fig. 35.9d). The intense rain percentiles (at and above the 80th percentile) also have a significant positive trend, with the 95th percentile (about 10.5 mm day^{-1}) having a trend of 2%/decade, more than twice as big as the increase in the mean rainfall. For the intermediate percentiles (65th to 20th) the trend is negative, with the trend for 30th and 40th percentiles being significant. The slope for the 30th percentile (about 0.5 mm day^{-1}) is −4%/decade. At the very low end (fifth–tenth percentiles) the trend is again positive, but not significant. Therefore, the GPCP observational analysis, at the monthly time scale, shows intense rainfall in the tropics increasing, intermediate rainfall amounts decreasing, and indeterminate changes at the lowest rainfall amounts. Although we are analyzing trends, probably driven by global warming, some of the plots in Fig. 35.9 indicate a shift around 1998 (e.g., 95th and 30th percentile plots), which is coincident with a PDO phase reversal mentioned earlier. Again, we may have a mix of long-term global warming effects and an inter-decadal forcing to produce the observed statistics. The intensity characteristics shown are for ocean and land combined, but are more pronounced over ocean. Over land alone similar trends are noted, but with weaker results in terms of significance tests, possibly due to the relatively small land area in the tropics and the short time period.

Comparing variations of observations such as GPCP to climate models certainly contributes to the validation of the models as in the previous section, but can also be useful to help understand the observational results. Figure 35.10 compares the tropical mean and selected intensity percentiles with two types of climate models (see Gu and Adler 2018 for details). From the CMIP5 model inter-comparison effort the curves in Fig. 35.10 labeled as such are from an ensemble of "free-running" ocean-atmosphere coupled models driven by time histories of CO_2, aerosols and

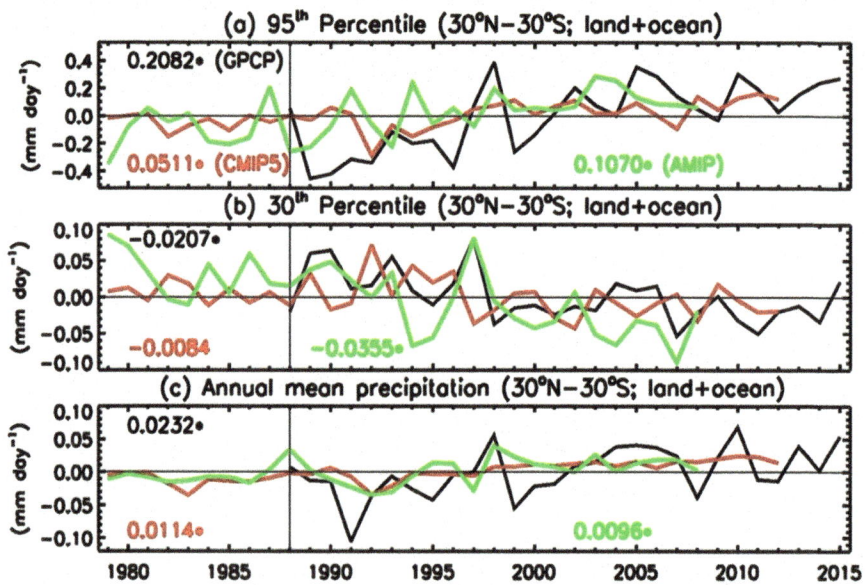

Fig. 35.10 Annual anomalies of the 95th, 30th precipitation percentiles and mean values between 30°N and 30°S for GPCP (black), AMIP (green), and Hist-Full (red) monthly rain rates. Also shown are their corresponding linear trends (mm day^{-1} per decade), and those followed by "*" are statistically significant

other variables. The AMIP-labeled curves are from an ensemble of models driven by observed SSTs and other variables. The AMIP results should reflect the effect on model precipitation caused by ENSO and PDO through the forcing related to the observed SSTs. The CMIP results will not reproduce particular ENSO or PDO features during the period, and since the results are from an ensemble, the plot essentially represents long-term simulated changes, i.e., those related mainly to global warming. The difference between the model types is therefore useful to aid in understanding the effect of the inter-decadal forcing (e.g., PDO) on the observations.

As can be seen in the bottom panel of Fig. 35.10 both GPCP and the two models all show a significant, positive trend in mean rainfall, with the models having a smaller magnitude than the observations when comparing the overlapping, but not identical periods. For the 95th and 30th percentiles (top and middle panel, respectively) the models again have the same sign of the trend as GPCP, but with magnitudes less than the observations. So, the GPCP intensity results confirm the climate model results in general, with rainfall intensity increasing in rainy areas and intermediate percentile rainrates decreasing. This decrease can be interpreted as dry areas are expanding, and this was confirmed by tracking trends in dry area defined by monthly rainrate thresholds (Gu and Adler 2018). The result that the percentile trends in the AMIP (SST driven) runs are greater than the "free-running" CMIP model results, indicate that the inter-decadal (PDO) effect may be present and in the same direction as the global warming signal. Therefore, the observed GPCP intensity trends may be an overestimate of what would be occurring with just global warming. However, the GPCP intensity trends are larger than even the AMIP runs and that can be interpreted that the models may be underestimating the effect of global warming on precipitation intensity changes. To understand this better there needs to be much more analysis of the models, including individual elements of the ensembles, and comparison with observations. A lengthening record of satellite precipitation is critical for this validation of the climate models.

35.7 Summary

Satellite observations over the last 36 or so years have allowed for a much more accurate estimation of precipitation over the entire globe. This satellite era has also allowed for a close study of variations and trends during this period to better understand the regional and global changes in this particularly important variable. The current GPCP climatological map shows the well-known precipitation features of both the Tropics and the middle and high latitudes, but now gives what might be termed as mature estimates of mean values of these regional features. In terms of mean annual precipitation over the entire globe the GPCP number is 2.69 mm day^{-1} with an estimated error of approximately $\pm 7\%$. The ocean and land totals when separated are 2.89 and 2.24 mm day^{-1}, respectively. The global number (and the ocean and land components) fit reasonably into large-scale water budgets using estimates of other branches of the water cycle (e.g., evaporation, transport), with perhaps an indication of a small underestimation over ocean. A similar, small value of possible underestimation over global oceans is obtained from examination of newer satellite-based estimates over a shorter period. Although there may be some "missed precipitation" on the light end of the intensity spectrum, the likely source of most of any underestimation probably comes from the underestimation of events or periods of heavy rainfall over tropical oceans. These absolute magnitudes will be refined as further analyses are done using data from the GPM mission and other new measurements and careful comparisons with ground measurements.

During the satellite era the global surface temperature has increased, although not at a constant rate, and atmospheric water vapor has also increased. However, the global precipitation value has shown a small, but not statistically significant positive trend over the period, although the tropical increase is statistically significant. Variations of precipitation are evident on even the largest scale, for the entire globe, and the separate ocean and land areas. ENSO, even though focused in the tropical Pacific Ocean, affects the global mean surface temperature, water vapor and precipitation magnitudes. During El Niños, global values of all three variables increase, with about a 2% increase in precipitation. La Niña events produce a smaller, negative temperature and precipitation signal for global temperature and precipitation. The two volcanic events during the period also produce temperature, water vapor and precipitation signals, all with a negative sign, with about a 3% dip in global precipitation.

In addition to affecting global precipitation, ENSO events show a very strong signal in the patterns of precipitation anomalies across the tropics and into middle and high latitudes. Composite results show the alternating areas of positive and negative rainfall anomalies along the Equator stretching east and west from the largest anomaly in the central Pacific and becoming weaker toward the west in Africa and toward the east in the Atlantic. But also obvious are extensions from the tropical features into middle and even high latitudes, indicating the extensive impact of this phenomenon.

Although the global total precipitation shows only a very small trend, the pattern or global map of observed trends shows a very distinctive pattern of positive and negative precipitation changes, with some linkages to the trend patterns for surface temperature and water vapor. Increasing rainfall is dominant in the western Pacific and Indian Ocean, in a narrow belt along the ITCZ location in the central and eastern Pacific and in the SPCZ. Areas of reduction are noted on either side of the ITCZ extending into the eastern Pacific and onto land (e.g., across the southern US). In general, the tropics show a significant positive trend, with mid-latitudes having zones of decrease in both hemispheres. The pattern of change is related primarily to the effect of global warming, although inter-decadal variations related to the PDO and AMO provide a second-level change that is important to understand when dealing with the relatively short satellite era.

The trends and changes in precipitation extend beyond means to changes in intensity. At the monthly level, in the tropics, there is shown to be a positive trend in intense rainfall at a rate faster than the mean rainfall rate. At an intermediate monthly rain rate, which roughly determines the boundary of dry areas, the rain rate decreases. These and other results indicate that regions of normally more intense rainfall climatologically are seeing an increase in rainrates, while climatologically dry areas are expanding. These changes in intensity are seemingly better defined than changes in mean rainfall and should be a focus of monitoring. These observed trends in mean values and intensity percentiles of precipitation are also very valuable for validating climate models and for using both the observations and the models to better understand the changes affecting this component of the water cycle.

All these results indicate the utility of the GPCP analyses and the value of this type of globally-complete, composite analysis of observations, including satellite and ground information. There is still much to be done in terms of utilizing new data sets and integrating them into the analysis without causing significant inhomogeneities to the analysis record. Use of these new estimates, for example from TRMM, CloudSat and GPM, are underway and will help in many ways to improve the analyses and allow for some adjustment of earlier periods. Areas of needed research emphasis include intense tropical rainfall where even the TRMM and GPM radars reach saturation, and light precipitation over both land and ocean, especially in middle and high latitudes. With this satellite era record extending as time moves on, its usefulness for comparison with models and in interpreting variations in terms of longer changes at the inter-decadal and trend scales will improve. However, the long-term record necessarily uses different satellites with different instruments of varying capabilities. Piecing these records together in time and in space (e.g., tropical and high latitudes) may require an increased focus on GPCP-type analyses. Methods to extend the analysis back in time before the satellite era will be limited, of course, by the lack of sufficient direct ocean estimates, but can be used with over-land gauge estimates and both numerical and statistical models. There is no doubt that a better understanding of all these scales of variations of precipitation on our planet will require continued extension and improvement of global precipitation analyses.

References

Adler, R. F., Huffman, G. J., Chang, A., Ferraro, R., Xie, P., Janowiak, J., Rudolf, B., Schneider, U., Curtis, S., Bolvin, D., Gruber, A., Susskind, J., Arkin, P., & Nelkin, E. (2003). The version 2 global precipitation climatology project (GPCP) monthly precipitation analysis (1979–present). *Journal of Hydrometeorology, 4*, 1147–1167. https://doi.org/10.1175/1525-7541(2003) 004<1147:TVGPCP>2.0.CO;2.

Adler, R. F., Wang, J.-J., Gu, G., & Huffman, G. J. (2009). A ten-year tropical rainfall climatology based on a composite of TRMM products. *Journal of the Meteorological Society of Japan, 87A*, 281–293. https://doi.org/10.2151/jmsj.87A.281.

Adler, R. F., Gu, G., & Huffman, G. J. (2012). Estimating climatological bias errors for the global precipitation climatology project (GPCP). *Journal of Applied Meteorology and Climatology, 51*, 84–99. https://doi.org/10.1175/JAMC-D-11-052.1.

Adler, R., Gu, G., Sapiano, M., Wang, J., & Huffman, G. (2017). Global precipitation: Means, variations and trends during the satellite era (1979–2014). *Surveys in Geophysics, 38*, 679–699. https://doi.org/10.1007/s10712-017-9416-4.

Adler, R. F., Sapiano, M., Huffman, G. J., Wang, J., Gu, G., Bolvin, D., Chiu, L., Schneider, U., Becker, A., Nelkin, E., Xie, P., Ferraro, R., & Shin, D.-B. (2018). The global precipitation climatology project (GPCP) monthly analysis (new version 2.3) and a review of 2017 global precipitation. *Atmosphere, 9,* 138. https://doi.org/10.3390/atmos9040138.

Allen, M. R., & Ingram, W. J. (2002). Constraints on future changes in climate and the hydrologic cycle. *Nature, 419,* 224–232. https://doi.org/10.1038/nature01092.

Arkin, P. A., & Xie, P. (1994). The global precipitation climatology project: First algorithm intercomparison project. *Bulletin of the American Meteorological Society, 75,* 401–419. https://doi.org/10.1175/1520-0477(1994)075<0401:TGPCPF>2.0.CO;2.

Behrangi, A., Stephens, G., Adler, R. F., Huffman, G. J., Lambrigtsen, B., & Lebsock, M. (2014). An update on the oceanic precipitation rate and its zonal distribution in light of advanced observations from space. *Journal of Climate, 27,* 3957–3965. https://doi.org/10.1175/JCLI-D-13-00679.1.

Behrangi, A., Nguyen, H., Lambrigtsen, B., Schreier, M., & Dang, V. (2015). Investigating the role of multi-spectral and near surface temperature and humidity data to improve precipitation detection at high latitudes. *Atmospheric Research, 163,* 2–12. https://doi.org/10.1061/j.atmospheres.2014.10.019.

Curtis, S., & Adler, R. F. (2003). Evolution of El Niño-precipitation relationships from satellites and gauges. *Journal of Geophysical Research, 108*(D4), 4153. https://doi.org/10.1029/2002JD002690.

Dai, A., & Wigley, T. M. L. (2000). Global patterns of ENSO-induced precipitation. *Geophysical Research Letters, 27,* 1283–1286. https://doi.org/10.1029/1999GL011140.

Gu, G., & Adler, R. F. (2011). Precipitation and temperature variations on the interannual time scale: Assessing the impact of ENSO and volcanic eruptions. *Journal of Climate, 24,* 2258–2270. https://doi.org/10.1175/2010JCLI3727.1.

Gu, G., & Adler, R. F. (2013). Interdecadal variability/long-term changes in global precipitation patterns during the past three decades: Global warming and/or Pacific decadal variability? *Climate Dynamics, 40,* 3009–3022. https://doi.org/10.1007/s00382-012-1443-8.

Gu, G., & Adler, R. F. (2018). Precipitation intensity changes in the tropics from observations and models. *Journal of Climate, 31,* 4775–4790. https://doi.org/10.1175/JCLI-D-17-0550.1.

Gu, G., Adler, R. F., Huffman, G. J., & Curtis, S. (2007). Tropical rainfall variability on interannual-to-interdecadal/longer-time scales derived from the GPCP monthly product. *Journal of Climate, 20,* 4033–4046. https://doi.org/10.1175/JCLI4227.1.

Gu, G., Adler, R. F., & Huffman, G. J. (2016). Long-term changes/trends in surface temperature and precipitation during the satellite era (1979–2012). *Climate Dynamics, 46,* 1091–1105. https://doi.org/10.1007/s00382-015-2634-x.

Hansen, J., Ruedy, R., Glascoe, J., & Sato, M. (1999). GISS analysis of surface temperature change. *Journal of Geophysical Research, 104,* 30997–31022. https://doi.org/10.1029/1999JD900835.

Held, I. M., & Soden, B. J. (2006). Robust responses of the hydrological cycle to global warming. *Journal of Climate, 19,* 5686–5699. https://doi.org/10.1175/JCLI3990.1.

Hou, A. Y., Kakar, R. K., Neeck, S., Azarbarzin, A. A., Kummerow, C. D., Kojima, M., Oki, R., Nakamura, K., & Iguchi, T. (2014). The global precipitation measurement mission. *Bulletin of the American Meteorological Society, 95,* 701–722. https://doi.org/10.1175/BAMS-D-13-00164.1.

Huffman, G. J., Adler, R. F., Arkin, P., Chang, A., Ferraro, R., Gruber, A., Janowiak, J., McNab, A., Rudolf, B., & Schneider, U. (1997). The global precipitation climatology project (GPCP) combined precipitation dataset. *Bulletin of the American Meteorological Society, 78,* 5–20. https://doi.org/10.1175/1520-0477(1997)078<0005:TGPCPG>2.0.CO;2.

Huffman, G. J., Adler, R. F., Morrissey, M., Bolvin, D. T., Curtis, S., Joyce, R., McGavock, B., & Susskind, J. (2001). Global precipitation at one-degree daily resolution from multi-satellite observations. *Journal of Hydrometeorology, 2,* 36–50. https://doi.org/10.1175/1525-7541(2001)002<0036:GPAODD>2.0.CO;2.

Huffman, G. J., Adler, R. F., Bolvin, D. T., & Gu, G. (2009). Improvements in the GPCP global precipitation record: GPCP version 2.1. *Geophysical Research Letters, 36*, L17808. https://doi.org/10.1029/2009GL040000.

Jaeger, L. (1983). In A. Street-Perrott, M. Beran, & R. Ratcliffe (Eds.), *Monthly and areal patterns of mean global precipitation. Variations in the global water budget* (pp. 129–140). Dordrecht: Springer. https://doi.org/10.1007/978-94-009-6954-4.

Kummerow, C. D., Simpson, J., Thiele, O., Barnes, W., Chang, A. T. C., Stocker, E., Adler, R. F., Hou, A. Y., Kakar, R., Wentz, F., Ashcroft, P., Kozu, T., Hong, Y., Okamoto, K., Iguchi, T., Kuroiwa, H., Im, E., Haddad, Z. S., Huffman, G. J., Ferrier, B., Olson, W. S., Zipser, E., Smith, E. A., Wilheit, T. T., North, G., Krishnamurti, T., & Nakamura, K. (2000). The status of the tropical rainfall measuring Mission (TRMM) after 2 years in orbit. *Journal of Applied Meteorology, 39*, 1965–1982. https://doi.org/10.1175/1520-0450(2001)040<1965:TSOTTR>2.0.CO;2.

L'Ecuyer, T., Beaudoing, H., Rodell, M., Olson, W., Lin, B., Kato, S., Clayson, C., Wood, E., Sheffield, J., Adler, R., Huffman, G., Bosilovich, M., Gu, G., Robertson, F., Houser, P., Chambers, D., Famiglietti, J., Fetzer, E., Liu, W., Gao, X., Schlosser, C., Clark, E., Lettenmaier, D., & Hilburn, K. (2015). The observed state of the energy budget in the early twenty-first century. *Journal of Climate, 28*, 8319–8346. https://doi.org/10.1175/JCLI-D-14-00556.1.

Park, K., Yoshimura, K., Kim, H., & Oki, T. (2017). Chronological development of terrestrial precipitation. *Bulletin of the American Meteorological Society, 98*, 2411–2427. https://doi.org/10.1175/BAMS-D-16-0005.1.

Rodell, M., Beaudoing, H. K., L'Ecuyer, T., Olson, W., Famiglietti, J. S., Houser, P. R., Adler, R., Bosilovich, M., Clayson, C. A., Chambers, D., Clark, E., Fetzer, E., Gao, X., Gu, G., Hilburn, K., Huffman, G., Lettenmaier, D. P., Liu, W. T., Robertson, F. R., Schlosser, C. A., Sheffield, J., & Wood, E. F. (2015). The observed state of the water cycle in the early 21st century. *Journal of Climate, 28*, 8289–8318. https://doi.org/10.1175/JCLI-D-14-00555.1.

Schneider, U., Becker, A., Finger, P., Meyer-Christoffer, A., Ziese, M., & Rudolf, B. (2014). GPCC's new land surface precipitation climatology based on quality-controlled in situ data and its role in quantifying the global water cycle. *Theoretical and Applied Climatology, 115*, 15–40. https://doi.org/10.1007/s00704-013-0860-x.

Skofronick-Jackson, G., Petersen, W. A., Berg, W., Kidd, C., Stocker, E. F., Kirschbaum, D. B., Kakar, R., Braun, S. A., Huffman, G. J., Iguchi, T., Kirstetter, P. E., Kummerow, C., Meneghini, R., Oki, R., Olson, W. S., Takayabu, Y. N., Furukawa, K., & Wilheit, T. (2017). The global precipitation measurement (GPM) mission for science and society. *Bulletin of the American Meteorological Society, 98*, 1657–1672. https://doi.org/10.1175/BAMS-D-15-00306.1.

Stephens, G. L., Vane, D. G., Boain, R. J., Mace, G. G., Sassen, K., Wang, Z., Illingworth, A. J., O'Connor, E. J., Rossow, W. B., Durden, S. L., Miller, S. D., Austin, R. T., Benedetti, A., Mitrescu, C., & the CloudSat Science Team. (2002). The CloudSat mission and the A-TRAIN: A new dimension to space-based observations of clouds and precipitation. *Bulletin of the American Meteorological Society, 83*, 1771–1790. https://doi.org/10.1175/BAMS-83-12-1771.

Sun, Y., Solomon, S., Dai, A., & Portmann, R. (2007). How often will it rain? *Journal of Climate, 20*, 4801–4818. https://doi.org/10.1175/JCLI4263.1.

Wang, J.-J., Adler, R. F., Huffman, G. J., & Bolvin, D. (2014). An updated TRMM composite climatology of tropical rainfall and its validation. *Journal of Climate, 27*, 273–284. https://doi.org/10.1175/JCLI-D-13-00331.1.

WCRP. (1986). Report of the First Session of the International Working Group on Data Management for the Global Precipitation Climatology Project. WCP-132, WMO/TD No. 171, 28 pp.

Xie, P., Janowiak, J. E., Arkin, P. A., Adler, R., Gruber, A., Ferraro, R., Huffman, G. J., & Curtis, S. (2003). GPCP pentad precipitation analyses: An experimental data set based on gauge observations and satellite estimates. *Journal of Climate, 16*, 2197–2214. https://doi.org/10.1175/2769.1.

Xie, S.-P., Deser, C., Vecchi, G. A., Ma, J., Teng, H., & Wittenberg, A. T. (2010). Global warming pattern formation: Sea surface temperature and rainfall. *Journal of Climate, 23*, 966–986. https://doi.org/10.1175/2009JCLI3329.1.

Chapter 36
Global Snowfall Detection and Measurement

Mark S. Kulie, Lisa Milani, Norman B. Wood, and Tristan S. L'Ecuyer

Abstract This chapter highlights the CloudSat Cloud Profiling Radar and its contribution to global snowfall research. Brief summaries of recent snowfall-related research using CloudSat snowfall products and near-global snowfall datasets are provided. Global snowfall detection statistics and quantitative snowfall rate estimates from CloudSat are also presented, including global mean zonal precipitation phase and snowfall-only statistics. A simple quality control methodology is also presented to alleviate ground clutter that can affect quantitative snowfall estimates. Gridded spatial analyses also provide valuable information on snowfall prone locations and corresponding annual mean snowfall accumulation estimates, including a monthly hemispheric snowfall variability analysis. Future research pathways to use and improve CloudSat snowfall products are also discussed.

Keywords Precipitation · Snowfall · CloudSat · CPR · A-train · TRMM · GPM · DPR · GMI · EarthCARE · Microwave · Radar · QPE · NEXRAD · Cloud processes · Microphysics · Antarctica · Greenland · Alaska · Himalaya · ERA-interim

M. S. Kulie (✉)
NOAA/NESDIS/Center for Satellite Applications and Research/Advanced Satellite Products Branch, Madison, WI, USA
e-mail: Mark.Kulie@noaa.gov

L. Milani
Goddard Space Flight Center, University of Maryland and NASA, Greenbelt, MD, USA

N. B. Wood
Cooperative Institute for Meteorological Satellite Studies, University of Wisconsin-Madison, Madison, WI, USA

T. S. L'Ecuyer
Department of Atmospheric and Oceanic Sciences, University of Wisconsin-Madison, Madison, WI, USA

© Springer Nature Switzerland AG 2020
V. Levizzani et al. (eds.), *Satellite Precipitation Measurement*, Advances in Global Change Research 69, https://doi.org/10.1007/978-3-030-35798-6_12

36.1 Introduction

The spaceborne precipitation remote sensing community has steadily evolved over the past few decades, first using passive microwave and/or infrared techniques to estimate precipitation over regions devoid of ground-based remote sensing or in situ observations. Spaceborne radars, which advantageously receive a direct signal from hydrometeors, have also contributed significantly to our understanding of global precipitation in recent decades. The Tropical Rainfall Measuring Mission (TRMM) accelerated the development of precipitation remote sensing tools preceding its 1997 launch and during its lengthy mission lifetime (Kummerow et al. 1998). TRMM era precipitation retrievals leveraged the first spaceborne radar (Ku-band Precipitation Radar) and highly calibrated microwave radiometer observations, including low latency products using blended satellite sensor and ground-based rain gauge synergy (e.g., Huffman et al. 2007). CloudSat followed as the second spaceborne radar in 2006, albeit with a higher radar frequency (W-band) than TRMM that aligned with its cloud remote sensing mandate (Stephens et al. 2002, also Chap. 2 of volume 1). CloudSat, however, is also uniquely configured for global precipitation studies. The Global Precipitation Measurement (GPM) mission – the TRMM flagship precipitation mission successor – provides enhanced radar (dual Ku/Ka-band frequencies and better radar sensitivity) and radiometer (higher frequency scattering sensitive channels) capabilities than TRMM (Hou et al. 2014; Skofronick-Jackson et al. 2017, also Chap. 1 of volume 1). GPM, launched in 2014, provides precipitation estimates in both tropical and higher latitudes. This higher latitude coverage, however, coincides with the added complexity to detect and quantitatively estimate solid and/or light precipitation. The CloudSat and GPM missions have initiated significant efforts to develop and refine snowfall detection and quantitative precipitation estimation (QPE) techniques using spaceborne assets.

Since the TRMM Precipitation Radar was limited to a predominantly tropical latitude orbit that did not frequently sample surface snowfall events, CloudSat truly ushered in a new era of snowfall remote sensing. CloudSat's 94 GHz Cloud Profiling Radar (CPR; Tanelli et al. 2008), combined with its Afternoon Constellation (A-Train) platform partners (L'Ecuyer and Jiang 2010; Stephens et al. 2018), was designed to sample global cloud distributions and cloud macro- and microphysical properties on an unprecedented level (Stephens et al. 2002, 2018). CloudSat's snowfall research potential, however, was realized relatively early in its mission lifetime (e.g., Liu 2008; Matrosov et al. 2008; Hudak et al. 2008; Kulie and Bennartz 2009). The CPR's ample sensitivity ($\sim$−28 to −30 dBZ) and orbital characteristics (sampling from 82°N–82°S latitudes) allowed it to be employed successfully for global snowfall research. CloudSat provided the first observational evidence to address the following fundamental global snowfall questions on a global, regional, and seasonal basis:

- Snowfall occurrence: Where does snow fall?
- Snowfall QPE: How much snow falls?

- Snowfall morphology: What fraction of snowfall (occurrence or QPE) originates from different snowfall types?
- Model evaluation: How consistent are CloudSat and model-derived snowfall datasets?
- Snow-related cloud processes: What cloud processes operate within snow-producing clouds, and are these processes adequately represented by numerical models?

In addition to valuable snowfall-related science, snowfall research initiated by CloudSat has also provided valuable independent information to assess GPM-generated snowfall statistics and refine GPM snowfall retrievals early in its mission.

This chapter primarily focuses on CloudSat snowfall applications. Recent global snowfall studies and discoveries enabled by CloudSat observations are first highlighted. The theoretical basis for CloudSat snowfall detection and QPE is then briefly described. Next, global snowfall characteristics derived from the CloudSat data record are presented, including snowfall detection statistics and QPE. Finally, concluding remarks and future work are discussed.

36.2   Global Snowfall: CloudSat's Recent Contributions

Early studies during the first few mission years illustrated the unique global snowfall perspective offered by CloudSat observations, including the first global snowfall detection and QPE maps from observational datasets (Liu 2008). Further studies using limited CloudSat datasets showed that light snowfall dominates the global snowfall rate distribution and highlighted microphysical variability (e.g., snowflake habit scattering properties and snow size distributions) that causes radar reflectivity (Z) to snowfall rate (S) conversion uncertainties (Liu 2008; Kulie and Bennartz 2009; Hiley et al. 2011). Early CPR comparisons to ground-based radar measurements were also undertaken, showing similar structural features, snowfall detection statistics, and independent QPE within reasonable uncertainty bounds for each instrument (Hudak et al. 2008; Matrosov et al. 2008). CloudSat annual snowfall estimates were also compared against gauge accumulations over Canada with varying degrees of accuracy based on geographic location (Liu 2008; Hiley et al. 2011), but overall consistency was demonstrated between the datasets. Numerous recent studies compared CloudSat snowfall products with QPE from ground-based radar networks. CloudSat's ability to detect light precipitation, combined with scanning radar issues (e.g., beam blockage in mountainous regions and shallow features not effectively observed at longer distances from radar sites), enabled CloudSat to observe snowfall more effectively than gauge-adjusted United States Next-Generation Weather Doppler Radar (NEXRAD) merged datasets (Smalley et al. 2014), thus highlighting the important role which CloudSat can play in snowfall-sensitive regions like the United States Intermountain West. Other studies also tout CloudSat's ability to detect lighter snowfall events compared to NEXRAD datasets, although NEXRAD might produce more realistic QPE for intense snowfall events due to possible CPR W-band attenuation issues (Cao et al. 2014; Chen et al. 2016). Finally, CloudSat and Sweden scanning radar network snowfall QPE were shown to agree extremely well below the ~1.0 mm h^{-1} threshold, but CloudSat also underestimates snowfall rates at higher values (Norin et al. 2015). These seminal studies provided evidence that CloudSat snowfall estimates compare favorably with independent retrievals from ground-based radars, while also highlighting strengths and weaknesses of the respective sensors and associated QPE uncertainties.

CloudSat's scientific worth has been demonstrated over remote regions like Antarctica and Greenland, where reanalysis or climate model datasets were previously the only available source of snowfall estimates over extended periods. Focused studies have characterized snowfall cycles and falling snow's contribution to the surface mass balance over the world's major ice sheets (Boening et al. 2012). CloudSat snowfall estimates have also been compared to reanalysis datasets. ERA-Interim reanalysis and CloudSat snowfall estimates over Antarctica demonstrate encouraging consistency (Palerme et al. 2014, 2017; Milani et al. 2018), while other reanalysis datasets capture seasonal and interannual variability in a similar fashion as CloudSat observations despite larger QPE discrepancies (Palerme et al. 2017). Larger ERA-Interim and CloudSat QPE differences also occur in surrounding Southern Ocean environments (Milani et al. 2018). Further studies have used CloudSat as an important climate model constraint. Climate models that optimally match CloudSat Antarctica snowfall output have been identified, thus allowing more confident predictions of future snowfall patterns over this climate sensitive region (Palerme et al. 2017).

CloudSat observations have also allowed different snowfall modes to be studied on a global basis. A CPR reflectivity profile analysis using the first year of CPR observations hinted at a variety of snowfall regimes with sometimes distinctive vertical reflectivity structures (Liu 2008). CloudSat snowfall events were also partitioned into isolated and horizontally extensive shallow or deep categories to study liquid water path trends in each category (Wang et al. 2013). Further snowfall partitioning studies were undertaken to classify snowfall events into "cumuliform" (convective) or "nimbostratus" categories using CloudSat cloud classification products (Kulie et al. 2016; Kulie and Milani 2018). While nimbostratus snowfall events produce a disproportionate amount of estimated global snowfall, shallow cumuliform snowfall is regionally important (>50% occurrence and/or mean annual accumulation). Oceanic convective snow also displays strong seasonal signals linked to sea ice coverage and air-sea interactions that force boundary layer convection. CloudSat observations have also been used to study snowfall characteristics and variability due to different weather states over Scandinavia, highlighting distinctive snowfall characteristics depending on prevailing winds and concomitant meteorological conditions (Norin et al. 2015).

CloudSat and coincident A-Train observations have also offered insights into important snow-producing cloud processes. Studies have highlighted the prevalence of supercooled cloud liquid water associated with snowfall events (Hiley et al. 2011; Battaglia and Delanoë 2013; Wang et al. 2013). More recent studies have illustrated the importance of snowfall in scavenging supercooled liquid water in polar regions and the corresponding overaggressive representation of the Wegener-Bergeron-Findeisen ice growth process in climate models compared to CloudSat and A-Train observations (McIlhattan et al. 2017).

Finally, CloudSat has been combined with GPM for synergistic global snowfall studies. These studies have exploited CloudSat's lighter snowfall sensitivity to evaluate GPM Dual Frequency Precipitation Radar (DPR) and Microwave Imager (GMI) snowfall detection capabilities (Casella et al. 2017; Panegrossi et al. 2017). Other studies have attempted to quantify global snowfall and higher latitude precipitation by combining QPE from both (and other) sensors by leveraging CloudSat (GPM DPR) products at lower (higher) snowfall rates (e.g., Behrangi et al. 2016; Adhikari et al. 2018).

36.3   CloudSat Snowfall Detection and QPE Algorithm

The CloudSat 2C-SNOW-PROFILE (2CSP) product provides surface and profile snowfall rate, snow water content, and snow particle size distribution retrievals for any CPR observation associated with a likely surface snowfall event (Wood et al. 2013). While the native CPR observations are collected using ~480 m range gates, instrument oversampling enables ~240 m vertical bin size in CloudSat profile products.

2CSP snowfall rates are created using a multi-step procedure that first classifies the precipitation phase as a likely surface snowfall event, then estimates a snowfall rate from the first usable CPR bin located above ground clutter (the "near-surface bin"). The third (fifth) bin above ground level is defined as the near-surface bin for ocean (land) surfaces, where ground elevations are determined by a global digital elevation model (DEM). Adopting the near-surface bin creates a blind-zone in the lowest ~0.8–1.0 km, whereby CloudSat surface snowfall rate estimates are defined by the near-surface bin reflectivity and corresponding conversion to snowfall rate. The 2C-PRECIP-COLUMN (2CPC) product (Haynes et al. 2009) indicates likely surface precipitation phase by incorporating ECMWF temperature profiles to approximate the melting level and hydrometeor melt fraction for each CPR profile. The 2CSP algorithm is applied to 2CPC profiles that are classified as "snow possible" or "snow certain". A minimum near-surface bin reflectivity threshold of -15 dBZ is used to identify likely surface snowfall events.

The 2CSP retrieval uses a radar reflectivity forward model in an optimal estimation framework, whereby measured snow microphysics properties are included as *a priori* knowledge to convert radar reflectivity (Z_e) into snowfall rates (S) in a physically consistent manner (Wood et al. 2014, 2015). Uncertainties induced by microphysical variability, including disparate scattering properties produced by modeled snowflake structures, are also generated by the 2CSP retrieval. These uncertainties can be quite large for single-frequency, W-band radar retrievals without added observational constraints (see Fig. 36.5 and Chaps. 13, 15, and 16 of volume 1). Unlike CloudSat studies that apply static W-band Z-S relationships based on different snow microphysics and microwave radiation scattering assumptions (e.g., Liu 2008; Matrosov et al. 2008; Kulie and Bennartz 2009; Hiley et al. 2011), 2CSP produces a range of possible Z-S relationships driven largely by temperature-dependent snow microphysical properties. These dynamic Z-S relationships have been reported in recent studies over Antarctica and the Southern Ocean (Milani et al. 2018).

All data products used for the current study are version R04 products acquired from the CloudSat Data Processing Center (DPC). Version R05 products have been recently released by the CloudSat DPC but were not available for the analyses presented herein.

36.4 Global Snowfall Characteristics

This section highlights global snowfall characteristics to illustrate CloudSat's capabilities and valuable knowledge gained over the first decade-plus of its mission lifetime.

36.4.1 CloudSat Sampling

The CloudSat CPR is a nadir-pointing radar with a 16-day repeat cycle. Despite being a non-scanning radar, its 82° latitude orbital maxima and associated orbital characteristics allow it to effectively observe high latitude snowfall events. Figure 36.1 shows both spatial and latitudinal views of CloudSat observations. Figure 36.1 (left panel) highlights the number of CloudSat overpasses located in high resolution grid boxes (0.1°) centered over Alaska to illustrate both the CloudSat sampling strategy and its increased sampling frequency at higher latitudes. Similar CloudSat overpass figures are shown in Kulie et al. (2016) and McIlhattan et al. (2017). The total zonal number of CloudSat observations as a function of latitude is also shown in Fig. 36.1 (right panel). The highest latitudes contain over 10^7 CloudSat zonal observations over the decade-plus period included in this study. CloudSat sampled both hemispheres symmetrically during the 2006 through early 2011 period, but hemispheric sampling asymmetry occurred in the late 2011 through 2016 period. The reduced Southern Hemisphere (SH) sampling is due to daylight-only (DO) operations that were initiated after CPR battery issues occurred in April, 2011. This SH sampling deficiency is accentuated during the peak of austral winter. Northern Hemisphere (NH) sampling during Daylight Only Operations (DO-Op),

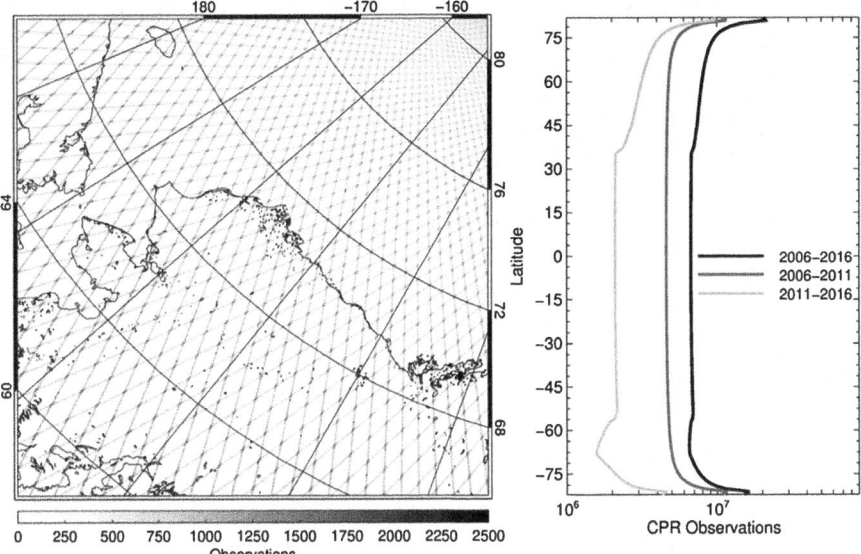

Fig. 36.1 CloudSat observations per 0.1° latitude/longitude grid box for the 2006–2016 time period (left panel). Zonally-averaged (1° latitude bins) number of CloudSat observations for the 2006–2016 (black), July 2006–April 2011 (dark gray), and December 2011–December 2016 (light gray) time periods (right panel)

while reduced compared to the 2006 through early 2011 period, is higher than the SH and approaches 2006–2011 values at the highest NH latitudes.

36.4.2   Snowfall Occurrence Statistics

CloudSat's orbital sampling shown in Fig. 36.1 fosters compelling high latitude precipitation science applications from an active sensor that were not previously possible. For instance, Fig. 36.2 shows latitudinally-averaged conditional occurrence fractions for different precipitation types (snow, rain, and possible mixed precipitation) identified by the CloudSat 2C-PRECIP-COLUMN (2CPC) product (Haynes et al. 2009, 2013). Unconditional snowfall occurrence fractions (i.e., the number of snow events in each latitude bin divided by the *total* number of CloudSat observations) are also shown. CloudSat conditional snowfall fraction estimates exceed 50% in both hemispheres near 60° latitude. The peak conditional snowfall fraction is slightly above 80% in the highest NH latitudes, while snow comprises 100% of all precipitation events in the SH over Antarctica. The SH displays a smooth latitudinal transition between rain and snow dominant regions, while the NH transition is more complex due to large land masses and the influence of the Himalaya Mountains at lower latitudes. Mixed precipitation fraction maximum

Fig. 36.2 Zonally-averaged conditional snow (black), rain (dark gray) and mixed precipitation (dot-dash) occurrence fractions from the 2C-PRECIP-COLUMN product. Unconditional snowfall occurrence fractions (light solid gray) are also shown. All results shown are averaged over 1° latitudinal belts for the 2006–2016 period

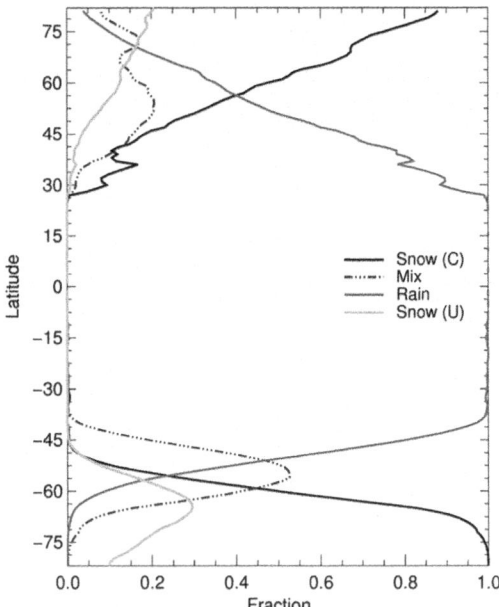

magnitudes are also much higher in the SH (~50%) versus the NH (~25%). Mixed precipitation fractions peak near 50–55° latitude in both hemispheres, although a small secondary NH peak exists near 70°. The mixed category is highly relevant to the snowfall community, as this cold season precipitation type naturally represents a source of precipitation classification uncertainty in the ~0–4 °C near-surface temperature range. Zonally-averaged unconditional snowfall occurrence fractions approach 30% in the SH peak near 65° latitude, then decrease at higher latitudes. NH unconditional snowfall occurrence fractions increase monotonically with latitude and display a peak zonal average near 20% at the highest latitudes.

As shown in Fig. 36.3, corresponding 1° latitude/longitude gridded conditional and unconditional snowfall occurrence fractions highlight further details that are embedded within the zonal mean results shown in Fig. 36.2. The interior of Greenland and virtually the entire Antarctic land mass (except for some coastal regions) experience 100% conditional snowfall occurrence fractions. SH latitudinal stratification in the conditional snowfall fraction field is also an obvious feature, as is the NH pattern produced by prominent storm tracks that are governed by large-scale atmospheric circulations. Conditional snowfall percentages exceed 50% over most NH land masses located higher than ~45°, but oceanic regions exceed that threshold at much higher latitudes. Similar to the zonal results (Fig. 36.2), the SH shows 50% conditional occurrence values south of about 60°S with much less latitudinal variability. Unconditional snowfall fractions within the 30–50% range commonly occur in the Southern Ocean. NH unconditional fractions are much lower, except for values exceeding 30% in various North Atlantic Ocean regions, the southern half of Greenland, inner Russia, the western Pacific Ocean, and the Himalaya and Alaska/ western Canada Pacific coastal mountain ranges.

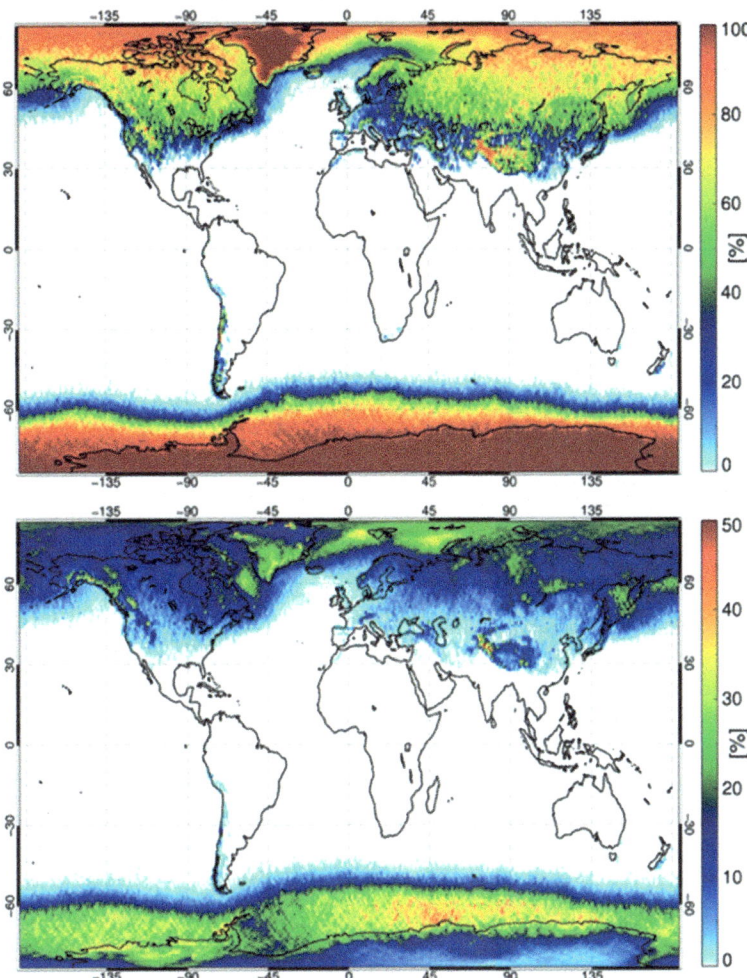

Fig. 36.3 Conditional (top) and unconditional (bottom) snow occurrence fractions (1° latitude/longitude grid boxes) from the 2C-PRECIP-COLUMN product for the 2006–2016 period

36.4.3   CPR Reflectivity Variability: Snowfall Events

The first global snowfall studies using temporally limited CloudSat datasets confirmed light snowfall dominance from a frequency of occurrence perspective (e.g., Liu 2008; Kulie and Bennartz 2009; Hiley et al. 2011). The CPR near-surface radar reflectivity normalized distributions shown in Fig. 36.4 reaffirm this fact using the 2006–2016 CloudSat snowfall database. These reflectivity distributions are further partitioned into nimbostratus and shallow cumuliform categories – another unique global snowfall aspect made possible by the CloudSat product suite. Snowfall events

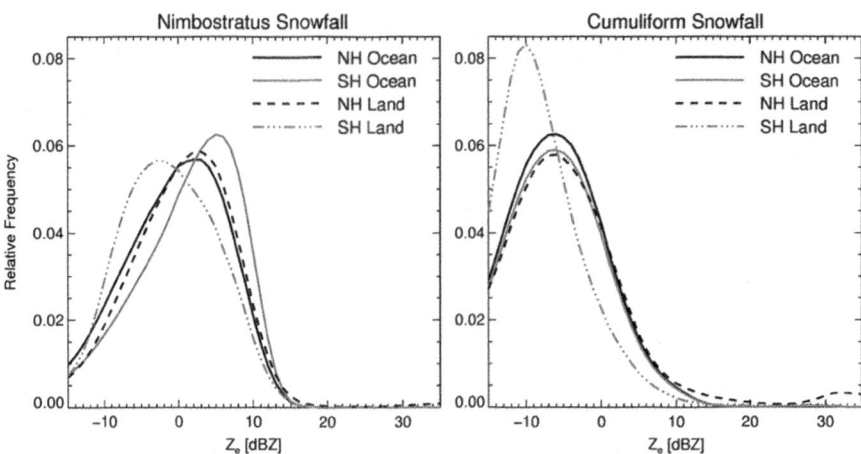

Fig. 36.4 Near-surface bin CPR normalized reflectivity distributions for nimbostratus (left) and cumuliform (right) snowfall events. Distributions are partitioned between Northern (NH) and Southern Hemisphere (SH) land and oceanic surfaces

are partitioned using cloud classifications from the 2B-CLDCLASS CloudSat product (Sassen and Wang 2008) and methodology described in Kulie et al. (2016). Figure 36.4 shows hemispheric, land/ocean, and snowfall mode reflectivity distribution differences. Nimbostratus snowfall events that are typically associated with deeper cloud structures peak at much higher reflectivity values compared to cumuliform snowfall. Further differences, however, are apparent. For instance, the SH ocean nimbostratus reflectivity distribution peaks near 6 dBZ – a much higher reflectivity value than the NH distribution (~2 dBZ peak) – and contain a larger fraction of events at higher reflectivity values, while the opposite trend exists for land events. SH ocean nimbostratus events are also much more intense than SH land events that are dominated by light Antarctic snowfall. Conversely, a NH land/ocean nimbostratus disparity exists (land snowfall events more intense), but this trend is not very pronounced.

Cumuliform snowfall reflectivity distribution peaks shown in Fig. 36.4 are quite similar, with two notable features. First, NH land cumuliform events exhibit higher relative counts than all other cumuliform and nimbostratus categories above the ~10 BZ threshold. Second, SH land cumuliform snowfall that occurs almost exclusively over Antarctica is extremely light. The first cumuliform snowfall feature, especially the 10–20 dBZ range, may indicate land enhancement by surface frictional convergence or orographic effects – a feature first highlighted in Kulie et al. (2016). A clear secondary cumuliform snowfall relative frequency increase also appears above the ~25 dBZ value. This feature is most likely related to ground clutter erroneously identified as shallow cumuliform cloud structures, since these extremely high W-band reflectivities are theoretically unrealistic for snowfall events.

36.4.4 Snowfall QPE

Snowfall QPE is the most useful hydrologic parameter produced by the 2CSP algorithm. Snow QPE, however, can be adversely affected by ground clutter, even though possible clutter comprises a very small fraction of snowfall events in Fig. 36.4. Ground clutter is typically avoided by using a near-surface bin located safely above intense surface radar signatures, but complex terrain and/or DEM inconsistencies can complicate the near-surface bin clutter avoidance procedure. Previous research has illustrated ground clutter QPE effects and offered ways to mitigate clutter (Kulie and Bennartz 2009), but these clutter quality control (QC) methodologies are most likely too harsh and remove legitimate snowfall populations (Hiley et al. 2011; Milani et al. 2018). Motivated by the cumuliform reflectivity distributions shown in Fig. 36.4, QC steps are applied to the 2CSP dataset in the current study. This QC procedure simply identifies and rejects any land cumuliform snowfall events that are associated with near-surface radar reflectivities equal to or greater than 25 dBZ (the land cumuliform reflectivity inflection point identified in Fig. 36.4). The 2CSP snow retrieval confidence flag also must indicate "moderate" or "high" confidence to be included in the final dataset.

Figure 36.5 shows the QC procedure effect on CloudSat zonally averaged unconditional snowfall fractions (left panel) and QPE (right panel) for the

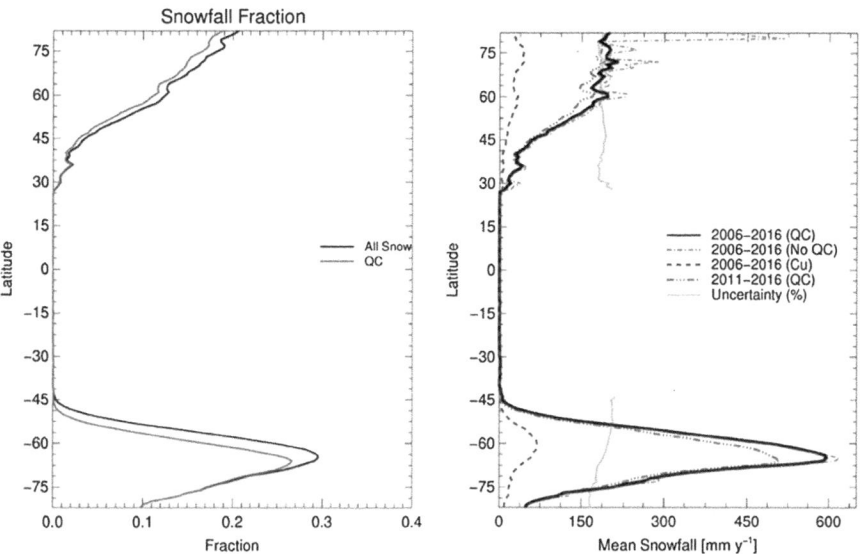

Fig. 36.5 Left panel: unconditional snowfall occurrence fraction with (gray) and without (black) ground clutter quality control applied. Right panel: mean annual snowfall rates with (solid black) and without (gray dash-dot) ground clutter quality control applied for all snowfall events over the 2006–2016 and 2011–2016 (dark gray dash-dot) periods. Cumuliform snowfall events with quality control applied for the 2006–2016 (gray dash) and mean fractional snowfall rate uncertainty (light gray solid) also shown

2006–2016 dataset. Snowfall fractions are reduced in both hemispheres by a few percent. Most events removed from the initial CloudSat dataset are oceanic light, shallow cumuliform events (due to the confidence flag threshold) and/or potential ground clutter events over rugged topography in Greenland, parts of northeastern Canada, and a few other mountainous regions (due to land cumuliform snow reflectivity threshold). The QC versus no QC QPE differences are extremely large over some NH latitudes. Substantial reductions in mean annual snowfall rate result from excluding the small population of extremely large near-surface reflectivities exceeding 25 dBZ, especially over the highest latitude bins.

Mean zonal NH snowfall rates are near 200 mm y^{-1} at latitudes higher than 60°, but steadily decrease to under 50 mm y^{-1} at about 40°N. The Himalayan effect slightly increases zonal mean values between 30-40°N. SH values are virtually unchanged due to the QC procedure. This QPE insensitivity indicates that the SH snowfall events removed due to the confidence threshold QC step are very light in intensity. SH mean zonal snowfall rates are much higher than the NH, peaking at about 600 mm y^{-1} near 60°S latitude. Mean annual cumuliform snowfall rates are also shown in Fig. 36.5 and indicate that cumuliform snowfall comprises ~15–20% of the mean annual snow in the NH. SH cumuliform snow, while higher in magnitude compared to the NH, is about 10–15% in the snowiest SH latitude belts. These zonal statistics, however, mask considerable regional variability with much higher cumuliform snowfall percentages (Kulie et al. 2016; Kulie and Milani 2018).

Two additional features are indicated in Fig. 36.5. First, the 2011–2016 DO-Op period displays ~10–15% lower zonal mean QPE compared to the entire dataset period and even larger relative differences compared to the 2006–2011 day-night operational period (not shown). Second, mean fractional 2CSP snowfall uncertainty values are consistently between ~160 and 200%. The fractional uncertainty is defined as the instantaneous 2CSP snowfall rate uncertainty divided by the retrieved instantaneous snowfall rate (Palerme et al. 2014). These large mean uncertainty values highlight the unconstrained nature of snowfall retrievals using single frequency W-band radar observations, yet represent the current state-of-the-art spaceborne snowfall retrieval capabilities.

Gridded 2CSP mean annual snowfall rates for the entire 2006–2016 period are shown in Fig. 36.6 and Fig. 36.7 using the quality-controlled 2CSP dataset. The gridded spatial fields shown in Fig. 36.6 highlight the Southern Ocean snowfall maxima that was described in previous snowfall occurrence and zonal QPE figures, where grid-averaged snowfall rates are consistently between 500 and 1000 mm y^{-1}. Figure 36.6 also highlights other prominent regions as among the snowiest in the world: Antarctic Peninsula, Alaska/Canada Pacific Coastal Range, southeast coastal Greenland, Himalayas. The southeast Greenland maxima shows consistent estimated values approaching 1000 mm y^{-1}. Oceanic NH maxima (~500 mm y^{-1}) are also prevalent in areas that receive ample convective snow such as near Svalbard, the Barents Sea, Western Pacific Ocean near the Kamchatka Peninsula. These regions have been highlighted in previous CloudSat-based cumuliform snowfall studies (Kulie et al. 2016; Kulie and Milani 2018).

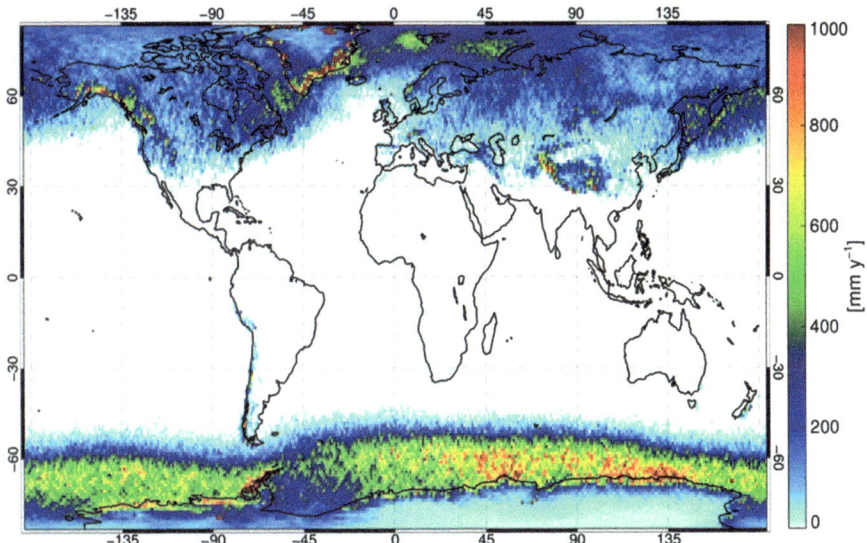

Fig. 36.6 2C-SNOW-PROFILE mean annual snowfall rate in 1° grid boxes for the 2006–2016 period

Hemispheric monthly snowfall rates and fractional occurrences, plus corresponding monthly variability statistics calculated over a multi-year period, are highlighted in Fig. 36.7. The NH shows large seasonal variability in both monthly snowfall rate and fraction fields. NH snowfall rates rapidly increase between September ($\sim$0.21 mm d^{-1}) and October ($\sim$0.38 mm d^{-1}), with a similarly large increase in snowfall fraction ($\sim$0.05–0.10). December and February receive the most NH snowfall ($\sim$0.43 mm d^{-1}), with higher (lower) variability over the 2006–2016 time period evident in February (December). Snowfall fractional occurrences remain above the $\sim$0.11 threshold during the peak boreal winter months (November through March). Boreal summer months experience substantially reduced snowfall fractions and rates, with snowfall fraction (rate) values falling to near 0.01 (0.05 mm d^{-1}) in July. Conversely, the seasonal SH snowfall cycle is more attenuated than the NH, especially SH snowfall fractions. SH snowfall occurrence fractions exhibit a relatively stable range between $\sim$0.07 and 0.09. The SH snowfall fraction stability is especially apparent between April and September. SH monthly snowfall rate and fractional occurrence variability is also reduced compared to the NH (i.e., generally less 25–75 percentile range variability compared to NH results). SH monthly snowfall rates are highest in austral winter and approach the peak NH winter median monthly magnitudes (slightly above 0.4 mm d^{-1}), while austral summer values reach a $\sim$0.23 mm d^{-1} minimum value – much higher than the corresponding NH minimum value. SH values shown in Fig. 36.7 are derived from the 2006 through early 2011 period, as reduced SH sampling during the CloudSat DO operations period (late 2011 through 2016) in austral winter causes extremely large variability, substantially reduces monthly median values, and eliminates the seasonal cycle signal evident in the 2006–2011 period.

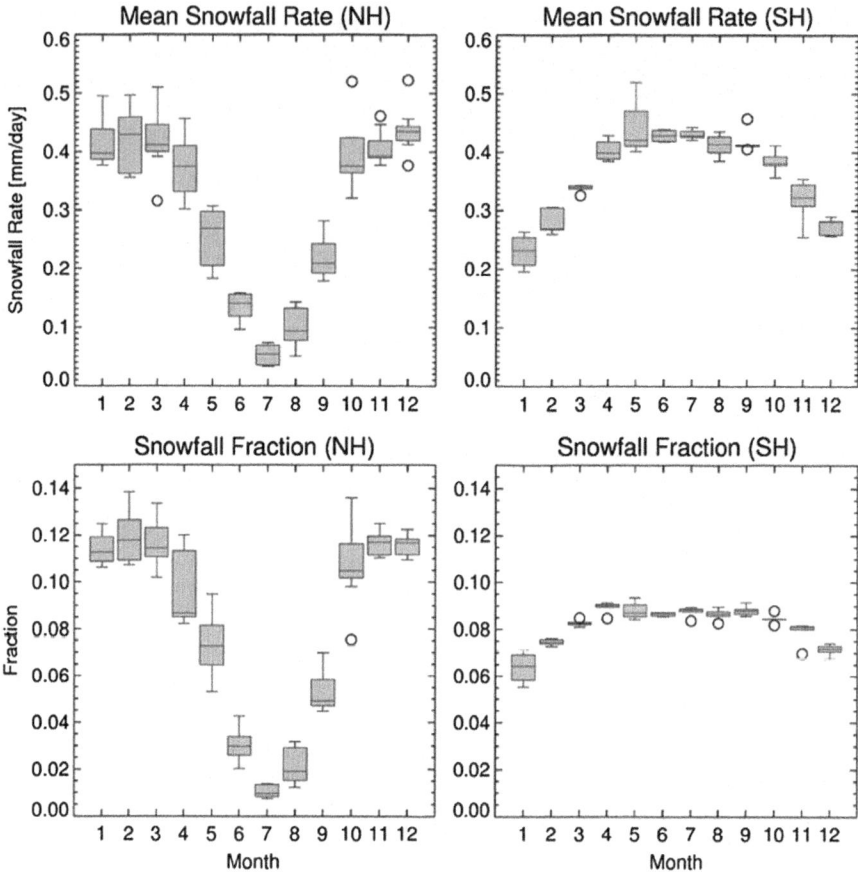

Fig. 36.7 Monthly snowfall rate [mm d^{-1}] (top panels) and unconditional snowfall fraction (bottom panels) for the Northern (left) and Southern (right) Hemispheres. Median (line), 25–75 percentile range (gray shaded), and minimum/maximum values extending up to 1.5 times the 25–75 interquartile range (thin lines) are shown. Outlier data points are indicated by circles. The Southern Hemisphere panels are restricted to the 2006–2011 time period, while the Northern Hemisphere panels use the entire 2006–2016 CloudSat dataset

The CloudSat snowfall QPE renditions shown in Figs. 36.6 and 36.7 represent a significant global snowfall monitoring achievement using an observationally-based dataset to address the fundamental global snowfall questions outlined in Sect. 36.1.

36.5 Concluding Remarks

CloudSat observations and products are an extremely valuable global snowfall research commodity and have initiated innovative snowfall-related research throughout the first decade-plus of the CloudSat mission. CloudSat's CPR offers the first quasi-global snowfall *observations* that augment previous global snowfall studies developed by model/reanalysis datasets. CloudSat results confirm that snowfall is prevalent and hydrologically important at higher latitudes, while simultaneously providing beneficial quantitative precipitation estimates. While still hampered by inherent uncertainties, CloudSat snowfall retrievals provide an available observational dataset that can help answer global hydrology challenges like the recent Global Energy and Water EXchanges (GEWEX) initiative to provide observational snowfall estimates across the major mountain basins around the globe. The extensive CloudSat snowfall dataset further offers unique observational insights and acts as an important independent model evaluation and constraint tool, especially in climate sensitive regions like Antarctica, Greenland, and other remote regions.

Future high priority research pathways using CloudSat snowfall datasets cover a wide possible spectrum of topics. Regional hydrology and specific snowfall regimes (e.g., cumuliform snow studies, orographic snow, etc.) will continue to garner attention, especially intercomparisons with other remote sensing and model-based datasets. CloudSat and A-Train observations will also be used for focused snow-producing cloud process studies, with ample possibilities to assess these processes within model microphysics schemes. The 2CSP microphysics retrievals (e.g., particle size distribution parameters) have been a relatively underexplored resource, but offer unique microphysics insights that will undoubtedly be studied. 2CSP product improvements will ultimately arise from rigorous comparisons against ground evaluation datasets. This comparison task is perpetually complicated by the difficulty of collecting spatially representative ground-based snowfall measurements with a high degree of accuracy. Further evaluation studies, however, will be required to understand 2CSP regional biases that may exist or offer ways to improve microphysical assumptions used in the 2CSP algorithm. Developing clutter mitigation schemes in areas prone to ground clutter are also desired, although updated DEM contained in R05 products may improve the embedded clutter problem in areas like Greenland. Finally, synergistic studies with other precipitation remote sensing datasets like GPM will offer further valuable insights, especially to build long-term global snowfall datasets. CloudSat also represents a natural bridge to the Earth Clouds, Aerosols and Radiation Explorer (EarthCARE) mission that will also operate a 94 GHz spaceborne radar and continue the W-band high latitude precipitation data record. Further global snowfall comparisons between spaceborne remote sensing datasets (e.g., GPM) will be undertaken, but recognizing and correcting for inherent algorithm differences to create equitable snowfall comparisons between disparate mission products will be needed (e.g., Skofronick-Jackson et al. 2018). This lengthy list of future research topics illustrates both the importance of the CloudSat snowfall data record and its continued role in guiding global snowfall research in future years.

Acknowledgments Partial support for this work from NASA grants NNX12AQ76G, NNX14AB22G, and NNX16AE21G is gratefully acknowledged. Data product procurement support from the CloudSat Data Processing Center is also appreciated.

References

Adhikari, A., Liu, C., & Kulie, M. S. (2018). Global distribution of snow precipitation features and their properties from 3 years of GPM observations. *Journal of Climate, 31*, 3731–3754. https://doi.org/10.1175/JCLI-D-17-0012.1.

Battaglia, A., & Delanoë, J. (2013). Synergies and complementarities of CloudSat-CALIPSO snow observations. *Journal of Geophysical Research, 118*, 721–731. https://doi.org/10.1029/2012JD018092.

Behrangi, A., Christensen, M., Richardson, M., Lebsock, M., Stephens, G., Huffman, G. J., Bolvin, D., Adler, R. F., Gardner, A., Lambrigtsen, B., & Fetzer, E. (2016). Status of high-latitude precipitation estimates from observations and reanalyses. *Journal of Geophysical Research, 121*, 4468–4486. https://doi.org/10.1002/2015JD024546.

Boening, C., Lebsock, M., Landerer, F., & Stephens, G. (2012). Snowfall-driven mass change on the East Antarctic ice sheet. *Geophysical Research Letters, 39*(21), L21501. https://doi.org/10.1029/2012GL053316.

Cao, Q., Hong, Y., Chen, S., Gourley, J. J., Zhang, J., & Kirstetter, P.-E. (2014). Snowfall detectability of NASA's CloudSat: The first cross-investigation of its 2c-snow-profile product and National Multi-Sensor Mosaic QPE (NMQ) snowfall data. *Progress In Electromagnetics Research, 148*, 55–61. https://doi.org/10.2528/PIER14030405.

Casella, D., Panegrossi, G., Sanò, P., Marra, A. C., Dietrich, S., Johnson, B. T., & Kulie, M. S. (2017). Evaluation of the GPM-DPR snowfall detection capability: Comparison with CloudSat-CPR. *Atmospheric Research, 197*, 64–75. https://doi.org/10.1016/j.atmosres.2017.06.018.

Chen, S., Hong, Y., Kulie, M., Behrangi, A., Stepanian, P. M., Cao, Q., You, Y., Zhang, J., Hu, J., & Zhang, X. (2016). Comparison of snowfall estimates from the NASA CloudSat Cloud Profiling Radar and NOAA NSSL Multi-Radar Multi-Sensor System. *Journal of Hydrology, 541*, 862–872. https://doi.org/10.1016/j.jhydrol.2016.07.047.

Haynes, J. M., L'Ecuyer, T. S., Stephens, G. L., Miller, S. D., Mitrescu, C., Wood, N. B., & Tanelli, S. (2009). Rainfall retrieval over the ocean with spaceborne W-band radar. *Journal of Geophysical Research, 114*(D8), D00A22. https://doi.org/10.1029/2008JD009973.

Haynes, J. M., L'Ecuyer, T. S., Vane, D., Stephens, G. L., & Reinke, D. (2013). Level 2-C precipitation column algorithm product process description and interface control document. Version P2_R04, CloudSat Project Doc., 17 pp. Available at http://www.cloudsat.cira.colostate.edu/sites/default/files/products/files/2C-PRECIP-COLUMN_PDICD.P2_R04.20130124.pdf. Last accessed 20 Nov 2018.

Hiley, M. J., Kulie, M. S., & Bennartz, R. (2011). Uncertainty analysis for CloudSat retrievals. *Journal of Applied Meteorology and Climatology, 50*, 399–418. https://doi.org/10.1175/2010JAMC2505.1.

Hou, A. Y., Kakar, R. K., Neeck, S., Azarbarzin, A. A., Kummerow, C. D., Kojima, M., Oki, R., Nakamura, K., & Iguchi, T. (2014). The global precipitation measurement (GPM) mission. *Bulletin of the American Meteorological Society, 95*, 701–722. https://doi.org/10.1175/BAMS-D-13-00164.1.

Hudak, D., Rodriguez, P., & Donaldson, N. (2008). Validation of the CloudSat precipitation occurrence algorithm using the Canadian C band radar network. *Journal of Geophysical Research, 113*, D00A07. https://doi.org/10.1029/2008JD009992.

Huffman, G. J., Bolvin, D. T., Nelkin, E. J., Wolff, D. B., Adler, R. F., Gu, G., Hong, Y., Bowman, K. P., & Stocker, E. F. (2007). The TRMM multi-satellite precipitation analysis: Quasi- global, multi-year, combined-sensor precipitation estimates at fine scale. *Journal of Hydrometeorology, 8*, 38–55. https://doi.org/10.1175/JHM560.1.

Kulie, M. S., & Bennartz, R. (2009). Utilizing spaceborne radars to retrieve dry snowfall. *Journal of Applied Meteorology and Climatology, 48*, 2564–2580. https://doi.org/10.1175/2009JAMC2193.1.

Kulie, M. S., & Milani, L. (2018). Seasonal variability of shallow cumuliform snowfall: A CloudSat perspective. *Quarterly Journal of the Royal Meteorological Society, 144*(S1), 329–343. https://doi.org/10.1002/qj.3222.

Kulie, M. S., Milani, L., Wood, N., Tushaus, S., Bennartz, R., & L'Ecuyer, T. (2016). A shallow cumuliform snowfall census using spaceborne radar. *Journal of Hydrometeorology, 17,* 1261–1279. https://doi.org/10.1175/JHM-D-15-0123.1.

Kummerow, C., Barnes, W., Kozu, T., Shiue, J., & Simpson, J. (1998). The tropical rainfall measuring Mission (TRMM) sensor package. *Journal of Atmospheric and Oceanic Technology, 15,* 809–817. https://doi.org/10.1175/1520-0426(1998)015<0809:TTRMMT>2.0.CO;2.

L'Ecuyer, T. S., & Jiang, J. H. (2010). Touring the atmosphere aboard the A-train. *Physics Today, 63*(7), 36–41. https://doi.org/10.1063/1.3463626.

Liu, G. (2008). Deriving snow cloud characteristics from CloudSat observations. *Journal of Geophysical Research, 113,* D00A09. https://doi.org/10.1029/2007JD009766.

Matrosov, S. Y., Shupe, M. D., & Djalalova, I. V. (2008). Snowfall retrievals using millimeter-wavelength cloud radars. *Journal of Applied Meteorology and Climatology, 47,* 769–777. https://doi.org/10.1175/2007JAMC1768.1.

McIlhattan, E. A., L'Ecuyer, T. S., & Miller, N. B. (2017). Observational evidence linking Arctic supercooled liquid cloud biases in CESM to snowfall processes. *Journal of Climate, 30,* 4477–4495. https://doi.org/10.1175/JCLI-D-16-0666.1.

Milani, L., Kulie, M. S., Casella, D., Dietrich, S., L'Ecuyer, T. S., Panegrossi, G., Porcù, F., Sanò, P., & Wood, N. B. (2018). CloudSat snowfall estimates over Antarctica and the Southern Ocean: An assessment of independent retrieval methodologies and multi-year snowfall analysis. *Atmospheric Research, 213,* 121–135. https://doi.org/10.1016/j.atmosres.2018.05.015.

Norin, L., Devasthale, A., L'Ecuyer, T. S., Wood, N. B., & Smalley, M. (2015). Intercomparison of snowfall estimates derived from the CloudSat cloud profiling radar and the ground-based weather radar network over Sweden. *Atmospheric Measurement Techniques, 8,* 5009–5021. https://doi.org/10.5194/amt-8-5009-2015.

Palerme, C., Kay, J. E., Genthon, C., L'Ecuyer, T. S., Wood, N. B., & Claud, C. (2014). How much snow falls on the Antarctic ice sheet? *The Cryosphere, 8*(4), 1577–1587. https://doi.org/10.5194/tc-8-1577-2014.

Palerme, C., Genton, C., Claud, C., Kay, J. E., Wood, N. B., & L'Ecuyer, T. S. (2017). Evaluation of current and projected Antarctic precipitation in CMIP5 models. *Climate Dynamics, 48,* 225–239. https://doi.org/10.1007/s00382-016-3071-1.

Panegrossi, G., Rysman, J.-F., Casella, D., Marra, A. C., Sanò, P., & Kulie, M. S. (2017). CloudSat-based assessment of GPM microwave imager snowfall observation capabilities. *Remote Sensing, 9*(12), 1263. https://doi.org/10.3390/rs9121263.

Sassen, K., & Wang, Z. (2008). Classifying clouds around the globe with the CloudSat radar: 1-year of results. *Geophysical Research Letters, 35,* L04805. https://doi.org/10.1029/2007GL032591.

Skofronick-Jackson, G., Petersen, W. A., Berg, W., Kidd, C., Stocker, E. F., Kirschbaum, D. B., Kakar, R., Braun, S. A., Huffman, G. J., Iguchi, T., & Kirstetter, P.-E. (2017). The global precipitation measurement (GPM) mission for science and society. *Bulletin of the American Meteorological Society, 98,* 1679–1695. https://doi.org/10.1175/BAMS-D-15-00306.1.

Skofronick-Jackson, G., Kulie, M., Milani, L., Munchak, S. J., Wood, N. B., & Levizzani, V. (2018). Satellite estimation of falling snow: A Global Precipitation Measurement (GPM) core observatory perspective. *Journal of Applied Meteorology and Climatology, 58,* 1429–1448. https://doi.org/10.1175/JAMC-D-18-0124.1.

Smalley, M., L'Ecuyer, T. S., Lebsock, M., & Haynes, J. (2014). A comparison of precipitation occurrence from the NCEP stage IV QPE product and the CloudSat cloud profiling radar. *Journal of Hydrometeorology, 15,* 444–458. https://doi.org/10.1175/JHM-D-13-048.1.

Stephens, G. L., Vane, D. G., Boain, R. J., Mace, G. G., Sassen, K., Wang, Z., Illingworth, A. J., O'Connor, E. J., Rossow, W. B., Durden, S. L., Miller, S. D., Austin, R. T., Benedetti, A., Mitrescu, C., & the CloudSat Science Team. (2002). The CloudSat mission and the A-train.

Bulletin of the American Meteorological Society, 83, 1771–1790. https://doi.org/10.1175/BAMS-83-12-1771.

Stephens, G., Winker, D., Pelon, J., Trepte, C., Vane, D., Yuhas, C., L'Ecuyer, T., & Lebsock, M. (2018). CloudSat and CALIPSO within the A-train: Ten years of actively observing the earth system. *Bulletin of the American Meteorological Society, 99*, 569–581. https://doi.org/10.1175/BAMS-D-16-0324.1.

Tanelli, S., Durden, S. L., Im, E., Pak, K. S., Reinke, D. G., Partain, P., Haynes, J. M., & Marchand, R. T. (2008). CloudSat's cloud profiling radar after two years in orbit: Performance, calibration, and processing. *IEEE Transactions on Geoscience and Remote Sensing, 46*, 3560–3573. https://doi.org/10.1109/TGRS.2008.2002030.

Wang, Y., Liu, G., Seo, E.-K., & Fu, Y. (2013). Liquid water in snowing clouds: Implications for satellite remote sensing of snowfall. *Atmospheric Research, 131*, 60–72. https://doi.org/10.1016/j.atmosres.2012.06.008.

Wood, N. B., L'Ecuyer, T. S., Vane, D., Stephens, G. L., & Partain, P. (2013). Level 2C snow profile process description and interface control document, V. P_R04, 21 pp. Available at http://www.cloudsat.cira.colostate.edu/sites/default/files/products/files/2C-SNOW-PROFILE_PDICD.P_R04.20130210.pdf. Last accessed 20 Nov 2018.

Wood, N. B., L'Ecuyer, T. S., Heymsfield, A. J., Stephens, G. L., Hudak, D. R., & Rodriguez, P. (2014). Estimating snow microphysical properties using collocated multisensor observations. *Journal of Geophysical Research, 119*, 8941–8961. https://doi.org/10.1002/2013JD021303.

Wood, N. B., L'Ecuyer, T. S., Heymsfield, A. J., & Stephens, G. L. (2015). Microphysical constraints on millimeter-wavelength scattering properties of snow particles. *Journal of Applied Meteorology and Climatology, 54*, 909–931. https://doi.org/10.1175/JAMC-D-14-0137.1.

Chapter 37
Snowfall Detection by Spaceborne Radars

Atsushi Hamada, Toshio Iguchi, and Yukari N. Takayabu

Abstract Algorithms to determine the thermodynamic phase state of precipitation observed from spaceborne radar are provided in this section. After briefly describing the classical methods to determine the thermodynamic phase of precipitation at the surface, some advanced methods to separate solid precipitation regions from liquid precipitation regions in the vertical profiles of radar measurements are described.

Keywords Precipitation · Snowfall · CloudSat · CPR · GPM · TRMM · DPR · PR · Microwave · Radar · Reflectivity · Dual-frequency · Mixed phase hydrometeors · Supercooled water · Graupel · Melting layer

37.1 Introduction

Precipitation plays a crucial role in the Earth's hydrological and energy cycles. Major types of the precipitation are rain and snow, and both are crucial components of the cycles over different regions and seasons. At the surface, snowfall basically occurs in the cold climate regions and highlands, whereas over the ocean in the tropics falling snow is observed in the upper troposphere associated with stratiform precipitation. These facts underscore the need to correctly measure solid precipitation, including snow, three-dimensionally over the entire globe.

Ground-based measurements obtained from meteorological stations and weather radars provide a quantitative information on snowfall. However, their observations are quite limited spatio-temporally, especially over open oceans. Recent spaceborne

A. Hamada (✉)
Faculty of Sustainable Design, University of Toyama, Toyama, Japan
e-mail: hamada@sus.u-toyama.ac.jp

T. Iguchi
National Institute of Information and Communications Technology (NICT), Koganei, Japan

Y. N. Takayabu
Atmosphere and Ocean Research Institute, The University of Tokyo, Chiba, Japan

© Springer Nature Switzerland AG 2020
V. Levizzani et al. (eds.), *Satellite Precipitation Measurement*, Advances in Global Change Research 69, https://doi.org/10.1007/978-3-030-35798-6_13

sensors onboard Earth-observing satellites, both active and passive sensors such as the Cloud Profiling Radar (CPR, Tanelli et al. 2008) onboard the CloudSat satellite (Stephens et al. 2008) and the Microwave Humidity Sounder (Bonsignori 2007) onboard the Metop satellites, can observe ice and snow particles in the atmosphere. They utilize high-frequency (higher than ~85 GHz) microwave channels that are sensitive to the scattering from ice and snow particles.

The use of recent spaceborne radars has produced a global climatology of precipitation according to its thermodynamic phase among liquid, solid, and mixed (Behrangi et al. 2012, 2014; Mülmenstädt et al. 2015; Kulie et al. 2016). Behrangi et al. (2012, 2014) used 3 years of CloudSat/CPR measurements to demonstrate the CPR superior capability in detecting light rainfall and snowfall both over land and ocean. The authors also derived the latitudinal distribution of the fraction of solid and mixed-phase precipitation to the total precipitation, where solid phase precipitation becomes dominant poleward of 45°–55° latitudes. Field and Heymsfield (2015) extended the analysis vertically by using vertical profiles of CloudSat measurements and indicated the importance of snow in the precipitation formation process, even in the tropics.

In this section, we focus on detection capability of snowfall from spaceborne precipitation radars and provide a summary of algorithms to determine the thermo-dynamic phase of precipitation and to separate solid precipitation regions from liquid precipitation regions in the vertical profiles of radar echoes.

37.2   Classical Methods to Determine Surface Precipitation Types

A simple and popular method to determine the phase of the hydrometeors at the surface is to use some ancillary information such as observations at surface meteo-rological stations or reanalysis datasets. The method dwells on the fact that the phase at the surface, rain or snow, basically depends on temperature and humidity near the ground. A more refined classification method has recently been proposed by Sims and Liu (2015) who suggested to incorporate the low-level vertical lapse rate and surface skin temperature as well as the wet-bulb temperature, which contains information both on air temperature and moisture. The method based on temperature and humidity has a long history in the development of global/regional rain-gauge-based precipitation datasets (Legates and Willmott 1990, Fuchs et al. 2001, Schneider et al. 2014, Yasutomi et al. 2011, Yatagai et al. 2012). Many datasets based on satellite passive microwave observations dwell on this simple method to discriminate between surface rainfall and snowfall. For example, the Global Satellite Mapping of Precipitation (GSMaP) (see Chap. 20 of volume 1 and Aonashi et al. 2009) product implements a technique based on the method by Sims and Liu (2015) to derive a probability of snowfall at the surface. The method performs quite well in discriminating between the regions where liquid or solid precipitation dominates.

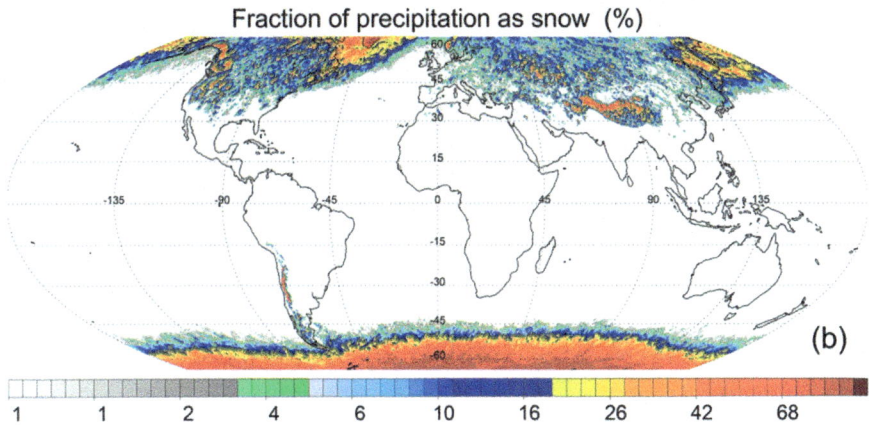

Fig. 37.1 Geographical distribution of the percentage of snow in the total precipitation. The grid box is 0.25 ° × 0.25 °. (Adapted of Adhikari et al. 2018)

This simple discrimination method using near-surface temperature and humidity can be easily applied to spaceborne radar precipitation datasets. Using the Precipitation Features (PFs) database (Liu et al. 2008) based on the dual-frequency precipitation radar (DPR) onboard the Global Precipitation Measurement (GPM) core observatory (Hou et al. 2014, Skofronick-Jackson et al. 2013) that was launched at the end of February 2014, Adhikari et al. (2018) derived a nearly global (65°S–65°N) distribution of near-surface (generally 0.5–1 km above the surface over the oceans; up to 2 km over land) snowfall rate and their fractions to the total precipitation (Fig. 37.1). They further compared the results with observations from the cloud profiling radar (CPR) onboard the CloudSat satellite to quantify the amount of snow missed by the Ku-band precipitation radar (KuPR) that is a component of the DPR. The GPM/KuPR is found to miss about a half of the global total snowfall. Casella et al. (2017) also conducted a similar analysis but more directly using coincident observations of GPM/DPR and CloudSat/CPR, and concluded that the fraction of snowfall amount detected by GPM/DPR is only 29–34% of the total.

37.3   Vertical Phase Distribution from Spaceborne Radars

Spaceborne precipitation radars can observe the vertical distribution of cloud and precipitation particles. Since radar echoes from solid precipitation are generally weaker and exhibit lower attenuation of electromagnetic waves than liquid precipitation, radars operating at a higher frequency, such as the cloud radar onboard CloudSat satellite, are more effective to detect solid precipitation. However, current spaceborne cloud radars are nadir-pointing and this imposes significant limitations in the sampling frequency. Precipitation radars such as the TRMM/PR and the GPM/DPR are less sensitive to solid precipitation, but still play a major role in describing the global distribution of solid precipitation. A widely accepted minimum detectable reflectivity of the TRMM/PR is around 18 dBZ, which is not sufficient to detect solid precipitation. In contrast, the minimum detectable reflectivity of the GPM/DPR is around 12 dBZ as a nominal value in a high-sensitivity scan mode (Hou et al. 2014; Toyoshima et al. 2015).

Several analyses using orbital data from the GPM/KuPR show that the GPM/KuPR may actually have higher sensitivity, around 12 dBZ, than the nominal value (Hamada and Takayabu 2016). Such a sensitivity enhancement has a significant impact on the detection of solid precipitation by spaceborne radars. Figure 37.2 shows a three-dimensional snapshot of mesoscale convective systems observed by GPM/KuPR near Costa Rica. Two panels show the same observation, but applying different threshold values and values below the specified thresholds are artificially masked out to demonstrate the impact of the enhancement in light precipitation detectability by GPM/KuPR compared to TRMM/PR. Bluish echoes in the middle to upper troposphere (Fig. 37.2a) refer to anvil clouds generally formed by ice-cloud

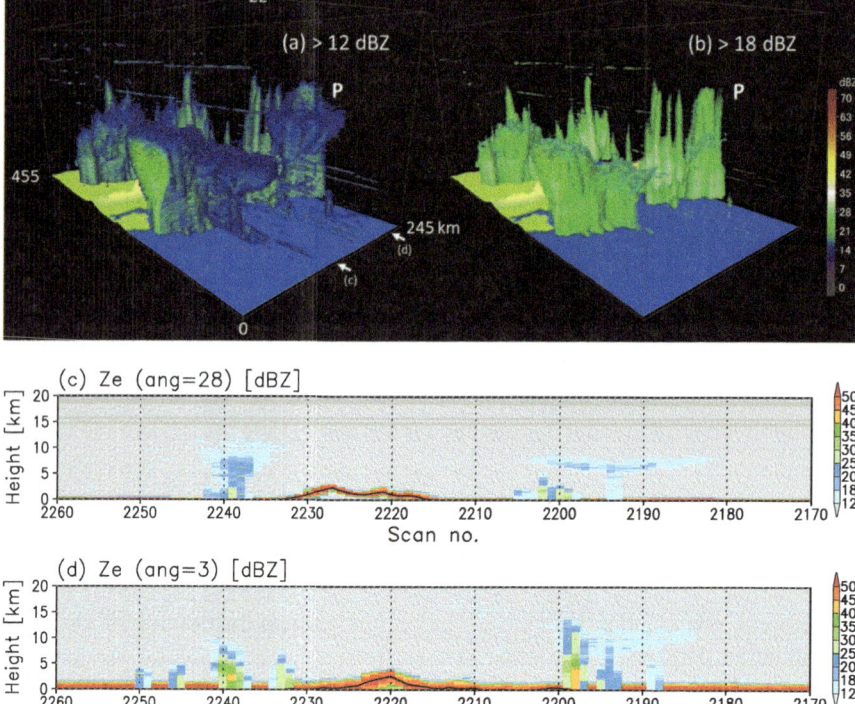

Fig. 37.2 Three-dimensional snapshot of GPM DPR observations near Costa Rica, 1113 UTC 1 Jul 2014, by setting different minimum detectable reflectivities: (**a**) 12 and (**b**) 18 dBZ. (**c**) Along-track cross section of reflectivity at the 28th angle bin [indicated by a white arrow in (**a**)]. The thin black line shows the topography contour. (**d**) Same as in (**c**), but at the third angle bin. (Adapted from Hamada and Takayabu 2016)

outflows from cumulonimbus (Houze 1993), containing mainly snow and cloud ice particles. When the minimum detectability is set to 18 dBZ, corresponding to the minimum detectable reflectivity of the TRMM/PR, only a gathering of isolated convective clouds is observed and the mesoscale organization is mostly masked out (Fig. 37.2b).

Another example, shown already in Fig. 1.2 in Chap. 1 of volume 1, is a case of snow bands over the Japan Sea. These latter typically form when strong cold air outbreaks from the Siberian Highs spread across the Japan Sea toward the coastal area of Japan. The formation process of these snow bands is similar to that of the so-called "lake-effect snow." These examples indicate that the GPM/DPR can effectively observe three-dimensional structure of various kinds of snowfall systems.

To retrieve vertical profiles of radar reflectivity and precipitation rates for the cases shown in Figs. 37.2 and 1.2 of volume 1, we have to consider the phase of precipitating particles at each vertical level. A simple way to separate the solid precipitation range from the liquid precipitation range is to use the vertical air temperature profile from an ancillary dataset (e.g., global reanalysis) and/or the thermodynamic phase of precipitation at the surface. For example, the GPM/DPR precipitation retrieval algorithm uses global objective analysis data to derive the 0 °C isotherm and to set physical variables such as density, dielectric constants, and falling velocity of falling hydrometeors (Iguchi et al. 2017).

The correct separation between the solid and liquid precipitation ranges in the observed vertical profiles is crucial, as there is a large difference in the scattering properties of electromagnetic waves between liquid and solid hydrometeors. Since the backscattering radar cross section of ice particle is less than one fourth of that of liquid water particle for the same diameter, a misclassification of the precipitation phase may introduce an order of magnitude errors in the estimated precipitation rates (Sims and Liu 2015). If a mixed-phase region such as a layer containing super-cooled water droplets in the updraft core of convective clouds exists, it may cause further problems.

Reanalysis datasets make available global vertical temperature and humidity profiles, but their reliability varies greatly depending on the availability of observations behind the reanalysis. In some regions such as open oceans, the errors in the 0 °C level derived from reanalysis datasets may be larger than the typical vertical resolutions of current spaceborne radars (range resolutions are 250 m for the GPM/DPR and 480 m for the CloudSat/CPR, and the corresponding vertical resolutions in the Level-2 dataset are 125 and 240 m, respectively). Therefore, a method that determines solid, liquid, and mixed-phase precipitation regions only from the radar measurements themselves is needed. However, the measurements from current spaceborne radars do not have direct information on whether the observed cloud and precipitation particles are liquid or solid.

If a bright band is observed in the vertical profile, we may easily separate between the solid and liquid precipitation ranges, because the bright band is typically located just below the 0 °C isotherm. However, convective precipitation normally has no bright band, and stratiform precipitation does not always develop a bright band.

Moreover, in the case of spaceborne radar measurements, bright bands located near the surface cannot be observed due to main lobe (surface) clutter (e.g., Liu 2008).

37.4 Use of Dual-Frequency Observations

Simultaneous radar measurements at multiple frequencies can improve the accuracy of the determination of solid precipitation ranges in the vertical profiles. Here we briefly account for algorithms developed for the GPM/DPR dual-frequency measurements. The GPM DPR consists of the two radars: the Ku-band radar (KuPR) and the Ka-band radar (KaPR). The KuPR radar has very similar specifications to the precipitation radar (PR) onboard the Tropical Rainfall Measuring Mission (TRMM) satellite, but it has a higher precipitation detectability threshold (Kojima et al. 2012, Hamada and Takayabu 2016). The KaPR was added to improve the retrieval of precipitation rates by increasing the precipitation detectability and by estimating the particle size more accurately than at single frequency.

By utilizing the difference in the scattering and attenuation properties of liquid and solid particles between Ku- and Ka-band electromagnetic waves, it is possible to estimate the mean diameter of precipitation particles provided an appropriate particle size distribution model is chosen. Since non-Rayleigh scattering is dominant at the higher-frequency for a given particle size, the difference in the equivalent radar reflectivity factor at the two frequencies can give information on particle size. The difference, which is called the dual-frequency ratio (DFR), is defined as follows:

$$\text{DFR} = Z_e(\text{Ku}) - Z_e(\text{Ka}) \propto \log_{10}\left[\frac{\int \sigma_{\text{Ku}}(D)N(D)dD}{\int \sigma_{\text{Ka}}(D)N(D)dD}\right] \qquad (37.1)$$

where Z_e is the equivalent radar reflectivity factor expressed in decibels (dBZ), $\sigma_{\text{Ku}}(D)$ and $\sigma_{\text{Ka}}(D)$ are back-scattering cross sections at Ku- and Ka-band, respectively, and $N(D)$ is a particle-size distribution function in terms of particle diameter D. Since snowflakes are generally larger than raindrops, DFR becomes much higher for ice or snow precipitation than for rain for a given value of $Z_e(\text{Ku})$ (Liao and Meneghini 2011). DFR measurements from the GPM/DPR has been used to detect the melting layer in the observed vertical echo profiles and has been implemented as one of the classification methods of precipitation type in the current GPM/DPR level 2 algorithm (Le and Chandrasekar 2013, Iguchi et al. 2017).

To obtain a reliable DFR value, we need to estimate Z_e accurately from the corresponding measured apparent radar reflectivity factors (Z_m) at the two bands. This requires highly accurate attenuation correction, which is generally very difficult to obtain, especially in the case of spaceborne radars. This is because we need to know the phase (solid or liquid) of the precipitating particles at each vertical level, which in turn is an information not available a priori. Another reason is the horizontal and vertical nonuniformity of precipitation within the radar footprint. Since the footprint size of TRMM/PR or GPM/DPR, around 5 km in diameter, is larger than the typical horizontal dimensions of a convective cloud, the effective attenuation may change drastically with different sub-pixel scale precipitation distributions by a non-linear effect, even if the measured Z_m profiles are identical. Considering all this framework, the current precipitation type classification method for GPM/DPR uses only the measured reflectivity factors, Z_m, and calculates a quantity called measured dual-frequency ratio (DFR_m):

$$DFR_m = Z_m(\text{Ku}) - Z_m(\text{Ka}) \text{ [dB]} \qquad (37.2)$$

Unlike DFR, DFR_m is affected by the differences in both scattering and attenuation properties of precipitation particles between the two frequencies, although the effect of attenuation is basically negligible for solid precipitation. Therefore, DFR_m may also take a large value when the attenuation in higher frequency echoes becomes significant such as in rainfall regions. Figure 37.3 shows vertical profiles

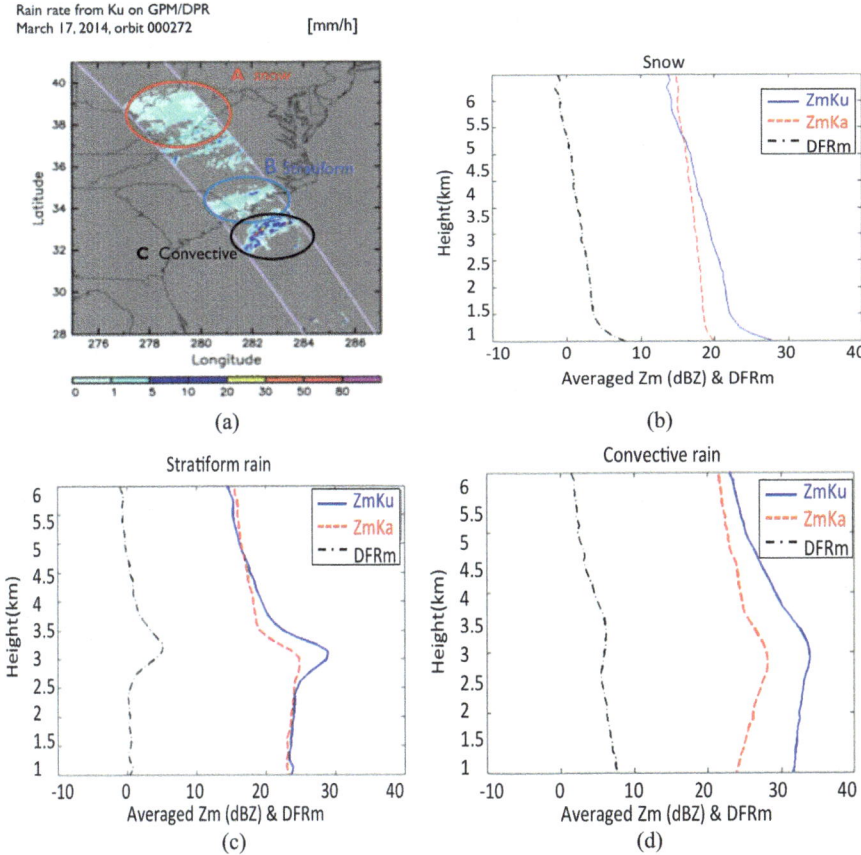

Fig. 37.3 (a) GPM DPR observation of rainfall rate on 17 March 2014 (orbit #000272). Circled areas A, B, and C represent snow, stratiform rain, and convective rain, respectively. (b) Averaged reflectivity profiles as well as dual-frequency ratio profile for snow. (c) Same as (b) for stratiform rain. (d) Same as (b) for convective rain. (Adapted from Le et al. 2017)

of Z_m at Ku- and Ka-band observed from GPM/DPR, and calculated DFR$_m$, for several types of surface precipitation (Le et al. 2017). For snow (Fig. 37.3a), the DFR$_m$ is several dBs, even when the Z_m at Ku-band remains relatively small, below 25 dBZ. The DFR$_m$ values increase toward the surface due to the size and accumulated effects of the scattering by the snow particles. Similar features can be seen in the vertical profiles for convective rainfall (Fig. 37.3d), but the Z_m values for both Ku- and Ka bands decrease downward below the melting layer (around 3 km) due to the attenuation by liquid precipitation particles. The DFR$_m$ values for stratiform precipitation show a downward increase above the melting layer, while they are near zero with slight downward increase below the melting layer because of the weak attenuation by light precipitation. These facts clearly indicate the usefulness of DFR$_m$ observations to detect solid precipitation regions at the surface. The information on surface precipitation types, called flagSurfaceSnowfall, has been implemented in the DPR level 2 products.

Iguchi et al. (2018) expand the use of the DFR$_m$ characteristics to detect the GPM/DPR footprints that contain intense ice precipitation including snow above -10 °C level. Their algorithm only considers the range bins with DFR$_m > 7$ dB and Z_m(Ku) > 27 dBZ to avoid the high uncertainty in the calculated Z_m around its minimum detectable value by GPM/KaPR reverberating into the DFR$_m$ values. Even though the algorithm is simple, the results correctly indicate the solid precipitation regions. A flag created by their algorithm, the flagHeavyIcePrecip, has been implemented in the DPR level 2 products since version 5. Figure 37.4a shows the percentage of observation pixels in which flagHeavyIcePrecip is set. Figure 37.4b shows the corresponding percentage in which the maximum value of Z_m exceeds 40 dBZ along the vertical above -10 °C level. This condition is generally met when there are enough graupels and cloud ice particles within strong updraft and implies lightning activity. The fact that the occurrence frequencies shown in Fig. 37.4b are much lower than those in Fig. 37.4a indicates that the method flagHeavyIcePrecip has a higher sensitivity to intense ice precipitation. Figure 37.4c is the same as Fig. 37.4a, but includes only the cases in which the surface air temperature is below ~1 °C. The distribution shows basically a similar pattern with that in Fig. 37.1, but Fig. 37.4c shows the frequencies of intense ice precipitation. (Note that the percentages shown in Fig. 37.4 are much lower than those shown in Fig. 37.1, because the former includes non-precipitating pixels in the calculation.) A relatively high occurrence is observed over the Atlantic Ocean both for mean and intense precipitation, whereas relatively lower occurrence is observed over the Antarctic Ocean.

$\longrightarrow$

Fig. 37.4 (continued) in which flagHeavyIcePrecip is set by condition A and the surface air temperature is below 1 °C. Percentages are calculated relative to the total number of the measured pixels that include no rain pixels. (Adapted from Iguchi et al. 2018)

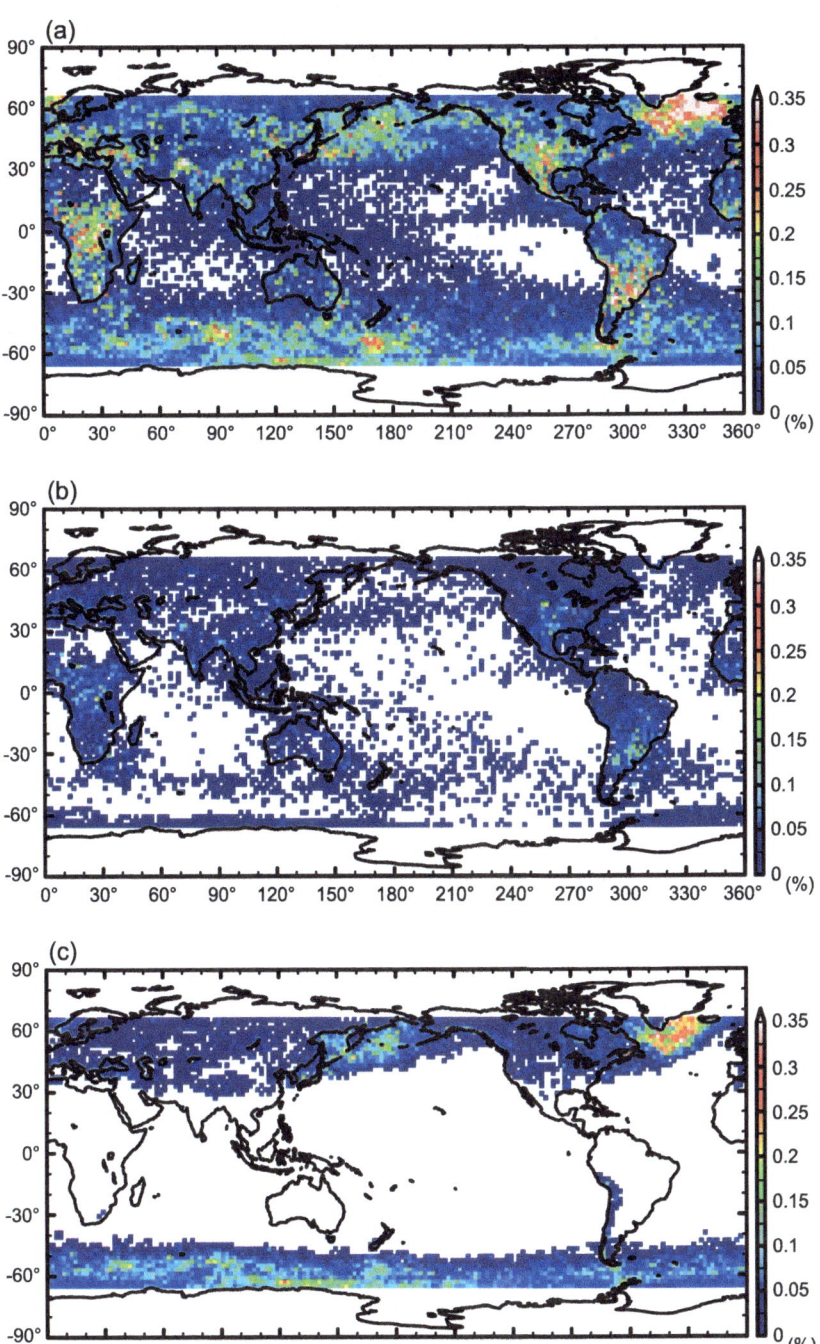

Fig. 37.4 Frequencies at which flagHeavyIcePrecip was set with different conditions over $2° \times 2°$ boxes in 2015. (**a**) Percentage of occurrences in which flagHeavyIcePrecip is set by "condition A [$DFR_m > 7$ dB, $Z_m(Ku) > 27$ dBZ above $-10°C$ level]". (**b**) Percentage of occurrences in which the maximum of Z_m of KuPR above the $-10°C$ height exceeds 40 dBZ. (**c**) Percentage of occurrences

37.5   Future Directions

Among the current spaceborne radars, the CloudSat/CPR seems to have the highest capability to detect solid precipitation, as the solid precipitation mostly consists of low-intensity snowfall. However, when the precipitation is intense or precipitating particles are large, the CloudSat/CPR can hardly quantify the precipitation amount, because non-Rayleigh scattering and multiple scattering pose problems at the operating frequency of the CPR (W-band; ~95 GHz). The GPM/DPR cannot observe very light precipitation including light snowfall, whereas it can quantify moderate to heavy precipitation and effectively detects intense ice precipitation. These results mean that spaceborne precipitation radars such as the GPM/DPR and the CloudSat/CPR are complementary and powerful instruments for the detection of solid precipitation, as well as liquid precipitation. A merging of the three-dimensional measurements of solid precipitation from the GPM/DPR and CloudSat/CPR can provide a unique and very useful dataset to study the ice microphysics such as precipitation formation processes and to improve the identification and quantification of solid precipitation by microwave radiometers.

References

Adhikari, A., Liu, C., & Kulie, M. S. (2018). Global distribution of snow precipitation features and their properties from 3 years of GPM observations. *Journal of Climate, 31*, 3731–3754. https://doi.org/10.1175/JCLI-D-17-0012.1.

Aonashi, K., Awaka, J., Hirose, M., Kozu, T., Kubota, T., Liu, G., Shige, S., Kida, S., Seto, S., Takahashi, N., & Takayabu, Y. N. (2009). GSMaP passive microwave precipitation retrieval algorithm: Algorithm description and validation. *Journal of the Meteorological Society of Japan, 87A*, 119–136. https://doi.org/10.2151/jmsj.87A.119.

Behrangi, A., Lebsock, M., Wong, S., & Lambrigtsen, B. (2012). On the quantification of oceanic rainfall using spaceborne sensors. *Journal of Geophysical Research, 117*, D20105. https://doi.org/10.1029/2012JD017979.

Behrangi, A., Tian, Y., Lambrigtsen, B. H., & Stephens, G. L. (2014). What does CloudSat reveal about global land precipitation detection by other spaceborne sensors? *Water Resources Research, 50*, 4893–4905. https://doi.org/10.1002/2013WR014566.

Bonsignori, R. (2007). The microwave humidity sounder (MHS): In-orbit performance assessment. In S. Habib et al. (Eds.), *Sensors, systems, and next-generation satellites XI* (International Society for Optical Engineering (SPIE proceedings)) (Vol. 6744, p. 67440A). https://doi.org/10.1117/12.737986.

Casella, D., Panegrossi, G., Sanò, P., Marra, A. C., Dietrich, S., Johnson, B. T., & Kulie, M. S. (2017). Evaluation of the GPM-DPR snowfall detection capability: Comparison with CloudSat-CPR. *Atmospheric Research, 197*, 64–75. https://doi.org/10.1016/j.atmosres.2017.06.018.

Field, P. R., & Heymsfield, A. J. (2015). Importance of snow to global precipitation. *Geophysical Research Letters, 42*, 9512–9520. https://doi.org/10.1002/2015GL065497.

Fuchs, T., Rapp, J., Rubel, F., & Rudolf, B. (2001). Correction of synoptic precipitation observations due to systematic measuring errors with special regard to precipitation phases. *Physics and Chemistry of the Earth (B), 26*, 689–693. https://doi.org/10.1016/S1464-1909(01)00070-3.

Hamada, A., & Takayabu, Y. N. (2016). Improvements in detection of light precipitation with the global precipitation measurement dual-frequency precipitation radar (GPM DPR). *Journal of*

Atmospheric and Oceanic Technology, 33, 653–667. https://doi.org/10.1175/JTECH-D-15-0097.1.

Hou, A. Y., Kakar, R. K., Neeck, S., Azarbarzin, A. A., Kummerow, C. D., Kojima, M., Oki, R., Nakamura, K., & Iguchi, T. (2014). The global precipitation measurement Mission. *Bulletin of the American Meteorological Society, 95*, 701–722. https://doi.org/10.1175/BAMS-D-13-00164.1.

Houze, R. A., Jr. (1993). *Cloud dynamics* (International geophysics series) (Vol. 53, 2nd ed.). Amsterdam: Academic. 496 pp, ISBN: 9780080921464.

Iguchi, T., Seto, S., Meneghini, R., Yoshida, N., Awaka, J., Le, M., Chandrasekar, V., & Kubota, T. (2017). *GPM/DPR level-2 algorithm theoretical basis document*, 68 pp. Available at http: // www.eorc.jaxa.jp/GPM/doc/algorithm/ATBD_DPR_201708_whole_1.pdf. Last accessed 6 Dec 2018.

Iguchi, T., Kawamoto, N., & Oki, R. (2018). Detection of intense ice precipitation with GPM/DPR. *Journal of Atmospheric and Oceanic Technology, 35*, 491–502. https://doi.org/10.1175/JTECH-D-17-0120.1.

Kojima, M., Miura, T., Furukawa, K., Hyakusoku, Y., Ishikiri, T., Kai, H., Iguchi, T., Hanado, H., & Nakagawa, K. (2012). Dual-frequency precipitation radar (DPR) development on the global precipitation measurement (GPM) core observatory. In H. Shimoda et al. (Eds.), *Earth observing missions and sensors: Development, implementation, and characterization II* (International Society for Optical Engineering (SPIE Proceedings)) (Vol. 8528, p. 85281A). https://doi.org/10.1117/12.976823.

Kulie, M. S., Milani, L., Wood, N. B., Tushaus, S. A., Bennartz, R., & L'Ecuyer, T. S. (2016). A shallow cumuliform snowfall census using spaceborne radar. *Journal of Hydrometeorology, 17*, 1261–1279. https://doi.org/10.1175/JHM-D-15-0123.1.

Le, M., & Chandrasekar, V. (2013). Hydrometeor profile characterization method for dual-frequency precipitation radar onboard the GPM. *IEEE Transactions on Geoscience and Remote Sensing, 51*, 3648–3658. https://doi.org/10.1109/TGRS.2012.2224352.

Le, M., Chandrasekar, V., & Biswas, S. (2017). An algorithm to identify surface snowfall from GPM DPR observations. *IEEE Transactions on Geoscience and Remote Sensing, 55*, 4059–4071. https://doi.org/10.1109/TGRS.2017.2687420.

Legates, D. R., & Willmott, C. J. (1990). Mean seasonal and spatial variability in gauge-corrected, global precipitation. *Quarterly Journal of the Royal Meteorological Society, 10*, 111–127. https://doi.org/10.1002/joc.3370100202.

Liao, L., & Meneghini, R. (2011). A study on the feasibility of dual wavelength radar for identification of hydrometeor phases. *Journal of Applied Meteorology and Climatology, 50*, 449–294. https://doi.org/10.1175/2010JAMC2499.1.

Liu, G. (2008). Deriving snow cloud characteristics from CloudSat observations. *Journal of Geophysical Research, 113*, D00A09. https://doi.org/10.1029/2007JD009766.

Liu, C., Zipser, E. J., Cecil, D. J., Nesbitt, S. W., & Sherwood, S. (2008). A cloud and precipitation feature database from nine years of TRMM observations. *Journal of Applied Meteorology and Climatology, 47*, 2712–2728. https://doi.org/10.1175/2008JAMC1890.1.

Mülmenstädt, J., Sourdeval, O., Delanoë, J., & Quaas, J. (2015). Frequency of occurrence of rain from liquid-, mixed-, and ice-phase clouds derived from A-train satellite retrievals. *Geophysical Research Letters, 42*, 6502–6509. https://doi.org/10.1002/2015GL064604.

Schneider, U., Becker, A., Finger, P., Meyer-Christoffer, A., Ziese, M., & Rudolf, B. (2014). GPCC's new land surface precipitation climatology based on quality-controlled in situ data and its role in quantifying the global water cycle. *Theoretical and Applied Climatology, 115*, 15–40. https://doi.org/10.1007/s00704-013-0860-x.

Sims, E. M., & Liu, G. (2015). A parameterization of the probability of snow-rain transition. *Journal of Hydrometeorology, 16*, 1466–1477. https://doi.org/10.1175/JHM-D-14-0211.1.

Skofronick-Jackson, G., Johnson, B. T., & Munchak, S. J. (2013). Detection thresholds of falling snow from satellite-borne active and passive sensors. *IEEE Transactions on Geoscience and Remote Sensing, 51*, 4177–4189. https://doi.org/10.1109/TGRS.2012.2227763.

Stephens, G. L., Vane, D. G., Tanelli, S., Im, E., Rokey, M., Reinke, D., Partain, P., Mace, G. G., Austin, R., L'Ecuyer, J. T., Haynes, M., Lebsock, K., Suzuki, D., Waliser, D. W., Kay, J., Gettelman, A., Wang, Z., & Marchand, R. (2008). The CloudSat mission: Performance and early science after the first year of operation. *Journal of Geophysical Research, 113*, D00A18. https://doi.org/10.1029/2008JD009982.

Tanelli, S., Durden, S. L., Im, E., Pak, K. S., Reinke, D. G., Partain, P., Haynes, J. M., & Marchand, R. T. (2008). CloudSat's cloud profiling radar after two years in orbit: Performance, calibration, and processing. *IEEE Transactions on Geoscience and Remote Sensing, 46*, 3560–3573. https://doi.org/10.1109/TGRS.2008.2002030.

Toyoshima, K., Masunaga, H., & Furuzawa, F. A. (2015). Early evaluation of Ku- and Ka-band sensitivities for the global precipitation measurement (GPM) dual-frequency precipitation radar (DPR). *SOLA, 11*, 14–17. https://doi.org/10.2151/sola.2015-004.

Yasutomi, N., Hamada, A., & Yatagai, A. (2011). Development of a long-term daily gridded temperature dataset and its application to rain/snow discrimination of daily precipitation. *Global Environmental Research, 15*, 165–172. Available at http: //www.airies.or.jp/attach.php/ 6a6f75726e616c5f31352d32656e67/save/0/0/15_2-10.pdf. Last accessed 5 Apr 2019.

Yatagai, A., Kamiguchi, K., Arakawa, O., Hamada, A., Yasutomi, N., & Kitoh, A. (2012). APHRODITE: Constructing a long-term daily gridded precipitation dataset for Asia based on a dense network of rain gauges. *Bulletin of the American Meteorological Society, 93*, 1401–1415. https://doi.org/10.1175/BAMS-D-11-00122.1.

Chapter 38
On the Duration and Life Cycle of Precipitation Systems in the Tropics

Rémy Roca, Dominique Bouniol, and Thomas Fiolleau

Abstract After decades of active research, knowledge of the precipitating meso-scale convective systems has made strong progress based on many regional inves-tigations. Yet a tropical-wide perspective on these systems is only recently emerging thanks to satellite observations. This chapter is dedicated to this tropical view with an emphasis on the duration and the life cycle of precipitating system that are physically sound characteristics of the organization of the system. The more organized systems are dominating the contribution to the precipitation totals despite their low occur-rence. It is shown that the systems lasting more than 12 h correspond up to 70% of the precipitation. At the system scale, the recent satellite observations dedicated to precipitation allow for a new appraisal of the conceptual model of mesoscale convective system (MCS) initiation, mature stage and dissipation. A preliminary quantification of the precipitation, latent heating and radiative cooling evolution within the life cycle of the MCS suggests the importance of the latent heating process in the first half of the system lifetime. In the second part, the radiative cooling needs to be accounted for to draw a complete picture of the energy budget of the systems.

Keywords Precipitation · Tropics · MCS · Water vapor · Convection · Deep clouds · GEO · LEO · TRMM · PR · CloudSat · CPR · Megha-Tropiques · GPM · Infrared · Microwave · Brightness temperature · TOOCAN · Climatology · ITCZ

R. Roca (✉) · T. Fiolleau
CNRS/LEGOS, Toulouse, France
e-mail: remy.roca@legos.obs-mip.fr

D. Bouniol
Météo France, CNRM/GMAP/OBS, Toulouse, France

© Springer Nature Switzerland AG 2020   729
V. Levizzani et al. (eds.), *Satellite Precipitation Measurement*, Advances in Global
Change Research 69, https://doi.org/10.1007/978-3-030-35798-6_14

38.1 Introduction

Precipitation at the surface results from the net condensation of the water vapor in the atmospheric column. The local condensation depends upon the thermodynamic and dynamic structure of the atmosphere. At the global scale, the conservation laws of water and energy impose a strong constraint on the future evolution of global precipitation (Stephens and Ellis 2008, O'Gorman et al. 2012). At the regional scales, the energy constraints are overwhelmed by the physics of the precipitation formation (Biasutti et al. 2018). In the tropics, atmospheric deep convection is the main source of local condensation.

Indeed, in the Tropics, deep convection is ubiquitous and spans a wide range of organization types from isolated deep convective cells to deep cells embedded within mesoscale structures as revealed by the earlier satellite imagery of the late 60's. The conceptual model of the individual convective cell life cycle and of the resulting precipitation has served as the basis for that of larger, organized systems (Redelsperger 1997). Decades of research have consolidated the broad concept of mesoscale convective systems (MCS) as an overarching model for all sorts of organized convective clouds (Houze 2018). The model of the life cycle and precipitation structure breaks down in three phases (e.g., Houze 1982). The initiation phase corresponds to the early part of the system life and is associated with a small extension of deep convective cells aggregating together. Precipitation at this stage is of convective nature. In the mature phase, the system grows in dimension and the complex interaction between old convective cells and precipitation with the thermodynamics and mesoscale circulation shapes a large, well organized system. The convective precipitation is still intense over a small area and is completed by a large stratiform precipitating zone as well as a non-precipitating part, both forming the anvil clouds. Finally, the system fades out and deep convection stops, cutting the feeding of the stratiform rainfall that also eventually vanishes. The non-precipitating clouds also dissipate but at a slower pace.

While a strong conceptual understanding of these precipitating systems has been acquired over the years mainly from ground-based radars, field campaigns and mesoscale modelling, a tropical wide perspective emerged only recently, mainly thanks to the improvements of the space-based observations of deep clouds and precipitation (Houze et al. 2015) and, in parallel, thanks to novel processing and retrieval algorithms.

In this chapter, we review the precipitating systems in the tropics with emphasis on the importance of duration and life cycle. After a brief technical introduction, the progress of understanding the contribution to the tropical water budget at the tropics scale is summarized. Then we explore recent efforts based on merging microwave (active and passive) data and infrared (IR) observations for revisiting the life cycle conceptual model with a tropical-wide dimension.

38.2   The Climatology and Morphology of the MCS

38.2.1   *Background and Robust Features*

The first step in assessing the structure of precipitating systems consists in the elaboration of their climatology and related precipitation characteristics. A review of the various techniques developed for the identification and tracking of the MCS is not relevant for this chapter. Instead we briefly recall the principles and review the existing tropical wide datasets. We further detail one of the approaches that has been developed and used by the Megha-Tropiques team and can serve as the basis of the chapter.

The identification of the cluster that composes the MCS and the tracking of the clusters in time throughout their life cycle are usually performed in two separate steps using any temporally resolved dataset. The most used is IR imagery from geostationary (GEO) platforms. A threshold or a combination of thresholds applied to brightness temperature (e.g., Maddox 1980, Boer and Ramanathan 1997) is used to delineate the cluster at one time step. Then the tracking step relies on an overlap assumption between consecutive time steps (e.g., Williams and Houze 1987). A review of the various IR-based techniques can be found in Fiolleau (2010). The recent development of the TOOCAN algorithm (Fiolleau and Roca 2013a) unifies the two steps into a single segmentation approach that combines time and space. In particular, the split/merge artefacts common to the overlap-based methods is removed.

An alternative description of convective clouds based on a statistical approach ("weather regimes") has also been developed to link the degree of organization of convection to precipitation (Lee et al. 2013). More recently a tropical wide analysis of space/time precipitation features was extracted, using a different set of assumptions and definitions, from gridded satellite and reanalysis precipitation products (White et al. 2017) to gridded low resolution IR archive (Dias et al. 2012).

However, the patterns of the geographical distribution of the MCS are robust with respect to used method and data and are associated with the large scale tropical well-known precipitation dynamic and precipitation features (ITCZ, warm-pool, Amazon basin, monsoon, etc.) as illustrated in Fig. 38.1.

In spite of different techniques (with various implementation and input observations), of the large spectrum of definitions and of a significant corpus of regional studies, robust climatological features nevertheless emerge at the tropical-wide scale:

- Short lived systems are more frequent than long lived ones, although their contribution to the total tropical precipitation is not dominant.
- Systems can last from a couple of hours up to a couple of days.
- Systems last longer more frequently over ocean than over land.
- The initiation of continental systems exhibits a strong diurnal cycle and is often linked to orography.

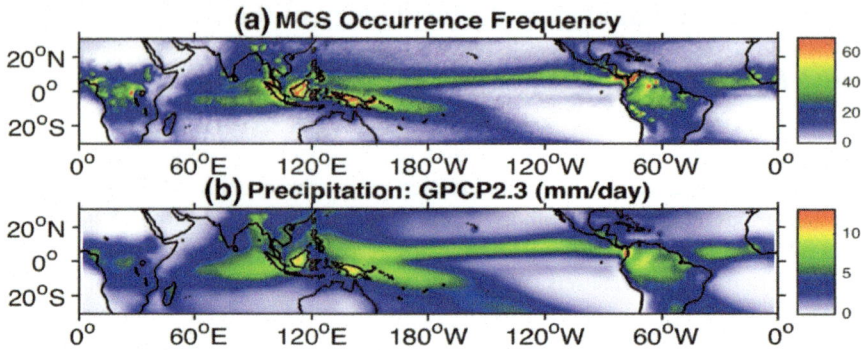

Fig. 38.1 (a) Climatology of occurrence of MCS (b) Climatology of precipitation. (Adapted from Huang et al. 2018)

These robust features have led, alongside specific studies, to the broad concept of sustainability of the explanation of the climatological distribution of the system duration. This energetic perspective suggests that over the ocean, where the moist and warm boundary layer provides a large source of energy, size growth and duration are not limited. Over land, the diurnal cycle of the boundary layer energetics prevents systems from persisting during night time, yielding to an overall shorter duration distribution (Houze 2004).

Owing to the various definition of the MCS, more quantitative aspects of the climatology and concept need to be based on the selection of a given database. In the following, quantitative aspects are illustrated using the results of the TOOCAN algorithm.

38.2.2 TOOCAN Specific Features of Tropical MCS

The TOOCAN algorithm operates on a time sequence of IR images to identify and track convective systems. The algorithm does not use the traditional detection and tracking steps, but a single three-dimensional segmentation step (space and time). The objective is to associate the convective core of an MCS to its anvil cloud in the spatiotemporal domain and to decompose the cold cloud shield in several convective systems in the three-dimensional domain. The algorithm relies on an image processing technique called region growing, which progresses from the convective core to the cloud edges. Adjacent pixels in a time series of IR images are assumed belonging to the same system, and that the optical depth of the cloudiness decreases away from the convective core to the cloud edges in the spatiotemporal domain. TOOCAN applies an iterative process of detection and growth identification of convective seeds at different temperature thresholds in the spatiotemporal domain, so that convective systems with various depths can be identified and characterized during their life cycles. This multistep, multi-threshold technique can be seen as a 3D extension of the detection and spread approach (Boer and Ramanathan 1997). The iterative process can be decomposed in three stages. Individual seeds in the three-dimensional space are first detected at a low brightness temperature threshold. Then, an intermediate cold cloud shield mask is identified at a 5 K warmer threshold. A stepwise growth of the convective seeds is finally performed until reaching the intermediate cold cloud shield boundaries. This step consists in adding edge pixels belonging to the cold cloud shield mask to all already identified convective seeds. The growth step is performed via successive expansion operations thanks to a 10-connectivity operator (8 pixels in space and 2 in time). This iterative process of detection and growing starts with a detection at a 190 K threshold and is repeated until each of the pixels colder than 235 K is associated to an individual system. Note that the segmentation of the convective systems does not depend on a given brightness temperature threshold, on a minimum area threshold.

As a result, the TOOCAN methodology overcomes the main issues of the "area-overlapping" methods by suppressing the split/merge artifacts implying a smoother evolution of the size of the cloud shield of the system during its life cycle. All geostationary images obtained by the operational meteorological geostationary satellite fleet at full space and time resolution over the 30°S–30°N latitude belt in the boreal summer months (June–September) for the 3 years 2012, 2013, and 2014 have been processed (Roca et al. 2017). Note that an extended database aiming to enhance the homogenization of the various geostationary platforms is under development.

Figure 38.2 shows the distribution of the MCS duration documenting a mode of the PDF occurring between 4 and 5 h both for oceanic and continental systems. In spite of the fact that short duration systems are the most frequent (about 50% of the MCS lasting less than 5 h), they only contribute about 10% to the cold cloud amount. In contrast, about 70% of cold clouds is explained by systems lasting longer than 10 h over land and 8 h over the ocean (Fig. 38.2c). MCSs can reach sizes up to 1.4×10^6 km^2 but a wide majority of the systems has sizes between 1×10^3 and 2×10^5 km^2.

38.3   Precipitation and MCS Duration

38.3.1   The Precipitation Totals and the Degree of Organization of Convection

A large number of studies have attempted to relate various metrics of the degree of organization of deep convection to the rainfall distribution, either adopting a static, instantaneous perspective (Nesbitt et al. 2006, Liu 2011, Liu and Zipser 2015, Yuan and Houze 2010) or a weather state approach (Jakob and Schumacher 2008, Lee et al. 2013, Duncan et al. 2014). The robust features among these various analyses is that the most organized systems, while having a relatively low occurrence, do

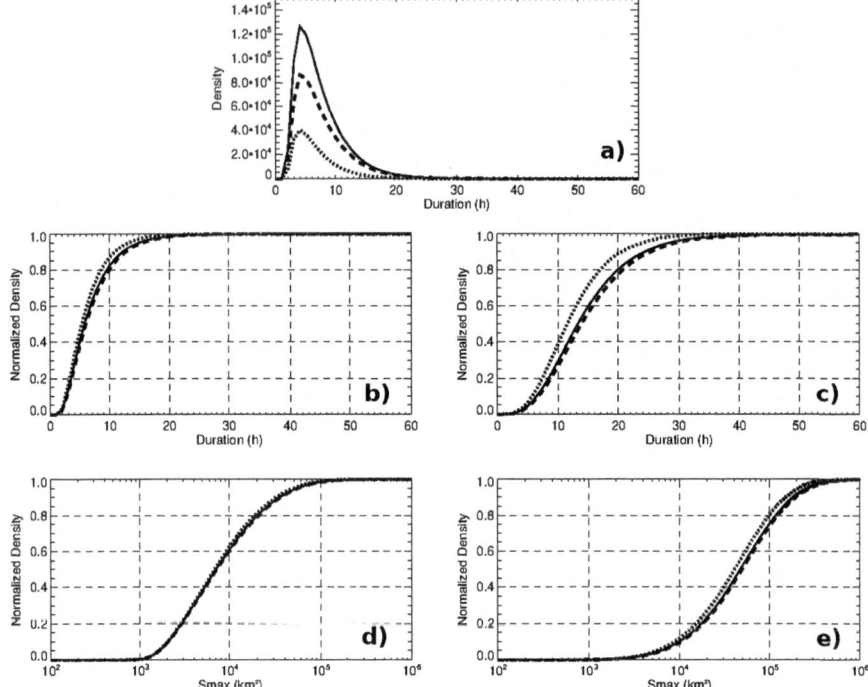

Fig. 38.2 Duration statistics for the 2012–2014 summer seasons. (**a**) Duration PDF, (**b**) duration CDF, and (**c**) duration CDF weighted by the MCS cold cloudiness. Maximum extent (**d**) CDF and (**e**) CDF weighted by the MCS cold cloudiness. Shown are the entire tropical belt (solid), continent (short dashed), and ocean (long dashed). (Adapted from Roca et al. 2017)

contribute to a large fraction of the total tropical precipitation. This key statement appears not only to characterize the integrated distribution of precipitation but also its extremes (Rossow et al. 2013) as well as the recent climatic observed precipitation trends (Tan et al. 2015).

The importance of the time dimension of the system has been relatively less investigated. Again, while the importance of the MCS to the tropical-wide precipitation is well established, a quantitative relationship to the morphology of the systems has only been recently explored.

38.3.2 The Precipitation Totals and the System Duration

The TOOCAN algorithm results are further used to quantify the role of the duration in determining the precipitation totals. The systems lasting up to 12 h are responsible for up to 20–30% of the precipitation totals (Fig. 38.3). In line with earlier generic properties of the systems, this contribution is higher over land (~40%); systems lasting up to 1 day reach in this way an 80% contribution mark. This confirms the

Fig. 38.3 (top) CDF of the tropical rainfall as a function of duration, and (bottom) as a function of duration. Solid lines are for all the tropics, long dashed for land and short dashed for ocean conditions. Black refers to the TAPEER 1.0 product (Chambon et al. 2013), blue to the 3B42v7 (Huffman et al. 2007) and red to the GPCP 1DD v1.2 (Huffman et al. 2001). (Adapted from Roca et al. 2014)

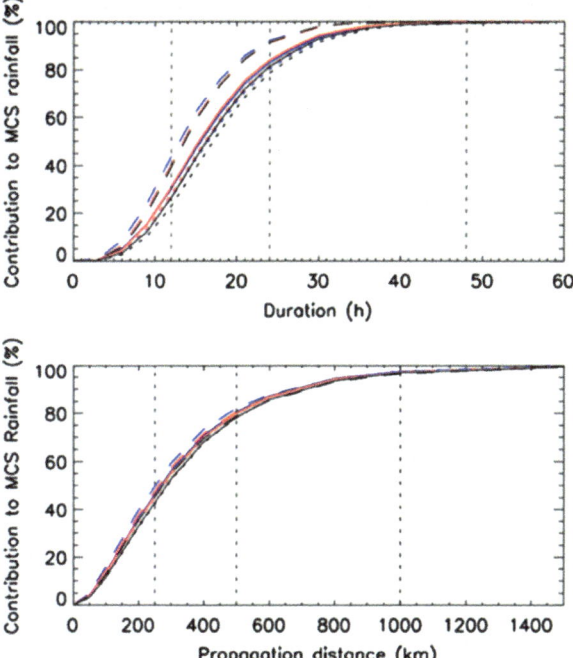

importance of the contribution of the low-occurrence/long-duration systems to the water budget of the tropics and provides a physically clear and sound metric (the duration) for the quantification and completes the static statistical perspective mentioned earlier. Note also that these quantitative results are robust with respect to the selection of the gridded precipitation product.

Another physically defined metric of organized convection linked to the life cycle is the propagative nature of the systems that is also illustrated in Fig. 38.3. The systems travelling more than 250 km contribute to 50% of the precipitation totals and the 80% mark is reached for systems propagating up to 500 km. Note that no contrast between land and ocean is observed in this case.

The duration/propagation properties of the MCSs in the tropics are well related to the water budget and climate models do not yet represent explicitly any of these important physical processes (Mapes and Neale 2011). This long-lasting problem should benefit from this recent enhanced depiction of the role of organized convection in the water cycle.

38.4   Precipitation and the MCS Life Cycle

MCS and their associated ice cloud anvils dominate rainfall and cloudiness over much of the Tropics, and even though the importance of MCSs is well understood, there is much left to learn about their growth, development and decay phases.

Indeed, active and passive instruments on low Earth orbiting (LEO) satellites allow for a comprehensive identification of the involved processes from space. It must be considered that these sensors provide a very limited number of instantaneous observations of each convective system, generally one, but up to six for the Megha-Tropiques satellite (Fiolleau and Roca 2013b). This fact prevents studies of the evolution of individual convective events. However, by matching such observations with life cycle information from a database of tracked MCS, it is possible to build up a composite picture of the evolution of the vertical structure of the MCS and its associated properties (surface rain, radiation, flash rate, latent and radiative heating, etc.).

Prior to merging the LEO information to build a composite perspective, it is important to clarify the various life stages of the cold cloud shield forming the MCS. Several approaches have been conceived to build composite life cycles (McAnelly and Cotton 1989, Futyan and Del Genio 2007, Mapes et al. 2009). In the following, the results of the TOOCAN algorithm are once more used to explore in detail the precipitation during the life cycle of the MCS.

38.4.1 The Linear Growth/Decay Model for the Cold Cloud Shield Life Cycle

The analysis of the TOOCAN statistics has revealed that, in spite of the apparent complexity, the life cycle of an MCS cold cloud shield can be simplified to a few degrees of freedom (Roca et al. 2017). In particular most of the systems are characterized by a two-stage growth/decay model for their life cycle, with the maximum extent reached at the middle of their life. This growth/decay can further be assumed to be linear within the life cycle.

Figure 38.4 shows an example of such cold shield evolution over its life cycle. This background time evolution of the cold cloudiness is important to set the context for the composite results. The inspection of this model for various MCS types shows that the simple description of the time evolution of the MCS shield explains up to 80% of the tropical cold cloudiness. Furthermore, the suitability of this growth/decay model is independent of geography, duration, maximum extent and propagation distance of the systems, making it a ubiquitous feature of tropical MCSs.

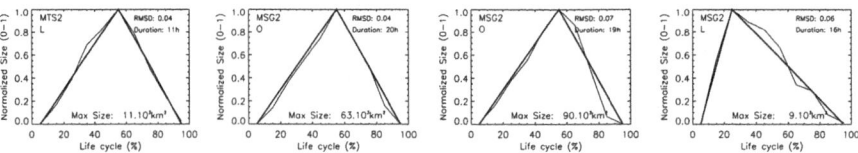

Fig. 38.4 Four examples of convective cold shield time evolution. (Adapted from Roca et al. 2017)

38.4.2   Compositing GEO and LEO Along the Life Cycle

Several studies make use of composite approaches for segmenting the MCS sample according to life duration or using normalized life cycle for accumulating information on different MCS properties. Jirak et al. (2003), Futyan and Del Genio (2007), Kondo et al. (2006) and Fiolleau and Roca (2013b) merged instantaneous surface rain estimation with tracked MCS life cycle. They found that the average rainfall rate reaches a maximum very early in the life cycle with the presence of convective cells. As the system continues to grow, the average rainfall rate decreases implying an increase in stratiform precipitation. However, Futyan and Del Genio (2007) note that over ocean the life cycle is less well defined and the stratiform rain fraction is essentially constant throughout its life cycle. Fiolleau and Roca (2013b) analyze the convective fraction and find twice as much convective rain in the continental systems than in their oceanic counterparts. Schumacher and Houze (2003, 2006) suggested that the higher stratiform rain fraction observed over oceanic regions is due to a greater convective sustainability. McAnelly and Cotton (1989) found that the smaller systems tend to be the least rainy.

The time evolution of the intensity, height penetration and mass flux of deep convection, the transport of ice particles from the convective to the stratiform region, and the subsequent microphysical processes are important characteristics of the internal MCS dynamics. One way to better understand the processes explaining differences in rain evolution with the life cycle and the physics behind the symmetrical evolution of the cold cloud shield is to make use of radar observations from TRMM and CloudSat (Futyan and Del Genio 2007, Mapes et al. 2009, Inoue et al. 2009, Bouniol et al. 2016, Imaoka and Nakamura 2012). Imaoka and Nakamura (2012) document the evolution of the TRMM PR reflectivity profile for MCSs over the Maritime Continent, focusing on MCSs lasting less than 5 h to incorporate in the statistics only MCSs completely enclosed in the radar swath. They found that the longer lasting systems tend to have higher maximum reflectivity values. Futyan and Del Genio (2007) examined TRMM PR reflectivity profiles over the African continent and the adjacent Atlantic Ocean. They obtained similar radar reflectivity profiles for convective cells during the mature stages for small and large MCS suggesting that the depth of convection does not directly control the horizontal extent of the system.

The use of the 94 GHz measurements of the CloudSat CPR, even if unable to determine the surface rain because of attenuation and multiple scattering in deep convective clouds, helps greatly in understanding the microphysical processes at the various stages of the life cycle of the different MCS sub-regions. Figure 38.5 shows the 94 GHz reflectivity profiles aggregated as a function of the life step and MCS sub-region for the same regions as those of Futyan and Del Genio (2007). The inter-quartile range is also documented in order to characterize the variability of the reflectivity profile at a given step/altitude/region corresponding to a mix of either different hydrometeor types and/or particle size distributions. A decrease in the magnitude of reflectivity is observed from the convective sub-region up to the cirriform anvil. Strong updrafts at the beginning of the life cycle are able to generate large hydrometeors that are lifted up to a high altitude and even detrained to the

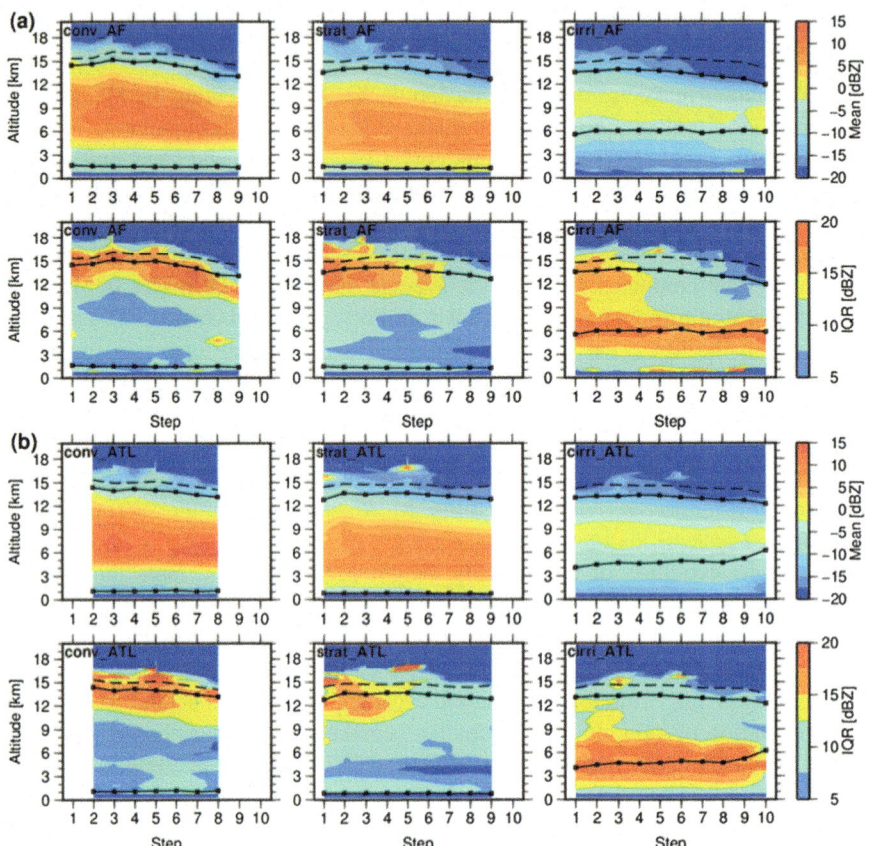

Fig. 38.5 Evolution of the mean profile of reflectivity (first row) and of the reflectivity interquartile range (second row) for (**a**) continental West Africa (AF) and (**b**) Atlantic Ocean (ATL). Solid upper line with squares and dashed line are the mean cloud-top height obtained from radar and radar–lidar, respectively. The lower line with squares is the mean cloud base. (Adapted from Bouniol et al. 2016)

stratiform sub-region, and a signature is also found in the nonprecipitating anvil. However, the convective intensity weakens after the first half of the life cycle. In contrast, for oceanic MCSs, the reflectivity is not as high in altitude, but its magnitude remains in the same range throughout the entire life cycle. Futyan and Del Genio (2007) obtained the same result for the mature stage of MCS using the TRMM PR reflectivity profile. The signature of detrainment from the convective towers is found only in the stratiform region for oceanic MCS.

The surface rain of an MCS is proportional to the net latent heat into the atmosphere and the radiative heating controlled by the ice cloud in the upper troposphere (Houze 2018). The diabatic heating profiles and their variability play an important role in the organization of MCSs through the generation of vorticity (Hagos and Zhang 2010). Figure 38.6 shows the evolution of convective and

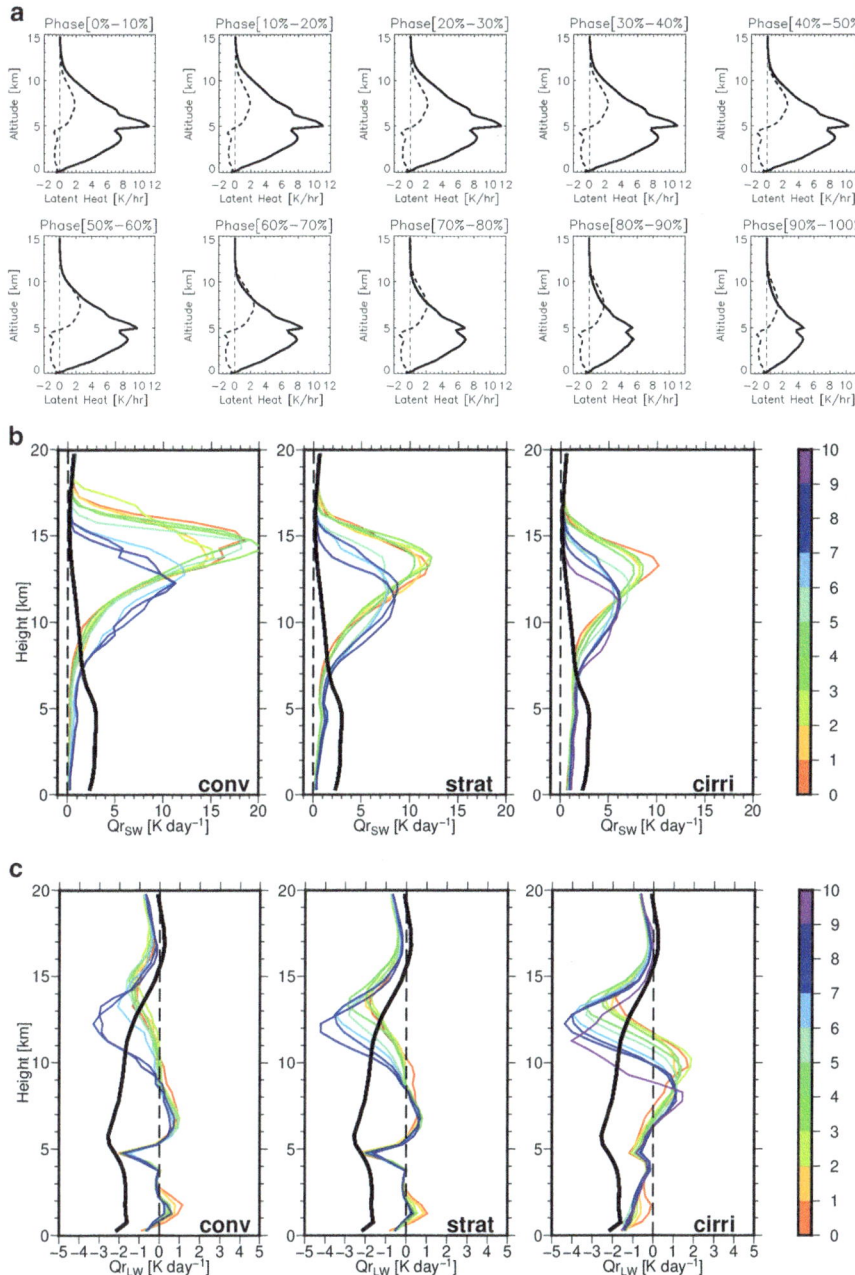

Fig. 38.6 Evolution of (**a**) latent heating profile in K h^{-1}, (**b**) shortwave radiative heating profiles at 0130 LT in K day^{-1} and (**c**) longwave radiative heating profiles in K day^{-1} for the different MCS sub-regions over the life cycle of a continental MCS. Each panel in (**a**) corresponds to a life step with convective heating in solid line and stratiform heating in dashed line. In (**b**) and (**c**) each panel corresponds to an MCS sub-region (convective, stratiform and cirriform), the color level showing the life stages. The black line shows the clear sky radiative heating profile

stratiform latent heat profiles of Shige et al. (2004, 2007) and the radiative heating profile from Henderson et al. (2013) composited according to the life cycle of African continental MCS for the different MCS sub-regions. For the convective sub-region, the latent heating profile is shaped by condensation and sublimation of water vapor in the intense updrafts leading to a positive "bottom-heavy" profile (Schumacher et al. 2004). Maximum heating rates appear at the early stages and decrease in the second part of the life cycle. In the stratiform sub-region, the latent heating profile remains positive above the melting layer and becomes negative below ("top heavy") because of stratiform precipitation evaporation processes. The cooling below the stratiform regions tends to feed cold pools that are important contributors for maintaining the MCS and triggering new convective cells. Compared with temporal variations of convective profiles, those in stratiform heating profiles are smaller.

The corresponding radiative heating profiles are shown in Fig. 38.6b, c for the shortwave (at 0130 LT) and longwave domain, respectively. In the shortwave domain the radiative heating is positive near cloud top. In particular, for stratiform and cirriform profiles this tends to reinforce the "top heavy shape". Note that such composite is obtained from a 0130 LT sampling and that this contribution of the shortwave radiative heating has to be scaled with the diurnal cycle. In the longwave domain, the radiative heating is negative at the top of the denser part of the cloud (Fig. 38.5) with a cooling effect that increases as the reflectivity decreases. Close to the anvil base, a positive contribution is found. This dipole of cooling and heating in the anvil may help in maintaining the mesoscale circulation (in particular during night time) explaining its longevity (Webster and Stephens 1980, Jensen and Del Genio 2003). Li et al. (2013) show that the radiative heating is an order of magnitude smaller than the latent heating at the regional scale. Figure 38.6 shows that this difference holds for heating profiles associated with an MCS. However, non-precipitating anvil clouds associated with an MCS have a lifespan far longer than the intense convective processes and therefore cannot be neglected even at the regional scale. The total heating profiles and their temporal variations are considerably affected by the area ratio of convective, stratiform and cirriform sub-region size and its temporal changes. Muller and Bony (2015) also demonstrated how the contrast between the MCS radiative heating and its dry environment is necessary to maintain aggregation.

38.5 Conclusions

After decades of progress in understanding the mechanisms of the tropical precipitation systems with field campaigns, mesoscale modelling and a large amount of regional observational studies, a tropical-wide perspective on these systems has emerged recently. Thanks to LEO satellite observations (TRMM, CloudSat, Megha-Tropiques, GPM, etc.), GEO imagery and renewed algorithms, the established weight of the organized systems in the precipitation totals and extremes, and within the recent climatic trends of tropical precipitation are now quantified. The overwhelming role of the systems lasting longer than 12 h is a strong outcome of these analyses that should guide climate model development to include parameterizations of the life cycle relevant variables, which are now absent in the models (Tapiador et al. 2018).

At the scale of the life cycle of the MCS, compositing LEO and GEO observations also prompts for revisiting the conceptual model of the system evolution. These studies suggest that both surface precipitation and latent heating distribution are dominant for the first half of the system lifetime and then radiation is contributed significantly to the energy budget of the MCS.

The future is promising in terms of extending these precipitation-related features to the whole energy and water budget elements and further linking the mesoscale morphological characteristics of the systems to the global climate energetics.

References

Biasutti, M., Voigt, A., Boos, W. R., Braconnot, P., Hargreaves, J. C., Harrison, S. P., Kang, S. M., Mapes, B. E., Scheff, J., Schumacher, C., Sobel, A. H., & Xie, S.-P. (2018). Global energetics and local physics as drivers of past, present and future monsoons. *Nature Geoscience, 11*, 392–400. https://doi.org/10.1038/s41561-018-0137-1.

Boer, E. R., & Ramanathan, V. (1997). Lagrangian approach for deriving cloud characteristics from satellite observations and its implications to cloud parameterization. *Journal of Geophysical Research, 102*, 21383–21399. https://doi.org/10.1029/97JD00930.

Bouniol, D., Roca, R., Fiolleau, T., & Poan, D. E. (2016). Macrophysical, microphysical, and radiative properties of tropical mesoscale convective systems over their life cycle. *Journal of Climate, 29*, 3353–3371. https://doi.org/10.1175/JCLI-D-15-0551.1.

Chambon, P., Jobard, I., Roca, R., & Viltard, N. (2013). An investigation of the error budget of tropical rainfall accumulation derived from merged passive microwave and infrared satellite measurements. *Quarterly Journal of the Royal Meteorological Society, 139*, 879–893. https://doi.org/10.1002/qj.1907.

Dias, J., Tulich, S. N., & Kiladis, G. N. (2012). An object-based approach to assessing the organization of tropical convection. *Journal of the Atmospheric Sciences, 69*, 2488–2504. https://doi.org/10.1175/JAS-D-11-0293.1.

Duncan, D. I., Kummerow, C. D., & Elsaesser, G. S. (2014). A Lagrangian analysis of deep convective systems and their local environmental effects. *Journal of Climate, 27*, 2072–2086. https://doi.org/10.1175/JCLI-D-13-00285.1.

Fiolleau, T. (2010). *Cycle de vie des systèmes convectifs de mousson dans les régions tropicales: Préparation à la mission Megha-Tropiques*. Thèse de Doctorat de l'École Polytechnique. Available at https://pastel.archives-ouvertes.fr/pastel-00576870/. Last accessed 20 Nov. 2018.

Fiolleau, T., & Roca, R. (2013a). An algorithm for the detection and tracking of tropical mesoscale convective systems using infrared images from geostationary satellite. *IEEE Transactions on Geoscience and Remote Sensing, 51*, 4302–4315. https://doi.org/10.1109/TGRS.2012.2227762.

Fiolleau, T., & Roca, R. (2013b). Composite life cycle of tropical mesoscale convective systems from geostationary and low earth orbit satellite observations: Method and sampling considerations. *Quarterly Journal of the Royal Meteorological Society, 139*, 941–953. https://doi.org/10.1002/qj.2174.

Futyan, J. M., & Del Genio, A. D. (2007). Deep convective system evolution over Africa and the tropical Atlantic. *Journal of Climate, 20*, 5041–5060. https://doi.org/10.1175/JCLI4297.1.

Hagos, S., & Zhang, C. (2010). Diabatic heating, divergent circulation and moisture transport in the African monsoon system. *Quarterly Journal of the Royal Meteorological Society, 136*(S1), 411–425. https://doi.org/10.1002/qj.538.

Henderson, D. S., L'Ecuyer, T. S., Stephens, G. L., Partain, P., & Sekiguchi, M. (2013). A multisensor perspective on the radiative impacts of clouds and aerosols. *Journal of Applied Meteorology and Climatology, 52*, 853–871. https://doi.org/10.1175/JAMC-D-12-025.1.

Houze, R. A., Jr. (1982). Cloud clusters and large-scale vertical motions in the tropics. *Journal of the Meteorological Society of Japan, 60*, 396–410. https://doi.org/10.2151/jmsj1965.60.1_396.

Houze, R. A., Jr. (2004). Mesoscale convective systems. *Reviews of Geophysics, 42*, RG4003. https://doi.org/10.1029/2004rg000150.

Houze, R. A., Jr. (2018). 100 years of research on mesoscale convective systems. In G. McFarquhar (Ed.), *A century of progress in atmospheric and related sciences: Celebrating the American Meteorological Society centennial* (Meteorological monographs). Boston: AMS. https://doi.org/10.1175/AMSMONOGRAPHS-D-18-0001.1.

Houze, R. A., Jr., Rasmussen, K. L., Zuluaga, M. D., & Brodzik, S. R. (2015). The variable nature of convection in the tropics and subtropics: A legacy of 16 years of the Tropical Rainfall Measuring Mission satellite. *Reviews of Geophysics, 53*, 994–1021. https://doi.org/10.1002/2015RG000488.

Huang, X., Hu, C., Huang, X., Chu, Y., Tseng, Y., Zhang, G. J., & Lin, Y. (2018). A long-term tropical mesoscale convective systems dataset based on a novel objective automatic tracking algorithm. *Climate Dynamics, 51*, 3145–3159. https://doi.org/10.1007/s00382-018-4071-0.

Huffman, G. J., Adler, R. F., Morrissey, M. M., Bolvin, D. T., Curtis, S., Joyce, R., McGavock, B., & Susskind, J. (2001). Global precipitation at one-degree daily resolution from multisatellite observations. *Journal of Hydrometeorology, 2*, 36–50. https://doi.org/10.1175/1525-7541(2001)002<0036:GPAODD>2.0.CO;2.

Huffman, G. J., Bolvin, D. T., Nelkin, E. J., Wolff, D. B., Adler, R. F., Gu, G., Hong, Y., Bowman, K. P., & Stocker, E. F. (2007). The TRMM multi-satellite precipitation analysis: Quasi- global, multi-year, combined-sensor precipitation estimates at fine scale. *Journal of Hydrometeorology, 8*, 38–55. https://doi.org/10.1175/JHM560.1.

Imaoka, K., & Nakamura, K. (2012). Statistical analysis of the life cycle of isolated tropical cold cloud systems using MTSAT-1R and TRMM data. *Monthly Weather Review, 140*, 3552–3572. https://doi.org/10.1175/MWR-D-11-00364.1.

Inoue, T., Vila, D., Rajendran, K., Hamada, A., Wu, X., & Machado, L. A. T. (2009). Life cycle of deep convective systems over the eastern tropical Pacific observed by TRMM and GOES-W. *Journal of the Meteorological Society of Japan, 87A*, 381–391. https://doi.org/10.2151/jmsj.87A.381.

Jakob, C., & Schumacher, C. (2008). Precipitation and latent heating characteristics of the major tropical western Pacific cloud regimes. *Journal of Climate, 21*, 4348–4364. https://doi.org/10.1175/2008JCLI2122.1.

Jensen, M. P., & Del Genio, A. D. (2003). Radiative and microphysical characteristics of deep convective systems in the tropical Western Pacific. *Journal of Applied Meteorology, 42*, 1234–1254. https://doi.org/10.1175/1520-0450(2003)042<1234:RAMCOD>2.0.CO;2.

Jirak, I. L., Cotton, W. R., & McAnelly, R. I. (2003). Satellite and radar survey of mesoscale convective system development. *Monthly Weather Review, 131*, 2428–2449. https://doi.org/10.1175/1520-0493(2003)131<2428:SARSOM>2.0.CO;2.

Kondo, Y., Higuchi, A., & Nakamura, K. (2006). Small-scale cloud activity over the maritime continent and the western Pacific as revealed by satellite data. *Monthly Weather Review, 134*, 1581–1599. https://doi.org/10.1175/MWR3132.1.

Lee, D., Oreopoulos, L., Huffman, G. J., Rossow, W. B., & Kang, I. S. (2013). The precipitation characteristics of ISCCP tropical weather states. *Journal of Climate, 26*, 772–788. https://doi.org/10.1175/JCLI-D-11-00718.1.

Li, W., Schumacher, C., & McFarlane, S. A. (2013). Radiative heating of the ISCCP upper level cloud regimes and its impact on the large-scale tropical circulation. *Journal of Geophysical Research, 118*, 592–604. https://doi.org/10.1002/jgrd50114.

Liu, C. (2011). Rainfall contributions from precipitation systems with different sizes, convective intensities, and durations over the tropics and subtropics. *Journal of Hydrometeorology, 12*, 394–412. https://doi.org/10.1175/2010JHM1320.1.

Liu, C., & Zipser, E. J. (2015). The global distribution of largest, deepest, and most intense precipitation systems. *Geophysical Research Letters, 42*, 3591–3595. https://doi.org/10.1002/2015GL063776.

Maddox, R. A. (1980). Mesoscale convective complexes. *Bulletin of the American Meteorological Society, 61*, 1374–1387. https://doi.org/10.1175/1520-0477(1980)061<1374:MCC>2.0.CO;2.

Mapes, B., & Neale, R. (2011). Parameterizing convective organization to escape the entrainment dilemma. *Journal of Advances in Modeling Earth Systems, 3*, 1–20. https://doi.org/10.1029/2011MS000042.

Mapes, B., Milliff, R., & Morzel, J. (2009). Composite life cycle of maritime tropical mesoscale convective systems in scatterometer and microwave satellite observations. *Journal of the Atmospheric Sciences, 66*, 199–208. https://doi.org/10.1175/2008JAS2746.1.

McAnelly, R. L., & Cotton, W. R. (1989). The precipitation life cycle of mesoscale convective complexes over the Central United States. *Monthly Weather Review, 117*, 784–808. https://doi.org/10.1175/1520-0493(1989)117<0784:TPLCOM>2.0.CO;2.

Muller, C., & Bony, S. (2015). What favors convective aggregation and why? *Geophysical Research Letters, 42*, 5626–5634. https://doi.org/10.1002/2015GL064260.

Nesbitt, S. W., Cifelli, R., & Rutledge, S. A. (2006). Storm morphology and rainfall characteristics of TRMM precipitation features. *Monthly Weather Review, 134*, 2702–2721. https://doi.org/10.1175/MWR3200.1.

O'Gorman, P. A., Allan, R. P., Byrne, M. P., & Previdi, M. (2012). Energetic constraints on precipitation under climate change. *Surveys in Geophysics, 33*, 585–608. https://doi.org/10.1007/s10712-011-9159-6.

Redelsperger, J. L. (1997). The mesoscale organization of deep convection. In R. K. S. Convection (Ed.), *The physics and parameterization of moist atmospheric* (pp. 159–160). London: Springer. ISBN: 978-94-015-8828-7.

Roca, R., Aublanc, J., Chambon, P., Fiolleau, T., & Viltard, N. (2014). Robust observational quantification of the contribution of mesoscale convective systems to rainfall in the tropics. *Journal of Climate, 27*, 4952–4958. https://doi.org/10.1175/JCLI-D-13-00628.1.

Roca, R., Fiolleau, T., & Bouniol, D. (2017). A simple model of the life cycle of mesoscale convective systems cloud shield in the tropics. *Journal of Climate, 30*, 4283–4298. https://doi.org/10.1175/JCLI-D-16-0556.1.

Rossow, W. B., Mekonnen, A., Pearl, C., & Goncalves, W. (2013). Tropical precipitation extremes. *Journal of Climate, 26*, 1457–1466. https://doi.org/10.1175/JCLI-D-11-00725.1.

Schumacher, C., & Houze, R. A., Jr. (2003). Stratiform rain in the tropics as seen by the TRMM precipitation radar. *Journal of Climate, 16*, 1739–1756. https://doi.org/10.1175/1520-0442(2003)016<1739:SRITTA>2.0.CO;2.

Schumacher, C., & Houze, R. A., Jr. (2006). Stratiform precipitation production over sub-Saharan Africa and the tropical East Atlantic as observed by TRMM. *Quarterly Journal of the Royal Meteorological Society, 132*, 2235–2255. https://doi.org/10.1256/qj.05.121.

Schumacher, C., Houze, R. A., Jr., & Kraucunas, I. (2004). The tropical dynamical response to latent heating estimates derived from the TRMM precipitation radar. *Journal of the Atmospheric Sciences, 61*, 1341–1358. https://doi.org/10.1175/1520-0469(2004)061<1341:TTDRTL>2.0.CO;2.

Shige, S., Takayabu, Y. N., Tao, W.-K., & Johnson, D. E. (2004). Spectral retrieval of latent heating profiles from TRMM PR data. Part I: Development of a model-based algorithm. *Journal of Applied Meteorology, 43*, 1095–1113. https://doi.org/10.1175/1520-0450(2004)043<1095:SROLHP>2.0.CO;2.

Shige, S., Takayabu, Y. N., Tao, W.-K., & Shie, C.-L. (2007). Spectral retrieval of latent heating profiles from TRMM PR data. Part II: Algorithm improvement and heating estimates over tropical ocean regions. *Journal of Applied Meteorology, 46*, 1098–1124. https://doi.org/10.1175/JAM2510.1.

Stephens, G. L., & Ellis, T. D. (2008). Controls of global-mean precipitation increases in global warming GCM experiments. *Journal of Climate, 21*, 6141–6155. https://doi.org/10.1175/2008JCLI2144.1.

Tan, J., Jakob, C., Rossow, W. B., & Tselioudis, G. (2015). Increases in tropical rainfall driven by changes in frequency of organized deep convection. *Nature, 519*, 451–454. https://doi.org/10.1038/nature14339.

Tapiador, F. J., Roca, R., Del Genio, A., Dewitte, B., Petersen, W. A., & Zhang, F. (2018). Is precipitation a good metric for model performance? *Bulletin of the American Meteorological Society, 100*, 223–233. https://doi.org/10.1175/BAMS-D-17-0218.1.

Webster, P. J., & Stephens, G. L. (1980). Tropical upper-tropospheric extended clouds: Inferences from winter MONEX. *Journal of the Atmospheric Sciences, 37*, 1521–1541. https://doi.org/10.1175/1520-0469(1980)037<1521:TUTECI>2.0.CO;2.

White, R. H., Battisti, D. S., & Skok, G. (2017). Tracking precipitation events in time and space in gridded observational data. *Geophysical Research Letters, 44*, 8637–8646. https://doi.org/10.1002/2017GL074011.

Williams, M., & Houze, R. A., Jr. (1987). Satellite-observed characteristics of winter monsoon cloud clusters. *Monthly Weather Review, 115*, 505–519. https://doi.org/10.1175/1520-0493(1987)115<0505:SOCOWM>2.0.CO;2.

Yuan, J., & Houze, R. A., Jr. (2010). Global variability of mesoscale convective system anvil structure from A-train satellite data. *Journal of Climate, 23*, 5864–5888. https://doi.org/10.1175/2010JCLI3671.1.

Chapter 39
Observational Characteristics of Warm-Type Heavy Rainfall

Byung-Ju Sohn, Geun-Hyeok Ryu, and Hwan-Jin Song

Abstract It has been shown that heavy rainfall can be produced from clouds whose cloud top is relatively low (or warm) with relatively few ice crystals if the atmospheric environment is very humid and there is continuous water vapor supply. This process was named "warm-type heavy rainfall" contrasting to the generally known heavy precipitation mechanism from vigorous and tall convective systems (i.e., cold-type rainfall). The water vapor supply in moist-adiabatically near neutral conditions results in a gentle upward motion and thus relatively lower cloud top. However, warm rains can be heavy because of the main growth of rain drops through the collision and coalescence processes below the melting layer.

Keywords Precipitation · Tropics · Warm rain · Convection · Heavy rain · TRMM · TMI · PR · GPROF · TMPA · CMORPH · PERSIANN · NRL-blended · Microwave · PCT · CAPE · ERA-Interim · East Asian monsoon · Microphysical processes · WRF · Collision · Coalescence · Total precipitable water

39.1 Introduction

Heavy rain events are particularly relevant because in many parts of the world they may cause serious floods costing losses of lives and properties. This is why continuous efforts are spent on investigating deep convection-related structures, cloud

B.-J. Sohn (✉)
School of Earth and Environmental Sciences, Seoul National University, Seoul, South Korea
e-mail: sohn@snu.ac.kr

G.-H. Ryu
National Meteorological Satellite Center, Korea Meteorological Administration, Seoul,
South Korea

H.-J. Song
National Institute of Meteorological Sciences, Korea Meteorological Administration, Seoul,
South Korea

© Springer Nature Switzerland AG 2020
V. Levizzani et al. (eds.), *Satellite Precipitation Measurement*, Advances in Global
Change Research 69, https://doi.org/10.1007/978-3-030-35798-6_15

microphysics, satellite rainfall monitoring, and relationship of heavy rainfall to large-scale synoptic set up, from which better weather forecasting can be achieved.

It is generally accepted that heavy rains are produced from convective clouds growing to significant heights with vigorous updrafts, along with lightning and hail produced during the convective development. A typical convective cell has a life cycle comprising three stages: cumulus stage, mature stage, and dissipation stage. In the period of evolving into the mature stage, with the continued strong upward motion, a great amount of water condenses and falls as a heavy precipitation. In this stage, cloud top often reaches to the tropopause level, lifting larger ice particles high into the storm, to result in 20-dBZ radar echo tops exceeding 14 km (Liu et al. 2007). Satellite-based rain estimation methods such as TRMM GPROF over land assume that heavy precipitation originates from deep convective clouds with abundant ice particles in the upper layer. However, studies of Sohn et al. (2013) and Song and Sohn (2015) noted that deep convection is not always necessary for producing heavy rain in cases characterized by a continuous water vapor supply into an already existing humid environment, such as the East Asian monsoon area during summer.

Satellite summer time (June, July and August) precipitation products (TMPA, CMORPH, PERSIANN, and NRL-blended) based on infrared (IR) data were validated against rain gauge data collected over the Korean peninsula (Sohn et al. 2010), and results surprisingly demonstrated that all satellite products significantly underestimated rain intensity. It was noted that underestimates are due to deficiencies of the retrievals based on passive microwave (PMW) data because the IR algorithms are generally tuned against PMW estimates. This fact suggests that PMW-based rain products also show severe underestimates of the summer rainfall over the Korean peninsula. In order to understand further why PMW-based algorithms partially fail over the Korean peninsula, TMI polarization-corrected temperatures at 85 GHz (TMI PCT85) are paired against TRMM Precipitation Radar (PR) based rain rates over Oklahoma and Korea, whose latitudes are nearly the same between 34° and 36°N (Ryu et al. 2012). The TMI GPROF algorithm over land has been tuned over the Oklahoma region (Olson et al. 2006). We hypothesized that the cloud-rain system over Korea is considerably different from that over Oklahoma, and the TMI algorithm may not work over Korea because of the different cloud-rain system. Figure 39.1 shows that the general relationship of more rainfall for lower TMI PCT85 is evident over Oklahoma − see the sector showing rain rate >20 mm h^{-1} and PCT85 < 200 K. In contrast, many heavy rain events over Korea are found under higher PCT85 conditions; a cluster of rain rates >20 mm h^{-1} is found between 200 and 250 K. An explanation of higher temperatures resulting in heavy rainfall is the low amount of ice crystals (inducing higher temperatures because of less scattering), but the system nevertheless produces heavy rainfall. However, this concept is considerably less intuitive compared to the concept that deep convective clouds cause heavy rainfall over places such as the US Great Plains. In this chapter, we review and document features of such warm-type rainfall and microphysical processes that may be responsible for producing heavy rainfall in spite of higher cloud top temperature.

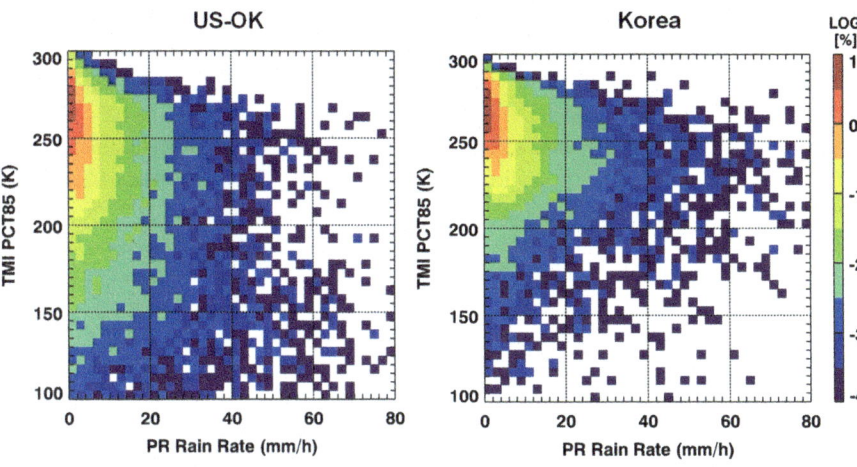

Fig. 39.1 Scatterplots of PR rain rate and TMI polarization corrected temperature at 85 GHz (PCT85) over Oklahoma (US-OK, left) and the Korean Peninsula (right)

39.2 Data and Analysis Method

This study uses TRMM 2A25 (PR near-surface rain rates and attenuation-corrected radar reflectivity profiles), TRMM 2A23 (PR storm heights), and TRMM 1B11 (TMI vertically/horizontally polarized brightness temperatures at 85 GHz) data (version 7). The PR is a 13.8 GHz radar with 5 km horizontal resolution and 0.25 km vertical resolution over the scan swath of 247 km, depending on the orbit. The spatial resolution and scan swath for the TMI sensor are 5.1 and 878 km, respectively. The PR reflectivity observed near the surface is converted into the near-surface rain rate based on the Z-R relationship, and the storm height is defined by the 15-dBZ reflectivity level. The TMI is a conically-scanning microwave radiometer with five frequencies. Here, the polarization-corrected temperature at 85.5 GHz (PCT85) is used in the analysis as an index of cloud ice scattering (Spencer et al. 1989). The PCT85 is defined as the combination of vertical (V) and horizontal (H) polarization components of the 85.5 GHz brightness temperature (i.e., $1.818 \times TB85V - 0.818 \times TB85H$). PCT85 has an advantage of alleviating the difference of surface emissivity between vertical and horizontal polarizations.

All instantaneous data are collected for the summer months (June–August) of 17 years (1998–2014) over Oklahoma (34–36.25°N, 100–96°W) and the southern part of the Korean peninsula (34–36.25°N, 126–130°E). Due to PR's northern coverage limit, 36.25°N is used as the northernmost boundary in this study. A total of 159,796/583,190 and 59,998/214,286 PR/TMI collocated pairs rain clouds are used in the analysis over US-OK and Korea, respectively. Note that 63% of the data are removed during the collocation process of PR and TMI measurements. This study investigates the relationship between the rain rate and PCT85, the frequency distribution of storm height and the PCT85 for light (<10 mm h^{-1}) and heavy

(>10 mm h^{-1}) rain, and the contoured frequency by altitude diagram (CFAD) of radar reflectivity for light and heavy rain cases.

The *K*-means clustering analysis is performed for collected CFADs at each $5° \times 5°$ grid box over the East Asian summer monsoon domain (10°–36.25°N, 100°–150°E) when heavy rains (>10 mm h^{-1}) are detected, in order to examine the rain types (for a detailed explanation see Song and Sohn 2015). A total of 2,811,327 reflectivity profiles for 77,066 heavy rain events (i.e., total number of selected $5° \times 5°$ grids) are collected in the form of CFADs. Although three types of heavy rains are classified over the East Asian domain, we focus only on two major types. The vertical structure, the spatial distribution of occurrence frequency, and the diurnal cycle of occurrence frequency for the two heavy rain types are examined. Furthermore, 6-hourly ERA-Interim reanalysis data (Dee et al. 2011) in a $1° \times 1°$ grid format are analyzed, in order to examine the environmental conditions associated with the two heavy rain types. For temporal match, ERA-Interim data are compared with TRMM instantaneous measurements by linearly interpolating 6-hourly reanalysis data into the TRMM scan time. This study examines 17-year mean composites of three environmental variables from ERA-Interim data: convective available potential energy (CAPE), total precipitable water (TPW), and vertically-integrated moisture flux convergence (ConQ). The 17-year mean field was constructed by weight averaging temporally-varying heavy rain areas at a given $5° \times 5°$ grid for each heavy rain type.

39.3   Comparison Between Korea and Oklahoma

In order to examine how the characteristics of rain clouds are different between Korea and Oklahoma, probability distribution functions (PDFs) obtained from PCT85 and storm height are computed for the two regions. Then the PDFs are classified depending on rain intensities in two classes: RR < 10 mm h^{-1} (light) and RR > 10 mm h^{-1} (heavy). Results are presented in Fig. 39.2. Note that light rain is dominant and covers around 90% of total rain cases in both regions. Although the frequency is higher for Korea when PCT85 > 250 K, the general distribution pattern is quite similar in the two regions (Fig. 39.2a). The frequency of PCT85 shows a very sharp curve peaking around 270 K and decreasing rapidly with increasing PCT85. The corresponding PDF distribution of storm height also has similar features with sharp peaks around 5–6 km altitude (Fig. 39.2c). In Korea, the frequency of the storms higher than 6 km decreases sharply and most of rain storm heights are located below 10 km altitude. In contrast, US-OK shows a weak peak at 5 km and slowly decreasing frequency along with increasing storm height. The PDFs for two regions are reversed at around 7 km height. The heavy rain corresponds to about 10% of the frequency in both regions, and distribution patterns for PCT85 and storm height are quite different. In Korea, there is a relatively high but broad peak around the 230–270 K interval, and then the frequency decreases rapidly with decreasing PCT85 (Fig. 39.2b). However, in Oklahoma the frequency decreases slowly with

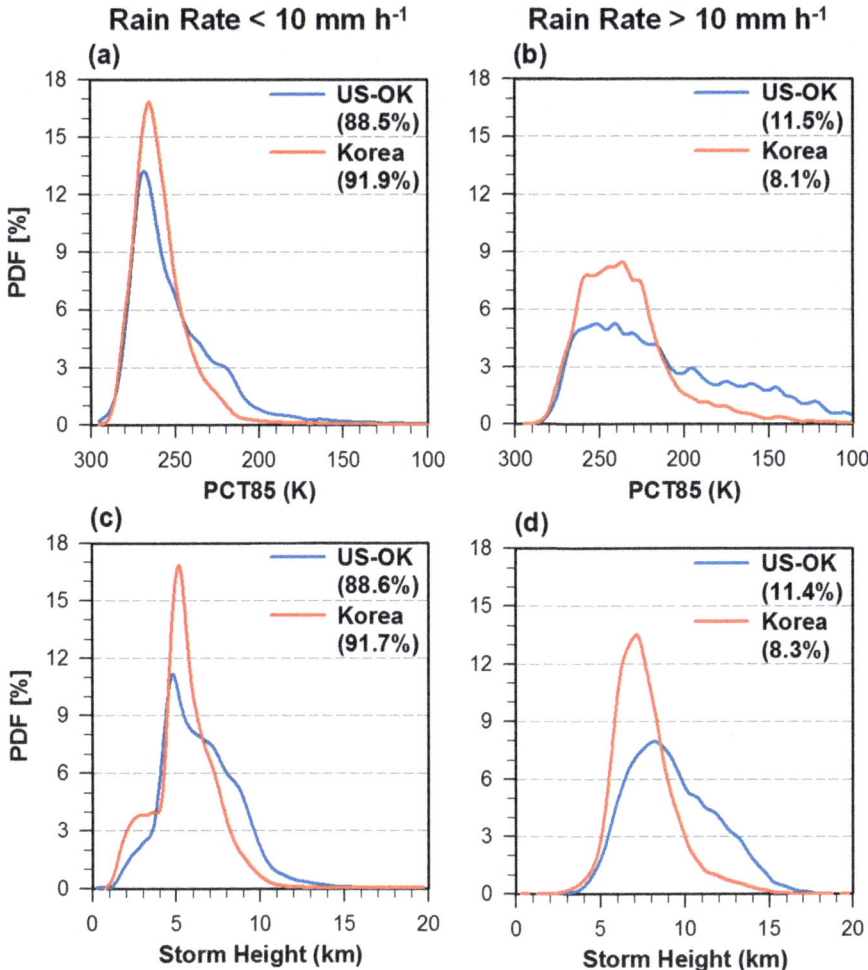

Fig. 39.2 Probability distribution functions (PDFs) of (**a**, **b**) TMI polarization corrected temperature at 85GHz (PCT85) and (**c**, **d**) PR storm height classified by rain rate intensity for Oklahoma (US-OK, blue line) and Korea (red line)

decreasing PCT85 if PCT85 < 270 K, indicating that more heavy rains occur when PCT85 is lower, compared with the Korea region. Conversely, heavy rains over Korea frequently occur under relatively higher PCT85 conditions. This pattern is also clearly shown in the PDF distribution of storm height (Fig. 39.2d). Comparing to light rain cases, the maximum frequency of storm height slightly increases from 5–6 km to 7 km over Korea, and the frequency decreases rapidly as shown in light rain cases. However, in the US-OK region, the maximum frequency of storm height is located at higher altitude (7–8 km) and there are much more storms higher than 10 km, compared to cases in Korea. In summary, heavy rains over Korea appear to

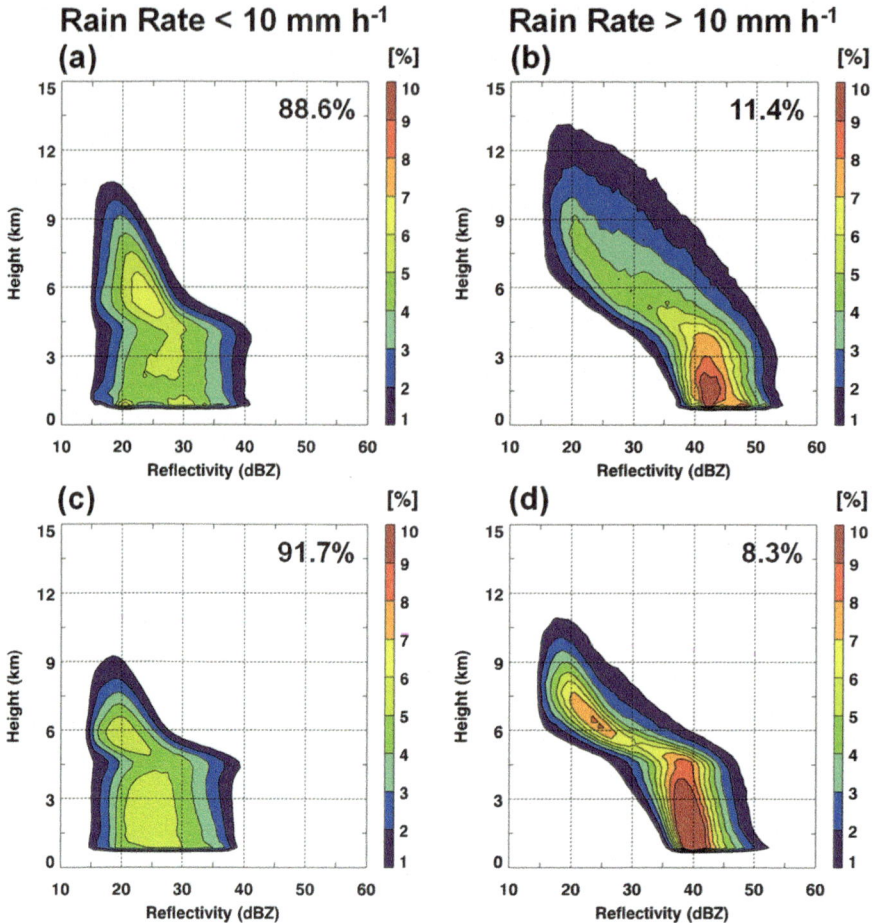

Fig. 39.3 The Contoured Frequency by Altitude Diagrams (CFADs) of the PR reflectivity classified by rain rate intensity for (**a, b**) Oklahoma (US-OK) and (**c, d**) Korea. Percentages refer to the occurrence of each rain class

occur more with clouds whose storm heights are much lower than over the US-OK region. This is consistent with the finding that heavy rainfall over Korea is often associated with relatively higher PCT85 (Fig. 39.1).

In order to examine how the vertical structure of rain clouds in Korea is different from that in US-OK, CFADs of PR reflectivity are constructed for two rain rate classes, light and heavy (see above). In the CFADs for the light rain in both regions (Fig. 39.3a, c), the vertical structure seems to be quite similar in both regions although the US-OK region seems to have higher storm heights. A relatively high reflectivity peak is found around 5 km altitude seemingly associated with the melting layer, and thus the maximum frequency region above the melting layer is located around 6 km in both regions. Although the storms producing light rain in US-OK are taller than in Korea, it seems that in both regions light rains are associated with stratiform-type precipitation.

The CFADs for heavy rains display different patterns when the two regions are compared (Fig. 39.3b, d). The most contrasting pattern is the reflectivity distribution above the melting layer. Above 5 km, while in Korea a sharp decrease of the reflectivity is detected, the decrease is slower in US-OK and the vertical extent of reflectivity reaches higher altitudes as well. At 9 km altitude, the reflectivity in US-OK reaches about 40 dBZ while the highest reflectivity in Korea at 9 km is smaller (30 dBZ). US-OK shows a vertically aligned pattern, suggesting that rain drops producing heavy rainfall at the surface may not grow below 5 km, being likely produced above the melting layer. On the other hand, there is an increasing tendency of reflectivity in CFADs for Korea (Fig. 39.3d). Since the root difference of reflectivity between two levels represents the growth of rain drops, the slope in reflectivity below the melting layer strongly suggests that rain drops grow rapidly below the melting layer and form heavy rainfall when they arrive at the surface probably due to collision and coalescence processes.

39.4   General Features Found Over the East Asian Monsoon Area

Figure 39.4 represents the vertical structures and spatial frequency distributions of two heavy rain types derived from the K-means clustering analysis. The 17-year climatological fields of geopotential height and moisture flux at 850 hPa are also displayed in Fig. 39.4d. Two major rain types covering around 74% of total rain cases emerged from the analysis. The first type (cold-type heavy rain) exhibits well-developed convective systems with large reflectivity in the upper layer (i.e., abundant ice water content) and high storm height (Fig. 39.4a). The cold-type heavy rain is mostly found over inland China (Fig. 39.4c). The second type (warm-type heavy rain) is characterized by more frequent occurrence (43.5%) and larger heavy rain area (996 km^2 within a given $5° \times 5°$ grid box) compared with the cold-type (30.0% and 818 km^2), and thus producing 77% more volumetric heavy rainfall (i.e., occurrence $\times$ area) than the cold-type (Fig. 39.4b). The warm-type heavy rain shows a relatively small reflectivity in the upper level (i.e., less ice water content) and lower storm height compared to the cold-type. Instead of ice microphysical process in the upper layer, growth of raindrops below the melting layer appears to be the main factor in producing the warm-type heavy rainfall. It will be discussed in Sect. 39.5 in details. The warm-type heavy rain commonly occurs over the tropical ocean region (10°–20°N) and over the strong water vapor flux region (Fig. 39.4d) along the western boundary of the North Pacific high (e.g., Taiwan, Korea, Japan, and adjacent oceans), indicating that the warm-type mainly develops in a humid environment.

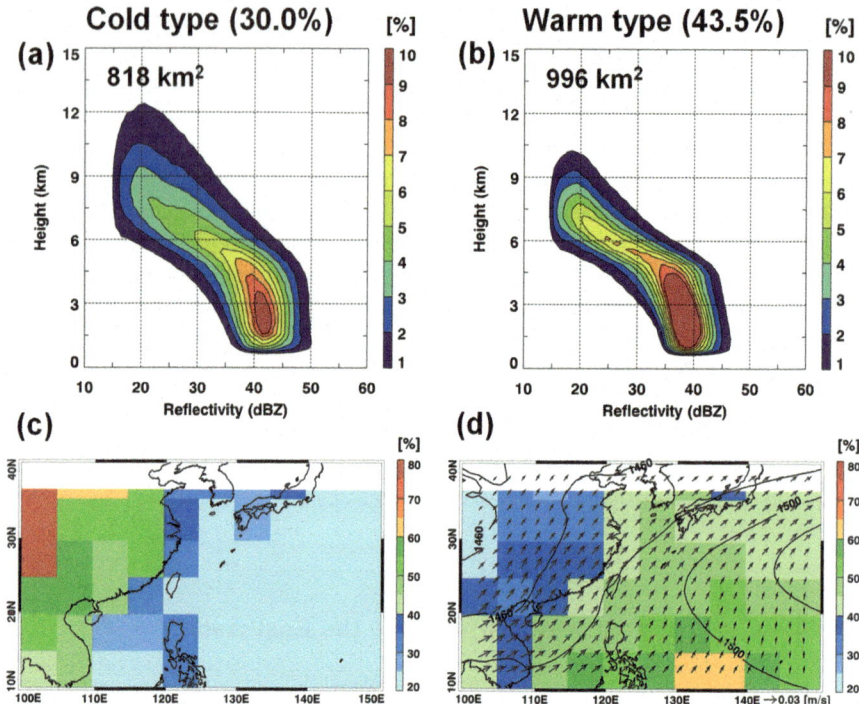

Fig. 39.4 Vertical structure (**a**, **b**) and occurrence frequency distribution (**c**, **d**) of the cold-type and warm-type heavy rain classified by K-means clustering analysis for the period June–August 1998–2014. The percentages indicate the total occurrence of each type. The area displayed in the top left of the upper figures represents the mean heavy rain area at a 5° × 5° grid over the entire domain. The 17-year summer mean geopotential height (gpm, solid lines) and moisture flux (m s^{-1}, arrows) at 850 hPa are shown in the bottom-right figure

Results from Fig. 39.4 are consistent with typical structures of precipitating clouds over the continental and monsoon (or oceanic) regions (Liu et al. 2008, Xu and Zipser 2012). Therefore, the mechanisms forming cold-type and warm-type heavy rain can be understood in the context of continental and oceanic rain regimes. Characteristics of continental and oceanic rain regimes are well represented by the diurnal variation of precipitation. In relation to solar activity and subsequent radiation processes, afternoon and early morning peaks of rainfall are well-known features for continental and oceanic rain regimes, respectively (Nesbitt and Zipser 2003). Diurnal variations of the cold-type and warm-type heavy rain cases exhibit local maxima around 1400–1900 LT and 0400–1100 LT, respectively (Fig. 39.5a), consistent with diurnal variations for land and ocean separately. Note also that the early morning maximum and the weak afternoon-night minimum are more evident in the warm-type case for latitudes north of 30°N than south of 30°N (Fig. 39.5b). The result implies that the rain system over the mid-latitudes of the domain is different from that from the subtropical/tropical oceans even if the whole ocean domain shows a dominant warm-type heavy rain.

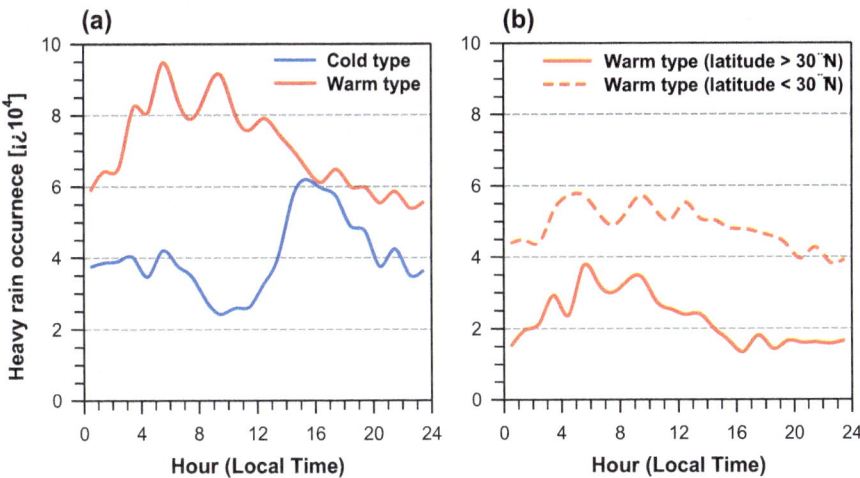

Fig. 39.5 (a) Diurnal variation of cold-type (blue line) and warm-type (red line) heavy rain (>10 mm h^{-1}) occurrence. (b) Diurnal variations of the warm-type over the oceanic area north of 30°N (solid line) and south of 30°N (dashed line)

In contrast to the diagnostic view in terms of continental and oceanic rain regime at a given location (e.g., Xu and Zipser 2012, Liu et al. 2008), the approach employed in this study (i.e., classified heavy rain types using the K-means clustering) has an advantage that can provide temporally-varying weather environments related to rain types.

Using the above concept, we try to understand which atmospheric environment is responsible for the two rain types. Figure 39.6 shows 17-year mean composites of CAPE, TPW, and ConQ for events when a certain rain type (cold or warm) occurs in a given 5° × 5° grid. As introduced in the methodology section, ERA-Interim data are interpolated into the TRMM observation time and data are averaged with the fraction of heavy rain (>10 mm h^{-1}) area as a weight. In Fig. 39.6, regions having the occurrence frequency lower than 20% are excluded from the analysis. It is obvious that cold-type heavy rains occur under relatively larger CAPE conditions, compared with warm-type conditions (Fig. 39.6a, b). The convective instability with high-CAPE conditions facilitates the development of cold-type heavy rain, likely inducing higher vertical extent of cloud systems with vigorous updrafts and abundant ice water content. On the other hand, warm-type heavy rain is thought to be associated with a more humid environment and strong water vapor convergence conditions compared to the cold-type (Fig. 39.6c–f). Under such conditions, deep convection with abundant ice water content seems not always necessary for producing heavy rainfall.

It is also worth noting that CAPEs for the warm type exhibit a strong contrast between the regions north and south of 25°N. Although both areas are subject to the warm-type heavy rain, we expect different mechanisms producing heavy rainfall north and south of 25°N. The results of this study mainly deal with the vertical

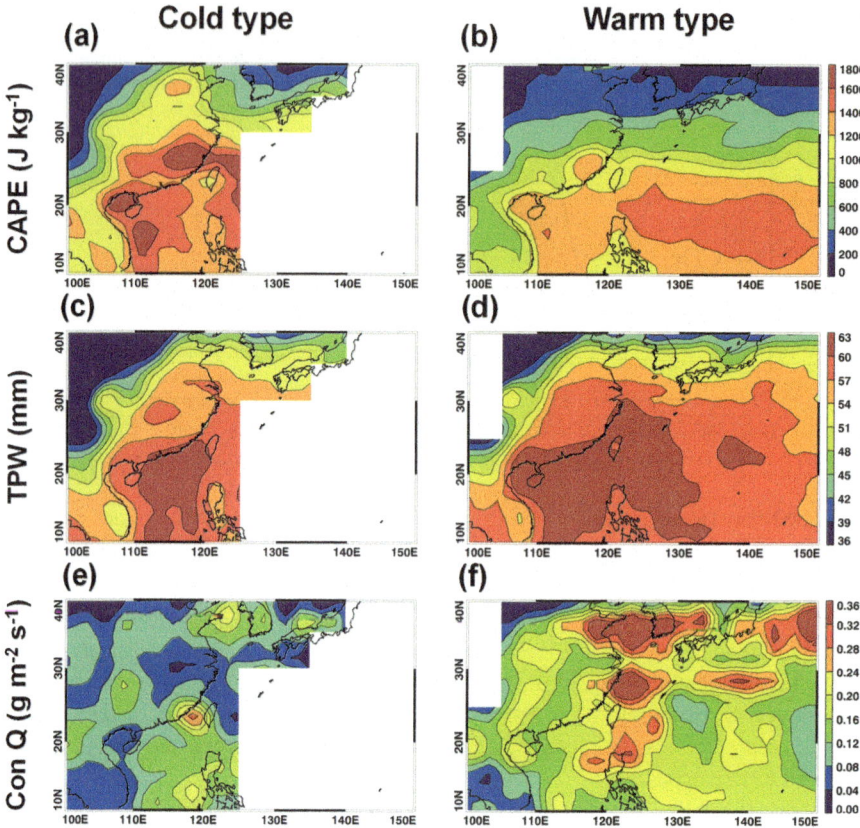

Fig. 39.6 Spatial distributions of 17-year summer mean (**a**, **b**) convective available potential energy (CAPE), (**c**, **d**) total precipitable water (TPW), and (**e**, **f**) vertically-integrated water vapor flux convergence (ConQ) for cold-type and warm-type heavy rain events. Regions showing occurrence frequency lower than 20% are excluded from the analysis

structure of precipitating clouds below the 10 km storm height where the radar reflectivity is greater than 15 dBZ (Fig. 39.4b). However, the precipitating clouds can often develop to the altitude of 15–16 km (approximate tropopause height). Therefore, the microphysical process within the cloud ice layer between 10 and 16 km (especially with the warm-type heavy rain) needs to be further studied in the future.

The thermodynamic structure inducing the warm type heavy rainfall is interesting although it is the water vapor convergence onto the region that forms the rain. In relation to this, the vertical distribution of averaged meteorological variables is obtained over the Korean peninsula domain (the same domain used for Figs. 39.2 and 39.3). Thus, the time series of composites shown in Fig. 39.7 refer to the southern part of the Korean peninsula (34–36.25°N, 126–130°E). In the figure, the equivalent potential temperature and relative humidity (RH) profiles are also

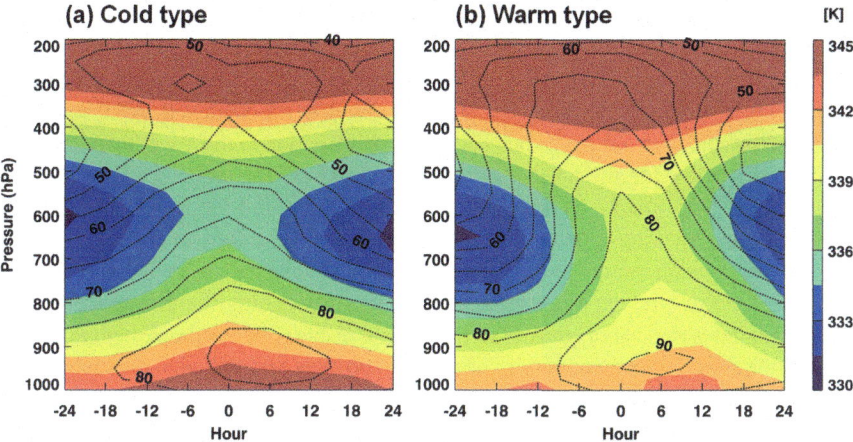

Fig. 39.7 The summer (June–August) mean equivalent potential temperature (color, K) and relative humidity (dotted line, %) profiles for (**a**) cold-type and (**b**) warm-type heavy rain events over the Korean peninsula. (Adapted from Song and Sohn 2015)

displayed. The equivalent potential temperature of the cold-type exhibits a near symmetrical pattern centered at 0 h when heavy rainfall events are observed by the TRMM satellite (Fig. 39.7a).

In contrast, the warm-type shows asymmetric equivalent potential temperature and RH distributions (Fig. 39.7b) quite different from those of the cold type. Much weaker vertical gradients of both equivalent potential temperature and RH are found for the warm type, suggesting that convective instability cannot fully explain the production of warm-type heavy rainfall, consistent with much lower CAPE conditions over the Korean peninsula shown in Fig. 39.6b. Furthermore, the duration time appears much longer for the warm type. It is important to emphasize that the conditions for the warm type generally satisfy moist-adiabatically near neutral condition in the middle-lower layer. In addition, the strong moisture flux convergence in the lower layer (Fig. 39.6f) keeps releasing latent heat with the generation of cloud liquid water. However, the air mass moving upward may have weak buoyancy under a near-neutral thermodynamic condition, which may be the main reason why the warm-type heavy rain stems from a generally lower cloud system with lack of ice crystals.

39.5 Role of Collision and Coalescence Processes

As discussed above, the development of warm-type heavy rain appears to be related with weak convection under very humid conditions. We have suggested in Sect. 39.3 that this may be mainly due to rain drop growth from the collision and coalescence processes. In order to verify this hypothesis and examine the associated microphysical process, we conduct 2-dimensional idealized experiments using the Weather Research and Forecasting (WRF) model version 3.6.1 (Skamarock et al. 2008). The horizontal 201 grids at 250-m interval, 80 vertical layers, and 60-min time period with a 3-sec time step are considered in the idealized WRF model. A thermal bubble with a 4-km radius and a 3-K perturbation in the lower troposphere is included for the forcing in the idealized simulation. The WRF Double-Moment 6-class (WDM6) microphysics parameterization (Lim and Hong 2010) is used to simulate precipitating clouds. The WDM6 scheme is known to produce realistic structures of heavy rain over the Korean peninsula (Song and Sohn 2018). The cold-type and warm-type experiments are constructed by adding the difference in temperature and water vapor profiles at 0 h shown in Fig. 39.7 to the default sounding condition. Here, four times larger perturbation profiles are applied to the temperature and moisture soundings since the result in Fig. 39.7 is produced from the 10-year composite of ERA-Interim data with a low resolution. As a result, convectively more unstable condition is prescribed (i.e., high CAPE and low TPW) in the cold-type experiment, whereas the more humid environment with moist-adiabatically neutral condition (i.e., low CAPE and high TPW) is considered in the warm-type experiment.

Figure 39.8 shows evolutionary features of radar reflectivity (Z), raindrop diameter (D_R) and raindrop number concentration (N_R) profiles for the cold-type and warm-type heavy rain experiments (Song et al. 2017). The main growth of radar reflectivity in the cold-type experiment is found in the middle to upper layers (Fig. 39.8a), indicating the importance of ice microphysics for the cold-type heavy rain. The growth of cold-type heavy rain may be interrupted by higher evaporation process under relatively dry conditions, shown by the low reflectivity below the freezing level in Fig. 39.8a. In contrast to the cold-type, the reflectivity in the warm-type experiment sharply increases with decreasing altitude below the melting layer (Fig. 39.8b). The main growth of warm-type heavy rain occurs in the lower part (related with liquid drop microphysics), not in the upper layer. It is evident that the increase of reflectivity below the melting layer for the warm-type experiment is closely linked with the increase of D_R (Fig. 39.8d) and the decrease of N_R (Fig. 39.8f) with decreasing height. Although the increase of D_R and the decrease of N_R are also found in the cold-type experiment (Fig. 39.8c, e), more rapidly increasing D_R and decreasing N_R features strongly suggest that the process of collision and coalescence of raindrops is mostly responsible for the warm-type heavy rainfall.

39.6 Conclusions

We first gave an attempt to answer why the TRMM rainfall estimate, based on the use of scattering signatures over the Korean peninsula, results in severe underestimation compared to surface rain gauge measurements. It was noted that the rain-cloud system over the Korean peninsula is very different from that found in

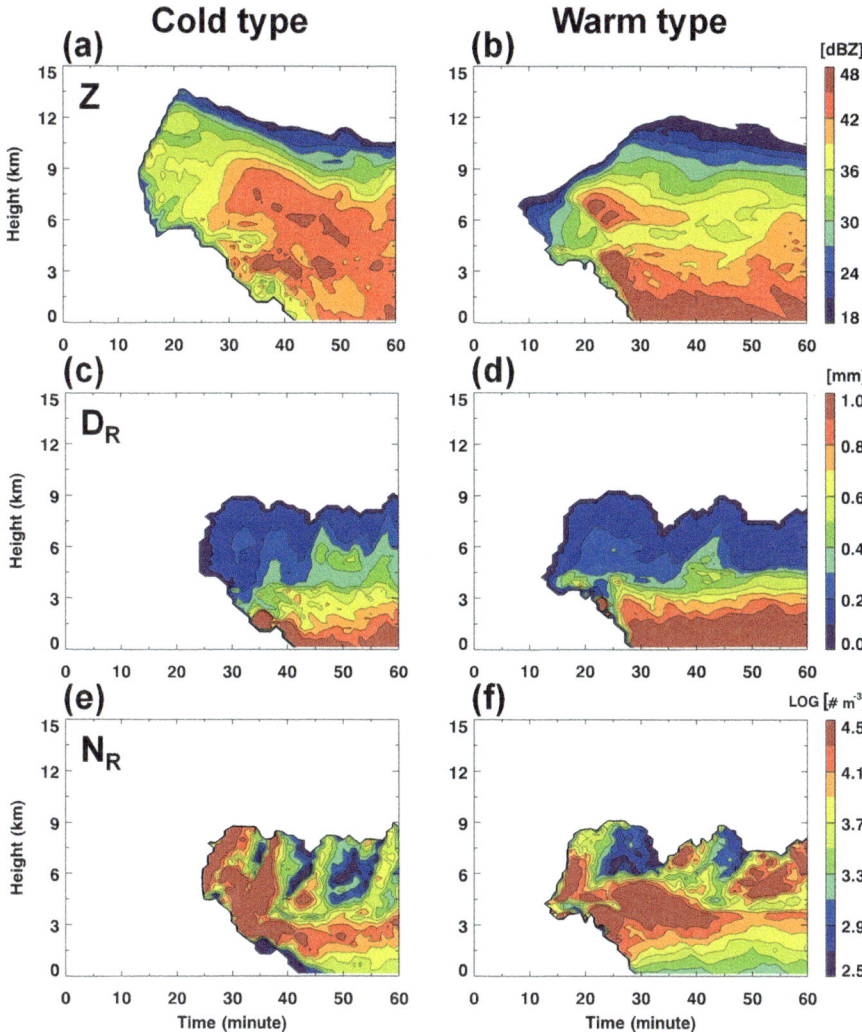

Fig. 39.8 Temporal evolution of (**a**, **b**) radar reflectivity (Z), (**c**, **d**) raindrop mean-mass diameter (D_R), and (**e**, **f**) raindrop number concentration (N_R) profiles for the cold-type and warm-type heavy rain. (Adapted from Song et al. 2017)

Oklahoma; clouds even producing heavy rainfall over Korca are substantially lower in height and are characterized by less available ice crystals even for the heavy rains. These cloud and rain features are likely the main reason for significant underestimation of rain retrievals by passive microwave measurements over the Korean peninsula. They suggest that deep convection may not always be necessary to produce heavy rainfall in very humid environment such as over the Korean peninsula if moisture convergence is prevalent with continuous supply into the region of interest. Similar conclusions were drawn from the study of precipitation character-istics of extreme events over Japan using TRMM observations (Hamada and Takayabu 2018).

We further explored whether the heavy rainfall associated with warm-type clouds is a general feature even for other regions. In doing so, analysis was extended into the East Asian monsoon area. Two dominant types of heavy rainfall emerged over the East Asian domain: a cold type showing well-developed deep convective system situated predominantly over mainland China, and a warm type mainly situated over the ocean with a moderate cloud depth. The warm type heavy rainfall, coexisting with the conventional cold-type heavy rainfall, is found to be a characteristic feature representing the East Asian summer monsoon rainfall. The East Asian monsoon area is very humid during the summer so that its environment seems to hold water vapor maintaining near moist-adiabatically neutral conditions. If there is additional large water vapor flux convergence, the accumulated water vapor may not be easily used for generating vigorous upward motions, even if latent heat is released in the condensation process, because of the near-neutral conditions. Consequently, the water vapor excess yields abundant cloud liquid water with lack of ice particles. Then the formed precipitation particles would grow continuously after melting into larger hydrometeors through dominant collision and coalescence processes until they fall, causing heavy rainfall at the surface, as revealed in the composites of radar reflectivity profiles. The concept of collision and coalescence of cloud droplets with falling rain drops to cause the warm-type heavy rainfall are proved to be valid from WRF model simulations.

This warm-type rainfall concept is not only applicable for the East Asian mon-soon area, as shown by the weak connection between extreme rainfall and deep convection in the global average sense (Hamada et al. 2015), suggesting that the warm-type rainfall discussed in this chapter may be a world-wide phenomenon.

Acknowledgments We appreciate Prof. Kenji Nakamura for his valuable comments. This work was supported by the Korea Meteorological Administration Research and Development Program under Grant KMIPA KM2018-06910.

References

Dee, D. P., Uppala, S. M., Simmons, A. J., Berrisford, P., Poli, P., Kobayashi, S., Andrae, U., Balmaseda, M. A., Balsamo, G., Bauer, P., Bechtold, P., Beljaars, A. C. M., van de Berg, L., Bidot, J., Bormann, N., Delsol, C., Dragani, R., Fuentes, M., Geer, A. J., Haimberger, L., Healy, S. B., Hersbach, H., Hólm, E. V., Isaksen, L., Kallberg, P., Köhler, M., Matricardi, M., McNally, A. P., Monge-Sanz, B. M., Morcrette, J.-J., Park, B.-K., Peubey, C., de Rosnay, P., Tavolato, C., Thépaut, J.-N., & Vitart, F. (2011). The ERA-interim reanalysis: Configuration and performance of the data assimilation system. *Quarterly Journal of the Royal Meteorological Society, 137*, 553–597. https://doi.org/10.1002/qj.828.

Hamada, A., & Takayabu, Y. N. (2018). Large-scale environmental conditions related to midsum-mer extreme rainfall events around Japan in the TRMM region. *Journal of Climate, 31*, 6933–6945. https://doi.org/10.1175/JCLI-D-17-0632.1.

Hamada, A., Takayabu, Y. N., Liu, C., & Zipser, E. J. (2015). Weak linkage between the heaviest rainfall and tallest storms. *Nature Communications, 6*, 6213. https://doi.org/10.1038/ncomms7213.

Lim, K.-S. S., & Hong, S.-Y. (2010). Development of an effective double-moment cloud microphysics scheme with prognostic cloud condensation nuclei (CCN) for weather and climate models. *Monthly Weather Review, 138*, 1587–1612. https://doi.org/10.1175/2009MWR2968.1.

Liu, C., Zipser, E. J., & Nesbitt, S. W. (2007). Global distribution of tropical deep convection: Different perspectives form TRMM infrared and radar data. *Journal of Climate, 20*, 489–503. https://doi.org/10.1175/JCLI4023.1.

Liu, C., Zipser, E. J., Cecil, D. J., Nesbitt, S. W., & Sherwood, S. (2008). A cloud and precipitation feature database from nine years of TRMM observations. *Journal of Applied Meteorology and Climatology, 47*, 2712–2728. https://doi.org/10.1175/2008JAMC1890.1.

Nesbitt, S. W., & Zipser, E. J. (2003). The diurnal cycle of rainfall and convective intensity according to three years of TRMM measurements. *Journal of Climate, 16*, 1456–1475. https://doi.org/10.1175/1520-0442-16.10.1456.

Olson, W. S., Kummerow, C., Yang, S., Petty, G. W., Tao, W.-K., Bell, T. L., Braun, S. A., Wang, Y., Lang, S. E., Johnson, D. E., & Chiu, C. (2006). Precipitation and latent heating distributions from satellite passive microwave radiometry. Part I: Improved method and uncertainties. *Journal of Applied Meteorology and Climatology, 45*, 702–720. https://doi.org/10.1175/JAM2369.1.

Ryu, G.-H., Sohn, B. J., Kummerow, C. D., Seo, E.-K., & Tripoli, G. J. (2012). Rain rate characteristics over the Korean peninsula and improvement of the Goddard Profiling (GPROF) database for TMI rainfall retrievals. *Journal of Applied Meteorology and Climatology, 51*, 786–798. https://doi.org/10.1175/JAMC-D-11-094.1.

Skamarock, W. C., Klemp, J. B., Dudhia, J., Gill, D. O., Barker, D. M., Duda, M. G., Huang, X.-Y., Wang, W., & Powers, J. G. (2008). *A description of the Advanced Research WRF Version 3*. NCAR Tech. Note TN-475_STR, USA, 113 pp. Available online at http://www2.mmm.ucar.edu/wrf/users/docs/arw_v3.pdf

Sohn, B. J., Han, H.-J., & Seo, E.-K. (2010). Validation of satellite-based high-resolution rainfall products over the Korean peninsula using data from a dense rain gauge network. *Journal of Applied Meteorology and Climatology, 49*, 701–714. https://doi.org/10.1175/2009JAMC2266.1.

Sohn, B. J., Ryu, G.-H., Song, H.-J., & Ou, M.-L. (2013). Characteristic features of warm-type rain producing heavy rainfall over the Korean peninsula inferred from TRMM measurements. *Monthly Weather Review, 141*, 3873–3888. https://doi.org/10.1175/MWR-D-13-00075.1.

Song, H.-J., & Sohn, B. J. (2015). Two heavy rainfall types over the Korean peninsula in the humid east Asian summer environment: A satellite observation study. *Monthly Weather Review, 143*, 363–382. https://doi.org/10.1175/MWR-D-14-00184.1.

Song, H.-J., & Sohn, B. J. (2018). An evaluation of WRF microphysics schemes for simulating the warm-type heavy rain over the Korean peninsula. *Asia-Pacific Journal of Atmospheric Sciences, 54*, 1–12. https://doi.org/10.1007/s13143-018-0006-2.

Song, H.-J., Sohn, B. J., Hong, S.-Y., & Hashino, T. (2017). Idealized numerical experiments on the microphysical evolution of warm-type heavy rainfall. *Journal of Geophysical Research, 122*, 1685–1699. https://doi.org/10.1002/2016JD025637.

Spencer, R. W., Goodman, H. M., & Hood, R. E. (1989). Precipitation retrieval over land and ocean with the SSM/I. Part I: Identification and characteristics of the scattering signal. *Journal of Atmospheric and Oceanic Technology, 6*, 254–273. https://doi.org/10.1175/1520-0426(1989) 006<0254:PROLAO>2.0.CO;2.

Xu, W., & Zipser, E. J. (2012). Properties of deep convection in tropical continental, monsoon, and oceanic rainfall regimes. *Geophysical Research Letters, 39*, L07802. https://doi.org/10.1029/2012GL051242.

Chapter 40
Satellite Precipitation Measurement and Extreme Rainfall

Olivier P. Prat and Brian R. Nelson

Abstract Extreme Rainfall can be defined by percentiles of the overall rainfall distribution at a given temporal scale. Generally, the 90th percentile value and above can be considered extreme. While satellite measurements of rainfall provide estimates at global scales and for ever increasing spatial resolution (currently 0.1 km^2), the many flavors of satellite precipitation measurement vary widely from algorithm to algorithm especially at the 90th percentile and above. Satellite estimates of extreme rainfall also vary widely based on certain conditions such as over land versus over ocean, over mountainous terrain versus flat terrain, and high latitudes versus the tropics. Finally, all satellite quantitative precipitation estimates are based on algorithms which ultimately include blending and adjustment using rain gauges. Therefore, satellite estimates of extreme rainfall can vary widely based on the algorithm. This chapter provides an overview of extreme rainfall characterization derived from satellite measurements.

Keywords Precipitation · Extreme rainfall · QPE · Microwave · Infrared · Radar · TRMM · GPM · IMERG · GPCP · CMAP · CMORPH · PERSIANN · TMPA · GHCN · USCRN · ITCZ · WMMD · Tropical cyclone · IDF curve

O. P. Prat (✉)
Cooperative Institute for Satellite Earth System Studies (CISESS), North Carolina State University, Asheville, NC, USA
e-mail: opprat@ncsu.edu

B. R. Nelson
NOAA/NESDIS, National Centers for Environmental Information (NCEI), Asheville, NC, USA

© Springer Nature Switzerland AG 2020
V. Levizzani et al. (eds.), *Satellite Precipitation Measurement*, Advances in Global Change Research 69, https://doi.org/10.1007/978-3-030-35798-6_16

40.1   Introduction

40.1.1   Physical Principles of Satellite Rainfall Measurements

Basically, there are three principles of precipitation retrieval from remote sensing platforms. Those principles are based on Infrared (IR), Passive Microwave (PMW), and Active Microwave (AMW) sensors. A fourth method includes Multi-sensor Techniques, which combine remotely sensed data from different observation platforms and rely on the advantage of each technique and platform in an attempt to provide optimal Quantitative Precipitation Estimates (QPEs). A more detailed description of the different measurement methods and the associated remotely sensed observation platforms can be found for instance in Kidd et al. (2010). Table 40.1 summarizes the different measurement methods and physical principles along with their pros and cons and the different precipitation products associated with each technique.

Satellite precipitation data have been used to study rainfall extremes. The instruments on board TRMM launched in 1997, provided an unprecedented way to look into extreme rainfall events. Numerous studies have used the combination of instruments (PR, TMI) on board of TRMM to quantify extreme events in relation with convective activity. They provided an accurate mapping of the occurrence, frequency, and seasonal and diurnal cycles of those extreme convective events (Nesbitt and Zipser 2003; Zipser et al. 2006; Hirose et al. 2009; Romatschke et al. 2010; Rasmussen and Houze 2011; Rasmussen et al. 2013; Rapp et al. 2014). Capturing those intense events require that they occur at the time the satellite overpasses (Prat and Barros 2010). The length of the satellite record (since 1998 for TRMM) provides enough statistical robustness at the sub-daily scale to derive the full diurnal cycle from individual overpasses and match gridded precipitation datasets (Nesbitt and Anders 2009; Prat and Nelson 2014). While there are many possible applications of satellite QPEs to characterize extreme precipitation events, in this chapter we will mostly focus on the measurement of extreme rainfall derived from multi-sensor techniques as well as on long-term extreme events analysis using gridded satellite products (Table 40.1).

40.1.2   Selected Satellite QPE Products

Table 40.1 summarizes the characteristics of some of the existing remotely sensed precipitation products. A brief description of some of the products mentioned in this chapter is provided here. The Global Precipitation Climatology Project (GPCP) has a suite of products spanning various time and space scales (Huffman et al. 2001). The GPCP monthly product, starting in 1979, is considered the best global precipitation product for climate studies and is used extensively in climate assessments. The GPCP daily product provides global precipitation estimates at a spatial resolution

Table 40.1 Physical principles of satellite precipitation retrieval methods along with advantages and known caveats along with a list of corresponding products

Methods	Physical principles	Advantages	Caveats	Products
Visible/ Infrared (VIS/IR)	Thermal IR measures emission from objects. Heavier rainfall associated with larger, taller clouds with colder top. Derive rainfall estimates by observing cloud top temperatures.	Simplicity Availability of data.	Indirect relationship between cloud top temperature and rainfall. Variations in the relationship occurring during the lifetime of an event, between rainfall systems, and climatological regimes.	GEOS-GPI AE GMSRA PERSIANN-MSA
Passive Microwave (PMW)	Earth MW radiation is attenueated by hydrometeors. Emission from raidrops (increase in MW radiation) and scattering by ice particles (decrease in MW radiation). Use of empirical techniques (calibrated against surface data) and physical techniques (radiative transfer)	Direct method. High-frequency PMW obs. are more sensitive to light rainfall and cloud properties. It is also relatively insensitive to surface and allows retrieval over problematic surfaces (AMSU-B).	Observations only available from LEO satellites (two observations/day). Emission from drops is low in comparison with background emissivity over land. Precipitation estimation over coastal areas.	TMI (GPROF) SSM/I AMSR-E AMSU-B GMI
Active Microwave (AMW)	Backscattered energy linked with precipitation. Relationship of precipitation intensity and backscatter depends on size distribution of particles.	Provides high quality rainfall estimates. Dual-Pol allows retrieving weak rain and snow (GPM).	Use of inadequate Z-R relationships (conv/strat). Attenuation effects can be important in case of non-uniform rain distribution within the rain cell.	TRMM PR GMI DPR
Multisensor techniques	TRMM TMPA: combines PMW and IR obs. (with/without RG correction). CMORPH: combines PMW and IR obs. With an advection scheme. PERSIANN: combines PMW and IR obs. With neural network technique. GPCP: same as TMPA + surface obs.	Overcome individual satellite deficiencies by exploiting the synergy between different satellite observations.	Depends of each product. Performance depends on the core physical principle (IR, PMW) and retrieval algorithm.	GPCP TRMM TMPA CMORPH PERSIANN IMERG CMAP, GSMaP NRL, MIRA

Gridded products combine estimates from different type of sensors

of 1°-daily from October 1996 to the present. The GPCP suite of products dwells on a satellite (IR and PMW) and in-situ product merging strategy. The Climate Prediction Center (CPC) MORPHing (CMORPH) technique provides daily/0.25° × 0.25° precipitation estimates spanning 60°S-60°N (Joyce et al. 2004). CMORPH combines precipitation estimates derived from PMW sensors with the advection of cloud features from more frequently available IR measurements. The bias-adjusted version of the product (i.e., CMORPH-CDR, with CDR standing for Climate Data Record) merges satellite observations and surface gauge analysis. For details regarding the CMORPH-CDR algorithm and the products generated see Xie et al. (2017). Comparable to GPCP and CMORPH are the precipitation estimates from the Precipitation Estimation from Remotely Sensed Information using Artificial Neural Networks, PERSIANN-CDR (Ashouri et al. 2015). PERSIANN-CDR combines IR satellite precipitation estimates and bias correction from in-situ data through GPCP and provides daily/0.25° × 0.25° precipitation estimates for the domain 60°S-60°N. The PERSIANN-CDR is available from 1983 to present. The TRMM Multi-satellite Precipitation Analysis (TMPA 3B42 version 7) is a combination of different remotely sensed PMW (TMI, SSM/I, AMSR, AMSU) and calibrated IR estimates with rain gauge corrected monthly accumulation (Huffman et al. 2007). The TMPA 3B42 provides 3-h/0.25° × 0.25° precipitation estimates for the domain between 50°S- 50°N, from which we will be computing rainfall accumulation at various time scales from daily to multiannual. We will be using the Research (3B42) and the Real-Time (3B42 RT) versions of TMPA. TMPA 3B42 is available from the Tropical Rainfall Measuring Mission (TRMM)'s inception in 1998 to the present time and until production is halted. TMPA 3B42 is currently gradually superseded by precipitation estimates from the Global Precipitation Measurement (GPM) mission which started in February 2014 (Hou et al. 2014). The newly available Integrated Multi-satellitE Retrievals for GPM (IMERG) provides an improvement both in terms of spatial coverage from 60°S-60°N, and spatial and temporal resolutions at 30-min/0.1° × 0.1° (Huffman et al. 2018).

40.2   Statistical Distribution of Rainfall Extremes

40.2.1   Average and Maximum Rainfall

Figure 40.1 presents the average daily precipitation (Fig. 40.1a, c, e) and the maximum daily precipitation (Fig. 40.1b, d, f) derived from PERSIANN-CDR (Fig. 40.1a, b), CMORPH (Fig. 40.1c, d), and GPCP (Fig. 40.1e, f). While average global precipitation patterns are relatively similar (Fig. 40.1a, c, e), maximum daily precipitation values exhibit large differences (Fig. 40.1b, d, f). The CMORPH bias-adjusted product displays higher values for the daily maximum rainfall while the GPCP product gives much lower values for the maximum daily rainfall and PERSIANN-CDR laying in between. The plates illustrate the large differences between remotely sensed products; differences that can be explained by the retrieval

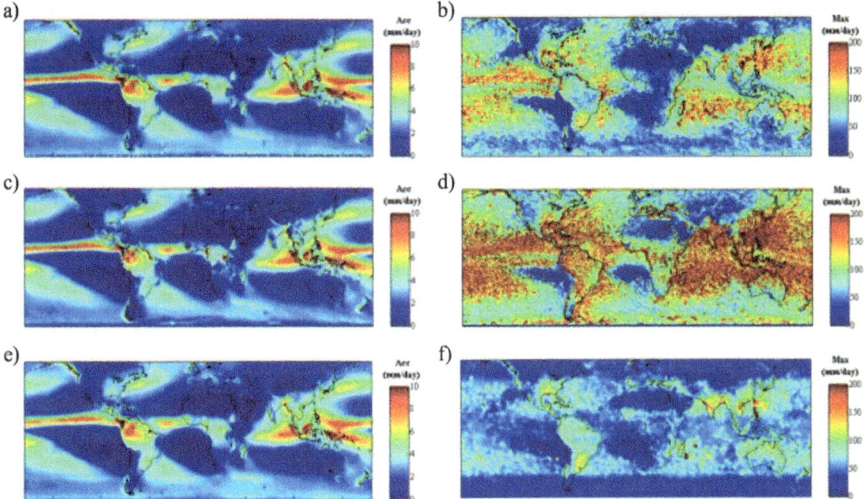

Fig. 40.1 Average precipitation (first column) and maximum daily precipitation (second column) for PERSIANN (**a**, **b**), CMORPH (**c**, **d**), and GPCP (**e**, **f**). (Adapted from Prat et al. 2017)

methods used (i.e., IR vs. PMW), the bias adjustment methods (i.e., monthly totals vs. daily adjustment using a moving window centered over a given date), the spatial resolution ($0.25° \times 0.25°$ for CMORPH and PERSIANN vs. $1° \times 1°$ for GPCP), and the temporal resolution (daily for CMORPH and PERSIANN vs. monthly for GPCP for the earlier part of the full period of record).

Gridded precipitation datasets are comparable in terms of daily average precipitation patterns in the sense that they are able to identify the areas of the most intense precipitation around the world; areas that correspond to well-established weather pattern of extreme convective activity. For instance, the highest daily average rainfall derived from TMPA 3B42 was found to match loosely the locations of two of the wettest places on earth (Prat and Nelson 2013b). Those maxima located near Quibdo (Colombia) at the convergence of the ITCZ, and near Mawsynram (Khasi Hills in Northeastern India) in the Khasi Hills, area of extreme convection and moisture convergence during the monsoon period (Romatschke et al. 2010), were however lower (half of the total for the second location) than reported by surface stations. Qualitatively, satellites accurately capture precipitation patterns and identify the wettest areas around the globe. Quantitatively, important differences are found when looking at maximum precipitation regardless of the accumulation duration. A comparison of maximum precipitation records across a variety of temporal scales (from 3-h to 2-year), shows that that TMPA 3B42 maximum rainfall fall well below maximum rainfall obtained from historical gauge records (Breña-Naranjo et al. 2015). While historical gauge records gathered from location around the world follow closely the empirical equation for the envelope of maximum precipitation ($R = aD^b$; with D is the duration in hour, R is the accumulation in mm, and the parameters $a = 422$, and $b = 0.4765$; see also National Research

Council (1994) and World Meteorological Organization (2009) for Probable Maximum Precipitation (PMP) methodology and expressions), the satellite maximum precipitation is found between 2 and 3.5 times lower than gauge records. Different lengths of records and observational footprints (microscale for the gauge vs. mesoscale for the satellite a $0.25° \times 0.25°$ i.e. 625 km^2) explain those discrepancies between satellite and surface stations observations. In particular, the satellite footprint being often larger than the size of convective cells, differences can be important during heavy rain events and peak rainfall periods (Wolff and Fisher 2008).

40.2.2 Thresholds, Percentiles and Spatial Distribution of Daily Extremes

The cumulative distribution of the daily rainfall events for the period 1998–2013 (Fig. 40.2) shows that the differences between the three products increase with increasing percentile values. The values for the daily rain rate differences at the 70th percentile range from 5 mm day^{-1} for GPCP to 7 mm day^{-1} for PERSIANN that is a 40% difference. This difference increases to 67% at the 90th percentile (12 mm day^{-1} for GPCP vs. 20 mm day^{-1} for CMORPH), and up to 100% at the 99th percentile (30 mm day^{-1} for GPCP vs. 60 mm day^{-1} for CMORPH). The

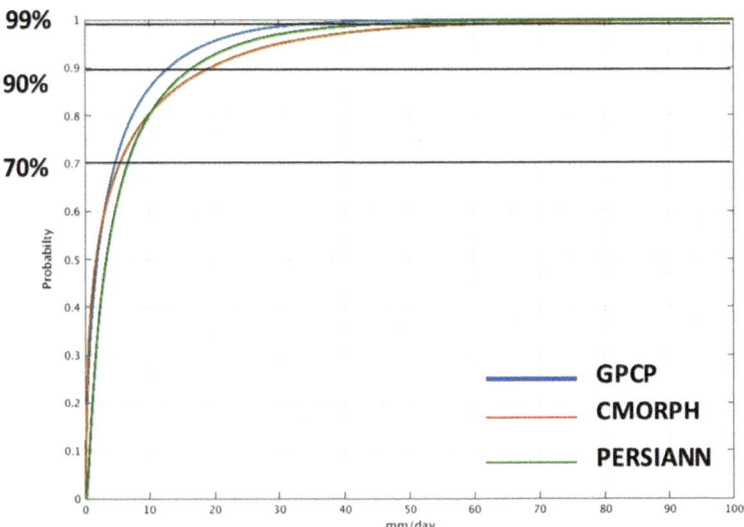

Fig. 40.2 Cumulative distribution function of daily rainfall derived from three satellite products GPCP, CMORPH, and PERSIANN-CDR for the period 1998–2013. (Adapted from Prat et al. 2017)

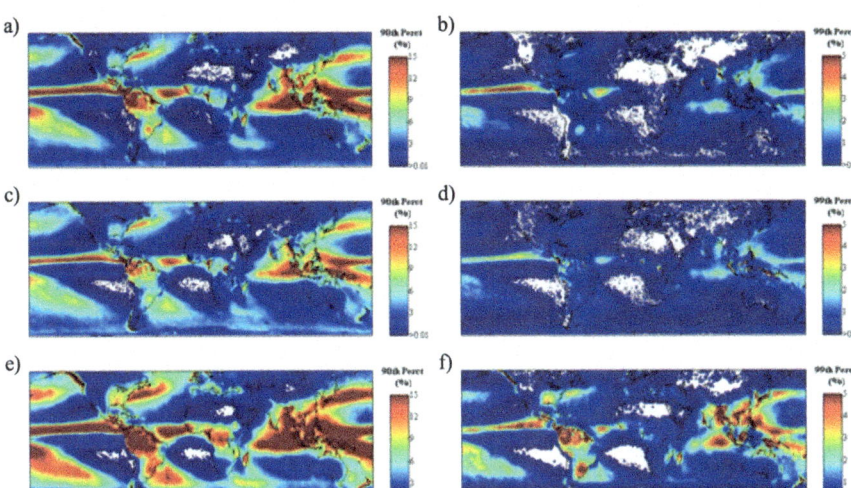

Fig. 40.3 Spatial distribution of daily rainfall above the 90th (first column) and 99th (second column) percentiles for PERSIANN (**a, b**), CMORPH (**c, d**), and GPCP (**e, f**). Values of the percentiles of each satellite product are defined according to the distribution presented in Fig. 40.2. (Adapted from Prat et al. 2017)

magnitude of such differences illustrates the impact of the different observing platforms, retrieval algorithms and bias adjustment techniques.

Figure 40.3 shows the spatial distribution of daily rainfall above the 90th (Fig. 40.3a, c, e) and 99th (Fig. 40.3b, d, f) percentiles with respect for the multi-sensor precipitation products PERSIANN (Fig. 40.3a, b), CMORPH (Fig. 40.3c, d), and GPCP (Fig. 40.3e, f). The geographical extent of daily rainfall above the 90th percentile is more widespread for GPCP (Fig. 40.3e), than for PERSIANN (Fig. 40.3a), and CMORPH (Fig. 40.3c) for which extremes are preferably localized over locations experiencing intense convective activity, that is along the ITCZ, the Gulf Stream in the Atlantic Ocean and the Kuroshio Current along the Pacific Coast of Japan over ocean, and the Amazon, Southeast Asia, and the Southeastern United States over land. With increasing values of the threshold that is for daily rainfall corresponding to the 99th percentile, the distribution of daily precipitation above the threshold is further reduced to mostly ocean locations for CMORPH (Fig. 40.3d), ocean and selected land locations (i.e., at the convergence of the ITCZ in Columbia) for PERSIANN (Fig. 40.3b), while GPCP displays both ocean and land widespread distributions of daily rainfall above the 99th percentile (Fig. 40.3f). Such differences clearly show the challenge in defining extreme precipitation from different observation platforms as well as the difficulty in quantifying those daily extremes.

Recently, Sun et al. (2018) have provided comparable information for the spatial distribution of the 90th percentile for 15 satellite, reanalysis, and in-situ global precipitation datasets (see Sun et al. 2018, their Figure 13). Like here, they found large differences overland and over ocean. The spread for the value of the 90th percentile is higher between the different type of products (reanalysis, in-situ for overland, and satellite) than between the different satellite products (CMAP, CMORPH, GPCP, MSWEP, PERSIANN-CCS, PERSIANN-CDR, TMPA 3B42). The most important differences are located along the tropical belt, both land and ocean, and over mountainous areas that correspond to the zones of strong convective and orographic activity (Sun et al. 2018). In Table 40.2 is provided a selection of recent studies of long-term evaluation of satellite precipitation products (SPPs) with respect to extreme rainfall. Those studies use slightly different definitions of extreme rainfall: percentiles (90th, 95th, or 99th and above: AghaKouchak et al. 2011; Lockhoff et al. 2014; Bharti et al. 2016; Jiang et al. 2017; Katiraie-Boroujerdy et al. 2017a, b; Nguyen et al. 2018; Sun et al. 2018), or thresholds for daily rainfall (>40-, 50-, or 100-mm day^{-1}: Gao and Liu 2013; Gosset et al. 2013; Gebremichael et al. 2014; Yong et al. 2014; Prat and Nelson 2015; Tan and Santo 2018). Apart from rare examples (Gosset et al. 2013; Bharti et al. 2016), the general consensus that arises from those studies is that SPPs tend to underestimate extreme precipitation (AghaKouchak et al. 2011; Gao and Liu 2013; Gebremichael et al. 2014; Lockhoff et al. 2014; Yong et al. 2014; Brena-Naranjo et al. 2015; Prat and Nelson 2015; Jiang et al. 2017; Katiraie-Boroujerdy et al. 2017a, b; Nguyen et al. 2018; Tan and Santo 2018). Furthermore, the performance of SPPs is decreasing with increasing threshold or percentiles (AghaKouchak et al. 2011; Lockhoff et al. 2014; Prat and Nelson 2015) and bias-adjusted products (TMPA 3B42, PERSIANN-CDR, IMERG_F) perform better than their unadjusted counterpart or other SPPs (Yong et al. 2014; Prat and Nelson 2015; Nguyen et al. 2018). Furthermore, SPPs performance depend on the choice of the metrics used for assessment (Bias, Correlation Coefficient (CC), Probability of Detection (POD), False Alarm ratio (FAR), Root Mean Square Error (RMSE) among others), the location where this evaluation is conducted (mountainous areas, areas with strong convective activity for instance), the season considered (cold versus warm precipitation), and the underlying physical principles (IR, PMW) of the algorithm used for rainfall retrieval and the bias-adjustment procedure if any (Table 40.1). Typically, there are no ideal SPPs when considering precipitation extremes. A key element in the success of using SPPs for extreme event identification, monitoring, and applications is to evaluate the limitations, strengths and weaknesses of each product.

40.3 Application of Satellite QPE to Extreme Events

40.3.1 Rainfall Extremes at the Daily Scale

In this section we provide some examples of applications of satellite data for extreme rainfall quantification. We also emphasize the differences between real-time and bias-adjusted satellite precipitation data. After investigating the ability of the different datasets to describe precipitation patterns, this section investigates their ability to capture intense and extreme precipitation at the daily scale. A conditional analysis

Table 40.2 Long term evaluation of gridded (level III) satellite precipitation products (SPPs) in relationship with extreme precipitation

Products	Location	Duration	Reference	Extreme definition	Main result when it comes to extreme rainfall	Authors
CMORPH, PERSIANN, TMPA (RT, 3B42)	Central US	4 year (2005–2008)	Radar (Stage IV)	Percentiles (90th, 95th)	SPPs performance varies as a function of the metric. SPP performance worsen with increasing precipitation threshold. No ideal SPP to capture rainfall extremes.	AghaKouchak et al. (2011)
CMORPH, PERSIANN, TMPA (RT, 3B42)	Tibetan Plateau	6 year (2004–2009)	Gauges	Threshold (>10 mm day^{-1})	Biases of PERSIANN and TMPA RT show dependence with topography. SPPs underestimate heavy rainfall (>10 mm day^{-1})	Gao and Liu (2013)
CMORPH, EPSAT, GPCP, GSMaP, PERSIANN, RFE, TMPA (RT, 3B42)	West Africa	8 yr. (JJAS 2003–2010)	Gauges	Threshold (>40 mm day^{-1})	SPPs (satellite only) tend to overestimate heavy rainfall (>40 mm day^{-1})	Gosset et al. (2013)
CMORPH, TMPA (RT, 3B42)	Blue Nile Basin	2 year (summers 2012–2013)	Gauges	Threshold (>20 mm day^{-1})	SPPs biases varying with respect to topography. SPPs underestimate heavy rain rates (>20 mm day^{-1})	Gebremichael et al. (2014)
GPCP 1DD	Europe	11 year (1998–2008)	Gauges (E-OBS)	Percentiles (90th, 95th)	POD (FAR) of extreme events drops (increases) with increasing percentile. SPP presents poor skill in the detection of extreme events. SPP is better in capturing MCS (JJA) and frontal systems (DJF) than thunderstorms.	Lockhoff et al. (2014)

(continued)

Table 40.2 (continued)

Products	Location	Duration	Reference	Extreme definition	Main result when it comes to extreme rainfall	Authors
TMPA (RT, 3B42:V6 and V7)	China (basin)	7 year (2003–2009)	Gauges	Threshold (>100 mm day^{-1})	Research SPPs are improved when compared with RT SPPs. TMPA V7 SPP RT statistics are close to research SPPs. All SPPs severely underestimate string events (128 mm day^{-1}). Caution when using SPP for monitoring precipitation extremes.	Yong et al. (2014)
TMPA 3B42	Global	16 year (1998–2013)	Gauges	Max precipitation (30 min to 2 year)	Underestimation (ratio 2 and 3.5) for max precipitation when compared to historical gauges.	Breña-Naranjo et al. (2015)
TMPA (RT, 3B42)	CONUS	11 year (2002–2012)	Gauges Radar (Stage IV)	Thresholds (>50, 100 mm day^{-1})	TMPA RT displays poor skills in capturing extreme rainfall (>100 mm day^{-1}) (NW US). TMPA 3B42 presents an improvement with respect to TMPA RT. TMPA 3B42 biases remain important for higher daily average rainfall (>5 mm day^{-1}).	Prat and Nelson (2015)
TMPA 3B42	Himalayas	16 year (1998–2013)	Gauges (gridded ¼ degree)	Percentiles (98th, 99th, 99.99th)	Important differences between SPP and gauges in 80% of the extreme events (>98th). For eight events (out of ten above the 99.99th percentile), SPP indicates higher rainfall. Differences might be due to the gauge locations (foothills) that miss orographic effect.	Bharti et al. (2016)

TMPA 3B42, IMERG	India	17 year (1998–2014) 3 month (JJA 2014: IMERG)	Gauges (gridded ¼ degree)	Percentile (75th)	TMPA over detects heavy rainfall events (>75th percentile). IMERG shows improvement in capturing heavy rainfall (limited assessment).	Prakash et al. (2016)
TMPA 3B42	China (basin)	15 year (1998–2012)	Gauges	Percentile (95th) Threshold (>50 mm day^{-1})	Underestimation of intense rainfall (>50 mm day^{-1}). Slight overestimation of extreme rainfall over the 95th percentile of daily rainfall.	Jiang et al. (2017)
PERSIANN-CDR (a, b) TMPA 3B42 (a)	Iran	(a) 10 year (1998–2007) (b) 30 year (1983–2012)	Gauges (gridded ¼ degree)	Percentile (90th) Threshold (>20 mm day^{-1})	(a) TMPA 3B42 captures extreme indices better than PERSIANN-CDR. (b) PERSIANN-CDR underestimates precipitation over the mountainous areas and the intensity of the 90th percentile.	Katiraie-Boroujerdy et al. (2017a, b)
PERSIANN, PERSIANN-CCS, PERSIANN-CDR	CONUS	13 year (2003–2015)	Gauges (gridded ¼ degree CPC)	Percentiles (95th, 99th) Threshold (>10 mm day^{-1})	PERSIANN-CDR (bias adjusted) outperforms the other two products. PERSIANN-CDR exhibits underestimation of extrems (> 95th, 99th percentiles).	Nguyen et al. (2018)
CMAP, CMORPH, GPCP, GSMaP, MSWEP, PERSIANN (CCS, CDR), TMPA 3B42	Global	8 year (2003–2010)	Gauges (gridded ¼ degree GPCC 1DD, CPC)	Percentile (90th)	Differences between SPPs depend on the location, topography, and the underlying algorithm for rainfall estimation (IR, PMW).	Sun et al. (2018)
TMPA (RT, 3B42), IMERG (_E, _L, _F)	Malaysia	2 year (2014–2016)	Gauges	Threshold (>50 mm day^{-1})	Good performance at the annual and monthly scale for all SPPs. All SPPs underestimate light (<1 mm day-1) and extreme (>50 mm day-1) rainfall.	Tan and Santo (2018)

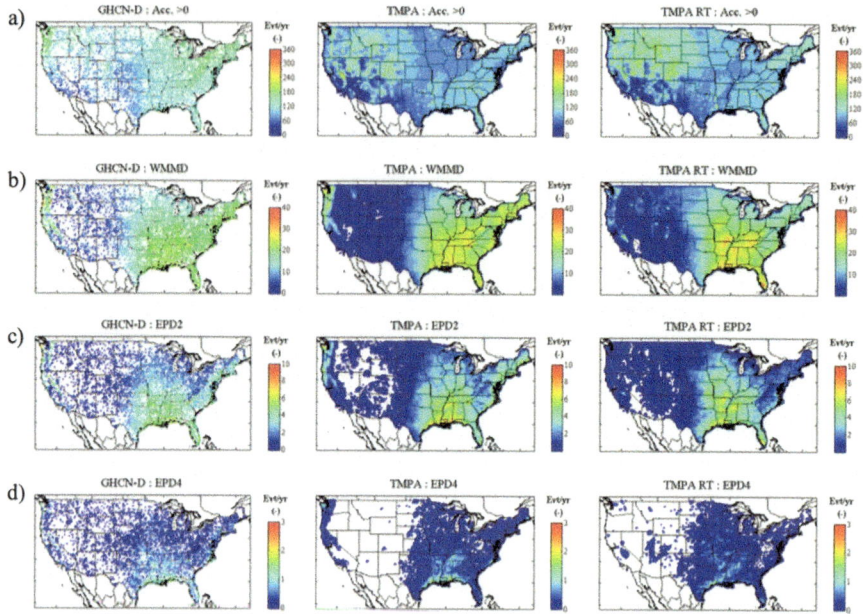

Fig. 40.4 (**a**) Number of rainy days (R > 0 mm day^{-1}), (**b**) Wet Millimeter Days (R > 17.8 mm day^{-1}: WMMD), (**c**) Rainy Days with accumulation greater than 2 in day^{-1} (R > 50.8 mm day^{-1}: EPD2), and (**d**) Rainy Days with accumulation greater than 4 in day^{-1} (R > 101.6 mm day^{-1}: EPD4) for GHCN-D (first column), TMPA 3B42 (second column), and TMPA 3B42RT (third column). (Adapted from Prat and Nelson 2015)

was conducted using different thresholds for the daily accumulation (Fig. 40.4). Figure 40.4a displays the average number of rainy days per year derived from GHCN-D (first column), 3B42 (second column), and 3B42RT (third column). For TMPA 3B42, and TMPA 3B42RT the daily accumulation is computed 1200Z-1200Z. For GHCN-D, the daily accumulation computed depends on the local time and is 0700 LST-0700 LST for most of the locations, which corresponds to 1200Z-1200Z over Eastern US. Therefore, some uncertainties could arise from computing daily accumulation over a slightly different time period. Although the number of rainy days appears consistent in terms of magnitude between in-situ data and satellite measurements, there are noticeable differences over specific areas. Despite a problematic visual comparison between point data measurements from GHCN-D and satellite gridded estimates due to the scarcity of station coverage, all products present a very similar pattern and a comparable number of rainy days throughout CONUS. When compared to GHCN-D, 3B42 displays a lower number of rainy days over the Northeast, the Middle-Atlantic, the Central Plains over the Ohio River Basin, and the Northwestern Coast (Fig. 40.4a). On the other hand, 3B42 has a higher number of rainy days over the Rockies encompassing part or all of the Missouri River Basin, the Colorado River Basin, and the Southwest over California and Nevada when compared to GHCN-D. Different sensitivity for light rainfall detection thresholds for each sensor, the ability to retrieve snow/frozen precipitation, the beam blockage over the Rockies, and/or the influence of temporal and spatial resolution can explain the differences. Overall, the GHCN-D stations and the satellite product TMPA 3B42 exhibit a satisfying visual agreement over CONUS despite the local differences mentioned above. More important differences are observed with the real-time TMPA 3B42RT dataset. Differences between the rain gauge-adjusted satellite dataset 3B42 and 3B42RT are particularly relevant over the Western US (Rocky Mountains) and at higher latitudes with more rainy days for 3B42RT. For daily accumulation greater than the Wet Millimeter Days (WMMD: $R > 17.8$ mm day^{-1}), significant differences are found over the Northwest and the Southeastern US (Fig. 40.4b). The Wet Millimeter Day threshold corresponds to the precipitation days that exceed the highest daily average over the area considered (Shepherd et al. 2007). For North America, this maximum daily average (17.8 mm day^{-1}) is recorded in Henderson Lake (British Columbia) (source National Centers for Environmental Information, NCEI). TMPA 3B42 shows similar distribution patterns of WMMD than in-situ data GHCN-D. The most important differences are found over the Northwest (Fig. 40.4b). The biggest differences are observed with 3B42RT that shows a much lower number of rainy days greater than WMMD when compared to the bias-adjusted 3B42 (Fig. 40.4b). This is consistent with the underestimation observed for the 3B42RT-based daily averages and to a lesser extent for 3B42 (Prat and Nelson 2015). The bias-adjustment increases the number of WMMD of 3B42 closer to the values obtained with the GHCN-D in-situ data. Conversely, 3B42 and 3B42RT that displayed less rainy days than GHCN-D (Fig. 40.4a), present a higher occurrence of WMMD over the Northeast, the Upper and Lower Mississippi River Basins (Fig. 40.4b). For a daily accumulation greater than 2 in day^{-1} (50.8 mm day^{-1}) (hereafter EPD2), GHCN-D and Stage IV display comparable counts for EPD2 over the Eastern US where rain gauge coverage is denser (Fig. 40.4c). Over the Lower Mississippi and the Southeast US, 3B42 and 3B42RT show a higher number of days with rainfall above 2 in day^{-1} as compared to the GHCN-D in-situ data. Daily precipitation greater than 4 in day^{-1} (101.6 mm day^{-1}) (hereafter EPD4) is limited to the Pacific Coast and East of 100°W, a domain regularly impacted by tropical cyclones (Prat and Nelson 2013a,b, 2014) (Fig. 40.4d). These EPD4 events are relatively infrequent (3 counts or less per year) and roughly correspond to the 0.1–0.5% top daily events regardless of the River Forecast Center (RFC) considered. The bias-adjusted 3B42 and real-time 3B42RT exhibit a comparable number of EPD4 events over the Southeastern US. The maximum occurrences are observed over the Lower Mississippi River Basin and are higher than the daily counts retrieved from the GHCN-D station data. Over the Northwest, the bias-adjusted (3B42) is able to better capture such extreme daily accumulation events (EPD4) with respect to 3B42RT that detects almost no days with rainfall above 4 in day^{-1} (Fig. 40.4d).

40.3.2 Daily Rainfall Extremes in Relation to Cyclonic Activity

Due to their global coverage, satellite precipitation products have gained popularity in the evaluation of the global distribution of precipitation by providing homogenous data regardless of the location. Among the numerous applications, we can cite the quantification of precipitation extremes in relation to tropical cyclone activity (Shepherd et al. 2007; Jiang and Zipser 2010; Chen et al. 2013; Prat and Nelson 2013a,b, 2016). For instance, the TMPA 3B42 dataset, which is available at the sub-daily temporal resolution of 3-h, allows for monitoring tropical cyclone (TC) activity from genesis to decay and from the tropical belt to higher latitudes. Figure 40.5 summarizes a long-term analysis of extreme daily precipitation events in relation to global tropical cyclone activity using the TC average annual rainfall (Fig. 40.5a) and the TC contribution to rainfall totals (Fig. 40.5b). The use of a global satellite precipitation product allows to directly compare the TC rainfall contribution and related extreme daily precipitation contribution independently of sensor type or measurement methods. It also allows for a global assessment in places where in-situ data are not always available, that is over sparsely populated areas and/or over developing countries. Results show that the maximum TC contribution over land is 50% (Southern Baja California), 37% (Southeast Taiwan), 45% (Southern Oman), 40% (Western Australia), and 29% (Southwestern Madagascar) for each of the five basins identified as (1) North and Central America (NCA), (2) Southeast and East Asia (SEA), (3) South and West Asia (SWA), (4) Oceania (OCE), and

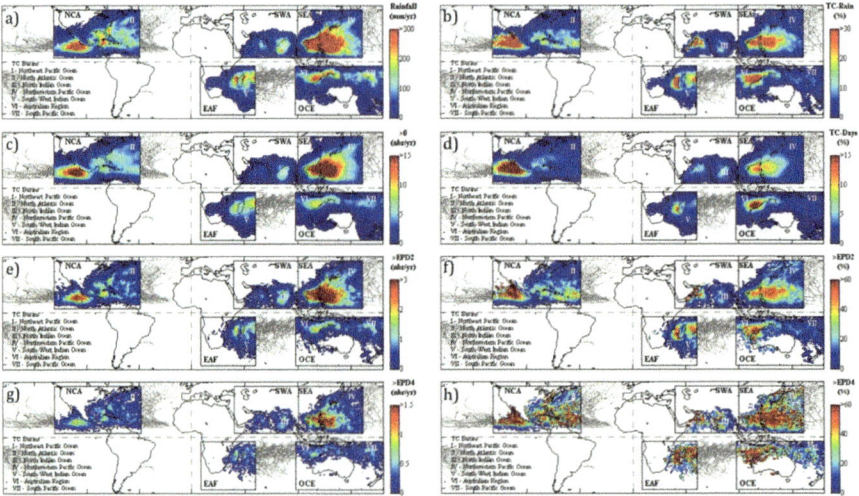

Fig. 40.5 (**a**) Average annual tropical cyclone rainfall, and (**b**) annual tropical cyclone rainfall contribution. (**c**) Average annual tropical cyclone number of rainy days with rainfall R > 0, and (**d**) tropical cyclone contribution to daily rainfall R > 0. (**e**) and (**f**) same as (**c**) and (**d**) for EPD2. (**g**) and (**h**) same as (**c**) and (**d**) for EPD4. (Adapted from Prat and Nelson 2016)

(5) East Africa (EAF) (Fig. 40.5b), respectively. Some of the largest TC contributions are observed over arid areas (Southern Baja California, Western Australia, Arabic Peninsula). While the first two regions experience relatively regular cyclonic activity, the Arabic Peninsula, despite exposure to a lesser cyclonic activity from the North Indian Ocean Basin ($\approx$1/4 of NCA), presents a higher contribution (40%), which is principally associated with Cyclone Keila (2010). Globally, we find a similar spatial distribution between the number of TC events (Fig. 40.5c) and the TC average annual rainfall (Fig. 40.5a), and between the TC-contribution (Fig. 40.5d) and the percentage of days associated with cyclonic activity (Fig. 40.5b). Because tropical storms are generally associated with more intense precipitation, the TC contribution is higher than the corresponding proportion of TC-days. For NCA, while tropical cyclones account locally for between 2% (East Coast, Gulf of Mexico) and 20% (Baja California, Mexico Pacific Coast) of precipitation days (Fig. 40.5d), they represent from 15% (Southeastern US, Atlantic Coast, Caribbean) to over 80% of EPD2 for Baja California (Fig. 40.5f). In addition, TCs cover more than 50% of EPD4 over Florida and Coastal Carolinas and account for almost all of EPD4 for Baja California (Fig. 40.5h). Furthermore, even over areas where TCs are infrequent like New England (Keim et al. 2007), they can be responsible for more than 70% of EPD4 (Fig. 40.5h). On average, TCs represent 2.7% (SWA), 3.1% (NCA), 3.7% (SEA), 4.1% (OCE), and 4.4% (EAF) of the total number of rainy days (Fig. 40.5d). Regardless of the domain, the proportion of TC-days remains relatively small but the proportion of extreme events linked to TC-activity increases with increasing daily rainfall threshold (EPD2, EPD4). More specifically, extreme rainfall linked with TC activity accounts for between 9–18%, and 13–30% for EPD2 (Fig. 40.5e, f), and EPD4 (Fig. 40.5g, h), respectively for the five domains. There is a 5-to-8-fold increase between the proportion of daily TC counts (2.7–4.4%) and the EPD4 ratio (13–31%) associated with TCs. One of the benefits of using global satellite precipitation data by contrast with in situ data, is that it allows to capture extreme events patterns at the local scale. For SEA, SWA, OCE, and EAF, the maximum proportion of TC-days is locally about 17%, 8%, 20%, and 18% (Fig. 40.5d), respectively. Locations experiencing marginal TC-activity such as the arid Arabian Peninsula (SWA) present clusters of TC extremes associated with individual storms. The maxima for EPD2 (Fig. 40.5f) and EPD4 (Fig. 40.5h) localized near Oman's easternmost point and over Southern Oman bear the signature of cyclones Gonu in 2007 (Dube et al. 2009; Abdalla and Al-Abri 2011) and Keila in 2010, respectively. Locally, TCs account for more than 80% of EPD4 events even for higher latitude areas which seldom experience TCs (Northwest China) or over areas farther inland in central India along the Satpura Range (Fig. 40.5h). A similar situation is found for OCE with an important ratio of TC-related EPD4 in Northern and Western Australia. We found that EPD4 associated with TCs accounted for more than 60% over northwestern Australia (Fig. 40.5h). Although each domain displays comparable trends and similar quantitative results for TC contribution to extreme precipitation events, the spatial distribution of extreme events is intertwined with TC activity and local climatological and geographical characteristics (arid, tropical, specific precipitation regimes).

Typically satellite precipitation products tend to underestimate rainfall associated with landfalling tropical cyclones (Villarini et al. 2011). While the pattern and quantitative value of the rainfall contribution of TCs were found to agree relatively well for estimates derived from TRMM 3B42 with respect to surface stations (Prat and Nelson 2013a), differences can be important on an event-to-event basis (Villarini et al. 2011). In the case of the 2009 Typhoon Morakot over Taiwan, all the satellite precipitation products tended to underestimate rainfall ranging from -19 to -61% for 3B42RT, PERSIANN-CCS, 3B42V6, and CMORPH (Chen et al. 2013).

40.3.3   Other Applications of SPPs with Respect to Rainfall Extremes

In addition to the examples provided before, there are numerous other applications of SPPs are centered around precipitation extremes. A selection is presented in Table 40.3.

In addition to the application to extreme events associated with tropical cyclones (Jiang and Zipser 2010; Prakash et al. 2012; Chen et al. 2013; Villarini et al. 2011; Prat and Nelson 2013a, b, 2016), other studies have focused on extreme precipitation associated with atmospheric rivers (Behrangi et al. 2016; Nayak and Villarini 2018), Indian monsoon (Bharti et al. 2016; Prakash et al. 2016; Prasanna 2016), the El Niño-Southern Oscillation (Curtis et al. 2007; Ricko et al. 2016), and convective systems (Zipser et al. 2006; Hirose et al. 2009, Rasmussen and Houze 2011; Levizzani et al. 2013; Rapp et al. 2014; Liu and Zipser 2015) to name a few.

SPPs have also been used with some encouraging success for the modeling of hydrological extremes including landslide (Kirschbaum et al. 2012), floods early warning and monitoring (Kachi et al. 2006; Zhang et al. 2015; Koriche and Rientjes 2016; Libertino et al. 2016; Tekeli and Fouli 2016; Tan and Santo 2018), and extreme rainfall and streamflow predictions (Su et al. 2017). With their period of record increasing, SPPs become more and more suitable for applications that require longer historical databases.

In recent years, SPPs have been used in the context of long-term event analysis for the determination of probable maximum precipitation, annual maximum series (AMS), and the computation of intensity–duration–frequency (IDF) curves (Endreny and Imbeah 2009; Pombo and de Oliveira 2015; Zhou et al. 2015; Marra et al. 2017; Taylor et al. 2017; Yang et al. 2018). However, although promising for the future (Marra et al. 2017) in particular for poorly gauged areas, the use of satellite precipitation estimates for hydrological design require extreme caution due to the underestimation of extreme rainfall. A thorough evaluation of the SPP biases in the upper percentiles and the development of appropriate adjustment methods (Gado et al. 2017; Faridzad et al. 2018; Ombadi et al. 2018), would allow SPPs to fill the observational gap in places where in-situ records are lacking or questionable. Furthermore, the next generation of SPPs such as GPM IMERG, for which

Table 40.3 Examples of recent studies using gridded (level III) satellite precipitation products (SPPs) for applications in relationship with extreme precipitation

Products	Location	Duration	Application	Conclusions	Authors
CMORPH, PERSIANN (PERSIANN, CCS), TMPA (RT, 3B42)	Western US	10 year (2003–2012)	Atmospheric rivers (AR)	SPPs significantly underestimate precipitation and barely capture orographic precipitation with difficulties over snow/frozen surfaces. Concerns regarding near-real time SPPs monitoring of ARs.	Behrangi et al. (2016)
CMORPH, PERSIANN, TMPA (RT, 3B42)	Central US	12 year (2003–2014)	Atmospheric rivers (AR)	TMPA 3B42 is found to be the best SPP for accumulation and rain rates associated with ARs. ARs contribute to about 35% of annual rainfall.	Nayak and Villarini (2018)
TMPA 3B42	(a) Southeastern US, (b) Global	12 year (1998–2009)	Tropical cyclones (TC)	The percentage of rainfall associated with TC increases with increasinf rain intensity and represents about 20% of heavy rainfall (> mm h^{-1}) (a). Globally, TCs contribute to 5–10% of annual precipitation for basins around the world (b).	Prat and Nelson (2013a, b)
TMPA 3B42	Global	15 year (1998–2012)	Tropical cyclones (TC)	TCs account for 3.5 ± 1% of the totalò number of rainy days over TC basins. TC days represent between 13% and 31% of daily extremes (>100 mm day^{-1}).	Prat and Nelson (2016)
GPCP-monthly (a), TMPA 3B42 (b)	Global	(a) 27 year (1979–2005) (b) 8 year (1998–2005)	ENSO	Monthyl and daily precipitation extremes in relation to ENSO. Frequency of intense rain rates (>20, 50 mm day^{-1}) show a relationship with ENSO.	Curtis et al. (2007)
TMPA 3B42	Ghana	7 year (1998–2006)	IDF curves	SPP useful to develop IDF curves for short gauge records or poorly gauged areas. Limitation of SPP IDF curves to durations of 3 h or higher.	Endreny and Imbeah (2009)

(continued)

Table 40.3 (continued)

Products	Location	Duration	Application	Conclusions	Authors
CMORPH	Eastern Mediterranean	16 year (1998–2013)	IDF curves	Good agreement between SPP and radar IDF curves for a range of varying climates. Potential for using SPP IDF curves in ungauged areas.	Marra et al. (2017)
PERSIANN-CDR	CONUS (river basin)	33 year (1983–2015)	IDF curves	Adjustment of annual maximum time series of SPP prior to derive IDF curves. Method improves Annual Maximum Series (AMS) in particular at high elevation. SPP IDF curves fall within Atlas 14 IDF curves uncertainties.	Faridzad et al. (2018)
PERSIANN-CDR	CONUS	35 year (1983–2017)	IDF curves	Method to develop IDF curves from SPP. SPP IDFs show considerable underestimation before adjustment. Extensive assessment of SPP uncertainties prior to computation of IDF curves.	Ombadi et al. (2018)
TMPA 3B42	Angola	16 year (1998–2013)	Annual max daily rainfall	TMPA 3B42 slightly underestimates annual maximum daily precipitation. SPP useful for estimating extreme precipitation values for different return periods.	Pombo and de Oliveira (2015)
TMPA RT	Global	15 year (1998–2012)	Extreme rainfall frequency	Provides useful early warning information for potentially extreme events as a complement to surface based data. Large uncertainties in Average Recurrence Interval (ARI) for regions with complex topography.	Zhou et al. (2015)
PERSIANN-CDR	Western US (CA, CO)	35 year (1983–2017)	Extreme rainfall frequency	Correction of SPP annual Maximum Series (AMS) to match gauge data. The method allows using SPP for extreme frequency analysis in ungauged areas.	Gado et al. (2017)
CMORPH, PERSIANN-CDR, TMPA 3B42	China (basin)	16 year (1998–2013)	Probable Maximum Precipitation (PMP)	CMORPH and TMPA 3B42 agree well with gauge data over complex terrain (correlation, 24 h PMP) and can be used for PMP estimation in ungauged regions.	Yang et al. (2018)

TMPA 3B42 (a) IMERG (b)	Italy	(a) 16 year (1998–2013) (b) 1 year (2014–2015)	Hydrologic design	TMPA 3B42 underestimates rainfall for deep convection systems. Preliminary analysis using IMERG shows significant improvement.	Libertino et al. (2016)
CMORPH	Ethiopia (river basin)	3 year (JJA: 2007–2009)	Flood early warning	Development of SPP based flood index (rainfall + DEM) for flood early warning. Effectiveness of SPPs for flood early warning.	Koriche and Rientjes (2016)
TMPA RT	Saudi Arabia	14 year (2000–2013)	Flood forecasting	Flood forecasting indexes derived from SPP capture high rain rates, daily, and seasonal variations of extreme events.	Tekeli and Fouli (2016)
TMPA 3B42	Global	13 year (1998–2010)	Landslides	SPP rainfall variability significantly correlates with increase in landslide activity. Use of SPPs for developing a global rainfall-triggered landslide climatology.	Kirschbaum et al. (2012)

preliminary assessments indicate improvements over existing TRMM products (Libertino et al. 2016; Prakash et al. 2016), open some promising perspectives for the use of SPP to hydrological design. In particular, shorter time scales (sub-hourly when compared to 3-hourly) would allow accounting for short term and extreme flash flood situations.

40.4   Verification of Satellite QPE Extremes with Respect to In-Situ Rain Gauge Observations

One of the challenges in assessing the ability of the different satellite products to capture extreme precipitation is the availability of quality controlled in-situ data. Furthermore, most of the bias-adjusted multi-sensor products use the same in-situ data and therefore are not fully independent. In addition to instrumental limitation, another limitation comes from the fact that we are comparing surface area rainfall (size of a sensor pixel scan for un-gridded data or size of a pixel for multisensory satellite products) to point measurements (gauge footprint). To compare the different satellite estimates with in-situ data, we assume that each rain gauge represents with sufficient accuracy the area-averaged rainfall over the native resolution of the different satellite products evaluated (i.e., $1° \times 1°$ for GPCP, $0.25° \times 0.25°$ for TMPA, CMORPH, PERSIANN, and $0.1° \times 0.1°$ for IMERG). While, there are well known limitations of using rain gauge point measurements to evaluate area-averaged rainfall retrieved from sensors with coarser spatial resolution (Ciach and Krajewski 1999; Ciach et al. 2003, 2007; Habib et al. 2004), the random sampling errors due to differing resolutions are mostly dominant at the sub-daily scales (Ciach and Krajewski 1999). For accumulation periods of several days the correlation distance (maximum distance between stations beyond which the correlations become insignificant, i.e. defined as the distance at which the pixel-to-pixel correlation decreases below $1/e \approx 0.37$) is of the order of several hundred kilometers (Gutowski et al. 2003). Those distances are several orders of magnitude greater than the satellite products spatial resolution. For unconditional rain rates, the correlation distance depends on the time scale considered. When it comes to extreme precipitation events (i.e., conditional analysis), this distance depends on the threshold considered (value or percentiles). Typically, over CONUS and for the daily rainfall value corresponding to the 90th percentile, the average correlation distance (at $1/e$) is about 30–80 km (Prat and Nelson 2015). Typically, the average correlation distance is greater than the representative footprint of each remotely sensed observation platform that varies from about 5 km (GPM DPR and TRMM PR), to 10 km for IMERG and 25 km for TMPA, CMORPH, and PERSIANN, and up to 100-km for GPCP. For such extreme daily events, the average correlation distance of about 30–80-km is then comparable with the satellite footprint. Another issue in assessing the ability of satellite products to capture extreme precipitation events stems from the fact that gauge daily accumulation reporting time (in-situ networks like GHCN-D

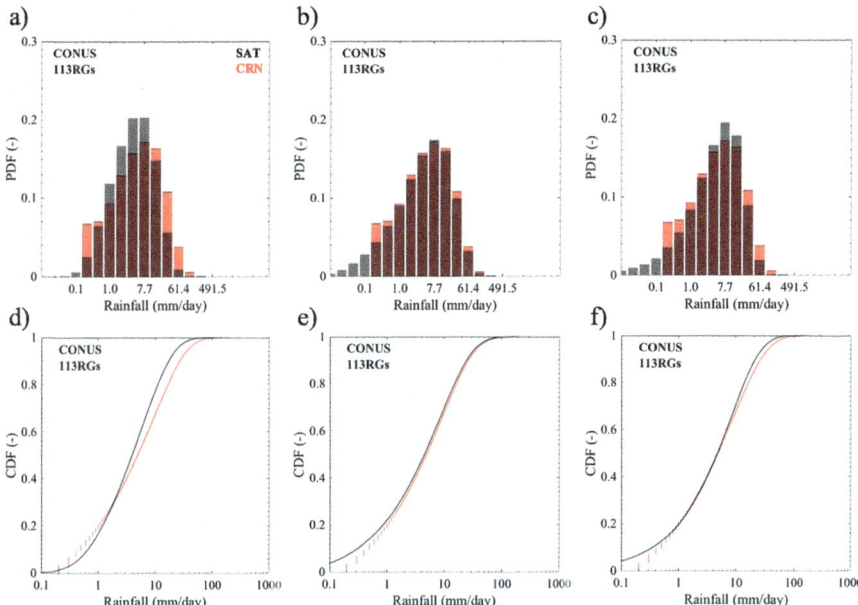

Fig. 40.6 (**a–c**) Distribution of daily rainfall for events observed simultaneously by USCRN stations and satellites. (**a**) PERSIANN-CDR, (**b**) CMORPH, and (**c**) GPCP. (**d–f**) Same as (**a–c**) but for the cumulative distribution function of USCRN stations and (**d**) PERSIANN-CDR, (**e**) CMORPH, and (**f**) GPCP for the period 2007–2015. (Adapted from Prat et al. 2017)

report rainfall totals on a 24-h accumulation: Menne et al. 2012) might not match satellite products daily accumulation that are typically 0000Z-0000Z. For satellite products available at a sub-daily temporal resolution (CMORPH, TMPA, IMERG), the daily accumulation should match (or at least come close to) the in-situ data daily accumulation. The US Climate Reference Network (USCRN, Diamond et al. 2013) operates 114 stations over CONUS (plus 21 over Alaska and Hawaii) and provides 5-min, 1-h, 1-day (local time) rain accumulations, which allows to compute any daily total to match the satellite window (0000Z-0000Z). Figure 40.6 shows a comparison of the daily rainfall distribution for each satellite product (PERSIANN, CMORPH, GPCP) compared with the USCRN measurements for the same period 0000Z-0000Z and for concurrent events; that is when rain is detected by the satellite and the surface station.

For PERSIANN (Fig. 40.6a) both distributions have the same width, while for CMORPH (Fig. 40.6b) and GPCP (Fig. 40.6c) the daily rainfall derived from satellite measurements tends to differ more on the left-hand side of the spectra, with the satellite indicating some daily accumulations below those recorded by the gauges. For the right-hand side of the spectra, toward higher intensity events, CMORPH presents a better agreement with the in-situ records (Fig. 40.6b), while GPCP (Fig. 40.6c) and PERSIANN (Fig. 40.6a) underestimate EPD2 extreme events. A comparison of the cumulative distribution from satellite and USCRN

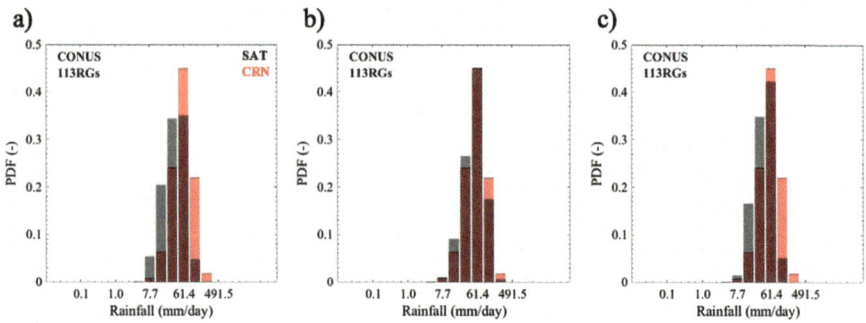

Fig. 40.7 Distribution of annual maximum daily rainfall for events observed simultaneously by USCRN stations and satellites. (**a**) PERSIANN-CDR, (**b**) CMORPH, and (**c**) GPCP for the period 2007–2015. (Adapted from Prat et al. 2017)

shows that for concurrent events CMORPH is in very good agreement with the USCRN (Fig. 40.6e), with GPCP coming in second with satellite-observed events above the 50th percentile found lower than USCRN events (Fig. 40.6f). Finally, PERSIANN tends to underestimate rainfall events located above the 35th percentile (Fig. 40.6d).

A similar comparison is presented for the values of the annual maximum daily rainfall recorded at every CONUS USCRN site and at the corresponding satellite pixel (Fig. 40.7). Results indicate that CMORPH (Fig. 40.7b) tends to better capture extreme precipitation than its counterparts GPCP (Fig. 40.7c) and PERSIANN (Fig. 40.7a). Being limited to the CONUS, this result is not for generalization. Several studies that investigated the performance of different satellite products with respect to surface observations indicated that these latter vary as a function of location and season. As a physically based retrieval, for which the measurement is directly affected by the size and the distribution of hydrometeors, PMW products typically have better performances than IR products that assume a direct relationship between a cloud-top brightness temperature and rain rate. The evaluation of satellite precipitation estimates is already challenging due to the lack of available uniformly deployed and availability of independent in-situ networks allowing us to compute rainfall accumulation for the same time period when the satellite products are generated. This challenge becomes even more complicated when trying to evaluates the performance of the satellite products for the infrequent extreme events.

The values of the 99th percentile of daily rainfall accumulation (Fig. 40.8) are presented for the three satellite products and USCRN in-situ data at each station sites. Among the three satellite products, CMORPH presents the highest values of the 99th percentile of daily rainfall when compared to PERSIANN and GPCP. Despite of this fact, such values still remain lower than the 99th percentile values retrieved from in-situ USCRN stations.

The results presented above that uses an independent in-situ dataset for evaluation (USCRN), confirm the results from numerous studies (see for instance the references cited in Table 40.2). All SPPs tend to underestimate extreme precipitation events and

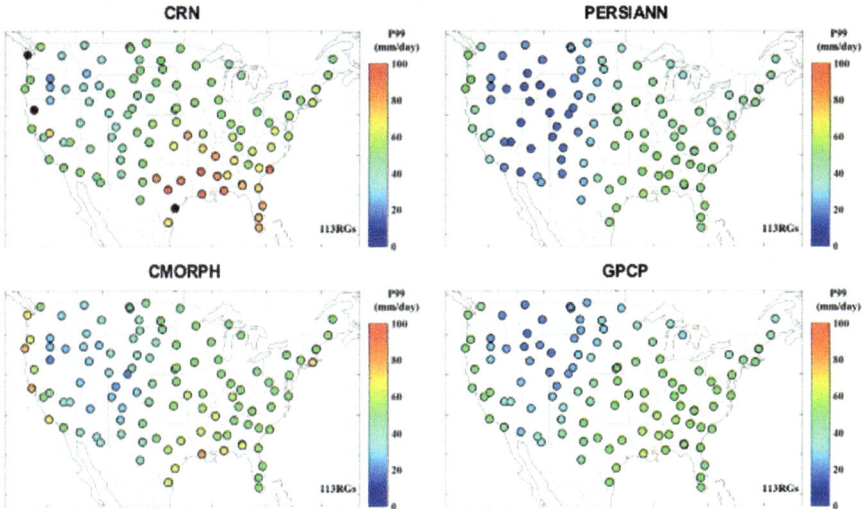

Fig. 40.8 Value of the 99th percentile daily rainfall retrieved from in-situ data (USCRN), and satellite products (PERSIANN-CDR, CMORPH, GPCP) for the period 2007–2015

that SPPs performance declines with increasing value of the percentile (90th percentile and higher) or threshold (above 50–100 mm day^{-1}).

40.5 Conclusions and Future Directions in Satellite QPEs for Extreme Events Measurements

The multitude of studies that evaluate the ability of satellite products to capture extreme precipitation, indicate that there are no products that are systematically better than the other to capture precipitation extremes, and that the performance of each product depends on the location, the season, and the time scale considered, as well as the metric used for the evaluation against in situ data (AghaKouchak et al. 2011; Nastos et al. 2013; Gebremichael et al. 2014; Su et al. 2018). Most gridded SPPs provide comparable global precipitation patterns and average rainfall. The performance of SPPs typically declines with increasing percentiles or precipitation thresholds. Except for a few situations and studies, SPPs tend to underestimate extreme precipitation (90th percentile and higher).

Recently, a new generation of satellite precipitation products has been made available. The launch of the GPM satellite in 2014 opened a new perspective in terms of increased temporal resolution as well as in terms of measurement accuracy. Building upon the legacy of the TRMM satellite and derived products in operation since 1998, the new generation of products is expected to bring significant improvements in terms of monitoring extreme events almost in real time. Figure 40.9 shows

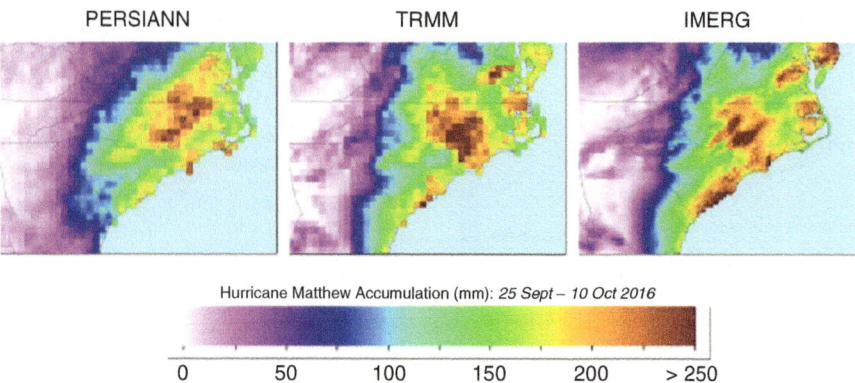

Fig. 40.9 Accumulated rainfall (mm) derived from PERSIANN-CDR, TRMM TMPA, and GPM IMERG during Hurricane Matthew (25 Sept.-10 Oct., 2016). (Figure courtesy of Andrew Shannon)

an example of the improvement provided by precipitation estimates from products of different platforms and algorithms. The resolution improvement from PERSIANN-CDR and TRMM TMPA 3B42 both at 0.25° resolution to GPM IMERG at 0.1° resolution clearly appears in the figure in which the rainfall accumulation associated with Hurricane Matthew provides a more detailed picture of the rainfall spatial distribution.

The improvement of SPPs both in terms of quality of the estimate as well as the reduced lag-time, will allow for the new generation of satellite precipitation products to be used for various applications from the local to the global scale. Those applications include but aren't limited to floods early warning, extreme events associated with tropical cyclones, atmospheric rivers, Indian monsoon, the El Niño-Southern Oscillation phenomenon, and convective systems for instance (Skofronick-Jackson et al. 2017). SPPs can adequately fill the observational gap due to sparse gauge coverage in places where few in-situ observations exist. The higher sampling frequency that ensures a 3-hourly global coverage by microwave sensors on board of a constellation of LEO satellites, could help monitoring the evolution of precipitation systems over regions prone to extreme rainfall including locations with complex topography (Panegrossi et al. 2016). The sensors on board of the GPM – Core Observatory (GPM-CO); the Microwave Imager (GMI) and the dualfrequency (Ka-/Ku-band) precipitation radar (DPR), provide information on the vertical structure of storm systems with an unprecedented definition. An idea of the progress in terms of observational capabilities when it comes to extreme rainfall events between the beginning of the TRMM era and the transition to GPM, can be found in studies that have looked at the global distributions of the most intense precipitation systems using TRMM PR (Zipser et al. 2006) and GPM Ku (Liu and Zipser 2015). With an increased coverage from 36°S-36°N to 65°S-65°N, it is possible to quantify to contribution of those extreme precipitation systems over higher latitude for the first time (Liu and Zipser 2015). Thanks to the retrospective processing, longer period of records of GPM derived SPPs will be available back to the beginning of the TRMM era. With an increase in both temporal and spatial resolution as well as an improvement in rainfall retrieval methods associated with remotely sensed observing platforms such as GPM and derived SPPs, new opportunities are now available for quantifying and monitoring precipitation extremes globally as well as a wide range of applications.

References

Abdalla, O., & Al-Abri, R. Y. (2011). Groundwater recharge in arid areas induced by tropical cyclones: Lessons learned from Gonu 2007 in Sultanate of Oman. *Environment and Earth Science, 63*, 229–239. https://doi.org/10.1007/s12665-010-0688-y.

AghaKouchak, A., Behrangi, A., Sorooshian, S., Hsu, K.-L., & Amitai, E. (2011). Evaluation of satellite-retrieved extreme precipitation rates across the Central United States. *Journal of Geophysical Research, 116*, D02115. https://doi.org/10.1029/2010JD014741.

Ashouri, H., Hsu, K.-L., Sorooshian, S., Braithwaite, D. K., Knapp, K. R., Cecil, L. D., Nelson, B. R., & Prat, O. P. (2015). PERSIANN-CDR daily precipitation climate data record from multisatellite observations for hydrological and climate studies. *Bulletin of the American Meteorological Society, 96*, 69–83. https://doi.org/10.1175/BAMS-D-13-00068.1.

Behrangi, A., Guan, B., Neiman, P. J., Schreier, M., & Lambrigtsen, B. (2016). On the quantification of atmospheric rivers precipitation from space: Composite assessments and case studies over the Eastern North Pacific Ocean and the Western United States. *Journal of Hydrometeorology, 17*, 369–382. https://doi.org/10.1175/JHM-D-15-0061.1.

Bharti, V., Singh, C., Ettema, J., & Turkington, T. A. R. (2016). Spatiotemporal characteristics of extreme rainfall events over the Northwest Himalaya using satellite data. *International Journal of Climatology, 36*, 3949–3962. https://doi.org/10.1002/joc.4605.

Breña-Naranjo, J. A., Pedrozo-Acuna, A., & Rico-Ramirez, M. A. (2015). World's greatest rainfall intensities observed by satellites. *Atmospheric Science Letters, 16*, 420–424. https://doi.org/10.1002/asl2.546.

Chen, S., Hong, Y., Cao, Q., Kirstetter, P.-E., Gourley, J. J., Qi, Y. C., Zhang, J., Howard, K., Hu, J. J., & Wang, J. (2013). Performance evaluation of radar and satellite rainfalls for Typhoon Morakot over Taiwan: Are remote-sensing products ready for gauge denial scenario of extreme events? *Journal of Hydrology, 506*, 4–13. https://doi.org/10.1016/j.jhydrol.2012.12.026.

Ciach, G. J., & Krajewski, W. F. (1999). On the estimation of rainfall error variance. *Advances in Water Resources, 22*, 585–595. https://doi.org/10.1016/S0309-1708(98)00043-8.

Ciach, G. J., Habib, E., & Krajewski, W. F. (2003). Zero-covariance hypothesis in the error variance separation method of radar rainfall verification. *Advances in Water Resources, 26*, 573–580. https://doi.org/10.1016/S0309-1708(02)00163-X.

Ciach, G. J., Krajewski, W. F., & Villarini, G. (2007). Product-error-driven uncertainty model for probabilistic quantitative precipitation estimation with NEXRAD data. *Journal of Hydrometeorology, 8*, 1325–1347. https://doi.org/10.1175/2007JHM814.1.

Curtis, S., Salahuddin, A., Adler, R. F., Huffman, G. J., Gu, G. J., & Hong, Y. (2007). Precipitation extremes estimated by GPCP and TRMM: ENSO relationships. *Journal of Hydrometeorology, 8*, 678–689. https://doi.org/10.1175/JHM601.1.

Diamond, H. J., Karl, T. R., Palecki, M. A., Baker, C. B., Bell, J. E., Leeper, R. D., Easterling, D. R., Lawrimore, J. H., Meyers, T. P., Helfert, M. R., Goodge, G., & Thorne, P. W. (2013). U.S. Climate Reference Network after one decade of operations: Status and assessment. *Bulletin of the American Meteorological Society, 94*, 485–498. https://doi.org/10.1175/BAMS-D-12-00170.1.

Dube, S. K., Jain, I., Rao, A. D., & Murty, T. S. (2009). Storm surge modelling for the Bay of Bengal and Arabian Sea. *Natural Hazards, 51*, 3–27. https://doi.org/10.1007/s11069-009-9397-9.

Endreny, T. A., & Imbeah, N. (2009). Generating robust rainfall intensity-duration-frequency estimates with short-record satellite data. *Journal of Hydrology, 371*, 182–191. https://doi.org/10.1016/j.jhydrol.2009.03.027.

Faridzad, M., Yang, T., Hsu, K.-L., Sorooshian, S., & Xiao, C. (2018). Rainfall frequency analysis for ungauged regions using remotely sensed precipitation information. *Journal of Hydrology, 563*, 123–142. https://doi.org/10.1016/j.jhydrol.2018.05.071.

Gado, T. A., Hsu, K.-L., & Sorooshian, S. (2017). Rainfall frequency analysis for ungauged sites using satellite precipitation products. *Journal of Hydrology, 554*, 646–665. https://doi.org/10.1016/j.jhydrol.2017.09.043.

Gao, Y. C., & Liu, M. F. (2013). Evaluation of high-resolution satellite precipitation products using rain gauge observations over the Tibetan Plateau. *Hydrology and Earth System Sciences, 17*, 837–849. https://doi.org/10.5194/hess-17-837-2013,2013.

Gebremichael, M., Bitew, M. M., Hirpa, F. A., & Tesfay, G. N. (2014). Accuracy of satellite rainfall estimates in the Blue Nile Basin: Lowland plain versus highland mountain. *Water Resources Research, 50*, 8775–8790. https://doi.org/10.1002/2013WR014500.

Gosset, M., Viarre, J., Quantin, G., & Alcoba, M. (2013). Evaluation of several rainfall products used for hydrological applications over West Africa using two high-resolution gauge networks. *Quarterly Journal of the Royal Meteorological Society, 139*, 923–940. https://doi.org/10.1002/qj.2130.

Gutowski, W. J., Jr., Decker, S. G., Donavon, R. A., Pan, Z., Arritt, R. W., & Takle, E. S. (2003). Temporal–spatial scales of observed and simulated precipitation in central U.S. climate. *J. Climate, 16*, 3841–3847. https://doi.org/10.1175/1520-0442(2003)016<3841:TSOOAS>2.0.CO;2.

Habib, E., Ciach, G. J., & Krajewski, W. F. (2004). A method for filtering out raingauge representativeness errors from the verification distributions of radar and raingauge rainfall. *Advances in Water Resources, 27*, 967–980. https://doi.org/10.1016/j.advwatres.2004.08.003.

Hirose, M., Oki, R., Short, D. A., & Nakamura, K. (2009). Regional characteristics of scale-based precipitation systems from ten years of TRMM PR data. *Journal of the Meteorological Society of Japan, 87A*, 353–368. https://doi.org/10.2151/jmsj.87A.353.

Hou, A. Y., Kakar, R. K., Neeck, S., Azarbarzin, A. A., Kummerow, C. D., Kojima, M., Oki, R., Nakamura, K., & Iguchi, T. (2014). The global precipitation measurement mission. *Bulletin of the American Meteorological Society, 95*, 701–722. https://doi.org/10.1175/BAMS-D-13-00164.1.

Huffman, G. J., Adler, R. F., Morrissey, M., Bolvin, D. T., Curtis, S., Joyce, R., McGavock, B., & Susskind, J. (2001). Global precipitation at one-degree daily resolution from multi-satellite observations. *Journal of Hydrometeorology, 2*, 36–50. https://doi.org/10.1175/1525-7541(2001)002<0036:GPAODD>2.0.CO;2.

Huffman, G. J., Adler, R. F., Bolvin, D. T., Gu, G., Nelkin, E. J., Bowman, K., Hong, Y., Stocker, E. F., & Wolf, D. B. (2007). The TRMM Multisatellite Precipitation Analysis (TMPA): Quasi-global, multiyear, combined-sensor precipitation estimates at fine scales. *Journal of Hydrometeorology, 8*, 38–55. https://doi.org/10.1175/JHM560.1.

Huffman, G. J., Bolvin, D. T, Braithwaite, D., Hsu, K., Joyce, R., Kidd, C., Nelkin, E. J., Sorooshian, S., Tan, J., & Xie, P. (2018, February 2018). NASA Global Precipitation Measurement (GPM) Integrated MultisatellitE Retrievals for GPM (IMERG). IMERG Algorithm Theoretical Basis Document (ATBD), version 5.2, 35 pp. Available at https://pmm.nasa.gov/sites/default/files/document_files/IMERG_ATBD_V5.2_0.pdf. Last Accessed 21 Nov 2018.

Jiang, H., & Zipser, E. D. (2010). Contribution of tropical cyclones to global precipitation from eight seasons of TRMM data: Regional, seasonal, and interannual variations. *Journal of Climate, 23*, 1526–1543. https://doi.org/10.1175/2009JCLI3303.1.

Jiang, D., Zhang, H., & Li, R. (2017). Performance evaluation of TMPA version 7 estimates for precipitation and its extremes in Circum-Bohai-Sea region, China. *Theoretical and Applied Climatology, 130*, 1021–1033. https://doi.org/10.1007/s00704-016-1929-0.

Joyce, R., Janowiak, J., Arkin, P., & Xie, P. (2004). CMORPH: A method that produces global precipitation estimates from passive microwave and infrared data at high spatial and temporal resolution. *Journal of Hydrometeorology, 5*, 487–503. https://doi.org/10.1175/1525-7541 (2004)005<0487:CAMTPG>2.0.CO;2.

Kachi, M., Oki, R., Shimizu, S., & Kojima, M. (2006, November 13–14). Global Precipitation Measurement (GPM) mission and its application for flood monitoring. In *Proceedings of the SPIE, conference on GEOSS and next-generation sensors and missions*, Goa, India, 6407, 64070A. https://doi.org/10.1117/12.694043.

Katiraie-Boroujerdy, P.-S., Asanjan, A. A., Hsu, K.-L., & Sorooshian, S. (2017a). Intercomparison of PERSIANN-CDR and TRMM-3B42V7 precipitation estimates at monthly and daily time scales. *Atmospheric Research, 193*, 36–49. https://doi.org/10.1016/j.atmosres.2017.04.005.

Katiraie-Boroujerdy, P.-S., Ashouri, H., Hsu, K.-L., & Sorooshian, S. (2017b). Trends of precipitation extreme indices over a subtropical semi-arid area using PERSIANN-CDR. *Theoretical and Applied Climatology, 130*, 249–260. https://doi.org/10.1007/s00704-016-1884-9.

Keim, B. D., Muller, R. A., & Stone, G. W. (2007). Spatiotemporal patterns and return periods of tropical storm and hurricane strikes from Texas to Maine. *Journal of Climate, 20*, 3498–3509. https://doi.org/10.1175/JCLI4187.1.

Kidd, C., Levizzani, V., & Laviola, S. (2010). Quantitative precipitation estimation from earth observation satellites. In F. Y. Testik, & M. Gebremichael (Eds.), *Rainfall: State of the science* (Geophysical monograph, Vol. 161, pp. 127–158). American Geophysical Union. https://doi. org/10.1029/2009GM000920.

Kirschbaum, D., Adler, R. F., Adler, D., Peters-Lidard, C., & Huffman, G. J. (2012). Global distribution of extreme precipitation and high-impact landslides in 2010 relative to previous years. *Journal of Hydrometeorology, 13*, 1536–1551. https://doi.org/10.1175/JHM-D-12-02.1.

Koriche, S. A., & Rientjes, T. H. M. (2016). Application of satellite products and hydrological modelling for flood early warning. *Physics and Chemistry of the Earth, 93*, 12–23. https://doi. org/10.1016/j.pce.2016.03.007.

Levizzani, V., Laviola, S., Cattani, E., & Costa, M. J. (2013). Extreme precipitation on the island of Madeira on 20 February 2010 as seen by satellite passive microwave sounders. *European Journal of Remote Sensing, 46*, 475–489. https://doi.org/10.5721/EuJRS20134628.

Libertino, A., Sharma, A., Lakshmi, V., & Claps, P. (2016). Global assessment of the timing of extreme rainfall from TRMM and GPM for improving hydrologic design. *Environmental Research Letters, 11*, 054003. https://doi.org/10.1088/1748-9326/11/5/054003.

Liu, C. T., & Zipser, E. J. (2015). The global distribution of largest, deepest, and most intense precipitation systems. *Geophysical Research Letters, 42*, 3591–3595. https://doi.org/10.1002/ 2015GL063776.

Lockhoff, M., Zolina, O., Simmer, C., & Schulz, J. (2014). Evaluation of satellite-retrieved extreme precipitation over Europe using gauge observations. *Journal of Climate, 27*(17), 607–623. https://doi.org/10.1175/JCLI-D-13-00194.1.

Marra, F., Morin, E., Peleg, N., Mei, Y. W., & Anagnostou, E. N. (2017). Intensity-duration-frequency curves from remote sensing rainfall estimates: Comparing satellite and weather radar over the eastern Mediterranean. *Hydrology and Earth System Sciences, 21*, 2389–2404. https:// doi.org/10.5194/hess-21-2389-2017.

Menne, M. J., Durre, I., Vose, S., Gleason, B. E., & Houston, T. G. (2012). An overview of the global historical climatology network-daily database. *Journal of Atmospheric and Oceanic Technology, 29*, 897–910. https://doi.org/10.1175/JTECH-D-11-00103.1.

Nastos, P. T., Kapsomenakis, J., & Douvis, K. C. (2013). Analysis of precipitation extremes based on satellite and high-resolution gridded data set over Mediterranean basin. *Atmospheric Research, 131*, 46–59. https://doi.org/10.1016/j.atmosres.2013.04.009.

National Research Council (NRC). (1994). *Estimating bounds on extreme precipitation events – A brief assessment* (p. 30). Washington, DC: National Academy Press. https://doi.org/10.17226/ 9195.

Nayak, M. A., & Villarini, G. (2018). Remote sensing-based characterization of rainfall during atmospheric rivers over the Central United States. *Journal of Hydrology, 556*, 1038–1049. https://doi.org/10.1016/j.jhydrol.2016.09.039.

Nesbitt, S. W., & Zipser, E. J. (2003). The diurnal cycle of rainfall and convective intensity according to three years of TRMM measurements. *Journal of Climate, 16*, 1456–1475. https://doi.org/10.1175/1520-0442-16.10.1456.

Nesbitt, S. W., & Anders, A. M. (2009). Very high-resolution precipitation climatologies from the Tropical Rainfall Measuring Mission precipitation radar. *Geophysical Research Letters, 36*, L15815. https://doi.org/10.1029/2009GL038026.

Nguyen, P., Ombadi, M., Sorooshian, S., Hsu, K.-L., AghaKouchak, A., Braithwaite, D., Ashouri, H., & Thorstensen, A. R. (2018). The PERSIANN family of global satellite precipitation data: A review and evaluation of products. *Hydrology and Earth System Sciences, 22*, 5801–5816. https://doi.org/10.5194/hess-22-58-1-2018.

Ombadi, M., Nguyen, P., Sorooshian, S., & Hsu, K.-L. (2018). Developing Intensity-Duration-Frequency (IDF) curves from satellite-based precipitation: Methodology and evaluation. *Water Resources Research, 54*, 7752–7766. https://doi.org/10.1029/2018WR022929.

Panegrossi, G., Casella, D., Dietrich, S., Marra, A. C., Sanò, P., Mugnai, A., Baldini, L., Roberto, N., Adirosi, E., Cremonini, R., Bechini, R., Vulpiani, G., Petracca, M., & Porcù, F. (2016). Use of the GPM constellation for monitoring heavy precipitation events over the Mediterranean region. *IEEE Journal of Selected Topics in Applied Earth Observations and Remote Sensing, 9*, 2733–2753. https://doi.org/10.1109/JSTARS.2016.2520660.

Pombo, S., & de Oliveira, R. P. (2015). Evaluation of extreme precipitation estimates from TRMM in Angola. *Journal of Hydrology, 523*, 663–679. https://doi.org/10.1016/j.jhydrol.2015.02.014.

Prakash, S., Mahesh, C., Gairola, R. M., & Pal, P. K. (2012). Comparison of high-resolution TRMM-based precipitation products during tropical cyclones in the North Indian Ocean. *Natural Hazards, 61*, 689–701. https://doi.org/10.1007/s11069-011-0055-7.

Prakash, S., Mitra, A. K., Pai, D. S., & AghaKouchak, A. (2016). From TRMM to GPM: How well can heavy rainfall be detected from space? *Advances in Water Resources, 88*, 1–7. https://doi.org/10.1016/j.advwatres.2015.11.008.

Prasanna, V. (2016). Heavy precipitation characteristics over India during the summer monsoon season using rain gauge, satellite and reanalysis products. *Natural Hazards, 83*, 253–292. https://doi.org/10.1007/s11069-016-2315-z.

Prat, O. P., & Barros, A. P. (2010). Assessing satellite-based precipitation estimates in the Southern Appalachian Mountains using rain gauges and TRMM PR. *Advances in Geosciences, 25*, 143–153. https://doi.org/10.5194/adgeo-25-143-2010.

Prat, O. P., & Nelson, B. R. (2013a). Precipitation contribution of tropical cyclones in the Southeastern United States from 1998 to 2009 using TRMM satellite data. *Journal of Climate, 26*, 1047–1062. https://doi.org/10.1175/JCLI-D-11-00736.1.

Prat, O. P., & Nelson, B. R. (2013b). Mapping the world's tropical cyclone rainfall contribution over land using the TRMM multi-satellite precipitation analysis. *Water Resources Research, 49*, 7236–7254. https://doi.org/10.1002/wrcr.20527.

Prat, O. P., & Nelson, B. R. (2014). Characteristics of annual, seasonal, and diurnal precipitation in the Southeastern United States derived from long-term remotely sensed data. *Atmospheric Research, 144*, 4–20. https://doi.org/10.1016/j.atmosres.2013.07.022.

Prat, O. P., & Nelson, B. R. (2015). Evaluation of precipitation estimates over CONUS derived from satellite, radar, and rain gauge data sets at daily to annual scales (2002–2012). *Hydrology and Earth System Sciences, 19*, 2037–2056. https://doi.org/10.5194/hess-19-2037-2015.

Prat, O. P., & Nelson, B. R. (2016). On the link between tropical cyclones and daily rainfall extremes derived from global satellite observations. *Journal of Climate, 29*, 6127–6135. https://doi.org/10.1175/JCLI-D-16-0289.1.

Prat, O. P., Nelson, B. R., Nick, E., & Ferraro, R. R. (2017, December 11–15). *Evaluation of daily extreme precipitation derived from long-term global satellite Quantitative Precipitation Estimates (QPEs)*. Abstract H53Q-06. 2017, AGU Fall Meeting, New Orleans, LA, USA.

Rapp, A. D., Peterson, A. G., Frauenfeld, O. W., Quiring, S. M., & Roark, E. B. (2014). Climatology of storm characteristics in Costa Rica using the TRMM precipitation radar. *Journal of Hydrometeorology, 15*, 2615–2633. https://doi.org/10.1175/JHM-D-13-0174.1.

Rasmussen, K. L., & Houze, R. A. (2011). Orogenic convection in subtropical South America as seen by the TRMM satellite. *Monthly Weather Review, 139*, 2399–2420. https://doi.org/10.1175/MWR-D-10-05006.1.

Rasmussen, K. L., Choi, S. L., Zuluaga, M. D., & Houze, R. A. (2013). TRMM precipitation bias in extreme storms in South America. *Geophysical Research Letters, 40*, 3457–3461. https://doi.org/10.1002/grl.50651.

Ricko, M., Adler, R. F., & Huffman, G. J. (2016). Climatology and interannual variability of quasi-global intense precipitation using satellite observations. *Journal of Climate, 29*, 5447–5468. https://doi.org/10.1175/JCLI-D-15-0662.1.

Romatschke, U., Medina, S., & Houze, R. A. (2010). Regional, seasonal, and diurnal variations of extreme convection in the South Asian region. *Journal of Climate, 23*, 419–439. https://doi.org/10.1175/2009JCLI3140.1.

Shepherd, J. M., Grundstein, A., & Mote, T. L. (2007). Quantifying the contribution of tropical cyclones to extreme rainfall along the coastal southeastern United States. *Geophysical Research Letters, 34*, L23810. https://doi.org/10.1029/2007GL031694.

Skofronick-Jackson, G., Petersen, W. A., Berg, W., Kidd, C., Stocker, E. F., Kirschbaum, D. B., Kakar, R., Braun, S. A., Huffman, G. J., Iguchi, T., Kirstetter, P. E., Kummerow, C., Meneghini, R., Oki, R., Olson, W. S., Takayabu, Y. N., Furukawa, K., & Wilheit, T. (2017). The global precipitation measurement (GPM) mission for science and society. *Bulletin of the American Meteorological Society, 98*, 1679–1695. https://doi.org/10.1175/BAMS-D-1500306.1.

Su, J. B., Lu, H. S., Wang, J. Q., Sadeghi, A. M., & Zhu, Y. H. (2017). Evaluating the applicability of four latest satellite-gauge combined precipitation estimates for extreme precipitation and streamflow predictions over the upper Yellow River basins in China. *Remote Sensing, 9*, 1176. https://doi.org/10.3390/rs9111176.

Su, X. L., Shum, C. K., & Luo, Z. C. (2018). Evaluating IMERG V04 final run for monitoring three heavy rain events over mainland China in 2016. *IEEE Geoscience and Remote Sensing Letters, 15*, 444–448. https://doi.org/10.1109/LGRS.2018.2793897.

Sun, Q., Miao, C., Duan, Q., Ashouri, H., Sorooshian, S., & Hsu, K.-L. (2018). A review of global precipitation data sets: Data sources, estimation, and intercomparisons. *Reviews of Geophysics, 56*, 79–107. https://doi.org/10.1002/2017RG000574.

Tan, M. L., & Santo, H. (2018). Comparison of GPM IMERG, TMPA 3B42 and PERSIANN-CDR satellite precipitation products over Malaysia. *Atmospheric Research, 202*, 63–76. https://doi.org/10.1016/j.atmosres.2017.11.006.

Taylor, C. M., Belusic, D., Guichard, F., Arker, D. J. P., Vischel, T., Bock, O., Harris, P. P., Janicot, S., Klein, C., & Panthou, G. (2017). Frequency of extreme Sahelian storms tripled since 1982 in satellite observations. *Nature, 544*, 475–478. https://doi.org/10.1038/nature22069.

Tekeli, A. E., & Fouli, H. (2016). Evaluation of TRMM satellite-based precipitation indexes for flood forecasting over Riyadh City, Saudi Arabia. *Journal of Hydrology, 541*, 471–479. https://doi.org/10.1016/j.jhydrol.2016.01.014.

Villarini, G., Smith, J. A., Baeck, M. L., Marchok, T., & Vecchi, G. A. (2011). Characterization of rainfall distribution and flooding associated with U.S. landfalling tropical cyclones: Analyses of Hurricanes Frances, Ivan, and Jeanne (2004). *Journal of Geophysical Research, 116*, D23116. https://doi.org/10.1029/2011JD016175.

Wolff, D. B., & Fisher, B. L. (2008). Comparisons of instantaneous TRMM ground validation and satellite rain-rate estimates at different spatial scales. *Journal of Applied Meteorology and Climatology, 47*(8), 2215–2237. https://doi.org/10.1175/2008JAMC1875.1.

World Meteorological Organization (WMO). (2009). *Manual on estimation of Probable Maximum Precipitation (PMP)* (3rd edn.), WMO-No. 1045, Geneva, Switzerland, 291 pp. Available at https://library.wmo.int/pmb_ged/wmo_1045_en.pdf

Xie, P., Joyce, R., Wu, S., Yoo, S.-H., Yarosh, Y., Sun, F., & Lin, R. (2017). Reprocessed, bias-corrected CMORPH global high-resolution precipitation estimates from 1998. *Journal of Hydrometeorology, 18*, 1617–1641. https://doi.org/10.1175/JHM-D-16-0168.1.

Yang, Y., Tang, G. Q., Lei, X. H., Hong, Y., & Yang, N. (2018). Can satellite precipitation products estimate probable maximum precipitation: A comparative investigation with gauge data in the Dadu River basin. *Remote Sensing, 10*, 41. https://doi.org/10.3390/rs10010041.

Yong, B., Chen, B., Gourley, J. J., Ren, L., Hong, Y., Chen, X., Wang, W., Chen, S., & Gong, L. (2014). Intercomparison of the Version-6 and Version-7 TMPA precipitation products over high and low latitudes basins with independent gauge networks: Is the newer version better in both real-time and post-real-time analysis for water resources and hydrologic extremes? *Journal of Hydrology, 508*, 77–87. https://doi.org/10.1016/j.jhydrol.2013.10.050.

Zhang, Y., Hong, Y., Wang, X. G., Gourley, J. J., Xue, X. W., Saharia, M., Ni, G. H., Wang, G. L., Huang, Y., Chen, S., & Tang, G. Q. (2015). Hydrometeorological analysis and remote sensing of extremes: Was the July 2012 Beijing flood event detectable and predictable by global satellite observing and global weather modeling systems? *Journal of Hydrometeorology, 16*, 381–395. https://doi.org/10.1175/JHM-D-14-0048.1.

Zhou, Y. P., Lau, W. K. M., & Huffman, G. J. (2015). Mapping TRMM TMPA into average recurrence interval for monitoring extreme precipitation events. *Journal of Applied Meteorology and Climatology, 54*, 979–995. https://doi.org/10.1175/JAMC-D-14-0269.1.

Zipser, E. J., Cecil, D. J., Liu, C., Nesbitt, S. W., & Yorty, D. P. (2006). Where are the most intense thunderstorms on earth? *Bulletin of the American Meteorological Society, 87*(8), 1057–1071. https://doi.org/10.1175/BAMS-87-8-1057.

Chapter 41
Rainfall Trends in East Africa from an Ensemble of IR-Based Satellite Products

Elsa Cattani, Andrés Merino, and Vincenzo Levizzani

Abstract The Africa Rainfall Climatology v2 (ARC2), Climate Hazards Group InfraRed Precipitation with Stations (CHIRPS) v2, and Tropical Applications of Meteorology using SATellite (TAMSAT) African Rainfall Climatology And Time Series v3 (TARCAT3) satellite rainfall products are exploited to study the spatial and temporal variability of East Africa (EA) rainfall between 1983 and 2017 through the time series of selected rainfall indices from the joint CCI/CLIVAR/JCOMM Expert Team on Climate Change Detection and Indices (ETCCDI). The indices total rainfall amount (PRCPTOT), Simple Daily Intensity (SDII), number of precipitating days (R1), maximum number of consecutive dry and wet days (CDD and CWD), and number of very heavy precipitating days (R20) were analyzed. The scope of the work is to draw the attention on the rainfall trend and variability identifying significant trend patterns regardless of the single satellite product, and also estimating the trend rate variability stemming from the multiplicity of the satellite products. The trend spatial patterns are recognized through the Mann-Kendall technique, considering the time series of the ensemble mean of the three satellite products and the corresponding time series of the standard deviations, which are interpreted as error bars associated with the ensemble mean time series. Indications on rainfall trends were extracted at annual and seasonal scales and the regions that more frequently exhibit statistically significant trends are located in eastern Kenya, Somalia at the border with eastern Ethiopia, northern Tanzania, and limited areas of South Sudan. At the seasonal scale increasing trends were identified for the October-November-December PRCPTOT, SDII, and R20 indices over eastern EA, with the exception of central Kenya, where negative trends with limited areas of significance stand out for R1 and CWD, distinguishable also at the yearly scale. In March-April-May rainfall decline is perceivable only through R1 and CWD in particular over the

E. Cattani (✉) · V. Levizzani
National Research Council of Italy, Institute of Atmospheric Sciences and Climate
(CNR-ISAC), Bologna, Italy
e-mail: e.cattani@isac.cnr.it

A. Merino
Department of Chemistry and Applied Physics, University of León, León, Spain

© Springer Nature Switzerland AG 2020
V. Levizzani et al. (eds.), *Satellite Precipitation Measurement*, Advances in Global
Change Research 69, https://doi.org/10.1007/978-3-030-35798-6_17

eastern EA region, whereas **PRCPTOT**, even though associated with negative trends, does not present any high confidence areas.

Keywords Precipitation · Satellite · East Africa · ETCCDI · Trend analysis

41.1   Introduction

The chapter focuses on the interannual variability and trends in the East Africa (EA, 5S–20N and 28–52E) rainfall during the last three decades from 1983 to 2017. Monitoring the variability of rainfall in EA is a necessity due to the extreme rainfall events that characterise this area (i.e., floods and droughts) and severely impact human life, water availability, and food security (FAO 2015; Adhikari et al. 2015; Brown et al. 2017). A further aggravating factor recently complicates the already difficult situation: extreme EA rainfall events are characterised by a prolonged duration impacting more consecutive wet seasons (Nicholson 2016). Examples of this kind of events are the 2011 drought with the consecutive failures of the October–November–December (OND) 2010 and March–April–May (MAM) 2011 wet seasons (Masih et al. 2014), and the more recent episodes of alternating floods and droughts from 2015 to the present days. The 2015 OND season was characterised by rains above average causing floods in localized areas of southern Somalia, Kenya, and Uganda supported by a very strong El Niño event; reduced rainfall followed during the 2016 MAM and OND in the eastern part of EA (FEWS NET 2017), sustained by a strong negative Indian Ocean Dipole (IOD) during 2016 boreal summer and fall (Lu et al. 2017) and a La Niña event; again OND rainfall was below average over east EA in 2017; and finally flood episodes characterised the 2018 boreal spring in Kenya, southern Somalia, and coastal Tanzania (source: Famine Early Warning Systems Network – FEWS NET – http://fews.net/east-africa, last accessed 23 Oct. 2018).

   In this frame satellite rainfall estimates represent a viable source of data providing reasonably consistent time series with appropriate spatial resolution and temporal duration, which represent an alternative to the rain gauge network measurements (Dinku et al. 2018, Ayehu et al. 2018; Cattani et al. 2016). Very often the ground-based networks in EA have not the necessary spatial coverage and density required to document the rainfall field that is very variable in terms of intensity and seasonality. Moreover, the public availability of the rain gauge data is a problem, considering that in many cases publicly available rain gauges represent only a limited sub-set of the station deployed over the territory (Dinku et al. 2018). Three satellite daily products are selected for trend monitoring due to their spatial resolution, temporal coverage, and a consolidated exploitation in African rainfall monitoring applications such as famine early warning and agricultural decision support: the Climate Prediction Centre (CPC) Africa Rainfall Climatology version 2.0 (ARC2), the Climate Hazards Group InfraRed Precipitation with Stations (CHIRPS) version 2.0, and the recently released version 3.0 of the Tropical Applications of

Meteorology using SATellite (TAMSAT) African Rainfall Climatology And Time Series (TARCAT3). All these products are based on infrared (IR) brightness temperatures from geostationary satellites for the identification of precipitating clouds, but they exploit rain gauge data for calibration and/or merging with the satellite-only estimates according to different methodologies. This means that the rainfall intensity estimation can be quite indirect with respect to microwave-based algorithms, with documented difficulties in the identification of warm rain occurring over mountainous and coastal areas and rainfall overestimation over arid and semi-arid regions due to sub-cloud evaporation (Dinku et al. 2011, 2018). Nevertheless, the requirement of at least 30 years of data to carry on a meaningful trend analysis is an obstacle since microwave-based rainfall products presently cannot provide 30+ years of daily data.

The variability and trend of the EA rainfall from the observational perspective have been already discussed in the literature (Lyon 2014; Liebmann et al. 2014; Seleshi and Zanke 2004; Cheung et al. 2008; Jury and Funk 2013; Maidment et al. 2015; Lyon and DeWitt 2012; Williams and Funk 2011; Yang et al. 2014; Asfaw et al. 2018; Gitau et al. 2018) focussing on the different wet seasons characterising the territory: the short rains season (OND) and the long-rains season (MAM) for the equatorial area (Kenya, Uganda, northern Tanzania, and Somalia); and the Kiremt (June-July-August-September, JJAS) and Belg (MAM) rain seasons for western Ethiopia and South Sudan. The emerging picture includes rainfall declines in the Kiremt and the long-rains season, whereas the short rains season is characterized by increasing trends. Cattani et al. (2018) extracted similar indications from the analysis of the ARC2, CHIRPS, and TARCAT version 2.0 data sets for the period 1983–2015, but more importantly they pointed out how trend spatial patterns can be different from product to product, making it difficult to derive a univocal trend mapping over the territory. Thus, the scope of the present work is to draw the attention on the trend variability stemming from the exploitation of the three satellite products in an ensemble configuration. Time series of the joint CCI/CLIVAR/ JCOMM Expert Team on Climate Change Detection and Indices (ETCCDI, http:// etccdi.pacificclimate.org/index.shtml, last accessed 23 Oct. 2018) computed with the ensemble mean approach are analyzed at yearly and seasonal scales. The Mann-Kendall technique is applied according to Soderberg and Hennet (2007) to extract trend spatial patterns considering the error bars (standard deviations) associated to each element of the ensemble time series.

The satellite data and the methodologies adopted for the investigations will be illustrated in Sect. 41.2, followed by the presentation of the interannual variability of the ETCCDI indices in Sect. 41.3. The trend analysis results at yearly and seasonal scales will be described in Sect. 41.4. Conclusions will be drawn in Sect. 41.5.

41.2   Data and Methods

41.2.1   Satellite Products

The rainfall product selection was driven by the availability of more than 30 years of data starting from 1983 in the case of TARCAT3 and ARC2 and 1981 for CHIRPS together with a high spatial resolution (0.0375° TARCAT3, 0.05° CHIRPS, and 0.1° ARC2) and a proven applicability in operational rainfall monitoring activities with feedback on agriculture and famine early warning. The three datasets are all based on IR brightness temperatures (BT) from geostationary platforms (Meteosat First and Second Generation), but they are substantially different in the IR BT calibration procedures and use of the rain gauge measurements.

TARCAT3 is the most recent released version of the University of Reading rainfall product centred over Africa (Maidment et al. 2017). First TARCAT3 validation results over EA were presented in Dinku et al. (2018) and Ayehu et al. (2018), but it has not yet used in trend studies to date. As the previous TARCAT version it relies on the cold cloud duration (CCD, i.e. the time period during which IR BTs are lower than specific temperature threshold values), now computed on a pentadal basis unlike TARCAT v2 that used a decadal time step. CCDs are used to identify precipitating clouds and are related to rainfall amount through linear relationships, whose coefficients and BT threshold values are determined by calibration with an historical rain gauge data set. The calibration process is presently performed at 1° spatial resolution with respect to the previous rectangular calibration zones. The improved calibration is designed to better capture local variations in the rainfall field reducing unrealistic spatial artefacts and rainfall underestimation (dry bias, Maidment et al. 2017) typically affecting TARCAT v2.

CHIRPS is a quasi-global land-only (50S-50N) satellite product (Funk et al. 2015). CCDs computed with a fixed IR BT threshold (235 K) represent the main input to the algorithm, with a calibration process making use of the Tropical Rainfall Measuring Mission (TRMM) Multi-satellite Precipitation Analysis (TMPA) 3B42 v7. The scheme of the algorithm includes the following steps: (1) IR-based rainfall estimates are extracted using the local (at 0.25° grid box) regressions between 3B42 v7 and CCDs; (2) these estimates are converted to percent anomalies and subsequently multiplied by the global monthly precipitation climatology CHPclim to produce rainfall estimates (CHIRP product) characterised by a reduced systematic bias; finally (3) CHIRP estimates are merged with rain gauge measurements from a wide data set including public archives and data provided by National Meteorological Agencies.

Three-hourly geostationary IR BTs and quality-controlled gauge measurements from the Global Telecommunication System (GTS) are the basis of the generation of the climatological data set ARC2 centred over Africa (40 S–40 N and 20 W–55 E) (Novella and Thiaw 2013). The Geostationary Operational Environmental Satellite (GOES) Precipitation Index (GPI) with a fixed threshold temperature (235 K) for the identification of the precipitating systems is exploited to compute the IR rainfall estimates, which are subsequently combined in a two-step merging approach with the GTS data. Similarly to CHIRPS, ARC2 makes use of real-time rain gauge data for satellite estimate bias correction. Nevertheless, it must be considered that the GTS network represents only a sub-set of the rain gauge stations used in CHIRPS.

Daily estimates from these three satellite products for the period 1983–2017 are used for the computation of the rainfall indices time series. All satellite data were projected on a common grid at 0.25°, using a local averaging approach to interpolate data from their original spatial resolution to the 0.25° grid (the National Center for Atmospheric Research – NCAR Command Language – NCL function "area_hi2lores"). Temporal gaps in the satellite time series were filled according to their duration, that is by interpolation for gaps up to 2 days, using long-term averages for longer gaps.

41.2.2   Rainfall Indices and Trend Analysis Methodology

Several authors exploited rainfall indices to characterize rainfall patterns (e.g., López-Moreno et al. 2010, Rodrigo and Trigo 2007). In this study, a set of six climate indices related to precipitation were selected, whose definitions are reported in Table 41.1, recommended by the joint CCI/CLIVAR/JCOMM Expert Team (ET) on Climate Change Detection and Indices (ETCCDI) (http://etccdi. pacificclimate.org/indices_def.shtml, last accessed 23 Oct. 2018). These indices have been widely used in studies of climate variability at global scale or regional scale (Aguilar et al. 2005; Zolina et al. 2008). The indices time series of the three datasets of daily precipitation described above were computed at each grid cell at annual and seasonal scales, i.e. winter (January–February, JF), spring (MAM), summer (JJAS) and autumn (OND).

The nonparametric Mann-Kendall test was applied to the rainfall index time series (Table 41.1) to evaluate the presence of significant trends and map their spatial distribution at annual and seasonal scales. The nonparametric Mann-Kendall test

Table 41.1 List of the rainfall indices exploited in the analysis

Index	Index name	Index definition	Unit
R1	Number of precipitating days	Days with rain rate (RR) ≥ 1 mm	# of days
R20	Number of heavy precipitating days	Days with RR ≥ 20 mm	# of days
CWD	Consecutive Wet Days	Maximum number of consecutive days with RR ≥ 1 mm	# of days
CDD	Consecutive Dry Days	Maximum number of consecutive days with RR <1 mm	# of days
PRCPTOT	Total precipitation	Total precipitation amount from days with RR ≥ 1 mm	mm
SDII	Simple Day Intensity Index	Ratio of PRCPTOT to number of wet days (RR ≥ 1 mm)	mm day^{-1}

(Kendall 1976) is perhaps the most commonly used test for trend evaluation in environmental sciences and it is widely used in climatic studies to determine trends of time series (Merino et al. 2016; Tian et al. 2016; Santos and Fragoso 2013; Cattani et al. 2018). The null hypothesis assumes that the sequence of values comes from an identically and independently distributed random variable. The alternative hypothesis assumes that the data are distributed according to an increasing or decreasing trend (Gallego et al. 2011). The expression of the statistic of the test, τ (Kendall), is given by:

$$\tau = \sum_{i=1}^{n-1} \sum_{j=i+1}^{n} sgn\left(X_j - X_i\right) \tag{41.1}$$

where X_j and X_i represent data values at the time step j and i; $sgn(X)$ is the sign function, with values equal to -1, 0, or 1 depending on whether the argument is respectively, negative, zero, or positive. The exact distribution of τ can be evaluated and, for $n > 10$, it approaches a normal distribution, especially if the correction $\tau' = \tau - sgn(\tau)$ is applied. If one considers the normalized variable:

$$Z = \frac{\tau'}{\sqrt{\frac{n(n-1)(2n+5)}{18}}} \tag{41.2}$$

then Z is normally distributed with mean 0 and variance 1. The Z value determines the sign of the trend; the greater the deviation from zero, the stronger the data trend (Yenilmez et al. 2011). Positive (negative) Z values indicate an upward (downward) trend.

The major drawback of the Mann-Kendall test occurs in presence of serial data correlation in space and time to which the test is very sensitive. In addition, climatological data do not normally satisfy the condition of independent and identically distributed samples, and thus the existence of positive serial correlation increases the probability that the test detects a trend when there is none (false positive). This would lead to rejection of the null hypothesis when it is actually true. A negative serial correlation, however, decreases the possibility of null hypothesis rejection (Yue and Wang 2004). For this reason, the modified Mann–Kendall trend test (Hamed and Rao 1998) has been used for trend analysis, taking into account the data serial correlation in space and time.

Finally, to estimate the magnitude of the change per time unit in the linear trend, the Sen slope estimator B was used (Theil 1950; Sen 1968), whose expression is:

$$B = median\left(\frac{X_i - X_j}{i - j}\right) \tag{41.3}$$

The Sen slope is related to the Kendall statistic τ' as follows: if $\tau' > 0$, then $B \geq 0$.

In the present work the trend analysis was performed considering the time series of the ensemble mean of the three satellite products and the corresponding time series of the standard deviations, which were interpreted as error bars associated with the ensemble mean time series:

$$X_i^{ens.\ mean} = \frac{x_i^{ARC2} + x_i^{TARCAT3} + x_i^{CHIRPS}}{3} \quad i = 1, \ldots, 35 \quad (41.4)$$

$$\sigma_i^{ens.\ mean} = \sqrt{1/2 * \sum_{m=1}^{3} \left(x_i^m - X_i^{ens.\ mean}\right)^2} \quad i = 1, \ldots, 35 \quad (41.5)$$

Each time series included 35 elements as the number of years considered in the study, i.e., from 1983 to 2017. The idea is to identify trend patterns, which can be significant regardless of the single satellite product, and also estimate the trend slope variability stemming from the multiplicity of the rainfall data sets. As typically applied, the Mann-Kendall test does not consider the uncertainty related to error bar for individual data value in the time series, thus to include the uncertainties the methodology proposed by Soderberg and Hennet (2007) was applied according to the following scheme:

1. Compute the rainfall index time series for each grid cell and satellite product;
2. Compute for each grid cell the ensemble mean and the corresponding error bar time series as in eq. (4) and (5);
3. Randomly generate 1000 time series, whose elements are within the interval $[X_i^{ens.\ mean} - \sigma_i^{ens.\ mean}, X_i^{ens.\ mean} + \sigma_i^{ens.\ mean}]$;
4. Apply the Mann-Kendall test and Sen slope estimator to each of the 1000 time series, and estimate the associated significance level;
5. Compute the mean Sen slope value and the corresponding standard deviation to account for the trend rate variability;
6. Estimate the percentage of time series that exhibit trends at the 0.05 significance level to extract the areas with trends with a high level of confidence.

41.3 Rainfall Seasonality and Variability

Rainfall over EA is characterized by a high spatial variability in terms of rainfall amount and seasonality due to the superimposition of various influencing factors. As recently summarized by Nicholson (2017) and references therein, rainfall variability can be ascribed to local geographic factors (i.e., complex topography and presence of large water bodies as Lake Victoria), coastal influences (e.g., sea breeze circulation), existence of remote factors such as the sea surface temperature (SST) in the tropical Pacific (El Niño Southern Oscillation - ENSO) and IOD and the Walker circulation,

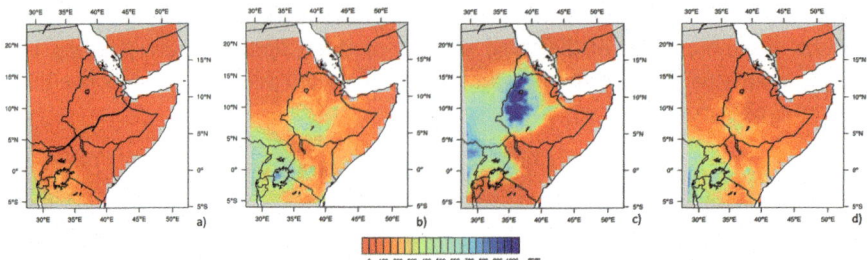

Fig. 41.1 Maps of the seasonal climatology of PRCPTOT: JF (**a**); MAM (**b**); JJAS (c); OND (**d**). For each satellite product the seasonal PRCPTOT climatology was firstly computed, followed by the average over the three products (ensemble mean). The black line in (**a**) identifies two EA regions (Western East Africa – WEA and Eastern East Africa – EEA) characterized by different rainfall annual cycles, monomodal with a single wet season – WEA, and bimodal with two distinct wet seasons – EEA, respectively. (Adapted from Cattani et al. 2016)

and elements of the regional circulation strongly interlinked with the geographic and remote factors.

Two are the areas with the higher amounts of rainfall: the Ethiopian Highlands, where rainfall amount ranges from 900 to 1800 mm year^{-1} according to the annual PRCPTOT climatology obtained from the ensemble mean of the three satellite products, and the region of Lake Victoria (Democratic Republic of Congo-DRC, Uganda, and northern Tanzania) with annual PRCPTOT values up to 2400 mm year^{-1}. The Lake Victoria region is also the only area where rain falls throughout the entire year, as shown in Fig. 41.1, unlike the rest of EA territory, where rain falls only during specific seasons. Arid and semi-arid areas are located in northern South Sudan (15–20N) and the Greater Horn of Africa (GHA, east Ethiopia, Somalia and eastern Kenya), where annual rainfall does not exceed 300 mm year^{-1}.

Seasonality is a pronounced characteristic of the EA rainfall. Wet seasons oscillate south-north with a latitudinal annual displacement (Fig. 41.1). In JF rainfall is confined over northern Tanzania (up to 300 mm cumulated in the two months). In MAM, the so-called "long rains" season, rainfall becomes more intense over DRC, Uganda, and Lake Victoria (MAM PRCPTOT up to 800 mm accumulated in the period), extending also to the eastern sector (Kenya, Somalia, and central and eastern Ethiopia) even though with lower amounts (up to 300 mm for the eastern EA and 500–600 mm for central Ethiopia). Rainfall reaches its maximum during the boreal summer months (JJAS) over western EA (West Ethiopia, southern South Sudan, DRC, and northern Uganda), giving rise to the Kiremt season, the longest and more intense wet season of EA. Rainfall is practically absent over the eastern sector of EA. In OND rainfall returns over south-eastern EA for a second but less intense wet season, also identified as the "short rains" season.

The seasonal migration toward and away from the equator of the Inter-Tropical Convergence Zone (ITCZ) with a twice-annual equatorial passage was usually invoked as the key factor controlling the rainfall seasonal cycle over equatorial Africa. However, recently this concept has been revised and new driving factors were proposed. Yang et al. (2015) suggested that the semiarid/arid climate in EA and its bimodal annual rainfall cycle can be explained by the ventilation mechanism, according to which the atmospheric convective stability is controlled by the inflow of air with low moist static energy from the relatively cool Indian Ocean. However, during the rainy seasons, the off-coast SST increases (and is highest during the long rains season) and consequently the incoming air in EA becomes less stable. Nicholson (2018) further questioned the link between ITCZ and the rain belt proposing the low-level Turkana jet as an important factor in creating the EA seasonal cycle.

According to the described spatial distribution of rainfall seasonality the EA territory was divided into two regions (black curved line in Fig. 41.1a), the northwest area including west Ethiopia and the Rift Valley, South Sudan, and northern Uganda, characterized by a single long wet season peaking in July-August, and the south east area with a bimodal rainfall annual cycle reaching its maximum during April-May and October-November (Cattani et al. 2016). This seasonality spatial scheme was adopted to analyse the temporal rainfall variability by averaging the rainfall index time series of each satellite product over the two areas (Figs. 41.2, 41.3 and 41.4).

Through the analysis of the rainfall index time series of the three satellite products it is possible to identify some of the major events, which hit the region in the period 1983–2017. The impact of the 2010–2011 drought episode is visible as a decrease in 2010 OND PRCPTOT (mean decrease over the EEA from -37 mm for CHIRPS to -62 mm for ARC2 in the season with respect to the mean values of the period 1983–2017), and in R1 (on average 3–4 precipitating days less), together with an increase in the maximum duration of the dry periods CDD from 3 to 15 days depending on the specific satellite product. These deficits in 2010 OND rainfall were followed by similar declines in the 2011 MAM season (a decrease of about 43–71 mm of total rainfall in the season, 4–5 days less of R1, and an extension of the CDD from 4 to 10 consecutive days). On the opposite side, flood occurred in 1997 stands out from the OND index time series with increases of about 240 mm in PRCPTOT, 10–18 days in R1, and 3–4 days in R20 (not shown), and a decrease in CDD from 12 to 25 days. In the last years after the flood episode in 2015, evident in the recovery OND PRCPTOT and R1 with respect to their mean values, rainfall decreased over EEA especially during 2016 MAM and OND, two consecutive drought seasons characterized by a decrease in PRCPTOT and R1. A different situation appears from the JJAS time series over WEA, where rainfall seems to have entered an increasing phase.

This marked interannual rainfall variability has been linked in recent years to remote driver variations according to the season and specific area considered (see Nicholson 2017 and references therein for a complete review of this topic). OND is the wet season with the highest interannual variability with respect to the other seasons. Conditions over the Indian Ocean have been identified as fundamental for the mechanism modulating the OND rainfall through the IOD (Wenhaji Ndomeni et al. 2018). Rainfall increases during the positive IOD phase (IOD+), when positive SST anomalies are present in the western Indian Ocean together with negative anomalies in the eastern part, whereas it decreases when the opposite SST

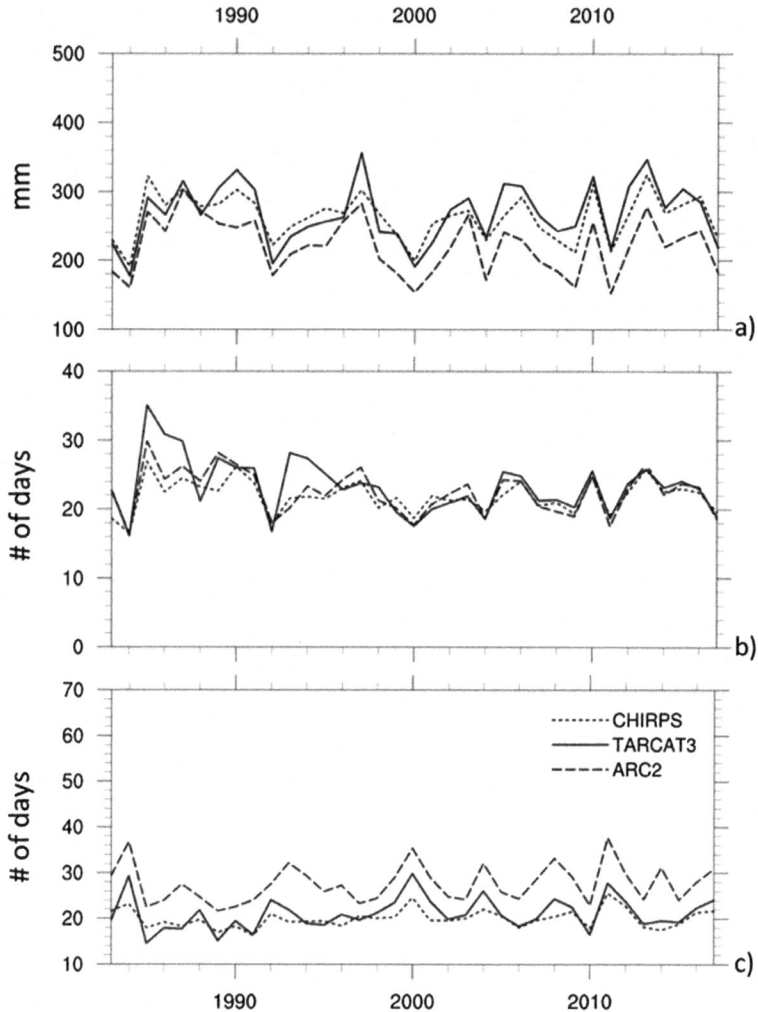

Fig. 41.2 MAM rainfall index time series for the three satellite products, ARC2 (long dashed), TARCAT3 (solid), and CHIRPS (short dashed): PRCPTOT (**a**); R1 (**b**); CDD (**c**). Time series were averaged over EEA

configuration (IOD-) occurs. On the other hand, there is also a strong evidence of the links among OND rain, the zonal vertical circulation cell in the central equatorial Indian Ocean (the Walker circulation) and ENSO.

A reinforcing effect on rainfall occurs in particular during the years when both IOD and ENSO are in their positive intensity phase (El Niño years). In these cases, anomalous low-level easterlies in the equatorial IO and the corresponding upper-level westerlies are intensified with a simultaneous weakening of the Walker circulation, and the moisture flux convergence over EA is enhanced. Drought conditions

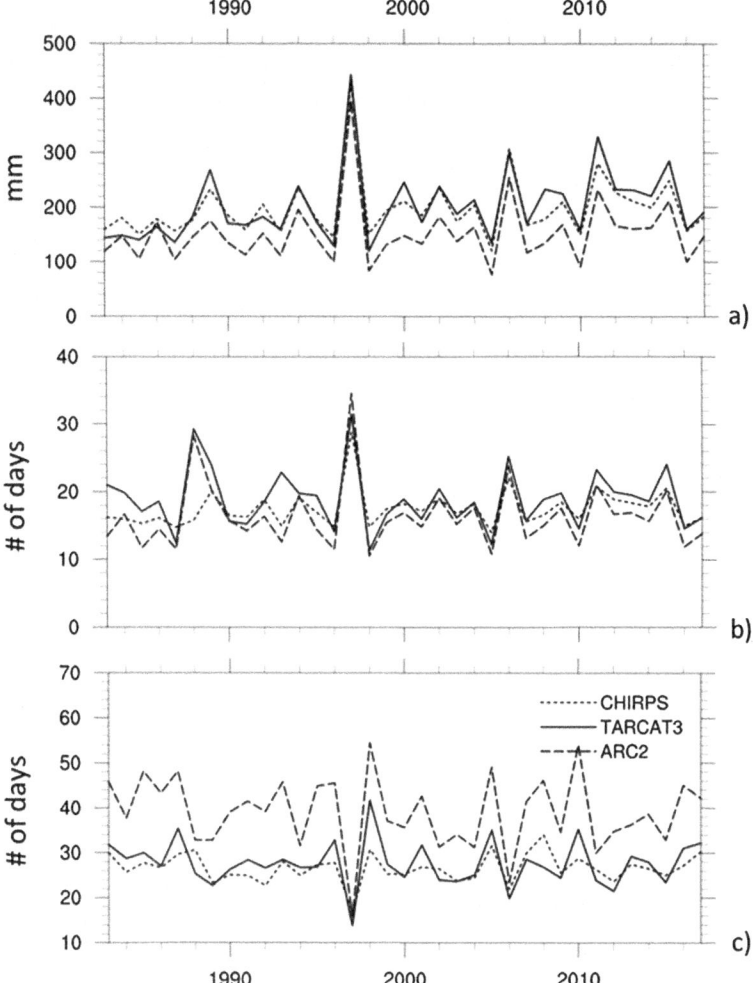

Fig. 41.3 Same as in Fig. 41.2 but for OND

characterise the whole EA territory during years with the negative phases of the IOD and/or ENSO, when the surface westerlies associated with upper level easterlies dominate the equatorial IO intensifying the Walker circulation.

Although the MAM rain season is characterised by a higher rainfall amount than the previous season (the mean values from the PRCPTOT time series in Fig. 41.2 range from 221.09 mm for ARC2 to 268.09 mm for TAMSAT3 against the corresponding values of 151.55 and 202.12 mm for the OND PRCPTOT), its interannual variability is less pronounced (e.g., the coefficients of variation of the OND PRCPTOT time series are 0.39, 0.33, and 0.28 for ARC2, TAMSAT3, and CHIRPS, whereas in the MAM season their values are 0.18, 0.17, and 0.13).

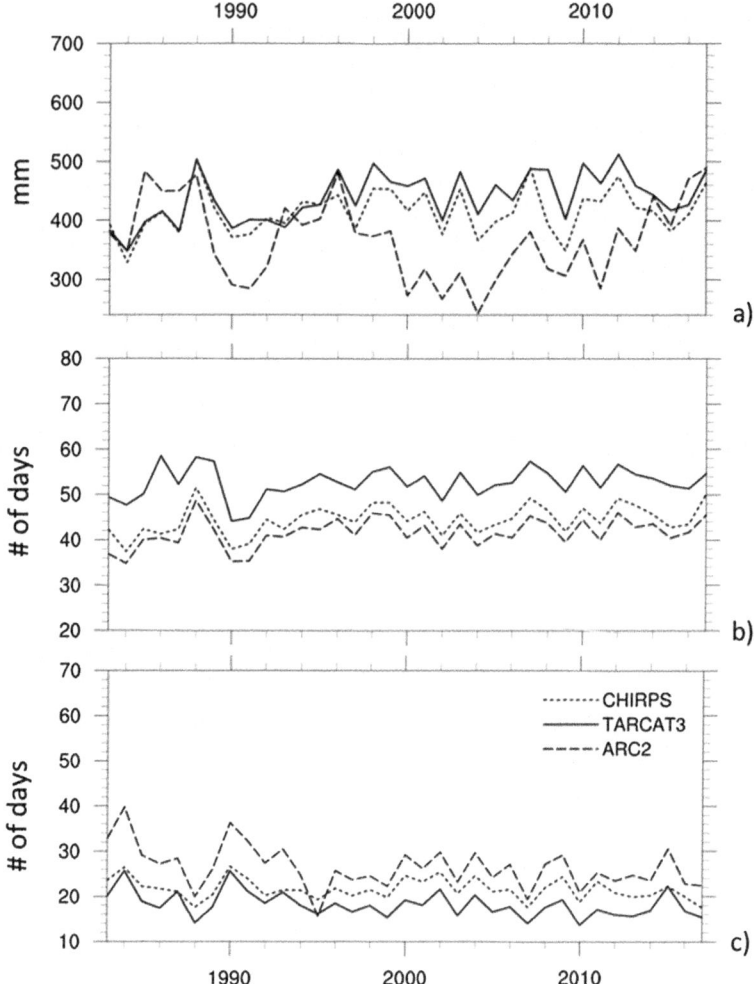

Fig. 41.4 Same as in Fig. 41.2 but for JJAS. Time series were averaged over WEA

Recently Degefu et al. (2017) examined the spatial patterns of the correlation among EA seasonal rainfall and various SST-based indices, demonstrating that the teleconnections with global SSTs are weak and statistically insignificant during this season. The Madden-Julian Oscillation was identified as a major factor for MAM rainfall variability in the region by Pohl and Camberlin (2006). JJAS rainfall variability is usually associated with the airflow from the Congo Basin and the Atlantic Ocean. However, the influence of the Pacific and Indian Oceans were recently discovered, with strong and statistically significant negative correlations between JJAS rainfall and the Niño3.4, IOD and the Central Indian Ocean index (CIndO, SST average over 55°–95°E and 25°S–10°N), being the effects of the IOD and CIndO largely restricted to the western part of Ethiopia along the Ethiopian-Sudanese border (Degefu et al. 2017).

Differences among the satellite products emerge from Figs. 41.2, 41.3, and 41.4. The better agreement stands out from the R1 time series, especially for the EEA area during MAM and OND. Differences among satellite products are more evident in JJAS (in the WEA area) for all rainfall indices including R1, where TARCAT3 exhibits a systematically higher number of precipitating days. This R1 overestimation of TARCAT3 is accompanied by an underestimation of the CDD index in JJAS. It has to be considered that WEA includes the Ethiopian Highlands and the Rift Valley, an area characterized by a very complex orography, which presents well recognised difficulties to the rainfall estimation from satellite (e.g., Dinku et al. 2011). In case of IR-based retrieval algorithms it is possible that cases of warm orographic rainfall were missed in particular by CHIRPS and ARC2, which use a fixed temperature threshold (235 K) for the identification of precipitating clouds. In this respect TARCAT3 has a more articulated approach, with seasonally and spatially varying temperature thresholds. Moreover, these thresholds are optimally selected in a range of values as the temperatures under which the highest level of agreement between rainfall from the historical rain gauge data set and CCD is obtained for the specific calibration cell (Maidment et al. 2014, their Appendix A2). As for the rainfall amount (PRCPTOT) the agreement between CHIRPS and TARCAT3 is substantial especially for MAM and OND. This similarity can be traced back to the adjustment of the rainfall calibration coefficients performed in the new version of the TARCAT product through CHPclim pentadal fields, aimed to the mean bias reduction and improvement of the geographical details in the rainfall estimates (Dinku et al. 2018). Note also that an attempt to reduce the dry bias affecting the previous version of TARCAT has been made by calibrating CCD against mean gauge rainfall as opposed to median rainfall. Probably this could have caused the higher PRCPTOT values that occasionally show up in the TARCAT3 MAM and JJAS time series. In this frame ARC2 is the product with systematically lower PRCPTOT values especially in WEA and during the boreal summer season, situation already described in Novella and Thiaw (2013). ARC2 is also characterized by the highest CDD values more evident in MAM and OND.

41.4   Trend Results

The trend analysis at annual and seasonal level (MAM, OND and JJAS) are presented by means of maps of the mean values of the Sen slope estimator (mean values obtained by averaging the 1000 Sen slope values for each grid cells from the randomly generated rainfall index time series), percentage of the 1000 randomly generated time series providing significant trend at confidence level ≥95% (PVAL %), and Sen slope estimator standard deviation (SD). The significance of the trends for each grid cell is extracted from PVAL%; the higher the value the higher will be the confidence associated with the trend. The aim is to identify trends resistant to time series variability induced by the three satellite products.

41.4.1   Trend Analysis of Annual Rainfall Indices

At the annual scale all rainfall indices except CDD show regions with significant trends (PVAL% $\geq$ 70%) (Fig. 41.5). There are three significant areas, central-eastern Kenya, eastern Ethiopia and Somalia at the border with Kenya, and northern Tanzania with Rwanda and Burundi. Note that each area has its own specific set of rainfall indices exhibiting significant trends.

The maximum duration of the wet periods (CWD) is the rainfall index with the highest PVAL% values for the central-eastern Kenya area, showing negative trends even though weak rates stand out (> −0.18 days per year; Fig. 41.6b) with a SD lower than 0.06 days per year (Fig. 41.7b). Other signals of decreasing rainfall come from R1 (not shown), which decreases at a mean rate $\leq$ −0.35 precipitating days per year associated with SD values within 0.25 days per year. PRCPTOT provides a substantial negative trend indication, but this is not supported by significance (Figs. 41.5a, 41.6a). On the contrary, the daily precipitation (SDII) increases in the area reaching mean trend rates up to 0.13 mm day^{-1}per year and SD in the range [0.02, 0.04] mm day^{-1} per year as result of the R1 negative trends (Figs. 41.6c and 41.7c).

For the region including eastern Ethiopia and Somalia at the border with Kenya the significant trends are those of SDII (largest PVAL% covering a wider area, Fig. 41.5c), followed by R20 and PRCPTOT, the latter with the lower PVAL% values (Fig. 41.5d and Fig. 41.5a, respectively). All the indices pointing to precipitation amount (SDII and PRCPTOT) entered an increasing phase in the period 1983–2017 accompanied also by an increasing tendency in the number of days with heavy rainfall (R20) (Fig. 41.6a, c, d). Rates for SDII include values from 0.03 up to 0.1 mm day^{-1} per year and SD in the range [0.02, 0.04] mm day^{-1} per year.

Weak positive trends of SDII (rates up to 0.06 mm day^{-1} per year with SD up to 0.03 mm day^{-1} per year) and negative of CWD (from −0.09 up to −0.21 days per year with SD up to 0.06 days per year) characterize the northern Tanzania area surrounding Lake Victoria (Figs. 41.6b, c and 41.7b, c).

41.4.2   Trend Analysis of Seasonal Rainfall Indices

41.4.2.1   January–February (JF) Season

Negative and significant trends characterize R1 over Tanzania and the region of Lake Victoria, the area where rainfall is manly located during this season (Fig. 41.8). The decreasing rates in the area reach mean values up to -0.40 days per season with

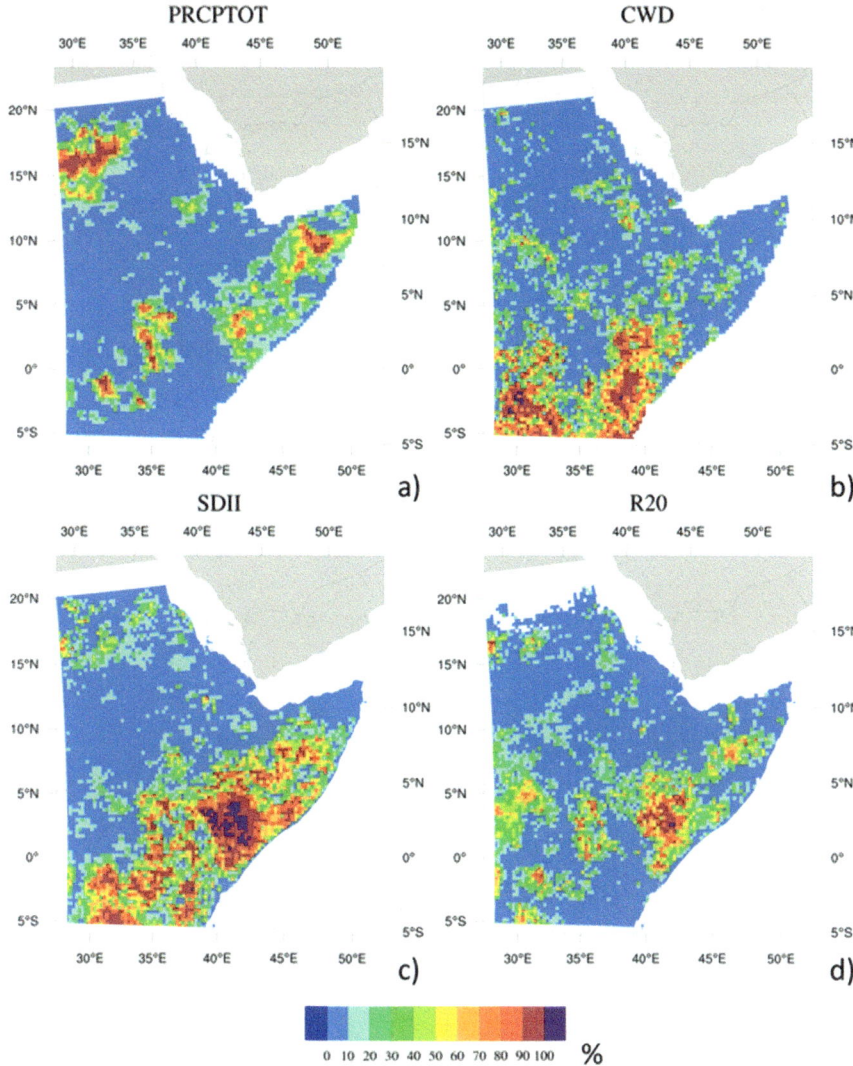

Fig. 41.5 Percentage of the 1000 randomly generated time series providing significant trend at confidence level ≥95% (PVAL%) for PRCPTOT (**a**), CWD (**b**), SDII (**c**), and R20 (**d**)

SD < 0.1 days per season. This area exhibits also a reduction of CWD and a corresponding increase of CDD, but in both cases the trends are not supported by PVAL% spatial patterns similar to those of R1 (not shown).

Another area shows a JF R1 trend significance, central Ethiopia (Rift Valley). Also in this case R1 has a decreasing tendency but with lower rates with respect to Tanzania and Lake Victoria (mean rates ≥-0.15 days per season with SD within 0.05 days per season).

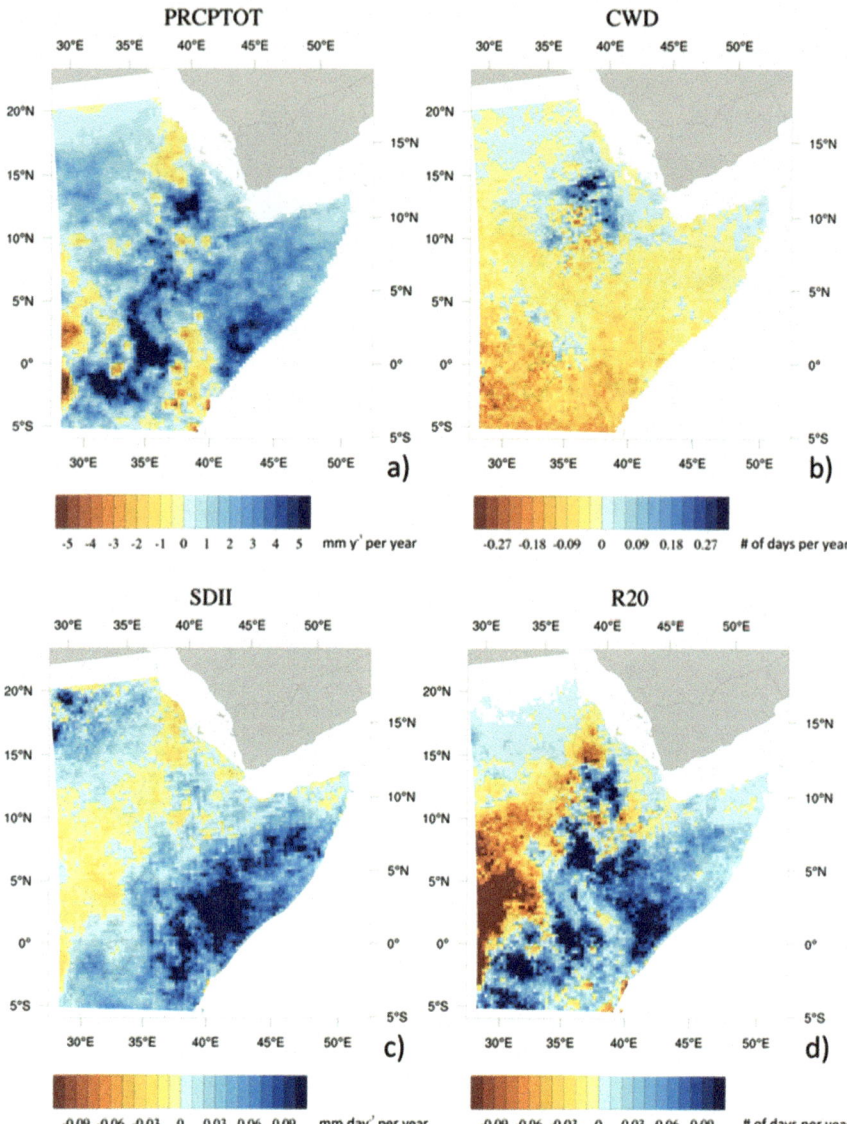

Fig. 41.6 Mean values of the Sen slope estimator for PRCPTOT (**a**), CWD (**b**), SDII (**c**), and R20 (**d**)

41.4.2.2 March–April–May (MAM) Season

Many studies describe the decline affecting MAM precipitation over EA (Williams and Funk 2011; Lyon and Dewitt 2012; Yang et al. 2014). Nicholson (2017) demonstrated through the Centennial Trends precipitation data sets (Funk et al.

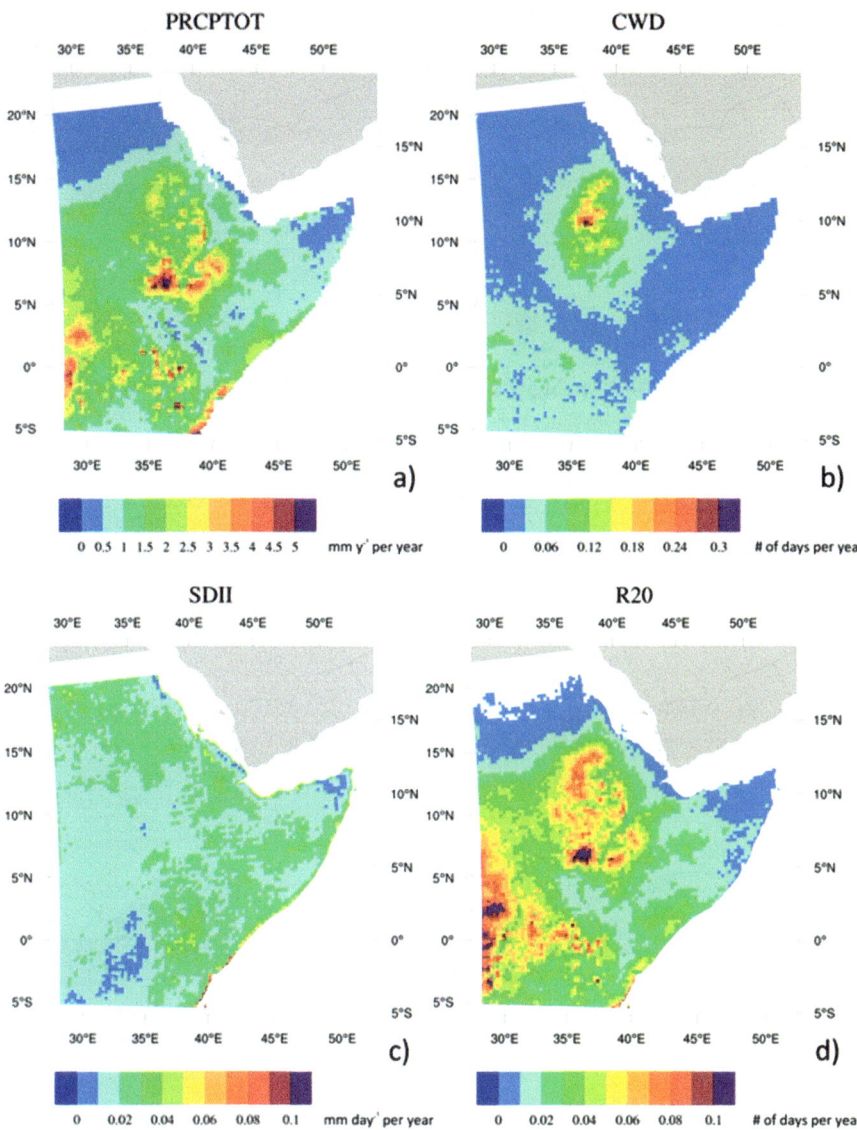

Fig. 41.7 Same as in Fig. 41.6 but for the standard deviation associated with the distribution of the 1000 Sen slope values computed for each grid cell

2015) that this is apparent in both WEA and EEA regions with different modalities, recognizable also from the mean PRCPTOT anomaly time series from the three satellite products (Fig. 41.9). Over WEA an abrupt shift occurred at the end of the '90s, followed by a persistent series of negative anomalies for the period 1998–2012, and a signal of recovery in the last years (Fig. 41.9a). A greater incidence of negative anomalies is apparent in EEA starting from 1998 with a predominance of negative

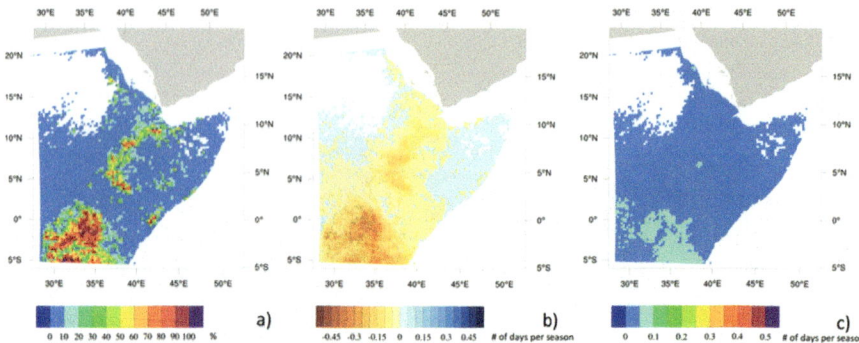

Fig. 41.8 Maps of PVAL% (**a**), Mean Sen slope estimator (**b**), and Sen slope SD (**c**) relative to the JF R1 index

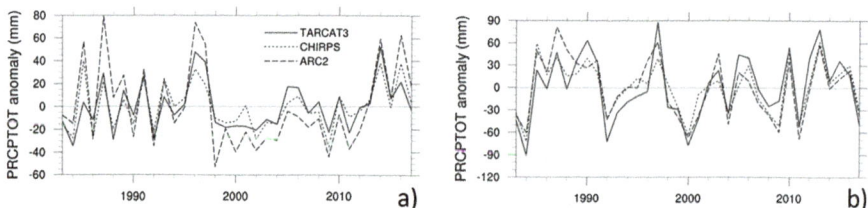

Fig. 41.9 MAM PRCPTOT mean anomalies time series over WEA (**a**), and EEA (**b**). Anomalies are computed from the mean MAM PRCPTOT time series of the two regions

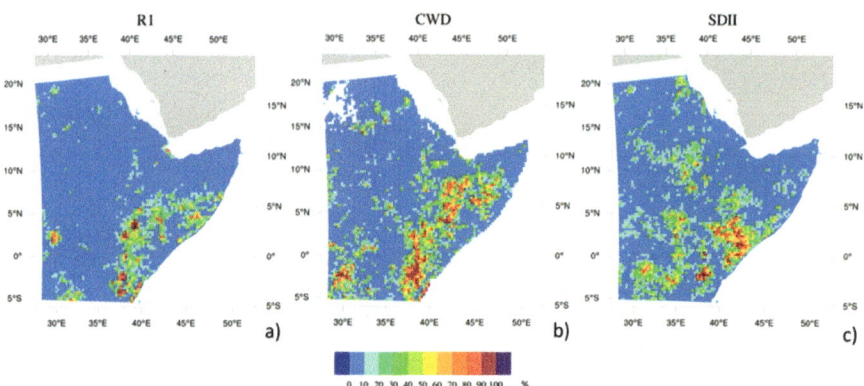

Fig. 41.10 PVAL% values relative to MAM R1(**a**), CWD (**b**), and SDII (**c**)

anomalies in the period 1998–2012, and a further decline in the last years (Fig. 41.9b). Nevertheless, with respect to the trend analysis the decline is perceivable only through R1 and CWD, which show a decreasing tendency in particular over EEA, where the areas with significant trends are located covering eastern Kenya and Somalia at the border with Ethiopia (Figs. 41.10 and 41.11). The mean trend

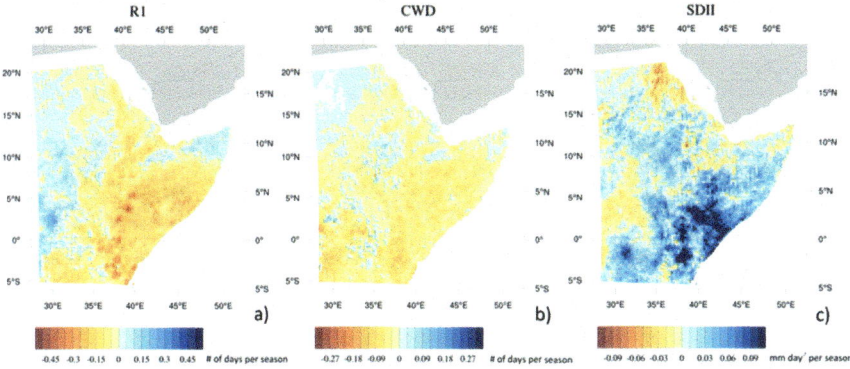

Fig. 41.11 MAM mean values of the Sen slope estimator for R1(**a**), CWD (**b**), and SDII (**c**)

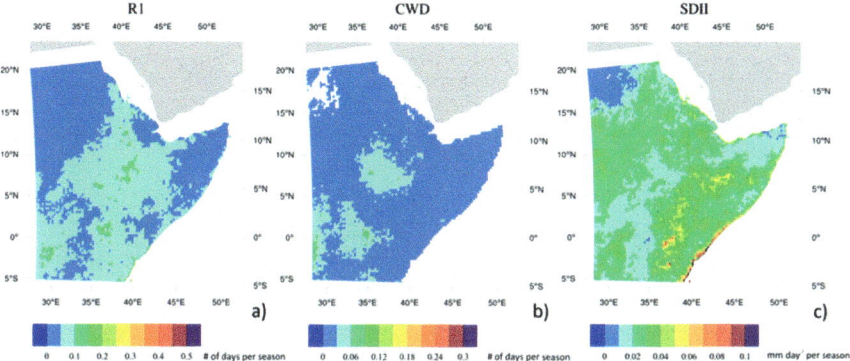

Fig. 41.12 MAM Sen slope SD values for R1(**a**), CWD (**b**), and SDII (**c**)

rates for the areas are generally within -0.31 and -0.44 days per season for R1 and -0.09 days per season for CWD (R1's SD $\leq$ 0.1 days per season and CWD's SD $\leq$ 0.03 days per season, Fig. 41.12). No significance is associated with the PRCPTOT, although the previously mentioned area shows the signature of PRCPTOT negative trends. Significant increasing trends are associated with the SDII index. The WEA region results completely lacking of high level confidence trends.

41.4.2.3 June–July–August–September (JJAS) Season

The three rainfall indices with significant trends are R1, PRCPTOT, and CDD with a general increasing tendency for the first two rainfall indices and an opposite tendency for CDD. These trend patterns are only locally interrupted, but without significance. The areas of higher confidence are several and not always shared among the three indices. They are located over the Ethiopian Rift Valley, South

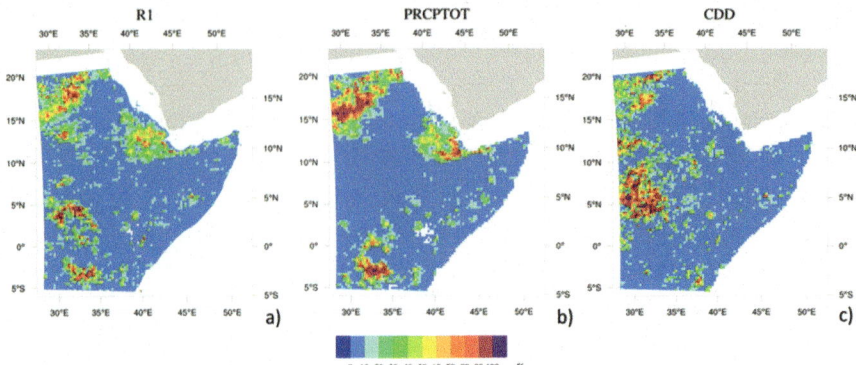

Fig. 41.13 JJAS PVAL% values for R1(**a**), PRCPTOT (**b**), and CDD (**c**)

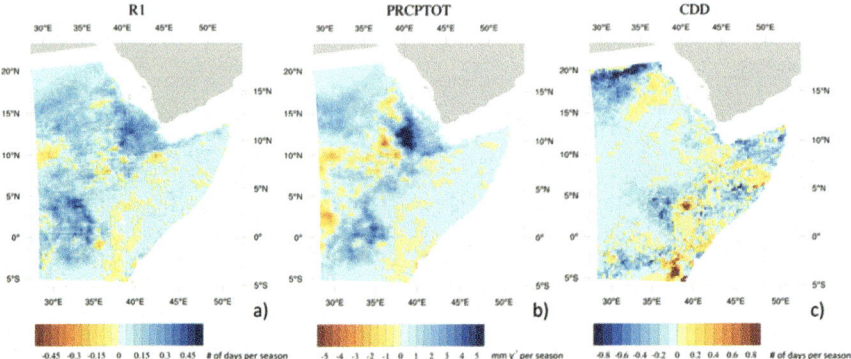

Fig. 41.14 JJAS mean values of the Sen slope estimator for R1(**a**), PRCPTOT (**b**), and CDD (**c**)

Sudan (15°–20°N and 5°–10°N), and northern Tanzania surrounding Lake Victoria (Fig. 41.13). In all these areas the increasing trend mean rates for PRCPTOT are within 2.5 mm year^{-1} per season (SD $\leq$ 0.5 mm year^{-1} per season) with values up to 6.9 mm year^{-1} per season in the Ethiopian Rift Valley, and reach peak values of 0.4–0.45 days per season over the Rift Valley for R1 (SD $\leq$ 0.1 days per season) (Figs. 41.14 and 41.15). As for CDD the area with the highest PVAL% values (South Sudan 5°–10°N) is associated with mean trend rate values up to -0.2 days per season and relatively high SD values. JJAS is recognized as a declining rainfall season similarly to MAM, in particular over WEA (Williams et al. 2012; Funk et al. 2015; Cheung et al. 2008). Nevertheless, some exceptions to these trends have been noted as summarized in Nicholson (2017) and confirmed also by the present results. Moreover, these results match those obtained by Cattani et al. (2018) by computing rainfall index trends for ARC2, CHIRPS, and the previous version of the TARCAT product and the time period 1983–2015.

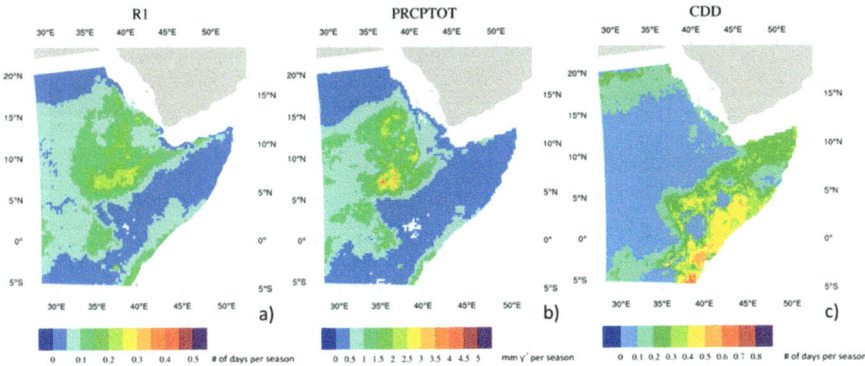

Fig. 41.15 JJAS Sen slope SD values for R1(**a**), PRCPTOT (**b**), and CDD (**c**)

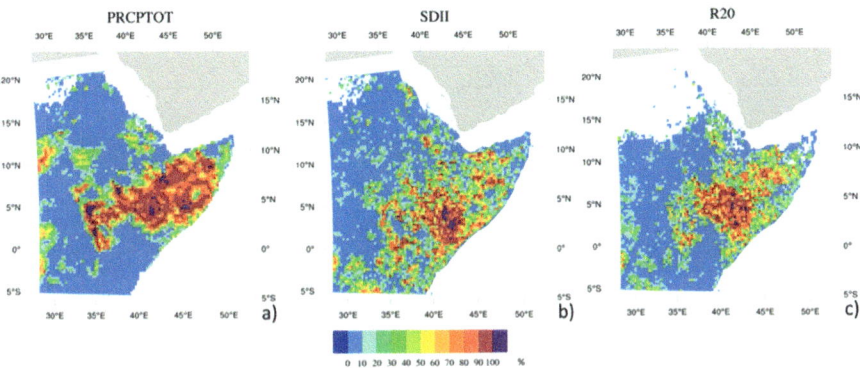

Fig. 41.16 OND PVAL% values for PRCPTOT (**a**), SDII (**b**), and R20 (**c**)

41.4.2.4   October–November–December (OND) season

OND is the season with the strongest signals in terms of extension of the area associated with high confidence trends. As shown in Figs. 41.16 and 41.17, a wide area covering southern Somalia and eastern Ethiopia exhibits increasing and significant trends of PRCPTOT, SDII, and R20. For PRCPTOT the mean rates mostly range from 2 up to 5 mm per season with SD values up to 1 mm per season (Fig. 41.18a), whereas for SDII the mean rates are within 0.29 mm day^{-1} per season with SD peak values of 0.06 mm day^{-1} per season. Mean rates from 0.03 up to 0.11 days per season with SD values between 0.01 and 0.02 days per season characterize the weak increasing trend of R20 over the area. Indications for increasing trends come also from R1 even though with lower (<70%, not shown) PVAL% values. In this picture of increasing rainfall over EEA noteworthy is the situation of central Kenya, where negative trends stand out from trend patterns of PRCTOT without significance, and with limited areas of significance in case of R1 and CWD (not

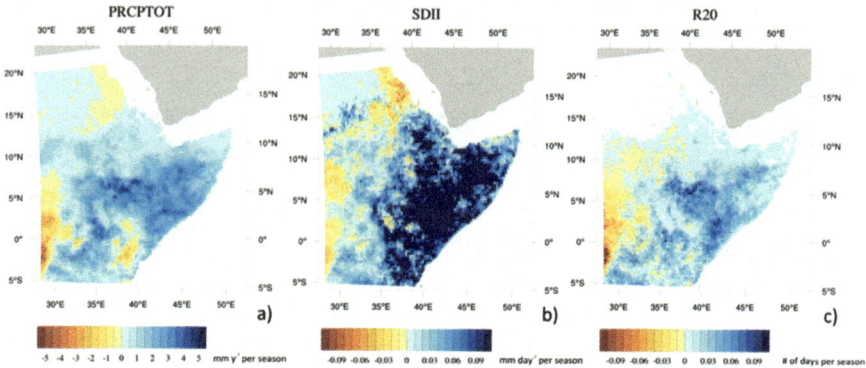

Fig. 41.17 OND mean values of the Sen slope estimator for PRCPTOT (**a**), SDII (**b**), and R20 (**c**)

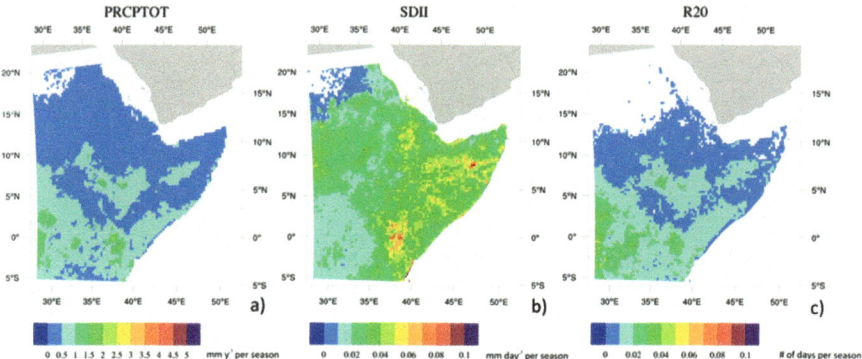

Fig. 41.18 OND Sen slope SD values for PRCPTOT (**a**), SDII (**b**), and R20 (**c**)

shown). This pattern, already highlighted in Cattani et al. (2018) with TARCAT
version 2 instead of TARCAT3, seems connected to what was found for the same
zone at the annual scale.

41.5 Conclusions

The time series of six ETCCDI rainfall indices, PRCPTOT, R1, CWD, CDD, SDII,
and R20, were computed from the daily rainfall estimates of three satellite products,
ARC2, TAMSAT3, and CHIRPS, for the period 1983–2017 to analyse the index
trend spatial patterns over East Africa at the annual and seasonal scales. This work
stems from the findings of Cattani et al. (2018) who extracted rainfall index trends
from ARC2, TARCAT version 2, and CHIRPS separately for the period 1983–2015,
demonstrating that the spatial trend patterns can differ among the rainfall products
and not always all satellite products show statistically significant trends of a specific

rainfall index over the same geographic areas. These results prompt for a trend analysis following a completely different methodology to extract statistically significant trend patterns independent of the satellite product variability, evaluating at the same time the trend rate ranges stemming from the three rainfall products. For this reason, the Mann-Kendall technique for trend assessment was applied considering for each grid cell at 0.25° the mean rainfall index time series (average among the three satellite products) and the corresponding error bars (standard deviations) for the three products. The transition from TARCAT version 2 to TARCAT3 partially reduced the differences among the satellite products, in particular those between CHIRPS and TARCAT3 as demonstrated also in Figs. 41.2, 41.3, and 41.4 due to the fact that the new version of TARCAT makes use of the CHPclim data set for TARCAT bias correction. Nevertheless, differences still remain with respect to the ARC2 product, as demonstrated for example in Fig. 41.4a for the JJAS over WEA, for which a dry bias is detected. This is the reason why it is important to explore trend using different satellite products.

At the annual scale statistically-significant trends occur mainly over three areas: central eastern Kenya, eastern Ethiopia and Somalia at the border with Kenya, and northern Tanzania surrounding Lake Victoria. Except CDD, all rainfall indices present significant trends even though not over the three previously mentioned areas at the same time. Over Kenya the signal of decreasing rainfall is revealed by negative trends of CWD and R1 with significance, whereas PRCPTOT does not go beyond a simple indication of a decrease not supported by significance. Unlike the central eastern Kenya, positive signals characterize the region of eastern Ethiopia and Somalia with increasing trends of SDII, R20, and to a lesser extent PRCPTOT. Only weak trends characterize northern Tanzania although statistically significant, positive for SDII and negative for CWD.

At the seasonal scale OND is the season with the most evident results in terms of the extension of the trend significance regions. Rising trends are found for PRCPTOT, SDII, and R20 over a wide region encompassing southern Somalia and eastern Ethiopia. An exception to this increasing tendency is represented by a portion of eastern Kenya, where R1 and CWD with limited areas of significance and PRCPTOT without significance show negative trends. These results for Kenya seem to confirm the findings at the annual scale.

As for MAM the significant trends are limited to an area encompassing eastern Kenya and Somalia at the border with Ethiopia with declining trends of R1 and CWD, which are confirmed also by PRCPTOT but without significance. WEA lacks completely of high confidence trends, although the analysis of the PRCPTOT anomalies show an abrupt shift from positive to negative values occurring at the end of the '90s followed by a series of negative anomalies for the period 1998–2012.

In contrast with other works, where JJAS is described as a declining rainfall season, our trend analysis results would seem more oriented to a rainfall increase. PRCPTOT and R1 increasing trends emerge for this season with high level confidence over the Ethiopian Rift Valley, South Sudan, and northern Tanzania surrounding Lake Victoria. These rising trends are supported by a statistically significant decrease in CDD only on South Sudan. Moreover, this is the season where the SDs associated with the mean trend rates reach the greatest values, for PRCPTOT and R1 in particular, denoting the more pronounced variability among the satellite products in this season.

Finally, the negative and significant trends of R1 over northern Tanzania and the region of Lake Victoria are the most evident results for JF.

Acknowledgments This study was supported by the European Union's Seventh Programme for research, technological development, and demonstration under Grant Agreement 603608 (eartH2Observe). The authors acknowledge NOAA/CPC; the Climate Hazards Group (CHG) of the University of California, Santa Barbara; the Dept. of Meteorology of the University of Reading for producing and providing full access to the precipitation datasets exploited in this article. Datasets can be accessed at the following web sites:

ARC2 (ftp://ftp.cpc.ncep.noaa.gov/fews/fewsdata/africa/arc2/bin)
CHIRPS (ftp://ftp.chg.ucsb.edu/pub/org/chg/products/CHIRPS-2.0)
TARCAT3 (https://www.tamsat.org.uk/data/archive).
NCAR is acknowledged for the software NCAR Command Language (NCL) version 6.4.0 (2017, Boulder, Colorado: UCAR/NCAR/CISL/TDD, https://doi.org/10.5065/D6WD3XH5). (All links last accessed 13 Dec. 2018)

References

Adhikari, U., Nejadhashemi, A. P., & Woznicki, S. A. (2015). Climate change and eastern Africa: a review of impact on major crops. *Food and Energy Security, 4*, 110–132. https://doi.org/10.1002/fes3.61.

Aguilar, E., Peterson, T. C., Obando, P. R., Frutos, R., Retana, J. A., Solera, M., Soley, J., García, I. G., Araujo, R. M., Santos, A. R., Valle, V. E., Brunet, M., Aguilar, L., Álvarez, L., Bautista, M., Castañón, C., Herrera, L., Ruano, E., Sinay, J. J., Sánchez, E., Oviedo, G. I. H., Obed, F., Salgado, J. E., Vázquez, J. L., Baca, M., Gutiérrez, M., Centella, C., Espinosa, J., Martínez, D., Olmedo, B., Espinoza, C. E. O., Núñez, R., Haylock, M., Benavides, H., & Mayorga, R. (2005). Changes in precipitation and temperature extremes in Central America and northern South America, 1961–2003. *Journal of Geophysical Research, 110*, D23107. https://doi.org/10.1029/2005JD006119.

Asfaw, A., Simane, B., Hassen, A., & Bantider, A. (2018). Variability and time series trend analysis of rainfall and temperature in northcentral Ethiopia: A case study in Woleca sub-basin. *Weather and Climate Extremes, 19*, 29–41. https://doi.org/10.1016/j.wace.2017.12.002.

Ayehu, G. T., Tadesse, T., Gessesse, B., & Dinku, T. (2018). Validation of new satellite rainfall products over the Upper Blue Nile Basin, Ethiopia. *Atmospheric Measurement Techniques, 11*, 1921–1936. https://doi.org/10.5194/amt-11-1921-2018.

Brown, M. E., Funk, C., Pedreros, D., Korecha, D., Lemma, M., Rowland, J., Williams, E., & Verdin, J. (2017). A climate trend analysis of Ethiopia: examining subseasonal climate impacts on crops and pasture conditions. *Climatic Change, 142*, 169–182. https://doi.org/10.1007/s10584-017-1948-6.

Cattani, E., Merino, A., & Levizzani, V. (2016). Evaluation of monthly satellite-derived precipitation products over East Africa. *Journal of Hydrometeorology, 17*, 2555–2573. https://doi.org/10.1175/JHM-D-15-0042.1.

Cattani, E., Merino, A., Guijarro, J. A., & Levizzani, V. (2018). East Africa rainfall trends and variability 1983-2015 using three long-term satellite products. *Remote Sensing, 10*, 931. https://doi.org/10.3390/rs10060931.

Cheung, W. H., Senay, G. B., & Singh, A. (2008). Trends and spatial distribution of annual and seasonal rainfall in Ethiopia. *International Journal of Climatology, 28*, 1723–1734. https://doi.org/10.1002/joc.1623.

Degefu, M. A., Rowell, D. P., & Bewket, W. (2017). Teleconnections between Ethiopian rainfall variability and global SSTs: Observations and methods for model evaluation. *Meteorology and Atmospheric Physics, 129*, 173–186. https://doi.org/10.1007/s00703-016-0466-9.

Dinku, T., Ceccato, P., & Connor, S. J. (2011). Challenges of satellite rainfall estimation over mountainous and arid parts of east Africa. *International Journal of Remote Sensing, 32*, 5965–5979. https://doi.org/10.1080/01431161.2010.499381.

Dinku, T., Funk, C., Peterson, P., Maidment, R., Tadesse, T., Gadain, H., & Ceccato, P. (2018). Validation of the CHIRPS satellite rainfall estimates over eastern of Africa. *Quarterly Journal of the Royal Meteorological Society, 144*(S1), 292–2312. https://doi.org/10.1002/qj.3244.

FAO. (2015). *The impact of disasters on agriculture and food security* (FAO report). ISBN:978-92-5-108962-0, 54 pp. Available at http://www.fao.org/resilience/resources/resources-detail/en/c/346258/. Last Accessed 24 Oct 2018.

FEWS NET. (2017). *Illustrating the extent and severity of the 2016 Horn of Africa drought, East Africa Special Report*. Available at http://fews.net/east-africa/special-report/february-3-2017. Last Accessed 24 Oct 2018.

Funk, C., Peterson, P., Landsfeld, M., Pedreros, D., Verdin, J., Shukla, S., Husak, G., Rowland, J., Harrison, L., Hoell, A., & Michaelsen, J. (2015). The climate hazards infrared precipitation with stations – A new environmental record for monitoring extremes. *Scientific Data, 2*, 150066. https://doi.org/10.1038/sdata.2015.66.

Gallego, M. C., Trigo, R. M., Vaquero, J. M., Brunet, M., García, J. A., Sigró, J., & Valente, M. A. (2011). Trends in frequency indices of daily precipitation over the Iberian Peninsula during the last century. *Journal of Geophysical Research, 116*, D02109. https://doi.org/10.1029/2010JD014255.

Gitau, W., Camberlin, P., Ogallo, L., & Bosire, E. (2018). Trends of intraseasonal descriptors of wet and dry spells over equatorial eastern Africa. *International Journal of Climatology, 38*, 1189–1200. https://doi.org/10.1002/joc.5234.

Hamed, K. H., & Rao, A. R. (1998). A modified Mann Kendall trend test for autocorrelated data. *Journal of Hydrology, 204*, 182–196. https://doi.org/10.1016/S0022-1694(97)00125-X.

Jury, M. R., & Funk, C. (2013). Climatic trends over Ethiopia: Regional signals and drivers. *International Journal of Climatology, 33*, 1924–1935. https://doi.org/10.1002/joc.3560.

Kendall, S. (1976). *Time series* (2nd ed., p. 195). London: Charles Griffin and Co. Ltd. ISBN:0852642415 9780852642412.

Liebmann, B., Hoerling, M. P., Funk, C., Bladé, I., Dole, R. M., Allured, D., Quan, X., Pegion, P., & Eischeid, J. K. (2014). Understanding recent eastern Horn of Africa rainfall variability and change. *Journal of Climate, 27*, 8630–8645. https://doi.org/10.1175/JCLI-D-13-00714.1.

López-Moreno, J. I., Vicente-Serrano, S. M., Angulo-Martínez, M., Beguería, S., & Kenawy, A. (2010). Trends in daily precipitation on the northeastern Iberian Peninsula, 1955–2006. *International Journal of Climatology, 30*, 1026–1041. https://doi.org/10.1002/joc.1945.

Lu, B., Ren, H.-L., Scaife, A. A., Wu, J., Dunstone, N., Smith, D., Wan, J., Eade, R., MacLachlan, C., & Gordon, M. (2017). An extreme negative Indian Ocean Dipole event in 2016: Dynamics and predictability. *Climate Dynamics, 51*, 89–100. https://doi.org/10.1007/s00382-017-3908-2.

Lyon, B. (2014). Seasonal drought in the Greater Horn of Africa and its recent increase during the March-May long rains. *Journal of Climate, 27*, 7953–7975. https://doi.org/10.1175/JCLI-D-13-77000459.1.

Lyon, B., & DeWitt, D. G. (2012). A recent and abrupt decline in the East African long rains. *Geophysical Research Letters, 39*, L02702. https://doi.org/10.1029/2011GL050337.

Maidment, R., Grimes, D., Allan, R. P., Tarnavsky, E., Stringer, M., Hewison, T., Roebeling, R., & Black, E. (2014). The 30 year TAMSAT African Rainfall Climatology And Time series (TARCAT) data set. *Journal of Geophysical Research, 119*. https://doi.org/10.1002/2014/JD021927.

Maidment, R., Allan, R. P., & Black, E. (2015). Recent observed and simulated changes in precipitation over Africa. *Geophysical Research Letters, 42*, 8155–8164. https://doi.org/10.1002/2015GL065765.

Maidment, R., Grimes, D., Black, E., Tarnavsky, E., Young, M., Greatrex, H., Allan, R. P., Stein, T., Nkonde, E., Senkunda, S., & Uribe Alcántara, E. (2017). A new, long-term daily satellite-based rainfall dataset for operational monitoring in Africa. *Scientific Data, 4*, 170063. https://doi.org/10.1038/sdata.2017.63.

Masih, I., Maskey, S., Mussá, F. E. F., & Trambauer, P. (2014). A review of droughts on the African continent – A geospatial and long-term perspective. *Hydrology and Earth System Sciences, 18*, 3635–3649. https://doi.org/10.5194/hess-18-3635-2014.

Merino, A., Fernández-Vaquero, M., López, L., Fernández-González, S., Hermida, L., Sánchez, J. L., García-Ortega, E., & Gascón, E. (2016). Large-scale patterns of daily precipitation extremes on the Iberian Peninsula. *International Journal of Climatology, 36*, 3873–3891. https://doi.org/10.1002/joc.4601.

Nicholson, S. E. (2016). An analysis of recent rainfall conditions in eastern Africa. *International Journal of Climatology, 36*, 526–532. https://doi.org/10.1002/joc.4358.

Nicholson, S. E. (2017). Climate and climatic variability of rainfall over eastern Africa. *Reviews of Geophysics, 55*, 590–635. https://doi.org/10.1002/2016RG000544.

Nicholson, S. E. (2018). The ITCZ and the seasonal cycle over equatorial Africa. *Bulletin of the American Meteorological Society, 99*, 337–348. https://doi.org/10.1175/BAMS-D-16-0287.1.

Novella, N. S., & Thiaw, W. M. (2013). African rainfall climatology version 2 for Famine Early Warning Systems. *Journal of Applied Meteorology and Climatology, 52*, 588–606. https://doi.org/10.1175/JAMC-D-11-0238.1.

Pohl, B., & Camberlin, P. (2006). Influence of the Madden-Julian Oscillation on East African rainfall: II. March–May season extremes and interannual variability. *Quarterly Journal of the Royal Meteorological Society, 132*, 2541–2558. https://doi.org/10.1256/qj.05.223.

Rodrigo, F. S., & Trigo, R. M. (2007). Trends in daily rainfall in the Iberian Peninsula from 1951 to 2002. *International Journal of Climatology, 27*, 513–529. https://doi.org/10.1002/joc.1409.

Santos, M., & Fragoso, M. (2013). Precipitation variability in Northern Portugal: Data homogeneity assessment and trends in extreme precipitation indices. *Atmospheric Research, 131*, 34–45. https://doi.org/10.1016/j.atmosres.2013.04.008.

Seleshi, Y., & Zanke, U. (2004). Recent changes in rainfall and rainy days in Ethiopia. *International Journal of Climatology, 24*, 973–983. https://doi.org/10.1002/joc.1052.

Sen, P. K. (1968). Estimates of the regression coefficient based on Kendall's tau. *Journal of the American Statistical Association, 63*(1–2), 1379–1389. https://doi.org/10.2307/2285891.

Soderberg, K., & Hennet, R. J.-C. (2007). Uncertainty and trend analysis – Radium in ground water and drinking water. *Ground Water Monitoring & Remediation, 27*, 122–129. https://doi.org/10.1111/j.1745-6592.2007.00167.x.

Theil, H. (1950). A rank-invariant method of linear and polynomial regression analysis: I. *Proceedings of the Koninklijke Nederlandse Akademie Wetenschappen, 53*, 386–392. [Available at http://www.dwc.knaw.nl/DL/publications/PU00018789.pdf. Last Accessed 12 Dec 2018].

Tian, J., Liu, J., Wang, J., Li, C., Nie, H., & Yu, F. (2016). Trend analysis of temperature and precipitation extremes in major grain producing area of China. *International Journal of Climatology, 37*, 672–687. https://doi.org/10.1002/joc.4732.

Wenhaji Ndomeni, C., Cattani, E., Merino, A., & Levizzani, V. (2018). An observational study of the variability of East African rainfall with respect to sea surface temperature and soil moisture. *Quarterly Journal of the Royal Meteorological Society, 144*(S1), 384–404. https://doi.org/10.1002/qj.3255.

Williams, A. P., & Funk, C. (2011). A westward extension of the warm pool leads to a westward extension of the Walker circulation, drying eastern Africa. *Climate Dynamics, 37*, 2147–2435. https://doi.org/10.1007/s00382-010-0984-y.

Williams, A. P., Funk, C., Michaelsen, J., Rauscher, S. A., Robertson, I., Wils, T. H. G., Koprowski, M., Eshetu, Z., & Loader, N. J. (2012). Recent summer precipitation trends in the Greater Horn of Africa and the emerging role of the Indian Ocean sea surface temperature. *Climate Dynamics, 39*, 2307–2328. https://doi.org/10.1007/s00382-011-1222-y.

Yang, W., Seager, R., & Cane, M. A. (2014). The East African long rains in observations and models. *Journal of Climate, 27*, 7185–7202. https://doi.org/10.1175/JCLI-D-13-00447.1.

Yang, W., Seager, R., & Cane, M. A. (2015). The annual cycle of the East Africa precipitation. *Journal of Climate, 28*, 2385–2404. https://doi.org/10.1175/JCLI-D-14-00484.1.

Yenilmez, F., Keskin, F., & Aksoy, A. (2011). Water quality trend analysis in Eymir Lake, Ankara. *Physics and Chemistry of the Earth, 36*, 135–140. https://doi.org/10.1016/j.pce.2010.05.005.

Yue, S., & Wang, C. Y. (2004). The Mann-Kendall test modified by effective sample size to detect trend in serially correlated hydrological series. *Water Resources Management, 18*, 201–218. https://doi.org/10.1023/B:WARM.0000043140.61082.60.

Zolina, O., Simmer, C., Kapala, A., Bachner, S., Gulev, S., & Maechel, H. (2008). Seasonally dependent changes of precipitation extremes over Germany since 1950 from a very dense observational network. *Journal of Geophysical Research, 113*, D06110. https://doi.org/10.1029/2007JD008393.

Chapter 42
Heavy Precipitation Systems in the Mediterranean Area: The Role of GPM

Giulia Panegrossi, Anna Cinzia Marra, Paolo Sanò, Luca Baldini, Daniele Casella, and Federico Porcù

Abstract Heavy precipitation systems typical of the Mediterranean area and often devastating its coastal regions, are described and analyzed here by exploiting active and passive microwave measurements and state-of-the-art precipitation products available in the Global Precipitation Measurement (GPM) mission era. The GPM is boosting its key role in integrating the established observational ground-based and satellite-borne tools not only for precipitation monitoring, but also for understanding and characterizing severe weather in the Mediterranean area. In this chapter, we present three events that have recently challenged observational and forecasting capabilities, and caused damages at the ground. Making use of ground based and satellite-borne instruments, we address the problem of estimating precipitation of a small-scale and short-living intense thunderstorm, the capability to render the 3-D structure of a mesoscale organized convective system, and the key role of spaceborne microwave sensors in the characterization and monitoring of a tropical-like cyclone. To this end, we exploited satellite measurements probably beyond the role they have been designed for, showing few strategies to blend satellite data and products with conventional meteorological data, with the aim to increase the knowledge of severe systems in the Mediterranean area and to support operational forecasting activities in a climate change perspective.

Keywords Precipitation · Mediterranean · Severe weather · Tropical-like cyclone · MCS · GPM · TRMM · DPR · GMI · Coastal regions · Raingauges · Radar · Visible · Infrared · Microwave · LINET · GEO · LEO · SEVIRI · H SAF · EPS-SG · AMSU · MHS · SSMIS · AMSR-2 · ATMS · MCS · Medicane · GPROF · IMERG · PNPR ·

G. Panegrossi (✉) · A. C. Marra · P. Sanò · D. Casella
National Research Council, Institute of Atmospheric Sciences and Climate (CNR-ISAC), Rome, Italy
e-mail: g.panegrossi@isac.cnr.it

L. Baldini
Institute of Atmospheric Sciences and Climate (ISAC), National Research Council (CNR), Roma, Italy

F. Porcù
Department of Physics and Astronomy, University of Bologna, Bologna, Italy

© Springer Nature Switzerland AG 2020
V. Levizzani et al. (eds.), *Satellite Precipitation Measurement*, Advances in Global Change Research 69, https://doi.org/10.1007/978-3-030-35798-6_18

Hydrometeors · Graupel · Hail · Cloud-to-ground lightning · Intra cloud lightning · Reflectivity · PCT

42.1 Introduction

The Mediterranean area, recognized as one of the major climate hot-spots in the world (Giorgi 2006), is a unique meteorological environment and a weather fore-casting challenge. Severe weather events of different nature are often observed in this region: deep convective systems (Manzato et al. 2014; Romero et al. 2015; Panegrossi et al. 2016; Marra et al. 2017), cut-off lows (Kotroni et al. 2006; Porcù et al. 2007), intense extratropical (Tripoli et al. 2005; Flaounas et al. 2015) and tropical-like cyclones (Reale and Atlas 2001; Emanuel 2005; Miglietta et al. 2011, 2013), tornadic supercells (Miglietta and Rotunno 2016; Miglietta et al. 2017). These systems often originate over the relatively warm sea and develop to hit coastal areas, mostly rich in assets and densely populated, causing major damages and casualties (Porcù and Carrassi 2009; Llasat et al. 2013). Several studies have recently been devoted to severe weather events over the Mediterranean area, both from the observational (Di Paola et al. 2014; Bech et al. 2015; Roberto et al. 2016; Panegrossi et al. 2016; Marra et al. 2017) and numerical modelling points of view (Cohuet et al. 2011; Buzzi et al. 2014; Davolio et al. 2017; Miglietta et al. 2017), or their combination (e.g., Laviola et al. 2011; Miglietta et al. 2013), addressing also their impact on the ground. In addition, over the last decade several coordinated interna-tional efforts have focused on the observation and forecasting of severe weather over the Mediterranean area: European projects such as EURAINSAT (Levizzani et al. 2007), MEDEX (Jansa et al. 2014), RiskMed (Bartzokas et al. 2010), FLASH (Price et al. 2011), or experiments such as HyMeX (Drobinski et al. 2014), among others.

The complex orography of the Mediterranean coastal regions and the need to monitor severe events during their offshore development make the use of conven-tional ground-based instruments (e.g., raingauges and weather radar networks) insufficient to provide the adequate coverage necessary to monitor and forecast severe weather in terms of time, location, and strength. Satellite measurements are a valuable tool to study and monitor severe events, especially over the sea and large part of the coastal regions, thanks to the capability of remote sensors to estimate precipitation and to increase the knowledge of structure and dynamics of such cloud systems.

The visible-infrared (VIS-IR) sensors available in the last decade at geostationary orbit with relatively high temporal sampling (15 min and less) and spatial resolution (3–5 km at nadir) and with constantly improving radiometric capabilities provide unprecedented data to study cloud upper level structure and dynamics, and convec-tive cloud properties (Mecikalski et al. 2010; Manzato et al. 2014). On the other hand, Passive Microwave (PMW) radiometers on board Low Earth Orbit (LEO) satellites have been largely used to extract quantitative information on precipitation and bulk cloud structure. While the launch of the first precipitation radar on board the

Tropical Rainfall Measuring Mission (TRMM, Kummerow et al. 2000) opened new possibilities to study tropical cloud structures and precipitation, the advent of the NASA/JAXA Global Precipitation Measurement (GPM) mission (Hou et al. 2014; Skofronick-Jackson et al. 2017) extended such possibilities to higher latitudes, including the Mediterranean area. The GPM Core Observatory (GPM-CO), equipped with the Ku/Ka-band Dual-frequency Precipitation Radar (DPR) and the GPM Microwave Imager (GMI), a high resolution, conically scanning multichannel PMW radiometer, is currently the reference satellite-borne platform for the quantitative estimation of precipitation and precipitation microphysics characterization from space. The GPM constellation of satellites carrying PMW radiometers ensures 1-hourly coverage (on average) over the Mediterranean area. Recently, Panegrossi et al. (2016) illustrated the potential of the GPM constellation for monitoring heavy precipitation events, while Marra et al. (2017) carried out an observational analysis of an extremely severe hailstorm, developed over the Mediterranean Sea, based on the GPM-CO.

The improving quality and capabilities of operational and research PMW sensors (including those on board the next EPS-SG satellites) are seeing the steady improvement of satellite-based retrieval techniques worldwide. In Europe one important effort has been carried out in this direction within the EUMETSAT Satellite Application Facility on Support to Operational Hydrology and Water Management (H SAF) (Mugnai et al. 2013b) designed to deliver near real-time (NRT) satellite products of hydrological interest (precipitation, soil moisture and snow parameters) mainly for operational hydrology. PMW products optimized for the Meteosat Second Generation (MSG) full disk area (Europe and African regions) for all the radiometers in the GPM constellation have been recently released, while the development of global products for the next generation of microwave (MW) radiometers on board the EPS-SG satellites is under way. The well-established scientific collaboration between H SAF and the GPM program, has contributed significantly to a prolific development and advancements in retrieval algorithm techniques (e.g., Casella et al. 2017; Sanò et al. 2015, 2016, 2018; Rysman et al. 2018), as well as in validation studies over the Mediterranean area (Petracca et al. 2018).

In the present chapter we focus on the analysis of specific cases typical of the Mediterranean area exploiting active and passive MW measurements and precipitation products available in the GPM era (both from the GPM-CO and constellation members). Three examples of precipitation systems that periodically devastate the Mediterranean coastal regions are analyzed here: small-scale isolated thunderstorms causing extensive floods, Mesoscale Convective Systems, and Mediterranean hurricanes (Medicanes). We illustrate the GPM capabilities and limitations towards the characterization and monitoring of these different types of heavy precipitation systems and show how GPM can integrate or replace the established observational tools.

42.2   Satellite Precipitation Products

The active and passive MW sensors currently on board the GPM constellation
satellites covering the Mediterranean area and considered in this study are: the
GPM-CO DPR and GMI, the GCOMW1 AMSR2, the DMSP (F16, F17, and F18)
SSMIS, the NOAA (18 and 19), the MetOp (A and B) AMSU/MHS, and the Suomi-
NPP ATMS. The PMW precipitation products developed at CNR-ISAC within the
EUMETSAT H SAF program, recently developed for all the GPM constellation
radiometers, are used in the analysis of the precipitation associated with the different
systems: the products based on the Cloud Dynamics and Radiation Database
(CDRD) Bayesian algorithm for conically scanning radiometers SSMIS (Casella
et al. 2013; Sanò et al. 2013; Mugnai et al. 2013a), and AMSR2 (Casella et al. 2017);
the products based on the Passive microwave Neural network Precipitation Retrieval
(PNPR) algorithm, initially designed for cross-track scanning radiometers, AMSU/
MHS (Sanò et al. 2015; Mugnai et al. 2013a) and ATMS (Sanò et al. 2016), and
recently developed for the GMI (hereafter, PNPR-GMI) and described by Sanò et al.
(2018). All H SAF products, except the global PNPR-GMI, are optimized for the
Mediterranean area. The most recent version (V05) of the NASA/JAXA GPM MW
products is also used: the surface rainfall rate available from GPROF-2A for each of
the GPM constellation radiometers (Kummerow et al. 2015) and the DPR products
(Iguchi et al. 2017, 2018), mainly for the analysis of the 3-D structure of the
precipitation systems. Moreover, combined MW/IR NRT products H SAF H03
and GPM IMERG are also included in the analysis. These products are usually
affected by larger uncertainties, but they are very effective for monitoring rapidly
evolving and localized convection because of their high spatial and temporal resolution. The H SAF H03 product, based on the histogram matching technique
combining H SAF PMW precipitation rate estimates from SSMIS and AMSU/
MHS with MSG SEVIRI IR observations (Mugnai et al. 2013b), provides rainfall
rate estimates at the SEVIRI IR grid resolution every 15 min (with 15 min latency).
The IMERG (Early run, V05) product (Huffman et al. 2015) provides surface rainfall
rate on a $0.1° \times 0.1°$ regular grid every half hour.

42.3   Isolated Deep-Convective Systems

Most flash floods that hit the coastal regions in the Mediterranean area are generated
by extremely localized convective systems, often not well represented and predicted
in Numerical Weather Prediction (NWP) models (Hally et al. 2015). It is therefore
crucial to rely on remote sensing data to monitor their development and evolution.
Panegrossi et al. (2016), who analyzed different precipitation systems over the
Mediterranean area hitting coastal regions, have obtained precipitation estimates in
good agreement with ground-based reference (raingauges and/or radar) by
exploiting the available overpasses of the GPM constellation. Such agreement is

subject to the availability of frequent LEO satellite overpasses, to the quality and consistency of the precipitation rate estimates derived from different sensors, and to the nature of the precipitation system, in particular its persistency and duration. Here we want to analyze the potential of state-of-the-art satellite precipitation products (both PMW and MW/IR) for quantification and monitoring of precipitation associated with a recent flash-flood case occurred on the coast of Tuscany, Italy. We also use ground-based observations from the Italian operational networks of raingauge and weather radars (Puca et al. 2014). The raingauge network consists of about 4000 raingauges with a mean density of 1 gauge per 75 km^2. The spatialization of raingauge measurements is provided operationally through a specific quality control and kriging method called GRISO (Rainfall Generator of Spatial Interpolation from Observation) on a 5×5 km^2 grid (Pignone et al. 2018). The Italian weather radar network, covering almost all the Italian territory, federates systems (most of them dual polarization) owned and operated by a variety of institutions. The Department of Civil Protection of Italy (DPC) coordinates the network and delivers products obtained by mosaicking data from all the radars. Among the available products, the Surface Rainfall Intensity (SRI, in mm h^{-1}) is made available every 10 min on a 1×1 km^2 grid.

In the night between 9 and 10 Sept. 2017 the convergence of a flow of warm, moist air of North Atlantic origin from the Channel of Sicily towards Tuscany with cooler winds from the west, gave rise to a condition of great instability, with very heavy rainfall enhanced by the physiography of the area. A flash flood hit the coastal city of Livorno (43.5°N, 10.3°E) between 0000 and 0400 UTC of 10 Sept., causing damages and casualties. Figure 42.1 shows the 24-h cumulated precipitation

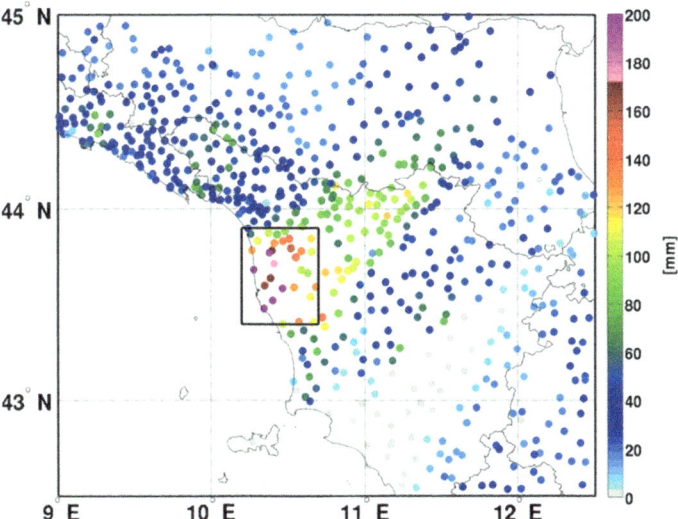

Fig. 42.1 Livorno flash-flood: cumulated precipitation measured in 24 h by the Italian raingauge network between 1200 UTC 9 Sept. and 1200 UTC 10 Sept., 2017

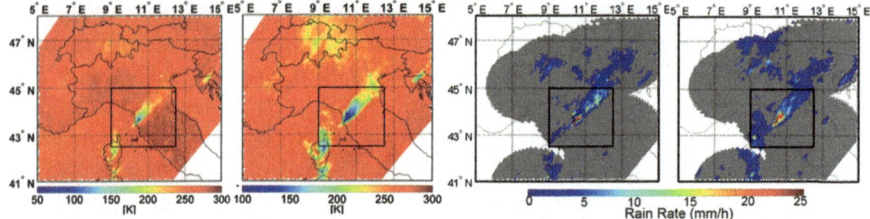

Fig. 42.2 Livorno flash-flood: GPM-CO overpass at 0117 UTC 10 Sept. GMI TBs at 89 GHz (V-pol) (first panel on the left) and 166 GHz (V-pol) (second panel), DPC radar SRI product at 0120 UTC (third panel), and PNPR-GMI rainfall rate estimate (last panel on the right) over the area covered by the radar. The black-box corresponds to the whole area shown in Fig. 42.1

measured by the operational raingauge network between 1200 UTC 9 Sept. and 1200 UTC 10 Sept. The black box in the figure indicates the area around the city of Livorno where most of the precipitation was registered: three stations measured over 230 mm of cumulated precipitation between 0000 and 0600 UTC of 10 Sept., with peaks of 75 mm in 30 min (150 mm h^{-1}) registered between 0100 and 0300 UTC.

There are 12 overpasses of the MW radiometers of the GPM constellation (6 AMSU/MHS, 2 GMI, 2 SSMIS, 1 AMSR2, 1 ATMS) that captured the storm throughout its transition over the selected area, the first at 1718 UTC on 9 Sept., and the last one at 0944 UTC on 10 Sept. One GPM-CO overpass captured the storm on 10 Sept. 0117 UTC. Unfortunately, the DPR swath did not cover the storm, but GMI provided unique multichannel images of the mature cell over the city of Livorno. Figure 42.2 shows the map of the 89 and 166 GHz (V-pol) GMI brightness temperatures (TBs) measured at the time of the GPM overpass when the storm hit the Livorno area. It is worth noting the different response of these two GMI channels to the different rainfall regimes. The 89 GHz is able to capture very well the small-scale, deep convective core of the storm (min TB at 95.9 K), but it is not sensitive to the light precipitation, while the 166 GHz saturates in correspondence of the deep convection (min TB 108.9 K), but shows high sensitivity to the moderate/light precipitation. The response of the different GMI channels to the precipitation, along with the high spatial resolution (up to ∼5 km), outlines the unique GMI capabilities for rainfall retrieval. In the two right panels, the corresponding PNPR-GMI rainfall rate is compared to the radar mosaic SRI, averaged to match the PNPR-GMI spatial resolution. The two rainfall maps show an overall agreement between the rainfall patterns (in terms of detection and variability), except for a more extended area affected by the intense precipitation (around 44.5°N 11.5°E), present in the PNPR-GMI map, with some underestimation (the radar peak at 41 mm h^{-1} is underestimated by 30%). As reported by Sanò et al. (2018), who analyzed this GMI overpass in more detail, the PNPR-GMI rainfall rate estimates are in good agreement with GPROF V05.

Figure 42.3 shows the 24-h cumulated precipitation obtained from the ground-based measurements (raingauges and radar), from the PMW precipitation products

(H SAF and GPM), and MW/IR products (H SAF H03 and GPM IMERG Early V05). A parallax-effect correction has been applied to the operational H03 product.

We show only NRT products not calibrated to ground-based data, and regridded at $0.1° \times 0.1°$ resolution, consistently with IMERG. The raingauge map was obtained cumulating half-hourly GRISO rainfall maps. Raingauge interpolation in coastal areas could be less accurate because of the asymmetrical distribution of nearby stations around the coastal stations. The radar map was obtained integrating the SRI available every 10 min, while the PMW cumulated precipitation (H SAF and GPM GPROF) was obtained from the regridded rainfall rate retrievals for all the overpasses (SSMIS, AMSR2, GMI, AMSU/MHS, and ATMS) available in the 24 h considered, assuming that the rain rate remains constant between two subsequent overpasses (as in Panegrossi et al. 2016). Finally, the H03 and IMERG cumulated precipitation was obtained by integrating the (mean) instantaneous precipitation rate estimates available every 15 and 30 min, respectively.

The two top panels in Fig. 42.3 show that the raingauges and the radar disagree in the region affected by the flood (black box). The radar underestimates the rainfall with respect to the raingauges that registered peaks of 120 and 150 mm h^{-1} (with over 256 mm cumulated between 0000 and 0600 UTC). The area of the peak of precipitation is located between 50 and 60 km south of the closest radar site and there are no natural obstacles between the radar and the precipitating cell. The underestimation likely arises from the combined effect of several causes, such as the use of a non-optimal Z-R relation for such intense precipitation and a residual attenuation that was not properly corrected. This is somewhat confirmed by the SRI product at the original 1×1 km^2 resolution (not shown). It reveals the presence of an intense cell of limited extension close to the radar, corresponding to the peak of cumulated precipitation at 160 mm in the radar map in Fig. 42.3, in good agreement with the gauges (20% underestimation). This cell likely aggravates the attenuation, and, as a result, even the spatial extension of the area of heavier precipitation is underestimated by the radar. It is worth noting that the regridding process tends to slightly smooth the peak observed by the radar.

There is a striking difference between the 24-h precipitation maps obtained from H SAF (PMW and MW/IR) and NASA (GPROF and IMERG) products. H SAF PMW products delineate quite well the area of intense precipitation, even though the peak of the precipitation is underestimated by 20% and is shifted to the north-east compared to the raingauges. The H SAF combined MW/IR product (H03) has a similar behavior, with a well-defined area of intense precipitation (slightly underestimated). H03 also shows overestimation of the light to moderate precipitation and a tendency to falsely detect light precipitation over the whole domain. The NASA products, both GPROF and IMERG, show significant underestimation for the precipitation peak compared to the raingauges, but they are able to depict the light precipitation areas better than the H SAF products as already documented by Panegrossi et al. (2016) for GPROF (V03) in the Genoa flood case on 9–11 Oct. 2014. Note that, while the NASA products are global products, all the H SAF products (except the PNPR-GMI which is a global product) are optimized for the Mediterranean area.

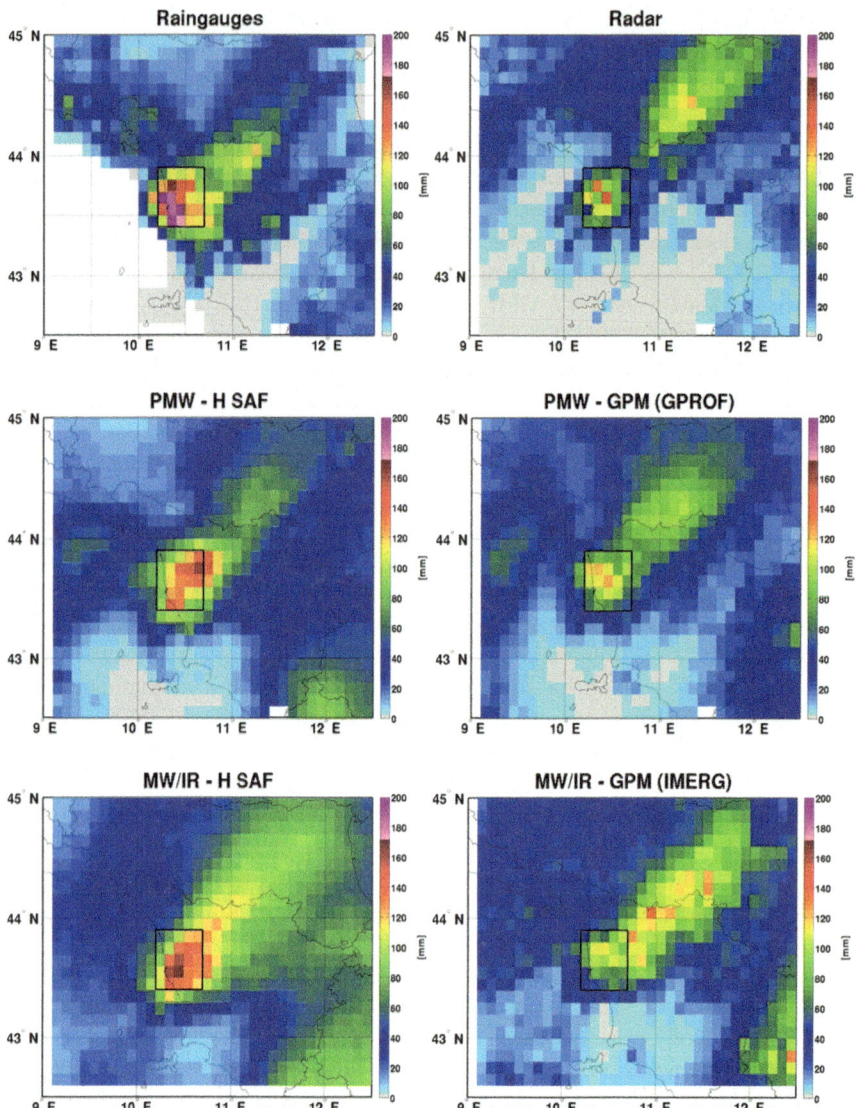

Fig. 42.3 24-h cumulated precipitation from 1200 UTC 9 Sept. to 1200 UTC 10 Sept. 2017 (on a regular 0.1° × 0.1° grid): raingauges and radar (top panels), PMW products (middle panels), and MW/IR products (see text for details)

Similarly to Panegrossi et al. (2016), the temporal evolution between 1600 UTC 9 Sept. and 1200 UTC 10 Sept. of the mean instantaneous and cumulated precipitation in the area affected by the flood (black box area in Figs. 42.1 and 42.3) is reported in Fig. 42.4. The mean instantaneous precipitation rates available from the PMW products (H SAF and GPROF) for all radiometers are shown in Fig. 42.4 (top

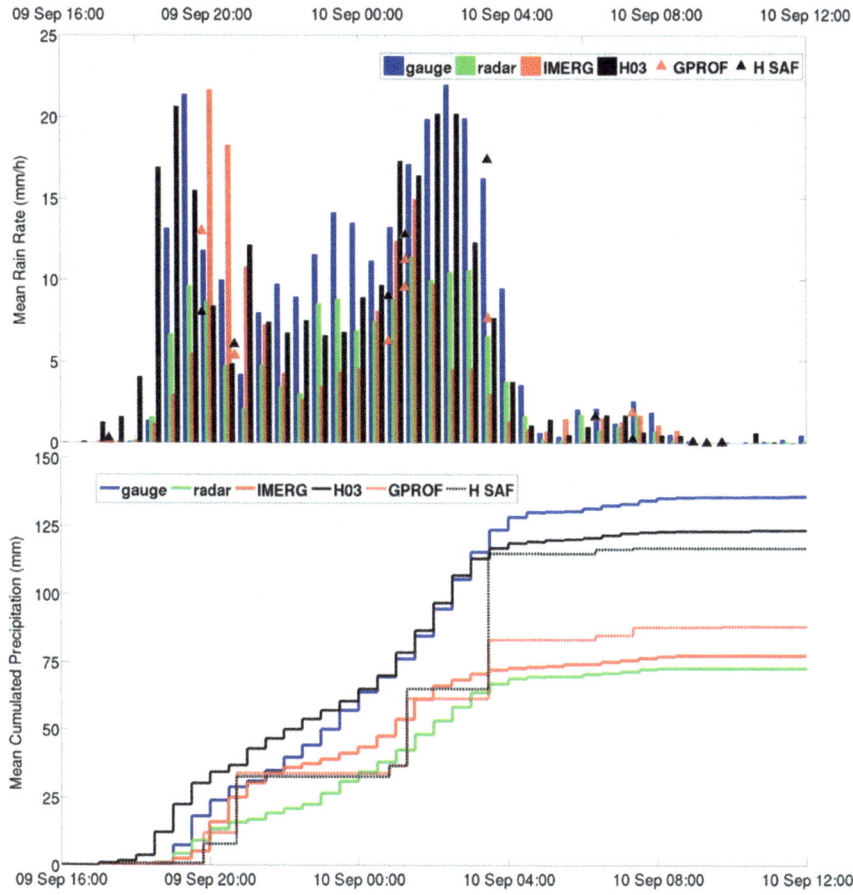

Fig. 42.4 Temporal evolution of the mean instantaneous precipitation (top panel) and mean cumulated precipitation (bottom panel), computed over the Livorno area (black box in Fig. 42.3) (see text for details)

panel) along with the half-hourly rainfall rates available from the MW/IR products (H03 and IMERG), and from raingauges and radar, while the bottom panel shows the corresponding curves of the mean cumulated precipitation.

The results confirm the significant disagreement between the two reference datasets observed already in Fig. 42.3, with an underestimation of the radar with respect to the raingauges. Until 0100 UTC the radar and the gauges are in better agreement (even though it is evident some underestimation especially in correspondence of the peak around 1900 UTC 9 Sept.). Afterwards the radar seems to miss the peak of the precipitation occurring around 0300 UTC (raingauge mean rainfall rate around 125 mm h^{-1} while the radar does not exceed 60 mm h^{-1}). The mean cumulated precipitation estimated by the radar is roughly 50% lower than the raingauges. These results underlie the difficulties in getting reliable reference rainfall maps in case of small-scale, and rapidly evolving intense events. In general, the PMW products overestimate the intensity of the rainfall rate peaks, while light and moderate rain is underestimated. H SAF and GPM products show an overall agreement in terms of instantaneous precipitation except at 0328 UTC 10 Sept. (F16 SSMIS overpass), where H SAF estimate is almost twice that from GPROF and radar, but is in good agreement with the raingauges (such discrepancy may be related to the missing high-frequency channels in F16 SSMIS data used as input by GPROF). While H03 instantaneous precipitation follows quite closely the temporal evolution of the precipitation and is in good agreement with the raingauges, IMERG seems to either follow or anticipate the peaks, and is in better agreement with the radar.

Several sources of uncertainties may concur to the discrepancies between ground reference and satellite estimates, e.g., the differences in the data acquisition and the temporal and spatial characteristics of the measuring systems, enhanced by the rapidly varying structures of the precipitation patterns. Finally, we remark that this event (as well as other isolated convective systems leading to devastating floods on the coastal regions of the Mediterranean area) is extremely challenging for all the observing systems considered in this study (and available to operational meteorology), given the small scales of the phenomenon and the high horizontal gradients of the observables. Despite the relatively poor performance of some of the products, however, we highlight the presence of a (at least qualitative) signature of the severity of the event in some of the satellite signals, making spaceborne sensors (not primarily designed for this application) useful in studying and monitoring such small scale/high impact events, whose occurrence at mid-latitude seems to increase in a climate change scenario.

42.4 Mesoscale Convective Systems

The event presented in the previous section is quite rare since convective cells often organize at the mesoscale, originating convective clusters that last many hours that constantly regenerate, and are able to impact wide regions with strong winds, heavy rainfall, hailstorms, and lightning. Marra et al. (2017) analyzed one of these clusters that developed over the Tyrrhenian Sea and hit the Gulf and the city of Naples in Italy on 5 Sept. 2015 (hereinafter referred to as the Naples hailstorm) using GPM-CO observations in conjunction with other satellite and ground-based measurements. A maximum cooling rate at cloud top around 4.5 K min^{-1} in the growth phase was found by means of MSG SEVIRI analyses (with a cloud top minimum temperature of 198 K), while the estimated peak updraft speed was compatible with 8–10 cm size hailstones at the ground. The maximum stroke rate (as measured by the LINET network, Betz et al. 2009) was around 300 min^{-1} and a total of 20,846 strokes were registered in 2 h (0700–0900 UTC) during the most intense phase of the storm.

One GPM-CO overpass captured the mature phase of the storm (0847 UTC) and Marra et al. (2017) evidenced both GMI and DPR unprecedented capabilities in

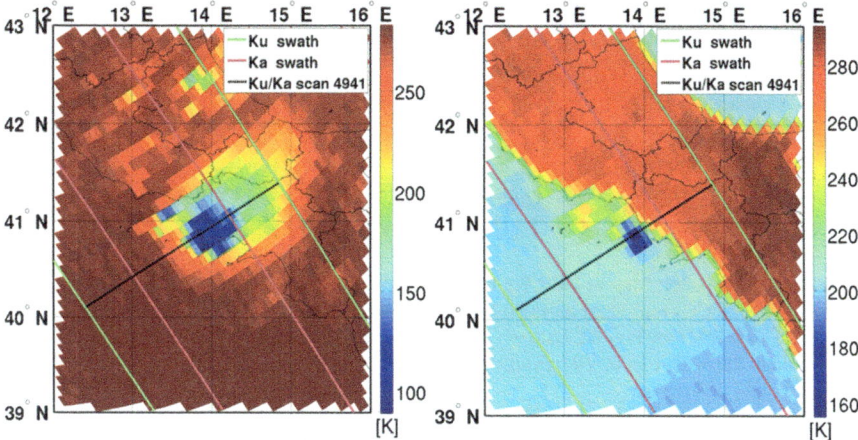

Fig. 42.5 Naples hailstorm on 5 Sept. 2015 0847 UTC: TBs at 166 GHz (V-pol) (left panel) and at 19 GHz (V-pol) (right panel) for the GMI orbit 8630. Ku- and Ka-band DPR swaths are evidenced in green and magenta, respectively. The position of the DPR-Ku scan 4941 (cross-section shown in Fig. 42.7) is shown in black

providing unique spaceborne measurements of the 3-D structure of precipitation, evidencing extremely rare features of the Naples hailstorm. Here we present some of the analysis carried out by Marra et al. (2017), but based on the latest version (V05) of all the GPM products.

Figure 42.5 shows the map of TBs at 166 GHz (V-pol), and at 19 GHz (V-pol). At 166 GHz the GMI measurements provide useful information on the different nature of the ice hydrometeors in the outflow region (low-density ice hydrometeors such as snowflakes and aggregates) and in the convective core (graupel or hail). The convective core shows a deep TB depression at 19 GHz and that the minimum TBs at 166 GHz are 20 K higher than at 89 GHz (not shown). This is a clear indication of the presence of large, high-density ice particles at different levels of the updraft region. Marra et al. (2017) evidenced also the role of the newly available 166 GHz polarization difference of GMI, indicating a complex physical structure in the upper cloud layers, with different hydrometeor characteristics: randomly oriented/tumbling ice particles brought to the upper levels by the strong updraft in the convective core, and non-spherical ice crystals, with preferential horizontal orientation, in the convective outflow region (likely horizontally oriented oblate ice particles).

DPR measurements provide further support to the GMI analysis. Figure 42.6 shows a map of DPR Ku-band median corrected reflectivity factor (Z_c) and strokes, both Intra Cloud (IC) and Cloud-to- Ground (CG), registered during a 4 min time interval around 0847 UTC (left panel). A hook echo associated to a mesocyclone is noticeable in the reflectivity map (it is very well defined in the ground-based C-band radar image shown in Marra et al. 2017). Moreover, most of the strokes are concentrated within the main cell identified by the area of maximum radar

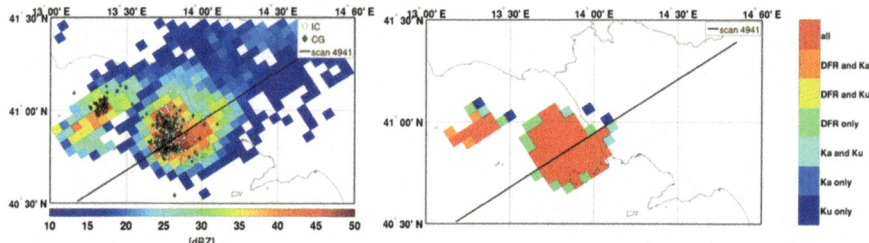

Fig. 42.6 Left panel: Map of DPR Ku-band median corrected reflectivity factor (Z$_c$) and LINET strokes (IC and CG) registered during a 4 min time interval around 0847 UTC with the position of DPR-Ku scan 4941 of Fig. 42.7 shown as a black line. Right panel: heavy ice precipitation flag available in the JAXA DPR V05 products determined with different criteria

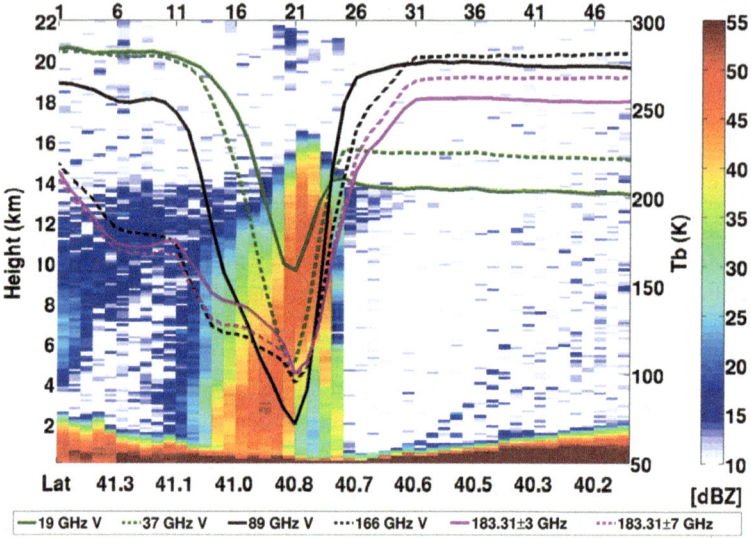

Fig. 42.7 DPR across-track scan 4941 of Ku measured reflectivity (Z$_m$) (DPR-NS V05). GMI TBs measured at four window channels (V-pol) and 183.31 GHz channels, corrected for parallax effects, are superimposed. On top the DPR Ku-band ray number along the cross-section is indicated. See Figs. 42.5 and 42.6 as reference for the position of the section

reflectivity (which is also coincident with the area of minimum TBs at 19 GHz). Here, the high values of the Z$_c$ (around 50 dBZ) are clear indications of the presence of hail. The DPR V05 products also provide a flag indicating the presence of heavy ice precipitation (Iguchi et al. 2018). The flag is defined by different methods and, in this case, all of them detect the occurrence of heavy ice precipitation within the cell core. This area includes the area of minimum TB at 19 GHz (V-pol) shown in Fig. 42.5.

Figure 42.7 shows the cross-section along scan 4941 of the reflectivity factor measured at Ku-band (Z$_m$) and the GMI TBs measured along the scan. The DPR shows a slanted structure of the main updraft and an overshooting top height at

Table 42.1 Analysis of Precipitation Features (PFs) found in 49 months of global observations (03/2014–03/2018, http://atmos.tamucc.edu/trmm/data/gpm, last accessed 27 Oct. 2018) based on minimum TB and PCT values

GMI TB	Rank	TRMM area (no CONUS)	Mediterranean area	CONUS	Other regions
19 GHz V-pol	2	–	1 (50%)	–	1 (50%)
23 GHz	4	–	1 (25%)	2 (50%)	1 (25%)
37 GHz PCT	7	–	1 (14%)	4 (57%)	2 (29%)
89 GHz PCT	161	138 (85.71%)	3(1.86%)	16 (9.94%)	4 (2.48%)

The second column shows the ranking of Naples hailstorm (for each TB or PCT) with respect to over 28 million (28,204,150) global PFs. Third to sixth column show the geographical distribution (in terms of number and %) of the PFs with TB and PCT minimum values equal to or lower than those found for the Naples hailstorm

16.25 km a.s.l. The slanted vertical structure, enhanced by the effect of the attenuation below 8 km, is a sign of vertical wind shear and storm severity, and is associated with the hook-echo in Fig. 42.6. The Z_m shows a well-defined structure of the storm, including the anvil and the main deep convective core, where the TBs reach extremely low values (158, 97, 67, and 87 K at 19, 37, 89 and 166 GHz, respectively). The high values of Z_m (~55 dBZ at 10 km, 40 dBZ at 14–15 km) indicate the presence of a very intense core of high-density ice particles (hail) in the upper levels maintained by a very strong updraft. A mirror echo above 20 km is present in correspondence of the updraft region, where the return signal from the surface is absent (Li and Nakamura 2002). The behavior of Ku and Ka-band (not shown) and the typical "knee" feature of the Ku/Ka Dual Wavelength Ratio (Battaglia et al. 2015) in the region of highest attenuation are typical of very intense convective cores characterized by large, high-density ice particles and strong updrafts.

GPM measurements were also used by Marra et al. (2017) to classify the severity of the Naples storm. A large database of GPM Precipitation Features (hereinafter PFs), i.e. contiguous pixels clustered on the basis of some specific criteria (Liu and Zipser 2015) was used to rank the Naples hailstorm in terms of minimum TBs (or Polarization Corrected Temperature, PCT) and 40 dBZ echo top height. The analysis carried out by Marra et al. (2017) is updated here considering 49 months of GPM global observations (03/2014–03/2018), and results are reported in Table 42.1. The exceptionality of the Naples hailstorm is confirmed, being characterized by the lowest 19 GHz (V-pol) TB minimum (158 K) registered in the Northern Hemisphere for the 49-month period analyzed (second to the lowest minimum globally registered for one hailstorm occurred in Argentina on 22 December 2017). The storm ranked first in the Mediterranean area also in terms of TB minimum at 23 GHz and PCT minimum at 37 GHz, and 4th and 7th globally with the other events found in the Continental US (CONUS) and in the tropics. Specific studies were carried out to find a relationship between minimum TBs at multiple frequencies and hail (e.g., Cecil 2009; Ferraro et al. 2015) and recently Mroz et al. (2017) found that the GMI PCT at 19 GHz shows the greatest potential for hail detection. All these studies confirm that the remarkable low TBs at low frequencies (noteworthy at 19 GHz), found for the Naples hailstorm, are linked to the large size of the convective core of the storm and to the presence of massive ice particles sustained at upper levels by very strong updrafts.

42.5 Medicanes

This section analyzes a Mediterranean hurricane (Medicane), or Tropical-Like Cyclone (TLC), a typical precipitation system of the Mediterranean area (e.g., Miglietta et al. 2013). Panegrossi et al. (2016) evidenced the role of GPM constellation radiometers in closely monitoring the evolution of the precipitation associated to Medicanes throughout their evolution, especially during their offshore development, when no other ground-based observations are available. The authors also outlined the improved observational capabilities of the GMI compared to other radiometers in depicting the precipitation structure of Medicane Qendresa, 7–8 Nov. 2014.

Here we focus on one storm called Numa that was classified as hybrid storm with both tropical and sub-tropical characteristics by NOAA/NESDIS, and as a Medicane by EUMETSAT. The sequence of Meteosat-10 Airmass RGBs, and ECMWF Mean Sea Level Pressure (MSLP) analyses, reveals that the depression started over the Strait of Sicily on 15 Nov. 2017 and deepened moving eastward on 16 Nov., accompanied by thunderstorms on a circular flow. Then it moved across the relatively warm Ionian Sea towards the Southern Apulia region in Italy, and between 17 and 18 Nov. revealed the well-defined, quasi cloud-free, eye structure surrounded by a whirl of clouds. It persisted over the coast of this region maintaining its tropical-like features for 24 h, causing extensive floods and damages. Then it moved eastward, and completely dissipated over Greece on 19 Nov. Compared to other Medicanes (e.g., Qendresa) for which a 986 hPa MSLP is inferred by ECWMF analysis, Numa does not show extremely low MSLP values (between 1006 and 1003 hPa). However, Numa is an interesting case to analyze in this context because, besides several overpasses of GPM constellation radiometers, one DPR overpass is also available. In particular, there are 53 PMW radiometer overpasses available between 1200 UTC 15 Nov. and 1200 UTC 19 Nov., and 15 of them occurred during the TLC phase (1200 UTC 17 Nov. – 1200 UTC 18 Nov.), when the storm persisted on the Apulia region coast. Two GPM-CO overpasses are available, one during Numa's development phase on 16 Nov. 1349 UTC, when the storm was captured by both GMI and DPR, and one during its mature phase on 18 Nov. 0359 UTC, when it was observed only by GMI.

Figure 42.8 shows the GPROF rainfall rate estimates for four Numa overpasses by AMSR2 and GMI. These two radiometers provide MW measurements at the

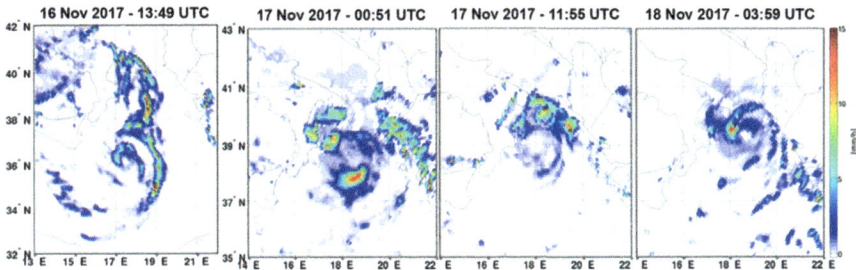

Fig. 42.8 Surface precipitation rate estimated by the GPROF V05 algorithm for four Numa overpasses by AMSR2 (middle panels) and GMI (left and right panels)

highest spatial resolution among the constellation radiometers (roughly three times higher than SSMIS). It is worth noting that only MW radiometers are able to provide such detail in the precipitation structure derived thanks to the multichannel response to precipitation microphysics, and this is particularly valuable over the sea where no other observations are available. The sequence shows how the precipitation evolves during the different stages of the storm development. During the initial phase, characterized by a less organized structure, regions of lighter and heavier (maximum estimated rainfall rate around 15 mm h^{-1}) rainfall coexist, while a well-defined eye and rainband structure can be observed during the mature phase. It is important to note that in the Southern Apulia region the storm caused floods and damages (with raingauge peaks of 165 mm in 24 h), relatively well predicted by NWP model simulations. The damages were not due to the rainfall intensity (mostly between 5 and 10 mm h^{-1}), but to the persistency of the Medicane over the area. The low rainfall intensity and the persistency are two important aspects of Medicanes evidenced in other studies (e.g., Miglietta et al. 2013), and specifically for Medicane Qendresa by Panegrossi et al. (2016).

The differences in the rainbands' morphology between the development and the mature phase of Numa are clearly visible in Fig. 42.9, showing the TBs measured at 37 and 89 GHz (V-pol) during the two GPM-CO overpasses. For the first overpass, the Storm Top Height (STH), estimated by the DPR Ku-band reflectivity factor, is also shown along with the LINET strokes registered in 1 h around the time of the overpass. At 37 GHz the rainbands appear very distinct, with higher TBs compared to the radiatively cold sea surface, and the pattern is well correlated to the surface rainfall rate map shown in Fig. 42.8. The absence of scattering signal at this frequency (often visible in presence of deep convection in the areas of lower 89 GHz TBs) evidences moderate (shallow) convective activity. The GMI (and AMSR2) 37 GHz channel (available at ~10 km spatial resolution on both radiometers) is able to delineate very well the fine rainfall structure in the rainbands, and evidences differences in the rainfall pattern and intensity between the development phase and the TLC phase of the storm. It is worth noting, for example, the high TBs at 37 GHz in correspondence of the rainband crossing the Southern Apulia region

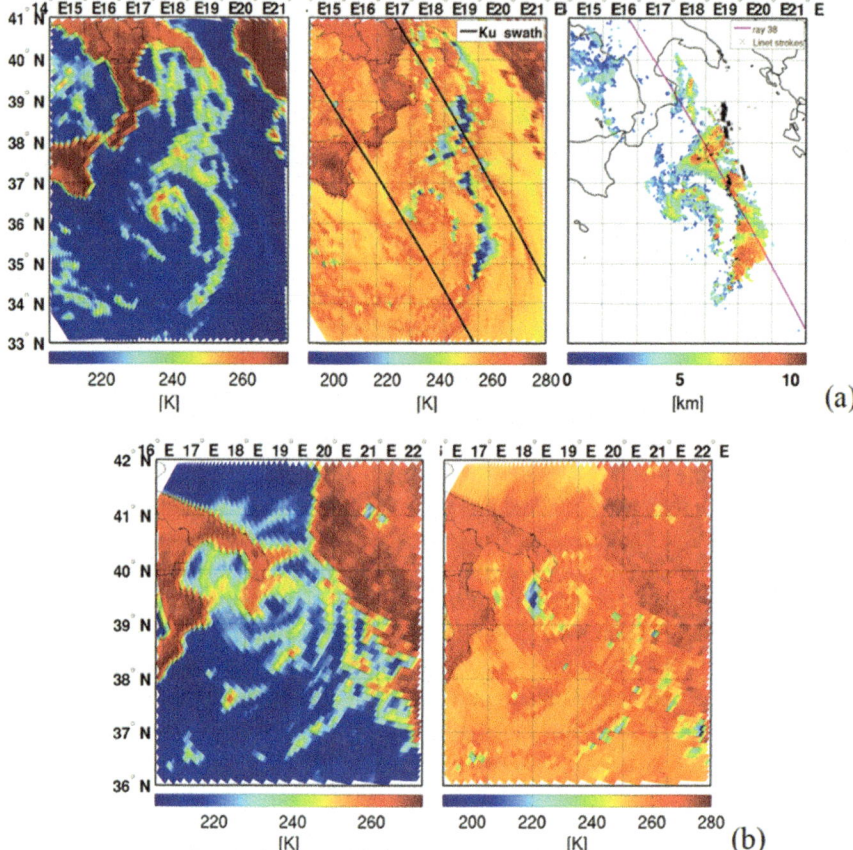

Fig. 42.9 (**a**) GPM-CO overpass on 16 Nov. 1350 UTC. Left panel: GMI TB at 37 GHz. Middle panel: GMI TB at 89 GHz (Ku-band swath is delimited by black lines). Right panel: Cloud Top Height (km), with LINET strokes, registered in 1 h around the time of the overpass, shown by black crosses and the position of Ray 38 (Fig. 42.10) shown in pink. (**b**) GPM-CO overpass on 18 Nov. 0359 UTC. Left panel: GMI TB at 37 GHz. Right panel: GMI TB at 89 GHz

during the mature phase, when the rotating storm, maintaining its position for more than 24 h, caused extensive floods.

The weak scattering signal at 89 GHz in correspondence of the main rainbands is a further evidence of the presence of low clouds, likely consistent with a slanted eyewall (as reported by Miglietta et al. 2013). The lowest TBs at 89 GHz (around 190 K) are found in correspondence of the convective regions. The TB pattern clearly shows that the convective areas characterize most part of the rainbands during the development phase, and they are more intense than in the mature phase, when the convection activity is limited to a small portion of the eyewall and is weaker. The DPR STH, as reconstructed from the Ku-band Z_m, shows that the maximum height during the development phase ranges between 6 and 10 km, with

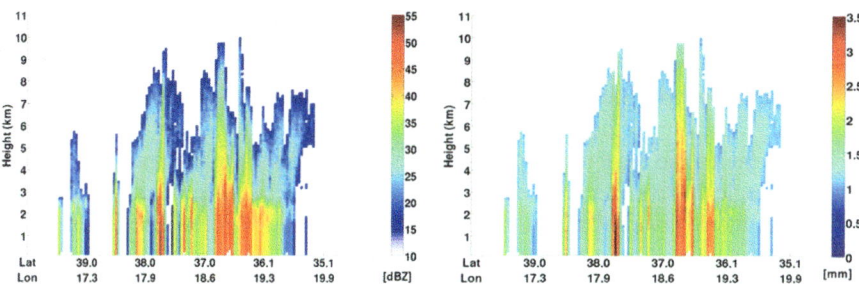

Fig. 42.10 Cross section along Ray 38 of DPR Ku-band Z_c (left panel) and DPR Ku particle size distribution D_m (right panel)

taller clouds mostly located on the eastern side of the cyclone center, confirming what was inferred from GMI TBs. The spatial distribution of the LINET strokes evidences that the electrical activity is concentrated in the areas of higher STH (and minimum TBs at 89 GHz). The lightning activity is more intense in the initial stage, with 177 LINET strokes registered in 1 h within 200 km from the MSLP minimum, than in the mature stage of the storm (8 strokes in 1 h), confirming what was found by Miglietta et al. (2013).

To our knowledge, Numa is the first Medicane that has been observed by the GPM DPR. DPR products provide several microphysics parameters (for example drop size distribution parameters, precipitation rate, liquid and ice water content). An example is shown in Fig. 42.10. The cross-section of DPR Ku Z_c along ray 38, where the most intense portions of the cyclone rainbands are found, shows the predominance of convective regions (e.g., 37.7°N-18.1°E; 36.7°N-18.8°E), with Z_c around 50 dBZ up to 4 km, confirming the analysis of GMI TBs. The region with higher concentration of LINET strokes (shown in Fig. 42.9) is found in correspondence of the convective core (at 36.7°N 18.8°E) where large values of mean mass-weighted drop diameter (D_m), exceeding 3 mm, are found in the upper cloud layers (up to 6–7 km). A stratiform region (36.1°N 19.3°E) is delimited by a clear signal of bright-band (peak of Z_c) in close correspondence with the freezing level height (~2400 m) as inferred from the ECMWF analysis. Lower (up to 2 mm) and nearly constant D_m values are found in this region. A further analysis of DPR (and GMI) measurements and products is needed to fully characterize Numa 3-D precipitation microphysics and structure during its development phase. It is worth noting here, however, that the analysis of GPM-CO overpasses of other Medicanes observed at different phases of their evolution over the sea will significantly contribute to our understanding and characterization of their differences and common features.

42.6   Final Remarks

The Mediterranean region is a challenging environment for meteorology from both the observational and the modelling point of view. The complex interplay between synoptic fluxes and local forcing (at all scales) makes the study and forecast of severe meteorological events a cumbersome task. To improve the observational capability in one of the regions more densely covered by ground networks of meteorological instruments, spaceborne sensors are increasingly used with novel multiplatform strategies. In the present chapter, the unprecedented GPM mission observations and products are used not only for the retrieval of precipitation rates, but also for the analysis of the precipitation structure, and for monitoring the evolution of extreme events which periodically devastate the Mediterranean coastal regions. Three examples of the possible role of GPM observations of severe weather are documented.

We first tested the capabilities of advanced satellite precipitation products to quantitatively estimate precipitation at the ground for a very intense and localized single cell that caused a flash flood in Italy. In this case, when large discrepancies between ground radar and raingauge measurements are found, the satellite products offer a wide spectrum of results, showing that better performances come from algorithms tailored for the specific region, while algorithms designed for global application are less sensitive to such small-scale thunderstorms.

The second case illustrates the use of the satellite data to study a mesoscale multicellular storm developed offshore and landed on the Italian coast, with widespread large hail (8–10 cm diameter) and unprecedented lightning records. The detailed 3-D view of the storm structure obtained by DPR, and the cloud upper layer features provided by PMW sensors, integrated the information from ground-based lightning network and weather radars, to provide a multi-sensor view of one of the most severe thunderstorm systems occurred worldwide in the recent years.

A class of events deserving attention in the last decade is analyzed as a third playground for satellite data exploitation: Medicanes. Comparative analysis between different GPM overpasses at two different stages of Medicane Numa evolution, and the DPR measurements available during its development phase in southern Mediterranean, reveal for the first time details about the rainband structure, convective vs. stratiform precipitation evolution, rainfall intensity, and precipitation microphysics of these interesting weather phenomena.

To fully exploit satellite data and products for studying and monitoring severe weather in the Mediterranean area, and, ultimately to improve forecasting (and nowcasting) capabilities in this complex region, efforts should be undertaken to provide products tailored for the area (working on algorithm calibration), and to provide the error structure of the products (working on validation). This would enable users to properly apply the most suitable products for the specific need, including data assimilation of precipitation-related fields over the sea where these systems often initiate. Moreover, long data records of satellites precipitation-related measurements and reliable products would help assessing climate change signatures in the Mediterranean area, where such severe events are becoming more and more frequent.

Acknowledgements Financial support by EUMETSAT's "Satellite Application Facility on Support to Operational Hydrology and Water Management" (H SAF) and by the Italian Research Project of National Interest 2015 (PRIN 2015) 4WX5NA is acknowledged. The authors would like to thank H SAF Management (in particular Dr. Stefano Dietrich at CNR-ISAC), the NASA PMM Research Program for supporting scientific collaborations with H SAF and HyMeX, the NASA PPS for providing GPM products, and the Italian Department for Civil Protection for providing ground-based radar and raingauge data. LINET data are provided by Nowcast GmhB (https://www.nowcast.de/, last accessed 27 Oct. 2018) at CNR ISAC-Rome within a scientific agreement lead by Dr. Stefano Dietrich. The authors express their sincere gratitude to Dr. Alberto Mugnai for his scientific work at CNR on microwave remote sensing of precipitation and for his guidance and support during the development of the H SAF precipitation products presented in this study.

References

Bartzokas, A., Azzopardi, J., Bertotti, L., Buzzi, A., Cavaleri, L., Conte, D., Davolio, S., Dietrich, S., Drago, A., Drofa, O., Gkikas, A., Kotroni, V., Lagouvardos, K., Lolis, C. J., Michaelides, S., Miglietta, M., Mugnai, A., Music, S., Nikolaides, K., Porcù, F., Savvidou, K., & Tsirogianni, M. I. (2010). The RISKMED project: Philosophy, methods and products. *Natural Hazards and Earth System Sciences, 10*, 1393–1401. https://doi.org/10.5194/nhess-10-1393-2010.

Battaglia, A., Tanelli, S., Mroz, K., & Tridon, F. (2015). Multiple scattering in observations of the GPM dual-frequency precipitation radar: Evidence and impact on retrievals. *Journal of Geophysical Research, 120*, 4090–4101. https://doi.org/10.1002/2014JD022866.

Bech, J., Arús, J., Castán, S., Pineda, N., Rigo, T., Montanyà, J., & van der Velde, O. (2015). A study of the 21 March 2012 tornadic quasi linear convective system in Catalonia. *Atmospheric Research, 158*, 192–209. https://doi.org/10.1016/j.atmosres.2014.08.009.

Betz, H.-D., Schmidt, K., Laroche, P., Blanchet, P., Oettinger, W. P., Defer, E., Dziewit, Z., & Konarski, J. (2009). LINET – An international lightning detection network in Europe. *Atmospheric Research, 91*, 564–573. https://doi.org/10.1016/j.atmosres.2008.06.012.

Buzzi, A., Davolio, S., Malguzzi, P., Drofa, O., & Mastrangelo, D. (2014). Heavy rainfall episodes over Liguria in autumn 2011: Numerical forecasting experiments. *Natural Hazards and Earth System Sciences, 14*, 1325–1340. https://doi.org/10.5194/nhess-14-1325-2014.

Casella, D., Panegrossi, G., Sanò, P., Mugnai, A., Smith, E. A., Tripoli, G. J., Dietrich, S., Formenton, M., Leung, W. Y., & Mehta, A. (2013). Transitioning from CRD to CDRD in Bayesian retrieval of rainfall from satellite passive microwave measurements: Part 2. Overcoming database profile selection ambiguity by consideration of meteorological control on microphysics. *IEEE Transactions on Geoscience and Remote Sensing, 51*, 4650–4671. https://doi.org/10.1109/TGRS.2013.2258161.

Casella, D., do Amaral, L. M. C., Dietrich, S., Marra, A. C., Sanò, P., & Panegrossi, G. (2017). The Cloud Dynamics and Radiation Database algorithm for AMSR2: Exploitation of the GPM observational dataset for operational applications. *IEEE Journal of Selected Topics in Applied Earth Observations and Remote Sensing, 10*, 3985–4001. https://doi.org/10.1109/JSTARS.2017.2713485.

Cecil, D. J. (2009). Passive microwave brightness temperatures as proxies for hailstorms. *Journal of Applied Meteorology and Climatology, 48*, 1281–1286. https://doi.org/10.1175/2009JAMC2125.1.

Cohuet, J. B., Romero, R., Homar, V., Ducrocq, V., & Ramis, C. (2011). Initiation of a severe thunderstorm over the Mediterranean Sea. *Atmospheric Research, 100*, 603–620. https://doi.org/10.1016/j.atmosres.2010.11.002.

Davolio, S., Silvestro, F., & Gastaldo, T. (2017). Impact of rainfall assimilation on high-resolution hydrometeorological forecasts over Liguria, Italy. *Journal of Hydrometeorology, 18*, 2659–2680. https://doi.org/10.1175/JHMD-17-0073.1.

Di Paola, F., Ricciardelli, E., Cimini, D., Romano, F., Viggiano, M., & Cuomo, V. (2014). Analysis of Catania flash flood case study by using combined microwave and infrared technique. *Journal of Hydrometeorology, 15*, 1989–1998. https://doi.org/10.1175/JHM-D-13-092.1.

Drobinski, P., Ducrocq, V., Alpert, P., Anagnostou, E., Béranger, K., Borga, M., Braud, I., Chanzy, A., Davolio, S., Delrieu, G., Estournel, C., Filali Boubrahmi, N., Font, J., Grubišić, V., Gualdi, S., Homar, V., Ivančan-Picek, B., Kottmeier, C., Kotroni, V., Lagouvardos, K., Lionello, P., Llasat, M. C., Ludwig, W., Lutoff, C., Mariotti, A., Richard, E., Romero, R., Rotunno, R., Roussot, O., Ruin, I., Somot, S., Taupier-Letage, I., Tintore, J., Uijlenhoet, R., & Wernli, H. (2014). HyMeX: A 10-year multidisciplinary program on the Mediterranean water cycle. *Bulletin of the American Meteorological Society, 95*, 1063–1082. https://doi.org/10.1175/BAMS-D-12-00242.1.

Emanuel, K. (2005). Genesis and maintenance of "Mediterranean hurricanes". *Advances in Geosciences, 2*, 217–220. https://doi.org/10.5194/adgeo-2-217-2005.

Ferraro, R., Beauchamp, J., Cecil, D., & Heymsfield, G. (2015). A prototype hail detection algorithm and hail climatology developed with the advanced microwave sounding unit (AMSU). *Atmospheric Research, 163*, 24–35. https://doi.org/10.1016/j.atmosres.2014.08.010.

Flaounas, E., Lagouvardos, K., Kotroni, V., Claud, C., Delanoë, J., Flamant, C., Madonna, E., & Wernli, H. (2015). Processes leading to heavy precipitation associated with two Mediterranean cyclones observed during the HyMeX SOP1. *Quarterly Journal of the Royal Meteorological Society, 142*(S1), 275–286. https://doi.org/10.1002/qj.2618.

Giorgi, F. (2006). Climate change hot-spots. *Geophysical Research Letters, 33*, L08707. https://doi.org/10.1029/2006GL025734.

Hally, A., Caumont, O., Garrote, L., Richard, E., Weerts, A., Delogu, F., Fiori, E., Rebora, N., Parodi, A., Mihalović, A., Ivković, M., Dekić, L., van Verseveld, W., Nuissier, O., Ducrocq, V., D'Agostino, D., Galizia, A., Danovaro, E., & Clematis, A. (2015). Hydrometeorological multi-model ensemble simulations of the 4 November 2011 flash flood event in Genoa, Italy, in the framework of the DRIHM project. *Natural Hazards and Earth System Sciences, 15*, 537–555. https://doi.org/10.5194/nhess-15-537-2015.

Hou, A. Y., Kakar, R. K., Neeck, S., Azarbarzin, A. A., Kummerow, C. D., Kojima, M., Oki, R., Nakamura, K., & Iguchi, T. (2014). The global precipitation measurement mission. *Bulletin of the American Meteorological Society, 95*, 701–722. https://doi.org/10.1175/BAMS-D-13-00164.1.

Huffman, G. J., Bolvin, D. T., Braithwaite, D., Hsu, K., Joyce, R., Kidd, C., Nelkin, E. J., & Xie, P. (2015). *NASA Global Precipitation Measurement (GPM) Integrated Multi-satellitE Retrievals for GPM (IMERG)*. Algorithm theoretical basis doc., v4.5, 26 pp. Available at https://pmm.nasa.gov/sites/default/files/document_files/IMERG_ATBD_V4.5.pdf. Last accessed 27 Oct 2018

Iguchi, T., Seto, S., Meneghini, R., Yoshida, N., Awaka, J., Le, M., Chandrasekar, V., & Kubota, T. (2017). *GPM/DPR Level-2. Algorithm theoretical basis Doc*. Available at https://pps.gsfc.nasa.gov/Documents/ATBD_DPR_201708_whole_1.pdf. Last accessed 27 Oct 2018

Iguchi, T., Kawamoto, N., & Oki, R. (2018). Detection of intense ice precipitation with GPM/DPR. *Journal of Atmospheric and Oceanic Technology, 35*, 491–502. https://doi.org/10.1175/JTECH-D-17-0120.1.

Jansa, A., Alpert, P., Arbogast, P., Buzzi, A., Ivancan-Picek, B., Kotroni, V., Llasat, M. C., Ramis, C., Richard, E., Romero, R., & Speranza, A. (2014). MEDEX: A general overview. *Natural Hazards and Earth System Sciences, 14*, 1965–1984. https://doi.org/10.5194/nhess-14-1965-2014.

Kotroni, V., Lagouvardos, K., Defer, E., Dietrich, S., Porcù, F., Medaglia, C. M., & Demirtas, M. (2006). The Antalya 5 December 2002 storm: Observations and model analysis. *Journal of Applied Meteorology, 45*, 576–590. https://doi.org/10.1175/JAM2347.1.

Kummerow, C. D., Simpson, J., Thiele, O., Barnes, W., Chang, A. T. C., Stocker, E., Adler, R. F., Hou, A., Kakar, R., Wentz, F., Ashcroft, P., Kozu, T., Hong, Y., Okamoto, K., Iguchi, T., Kuroiwa, H., Im, E., Haddad, Z., Huffman, G., Ferrier, B., Olson, W. S., Zipser, E., Smith, E. A., Wilheit, T. T., North, G., Krishnamurti, T., & Nakamura, K. (2000). The status of the Tropical Rainfall Measuring Mission (TRMM) after two years in orbit. *Journal of Applied Meteorology, 39*, 1965–1982. https://doi.org/10.1175/1520-0450(2001)040<1965:TSOTTR>2.0.CO;2.

Kummerow, C. D., Randel, D. L., Kulie, M., Wang, N.-Y., Ferraro, R., Munchak, S. J., & Petkovic, V. (2015). The evolution of the Goddard Profiling Algorithm to a fully parametric scheme. *Journal of Atmospheric and Oceanic Technology, 32*, 2265–2280. https://doi.org/10.1175/JTECH-D-15-0039.1.

Laviola, S., Moscatello, A., Miglietta, M. M., Cattani, E., & Levizzani, V. (2011). Satellite and numerical model investigation of two heavy rain events over the Central Mediterranean. *Journal of Hydrometeorology, 12*, 634–649. https://doi.org/10.1175/2011JHM1257.1.

Levizzani, V., Bauer, P., & Turk, F. J. (Eds.). (2007). *Measuring precipitation from space: EURAINSAT and the future*. Dordrecht: Springer, 748 pp., ISBN: 1-4020-5834-9.

Li, J., & Nakamura, K. (2002). Characteristics of the mirror image of precipitation observed by the TRMM Precipitation Radar. *Journal of Atmospheric and Oceanic Technology, 19*, 145–158. https://doi.org/10.1175/1520-0426.

Liu, C., & Zipser, E. J. (2015). The global distribution of largest, deepest, and most intense precipitation systems. *Geophysical Research Letters, 42*, 3591–3595. https://doi.org/10.1002/2015GL063776.

Llasat, M. C., Llasat-Botija, M., Petrucci, O., Pasqua, A. A., Rosselló, J., Vinet, F., & Boissier, L. (2013). Towards a database on societal impact of Mediterranean floods within the framework of the HYMEX project. *Natural Hazards and Earth System Sciences, 13*, 1337–1350. https://doi.org/10.5194/nhess-13-1337-2013.

Manzato, A., Davolio, S., Miglietta, M. M., Pucillo, A., & Setvák, M. (2014). 12 September 2012: A supercell outbreak in NE Italy? *Atmospheric Research, 153*, 98–118. https://doi.org/10.1016/j.atmosres.2014.07.019.

Marra, A. C., Porcù, F., Baldini, L., Petracca, M., Casella, D., Dietrich, S., Mugnai, A., Sanò, P., Vulpiani, G., & Panegrossi, G. (2017). Observational analysis of an exceptionally intense hailstorm over the Mediterranean area: Role of the GPM Core Observatory. *Atmospheric Research, 192*, 72–90. https://doi.org/10.1016/j.atmosres.2017.03.019.

Mecikalski, J. R., MacKenzie, W. M., Jr., König, M., & Muller, S. (2010). Cloud top properties of growing cumulus prior to convective initiation as measured by Meteosat Second Generation. Part I: Infrared fields. *Journal of Applied Meteorology and Climatology, 49*, 521–534. https://doi.org/10.1175/2009JAMC2344.1.

Miglietta, M. M., & Rotunno, R. (2016). An EF3 multivortex tornado over the Ionian Region: Is it time for a dedicated warning system over Italy? *Bulletin of the American Meteorological Society, 97*, 337–344. https://doi.org/10.1175/BAMS-D-14-00227.1.

Miglietta, M. M., Moscatello, A., Conte, D., Mannarini, G., Lacorata, G., & Rotunno, R. (2011). Numerical analysis of a Mediterranean 'hurricane' over south-eastern Italy: Sensitivity experiments to sea surface temperature. *Atmospheric Research, 101*, 412–426. https://doi.org/10.5194/adgeo-2-217-2005.

Miglietta, M. M., Laviola, S., Malvaldi, A., Conte, D., Levizzani, V., & Price, C. (2013). Analysis of tropical-like cyclones over the Mediterranean Sea through a combined modelling and satellite approach. *Geophysical Research Letters, 40*, 2400–2405. https://doi.org/10.1002/grl.50432.

Miglietta, M. M., Mazon, J., & Rotunno, R. (2017). Numerical simulations of a tornadic supercell over the Mediterranean. *Weather and Forecasting, 32*, 1209–1226. https://doi.org/10.1175/WAF-D-16-0223.1.

Mroz, K., Battaglia, A., Lang, T. J., Cecil, D. J., Tanelli, S., & Tridon, F. (2017). Hail-detection algorithm for the GPM Core Observatory satellite sensors. *Journal of Applied Meteorology and Climatology, 56*, 1939–1957. https://doi.org/10.1175/JAMC-D-16-0368.1.

Mugnai, A., Smith, E. A., Tripoli, G. J., Bizzarri, B., Casella, D., Dietrich, S., Di Paola, F., Panegrossi, G., & Sanò, P. (2013a). CDRD and PNPR satellite passive microwave precipitation retrieval algorithms: EuroTRMM/EURAINSAT origins and H-SAF operations. *Natural Hazards and Earth System Sciences, 13*, 887–912. https://doi.org/10.5194/nhess-13-887-2013.

Mugnai, A., Casella, D., Cattani, E., Dietrich, S., Laviola, S., Levizzani, V., Panegrossi, G., Petracca, M., Sanò, P., Di Paola, F., Biron, D., De Leonibus, L., Melfi, D., Rosci, P., Vocino, A., Zauli, F., Pagliara, P., Puca, S., Rinollo, A., Milani, L., Porcù, F., & Gattari, F. (2013b). Precipitation products from the hydrology SAF. *Natural Hazards and Earth System Sciences, 13*, 1959–1981. https://doi.org/10.5194/nhess-13-1959-2013.

Panegrossi, G., Casella, D., Dietrich, S., Marra, A. C., Petracca, M., Sanò, P., Mugnai, A., Baldini, L., Roberto, N., Adirosi, E., Cremonini, R., Bechini, R., Vulpiani, G., & Porcù, F. (2016). Use of the GPM constellation for monitoring heavy precipitation events over the Mediterranean region. *IEEE Journal of Selected Topics in Applied Earth Observations and Remote Sensing, 9*, 2733–2753. https://doi.org/10.1109/JSTARS.2016.2520660.

Petracca, M., D'Adderio, L. P., Porcù, F., Vulpiani, G., Sebastianelli, S., & Puca, S. (2018). Validation of GPM Dual-Frequency Precipitation Radar (DPR) rainfall products over Italy. *Journal of Hydrometeorology, 19*, 907–925. https://doi.org/10.1175/JHM-D-17-0144.1.

Pignone, F., Rebora, N., Silvestro, F., & Castelli, F. (2018). GRISO: A new interpolator of raingauge observations. *Journal of Atmospheric and Oceanic Technology* (under review).

Porcù, F., & Carrassi, A. (2009). Toward an estimation of the relationship between cyclonic structures and damages at the ground in Europe. *Natural Hazards and Earth System Sciences, 9*, 823–829. https://doi.org/10.5194/nhess-9-823-2009.

Porcù, F., Carrassi, A., Medaglia, C. M., Prodi, F., & Mugnai, A. (2007). A study on cut-off low vertical structure and precipitation in the Mediterranean region. *Meteorology and Atmospheric Physics, 96*, 121–140. https://doi.org/10.1007/s00703-006-0224-5.

Price, C., Yair, Y., Mugnai, A., Lagouvardos, K., Llasat, M. C., Michaelides, S., Dayan, U., Dietrich, S., Di Paola, F., Galanti, E., Garrote, L., Harats, N., Katsanos, D., Kohn, M., Kotroni, V., Llasat-Botija, M., Lynn, B., Mediero, L., Morin, E., Nicolaides, K., Rozalis, S., Savvidou, K., & Ziv, B. (2011). The FLASH project: Using lightning data to better understand and predict flash floods. *Environmental Science & Policy, 14*, 898–911. https://doi.org/10.1007/s10712-011-9146-y.

Puca, S., Porcu, F., Rinollo, A., Vulpiani, G., Baguis, P., Balabanova, S., Campione, E., Ertürk, A., Gabellani, S., Iwanski, R., Jurašek, M., Kaňák, J., Kerényi, J., Koshinchanov, G., Kozinarova, G., Krahe, P., Lapeta, B., Lábó, E., Milani, L., Okon, L.'., Öztopal, A., Pagliara, P., Pignone, F., Rachimow, C., Rebora, N., Roulin, E., Sönmez, I., Toniazzo, A., Biron, D., Casella, D., Cattani, E., Dietrich, S., Di Paola, F., Laviola, S., Levizzani, V., Melfi, D., Mugnai, A., Panegrossi, G., Petracca, M., Sanò, P., Zauli, F., Rosci, P., De Leonibus, L., Agosta, E., & Gattari, F. (2014). The validation service of the hydrological SAF geostationary and polar satellite precipitation products. *Natural Hazards and Earth System Sciences, 14*, 871–889. https://doi.org/10.5194/nhess-14-871-2014.

Reale, O., & Atlas, R. (2001). Tropical cyclone–like vortices in the extratropics: Observational evidence and synoptic analysis. *Weather and Forecasting, 16*, 7–34. https://doi.org/10.1175/1520-0434.

Roberto, N., Adirosi, E., Baldini, L., Casella, D., Dietrich, S., Gatlin, P., Panegrossi, G., Petracca, M., Sanò, P., & Tokay, A. (2016). Multi-sensor analysis of convective activity in central Italy during the HyMeX SOP 1.1. *Atmospheric Measurement Techniques, 9*, 535–552. https://doi.org/10.5194/amt-9-535-2016.

Romero, R., Ramis, C., & Homar, V. (2015). On the severe convective storm of 29 October 2013 in the Balearic Islands: Observational and numerical study. *Quarterly Journal of the Royal Meteorological Society, 141*, 1208–1222. https://doi.org/10.1002/qj.2429.

Rysman, J.-F., Panegrossi, G., Sanò, P., Marra, A. C., Dietrich, S., Milani, L., & Kulie, M. S. (2018). SLALOM: An allsurface snow water path retrieval algorithm for the GPM Microwave Imager. *Remote Sensing, 10*, 1278. https://doi.org/10.3390/rs10081278.

Sanò, P., Casella, D., Mugnai, A., Schiavon, G., Smith, E. A., & Tripoli, G. J. (2013). Transitioning from CRD to CDRD in Bayesian retrieval of rainfall from satellite passive microwave measurements: Part 1. Algorithm description and testing. *IEEE Transactions on Geoscience and Remote Sensing, 51*, 4119–4143. https://doi.org/10.1109/TGRS.2012.2227332.

Sanò, P., Panegrossi, G., Casella, D., Di Paola, F., Milani, L., Mugnai, A., Petracca, M., & Dietrich, S. (2015). The Passive microwave Neural network Precipitation Retrieval (PNPR) algorithm for AMSU/MHS observations: Description and application to European case studies. *Atmospheric Measurement Techniques, 8*, 837–857. https://doi.org/10.5194/amt-8-837-2015.

Sanò, P., Panegrossi, G., Casella, D., Marra, A. C., Di Paola, F., & Dietrich, S. (2016). The new Passive microwave Neural network Precipitation Retrieval (PNPR) algorithm for the cross-track scanning ATMS radiometer: Description and verification study over Europe and Africa using GPM and TRMM spaceborne radars. *Atmospheric Measurement Techniques, 9*, 5441–5460. https://doi.org/10.5194/amt-9-5441-2016.

Sanò, P., Panegrossi, G., Casella, D., Marra, A. C., D'Adderio, L. P., Rysman, J.-F., & Dietrich, S. (2018). The Passive Microwave Neural Network Precipitation Retrieval (PNPR) algorithm for the CONICAL scanning Global Microwave Imager (GMI) radiometer. *Remote Sensing, 10*, 1122. https://doi.org/10.3390/rs10071122.

Skofronick-Jackson, G., Petersen, W. A., Berg, W., Kidd, C., Stocker, E. F., Kirschbaum, D. B., Kakar, R., Braun, S. A., Huffman, G. J., Iguchi, T., Kirstetter, P. E., Kummerow, C., Meneghini, R., Oki, R., Olson, W. S., Takayabu, Y. N., Furukawa, K., & Wilheit, T. (2017). The Global Precipitation Measurement (GPM) mission for science and society. *Bulletin of the American Meteorological Society, 98*, 1679–1695. https://doi.org/10.1175/BAMS-D-15-00306.1.

Tripoli, G. J., Medaglia, C. M., Dietrich, S., Mugnai, A., Panegrossi, G., Pinori, S., & Smith, E. A. (2005). The 9–10 November 2001 Algerian flood: A numerical study. *Bulletin of the American Meteorological Society, 86*, 1229–1235. https://doi.org/10.1175/BAMS-86-9-1229.

Chapter 43
Dryland Precipitation Climatology from Satellite Observations

Efrat Morin, Francesco Marra, and Moshe Armon

Abstract Dryland regions cover a wide portion of the Earth's land area. However, an accurate climatic characterization of precipitation properties is lacking due to the limited ground monitoring systems typical of these regions. Precipitation estimation from satellite observations provides the opportunity to improve our knowledge of precipitation climatology in these regions. In this chapter, we summarize the current research utilizing and assessing satellite precipitation estimates in the drylands, and we conduct a quasi-global analysis of dryland precipitation climatology based on the TRMM (TMPA) data set, with the aim of highlighting potential and limitations of satellite estimates for such areas. Distinct climatological features are observed over regions characterized by different levels of aridity. Dryer areas are associated with smaller yearly precipitation amounts and larger coefficient of variation, lower number of rainy days, heavier tail of the distribution of the extremes, smaller fraction of rainy area and shorter correlation distance. We conclude providing some recommendations for future applications on dryland precipitation estimates and climatology based on satellite products.

Keywords Precipitation · Dryland · Desert · Hyperarid · Arid · Semi-arid · Aridity index · TRMM · Rain cells · Autocorrelation · Frequency analysis · Climatology · Satellite · Radar

E. Morin (✉) · M. Armon
The Fredy & Nadine Herrmann Institute of Earth Sciences, Hebrew University of Jerusalem, Jerusalem, Israel
e-mail: efrat.morin@mail.huji.ac.il

F. Marra
The Fredy & Nadine Herrmann Institute of Earth Sciences, Hebrew University of Jerusalem, Jerusalem, Israel

National Research Council of Italy, Institute of Atmospheric Sciences and Climate, CNR-ISAC, Bologna, Italy

© Springer Nature Switzerland AG 2020   843
V. Levizzani et al. (eds.), *Satellite Precipitation Measurement*, Advances in Global Change Research 69, https://doi.org/10.1007/978-3-030-35798-6_19

43.1 Introduction

Drylands comprise more than a third of the Earth's land (Fig. 43.1a). They are found in all of the six inhabited continents, and host about 20% of the world's population (Safriel et al. 2005). In this environment, precipitation is the main source of fresh water, aquifer recharge, runoff and soil moisture, all crucial for human populations and ecological systems living therein; yet it is typically meager, irregular and highly variable (Goodrich et al. 1995; Ahrens 2003). Due to the bare surfaces and low infiltration rates of the arid soils, storms showering rather small amounts of rainfall can, when featuring high precipitation rates, lead to large and sudden floods that pose risks to the local population. Precipitation measurements in dryland areas are severely limited by the scarcity of ground monitoring systems, including rain gauge networks and weather radars. In fact, the sparse population density and the harsh conditions make installation and maintenance of measurement devices difficult. Quantitative estimation of precipitation from satellite observations provides a unique opportunity to better specify the climatology of precipitation in the drylands, characterize the unique properties of storm events, assess and manage water resources, and improve flood risk management. In this chapter we depict the state of the art in the use of satellite observations for dryland climatological studies (Sect. 19.4), and we present a new, semi-global climatological analysis based on the TRMM Multi-satellite Precipitation Analysis (TMPA; Huffman et al. 2007, 2010) (Sect. 19.5). We conclude providing recommendations for further research and scientific efforts.

43.2 Background on Dryland Climate

There is no single common definition for dryland climate (Nicholson 2011). Simple definitions use thresholds on the mean annual precipitation (e.g., 250 mm), while more accurate definitions rely on the balance between potential water loss and precipitation amount. Indices like the ratio between precipitation and potential evapotranspiration (UNEP 1992, Fig. 43.1a; e.g., Cherlet et al. 2018), or the Budyko index, i.e. the ratio between net energy and precipitation (Budyko 1958), enable quantifying and ranking the degree of aridity (Warner 2004).

Figure 43.1a shows an aridity index map based on the UNEP (1992) definition. Average sea-level pressure contours manifest semi-permanent high-pressure centers ("H") related to subsidence of air and low rainfall amounts. Although most drylands are located in the sub-tropical "desert belt" (roughly latitudes of 20°-30°), some of them extend even up to ~50° latitude. Their distribution is driven by different climatic mechanisms causing aridity. In general, deserts may be classified as warm, cold, or foggy/coastal deserts (Shmida 1985). Most warm deserts are caused by subsidence from planetary scale circulation (e.g., the polar side of Hadley cells) and are situated in the "desert belt". Examples include the Sahara in Africa, the Arabian Desert in south-east Asia and the Chihuahua Desert in North America. Most cold deserts extend the dryland regions poleward, and result either from the distance from moisture sources, or from orographic effects (rain shadow; Warner 2004). Examples include the central Asia deserts (e.g., Gobi and Turkestan), the Patagonian and the North-American deserts. Coastal deserts result mainly from cold ocean currents (Fig. 43.1a), like the Benguela Current (Namib desert) and the Humboldt current (Atacama desert).

From the meteorological point of view, most of the precipitation in dryland areas is caused by the occasional deflection of tropical (originated equatorward, in lower latitudes, and focused mostly in summer months) or extra-tropical (poleward, wintertime) weather systems from their regular routes, that cause them to precipitate over normally much drier regions (Nicholson 2011). Tropical systems affecting the drylands are more common at low latitudes, as seen in the southern Sahara Desert (Fig. 43.1b). These include tropical depressions and lows, easterly waves, or mesoscale convective systems, and are sometimes associated with large-scale monsoons. These systems may cause high rainfall rates and flash-floods (e.g., Hirschboeck 1988, Sheppard et al. 2002). Western Sahara is characterized by maximal rainfall in autumn, caused by south-penetrating extra-tropical lows and tropical-extra tropical interactions, such as Tropical Plumes. Central Sahara is characterized by negligible rainfall amounts peaking either in spring or having no preferential seasonality.

Extra-tropical systems generate rainfall mainly in the higher latitude drylands, particularly in proximity of Mediterranean climate regions. For example, the northern Sahara, Arabian and Sinai-Negev deserts, situated downwind of the Mediterranean, are all affected by winter precipitation generated by Mediterranean lows that occasionally divert south/eastward from their usual routes (Kahana et al. 2002; Barth and Steinkohl 2004; Dayan and Morin 2006; Kushnir et al. 2017; Armon et al. 2018) (Fig. 43.1b). These systems commonly consist of lower rainfall rates, and tend to have faster-moving rain-cells (Belachsen et al. 2017).

Tropical – extra-tropical interactions may as well cause desert rainfall. They are generated from a mid to upper-tropospheric jet that draws moisture mainly from low latitudes into the subtropics and generates substantial amounts of rainfall over broad regions, such as the western Sahara (Fig. 43.1b), but could even extend far away from any ocean (e.g., Ziv 2001; Knippertz and Wernli 2010). These include various – sometimes overlapping – titles, such as diagonal cloud bands, tropical plumes, atmospheric rivers, tropical moisture exports, etc. (Knippertz 2003; Rubin et al. 2007; Nicholson 2011; Ralph et al. 2017; Tubi et al. 2017). In this type of systems, stratiform precipitation can be widespread and prolonged with low intensities and even virgae, and may enclose convective elements exhibiting high intensity and causing floods (Geerts and Dejene 2005; Muller et al. 2008; Tubi and Dayan 2014; Armon et al. 2018).

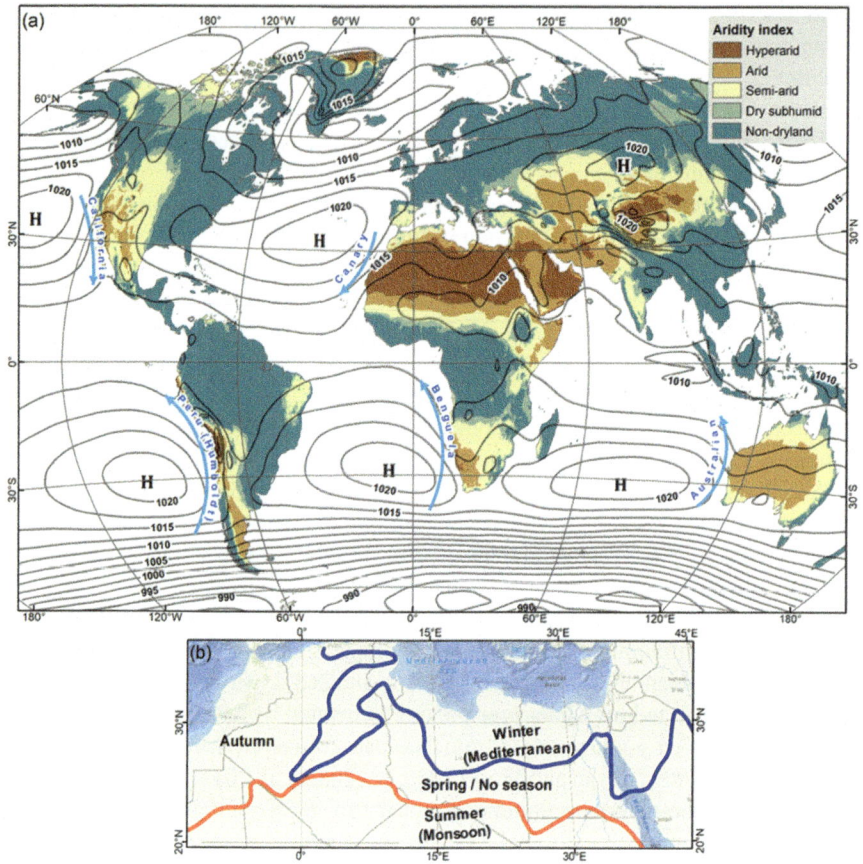

Fig. 43.1 (**a**) Aridity index map based on the UNEP (1992) definition. (**b**) Magnification of the northern Sahara with (blue) the southernmost line of winter precipitation (i.e., with maximal monthly rainfall occurring in winter) and (red) the northernmost line of summer precipitation. Aridity index: Climatic Research Unit of University of East Anglia (New et al. 2002); Pressure: NCEP Reanalysis 1948–2018 [NOAA/OAR/ESRL PSD, Boulder, Colorado, USA, https://www.esrl.noaa.gov/psd/, last accessed 3 Apr. 2019 (Kalnay et al. 1996)]; Precipitation: Tropical Rainfall Measuring Mission [TRMM (TMPA/3B43) Rainfall Estimate L3 1 month 0.25° × 0.25° V7, NASA Goddard Earth Sciences Data and Information Services Center, Accessed 06/2018 https://doi.org/10.5067/TRMM/TMPA/MONTH/7]

43.3 Accuracy of Satellite Precipitation Estimates for Dryland Regions

Despite the number of evaluation studies covering different regions around the globe, only a few of them explicitly focused on satellite precipitation products in dryland areas. This is likely due to the data scarcity characterizing these regions. Varying degrees of accuracy are observed, including large under- and over-estimation, depending on the examined regions and products and with different products alternatively performing better or worse than others. Often overestimation of light precipitation and false precipitation signals due to raindrops evaporation are reported (e.g., Dinku et al. 2010; Habib et al. 2012; Yang and Luo 2014; Ghajarnia et al. 2015; Milewski et al. 2015; Serrat-Capdevila et al. 2016; Mahmoud et al. 2018). However, when dealing with poor density of the reference ground data, the accuracy of satellite precipitation products is likely underestimated, mostly due to spatiotemporal scale mismatch between satellite products (spatial aggregation and temporal sampling) and rain gauges (spatial sampling and temporal aggregation) (Amitai et al. 2012; Maggioni et al. 2017; Tang et al. 2018; Tian et al. 2017). This mismatch is likely enhanced by the small-scale precipitation variability and by the ground network sparseness characterizing these areas. This needs to be taken into account when utilizing satellite precipitation products over dryland regions, since a proper separation between errors related to the satellite retrieval algorithms and to the sampling/aggregation scale mismatch is still missing. Amitai et al. (2012) found "very good agreement" between the instantaneous estimates (thus overcoming the temporal sampling limitation) of the TRMM precipitation radar and a uniquely high density rain gauge network (thus smoothing the spatial sampling limitation of rain gauges) over a semiarid-arid area in the United States. Hydrological evaluation methods may provide improved information on the quality of satellite precipitation products by smoothing some of the issues related to the scales mismatch, even if additional uncertainty arises from the choice of the hydrological model and of its parametrization (e.g., Bitew and Gebremichael 2011; Knoche et al. 2014).

Bias correction techniques were reported not performing well in dryland areas so that satellite products are commonly trusted only in areas where very poor ground data is available (Guo et al. 2015; Tekeli and Fouli 2016). Despite this, the error variance of a given satellite precipitation product was shown to be reasonably transferrable within climatic regions, e.g., using the Köppen-Geiger classification (Peel et al. 2007), permitting to extract quantitative information on the error variance even in absence of in situ ground measurements (Tang and Hossain 2012, Gebregiorgis and Hossain 2015). Even though this quantification is likely influenced by the ground validation network density, it permits important assumptions on the satellite estimation error for ungauged areas of the Earth. At this concern, it should be pointed out that topography is, with climate, one of the main factors driving the accuracy of satellite precipitation products, possibly causing further finer-scale variations in the performance (Derin et al. 2016). Data-driven clustering techniques represent a potential approach for identifying homogeneous regions in precipitation climatology and obtain improved climatic information for data scarce areas such as the drylands (Mantas et al. 2015; Demirdjian et al. 2018).

43.4 Satellite-Derived Climatological Properties of Dryland Precipitation

Precipitation in dryland regions are often very local and short living. Recent studies, conducted in the dryland areas of the eastern Mediterranean using high-resolution estimates obtained from X-Band and C-Band weather radars, derived information on

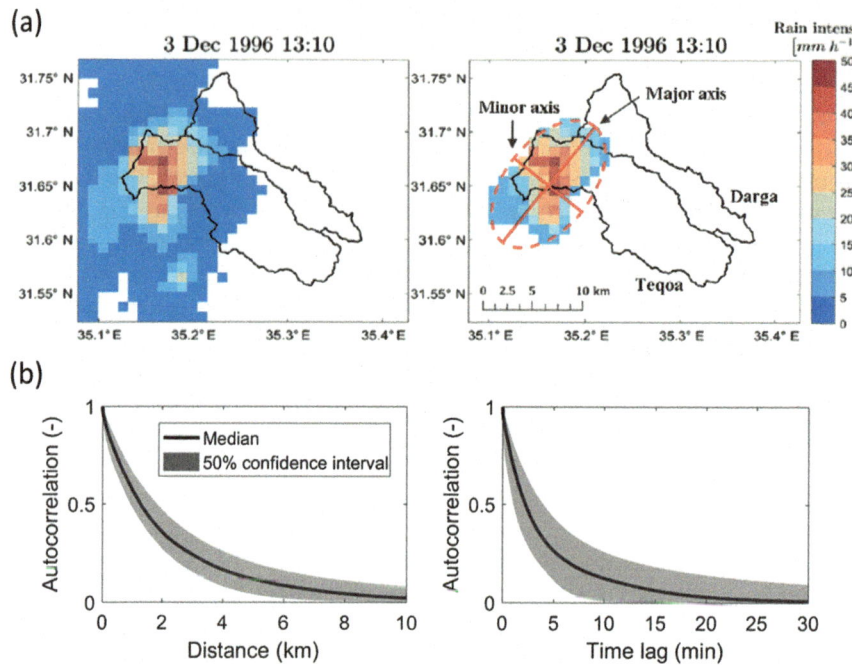

Fig. 43.2 (**a**) An example of a convective rain-cell derived from a C-Band weather radar and a fitted ellipse (Belachsen et al. 2017). (**b**) Spatial and temporal autocorrelation of convective rainfall fields in an arid region of the eastern Mediterranean as observed from 60 m, 1 min resolution X-Band weather radar estimates. (Adapted from Marra and Morin 2018)

the fine scale structure of dryland precipitation at high spatiotemporal resolution. For instance, the mean size of convective rain cells (connected regions exceeding 5 mm h^{-1} intensity in 5-min radar scans) in the dry region draining to the Dead Sea is 90 km^2, and their mean life time is 13 min (Fig. 43.2a, Belachsen et al. 2017). Spatial and temporal correlation distances of the rainfall fields are 1.5–2.8 km and ~1.8–6.4 min, respectively (Fig. 43.2b, Sharon 1979; Marra and Morin 2018; Peleg et al. 2018). As regards the extremes, precipitation rates for short durations and long return periods (i.e., the most extreme events) are higher in dry areas, as consistently shown by radar-based and rain gauge-based analyses (Fig. 43.3a, Marra and Morin 2015). Satellite products, owing to their resolution, cannot resolve the scales typical of dryland precipitation. However, since rain gauge and weather radar estimates suffer from the sparse and uneven distribution of the measurement networks typical of dryland regions (Michaelides et al. 2009), the combination of satellite-based estimates can provide improved, and sometimes unique, information (e.g., Huffman et al. 1997). Despite this, only a few studies aimed at deriving the climatology of dryland regions using satellite-based precipitation estimates.

Precipitation seasonality was shown to be well represented by satellite data. For example, Cattani et al. (2016) showed that satellite precipitation products are able to fully reproduce the monthly climatology of East Africa also in the drier areas, even if

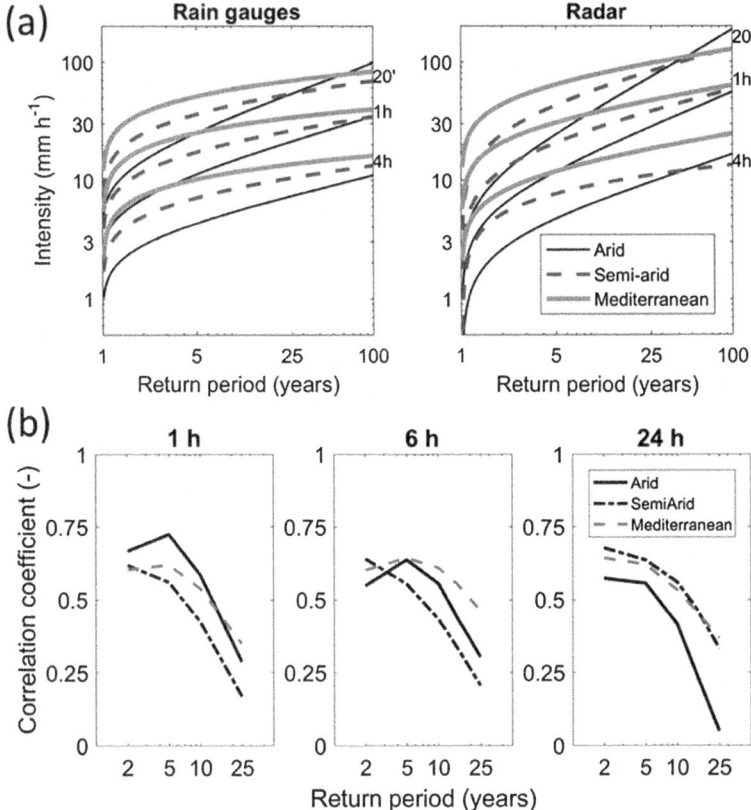

Fig. 43.3 (**a**) Average relation between rain intensity and return period for arid, semi-arid and Mediterranean climates as observed from rain gauges and a C-Band weather radar in the eastern Mediterranean (adapted from Marra and Morin 2015). (**b**) Spatial correlation coefficient between quantiles derived from C-Band weather radar and satellite-based estimates (high-resolution CMORPH) over arid, semi-arid and Mediterranean climates in the eastern Mediterranean. (Adapted from Marra et al. 2017)

the well-known estimation issues related to complex terrain were confirmed. Kelley (2014) examined the seasonal precipitation patterns in the Sahara Desert using TRMM observations and, contrasting the common idea that rainfall in the area is highly unpredictable, identified four distinct rainy seasons separated by drier months and characterized by peculiar precipitation properties. The region between 20–27°N and 22–32°E was identified as the driest portion of the desert, with less than 5 mm average yearly precipitation.

Satellite observations were also used to get information about the vertical structure of precipitation. Using the TRMM precipitation radar over Africa, Geerts and Dejene (2005) observed that in drier areas, such as the Sahel and the northern Savannah regions, precipitating storms are characterized by high echo tops, high hydrometeor loading aloft accompanied by low-level evaporation and little indication of radar bright band at the freezing level.

In recent years, owing to the growing length of data records, some studies attempted to derive information on the frequency of extreme precipitation from satellite-based estimates (Zhou et al. 2015; Demirdjian et al. 2018). Interestingly, probably due to the lack of ground-based estimates able to provide the sought information, many of these studies focused on dryland areas. The problem was approached either exploiting the relatively high temporal resolution of satellite products to downscale daily information from rain gauges (Awadallah et al. 2011; Awadallah and Awadallah 2013), or directly quantifying the frequency of extreme precipitation from the satellite based archives, with the objective of hydrological design or heavy rainfall monitoring/early warning (Endreny and Imbeah 2009; Gado et al. 2017; Marra et al. 2017; Faridzad et al. 2018). Focusing on the eastern Mediterranean region, Marra et al. (2017) compared the analyses based on the high-resolution CMORPH (Joyce et al. 2004; Xie et al. 2017) to the ones from a long weather radar archive (Marra and Morin 2015) finding good spatial correlation between the quantiles derived from different sensors on semiarid and arid climates (Fig. 43.3b). In general, drier climates were found to be better described by thicker-tailed distributions of the rainfall extremes, even if important uncertainties arise from the small number of rainfall events occurring in these regions.

43.5 Quasi-Global Dryland Precipitation Climatology from TRMM Observations

The above studies demonstrate the large potential of satellite precipitation products in resolving climatological properties of precipitation in dryland regions with scarce ground data. In this section we present a semi-global climatological analysis based on Goddard Earth Sciences Data and Information Services Center (2016), TRMM (TMPA) Precipitation L3 1 day $0.25° \times 0.25°$ V7 (Huffman et al. 2007, 2010), which covers the globe between 50°S and 50°N latitude with data available from 1998 to present. We analyzed the data from 1 September 1998 to 31 August 2017 for a total of 19 hydrological years. The mean annual precipitation map is shown in Fig. 43.4a, and some artifacts caused by the retrieval algorithms are highlighted (Fig. 43.4b, c). Given the small precipitation amounts characterizing the drylands, such artifacts are expected to weigh more, with respect to other climatic regions, in affecting the accuracy of satellite precipitation estimates.

Drylands were grouped into 4 classes based on the UNEP (1992) aridity index (AI) (Fig. 43.1a). AI ≤ 0.05 refers to hyper-arid, $0.05 < $ AI ≤ 0.2 to arid, $0.2 < $ AI ≤ 0.5 to semi-arid, and, $0.5 < $ AI ≤ 0.65 to dry-subhumid regions (e.g., Cherlet et al. 2018). Higher AI represent non-dryland regions, grouped apart in a fifth class. The fraction of analyzed TMPA pixels belonging to each class is given in the second column of Table 43.1. Figure 43.5 presents the distribution of mean annual precipitation (a), the coefficient of variation of the mean annual precipitation (b), and the mean number of rainy days (> 0.1 mm day^{-1}) per year (c), for the examined climatic classes. A strong dependence of these properties on the climatic class emerges.

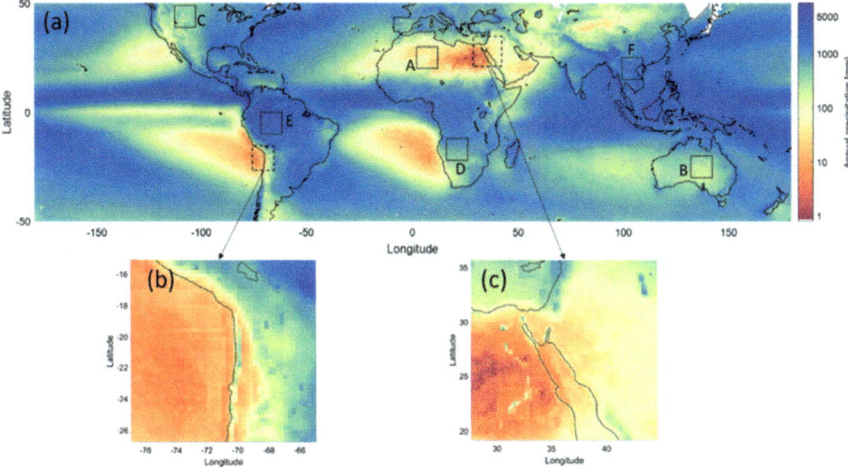

Fig. 43.4 (**a**) Mean annual precipitation computed for 19 hydrological years (1 Sept. 1998–31 Aug. 2017). Colors are mapped in log10 scale to better emphasize dryland regions. Solid line labeled squares delimit regions used for autocorrelation analysis (see Fig. 43.6). Precipitation estimation artifacts (dashed line squares) are spotted near the shoreline of the Atacama Desert (**b**) and the Nile and the Red Sea (**c**). Precipitation data are based on TMPA Precipitation L3 1 day – 0.25° × 0.25° V7, Edited by Andrey Savtchenko, Goddard Earth Sciences Data and Information Services Center, Accessed 06/2018 https://doi.org/10.5067/TRMM/TMPA/DAY/7

Table 43.1 Spatial statistical properties of daily precipitation

	Fraction of analyzed pixel number	Fraction of rainy pixels	Fraction of rainy pixels during the wettest month	Spatial autocorrelation distance for two selected regions (°)
Hyper-arid	0.11	0.04	0.08	0.22,0.34
Arid	0.18	0.11	0.20	
Semi-arid	0.21	0.22	0.37	0.60,0.67
Dry-subhumid	0.08	0.27	0.46	
Non-dryland	0.42	0.38	0.59	0.86,1.87

As expected, dryer climates are characterized by lower mean annual amounts, accompanied by larger inter-year variability of these amounts, and smaller number of rainy days occurring every year (Fig. 43.5, Table 43.1).

As highlighted in the review above, an important feature of dryland precipitation is the spatial distribution, generally characterized by small area coverage and large variability. The area coverage is estimated as the fraction of rainy pixels, computed for all days and pixels (Table 43.1, third column), as well as focusing on the wettest month of each pixel (Table 43.1, fourth column). While in the non-dryland regions precipitation area coverage is close to 40%, in the drylands this fraction drops as

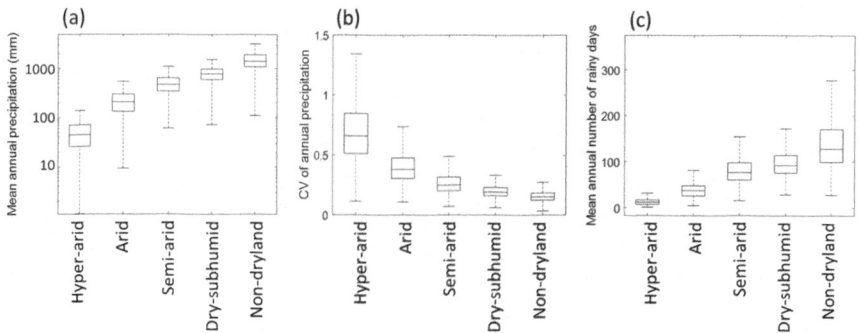

Fig. 43.5 Box plots of mean annual precipitation (**a**), coefficient of variation (CV) of annual precipitation (**b**), and, mean number of rainy days (> 0.1 mm day^{-1}) per year, comparing five climatological classes: hyper-arid, arid, semi-arid, dry-subhumid and non-dryland. Each box shows statistics of all pixels of a given climatological classification. Outliers are not shown

aridity increases, reaching 4% in the hyper-arid regions. When only the wettest month is considered, similar trends are found, with the fraction of rainy pixels dropping from ~60% of non-drylands to 8% of hyper-arid climates. To better quantify the spatial variability of dryland precipitation, we compute the spatial autocorrelation of the daily precipitation within six 10° × 10° windows characterized by different climates (2 for hyper-arid/arid climates, 2 for semi-arid/dry-subhumid, and 2 for non-dryland, see Fig. 43.4a). The 2-dimensional autocorrelation function is calculated using the Wiener-Khintchine theorem and the fast Fourier transform (Nerini et al. 2017; Marra and Morin 2018), and an exponential function is fitted to the median of all the daily autocorrelation fields. The resulting curves are shown in Fig. 43.6. The autocorrelation drops faster in drier areas, with a correlation distance (i.e., the distance at which autocorrelation decreases by e^{-1}) decreasing from 0.86–1.87° for the non-dryland regions, to 0.6–0.67° and 0.22–0.34° for the semi-arid/dry-subhumid and hyper-arid/arid regions, respectively (Fig. 43.6, Table 43.1).

Understanding the properties of extreme precipitation is fundamental for flood risk management in dryland regions. Typically, daily precipitation distributions are highly skewed, leading to heavy tails in the distribution of extreme precipitation amounts (Papalexiou and Koutsoyiannis 2013). Previous studies showed that drier climates seem to be characterized by heavier tails in the distribution of extremes, particularly for short (sub-daily) durations (e.g., Marra and Morin 2015), suggesting that, given a common low probability of occurrence, higher precipitation amounts should be expected in drier areas at these temporal scales. However, owing to the peculiar characteristics of dryland precipitation (i.e., the low number of rain events per year), it is not completely understood how much this signal is influenced by the chronic data scarcity characterizing dryland areas.

Recent methodological developments (Marani and Ignaccolo 2015; Zorzetto et al. 2016) in the analysis of extreme precipitation suggest that examining the distribution

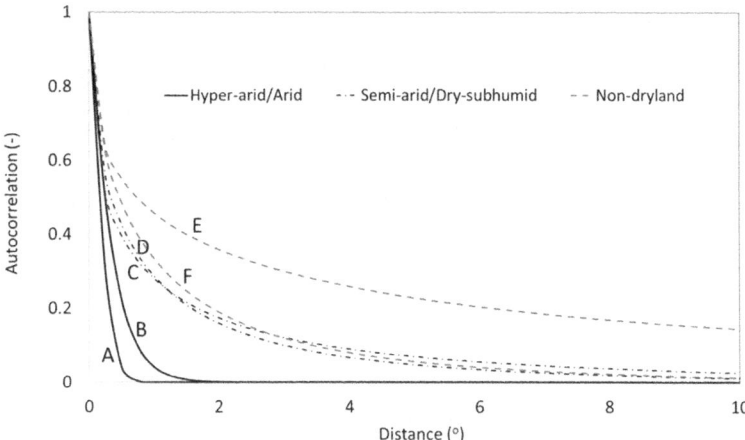

Fig. 43.6 Spatial autocorrelation curves, using an exponential function, of daily precipitation over $10° \times 10°$ windows (shown in Fig. 43.4a with matching labels) for three climatic categories: hyper-arid/arid, semi-arid/dry-subhumid, and, non-dryland. Two curves for each climatic class

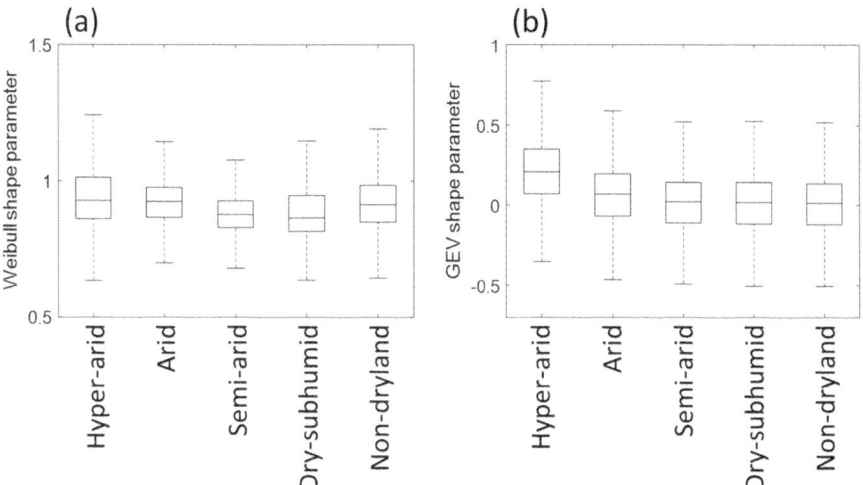

Fig. 43.7 Same as Fig. 43.5 but for the shape parameters of: a) the Weibull distribution for all the non-zero daily precipitation, and b) the GEV distribution for annual daily maxima

of the ordinary rain events could help shedding light on the distribution of the extremes. Theoretical reasoning (Wilson and Toumi 2005) confirmed by observations (Marani and Ignaccolo 2015; Zorzetto et al. 2016) points to the Weibull distribution as adequately describing the distribution of the non-zero daily precipitation amounts. Figure 43.7a presents the shape parameter of the Weibull distribution fitted to all the non-zero daily (ordinary) precipitation events. At the same time, Fig. 43.7b shows the shape parameter of a Generalized Extreme Value (GEV) distribution fitted to annual daily maxima (Coles 2001), which represents a standard approach to the analysis of extremes. As anticipated, the GEV shape parameters tend to decrease when moving from the driest regions to less dry areas (it should be noted that the large spread is caused by the use of few years of annual maxima to derive the distribution parameters). As shown by Marani and Ignaccolo (2015), the combination of the distribution of the ordinary events and the average number of occurrences per year shape the distribution of the extremes. The shape parameter of the Weibull distribution (Fig. 43.7a) shows a mild decrease when moving from hyper-arid to semi-arid climates, but the climatic signal is generally less well defined, suggesting that in dryland areas the number of events per year dominates the characteristics of the distribution of the extremes.

43.6 Conclusions and Recommendations

Satellite observations provide opportunities to improve our knowledge of precipitation climatology in dryland regions, where traditional monitoring systems are often lacking. However, dryland regions pose also unique challenges. In this chapter, we presented a quasi-global analysis based on the TMPA data set, showing how the aridity level is associated with unique characteristics of precipitation climatology.

As aridity increases, precipitation is characterized by smaller yearly amounts, higher coefficient of variation, lower number of rainy days, heavier tail of the distribution of the extremes, smaller fraction of rainy area and lower correlation distance. All these characteristics concur making the measurement of dryland precipitation from satellite-borne sensors a rather challenging task. High space-time resolutions are required to adequately capture the precipitation structures. Presently, the resolutions of satellite products are still coarse compared to the typical scale of dryland precipitation. This arises from the low spatial resolution and the low number of overpasses of the low Earth orbiting satellites carrying passive and active microwave sensors. Some of these errors can be reduced by combining data in multi-satellite multi-sensor products and by adopting proper statistical analyses, but a thorough quantification of the uncertainties of these products over dryland regions is still lacking. Standard methodologies for the assessment of satellite-based precipitation products are difficult to apply, as well as standard bias adjustment techniques, since the combination of high spatial variability and low network density leads to systematic under-sampling of the gauges.

Future research should focus on these validation and adjustment issues, by including information from dense gauge networks and distributed information from ground-based weather radar systems. Clustering and other data-driven methods could also be explored to improve the representativeness of the precipitation reference.

References

Ahrens, C. D. (2003). *Meteorology today: An introduction to weather, climate, and the environment* (7th ed., p. 624). Pacific Grove: Thomson/Brooks/Cole. ISBN:0534397719.

Amitai, E., Unkrich, C. L., Goodrich, D. C., Habib, E., & Thill, B. (2012). Assessing satellite-based rainfall estimates in semiarid watersheds using the USDA-ARS walnut gulch gauge network and TRMM PR. *Journal of Hydrometeorology, 13*, 1579–1588. https://doi.org/10.1175/JHM-D-12-016.1.

Armon, M., Dente, E., Smith, J. A., Enzel, Y., & Morin, E. (2018). Synoptic-scale control over modern rainfall and flood patterns in the Levant drylands with implications for past climates. *Journal of Hydrometeorology, 19*, 1077–1096. https://doi.org/10.1175/JHM-D-18-0013.1.

Awadallah, A. G., & Awadallah, N. A. (2013). A novel approach for the joint use of rainfall monthly and daily ground station data with TRMM data to generate IDF estimates in a poorly gauged arid region. *Open Journal of Modern Hydrology, 2013*, 1–7. https://doi.org/10.4236/ojmh.2013.31001.

Awadallah, A. G., ElGamal, M., ElMostafa, A., & ElBadry, H. (2011). Developing intensity-duration-frequency curves in scarce data region: An approach using regional analysis and satellite data. *Engineering, 03*, 215–226. https://doi.org/10.4236/eng.2011.33025.

Barth, H.-J., & Steinkohl, F. (2004). Origin of winter precipitation in the central coastal lowlands of Saudi Arabia. *Journal of Arid Environments, 57*, 101–115. https://doi.org/10.1016/S0140-1963(03)00091-0.

Belachsen, I., Marra, F., Peleg, N., & Morin, E. (2017). Convective rainfall in a dry climate: Relations with synoptic systems and flash-flood generation in the Dead Sea region. *Hydrology and Earth System Sciences, 21*, 5165–5180. https://doi.org/10.5194/hess-21-5165-2017.

Bitew, M. M., & Gebremichael, M. (2011). Evaluation of satellite rainfall products through hydrologic simulation in a fully distributed hydrologic model. *Water Resources Research, 47*, 1–11. https://doi.org/10.1029/2010WR009917.

Budyko, M. I. (1958). *The heat balance of the Earth's surface* (259 pp). Translated. U.S. Department of Commerce, Washington, DC. Available at https://catalog.hathitrust.org/Record/001485928. Last accessed 13 Dec 2018.

Cattani, E., Merino, A., & Levizzani, V. (2016). Evaluation of monthly satellite-derived precipitation products over East Africa. *Journal of Hydrometeorology, 17*, 2555–2573. https://doi.org/10.1175/JHM-D-15-0042.1.

Cherlet, M., Hutchinson, C., Reynolds, J. N., Hill, J., Sommer, S., & von Maltitz, G. (2018). *World atlas of desertification* (3rd ed., p. 295). Luxembourg: Publication Office of the European Union. https://doi.org/10.2760/06292.

Coles, S. (2001). *An introduction to statistical modeling of extreme values* (p. 183). London: Springer. ISBN:978-1-4471-3675-0.

Dayan, U., & Morin, E. (2006). Flash flood – Producing rainstorms over the Dead Sea: A review. *New Front. Dead Sea Paleoenviron. Res. Geological Society of America Special Papers, 401*, 53–62. https://doi.org/10.1130/2006.2401(04).

Demirdjian, L., Zhou, Y., & Huffman, G. J. (2018). Statistical modeling of extreme precipitation with TRMM data. *Journal of Applied Meteorology and Climatology, 57*, 15–30. https://doi.org/10.1175/JAMC-D-17-0023.1.

Derin, Y., Anagnostou, E. N., Berne, A., Borga, M., Boudevillain, B., Buytaert, W., Chang, C.-H., Delrieu, G., Hong, Y., Hsui, Y. C., Lavado-Casimiro, W., Manze, B., Moges, S., Nikolopoulos, E. I., Sahlu, D., Salerno, F., Rodríguez-Sánchez, J.-P., Vergara, H. J., & Yilmaz, K. K. (2016). Multiregional satellite precipitation products evaluation over complex terrain. *Journal of Hydrometeorology, 17*, 1817–1836. https://doi.org/10.1175/JHM-D-15-0197.1.

Dinku, T., Connor, S. J., & Ceccato, P. (2010). Comparison of CMORPH and TRMM-3B42 over mountainous regions of Africa and South America. In M. Gebremichael & F. Hossain (Eds.), *Satellite rainfall applications for surface hydrology* (pp. 193–204). Dordrecht: Springer. ISBN:978-90-481-2915-7.

Endreny, T. A., & Imbeah, N. (2009). Generating robust rainfall intensity-duration-frequency estimates with short-record satellite data. *Journal of Hydrology, 371*, 182–191. https://doi.org/10.1016/j.jhydrol.2009.03.027.

Faridzad, M., Yang, T., Hsu, K.-L., Sorooshian, S., & Xiao, C. (2018). Rainfall frequency analysis for ungauged regions using remotely sensed precipitation information. *Journal of Hydrology, 563*, 123–142. https://doi.org/10.1016/j.jhydrol.2018.05.071.

Gado, T. A., Hsu, K.-L., & Sorooshian, S. (2017). Rainfall frequency analysis for ungauged sites using satellite precipitation products. *Journal of Hydrology, 554*, 646–655. https://doi.org/10.1016/j.jhydrol.2017.09.043.

Gebregiorgis, A. S., & Hossain, F. (2015). How well can we estimate error variance of satellite precipitation data around the world? *Atmospheric Research, 154*, 39–59. https://doi.org/10.1016/j.atmosres.2014.11.005.

Geerts, B., & Dejene, T. (2005). Regional and diurnal variability of the vertical structure of precipitation systems in Africa based one spaceborne radar data. *Journal of Climate, 18*, 893–916. https://doi.org/10.1175/JCLI-3316.1.

Ghajarnia, N., Liaghat, A., & Daneshkar Arasteh, P. (2015). Comparison and evaluation of high resolution precipitation estimation products in Urmia Basin-Iran. *Atmospheric Research, 158–159*, 50–65. https://doi.org/10.1016/j.atmosres.2015.02.010.

Goodrich, D. C., Faurès, J. M., Woolhiser, D. A., Lane, L. J., & Sorooshian, S. (1995). Measurement and analysis of small-scale convective storm rainfall variability. *Journal of Hydrology, 173*, 283–308. https://doi.org/10.1016/0022-1694(95)02703-R.

Guo, H., Chen, S., Bao, A., Hu, J., Gebregiorgis, A. S., Xue, X., & Zhang, X. (2015). Intercomparison of high-resolution satellite precipitation products over Central Asia. *Remote Sensing, 7*, 7181–7211. https://doi.org/10.3390/rs70607181.

Habib, E., Elsaadani, M., & Haile, A. T. (2012). Climatology-focused evaluation of CMORPH and TMPA satellite rainfall products over the Nile Basin. *Journal of Applied Meteorology and Climatology, 51*, 2105–2121. https://doi.org/10.1175/JAMC-D-11-0252.1.

Hirschboeck, K. K. (1988). Flood hydroclimatology. In V. R. Baker (Ed.), *Flood geomorphology* (pp. 27–49). Wiley-Interscience. ISBN: 978-0-471-62558-2.

Huffman, G. J., Adler, R. F., Arkin, P., Chang, A., Ferraro, R., Gruber, A., Janowiak, J., McNab, A., Rudolf, B., & Schneider, U. (1997). The global precipitation climatology project (GPCP) combined precipitation dataset. *Bulletin of the American Meteorological Society, 78*, 5–20. https://doi.org/10.1175/1520-0477(1997)078<0005:TGPCPG>2.0.CO;2.

Huffman, G. J., Bolvin, D. T., Nelkin, E. J., & Wolff, D. B. (2007). The TRMM multisatellite precipitation analysis (TMPA): Quasi-global, multiyear, combined-sensor precipitation stimates at fine scales. *Journal of Hydrometeorology, 8*, 38–55. https://doi.org/10.1175/JHM560.1.

Huffman, G. J., Adler, R. F., Bolvin, D. T., & Nelkin, E. J. (2010). The TRMM multi-satellite precipitation analysis (TMPA). In M. Gebremichael & F. Hossain (Eds.), *Satellite rainfall applications for surface hydrology* (pp. 3–22). Dordrecht: Springer. ISBN:978-90-481-2915-7.

Joyce, R. J., Janowiak, J. E., Arkin, P. A., & Xie, P. (2004). CMORPH: A method that produces global precipitation estimates from passive microwave and infrared data at high spatial and temporal resolution. *Journal of Hydrometeorology, 5*, 487–503. https://doi.org/10.1175/1525-7541(2004)005<0487:CAMTPG>2.0.CO;2.

Kahana, R., Ziv, B., Enzel, Y., & Dayan, U. (2002). Synoptic climatology of major floods in the Negev Desert, Israel. *International Journal of Climatology, 22*, 867–882. https://doi.org/10.1002/joc.766.

Kalnay, E., Kanamitsu, M., Kistler, R., Collins, W., Deaven, D., Gandin, L., Iredell, M., Saha, S., White, G., Woollen, J., Zhu, Y., Chelliah, M., Ebisuzaki, W., Higgins, W., Janowiak, J., Mo, K. C., Ropelewski, C., Wang, J., Leetmaa, A., Reynolds, R., Jenne, R., & Joseph, D. (1996). The NCEP/NCAR 40-year reanalysis project. *Bulletin of the American Meteorological Society, 77*, 437–472. https://doi.org/10.1175/1520-0477(1996)077<0437:TNYRP>2.0.CO;2.

Kelley, O. A. (2014). Where the least rainfall occurs in the Sahara Desert, the TRMM radar reveals a different pattern of rainfall each season. *Journal of Climate, 27*, 6919–6939. https://doi.org/10.1175/JCLI-D-14-00145.1.

Knippertz, P. (2003). Tropical–extratropical interactions causing precipitation in Northwest Africa: Statistical analysis and seasonal variations. *Monthly Weather Review, 131*, 3069–3076. https://doi.org/10.1175/1520-0493(2003)131<3069:TICPIN>2.0.CO;2.

Knippertz, P., & Wernli, H. (2010). A Lagrangian climatology of tropical moisture exports to the northern hemispheric extratropics. *Journal of Climate, 23*, 987–1003. https://doi.org/10.1175/2009JCLI3333.1.

Knoche, M., Fischer, C., Pohl, E., Krause, P., & Merz, R. (2014). Combined uncertainty of hydrological model complexity and satellite-based forcing data evaluated in two data-scarce semi-arid catchments in Ethiopia. *Journal of Hydrology, 519*, 2049–2066. https://doi.org/10.1016/j.jhydrol.2014.10.003.

Kushnir, Y., Dayan, U., Ziv, B., Morin, E., & Enzel, Y. (2017). Climate of the Levant: Phenomena and mechanisms. In Y. Enzel & B.-Y. Ofer (Eds.), *Quaternary of the Levant: Environments, climate change, and humans* (pp. 31–44). Cambridge: Cambridge University Press. https://doi.org/10.1017/9781316106754.

Maggioni, V., Nikolopoulos, E. I., Anagnostou, E. N., & Borga, M. (2017). Modeling satellite precipitation errors over mountainous terrain: The influence of gauge density, seasonality, and temporal resolution. *IEEE Transactions on Geoscience and Remote Sensing, 55*, 4130–4140. https://doi.org/10.1109/TGRS.2017.2688998.

Mahmoud, M. T., Al-Zahrani, M. A., & Sharif, H. O. (2018). Assessment of global precipitation measurement satellite products over Saudi Arabia. *Journal of Hydrology, 559*, 1–12. https://doi.org/10.1016/j.jhydrol.2018.02.015.

Mantas, V. M., Liu, Z., Caro, C., & Pereira, A. J. S. C. (2015). Validation of TRMM multi-satellite precipitation analysis (TMPA) products in the Peruvian Andes. *Atmospheric Research, 163*, 132–145. https://doi.org/10.1016/j.atmosres.2014.11.012.

Marani, M., & Ignaccolo, M. (2015). A metastatistical approach to rainfall extremes. *Advances in Water Resources, 79*, 121–126. https://doi.org/10.1016/j.advwatres.2015.03.001.

Marra, F., & Morin, E. (2015). Use of radar QPE for the derivation of intensity–duration–frequency curves in a range of climatic regimes. *Journal of Hydrology, 531*, 427–440. https://doi.org/10.1016/j.jhydrol.2015.08.064.

Marra, F., & Morin, E. (2018). Autocorrelation structure of convective rainfall in semiarid-arid climate derived from high-resolution X-band radar estimates. *Atmospheric Research, 200*, 126–138. https://doi.org/10.1016/j.atmosres.2017.09.020.

Marra, F., Morin, E., Peleg, N., Mei, Y., & Anagnostou, E. N. (2017). Intensity-duration-frequency curves from remote sensing rainfall estimates: Comparing satellite and weather radar over the eastern Mediterranean. *Hydrology and Earth System Sciences, 21*, 2389–2404. https://doi.org/10.5194/hess-21-2389-2017.

Michaelides, S., Levizzani, V., Anagnostou, E. N., Bauer, P., Kasparis, T., & Lane, J. E. (2009). Precipitation: Measurement, remote sensing, climatology and modeling. *Atmospheric Research, 94*, 512–533. https://doi.org/10.1016/j.atmosres.2009.08.017.

Milewski, A., Elkadiri, R., & Durham, M. (2015). Assessment and comparison of TMPA satellite precipitation products in varying climatic and topographic regimes in Morocco. *Remote Sensing, 7*, 5697–5717. https://doi.org/10.3390/rs70505697.

Muller, A., Reason, C. J. C., & Fauchereau, N. (2008). Extreme rainfall in the Namib desert during late summer 2006 and influences of regional ocean variability. *International Journal of Climatology, 28*, 1061–1070. https://doi.org/10.1002/joc.1603.

Nerini, D., Besic, N., Sideris, I., Germann, U., & Foresti, L. (2017). A non-stationary stochastic ensemble generator for radar rainfall fields based on the short-space Fourier transform. *Hydrology and Earth System Sciences, 21*, 2777–2797. https://doi.org/10.5194/hess-21-2777-2017.

New, M., Lister, D., Hulme, M., & Makin, I. (2002). A high-resolution data set of surface climate over global land areas. *Climate Research, 21*, 1–25. https://doi.org/10.3354/cr021001.

Nicholson, S. E. (2011). *Dryland climatology*. New York., 516 pp: Cambridge University Press. https://doi.org/10.1017/CBO9780511973840.

Papalexiou, S. M., & Koutsoyiannis, D. (2013). Battle of extreme value distributions : A global survey on extreme daily rainfall. *Water Resources Research, 49*, 187–201. https://doi.org/10.1029/2012WR012557.

Peel, M. C., Finlayson, B. L., & McMahon, T. A. (2007). Updated world map of the Köppen-Geiger climate classification. *Hydrology and Earth System Sciences, 11*, 1633–1644. https://doi.org/10.5194/hess-11-1633-2007.

Peleg, N., Marra, F., Fatichi, S., Molnar, P., Morin, E., Sharma, A., & Burlando, P. (2018). Intensification of convective rain cells at warmer temperatures observed from high-resolution weather radar data. *Journal of Hydrometeorology, 19*, 715–726. https://doi.org/10.1175/JHM-D-17-0158.1.

Ralph, F. M., Dettinger, M., Lavers, D., Gorodetskaya, I. V., Martin, A., Viale, M., White, A. B., Oakley, N., Rutz, J., Spackman, J. R., Wernli, H., & Cordeira, J. (2017). Atmospheric rivers emerge as a global science and applications focus. *Bulletin of the American Meteorological Society, 98*, 1969–1973. https://doi.org/10.1175/BAMS-D-16-0262.1.

Rubin, S., Ziv, B., & Paldor, N. (2007). Tropical plumes over eastern North Africa as a source of rain in the Middle East. *Monthly Weather Review, 135*, 4135–4148. https://doi.org/10.1175/2007MWR1919.1.

Safriel, U., & Coauthors. (2005). Dryland Systems. In R. Hassan, R. Scholes, & N. Ash (Eds.), *Ecosystems and human Well-being: Current state and trends, volume 1* (pp. 623–662). Washington, DC: Island Press. ISBN:9781559632270.

Serrat-Capdevila, A., Merino, M., Valdes, J. B., & Durcik, M. (2016). Evaluation of the performance of three satellite precipitation products over Africa. *Remote Sensing, 8*. https://doi.org/10.3390/rs8100836.

Sharon, D. (1979). Correlation analysis of the Jordan valey rainfall field. *Monthly Weather Review, 107*, 1042–1047. https://doi.org/10.1175/1520-0493(1979)107<1042:CAOTJV>2.0.CO;2.

Sheppard, P. R., Comrie, A. C., Packin, G. D., Angersbach, K., & Hughes, M. K. (2002). The climate of the US southwest. *Climate Research, 21*, 219–238. https://doi.org/10.3354/cr021219.

Shmida, A. (1985). Biogeography of the desert floras of the world. In M. Evenari, I. Noy-Meir, & D. W. Goodall (Eds.), *Hot deserts and arid shrublands* (pp. 23–77). Amsterdam: Elsevier. ISBN:044442296X.

Tang, L., & Hossain, F. (2012). Investigating the similarity of satellite rainfall error metrics as a function of Köppen climate classification. *Atmospheric Research, 104–105*, 182–192. https://doi.org/10.1016/j.atmosres.2011.10.006.

Tang, G., Behrangi, A., Long, D., Li, C., & Hong, Y. (2018). Accounting for spatiotemporal errors of gauges: A critical step to evaluate gridded precipitation products. *Journal of Hydrology, 559*, 294–306. https://doi.org/10.1016/j.jhydrol.2018.02.057.

Tekeli, A. E., & Fouli, H. (2016). Evaluation of TRMM satellite-based precipitation indexes for flood forecasting over Riyadh City, Saudi Arabia. *Journal of Hydrology, 541*, 471–479. https://doi.org/10.1016/j.jhydrol.2016.01.014.

Tian, F., Hou, S., Yang, L., Hu, H., & Hou, A. (2017). How does the evaluation of GPM IMERG rainfall product depend on gauge density and rainfall intensity? *Journal of Hydrometeorology, 19*, 339–349. https://doi.org/10.1175/JHM-D-17-0161.1.

Tubi, A., & Dayan, U. (2014). Tropical plumes over the Middle East: Climatology and synoptic conditions. *Atmospheric Research, 145–146*, 168–181. https://doi.org/10.1016/j.atmosres.2014.03.028.

Tubi, A., Dayan, U., & Lensky, I. M. (2017). Moisture transport by tropical plumes over the Middle East: A 30-year climatology. *Quarterly Journal of the Royal Meteorological Society, 143*, 3165–3176. https://doi.org/10.1002/qj.3170.

UNEP. (1992). *World atlas of desertification* (p. 69). London: Edward Arnold. ISBN:0-340-55512-2.

Warner, T. T. (2004). *Desert meteorology* (p. 595). New York: Cambridge University Press. https://doi.org/10.1017/CBO9780511535789.

Wilson, P. S., & Toumi, R. (2005). A fundamental probability distribution for heavy rainfall. *Geophysical Research Letters, 32*, 1–4. https://doi.org/10.1029/2005GL022465.

Xie, P., Joyce, R., Wu, S., Yoo, S.-H., Yarosh, Y., Sun, F., & Lin, R. (2017). Reprocessed, bias-corrected CMORPH global high-resolution precipitation estimates from 1998. *Journal of Hydrometeorology, 18*, 1617–1641. https://doi.org/10.1175/JHM-D-16-0168.1.

Yang, Y., & Luo, Y. (2014). Evaluating the performance of remote sensing precipitation products CMORPH, PERSIANN, and TMPA, in the arid region of Northwest China. *Theoretical and Applied Climatology, 118*, 429–445. https://doi.org/10.1007/s00704-013-1072-0.

Zhou, Y., Lau, W. K. M., & Huffman, G. J. (2015). Mapping TRMM TMPA into average recurrence interval for monitoring extreme precipitation events. *Journal of Applied Meteorology and Climatology, 54*, 979–995. https://doi.org/10.1175/JAMC-D-14-0269.1.

Ziv, B. (2001). A subtropical rainstorm associated with a tropical plume over Africa and the middle-east. *Theoretical and Applied Climatology, 102*, 91–102. https://doi.org/10.1007/s007040170037.

Zorzetto, E., Botter, G., & Marani, M. (2016). On the emergence of rainfall extremes from ordinary events. *Geophysical Research Letters, 43*, 8076–8082. https://doi.org/10.1002/2016GL069445.

Chapter 44
Hailfall Detection

Ralph R. Ferraro, Daniel Cecil, and Sante Laviola

Abstract The identification of hailfall from passive microwave sensors is based on the detection of scattering signatures associated with large ice particles at various microwave observation frequencies, depending upon the sensor being used. The larger the ice size and ice volume within a satellite footprint (FOV), the greater the brightness temperature (TB) depression will be. For microwave imagers, 37 or 85 GHz is frequently used. Microwave sounders, although typically with larger FOV sizes, have an added advantage of observation channels at 150 and 183 GHz, the latter of which are sensitive to water vapor at various altitudes in the atmosphere. Thus, TB depressions at the highest peaking channels (e.g., 183 ± 1 GHz) are indicative of strong cloud updrafts associated with hail. In addition, recent work with precipitation radars from TRMM and GPM has shown skill in determining regions of hail, in particular, the three-dimensional structure of hail systems from space. This chapter will review various approaches to hail detection that have been developed and used to map regions of hail. Strengths and weaknesses of each approach will be described, and application examples will be presented.

Keywords Precipitation · Hail · Hailfall · Hailstone · Hailstorm · Tornado · Graupel · Convection · Updraft · Brightness temperature · Microwave · Sounder · infrared · TRMM · GPM · GMI · TMI · AMSR-E · AMSU-B · MHS · ATMS · Scattering

R. R. Ferraro (✉)
NOAA/NESDIS/Center for Satellite Applications and Research, College Park, MD, USA
e-mail: Ralph.R.Ferraro@noaa.gov

D. Cecil
NASA, Marshall Space Flight Center, Huntsville, AL, USA

S. Laviola
National Research Council, Institute of Atmospheric Science and Climate (CNR-ISAC), Bologna, Italy

44.1 Physical Basis

Nearly 40 years ago, passive microwave (MW) measurements from research satellites were shown to be related to severe weather occurrences over land (Spencer et al. 1983, 1987; Spencer and Santek 1985). They utilized the physical principle of scattering by millimeter or larger ice particles related to precipitation, and even hail, at 37 GHz, as the main indicator of severe weather. Figure 44.1 presents a schematic illustrating the combined effect of ice particle size and its three-dimensional distribution within the cloud, typically representing only a fraction of the larger satellite FOV. It should be noted that there are various types of ice that can cause scattering, including graupel (up to 5 mm in diameter), hail (>5 mm) and large hail (>5 cm).

Very little research took place beyond that original ground-breaking research, until Cecil (2009, 2011) and Cecil and Blankenship (2012) demonstrated a strong relationship between the hail occurrence and the MW brightness temperatures (TB), primarily at 37 GHz, but also at 85 GHz. These studies were performed with the Tropical Rainfall Measuring Mission (TRMM) Microwave Imager (TMI) and the Earth Observation System (EOS) Aqua Advanced Microwave Scanning Radiometer (AMSR-E). The hail climatologies derived from these sensors were consistent with those observed from surface observations. Limitations to these data include the 35°S to 35°N spatial domain of the TMI and the 1300 am/pm local observation time of the AMSR-E sensor, both inhibiting the development of a true global land hail climatology that sampled the complete diurnal cycle.

Fig. 44.1 Hail storm schematic illustrating the combined effect of hail location, size distribution and vertical extent which impact the satellite measurement. Courtesy of http://agatelady.blogspot.com/2013/08/all-about-hail.html (last accessed 26 Oct. 2018)

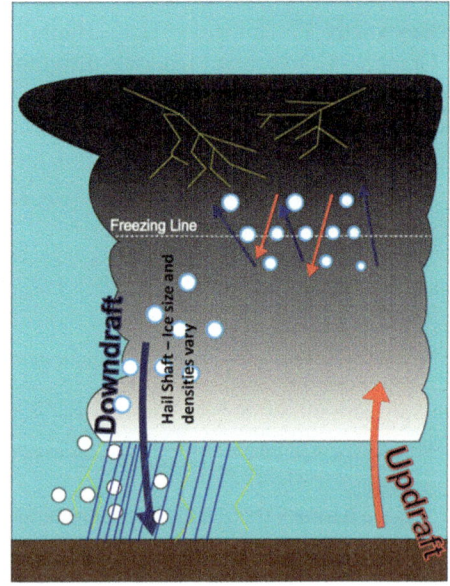

Similar work was done using with passive MW sounders flown by NOAA and EUMETSAT such as the Advanced Microwave Sounding Unit (AMSU) 'A' and 'B' sensors (and the Microwave Humidity Sounder – MHS – which replaced AMSU-B). In convective precipitation with strong vertical updrafts, the reduction in TB has been noted to occur progressively higher in the 183 GHz bands, with TB reductions of 100 K or more even at 183 ± 1 GHz, which has a peak in its water vapor weighting function at approximately 7 km (Hong and Heygster 2005a, b; Ferraro et al. 2005). The magnitude of the TB reduction is the net effect of the volume of ice in the satellite FOV, its size distribution (recall, it is the mm-sized ice that is important for the Mie scattering), and the vertical extent of the ice. The AMSU-A FOV's are 48 km at nadir and increase to nearly 100 km at the very edge of the scan; MHS FOV's are 16 km at nadir and increase to approximately 50 km at the edge of the scan. Subsequent to these studies, Laviola and Levizzani (2011) and Ferraro et al. (2015) were able to develop hail detection algorithms with AMSU/MHS. The advantage of these methods is that for the past decade, on the order of three satellites, spaced approximately 4 h apart in overpass time, have been in operation. This allows for a much better opportunity to develop a hail climatology due to a better sampling of the diurnal cycle.

44.2 Microwave Imager and Active Radar Retrievals

NASA's TRMM and Global Precipitation Mission (GPM) have enabled recent work aimed at using satellite-borne microwave remote sensors to detect hail. Both the TRMM and GPM satellites carried passive MW imagers (with frequencies near 10, 19, 37, and 85 GHZ) and active MW radars (Ku-band near 13.8 GHz). The GPM Microwave Imager (GMI) has additional higher frequency passive MW channels (166 and 183 GHz), and a second active radar frequency (Ka-band at 36 GHz). Other passive MW radiometers measure similar frequencies from satellites that are broadly referred to as the "GPM Constellation", even including satellites from other agencies and other nations that pre-dated GPM by several years.

As was described earlier, much of the feasibility to detect hail from these sensors dates back over 30 years ago. The TMI achieved significantly improved spatial resolution for 37 GHz measurements (16×9 km^2 footprint, Kummerow et al. 1998), as did the AMSR-E (14×8 km^2, JAXA 2006) and GMI (15×9 km^2, Hou et al. 2014). Cecil (2009) used TMI data to compute the likelihood of a large hail (19 mm diameter) report in the US as a function of TB (Fig. 44.2). Those likelihoods increased rapidly with decreasing TB for the 19 and 37 GHz channels; for the higher resolution 85 GHz channel, the likelihood leveled off at around 50% for the lowest TB values. As pointed out by Mroz et al. (2017, Fig. 44.3), extinction coefficients for idealized spherical ice have strong peaks at diameters of only a few mm for the highest frequency channels on modern radiometers. That is, large graupel can be more effective than hail at generating a signal in the 85 GHz (or higher) channels. Low TB in the 19-GHz channel gave the lowest false alarm rates, but its coarse

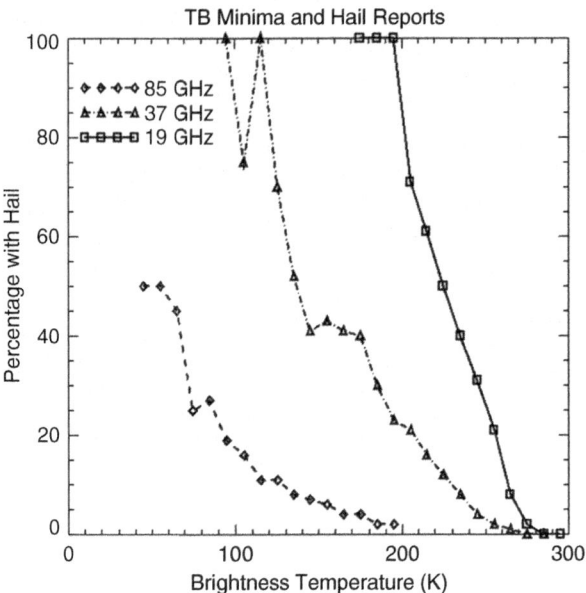

Fig. 44.2 Percentage of TMI brightness temperature local minima associated with large (at least 2 cm diameter) hail reports in the US. (Adapted from Cecil 2009)

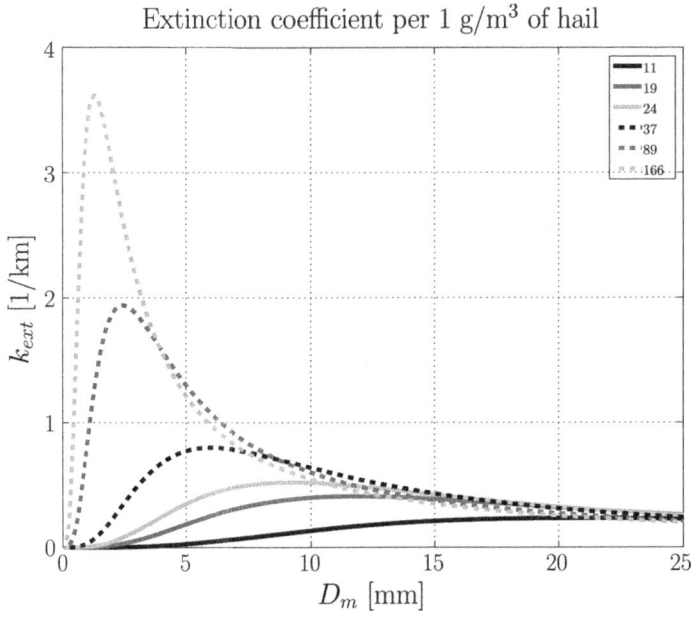

Fig. 44.3 Extinction coefficient per 1 g m^{-3} concentration of spherical hailstones as a function of diameter, at selected GMI frequencies. (Adapted from Mroz et al. 2017)

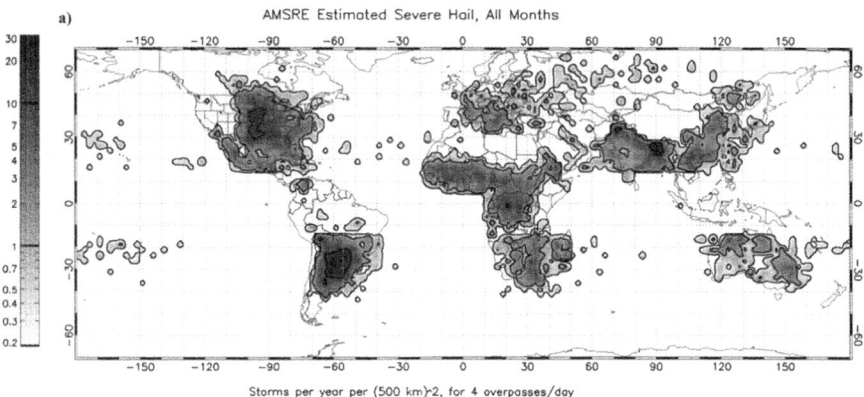

Fig. 44.4 Hailstorm frequency of occurrence estimated from AMSR-E at 36-GHz. (Adapted from Cecil and Blankenship 2012)

horizontal resolution (30 × 18 km^2 from TRMM) was seen as a drawback in comparison to the 37 GHz channel. The TRMM orbit only extended to ±35° latitude, so Cecil and Blankenship (2012) applied the TMI 37-GHz hail probabilities to AMSR-E data with some modifications. They assigned a fractional weight based on the hail probabilities to each event seen by AMSR-E that satisfied threshold TB values – the lower the 37 GHz TB, the greater the weight for that event. This led to a global hail climatology estimated from the 2003–2010 record of AMSR-E (Fig. 44.4). Their hail climatology is essentially an accumulation of probabilistic assessments of individual scenes.

The initial studies relating satellite-borne passive MW data to severe weather relied upon reports in the US of severe weather observed by storm spotters at ground level [tornado, 26 m s^{-1} wind, or 19 mm hail for Spencer et al. 1987; 19 mm hail for Cecil 2009; 25 mm hail for Cecil and Blankenship 2012]. Questions naturally arise about how well those relationships built from US data apply to weather regimes in other parts of the world, and how results would vary if considering smaller hail sizes, or hail aloft that does not necessarily reach the ground before melting. Cecil and Blankenship (2012) did attempt to account for some regional differences, by applying different scalings (based on observed radar reflectivity profiles) for land versus ocean and for tropics versus mid-latitudes. Ni et al. (2016) compared TRMM properties of storms with reported hail in China to those in the US. Large differences in TRMM properties between the two regions had to do with how hail events are reported. The Chinese database is dominated by graupel and small hail, and limited to events at fixed observing sites (as opposed to storm spotters from the general public reporting "severe", i.e., 25 mm and larger, hail in the US). When comparing properties associated with specific hail sizes, the satellite-based remote sensing properties were similar for both countries. This gives some confidence for applying the US-based results more broadly, although it does not address uncertainties related to either very high or very low latitudes. The small hail that dominates the Chinese

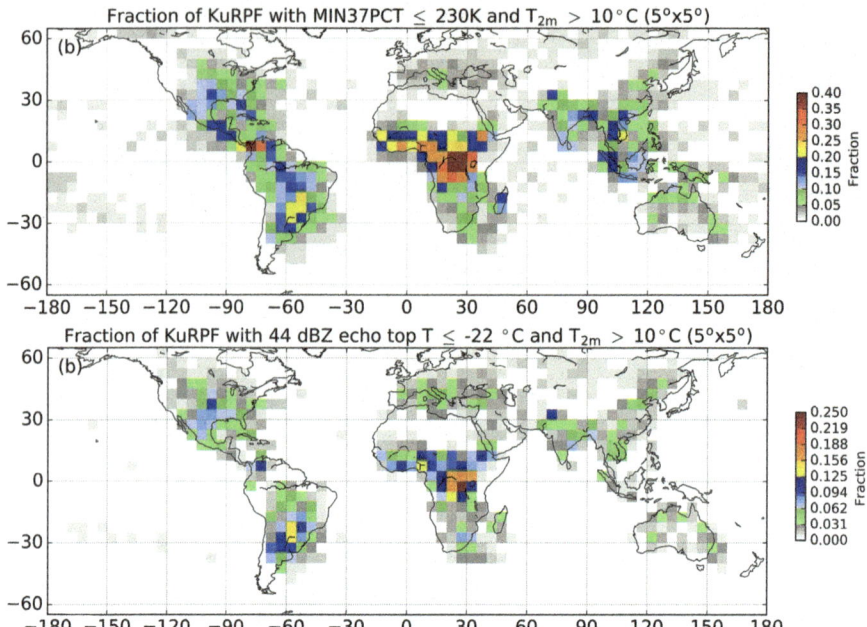

Fig. 44.5 GPM cases from Mar 2014 – Feb 2017 satisfying the large hail thresholds from Ni et al. (2017). Cases with near-surface temperature below 10 °C are screened out to avoid snow and ice contamination. (Adapted from Ni et al. 2017)

database is predominantly from high terrain, and those events had much weaker signatures in both the passive and active MW measurements.

While Cecil (2009) and Cecil and Blankenship (2012) approached hail detection in a probabilistic sense, others have taken a deterministic approach and sought measurement thresholds that optimize skill scores or that match typical (mean or median) values from hail cases. Ni et al. (2017) used TRMM data together with US reports of large (19 mm or larger) hail (similar to Cecil 2009) to optimize skill scores for passive 37 GHz brightness temperatures and active Ku-band radar reflectivities. This led to thresholds of 230 K for 37 GHz TB or 44 dBZ radar reflectivity at the −22 °C temperature level. Ni et al. (2017) mapped the fraction of cases satisfying those thresholds in 3 years of GPM data (Fig. 44.5). While the general patterns are similar using either threshold, the 37 GHz passive MW approach yields many more hail events in the tropics than the active radar approach. Cases satisfying the 230-K threshold at 37 GHz in the tropics tend to have lower reflectivity values between about −10 °C and − 40 °C, but higher overall echo tops, than cases from the subtropics that also satisfy the threshold (Fig. 44.6). This difference in reflectivity profiles between tropics and subtropics led Cecil and Blankenship (2012) to impose a regional scaling on their AMSR data. Without that scaling, Fig. 44.4 would have had much higher values equatorward of 15°.

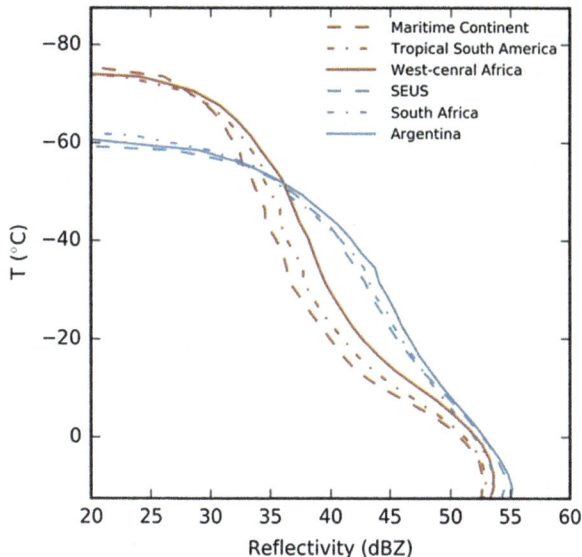

Fig. 44.6 Median profiles of maximum Ku-band radar reflectivity from TRMM cases that satisfy the 230 K threshold at 37 GHz. Red lines are for tropical regions (Maritime Continent, tropical South America, and west-central Africa). Blue lines are for subtropical regions (southeastern US, South Africa and northern Argentina)

The aforementioned studies have focused mostly on large hail reported at the ground, using the US National Weather Service criterion for a severe thunderstorm (at least 0.75″ – 19 mm diameter before 2010, and at least 1″ – 25 mm diameter since 2010). Mroz et al. (2017) instead used hydrometeor type retrievals from polarimetric radar data in the US to identify hail, and determined thresholds for GPM measurements that optimize skill scores. The polarimetric radar retrievals of hail do not specify a minimum size, and do not require that the hail reaches the ground without first melting. Mroz et al. (2017) also departed from the previous studies by testing all the GMI frequencies (from 10.7 to 183.3 GHz), both Ku- and Ka-band active radar measurements, and dual-variable proxies that can combine multiple frequencies or combine passive and active measurements. Their highest skill score resulted from a combination of Ku- and Ka-band radar measurements in the mixed phase region (averaged over the 4 km immediately above the $-10\,°\mathrm{C}$ isotherm), but that scored only slightly higher than using Ku-band radar by itself. The 19 GHz passive MW measurements yielded slightly better skill than 37 GHz, and the most effective radar-radiometer combination involved 10.7 GHz passive together with mixed-phase layer radar reflectivity from the Ku-band. Since the dual-variable proxies were only slightly more skillful than the Ku-band radar-only proxy, Mroz et al. (2017) mapped the fraction of GPM radar profiles satisfying the mixed-phase reflectivity >40.42 dBZ threshold (Fig. 44.7).

The different approaches to hailstorm identification lead to broadly similar results in Figs. 44.4, 44.5 and 44.7. They all highlight the same hotspots in central and western tropical Africa, the central US, northern Argentina through southern Brazil, Pakistan, and eastern India/Bangladesh. They all show enhancements across southern Europe and southern Africa, with a smattering of storms extending across the

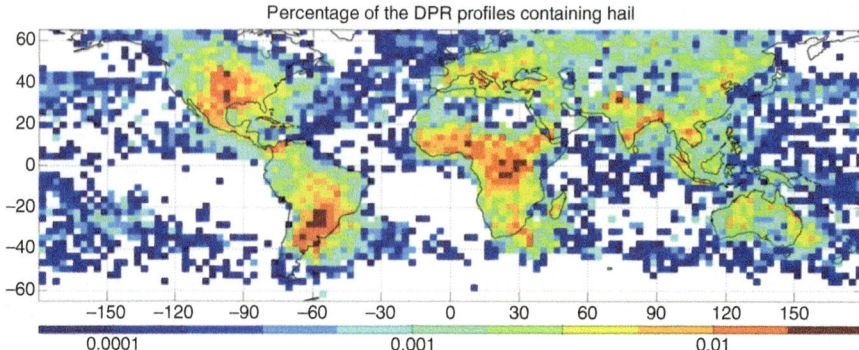

Fig. 44.7 Fraction of GPM radar profiles with mean mixed-phase Ku-band reflectivity >40.42 dBZ, Apr 2014 – Mar 2016. (Adapted from Mroz et al. 2017)

high latitudes of Canada and Russia. Key differences in Figs. 44.4, 44.5, and 44.7 mostly involve uncertainty over how often hail occurs in the deep tropics. It is quite reasonable to expect that hail is common aloft in the strong thunderstorms of tropical Africa. Since Fig. 44.6 uses an algorithm trained on retrievals of hail anywhere in the vertical, its representation of the tropics is entirely plausible. Figure 44.4 and 44.5 instead are intended to represent large hail that reaches the surface. Whether this happens often in the tropics is debatable. Figure 44.6 and Cecil (2011) show that radar profiles appear a bit weaker for storms in the tropics than one would expect based solely on their passive MW signatures. The storms there may tend to be dominated by deep layers and large concentrations of small hail, which could melt before reaching the surface. If/when large hail stones are generated, they would be expected to reach the surface, since their large fall speeds would limit melting upon descent. There are some reports of hail reaching the ground, but not enough to support the bullseye in central Africa especially seen in Fig. 44.5. On the other hand, the lack of more reports from the ground could be partially explained by combinations of low population density and limited institutional infrastructure for collecting and conveying storm reports.

These satellite-based approaches have much better agreement with reanalysis-derived maps of severe weather environments than with maps of ground-based hail reports (Williams 1973; as reprinted in Barnes 2001, and Cecil and Blankenship 2012). Ground-based hail reports are difficult to meaningfully compare across national boundaries, and across eras, due to varying reporting standards. That is a key motivation for using satellite-borne instruments to map hailstorms. Maps of ground-based reports are also skewed toward mountain ranges, likely with contamination by events that are truly graupel instead of hail. On the other hand, the satellite-based approaches likely struggle to detect cases with very small hail (they are mostly trained using reports of larger hail) or with horizontally small storm sizes (due to satellite footprints being on the order of 10 km).

44.3 Microwave Sounder Retrievals

There are two primary hail detection algorithms that have been developed for use with MW sounders – MWCC and "NOAA". These are described in the following two sections.

44.3.1 MicroWave Cloud Classification (MWCC)

Radiometric measurements in the MW regime commonly sense thick clouds and precipitation by registering a general variation of the nominal signal as observed in clear sky conditions. For frequencies higher than 85 GHz, the atmosphere is not completely transparent due to increasing water vapor absorption as the observation frequency increases. The perturbation in the radiation field induced by thick clouds is detected by satellite sensors depending on the observation frequency, the channel weighting function and the cloud structure in terms of vertical development, liquid and ice content, phase and size distribution of cloud droplets. Frozen aggregates on top of clouds largely depress frequencies of 150 GHz and above. Several studies and modelling in the past have examined the effects of clouds on MW high frequencies in the range 90–190 GHz (Skofronick-Jackson et al. 2002; Bennartz and Bauer 2003; Hong and Heygster 2005a, b) showing, for example, that the water vapor absorption lines around 183.31 GHz can be potentially used for delineating the distribution of hydrometeors in clouds (Burns et al. 1997).

The high sensitivity of MW TB in the range 90–190 GHz has been used in characterizing the cloud properties and delineating the cloud field by distinguishing the type of clouds and their top altitude. The MicroWave Cloud Classification (MWCC) method originally developed for AMSU-B/MHS and now fitted for the Suomi National Polar Partnership (S-NPP) Advanced Technology Microwave Sounder (ATMS), DMSP-Special Sensor Microwave Imager/Sounder (SSMIS), and for the multi-channel conical-scanning radiometer GMI, exploits the properties of the high frequency channels to classify the cloud type (stratiform/convection) by identifying the cloud top altitude and the phase of hydrometeors.

The retrieval scheme first described in Laviola and Levizzani (2009) and further enhanced in Levizzani et al. (2013) is physically based on the impact of clouds on the MW signal at 183.31 GHz with respect to the clear sky condition. The basic concept of the method is that different stages of cloud formation have different impacts on the water vapor channel response at 183.31 GHz and one can delineate the vertical distribution of clouds by evaluating their vertical penetration in the atmosphere on the basis of the signal depression measured in each channel. A quantitative evaluation of the signal perturbation with respect to the unperturbed field is instrumental to derive cloud information. The signal perturbation on a certain frequency around 183.31 GHz is quantified as the percentage variation with respect to the unperturbed field (background, TB^i_{max}) which is dynamically calculated as the

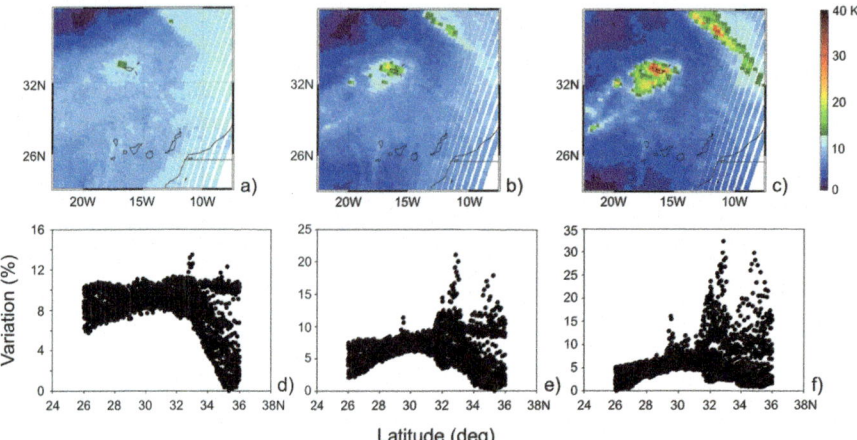

Fig. 44.8 MetOp-A MHS, 20 February 2010 1220 UTC. Top: signal perturbation (K) with respect to clear sky conditions induced by scattering hydrometeors on the MHS brightness temperatures in the channels at (**a**) 184, (**b**) 186 and (**c**) 190 GHz. Bottom: diagrams corresponding to the computed percent variation of the signal at the above frequencies. Values lower than 10% are generally associated with water vapor absorption while the spikes identify the scattering due to precipitating clouds. Values close to zero correspond to low water vapor absorption in the upper, middle and lower atmosphere at (**d**) 184, (**e**) 186 and (**f**) 190 GHz, respectively

absolute maximum of the observed brightness temperatures. For each channel an offset has been calculated to reduce the absorption of water vapor, which is substantial when a stratiform cloud is detected. Thus, all perturbations of different water vapor frequencies are combined in a cascade test, which computes the scalar P_{MWCC} indicating the type of clouds for each observed pixel. When a cloud is categorized a detector for ice aggregates identifies the frozen hydrometeors in the stratiform cloud type.

An example showing the concept at the basis of the MWCC can be seen in Fig. 44.8 referring to a cloud system over the Island of Madeira that perturbs the three water vapor channels of the MHS on board the MetOp-A satellite. The impact of clouds on the sounding frequency is shown in absolute terms (K) and in terms of percentage variation of the radiance field with respect to the background TB. The top plates in Fig. 44.8 show that the perturbation induced by clouds affects less the 184 GHz channel, whose weighting function peaks in the upper troposphere. Thus, the contribution of that frequency in the computational scheme of the MWCC is low with respect to other channels where the impact of the cloud system is much higher. The scatterplots in the bottom plates of Fig. 44.8 quantify this perturbation showing very modest spikes higher than 10% (off-set threshold) at 184 GHz referring to the small impact of deep convective clouds. The marked spikes evaluated around 22% and 35% at 186 and 190 GHz, respectively, are in turn due to the scattering by ice hydrometeors depressing the radiation at highest sounding frequencies. The signal reduction is thus a measure of the cloud vertical development impacting on the

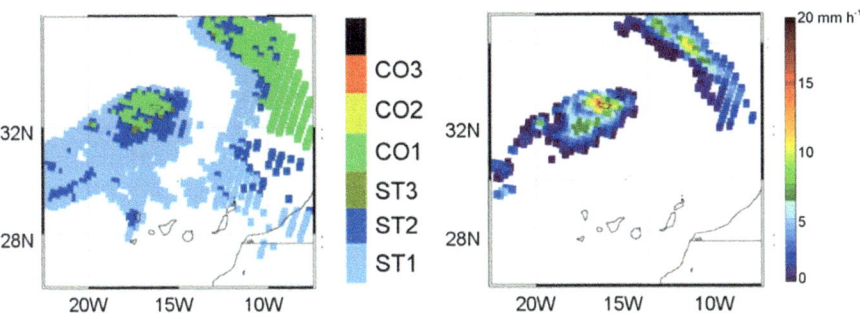

Fig. 44.9 20 February 2010 1220 UTC. The MWCC product (left) identifies as shallow convection (CO1) the main core of precipitation as retrieved by the 183-WSL (right). Stratiform clouds (ST1 and ST2) surround the precipitation with low rain intensity to no-rain

considered frequency. Finally, a background percentage variation due to the pure absorption of water vapor at frequencies 184, 186, and 190 GHz has been statistically assessed in order to establish a series of frequency-dependent offsets for better isolating the cloud signals. This has been conceived to avoid the shadowing effect due to the absorption in the 183.31 GHz band by water vapor surrounding the clouds. Therefore, the retrieval scheme of the MWCC has been conceived with a high-pass threshold filtering the background signals and activating any conditional steps superintending the computational chain when a specific threshold value is exceeded.

Figure 44.9 shows the MWCC product compared with the rain rate retrieval of the 183-WSL algorithm (Laviola and Levizzani 2011; Laviola et al. 2013) for the case study described in Fig. 44.8. The main module of the MWCC classification provides three different classes for each cloud category. Cloud type is assessed in terms of cloud top height divided into six categories, 3 stratiform (ST) and 3 convective (CO), with height increasing from the ST1 (lowest top) to the CO3 (highest top) class. The stratiform cloud categories are: ST1 (1–3 km), ST2 (3–5 km), and ST3 (5–6 km). The convective clouds categories are: CO1 (6–7 km), CO2 (7–9 km), and CO3 (>9 km). The precipitation affecting the island was mainly generated by a convective system classified by the algorithm as shallow convection CO1 reaching around 6–7 km altitude. The comparison with the 183-WSL rain map demonstrates that stratiform clouds of type ST1 and ST2, with low rain intensities decreasing from 5–6 mm h^{-1} to no-rain, surround the precipitating convective cores, corresponding to rain intensities of 10–15 mm h^{-1}. Note the almost complete absence of the ST3 cloud type, which is commonly observed in the transition region between stratiform and shallow convection development. This absence of high-level stratiform clouds confirms the shallow convection type of the cloud system, which thus most probably developed as isolated towers embedded in a preexisting low-layer of oceanic stratiform rain clouds.

44.3.2 Hail Detection with MWCC Method

New investigations originally conceived for improving the MWCC computational scheme during severe thunderstorms revealed the high correlation between the occurrence of hail and the TB depression in deep convective clouds classified by the MWCC method. The increase of the ice mass during the towering stage of convections tends to progressively reduce the signal measured by high frequency channels in the range 90–190 GHz. However, the formation of graupel or hailstones due to the strong updrafts seems to be identified by a marked reduction of the signal increasing with increasing hydrometeor size (Cecil 2009). Thus, because the goal is to quantify and isolate the scattering signature by hailstones, we evaluate the total scattering by cloud ice hydrometeors in terms of TB depression as follows

$$\Delta TB^i = TB^i_{max} - TB^i_{sat} \qquad i = 1, 2, 3, 4, 5 \qquad (44.1)$$

where TB^i_{max} is the background TB for clear-sky condition and TB^i_{sat} the satellite TB measured by the i channel.

Figure 44.10 shows the TB measured by MHS on NOAA-19 during the hailstorm over Italy on 04 September 2015 at 1253 UTC. An intense hailstorm is approaching Western Sardinia meanwhile Central Italy is affected by localized deep convections. The observed TB_{max} used as background brightness temperature in Eq. (44.1) are, respectively: $TB^1_{max} = 307.23$ K, $TB^2_{max} = 305.57$ K, $TB^3_{max} = 273.78$ K, $TB^4_{max} = 279.05$ K, $TB^5_{max} = 288.73$ K. Due to the intense hailstorm, the radiation field at 89 GHz is strongly perturbed by very large hail (black circle in Fig. 44.10) reaching the minimum value $TB^1_{sat} = 131.5$ K corresponding to the total signal variation $\Delta TB^1 = 175.7$ K. Lower impact from other ice hydrometeors is registered at this frequency. The higher frequencies show in turn high sensitivities to the scattering both by the smallest and largest hailstones. The scattering signature at 157 GHz for very large hail matches value of $\Delta TB^2 = 185.1$ K whereas values of 166.3, 140.5 and 160.9 K were found, respectively, for channels 3 to 5. The scattering by ice in deep convection (light green arrow) over Central Italy greatly increases the signal reduction at higher frequencies, but at 89 GHz just a few cold spots where the ice mass is largest can be observed. Furthermore, applying eq. (44.1) to the region of the storms (dashed rectangle in Fig. 44.10) we can isolate the contribution of scattering by hail with respect to the total extinction of radiation.

The advantage of measurements at 183.31 GHz is that the surface emissivity has a negligible impact on the observation and the scattering signature by ice results from a background level marked by the absorption of water vapor at each level of the weighting functions.

The hail detector prototype developed for the MWCC method (hereafter MWCC-H) is based on a logistic model of growth (Verhulst 1845), improved with the dynamic carrying capacity (Mayer and Ausbel 1999). The dynamics of the hail model expects that the model accuracy starts for probability values >0.45 where large hailstone formation typically occurs. Lowest values are usually associated with

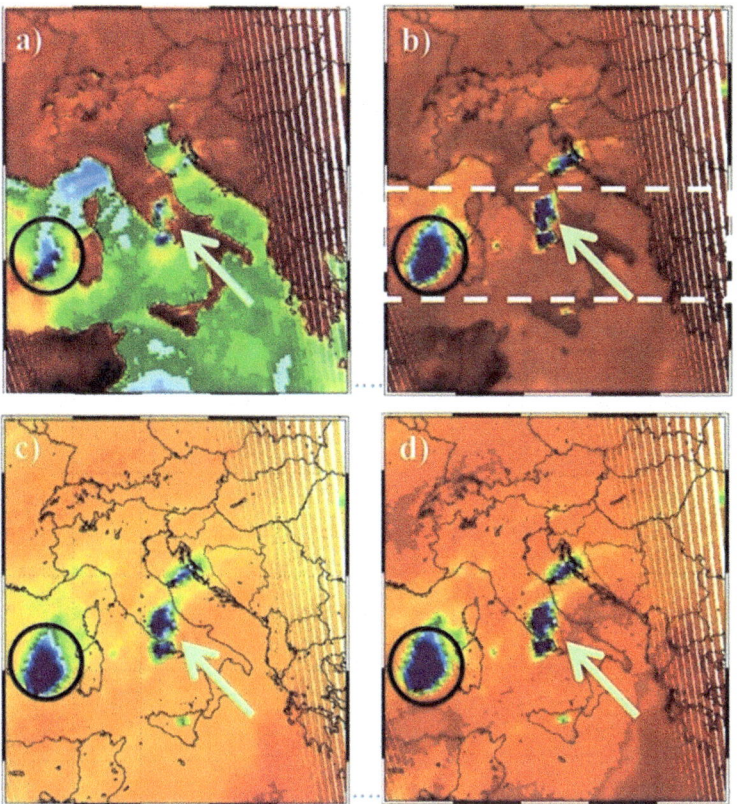

Fig. 44.10 4 September 2015 1253 UTC. Brightness temperatures during the hailstorm over Sardinia and Central Italy. From (**a**) to (**d**), the MHS-N19 measurements at 89, 157, 186 and 190 GHz, respectively. The black circle indicates the hailstorm; the light green arrow marks the deep convection that evolved in a strong hailstorm during the successive hours. The dashed rectangle in (**b**) groups the longitudes where the main cores of the storm formed

small hailstones or supercooled water (radiometrically difficult to be distinguished) while values >0.60 are associated with extra-large hailstones.

Figure 44.11 shows an example of MWCC-H application during the intense hailstorm over Western Sardinia and Central Italy on 04 and 05 September 2015. The Meteosat-10 infrared image (Fig. 44.11a) shows the V-shaped feature characterizing the intense hailstorm over the Tyrrhenian coast of Central Italy. Ground reports confirmed the existence of very large hail hitting the Gulf of Naples and inland. Figure 44.11b–d show the life cycle of storm using the MWCC-H method applied to MHS observations. Clusters of large and extra-large hail are nested in a well-organized convective system travelling from Western Sardinia to Central Italy. The MWCC-H retrieval sequence document that the system starts off the coast of Sardinia with retrieved extra-large hail (Fig. 44.11b), decreases during the night (Fig. 44.11c) and triggers a new hailstorm in the early morning over Central Italy

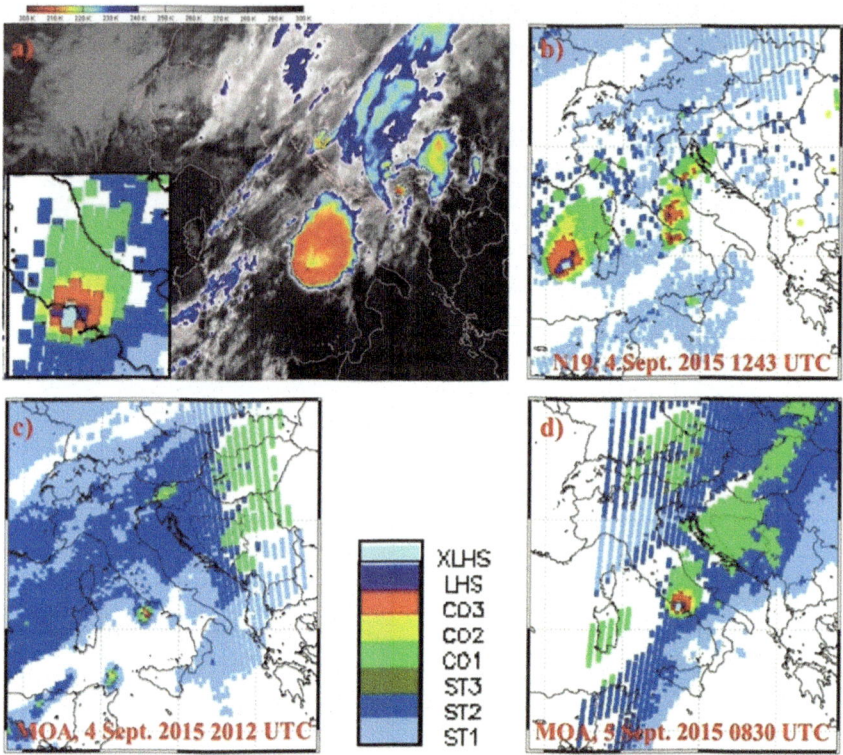

Fig. 44.11 Severe hailstorm over Western Sardinia and Central Italy, September 04–052015. (**a**) V-shaped supercell over Central Italy as seen from the Meteosat-10 at IR10.8 on 5 Sept 0900 UTC. (**b–d**) Hailstorm evolution as retrieved by the MWCC-H algorithm. Large and extra-large hailstones (violet and cyan, respectively) are nested in a mesoscale convective system (green, yellow, and red for CO1, CO2, and CO3, respectively) travelling from Western Sardinia. Stratiform clouds from low to middle atmospheric levels are ST1, ST2, and ST3, respectively

(Fig. 44.11d). The Meteosat image in Fig. 44.11a shows the main core of the hailstorm, which is classified as extra-large hail by the MWCC-H (inner small box). This region corresponds to the cloud top where brightness temperatures are higher than those of the surrounding anvil, indicating that convection intruded the stratosphere above the anvil top (Melani et al. 2003).

44.3.3 NOAA AMSU-b/MHS

The NOAA hail method (Ferraro et al. 2015) exploits the same physical basis that was described with the MWCC method. Using colocations of AMSU/MHS TB's and published spotter reports of hail (e.g., within 5 min and 25 km of each other), Ferraro et al. (2015) developed a threshold-based AMSU hail detection algorithm that uses AMSU-B/MHS observations; it can be extended to other sounders with similar sensor characteristics. It was developed on a 10-year (2000–2009) training data set that limited the matchups to zenith angles within 25 degrees of nadir (to avoid FOV size changes along the scan). Because the interest is in mapping hail associated with severe weather (i.e., 25 mm or larger) this reduced the colocation data set down to about 25% of all data, or approximately 1700 points. It should be noted that the training data set represents an extremely small fraction of all possible AMSU observations because of the sparse nature of the surface hail reports, i.e., these are based on public reports and are limited to where people actual live and when they observe the hail.

Based on the mean TB that coincided with hail of 25 mm in diameter, the "NOAA" algorithm detects the presence of hail when all of the following conditions are met:

$$TB89 < 228.2\ K$$
$$TB150/157 < 206.9\ K$$
$$TB183 + 1 < 211.1\ K \tag{44.2}$$
$$TB183 + 3 < 204.6\ K$$
$$TB183 + 7/190.3 < 200.5\ K$$

Although the hail detection algorithm used all five AMSU/MHS channels, the contribution of each individual AMSU channel was examined. It was found that although a single channel could adequately detect hail, the false alarm rate was substantially higher, presumably due to the effect of surface type (Ferraro et al. 2005). Thus, the five-channel combination performed the best when weighting the probability of detection (POD) and false alarm rate (FAR).

Figure 44.12 shows the AMSU/MHS derived climatology from all existing NOAA and EUMETSAT satellites operating during the period 2000–2011 based on the study of Ferraro et al. (2015). It clearly indicates the maximum period of hail during the month of May across the central United States. Qualitatively, this agrees well with the storm spotter results. Grouping all occurrences of hail (March – September) for the 12-year period yielded a remarkable correlation of 0.82.

Drawbacks of this threshold technique is that it does not necessarily account for the ambient background TB, thus, it implicitly assumes that the thresholds are for some "mean" surface and atmospheric condition. Because the data set comes from a diverse set of atmospheric and surface types, it might be desirable to look at the TB depressions to isolate the effect of the hail. In addition, the technique does not account for variations in freezing level height, thus, when it is applied outside of the US or in cases with a high freezing level height, false alarms can increase because of the hail melting before it reaches the ground.

Figure 44.13 shows an example of this technique applied to a hail event over the middle Atlantic states of the US on 23 June 2015. As can be seen, the MHS detected hail in regions nearby where there were surface reports of hail. As previously

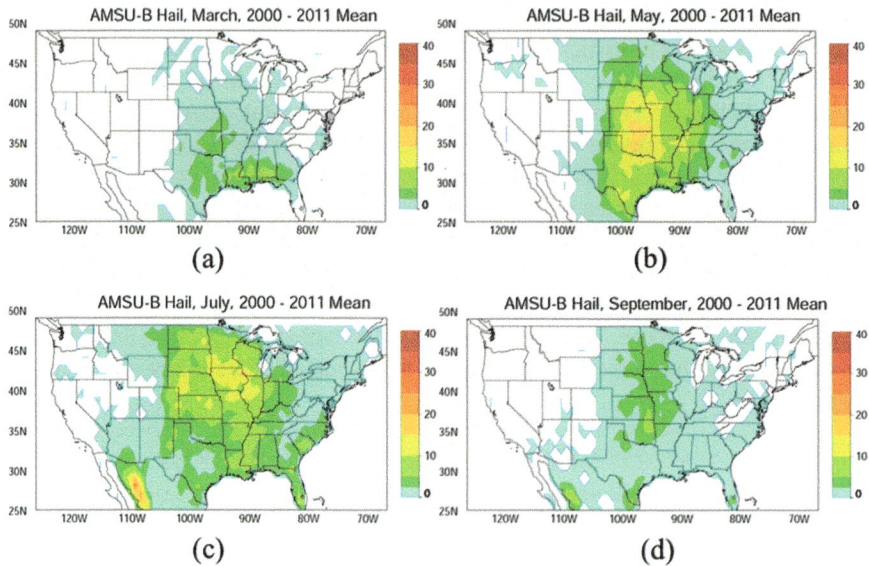

Fig. 44.12 Twelve-year (2000–11) mean AMSU derived hail occurrences for (**a**) March, (**b**) May, (**c**) July and (**d**) September. Values are number of occurrences per month within the 1° grid box. (Adapted from Ferraro et al. 2015)

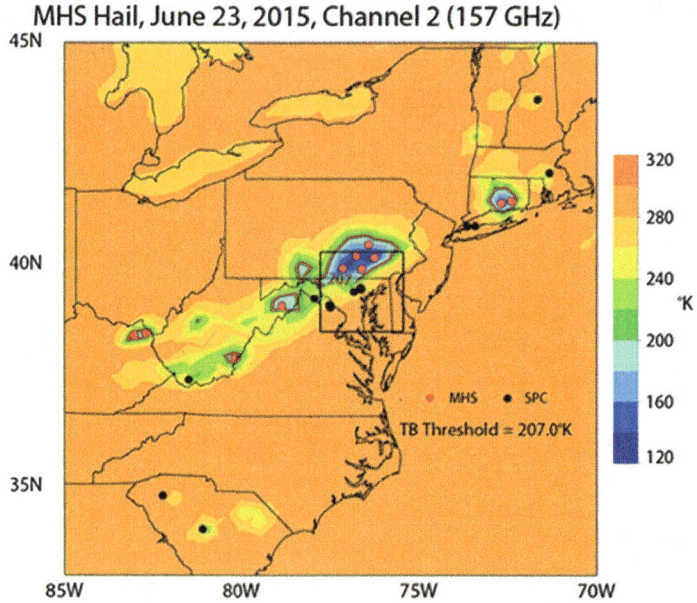

Fig. 44.13 MetOp-A MHS 157 GHz TB (K) for 23 June 2015 2230 local time. A 207 K contour (red line) is shown which is the hail threshold at this channel for the NOAA algorithm. Red dots indicate the NOAA hail detection (using all five channels) whereas the black dots indicate surface reports of hail. The region within the box reported hail up to 100 mm in diameter

mentioned, one limitation is that the spotter surface reports are where people live and report and not necessarily where the hail actually occurred.

By exploiting the diurnal sampling of the NOAA and EUMETSAT satellites, a 6-hourly climatology was also developed, which had relatively good spatial agreement with the storm reports. Interestingly, the biggest disparity was from midnight to 0600 am local time; a period where the spotter reports were most likely to be the least reliable due to darkness and at a time of the day when people were most likely asleep.

44.4 Summary

Passive MW sensors have been exploited to detect and map regions of hail. This can be important for many applications, including near real time monitoring for public safety and transportation concerns, as well as longer term climate studies. Both imagers and sounders have advantages and disadvantages related to this topic, with the former providing better spatial resolution while the latter takes advantage of the higher frequencies at 183 GHz and better diurnal sampling. Synergistic use of the active precipitation radar on TRMM and GPM also provide useful information on the vertical structure of the cloud system, helping delineate freezing level heights and whether the hail is actually reaching the surface. This is a limitation of these techniques, especially in warmer climates where large hail can be detected aloft but rarely reaches the surface.

References

Barnes, G. (2001) Severe local storms in the tropics. *Severe Convective Storms, Meteorological Monograph, 50*, 359–432, American Meteorological Society. https://doi.org/10.1175/0065-9401-28.50.359.

Bennartz, R., & Bauer, P. (2003). Sensitivity of microwave radiances at 85 –183 GHz to precipitating ice particles. *Radio Science, 38*(4), 8075. https://doi.org/10.1029/2002RS002626.

Burns, B. A., Wu, X., & Diak, G. R. (1997). Effects of precipitation and cloud ice on brightness temperatures in AMSU moisture channels. *IEEE Transactions on Geoscience and Remote Sensing, 35*, 1429–1437. https://doi.org/10.1109/36.649797.

Cecil, D. (2009). Passive microwave brightness temperatures as proxies for hailstorms. *Journal of Applied Meteorology and Climatology, 48*, 1281–1286. https://doi.org/10.1175/2009JAMC2125.1.

Cecil, D. (2011). Relating passive 37-GHz scattering to radar profiles in strong convection. *Journal of Applied Meteorology and Climatology, 50*, 233–240. https://doi.org/10.1175/2010JAMC2506.1.

Cecil, D., & Blankenship, C. B. (2012). Toward a global climatology of severe hailstorms as estimated by satellite passive microwave imagers. *Journal of Climate, 25*, 687–703. https://doi.org/10.1175/JCLI-D-11-00130.1.

R. R. Ferraro et al.

Ferraro, R., Beauchamp, J., Cecil, D., & Heymsfield, G. (2015). A prototype hail detection algorithm and climatology developed with the Advanced Microwave Sounding Unit (AMSU). *Atmospheric Research, 163*, 24–35. https://doi.org/10.1016/j.atmosres.2014.08.010.

Ferraro, R. R., Weng, F., Grody, N., Zhao, L., Meng, H., Kongoli, C., Pellegrino, P., Qiu, S., & Dean, C. (2005). NOAA operational hydrological products derived from the AMSU. *IEEE Transactions on Geoscience and Remote Sensing, 43*, 1036–1049. https://doi.org/10.1109/TGRS.2004.843249.

Hong, G., & Heygster, G. (2005a). Detection of deep convective clouds from AMSU-B water vapor channels measurements. *Journal of Geophysical Research, 110*, D05205. https://doi.org/10.1029/2004JD004949.

Hong, G., & Heygster, G. (2005b). Sensitivity of microwave brightness temperatures to hydrometeors in a tropical deep convective cloud system at 89–190 GHz. *Radio Science, 40*, RS4003. https://doi.org/10.1029/2004RS003129.

Hou, A. Y., Kakar, R. K., Neeck, S., Azarbarzin, A. A., Kummerow, C. D., Kojima, M., Oki, R., Nakamura, K., & Iguchi, T. (2014). The Global Precipitation Measurement mission. *Bulletin of the American Meteorological Society, 95*, 701–722. https://doi.org/10.1175/BAMS-D-13-00164.1.

JAXA. (2006). *AMSR-E data users handbook* (4th edn.). Japan Aerospace Exploration Agency. Available from http://www.eorc.jaxa.jp/en/hatoyama/amsr-e/amsr-e_handbook_e.pdf. Last accessed 26 Oct 2018.

Kummerow, C., Barnes, W., Kozu, T., Shiue, J., & Simpson, J. (1998). The Tropical Rainfall Measuring Mission (TRMM) sensor package. *Journal of Atmospheric and Oceanic Technology, 15*, 809–817. https://doi.org/10.1175/1520-0426(1998)015<0809:TTRMMT>2.0.CO;2.

Laviola, S., & Levizzani, V. (2009). Observing precipitation by means of water vapor absorption lines: A first check of the retrieval capabilities of the 183-WSL rain retrieval method. *Italian Journal of Remote Sensing, 41*, 39–49.

Laviola, S., & Levizzani, V. (2011). The 183-WSL fast rain rate retrieval algorithm. Part I: Retrieval design. *Atmospheric Research, 99*, 443–461. https://doi.org/10.1016/j.atmosres.2010.11.013.

Laviola, S., Levizzani, V., Cattani, E., & Kidd, C. (2013). The 183-WSL fast rain rate retrieval algorithm. Part II: Validation using ground radar measurements. *Atmospheric Research, 134*, 77–86. https://doi.org/10.1016/j.atmosres.2013.07.013.

Levizzani, V., Laviola, S., Cattani, E., & Costa, M. J. (2013). Extreme precipitation on the Island of Madeira on 20 February 2010 as seen by satellite passive microwave sounders. *European Journal of Remote Sensing, 46*, 475–489. https://doi.org/10.5721/EuJRS20134628.

Mayer, P. S., & Ausbel, J. H. (1999). Carrying capacity: A model with logistically varying limits. *Technological Forecasting Social Change, 61*, 209–214. https://doi.org/10.1016/S0040-1625(99)00022-0.

Melani, S., Cattani, E., Torricella, F., & Levizzani, V. (2003). Characterization of plumes on top of a deep convective storm using AVHRR imagery and radiative transfer simulations. *Atmospheric Research, 67–68*, 485–499. https://doi.org/10.1016/S0169-8095(03)00061-9.

Mroz, K., Battaglia, A., Lang, T. J., Cecil, D. J., Tanelli, S., & Tridon, F. (2017). Hail detection algorithm for the GPM core satellite sensors. *Journal of Applied Meteorology and Climatology, 56*, 1939–1957. https://doi.org/10.1175/JAMC-D-16-0368.1.

Ni, X., Liu, C., Zhang, Q., & Cecil, D. J. (2016). Properties of hail storms over China and the United States from the Tropical Rainfall Measuring Mission. *Journal of Geophysical Research, 121*, 12031–12044. https://doi.org/10.1002/2016JD025600.

Ni, X., Liu, C., Cecil, D. J., & Zhang, Q. (2017). On the hail detection using satellite passive microwave radiometers and precipitation radar. *Journal of Applied Meteorology and Climatology, 56*, 2693–2709. https://doi.org/10.1175/JAMC-D-17-0065.1.

Skofronick-Jackson, G., Gasiewski, A. J., & Wang, J. R. (2002). Influence of microphysical cloud parameterizations on microwave brightness temperatures. *IEEE Transactions on Geoscience and Remote Sensing, 40*, 187–196. https://doi.org/10.1109/36.981360.

Spencer, R., & Santek, D. (1985). Measuring the global distribution of intense convection over land with passive microwave radiometry. *Journal of Climate and Applied Meteorology, 24*, 860–864. https://doi.org/10.1175/1520-0450(1985)024<0860:MTGDOI>2.0.CO;2.

Spencer, R., Olson, W. S., Rongzhang, W., Martin, D., Weinman, J., & Santek, D. (1983). Heavy thunderstorms observed over land by the Nimbus-7 scanning multichannel microwave radiometer. *Journal of Climate and Applied Meteorology, 22*, 1041–1046. https://doi.org/10.1175/1520-0450(1983)022<1041:HTOOLB>2.0.CO;2.

Spencer, R., Howland, M., & Santek, D. (1987). Severe storm identification with satellite microwave radiometry: An initial investigation with Nimbus-7 SMMR data. *Journal of Climate and Applied Meteorology, 26*, 749–754. https://doi.org/10.1175/1520-0450(1987)026<0749:SSIWSM>2.0.CO;2.

Verhulst, P. F. (1845). *Recherches mathématiques sur la loi d'accroissement de la population* [Mathematical Researches into the Law of Population Growth Increase]. *Nouveaux Mémoires de l'Académie Royale des Sciences et Belles-Lettres de Bruxelles, 18*, 1–42.

Williams, L. (1973) *Hail and its distribution*. Studies of the Army Aviation (V/STOL Environment), Army Engineer Topographic Laboratories Rep. 8, ETL-SR-73-3, 27 pp.

Chapter 45
Improving High-Latitude and Cold Region Precipitation Analysis

Ali Behrangi

Abstract Accurate estimation of precipitation is important for science and application. In high latitudes precipitation estimation is difficult due to several challenges in remote sensing of precipitation and sparseness of in situ observations. Furthermore, in situ observations can also have large errors, especially for measurement of snowfall that occurs frequently in high latitudes. Here, we show how CloudSat and the Gravity Recovery and Climate Experiment (GRACE) can provide additional information to refine our quantification of high latitude precipitation. We show this through case studies over ocean, Eurasia, Tibetan Plateau, and arctic basins. The results suggest that combination of CloudSat and GRACE can provide valuable information on quantifying snowfall accumulation that can also be used to assess gauge undercatch correction methods. Together with ongoing efforts under GPM to advance Level 2 precipitation retrievals and their combination, great opportunities exist to advance precipitation retrieval in high latitudes.

Keywords Precipitation · High latitudes · GPM · IMERG · CloudSat · GRACE · GPCC · GPCP

45.1 Introduction

The observed changes in the hydrologic regime of the high latitude regions and its expected future amplifications (Solomon et al. 2007) has motivated studies of the water cycle and its acceleration in a more spatially and temporally-comprehensive way than that has been previously possible. The reported changes in high latitude hydrologic cycle include decreasing snow cover, mountain glacier area, permafrost extent and lake area, and increasing freshwater discharge (e.g., Smith et al. 2005; Alley et al. 2007; Yang et al. 2002; Meier and Carter 2006; McClelland et al. 2006), all requiring good

A. Behrangi (✉)
Department of Hydrology and Atmospheric Sciences, University of Arizona, Tucson, AZ, USA
e-mail: behrangi@email.arizona.edu

© Springer Nature Switzerland AG 2020   881
V. Levizzani et al. (eds.), *Satellite Precipitation Measurement*, Advances in Global Change Research 69, https://doi.org/10.1007/978-3-030-35798-6_21

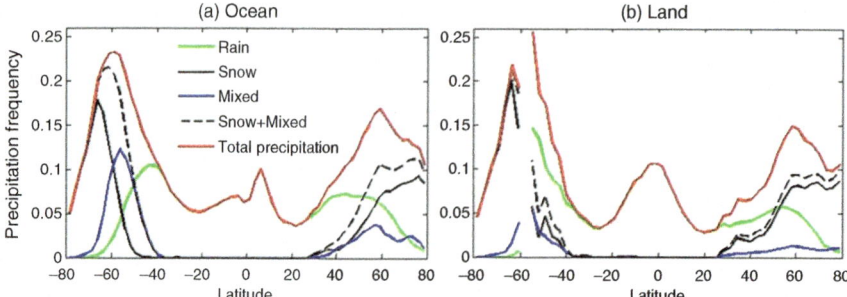

Fig. 45.1 Zonal distribution of precipitation occurrence and its phase over (**a**) ocean and (**b**) land from CloudSat (footprint size ~1.4 × 1.7 km^2) 2C-RAIN-COLUMN product (Haynes et al. 2009). The plots are adapted from Behrangi et al. (2014a, b) and are constructed using 3 years (2007–2009) of "certain" precipitation occurrences

quantitative knowledge on precipitation amount and its distribution. Therefore, improved quantification of precipitation is critical for the understanding and quantification of the climate change impact on hydrology of the high latitude regions, though often not obtainable from gauges because of their sparseness at higher latitudes.

Because of the sparseness of rain gauges and of the need for a comprehensive quantification of global precipitation, precipitation retrieval from satellites has been an important activity over the last few decades. With the operation of the TRMM microwave imager (TMI) and precipitation radar (PR), significant advancement in the science and retrieval of moderate and intense precipitation was obtained. However, due to its limited coverage (~35°S/N) and lack of sufficient sensitivity to take on the challenges of measuring light rain and falling snow, TRMM offered limited impact on improving the remote sensing of high latitude precipitation. Therefore, a major weakness in remote sensing of precipitation remained in high latitudes where precipitation occurs more frequently than in the tropics (Fig. 45.1) and light rain and snowfall are dominant precipitation types.

Large uncertainties in retrieval of high latitude precipitation over both ocean and land have been highlighted in previous studies (Adler et al. 2012; Behrangi et al. 2012, 2014a, b). For example, by analyzing precipitation products from infrared (IR) and multiple microwave (MW) sensors, Behrangi et al. (2012) showed that precipitation detection and estimation capabilities reduce sharply poleward of 20°S/N. Missed precipitation at high latitudes (e.g., at about 60° latitude) could add up to approximately 50% of the total precipitation amount and even more for precipitation occurrence (Behrangi et al. 2012, 2014a; Lin and Hou 2012). These shortcomings have been related to the limitations in retrieval techniques (e.g., scattering-based over land), sensor sensitivity, poor understanding of the precipitation microphysics and surface emissivity, and difficulties in distinguishing between light rain and cloud (Berg et al. 2006; Lebsock and L'Ecuyer 2011).

The GPM core instruments, the Ku/Ka (13.6/35.5 GHz)-band Dual-frequency Precipitation Radar (DPR) and GPM Microwave Imager (GMI), have geographical coverage of ~65°S/N (covering about 91% of the globe) which is about 40% more than that of TRMM (~35°S/N; covering ~51% of globe). Furthermore, the increased sensitivity of the DPR relative to the PR, together with the addition of four high-frequency channels on the GMI (e.g., at about 166 and 183 GHz), have given GPM new capabilities to advance measuring light rain and falling snow. DPR enables extracting quantitative information on microphysical properties such as the particle size distribution, important to reduce a priori assumptions in radar and radiometer retrieval algorithms, and thus providing more accurate precipitation estimates relative to the TRMM PR. If we also add the higher sensitivity of DPR to light rain and snowfall, DPR becomes a robust benchmark for GPM radiometric retrievals in middle and high latitudes and facilitates developing more physically based algorithms to retrieve precipitation signals from passive MW radiometers over land and ocean. Through construction of a common cloud/radiance database for passive MW radiometer retrievals, the combined DPR and GMI measurements is key to unifying the GPM constellation radiometers into one consistent framework to produce uniform global precipitation products. Together with the efforts to quantify surface emissivity over land (Ferraro et al. 2013; Norouzi et al. 2015), this has created unprecedented opportunities for advancing passive MW retrievals of light rain and snowfall rates, especially over frozen land surfaces where previous retrievals often had missing values.

The improvement in Level 2 GPM MW products should also improve the GPM merged multi-sensor products (e.g., Level 3 NASA IMERG, Huffman et al. 2018) compared to those developed during the TRMM era such as the TRMM Multisatellite Precipitation Analysis (TMPA, Huffman et al. 2007) that showed large errors over cold regions (e.g., Behrangi et al. 2014c; Wen et al. 2018).

Through a comparative study in winter over the mountainous Tuolumne and the adjacent Cherry/Eleanor basins in California, Behrangi et al. (2018) showed that IMERG-HQ V5, which is a merged MW product, has large underestimations (Fig. 45.2). However, IMERG-HQ was able to capture monthly snowfall patterns with high correlation (i.e., Corr = 0.88) compared to in situ observations. With bias correction, IMERG showed a significant improvement over the MW-only product (Fig. 45.2), in agreement with several previous studies (e.g., Behrangi et al. 2011)

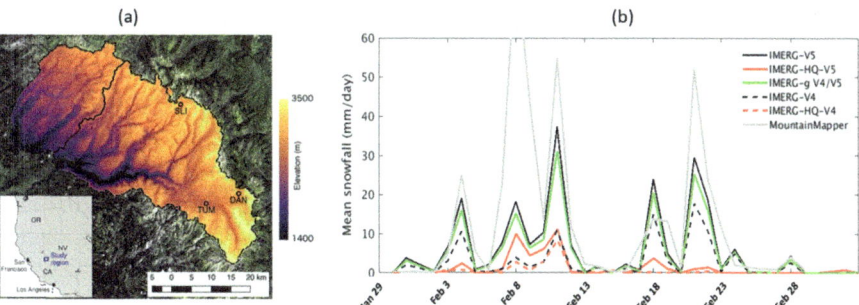

Fig. 45.2 (a) The mountainous Tuolumne and the adjacent Cherry/Eleanor basins in California, (b) Time series of daily mean snowfall rate, averaged over the snow grids. Mountain mapper (Schaake et al. 2004) is used as reference precipitation and IMERG-g represents IMERG with gauge correction. (Adapted from Behrangi et al. 2018, with modifications)

discussing the critical role of bias adjustment to advance precipitation products. However, there are several issues with using in situ data in cold and high latitude regions including: (1) the precipitation gauge network is often sparse and discontinuous in most regions—a substantial decrease in the number of high latitude precipitation stations since 1990 has exacerbated this limitation; (2) precipitation measurement must be bias corrected to account for wetting loss and gauge undercatch (Goodison et al. 1998; Yang et al. 2002) as the bias correction factors are largest for solid precipitation (as high as 300%, Fuchs et al. 2001) and depend on the choice of the correction method, and (3) in situ data are mainly missing over ocean.

In light of the recent advancement in observing systems and retrieval techniques, other alternatives are available to provide insights on cumulative precipitation over cold regions. Among those CloudSat (Stephens et al. 2002) and the Gravity Recovery and Climate Experiment (GRACE) (Tapley et al. 2004) have been found particularly useful.

45.2 Utilizing CloudSat to Assess High Latitude Precipitation

The 94 GHz (W band) CloudSat CPR (Stephens et al. 2002), with a minimum detectable signal of −28 dBZ, allows high-quality detection and estimation of snowfall, light rainfall and drizzle, dominant types of precipitation at high latitudes, that go largely undetected with other sensors (Stephens et al. 2008; Behrangi et al. 2012). Evaluation of CloudSat precipitation using ship data demonstrated that the CloudSat retrieval is consistent with ship observations well into high latitudes (Ellis et al. 2009). Good performance has also been reported over land (Smalley et al. 2014; Chen et al. 2016). Furthermore, focusing on high latitude regions enables collecting ample samples over both land and ocean. CloudSat estimates are particularly unique over ocean where ground observations barely exist.

45.2.1 Analysis of Zonal Precipitation Distribution Over Ocean

Behrangi et al. (2012, 2014b) merged precipitation estimates from CloudSat CPR, TRMM PR, and Aqua's Advanced Microwave Scanning Radiometer for Earth Observing System (AMSR-E) using a method described in Berg et al. (2010) and Behrangi et al. (2012). The complementary estimates from the Merged CloudSat, TRMM, and AMSR (on Aqua) (named MCTA) includes snowfall and rainfall, from drizzle to the most intense rain rates captured by TRMM. At higher latitudes, where TRMM PR does not exist, precipitation estimates from AMSR-E (preceding CloudSat by about 1 min) complemented CloudSat to capture intense precipitation rates that can be underestimated due to saturation of CloudSat signals in heavy

Fig. 45.3 Zonal
precipitation averages
(in mm day^{-1}) for the full
year 2015. The GPCP,
MCTA, and IMERG are
Version 2.3, Version 2, and
Version 03, respectively.
Other GPM products are
Version 04. (Adapted from
Skofronick-Jackson et al.
2017)

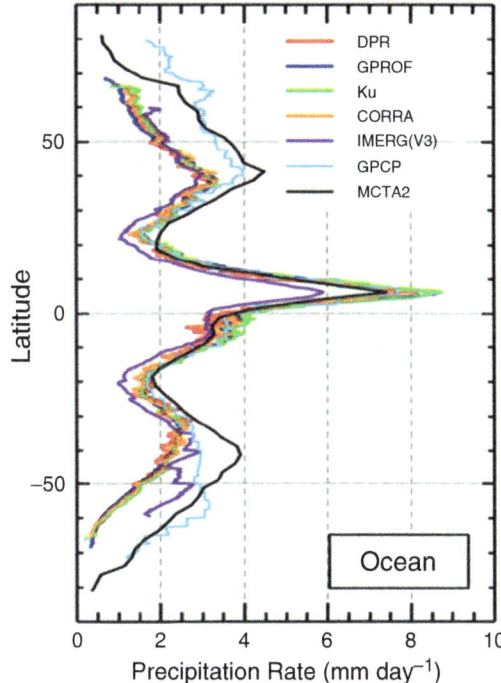

precipitation. MCTA confirmed that the Global Precipitation Climatology Project
(GPCP) likely presents a more realistic precipitation distribution in high latitudes
compared to other MW-based products. However, MCTA also showed that GPCP
contains large errors in estimating mean precipitation rates outside the tropics,
particularly over the Southern Ocean. The later version of MCTA (V2) includes
mixed phase precipitation that was missed in CloudSat estimates at the time of the
MCTA construction. Based on the MCTA guidance and other analysis, GPCP is
currently used to bias adjust the GPM IMERG over ocean at high latitudes, where
the GPM L2 products continue to underestimate (Fig. 45.3).

45.2.2 Regional Analysis and Precipitation Pattern

CloudSat data can also be used for seasonal and regional analysis. Behrangi et al.
(2016) compared seasonal mean precipitation intensity maps of GPCP with
CloudSat and a few reanalyses in high latitudes. Despite the general agreement in
their seasonal patterns, GPCP showed some large inconsistencies compared to other
products. For example, as can be seen in Fig. 45.4, over the Atlantic Ocean south of
70°N, CloudSat reports average precipitation during fall and winter >5 mm day^{-1} in
many grid boxes, similar to reanalyses, but very different from GPCP. Unlike other
products, GPCP does not present any noticeable precipitation gradient around 70°N

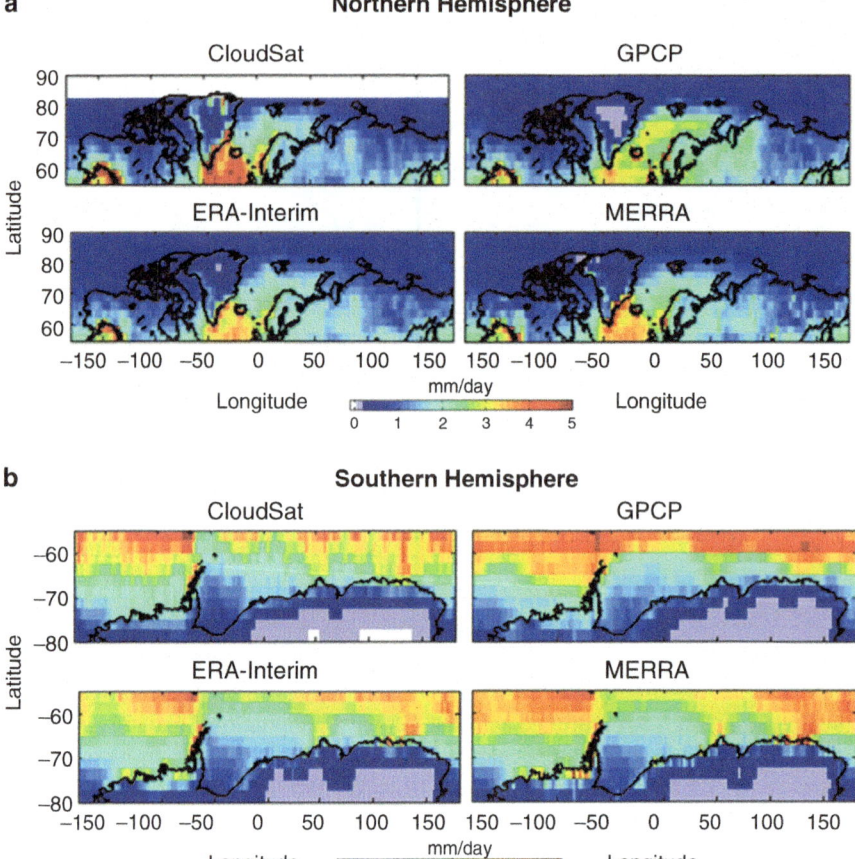

Fig. 45.4 Maps of precipitation mean intensity (2007–2009) from GPCP and a few other products poleward of latitude 55° over (**a**) Northern Hemisphere, and (**b**) Southern Hemisphere. (Adapted from Behrangi et al. 2016)

over the Atlantic Ocean. Further analyses showed that the AIRS/TOVS precipitation product used in GPCP captures the precipitation gradient around 70°N, but the issue seems to be related to the post-processing climatology adjustment utilized in GPCP. Furthermore, GPCP shows higher precipitation rates over Eurasia compared to other products (Fig. 45.4a). A similar analysis over the Southern Hemisphere showed that GPCP has a nearly continuous zonal band of intense precipitation over the ocean around 60°S, but this is not consistent with CloudSat and reanalysis observations. The zone near the 60°S band often hosts intense precipitation events that can saturate the CloudSat signal. This issue was mitigated by combining CloudSat and AMSR-E estimates in Behrangi et al. (2016), but the new capability offered by the GPM core observatory with 67°S-N coverage (versus 37°S-N in TRMM) can potentially help improve the analysis.

45.3 Utilizing GRACE to Assess Cold Region Precipitation

The GRACE mission has retrieved mass variations within the Earth with high accuracy (Tapley et al. 2004) since 2002. Upon its success, the GRACE follow on (GRACE-FO) mission was launched in May 2018 to continue the GRACE data record. Over land the hydrological cycle causes time-varying gravity changes and hence by combining GRACE Terrestrial Water Storage Anomaly (TWSA) estimates with independent data sets, terrestrial water storages such as snow, surface water, soil moisture and groundwater have been studied (e.g., Niu et al. 2007; Swenson and Wahr 2007; Swenson et al. 2008). Recent studies have shown that GRACE observations are valuable for precipitation estimation in cold regions (Seo et al. 2010; Behrangi et al. 2016, 2017).

During cold months, wide land areas at high latitudes experience air and surface temperature <0 °C and precipitation mainly falls as snow (Fig. 45.5). As in such conditions evapotranspiration and runoff are generally very small (Serreze et al. 2003), GRACE observations of TWSA are primarily due to the accumulated snowfall (Landerer et al. 2010). This enables gridded analysis of snow accumulation through construction of monthly and seasonal maps.

In cold regions, where satellite and ground measurements face the highest uncertainty, GRACE-based snowfall accumulation provides a completely independent technique, gravimetry versus radiometry with no need for empirical parameterizations. It also captures accumulated snowfall during cold-season and does not miss snowfall occurring between two satellite overpasses.

Figure 45.6a–c compares maps of snowfall accumulation estimates from GPCP, CloudSat and GRACE using datasets collected from three consecutive winters (2007–2010) and over grids that remain below 0 °C. The 3 years provide large

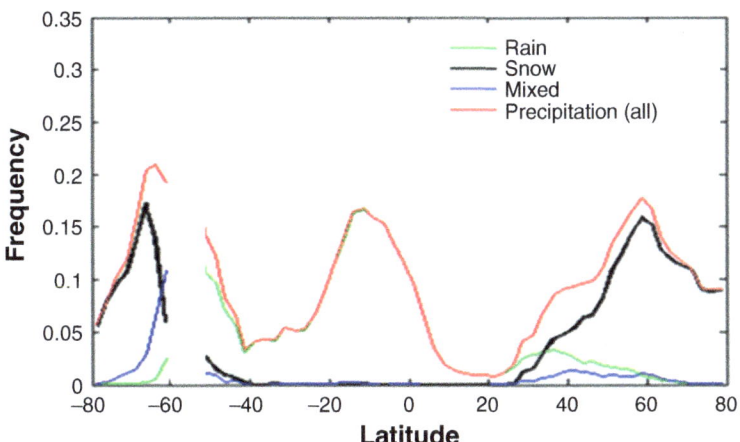

Fig. 45.5 Zonal distribution of precipitation frequency over land during boreal winter based on CloudSat

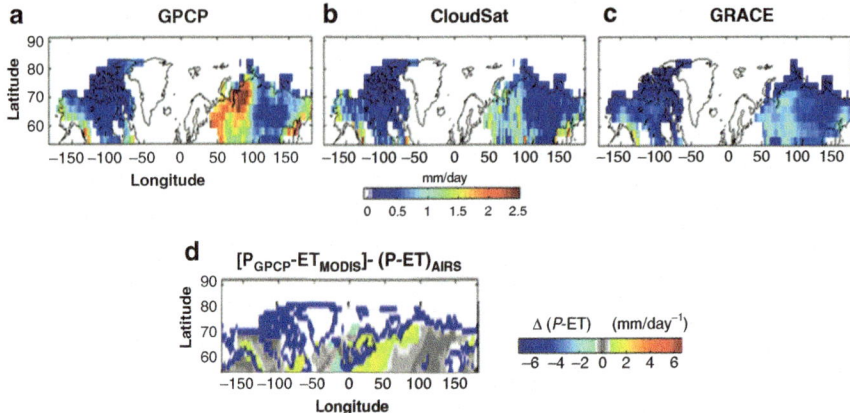

Fig. 45.6 Comparison of wintertime snowfall rate estimates from (**a**) GPCP, (**b**) CloudSat, and (**c**) GRACE. Only grids with T2 m < −1 °C across all days in winter are considered for this comparison. Panel (**d**) shows the difference between GPCP precipitation (P) minus MODIS evapotranspiration (ET) and P-ET calculated from water vapor convergence based on AIRS water vapor field and ERA wind vectors. (Adapted from Behrangi et al. 2016, 2014d, with modifications)

samples for stable grid analysis using CloudSat. GRACE does not have such sampling limitations. While GRACE and CloudSat are broadly consistent in producing the snowfall maps, the GPCP snowfall rates over Eurasia are almost twice those estimated by CloudSat and GRACE. This is consistent with previous findings of Behrangi et al. (2014d) who used water vapor convergence from reanalysis and AIRS to calculate seasonal precipitation accumulation (Fig. 45.6d). The various independent analyses suggest that GPCP is overestimating precipitation rate over Eurasia, likely related to GPCP's overcorrection for gauge undercatch issues.

GRACE, CloudSat, and the water vapor convergence approach using AIRS agree in showing that GPCP is likely overestimating precipitation accumulation over Eurasia. It is also possible to verify such findings via mass balance closure at basin scale through available streamflow observations. Figure 45.7b shows a multi-annual (2003–2010) monthly precipitation estimate from GRACE (using basin mass balance or water closure) and various precipitation products over the Ob river basin shown in Fig. 45.7a that drains into the Arctic ocean. The shaded area around the GRACE-based precipitation estimate represents uncertainty resulted from the spread among evapotranspiration (ET) products used to calculate precipitation rate from the water balance equation. GPCP applies correction factors, recommended by Legates and Willmott (1990), to GPCC for bias adjustment of satellite precipitation products (GPCC-L). Therefore, the similarity of GPCP to GPCC-L makes sense. Both GPCP and GPCC-L show larger precipitation over the Ob basin than that estimated from GRACE and GPCC-F (i.e., GPCC corrected by the method of Fuchs et al. 2001), in agreement with the results shown in Fig. 45.6.

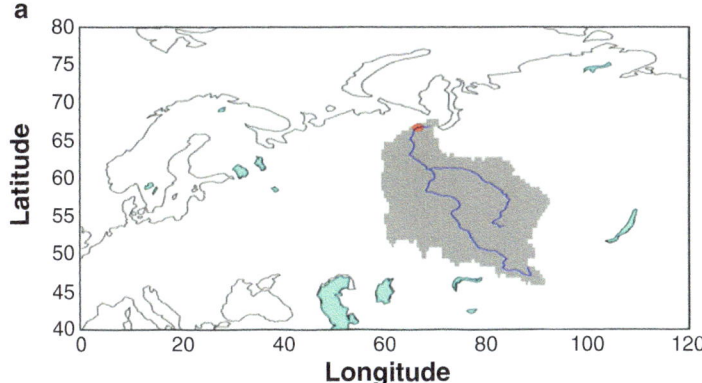

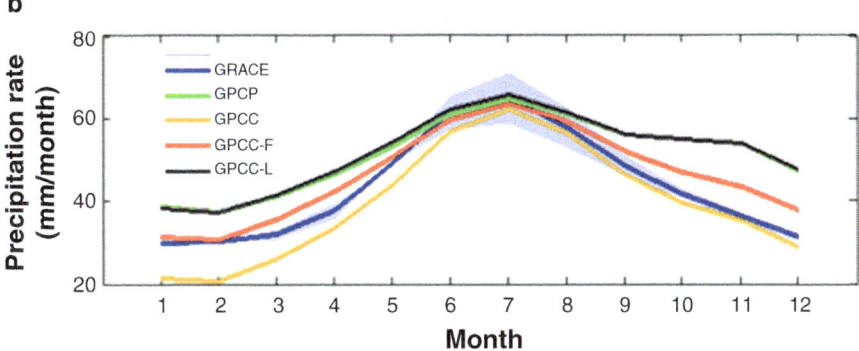

Fig. 45.7 (**a**) Location of the Ob basin and the streamflow gauge (the red dot), (**b**) Multi-annual (2003–2010) monthly precipitation estimate from GRACE and various precipitation products over the basin. Only regions above the outlet are used in calculating monthly average precipitation. GPCC has no gauge undercatch correction. GPCC-F and GPCC-L are GPCC corrected by Fuchs et al. (2001) and Legates and Willmott (1990) methods, respectively. GPCP uses Legates correction factor

While the mass balance approach at basin scale requires streamflow observations, streamflow observations may not be available in many regions. This is specially the case for mountainous and remote regions. However, one can eliminate this requirement over endorheic basins, defined as closed basin that retains water within the basin and allows no outflow. Behrangi et al. (2017) used this method to calculate the monthly precipitation accumulation over two large endorheic basins in the Tibetan Plateau. The location of these two basins are shown in Fig. 45.8a. Figure 45.8b shows the multi-annual (2003–2009) monthly precipitation estimate from GRACE and various precipitation products over Basin 1, with average elevation >5000 m above sea level.

The mass balance analysis over the Tibetan Plateau showed comparable results with other products in warm months, when satellite precipitation products are expected to perform reasonably. This adds credibility to the GRACE estimate over

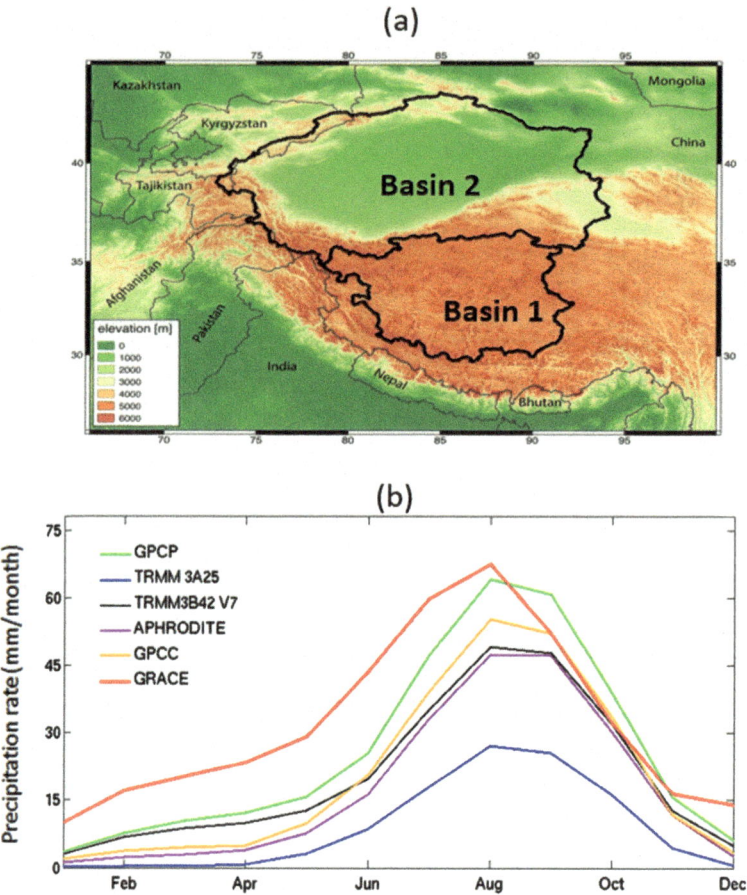

Fig. 45.8 (**a**) Location of two closed (not reaching ocean) basins plotted over a colormap of surface elevations with river networks and lakes shown in blue. (**b**) Multi-annual (2003–2009) monthly precipitation estimate from GRACE and various precipitation products over Basin 1. (Adapted from Behrangi et al. 2017 with modifications)

the region. However, during cold months all of the precipitation products show much less total precipitation accumulation than that estimated by GRACE, in agreement with our expectation for missed or underestimated precipitation amount by current precipitation products. GPCC also shows large underestimation as it is not corrected for gauge undercatch and is based on a handful of in situ observations, mainly located at the lower elevations. It is important to note that in calculation of mass balance over the Tibetan Plateau, sublimation from reanalysis is considered. While the sublimation amount is about an order of magnitude less than precipitation amount, refined estimates of sublimation can improve the results.

45.4   Concluding Remarks

A recent analysis of the latest GPM products at high latitudes has shown that the skill of GPM passive MW products has improved and now estimates are available over frozen surfaces. While light rain and snowfall continue to be underestimated, even if the precipitation pattern is well captured, other observational resources can help correct for bias. Currently, several merged precipitation products use gridded in situ data sets, such as GPCC, to reduce precipitation biases. However, at high latitudes and over mountainous regions, in-situ data are often sparse and erroneous; hence, there is a need for other observational sources to constrain total precipitation amount. Here we showed how CloudSat (Stephens et al. 2002) and GRACE (Tapley et al. 2004) can provide guidance to constrain precipitation amount over cold regions.

Besides CloudSat and GRACE, there are other emerging opportunities that can be utilized. Among these, the shipboard Ocean Rainfall And Ice-phase precipitation measurement Network (OceanRAIN), which is the only systematic long-term disdrometer-based shipboard precipitation data set over the global oceans (Klepp et al. 2018). There is also emerging opportunities via the WMO Solid Precipitation Intercomparison Experiment (SPICE) project (2013–2016) that developed a new method for bias correction of automatic measurements (Rasmussen et al. 2012; Kochendorfer et al. 2017) in various climate regimes.

With merged products (such as IMERG and CMORPH) extending their coverage to the poles, it is important to bring together and assess various observational alternatives that can help constrain precipitation amount at high latitudes and over cold regions. Reanalysis and regional models have also shown good skill in estimating precipitation amount and spatial pattern of precipitation at high latitudes. Investigating the precipitation patterns suggested by models and reanalysis is important to identify areas of major discrepancies between models and observations. Furthermore, in some areas there are known issues with satellite precipitation data, where model data may provide guidance for further analysis and diagnostic assessment of satellite products.

The recently launched GRACE-FO mission and the near future launch and operation of the Earth, Clouds, Aerosols, and Radiation Explorer mission (EarthCARE) (Illingworth et al. 2015) will provide continued observations similar to GRACE and CloudSat that help produce longer data record. Together with ongoing effort under GPM to advance Level 2 precipitation retrievals and their combination, and the improvement in accuracy and resolution of reanalysis and regional models, excellent opportunities are at hands to advance precipitation estimates at high latitudes and in cold regions.

References

Adler, R. F., Gu, G., & Huffman, G. J. (2012). Estimating climatological bias errors for the Global Precipitation Climatology Project (GPCP). *Journal of Applied Meteorology and Climatology, 51*, 84–99. https://doi.org/10.1175/JAMC-D-11-052.1.

Alley, R., Berntsen, T., Bindoff, N. L., Chen, Z., Chidthaisong, A., Friedlingstein, P., Gregory, J. M., Hegerl, G. C., Heimann, M., Hewitson, B., Hoskins, B. J., Joos, F., Jouzel, J., Kattsov, V., Lohmann, U., Manning, M., Matsuno, T., Molina, M., Nicholls, N., Overpeck, J., Qin, D., Raga, G., Ramaswamy, V., Ren, J., Rusticucci, M., Solomon, S., Somerville, R., Stocker, T. F., Stott, P. A., Stouffer, R. J., Whetton, P., Wood, R. A., & Wratt, D. (2007). Summary for policymakers. In S. Solomon et al. (Eds.), *Climate change 2007: The physical science basis* (Contribution of Working Group I to the 4th Assessment Report of the IPCC) (pp. 1–18). Cambridge University Press, Cambridge. Available at https: //www.ipcc.ch/pdf/assessment-report/ar4/wg1/ar4-wg1-spm.pdf. Last accessed 22 Nov. 2018.

Behrangi, A., Khakbaz, B., Jaw, T. C., AghaKouchak, A., Hsu, K.-L., & Sorooshian, S. (2011). Hydrologic evaluation of satellite precipitation products over a mid-size basin. *Journal of Hydrology, 397*, 225–237. https://doi.org/10.1016/j.jhydrol.2010.11.043.

Behrangi, A., Lebsock, M., Wong, S., & Lambrigtsen, B. (2012). On the quantification of oceanic rainfall using spaceborne sensors. *Journal of Geophysical Research, 117*, D20105. https://doi.org/10.1029/2012JD017979.

Behrangi, A., Tian, Y., Lambrigtsen, B. H., & Stephens, G. L. (2014a). What does CloudSat reveal about global land precipitation detection by other spaceborne sensors? *Water Resources Research, 50*, 4893–4905. https://doi.org/10.1002/2013WR014566.

Behrangi, A., Stephens, G. L., Adler, R. F., Huffman, G. J., Lambrigtsen, B. H., & Lebsock, M. (2014b). An update on the oceanic precipitation rate and its zonal distribution in light of advanced observations from space. *Journal of Climate, 27*, 3957–3965. https://doi.org/10.1175/JCLI-D-13-00679.1.

Behrangi, A., Andreadis, K., Fisher, J. B., Turk, F. J., Granger, S., Painter, T., & Das, N. (2014c). Satellite-based precipitation estimation and its application for streamflow prediction over mountainous Western U.S. Basins. *Journal of Applied Meteorology and Climatology, 53*, 2823–2842. https://doi.org/10.1175/JAMC-D-14-0056.1.

Behrangi, A., Wong, S., Mallick, K., & Fisher, J. B. (2014d). On the net surface water exchange rate estimated from remote-sensing observation and reanalysis. *International Journal of Remote Sensing, 35*, 2170–2185. https://doi.org/10.1080/01431161.2014.889866.

Behrangi, A., Christensen, M., Richardson, M., Lebsock, M., Stephens, G. L., Huffman, G. J., Bolvin, D. T., Adler, R. F., Gardner, A., Lambrigtsen, B. H., & Fetzer, E. (2016). Status of high-latitude precipitation estimates from observations and reanalyses. *Journal of Geophysical Research, 121*, 4468–4486. https://doi.org/10.1002/2015JD024546.

Behrangi, A., Gardner, A. S., Reager, J. T., & Fisher, J. B. (2017). Using GRACE to constrain precipitation amount over cold mountainous basins. *Geophysical Research Letters, 44*, 219–227. https://doi.org/10.1002/2016GL071832.

Behrangi, A., Bormann, K. J., & Painter, T. H. (2018). Using the airborne snow observatory to assess remotely sensed snowfall products in the California Sierra Nevada. *Water Resources Research, (10)*, 7331–7346. https://doi.org/10.1029/2018WR023108.

Berg, W., L'Ecuyer, T. S., & Kummerow, C. D. (2006). Rainfall climate regimes: The relationship of regional TRMM rainfall biases to the environment. *Journal of Applied Meteorology and Climatology, 45*, 434–454. https://doi.org/10.1175/JAM2331.1.

Berg, W., L'Ecuyer, T. S., & Haynes, J. M. (2010). The distribution of rainfall over oceans from spaceborne radars. *Journal of Applied Meteorology and Climatology, 49*, 535–543. https://doi.org/10.1175/2009JAMC2330.1.

Chen, S., Hong, Y., Kulie, M., Behrangi, A., Stepanian, P. M., Cao, Q., You, Y., Zhang, J., Huk, J., & Zhang, X. (2016). Comparison of snowfall estimates from the NASA CloudSat Cloud

Profiling Radar and NOAA/NSSL multi-radar multi-sensor system. *Journal of Hydrology, 541*, 862–872. https://doi.org/10.1016/j.jhydrol.2016.07.047.

Ellis, T. D., L'Ecuyer, T. S., Haynes, J. M., & Stephens, G. L. (2009). How often does it rain over the global oceans? The perspective from CloudSat. *Geophysical Research Letters, 36*, L03815. https://doi.org/10.1029/2008GL036728.

Ferraro, R. R., Peters-Lidard, C. D., Hernandez, C., Turk, F. J., Aires, F., Prigent, C., Lin, X., Boukabara, S.-A., Furuzawa, F. A., Gopalan, K., Harrison, K. W., Karbou, F., Li, L., Liu, C., Masunaga, H., Moy, L., Ringerud, S., Skofronick-Jackson, G. M., Tian, Y., & Wang, N.-Y. (2013). An evaluation of microwave land surface emissivities over the Continental United States to benefit GPM-era precipitation algorithms. *IEEE Transactions on Geoscience and Remote Sensing, 51*, 378–398. https://doi.org/10.1109/TGRS.2012.2199121.

Fuchs, T., Rapp, J., Rubel, F., & Rudolf, B. (2001). Correction of synoptic precipitation observations due to systematic measuring errors with special regard to precipitation phases. *Physics and Chemistry of the Earth, Part B, 26*, 689–693. https://doi.org/10.1016/S1464-1909(01)00070-3.

Goodison, B. E., Louie, P. Y. T., & Yang, D. (1998). *WMO solid precipitation measurement intercomparison* (Report 67, 212 pp). Geneva: WMO. Available at http://www.wmo.int/pages/prog/www/reports/WMOtd872.pdf. Last accessed 22 Nov. 2018.

Haynes, J. M., L'Ecuyer, T. S., Stephens, G. L., Miller, S. D., Mitrescu, C., Wood, N. B., & Tanelli, S. (2009). Rainfall retrieval over the ocean with spaceborne W-band radar. *Journal of Geophysical Research, 114*(D8). https://doi.org/10.1029/2008JD009973.

Huffman, G. J., Adler, R. F., Bolvin, D. T., Gu, G. J., Nelkin, E. J., Bowman, K. P., Hong, Y., Stocker, E. F., & Wolff, D. B. (2007). The TRMM multisatellite precipitation analysis (TMPA): Quasi-global, multiyear, combined-sensor precipitation estimates at fine scales. *Journal of Hydrometeorology, 8*, 38–55. https://doi.org/10.1175/JHM560.1.

Huffman, G. J., Bolvin, D., Braithwaite, D., Hsu, K.-L., Joyce, R., Kidd, C., Nelkin, E. J., Sorooshian, S., Tan, J., & Xie, P. (2018). *Integrated Multi-satellitE Retrievals for GPM (IMERG), ATBD, V. 5.2*, 31 pp. Available at https://pmm.nasa.gov/sites/default/files/document_files/IMERG_ATBD_V5.2.pdf. Last accessed 22 Nov 2018.

Illingworth, A. J., Barker, H. W., Beljaars, A., Ceccaldi, M., Chepfer, H., Clerbaux, N., Cole, J., Delanoë, J., Domenech, C., Donovan, D. P., Fukuda, S., Hirakata, M., Hogan, R. J., Huenerbein, A., Kollias, P., Kubota, T., Nakajima, T., Nakajima, T. Y., Nishizawa, T., Ohno, Y., Okamoto, H., Oki, R., Sato, K., Satoh, M., Shephard, M. W., Velázquez-Blázquez, A., Wandinger, U., Wehr, T., & van Zadelhoff, G.-J. (2015). The EarthCARE satellite: The next step forward in global measurements of clouds, aerosols, precipitation, and radiation. *Bulletin of the American Meteorological Society, 96*, 1311–1332. https://doi.org/10.1175/BAMS-D-12-00227.1.

Klepp, C., Michel, S., Protat, A., Burdanowitz, J., Albern, N., Dahl, A., Kähnert, M., Louf, V., Bakan, S., & Buehler, S. A. (2018). OceanRAIN, a new in-situ shipboard global ocean surface-reference dataset of all water cycle components. *Scientific Data, 5*, 180122. https://doi.org/10.1038/sdata.2018.122.

Kochendorfer, J., Nitu, R., Wolff, M., Mekis, E., Rasmussen, R. M., Baker, B., Earle, M. E., Reverdin, A., Wong, K., Smith, C. D., Yang, D., Roulet, Y.-A., Buisan, S., Laine, T., Lee, G., Aceituno, J. L. C., Alastrué, J., Isaksen, K., Meyers, T., Brækkan, R., Landolt, S., Jachcik, A., & Poikonen, A. (2017). Errors and adjustments for single-Alter shielded and unshielded weighing gauge precipitation measurements from WMO-SPICE. *Hydrology and Earth System Sciences, 21*, 3525–3542. https://doi.org/10.5194/hess-21-3525-2017.

Landerer, F. W., Dickey, J. O., & Güntner, A. (2010). Terrestrial water budget of the Eurasian pan-Arctic from GRACE satellite measurements during 2003–2009. *Journal of Geophysical Research, 115*, D23115. https://doi.org/10.1029/2010JD014584.

Lebsock, M. D., & L'Ecuyer, T. S. (2011). The retrieval of warm rain from CloudSat. *Journal of Geophysical Research, 116*, D20209. https://doi.org/10.1029/2011JD016076.

Legates, D. R., & Willmott, C. J. (1990). Mean seasonal and spatial variability in gauge-corrected, global precipitation. *International Journal of Climatology, 10,* 111–127. https://doi.org/10. 1002/joc.3370100202.

Lin, X., & Hou, A. Y. (2012). Estimation of rain intensity spectra over the Continental United States using ground radar–gauge measurements. *Journal of Climate, 25,* 1901–1915. https://doi.org/ 10.1175/JCLI-D-11-00151.1.

McClelland, J. W., Déry, S. J., Peterson, B. J., Holmes, R. M., & Wood, E. F. (2006). A Pan-Arctic evaluation of changes in river discharge during the latter half of the 20th century. *Geophysical Research Letters, 33,* L06715. https://doi.org/10.1029/2006GL025753.

Meier, M. F., & Carter, C. L. (2006). Observational evidence of increases in freshwater inflow to the Arctic Ocean. *Arctic, Antarctic, and Alpine Research, 36,* 117–122. https://doi.org/10.1657/ 1523-0430(2004)036[0117:OEOIIF]2.0.CO;2.

Niu, G.-Y., Seo, K.-W., Yang, Z.-L., Wilson, C., Su, H., Chen, J., & Rodell, M. (2007). Retrieving snow mass from GRACE terrestrial water storage change with a land surface model. *Geophysical Research Letters, 34,* L15704. https://doi.org/10.1029/2007GL030413.

Norouzi, H., Temimi, M., Prigent, C., Turk, F. J., Khanbilvardi, R., Tian, Y., Furuzawa, F., & Masunaga, H. (2015). Assessment of the consistency among global microwave land surface emissivity products. *Atmospheric Measurement Techniques, 8,* 1197–1205. https://doi.org/10. 5194/amt-8-1197-2015.

Rasmussen, R., Baker, B., Kochendorfer, J., Meyers, T., Landolt, S., Fischer, A. P., Black, J., Thériault, J. M., Kucera, P., Gochis, D., Smith, C., Nitu, R., Hall, M., Ikeda, K., & Gutmann, E. (2012). How well are we measuring snow: The NOAA/FAA/NCAR winter precipitation test bed. *Bulletin of the American Meteorological Society, 93,* 811–829. https://doi.org/10.1175/ BAMS-D-11-00052.1.

Schaake, J., Henkel, A., & Cong, S. (2004). Application of PRISM climatologies for hydrologic modeling and forecasting in the western U.S. In *18th conference on hydrology,* American Meteor Society, 5.3, Seattle, WA. Available at https: //ams.confex.com/ams/84Annual/ techprogram/paper_72159.htm. Last accessed 22 Nov 2018.

Seo, K.-W., Ryu, D., Kim, B.-M., Waliser, D. E., Tian, B., & Eom, J. (2010). GRACE and AMSR-E-based estimates of winter season solid precipitation accumulation in the Arctic drainage region. *Journal of Geophysical Research, 115,* D20117. https://doi.org/10.1029/ 2009JD013504.

Serreze, M. C., Bromwich, D. H., Clark, M. P., Etringer, A. J., Zhang, T., & Lammers, R. (2003). Large-scale hydro-climatology of the terrestrial Arctic drainage system. *Journal of Geophysical Research, 108,* 8160. https://doi.org/10.1029/2001JD000919.

Skofronick-Jackson, G., Petersen, W. A., Berg, W., Kidd, C., Stocker, E. F., Kirschbaum, D. B., Kakar, R., Braun, S. A., Huffman, G. J., Iguchi, T., Kirstetter, P.-E., Kummerow, C., Meneghini, R., Oki, R., Olson, W. S., Takayabu, Y. N., Furukawa, K., & Wilheit, T. (2017). The Global Precipitation Measurement (GPM) mission for science and society. *Bulletin of the American Meteorological Society, 98,* 1657–1672. https://doi.org/10.1175/BAMS-D-15-00306.1.

Smalley, M., L'Ecuyer, T., Lebsock, M., & Haynes, J. (2014). A comparison of precipitation occurrence from the NCEP Stage IV QPE product and the CloudSat Cloud Profiling Radar. *Journal of Hydrometeorology, 15,* 444–458. https://doi.org/10.1175/JHM-D-13-048.1.

Smith, L. C., Sheng, Y., MacDonald, G. M., & Hinzman, L. D. (2005). Disappearing Arctic lakes. *Science, 308*(5727), 1429. https://doi.org/10.1126/science.1108142.

Solomon, S., Qin, D., Manning, M., Alley, R. B., Berntsen, T., Bindoff, N. L., Chen, Z., Chidthaisong, A., Gregory, J. M., Hegerl, G. C., Heimann, M., Hewitson, B., Hoskins, B. J., Joos, F., Jouzel, J., Kattsov, V., Lohmann, U., Matsuno, T., Molina, M., Nicholls, N., Overpeck, J., Raga, G., Ramaswamy, V., Ren, J., Rusticucci, M., Somerville, R., Stocker, T. F., Whetton, P., Wood, R. A., & Wratt, D. (2007). Technical summary. In S. Solomon et al. (Eds.), *Climate change 2007: The physical science basis* (Contribution of Working Group I to the Fourth Assessment Report of the Intergovernmental Panel on Climate Change) (pp. 20–91).

Cambridge: Cambridge University Press. Available at https: //www.ipcc.ch/pdf/assessment-report/ar4/wg1/ar4-wg1-ts.pdf. Last accessed 22 Nov. 2018.

Stephens, G. L., L, G., Vane, D. G., Boain, R. J., Mace, G. G., Sassen, K., Wang, Z., Illingworth, A. J., J, E., O'Connor, W. B., Rossow, S. L., Durden, S. D., Miller, R. T., Austin, A., Benedetti, C. M., & The CloudSat Science Team. (2002). The CloudSat mission and the a-train – A new dimension of space-based observations of clouds and precipitation. *Bulletin of the American Meteorological Society, 83*, 1771–1790. https://doi.org/10.1175/BAMS-83-12-1771.

Stephens, D. G., Vane, S., Tanelli, S., Im, E., Durden, S., Rokey, M., Reinke, D., Partain, P., Mace, G. G., Austin, R., L'Ecuyer, T., Haynes, J., Lebsock, M., Suzuki, K., Waliser, D., Wu, D., Kay, J., Gettelman, A., Wang, Z., & Marchand, R. (2008). CloudSat mission: Performance and early science after the first year of operation. *Journal of Geophysical Research, 113*, D00A18. https://doi.org/10.1029/2008JD009982.

Swenson, S., & Wahr, J. (2007). Multi-sensor analysis of water storage variations of the Caspian Sea. *Geophysical Research Letters, 34*, L16401. https://doi.org/10.1029/2007GL030733.

Swenson, S., Famiglietti, J., Basara, J., & Wahr, J. C. W. (2008). Estimating profile soil moisture and groundwater variations using GRACE and Oklahoma Mesonet soil moisture data. *Water Resources Research, 44*. https://doi.org/10.1029/2007WR006057.

Tapley, B. D., Bettadpur, S., Ries, J. C., Thompson, P. F., & Watkins, M. M. (2004). GRACE measurements of mass variability in the Earth system. *Science, 305*, 503–505. https://doi.org/10.1126/science.1099192.

Wen, Y., Behrangi, A., Chen, H., & Lambrigtsen, B. (2018). How well the early 2017 California atmospheric river precipitation events were captured by satellite products and ground-based radars? *Quarterly Journal of the Royal Meteorological Society, 144*(S1), 344–359. https://doi.org/10.1002/qj.3253.

Yang, D., Kane, D., Hinzman, K. D., Zhang, X., Zhang, T., & Ye, H. (2002). Siberian Lena river hydrological regime and recent change. *Journal of Geophysical Research, 107*(D23), 4694. https://doi.org/10.1029/2002JD002542.

Chapter 46
Latent Heating Retrievals from Satellite Observations

Yukari N. Takayabu and Wei-Kuo Tao

Abstract With the launch of the Tropical Rainfall Measuring Mission satellite, equipped with high-functioning sensors to measure precipitation, four-dimensional latent heating retrievals from satellite observation have become available. With the launch of the Global Precipitation Measurement mission Core Observatory, the latent heating retrievals are now available at higher latitudes. Algorithms of two standard NASA/JAXA products, the Spectral Latent Heating, and the Convective Stratiform Heating and some results are introduced.

Keywords Precipitation · Latent heating · Spectral latent heating · Convective stratiform heating · TRMM · PR · GPM · DPR · GMI · NASA · Goddard Cumulus Ensemble model · GATE · TOGA-COARE · JMA · Local forecast model · NU-WRF · WRF-ARW

46.1 Introduction

Convective latent heating plays an essential role for tropical convection, generating mesoscale circulations (e.g., Houze et al. 1989). Moreover, vertical profiles of diabatic heating, to which convective latent heating contributes largely, affects the general circulation of the atmosphere (e.g., Hartmann et al. 1984). For mid-latitude systems, latent heating also contributes to generations of potential vorticities (PVs) resulting in, for example, explosive intensifications of mid-latitude storms (Boettcher and Wernli 2011).

Three-dimensional (3D) observations of precipitation from space started with the Tropical Rainfall Measuring Mission (TRMM) Precipitation Radar (PR) in 1997

Y. N. Takayabu (✉)
Atmosphere and Ocean Research Institute, The University of Tokyo, Chiba, Japan
e-mail: yukari@aori.u-tokyo.ac.jp

W.-K. Tao
NASA/Goddard Space Flight Center, Greenbelt, MD, USA

© Springer Nature Switzerland AG 2020
V. Levizzani et al. (eds.), *Satellite Precipitation Measurement*, Advances in Global Change Research 69, https://doi.org/10.1007/978-3-030-35798-6_22

paving the way to obtain 3D latent heating (LH) retrievals. The latent heating retrieval was actually one of the important objectives of TRMM (Simpson et al. 1988). However, what the PR can obtain is the information from precipitation particle size and density. Therefore, we cannot capture the direct instances of release of latent heating associated with phase changes of water, e.g. condensation/evaporation, freezing/melting, etc., directly from the PR observations. Still, efforts have been put to estimate the 3D latent heating distribution utilizing the radar observations from space as a very powerful constraint, with some aids of numerical cloud resolving models. In this paper, we describe how two latent heating products from JAXA and NASA, namely the Spectral Latent Heating (SLH) and the Convective and Stratiform Heating (CSH), respectively, are generated.

46.2   The Spectral Latent Heating (SLH) Algorithm

Late Professor Tsuyoshi Nitta was the TRMM Project Scientist in Japan at the timing of the successful launch of the satellite on 27 November 1997. As a tropical meteorologist, Nitta wished to obtain the latent heating profiles as a scientific breakthrough. Prof. Nitta used to mention to Y. N. Takayabu that we should utilize the 3D precipitation observation from the TRMM PR to obtain the convective latent heating profiles. To realize Nitta's wish, after some early analyses of 3D precipitation products obtained from the TRMM PR (e.g., Takayabu 2002), Y. N. Takayabu designed an algorithm to estimate the latent heating utilizing radar-observed precipitation profiles as constraints with an aid of a numerical cloud resolving model. Then, with the participation of S. Shige who was just visiting W.-K. Tao to study the Goddard Cumulus Ensemble Model (GCEM), the SLH algorithm was constructed from the Tropical Ocean and Global Atmosphere-Coupled Ocean-Atmosphere Response Experiments (TOGA-COARE) simulation utilizing the GCEM (Shige et al. 2004, 2007, etc.) for tropical and subtropical TRMM observation region (36°N-36°S).

With the launch of the GPM Core satellite on 28 February, 2014, the observation region of the space-borne precipitation radar extended to 65°N-65°S. The Core satellite carries the dual-frequency precipitation radar (DPR) and the GPM microwave radiometer (GMI). Since the precipitation systems are very different between the tropics and the mid latitudes, a new module was added to the SLH algorithm for mid-latitudes to make it applicable to the entire GPM observation region.

For the GPM SLH algorithm, we retrieve latent heating variables utilizing two separate algorithms for tropics and for mid-latitudes. To apply these algorithms, the location of each GPM KuPR pixel is assigned to either tropics or mid-latitudes, depending on monthly maps of precipitation regimes determined in a similar manner as described in Takayabu (2008). Then, we retrieve 3D convective latent heating, Q1-QR, and Q2, applying either tropical/mid-latitude algorithms to precipitation data observed from GPM DPR (KuPR). Here, Q1 and Q2 are apparent heat source and apparent moisture sink (Yanai et al. 1973), respectively, and QR is radiative heating of the atmosphere. In the following subsections, we briefly describe the SLH algorithms for tropical precipitation and for mid-latitude precipitation.

46.2.1 Latent Heating Retrieval for the Tropical Region

With the TRMM PR, we first developed the SLH algorithm for the tropics as summarized in Fig. 46.1. First, utilizing the GCEM, we simulated the TOGA-COARE precipitation, from which we made three spectral look-up tables (LUTs) of latent heating profiles in terms of precipitation top height (PTH) for convective and shallow stratiform rain, and melting level precipitation (MLP) rate for deep stratiform rain. Utilizing TRMM PR precipitation profiles and these tables, we estimated the latent heating associated with the precipitation.

Table 46.1 shows a brief summary of the evolution of the SLH algorithm (Shige et al. 2004, 2007, 2008, 2009). Deep stratiform rain is now further divided into two new categories: deep stratiform with decreasing precipitation from the melting level toward the surface (downward decreasing: DD), and deep stratiform with increasing precipitation from the melting level toward the surface (downward increasing: DI). The SLH algorithm computes deep stratiform cooling magnitudes as a function of Pm (melting level) − Ps (surface rain rate), assuming the evaporative cooling rate below the melting level in deep stratiform regions is proportional to the reduction in the precipitation profile toward the surface from the melting level (based on 1D water substance conservation). However, increasing precipitation profiles are found in some portions of stratiform regions, especially in regions adjacent to convective regions where 1D water substance conservation may be invalid. A LUT for deep stratiform with increasing precipitation toward the surface from the melting level (called "intermediate") is produced with the amplitude determined by Ps.

Utilizing these TRMM SLH, we found the existence of a clear congestus regime with relatively high sea surface temperature with suppressed deep convection associated with large-scale atmospheric subsidence (Fig. 46.2, Takayabu et al. 2010). Comparing the Q1-QR distributions at 7.5 and at 2.0 km in Fig. 46.3, several regions of congestus regime are noticed; around Hawaii in the north Pacific, to the east of the South Pacific Convergence Zone (SPCZ) along the equator, over the north and south Atlantic to the east of continents, and over the western Indian Ocean. In particular, it is interesting to note that while major deep heating shows SPCZ bending southward, the lower-level congestus heating distribution looks like the "double ITCZ" often found in numerical models. This finding was connected to unravel the double ITCZ problem that climate model developers have long been plagued by (Hirota et al. 2011).

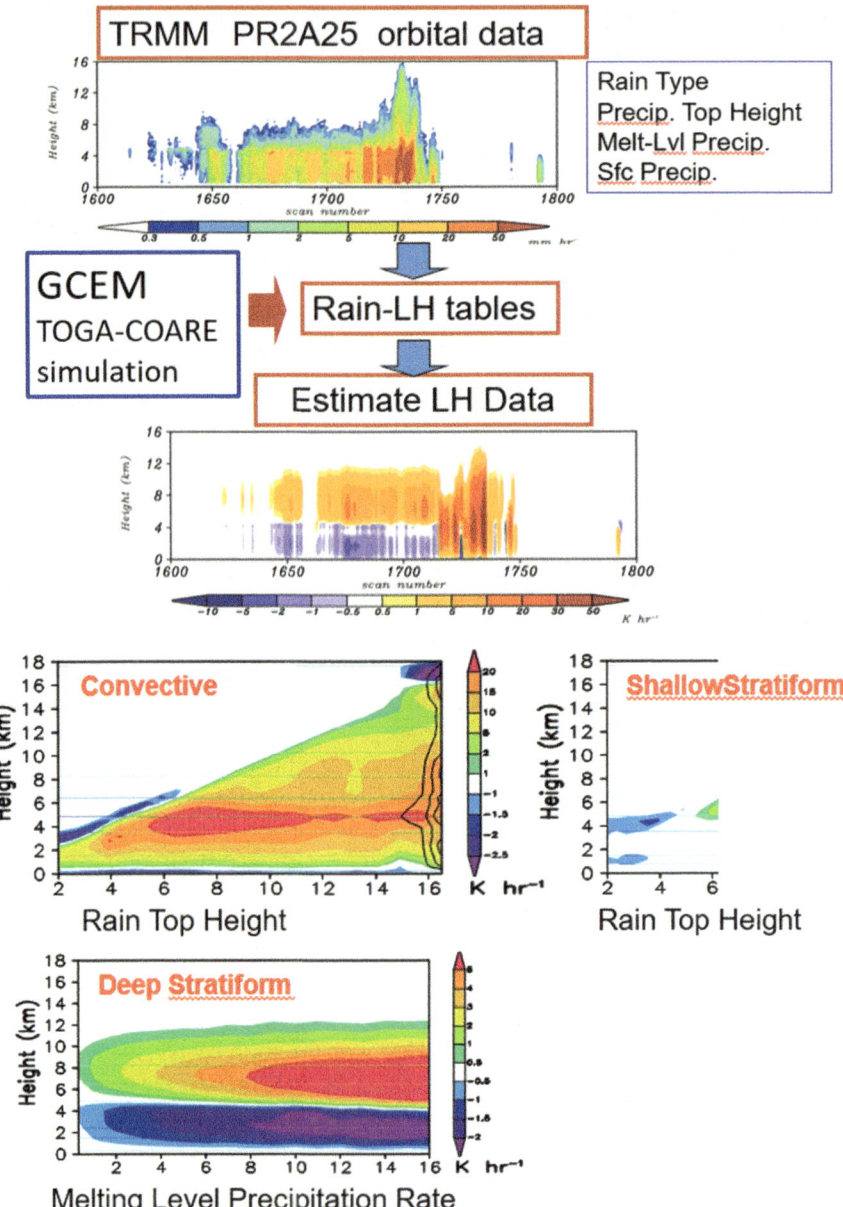

Fig. 46.1 Summary of the SLH for the tropics. Top panel shows the flow chart of the algorithm. Rain-LH lookup tables are constructed from the TOGA-COARE simulation with the GCEM. With the LUT, latent heating variables are estimated from the TRMM PR2A25 precipitation data. The bottom three panels show rain-LH spectral LUTs for convective, deep stratiform, and shallow stratiform precipitation

Table 46.1 Evolution of the SLH algorithm with a brief summary of references

Highlights	References
First paper on the SLH algorithm – index the LUTs according to vertical rain profile information: precipitation top height (PTH) for convective and shallow stratiform rain and melting-level rain intensity for anvil (deep stratiform with a PTH higher than the melting level).	Shige et al. (2004)
Improved SLH by separating convective heating retrieval into upper-level heating due to ice processes and lower-level heating due to liquid water processes, and shifted up or down the stratiform heating profile by matching the melting level in the LUT with the observed one. Good agreement between SLH estimates and diagnostic calculations for SCSMEX.	Shige et al. (2007)
Expanded SLH to the retrieval of Q_2. Good agreement between SLH estimates and diagnostic calculations for SCSMEX. Differences of SLH Q_2 estimates between the western Pacific Ocean and the Atlantic Ocean consistent with the results from the budget study.	Shige et al. (2008)
Comparisons of the LUTs from two- and three-dimensional CRM simulations. Less agreement between SLH estimates using the LUTs from three-dimensional CRM simulations and diagnostic calculations for SCSMEX using the LUTs from two-dimensional CRM simulations.	Shige et al. (2009)

Fig. 46.2 SLH-estimated Q1-QR profiles 30°N-30°S at all longitudes over ocean, stratified with pressure velocity at 500 hPa. Congestus regime is found in the large-scale subsidence region with Q1R peaking around 2 km. (Adapted from Takayabu et al. 2010)

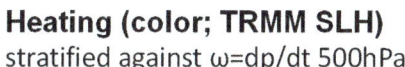

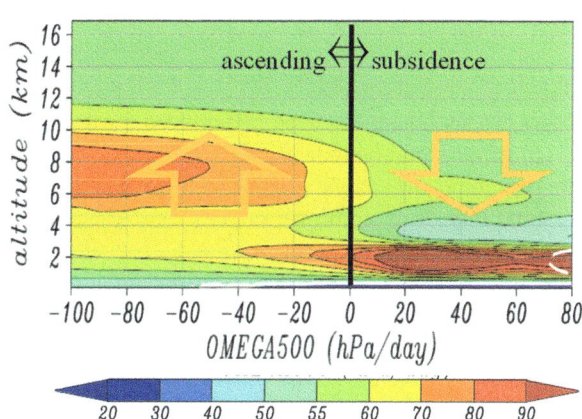

46.2.2 SLH Algorithm for Mid Latitudes

46.2.2.1 Construction of LUTs

For the GPM-SLH, we modified the SLH algorithm to be applicable also to mid-latitude precipitation systems. To this aim, separate LUTs are constructed for the mid-latitude precipitation using the Japan Meteorological Agency (JMA)'s high resolution (horizontally 2 km) local forecast model (LFM). With the collaboration of

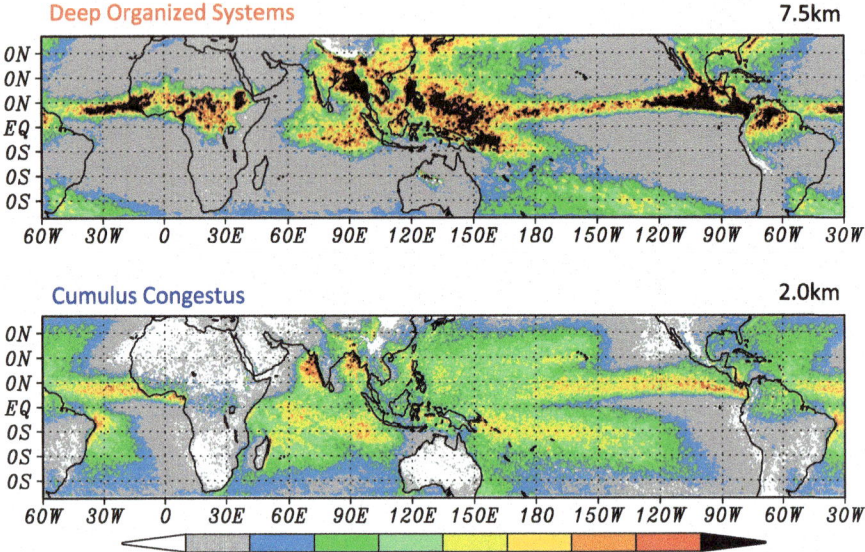

Fig. 46.3 SLH Q1-QR distributions at 7.5 and at 2.0 km averaged for JJA 1998–2007. (Adapted from Takayabu et al. 2010)

JMA's numerical forecast group, 3 and 4 h forecast data for 8 extratropical cyclone cases (16 shots) are collected and utilized for the LUTs.

In treating mid-latitude systems, in contrast to the tropical systems, we first have to overcome the "cloud base problem". For the tropical systems, we basically assume that deep stratiform rainfalls are associated with a mesoscale convective systems (MCS), and their cloud base coincides with the freezing level around 4.5 km. For mid latitude stratiform precipitation, however, this assumption does not hold, and we have to determine the cloud base levels, which can be either above the freezing level or below the freezing level, from the precipitation radar data.

LH is related to various phase changes of water. Figure 46.4 shows the LH profiles associated with various phase change processes. The left panel shows an example where the cloud base is below the freezing level. In this case, melting and condensation occur around similar levels below 0 °C, so that they cancel each other. On the other hand, when cloud base is above the freezing level (right panel in the same figure), evaporation and melting occur together and result in a large cooling. Therefore, the cloud base level information is necessary to retrieve the latent heating. Since we only observe radar reflectivity profiles from precipitation with GPM's DPR, we obtained the cloud base information from the precipitation profiles.

Assuming that precipitation evaporates once it abandons the cloud base, precipitation maximum level (Pmax) can represent the cloud base. When we draw a scatter plot of the precipitation maximum heights against cloud base heights for three types

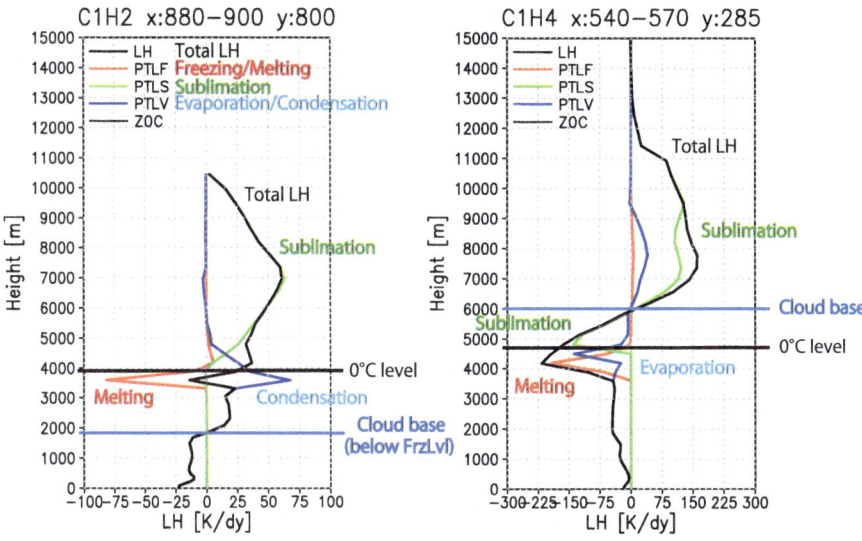

Fig. 46.4 Examples of latent heating profiles associated with cloud processes for cloud base level below the freezing level (left) and above the freezing level (right). Latent heating data are obtained from different grids of the LFM simulation of an extratropical cyclone

of precipitation (Downward Decreasing-DD stratiform precipitation from the 0 °C level, Downward Increasing-DI stratiform precipitation, and others), we can confirm that the assumption is valid (not shown).

Finally, six look-up tables (convective, shallow stratiform, 3 types of deep stratiform, and other) are constructed for LH, Q1-QR (Q1R), and Q2, respectively. Table 46.2 shows some details of the LUTs. Convective and shallow stratiform LH profiles are derived from the precipitation top heights. For the deep stratiform and other precipitation, maximum precipitation (Pmax) is chosen as the reference. In these LUTs, LH profiles are given for the standardized altitude with zero at the Pmax level, and +1.0 and −1.0 corresponding to the precipitation top height (PTH) and the precipitation bottom height (PBH), respectively.

Utilizing the 6 LUTs and precipitation profile data from the DPR Ku-band radar (KuPR) data, profiles of LH, Q1-QR and Q2 are retrieved. Precipitation type (convective/ stratiform/ other), PTH, PBH, Pmax, and Pnsfc are input values.

46.2.2.2   SLH Retrieval from GPM KuPR: Consistency Check and Performance

Figure 46.5 shows an example where GPM KuPR is utilized for the retrieval. The left panels show precipitation and precipitation types from GPM KuPR observed around 1800 UTC 16 April 2016. Note that there is no guarantee that the simulation corresponds perfectly to the observation along the satellite orbit. However, from the figure, we can confirm that the performance of the LFM simulation for this case is excellent.

Table 46.2 LUTs constructed from 8 extratropical cyclone cases simulated using the JMA LFM

LUT type	Reference variables [range, unit]	Retrieved variables [range, unit]	Profiles [range, unit]
Convective	PTH [0–12, km] Pnsfc [mm h^{-1}]	LH, Q1-QR, Q2 [−2.5 to 35, K h^{-1}]	Height [0–12, km]
Shallow Stratiform	PTH [0–3, km] Pnsfc [mm h^{-1}]	LH, Q1-QR, Q2 [−1.5 to 15, K h^{-1}]	Height [0–5, km]
Stratiform DD	Pmax,Pnsfc [mm h^{-1}] PTH, PBH [km]	LH, Q1-QR, Q2 [−2.5 to 15, K h^{-1}]	Pmax-relative standardized height [−2 to 2, nodim]
Stratiform DI	Pmax,Pnsfc [mm h^{-1}] PTH, PBH [km]	LH, Q1-QR, Q2 [−2.5 to 15, K h^{-1}]	Pmax-relative standardized height [−2 to 2, nodim]
Stratiform SubZero	Pmax,Pnsfc [mm h^{-1}] PTH, PBH [km]	LH, Q1-QR, Q2 [−2.5 to 15, K h^{-1}]	Pmax-relative standardized height [−2 to 2, nodim]
Other	Pmax,Pnsfc [mm h^{-1}] PTH, PBH [km]	LH, Q1-QR, Q2 [−2.5 to 15, K h^{-1}]	Pmax-relative standardized height [−2 to 2, nodim]

PTH and PBH stand for the precipitation top height and precipitation bottom height, respectively.
Pmax and Pnsfc are maximum precipitation and near-surface precipitation, respectively

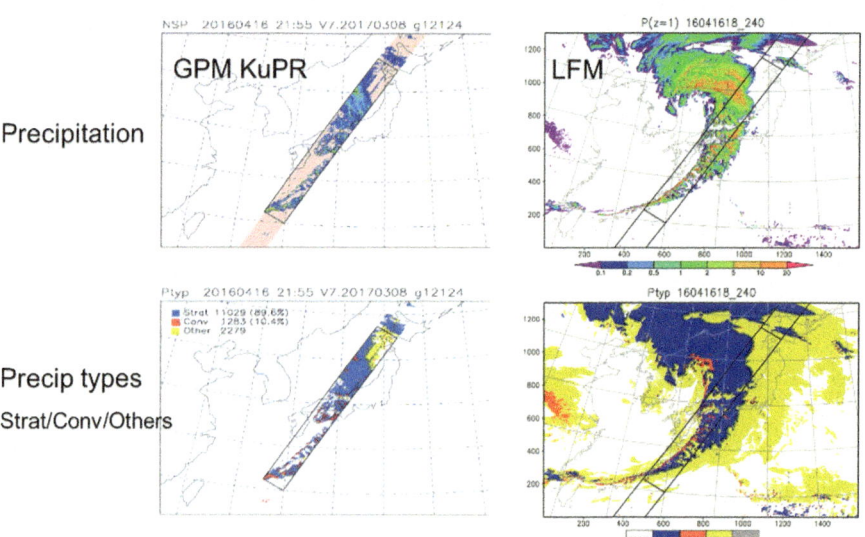

Fig. 46.5 Upper panels: GPM-observed precipitation and LFM-simulated precipitation at 1800
UTC 16 Apr. 2016 (4 h F). Lower panels: precipitation types

Then, the simulated and retrieved LH along the GPM orbit are compared in Fig. 46.6. Left panels show LFM-simulated precipitation and LH, middle panels show LFM precipitation and retrieved LH from LFM precipitation, and right panels show GPM-observed precipitation and retrieved LH with the GPM precipitation. Since precipitation simulation is confirmed to be very good, we can compare these cross sections. The retrieval looks quite successful. Average profiles of LH are compared also for eight cases utilized to construct the LUTs and their good consistency are confirmed (not shown).

Figure 46.7 shows the three-dimensional view of LH retrieved with GPM DPR for the above case. Northward (to the right) from the center, we can see an ascending

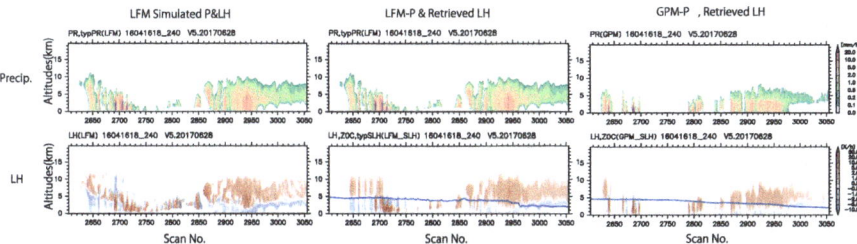

Fig. 46.6 Simulated and (LFM- and GPM-) retrieved LH along the GPM orbit shown in Fig. 46.5

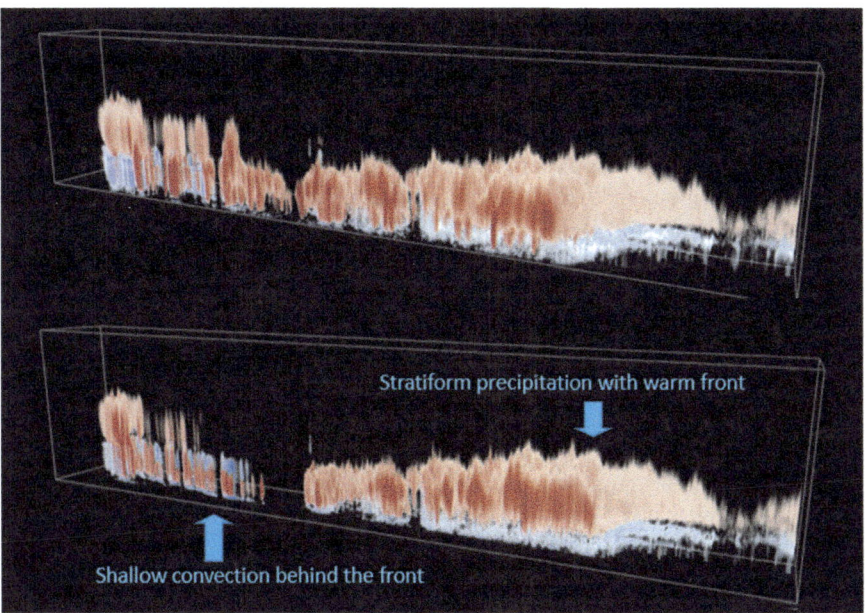

Fig. 46.7 Three-dimensional SLH latent heating structure derived from the GPM DPR observation case on 16 Apr. 2016, the same case as in Fig. 46.5. Northward is to the right and southward is to the left. Top and bottom panels show different cross sections of the cyclone. Reddish colors show the heating and bluish colors show the cooling

cloud base associated with the warm front. To the left, in the top panel, we can find deep convection at the cold front, while in the bottom panel showing the cross section just behind the cold front, we can find shallow convection behind the front, and very shallow heating with cooling atop. This shallow convection is also pointed out by Posselt et al. (2008) with Cloudsat observations.

To summarize, we developed a new Spectral LH algorithm for mid-latitude precipitation, utilizing 8 extratropical cyclone cases simulated with JMA LFM. To this end, cloud base vs freezing level relationship was an issue, for we utilize precipitation profiles observed from the GPM KuPR. After all, we found that precipitation maximum height can well represent the cloud base height. Six LUTs were constructed to retrieve LH from GPM KuPR. This new algorithm performed successfully: the consistency check and the retrieval results were quite satisfactory.

46.3   Convective and Stratiform Heating (CSH) Algorithm

Joanne Simpson, TRMM Project Scientist, suggested to use the GCEM (a cloud-resolving model, CRM) simulated LH structure and its associated cloud and precipitation properties to retrieve LH for TRMM. The first LH retrieval algorithm was able to estimate LH structures based on the GCEM simulated vertical profiles of cloud hydrometeors (cloud water, cloud ice, rain, snow and graupel) and surface rainfall. However, this LH algorithm requires artificial coefficients related to non-precipitating clouds (cloud and cloud ice) to estimate condensation and deposition heating. These coefficients could be determined based on the surface rainfall.

It is noted that the LH profiles are quite different between the convective and stratiform regions (Houze 1982; Johnson 1984). Positive LH heating via condensation and deposition (and freezing) dominates throughout the column in the convective region. On the other hand, LH heating occurs above the melting level through deposition while cooling from evaporation prevails beneath the melting level in the stratiform region. The melting (cooling) also occurrs in the stratiform region. Therefore, Tao et al. (1993) proposed a new LH algorithm known as the Convective-Stratiform Heating (CSH) algorithm. The CSH algorithm only needs to have derived satellite surface rainfall amount and its stratiform percentage to derive the LH structures. An LUT for LH profiles is also required for the CSH algorithm. Tao et al. (1993) used a simple LUT consisting of rain-normalized Q1 profiles for the convective and stratiform region composited for land and ocean from sounding budgets and a few GCE simulations. Tao et al. (2000, 2001) tested the performance of the CSH algorithm through self-consistency checking using GCEM-simulated cloud heating data as "truth", and the algorithm was used to retrieve LH structures using 1-month of TRMM rainfall products. Later, the CSH algorithm was redesigned and improved by using several GCEM simulations and differentiating the LUTs into many bins and into separate heating components (Tao et al. 2010).

The updated Goddard CSH algorithm (V6) used two different high-resolution CRMs to generate the LUTs. One is for tropical regions (TRMM domain) and the other for high altitude (GPM domain). The followings will describe the V6 CSH algorithm.

46.3.1 CSH Algorithm for Tropical Region

46.3.1.1 Cases for CSH Look-Up Table

The GCEM model is used with imposed with large-scale advective forcing in temperature and water vapor. This large-scale advective forcing is derived from extensive rawinsonde networks through a residual method (known as a diagnostic heating budget; Yanai et al. 1973; Houze 1982, 1997; Johnson 1984). This derived large-scale advective forcing includes many different geographic locations (tropical oceans, tropical continental and central US). For tropical ocean cases, they consist of the GATE (Global Atmospheric Research Program's Atlantic Tropical Experiment), SCSMEX/NESA (South China Sea Monsoon Experiment/Northern Enhanced Sounding Array), TOGA COARE (Tropical Ocean Global Atmosphere – Coupled Ocean Atmosphere Response Experiment), KWAJEX (Kwajalein Experiment), TWP-ICE (Tropical Warm Pool – International Cloud Experiment) and DYNAMO (Dynamics of the MJO). For continental cases, they consist of the MC3E (Midlatitude Continental Convective Clouds Experiment), GoAMAZON (Green Ocean Amazon Experiment), and the Department of Energy's Atmospheric Radiation Measurement Southern Great Plains site (1997 and 2002). Figure 46.8 shows the geographic locations of these field campaigns. The GCEM-simulated rainfall, Q1 (apparent heat source) and Q2 (apparent moist sink) are always in excellent agreement with the sounding estimated (Tao et al. 2010; Lang and Tao 2018). In all, GCEM-simulated LH structures have more than 355 days (122 days for land case and 233 days for oceanic cases) of model integrations.

46.3.1.2 Updated CSH

The Goddard CSH algorithm has been recently updated (Lang and Tao 2018). First, it also uses two new metrics for mapping the simulated LH structures to the satellite: echo top heights and low-level echo gradients. Secondly, the profiles are binned according to mean conditional surface rain rates obtained from GCEM sub-domains roughly consistent with the size of the CSH LH product grids, i.e., 32 (16) km model subdomains are used to build the LUTs for the TRMM (GPM) half (quarter) degree gridded products. This approach has several potential advantages regarding heating structures. Obviously, having many more profiles in the LUTs, it allows for the

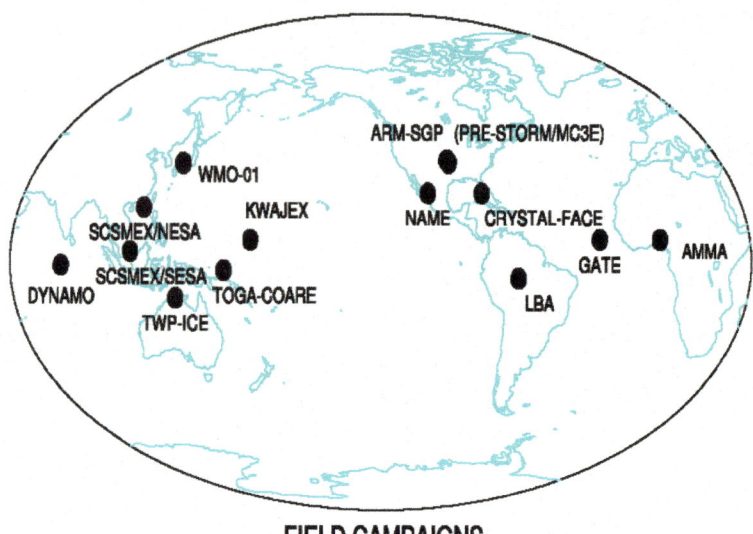

FIELD CAMPAIGNS

Fig. 46.8 Geographic locations of the twelve field campaigns that provide data to drive and evaluate CRM simulations. These campaigns are: ARM-SGP campaigns conducted in summer 1997, spring 2000 and summer 2002; GATE (1974); KWAJEX (1999); TOGA-COARE (1992 and 1993); TWP-ICE (2006); SCSMEX/NESA and SESA (1998); AMMA (2006; African Monsoon Multidisciplinary Analysis experiment); AMIE/DYNAMO (2011; ARM Madden-Julian Oscillation Experiment/Dynamics of the Madden-Julian Oscillation). (Adapted from Zeng et al. 2009)

possibility of having many more heating structures, and using conditional rain rates it allows for smaller areas of intense rain to be distinguished from larger areas of weak rain. In addition to conditional surface rain intensities and stratiform fraction, the land and ocean LUTs are further binned according to mean echo top heights (5 bins, 0–2, 2–4, 4–6, 6–8, and 8+ km) and low-level (0–2 km AGL) vertical reflectivity gradients (downward increasing or decreasing) to improve the depth of heating and intensity of low-level evaporative cooling in stratiform regions, respectively.

The GCEM was upgraded with an improved Goddard 4ICE (cloud ice, snow, graupel and hail) scheme (Lang et al. 2014; Tao et al. 2016a) cloud microphysics. This new 4ICE scheme was tested for an intense continental squall line and a moderate, less organized continental case; as well as long term model integration with imposed large-scale advective forcing (those discussed in the previous section). In addition to add hail, the bin microphysics-based rain evaporation correction has been implemented into the 4ICE scheme.

The horizontal resolution used in the GCEM-simulated LH is improved by using 200 m instead of 1000 m. This is because the previous CSH LUT might under-represent LH associated with shallow clouds. The model domain size is also increased to better simulate mesoscale convective systems. Another major modification is the convective and stratiform classification that is crucial for CSH

Table 46.3 Evolution of the Goddard CSH algorithm with key improvements and references

Highlights	References
First paper to use a CRM to develop a LH algorithm – LH structure estimated from vertical hydrometeor profiles.	Tao et al. (1990)
First paper on the CSH algorithm – composite both sounding estimated and CRM modeled convective and stratiform heating profiles into a simple look-up table (LUT).	Tao et al. (1993)
Examined the performance of the CSH algorithm using CRM simulated (consistency check), SSM/I and ship borne estimated rainfall and stratiform percentages. Retrieved LH sensitive to surface rainfall amount and stratiform percentage.	Tao et al. (2000)
First paper to retrieve LH based on 1-month of TRMM-estimated rainfall products.	Tao et al. (2001)
Improved CSH by using several CRM-simulations to build the LUTs, individual heating components retrieved separately, LUTs separated into many surface rain rate intensity bins and stratiform fractions.	Tao et al. (2010)
Improved CSH by using even more CRM simulations with better microphysics at finer resolution to build the LUTs and added new metrics for echo top heights and low-level dBZ gradients to the LUTs.	Lang and Tao (2018)
Expanded the retrieval of LH to higher latitudes and the cold season using NU-WRF simulations of synoptic storms to build separate LUTs.	Tao et al. (2019)
Review papers on LH algorithms, applications and evaluations.	Tao et al. (2006, 2016a)

Adapted from Tao et al. (2019)

algorithms. The modified separation method is now based on the Steiner et al. (1995) method that is consistent with the DPR method. Table 46.3 shows the highlights and references for the CSH algorithm evolution.

46.3.2 CSH for High Latitudes

46.3.2.1 NU-WRF and Cases

To support GPM, the CSH algorithm needs to include LH profiles associated with higher latitudes (poleward of 35°N/35°S) and the winter season in its LUTs. Large-scale advective forcing derived from sounding networks is unavailable for winter at high-latitudes weather events. Therefore, a regional scale numerical model such as the NASA Unified-Weather Research and Forecasting model (NU-WRF; Peters-Lidard et al. 2015) is used as it is better suited to handle precipitation systems associated with high latitude frontal and snow events and cyclones. Six synoptic-scale storm cases are selected for the NU-WRF simulations to generate the cold-season LH LUTs. Three cases are winter storms observed over the eastern continental US (CONUS) in 2014 and 2015. The other three cases are storms that developed over the eastern Pacific off the coast of California during the CalWater 2015 field campaign (Ralph et al. 2016).

NU-WRF version 8 is based on NCAR WRF-ARW version 3.7.1, together with the updated Goddard 4-ICE bulk microphysics scheme (Lang et al. 2014; Tao et al. 2016b) and the updated Goddard short- and longwave radiation schemes (Chou and Suarez 1999, 2001; Matsui and Jacob 2014) to conduct high-resolution (3-km inner nested domains) extratropical simulations. The initial and lateral boundary conditions of the outer domains and the initial conditions for the inner domains come from NCEP final (FNL) operational global analysis data (NCEP/National Weather Service/NOAA/US Department of Commerce) with a grid spacing of 1° in both latitude and longitude with 6-hourly intervals. Two nested domains are used at 9 and 3 km, respectively, with 61 vertical layers.

The first step is to identify and quantify basic relations between GPM (TRMM) observable structures/quantities such as radar and precipitation and LH profiles simulated from NU-WRF. Officially, the CSH algorithm uses retrieved precipitation properties from the TRMM/GPM combined algorithm as input. In addition to retrieve precipitation rates from the combined algorithm, corresponding information on echo top heights, freezing level heights and maximum radar reflectivity (dBZ) heights is available from the DPR (and re-analysis data). The relationships between LH profiles, radar reflectivity (i.e., maximum column intensity and its height), freezing height, echo top height and surface precipitation (i.e., intensity) must be established based on NU-WRF simulations.

46.3.2.2 NU-WRF and Look-Up Tables

There is a high correlation between precipitation rate and composite radar reflectivity. High (low) composite dBZs are always coincident with large (small) surface precipitation rates; composite radar reflectivity >40 dBZ corresponds with precipitation rates >12 mm h^{-1}. Also, the region with high precipitation rates has a high freezing level as it mainly occurred in the southern part of the system. High precipitation rates are also usually associated with maximum radar echoes at a low height. This may indicate that warm rain processes dominate the high precipitation rates.

These results suggest that all of the aforementioned NU-WRF simulated precipitation quantities should be used to build relationships with LH. Collectively, LH estimates can be made and refined according to the ambient reflectivity/precipitation structures and thus provide the basis by which the CSH algorithm and its LUTs can be expanded to colder environments. Table 46.4 shows the key parameters, i.e., surface precipitation rate (mm h^{-1}), maximum radar reflectivity height (m), freezing level height (m), echo top height (m), decreasing flag and maximum radar reflectivity intensity (dBZ), which are to be used for the new CSH cold season LH LUT. The bin ranges are based on the close examination of the results from the NU-WRF-simulations. Also note that the range of the surface precipitation intensity and other precipitation properties (radar reflectivity, freezing level height) are selected through detailed analysis of the NU-WRF simulations.

Table 46.4 Key parameters used for the new CSH cold season LH look-up table. The look-up table is built from NU-WRF simulated LH profiles associated with six synoptic storm events

Variables	Bin ranges
Surface rainfall rate (mm h − 1)	0., 0.178, 1.0, 1.78, 3.16, 5.62, 7.5, 10.0, 13.3, 17.8, 22.4, 27.0, 31.6, 44.0, 56.2, 70.0, 100.0, 999.0
Max dBZ height (m)	0, 500, 1000, 1500, 2000, 3000, 4000, 5000, 99,999
Freezing level height (m)	Negative, 0, 500, 1000, 1500, 2000, 2500, 3000, 3500, 4000, 4500, 5000, 5500, 99,999
Echo top height (m)	0, 1000, 2000, 3000, 4000, 5000, 6000,7000, 8000, 9000, 10,000, 99,999
Decreasing flag	"Decreasing", "non-decreasing"
Max dBZ intensity	−10, −9, −8, …78, 79, 80 (from −10 to 80 with an interval of 1)

46.3.2.3 Consistency Check

As with the development of the CSH algorithm, the first form of validation with the revised algorithm is via self-consistency checks (see Tao et al. 2000, 2006, 2010) wherein LUTs derived from one or more simulated synoptic events can be used with the new algorithm to retrieve heating for an independent synoptic event with the modeled LH fields for that event serving as truth. In this way, an optimal set of conditions, relations and LUTs can be identified based on biases and errors obtained via the self-consistency checks using the model as truth.

Two different self-consistency check methods are conducted for one CONUS and one CalWater case. The differences between these two types of self-consistency checks are included in the LUTs. The first self-consistency check uses a LUT produced at a specific model integration time. The second one uses a LUT produced from the entire model integration, except for the spin-up time. The retrieved LH structures from the first consistency check are in excellent agreement with the model "truth", i.e., the actual simulated LH structures for both cases (Fig. 46.9). The retrieved LH structures (level of maximum and minimum heating/cooling) from the second consistency check are quite similar to the "truth" for both cases, but the mean LH profile is slightly stronger than the "truth" profile.

46.4 Summary and Future Issues

We have developed two standard products, SLH and CSH described above, utilizing the GPM observation, and have released their latest Ver.6 in 2018. The primary modifications in their latest versions are their applications to the mid-latitude precipitation.

Utilizing the SLH and CSH latent heating products, we plan to quantitatively study the role of convective latent heating in the global water and energy budget as a next step. Further validations of latent heating products and improvements will be

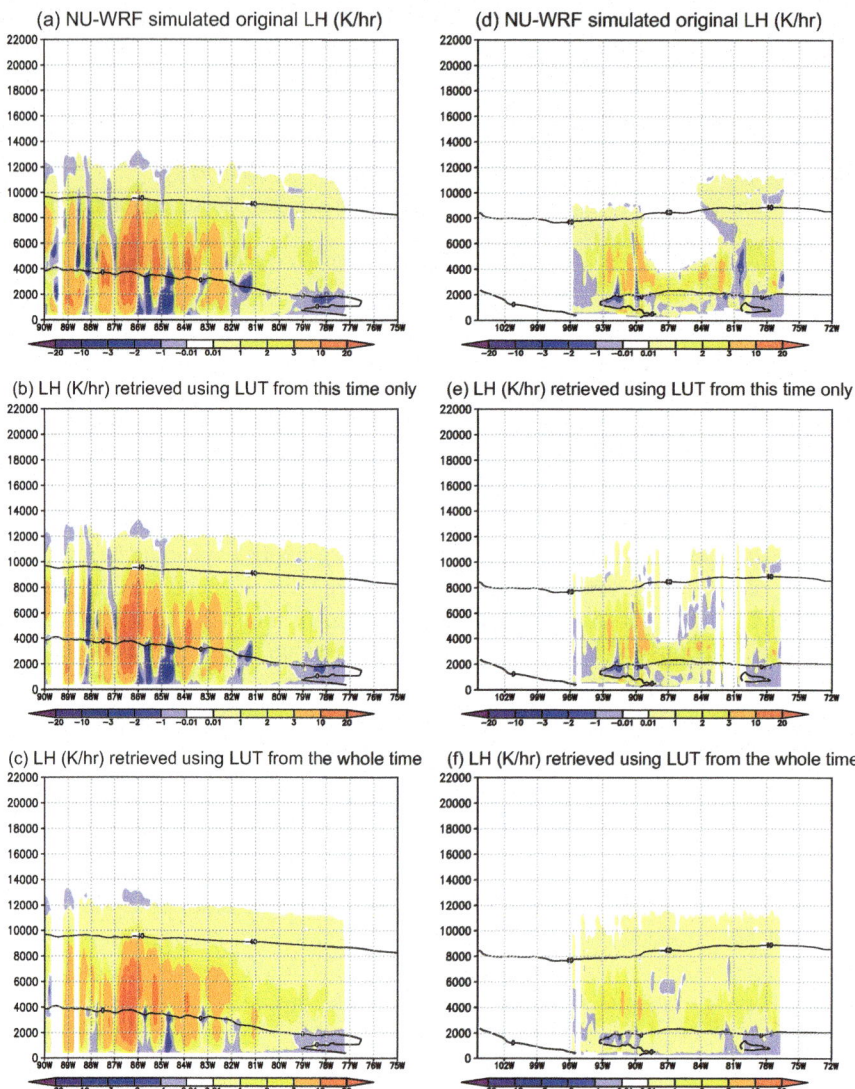

Fig. 46.9 Vertical cross sections of latent heating/cooling rates and air temperature (contour lines) along a SW-NE direction over eastern CONUS, (**a**) directly from the NU-WRF simulation at 1800 UTC 16 March 2014, (**b**) retrieved using a look-up table (LUT) based on the NU-WRF simulation results at this time only, and (**c**) retrieved using a LUT based on the NU-WRF simulation results from every 10 min for the whole 24 h, simulation period, except for the first 6 h. (**d–f**) Vertical cross sections along a W-E direction over eastern CONUS

expected for mid- and high-latitude precipitation, precipitation over land, and for some specific phenomena, such as extratropical and tropical cyclones. Quantifications of convective latent heating should shed light on the diabatic heating effects on the rapid intensifications of extratropical and tropical cyclones.

Acknowledgements The first author acknowledges Drs. Shoichi Shige, Atsushi Hamada, and Chie Yokoyama at the University of Tokyo, and Yasutaka Ikuta at the JMA for their collaborations in the development of the SLH algorithm. The second author acknowledges Steve Lang and Taka Iguchi for their collaborations in the development of the CSH algorithm. This research was supported by the JAXA Precipitation Measuring Mission (PMM) Research Announcement, and NASA Precipitation Measuring Mission (PMM) and NASA Energy and Water cycle Study (NEWS). The authors are grateful to Drs. M. Kojima and R. Oki at JAXA, and Drs. R. Kakar, G. S. Jackson and J. Entin at NASA headquarters for their support of this research. Acknowledgment is also made to the JAXA EORC and the NASA Goddard Space Flight Center and NASA Ames computing centers.

References

Boettcher, M., & Wernli, H. (2011). Life cycle study of a diabatic Rossby wave as a precursor to rapid cyclogenesis in the North Atlantic—Dynamics and forecast performance. *Monthly Weather Review, 139*, 1861–1878. https://doi.org/10.1175/2011MWR3504.1.

Chou, M.-D., & Suarez, M. J. (1999). *A solar radiation parameterization for atmospheric studies* (NASA technical reports). NASA/TM-1999-10460, Vol. 15, 38 pp. Available at https://ntrs.nasa.gov/search.jsp?R=19990060930. Last accessed 2 Dec 2018.

Chou, M.-D., & Suarez, M. J. (2001). *A thermal infrared radiation parameterization for atmospheric studies.* NASA/TM-2001-104606, Vol. 19, 55 pp. Available at https://ntrs.nasa.gov/search.jsp?R=20010072848. Last accessed 2 Dec 2018.

Hartmann, D. L., Hendon, H. H., & Houze, R. A., Jr. (1984). Some implications of the mesoscale circulations in tropical cloud clusters for large-scale dynamics and climate. *Journal of the Atmospheric Sciences, 41*, 113–121. https://doi.org/10.1175/1520-0469(1984)041<0113:SIOTMC>2.0.CO;2.

Hirota, N., Takayabu, Y. N., Watanabe, M., & Kimoto, M. (2011). Precipitation reproducibility over tropical oceans and its relationship to the double ITCZ problem in CMIP3 and MIROC5 climate models. *Journal of Climate, 7*, 4859–4873. https://doi.org/10.1175/2011JCLI4156.1.

Houze, R. A., Jr. (1982). Cloud clusters and large-scale vertical motions in the tropics. *Journal of the Meteorological Society of Japan, 60*, 396–409. https://doi.org/10.2151/jmsj1965.60.1_396.

Houze, R. A., Jr. (1997). Stratiform precipitation in regions of convection: A meteorological paradox? *Bulletin of the American Meteorological Society, 78*, 2179–2196. https://doi.org/10.1175/1520-0477(1997)078<2179:SPIROC>2.0.CO;2.

Houze, R. A., Jr., Rutledge, S. A., Biggerstaff, M. I., & Smull, B. F. (1989). Interpretation of Doppler weather radar displays of midlatitude mesoscale convective systems. *Bulletin of the American Meteorological Society, 70*, 608–619. https://doi.org/10.1175/1520-0477(1989)070<0608:IODWRD>2.0.CO;2.

Johnson, R. H. (1984). Partitioning tropical heat and moisture budgets into cumulus and mesoscale components: Implication for cumulus parameterization. *Monthly Weather Review, 112*, 1656–1665. https://doi.org/10.1175/1520-0493(1984)112<1590:PTHAMB>2.0.CO;2.

Lang, S. E., & Tao, W.-K. (2018). The next-generation Goddard convective-stratiform heating algorithm: New tropical and warm season retrievals for GPM. *Journal of Climate, 31*, 5997–6026. https://doi.org/10.1175/JCLI-D-17-0224.1.

Lang, S. E., Tao, W.-K., Chern, J.-D., Wu, D., & Li, X. (2014). Benefits of a fourth ice class in the simulated radar reflectivities of convective systems using a bulk microphysics scheme. *Journal of the Atmospheric Sciences, 71*, 3583–3612. https://doi.org/10.1175/JAS-D-13-0330.1.

Matsui, T., & Jacob, J. P. (2014). *Goddard radiation scheme for NU-WRF version 2014.* NASA Goddard Space Flight Center, 4 pp. Version 2016 available at https://modelingguru.nasa.gov/servlet/JiveServlet/previewBody/2056-102-16-6426/nuwrf_userguide_patch7.pdf.   Last accessed 2 Dec 2018.

Peters-Lidard, C. D., Kemp, E. M., Matsui, T., Santanello, J. A., Jr., Kumar, S. V., Jacob, J. P., Clune, T., Tao, W.-K., Chin, M., Hou, A. Y., Case, J. L., Kim, D., Kim, K.-M., Lau, W., Liu, Y., Shi, J.-J., Starr, D., Tan, Q., Tao, Z., Zaitchik, B. F., Zavodsky, B., Zhang, S. Q., & Zupanski, M. (2015). Integrated modeling of aerosol, cloud, precipitation and land processes at satellite-resolved scales. *Environmental Modelling and Software, 67*, 149–159. https://doi.org/10.1016/j.envsoft.2015.01.007.

Posselt, D. J., Stephens, G. L., & Miller, M. (2008). CloudSat adding a new dimension to a classical view of extratropical cyclones. *Bulletin of the American Meteorological Society, 89*, 599–610. https://doi.org/10.1175/BAMS-89-5-599.

Ralph, F. M., Prather, K. A., Cayan, D., Spackman, J. R., DeMott, P., Dettinger, M., Fairall, C., Leung, R., Rosenfeld, D., Rutledge, S., Waliser, D., White, A. B., Cordeira, J., Martin, A., Helly, J., & Intrieri, J. (2016). CalWater field studies designed to quantify the roles of atmospheric rivers and aerosols in modulating U.S. West Coast precipitation in a changing climate. *Bulletin of the American Meteorological Society, 97*, 1209–1228. https://doi.org/10.1175/BAMS-D-14-00043.1.

Shige, S., Takayabu, Y. N., Tao, W.-K., & Johnson, D. E. (2004). Spectral retrieval of latent heating profiles from TRMM PR data. Part I: Development of a model-based algorithm. *Journal of Applied Meteorology, 43*, 1095–1113. https://doi.org/10.1175/1520-0450(2004)043<1095:SROLHP>2.0.CO;2.

Shige, S., Takayabu, Y. N., Tao, W.-K., & Shie, C.-L. (2007). Spectral retrieval of latent heating profiles from TRMM PR data. Part II: Algorithm improvement and heating estimates over tropical ocean regions. *Journal of Applied Meteorology and Climatology, 46*, 1098–1124. https://doi.org/10.1175/JAM2510.1.

Shige, S., Takayabu, Y. N., & Tao, W.-K. (2008). Spectral retrieval of latent heating profiles from TRMM PR data. Part III: Estimating apparent moisture sink profiles over tropical oceans. *Journal of Applied Meteorology and Climatology, 47*, 620–640. https://doi.org/10.1175/2007JAMC1738.1.

Shige, S., Takayabu, Y. N., Kida, S., Tao, W.-K., Zeng, X., Yokoyama, C., & L'Ecuyer, T. (2009). Spectral retrieval of latent heating profiles from TRMM PR data. Part IV: Comparisons of lookup tables from two- and three-dimensional cloud-resolving model simulations. *Journal of Climate, 22*, 5577–5594. https://doi.org/10.1175/2009JCLI2919.1.

Simpson, J., Adler, R. F., & North, G. R. (1988). A proposed tropical rainfall measuring mission (TRMM) satellite. *Bulletin of the American Meteorological Society, 69*, 278–295. https://doi.org/10.1175/1520-0477(1988)069<0278:APTRMM>2.0.CO;2.

Steiner, M., Houze, R. A., Jr., & Yuter, S. E. (1995). Climatological characterization of three-dimensional storm structure from operational radar and rain gauge data. *Journal of Applied Meteorology, 34*, 1978–2007. https://doi.org/10.1175/1520-0450(1995)034<1978:CCOTDS>2.0.CO;2.

Takayabu, Y. N. (2002). Spectral representation of rain profiles and diurnal variations observed with TRMM PR data over the equatorial area. *Geophysical Research Letters, 29*, 25-1–25-4. https://doi.org/10.1029/2001GL014113.

Takayabu, Y. N. (2008). Observing rainfall regimes using TRMM PR and LIS Data. *GEWEX News*, 18, 2, 9–10. Available at https://www.gewex.org/gewex-content/files_mf/1432208504May2008.pdf Last accessed 2 Dec 2018.

Takayabu, Y. N., Shige, S., Tao, W.-K., & Hirota, N. (2010). Shallow and deep latent heating modes over tropical oceans observed with TRMM PR spectral latent heating data. *Journal of Climate, 23*, 2030–2046. https://doi.org/10.1175/2009JCLI3110.1.

Tao, W.-K., Simpson, J., Lang, S., McCumber, M., Adler, R. F., & Penc, R. (1990). An algorithm to estimate the heating budget from vertical hydrometeor profiles. *Journal of Applied Meteorology, 29*, 1232–1244. https://doi.org/10.1175/1520-0450(1990)029<1232:AATETH>2.0.CO;2.

Tao, W.-K., Lang, S., Simpson, J., & Adler, R. F. (1993). Retrieval algorithms for estimating the vertical profiles of latent heat release: Their applications for TRMM. *Journal of the Meteorological Society of Japan, 71*, 685–700. https://doi.org/10.2151/jmsj1965.71.6_685.

Tao, W.-K., Lang, S., Simpson, J., Olson, W. S., Johnson, D., Ferrier, B., Kummerow, C. D., & Adler, R. F. (2000). Retrieving vertical profiles of latent heat release in TOGA COARE convective systems using a cloud resolving model, SSM/I and radar data. *Journal of the Meteorological Society of Japan, 78*, 333–355. https://doi.org/10.2151/jmsj1965.78.4_333.

Tao, W.-K., Lang, S., Olson, W. S., Yang, S., Meneghini, R., Simpson, J., Kummerow, C. D., Smith, E. A., & Halverson, J. (2001). Retrieved vertical profiles of latent heating release using TRMM rainfall products for February 1998. *Journal of Applied Meteorology, 40*, 957–982. https://doi.org/10.1175/1520-0450(2001)040<0957:RVPOLH>2.0.CO;2.

Tao, W.-K., Smith, E. A., Adler, R. F., Haddad, Z. S., Hou, A. Y., Iguchi, T., Kakar, R., Krishnamurti, T. N., Kummerow, C. D., Lang, S., Meneghini, R., Nakamura, K., Nakazawa, T., Okamoto, K., Olson, W. S., Satoh, S., Shige, S., Simpson, J., Takayabu, Y. N., Tripoli, G. J., & Yang, S. (2006). Retrieval of latent heating from TRMM measurements. *Bulletin of the American Meteorological Society, 87*, 1555–1572. https://doi.org/10.1175/BAMS-87-11-1555.

Tao, W.-K., Lang, S., Zeng, X., Shige, S., & Takayabu, Y. N. (2010). Relating convective and stratiform rain to latent heating. *Journal of Climate, 23*, 1874–1893. https://doi.org/10.1175/2009JCLI3278.1.

Tao, W.-K., Takayabu, Y. N., Lang, S., Olson, W. S., Shige, S., Hou, A. Y., Jiang, X., Lau, W., Krishnamurti, T. N., Waliser, D., Zhang, C., Johnson, R., Houze, R., Ciesielski, P., Grecu, M., Hagos, S., Kakar, R., Nakamura, K., Braun, S., & Bhardwaj, A. (2016a). TRMM latent heating retrieval and comparison with field campaigns and large-scale analyses. In R. G. Fovell & W.-W. Tung (Eds.), *Multi-scale convection-coupled systems in the tropics* (AMS meteorological monographs). Boston. https://doi.org/10.1175/AMSMONOGRAPHS-D-15-0013.1.

Tao, W.-K., Wu, D., Lang, S., Chern, J.-D., Peters-Lidard, C., Fridlind, A., & Matsui, T. (2016b). High-resolution NU-WRF simulations of a deep convective-precipitation system during MC3E: Further improvements and comparisons between Goddard microphysics schemes and observations. *Journal of Geophysical Research, 121*, 1278–1305. https://doi.org/10.1002/2015JD023986.

Tao, W.-K., Iguchi, T., & Lang, S. (2019). Expanding the Goddard CSH algorithm for GPM: New extratropical retrievals. *Journal of Applied Meteorology and Climatology, 58*, 921–946. https://doi.org/10.1175/JAMC-D-18-0215.1.

Yanai, M., Esbensen, S., & Chu, J. (1973). Determination of average bulk properties of tropical cloud clusters from large-scale heat and moisture budgets. *Journal of the Atmospheric Sciences, 30*, 611–627. https://doi.org/10.1175/1520-0469(1973)030<0611:DOBPOT>2.0.CO;2.

Zeng, X., Tao, W.-K., Zhang, M.-H., Hou, A. Y., Xie, S., Lang, S., Li, X., Starr, D. O.'. C., & Li, X. (2009). A contribution by ice nuclei to global warming. *Quarterly Journal of the Royal Meteorological Society, 135*, 1614–1629. https://doi.org/10.1002/qj.449.

Part VI
Applications

Chapter 47
Operational Applications of Global Precipitation Measurement Observations

Anita LeRoy, Emily Berndt, Andrew Molthan, Bradley Zavodsky, Matthew Smith, Frank LaFontaine, Kevin McGrath, and Kevin Fuell

Abstract The NASA Global Precipitation Measurement Mission core satellite was launched in 2014, and the data observations provided by this mission are freely available to scientists around the globe in near real time. As part of the Early Adopter program, NASA's Short-term Prediction Research and Transition (SPoRT) center has worked to provide these data to National Weather Service forecasters and hydrologists to determine its value within the operational environment. Over the course of three formal assessments with targeted users, forecasters have provided feedback to SPoRT on what types of data and value-added products are relevant to the offices' primarily forecast challenges, how these GPM precipitation datasets can best be displayed, and what the products' value is for specific weather-related phenomena. Forecasters in operational weather forecasting and hydrology valued using the precipitation estimate data, particularly in heavy rain events in which the impacts could be dangerous. These type of events include areal flooding from atmospheric river events, flash flooding from convective events, and assessing precipitation for mudslide risks. The gridded precipitation dataset from GPM, called IMERG, provided the most value to hydrologists. IMERG also provided some value to forecasters in retrospective cases, due to its longer latency. Snowfall-related products derived from GPM helped forecasters assess the intensity and location of falling snow, supplementing other datasets like ground-based radar. The passive microwave imagery, including RGB composites of different polarizations of single

A. LeRoy (✉) · K. Fuell
Earth System Science Center (ESSC), University of Alabama in Huntsville, Huntsville, AL, USA
e-mail: anita.leroy@nasa.gov

E. Berndt · A. Molthan · B. Zavodsky
NASA, Marshall Space Flight Center (MSFC), Huntsville, AL, USA
e-mail: emily.b.berndt@nasa.gov

M. Smith
Information Technology and Systems Center, University of Alabama in Huntsville, Huntsville, AL, USA

F. LaFontaine · K. McGrath
Jacobs, ESSCA Group, NASA-Marshall Space Flight Center, Huntsville, AL, USA

© Springer Nature Switzerland AG 2020
V. Levizzani et al. (eds.), *Satellite Precipitation Measurement*, Advances in Global Change Research 69, https://doi.org/10.1007/978-3-030-35798-6_23

channels, helped tropical forecasters infer structure and intensity in tropical systems. The details and context of these results are provided herein to guide possible operational use of GPM's datasets for forecasting and nowcasting.

Keywords Precipitation · Rainfall · Snowfall · GPM · GMI · NASA · Short-term Prediction Research and Transition Center (SPoRT) · NOAA · National Weather Service · R2O/O2R · IMERG · GPROF · Rain gauges · AWIPS · Forecast · Landslide · Floods · Atmospheric river

47.1 Introduction

The launch of next-generation satellites with advanced sensors, such as Global Precipitation Measurement (GPM) Core Observatory with the GPM Microwave Imager (GMI) and Dual-frequency Precipitation Radar (DPR) onboard, provide a unique opportunity to develop new products, tools, and capabilities to enhance operational weather forecasting. Many of the next-generation weather satellites boast of capabilities that have never been used in the operational environment. The operational environment is often fast-paced and requires forecasters to examine and make decisions based on analyzing many different types of data very quickly, particularly in rapidly evolving weather threat situations. This is a stark contrast to the research world, where researchers use a limited number of datasets to answer very specific scientific questions, taking months or years to perform analyses and arrive at conclusions. The National Aeronautics and Space Administration (NASA) Short-term Prediction Research and Transition (SPoRT) Center strives to bridge the gap between researchers and the operational community of end users to facilitate the transition of research products and capabilities to an operational environment.

NASA SPoRT was established in 2002 to transition NASA satellite observations and capabilities to end users to improve short-term, regional weather forecasts (Goodman et al. 2004; Jedlovec 2013). SPoRT began providing NASA Moderate Resolution Imaging Spectroradiometer (MODIS) imagery to National Weather Service (NWS) Weather Forecast Offices (WFOs) as a high-resolution complement to basic satellite imagery to improve situational awareness and short-term forecasts (Jedlovec 2013). Since 2002 SPoRT has used a research to operations / operations to research (R2O/O2R) paradigm to transition both NASA and National Oceanic and Atmospheric Administration (NOAA) satellite observations and research capabilities to operational end users throughout all NWS regions. Successful transition of research products and capabilities to operations can be facilitated through interactive partnerships between end users and product or algorithm developers through an iterative product development and assessment process. SPoRT conducted a series of assessments of GPM rain rate products with NWS weather forecasters and hydrologists, with the goal of determining the utility of GPM products in operations and providing feedback to product and algorithm developers. Herein, Sect. 47.2 will describe the background and methods while Sect. 47.3 will highlight operational applications of GPM data with a primary focus on rain rate, but other relevant applications related to supporting high impact weather events, snowfall, and hurricane forecasting are briefly highlighted.

47.2 Background and Methods

47.2.1 GPM

The Global Precipitation Measurement (GPM) is a joint mission with NASA and the Japan Aerospace Exploration Agency (JAXA). GPM officially began with the launch of its Core Observatory platform in February 2014. The mission, however, includes data from an international constellation of 12 satellites all with similar passive microwave instruments. GPM's goal is to explore and build a better understanding of Earth's water and energy cycles (Hou et al. 2014). The Core Observatory's microwave imager, GPM Microwave Imager (GMI), and its Dual-frequency Precipitation Radar (DPR) are being used to intercalibrate GPM products using an algorithm called the Goddard Profiling algorithm (GPROF; Kummerow et al. 2001). GPROF compares the passive microwave retrievals to a database of profiles which contains radar estimates, environmental parameters, and in some cases gauge estimates, to create rain rate estimates.

47.2.2 Data

Passive microwave GPM Constellation data and products highlighted herein were obtained from NASA's Precipitation Processing System (PPS; GPM Science Team 2014a, b, c). GPM Constellation data are available from GMI, MHS (Microwave Humidity Sounder), AMSR2 (Advanced Microwave Scanning Radiometer-2), SSMIS (Special Sensor Microwave Imager Sounder), and ATMS (Advanced Technology Microwave Sounder) and utilized in near-real time production and distribution of experimental, value-added products. SPoRT has introduced GPM Constellation brightness temperatures from the 36–37 GHz and 88–89 GHz channels (e.g., Level 1C intercalibrated brightness temperatures) as well as multispectral or Red Green Blue (RGB) imagery derived from 37 and 89 GHz channel vertical and horizontal polarizations following Lee et al. (2002). SPoRT has also closely collaborated with the NOAA National Environmental Satellite, Data, and Information Service (NESDIS) on activities related to passive microwave swath snowfall rate products (Meng et al. 2017a, b; Ferraro et al. 2018). The snowfall rate product is directly obtained from NESDIS algorithm developers and SPoRT tailors the data for application in the operational environment. The NESDIS Snowfall Rate product is derived from polar-orbiting sensors onboard Suomi-National Polar-orbiting Partnership (S-NPP), GPM, NOAA, MetOp, and Defense Meteorological (DMSP) satellites

(Meng et al. 2017a, b; Kongoli et al. 2015, 2018). The product is useful for anticipating or validating snowfall in radar-deprived regions and tracking snowfall maxima.

The primary focus of this paper is application of the GPM Level 2 swath rain rate estimates that are available in near real time (NRT, <1 h from satellite valid time) and the gridded Level 3 calibrated precipitation rain rate field (precipitationCal) IMERG (Integrated Multi-satellitE Retrievals for GPM; Huffman et al. 2015, 2018). The Level 2 rain rates utilize the GPM Constellation of satellites, as processed through the GPROF algorithm to estimate the rain rates that are available in the native resolution and swath width of the instrument. The Level 3 IMERG rain rate product is a $0.1° \times 0.1°$ gridded precipitation estimate with nearly global coverage ($60°N – 60°S$). IMERG rain rate estimates also utilize GPROF to produce precipitation estimates every 30 min. The Early run of IMERG is available several hours sooner than the Late run, and the Late run uses more advanced techniques to improve the accuracy of precipitation estimates such as both forward and backward morphing. IMERG produces an estimate for precipitation as well as a likelihood of precipitation phase.

SPoRT introduces "value-added" experimental products to NWS operational forecasters when features or functionality are added that does not exist in the initial dataset. For example, based on prior assessments of the satellite-based Quantitative Precipitation Estimate (QPE; Kuligowski 2002) products' utility in operations, SPoRT developed cumulative precipitation products from IMERG at synoptic times (1-, 3-, 6-, 12-, and 24-h; Smith et al. 2016). GPM datasets were obtained in near-real time from NASA's PPS and formatted to be compatible with the Advanced Weather Interactive Processing System (AWIPS) for integration in the operational environment with a latency of about 40 min for the swath rain rate, 6 h for the Early IMERG, and 14 h for the Late IMERG, with latency almost entirely upstream of SPoRT's initial acquisition of the data. Products are ultimately distributed to NWS end users by SPoRT via the Unidata Local Data Manager (LDM; Unidata 2018).

47.2.3 Motivation

Accurately measuring precipitation rate and accumulation is an integral part of the operational forecasting process and assessing potential hazards such as drought, flash flooding, and landslides/mudslides. Although rain is the most common weather event and has been recorded, along with temperature, in the most basic of climate records, precipitation is actually fiendishly difficult to measure accurately. Forecasters rely on ground-based radars, using different mathematical relationships and algorithms, to estimate rain rates or amounts. Leaving aside all arguments for or against the various methods of calculating a precipitation estimate from radar reflectivity, radars must be able to observe the precipitation in question; the rain must be falling near a ground-based radar that is not beam-blocked, which largely excludes mountainous areas, large bodies of water, and any global location with limited weather observing infrastructure. Therefore, estimating precipitation with ground-based radar is a challenge for NWS forecasters in the mountainous regions of the western United States and Alaska. A rain gauge offers a straight-forward measurement of rainfall, but precipitation is highly spatially variable. Even in dense rain gauge networks, statistically significant differences in rainfall can be recorded. It would be impractical to have a rain gauge network dense enough to confidently measure global rain rates, and again mountainous terrain would be difficult to cover. And as already alluded to, these methods are relevant to precipitation falling on land surfaces. Precipitation from atmospheric rivers originating in tropical oceans accounts for a large proportion of terrestrial rainfall and high impact events at a large range of latitudes. These types of events pose a unique challenge since the precipitation may develop over oceanic regions that lack radar and gauge observations and move inland, creating high impact events due to heavy rainfall.

Space-based sensors can be utilized to address the challenges and limitations with observing precipitation rate and accumulation with ground-based radars and gauges. The best space-based estimates of rain rate come from radar and passive microwave instruments on satellites with Low Earth Orbits, providing global observations and filling in the observational gaps where ground-based observations are missing (i.e., mountainous and oceanic regions). As with any dataset, limitations exist. The swath width of the satellite observations is narrow (100s to 1000s of km) and the orbital dynamics limit these observations to a few overpasses a day over any given location on earth, making it difficult to continuously track a single storm. Methods for getting from a brightness temperature to a rain rate vary significantly, from incorporating infrared and passive microwave together with empirical relationships (e.g., Self-Calibrating Multivariate Precipitation Retrieval, SCaMPR; Kuligowski 2002), to integrating multiple instruments as close to near-real time as possible (e.g., Microwave Integrated Retrieval System, MiRS; Boukabara et al. 2011), to utilizing an a priori database of precipitation profiles to statistically estimate rainfall (GPROF; Kummerow et al. 2001). The nominal performance requirements for these methods and algorithms, including their latency, are targeted to satisfy research requirements rather than operational requirements. Methods vary over land and oceans, and the resolution of these instruments (5 km to 10s of km) would negate the possibility for point observations.

While considering all these different strengths and weaknesses, an interactive collaboration between research scientists, NWS forecasters, and SPoRT was used to determine whether the GPROF satellite-based measurements of precipitation rate are an effective tool for operational weather forecasters and hydrologists. Feedback from operational end users led to the development of value-added data products from next-generation sensors to address forecasting challenges unique to the operational environment. Within the context of its R2O/O2R paradigm and testbed environment, SPoRT conducted a series of assessments of experimental GPM rain rate products with NWS weather forecasters and hydrologists, with the goal of determining the utility of GPM products in operations and providing feedback to product and algorithm developers.

47.2.4   End User Interaction

SPoRT adheres to an R2O/O2R paradigm (Fig. 47.1) to introduce experimental products to end users through an iterative process.

Note that throughout the iterative process, the specific operational environment and forecasters' needs were the driving force behind product adjustments and changes; successful R2O/O2R transitions have kept the end user at the forefront of the activity. An integral part of the paradigm is integrating the new product or capability into the analysis tools in the end users' operational environment such as AWIPS, the processing, display, and telecommunications system for the NWS. The ability to use the new product within and alongside existing tools and datasets makes it easier for end users to adopt a new product in the fast-paced operational environment. Forecasters can also integrate new datasets more easily if they receive adequate applications-focused training on the new product. SPoRT-developed training materials demonstrate how the new product addresses an existing forecast challenge within a focused case example. The training materials also show other products that a forecaster would routinely utilize to address that challenge. This link from the familiar products to the unfamiliar product improves forecaster confidence and understanding of the new product, helping them integrate it into their existing process. In order to further facilitate the successful implementation of new data products into NWS operations, including those for precipitation, focused operational

Fig. 47.1 SPoRT research-to-operations/operations-to-research paradigm. (Adapted from Jedlovec 2013)

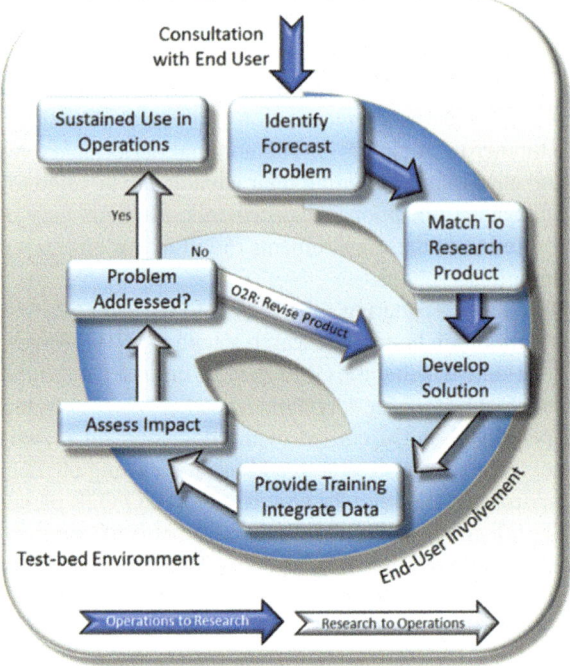

assessments are conducted. Product assessment activities provide the opportunity to assess the utility and impact of the product in the operational environment where the end user provides actionable feedback to the developers to improve the product to better address their needs. To garner feedback, end users are asked to respond to Likert-type questions (Likert 1932) to rate the impact of the new product on the forecasting and decision process. As SPoRT facilitates feedback between end users and product and/or algorithm developers, a new product can undergo several iterations through the development and assessment cycles of the R2O/O2R paradigm until it is successfully transitioned into operations.

Since 2015 SPoRT has conducted three assessments of GPM Level 2 swath rain rate and the IMERG Early and Late products with NWS forecasters and hydrologists to determine the utility of satellite derived rain rate products in the operational weather forecasting environment. SPoRT selected rain rate products from the GPM datasets for experimental demonstration of GPM observations based on the unique needs and scopes of NWS WFOs and River Forecast Centers (RFCs). In addition to issuing short- and long-term weather forecasts, WFOs are responsible for issuing forecast products regarding areal flooding, flash flooding, and other highly localized, quickly evolving (often minutes to hours) precipitation events. RFCs have regional areas of responsibility and focus on modeling streamflow through river basins and making gridded products of observed and modeled precipitation, typically at scales of hours to days. As a GPM Early Adopter SPoRT was able to obtain initial access to the swath rain rate and IMERG datasets in April 2015.

The goal of the initial assessment activity, during July and August 2015, was to determine whether the GPM Level 2 swath rain rates and Level 3 IMERG calibrated rain rate products have utility in NWS forecasting operations. Select partner WFOs and RFCs from Alaska, the Southwestern US, and the Southeastern US were asked to compare the GPM products to gauges and other satellite products available operationally. The Level 2 swath rain rates were found to have utility primarily at Alaska WFOs where there are gaps in radar coverage due to the complex terrain (Smith et al. 2016). The product was deemed most useful for identifying rainfall location and intensity used in near-real time operations. Forecasters recommended that improved detection efficiency of low rain rates and improved accuracy of higher rain rates would improve the operational utility of the GPM products. The utility of near-real time operations of IMERG were considerably lower; in about 66% of evaluated events, respondents stated that IMERG had a "very small" impact on their operations, typically citing long latency as the main limitation, meaning that the IMERG data were not suitable for NRT operations in WFOs (Smith et al. 2016). Despite these limitations, forecasters found value in applying the IMERG data to post-event analysis of high impact landslide and flooding events, which are described in Sect. 47.3.

The second assessment during August and September 2016 focused on the use of Version 4a GPM Level 2 swath rain rates and Version 3b IMERG datasets (LeRoy et al. 2017a). The primary participants were again from Alaska and the Southwestern U.S. and the focus was to examine events related to precipitation from the Southwest monsoon, atmospheric rivers, and more commonplace precipitation activity commensurate with summer season instability. Because the participating WFOs are in areas largely not well-represented by rain gauges or are in radar-void regions with beam-blockage due to complex terrain, forecasters also used this dataset to validate other satellite data, such as single channel infrared imagery. In 53% of events evaluated, the GPM data were used to provide information where ground-based precipitation products did not provide detailed information (LeRoy et al. 2017a). This assessment uncovered a data error that resulted in clear air returns and marine layer returns of very low rain rates that can more accurately be described as statistical noise (LeRoy et al. 2017a). The noise is a side effect of the GPROF algorithm assigning a probability of precipitation to each pixel, along with a rain rate estimate in versions 4 and earlier (G. Huffman 2016, personal communication). As an example of O2R interactions, forecasters and SPoRT personnel were able to alert developers of the frequency of data artifacts in higher latitudes. As a result, the end users, SPoRT, and the GPM team were able to develop a solution to filter the raw rain rate data to eliminate or substantially decrease the clear air returns so that the product better suited the operational forecasting needs and developers choose to prioritize corrections to the data in subsequent algorithm versions. Despite the clear air returns, forecasters stated that the Level 2 swath rain rates had a "high" impact on the analysis of 57% of events evaluated during the assessment (LeRoy et al. 2017a). Related to IMERG, forecasters at WFOs requested post-event data after high-impact events, for similar applications as the previous assessment, to help supplement rain rate information in otherwise data-sparse locations and as an additional observational tool to be included in reports. For post-event analysis and for incoming events, IMERG provided reliable qualitative information about the location and relative intensity of rain in otherwise data-void locations, particularly in mountainous and offshore locations. IMERG was used for post-event analysis in approximately 67% of events evaluated during this assessment (LeRoy et al. 2017a). Another limitation of IMERG noted by the forecasters was the coverage area. The latency of the product makes it much more viable in RFCs, in which the hydrologic forecasting is longer-term and basin response can be delayed, and the RFC hydrologists considered incorporating the gridded dataset into hydrologic models; however, because IMERG did not extend beyond 60°N, hydrologists did not incorporate the dataset in their models for operational activities.

Once the next version of GPROF was available, SPoRT conducted the third assessment with Alaska region forecasters and hydrologists during late summer 2017 (LeRoy et al. 2017b). Version 5 of the swath rain rates showed improved accuracy and detection efficiency in both stratiform and convective rain events in field studies, and also featured a correction to the spurious clear air returns observed in the prior assessment. IMERG Version 4b showed improved accuracy near coastlines and incorporated other algorithm changes that were expected to improve the accuracy of rain rates in general. GPM algorithm developers also indicated that some processing activities were streamlined, reducing Early IMERG latencies to about 4 h. To address limited latitudinal extent of IMERG that limited its operational utility in the previous assessment with Alaska forecasters and hydrologists, SPoRT collaborated with the GPM Science Team on a solution and identified the Level

3 High Quality Precipitation (HQPrecipitation, hereafter referred to as HQPrecip) field to potentially meet the operational need (LeRoy et al. 2017b). The HQPrecip combined swath rain rate information with the gridded and morphed rain rates in IMERG (Huffman et al. 2015, 2018). SPoRT began providing the HQPrecip product in AWIPS format to NWS partners, primarily the Alaska Pacific RFC, which increased GPM product utility for Alaska Region forecasters by giving insight into precipitation north of IMERG's normal data extent (60°N). Forecasters stated that it provided some additional continuity particularly when precipitation systems like atmospheric rivers would transition from the Gulf of Alaska to the mainland, impacting the interior of the state well beyond 60°N. This product mainly provided a gap-fill solution while anticipating a "global" IMERG product in the future. To summarize the assessment results, the accuracy of the swath rain rate product was viewed favorably when compared to most other precipitation data sources by the forecasters (Fig. 47.2) and the GPM data had "some" or "high" impact to operations in 64% of evaluated cases (LeRoy et al. 2017b). With the slight improvement of latency of the IMERG product to 4 h, forecasters began to use the product not only for post-event analysis but to also increase situational awareness of incoming high-impact precipitation events and especially for the assessment of precipitation moving onshore from the data-void oceanic regions and radar-beam blocked mountainous regions (LeRoy et al. 2017b).

This is the purpose of the SPoRT paradigm: a NASA research product (s) underwent iterative improvements and value-added adjustments based on feedback from end users who were able to use the product(s) seamlessly within their operations environment. Forecaster feedback was communicated to algorithm and product developers and each iterative cycle of the R2O/O2R paradigm unveiled increased utility and impact on decision making in the operational environment. As

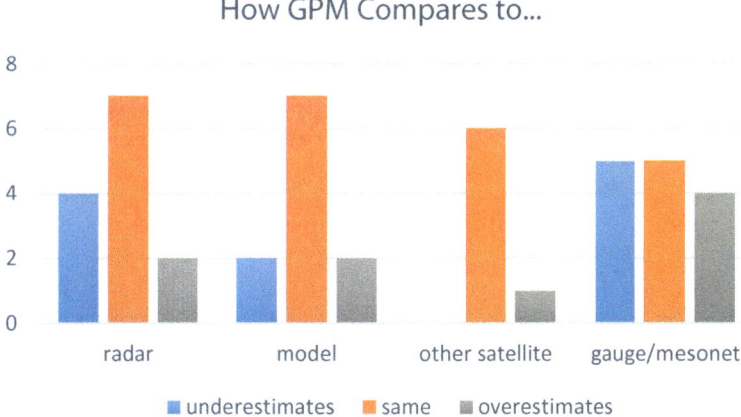

Fig. 47.2 GPM Level 2 swath rain rates were typically estimated by forecasters to be the same as radar, model estimated rain, other satellite products, and gauge data for the events they evaluated. Number of events in which comparisons were performed is shown on the y-axis. (Adapted from LeRoy et al. 2017b)

SPoRT engaged with end users to assess GPM data in operations, limitations (e.g., detection efficiency, noise, spatial coverage, and latency) were identified and brought to the attention of the GPM Science Team. As a result, the GPM Science Team were able to prioritize addressing these limitations in subsequent algorithm versions earlier than planned. For example, including high-latitude data was prioritized 1.5 versions before the notional plan called for it (G. J. Huffman 2018, personal communication). This R2O/O2R interaction helped accelerate the usability of GPM data in NWS operations and SPoRT plays a key role in providing data and end user support beyond the scope of the GPM Science Team's priorities (G. J. Huffman 2018, personal communication). These assessments have been an avenue to discover novel applications of GPM data as a direct result of forecasters utilizing the data in the operational environment. This paper will highlight novel application examples of GPM products related to anticipating hazards due to heavy precipitation such as landslides and flooding, atmospheric river events, and rain gauge verification, as well as briefly highlight other applications related activities including supporting high impact events, anticipating snowfall, and assessing hurricane structure and intensity.

47.3   Applications

47.3.1   Landslides/Mudslides

Alaska is highly susceptible to landslides and mudslides due to a variety of factors related to its steep slopes, silty soils (Dai et al. 2002), and retreating glaciers (Kääb et al. 2003), and exposure to triggers such as high winds (Buma and Johnson 2015) and heavy rains in its temperate rainforest climate (Hong et al. 2007). Because of this and the NWS mission to protect life and property, forecasters in Alaska Region routinely assess precipitation to ascertain landslide risk and issue public products to alert the public of such landslide potential. The introduction of GPM datasets to Alaska Region forecasters provided a value-added dataset to assess the impact of precipitation on landslide risk allowing for the ability to assess heavy precipitation potential where observations and radar coverage lack. During our interaction with Alaska Region forecasters, two events occurred where GPM data provided the forecasters with more information for post and pre-event analysis.

For example, the trial assessment which occurred in 2015 was the first time forecasters in Alaska used the precipitation estimates from GPM. Because these products are derived from a NASA research mission and are considered experimental to NWS, forecasters were evaluating the impact these products could have in operations, and whether there were specific benefits and limitations to their use. A major benefit of these products was that they provide information in data-sparse locations. One limitation identified was the latency, particularly in IMERG; recall that the early versions of these products (V03 in 2015) included processing latencies of about 6 h for the early run of IMERG. The trial assessment partly aimed to determine if forecasters would find value in IMERG despite its latency, and whether they had confidence in these products. A tragic landslide, occurred in Southeast Alaska on 18 August 2015, would shed light on these questions.

At about 10 AM Alaska Daylight Time (AKDT) on 18 August, a landslide of mud, trees, and other entrained debris swept down a hillside on Harbor Mountain and tore through a housing development in Sitka, Alaska, destroying a home and killing three people. In the aftermath, forecasters at the NWS WFO Juneau (AJK) documented the weather events leading up to the landslide and requested some additional imagery from IMERG and GPM swath rain rate estimates to help supplement their post-event analysis. Imagery like Fig. 47.3, a 6-h precipitation accumulation from IMERG, helped forecasters determine where the heaviest rain fell just prior to the landslide event, supplementing the radar information from the Baranof Island Weather Surveillance Radar, 1988, Doppler (WSR-88D) radar, which is partially beam-blocked by the terrain. Forecasters also requested imagery from the DPR on GPM. Despite the experimental nature of these data products, they were able to effectively utilize them for the post-event analysis and official report.

In subsequent assessments, forecasters had more confidence and familiarity with the GPM rain rate estimates and integrated them into their workflow with the ability to quickly synthesize the new dataset with existing observations and tools. For example, as a forecaster stated, "Comparisons between the sparse rain gauge

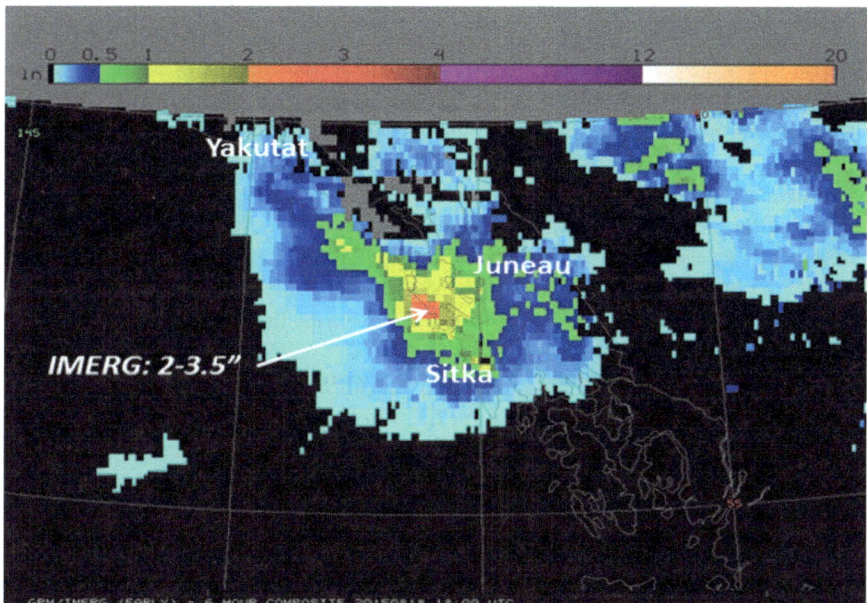

Fig. 47.3 IMERG 6 h precipitation accumulation (inches) 1200–1800 UTC 18 August 2015. IMERG imagery gave forecasters insight into the heaviest rain rates outside of radar coverage. IMERG was used by forecasters to analyze an event that resulted in a mudslide in the Juneau WFO's area of responsibility on 18 August 2015 (annotations by Aaron Jacobs, NWS WFO AJK)

network over SE [Southeast] Alaska and the GPM product revealed generally good agreement. This gave better confidence in using this product to estimate rainfall in data sparse areas." As a result, forecasters utilized the GPM rain rate estimates for another landslide event, this time during the assessment that took place late summer 2017. On 4 September 2017, forecasters observed moderately heavy rains offshore in the GPM swath rain rate estimates prior to a landslide near Sitka, which fortunately caused no fatalities or major damage. With experience from the previous assessments, forecasters were able to utilize the GPM data to assess the landslide risk rather than utilizing the GPM data solely for post-event analysis.

47.3.2   Flooding

Another post-event application of the product was the assessment of a flash flood event at Superman Canyon near Gallup, NM. GPM IMERG was used for a post-event report of a local canyon flash flood fatality by the NWS WFO Albuquerque, NM (ABQ). Operational forecasters augmented the official National Climate Data Center (now known as National Center for Environmental Information, NCEI) Storm Events Database with IMERG 3-h and 6-h accumulated precipitation products. Forecasters gave positive feedback: "This data helped me supplement the info provided by Navajo Police along with the AHPS [Advanced Hydrologic Prediction Service] data to determine that the event occurred on 13 July." The gridded IMERG Early and Late products helped to supplement existing information, like the WSR-88D radar near Albuquerque, which has compromised low-level radar coverage toward northwestern NM, and the accumulation products developed from IMERG, shown in Fig. 47.4, helped forecasters better understand the timing of this event. The daily precipitation from IMERG (Fig. 47.4) provided additional information about how much rain had fallen near Gallop, NM, where the complex terrain blocks the radar and limits surface precipitation estimates. The forecasters already had some experience and confidence in the IMERG dataset and were able to use it as an additional reliable data source to write the post-event analysis and determine the time the event occurred.

In later assessments, forecasters used the GPM swath rain rate estimates to forecast flooding events. In a case from 2016 from NWS WFO AJK, the forecasters had observed several swath rain rate overpasses over the system offshore. The forecasters eventually issued flood advisories and a Special Weather Statement (SPS) discussing the observations from the swath rain rates and the expected impact of the heavy rain in downstream locations. As the event moved onshore, GPM swath rain rate data were very comparable to the available gauge estimates. Seeing the agreement between the satellite estimates and the available gauge data gave the forecasters confidence in the swath rain rates as the event progressed, helping forecasters go from using the datasets in post-event analysis, as shown above in the landslide application, to integrating the GPM data in the real-time forecasting process to assess timing and intensity of rainfall and associated hazards to influence public forecasts and text products as shown here and in the following examples.

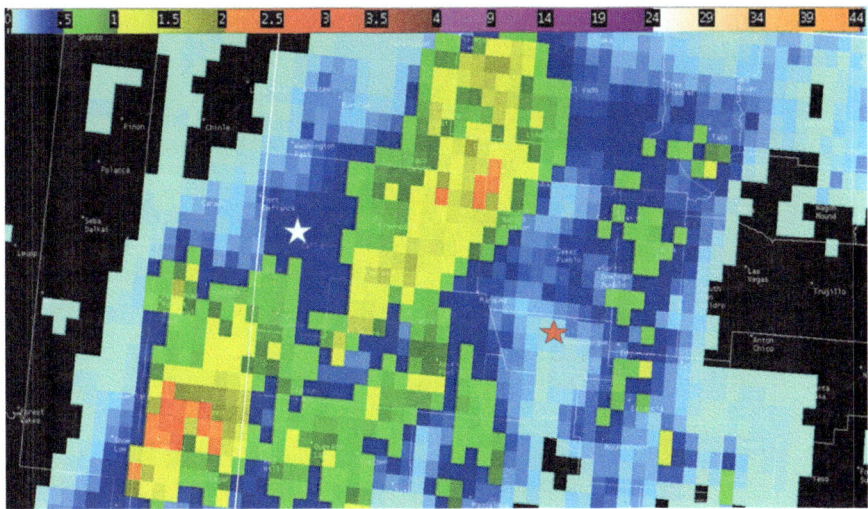

Fig. 47.4 Daily precipitation (inches) from Early IMERG for precipitation ending 1200 UTC 13 July 2015. Gallup, NM, is labeled with a white star, and the ABQ radar is labeled with a red star

47.3.3 Atmospheric Rivers

The GPM precipitation estimate products were also used by forecasters to assess the potential rain rate of incoming systems called atmospheric rivers (ARs). ARs are plumes of water vapor transport that originate in the tropics and are often responsible for flooding events when they reach land, especially in mountainous terrain (Zhu and Newell 1998; Ralph and Dettinger 2011). One challenge of forecasting the effects of an AR is that they are well beyond the reach of ground-based radar, leaving forecasters to observe them using satellite products like the GPM rain rate estimates or use model data to estimate the extent and intensity of the moisture. In fact, AR events are among the cases in which forecasters stated that GPM had the most impact on their operational decisions.

For example, on 20 August 2017, forecasters were tracking the development of an atmospheric river in the North Pacific Ocean, specifically working to target when and where the most intense rainfall was likely to occur. Through an anonymous survey, forecasters stated before the event reached their area of responsibility that, "the [GPM Level 2 swath] rain rates showed a [maximum] about where the second plume / or impulse would be based on satellite vapor imagery and IR [infrared] shots. This [plume] is due into SEAK [Southeast Alaska] Monday night and added confidence to the forecast that the second wave was going to a significant feature." Additional feedback submitted during this event supported the utility and accuracy of the swath rain rates, which they examined alongside other products within their AWIPS system. The need for accurate datasets for atmospheric rivers is so great that, in another AR event, forecasters stated that even the higher latency IMERG product

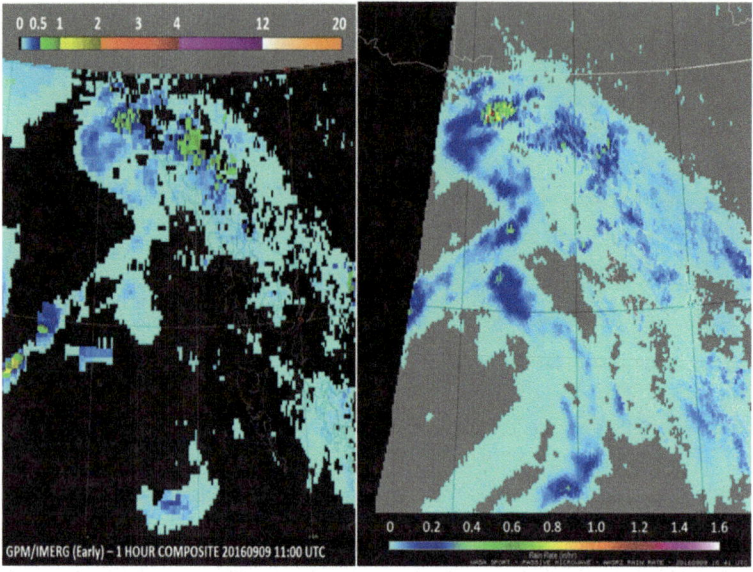

Fig. 47.5 (**a**) IMERG 1-h precipitation accumulation (inches) 1100 UTC 9 September 2016, and (**b**) swath rain rate estimates (in h^{-1}) 1041 UTC 9 September 2016 offshore of southeast Alaska helped a forecaster issue a flood advisory by showing rain rate intensity produced by the system prior to landfall

had a high impact on operations. In a case example from 9 September 2016 (Fig. 47.5), forecasters used the GPM products to issue a flood advisory – anticipating the impact of the precipitation associated with the AR. Despite the 4-h latency of IMERG, forecasters found value in the GPM product as a tool to increase situational awareness of the location, movement, and intensity of offshore precipitation; a forecaster indicated that, "[...] we can look at what rain rates the system was producing over the Gulf and then take those values and move them over the panhandle." Once the systems reach land, forecasters continue to use the satellite data to assess rainfall and flood potential and modify public messaging, particularly in that these datasets help supplement gauge information and the relatively sparse radar available in Southeast Alaska.

47.3.4   Rain Gauge Verification

During the 2016 summer assessment, a forecaster at the APRFC demonstrated a specific use of the GPM precipitation products in their environment. RFCs use precipitation and river gauge data along with other available rainfall estimates to model streamflow in river basins, and in Alaska, gauges are sparse and occasionally faulty. Gauge data is also used in hydrologic models and must be "quality controlled" to prevent faulty data from being passed to the model framework. Therefore, additional datasets are desired to verify the gauge data and supplement spatial gaps where no gauges are present.

An APRFC hydrologist submitted a case focusing on the forecast challenge in the Taiya river basin. The Taiya basin is in steep terrain and is not well represented by the few gauges in the area. A weak atmospheric river reached Southeast Alaska and produced locally heavy rainfall, resulting in the Taiya River reaching moderate flood stage. The anonymous feedback provided for this flooding event included the forecaster showing IMERG data over this challenging location, stating that the rainfall amounts in IMERG appeared to match the known rainfall. Indeed, the IMERG gridded datasets provided value in this and similar cases in which there were data challenges in Southeast Alaska. Recall that IMERG coverage does not go beyond 60°N latitude, covering very little of the Alaskan mainland.

Through collaboration with the GPM Science Team, SPoRT was able to introduce another experimental GPM product to the forecasters from the HQPrecip field which incorporates the high-quality calibrated swath data with the gridded, half-hourly IMERG data. The GPM Science Team was able to prioritize including high-latitude data 1.5 algorithm versions before planned inclusion (G. J. Huffman 2018, personal communication). The resulting product showed Level 3 precipitation estimates wherever there was Level 2 swath data, giving forecasters a gridded dataset that could be incorporated into their existing decision support system highlighting precipitation over portions of the Alaskan interior. Figure 47.6 is an example of the additional higher latitudinal coverage of HQPrecip (Fig. 47.6a) compared to standard IMERG (Fig. 47.6b). The product was launched in time for an assessment activity planned for the summer of 2017, less than a year from the prior assessment in which end users articulated the need for additional IMERG coverage. By providing actionable feedback, forecasters helped spur the development of a new experimental product, demonstrating an example of the R2O/O2R paradigm used by SPoRT.

This new product was put to use in the 2017 assessment series. In one example, hydrologists at the APRFC used HQPrecip to examine the impact of a system reaching landfall on the far western coast of Alaska. In this case, the forecaster used the product to observe a system coming in off the ocean, and was able to follow up with the impact it was having on land in a region with a sparse gauge network, providing an estimate of rainfall there. In another event, the forecaster used the HQPrecip to compare to gauge data, specifically because two gauges were showing much higher rain rates than the other gauges in the vicininty. These applications further highlight the use of this product to improve the continuity of observations available and to help verify or supplement gauge observations.

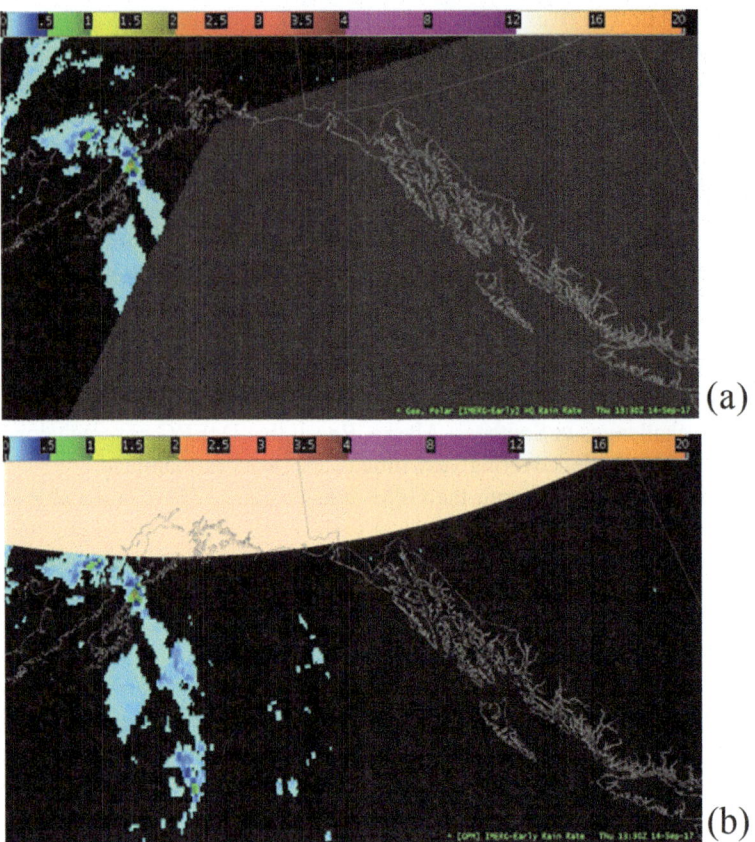

Fig. 47.6 (**a**) HQPrecip precipitation (inches) and (**b**) IMERG-Early precipitation (inches) for 0800 UTC on 07 February 2019 over Alaska

47.3.5 Additional Applications

47.3.5.1 Supporting High Impact Events

The hurricane season of 2017 brought several significant storms to the Atlantic basin including three major hurricanes with significant impacts to the United States: Hurricanes Harvey, Irma, and Maria. Hurricane Harvey produced record-setting rainfall and flooding within the Houston metro area. NASA remote sensing including observations from GPM and Soil Moisture Active Passive (SMAP) captured torrential rains and lingering, high soil moisture content with regions prone to flooding well after the event. SPoRT operates the NASA Land Information System in collaboration with NASA Goddard, and the integration of NOAA radar-estimated rainfall and atmospheric forcing captured the signature of Harvey's rainfall in southeastern Texas as saturated soils persisted in the weeks and months following the storm. Hurricane Irma's faster movement brought different impacts to Florida, and scientists within NASA's Earth Science programs provided flood mapping, damage mapping, and tracking of the loss or restoration of lights and power through various optical and synthetic aperture radar remote sensing techniques. Many of these same techniques were applied to monitoring the impacts from Hurricane Maria in Puerto Rico and the US Virgin Islands, which experienced the brunt of the storm, and helped to document and monitor the long-lasting and continuing impacts of Maria on Puerto Rico's infrastructure. Prior to Hurricane Maria's landfall, SPoRT responded to a request from the NWS Southeast RFC for satellite-based QPE over Puerto Rico due to fears that the island's two Doppler radars would be damaged by the storm. SPoRT quickly provided GPM IMERG data to both the RFC and the Puerto Rico WFO who were then able to access the GPM observations online and within AWIPS. In fact, During Maria, both radars sustained significant damage leaving hydrologic forecasters without a way to determine rainfall amounts for flood forecasting. Therefore, GPM played a critical role in helping NWS forecasters map heavy rainfall and other impacts following damage to and loss of the WSR-88D weather radar. Data products generated by GPM and delivered to NWS partners in Puerto Rico by SPoRT were commented on by NWS staff for providing crucial information on rainfall amounts during the radar outage, supplemented with other agency information including United States Geological Survey (USGS) stream gauges and other satellite products from the GOES series.

47.3.5.2 Precipitation – Applications Related to Snowfall

SPoRT has closely collaborated with NESDIS since 2014 to engage with NWS forecasters to assess the application of the passive-microwave derived NESDIS Snowfall Rate product (Meng et al. 2017a) in NWS operations (Meng et al. 2017b; Ferraro et al. 2018). The first assessment was conducted during the 2014 winter season to determine the operational utility of the product as it relates to radar gaps, beam blockage, and tracking of snowfall rate maxima. In 2014, the initial snowfall rate product was derived from the Advanced Microwave Sounding Unit (AMSU) and MHS sensors onboard NOAA-18, -19 and MetOp-A, -B. The product was produced by the Satellite Climate Studies Branch at NOAA NESDIS Center for Satellite Applications and Research (STAR) and SPoRT converted it to an AWIPS and N-AWIPS (i.e., National Center for Environmental Prediction AWIPS) format for distribution to NWS WFOs and National Centers, developed applications-based training, and assessed the product with end users. At that time, up to 8 land-only snowfall rate retrievals were available per day, limited to regions with a surface temperature greater than $-6\,°C$ (22 °F) and maximum detectable snowfall rates of 50.8 mm h^{-1} (2 in h^{-1}) (for snow to liquid ratio of 10 to 1). Forecaster feedback from the first assessment was positive in that the product was useful for improving data coverage where radar was lacking and valuable for tracking snowfall maxima in combination with other satellite observations (Meng et al. 2017b). During the first assessment, forecasters discovered a novel application of the data as a short-term forecasting tool to increase confidence and situational awareness of snowfall development aloft and cloud seeding before snow is detected by ground based radar (Meng et al. 2017b).

At times the product was limited by the latency and temporal gaps between overpasses, so the product developers addressed forecaster concerns by utilizing Direct Broadcast data where it is possible to reduce latency to 30–60 min and developed a radar-merged snowfall rate product with a temporal frequency of 10 min. This means that the passive microwave swath is complemented by National Severe Storm Laboratory's Multi-Radar/Multi-Sensor (MRMS; Zhang et al. 2016) data to increase situational awareness of snowfall rate and fill in the gaps between polar-orbiter overpasses. As part of the R2O/O2R paradigm, the developers at STAR added S-NPP ATMS snowfall rate retrievals to the product and improved the algorithm for use in colder regimes in order to expand to an Alaska domain (Meng et al. 2017b). After these changes, the product was assessed again during the 2016 winter season during which forecasters found the swaths and merged snowfall rate products valuable for identifying snowfall in data-deprived regions. One forecaster example highlights use of the product to increase situational awareness of a rain to snow transition event. A forecaster from the NWS WFO Charleston, WV, indicated the utility of the product for a rain to snow transition event with this anonymous feedback: "Much of the precipitation across West Virginia was still in the form of rain…with an area of snow extending from northwest PA across Ohio into southwest portions of the state. There appears to be several observations of rain across Ohio with surface temperatures of 32–35 °F where the [SFR] Snowfall Rate product indicated snow in the clouds". Another example highlights use of the product during a widespread heavy snow event on 23 January 2016. A forecaster at the NWS WFO Sterling, VA, noted in a forecast discussion that the product was observing 1–2 inches of snow an hour, and warned of whiteout conditions in the impacted areas. In this event, the snowfall rate product compared well with ground observations of snowfall in both amounts and areal extent.

Prior to the 2018 winter assessment, additional observations were added from the SSMIS instruments onboard DMSP -F16, -F17, and -F18 as well as GMI onboard the NASA GPM Core Observatory and there were also improvements to the snowfall detection algorithm (Meng et al. 2017b). During the 2018 assessment, forecasters continued to provide feedback that the NESDIS Snowfall Rate product is valuable for use in radar-deprived regions, depicting where the heaviest snow is falling, and verifying model performance, anticipating rain to snow transitions, and identifying snowfall aloft before detection by WSR-88D radar. SPoRT and NESDIS have continued to collaborate on R2O/O2R activities since 2014 to transition and assess the NESDIS Snowfall Rate product in the operational environment. Targeted training, assessments, and deliberate interaction with forecasters and algorithm developers has led to product improvements and increased usability and applicability of the NESDIS Snowfall Rate product in operations (Meng et al. 2017b; Ferraro et al. 2018).

47.3.5.3   Hurricane Structure and Intensity

Shortly after the launch of the GPM Core Observatory in 2014, SPoRT began providing experimental passive microwave products to the National Hurricane Center (NHC) to supplement existing observations in the N-AWIPS display system. SPoRT provided GPM's intercalibrated brightness temperatures from the 37 and 89 GHz channels from GMI and the suite of GPM Constellation Satellites. In addition to providing the 37 and 89 GHz brightness temperatures from GPM, SPoRT developed a technique to provide passive microwave RGB imagery (Lee et al. 2002) to forecasters in N-AWIPS format. RGB imagery combines radiometer channels into the red, green, and blue color components of the data display to leverage the advantages of each channel to allow for quick identification of features without analyzing the single channels independently. The resulting RGB image provides a qualitative means to observe an emphatic view of extent and intensity of precipitation and locations of deep, ice-laden convection, both of which are beneficial for identifying hurricane eye location and eyewall structure. Passive microwave RGB imagery and single channel brightness temperatures are used by the NHC to identify the tropical storm or hurricane center ("center fix"), used in intensity measurements, to readily identify vigorous convective activity in tropical storms, and to infer locations of heavy precipitation. For example, during Hurricane Irma GMI data helped NHC forecasters recognize that the storm was undergoing an eyewall replacement during the overpass. The forecasters modified their intensity estimates accordingly and noted the eyewall changes, citing GMI in the publically issued text product. NHC continues to extensively utilize and cite these passive microwave products in operations leveraging the ability for observations in cloudy regions to detect features of interest to determine hurricane structure and intensity.

47.4   Conclusions

Since the launch of GPM in 2014, NASA SPoRT has collaborated with NWS forecasters to experimentally transition the research-quality datasets developed from the GPROF algorithm to the operational environment to assess their utility for a variety of applications. With a main focus on forecasting precipitation and related hazards as discussed in this paper, SPoRT has also collaborated with forecasters to provide GPM data to supplement the NWS Puerto Rico weather radar outage during the 2017 hurricane season, applications related to passive microwave-derived snowfall rate, and forecasting hurricane structure and intensity. SPoRT uses an interactive R2O/O2R paradigm that involves the end-user in the entire process to match research products to forecasting challenges, develop a solution, provide data in appropriate formats, and develop applications-based-training to demonstrate how the data can be analyzed in tandem with familiar products. These activities occur in a focused testbed environment in which forecasters have the opportunity to integrate and use the datasets in the operational environment and provide actionable feedback on how the dataset can be improved to meet their needs. This process results in incremental improvements to the dataset to benefit operational end users, tailoring the products to their environment and applications.

This paper highlights applications of GPM products that were a direct result of experimentally transitioning value-added GPM products to NWS forecasters and hydrologists. The novel applications highlighted in this paper related to GPM rain rate estimates include understanding and forecasting landslide threat, observing and responding to flooding incidents, forecasting the location and intensity of precipitation associated with atmospheric rivers, and supplementing and verifying rain gauge data. These activities highlighted are a testament to the interactive R2O/O2R process whereby SPoRT has bridged the gap between operational end users and mission scientists to expand the utility of research products in the operational environment for integration into weather forecasting applications. For example, by engaging with the end users and the GPM science team, it was determined that the Level 2 swath rain rates were most useful for identifying rainfall location and intensity in near-real time WFO operations while the higher latency IMERG data were more applicable in RFCs, in WFOs for post-event analysis, and in some cases, during offshore events to provide areal extent and intensity information in data-sparse regions. Forecasters also provided information on limitations such as detection efficiency, noise, spatial coverage, and latency and SPoRT collaborated with the GPM Science Team to either improve the products on the user side or prioritize solutions in the next algorithm update. This R2O/O2R interaction helped accelerate the usability of GPM data in NWS operations. As of this writing, passive microwave single channel imagery and IMERG precipitation type and rate are soon to be delivered to forecasters via the operational NOAA data stream (e.g., Satellite Broadcast Network). SPoRT continues to provide experimental data to operational end users and look toward the next-generation of NASA's precipitation missions in development now that have the potential to improve short-term forecasts on a regional scale.

Acknowledgments NASA SPoRT is appreciative of the NWS forecasters and hydrologists who have voluntarily used their time to evaluate GPM products in the often fast-paced operational environment and provide valuable feedback in the form of emails and feedback forms. SPoRT would also like to thank the GPM Scientists for valuable input and interaction regarding data products and algorithm updates. This work was supported by the NASA Research and Analysis Program as part of the Short-term Prediction Research and Transition Center at Marshall Space Flight Center. GPM Constellation Level 2 swath rain rate, Level 1C calibrated brightness temperature, and both Early and Late IMERG products are acquired by SPoRT in near real time from NASA's Precipitation Processing System. The NESDIS snowfall rate is provided to SPoRT in near-real time by Huan Meng and Jun Dong of the Satellite Climate Studies Branch at NOAA/NESDIS/STAR.

References

Boukabara, S. A., Garrett, K., Chen, W., Iturbide-Sanchez, F., Grassotti, C., Kongoli, C., Chen, R., Liu, Q., Yan, B., Weng, F., Ferraro, R. R., Kleespies, T. J., & Meng, H. (2011). MiRS: An all-weather 1DVAR satellite data assimilation and retrieval system. *IEEE Transactions on Geoscience and Remote Sensing, 49*, 3249–3272. https://doi.org/10.1109/TGRS.2011.2158438.

Buma, B., & Johnson, A. C. (2015). The role of windstorm exposure and yellow cedar decline on landslide susceptibility in southeast Alaskan temperate rainforests. *Geomorphology, 228*, 504–511. https://doi.org/10.1016/j.geomorph.2014.10.014.

Dai, F. C., Lee, C. F., & Ngai, Y. Y. (2002). Landslide risk assessment and management: An overview. *Engineering Geology, 64*, 65–87. https://doi.org/10.1016/S0013-7952(01)00093-X.

Ferraro, R., Meng, H., Zavodsky, B., Kusselson, S., Kann, D., Guyer, B., Jacobs, A., Perfater, S., Folmer, M., Dong, J., Kongoli, C., Yan, B., Wang, N.-Y., & Zhao, L. (2018). Snowfall rates from satellite data help weather forecasters. *Eos, 99*. https://doi.org/10.1029/2018EO096715.

Goodman, S. J., Lapenta, W., Jedlovec, G., Dodge, J., & Bradshaw, T. (2004). *The NASA short-term prediction research and transition (SPoRT) center: A Collaborative model for accelerating research into operations.* Available at https://ntrs.nasa.gov/search.jsp?R=20040028009. Last accessed 28 Nov 2018.

GPM Science Team. (2014a). *GPM constellation level 1C common calibrated brightness temperatures, version 2a, 4a, 5, NASA precipitation processing system, obtained real-time since April 2014 to present.* Available at ftp://jsimpson.pps.eosdis.nasa.gov/data/1C/. Last accessed 28 Nov 2018.

GPM Science Team. (2014b). *GPM constellation GPROF precipitation rate, version 3a, 4a, 5, NASA precipitation processing system, obtained real-time since April 2015 to present.* Available at ftp://jsimpson.pps.eosdis.nasa.gov/data/GPROF. Last accessed 28 Nov 2018.

GPM Science Team. (2014c). *GPM IMERG 3a, 3b, 4b, 5, NASA precipitation processing system, obtained real-time since April 2015 to present.* Available at ftp://jsimpson.pps.eosdis.nasa.gov/NRTPUB/imerg/. Last accessed 28 Nov 2018.

Hong, Y., Adler, R. F., & Huffman, G. J. (2007). Use of satellite remote sensing data in the mapping of global landslide susceptibility. *Natural Hazards, 43*, 245–256. https://doi.org/10.1007/s11069-006-9104-z.

Hou, A. Y., Kakar, R. K., Neeck, S., Azarbarzin, A. A., Kummerow, C. D., Kojima, M., Oki, R., Nakamura, K., & Iguchi, T. (2014). The Global Precipitation Measurement mission. *Bulletin of the American Meteorological Society, 95*, 701–722. https://doi.org/10.1175/BAMS-D-13-00164.1.

Huffman, G. J., Bolvin, D. T., Braithwaite, D., Hsu, K.-L., & Joyce, R. (2015). *Algorithm theoretical basis document (ATBD) version 4.5: NASA global precipitation measurement (GPM) Integrated Multi-satellitE Retrievals for GPM (IMERG).* Greenbelt: NASA. Available at https://pmm.nasa.gov/sites/default/files/document_files/IMERG_ATBD_V4.5.pdf. Last accessed 27 Nov 2018.

Huffman, G. J., Bolvin, D. T., Braithwaite, D., Hsu, K.-L., Joyce, R., Kidd, C., Nelkin, E. J., Sorooshian, S., Tan, J., & Xie, P. (2018). Algorithm Theoretical Basis Document (ATBD) Version 5.2 for NASA Global Precipitation Measurement (GPM) Integrated Multi-satellitE Retrievals for GPM (IMERG), NASA 34 pp. Available at https://docserver.gesdisc.eosdis.nasa.gov/public/project/GPM/IMERG_ATBD_V5.pdf. Last accessed 28 Nov 2018.

Jedlovec, G. (2013). Transitioning research satellite data to the operational weather community: The SPoRT paradigm. *IEEE Geoscience and Remote Sensing Magazine, 1*, 62–66. https://doi.org/10.1109/MGRS.2013.2244704.

Kääb, A., Wessels, R., Haeberli, W., Huggel, C., Kargel, J. S., & Khalsa, S. J. S. (2003). Rapid ASTER imaging facilitates timely assessment of glacier hazards and disasters. *Eos, 84*(13), 117–121. https://doi.org/10.1029/2003EO130001.

Kongoli, C., Meng, H., Dong, J., & Ferraro, R. (2015). A snowfall detection algorithm over land utilizing high-frequency passive microwave measurements – Application to ATMS. *Journal of Geophysical Research, 120*, 1918–1932. https://doi.org/10.1002/2014JD022427.

Kongoli, C., Meng, H., Dong, J., & Ferraro, R. (2018). A hybrid snowfall detection method from satellite passive microwave measurements and global forecast models. *Quarterly Journal of the Royal Meteorological Society, 144*(S1), 120–132. https://doi.org/10.1002/qj.3270.

Kuligowski, R. J. (2002). A self-calibrating real-time GOES rainfall algorithm for short-term rainfall estimates. *Journal of Hydrometeorology, 3*(2), 112–130. https://doi.org/10.1175/1525-7541(2002)003<0112:ASCRTG>2.0.CO;2.

Kummerow, C. D., Hong, Y., Olson, W. S., Yang, S., Adler, R. F., McCollum, J., Ferraro, R. R., Petty, G., Shin, D.-B., & Wilheit, T. T. (2001). The evolution of the Goddard Profiling Algorithm (GPROF) for rainfall estimation from passive microwave sensors. *Journal of Applied Meteorology, 40*(11), 1801–1820. https://doi.org/10.1175/1520-0450(2001)040<1801: TEOTGP>2.0.CO;2.

Lee, T. F., Turk, F. J., Hawkins, J., & Richardson, K. (2002). Interpretation of TRMM TMI images of tropical cyclones. *Earth Interactions, 6*, 1–17. https://doi.org/10.1175/1087-3562(2002) 006<0001:IOTTIO>2.0.CO;2.

LeRoy, A., Smith, M., Jacobs, A., Case, J. L., & Holloway, E. (2017a). Transition and assessment of GPM rain rate products to operations. In *Proceedings of the 7th conference on transition of research to operations*, 97th AMS annual meeting, Seattle, WA, 7.3. Available at https://ams. confex.com/ams/97Annual/webprogram/Paper308534.html. Last accessed 29 Nov 2018.

LeRoy, A., Smith, M., & Huffman, G. J. (2017b). *NASA SPoRT assessment and application of GPM rain rates in operational hydrology*. EUMETSAT Meteor. Satellite conference, Rome. Available at https://www.eumetsat.int/website/home/News/ConferencesandEvents/ PreviousEvents/DAT_3212307.html. Last accessed 29 Nov 2018.

Likert, R. (1932). *A technique for the measurement of attitudes*. Archives of psychology. Available at https://legacy.voteview.com/pdf/Likert_1932.pdf. Last accessed 29 Nov 2018.

Meng, H., Dong, J., Ferraro, R., Yan, B., Zhao, L., Kongoli, C., Wang, N.-Y., & Zavodsky, B. (2017a). A 1DVAR-based snowfall rate retrieval algorithm for passive microwave radiometers. *Journal of Geophysical Research, 122*, 6520–6540. https://doi.org/10.1002/ 2016JD026325.

Meng, H., Berndt, E., & White, K. (2017b). NESDIS snowfall rate product and its applications. *JPSS Science Seminar*. Available at http://www.jpss.noaa.gov/assets/pdfs/science_seminars/pre sentations/SnowfallRate_Meng_White_Berndt_2017.pdf. Last accessed 29 Nov 2018.

Ralph, F. M., & Dettinger, M. D. (2011). Storms, floods, and the science of atmospheric rivers. *Eos, 92*(32), 265–266. https://doi.org/10.1029/2011EO320001.

Smith, M., Leroy, A., Case, J., & Bolvin, D. T. (2016). Initial results of integrating GPM/IMERG precipitation data into operational forecasting environments. In *Proceedings of the 30th conference on hydrology*, 96th AMS annual meeting, New Orleans, LA, P9.6. Available at https:// ams.confex.com/ams/96Annual/webprogram/Paper288254.html. Last accessed 29 Nov 2018.

Unidata. (2018). *Local data manager [software]*. Boulder: UCAR/Unidata. Available at https://doi. org/10.5065/D64J0CT0. Last accessed 29 Nov 2018.

Zhang, J., Howard, K., Langston, C., Kaney, B., Qi, Y., Tang, L., Grams, H., Wang, Y., Cocks, S., Martinaitis, S., Arthur, A., Cooper, K., Brogden, J., & Kitzmiller, D. (2016). Multi-Radar Multi-Sensor (MRMS) quantitative precipitation estimation: Initial operating capabilities. *Bulletin of the American Meteorological Society, 97*, 621–638. https://doi.org/10.1175/BAMS-D-14-00174.1.

Zhu, Y., & Newell, R. E. (1998). A proposed algorithm for moisture fluxes from atmospheric rivers. *Monthly Weather Review, 126*, 725–735. https://doi.org/10.1175/1520-0493(1998)126<0725: APAFMF>2.0.CO;2.

Chapter 48
Assimilation of Precipitation Observations from Space into Numerical Weather Prediction (NWP)

Sid-Ahmed Boukabara, Erin Jones, Alan Geer, Masahiro Kazumori, Kevin Garrett, and Eric Maddy

Abstract This chapter focuses on one of the major applications of space observations impacted by precipitation: the use of these data in Numerical Weather Prediction (NWP) data assimilation (DA) systems. Applications in global systems will be the focus of this study. The assimilation of observations impacted by precipitation is done either directly or after geophysical data (generally precipitation amount) is inverted. Assimilating and inverting data using physical retrieval approaches are inherently intertwined and share the same objectives. After all, the purpose of retrieving precipitation from satellite data aims at extracting the precipitation signal from the measurements while accounting for non-precipitation signals in the same measurements. This is essentially the same objective as assimilating the precipitation-impacted data into NWP: the purpose being to extract the geophysical information from the measurements and use it to help generate a geophysical analysis. Different numerical weather prediction centers are adopting different approaches on how to handle the assimilation of precipitation data. Some are directly assimilating radiances and some are using pre-processed retrievals. The goal of this chapter is not to assess the pros and cons of the different methods, but instead to give the reader an overview of the approaches adopted or being explored. Assimilating precipitation-impacted measurements is an ongoing active effort in the research community, as well as in major operational NWP centers around the world. To remain concise, we will provide broad overviews of the current trends from multiple institutions, and do not pretend to be exhaustive in describing all approaches dealing with this issue. Instead, multiple and diverse references will be included to allow readers to obtain further details on a diverse set of methodologies being pursued. For

S.-A. Boukabara (✉) · E. Jones · K. Garrett · E. Maddy
NOAA/NESDIS/STAR, College Park, MD, USA
e-mail: sid.boukabara@noaa.gov

A. Geer
ECMWF, Reading, Berkshire, UK

M. Kazumori
Japan Meteorological Agency, Tokyo, Japan

© Springer Nature Switzerland AG 2020
V. Levizzani et al. (eds.), *Satellite Precipitation Measurement*, Advances in Global Change Research 69, https://doi.org/10.1007/978-3-030-35798-6_24

completeness, the chapter will be outlined in a way that allows the reader a gradual understanding of (1) the main motivations behind the significant effort being undertaken by the international community to tackle this problem, (2) the challenges facing us when handling cloud and rain-impacted space observations, (3) the methodologies adopted by different teams and their potential shortfalls when addressing some of those challenges, and (4) some of the major results when using these approaches in an NWP context.

Keywords Precipitation · Rainfall · Numerical weather prediction · Data assimilation · Satellite radiances · Particle size · Microphysics · Total precipitable water · Radiative transfer model · Field of view · CRTM · Rain water path · Graupel water path · Microwave · Brightness temperature · Bayesian technique · GPROF · Ensemble forecast · AMSR2 · AMSR-E · GPM · GMI · GCOM-W · SSMIS · MHS · Global spectral model

48.1 Introduction

In this chapter, we will examine a diverse set of meteorological and hydrological applications that benefit from satellite-based measurement of precipitation, with an important impact on large parts of society and industry. As will be described in detail later, we will focus primarily on the measurement of precipitation from space using passive microwave data. The choice of focusing on the passive microwave signal is driven by the advantage that the passive microwave signal is directly linked to the precipitation itself. The signal is also usually obtained from a large number of sensors, offering therefore significantly superior temporal and spatial coverages. As shown in previous chapters however, precipitation is also measured using active microwave as well as infrared sensors.

48.1.1 Understanding the Remote Sensing of Precipitation

Figure 48.1 presents an overview of the different mechanisms that play a role, with different degrees of impact, in the measurement of precipitation from space. The measurement from a spaceborne sensor is usually a combination of multiple geophysical signals (in the case of passive radiometers, usually called radiances or brightness temperatures). It is important to note that once this composite geophysical signal reaches the sensor's receptor, it will also be impacted by many sensor-specific factors before being registered as signal by the instrument. These factors include microwave waveguide efficiencies, receptor antenna patterns, gain factor, instrumental noise (NeDT), etc. The method of extracting the geophysical *composite* signal from the actual *noisy* instrument measurement relies on sophisticated radiometric models (calibration process), and is not the scope of this chapter. In the

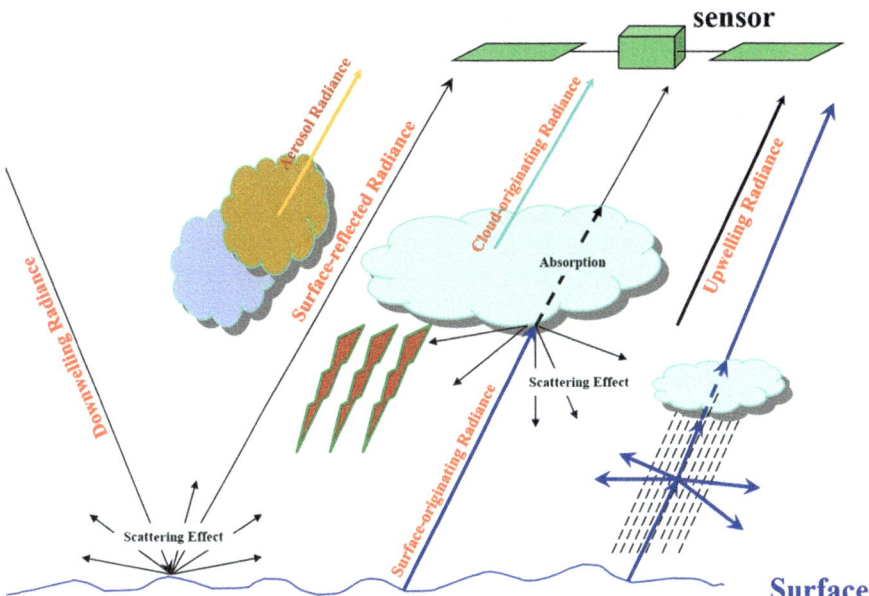

Fig. 48.1 Measuring the environmental state (including precipitation) from space. The signal measured at the sensor level is a composite signal that includes the effect of absorption and scattering from the surface, the dry atmosphere, cloud, and precipitation. The degree of sensitivity of the signal to these parameters depends heavily on the operating frequencies and other sensor characteristics such as viewing angle, polarization, etc. See text for more details

following, we will assume that the precipitation signal we are referring to is cleared of calibration and radiometric issues. Understanding fully the components of the geophysical signals, and how they vary spectrally, is important for handling this signal in a DA or physical inversion context.

The geophysical signal measured by spaceborne instruments is a composite of multiple signals -see Fig. 48.1. These include: (1) the upwelling radiance, or the signal emanating from the dry atmosphere (free of cloud and precipitation), which is generally sensitive to pressure, temperature, and moisture; (2) the cloud-originating radiance, or the signal from suspended cloud layers, transmitted through the upper part of the atmosphere and generally sensitive to the same parameters as the upwelling radiance as well as to cloud characteristics (cloud amount, particle size and density, vertical distribution, etc.); (3) the rain-originated radiance, or the signal from precipitating clouds that is transmitted through the atmosphere and sensitive to pressure, temperature, and moisture as well as precipitating cloud parameters (e.g., rain amount, particle size, shape, distribution, density, etc.); (4) the surface-originating radiance, or the signal that emanates from the surface itself and is transmitted through the dry atmosphere. This surface signal will vary with the surface type (ocean, land, cryosphere), each having specific emissivity and reflectivity characteristics. The composite precipitation signal could also include (5) the same surface-originating radiance but partially absorbed and/or scattered through the atmosphere (through precipitating and suspended clouds for instance), before reaching the sensor receptors; (6) the downwelling radiance: signal that is emanating from the atmosphere (all components) down to the surface before it gets reflected by the surface and transmitted through the atmosphere and reaching the sensor (either directly and/or after absorption/scattering by atmospheric particles). All these radiative effects of cloud particles depend on the spectrally-varying optical properties of these components (absorption and scattering/extinction coefficients). This means that some types of clouds and/or aerosols could have significant or very little to no impact on the space-based measurement, depending on the spectral range that the measurement is made at. The other spectrally-dependent features of the radiances come from the emittance and reflectance of the surface. Indeed, the spectrum of these parameters will drive the degree of impact the surface-originating signals will have on the overall signal. The emittance and reflectance depend on the emissivity and reflectivity of the surface as well as its temperature. The emissivity and reflectivity typically depend on the geometry of the surface (to what degree the surface is specular versus Lambertian), in some cases its temperature, and the intrinsic characteristics of the surface (e.g., type of surface). These varying sensitivities to different geophysical parameters at different frequencies drive the information content of the space-based observations. The purpose of remote sensing and satellite DA is to extract the information needed from the signal while accounting for the other parameters that also impact it. In DA and physically-based remote sensing, this usually requires the ability to simulate measurements using a forward operator (i.e., a radiative transfer, or RT, model). All these individual signals from different components of the environment (surface, atmosphere, hydrometeors, etc.) have to be accounted for in simulating the measurement using a forward operator. While the description given above of the sensor measurement composition as illustrated in Fig. 48.1, is simplified for didactic reasons, it is worth noting that there are many complicating factors that have to be accounted as well. For example, signal from the surface is sometimes scattered back (by clouds and precipitation) in different directions including towards the surface and toward other cloud layers. This reverberation of signals across the atmosphere and the surface is usually labeled multiple scattering, which is also modeled by RT models, usually at the expense of higher computation time. As another example of complicating factors, it is also worth noting that the cloud water phase plays a significant role in altering or modulating the signals. Indeed, cloud could be composed of precipitating or suspended water droplets in different phases (liquid, frozen or mixed-phase). This has, again depending on the spectral region in question, a dramatic impact on the optical properties and therefore on the overall signal. The physically-based remote sensing approaches and the DA methodologies all aspire to extract useful geophysical information from these complex space-based measurements of precipitation. This useful geophysical information could be either the precipitation information itself or, as importantly, information on the temperature and moisture profile, for example, in the presence of precipitation.

48.1.2 Applications of Space-Borne Precipitation Measurements

Precipitation measurements from space provide global coverage of both rain and snowfall, and therefore provide valuable information to a variety of environmental applications. Climatologies of global precipitation observations help to identify shifts in precipitation, regional, annual or inter-annual patterns, and overall global circulation patterns (Jones et al. 2004; Lyon and Barnston 2005). For environmental monitoring, precipitation observations are used to assess drought, or other risks to environment and health. Precipitation observations are vital for hydrological applications, specifically monitoring and forecasting stream flows and reservoir levels. And for weather prediction, observations are used to provide nowcasting guidance on flash flooding or other extreme weather threats, along with improving the initial conditions (analyses) for regional and global NWP models (Kirschbaum et al. 2017).

Weather prediction applications for space-borne precipitation measurements rely on the inference of precipitation coverage and intensity from radiometric observations. For example, a forecaster viewing cold brightness temperatures from a Geostationary satellite infrared imager window channel may infer convection and heavy precipitation. Based on correlations between precipitation and radiometric observations, various statistical or empirical approaches can be adopted by computer algorithms to infer rain or snowfall. These so-called retrievals have the ability to process large volumes of satellite observations, from a variety of types of platforms and sensors, and provide precipitation information in near-real-time to a variety of downstream applications in weather prediction. Examples of retrieval methodologies may be found for infrared radiances (Scofield and Kuligowski 2003), infrared and passive microwave radiances (Joyce et al. 2004), passive microwave radiances (Iturbide-Sanchez et al. 2011), and active microwave or combined active/passive sensors (Skofronick-Jackson et al. 2017).

Similar approaches may also be adopted by NWP DA systems for use of precipitation information in weather prediction models. However, in typical NWP DA applications, precipitation information in the form of radiances is often removed from the assimilation by quality control (QC) procedures, or the signal itself is removed from the satellite observation before it is assimilated, as is the case in cloud-clearing (Wang et al. 2017; Reale et al. 2018). Another approach is to assimilate the precipitation retrievals themselves (Mahfouf et al. 2005). For either case, in the construct of DA, the assumptions made about the hydrometeor type, phase, vertical distribution, size, density, non-linear occurrence in time and space, background error covariance, and mapping to optical properties (used by the forward operator) all pose challenges for producing an accurate and properly balanced analysis. This chapter will highlight further those challenges for assimilating precipitation observations in state-of-the-art NWP DA systems and some approaches to overcome them.

48.2 Precipitation Measurement from Space: Sensitivity Assessment

In order to fully exploit satellite measurements (through either DA or physical retrievals), we have to understand the composition of these measurements like we did in Sect. 48.1 but we also have to know the degree of sensitivity to these parameters. This will account for sensitivities in the assimilation process (or in the physical retrieval process). In other words, we have to understand what the essential parameters having an impact on the observations are, how to simulate their impacts on measurements, how to account for the natural correlations between them, how to describe their spatial and temporal characteristics, etc. For this reason, we will now highlight the major parameters that spaceborne observations of precipitation are sensitive to. This will inform us on the geophysical information content of hydrometeor-impacted data. Later on, we will also focus on their variability in time and space, since these two aspects play a critical role in determining what approach is best suited for the inversion of the observations into geophysical data. As mentioned previously, we will focus mainly on microwave observations since they have a direct linkage to hydrometeors. Data from other spectral regions such as infrared tend to be saturated quickly and at higher altitudes due to the strong impact of cloud in this part of the spectrum. Their linkages to actual precipitation on the ground are thus usually by inference.

48.2.1 Sensitivity to Microphysical Properties

In typical microwave channels used currently for satellite precipitation measurements, the most important parameters impacting the signal relate to macro and microphysical properties of the hydrometeors themselves. Figure 48.2 assesses the sensitivity of the signal using the Community Radiative Transfer Model (CRTM) (Chen et al. 2008; Ding et al. 2011; Liu and Boukabara 2014) to different hydrometeor properties.

Note that almost all microphysical properties have impact on the signal, especially in high microwave frequencies (above 80 GHz), but for simplicity, we focus on the most important ones: the water phase (liquid vs frozen), the total amount of precipitating water, and the effective radius of particle size. Figure 48.2 shows the simulated spectral variation of remotely sensed brightness temperatures (top panels) and its departure from clear-sky simulations (bottom panels) to illustrate the hydrometeor signal. These are shown for typical liquid precipitation (left panels, labeled Rain Water Path, RWP) and frozen precipitation of graupel-size particles (right panels, labeled Graupel Water Path, GWP). The different curves represent several variations of the amount and particle size combination. Solid lines represent simulations using the nominal particle size of 500 μm while the dashed lines represent

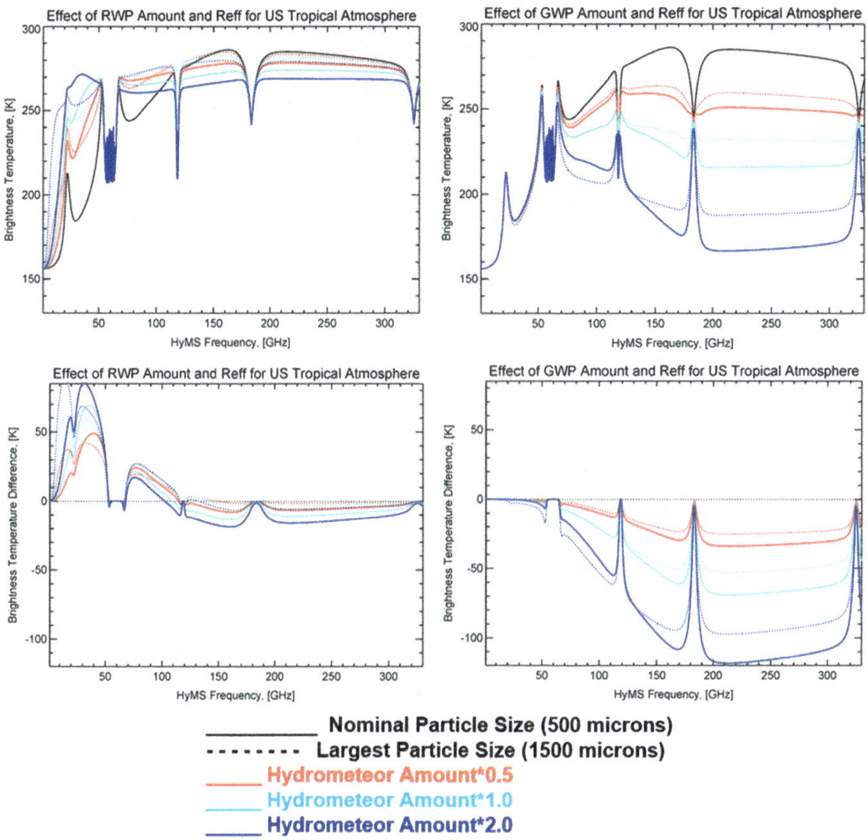

Fig. 48.2 Effect of rain and ice/graupel amounts and particle effective radius on the microwave spectrum from 3 to 330 GHz using a US Tropical Atmosphere. The top panels show the actual simulations of measurements for different scaling factors of hydrometeors amounts and two specific particle sizes (500 and 1500 μm), while the bottom panels show the departures of those simulations from clear-sky (hydrometeor-free) simulations. In the top panels, the black line represents the microwave spectrum in the absence of cloud or precipitation. HyMS represents a hypothetical hyperspectral microwave sensor with measurements ranging from 3 to 330 GHz with a 0.1 GHz spectral resolution. See text for more details

larger particle size of 1500 μm. The colors represent scaled total amounts, with scaling factors ranging from 0.5 (red) to 1.5 (dark blue).

The first striking feature is that the impacts of the water phase (frozen and/or liquid precipitation) on the signal are significantly different. Liquid phase tends to have a bigger impact in the low frequency range of the microwave spectrum, while ice tends to have a much bigger impact in the higher end of the spectrum. It can also be seen that the hydrometeors impact on brightness temperature is significant and spectrally-shaped. Reaching maxima of 90 K in the low frequencies for warm rain (absorption), and around 120 K (depression of measurements) from scattering by frozen precipitation in the higher frequency range. It is also evident that while the first order impact is generally due to the hydrometeor amount, the impact of the particle size is significant as well. For the same amount for example (same Fig. 48.2), the impact due to varying the particle size could reach 10–20 K in some window regions. It is however noticeable that at low frequencies, the impact of the precipitation particle size could be more significant than the impact of the amount itself. This is not unexpected since the impact of hydrometeors in the microwave region is only measurable in the resonance region. In other words, electromagnetic waves with a certain wavelength interact with particles of certain sizes – within resonance range. It is only within this range that the precipitation amount becomes important. Certain precipitation situations would therefore be transparent to the microwave signal at certain frequencies (outside the resonance region/range) if their particle sizes are too small to resonate with the electromagnetic signal. In this case, the hydrometeor amount would have little to no impact. This is why the particle size has a particularly important impact on the precipitation signal. It is worth noting that the hydrometeors (with their type, size, shape, density, etc.) all have specific optical properties (absorption, scattering, extinction) by which they contribute to the radiometric signal. This is why a liquid phase precipitation has dramatically different signatures than the frozen precipitation. Not shown here but this is also why ice particle shape has a significant impact on the signal, due to its dramatically different optical properties for different shapes.

The preceding sensitivity study of the three major hydrometeors properties (water phase, particle size and hydrometeor amount) indicates that in order to properly exploit satellite data impacted by precipitation for the purpose of assessing the amount of precipitation in the atmospheric column or to retrieving the underlying sounding data (temperature and moisture in particular), one must imperatively account for the effect of the phase of the water as well as the estimated particle size and amount. Failure to do so will result in large uncertainties in the inverted/ assimilated geophysical data. These uncertainties will indeed impact the inverted/ analyzed precipitation amount itself and/or the other surrounding environment parameters (such as temperature and moisture). In later sections, we will show that the water phase is sometimes accounted for in a simplified way by accounting for both frozen and liquid phases in the assimilation and retrieval algorithms. Attempts to account for the particle size when extracting precipitation information from the measurements are emerging but not yet widely used. Accounting for the hydrometeor amount alone is not optimal and will certainly carry large uncertainties due to the effect of water phase and particle size.

Other microphysical properties not studied here also have large impacts and should be considered as well. These properties include the density of the particles, their shapes, and their size distributions: effective radius is a simple metric to inform about particle size, but size distributions give a fuller picture of the many particle sizes that are along the line of sight in a volume of precipitating atmosphere.

48.2.2 Sensitivity to the Surrounding Environment

In addition to the impacts from the hydrometeors, the microwave signal is also strongly influenced by the surrounding environment. This includes the temperature and moisture profiles in the same line of sight containing precipitation, since they modulate the transmission above, within, and below the precipitating layers. It also includes the surface characteristics (emissivity/reflectivity and skin temperature), which have direct and indirect effects on the measurement, as described in Sect. 48.1. The detection of precipitation and the inference of its intensity relies largely on the contrast in signal between precipitating scenes and the ambient environment. Therefore, the degree to which the contrast is evident depends not only on the type of precipitation (e.g., stratiform vs. convective), but also on the environment the precipitation occurs in, which may help to either inhibit or enhance the signal. Antenna beam filling, for instance, illustrates the modulating role of the cloud/ precipitation fraction when cloud/precipitation is only present over a portion of the line of sight, or field of view (FOV). The presence of precipitation over part of a remotely sensed FOV, or the variation in precipitation rates over an FOV, has an impact on the observed radiance that may make it difficult to determine what conditions were actually present at the time of the observation, or what the true average precipitation rate was. In fact, this beam filling error has been found to be amongst the largest contributing factors to errors in retrieved rainfall rates (Short and North 1990), though the relationship between the observed radiance and the average precipitation rate can be estimated by modeling the distribution of the precipitation rate over the FOV (e.g., Short and North 1990; Bell 1987).

Figures 48.3 and 48.4 represent the results of the sensitivity assessment of the signal for two microwave frequencies (19 and 37 GHz respectively) as a function of a multitude of parameters (see figure captions for a full description). The figures represent simulated impacts (assessed based on a nominal precipitation simulation) as a function of parameters of the hydrometeors and of the surrounding environment and a scaling factor of rain and graupel amount for the hydrometeors. It is important to note that depending on the frequency, the impacts are significantly different. The dependencies at 37 GHz are significantly different (in amplitude and shape) from the ones at 19 GHz. They are also very different from those at other frequencies, such as 183 GHz (not shown). Overall, hydrometeor amounts are almost always a major factor impacting the signal, as described in the previous section, but this impact is heavily modulated by the surrounding environment (e.g., moisture in the atmosphere, etc.). Additionally, particle size also plays an important role in a remotely sensed signal of precipitation, as there is signal resonance at certain particle sizes as mentioned in the previous section. For instance, at 37 GHz the signal is at a maximum in the presence of rain particles with an effective radius around 600 µm (and a GWP scaling factor of 2), while this maximum occurs at 19 GHz when particles have an effective radius around 1400 µm. This resonance is itself modulated by the hydrometeor amount and other parameters of the surrounding environment. Another striking feature (as can be seen in Figs. 48.3 and 48.4), especially for surface-sensing channels, is the significant impact of surface emissivity on the

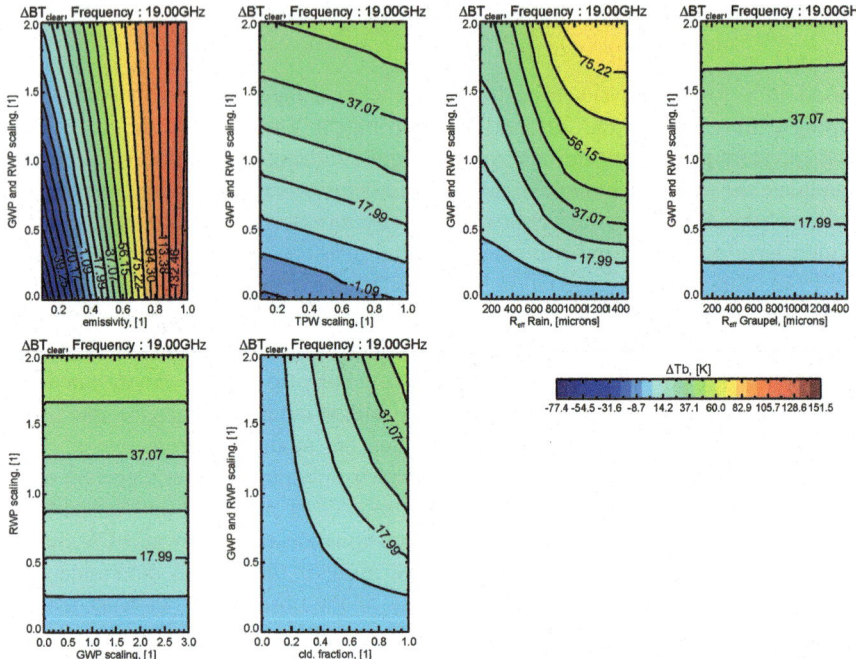

Fig. 48.3 Sensitivity of the precipitation-impacted measurement at 19 GHz, defined as the difference between clear-sky simulation and precipitation-based simulation, as a function of hydrometeors and multiple surrounding environment parameters. The hydrometeors and surrounding environment parameters include (clockwise, from top left) surface emissivity, total precipitable water (TPW), effective particle radius of liquid rain, effective particle radius of frozen graupel, amount of graupel (GWP), and cloud fraction. The x-axes on the plots represent parameters of hydrometeors and their surrounding environment, and the y-axes represent scaling factors for rain and graupel hydrometeor amounts (with 0 being no hydrometeors used in simulation and 2 being double the original amounts of liquid and frozen precipitation used). Contours represent signal amounts

signal. The impact of surface emissivity, however, is also not independent of other factors; at low emissivities (e.g., over water) in our 37 GHz example (Fig. 48.4), for instance, hydrometeor amount is important, while at higher emissivities (e.g., over land) the signal is almost independent of hydrometeor amount. Factors such as cloud fraction and humidity additionally influence the impact that surface emissivity will have on the signal, as they affect the degree to which the sensor can "see" the surface. This is particularly important over land, as the surface emissivity will tend to vary greatly in the presence of precipitation (e.g., from high values – around 0.9, to low values – around 0.4, depending on the frequency used: see Figs. 48.3 and 48.4). This suggests that emissivity is a crucial factor to account for when exploiting precipitation-impacted microwave measurements.

As the prior sensitivity assessment illustrates, a proper exploitation of the microwave signal at frequencies currently used in many satellites generally requires that hydrometeor parameters and the surrounding environment are accounted for when precipitation is present. Hydrometer amount, phase, fraction and particle size are

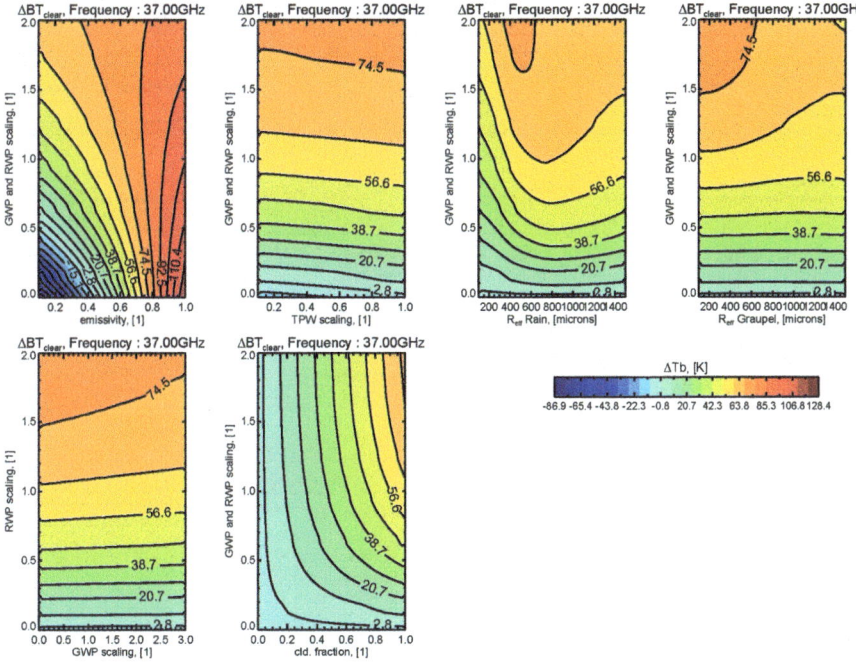

Fig. 48.4 Same as in Fig. 48.3, but for a 37 GHz simulation

particularly important to the microwave signal, as is emissivity. Though the precise impacts of hydrometeors and environmental conditions on the microwave signal will vary by frequency and from one location to another, the effects of emissivity and precipitation are so intertwined and significant that it is not practical to ignore them in the processes of inversion and assimilation without encountering uncertainties that are too high to make the inversion or assimilation of the data useful. In summary, exploiting the precipitation observation from space requires accounting for several critical factors (some of them inherent to the hydrometeors themselves and some of them depend on the surrounding environment) that impact the overall signal in a highly complex and varying fashion. This impact varies therefore pixel by pixel, and changes from one frequency to another.

48.3   Important Factors for the Physical Inversion and Data Assimilation of Precipitation Observations

In this section, we will discuss the different aspects that must be accounted for when handling precipitation observations, either in a physical retrieval or variational DA context, and briefly describe how these factors are typically handled. Figure 48.5 gives an overview of these driving factors, which, for simplicity, have been divided

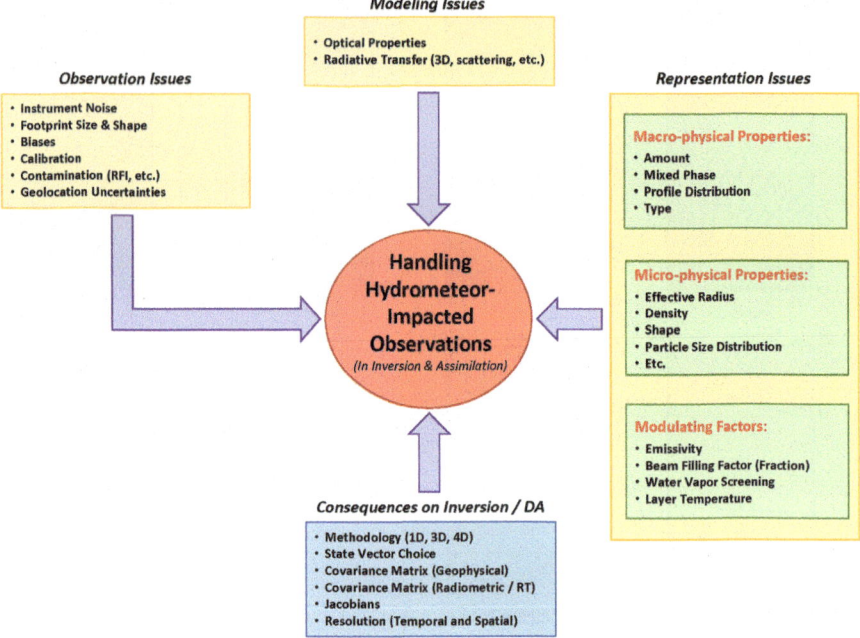

Fig. 48.5 Overview of the different driving factors impacting how the precipitation observations are handled in a physical retrieval and/or DA context. These driving factors include observations characteristics, modeling capabilities, and signal representation issues. These driving factors determine what approach and techniques are adopted for the assimilation and physical inversion of precipitation observations (what methodology to use, what to include in the state vector, etc.)

into three groups: (1) observations-related issues, such as noise and footprint shape/size; (2) modeling issues, like the prescription of hydrometeor optical properties in an RT model; and (3) signal-representation issues, or literally how precipitation is represented in the inversion or DA system. As will be described in this section, these factors will drive the technical implementation of the precipitation handling systems in inversion and DA (e.g., what to invert and/or assimilate, how to account for biases, etc.) and play a key role in what variables are chosen to be included in the system's *state vector*, or the set of parameters (like temperature, humidity, and emissivity) used to represent the state of the atmosphere. Instrumental issues are typically handled dynamically within the system by having (1) variable bias correction (commonly called VarBC; Zhu et al. 2015) mechanisms to assess the level of bias while assimilating the observations, as well as accounting for (2) variable noise error levels computed on the fly, that will dynamically modulate the weights allocated to the observations.

The effects of hydrometeors, including frozen and liquid precipitation, have to be modeled properly. This includes the optical properties modeling which in turn is key to the RT modeling that simulates the precipitation observations. Optical properties are pre-calculated and stored in look up tables for easy access by RT models (see Sect. 48.3.3 for more information on how they are computed). The signal-representation issues are perhaps the most important of all driving factors. These determine what parameters will be available for the users: what will be inverted or analyzed? By exploiting the precipitation observations, do we intend to make available only the precipitation amount? Or do we intend to also produce side-parameters of cloud particle size as well? What about the surrounding moisture and temperature in the atmosphere (below, above and within the precipitating layers)? These determine how to represent the precipitation and what parameters are analyzed or inverted. Is precipitation represented simply by a total water liquid amount, or instead, by a profile of liquid and/or frozen water, how about the particle size information, the density, and other microphysical parameters? Or do we even represent the precipitation when performing the assimilation and retrieval of precipitation observations? Indeed, some approaches do not analyze precipitation per se, but instead rely on the analysis of temperature and moisture, and simply generate the cloud and precipitation information based on cloud-resolving models, which are then fed to forward RT models to simulate and compare to precipitation observations.

The parameters chosen to be included will be part of what is called the assimilation *state vector* and will be part of the inversion or assimilation process and will therefore be of great importance to the users. Section 48.2 above described the sensitivity of the observations to two types of parameters: (1) those related to the hydrometeors themselves (labeled as micro- and macro-physical properties in Fig. 48.5) as well as (2) surrounding environment parameters that modulate the radiometric impact of the hydrometeors themselves. According to that sensitivity study, precipitation amount and type as well as corresponding particle size, plus the beam filling factor (cloud/precip fractions) are all crucial and should be considered to include in the state vector. And so are the surrounding environment parameters of emissivity as well as temperature and moisture.

Other fundamental factors play a crucial role as well and impact the high-level approach adopted in the assimilation of precipitation observations. These include (1) the limitations of the forecasting model itself used in the assimilation process, (2) the uncertainties in the forward modeling and Jacobians modeling, also used the assimilation and physical retrievals, as well as (3) the natural spatio-temporal variabilities of precipitation. These are three fundamental aspects that will be addressed in sub-sections below. Other factors include observation issues such as the noise level of the observations, footprint sizes and shapes, as well as the bias and calibration issues that might impact them. Other non-modeled (such as geolocation uncertainties) and sometimes anthropogenic impacts on the observations (such as contamination by radio frequency interference, or RFI) are usually handled by filtering them out. Several filtering algorithms exist but a first-order filtering usually exist, based on a simple departure-from-background threshold, called a *gross check* filter.

The following sub-sections will first introduce the theoretical basis for doing DA or physical retrieval, that will lead to the fundamental equations that govern the process of handling precipitation observations in those systems. This mathematical description will also allow to gain insight on the assumptions made for the system to be valid (such as local-linearity, Gaussianity of the errors distribution, independence of background and observation errors, etc.). This will be followed by a sub-section to describe the main limitation due to the natural spatio-temporal variability characteristics of the precipitation and cloud. The main objective being to assess the pros and cons of several approaches used in DA of precipitation observations, including 3D-Var, 4D-Var, 1D-Var and 1D + 4D-Var approaches. And in the particular, the handling of non-linearities. The use of ensembles will also be discussed briefly.

48.3.1   The Theoretical and Mathematical Basis for the Inversion of Satellite Radiances

Most retrieval and DA techniques share a common starting point in the Bayes theorem (e.g., Rodgers 2000). This shows how to calculate the probability density function (PDF, or just probability) of a state given some observations and an initial knowledge of the state:

$$P(x|y) = \frac{P(y|x)P(x)}{P(y)} = \frac{P(y|x)P(x)}{\int P(y|x)P(x)dx} \tag{48.1}$$

This conditional probability $P(x|y)$ is known as the *posterior* probability, i.e. the probability of a particular state given the observations. Bayes theorem shows how this can be computed from $P(x)$, the initial knowledge or *prior* PDF of the atmospheric state, and $P(y|x)$, the probability of getting the observations given any particular state. The denominator $P(y)$ is a normalizing factor and is just the integral of the numerator over all possible states, i.e. $\int P(y|x)P(x)dx$.

The Bayes theorem compactly summarizes the entire problem of inverting satellite radiances: the prior knowledge of the state $P(x)$ can be derived from a forecast model, an ensemble of forecast models, or possibly a representative database of atmospheric states. The probability of getting the observations $P(y|x)$ is computed using a forward model linking the observations to the state, defined as.

$$y = H(x) \tag{48.2}$$

When doing precipitation retrievals, the forward model $H(\ )$ is an RT model. When doing DA, the forward model $H(\ )$ may include not just the observation operator, but also an atmospheric forecast model, to allow the assimilation of observations valid at different times. The probability of getting the observations $P(y|x)$ must also account for the observation error. This observation error includes not just the measurement error but also the error in forward model and possibly the error of representation (which will be explained below).

A main difference between retrievals and DA is in the scope of the state x and in the number of observations y being used. In a retrieval, the aim is to derive the atmospheric state at the instantaneous time and location of a single observation. The atmospheric state may be described by variables as simple as the instantaneous precipitation rate and the total column water vapor, in order not to exceed the information content of the observations, though in some retrieval systems the state vector contains a more extended description of the atmosphere (like temperature, moisture, and hydrometeor profiles). The inversion process finds a state vector compatible with the information in the observations (see Chap. 7 of volume 1). In DA, the aim is to describe the full state of the atmosphere over a large (and often global) model domain, and to describe the evolution of this model through time, in order to make forecasts. Hence, the state typically contains the full thermodynamic description of the atmosphere along with cloud and precipitation hydrometeors. The state is represented on a model grid which means that the precipitation rate (for example) at any grid point represents a finite volume in space and time, covering the area of the grid box, which might be tens of kilometers, and the timestep of the model, which might be as much as 30 min. In order to derive the state of the atmosphere on a possibly global domain, all available observations are assimilated, whether from satellite or from in-situ measurements, and not just relating to precipitation but to all atmospheric variables.

In DA the state variables on the model grid do not exactly represent the instantaneous and local state seen by the observations; the resulting errors are known as the error of representation (e.g., Janjić et al. 2017). Precipitation is a rapid, localized process and hence it gives rise to very large errors of representation; these errors can be so large that they dominate any other source of observation error in the assimilation of precipitating observations (Geer and Bauer 2011). The error of representation means that DA and retrieval estimates of precipitation can be very different, but this is because they are aiming to do different things: a retrieval to characterize the exact state of the atmosphere that gave rise to a particular observation; DA to provide initial conditions for forecasts. In finding these initial conditions, the exact location and timing of the observed precipitation can be less important.

This leads to the other main difference between retrievals and DA, which is the *cycling* of information. The posterior probability $P(x_t|y_t)$ valid at a time t initializes a forecast model which is then used to provide the prior probability $P(x_{t+1})$ for assimilating observations at a later time, $t + 1$. In practice, there are close links between retrievals and DA. For example, global operational forecasts are widely used as a source of prior information, $P(x)$, for satellite retrievals.

The solution of Bayes theorem is done in many different ways, leading to diverse approaches in both DA and retrievals. Bayes does not define or require any specific shape of PDF, but in the majority of algorithms, all PDFs are represented by Gaussian distributions. A smaller class of solutions attempts to preserve the non-parametric nature of Bayes theorem. The 'Bayesian' approach to precipitation retrievals is used by, among others, the Goddard Profiling Algorithm (GPROF; Kummerow et al. 2001) where the prior probability $P(x)$ is represented by a database of atmospheric states x_i so that

$$P(x) \approx \frac{1}{N} \sum\nolimits_{i=1}^{N} \delta(x - x_i) \tag{48.3}$$

where $\delta()$ is the delta function. The database of states used in GPROF was originally derived from cloud resolving models but more recently it has come from a set of nearby radar retrievals – for example taking advantage of the combination of precipitation radar and passive microwave onboard GPM. In DA, the counterpart of the Bayesian retrieval is the particle filter (e.g., van Leeuwen 2010). Here, an ensemble of model forecasts provides the 'particles' that form the non-parametric representation of $P(x)$. The posterior $P(x|y)$ is computed by re-weighting the particles according to Bayes theorem. The additional challenge in DA is to somehow fit millions of observations simultaneously and to maintain an ensemble of forecasts (particles) that remotely represents the true atmosphere; hence this is not yet a practical technique in weather forecasting. However, given the non-Gaussian nature of precipitation PDFs, the particle filter will be a strong candidate to provide the DA component in future nowcasting applications. Many of the practical issues of applying to the particle filter to atmospheric DA are being solved (e.g., Poterjoy 2016). Note that in practice neither Bayesian retrievals nor particle filters are fully non-parametric since they typically represent the observation error as Gaussian.

If we do assume a Gaussian form for both the observation error and prior PDFs, the Bayes theorem can be rewritten in terms that are more familiar from the majority of retrieval and DA algorithms. For example (e.g., Rodgers 2000), if the probability of getting the observation is a Gaussian distribution centered on the observation itself, then we can write:

$$P(y|x) = \frac{1}{c_1} e^{-\frac{1}{2}(y - H(x))^T R^{-1}(y - H(x))} \tag{48.4}$$

Similarly, if the prior state distribution is centered on a best prior (or background) forecast x^f, it is given by

$$P(x) = \frac{1}{c_2} e^{-\frac{1}{2}(x - x^f)^T B^{-1}(x - x^f)} \tag{48.5}$$

Here R and B are the covariance matrices for the observation error and background error and c_1 and c_2 are normalizing constants. Substituting these into Bayes theorem and taking logarithms of both side (introducing a new constant c_3) we get the cost function:

$$-2\ln P(x|y) = (y - H(x))^T R^{-1}(y - H(x)) + (x - x^f)^T B^{-1}(x - x^f) + c_3 \tag{48.6}$$

Variational methods use a minimization algorithm to vary x in order to find the highest posterior probability $P(x|y)$ or equivalently the lowest value of $-2\ln P(x|y)$. This is often referred to as the maximum likelihood solution. Compared to the Bayesian approach, we are no longer looking for the full distribution $P(x|y)$ but only its best estimate. In DA, this best estimate is called the analysis (in retrievals it is of course simply the retrieval). The cost function shows that both retrievals and analyses are a weighted combination of information from the observations y and the background forecast x^f, with the relative weights controlled by the observation and background error, R and B. In DA, the background forecast is often of very high accuracy, so the impact of observations is a small correction to an already good forecast. By contrast, retrievals hope to minimize the contribution from the background, so that the retrieval is based mainly on observational information. Whether this is true or not can be judged by computing the *information content* of the retrieval (see Rodgers 2000).

To find the minimum of the cost function, variational algorithms will typically compute its gradient repeatedly at different values of x as the algorithm steps towards the minimum. The gradient is obtained by linearizing the cost function. This requires a linearized forward operator H, defined from the nonlinear operator by

$$H(x + dx) - H(x) \approx \mathbf{H}dx \tag{48.7}$$

This linearization is of particular importance when using observations of precipitation. First, the RT involved can be highly nonlinear, as is especially the case in precipitating conditions; this affects both retrievals and DA (though, as described in following sections, this RT non-linearity is, to a certain extent, mitigated by 1D-Var minimization techniques (e.g., the Levenberg-Marquardt algorithm; Levenberg 1944) that account for the local-linearity of the RT). Moreover, since in DA the forward model $H(\ )$ may also contain a forecast model, the dependence of precipitation on the other state variables (such as wind, humidity and temperature) is also highly nonlinear. An approach that has allowed successful assimilation of rain-affected observations at the European Centre for Medium-range Weather Forecasts (ECMWF) is the incremental variational method (Courtier et al. 1994) in which the solution of the nonlinear version of the cost function is approximated by a series of linearized cost functions. Here the linearization error reduces as x gets closer to the solution and dx gets smaller. In practice, this means that incremental variational assimilation can, for example, create convection in the analysis at a location where none existed in the background forecast (Bauer et al. 2010). The ECMWF algorithm is also a four-dimensional variational assimilation (4D-Var) meaning that $H(\ )$ does contain a forecast model. This enables the DA to fit observations across a time-window (at ECMWF typically 12 h). Hence the linearized forward operator H similarly contains a linearized version of the forecast model as well as of the observation operator. The linearized version of the forecast model (in its tangent-linear and adjoint versions) is used to map information forward and backward in time between the state x and the observations y, which enables observations to be assimilated from any time throughout the assimilation window. To derive such a linearized forecast model and to make it work for cloud and precipitation is a difficult and time-consuming task (e.g., Janisková and Lopez 2013) that has only been undertaken by a few weather centers worldwide and that also introduces errors into the assimilation process which should be accounted for. However, it is essential to assimilating precipitation observations in the standard variational assimilation framework, and the errors introduced by it may be addressed by inflating observation errors.

Over the past decades, weather forecasting centers have either superseded or enhanced pure variational algorithms using ensemble techniques. A variational DA algorithm only propagates forward the maximum likelihood of the state estimate. In contrast, an ensemble approach also propagates its error covariance, as represented by an ensemble of forecasts (e.g., Hunt et al. 2007). Similar to the Bayesian retrieval or particle filter approach, a set or *ensemble* of forecasts gives information on the PDF of the prior, $P(x)$. Each ensemble member can be updated using observations to form a posterior ensemble representing not just the maximum but the spread of $P(x|y)$. However, ensemble algorithms usually assume Gaussianity of these error distributions, so what the ensemble provides is the mean and error covariance. Variational assimilation algorithms can be improved by using an ensemble to provide dynamically evolving background error covariances, B. This is known as *hybrid* variational assimilation.

For the assimilation of precipitation observations, ensemble techniques can circumvent the difficult task of creating linearized versions of the forecast model. The ensemble-variational approach, such as the 4D-EnVar used operationally at the National Centers for Environmental Prediction (NCEP), uses the ensemble to help propagate information through the assimilation time window. This contrasts with the use of a linearized forecast model in normal variational assimilation. 4D-EnVar has allowed NCEP to introduce operational assimilation of all-sky microwave radiances (Zhu et al. 2016). A further benefit of ensembles is that introducing new variables into the analysis, such as precipitation, can be straightforward. This is because ensemble-based background error covariances implicitly include cross-covariances between different variables, such as between precipitation and vertical wind. In variational assimilation these cross-correlations need to be explicitly modelled, which requires the effort of developing a model for these correlations that may struggle to represent some very complex and variable atmospheric processes. See Bannister (2008) for more information on background error modeling in variational assimilation and Montmerle and Berre (2010) for the difficulties of modeling background error covariances in precipitating areas.

There are two potential main limitations of ensemble techniques, particularly when assimilating precipitation. First, although they do not need to explicitly model the linearized H, this linearity is assumed in the algorithm, which finds the minimum of the linearized cost function. Incremental variational assimilation has an advantage in being able to find the minimum of the nonlinear cost function. Second, the analysis can only be found within the space spanned by the prior ensemble members. In more practical terms, this means that ensemble techniques cannot create convective precipitation in the analysis at a location where it does not exist in at least one member of the background ensemble (at least, not without ad-hoc modifications). Geer et al. (2018) and Gustafsson et al. (2018) give a more detailed overview of the DA techniques in use around the world for the assimilation of cloud and precipitation observations. These cover respectively the assimilation of all-sky satellite radiances and the assimilation of observations for nowcasting and local-area forecasting, where in particular the assimilation of precipitation radar is becoming important.

One other technique combines retrievals and DA to avoid the problems of the nonlinearity of the forecast model in the $H(\ )$ operator. This technique is the combination of a prior retrieval, followed by the assimilation of that retrieval into a DA system. Where that is done variationally using the same prior ensemble in the retrieval as in the analysis, this technique is known as 1D + 4D-Var (e.g., Marécal and Mahfouf 2002). The atmospheric state in the 1D-Var retrieval can be extended to include precipitation variables, and the forward operator can include a 1D moist physics scheme to constrain the solution, but the aim is to keep the most nonlinear parts of the problem inside the 1D-Var part of the retrieval. Then, a pseudo-observation such as total column water vapour can be assimilated into the main 4D-Var assimilation system, without needing to extend that assimilation system to analyze precipitation variables directly. It is hoped that information on atmospheric water vapour will then help remove or add precipitation in the main analysis, so as to fit the observations. The 1D-Var stage is an optimal retrieval of information (including precipitation) from the satellite data and it is able to fit the observations very closely. More precisely, this is because it does not need to take the error of representation into account. The main limitation is that simply assimilating the pseudo-observation into 4D-Var is not optimal, i.e. not fully consistent with Bayes theorem. The 1D-Var retrieval introduces background information into the pseudo-observation, or more precisely it makes $P(y|x)$ partly dependent on the prior $P(x)$. However, techniques have been developed to remove this dependence on the prior, making it possible to assimilate retrievals optimally (see Migliorini 2012). Further, the 1D-Var retrieval can be replaced by a Bayesian retrieval based on an ensemble of local model grid points, which can be argued to have less dependence on the prior. This technique is operationally used to assimilate precipitation radar at Météo-France (Caumont et al. 2010), and is known in that context as 1D-Bay+3D-Var (because their main limited-area assimilation system uses 3D variational assimilation, or 3D-Var).

48.3.2 The Issue of Non-linearities When Assimilating Precipitation Measurements

As described in the previous section, in most DA (or physical retrieval) methodologies, the main objective is to seek an optimal geophysical solution that is consistent with the observations at hand. This is usually done by minimizing a cost function that will help the system fit the observations while at the same time maintaining geophysical consistency between all parameters that constitute the solution. This is, as shown previously, a highly non-linear and ill-posed problem. The variational approach attempts to handle this non-linearity using an iterative method which divides the path to the solution in small steps assumed to be locally linear: in other words, the path that will lead to the solution corresponding to the absolute minimum of the cost function. Non-linearity in DA and physical retrievals essentially comes from three main components of the cost function: the RT operator (as in 1D-Var retrievals and 3-4D-Var DA), the spatial distribution of precipitation (in the case of 3D-Var and 4D-Var assimilation which account for spatial variability), and the time evolution of precipitation (as in 4D-Var DA which accounts for temporal variability). Techniques employed in physical retrieval and DA usually divide the cost function into locally-linear regimes. Figure 48.6 shows an example of non-linearity due to RT, spatial evolution, and temporal evolution (see figure caption for a detailed description). The cost function is multi-dimensional and complicated to visualize, but this should give us some insight into the different components which create non-linearities in it. The high non-linearity of the RT is maximum in cloudy and precipitating conditions (as shown in Fig. 48.6), but this RT-based non-linearity is *decomposable* into local linear regimes, and therefore iterative or incremental variational inversion/assimilation would be appropriate for handling this type of non-linearity. Unlike RT-based non-linearities, however, the temporal and spatial evolutions of precipitation measurements, present highly non-linear features that are challenging to decompose iteratively into small linear regimes. This can be understood intuitively: when using regular DA cycles of 6 h (or even 1 h), a situation may arise where one temporal grid cell is clear sky and the following grid cell (1 or 6 h later) is not. This is highly non-linear (even discontinuous). Similarly, spatially, one clear-sky grid cell (of the order of 20 km, for instance) may be juxtaposed to a neighboring grid cell of intense precipitation. This is also spatially discontinuous and therefore highly non-linear. It is this high non-linearity (with first degree derivative discontinuity) that makes the assimilation of precipitation observations with 3D and 4D approaches challenging. Ideally, much more frequent cycles (on the order of minutes) and much higher spatial resolutions (sub-km) would be needed to use 3D and 4D approaches in precipitating conditions. Using a 1D-Var approach, though presenting challenges mentioned previously, offers one way to account for the non-linearities of precipitation observations before using spatially- and temporally-constrained DA systems.

One should recall at this point the previous section describing the mathematical basis of variational assimilation/inversion and the corresponding assumptions for it to be valid (in particular the assumption of local-linearity of the cost function). The cost function used for 4D and 3D variational approaches accounts for the temporal and spatial correlations of the different variables being analyzed, and therefore creates significant non-linearities in precipitating situations and for observations of precipitation, as temporal and spatial constraints/correlations are included in the minimization of the cost function. This, in turn, makes the use of simple formalism in assimilation difficult to implement. Indeed, in cloudy and precipitating conditions, it is not necessarily optimal to have points in time or space influence or be influenced by neighboring points through temporal or spatial correlations, as cloud and precipitation features can vary radically over time and space. This suggests that using 3D or 4D approaches for the exploitation of spaceborne observations might not be optimal

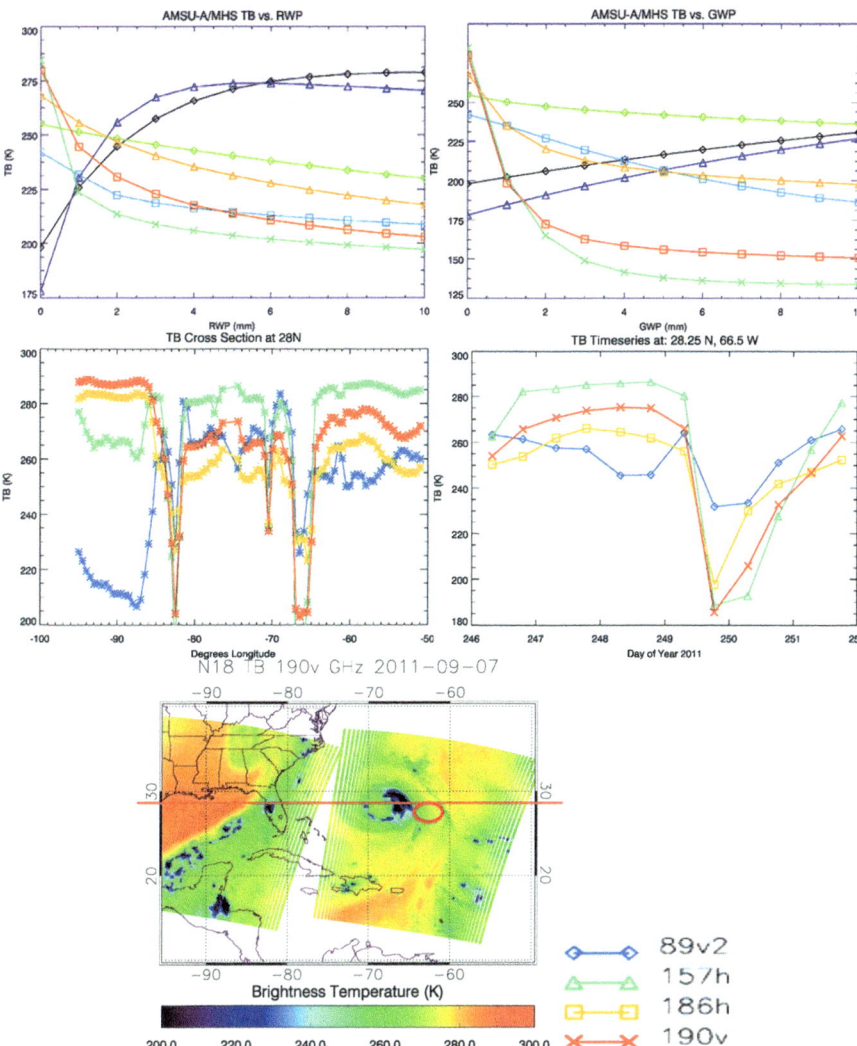

Fig. 48.6 Evolution of simulated brightness temperature as a function of hydrometeor amount (RWP – top left, and ice-graupel water amount – top right) highlighting the non-linearity (but incrementally locally-linear) of the RT. The different colors/symbols represent different channels with precipitation-sensitive frequencies in the AMSU/MHS sensor pair. Middle plots show the spatial evolution (left) and temporal evolution (right) of real NOAA-18 MHS data at different frequencies from 8 September 2011. A cross section at 28°N is highlighted (in red) in the left bottom map showing the field of 190 GHz brightness temperature. The temporal evolution is computed by averaging the area highlighted by a red circle on the map, using several passes of NOAA-18 over multiple days. The non-linearities due to spatial and temporal evolutions of space observations are highly non-linear and their linearization (incrementally) would be challenging due to abrupt variations with local first degree derivative discontinuities

in situations that are highly non-linear in nature, as is the case for precipitating situations, unless the time step and spatial resolution are significantly enhanced (from what is done currently) to allow local temporal and spatial linearizations.

48.3.3 Uncertainties in Radiative Transfer and Jacobians

The basis for inversion of satellite radiometric observations into geophysical parameters is the ability of the forward operator to accurately simulate both the observed signal and the sensitivities of the signal to perturbations in the geophysical state. The degree of accuracy depends on the RT model employed.

For operational DA systems, fast RT models such as the CRTM (Han et al. 2006; Ding et al. 2011) and Radiative Transfer for TOVS (RTTOV) (Saunders et al. 1999) are used to minimize the number of complex calculations needed to solve the RT equation for a desired wave number. In clear-sky, this mainly consists of training an absorption model from line-by-line calculations to provide atmospheric transmittances in the presence of various gaseous species (Clough et al. 2005). Additionally, when clouds and precipitation are present, this involves making assumptions about the hydrometeor phase, size, density, amount, vertical distribution, and correspondence to optical properties used in the RT model. For precipitation-sized hydrometeors, bulk-scattering optical properties such as the phase function, single-scattering albedo, and extinction cross-section are developed offline from either Mie theory (a spherical particle assumption) or assumed ice crystal habit distributions of complex shapes or roughened particles (Bi and Yang 2017; Baum et al. 2005; Liu 2008; Yi et al. 2016). The uncertainties in RT for various parts of the EM spectrum are due to these assumptions; for example, in a cold rain case with ice clouds, sensitivity in the solar or infrared portion of the spectrum is largely due to particle size, optical depth, and vertical extent (i.e. cloud top temperature and geometric thickness) of the cloud (Garrett et al. 2009). For the MW spectrum, sensitivity mainly exists for large ice particles and their amount, shape, size, and density. Typically, in DA, since the prior information is not available to capture these quantities, their assumptions lead to large uncertainties in the RT.

Figure 48.7 highlights the error space due to assumptions about the cloud microphysical properties and optical properties specified in the RT model. Shown are brightness temperature differences from 1–330 GHz for scenes with non-precipitating cloud (clwp), precipitating cloud (rwpp), and both clwp and rwp, with varying cloud and raindrop effective radii. The differences are computed relative to a clear-sky simulation. It is evident even for the simple case of rwpp only, that an assumption of effective radius for rain drops of 250 or 1000 μm leads to brightness temperature differences of up to 30 K in some spectral regions (23 GHz), or changes to the gradient (60–118 GHz).

More complex scenes involving frozen hydrometeors can add to the uncertainty in RT simulations. It is therefore important to develop approaches for reducing these uncertainties due to the various assumptions in addition to improving the RT itself.

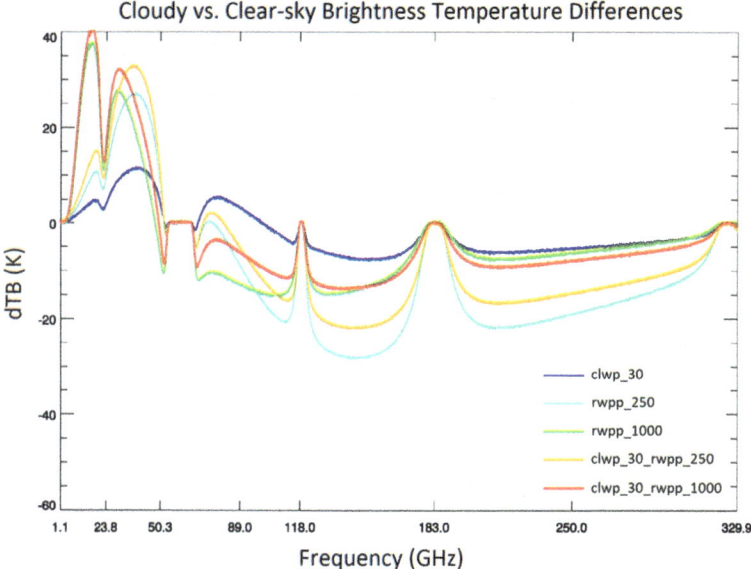

Fig. 48.7 Simulated brightness temperature differences from 1–330 GHz, between clear and cloudy/precipitating cases with varying assumed cloud and rain effective radii

48.3.4   Model-Related Limiting Factors in the Assimilation of Precipitation

The assimilation of precipitation is challenging simply by virtue of the fact that precipitation processes are difficult to model. As stated in Sects. 48.3.1 and 48.3.2, the non-linear processes of precipitation must be linearized to be handled by the variational framework of a DA system, contributing to error. Also mentioned in Sects. 48.3.1 and 48.3.2 is the fact that precipitation processes are quickly evolving and occur on scales that often do not match up with the grid of forecast and DA systems, making it difficult to accurately represent the true state of the atmosphere in precipitating conditions, additionally contributing to error. Even if these problems can be adequately addressed, challenges still arise owing to the fact that NWP models do not provide all of the information required to simulate radiances in precipitating conditions with an RT model, like particle shape and density and the particle size distribution; these properties must be estimated and assumed since they are not explicitly measured or observed.

Related to the difficulties of modeling precipitation processes is the ease with which observations of precipitation can disturb the balance of an NWP model. It is known that models can quickly deviate from their initial conditions, thereby making it difficult to assimilate observations from processes that are not easily represented with a forecast model, even if the initial state of a cycling DA system can accurately depict reality over the model domain. This harkens back to errors of representation;

the model's inability to forecast precipitation well on the spatial and temporal scales at which precipitation occurs consequently leads to situations in which future precipitating observations differ too greatly from future DA backgrounds for the observations to be used or weighted heavily in subsequent analyses without producing unnatural gradients and upsetting the stability of the system. This leads to many observations impacted by precipitation to be rejected altogether from the DA system.

48.4 The Assimilation of Precipitating Data: Different Methods and Perspectives

As mentioned earlier in this chapter, assimilating satellite data impacted by hydrometeors (suspended cloud, liquid, and frozen precipitation) is an active area of work in both NWP research and operations. The following sections will attempt to give the reader a summary of these activities. After a review of major efforts in this area that have taken place in the last couple of decades, examples of implementations at the ECMWF and the Japan Meteorological Agency (JMA) are shown. This is followed by a description of an ongoing research activity at NOAA's Center for Satellite Applications and Research (STAR) which combines 1D-Var pre-processing with DA for the handling of precipitation observations in NWP.

48.4.1 A Summary of Previous Efforts

There is a long history of research on the assimilation of precipitation-related information into weather-forecasting models. The early generation of DA methods could only ingest data in the form of the prognostic variables, which meant that direct assimilation of rain-affected satellite radiances or radar reflectivities was not possible. Hence, attempts were made to change the analyzed temperature and moisture fields in order to create (or remove) precipitation in the subsequent forecast. These attempts included 'physical initialization' (e.g., Krishnamurti et al. 1991) and the assimilation of heating rates inferred from satellite observations (e.g., Puri and Miller 1990). A final approach, latent heat nudging, was sufficiently promising enough that it became operational in a number of limited area models (e.g., Jones and Macpherson 1997; Stephan et al. 2008).

The advent of variational assimilation in the 1990s replaced the ad-hoc initialization precipitation initialization schemes with the mathematically rigorous formulation of the inverse problem (see Sect. 48.3.1). The new framework had a number of advantages for the assimilation of precipitation. First, it allowed any observation to be assimilated, as long as a forward model could be specified and assumptions such as linearity and Gaussianity could be met. Second, 4D variational assimilation (4D-Var) allowed a forecast model to become part of the forward model, giving a rigorous mapping between precipitation and the standard prognostic variables such as temperature, wind, and moisture. However, this required the development of linearized and regularized moist physics models (one of the main issues being to replace on-off convection switches with continuous functions) from which tangent linear and adjoint models could be derived (e.g., Zou et al. 1993). The possibility of assimilating precipitation quantities, such as 24-hour accumulated precipitation, was then explored by many different studies (e.g., Županski and Mesinger 1995; Tsuyuki 1997; Benedetti et al. 2003). Although the early studies helped demonstrate the possibility of 4D-Var rain and cloud assimilation, the research was generally not mature enough to be used in operational forecasting systems. Many of the early studies were identified as being incomplete, for example lacking detailed treatment of background error covariances, using near-zero observation error, or being based on idealized case studies (Errico et al. 2007). The goal was still to demonstrate benefits from variational assimilation of precipitation-related data in an operational forecasting system. This is much harder than in an idealized scenario because operational forecasting systems, even 20 years ago, were already very well-constrained by conventional observations and clear-sky satellite data, and already able to generate precipitation forecasts of reasonable quality.

Two early approaches did become operational at weather forecasting centers. In the JMA mesoscale assimilation system, a special cost function was added for assimilating precipitation retrievals from radar (Koizumi et al. 2005) and microwave imagers such as AMSR-E (Tauchi et al. 2004). These assimilation approaches became active in 2002 and 2004 respectively (see the survey of Bauer et al. 2011). At ECMWF, the 1D + 4D-Var approach was adopted as the preferred option for using rain-affected microwave radiances (Moreau et al. 2004) and it became operational in 2005 (Bauer et al. 2006a, b). In this approach a retrieval was made from rain-affected microwave imager radiances using a 1D-Var framework that incorporated a 1D moist physics model. The moist physics model helped constrain the solution and allowed the inference of humidity and temperature from precipitation. The retrieved total column water vapor was assimilated into the main 4D-Var analysis. The benefits of the 1D-Var step were to isolate the most nonlinear part of the process within the retrieval and to provide an additional layer of QC, as mentioned in previous sections.

It is difficult to assess the full impact of the ECMWF 1D + 4D-Var approach from the results published at the time, but there were clear improvements to the short-range forecasts of lower-tropospheric humidity (see Geer et al. 2017). The 1D-Var retrievals were able to fit the observations quite closely (as measured in observation space) and the precipitation retrievals were reasonably consistent with those from co-located precipitation radar precipitation retrievals, but information transmission into 4D-Var was more limited (Geer et al. 2008). There were also practical issues such as the asymmetric selection of data leading to an inbuilt sampling bias. Wang et al. (2012) assessed the ECMWF 1D-Var retrievals by comparison to Bayesian retrievals from GPROF; they showed that, for retrievals from the same observations, the ECMWF 1D-Var produced more cloud and less precipitation than the Bayesian retrievals. This difference in cloud-rain partitioning was explained by differences in the climatologies of the retrieval constraints: in the 1D-Var retrievals, the results had to be consistent with the ECMWF model moist physics; in the Bayesian retrieval they had to be consistent with the prior database of radar and passive rain retrievals. A further issue with the 1D-Var + 4D-Var approach, as with any assimilation of retrievals, is that it is not optimal without using special methods (e.g., Migliorini 2012, Sect. 48.3.1). Retrievals are a blend of a-priori and retrieved information; in the rainy 1D + 4D-Var their assimilation actually helped pull the analysis back towards the prior solution (Geer et al. 2010). These were some of the reasons that encouraged ECMWF to move to 'all-sky' direct radiance assimilation in 4D-Var (Bauer et al. 2010). In this approach, radiances are assimilated whether they are clear, cloudy or affected by precipitation, and 4D-Var is left free to infer the increments (in whatever part of the state) to fit the radiances best. The all-sky approach has now been developed to the point that microwave water vapor, cloud and precipitation sensitive radiances now provide over 20% of the observational constraint in the ECMWF operational system (Geer et al. 2017; Sect. 48.4.2).

Currently, at most operational forecasting centers the all-sky approach is either already in use or in development for the assimilation of microwave radiances directly in all weather conditions (Geer et al. 2018). Since 2010, however, Météo-France has successfully used a '1D-Bay+3D-Var' approach in their limited area forecasting (Caumont et al. 2010; Wattrelot et al. 2014); rather than a 1D-Var variational retrieval, this uses a Bayesian retrieval step similar to the GPROF approach, but with the a-priori database populated from the surrounding grid-points of the first guess forecast. Section 48.4.4 will also outline the development of a new 1D-Var + 4D-EnVar approach. Another trend is the move towards ensemble approaches, which use the ensemble-derived covariances to replace some of the more complex parts of a pure variational DA: avoiding the need to explicitly model covariances between the control variables (e.g., between precipitation and temperature) and the need to derive tangent-linear and adjoint moist physics models. For example, NCEP assimilate AMSU-A observations in cloudy (but not precipitating) areas with the help of the 4D-EnVar formulation (Zhu et al. 2016). Apart from operational global forecasting centers, the nowcasting and storm forecasting community is also working towards using similar tools, such as ensemble DA and direct radar reflectivity assimilation for convection (e.g., Sun et al. 2014).

48.4.2 Global Assimilation of Precipitation Affected Radiances at ECMWF

Precipitation-affected microwave observations are now routinely assimilated at ECMWF using an 'all-sky' approach. This assimilates the observations directly as radiances into the operational 4D-Var analysis, under all weather conditions (e.g., clear-sky, cloudy or precipitating). The observation operator is RTTOV-SCATT (Bauer et al. 2006c) which is able to simulate the effect of cloud and precipitation hydrometeors, including multiple scattering, with sufficient accuracy and speed for DA. The tangent-linear and adjoint of the forecast model includes moist physics operators. This allows the DA to update dynamical control variables at the beginning of the 4D-var assimilation window in order to fit the cloud and precipitation that is observed in the microwave radiances. This can be labelled the "generalized 4D-Var tracing" effect. It is well known that a 4D-Var DA can extract wind information from radiances sensitive to water vapor or other constituents using the adjoint of the tracer transport equations (e.g., Peubey and McNally 2009; Allen et al. 2013). In the same way, the forecast model embodies all the transport and diabatic processes that lead to the formation of cloud and precipitation, and its adjoint model can also be used to infer changes to the circulation (or more local changes if necessary) that improve the fit of the analysis to the precipitation-affected observations.

Geer et al. (2014) ran single-observation case studies to demonstrate the generalized 4D-Var tracing mechanism in practice. For example, in one case study, the intensity of a low-pressure system was reduced in the analysis so that less frontal precipitation was generated 11 h later, hence better fitting an all-sky observation made at that time. In the tropics, the adjustments are generally on more local scales – for example to add a mesoscale convection system in the analysis, 4D-Var adds moisture and sets up a local pattern of low-level convergence, upper-level divergence to generate or enhance deep convection. Since the main goal of ECMWF is medium-range prediction, the improved analysis of precipitation is not the main goal, but rather the improvement to the dynamical initial conditions. For example, the addition of a group of all-sky instruments was shown to improve day-5 geopotential height forecasts by 2% to 4% (Geer et al. 2017) which in the context of global DA is a very substantial improvement. Figure 48.8 shows the current impact of microwave water vapor, cloud and precipitation sensitive observations on the quality of the 24 h forecast, as measured by forecast sensitivity to observation impact (FSOI; Langland and Baker 2004). Due to the move to all-sky assimilation, these instruments are now one of the most important parts of the observing system at ECMWF, giving similar benefit to the microwave temperature-sensitive channels.

One way to look at the use of all-sky radiances at ECMWF is that it is simply extends the use of satellite data beyond the 'clear-sky' approach that was necessary until recently. The forecast benefit measured in Fig. 48.8 comes from the full range of clear-sky, cloudy and precipitating scenes. Some of the information in precipitating areas comes from the cloud and water vapor visible in the observations, not just from their precipitation content. Hence, Fig. 48.8 measures much more than just the impact of precipitation assimilation on the forecasts. However, going from clear-sky to all-sky assimilation roughly doubled the impact of microwave humidity sounding data (Geer et al. 2014). Further, Geer et al. (2017) broke down the FSOI statistics into the clear-sky and cloudy parts, showing a substantial part of the forecast impact does come from cloudy scenes. Figure 48.9 here uses the same data but breaks it down into clear, cloudy and precipitating scenes (based on a rough C37 threshold (Geer and Bauer 2011) of <0.05 for clear-sky, < 0.2 for cloud and > 0.2 for precipitation). It shows the results from the microwave imager AMSR2 (for more details of its assimilation at ECMWF see Kazumori et al. 2016). By number the

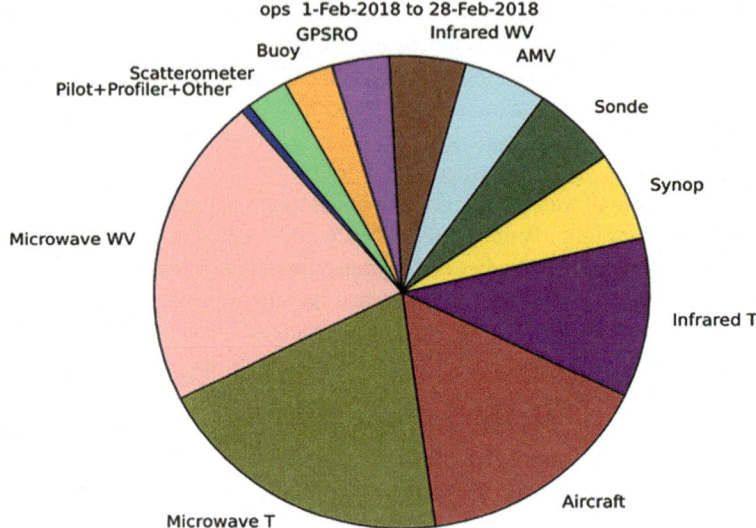

Fig. 48.8 FSOI in ECMWF operational forecasts, for different categories of observations, for February 2018. Microwave WV, cloud, and precipitation observations (the "Microwave WV" category) account for 21% of short-range forecast impact, with "moist" assimilation having slightly more impact than "dry" assimilation ("Microwave T")

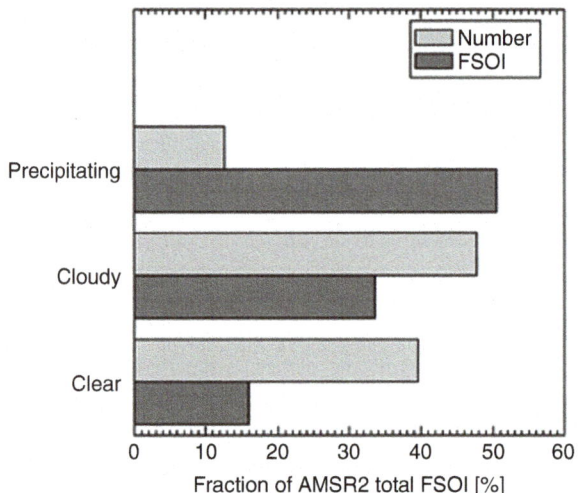

Fig. 48.9 Impact of all-sky AMSR2 radiances in the ECMWF operational system during July and August 2016 (see Geer et al. 2017), in terms of FSOI (forecast impact) and number of observations, all given as a percentage of the total (FSOI or number of observations) for AMSR2. Scenes have been characterized according to whether they are clear-sky, cloudy or precipitating. Although the number of precipitating scenes is relatively small, they have large impact on the 24 h forecast quality

precipitating sample is the smallest, but by forecast impact, it is the largest. This is typical of all-sky microwave imager assimilation at ECMWF. It is thought that, because precipitating scenes are usually found in the most active weather areas, and because these areas are generally where the greatest sensitivity to initial conditions lies (McNally 2002), and finally because these areas are typically less well supplied with other satellite observations, the precipitation-affected data has great scope to improve the analysis and subsequent forecast.

48.4.3 Global Assimilation of Precipitation Affected Radiances at JMA: Impacts on Tropical Cyclones

Development of an all-sky microwave radiance assimilation scheme for the JMA global DA system is in progress. In the JMA global NWP system (JMA 2013), the primary targets are accurate analysis and forecasting of tropical cyclones and predictions of severe precipitation events. The analysis system uses an incremental 4D-Var approach (Courtier et al. 1994) to produce initial conditions for JMA's global spectral model (GSM). The all-sky microwave radiance assimilation scheme uses the 4D-Var system and RTTOV-SCATT for the RT calculation. Cloud liquid water, cloud ice water, cloud fraction, rain and snow profiles are not control variables in the DA system, but they are provided by the JMA GSM. The microwave imager observations are averaged to match the model resolution used in the minimization and thinned in 150 km × 150 km boxes. The microwave humidity sounding radiance are thinned in 180 km. Situation dependent observation errors are set using a symmetric cloud amount parameter based on Geer and Bauer (2011).

Tropical cyclones are one of the severe meteorological phenomena of the Earth and its accurate prediction is crucial for disaster prevention and mitigation. The accurate analysis and forecasting of tropical cyclones are a scientifically challenging task. Water vapor, clouds and precipitation in the atmosphere have important roles for tropical cyclone formation and development. Space-based microwave radiance observations from various microwave radiometers onboard polar-orbiting satellites (e.g., AMSR2/GCOM-W, GMI/GPM, SSMIS/DMSP, MHS/NOAA and Metop) contain clouds and precipitation information. The microwave low frequency observations (from 19 to 37 GHz) have information on total column water vapor, cloud liquid water, precipitation and surface wind speed over oceans. Microwave water vapor sounding observations at 183 GHz water vapor absorption line have upper- and middle- tropospheric water vapor information and clouds and solid precipitation information. The hydrometeor information is crucial data source to produce initial fields for numerical weather predictions.

The assimilation of cloud and precipitation affected radiance can reduce errors in first-guess fields from numerical weather prediction models. Accurately analyzed temperature, water vapor, clouds, precipitation and wind fields under cloudy conditions should be used for the initial field of the NWP models.

During the lifetime of a tropical cyclone, cloudy areas are spatially spread around the center of the tropical cyclone in the genesis and the developing stages. All-sky microwave radiance assimilation can make use of cloud- and precipitation affected radiance to improve the analysis accuracy in the cloudy conditions. Furthermore, the cloud and precipitation phenomena exhibit nonlinear behavior in their formation and dissipation. To consider such nonlinearity in the incremental 4D-Var DA, outer-loop iteration is thought as an effective method. In the outer-loop iterations, the updates of the trajectory and the re-computations of the departures from the observations are performed in the minimizations of the 4D-Var DA.

To confirm the effect of the all-sky microwave radiance assimilation for tropical cyclone analysis and forecasting, DA experiments were performed by using the JMA global NWP system. Three DA experiments were conducted to investigate impacts of all-sky microwave radiance assimilation, and outer-loop introduction in the JMA global NWP system. Control run (CNTL) uses same configuration as the JMA operational system in November 2016. CNTL uses a clear sky microwave radiance assimilation configuration and no outer-loop iterations. TEST1 is the same as CNTL but single outer-loop iteration (single update of the trajectory in the minimization and re-computations of FG departure for all observation type) is introduced. TEST2 includes the all-sky microwave radiance assimilation and the single outer-loop iteration. The microwave radiance data from AMSR2, GMI, SSMIS and MHS were assimilated under all-sky conditions. In the TEST2, RTTOV-SCATT is used for all-sky microwave RT calculation to include scattering and emission effects from hydrometeors in atmosphere. Temperature, water vapor, cloud liquid water, cloud ice water, cloud fraction, rain and snow profiles are obtained from JMA global model for the RT calculation. The symmetric observation errors defined with the symmetric cloud amount are used in the TEST2 for the all-sky microwave radiance data.

From comparisons of the experiment results of TEST2 and CNTL for various tropical cyclone cases, it is found that the all-sky assimilation can provide clear positive impacts for tropical cyclone analysis and forecasting. The all-sky assimilation and the outer-loop introduction reduced the error of tropical cyclone track predictions for various ocean basins. The results are supported by the improvement of short-range forecast skill on temperature, specific humidity, and wind vector fields. Figure 48.10 shows one typical improved tropical cyclone central pressure predictions (Hurricane Jimena in 2015) from the DA experiments. As shown in the Fig. 48.10a, TEST2 case (all-sky assimilation plus outer-loop introduction) showed a realistic rapid intensification of the tropical cyclone compared to those of CNTL runs. The results from TEST1 (Fig. 48.10b) showed improvements for the tropical cyclone intensity predictions.

However, TEST2 showed much realistic predictions of the rapid intensification (NOAA best track data are shown with a black line). The improvements were found for the development stages in the tropical cyclone lifetime. The all-sky assimilation could provide observational information for cloudy areas around the tropical cyclone. However, minor improvements were found in the decaying stages for both cases. To clarify the mechanism of the improved tropical cyclone intensity prediction by the all-sky microwave radiance assimilation, analyzed water vapor fields in cloudy regions around the tropical cyclone were compared. Figure 48.11

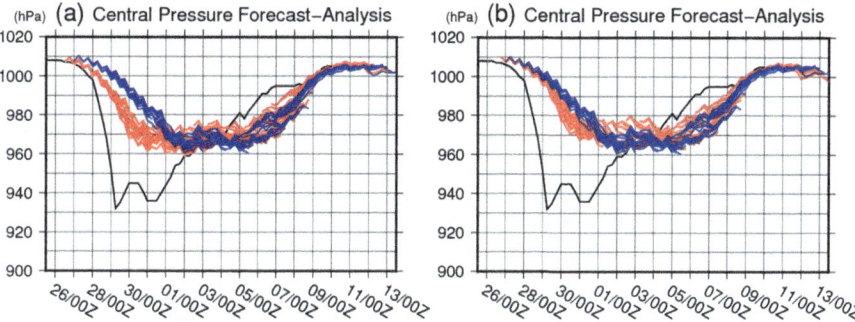

Fig. 48.10 Comparisons of tropical cyclone prediction (central pressure) for Hurricane Jimena in 2015. (**a**) Comparison of TEST2 (all-sky assimilation plus outer-loop iteration) and CNTL. (**b**) Comparison of TEST1 (outer-loop iteration) and CNTL. Red lines denote the predicted central pressure of TEST and blue lines denote those of CNTL. Black solid line is NOAA best track data. The period is from 25 August to 13 September 2015

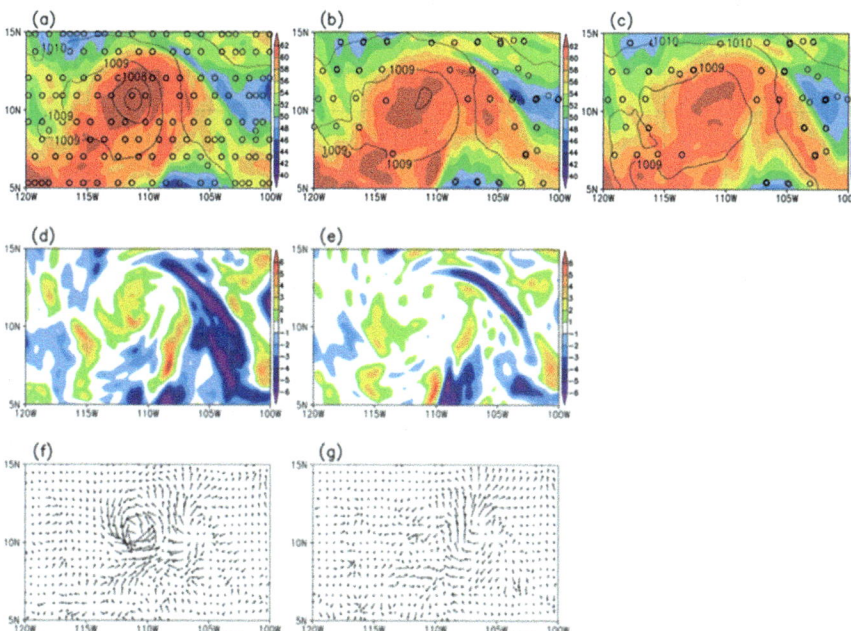

Fig. 48.11 Comparisons of analyzed TPW fields for Hurricane Jimena in 2015. The analysis is for 26 August 2015 1200 UTC. (**a**) analyzed TPW of TEST2, (**b**) TEST1, and (**c**) CNTL. The units are mm. The contour lines denote analyzed sea surface pressure in hPa. Small black circles indicate locations of assimilated microwave radiance data. (**d**) TPW difference of TEST2 and CNTL, (**e**) TPW difference TEST1 and CNTL. The units are mm. (**f**) surface wind vector difference of TEST2 and CNTL, (**g**) surface wind vector difference TEST1 and CNTL. The units are m s^{-1}

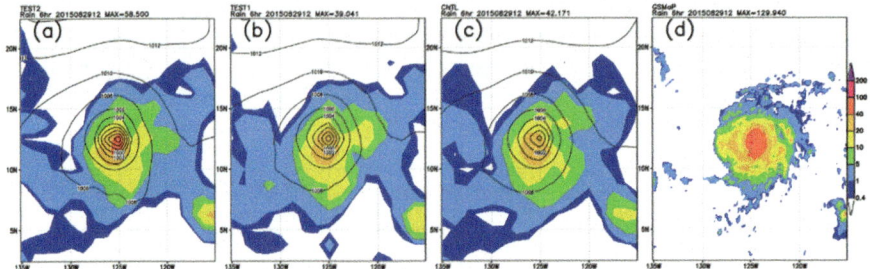

Fig. 48.12 Comparisons of 6 h accumulated precipitation forecasting from 0600 UTC 29 August 2015 initial time (tropical cyclone developing stage). (**a**) TEST2, (**b**) TEST1, (**c**) CNTL, and (**d**) GSMaP. The units are mm. Contours denote sea surface pressure in hPa

shows the comparison of total precipitable water (TPW) vapor analysis and their differences from CNTL for the early stage of the tropical cyclone (26 August 2015 1200 UTC). Difference of analyzed sea surface wind vectors are shown with arrows in the bottom panels. TEST2 result showed high TPW amount in the center of the tropical cyclone was analyzed in the all-sky assimilation and the outer-loop introduction. Black open circles indicate the locations of the assimilated microwave radiance data. TEST1 result showed increased TPW around the tropical cyclone center through the clear sky assimilation and outer-loop introduction. These results showed the all-sky assimilation plus the outer-loop introduction could change the water vapor and wind fields realistically under cloudy conditions (i.e., in the core of the tropical cyclone) and deeper central pressure of the tropical cyclone was analyzed in TEST2.

Moreover, precipitation forecasting of the tropical cyclone were compared with the Japanese Aerospace Exploration Agency's (JAXA) precipitation product, GSMaP. Figure 48.12 shows the 6 h-accumulated precipitation forecasts of the experiments at the developing stage (29 August 2015 1200 UTC) of the tropical cyclone. In the core of the tropical cyclone, TEST2 result showed strong precipitation, which is similar to GSMaP precipitation products. However, results of other experiments showed relatively weak precipitation. The all-sky microwave radiance assimilation together with the out-loop iteration use contributed to the realistic precipitation forecast of the tropical cyclone.

Generally, in the tropical cyclone developing stages, spatially spread cumulus clouds are widely distributed, and they are in a circulation. During those stages, the all-sky microwave radiance DA can provide water vapor information under the cloudy conditions and the assimilation leads to increases of water vapor concentration in and around the tropical cyclone center. They help the enhancement of the circulation in the lower troposphere through the model physics in 4D-Var. The increase of water vapor, deeper central pressure and enhanced water vapor flows toward the tropical cyclone center can be retained in cycling of the DA. However, in the decaying stages, the impact of all-sky radiance assimilation was relatively smaller compared to the early stage of the tropical cyclone lifetime. The reason is that the amount of the clouds in the decaying stages was decreased around the tropical cyclone and there is an intrusion of relatively dry atmosphere from high latitude areas. The atmosphere is under clear conditions and clear-sky microwave radiance data have already assimilated there.

48.4.4 Assimilation of Precipitation Affected Radiances Using a 1D-Var + 4D-EnVar

We have shown in the previous sections on precipitation signal sensitivity and on the important factors driving the assimilation and inversion of precipitation measurements that one of the approaches used to handle precipitation-induced non-linearities is to combine a 1D-Var pre-processor with a regular (3D or 4D) assimilation approach. This section aims to highlight the results of using this approach and to show some of its benefits. It is worth noting, as precipitation accounts for roughly 5% of cases, that the impact of assimilating precipitation observations is neither likely nor necessarily expected to be great on global statistics. We will therefore focus mainly on highlighting local/regional impacts, such as storms initialization, front placement, and the like.

As described in the section above, caution must be used when employing the 1D-Var as pre-processing to the DA step. Consistency between 1D-Var and the DA must be met, as to not introduce artifacts in the analysis. Using similar backgrounds, similar RT models, similar assumptions on some *hard-coded* values (such as density of hydrometeors, or optical properties), between the 1D-Var and the DA systems are few examples where consistency must be respected. 1D-Var pre-processing step should also be used, to the maximum extent possible, in a complementary fashion to the DA, and not be a redundant one. Employing the 1D-Var to process data that are otherwise filtered out in the DA, or implementing the 1D-Var and using its output as a first guess (and not as background information), are few ways to ensure the 1D-Var and the DA are complementary to each other. The most valuable benefit of using a 1D-Var pre-processing step prior to DA is that it allows for the ability to provide geophysical information that is consistent with the actual satellite observations, and therefore theoretically more accurate than a forecast. This information will be provided to the DA system, on a point by point basis, as a sort of pre-correction of the forecast background before observations are assimilated. This results, for instance, in the ability to generate precipitation in the background where it does not exist but measurements indicate that it should, and to remove precipitation in the background where measurements suggest its absence. In other words, a 1D-Var pre-processor has the ability to shift and adjust fronts and other geophysical features (provided the 1D-Var system can solve for them) in the background, if they happen to be misplaced.

Note that this applies not only to precipitation displacement correction but to other geophysical data as well (as long as the 1D-Var system is able to solve for them). Typically, fast moving atmospheric features or highly-contrasted backgrounds, such as precipitation, suspended cloud, and atmospheric moisture have the potential to benefit from a 1D-Var pre-processing correction when the background contains misplacement errors.

An added benefit, although not related to precipitation per se, is that this 1D-Var pre-processing stage can benefit the DA in other ways, such as dynamically providing parameters to the DA system that are not currently analyzed in it, thereby relieving the DA system from relying on inaccurate climatology-based information or fixed/gridded databases. An example of this could be the provision of dynamic CO_2 values for instance when pre-processing temperature-sensitive infrared channels before they are assimilated. It could also be the provision of dynamic land emissivity values for surface-sensitive microwave channels, relevant to the footprint size of the measurement, instead of relying on a gridded, averaged emissivity database. Additional benefits of using this 1D-Var preprocessing include: (1) providing a generic, theoretically more accurate, quality-control system to assess discrepancies between the measurement and the background (i.e., if the 1D-Var pre-processor cannot handle it, the DA system can likely not handle it either, and the observation can therefore be rejected); and (2) providing valuable information (like temperature profiles in precipitating conditions) that is simply not exploitable by the current DA. The 1D-Var pre-processor used in the following example is the Multi-Instrument Inversion and Data Assimilation Pre-processing System (MIIDAPS), developed by NOAA STAR. As MIIDAPS is an upgrade of the previously discussed MiRS algorithm (see Chap. 19 in volume 1 and Boukabara et al. 2011), we will not describe it in detail here other than to mention that MIIDAPS has been extended to process non-microwave sensors and to act as both a standalone inversion system as well as a DA pre-processing system.

In Fig. 48.13, an example of MIIDAPS pre-processing impact is highlighted. It shows the ability to correct potential displacements of meteorological front locations in the background forecasts, preceding the assimilation process. To make sure this is done in a controlled environment, the impact is assessed using the GEOS-5 Nature Run (G5NR) developed by the National Aeronautics and Space Administration (NASA; Gelaro et al. 2014). The Nature Run is obtained by free-running a forecast model, generating a large set of geophysical analyses over a long period of time. This is used here to test the impact of 1D-Var pre-processing but it is used for a multitude of applications such as observation system simulation experiments (OSSEs; Boukabara et al. 2016).

The top left panel in Fig. 48.13 shows the TPW field from the G5NR, over the southern coast of South Africa. On that panel is highlighted a trapezoid where SNPP/ATMS data were simulated to reproduce what the sensor would be seeing if it was measuring over that area. The purpose then is to undertake a DA experiment, with and without the 1D-Var pre-processing to assess the impact of the 1D + DA approach. In the panel labeled (a) is the field of differences between G5NR and the background (assumed to be past 6 h analysis from the G5NR). *Dipoles* of juxtaposed positive/negative differences from G5NR exist within the trapezoid, typical of displacement errors, in this case of moisture front. Indeed, when displacement occurs -when forecast misplaces the location of a front for example- the

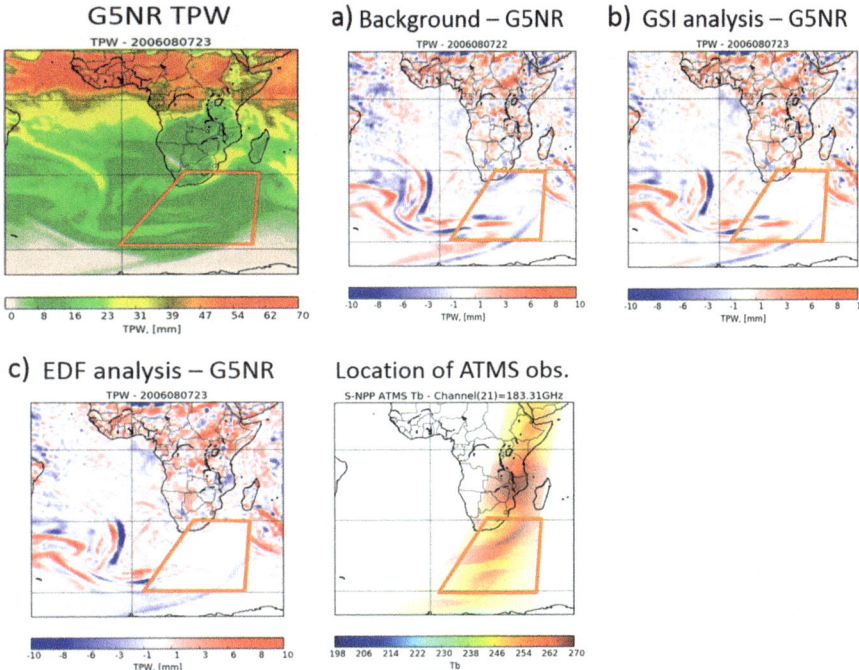

Fig. 48.13 Example Analysis Cycle at 2300 UTC 7 August 2006. (**a**) Background-G5NR shows large displacements (dipoles) in TPW field, (**b**) GSI analysis-G5NR reduces magnitude of dipoles slightly where SNPP ATMS data exists (red trapezoid), (**c**) Environmental Data Fusion (EDF) analysis through MIIDAPS-based background adjustment removes most of dipole feature and reduces TPW differences where SNPP ATMS data exists

difference between the forecasted and the actual field has a bimodal distribution: significant positive and negative differences coexist, next to each other. When ATMS data are assimilated using the Gridpoint Statistical Interpolation (GSI) system, the result of the analysis is shown in panel (b). Within the trapezoid area, where the simulated ATMS data was assimilated, the difference is reduced. Note that the points right where the dipoles occur, were not assimilated because they essentially didn't pass the *gross check* since they were above the threshold imposed in the GSI. So, all improvements (reduction in differences) in that area are due to the indirect effect of assimilating neighboring points. Panel (c) shows the same difference field when the MIIDAPS 1D-Var pre-processor is applied to correct the background before data are assimilated. The differences disappear entirely in the area where the ATMS were assimilated (within the trapezoid). This demonstrates one of the main advantages of the 1D-Var applied as pre-processing to DA: correcting misplacements of background information.

The test case shown in Fig. 48.13 was shown for TPW and in simulation (using G5NR), to ensure that we do not have uncertainties on the exact location of the moisture front. The 1D-Var has also been applied to real data, to correct

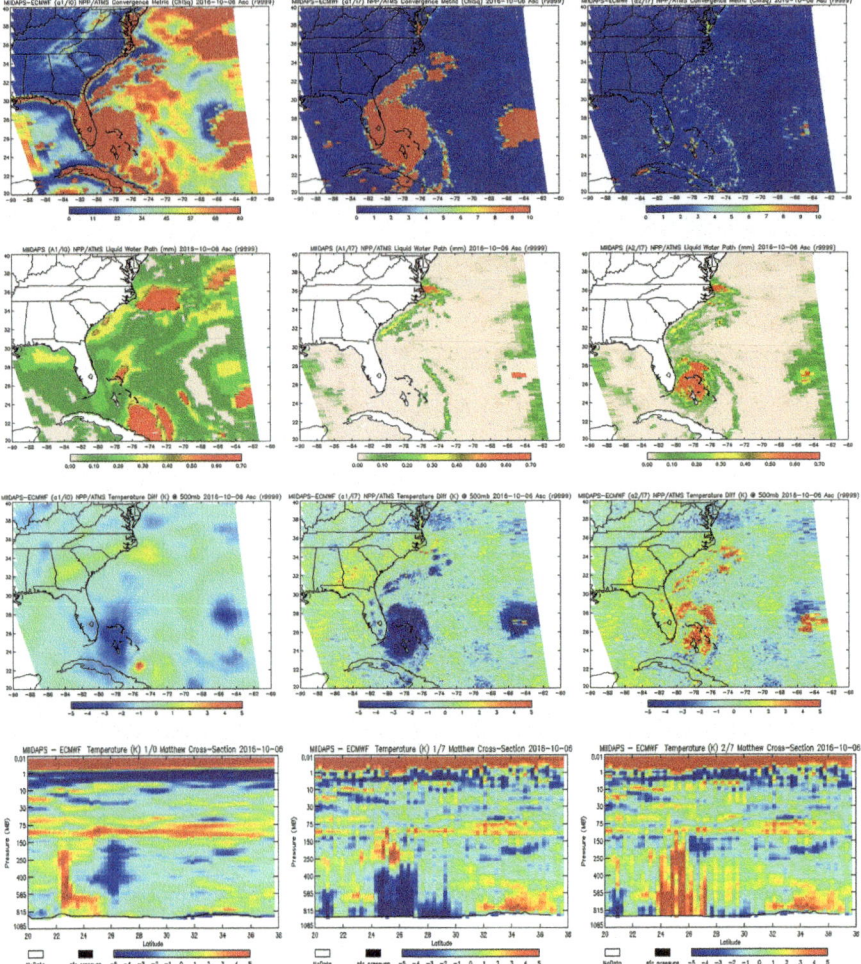

Fig. 48.14 Highlighting the impact of using a 1D-Var system as preprocessing to DA. One of the main features is the generation (when absent) and displacement correction of hydrometeors in the analysis (shown is the case of Hurricane Matthew over the US Floridian coast, from October 2016). Top panels show the evolution of the convergence metric (from iteration#0: background, to iteration 4 to the last iteration 7 in the 1D-Var preprocessor). The second row shows the evolution of the cloud field within the iterative process, again starting with the background to the final cloud field retrieved. The third and fourth rows show the evolution of temperature field at 500 hPa and the vertical cross-section of temperature respectively. Instead of the temperature itself, the departure from the reference (ECMWF analyses) is shown. Both spatial and vertical displacements are corrected. See text for more details

misplacements in forecast background of cloud and precipitation in addition to moisture and temperature (in and around precipitating conditions). Figure 48.14 shows an example of this for a number of parameters for the case of Hurricane Matthew. What can be seen in Fig. 48.14 is that using the pre-processor allows the cloud to be generated when it is absent in the background, while it reduces to zero the cloud in the background where there should be no cloud, at least according to the precipitation measurements (second row of Fig. 48.14). We can see that the cloud that was misplaced north of the Cuban island (left) was positioned further north (east of the Florida tip) in the right panel. The radiometric consistency -convergence- is achieved (top row) by the pre-processor through the iterative process. The middle panel are intermediate results during the iterative process. The bimodal spatial distribution of the temperature difference in the third row, left panel, is illustrated in the red dot showing a very positive difference, juxtaposed to a spread of blue, indicating a very negative difference. This bimodal difference (with respect to ECMWF) is also shown vertically in the last row, left panel. In the left panels, these bimodal differences disappear.

48.5 Final Thoughts on Precipitation Data and Data Assimilation for Numerical Weather Prediction

Observations of precipitation from space provide critical information for many applications from environmental monitoring to medium-range weather prediction. Their use is essential since precipitation is exactly "where the weather is," and satellites fill the critical gaps where surface-based or in-situ measurements are non-existent. For NWP applications, there are many challenges for using these data which have been discussed, and can be summarized by how best to approach modeling of the background a-priori and error covariance, projection of the background precipitation fields into radiometric space through RT, and characterization of observation errors and biases. To a large degree, the challenges of exploiting precipitating observations are due to the non-linearities of the problem: the non-linearity of hydrometeors spatially, the non-linearity of hydrometeors temporally, and the non-linear response to satellite observations (brightness temperatures) to changes in hydrometeor fields.

To overcome these challenges, various approaches have been adopted. A few of these were discussed; ECMWF directly assimilate cloudy and precipitating radiances into their 4D-Var system using the generalized 4D-Var tracing. This method allows for the forecast model moist physics, tangent-linear and adjoint to adjust variables which influence cloud formation and dissipation, in order to fit the affected satellite radiances at the analysis time. Similarly, JMA uses an incremental FJMA4D-Var system to directly assimilate precipitating radiances, with the GSM providing state variables for various cloud hydrometeor profiles along with cloud fraction for the forward simulations. At NESDIS, the focus is on using a 1D-Var preprocessor to analyze the cloud and precipitation fields to better constrain the 4D-Var analysis.

As these approaches evolve and new methodologies are developed, the focus will remain on improving the analysis fields within areas of clouds and precipitation.

These improvements rely upon the advancements in both RT modeling, forecast models, and their coupling within DA systems.

References

Allen, D. R., Hoppel, K. W., Nedoluha, G. E., Kuhl, D. D., Baker, N. L., Xu, L., & Rosmond, T. E. (2013). Limitations of wind extraction from 4D-Var assimilation of ozone. *Atmospheric Chemistry and Physics, 13*, 3501–3515. https://doi.org/10.5194/acp-13-3501-2013.

Bannister, R. N. (2008). A review of forecast error covariance statistics in atmospheric variational data assimilation. II: Modelling the forecast error covariance statistics. *Quarterly Journal of the Royal Meteorological Society, 134*, 1971–1996. https://doi.org/10.1002/qj.340.

Bauer, P., Auligné, T., Bell, W., Geer, A., Guidard, V., Heilliette, S., Kazumori, M., Kim, M.-J., Liu, E. H.-C., McNally, A. P., Macpherson, B., Okamoto, K., Renshaw, R., & Riishøjgaard, L.-P. (2011). Satellite cloud and precipitation assimilation at operational NWP centres. *Quarterly Journal of the Royal Meteorological Society, 137*, 1934–1951. https://doi.org/10.1002/qj.905.

Bauer, P., Geer, A. J., Lopez, P., & Salmond, D. (2010). Direct 4D-Var assimilation of all-sky radiances. Part I: Implementation. *Quarterly Journal of the Royal Meteorological Society, 136*, 1868–1885. https://doi.org/10.1002/qj.659.

Bauer, P., Lopez, P., Benedetti, A., Salmond, D., & Moreau, E. (2006a). Implementation of 1D+ 4D-Var assimilation of precipitation-affected microwave radiances at ECMWF. I: 1D-Var. *Quarterly Journal of the Royal Meteorological Society, 132*, 2277–2306. https://doi.org/10.1256/qj.05.189.

Bauer, P., Lopez, P., Salmond, D., Benedetti, A., Saarinen, S., & Bonazzola, M. (2006b). Implementation of 1D+ 4D-Var assimilation of precipitation-affected microwave radiances at ECMWF. II: 4D-Var. *Quarterly Journal of the Royal Meteorological Society, 132*, 2307–2332. https://doi.org/10.1256/qj.06.07.

Bauer, P., Moreau, E., Chevallier, F., & O'Keeffe, U. (2006c). Multiple-scattering microwave radiative transfer for data assimilation applications. *Quarterly Journal of the Royal Meteorological Society, 132*, 1259–1281. https://doi.org/10.1256/qj.05.153.

Baum, B. A., Yang, P., Heymsfield, A. J., Platnick, S., King, M. D., & Thomas, S. M. (2005). Bulk scattering properties for the remote sensing of ice clouds II: Narrowband models. *Journal of Applied Meteorology, 44*, 1896–1911. https://doi.org/10.1175/JAM2309.1.

Bell, T. L. (1987). A space-time stochastic model of rainfall for satellite remote-sensing studies. *Journal of Geophysical Research, 92*, 9631–9643. https://doi.org/10.1029/JD092iD08p09631.

Benedetti, A., Stephens, G. L., & Vukićević, T. (2003). Variational assimilation of radar reflectivities in a cirrus model. II: Optimal initialization and model bias estimation. *Quarterly Journal of the Royal Meteorological Society, 129*, 301–319. https://doi.org/10.1256/qj.02.64.

Bi, L., & Yang, P. (2017). Improved ice particle optical property simulations in the ultraviolet to far-infrared regime. *Journal of Quantitative Spectroscopy and Radiative Transfer, 189*, 228–237. https://doi.org/10.1016/j.jqsrt.2016.12.007.

Boukabara, S. A., Garrett, K., Chen, W., Iturbide-Sanchez, F., Grassotti, C., Kongoli, C., Chen, R., Liu, Q., Yan, B., Weng, F., Ferraro, R., Kleespies, T. J., & Meng, H. (2011). MiRS: An all-weather 1DVAR satellite data assimilation and retrieval system. *IEEE Transactions on Geoscience and Remote Sensing, 49*, 3249–3272. https://doi.org/10.1109/TGRS.2011.2158438.

Boukabara, S.-A., Moradi, I., Atlas, R., Casey, S. P. F., Cucurull, L., Hoffman, R. N., Ide, K., Kumar, V. K., Li, R., Li, Z., Masutani, M., Shahroudi, N., Woollen, J., & Zhou, Y. (2016). Community Global Observing System Simulation Experiment (OSSE) Package (CGOP): Description and usage. *Journal of Atmospheric and Oceanic Technology, 33*(8), 1759–1777. https://doi.org/10.1175/JTECH-D-16-0012.1.

Caumont, O., Ducrocq, V., Wattrelot, É., Jaubert, G., & Pradier-Vabre, S. (2010). 1D+3DVar assimilation of radar reflectivity data: A proof of concept. *Tellus A, 62*, 173–187. https://doi.org/10.1111/j.1600-0870.2009.00430.x.

Chen, Y., Weng, F., Han, Y., & Liu, Q. (2008). Validation of the community radiative transfer model (CRTM) by using CloudSat data. *Journal of Geophysical Research, 113*(D8), 2156–2202. https://doi.org/10.1029/2007JD009561.

Clough, S. A., Shephard, M. W., Mlawer, E. J., Delamere, J. S., Iacono, M. J., Cady-Pereira, K., Boukabara, S., & Brown, P. D. (2005). Atmospheric radiative transfer modeling: A summary of the AER codes. *Journal of Quantitative Spectroscopy and Radiative Transfer, 91*, 233–244. https://doi.org/10.1016/j.jqsrt.2004.05.058.

Courtier, P., Thépaut, J.-N., & Hollingsworth, A. (1994). A strategy for operational implementation of 4D-Var, using an incremental approach. *Quarterly Journal of the Royal Meteorological Society, 120*, 1367–1387. https://doi.org/10.1002/qj.49712051912.

Ding, S., Yang, P., Weng, F., Liu, Q., Han, Y., Van Delst, P., Li, J., & Baum, B. (2011). Validation of the community radiative transfer model. *Journal of Quantitative Spectroscopy and Radiative Transfer, 112*, 1050–1064. https://doi.org/10.1016/j.jqsrt.2010.11.009.

Errico, R. M., Bauer, P., & Mahfouf, J.-F. (2007). Issues regarding the assimilation of cloud and precipitation data. *Journal of the Atmospheric Sciences, 64*, 3785–3798. https://doi.org/10.1175/2006JAS2044.1.

Garrett, K. J., Yang, P., Nasiri, S. L., Yost, C. R., & Baum, B. A. (2009). Influence of cloud top height and geometric thickness on a MODIS infrared-based cirrus cloud retrieval. *Journal of Applied Meteorology and Climatology, 48*, 818–832. https://doi.org/10.1175/2008JAMC1915.1.

Geer, A. J., & Bauer, P. (2011). Observation errors in all-sky data assimilation. *Quarterly Journal of the Royal Meteorological Society, 137*, 2024–2037. https://doi.org/10.1002/qj.830.

Geer, A. J., Bauer, P., & Lopez, P. (2008). Lessons learnt from the operational 1D+4D-Var assimilation of rain-and cloud-affected SSM/I observations at ECMWF. *Quarterly Journal of the Royal Meteorological Society, 134*, 1513–1525. https://doi.org/10.1002/qj.304.

Geer, A. J., Bauer, P., & Lopez, P. (2010). Direct 4D-Var assimilation of all-sky radiances. Part II: Assessment. *Quarterly Journal of the Royal Meteorological Society, 136*, 1886–1905. https://doi.org/10.1002/qj.681.

Geer, A. J., Baordo, F., Bormann, N., & English, S. (2014). All-sky assimilation of microwave humidity sounders. Technical Memoranda, 741, ECMWF, 57 pp. Available at https://www.ecmwf.int/en/elibrary/9507-all-sky-assimilation-microwave-humidity-sounders, last accessed 23 Nov 2018.

Geer, A. J., Baordo, F., Bormann, N., Chambon, P., English, S. J., Kazumori, M., Lawrence, H., Lean, P., Lonitz, K., & Lupu, C. (2017). The growing impact of satellite observations sensitive to humidity, cloud and precipitation. *Quarterly Journal of the Royal Meteorological Society, 143*, 3189–3206. https://doi.org/10.1002/qj.3172.

Geer, A., Lonitz, K., Weston, P., Kazumori, M., Okamoto, K., Zhu, Y., Liu, E., Collard, A., Bell, W., Migliorini, S., Chambon, P., Fourrié, N., Kim, M., Köpken-Watts, C., & Schraff, C. (2018). All-sky satellite data assimilation at operational weather forecasting centres. *Quarterly Journal of the Royal Meteorological Society, 144*, 1191–1217. https://doi.org/10.1002/qj.3202.

Gelaro, R., Putman, W. M., Pawson, S., Draper, C., Molod, A., Norris, P. M., Ott, L., Prive, N., Reale, O., & Achutavarier, D. (2014). *Evaluation of the 7-km GEOS-5 nature run*. Tech. Rep. Series on Global Modeling and Data Assimilation, R. D. Koster, Ed., NASA, Greenbelt, MD, Goddard Space Flight Center, Vol. 36, 305 pp. TM-2014-104606v36. Available at https://ntrs.nasa.gov/search.jsp?R=20150011486, last accessed 23 Nov 2018.

Gustafsson, N., Janjić, T., Schraff, C., Leuenberger, D., Weissman, M., Reich, H., Brousseau, P., Montmerle, T., Wattrelot, E., Bučánek, A., Mile, M., Hamdi, R., Lindskog, M., Barkmeijer, J., Dahlbom, M., Macpherson, B., Ballard, S., Inverarity, G., Carley, J., Alexander, C., Dowell, D., Liu, S., Ikuta, Y., & Fujita, T. (2018). Survey of data assimilation methods for convective-scale numerical weather prediction at operational centres. *Quarterly Journal of the Royal Meteorological Society, 144*, 1218–1256. https://doi.org/10.1002/qj.3179.

Han, Y., van Delst, P., Liu, Q., Weng, F., Yan, B., Treadon, R., & Derber, J. (2006). JCSDA Community Radiative Transfer Model (CRTM) – Version 1 (NOAA Technical Report NESDIS 122), 33 pp. Available at https://repository.library.noaa.gov/view/noaa/1157, last accessed 23 Nov 2018.

Hunt, B. R., Kostelich, E. J., & Szunyogh, I. (2007). Efficient data assimilation for spatiotemporal chaos: A local ensemble transform Kalman filter. *Physica D: Nonlinear Phenomena, 230*, 112–126. https://doi.org/10.1016/j.physd.2006.11.008.

Iturbide-Sanchez, F., Boukabara, S. A., Chen, R., Garrett, K., Grassotti, C., Chen, W., & Weng, F. (2011). Assessment of a variational inversion system for rainfall rate over land and water surfaces. *IEEE Transactions on Geoscience and Remote Sensing, 49*, 3311–3333. https://doi.org/10.1109/TGRS.2011.2119375.

Janisková, M., & Lopez, P. (2013). Linearized physics for data assimilation at ECMWF. In S. K. Park & L. Xu (Eds.), *Data Assimilation for Atmospheric, Oceanic and Hydrologic Applications (Vol. II)* (pp. 251–286). Berlin/Heidelberg: Springer. ISBN:978-3-642-35088-7.

Janjić, T., Bormann, N., Bocquet, M., Carton, J. A., Cohn, S. E., Dance, S. L., Losa, S. N., Nichols, N., Potthast, R., Waller, J. A., & Weston, P. (2017). On the representation error in data assimilation. *Quarterly Journal of the Royal Meteorological Society, 144*, 1257–1278. https://doi.org/10.1002/qj.3130.

JMA. (2013). *Outline of the operational numerical weather prediction at the Japan Meteorological Agency.* Appendix to WMO Technical Progress Report on the global data-processing and forecasting system and numerical weather prediction. Available at http://www.jma.go.jp/jma/jma-eng/jma-center/nwp/outline2013-nwp/index.htm, last accessed 23 Nov 2018.

Jones, C. D., & Macpherson, B. (1997). A latent heat nudging scheme for the assimilation of precipitation data into an operational mesoscale model. *Meteorological Applications, 4*, 269–277. https://doi.org/10.1017/S1350482797000522.

Jones, C., Waliser, D. E., Lau, K. M., & Stern, W. (2004). Global occurrences of extreme precipitation and the Madden-Julian Oscillation: Observations and predictability. *Journal of Climate, 17*, 4575–4589. https://doi.org/10.1175/3238.1.

Joyce, R. J., Janowiak, J. E., Arkin, P. A., & Xie, P. (2004). CMORPH: A method that produces global precipitation estimates from passive microwave and infrared data at high spatial and temporal resolution. *Journal of Hydrometeorology, 5*, 487–503. https://doi.org/10.1175/1525-7541(2004)005<0487:CAMTPG>2.0.CO;2.

Kazumori, M., Geer, A. J., & English, S. J. (2016). Effects of all-sky assimilation of GCOM-W/AMSR2 radiances in the ECMWF numerical weather prediction system. *Quarterly Journal of the Royal Meteorological Society, 142*, 721–737. https://doi.org/10.1002/qj.2669.

Kirschbaum, D. B., Huffman, G. J., Adler, R. F., Braun, S., Garrett, K., Jones, E., McNally, A., Skofronick-Jackson, G., Stocker, E., Wu, H., & Zaitchik, B. F. (2017). NASA's remotely sensed precipitation: A reservoir for applications users. *Bulletin of the American Meteorological Society, 98*, 1169–1184. https://doi.org/10.1175/BAMS-D-15-00296.

Koizumi, K., Ishikawa, Y., & Tsuyuki, T. (2005). Assimilation of precipitation data to the JMA mesoscale model with a four-dimensional variational method and its impact on precipitation forecasts. *SOLA, 1*, 45–48. https://doi.org/10.2151/sola.2005-013.

Krishnamurti, T. N., Xue, J., Bedi, H. S., Ingles, K., & Oosterhof, D. (1991). Physical initialization for numerical weather prediction over the tropics. *Tellus A, 43*, 53–81. https://doi.org/10.3402/tellusa.v43i4.11938.

Kummerow, C. D., Hong, Y., Olson, W. S., Yang, S., Adler, R. F., McCollum, J., Ferraro, R. R., Petty, G., Shin, D.-B., & Wilheit, T. T. (2001). The evolution of the Goddard Profiling Algorithm (GPROF) for rainfall estimation from passive microwave sensors. *Journal of Applied Meteorology, 40*, 1801–1820. https://doi.org/10.1175/1520-0450(2001)040<1801:TEOTGP>2.0.CO;2.

Langland, R. H., & Baker, N. L. (2004). Estimation of observation impact using the NRL atmospheric variational data assimilation adjoint system. *Tellus A, 56,* 189–201. https://doi.org/10.3402/tellusa.v56i3.14413.

Levenberg, K. (1944). A method for the solution of certain non-linear problems in least squares. *Quarterly of Applied Mathematics, 2,* 164–168. Available at https://www.jstor.org/stable/43633451, last accessed 23 Nov 2018.

Liu, G. (2008). A database of microwave single-scattering properties for nonspherical ice particles. *Bulletin of the American Meteorological Society, 89,* 1563–1570. https://doi.org/10.1175/2008BAMS2486.1.

Liu, Q., & Boukabara, S.-A. (2014). Community Radiative Transfer Model (CRTM) applications in supporting the Suomi National Polar-orbiting Partnership (SNPP) mission validation and verification. *Remote Sensing of Environment, 140,* 744–754. https://doi.org/10.1016/j.rse.2013.10.011.

Lyon, B., & Barnston, A. G. (2005). ENSO and the spatial extent of interannual precipitation extremes in tropical land areas. *Journal of Climate, 18,* 5095–5109. https://doi.org/10.1175/JCLI3598.1.

Mahfouf, J.-F., Bauer, P., & Marécal, V. (2005). The assimilation of SSM/I and TMI rainfall rates in the ECMWF 4D-Var system. *Quarterly Journal of the Royal Meteorological Society, 131,* 437–458. https://doi.org/10.1256/qj.04.17.

Marécal, V., & Mahfouf, J.-F. (2002). Four-dimensional variational assimilation of total column water vapor in rainy areas. *Monthly Weather Review, 130,* 43–58. https://doi.org/10.1175/1520-0493(2002)130<0043:FDVAOT>2.0.CO;2.

McNally, A. P. (2002). A note on the occurrence of cloud in meteorologically sensitive areas and the implications for advanced infrared sounders. *Quarterly Journal of the Royal Meteorological Society, 128,* 2551–2556. https://doi.org/10.1256/qj.01.206.

Migliorini, S. (2012). On the equivalence between radiance and retrieval assimilation. *Monthly Weather Review, 140,* 258–265. https://doi.org/10.1175/MWR-D-10-05047.1.

Montmerle, T., & Berre, L. (2010). Diagnosis and formulation of heterogeneous background-error covariances at the mesoscale. *Quarterly Journal of the Royal Meteorological Society, 136,* 1408–1420. https://doi.org/10.1002/qj.655.

Moreau, E., Lopez, P., Bauer, P., Tompkins, A. M., Janisková, M., & Chevallier, F. (2004). Variational retrieval of temperature and humidity profiles using rain rates versus microwave brightness temperatures. *Quarterly Journal of the Royal Meteorological Society, 130,* 827–852. https://doi.org/10.1256/qj.03.118.

Peubey, C., & McNally, A. P. (2009). Characterization of the impact of geostationary clear-sky radiances on wind analyses in a 4D-Var context. *Quarterly Journal of the Royal Meteorological Society, 135,* 1863–1876. https://doi.org/10.1002/qj.500.

Poterjoy, J. (2016). A localized particle filter for high-dimensional nonlinear systems. *Monthly Weather Review, 144,* 59–76. https://doi.org/10.1175/MWR-D-15-0163.1.

Puri, K., & Miller, M. J. (1990). The use of satellite data in the specification of convective heating for diabatic initialization and moisture adjustment in numerical weather prediction models. *Monthly Weather Review, 118,* 67–93. https://doi.org/10.1175/1520-0493(1990)118<0067:TUOSDI>2.0.CO;2.

Reale, O., McGrath-Spangler, E. L., McCarty, W., Holdaway, D., & Gelaro, R. (2018). Impact of adaptively thinned AIRS cloud-cleared radiances on tropical cyclone representation in a global data assimilation and forecast system. *Weather and Forecasting, 33,* 909–931. https://doi.org/10.1175/WAF-D-17-0175.1.

Rodgers, C. D. (2000). *Inverse methods for atmospheric sounding: Theory and practice* (Vol. 2, 258 pp). Singapore: World Scientific. https://doi.org/10.1142/3171.

Saunders, R. W., Matricardi, M., & Brunel, P. (1999). An improved fast radiative transfer model for assimilation of satellite radiance observations. *Quarterly Journal of the Royal Meteorological Society, 125,* 1407–1425. https://doi.org/10.1002/qj.1999.49712555615.

Scofield, R. A., & Kuligowski, R. J. (2003). Status and outlook of operational satellite precipitation algorithms for extreme-precipitation events. *Monthly Weather Review, 18,* 1037–1051. https://doi.org/10.1175/1520-0434(2003)018<1037:SAOOOS>2.0.CO;2.

Short, D. A., & North, G. R. (1990). The beam filling error in the Nimbus 5 electronically scanning microwave radiometer observations of Global Atlantic Tropical Experiment rainfall. *Journal of Geophysical Research, 95,* 2187–2193. https://doi.org/10.1029/JD095iD03p02187.

Skofronick-Jackson, G., Petersen, W. A., Berg, W., Kidd, C., Stocker, E. F., Kirschbaum, D. B., Kakar, R., Braun, S. A., Huffman, G. J., Iguchi, T., Kirstetter, P. E., Kummerow, C., Meneghini, R., Oki, R., Olson, W. S., Takayabu, Y. N., Furukawa, K., & Wilheit, T. (2017). The Global Precipitation Measurement (GPM) mission for science and society. *Bulletin of the American Meteorological Society, 98,* 1679–1695. https://doi.org/10.1175/BAMS-D-15-00306.1.

Stephan, K., Klink, S., & Schraff, C. (2008). Assimilation of radar-derived rain rates into the convective-scale model COSMO-DE at DWD. *Quarterly Journal of the Royal Meteorological Society, 134,* 1315–1326. https://doi.org/10.1002/qj.269.

Sun, J., Xue, M., Wilson, J. W., Zawadzki, I., Ballard, S. P., Onvlee-Hooimeyer, J., Joe, P., Barker, D. M., Li, P. W., Golding, B., & Xu, M. (2014). Use of NWP for nowcasting convective precipitation: Recent progress and challenges. *Bulletin of the American Meteorological Society, 95*(3), 409–426. https://doi.org/10.1175/BAMS-D-11-00263.1.

Tauchi, T., Takeuchi, Y., & Sato, Y. (2004). Assimilation of the Aqua/AMSR-E data to numerical weather predictions. *IGARSS'04, 5,* 3199–3202. https://doi.org/10.1109/IGARSS.2004.1370381.

Tsuyuki, T. (1997). Variational data assimilation in the tropics using precipitation data. Part III: Assimilation of SSM/I precipitation rates. *Monthly Weather Review, 125,* 1447–1464. https://doi.org/10.1175/1520-0493(1997)125<1447:VDAITT>2.0.CO;2.

van Leeuwen, P. J. (2010). Nonlinear data assimilation in geosciences: An extremely efficient particle filter. *Quarterly Journal of the Royal Meteorological Society, 136,* 1991–1999. https://doi.org/10.1002/qj.699.

Wang, F., Kummerow, C. D., Geer, A. J., Bauer, P., & Elsaesser, G. (2012). Comparing rain retrievals from GPROF with ECMWF 1D-Var products. *Quarterly Journal of the Royal Meteorological Society, 138,* 1852–1866. https://doi.org/10.1002/qj.1931.

Wang, P., Li, J., Li, Z., Lin, A. H. N., Li, J., Schmit, T. J., & Goldberg, M. D. (2017). The impact of Cross-track Infrared Sounder (CrIS) cloud-cleared radiances on Hurricane Joaquin (2015) and Matthew (2016) forecasts. *Journal of Geophysical Research, 122,* 13201–13218. https://doi.org/10.1002/2017JD027515.

Wattrelot, E., Caumont, O., & Mahfouf, J.-F. (2014). Operational implementation of the 1D+ 3D-Var assimilation method of radar reflectivity data in the AROME model. *Monthly Weather Review, 142,* 1852–1873. https://doi.org/10.1175/MWR-D-13-00230.1.

Yi, B., Yang, P., Liu, Q., van Delst, P., Boukabara, S.-A., & Weng, F. (2016). Improvements on the ice cloud modeling capabilities of the Community Radiative Transfer Model. *Journal of Geophysical Research, 121,* 13577–13590. https://doi.org/10.1002/2016JD025207.

Zhu, Y., Derber, J. C., Purser, R. J., Ballish, B. A., & Whiting, J. (2015). Variational correction of aircraft temperature bias in the NCEP's GSI analysis system. *Monthly Weather Review, 143,* 3774–3803. https://doi.org/10.1175/MWR-D-14-00235.1.

Zhu, Y., Liu, E., Mahajan, R., Thomas, C., Groff, D., Van Delst, P., Collard, A., Kleist, D., Treadon, R., & Derber, J. C. (2016). All-sky microwave radiance assimilation in NCEP's GSI analysis system. *Monthly Weather Review, 144,* 4709–4735. https://doi.org/10.1175/MWR-D-15-0445.1.

Zou, X., Navon, I. M., & Sela, J. G. (1993). Variational data assimilation with moist threshold processes using the NMC spectral model. *Tellus A, 45,* 370–387. https://doi.org/10.3402/tellusa.v45i5.14900.

Županski, D., & Mesinger, F. (1995). Four-dimensional variational assimilation of precipitation data. *Monthly Weather Review, 123,* 1112–1127. https://doi.org/10.1175/1520-0493(1995)123<1112:FDVAOP>2.0.CO;2.

Chapter 49
Precipitation Ensemble Data Assimilation in NWP Models

Takemasa Miyoshi, Shunji Kotsuki, Koji Terasaki, Shigenori Otsuka, Guo-Yuan Lien, Hisashi Yashiro, Hirofumi Tomita, Masaki Satoh, and Eugenia Kalnay

Abstract This chapter describes the authors' effort on ensemble data assimilation of satellite precipitation measurements. The Local Ensemble Transform Kalman Filter (LETKF) was implemented with the Nonhydrostatic Icosahedral Atmospheric Model (NICAM), and JAXA's GSMaP (Global Satellite Mapping of Precipitation) data were assimilated at 112-km resolution.

Keywords Precipitation · Numerical weather prediction · Data assimilation · Local Ensemble Transform Kalman Filter (LETKF) · Nonhydrostatic Icosahedral Atmospheric Model (NICAM) · JAXA · RIKEN · GSMaP · TRMM · TMPA · NCEP

Data assimilation combines a numerical model simulation with actual measurements and brings synergy, meaning that it finds the optimal initial condition and model parameters and minimizes the misfit between the model simulation with measurements. In this chapter, we provide a brief summary of our research activities at

T. Miyoshi (✉) · K. Terasaki · S. Otsuka · H. Tomita
RIKEN Center for Computational Science, Kobe, Japan
e-mail: takemasa.miyoshi@riken.jp

S. Kotsuki
RIKEN Center for Computational Science, Kobe, Japan

Center for Environmental Remote Sensing, Chiba University, Chiba, Japan

G.-Y. Lien
Research and Development Center, Central Weather Bureau, Taipei, Taiwan

H. Yashiro
Satellite Observation Center, National Institute for Environmental Studies, Tsukuba, Japan

M. Satoh
Atmosphere and Ocean Research Institute (AORI), The University of Tokyo, Chiba, Japan

E. Kalnay
Department of Atmospheric and Oceanic Science, University of Maryland,
College Park, MD, USA

© Springer Nature Switzerland AG 2020
V. Levizzani et al. (eds.), *Satellite Precipitation Measurement*, Advances in Global Change Research 69, https://doi.org/10.1007/978-3-030-35798-6_25

RIKEN, the Japan's flagship research institute for all sciences, in collaboration with the University of Tokyo and the University of Maryland, on enhancing ensemble data assimilation of satellite precipitation measurements to improve global numerical weather prediction (NWP).

The project, funded by JAXA on ensemble data assimilation of satellite precipitation measurements, started in 2013. Motivated by the previous successes at the University of Maryland (Lien et al. 2013) with idealized simulation experiments to assimilate precipitation data using a low-resolution (T30/L7), simplified global model, the project aimed to improve global NWP beyond a few days using real-world satellite-derived precipitation measurements. In this project, we used the Nonhydrostatic Icosahedral Atmospheric Model (NICAM, Satoh et al. 2014) and assimilated JAXA's GSMaP (Global Satellite Mapping of Precipitation, Ushio et al. 2009). We had a close collaboration with a NASA-funded project at the University of Maryland (PI: Eugenia Kalnay) on assimilating TMPA (TRMM Multisatellite Precipitation Analysis, Huffman et al. 2007) data with the National Centers for Environmental Prediction (NCEP) Global Forecast System (GFS) (Lien et al. 2016a, b).

The project started from scratch. Namely, we had no data assimilation system implemented with NICAM at the beginning of the project. We first developed the Local Ensemble Transform Kalman Filter (LETKF, Hunt et al. 2007) with NICAM at relatively low 112-km resolution to assimilate only conventional observations from NCEP (known as the PREPBUFR data). As the first step, we started from implementing the existing LETKF system (Miyoshi 2005, 2011; Miyoshi and Yamane 2007; Miyoshi et al. 2007, 2010) with the longitude-latitude grid interface (LL-LETKF). Here, we applied horizontal interpolation between the longitude-latitude grid and the NICAM's native icosahedral grid at each data assimilation cycle. Terasaki et al. (2015) developed the LETKF with the native icosahedral grid interface (ICO-LETKF) and showed that ICO-LETKF outperformed LL-LETKF by removing the interpolation (Fig. 49.1).

Using the NICAM-LETKF system, Kotsuki et al. (2017a) developed a system to assimilate the GSMaP data and to improve forecasts beyond a few days, following the success of Lien et al. (2013) and Lien et al. (2016a, b). Here, applying the Gaussian Transformation (GT) is essential to assimilate precipitation data. The GT transforms the highly-skewed precipitation variable into a Gaussian variable (Fig. 49.2). This way, the observed variable agrees better with the underlying assumption of the Gaussian PDF in the LETKF or any other ensemble Kalman filter implementation. Here, the delta function at zero precipitation is transformed to a delta function whose area covers the probability of no precipitation. We need to decide the location of the transformed delta function, and Lien et al. (2016b) proposed and tested two different approaches. The results showed only a small difference between the two approaches, so that the location of the delta function has only a marginal impact on the analysis accuracy. The GT is constructed empirically with a historic record. Lien et al. (2016a, b) used 10 years of 3- to 9-h precipitation forecasts initialized from the NCEP's operational analysis data to construct the GT for model prediction. For observation, past 10-year observation data of TMPA were used. Kotsuki et al. (2017a) improved the algorithm to use the

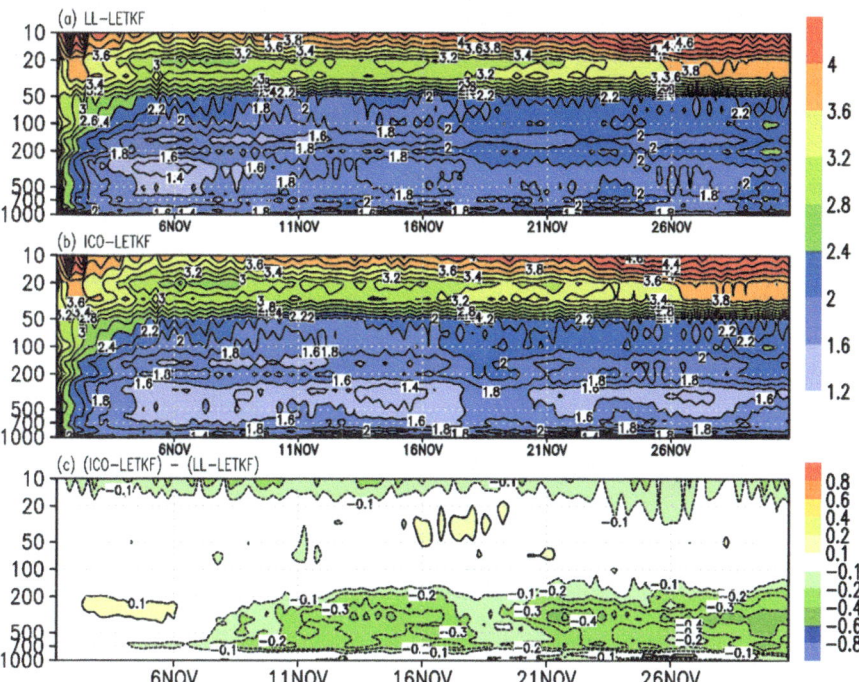

Fig. 49.1 Time-series of the global-mean root mean square differences (RMSD) for temperature (K) relative to the ERA-Interim. (**a**) LL-LETKF, (**b**) ICO-LETKF, and (**c**) the difference between LL-LETKF and ICOLETKF. The horizontal and vertical axes represent the date in 2011 and pressure level (hPa), respectively. Negative values of panel (**c**) corresponds to ICO-LETKF's advantage. (Adapted from Terasaki et al. 2015)

past record of only 30 days, rather than 10 years. This makes the method more applicable to various systems such as NICAM-LETKF with which no long-term analysis data exist. With the GT, the observation-minus-first-guess (a.k.a. O-B) departures show a histogram much closer to Gaussian (Fig. 49.3). This contributes to an improvement by assimilating GSMaP in up to 5-day forecasts (Fig. 49.4). Here, spatial thinning of GSMaP data was essential due to significant error correlations in space (Kotsuki et al. 2017a). We will describe our effort on treating the observation error correlations in data assimilation later in this chapter.

Kotsuki et al. (2017a) extended the use of the GT by employing the inverse of the empirical GT for making better precipitation analyses. NWP models usually have a bias, which can be mitigated by applying the inverse GT based on the observation data. Namely, the GT absorbs the bias, and we assume that the transformed Gaussian variable is unbiased. The precipitation analysis in the Gaussian variable can be transformed back to the precipitation variable by either model-based or observation-based inverse GT. If we use the observation-based (model-based) inverse GT, the precipitation field looks like observation (model) fields (Fig. 49.5). This is useful when we produce precipitation analysis data.

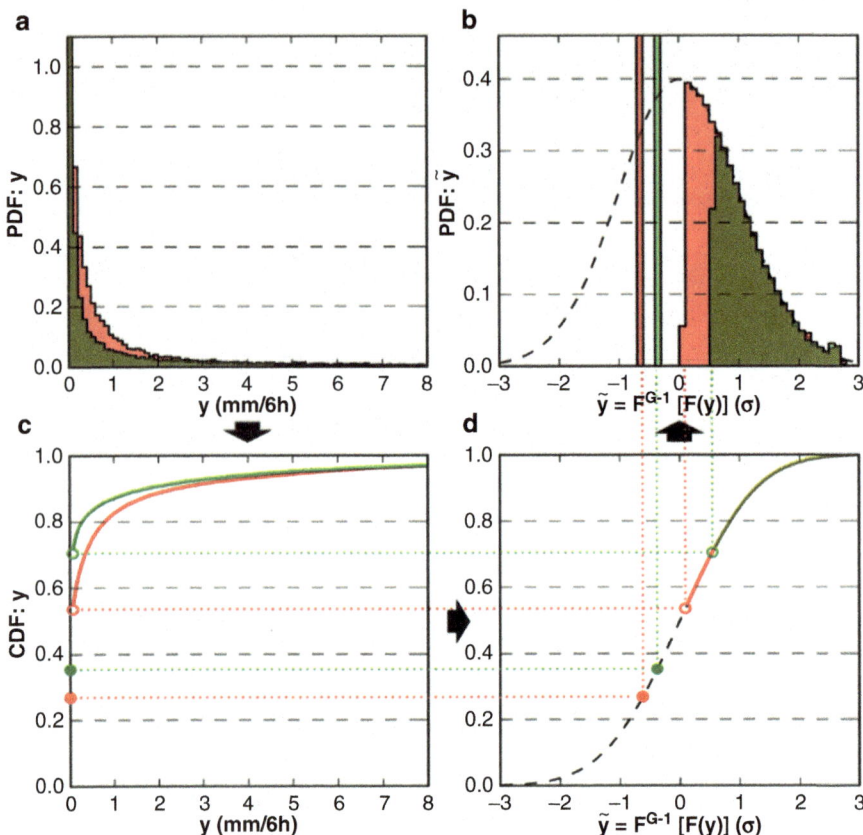

Fig. 49.2 The probability density function (PDF; **a** and **b**) and cumulative density function (CDF; **c** and **d**) of the original precipitation and the transformed precipitation based on the 10-year (2001–2010) model (red) and observation (green) climatologies at a grid point in the extratropics (39.08°N, 76.98°W; near Maryland). All plots correspond to the 11–20 Jan period. The procedure of the Gaussian transformation is indicated by the arrows, i.e., (**a**) to (**c**) to (**d**) to (**b**). The open circles correspond to zero precipitation probability and the solid circles correspond to the half value (median) of zero precipitation probability. (Adapted from Lien et al. 2016a)

Kotsuki et al. (2018) further extended the GSMaP data assimilation to estimate a NICAM's model parameter. NICAM at the relatively low resolution of 112 km uses a large-scale condensation scheme. Here we estimated a parameter which controls how fast the oversaturated vapor falls as precipitation. The parameter estimation system worked as designed. We assimilated GSMaP data to estimate the parameter that gives better precipitation analysis and forecast relative to the GSMaP data. The results showed that the precipitation analysis and forecast were closer to the GSMaP data as designed (Fig. 49.6). However, the cloud amount increased due to parameter estimation. The model with the default parameter overproduced weak precipitation in a broader area, and the parameter estimation adjusted the large-scale condensation to slow down the precipitation process. This makes oversaturated vapor remain in the atmosphere longer, with increased cloud amount, leading to a change in the

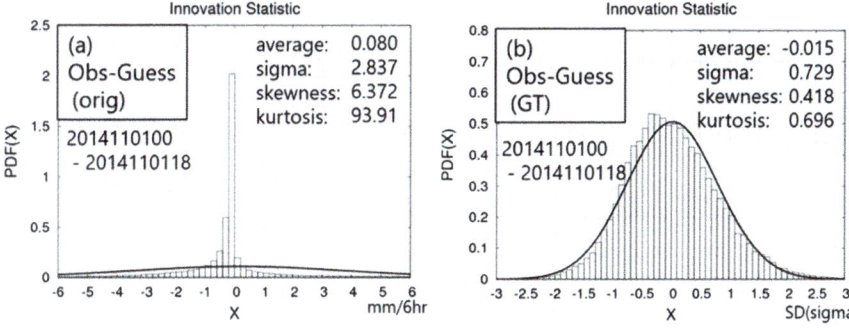

Fig. 49.3 Innovation statistics 0000–1800 UTC 1 November 2014 for (**a**) original precipitation data (mm/6 h) and (**b**) precipitation after applying the GT (standard deviation). Bold lines show the normal distributions computed by the mean and standard deviation of the innovation samples. (Adapted from Kotsuki et al. 2017a)

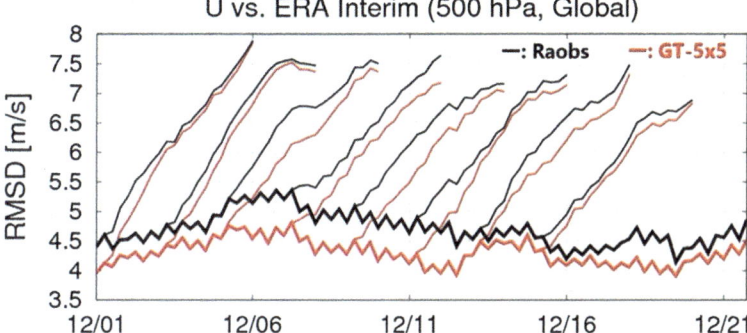

Fig. 49.4 Time series of RMSDs for zonal wind (m s^{-1}) at 500 hPa relative to the ERA Interim reanalysis. (black lines) control experiment without assimilating GSMaP and (red lines) test experiment with assimilation of GSMaP with the GT. Bold and thin lines denote the analyses and the 5-day forecasts from the eight initial times, respectively. The abscissa shows month/date in 2014. (Adapted from Kotsuki et al. 2017a)

radiation budget, i.e., global circulation. This causes overall degradation to the atmospheric analysis. Yet, the system worked as designed, and we obtained improved precipitation analysis and forecasts relative to the GSMaP data. The lesson learned here is that we should look at different variables simultaneously for overall improvement by parameter estimation.

After getting successful results when assimilating GSMaP data, we improved the NICAM-LETKF system in various ways. Yashiro et al. (2016) developed the NICAM-LETKF system for future high-performance computing (HPC) systems. Terasaki and Miyoshi (2017) implemented the radiative transfer model known as the RTTOV (Saunders et al. 2013) to assimilate satellite radiance data. The NICAM-LETKF system now assimilates the Advanced Microwave Sounding Unit (AMSU)-A radiances by default. The AMSU-A radiances provide information about mainly temperature profiles and have a large impact. Kotsuki et al. (2017b) developed the

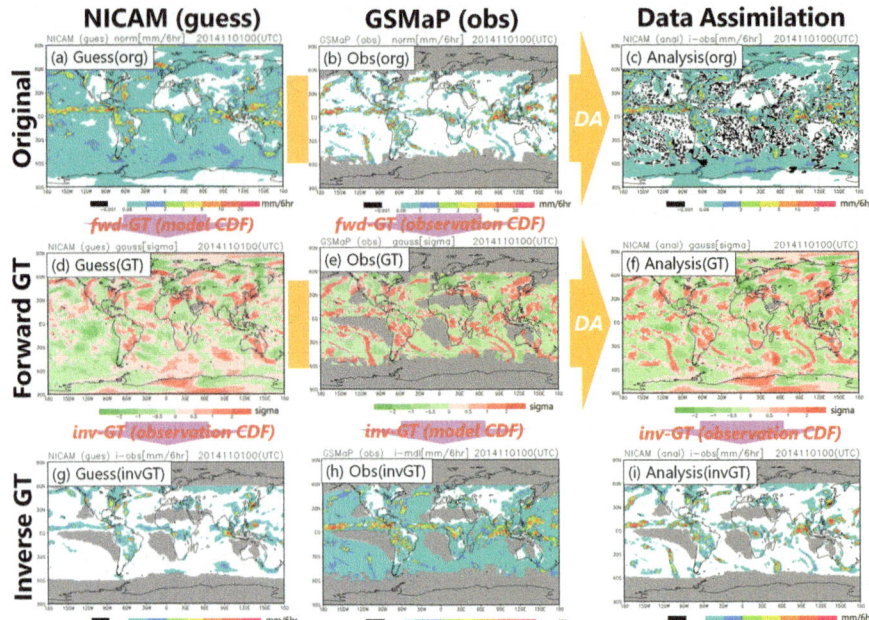

Fig. 49.5 Global precipitation patterns (mm/6 h) at the first data assimilation step (0000 UTC 1 November 2014). (**a**) original NICAM forecast, (**b**) original GSMaP, (**c**) analysis without applying GT, (**d**) GT'ed. NICAM forecast, (**e**) GT'ed. GSMaP and (**f**) analysis with applying the GT. Panels (**g–i**) demonstrate impacts of the inverse GT. Panels (**g**) and (**i**) are obtained by applying observation-CDF-based inverse GT to panels (**d**) and (**f**). Panel (**h**) is obtained by applying model-CDF-based inverse GT to panel (**h**). Gray and black colors show missing value and negative precipitation value, respectively. (Modified from Kotsuki et al. 2017a)

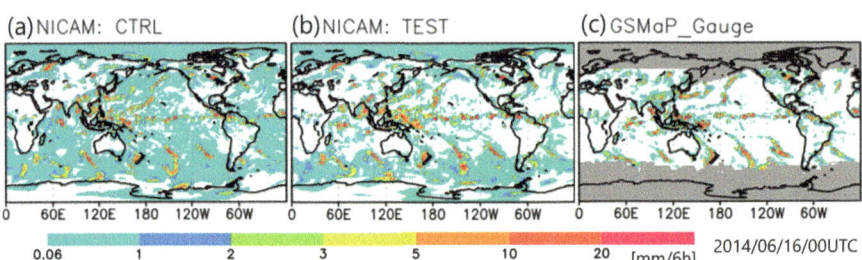

Fig. 49.6 Global precipitation forecasts (mm/6 h) at 0000 UTC 16 June 2014. (**a**) control experiment with the default model setting, (**b**) test experiment with the model parameter estimation, and (right) GSMaP_Gauge observation data, respectively. Overproduced precipitation over ocean in the control experiment is successfully mitigated by the model parameter estimation. (Adapted and modified from Kotsuki et al. 2018)

adaptive relaxation to prior perturbation (RTPP) and adaptive relaxation to prior spread (RTPS) algorithms to replace the adaptive multiplicative inflation method of Miyoshi (2011).

Satellite data generally provide a large coverage of scanning with a single sensor and tend to have correlated errors in the horizontal. Previous studies (Weston et al. 2014; Campbell et al. 2017) included inter-channel error correlations of satellite radiances in variational data assimilation systems. However, no study has investigated the impact of horizontal error correlations of satellite data. Kotsuki et al. (2017a) needed to thin the GSMaP data to avoid negative impacts from the horizontal error correlations. To use the dense satellite data more effectively, it would be important to include the horizontal error correlations explicitly in data assimilation. Following Miyoshi et al. (2013), Terasaki and Miyoshi (2014) investigated the impact of including the observation error correlations in data assimilation with a toy model. These theoretical studies suggest potential large improvements by including the observation error correlations. Our future research will focus on more realistic scenarios with the NICAM-LETKF.

We have also explored applying data assimilation to satellite-based global precipitation nowcasting. Otsuka et al. (2016) developed an optical-flow system with GSMaP data. Here, two successive images of GSMaP data are analyzed to find the flow field, with which the GSMaP precipitation image is extrapolated in space and time. This can provide useful prediction for a half day or so. Otsuka et al. (2016) applied the LETKF to improve the flow field and showed improvement of precipitation nowcasting. Nowcasting has an advantage of rapid processing. This system is now run routinely in real time at RIKEN and is made available to the public under the license of the Japanese Meteorological Service Act. Since the nowcasting accuracy drops quickly after 12 h, NWP with NICAM-LETKF is necessary for longer lead times. Optimally merging the nowcasting and NWP is an important subject of our future research.

So far, we described mainly the use of GSMaP data. Direct assimilation of GPM/DPR reflectivity is an important research topic to explore in order to more effectively use GPM satellite precipitation data. As the first step, Kotsuki et al. (2014) compared the NICAM cloud-permitting simulation at 3.5-km with actual GPM/DPR reflectivity. The results showed systematic differences between simulated and observed reflectivity. Our preliminary test with assimilating DPR reflectivity suggests that simply analyzing model states be not sufficient. Our future study will focus on estimating model microphysics parameters by assimilating DPR reflectivity.

References

Campbell, W. F., Satterfield, E., Ruston, B., & Baker, N. (2017). Accounting for correlated observation error in a dual formulation 4D-variational data assimilation system. *Monthly Weather Review, 145*, 1019–1032. https://doi.org/10.1175/MWR-D-16-0240.1.

Huffman, G. J., Bolvin, D. T., Nelkin, E. J., Wolff, D. B., Adler, R. F., Gu, G., Hong, Y., Bowman, K. P., & Stocker, E. F. (2007). The TRMM Multisatellite Precipitation Analysis (TMPA): Quasi-global, multiyear, combined-sensor precipitation estimates at fine scales. *Journal of Hydrometeorology, 8*, 38–55. https://doi.org/10.1175/JHM560.1.

Hunt, B. R., Kostelich, E. J., & Szunyogh, I. (2007). Efficient data assimilation for spatiotemporal chaos: A local ensemble transform Kalman filter. *Physica D: Nonlinear Phenomena, 230*, 112–126. https://doi.org/10.1016/j.physd.2006.11.008.

Kotsuki, S., Terasaki, K., & Miyoshi, T. (2014). GPM/DPR precipitation compared with a 3.5-km-resolution NICAM simulation. *SOLA, 10*, 204–209. https://doi.org/10.2151/sola.2014-043.

Kotsuki, S., Miyoshi, T., Terasaki, K., Lien, G.-Y., & Kalnay, E. (2017a). Assimilating the global satellite mapping of precipitation data with the Nonhydrostatic Icosahedral Atmospheric Model (NICAM). *Journal of Geophysical Research, 122*, 631–650. https://doi.org/10.1002/2016JD025355.

Kotsuki, S., Ota, Y., & Miyoshi, T. (2017b). Adaptive covariance relaxation methods for ensemble data assimilation: Experiments in the real atmosphere. *Quarterly Journal of the Royal Meteorological Society, 143*, 2001–2015. https://doi.org/10.1002/qj.3060.

Kotsuki, S., Terasaki, K., Yashiro, H., Tomita, H., Satoh, M., & Miyoshi, T. (2018). Online model parameter estimation with ensemble data assimilation in the real global atmosphere: A case with the Nonhydrostatic Icosahedral Atmospheric Model (NICAM) and the global satellite mapping of precipitation data. *Journal of Geophysical Research, 123*, 7375–7392. https://doi.org/10.1029/2017JD028092.

Lien, G.-Y., Kalnay, E., & Miyoshi, T. (2013). Effective assimilation of global precipitation: Simulation experiments. *Tellus A, 65*, 1–16. https://doi.org/10.3402/tellusa.v65i0.19915.

Lien, G.-Y., Kalnay, E., Miyoshi, T., & Huffman, G. J. (2016a). Statistical properties of global precipitation in the NCEP GFS model and TMPA observations for data assimilation. *Monthly Weather Review, 144*, 663–679. https://doi.org/10.1175/MWR-D-15-0150.1.

Lien, G.-Y., Miyoshi, T., & Kalnay, E. (2016b). Assimilation of TRMM multisatellite precipitation analysis with a low-resolution NCEP global forecasting system. *Monthly Weather Review, 144*, 643–661. https://doi.org/10.1175/MWR-D-15-0149.1.

Miyoshi, T. (2005). *Ensemble Kalman filter experiments with a primitive-equation global model.* Ph.D. dissertation, University of Maryland, College Park, 197 pp. Available at https://drum.lib.umd.edu/handle/1903/3046, last accessed 23 Nov 2018.

Miyoshi, T. (2011). The Gaussian approach to adaptive covariance inflation and its implementation with the local ensemble transform Kalman filter. *Monthly Weather Review, 139*, 1519–1535. https://doi.org/10.1175/2010MWR3570.1.

Miyoshi, T., & Yamane, S. (2007). Local ensemble transform Kalman filtering with an AGCM at a T159/L48 resolution. *Monthly Weather Review, 135*, 3841–3861. https://doi.org/10.1175/2007MWR1873.1.

Miyoshi, T., Yamane, S., & Enomoto, T. (2007). Localizing the error covariance by physical distances within a local ensemble transform Kalman filter (LETKF). *SOLA, 3*, 89–92. https://doi.org/10.2151/sola.2007-023.

Miyoshi, T., Sato, Y., & Kadowaki, T. (2010). Ensemble Kalman filter and 4D-Var intercomparison with the Japanese operational global analysis and prediction system. *Monthly Weather Review, 138*, 2846–2866. https://doi.org/10.1175/2010MWR3209.1.

Miyoshi, T., Kalnay, E., & Li, H. (2013). Estimating and including observation-error correlations in data assimilation. *Inverse Problems in Science and Engineering, 21*, 387–398. https://doi.org/10.1080/17415977.2012.712527.

Otsuka, S., Kotsuki, S., & Miyoshi, T. (2016). Nowcasting with data assimilation: A case of global satellite mapping of precipitation. *Weather and Forecasting, 31*, 1409–1416. https://doi.org/10.1175/WAF-D-16-0039.1.

Satoh, M., Tomita, H., Yashiro, H., Miura, H., Kodama, C., Seiki, T., Noda, A. T., Yamada, Y., Goto, D., Sawada, M., Miyoshi, T., Niwa, Y., Hara, M., Ohno, T., Iga, S., Arakawa, T., Inoue, T., & Kubokawa, H. (2014). The non-hydrostatic icosahedral atmospheric model: Description and development. *Progress in Earth and Planetary Science, 1*, 18. https://doi.org/10.1186/s40645-014-0018-1.

Saunders, R., Hocking, J., Rundle, D., Rayer, P., Matricardi, M., Geer, A., Lupu, C., Brunel, P., & Vidot, J. (2013). *RTTOV-11: Science and validation report.* NWP-SAF Rep., UK Met Office,

62 pp. Available at https://nwpsaf.eu/oldsite/deliverables/rtm/docs_rttov11/rttov11_svr.pdf, last accessed 23 Nov 2018.

Terasaki, K., & Miyoshi, T. (2014). Data assimilation with error-correlated and non-orthogonal observations: Experiments with the Lorenz-96 model. *SOLA, 10,* 210–213. https://doi.org/10.2151/sola.2014-044.

Terasaki, K., & Miyoshi, T. (2017). Assimilating AMSU-A radiances with the NICAM-LETKF. *Journal of the Meteorological Society of Japan, 95,* 433–446. https://doi.org/10.2151/jmsj.2017-028.

Terasaki, K., Sawada, M., & Miyoshi, T. (2015). Local ensemble transform Kalman filter experiments with the nonhydrostatic icosahedral atmospheric model NICAM. *SOLA, 11,* 23–26. https://doi.org/10.2151/sola.2015-006.

Ushio, T., Sasashige, K., Kubota, T., Shige, S., Okamoto, K., Aonashi, K., Inoue, T., Takahashi, N., Iguchi, T., Kachi, M., Oki, R., Morimoto, T., & Kawasaki, Z.-I. (2009). A Kalman filter approach to the global satellite mapping of precipitation (GSMaP) from combined passive microwave and infrared radiometric data. *Journal of the Meteorological Society of Japan, 87A,* 137–151. https://doi.org/10.2151/jmsj.87A.137.

Weston, P. P., Bell, W., & Eyre, J. R. (2014). Accounting for correlated error in the assimilation of high-resolution sounder data. *Quarterly Journal of the Royal Meteorological Society, 140,* 2420–2429. https://doi.org/10.1002/qj.2306.

Yashiro, H., Terasaki, K., Miyoshi, T., & Tomita, H. (2016). Performance evaluation of a throughput-aware framework for ensemble data assimilation: The case of NICAM-LETKF. *Geoscientific Model Development, 9,* 2293–2300. https://doi.org/10.5194/gmd-9-2293-2016.

Chapter 50
PERSIANN-CDR for Hydrology and Hydro-climatic Applications

Phu Nguyen, Hamed Ashouri, Mohammed Ombadi, Negin Hayatbini, Kuo-Lin Hsu, and Soroosh Sorooshian

Abstract Satellite-retrieved precipitation datasets represent a promising input data source to be utilized in hydroclimatic and hydrologic applications. Due to their characteristics of high spatiotemporal resolution, near real-time availability and quasi global coverage, satellite-retrieved precipitation datasets promise to provide a remedy for the long-standing issues associated with ground rainfall information. In this article, we shed light on the Precipitation Estimation from Remotely Sensed Information using Artificial Neural Networks – Climate Data Records (PERSIANN-CDR) dataset and its use in hydroclimatic and hydrologic applications. In particular, we highlight the use of PERSIANN-CDR for rainfall trend analysis, observation of extreme rainfall events such as Hurricanes, and evaluation of climate models' simulations of precipitation based on their historical performance. Regarding the use of PERSIANN-CDR for hydrologic applications, we show examples of utilizing the dataset in rainfall-runoff modeling as well as its use in rainfall frequency analysis and the development of intensity-duration-frequency (IDF) curves.

Keywords Precipitation · Rainfall · PERSIANN-CDR · PERSIANN-CCS · TMPA · CMORPH · GPCP · Hydrology · Hydroclimatology · Neural networks · Rain gauges · Brightness temperature · Mann-Kendall test · ETCCDI · CCI · JCOMM · CLIVAR · Extreme value theory · CMIP5 · IDF curves

50.1 Introduction

Precipitation is a vital component of water and energy cycle and is one of the most, if not the most, important meteorological input for hydrometeorological and climatic models (Sorooshian et al. 2011). Reliable long-term precipitation data on a global scale is essential for a wide range of hydrological, hydrometeorological, and

P. Nguyen (✉) · H. Ashouri · M. Ombadi · N. Hayatbini · K.-L. Hsu · S. Sorooshian
Center for Hydrometeorology & Remote Sensing (CHRS), University of California, Irvine, CA, USA
e-mail: ndphu@uci.edu

© Springer Nature Switzerland AG 2020
V. Levizzani et al. (eds.), *Satellite Precipitation Measurement*, Advances in Global Change Research 69, https://doi.org/10.1007/978-3-030-35798-6_26

climatological applications. Gauge-based data sets generally provide long-term, direct physical measurement of precipitation, however, their sparse point measurements as well as susceptibility to certain errors have elevated the importance of satellite-based rainfall estimates (Kidd et al. 2017; Rana et al. 2015; Xie and Arkin 1997).

Satellite observations make up for such deficiencies by providing coverage that is spatially more homogeneous and temporally complete globally (Kidd and Levizzani 2011; Xie et al. 2003). Over the past recent decades, the availability of satellite-based observations has motivated researchers to investigate several hydroclimatic processes and develop methods for the incorporation of these datasets in hydrologic applications. Sun et al. (2018) provided a summary of major satellite-related precipitation data sets that are currently available and Maggioni et al. (2016) presented a consolidated and detailed review of the algorithms used in satellite precipitation data sets. Some of the currently operationally available satellite-derived data sets are the Tropical Rainfall Measuring Mission (TRMM) Multisatellite Precipitation Analysis (TMPA) (Huffman et al. 2007), the Precipitation Estimation from Remotely Sensed Information using Artificial Neural Networks – Climate Data Records (PERSIANN-CDR) (Ashouri et al. 2015), and the Climate Prediction Center (CPC) morphing technique (CMORPH) (Joyce et al. 2004) products.

The focus of this chapter is on the applications of PERSIANN algorithms family, particularly PERSIANN Climate Data Record (PERSIANN-CDR, Ashouri et al. 2015). This algorithm provides long-term, high-resolution, satellite-based precipitation estimates for hydroclimatological applications. PERSIANN-CDR's daily (sub-daily; 3-hourly) and 0.25° precipitation data from 1983 to present makes it suitable to study the behavior of extreme precipitation patterns on a global scale over the past three decades (Hsu et al. 1997; Sorooshian et al. 2000). This satellite-based precipitation product provides daily precipitation estimates from the year 1983 to present at a resolution of 0.25° on the archive of Gridded Satellite (GridSat-B1) IR satellite data (Ashouri et al. 2015; Lee 2014). PERSIANN-CDR utilizes an artificial neural network to assign a surface rain rate based on brightness temperature retrievals of infrared information from geostationary Earth-Orbiting (GEO) satellites, specifically from the archive of Gridded Satellite (GridSat-B1). The artificial neural network is trained with stage IV hourly precipitation data from the National Centers for Environmental Prediction (NCEP). The high-resolution PERSIANN estimates are then adjusted by the GPCP (Global Precipitation Climatology Project) data at a resolution of 2.5° to downscale and remove the bias from the precipitation estimates (Ashouri et al. 2015). Toward this aim, the merged analysis (1979-present) from monthly GPCP global precipitation products is utilized for each month of the 30-year period at each 2.5° grid box of PERSIANN data. The corresponding 0.25° 3-hourly PERSIANN rain-rate estimates are aggregated to monthly scale after applying a proper threshold value to filter out pixels associated with no rain-rate. Then a correction factor based on the ratio of 2.5° monthly GPCP precipitation and PERSIANN rain-rate estimates at each 0.25° pixel is calculated and adjusted to be applied to 3-hourly PERSIANN estimate. GPCP version 2.2 (http://precip.gsfc.nasa.gov, last accessed 24 Nov. 2018), and GPCP 1° daily precipitation product

(Huffman et al. 2001) are used for correction and evaluation purposes respectively. For more detail regarding the process of adjusting PERSIANN data using monthly GPCP data refer to Ashouri et al. (2015).

The resulting final PERSIANN-CDR data is available through NOAA NCEI CDR program at https://www.ncdc.noaa.gov/cdr/atmospheric/precipitation-persiann-cdr (last accessed 24 Nov. 2018). PERSIANN-CDR has been widely used and validated by the scientific and user communities (Miao et al. 2015; Yang et al. 2016; Tan et al. 2015; Zhu et al. 2016; Duan et al. 2016, among many others). The availability of historical record (+35 years) at high spatiotemporal resolution of satellite-based data have enabled the investigation of trends and changes of many hydroclimatic processes. For example, the PERSIANN-CDR dataset has been used in investigating the response of precipitation regimes to the Amazonian deforestation (Khanna et al. 2017). An additional example is examining the increased frequency and intensity of floods in the Niger River basin during the last decade (Casse and Gosset 2015). The global PERSIANN Cloud Classification System (PERSIANN-CCS) is another member of the PERSIANN family of rainfall retrieval algorithms. It has a higher spatio-temporal resolution (0.04° and 30 min) compared to PERSIANN-CDR. It is derived from IR brightness temperature data from GEO satellites and uses PMW measurements from LEO satellites to update its parameters. This patch-based cloud classification and rainfall estimation algorithm uses histogram matching and exponential regression to fit curves to the plots of pixel brightness temperature versus rainfall rate (Hong et al. 2007).

The use of satellite-retrieved precipitation is promising mainly due to its characteristics of real-time information, near-global coverage and high spatiotemporal resolution.

Real-time (or near real-time) information provided by satellite-based precipitation proved to be invaluable in monitoring and capturing the development of cyclones since it has the ability of providing rainfall intensity information over the oceans before cyclones make landfall. In addition, it consistently provides rainfall estimates over land during such events without being affected by the high-speed wind and torrential rainfall. An example of real-time monitoring of cyclones using satellite-based precipitation is the tracking of Typhoon Haiyan (Nguyen et al. 2014) which struck Southeast Asia in the year 2013. In the study of Nguyen et al. (2014), PERSIANN-CCS dataset has been used in monitoring the overall propagation and estimating rainfall intensities during the event.

Overall, PERSIANN datasets have been used in a wide range of studies during the last decades; these studies can be classified in three categories. Firstly, hydro-meteorological studies such as investigating diurnal rainfall patterns (Sorooshian et al. 2002), drought monitoring (Zambrano et al. 2017) and evaluating climate model simulations of precipitation (Nguyen et al. 2018). Secondly, hydrologic applications such as examining the use of satellite-based precipitation in runoff prediction (e.g., Behrangi et al. 2011; Ashouri et al. 2016a; Liu et al. 2017; Hsu et al. 2013) and use of satellite-based precipitation in rainfall frequency analysis (e.g., Gado et al. 2017; Ombadi et al. 2018). Finally, these datasets have been integrated with other precipitation estimates from different sources including gauge, radar and satellite to compile comprehensive precipitation datasets (e.g., Chiang et al. 2007). In the subsequent sections of this chapter, some of these applications will be discussed in more detail.

50.2 Hydro-climatic Applications

50.2.1 RainSphere for Global Precipitation Analysis and Visualization

CHRS RainSphere is one of the applications derived from PERSIANN-CDR (Ashouri et al. 2015) precipitation estimates records to facilitate trend analysis in annual global precipitation and future precipitation projections studies. CHRS RainSphere interface includes search capabilities as well as the ability to automatically generate reports with basic statistics and summaries. It has several options for visualization of precipitation data along with additional spatial reference so users can quickly and easily extract meaningful information. (Nguyen et al. 2017a, 2018).

Figure 50.1 shows the map visualization for the accumulated total precipitation for 1 January to 31 December 2014 along with options to explore spatial patterns of

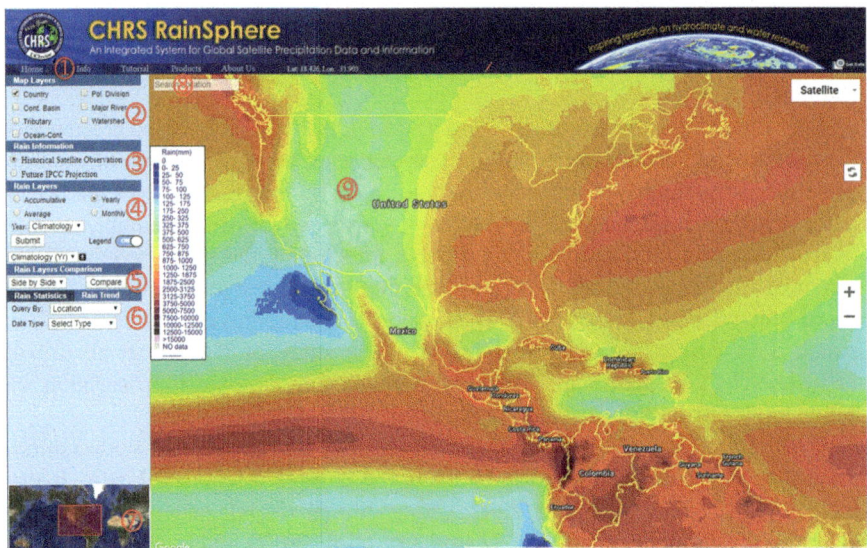

Fig. 50.1 CHRS RainSphere interface: (1) Navigation Bar, (2) Map Layers, (3) Rain Information, (4) Rain Layers, (5) Rain Comparison, (6) Rain Statistics, (7) Reference Map, (8) Search Location, (9) Map Canvas (URL: http://rainsphere.eng.uci.edu/, last accessed 24 Nov. 2018). (Nguyen et al. 2017a, 2018)

precipitation within a specified area. In this case the "Country" map layer is added for spatial reference, and additional options for spatial reference include political divisions (e.g., states and provinces), continental basins, major rivers, tributaries, and watersheds. Other than rain information and map layers, rain comparison, and rain statistics are further information that users are able to extract based on their requirements.

CHRS RainSphere also provides users with the general trends and average behavior for the selected area during the selected time span using a bar plot with precipitation amount in mm for each time step (day, month, or year). Corresponding mean and linear regression along with the Mann-Kendall test (Mann 1945; Kendall 1976) results automatically will be generated and reported as part of the statistics suite.

The Mann-Kendall test is used to statistically investigate whether to reject the null hypothesis of no trend in the data, with a p value equal to or greater than 0.05 in this calculation indicating acceptance of the null hypothesis. Below that value the alternative hypothesis is accepted with a smaller value of p indicating higher confidence that a trend exists. This test has been demonstrated as a useful tool for evaluating global climate trends (Damberg and AghaKouchak 2014). For illustration, Fig. 50.2 displays the statistical summary for yearly precipitation in the state of California from 1983 to 2015.

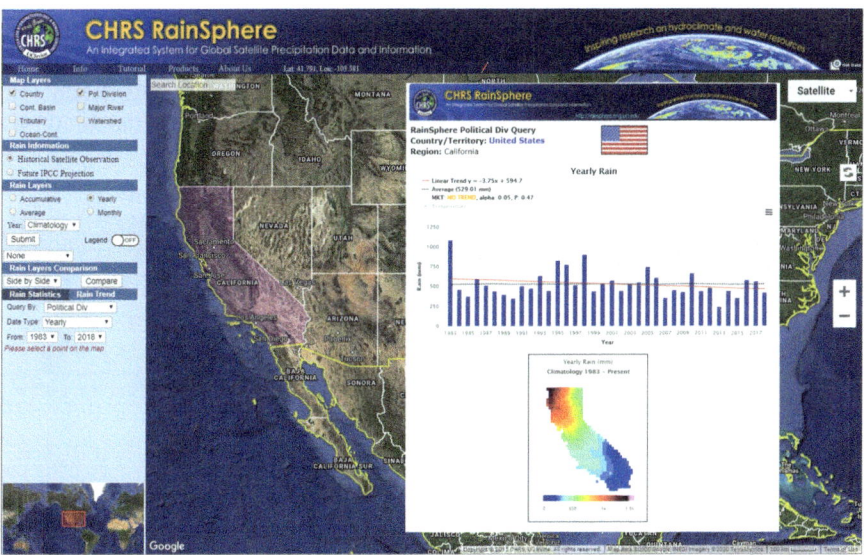

Fig. 50.2 Downloadable Rain Query Report including: Rain Linear Trend, Rain Average, Mann-Kendall Test. [http://rainsphere.eng.uci.edu/, last accessed 24 Nov. 2018]

50.2.2   Evaluation of PERSIANN-CDR on Extreme Events

One of the main applications of PERSIANN-CDR's long-term precipitation climate
data record is studying extreme precipitation events. As an example, we looked at
Hurricane Katrina, one of the five deadliest and the costliest hurricanes ever to strike
the US Katrina hit Southeast US in August 2005 and caused inflicted loss of lives
and economic damages in the region (Graumann and National Climatic Data Center
2006). As shown in Fig. 50.3, PERSIANN-CDR (a) shows similar precipitation
patterns to the radar data (b). Moreover, unlike Stage IV radar data which suffers
from blockages in mountainous regions or outages during a catastrophic event, the
spatial coverage provided by PERSIANN-CDR is very valuable and captures a wide
view of the precipitation and hurricane landfall. In order to investigate the performance of PERSIANN-CDR compared to other high-resolution satellite-based precipitation products, the TMPA V7 research version product is used. PERSIANN-
CDR and TMPA were each compared to Stage IV radar data. As shown in the
scatterplots in Fig. 50.3, PERSIANN-CDR shows a higher correlation coefficient
than TMPA however, the bias in TMPA is lower than that in PERSIANN-CDR. It is
noteworthy that the TMPA research version is also bias corrected with GPCP data at
monthly scale. The differences observed between TMPA and PERSIANN-CDR
performance mainly relate to 1) the difference in the algorithm of these two products,
and 2) the inputs to these algorithms. During validation and performance comparison, it is imperative to note that, gauge-information aside, PERSIANN-CDR only
uses IR data as its input whereas TMPA uses, passive microwave, IR, and radar data.

Expert Team on Climate Change Detection and Indices (ETCCDI) (Klein Tank
et al. 2009), sponsored jointly by the World Meteorological Organization (WMO)

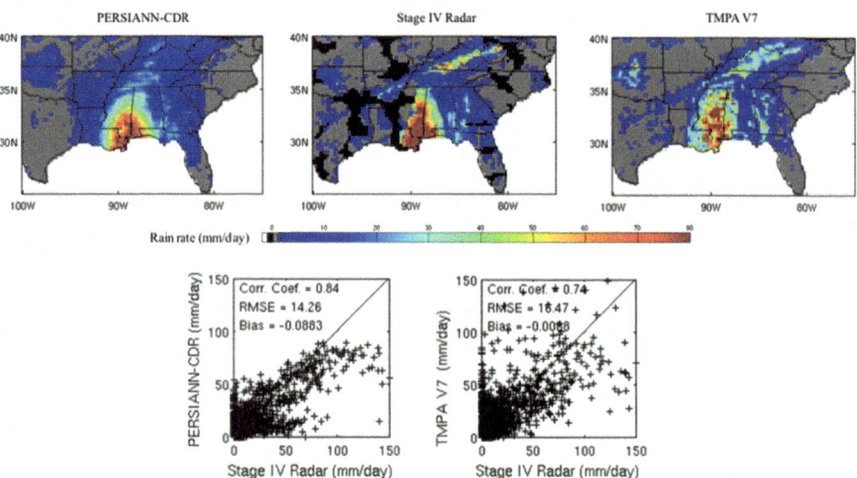

Fig. 50.3 Rainfall (mm day^{-1}) over land during Hurricane Katrina on 29 August 2005 from: (**a**)
PERSIANN-CDR, (**b**) Stage IV Radar (Lin and Mitchell 2005), and (**c**) TMPA v7 (Huffman et al.
2007). Black and gray pixels show radar blockages and zero precipitation, respectively. (Adapted
from Ashouri et al. 2015)

Commission for Climatology (CCl), the Joint Commission for Oceanography and Marine Meteorology (JCOMM), and the World Climate Research Program (WCRP) on Climate Variability and Predictability (CLIVAR), has defined various precipitation indices for studying extremes. Using two count related indices, the annual average count of days when rainfall ≥10 mm (R10mm), and the annual average count of days when rainfall ≥20 mm (R20mm), the performance of PERSIANN-CDR in reproducing the number of rainy days over the US for 1983–2011 is investigated against the 0.25° daily CPC Unified Gauge-Based Analysis of Precipitation data (Xie et al. 2010). As shown in Fig. 50.4, in general PERSIANN-CDR reproduces the same patterns as depicted in CPC. PERSIANN-CDR, however, underestimates R10mm and R20mm on the west cost of the US. The underestimation over the Sierra Nevada Mountains might be most likely due to 1) the type of precipitation in this region, being snow dominated rather than rain and/or 2) being orographic rain which satellite and radar have difficulties to fully capture.

The scatterplots and the statistics, Correlation Coefficient (Corr. Coef.), Root Mean Square Error (RMSE), and Bias of PERSIANN-CDR against CPC are shown in Fig. 50.5. As shown, correlation coefficient is high in both cases (> 0.9). PERSIANN-CDR tends to show a lower bias in R10mm than R20mm. Over mountainous regions like Sierra Nevada mountains, the agreement between PERSIANN-CDR and CPC degrade.

In another extreme precipitation study the performance of PERSIANN-CDR in capturing extreme rainfall events was validated against gauge observation in China. East Asia (EA, Xie et al. 2007) ground-based gridded daily precipitation data set is comprised of more than 1300 ground-based stations across China interpolated into 0.5° × 0.5° grid boxes using the Optimal Interpolation (OI) method.

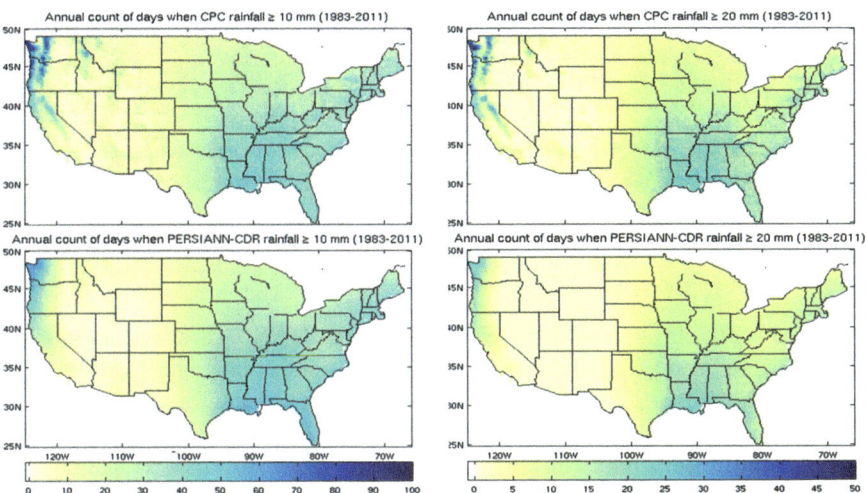

Fig. 50.4 Annual average count of days where rainfall ≥10 mm (left column) and rainfall ≥20 mm (right column) for CPC (top), and PERSIANN-CDR (bottom) for 1983–2011

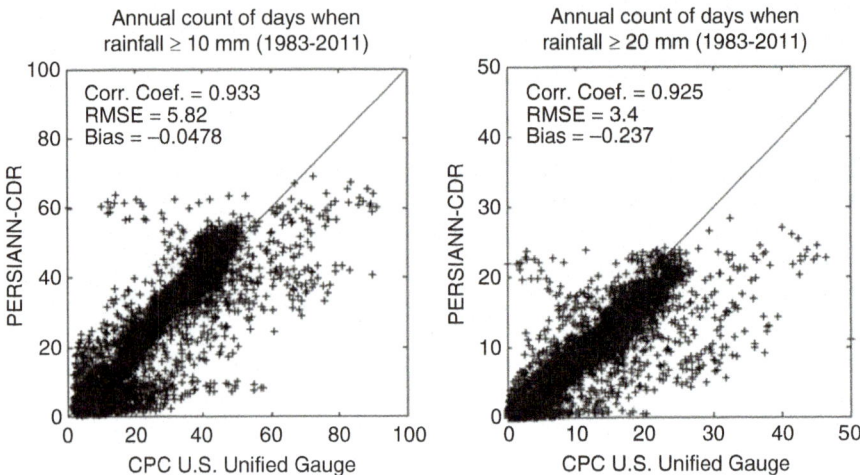

Fig. 50.5 Scatter plots of the annual average count of days where rainfall ≥10 mm (left column) and rainfall ≥20 mm (right) for PERSIANN-CDR against CPC. Correlation coefficient, RMSE, and Bias are shown on the plots

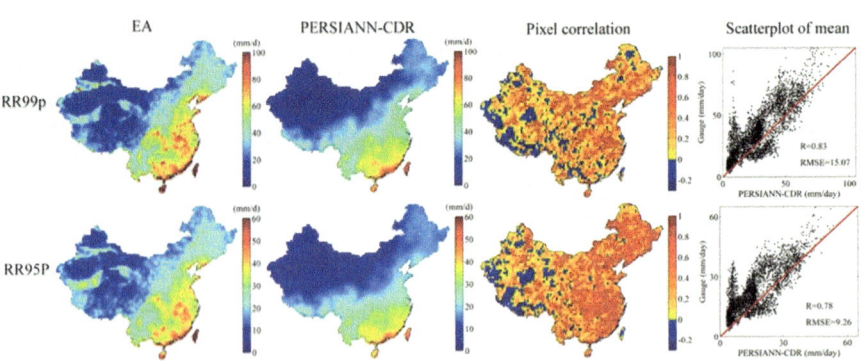

Fig. 50.6 The 99th and 95th percentile indices of extreme daily precipitation from the EA data set (first column), and PERSIANN-CDR (second column). The spatial correlation distribution and the scatterplots of the indices from the EA and PERSIANN-CDR data sets are shown in the third and fourth columns, respectively. The stippled areas in the third column show the significant correlation coefficient at the 95% level (Miao et al. 2015)

Figure 50.6 illustrates the performances of PERSIANN-CDR in capturing the 99th (RR99p) and 95th (RR95p) percentile indices of the daily precipitation during the period of 1983–2006. As shown, PERSIANN-CDR captures the spatial distribution of RR99p and RR95p similar to what the EA data set shows, with increasing RR99p and RR95p from North to South and from East to West. In addition, the scatterplots show high correlation coefficient between the percentile indices extracted from PERSIANN-CDR and EA datasets (Miao et al. 2015, for a complete analysis on all extreme precipitation indices). It is noteworthy the agreement between PERSIANN-CDR and EA dataset was closer in data rich regions in the west and south. As shown in Fig. 50.6, correlation coefficients in dry and arid regions in the western (Tibetan Plateau) and northwestern (Taklamakan Desert) China are relatively low. The main reason for this discrepancy is lack of enough gauge stations in these regions. With lesser and much sparse gauge stations in this region, the error and uncertainty that is introduced in the interpolated EA product could be significant.

In order to consider the effects of climate change, Ashouri et al. (2016a, b) developed non-stationary statistical models based on Extreme Value Theory (EVT) to investigating whether changes in our climate system have altered the probability distribution of climate extremes. The study – carried out over US – could identify regions where over the past three decades, the odds of record-setting extreme precipitation events have increased.

50.2.3 Evaluation of CMIP5 Model Precipitation

General circulation models (GCMs) are important tools for simulating the current state of the climate and projecting future changes of precipitation under different greenhouse gas emission scenarios. The predictive skills in precipitation simulations of the Coupled Model Intercomparison Project Phase 5 (CMIP5) models (Taylor et al. 2012), especially in capturing extreme precipitation events, are highly model dependent. PERSIANN-CDR was used as a tool to evaluate the ability of 32 CMIP5 models to capture the behavior of extreme precipitation estimates globally (Nguyen et al. 2017b). The work uniquely defines study regions by partitioning global land areas into 26 groups based on continent and climate zone type then uses PERSIANN-CDR as a baseline to investigate 8 extreme precipitation indices: (a) Total: R99pTOT – annual total precipitation when daily precipitation amount on a wet day >99 percentile, R95pTOT – annual total precipitation when daily precipitation amount on a wet day >95 percentile, R10mmTOT – annual total precipitation when daily precipitation amount $\geq$ 10 mm, and PRCPTOT – annual total precipitation in wet days; (b) Intensity: SDII – simple daily intensity index; (c) Frequency: R10mm – annual count of days when daily precipitation amount $\geq$ 10 mm; and d) Duration: CWD – annual maximum number of consecutive days with daily precipitation amount $\geq$ 1 mm, CDD – annual maximum number of consecutive days with daily precipitation amount < 1 mm from each of the 32 CMIP5 models. The extreme indices are recommended by the joint CCl/CLIVAR/JCOMM Expert Team (ET) on Climate Change Detection and Indices (ETCCDI).

A comprehensive assessment of each model's performance in each defined continent-climate zone group provides insight for users to select suitable models for their region of interest from a larger pool of models containing those with less skill. For instance, one can use the correlation and/or RMSE criteria in Fig. 50.7 to select a number of models among the 32 CMIP5 models for southeast China (CCZ6)

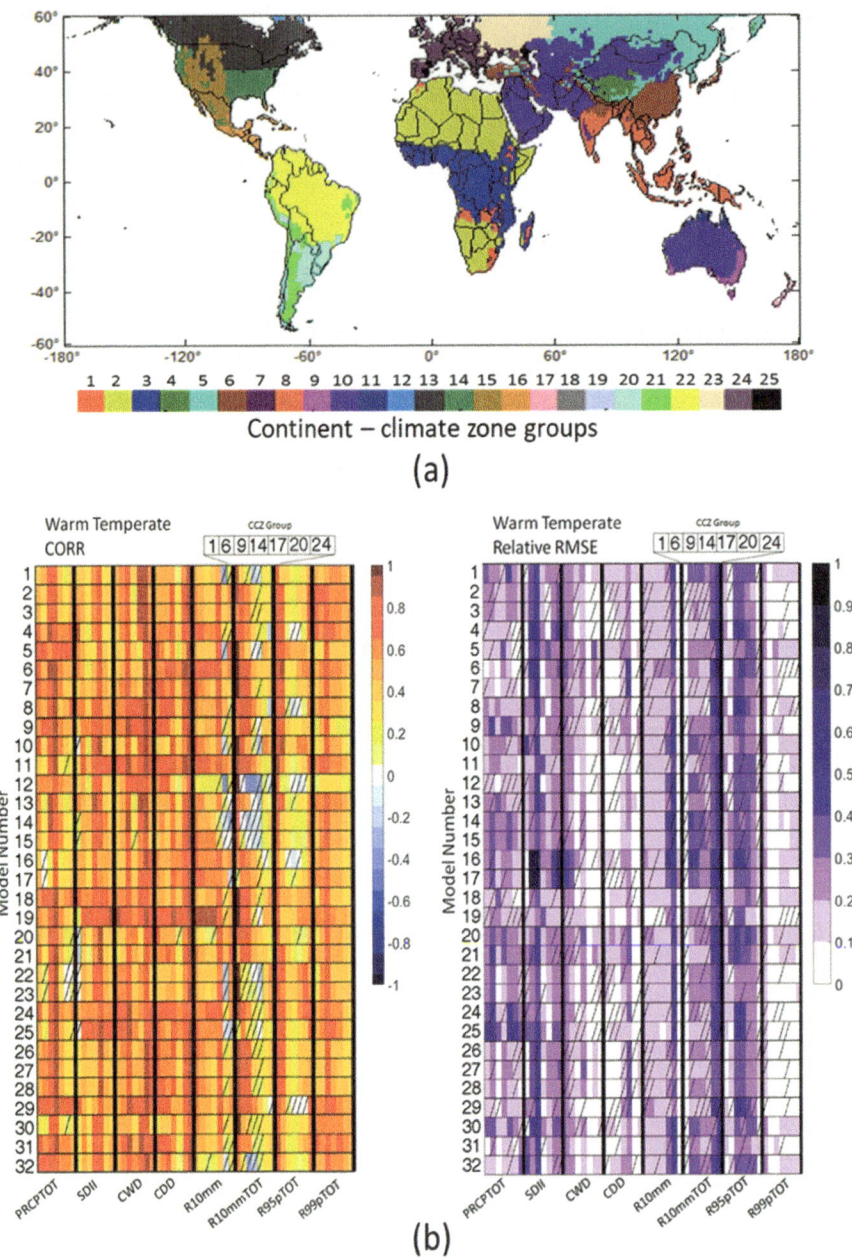

Fig. 50.7 (a) The 26 continent-climate zone groups, (b) Correlation (left) and relative RMSE (right) for precipitation indices in Warm Temperate continent-climate zone (CCZ) groups (statistical insignificance at 0.05 in hatched boxes). (Adapted from Nguyen et al. 2017b)

based on one or some of the 8 precipitation indices. If the selection is based on the simple daily intensity index (SDII) with the highest correlation, HadGEM2-ES, EC-EARTH, MIROC5 are the 3 best choices respectively. Such model selection can be adapted to any region globally depending on the phenomenon of interest, and can be combined with user-specific criteria.

50.3 Hydrology Applications

50.3.1 Hydrologic Modeling

Of more hydrological related applications of satellite-based precipitation products is rainfall-runoff modeling. With their continuous coverage both in space and time, satellite products bring invaluable information in modeling extreme events such as floods and droughts. In order to investigate the capability and accuracy of PERSIANN-CDR in modeling streamflow, PERSIANN-CDR's daily precipitation data are used in the NOAA's National Weather Service (NWS) Distributed Hydrologic Model Intercomparison Project – Phase 2 (DMIP2) test frame. Three test basins from Oklahoma are chosen. It is essential that as the very initial steps of conducting any study with satellite-products, the quality of satellite products be tested against ground-truth observations. The Taylor diagram in Fig. 50.8 illustrates

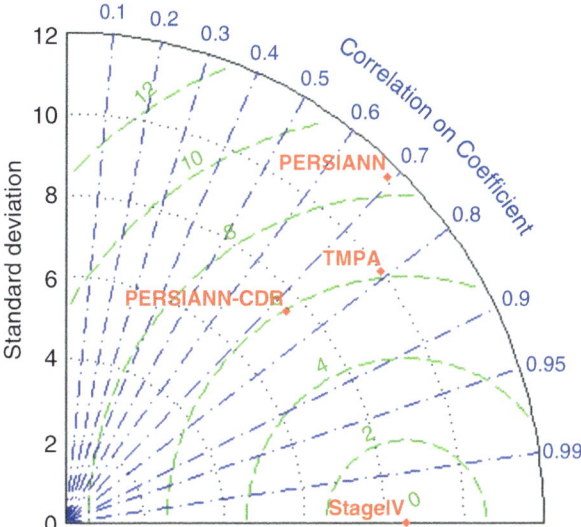

Fig. 50.8 Precipitation comparison plots SLOA4 basins between PERSIANN, PERSIANN-CDR, and TMPA against Stage IV gauge-adjusted radar data for 2003–2010. (Adapted from Ashouri et al. 2016a, b)

the evaluations of PERSIANN, PERSIANN-CDR, and TMPA precipitation products against Stage IV gauge-adjusted radar data as the reference dataset for SLOA4
basin. As shown PERSIANN-CDR and TMPA show close agreement with a
relatively higher correlation coefficient (values on the arch) for TMPA, lower
standard deviation (values on the y-axis) for PERSIANN-CDR, and almost the
same root mean square deviation (RMSD; values on dashed curved lines) for both
products.

With respect to the hydrological model, the widely used NOAA/NWS/Office
of Hydrologic Development's HL-RDHM (Koren et al. 2003, 2004, 2014) was
selected as the hydrological model to simulate the streamflow using the precipitation data products. Figure 50.9 illustrates the comparison between the
observed and simulated streamflow from Stage IV, TMPA, PERSIANN, and
PERSIANN-CDR over SLOA4 basin from 2003 to 2010. As can be seen, the
performance of PERSIANN-CDR is satisfying when compared to Stage IV
and TMPA products but PERSIANN-CDR has the capability to extend
streamflow simulation back to 1983 where other high-resolution products are
not available.

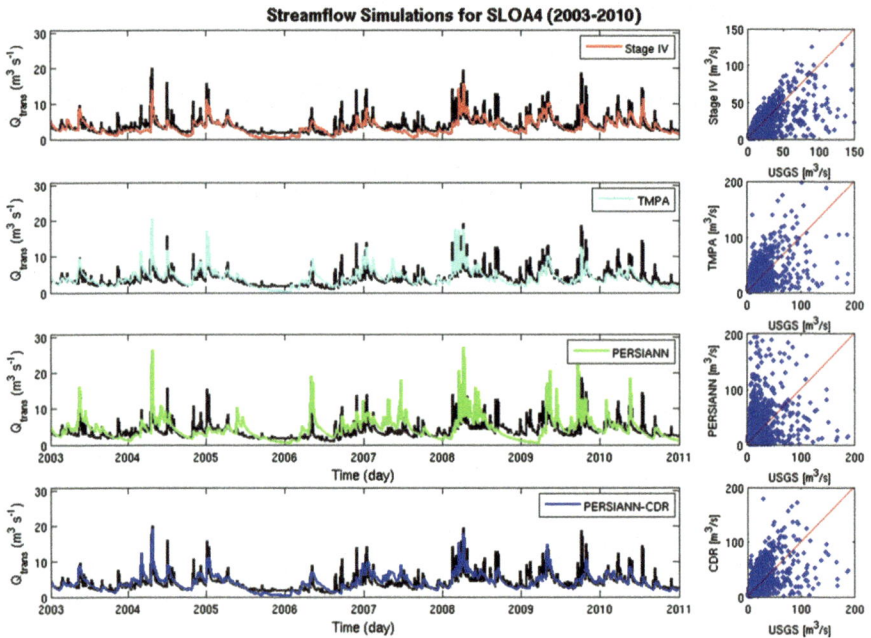

Fig. 50.9 Simulated and observed streamflow hydrographs and respective scatterplots at the outlet
of SAVOY basin using (from top to bottom) Stage IV, TMPA, PERSIANN, and PERSIANN-CDR
precipitation products. The solid black line shows the USGS observations. (Adapted from Ashouri
et al. 2016a, b)

50.3.2 Rainfall Frequency Analysis Using Satellite-Retrieved Precipitation

In recent years, a handful of studies examined utilizing satellite-retrieved precipitation in rainfall frequency analysis to develop Intensity-Duration-Frequency (IDF) curves (e.g., Endreny and Imbeah 2009; Awadallah et al. 2011; Marra et al. 2017; Ombadi et al. 2018). This application of satellite-based precipitation datasets is of utmost importance to the developing countries where in-situ rainfall measurement has sparse distribution, insufficient record length and poor data quality. Furthermore, even in regions with dense ground-based rainfall gauge networks, satellite-based precipitation can provide valuable information about the spatial distribution of rainfall. This is primarily because precipitation retrieval algorithms from satellites provide area-averaged rainfall estimates unlike ground gauge observation which represent point measurements.

Among the satellite-based precipitation products that have been used for rainfall frequency analysis is PERSIANN-CDR. It represents a unique dataset due to its long historical record (1983 – present) which provides sufficient sample size for frequency analysis. Recently, the potential of using PERSIANN-CDR to develop IDF curves has been investigated (Ombadi et al. 2018; Gado et al. 2017). Ombadi et al. (2018) developed a general framework for developing IDF curves from satellite-retrieved precipitation. It is based mainly on two steps, firstly, bias adjustment using a regression model that utilizes elevation as a predictor variable, secondly, transformation of areal rainfall to point rainfall. The parameters of the bias adjustment model for PERSIANN-CDR were estimated over the Contiguous United States (CONUS) using CPC Unified Gauge-Based Analysis of Daily Precipitation over CONUS. The transformation of areal-to-point rainfall is necessary to develop point IDF curves since satellite-based precipitation products estimate an areal average of rainfall over a grid cell. Ombadi et al. (2018) adopted an approach for rainfall transformation that is used for the reverse transformation (i.e., point-to-area); the method is based on the stochastic representation of rainfall fields in space and time (Sivapalan and Blöschl 1998).

The framework was evaluated by developing IDF curves over CONUS and the results were compared to NOAA Atlas 14 (Bonnin et al. 2006, Perica et al. 2013). Figure 50.10 shows the distribution of relative errors of IDF curves developed from PERSIANN-CDR in comparison to NOAA Atlas 14. The results highlight the potential of using satellite-based precipitation datasets in developing IDF curves, median relative errors are in the range of (17–22%), (6–12%) and (3–8%) for 1-, 2- and 3-days IDFs, respectively, and return periods in the 2–100-year range.

Also, in a recent study (Gado et al. 2017), PERSIANN-CDR dataset has been used as an alternative approach to implement regional frequency analysis. Regional frequency analysis is commonly implemented in rainfall frequency analysis either to estimate design rainfalls at ungauged sites or to improve statistical inference by providing a larger sample size. Traditional regional frequency analysis is implemented by firstly delineating homogenous groups of sites. This step can be

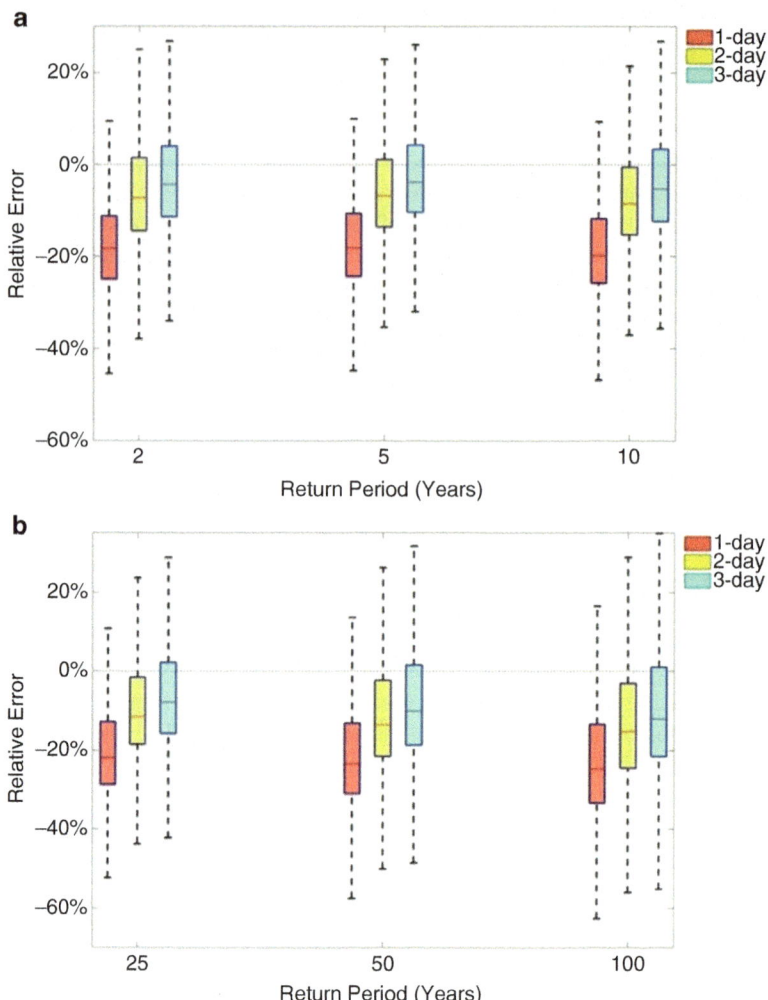

Fig. 50.10 Boxplots of satellite-based IDF relative error for durations of (1, 2 and 3) days and return periods of 25, 50 and 100 years. (Adapted from Ombadi et al. 2018)

performed using different methods such as cluster analysis (Tasker 1982), discriminant analysis (Wiltshire 1986), region of influence (Burn 1990) and the widely used method of discordancy measure (Hosking and Wallis 1993). The new approach is geared toward the estimation of design rainfalls at ungauged locations within a partially-gauged homogenous region and it consists of three steps. Firstly, satellite-based precipitation is adjusted for bias using the probability matching method (Calheiros and Zawadzki 1987). The distributions are constructed from satellite-based and in-situ rainfall measurements. Secondly, a relationship is constructed between original and corrected satellite-based precipitation. This

Table 50.1 Evaluation results of the two regional frequency approaches: Index Flood Method (IFM) and Regional Rainfall Frequency Analysis using Satellite Precipitation (RRFA-S)

RMSE (mm)			
Colorado		California	
IFM	RRFA-S	IFM	RRFA-S
<u>4.09</u>	5.95	6.13	<u>5.05</u>
<u>7.05</u>	8.45	7.77	<u>6.09</u>
13.42	<u>13.38</u>	10.33	<u>7.88</u>
20.36	<u>18.73</u>	12.67	<u>9.78</u>
29.49	<u>25.85</u>	15.43	<u>12.29</u>
41.28	<u>35.15</u>	18.68	<u>15.53</u>
82.44	<u>68.33</u>	28.51	<u>26.24</u>
37.86	<u>19.32</u>	15.88	<u>13.63</u>

Adapted from Gado et al. (2017)
Underlined values indicate the best result in each site

relationship is derived from gauged sites at a homogenous region and it is assumed to be valid for ungauged sites in the same region. Finally, this relationship is used to correct satellite-based precipitation at ungauged sites and derive IDF curves.

This method has been implemented to derive IDF curves over two regions in Colorado and California with 11 and 18 gauged sites respectively. These two regions have been identified as homogenous from previous studies (Sveinsson et al. 2002; Bonnin et al. 2011). In order to assess the performance of the methodology, the leave-one-out cross-validation was used to compare the accuracy of the methodology with the traditional index flood method. This is achieved by estimating the quantiles at a specific gauged site using the two approaches while neglecting data from that specific site. Then, the results are compared against at-site quantile estimates. The results in terms of the selected metrics, namely Bias, Relative Bias, Root Mean Square Error (RMSE) and relative RMSE indicated that the regional frequency analysis based on satellite-based precipitation provides more accurate results in most sites. In particular, it can be clearly seen that RMSE values, shown in Table 50.1, are considerably reduced using the approach based on satellite-retrieved precipitation.

50.4 Conclusions

There is a great zeal in current decades to investigate hydrologic and hydro-climatic processes with the availability and high reliability of ever-growing remotely sensed information. Characteristics such as high spatiotemporal resolution, real-time near-global coverage, make a wide range of applications viable through the incorporation of satellite-retrieved precipitation datasets such as PERSIANN-CDR.

Among the most important applications are the investigation of trends and changes of many hydroclimatic processes such as precipitation regimes response or the frequency and intensity of floods. Hurricane Katrina has been studied as an example of extreme precipitation event to compare the rainfall patterns to the radar data and high-resolution satellite-based precipitation products from TMPA v7. The results show a higher correlation coefficient for PERSIANN-CDR than TMPA compared to Stage IV radar data, however, the bias in TMPA is lower than that in PERSIANN-CDR. In terms of count related indices PERSIANN-CCS in general produces the same patterns as CPC Unified Gauge-Based Analysis of Precipitation data. Another extreme precipitation study over China is conducted to assess the performance of PERSIANN-CDR in capturing extreme rainfall events, and was validated against gauge observation. PERSIANN-CDR captures the spatial distribution of the 99th (RR99p) and 95th (RR95p) percentile indices of the daily precipitation during the period of 1983–2006 similar to what the East Asia ground-based gridded daily precipitation data set shows.

Two case studies have been highlighted as examples of utilizing satellite-based precipitation datasets for hydrologic applications. Firstly, runoff prediction by forcing hydrologic models with satellite-based precipitation datasets. In the case study of Ashouri et al. (2016a, b), PERSIANN-CDR dataset has been used to predict runoff in three sub-basins of the Illinois River basin. The results demonstrated that PERSIANN-CDR-derived streamflow simulations are comparable to USGS observations in terms of correlation coefficients, bias, and index of agreement criterion. Secondly, the use of satellite-based precipitation in rainfall frequency analysis. In the study of Gado et al. (2017), a new approach to regional frequency analysis was proposed using PERSIANN-CDR dataset. The methodology was evaluated in two homogenous regions in Colorado and California; results demonstrated the efficiency of the methodology.

In the coming years, it is expected that satellite-based precipitation will play a fundamental role in hydrometeorological research. Given the continuous advancement in satellite sensors technology and retrieval algorithms, satellite-based precipitation datasets will be available in an improved spatiotemporal resolution. This is of utmost importance to the observation of precipitation heterogeneous nature, thus, enabling comprehensive investigation of key hydrometeorological processes. Moreover, as records of satellite-based precipitation extend to cover more years, trend analysis studies will be more conceivable. Similarly, extended record length will facilitate the utilization of satellite-based precipitation in rainfall frequency analysis for infrastructure design.

Acknowledgements This research was partially supported by the NASA Precipitation Measurement Missions (award # NNX10AK07G), NASA MEaSURES (award # NNX13AM12G), NASA MIRO (award # NNX15AQ06A), the ICIWaRM of the US Army corps of Engineering, UNESCO's G-WADI program, Cooperative Institute for Climate and Satellites (CICS) program (NOAA prime award #NA14NES4320003, subaward # 2014-2913-03) for OHD-NWS student fellowship, Army Research Office (award # W911NF-11-1-0422), National Science Foundation (NSF award # 1331915), Department of Energy (DoE prime award # DE-IA0000018) and California Energy Commission (CEC Award # 300-15-005).

References

Ashouri, H., Hsu, K.-L., Sorooshian, S., Braithwaite, D. K., Knapp, K. R., Cecil, L. D., Nelson, B. R., & Prat, O. P. (2015). PERSIANN-CDR: Daily precipitation climate data record from multisatellite observations for hydrological and climate studies. *Bulletin of the American Meteorological Society, 96*, 69–83. https://doi.org/10.1175/BAMS-D-13-00068.1.

Ashouri, H., Nguyen, P., Thorstensen, A., Hsu, K.-L., Sorooshian, S., & Braithwaite, D. (2016a). Assessing the efficacy of high-resolution satellite-based PERSIANN-CDR precipitation product in simulating streamflow. *Journal of Hydrometeorology, 17*, 2061–2076. https://doi.org/10.1175/JHM-D-15-0192.1.

Ashouri, H., Sorooshian, S., Hsu, K.-L., Bosilovich, M. G., Lee, J., Wehner, M. F., & Collow, A. (2016b). Evaluation of NASA's MERRA precipitation product in reproducing the observed trend and distribution of extreme precipitation events in the United States. *Journal of Hydrometeorology, 17*, 693–711. https://doi.org/10.1175/JHM-D-15-0097.1.

Awadallah, G. A., El Gamal, M., El Mostafa, A., & El Badry, H. (2011). Developing intensity-duration-frequency curves in scarce data region: An approach using regional analysis and satellite data. *Engineering, 3*, 215–226. https://doi.org/10.4236/eng.2011.33025.

Behrangi, A., Khakbaz, B., Jaw, T. C., AghaKouchak, A., Hsu, K.-L., & Sorooshian, S. (2011). Hydrologic evaluation of satellite precipitation products over a mid-size basin. *Journal of Hydrology, 397*, 225–237. https://doi.org/10.1016/j.jhydrol.2010.11.043.

Bonnin, G. M., Martin, D., Lin, B., Parzybok, T., Yekta, M., & Riley, D. (2006). NOAA Atlas 14 Volume 2 Version 3.0, Precipitation-Frequency Atlas of the United States. Silver Spring, MD, NOAA, National Weather Service, 295 pp. Available at http://www.nws.noaa.gov/oh/hdsc/PF_documents/Atlas14_Volume2.pdf, last accessed 24 Nov 2018.

Bonnin, G. M., Martin, D., Lin, B., Parzybok, T., Yekta, M., & Riley, D. (2011). NOAA Atlas 14 Volume 1 Version 5.0, Precipitation-Frequency Atlas of the United States. Silver Spring, MD, NOAA, National Weather Service. Available at http://www.nws.noaa.gov/oh/hdsc/PF_documents/Atlas14_Volume1.pdf, last accessed 24 Nov 2018.

Burn, D. H. (1990). Evaluation of regional flood frequency analysis with a region of influence approach. *Water Resources Research, 26*, 2257–2265. https://doi.org/10.1029/WR026i010p02257.

Calheiros, R. V., & Zawadzki, I. (1987). Reflectivity-rain rate relationships for radar hydrology in Brazil. *Journal of Climate and Applied Meteorology, 26*, 118–132. https://doi.org/10.1175/1520-0450(1987)026<0118:RRRRFR>2.0.CO;2.

Casse, C., & Gosset, M. (2015). Analysis of hydrological changes and flood increase in Niamey based on the PERSIANN-CDR satellite rainfall estimate and hydrological simulations over the 1983–2013 period. *Proceedings of the International Association of Hydrological Sciences, 370*, 117–123. https://doi.org/10.5194/piahs-370-117-2015.

Chiang, Y. M., Hsu, K.-L., Chang, F.-J., Hong, Y., & Sorooshian, S. (2007). Merging multiple precipitation sources for flood forecasting. *Journal of Hydrology, 340*, 183–196. https://doi.org/10.1016/j.jhydrol.2007.04.007.

Damberg, L., & AghaKouchak, A. (2014). Global trends and patterns of drought from space. *Theoretical and Applied Climatology, 117*, 441–448. https://doi.org/10.1007/s00704-013-1019-5.

Duan, Z., Liu, J., Tuo, Y., Chiogna, G., & Disse, M. (2016). Evaluation of eight high spatial resolution gridded precipitation products in Adige Basin (Italy) at multiple temporal and spatial scales. *Science of the Total Environment, 573*, 1536–1553. https://doi.org/10.1016/j.scitotenv.2016.08.213.

Endreny, T. A., & Imbeah, N. (2009). Generating robust rainfall intensity–duration–frequency estimates with short-record satellite data. *Journal of Hydrology, 371*, 182–191. https://doi.org/10.1016/j.jhydrol.2009.03.027.

Gado, T. A., Hsu, K.-L., & Sorooshian, S. (2017). Rainfall frequency analysis for ungauged sites using satellite precipitation products. *Journal of Hydrology, 554*, 646–655. https://doi.org/10.1016/j.jhydrol.2017.09.043.

Graumann, A., and National Climatic Data Center (US). (2006). Hurricane Katrina: A climatological perspective: preliminary report, Asheville, NC, U.S. Dept. of Commerce, National Oceanic and Atmospheric Administration, National Environmental Satellite Data and Information Service, National Climatic Data Center, 27 pp. Available at https://iucat.iu.edu/iue/8006266, last accessed 24 Nov 2018.

Hong, Y., Gochis, D., Cheng, J. T., Hsu, K.-L., & Sorooshian, S. (2007). Evaluation of PERSIANN-CCS rainfall measurement using the NAME event rain gauge network. *Journal of Hydrometeorology, 8*, 469–482. https://doi.org/10.1175/JHM574.1.

Hosking, J. R. M., & Wallis, J. R. (1993). Some statistics useful in regional frequency analysis. *Water Resources Research, 29*, 271–281. https://doi.org/10.1029/92WR01980.

Hsu, K.-L., Gao, X., Sorooshian, S., & Gupta, H. V. (1997). Precipitation estimation from remotely sensed information using artificial neural networks. *Journal of Applied Meteorology, 36*, 1176–1190. https://doi.org/10.1175/1520-0450(1997)036<1176:PEFRSI>2.0.CO;2.

Hsu, K.-L., Sellars, S., Nguyen, P., Braithwaite, D., & Chu, W. (2013). G-WADI PERSIANN-CCS GeoServer for extreme event analysis. *Sciences in Cold and Arid Regions, 5*, 6–15. https://doi.org/10.3724/SP.J.1226.2013.00006.

Huffman, G. J., Adler, R. F., Morrissey, M. M., Bolvin, D. T., Curtis, S., Joyce, R., McGavock, B., & Susskind, J. (2001). Global precipitation at one-degree daily resolution from multisatellite observations. *Journal of Hydrometeorology, 2*, 36–50. https://doi.org/10.1175/1525-7541(2001)0022.0.CO;2.

Huffman, G. J., Bolvin, D. T., Nelkin, E. J., Wolff, D. B., Adler, R. F., Gu, G., Hong, Y., Bowman, K. P., & Stocker, E. F. (2007). The TRMM multisatellite precipitation analysis (TMPA): Quasi-global, multiyear, combined-sensor precipitation estimates at fine scales. *Journal of Hydrometeorology, 8*, 38–55. https://doi.org/10.1175/JHM560.1.

Joyce, R. J., Janowiak, J. E., Arkin, P. A., & Xie, P. (2004). CMORPH: A method that produces global precipitation estimates from passive microwave and infrared data at high spatial and temporal resolution. *Journal of Hydrometeorology, 5*, 487–503. https://doi.org/10.1175/1525-7541(2004)005<0487:CAMTPG>2.0.CO;2.

Kendall, M. (1976). *Rank Correlation Methods* (4th ed.). London: Griffin.

Khanna, J., Medvigy, D., Fueglistaler, S., & Walko, R. (2017). Regional dry-season climate changes due to three decades of Amazonian deforestation. *Nature Climate Change, 7*, 200–204. https://doi.org/10.1038/nclimate3226.

Kidd, C., & Levizzani, V. (2011). Status of satellite precipitation retrievals. *Hydrology and Earth System Sciences, 15*, 1109–1116. https://doi.org/10.5194/hess-15-1109-2011.

Kidd, C., Becker, A., Huffman, G. J., Muller, C. L., Joe, P., Skofronick-Jackson, G., & Kirschbaum, D. B. (2017). So, how much of the Earth's surface is covered by rain gauges? *Bulletin of the American Meteorological Society, 98*, 69–78. https://doi.org/10.1175/BAMS-D-14-00283.1.

Klein Tank, A. M. G., Zwiers, F. W., & Zhang, X. (2009). Guidelines on analysis of extremes in a changing climate in support of informed decisions for adaptation, climate data and monitoring. WCDMP-No 72. WMO-TD No 1500, 56 pp. Available at https://www.ecad.eu/documents/WCDMP_72_TD_1500_en_1.pdf, last accessed 24 Nov 2018.

Koren, V., Smith, M., & Duan, Q. (2003). Use of a priori parameter estimates in the derivation of spatially consistent parameter sets of rainfall-runoff models. *Calibration of Watershed Models, 6*, 239–254. https://doi.org/10.1002/9781118665671.ch18.

Koren, V., Reed, S., Smith, M., Zhang, Z., & Seo, D. J. (2004). Hydrology laboratory research modeling system (HL-RMS) of the US national weather service. *Journal of Hydrology, 291*, 297–318. https://doi.org/10.1016/j.jhydrol.2003.12.039.

Koren, V. I., Smith, M. B., Cui, Z., & Cosgrove, B. A. (2014). Physically-based modifications to the Sacramento Soil Moisture Accounting model. Part A: Modeling the effects of frozen ground on the runoff generation process. *Journal of Hydrology, 519*, 3475–3491. https://doi.org/10.1016/j.jhydrol.2014.03.004.

Lee, H. T. (2014). Climate algorithm theoretical basis document (C-ATBD): Outgoing longwave radiation (OLR)—Daily. NOAA's Climate Data Record (CDR) Program. Tech. Rep. CDRP-ATBD-0526, 46 pp. Available at https://www1.ncdc.noaa.gov/pub/data/sds/cdr/CDRs/

Outgoing%20Longwave%20Radiation%20-%20Daily/AlgorithmDescription.pdf, last accessed 24 Nov 2018.

Lin, Y., & Mitchell, K. E. (2005). 1.2 the NCEP stage II/IV hourly precipitation analyses: Development and applications. 19th conference on Hydrology. Available at https://ams. confex.com/ams/Annual2005/techprogram/paper_83847.htm, last accessed 24Nov 2018.

Liu, X., Yang, T., Hsu, K.-L., & Sorooshian, S. (2017). Evaluating the streamflow simulation capability of PERSIANN-CDR daily rainfall products in two river basins on the Tibetan Plateau. Hydrology and Earth System Sciences, 21, 169–181. https://doi.org/10.5194/hess-21-169-2017.

Maggioni, V., Meyers, P. C., & Robinson, M. D. (2016). A review of merged high-resolution satellite precipitation product accuracy during the Tropical Rainfall Measuring Mission (TRMM) era. Journal of Hydrometeorology, 17, 1101–1117. https://doi.org/10.1175/JHM-D-15-0190.1.

Mann, H. B. (1945). Nonparametric tests against trend. Econometrica: Journal of Econometric Society, 245–259. Available qat https://www.jstor.org/stable/1907187?seq=1#metadata_info_tab_contents, last accessed 24 Nov 2018.

Marra, F., Morin, E., Peleg, N., Mei, Y., & Anagnostou, E. N. (2017). Intensity–duration–frequency curves from remote sensing rainfall estimates: Comparing satellite and weather radar over the eastern Mediterranean. Hydrology and Earth System Sciences, 21, 2389–2404. https://doi.org/10.5194/hess-21-2389-2017.

Miao, C., Ashouri, H., Hsu, K., Sorooshian, S., & Duan, Q. (2015). Evaluation of the PERSIANN-CDR Daily Rainfall Estimates in Capturing the Behavior of Extreme Precipitation Events over China. Journal of Hydrometeorology, 16, 1387–1396. https://doi.org/10.1175/JHM-D-14-0174.1.

Nguyen, P., Sellars, S., Thorstensen, A., Tao, Y., Ashouri, H., Braithwaite, D., Hsu, K., & Sorooshian, S. (2014). Satellites track precipitation of Super Typhoon Haiyan. EOS Transactions, 95, 133–135. https://doi.org/10.1002/2014EO160002.

Nguyen, P., Sorooshian, S., Thorstensen, A., Tran, H., Huynh, P., Pham, T., Ashouri, H., Hsu, K.-L., AghaKouchak, A., & Braithwaite, D. (2017a). Exploring trends through "RainSphere": Research data transformed into public knowledge. Bulletin of the American Meteorological Society, 98, 653–658. https://doi.org/10.1175/BAMS-D-16-0036.1.

Nguyen, P., Thorstensen, A., Sorooshian, S., Zhu, Q., Tran, H., Ashouri, H., Miao, C., Hsu, K.-L., & Gao, X. (2017b). Evaluation of CMIP5 model precipitation using PERSIANN-CDR. Journal of Hydrometeorology, 18, 2313–2330. https://doi.org/10.1175/JHM-D-16-0201.1.

Nguyen, P., Thorstensen, A., Sorooshian, S., Hsu, K.-L., AghaKouchak, A., Ashouri, H., Tran, H., & Braithwaite, D. (2018). Global precipitation trends across spatial scales using satellite observations. Bulletin of the American Meteorological Society, 99, 689–697. https://doi.org/10.1175/BAMS-D-17-0065.1.

Ombadi, M., Nguyen, P., Sorooshian, S., & Hsu, K. (2018). Developing intensity-duration-frequency (IDF) curves from satellite-based precipitation: Methodology and evaluation. Water Resources Research, 54, 7752–7766. https://doi.org/10.1029/2018WR022929.

Perica, S., Martin, D., Pavlovic, S., Roy, I., Laurent, M. S., Trypaluk, C., et al. (2013). NOAA Atlas 14, Volume 9, Version 2, Precipitation-Frequency Atlas of the United States, Southeastern States (Vol. 18). Silver Spring, MD: NOAA, National Weather Service.

Rana, S., McGregor, J., & Renwick, J. (2015). Precipitation seasonality over the Indian subcontinent: An evaluation of gauge, reanalyses, and satellite retrievals. Journal of Hydrometeorology, 16, 631–651. https://doi.org/10.1175/JHM-D-14-0106.1.

Sivapalan, M., & Blöschl, G. (1998). Transformation of point rainfall to areal rainfall: Intensity-duration-frequency curves. Journal of Hydrology, 204, 150–167. https://doi.org/10.1016/S0022-1694(97)00117-0.

Sorooshian, S., Hsu, K.-L., Gao, X., Gupta, H. V., Imam, B., & Braithwaite, D. (2000). Evaluation of PERSIANN system satellite–based estimates of tropical rainfall. Bulletin of the American Meteorological Society, 81, 2035–2046. https://doi.org/10.1175/1520-0477(2000)081<2035:EOPSSE>2.3.CO;2.

Sorooshian, S., Gao, X., Hsu, K.-L., Maddox, R. A., Hong, Y., Gupta, H. V., & Imam, B. (2002). Diurnal variability of tropical rainfall retrieved from combined GOES and TRMM satellite information. *Journal of Climate, 15*, 983–1001. https://doi.org/10.1175/1520-0442(2002) 015<0983:DVOTRR>2.0.CO;2.

Sorooshian, S., AghaKouchak, A., Arkin, P., Eylander, J., Foufoula-Georgiou, E., Harmon, R., Hendrickx, J. M., Imam, B., Kuligowski, R., Skahill, B., & Skofronick-Jackson, G. (2011). Advancing the remote sensing of precipitation. *Bulletin of the American Meteorological Society, 92*, 1271–1272. https://doi.org/10.1175/BAMS-D-11-00116.1.

Sun, Q., Miao, C., Duan, Q., Ashouri, H., Sorooshian, S., & Hsu, K.-L. (2018). A review of global precipitation data sets: Data sources, estimation, and intercomparisons. *Reviews of Geophysics, 56*, 79–107. https://doi.org/10.1002/2017RG000574.

Sveinsson, O. G., Salas, J. D., & Boes, D. C. (2002). Regional frequency analysis of extreme precipitation in northeastern Colorado and Fort Collins flood of 1997. *Journal of Hydrologic Engineering, 7*, 49–63. https://doi.org/10.1061/(ASCE)1084-0699(2002)7:1(49).

Tan, M. L., Ibrahim, A. L., Duan, Z., Cracknell, A. P., & Chaplot, V. (2015). Evaluation of Six High-Resolution Satellite and Ground-Based Precipitation Products over Malaysia. *Remote Sensing, 7*, 1504–1528. https://doi.org/10.3390/rs70201504.

Tasker, G. D. (1982). Comparing methods of hydrologic regionalization. *Journal of the American Water Resources Association, 18*, 965–970. https://doi.org/10.1111/j.1752-1688.1982. tb00102.x.

Taylor, K. E., Stouffer, R. J., & Meehl, G. A. (2012). An overview of CMIP5 and the experiment design. *Bulletin of the American Meteorological Society, 93*, 485–498. https://doi.org/10.1175/ BAMS-D-11-00094.1.

Wiltshire, S. E. (1986). Regional flood frequency analysis I: Homogeneity statistics. *Hydrological Sciences Journal, 31*, 321–333. https://doi.org/10.1080/02626668609491051.

Xie, P., & Arkin, P. A. (1997). Global precipitation: A 17-year monthly analysis based on gauge observations, satellite estimates, and numerical model outputs. *Bulletin of the American Meteorological Society, 78*, 2539–2558. https://doi.org/10.1175/1520-0477(1997)078<2539: GPAYMA>2.0.CO;2.

Xie, P., Janowiak, J. E., Arkin, P. A., Adler, R. F., Gruber, A., Ferraro, R. R., Huffman, G. J., & Curtis, S. (2003). GPCP pentad precipitation analyses: An experimental dataset based on gauge observations and satellite estimates. *Journal of Climate, 16*, 2197–2214. https://doi.org/10. 1175/2769.1.

Xie, P., Chen, M., Yang, S., Yatagai, A., Hayasaka, T., Fukushima, Y., & Liu, C. (2007). A gauge-based analysis of daily precipitation over East Asia. *Journal of Hydrometeorology, 8*, 607–626. https://doi.org/10.1175/JHM583.1.

Xie, P., Chen, M., & Shi, W. (2010). CPC unified gauge-based analysis of global daily precipitation. Preprints, 24th Conference on Hydrology, Atlanta, GA, American Meteorological Society. (Vol. 2). Available at https://ams.confex.com/ams/90annual/techprogram/paper_163676.htm, last accessed 24 Nov 2018.

Yang, X., Yong, B., Yong, H., Chen, S., & Zhang, X. (2016). Error analysis of multi-satellite precipitation estimates with an independent raingauge observation network over a medium-sized humid basin. *Hydrological Sciences Journal, 61*, 1813–1830. https://doi.org/10.1080/ 02626667.2015.1040020.

Zambrano, F., Wardlow, B., Tadesse, T., Lillo-Saavedra, M., & Lagos, O. (2017). Evaluating satellite-derived long-term historical precipitation datasets for drought monitoring in Chile. *Atmospheric Research, 186*, 26–42. https://doi.org/10.1016/j.atmosres.2016.11.006.

Zhu, Q., Xuan, W., Liu, L., & Xu, Y. P. (2016). Evaluation and hydrological application of precipitation estimates derived from PERSIANN-CDR, TRMM 3B42V7, and NCEP-CFSR over humid regions in China. *Hydrocarbon Processing, 30*, 3061–3083. https://doi.org/10. 1002/hyp.10846.

Chapter 51
Soil Moisture and Precipitation: The SM2RAIN Algorithm for Rainfall Retrieval from Satellite Soil Moisture

Luca Ciabatta, Stefania Camici, Christian Massari, Paolo Filippucci, Sebastian Hahn, Wolfgang Wagner, and Luca Brocca

Abstract The standard approach for measuring instantaneous rainfall rates from space is based on the inversion of the atmospheric signals reflected or radiated by atmospheric hydrometeors, i.e., a "top-down" approach. Recently, a new "bottom-up" approach has been proposed that exploits satellite soil moisture observations for obtaining accumulated rainfall estimates. The approach, referred to as SM2RAIN, is based on the inversion of the hydrological water balance. In this chapter, after a short description of the SM2RAIN algorithm and its application to satellite soil moisture data, the two most recent satellite rainfall products obtained by the application of SM2RAIN to ESA-CCI (European Space Agency – Climate Change Initiative) and ASCAT (Advanced SCATterometer) soil moisture products are illustrated. Then, we have investigated the use of SM2RAIN-derived rainfall products, in comparison with "top-down" precipitation products, for improving flood forecasting over 600 basins in Europe. Finally, the limitations of the SM2RAIN algorithm and the future research and technological developments to address such limitations are provided.

Keywords Precipitation · Rainfall · Soil moisture · SM2RAIN · ESA-CCI · ASCAT · SMOS · AMSR · CYGNSS · TMPA · CMORPH · ERA5 reanalysis · GPCC · Floods

L. Ciabatta (✉) · S. Camici · C. Massari · P. Filippucci
National Research Council, Research Institute for Geo-Hydrological Protection (CNR-IRPI), Perugia, Italy
e-mail: l.ciabatta@irpi.cnr.it

S. Hahn · W. Wagner
Department of Geodesy and Geoinformation, Research Group Remote Sensing, TU Wien, Vienna, Austria

L. Brocca
Research Institute for Geo-Hydrological Protection (IRPI), National Research Council (CNR), Perugia, Italy

© Springer Nature Switzerland AG 2020   1013
V. Levizzani et al. (eds.), *Satellite Precipitation Measurement*, Advances in Global Change Research 69, https://doi.org/10.1007/978-3-030-35798-6_27

51.1 Introduction

In 2013, a new "bottom-up" approach was proposed by Brocca et al. (2013) for estimating rainfall from space. Instead of considering the interaction between microwave and infrared signal with hydrometeors, as is usually done in state-of-the-art "top-down" precipitation retrieval techniques, the bottom-up approach takes advantage of the capability of spaceborne microwave sensors for measuring soil moisture. The soil is considered as a natural raingauge (Brocca et al. 2014) and from the knowledge of the variation in time of soil moisture (SM), the accumulated rainfall between two satellite overpasses is computed.

The first studies highlighting the benefit in using satellite SM measurement for improving satellite-based rainfall estimates have been carried out by Crow et al. (2009, 2011) and Pellarin et al. (2008, 2013). In these and similar more recent studies, satellite SM data have been used for correcting rainfall that employ different approaches: Kalman filtering in Crow et al. (2011), particle filtering in Wanders et al. (2015) and Román-Cascón et al. (2017), and multiplicative factors in Pellarin et al. (2013). These studies demonstrated the largest improvements in areas where satellite SM data perform well, as in Australia by using the SMOS (Soil Moisture Ocean Salinity) SM product (Brocca et al. 2016b) and in Western United States by using the AMSR (Advanced Microwave Scanning Radiometer) SM product (Crow et al. 2011).

Instead of using SM observations for rainfall correction, Brocca et al. (2013) firstly proposed the use of SM for providing direct estimates of rainfall accumulations. This method, called SM2RAIN, is based on the inversion of the soil water balance equation. That is, it estimates the rainfall by using the change in time of the amount of water stored in the soil, thus considering it "as a natural raingauge". SM2RAIN has been applied both at a local (Brocca et al. 2013, 2015) and global scale (Brocca et al. 2014, 2017; Koster et al. 2016; Ciabatta et al. 2018) with ground and satellite SM data as input with satisfactory results in terms of rainfall estimation. Recently, the use of a constellation of satellite SM products has been also investigated in India and Italy by Brocca et al. (2016a) and Tarpanelli et al. (2017), who obtained significantly improved performance with respect to the use of a single SM product. A similar approach has been also proposed by Tian et al. (2014) for estimating snowfall from snow water equivalent observations obtained by passive microwave satellite sensors.

Massari et al. (2014), in a study over a small catchment in southern France, found that the correction of rainfall through SM2RAIN provides improvement in flood modelling when compared to the use of rain gauge observations only. Similar results are obtained in Ciabatta et al. (2016) for flood simulation in four basins in Italy and by Massari et al. (2018) and Camici et al. (2018) for 15 basins in the Mediterranean area. More recently, SM2RAIN-derived rainfall datasets have been used by Abera et al. (2017) for water budget assessment in the Upper Blue Nile (Ethiopia), by Brunetti et al. (2018) for landslide prediction throughout Italy, and by Thaler et al. (2018) for crop modelling in Austria. Therefore, the SM2RAIN approach is becoming a well-established method that has been found of benefit for hydrological (floods, water resources management), hydrogeological (landslides) and agricultural applications.

Bottom-up and top-down remote sensing approaches exhibit two major differences. Firstly, top-down approaches provide instantaneous rainfall measurements from which the 3-hourly and/or daily accumulated rainfall can be estimated (Fig. 51.1 upper panel). However, if the satellite sensors do not pass over when it rains, a significant underestimation of rainfall is expected from these algorithms. Differently, the bottom-up approach is based on the estimation of accumulated rainfall between two consecutive SM measurements (Fig. 51.1 lower panel). Therefore, the method keeps track of the total rainfall fallen between satellite overpasses, with an expected higher degree of accuracy for rainfall accumulation estimates. Secondly, top-down estimates are available over land and over sea, are not significantly affected by surface conditions, and exploit a constellation of satellite sensors also specifically dedicated to rainfall measurement (e.g., Huffman et al. 2007; Hou et al. 2014). Bottom-up rainfall estimates are only available over land, and are affected by surface conditions as frozen and saturated soils, snow, and vegetation density. These two points make the two approaches highly complementary. Ciabatta et al. (2017) and Chiaravalloti et al. (2018) have already demonstrated that the integration of bottom-up and top-down approaches is able to provide highly accurate rainfall estimates, better than using either of the parent approaches alone.

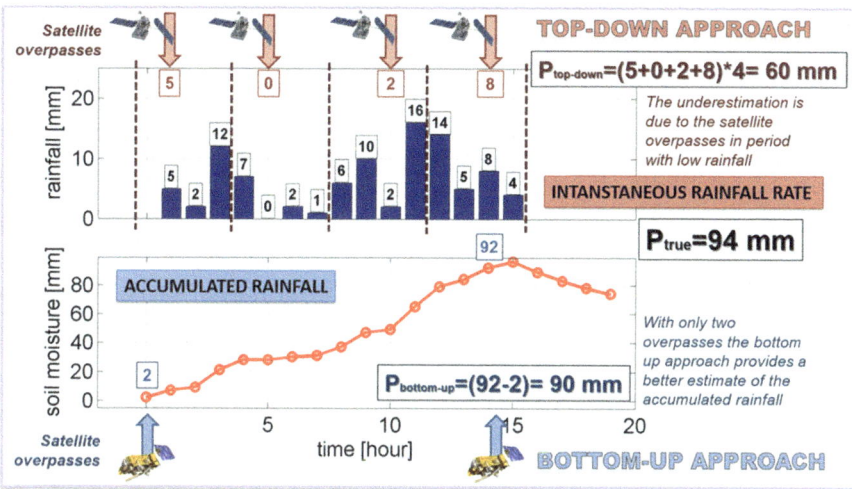

Fig. 51.1 Bottom-up vs. top-down perspective for rainfall retrieval from remote sensing assuming no error in the satellite measurements and in the retrieval algorithms. Due to the satellite overpass during low rainfall intensities, the "top down" method may fail in estimating the accumulated rainfall whereas the "bottom up" approach accurately reproduces the observations even with a lower number of overpasses

51.2   SM2RAIN Algorithm

The SM2RAIN algorithm is based on the inversion of the soil water balance equation and allows to estimate the amount of water entering into the soil by using as input SM information. SM2RAIN has been mostly used to retrieve rainfall from in situ and satellite SM data (e.g., Brocca et al. 2014, 2015, 2016a, b; Koster et al. 2016; Ciabatta et al. 2017; Massari et al. 2017a). Specifically, the soil water balance equation can be described by the following equation (over non-irrigated areas):

$$nZ\frac{dS(t)}{dt} = p(t) - g(t) - sr(t) - e(t) \tag{51.1}$$

where n [−] is the soil porosity, Z [mm] is the soil layer depth, $S(t)$ [−] is the relative saturation of the soil or relative SM, t [days] is the time, $p(t)$ [mm/day] is the rainfall rate, $g(t)$ [mm day^{-1}] is the drainage (deep percolation plus subsurface runoff) rate, $sr(t)$ [mm day^{-1}] is the surface runoff and $e(t)$ [mm day^{-1}] is the actual evapotranspiration. The drainage rate is related to the relative SM through a power law equation (Brocca et al. 2014):

$$g(t) = K_s S(t)^{3+\frac{2}{\lambda}} \tag{51.2}$$

where K_s [mm day^{-1}] is the saturated hydraulic conductivity and λ [−] is the pore size distribution index. The actual evapotranspiration rate is assumed to be linearly related to potential evapotranspiration, $ET_{pot}(t)$ [mm day^{-1}]:

$$e(t) = ET_{pot}(t)S(t) \tag{51.3}$$

The potential evapotranspiration can be computed through the empirical relation of Blaney and Criddle as modified by Doorenbos and Pruitt (1977):

$$ET_{pot}(t) = K_c\{-2 + 1.26[\xi\,(0.46T_a(t) + 8.13)]\} \tag{51.4}$$

where $T_a(t)$ [°C] is the air temperature, ξ [−] is the percentage of total daytime hours for the period used (daily or monthly) out of total daytime hours of the year (365 × 12), and K_c [−] is a correction factor for taking into account the empirical nature of eq. (51.4). By assuming that the rate of surface runoff is negligible, i.e., $sr(t) = 0$ (Brocca et al. 2015), eq. (51.1) is rewritten as:

$$p(t) = Z^*\frac{dS(t)}{dt} + K_s S(t)^{3+\frac{2}{\lambda}} + ET_{pot}(t)S(t) \tag{51.5}$$

where $Z* = Zn$ [mm] represents the water capacity of the soil layer.

Therefore, eq. (51.5) is used for estimating rainfall rate from SM, $S(t)$, and air temperature, $T_a(t)$ data. Four parameters ($Z*$, K_s, λ, K_c) need to be estimated by calibration using a reference rainfall dataset; note that in previous studies the parameter K_s is defined as a and the expression $\left(3 + \frac{2}{\lambda}\right)$ as b. In most of the previous applications of SM2RAIN, we have neglected the contribution of evapotranspiration (see Brocca et al. 2015 for more details) thus yielding to:

$$p(t) = Z^* \frac{dS(t)}{dt} + K_s S(t)^{3+\frac{2}{\lambda}} \tag{51.6}$$

where the knowledge of only SM observations is needed for estimating rainfall.

51.3   SM2RAIN-Derived Rainfall Products

In the last five years, we have developed and distributed several SM2RAIN-derived datasets by applying the algorithm to ASCAT (e.g., Brocca et al. 2017), SMOS (Brocca et al. 2016b), SMAP (Soil Moisture Active and Passive, Koster et al. 2016), RapidScat (Brocca et al. 2016a), AMSR (Brocca et al. 2014) and AMSR2 (Advanced Microwave Scanning Radiometer 2, Tarpanelli et al. 2017) SM products. Two of the most recent and relevant datasets are briefly illustrated below.

51.3.1   SM2RAIN-CCI

The SM2RAIN-CCI rainfall product has been developed in Ciabatta et al. (2018) and relies on the application of SM2RAIN to ESA-CCI (European Space Agency – Climate Change Initiative) SM product. The product is freely available at https://doi. org/10.5281/zenodo.846260 (last accessed 28 Oct. 2018) and is expected to be updated every year with the most recent ESA CCI SM dataset. The current version of the product is available from 1998 to 2015 (18 years), and it provides daily rainfall estimates over land sampled at 0.25-degree resolution. Ciabatta et al. (2018) have described in full details the procedure used for producing the SM2RAIN-CCI rainfall product. Moreover, the assessment of the product with global scale gauge-based rainfall products has been carried out, also in comparison with other satellite-based rainfall datasets, e.g., TMPA (Tropical Rainfall Measuring Mission Multi-satellite Precipitation Analysis) and CMORPH (Climate Prediction Center MORPHing). Results show good performance over Africa, Brazil, western US, India and Australia, in terms of both Pearson's correlation, R, and mean annual rainfall estimation.

51.3.2   SM2RAIN-ASCAT

The most recent product we have developed is based on the application of SM2RAIN to ASCAT SM (Wagner et al. 2013) product at full resolution (12.5 km sampling). In previous studies, the ASCAT SM product has been aggregated at 1° and 0.5° resolution in order to facilitate the computations. Recently, under the EUMETSAT (European Organisation for the Exploitation of Meteorological Satellites) project "Global-SM2RAIN", a full investigation of the capability of ASCAT SM product to provide rainfall estimates is being investigated. The current version of the rainfall product is available from 2007 to 2017 (11 years) at daily temporal resolution and sampled on an irregular grid with spacing of 12.5 km. The dataset is freely available and can be obtained by contacting the authors. A first assessment of the product is carried out by aggregating the data at 1-degree resolution and by comparing the data with gauge-based rainfall dataset from GPCC (Global Precipitation Climatology Center). The latter represents an independent dataset as SM2RAIN parameters have been calibrated on ERA5 reanalysis rainfall

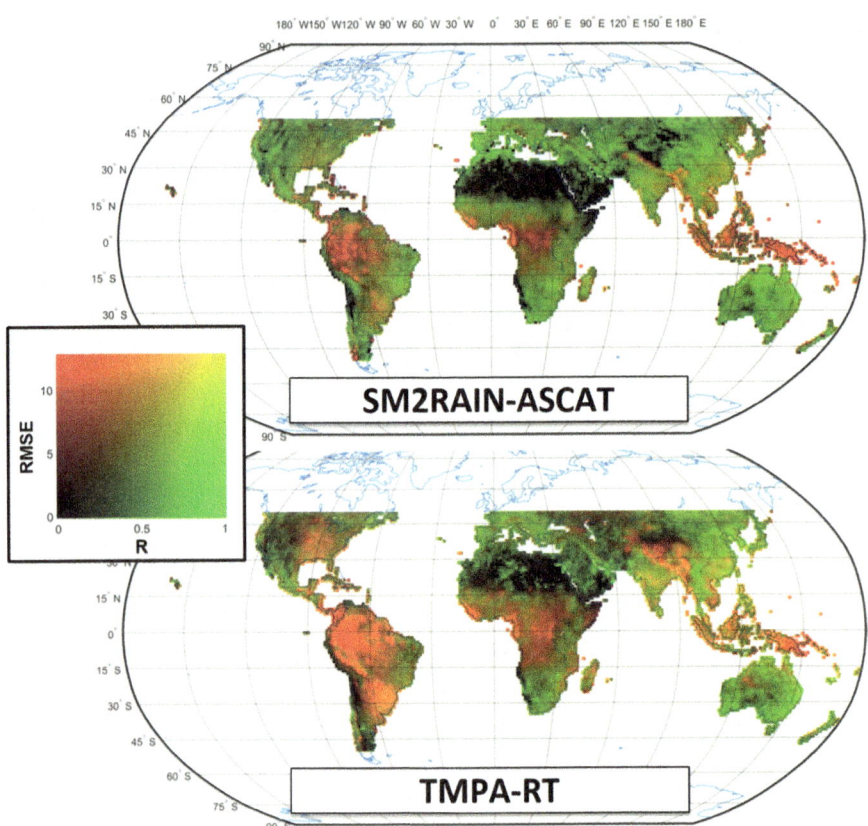

Fig. 51.2 Comparison in terms of Pearson's correlation, R, and root mean square error, RMSE, between SM2RAIN-ASCAT and TMPA-RT satellite-based rainfall products as compared with GPCC gauge-based dataset used as benchmark. The analysis is carried out at 1-day and 1° temporal spatial resolution. The maps clearly show the differences in the accuracy of the two products, with SM2RAIN-ASCAT performing well in the eastern US, Brazil, the Sahel, south-eastern Asia and Australia (green colors)

(from ECMWF). Figure 51.2 shows the results of the assessment for both SM2RAIN-ASCAT and TMPA-RT product for the period 2007–2017 at daily time scale. Maps show the results both in term of R and root mean square error, RMSE. Except tropical forests, Sahara Desert and high mountains (e.g., Himalaya), the SM2RAIN performs well with median R and RMSE equal to 0.51 and 3.89 mm day^{-1}, respectively. The accuracy of TMPA-RT is lower than SM2RAIN-ASCAT (median R/RMSE = 0.44/5.63 mm day^{-1}), particularly in eastern US, Brazil, the Sahel, south-eastern Asia and Australia. In these regions, we expect larger benefits in using SM2RAIN-ASCAT rainfall product for hydrological and climate applications.

51.4 Flood Modelling in Europe Through SM2RAIN-Derived Rainfall Products

The capability of SM2RAIN to keep track of the amount of water falling into the soil makes the derived rainfall products particularly useful for hydrological applications in which the accurate estimation of accumulated rainfall is needed, i.e. in cases where the accumulated rainfall amounts are more important than the knowledge of instantaneous rainfall rates. We have used SM2RAIN-ASCAT rainfall product, in combination with TMPA-RT and CMORPH, for flood simulations throughout Europe and the main results of this analysis are described here.

51.4.1 In Situ and Satellite Datasets

To perform a robust assessment of satellite-based rainfall products for flood modelling, a dataset of 600 basins throughout Europe has been compiled. For each basin, daily river discharge observations for a period of at least 1 year (average length of 5.6 years) between 2007 and 2013 are available. Daily precipitation and air temperature observations have been collected from in situ stations by using the E-OBS (Haylock et al. 2008) dataset. Three different satellite-based rainfall datasets have been used including: SM2RAIN-ASCAT, TMPA-RT and CMORPH (Joyce et al. 2004) real time versions. Additionally, the combination of TMPA-RT (and CMORPH) with SM2RAIN-ASCAT following the same merging approach proposed in Ciabatta et al. (2016) has been implemented. Therefore, for each basin, six river discharge simulations have been carried out, i.e., by using as precipitation input: (1) E-OBS, (2) CMORPH, (3) TMPA-RT, (4) SM2RAIN-ASCAT, (5) SM2RAIN-ASCAT+ CMORPH, and (6) SM2RAIN-ASCAT+ TMPA-RT.

51.4.2 MISDc Rainfall-Runoff Modelling

MISDc (Modello Idrologico Semi-Distribuito in continuo) is a continuous rainfall-runoff model developed by Brocca et al. (2011) for the operational forecasting of flood events in central Italy. In this chapter, a two-layer version of the model is used. With respect to the previous version, it includes a snow module and a different infiltration equation. The model uses as input daily rainfall and air temperature data and simulates the temporal evolution of river discharge and SM for a surface and a root-zone soil layer. Water is extracted from the first layer by evapotranspiration, which is calculated by a linear function between the potential evaporation and SM. A non-linear relation is used for computing the percolation from the surface to the root zone layer. The rainfall excess is calculated by a power law relationship as a function of the first layer SM while base flow is a non-linear function of the SM of the second layer. Full details on model equations are given in Brocca et al. (2011) and recent applications can be found in Camici et al. (2018) and Massari et al. (2018).

51.4.3 Results

The use of satellite-based rainfall products for flood modelling requires some pre-processing steps. We have employed the same approach as in Camici et al. (2018) in which MISDc model has been recalibrated in each simulation by maximizing the Kling-Gupta Efficiency (KGE, Gupta et al. 2009) with respect to observed daily discharge. The calibration of the merged products, i.e., SM2RAIN-ASCAT+ CMORPH and SM2RAIN-ASCAT+TMPA-RT, has been also carried out to maximize KGE. Therefore, for each rainfall product and basin, MISDc parameter values are calibrated and a validation period is not considered in this analysis. Certainly, this procedure does not allow for the evaluation of the satellite rainfall product in an operational context, but here we want to assess the best information that can be extracted from each product by using the maximum length of the available data.

The results in terms of KGE for all basins and products are shown in Fig. 51.3 as boxplots. As expected, very good performances are obtained with raingauge observations from E-OBS with a mean KGE equal to 0.77. Among the three satellite single rainfall products, SM2RAIN-ASCAT provides the best scores (mean KGE = 0.67) followed by TMPA-RT (mean KGE = 0.57), and CMORPH (mean KGE only 0.31). The percentages of basins in which SM2RAIN-ASCAT, TMPA-RT, and CMORPH provide the best KGE are equal to 77%, 21% and 4%, respectively. The integration of SM2RAIN-ASCAT with TMPA-RT and CMORPH provides improved performance close to the ones obtained with E-OBS. Specifically, for 27% (24%) of basins SM2RAIN-ASCAT+TMPA-RT (SM2RAIN-ASCAT +CMORPH) shows better results than E-OBS. It should be underlined that such satellite rainfall products are potentially available in near real-time, thus representing an important new data source for flood forecasting in Europe.

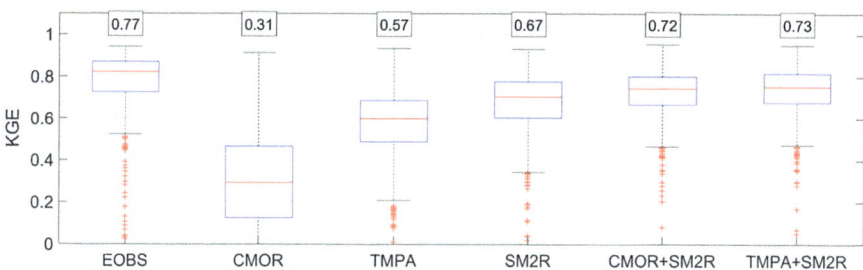

Fig. 51.3 Boxplot of KGE for the six investigated rainfall products and for the 600 basins. For each box, the red line represents the median values and the blue box represents the 25th and 75th percentile, the black dotted whiskers extend to the most extreme data points and cross symbols represent outliers. The numbers in the top boxes indicate the mean value for each rainfall product. The integration of SM2RAIN-ASCAT (SM2R) with TMPA-RT (TMPA) and CMORPH (CMOR) provides the best performance close to the ones obtained with high-quality raingauge observations (EOBS)

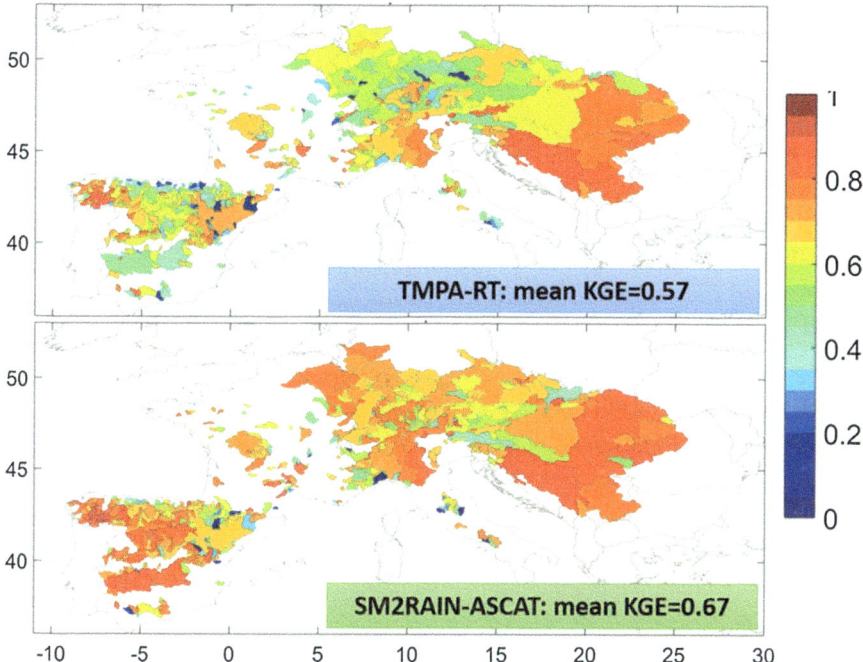

Fig. 51.4 Spatial distribution of KGE performance for TMPA-RT and SM2RAIN-ASCAT rainfall products over 600 basins in Europe. Red colours mean high KGE values and, hence, better performances. Overall, SM2RAIN-ASCAT is performing better over 77% of basins with TMPA-RT showing better results over some basins in Italy (mainly close to the western Alps), in central France and in South-Eastern Europe (Balkans)

To visualize the spatial distribution of the performances, Fig. 51.4 shows the KGE values for each basin and the two satellite products TMPA-RT and SM2RAIN-ASCAT as coloured polygons by mapping from the larger to the smaller basins (to visualize all basins and avoid overlapping). Overall, SM2RAIN-ASCAT is performing better than TMPA-RT over Central Europe and Spain while TMPA-RT is better over some basins in Italy (mainly close to the western Alps), in central France and in south-eastern Europe (Balkans).

As an example, Fig. 51.5 shows the simulation of river discharge for four randomly selected basins (Rhine at Köln, Danube at Pfelling, Duero at Almazan,

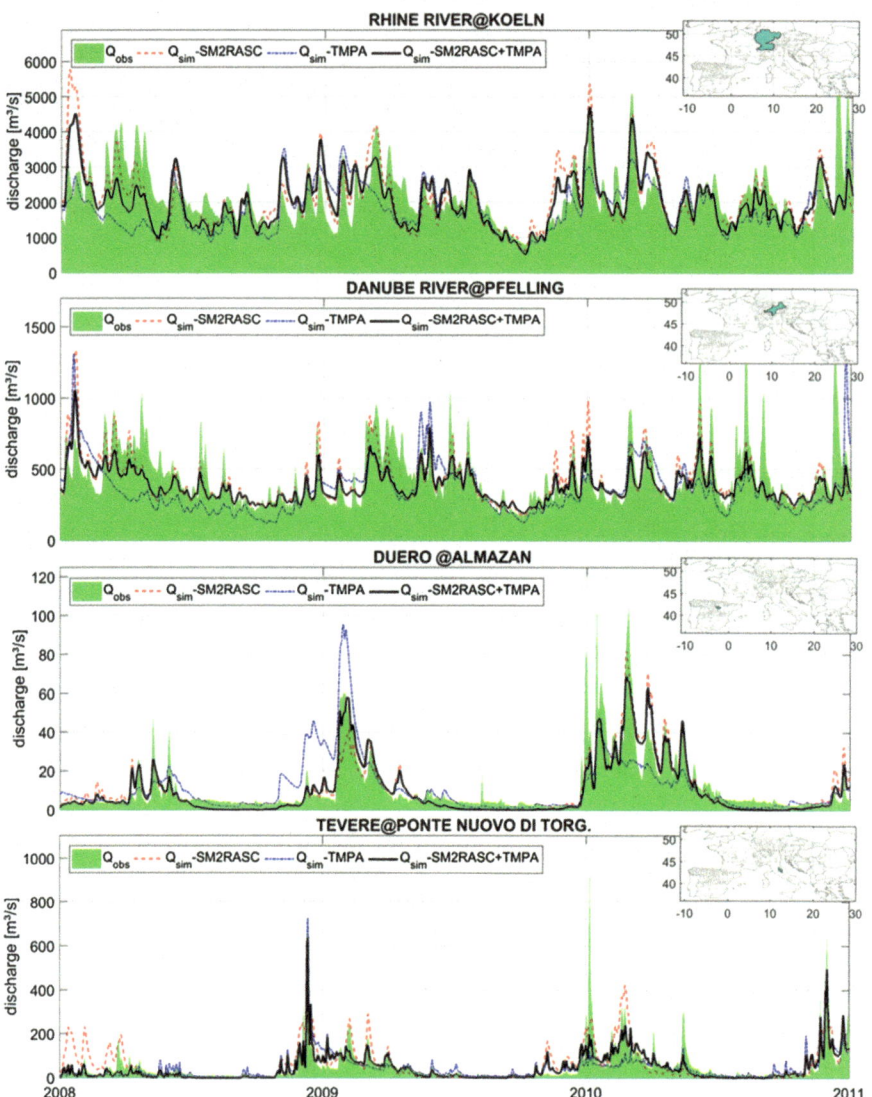

Fig. 51.5 Simulation of 3-year discharge at four basins across Europe by using SM2RAIN-ASCAT (red line), TMPA-RT (blue line) and SM2RAIN-ASCAT+TMPA-RT (black line) as input rainfall. The comparison with in situ discharge (green area) clearly underlined the benefit of integrating top-down and bottom-up approaches for flood simulation with a better reproduction of both high and low flows

Tevere at Ponte Nuovo di Torgiano) using SM2RAIN-ASCAT, TMPA-RT and TMPA-RT + SM2RAIN-ASCAT as input. The results highlight that the integration of top-down and bottom-up approaches is highly beneficial for reproducing river discharge observations in Europe, both for high and low flows. A nice example is visible for the Duero river at Almazan where the integrated product corrects the overestimation of TMPA-RT at the beginning of 2009 and the underestimation in 2010.

51.5 Limitations and Future Directions

In the previous paragraphs, and in the studies mentioned in the introduction, we have shown the good capability of SM2RAIN method to reproduce accumulated rainfall, with good performance when the rainfall product is applied in hydrological applications. However, we are also aware of the limitations of the method. Indeed, SM2RAIN may fail in reproducing rainfall when the soil is close to saturation (e.g., Brocca et al. 2013; Koster et al. 2016) and if the temporal resolution of the data is too coarse (e.g., > 2–3 days, see Brocca et al. 2016b). Moreover, the method is dependent on the accuracy of the original satellite SM dataset used as input. Therefore, high errors are expected over mountainous, urbanized and highly vegetated areas and during frozen/snow soil conditions (Brocca et al. 2014, 2017). The most important issue in using satellite surface SM data (layer depth < 5 cm) for rainfall estimation is due to the high temporal frequency fluctuations of satellite SM signals (due to noise and not to rainfall) that, when positive, are interpreted as rainfall by the SM2RAIN algorithm.

The SM2RAIN algorithm has been developed quite recently and a number of future steps is needed to fully exploit its potential for rainfall estimation. Two different kinds of development are envisaged: technological and methodological. For the technological developments, three future directions are foreseen:

1. An operational rainfall product from the SM2RAIN algorithm still has to be provided. Under the EUMETSAT H-SAF (Satellite Application Facility on Support to Operational Hydrology and Water Management, http://hsaf. meteoam.it, last accessed 28 Oct. 2018) project, an operational product integrating bottom-up and top-down approaches is being developed, and it is expected to be delivered in 2019. A comprehensive assessment of the product accuracy in time and space will be carried out, which will provide the estimation of rainfall uncertainty.
2. The integration of different satellite SM products, as initially demonstrated in Tarpanelli et al. (2017) for producing a higher quality rainfall product through SM2RAIN, is being investigated under the ESA SMOS+rainfall project.
3. The most recent satellite SM products characterized by higher spatial resolution are also being analysed to highlight the potential of obtaining high-resolution (1 km) rainfall products from space. Specifically, SM2RAIN has been applied to

the integration of ASCAT and Sentinel-1 SM product, i.e., SCATSAR, in Bauer-Marschallinger et al. (2018) for the Italian territory. Moreover, the application of SM2RAIN to NASA's Cyclone Global Navigation Satellite System (CYGNSS) SM products as obtained by Chew and Small (2018) is being investigated due to the low cost and high temporal frequency of the satellite SM product obtained through CYGNSS.

For the methodological developments, four future directions are foreseen:

1. The loss function in eq. (51.6) has been kept as simple as possible in all previous applications of SM2RAIN. However, as suggested in Koster et al. (2018), the use of a more complex loss function, taking into account of the evapotranspiration component, and fully exploiting the drying rate of SM time series, is expected to provide more robust and accurate results. More specifically, the analysis of the SM recession rate would allow the model to directly obtain the loss function and its parameterization from SM observations only, thus potentially enabling the derivation of a self-calibrated SM2RAIN formulation that does not need rainfall observations for its calibration.
2. The calibration of SM2RAIN parameter values is usually carried out pixel-by-pixel, thus not considering the space component. We want to develop more elaborated calibration procedures exploiting the expected continuity in space of rainfall and soil parameters, thus obtaining a more robust parameterization. The analysis of the temporal variability of model parameter values is also foreseen.
3. The integration between bottom-up and top-down approaches have been carried out by considering a simple linear weighting (e.g., Ciabatta et al. 2016; Brocca et al. 2016b). Even though the linear weighting has been found successful in obtaining a higher quality rainfall product, it does not consider the different error behavior of the two approaches. For instance, we are aware of the errors of SM2RAIN approach as a function of land cover and time (e.g., at saturation), and similar for the top-down approaches. We have recently developed a more elaborated merging technique that considers the time and space variability of the errors in the two approaches, under a Bayesian framework. First results have shown to be promising (Maggioni et al. 2017), and further investigations are being carried out.
4. The reduction of the noise in the SM signal is found to be mandatory to obtain reliable rainfall estimates. The current SM2RAIN approach considers the application of the exponential filter (Wagner et al. 1999), which may fail to correctly reproduce the timing of rainfall. More specific filters, such as those based on wavelet (e.g., Massari et al. 2017b), should be tested for improving the reduction of noise in SM signal and thus obtaining more accurate rainfall estimates.

For each of the points above, specific studies have been already started, or are going to be started in the future. Finally, we expect to apply SM2RAIN-derived rainfall in more applications over large areas and longer time periods to properly assess its potential. Apart from flood prediction, as shown in this chapter, SM2RAIN-derived rainfall is being applied for landslide prediction (Brunetti et al.

2018), water management (Abera et al. 2017), crop prediction (Thaler et al. 2018), and new applications such as bird migration and disease prevention.

Acknowledgements The authors gratefully acknowledge support from the EUMETSAT through the "Satellite Application Facility on Support to Operational Hydrology and Water Management (H SAF)" CDOP3 (EUM/C/85/16/DOC/15) and the Global SM2RAIN project (contract no. EUM/CO/17/4600001981/BBo), and from the ESA through SMOS+rainfall project (contract no. 4000114738/15/I-SBo) and the Climate Change Initiative, CCI (contract no. 4000104814/11/I-NB).

References

Abera, W., Formetta, G., Brocca, L., & Rigon, R. (2017). Modeling the water budget of the Upper Blue Nile basin using the JGrass-NewAge model system and satellite data. *Hydrology and Earth System Sciences, 21*, 3145–3165. https://doi.org/10.5194/hess-21-3145-2017.

Bauer-Marschallinger, B., Paulik, C., Hochstöger, S., Mistelbauer, T., Modanesi, S., Ciabatta, L., Massari, C., Brocca, L., & Wagner, W. (2018). Soil moisture from fusion of scatterometer and SAR: Closing the scale gap with temporal filtering. *Remote Sensing, 10*(7), 1030. https://doi.org/10.3390/rs10071030.

Brocca, L., Melone, F., & Moramarco, T. (2011). Distributed rainfall-runoff modelling for flood frequency estimation and flood forecasting. *Hydrological Processes, 25*(18), 2801–2813. https://doi.org/10.1002/hyp.8042.

Brocca, L., Melone, F., Moramarco, T., & Wagner, W. (2013). A new method for rainfall estimation through soil moisture observations. *Geophysical Research Letters, 40*(5), 853–858. https://doi.org/10.1002/grl.50173.

Brocca, L., Ciabatta, L., Massari, C., Moramarco, T., Hahn, S., Hasenauer, S., Kidd, R., Dorigo, W., Wagner, W., & Levizzani, V. (2014). Soil as a natural rain gauge: Estimating global rainfall from satellite soil moisture data. *Journal of Geophysical Research, 119*(9), 5128–5141. https://doi.org/10.1002/2014JD021489.

Brocca, L., Massari, C., Ciabatta, L., Moramarco, T., Penna, D., Zuecco, G., Pianezzola, L., Borga, M., Matgen, P., & Martínez-Fernández, J. (2015). Rainfall estimation from in situ soil moisture observations at several sites in Europe: An evaluation of SM2RAIN algorithm. *Journal of Hydrology and Hydromechanics, 63*(3), 201–209. https://doi.org/10.1515/johh-2015-0016.

Brocca, L., Massari, C., Ciabatta, L., Wagner, W., & Stoffelen, A. (2016a). Remote sensing of terrestrial rainfall from Ku-band scatterometers. *IEEE Journal of Selected Topics in Applied Earth Observations and Remote Sensing, 9*(1), 533–539. https://doi.org/10.1109/JSTARS.2015.2508065.

Brocca, L., Pellarin, T., Crow, W. T., Ciabatta, L., Massari, C., Ryu, D., Su, C.-H., Rudiger, C., & Kerr, Y. (2016b). Rainfall estimation by inverting SMOS soil moisture estimates: A comparison of different methods over Australia. *Journal of Geophysical Research, 121*(20), 12062–12079. https://doi.org/10.1002/2016JD025382.

Brocca, L., Crow, W. T., Ciabatta, L., Massari, C., de Rosnay, P., Enenkel, M., Hahn, S., Amarnath, G., Camici, S., Tarpanelli, A., & Wagner, W. (2017). A review of the applications of ASCAT soil moisture products. *IEEE Journal of Selected Topics in Applied Earth Observations and Remote Sensing, 10*(5), 2285–2306. https://doi.org/10.1109/JSTARS.2017.2651140.

Brunetti, M. T., Melillo, M., Peruccacci, S., Ciabatta, L., & Brocca, L. (2018). How far are we from the use of satellite rainfall products in landslide forecasting? *Remote Sensing of Environment, 210*, 65–75. https://doi.org/10.1016/j.rse.2018.03.016.

Camici, S., Ciabatta, L., Massari, C., & Brocca, L. (2018). How reliable are satellite precipitation estimates for driving hydrological models: A verification study over the Mediterranean area. *Journal of Hydrology, 563*, 950–961. https://doi.org/10.1016/j.jhydrol.2018.06.067.

Chiaravalloti, F., Brocca, L., Procopio, A., Massari, C., & Gabriele, S. (2018). Assessment of GPM and SM2RAIN-ASCAT rainfall products over complex terrain in southern Italy. *Atmospheric Research, 206*, 64–74. https://doi.org/10.1016/j.atmosres.2018.02.019.

Ciabatta, L., Brocca, L., Massari, C., Moramarco, T., Gabellani, S., Puca, S., & Wagner, W. (2016). Rainfall-runoff modelling by using SM2RAIN-derived and state-of-the-art satellite rainfall products over Italy. *International Journal of Applied Earth Observation and Geoinformation, 48*, 163–173. https://doi.org/10.1016/j.jag.2015.10.004.

Ciabatta, L., Marra, A. C., Panegrossi, G., Casella, D., Sanò, P., Dietrich, S., Massari, C., & Brocca, L. (2017). Daily precipitation estimation through different microwave sensors: Verification study over Italy. *Journal of Hydrology, 545*, 436–450. https://doi.org/10.1016/j.jhydrol.2016.12.057.

Ciabatta, L., Massari, C., Brocca, L., Gruber, A., Reimer, C., Hahn, S., Paulik, C., Dorigo, W., Kidd, R., & Wagner, W. (2018). SM2RAIN-CCI: A new global long-term rainfall data set derived from ESA CCI soil moisture. *Earth System Science Data, 10*, 267–280. https://doi.org/10.5194/essd-10-267-2018.

Chew, C. C., & Small, E. E. (2018). Soil moisture sensing using spaceborne GNSS reflections: Comparison of CYGNSS reflectivity to SMAP soil moisture. *Geophysical Research Letters, 45* (9), 4049–4057. https://doi.org/10.1029/2018GL077905.

Crow, W. T., Huffman, G. J., Bindlish, R., & Jackson, T. J. (2009). Improving satellite-based rainfall accumulation estimates using spaceborne surface soil moisture retrievals. *Journal of Hydrometeorology, 10*(1), 199–212. https://doi.org/10.1175/2008JHM986.1.

Crow, W. T., van Den Berg, M. J., Huffman, G. J., & Pellarin, T. (2011). Correcting rainfall using satellite-based surface soil moisture retrievals: The Soil Moisture Analysis Rainfall Tool (SMART). *Water Resources Research, 47*(8), W08521. https://doi.org/10.1029/2011WR010576.

Doorenbos, J., & Pruitt, W. O. (1977). *Background and development of methods to predict reference crop evapotranspiration (ETo).* FAO Irrigation and Drainage Paper No. 24, 108–119. Available at http://www.fao.org/3/a-f2430e.pdf, last accessed 29 Nov 2018.

Gupta, H. V., Kling, H., Yilmaz, K. K., & Martinez, G. F. (2009). Decomposition of the mean squared error and NSE performance criteria: Implications for improving hydrological modelling. *Journal of Hydrology, 377*(1), 80–91. https://doi.org/10.1016/j.jhydrol.2009.08.003.

Haylock, M. R., Hofstra, N., Klein Tank, A. M. G., Klok, E. J., Jones, P. D., & New, M. (2008). A European daily high-resolution gridded data set of surface temperature and precipitation for 1950-2006. *Journal of Geophysical Research, 113*(D20). https://doi.org/10.1029/2008JD010201.

Hou, A. Y., Kakar, R. K., Neeck, S., Azarbarzin, A. A., Kummerow, C. D., Kojima, M., Oki, R., Nakamura, K., & Iguchi, T. (2014). The global precipitation measurement mission. *Bulletin of the American Meteorological Society, 95*, 701–722. https://doi.org/10.1175/BAMS-D-13-00164.1.

Huffman, G. J., Bolvin, D. T., Nelkin, E. J., Wolff, D. B., Adler, R. F., Gu, G., Hong, Y., Bowman, K. P., & Stocker, E. F. (2007). The TRMM Multisatellite Precipitation Analysis (TMPA): Quasi-global, multiyear, combined-sensor precipitation estimates at fine scales. *Journal of Hydrometeorology, 8*(1), 38–55. https://doi.org/10.1175/JHM560.1.

Joyce, R. J., Janowiak, J. E., Arkin, P. A., & Xie, P. (2004). CMORPH: A method that produces global precipitation estimates from passive microwave and infrared data at high spatial and temporal resolution. *Journal of Hydrometeorology, 5*, 487–503. https://doi.org/10.1175/1525-7541(2004)005<0487:CAMTPG>2.0.CO;2.

Koster, R. D., Brocca, L., Crow, W. T., Burgin, M. S., & De Lannoy, G. J. M. (2016). Precipitation estimation using L-band and C-band soil moisture retrievals. *Water Resources Research, 52*(9), 7213–7225. https://doi.org/10.1002/2016WR019024.

Koster, R. D., Crow, W. T., Reichle, R. H., & Mahanama, S. P. (2018). Estimating basin-scale water budgets with SMAP soil moisture data. *Water Resources Research, 54*, 4228–4244. https://doi.org/10.1029/2018WR022669.

Maggioni, V., Massari, C., Barbetta, S., Camici, S., & Brocca, L. (2017). Merging satellite precipitation products for improved streamflow simulations. *AGU Fall Meeting 2017*, abstract #H54F-03.

Massari, C., Brocca, L., Moramarco, T., Tramblay, Y., & Didon Lescot, J.-F. (2014). Potential of soil moisture observations in flood modelling: Estimating initial conditions and correcting rainfall. *Advances in Water Resources, 74*, 44–53. https://doi.org/10.1016/j.advwatres.2014.08.004.

Massari, C., Crow, W. T., & Brocca, L. (2017a). An assessment of the accuracy of global rainfall estimates without ground-based observations. *Hydrology and Earth System Sciences, 21*, 4347–4361. https://doi.org/10.5194/hess-21-4347-2017.

Massari, C., Su, C.-H., Brocca, L., Sang, Y. F., Ciabatta, L., Ryu, D., & Wagner, W. (2017b). Near real time de-noising of satellite-based soil moisture retrievals: An intercomparison among three different techniques. *Remote Sensing of Environment, 198*, 17–29. https://doi.org/10.1016/j.rse.2017.05.037.

Massari, C., Camici, S., Ciabatta, L., & Brocca, L. (2018). Exploiting satellite-based surface soil moisture for flood forecasting in the Mediterranean area: State update versus rainfall correction. *Remote Sensing, 10*(2), 292. https://doi.org/10.3390/rs10020292.

Pellarin, T., Ali, A., Chopin, F., Jobard, I., & Bergès, J. C. (2008). Using spaceborne surface soil moisture to constrain satellite precipitation estimates over West Africa. *Geophysical Research Letters, 35*(2). https://doi.org/10.1029/2007GL032243.

Pellarin, T., Louvet, S., Gruhier, C., Quantin, G., & Legout, C. (2013). A simple and effective method for correcting soil moisture and precipitation estimates using AMSR-E measurements. *Remote Sensing of Environment, 136*, 28–36. https://doi.org/10.1016/j.rse.2013.04.011.

Román-Cascón, C., Pellarin, T., Gibon, F., Brocca, L., Cosme, E., Crow, W. T., Fernández, D., Kerr, Y., & Massari, C. (2017). Correcting satellite-based precipitation products through SMOS soil moisture data assimilation in two land-surface models of different complexity: API and SURFEX. *Remote Sensing of Environment, 200*, 295–310. https://doi.org/10.1016/j.rse.2017.08.022.

Tarpanelli, A., Massari, C., Ciabatta, L., Filippucci, P., Amarnath, G., & Brocca, L. (2017). Exploiting a constellation of satellite soil moisture sensors for accurate rainfall estimation. *Advances in Water Resources, 108*, 249–255. https://doi.org/10.1016/j.advwatres.2017.08.010.

Thaler, S., Brocca, L., Ciabatta, L., Eitzinger, J., Hahn, S., & Wagner, W. (2018). Effects of different spatial precipitation input data on crop model outputs under a Central European climate. *Atmosphere, 9*(8), 290. https://doi.org/10.3390/atmos9080290.

Tian, Y., Liu, Y., Arsenault, K. R., & Behrangi, A. (2014). A new approach to satellite-based estimation of precipitation over snow cover. *International Journal of Remote Sensing, 35*(13), 4940–4951. https://doi.org/10.1080/01431161.2014.930208.

Wagner, W., Lemoine, G., & Rott, H. (1999). A method for estimating soil moisture from ERS scatterometer and soil data. *Remote Sensing of Environment, 70*(2), 191–207. https://doi.org/10.1016/S0034-4257(99)00036-X.

Wagner, W., Hahn, S., Kidd, R., Melzer, T., Bartalis, Z., Hasenauer, S., Figa-Saldaña, J., de Rosnay, P., Jann, A., Schneider, S., Komma, J., Kubu, G., Brugger, K., Aubrecht, C., Züger, J., Gangkofner, U., Kienberger, S., Brocca, L., Wang, Y., Blöschl, G., Eitzinger, J., Steinnocher, K., Zeil, P., & Rubel, F. (2013). The ASCAT soil moisture product: A review of its specifications, validation results, and emerging applications. *Meteorologische Zeitschrift, 22*(1), 5–33. https://doi.org/10.1127/0941-2948/2013/0399.

Wanders, N., Pan, M., & Wood, E. F. (2015). Correction of real-time satellite precipitation with multi-sensor satellite observations of land surface variables. *Remote Sensing of Environment, 160*, 206–221. https://doi.org/10.1016/j.rse.2015.01.016.

Chapter 52
Drought Risk Management Using Satellite-Based Rainfall Estimates

Elena Tarnavsky and Rogerio Bonifacio

Abstract In this chapter, we present an overview of the role of satellite-based rainfall estimates (SREs) in drought risk management applications, ranging from simple anomaly and index-based approaches to cross-cutting drought early warning systems (EWS) and financial instruments such as weather index-based insurance (WII) schemes. We contend that meteorological, hydrological, agricultural, and socioeconomic are aspects – not types – of drought, and a universally acceptable drought definition is not a prerequisite for the effective and efficient assessment of the impacts of drought using SREs and other satellite-based datasets and/or models. This is illustrated through examples from the work of the co-authors, as well as the wider community. The chapter concludes with a synthesis of the challenges for SREs and the current trends in the development and application of SREs in drought risk management, including an outlook of the priorities for future research and applications.

Keywords Precipitation · Satellite rainfall estimates · Drought · Early warning system · Weather index-based insurance · Soil water deficit · Water stress · Famine · Agriculture · APHRODITE · ARC2 · CHIRPS · EPSAT-SG · GPI · GPROF · GSMaP · MERRA · PERSIANN · RFE2.0 · TAMSAT · TMPA · Palmer drought severity index · Standardized precipitation index · Water requirements satisfaction index · DSSAT · MARSOP · JRC · World Food Programme · Seasonal performance probability · Soil moisture · NDVI · Scaled drought condition index · Enhanced combined drought index

E. Tarnavsky (✉)
Department of Meteorology, University of Reading, Reading, UK
e-mail: e.tarnavsky@reading.ac.uk

R. Bonifacio
World Food Programme, Vulnerability Assessment and Mapping Unit, Rome, Italy

52.1 Introduction

Droughts occur under all climates and are generally associated with dry weather conditions – below-average precipitation and sustained high temperatures – that persist over a sufficiently long time period to impact on human activities. Unlike other weather-related disasters such as hurricanes, which are methodically identified and classified in terms of wind speeds, droughts have proved more difficult to define, as they are triggered by different factors under different climates and develop on a range of scales in space (localised or regional to quasi-continental) and time (from several days and intra-seasonal to inter-seasonal and multi-annual).

In terms of meteorology, droughts are most directly linked with precipitation, as rainfall and snowmelt are the main water inputs in the hydrological cycle. From a hydrological perspective, droughts and floods are the respective extremes of rainfall deficit and excess, while in agricultural terms, droughts can lead to food production shortage or wholesale crop failure, causing famine. Depending on the area of human activities impacted, droughts can trigger water, food, and/or energy crises, causing substantial economic and human life losses. These characteristics are reflected in the World Meteorology Organization's definitions of drought[1] and many of the over 150 definitions of drought reviewed by Wilhite and Glantz (1985). Despite this, a uniform drought definition has not emerged as a practical solution (Lloyd-Hughes 2014; Dinku et al. 2015) and some have identified this as an impediment to drought monitoring (Heim 2002; Enenkel et al. 2015). Functionally, however, and in relation to their impact on human activities, we contend that four key aspects – not types – of drought have emerged and have been successfully assessed with the use of satellite-based rainfall estimates (SREs) in drought risk management tools; these are:

- Meteorological (rainfall deficit, dry spells) specific to different regions/climates,
- Hydrological (water supply shortage, reservoir storage depletion) specific to different regions and catchment characteristics,
- Agricultural (soil water deficit, crop-yield shortfall) specific to different agro-climatic regions and crops, and
- Socio-economic (food shortages, famine) specific to different livelihood systems and policy and governance settings.

While the first three aspects are linked to the physical processes involved in drought, socio-economic drought aspects are focused on the two-way evaluation of

---

[1]Meteorological drought: "(1) Prolonged absence or marked deficiency of precipitation, (2) Period of abnormally dry weather sufficiently prolonged for the lack of precipitation to cause a serious hydrological imbalance"; Hydrological drought: "Period of abnormally dry weather sufficiently prolonged to give rise to a shortage of water as evidenced by below normal streamflow and lake levels and/or the depletion of soil moisture and a lowering of groundwater levels"; Agricultural drought: "...when rainfall amounts and distribution, soil water reserves and evaporation losses combine to cause crop or livestock yields to diminish markedly" (WMO 1992).

the balance between supply and demand of water, energy, and food resources, including qualitative and/or quantitative impact-based analyses under different socio-economic scenarios.

Here, we provide an overview of drought risk management approaches with respect to the role of SREs in addressing the four aspects of drought. First, we summarise the evolution of rainfall observation and estimation for drought monitoring, as well as the strengths and weaknesses of SRE products used in drought assessments, including evaluation and inter-comparisons of different SREs. This is followed by an outline of the progress with existing drought risk management approaches ranging from simple rainfall-based anomalies and indices to modelling, and cross-cutting approaches such as early warning system (EWS) analytical tools and weather-based insurance instruments (WII). The chapter concludes with a synthesis of trends drawn from research and practice, a summary of the challenges for SREs and drought risk management applications, and a roadmap for future priorities in this area.

52.2   Rainfall Monitoring for Drought Assessment

Rainfall affects a wide range of human activities and understanding its role in creating drought conditions requires detailed records over space and time. Due to technological advances that have defined rainfall monitoring over the years, data from ground-based rainfall observations and satellite-based rainfall estimates (SREs) have been the main input for many drought-related decision-support applications that necessarily include data on other variables related to the water cycle (e.g., runoff, soil moisture, reservoir storage, and crop water stress).

Since meteorological, hydrological, and agricultural drought aspects – all linked to rainfall deficit – are region-specific, monitoring rainfall entails collating quality information on both rainfall occurrence and amount in real time and over the long term, as well as across a wide range of space-time scales (local to global and sub-daily to monthly). Moreover, rainfall monitoring data need to provide comparable and reliable information on rainfall over space and time that is accessible at low operational cost. This is because errors in the input data impact the estimation of hydrological fluxes and agricultural productivity, and subsequently can compromise decision-making information. However, rainfall is difficult to measure due to its high spatial and temporal variability. Here, we trace the evolution of rainfall observation technology and common approaches to rainfall product evaluation.

52.2.1 Evolution of Rainfall Observation Technology: From Rainfall Observations to Rainfall Estimation

Since the nineteenth century, rainfall observations have been collected at ground-based synoptic stations equipped with rain gauges, albeit not continuously over time across the globe. Ground-based radar, invented in the nineteenth century, has been in use for operational rainfall monitoring since 1940s; however, it is expensive to deploy and requires substantial technical expertise to operate. Substantial progress in rainfall monitoring has been made with the launch of a series of geostationary (GOES in 1970s and Meteosat in 1980s) and polar-orbiting (NOAA in 1970s, TRMM in 1997, QuickScat in 1999, Metop-A in 2006, and GPM5 in 2014) meteorological satellites. The advantages and disadvantages of gauge-based rainfall observations and satellite-based rainfall estimates are summarised in Tables 52.1 and 52.2, respectively. The key strengths of SREs are their spatial and temporal continuity. This is further enhanced by diverse satellite-based observations on other drought-relevant variables such as vegetation, soil moisture, and surface and groundwater storage, allowing for the joint monitoring of the meteorological, hydrological, and agricultural aspects of drought.

Although SREs are widely used for a wide range of drought monitoring applications, ground-based rainfall observations have not been made obsolete for two reasons. Firstly, rain gauge data are required for indirect approaches of rainfall

Table 52.1 Advantages and disadvantages of gauge-based rainfall observations

Advantages	Disadvantages
Comparability Rainfall amounts can be compared over space and time with standardised gauges or the use of correction factors for differing gauge designs.	**Point-based measurement** Rain gauges measure rainfall at a point and reliable aerial observations are difficult to obtain as rainfall can vary greatly over short distances.
Long-term observations In many places observations date back to over 100 years.	**Gauge distribution** Gauges are read daily and thus, located in easily accessible places in cities and along roads, very sparse in remote agricultural areas; gauge distribution may not be representative of regionally-variable rainfall.
Simplicity Inexpensive, simple to manufacture and use, and easy to maintain instruments.	**Measurement error** Observers report amount of rainfall into the gauge, which can differ substantially from actual rainfall (e.g., wind losses, evaporation, side-wetting, splashing, inappropriate setting, etc); room for error in the transcription of gauge measurements.
	Time delay Typically read at 6:00 a.m. for previous day's observation, but weeks to months can elapse before data are assembled into useful datasets; telemetered gauges are increasingly affordable, but still expensive to install in many locations, more difficult to maintain, and prone to signal failures and vandalism.

Table 52.2 Advantages and disadvantages of satellite-based rainfall estimates

Advantages	Disadvantages
Comparability Rainfall amounts can be compared over space and time, provided a consistent approach is used for rainfall retrieval.	**Reliability** The approach for rainfall retrieval is indirect and thus, estimates may not be equally reliable in all cases and over all areas; hence, reliability is contingent on the interpretation of a skilled specialist and gauge data input for calibration.
Long-term observations Satellites have now been in orbit for over 35 years.	**Calibration** Rainfall estimation is reliant on calibration against ground-based observations and thus, satellite approaches still require incorporation of observations from rain gauges.
Real time availability Imagery are available in near-real time, typically within an hour or so from acquisition.	
Reliable area estimates Images cover large areas and represent areal rainfall better than gauge observations.	
Cost Imagery from a wide range of satelliteborne instruments are available to national meteorological agencies and other non-profit users free of charge.	

retrieval from satellite-based earth observations. Albeit somewhat inconsistently, the key point is that SREs serve as a critical data input in areas of sparse rain gauge networks, while they are not independent of gauge data and their performance is often assessed as poorer over sparsely-gauged regions. Secondly, most applications rely on a range of different methods of rainfall observation that work in conjunction in order to inform stakeholders on different aspects of rainfall. Last but not least, gauge observations are used for evaluation and inter-comparison of SREs.

52.2.2   Evaluations and Inter-comparisons of Satellite-Based Rainfall Estimates: Choosing a Rainfall Product

The main SREs for drought assessment include products at sub-daily, daily, pentadal (5-daily), dekadal (approx. 10-daily), monthly, and/or seasonal products time scales from the following sources: APHRODITE (Yatagai et al. 2012), ARC2 (Novella and Thiaw 2013), CHIRP(S) (Funk et al. 2014), EPSAT-SG (Bergés et al. 2010), FAO-RFE algorithm (Dinku et al. 2015), GPI (Arkin and Meisner 1987), GPROF 6.0 (Kummerow et al. 2001), GSMaP (Ushio et al. 2009), MERRA (Rienecker et al. 2011), PERSIANN (Nguyen et al. 2018), RFE2.0 (NOAA 2001), TAMSAT (Maidment et al. 2017), TRMM 3B43 and 3B42 (a.k.a. TMPA) products (Huffman et al. 2007), and the new IMERG product (see Chap. 19 of volume 1). Many of these products are used for assessment of both floods and droughts (Grimes et al. 1999;

Toté et al. 2015). Dekadal time step products have been particularly useful for agro-meteorological monitoring and modelling, as a daily time step makes modelling data-intensive without a proportional gain in information, while a monthly time step fails to capture important vegetation growth stages (Verdin and Klaver 2002). As defined by the World Meteorological Organisation (WMO), dekad 1 in each month corresponds to first to the tenth inclusive, dekad 2 – from the 11th to the 20th inclusive, and dekad 3 covers the remaining 8–11 days depending on the month (WMO 1992).

The general aim and key contribution of studies that evaluate SRE products relative to ground-based observations and/or through inter-comparison with other SREs is to provide users (decision-makers) with the information required to make an informed choice of optimal SRE product(s) for the intended application, including drought risk management. Comprehensive reviews of evaluation and inter-comparison studies over Africa are provided elsewhere (see Table 5 in Awange et al. 2015). Here, we highlight studies that assess the quality of SREs from different sources in a novel way and we present a collation of relevant evaluation and inter-comparison studies for further reference (Table 52.3). For example, errors in both gauge observations and satellite-based estimates were perhaps first considered by accounting for uncertainty in gauge observations using block kriging (Grimes et al. 1999). Later, uncertainty of SREs was assessed through the bias and the root mean square error of linear regression as part of testing and evaluating a new merging algorithm over West Africa (Roca et al. 2010).

Table 52.3 Selected evaluation and inter-comparison studies for satellite-based rainfall estimates (SREs)

Region/Country	SRE Products
Africa (Thiemig et al. 2012)	CMORPH, GPROF6.0, PERSIANN, RFE2.0, TMPA[a]
Africa (Awange et al. 2015)	ARC2, CMORPH, GSMaP, PERSIANN, TAMSAT, TMPA[a]
Africa (Serrat-Capdevila et al. 2016)	CMORPH, PERSIANN, TMPA
East Africa (Kimani et al. 2017)	CHIRPS, CMAP, CMORPH, GPCP, PERSIANN-CDR, TAMSAT, TRMM 3B43
Ghana (Amekudzi et al. 2016)	RFE2.0, TRMM
Indonesia (Vernimmen et al. 2012)	CMORPH, PERSIANN, TMPA
Iran (Katiraie-Boroujerdi et al. 2013)	APHDRODITE, MERRA
Mozambique (Toté et al. 2015)	CHIRPS, RFE2.0, TAMSAT
Nile Basin (Hessels 2015)	CHIRPS, CMORPH, TRMM 3B43
Uganda (Maidment et al. 2013)	GPCP-1DD, RFE 2.0, TAMSAT
Uganda (Diem et al. 2014)	ARC2, RFE2.0, TMPA[a]
USA (AghaKouchak et al. 2012)	CMORPH, PERSIANN, TMPA
West Africa (Lamptey 2008)	GPCP
West Africa (Roca et al. 2010)	EPSAT-SG, GSMaP, TMPA
West Africa (Jobard et al. 2011)	EPSAT-SG, TMPA[a], CMORPH, PERSIANN, GPI

[a]Original study refers to TRMM 3B42, TMPA is used for consistency here

The key challenges for evaluation and inter-comparison investigations have been: (a) declining number of reporting stations, especially across the African continent (Awange et al. 2015) where only about 25% of automated stations actually report in any given dekad (Tarnavsky et al. 2014), (b) poor quality and temporal extent of gauge data records, including the inherent uncertainty in ground-based rainfall observations, particularly for representing aerial rainfall, (c) the fundamental differences between ground-based observations and SREs that make evaluations and inter-comparisons challenging, and (d) lack of a systematic evaluation framework and tools to enable users/decision-makers to assess the optimal SRE products for the intended applications.

52.3   Drought Risk Management

Drought risk management applications include rainfall-based monitoring, hydrological modelling, crop (yield) monitoring and forecasting, as well as early warning systems (EWS) and risk management instruments such as weather index-based insurance (WII). These applications use one or more of the approaches illustrated in Fig. 52.1, i.e. simple satellite-based rainfall estimates summarised into anomalies or indices such as the Standardized Precipitation Index (SPI) or as part of multivariable anomalies/indices and together with water balance modelling into indices such as the Palmer Drought Severity Index (PDSI); hydrological drought indices such as the runoff ratio and streamflow index or dynamic models such as DryMOD (Tarnavsky et al. 2013); and a broad family of crop models, including simplified water stress models such as the Water Requirements Satisfaction Index (WRSI) and

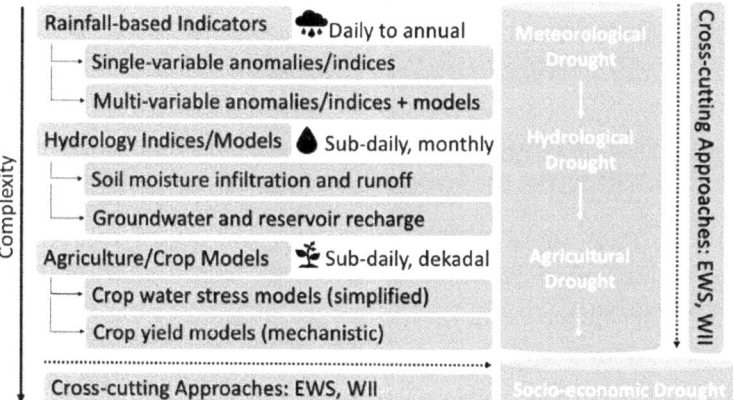

Fig. 52.1 Approaches to drought risk management using satellite-based rainfall estimates (*EWS* Early Warning Systems, *WII* Weather Index-based Insurance). Complexity refers to both the complexity of the modelling approach and its operational application over large areas at high spatial and temporal timescales

detailed mechanistic yield models such as the Decision Support System for Agro-technology Transfer (DSSAT). These approaches address respectively the meteoro-logical, hydrological, and agricultural aspects of drought. In contrast, cross-cutting applications such as EWS and WII are focused on impact analyses using either simple approaches such as rainfall anomalies or more complex methods based on the integration of more than one approach as illustrated in Fig. 52.1 and depending on the specific objectives and space-time scales. This is because EWS and WII appli-cations involve impact-based analyses that requires detailed understanding of socio-economic drought conditions, particularly in developing countries.

SREs are the first step toward drought assessment but also, an integral part of many approaches, and are the most variable input in both space and time. SREs are most commonly used to (i) identify drought risk zones, (ii) monitor drought condi-tions, and (iii) forecast drought events and their impacts. Generally, the first category involves climatological data analyses, the second category uses (near-)real time data, and the third category integrates past and present meteorological, hydrological, agricultural, and socio-economic data to develop future forecasts. All three catego-ries use the approaches summarised in Fig. 52.1.

Drought impacts on a range of human activities through diminished agricultural productivity, reduced potable water supply and hydropower capacity, as well as subsequently affecting economic activities/outputs and in extreme cases, causing famine, severe impacts on food security, and increased potential for civil unrest. The frequency, intensity, and duration of droughts is projected to increase under chang-ing climate conditions (Pozzi et al. 2013). This has promoted an increased interest in transitioning away from reactive responses to more proactive, anticipatory risk management approaches such as drought early warning systems that provide an associated impact analyses in terms of populations affected and costs of recovery and financial instruments for risk transfer such as weather index-based insurance.

52.3.1   Rainfall-Based Drought Indicators

Here, we provide an overview of rainfall-based and multi-dimensional indicators of drought, including studies that benchmark drought indicators, while comprehensive reviews are provided elsewhere (e.g., Smakhtin and Hughes 2004; Heim 2002; Zargar et al. 2011). While the first drought indicators were developed before the satellite era and relied on synoptic station data, our summary is focused on SRE-based indicators.

52.3.1.1   Single-Variable Indicators: Rainfall Anomalies/Indices

Rainfall anomalies, the difference between rainfall for a given time period relative to a long-term average or climatology, have been widely used to assess the magnitude of rainfall deficit under drought conditions as part of routine monitoring or to assess the extent and severity of drought in event-specific analyses. For example, the

Monitoring Agricultural ResourceS Operationally (MARSOP) programme of the European Commission's Joint Research Centre (JRC) in Ispra, Italy, made use of rainfall anomalies from the TAMSAT dataset operationally produced at the dekadal, monthly, and seasonal time steps (Tarnavsky et al. 2014). The World Food Programme's VAM Seasonal Explorer (available at http://dataviz.vam.wfp.org/Agroclimatic_Charts, last accessed 25 Nov. 2018) regularly provides an assessment of rainfall from the CHIRPS dataset relative to long-term climatology (see example in Fig. 52.2) with plots and data available at national and sub-national level. In a detailed, event-specific analysis of the Greater Horn of Africa drought of 2010/2011,

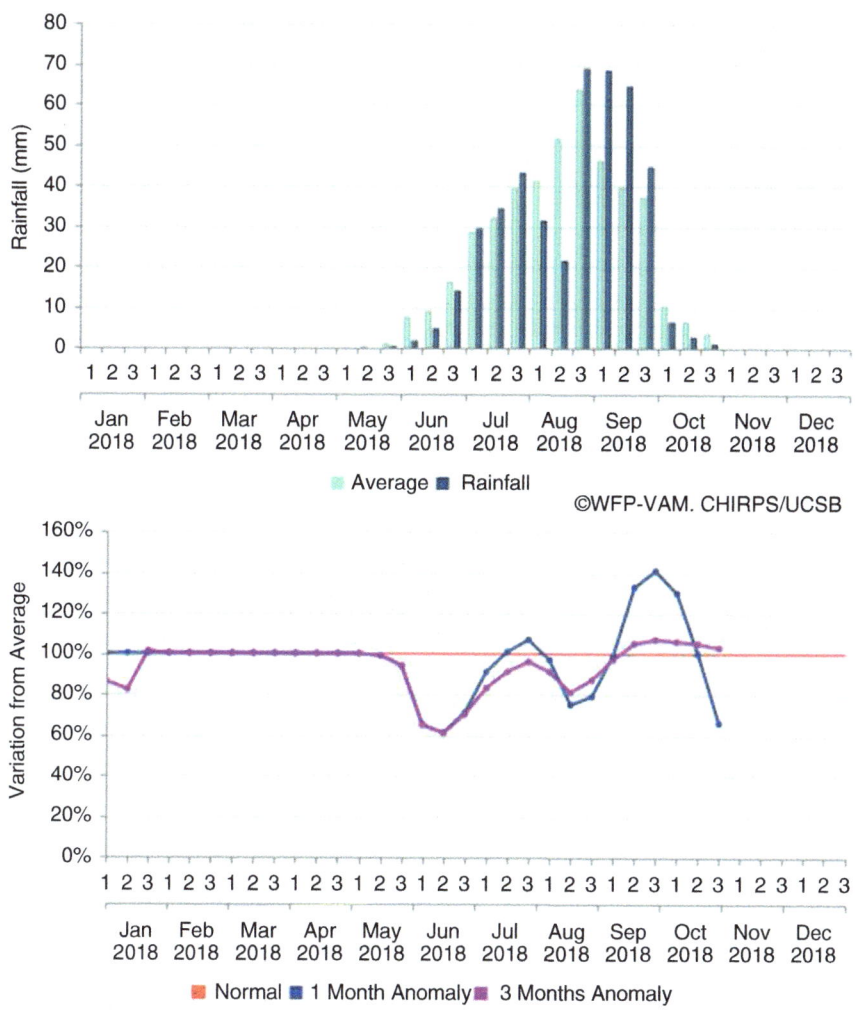

Fig. 52.2 Example of rainfall anomalies from the World Food Programme Vulnerability Assessment and Mapping (VAM) unit's Seasonal Explorer for the Matam region in Senegal

rainfall anomalies derived from **TRMM** 3B43 helped to illustrate that rainfall was at least 50–75% below average across approximately half of the drought-affected region (Nicholson 2014).

The Standardised Precipitation Index (SPI) is widely used to calculate a probability distribution of rainfall that is normalised to resolve wet and dry areas over time at multiple temporal ranges, i.e. 3 months, 6 months, etc. (Lessel et al. 2016; Senay et al. 2015). In the shorter term, SPI is related to soil moisture deficit, while in the long-run it provides an indication of groundwater and reservoir storage levels, although it does not account for changes in evapotranspiration.

In addition to SPI, the drought early warning system of the USAID Famine Early Warning Systems NETwork (FEWS NET) provides key rainfall-based metrics routinely calculated from SREs since the mid-1990s (Senay et al. 2015; Tadesse et al. 2008):

- Start of season (SOS) maps, a.k.a. rainfall onset maps,
- SOS anomalies,
- Dekadal rainfall anomalies,
- End of season (EOS) maps, a.k.a. rainfall cessation maps, and
- Short-term dryness indicators: number of rainy and non-rainy days over the previous 30 days, consecutive dry days, and days since a rain event over a specified time period (usually the previous 30 days).

In terms of probabilistic approaches for seasonal rainfall estimation, SPI has been integrated with short-range forecasts merged with long-term rainfall climatology (Dutra et al. 2014) or used as driving data input to the Seasonal Performance Probability (SPP) tool (Novella and Thiaw 2016). The key advantages of such approaches, particularly in terms of operational implementation, are the near-real time availability of the probabilistic SPI that is also stable (as it does not depend directly on gauge data input or a single SRE product) and the uncertainty estimated provided through the ensemble spread. The key contribution of rainfall-based indicators has been that they place rainfall deficit in region-specific context by exploiting climatological and near-real time records, thus enabling the characterisation of the meteorological aspects of drought, as well as its extent and severity.

52.3.1.2   Multi-variable Indicators: Anomalies/Indices and Models

Multi-variable indicators integrate rainfall input with other information such as land surface temperature (LST), vegetation indices (VI), soil moisture (SM) and/or water-balance modelling. For example, the Palmer Drought Severity Index (PDSI), an indicator of relative dryness used in the US (Palmer 1965), is based on rainfall and temperature data combined with a water balance model to calculate evapotranspiration, soil moisture, and runoff, as indicators of meteorological and hydrological drought. Since drought assessment with PDSI is carried out with a monthly time step, it has proved particularly suitable for detecting long-term droughts.

Other indicators blend SREs with VIs for example, to attribute changes in vegetation greenness to rainfall patterns over agricultural areas in Senegal (Li et al. 2004) or to support a new conceptual framework for assessing weather-related trends in rainfall and vegetation over Sub-Saharan Africa (Hoscilo et al. 2014). The Scaled Drought Condition Index (SDCI), a weighted combination of LST, the normalised difference vegetation index (NDVI), and TRMM 3B43 rainfall datasets at 1-km spatial resolution, generally showed agreement with the United States Drought Monitor (USDM) over arid and humid regions in the US, outperforming drought indicators based on VIs, the PDSI, 3- and 6-months SPI, and the Z-index (Rhee et al. 2010). In an evaluation over Uruguay, the Drought Severity Index (DSI), a modification of SDCI with climatological anomalies, calculated from TRMM and CMORPH SREs, and satellite-based LST and VI data showed very good correlations in the warm months (Oct-Mar), although it underperformed for the colder months (Apr-Sep) due to higher rainfall variability in the colder season (Lessel et al. 2016).

Commonly used drought indices such as SPI, the Standardized Precipitation Evapotranspiration Index (SPEI), and Soil Moisture Anomalies (SMA), calculated from TRMM 3B43, GPCP, and CMAP SREs, were assessed over four key river basins in Africa (Naumann et al. 2014). While for rainy season onset and the timing of drought and recovery all datasets showed good agreement, for the magnitude of the rainy season and extreme events, disagreements were noted, especially in the extent of the drought-affected area. TRMM 3B43, which combines TRMM with in situ data, was recommended for operational applications to drought monitoring due to its higher spatial resolution, noting that the main source of uncertainty in the drought index calculated is the uncertainty in rainfall.

SREs combined with SM, LST, and VI data into a new Enhanced Combined Drought Index (ECDI, Enenkel et al. 2016) helped identify droughts in Ethiopia, particularly in terms of anomalies related to the onset and end of the crop growing season. ECDI outperformed both SPEI and the self-calibrated PDSI (scPDSI) in terms of spatial and temporal resolution, although it correlated better with the SPEI than with the scPDSI. Taking into account satellite-based soil moisture in agricultural drought modelling is advantageous as it closes the gap between atmospheric process and land surface response by soil and vegetation, while rainfall is not accessible to plants due to runoff and/or increased evaporation rates.

The key strengths of multi-variable drought indicators are that they enable accounting for the linkages with land surface conditions such as temperature, vegetation, and soil moisture and their responses to rainfall deficit of varying extent and magnitude.

52.3.2   Hydrological Models for Drought Assessment

With regard to hydrological extremes, information on rainfall allows stakeholders to predict river flows, plan hydropower generation, mitigate against floods/droughts, plan water harvesting activities, anticipate shortfalls in the availability of fresh water, and estimate ground-water recharge. Relevant to freshwater resources, soil erosion, sedimentation, and crusting are heavy rainfall events in areas of poor soil quality, lack of vegetation cover, and steep topography, where rainfall can cause erosion, reservoir sedimentation, and soil crusting.

In the modelling of hydrological extremes – floods and droughts – SREs are typically combined with information on other variables such as topography, elevation, slope, catchment area, vegetation, and soil moisture to derive estimates of surface runoff, discharge, and groundwater recharge. This is because meteorological indices alone fail to capture the extent and severity of hydrological drought (Trambauer et al. 2014) and discharge is determined by key factors such as elevation, slope, and catchment area (Staudinger et al. 2015). Further water balance studies making use of SREs are summarised in Wang et al. (2014, their Table 1). Here, we provide a brief overview of deterministic and probabilistic approaches to hydrological modelling for assessment of the hydrological aspects of droughts at different space-time scales.

In terms of surface water availability, including runoff generation, data on topography can be combined with evapotranspiration and TRMM-based rainfall data in water balance modelling to anticipate anomalous conditions (Senay et al. 2015). For example, TRMM data together with MODIS evapotranspiration data and information on water storage change from the Gravity Recovery and Climate Experiment (GRACE) provided adequate inputs for water balance modelling in Australia with satellite-based data sources being more coherent at the seasonal and annual scale than at the monthly scale (Wang et al. 2014). As part of routine hydrological monitoring, the WFP VAM unit monitors rainfall anomalies at the river basin scale (available at http://dataviz.vam.wfp.org/seasonal_explorer/river_basins, last accessed 25 Nov. 2018) (Fig. 52.3) and work is in progress to develop this into a comprehensive decision-support tool using other satellite-based datasets and a user-tailored water balance model.

With regard to large-scale atmospheric processes and their influence on surface hydrology, the GPCP, CRU, and the University of Delaware (UDEL) rainfall products were successfully used to establish the relationship between hydrological extremes such as droughts and floods and the timing of El Niño/La Niña events (Zaroug et al. 2014). Daily Meteosat-based rainfall estimates and enhanced daily rainfall estimates with information from numerical weather prediction on African Easterly Wave phase and storm type outperformed rain gauge data alone in a river flow forecasting model for West Africa (Grimes and Diop 2003).

In terms of probabilistic applications, detailed hydrological modelling using SREs can help to develop strategies and policies that avoid agricultural and socio-economic drought, for example by exploiting opportunities such as runoff rainfall harvesting in semi-arid and arid regions. For example, a probabilistic approach can help assess the likelihood of rainfall being within a specific range for applications of improved SREs to drought risk management across the African continent, including interpretation of outputs from probabilistic climate forecasts (Husak et al. 2007). This methodology has allowed the generation of model input data on rainfall, providing decision-makers with outputs on the likelihood of various rainfall accumulations so that different scenarios can be evaluated. Elements of this technique (Husak et al. 2007) were further developed (Tarnavsky et al. 2012) to enable the stochastic simulation of rainfall at 1-km spatial scale using TRMM-based datasets.

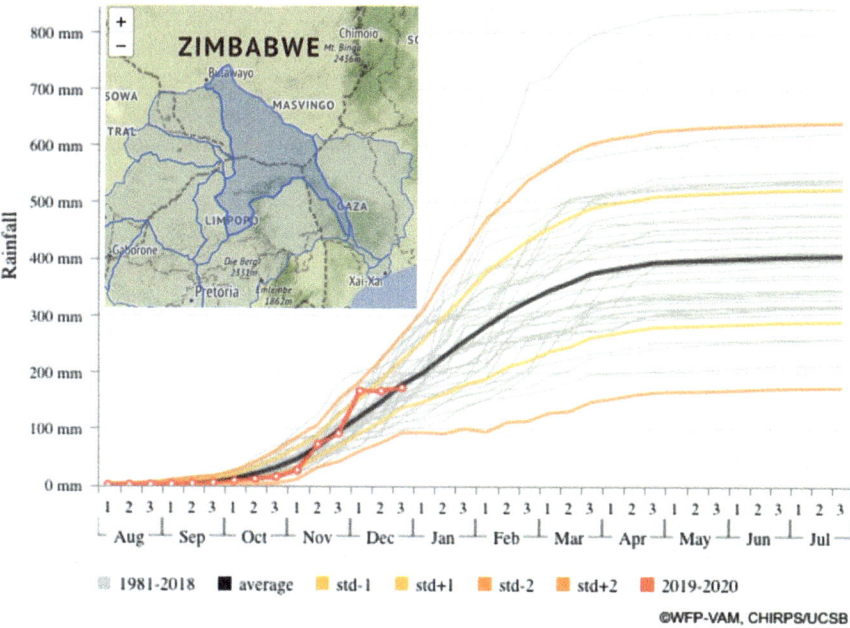

Fig. 52.3 Example of rainfall anomalies from the World Food Programme Vulnerability Assessment and Mapping (VAM) unit's Seasonal Explorer for part of the Limpopo River Basin. Note: std. is standard deviation

This was applied to identify potential areas and times of available surface runoff that can be harvested for agricultural uses in semi-arid regions through a new, Dryland hydrological MODel (DryMOD) with an hourly time step that was piloted in Tunisia and Senegal (Tarnavsky et al. 2013).

The key role of rainfall in assessing the hydrological aspects of drought are in that rainfall provides the main data input, while the modelling approach should be tailored for each application to the suitable time and space scales, levels of complexity for process representation, and how the propagation of error and uncertainty in the rainfall data input are treated within the chosen modelling framework.

52.3.3 Crop Water Stress and Crop Yield Modelling

In agricultural production monitoring, information on rainfall enables stakeholders to select crop varieties according to season's start and length, determine irrigation needs, predict likely harvest, anticipate yield shortfalls, forecast crop prices, and advise commodity trading. Crop/agro-climatology models are essential in decision support to assess and explain weather-related production variability over time and across large regions, as well as to monitor and predict crop responses to current and future climate conditions. This is useful for designing and implementing climate change coping strategies. Similarly to hydrological modelling, since errors in yield model outputs can be related to the accuracy of SREs and soil data (Thornton et al. 1997), such models rely on the accuracy of rainfall and/or soil moisture data inputs to estimate the impacts of moisture stress on crops during the growing season and crop yields at harvest time.

Simplified models focus on crop water stress, while more complex, deterministic crop models take into account the allocation of available water and energy to the different parts of crop plants to estimate yield. The latter can also represent land management strategies such as irrigation and nutrient inputs.

The crop water stress concept has been applied in the Water Requirements Satisfaction Index (WRSI) model developed by the UN's Food and Agriculture Organisation (FAO) and widely used for drought and famine early warning since the 1980s. WRSI of 95–100% indicates no water deficit (i.e., adequate rainfall and moisture availability or absence of yield reduction due to water deficit), values between 95% and 50% indicate varying degree of water stress and yield reduction due to inadequate water supply, and values below 50% indicate crop failure (Smith 1992 in Senay and Verdin 2002, 2003). Other crop water stress approaches consider the percentage available water (PAW) calculated from a soil water balance (SWB) model. PAW has been used as an indicator of plant stress: PAW below 50% indicates the beginning of plant stress, between 30 and 20% – moderate plant stress, and below 20% – severe plant stress (Lessel et al. 2016). Since crop-yield shortfalls in areas of rainfed agriculture (e.g., West and Sub-Saharan Africa) are often caused by rainfall shortages, parameters such as seasonal rainfall, season onset and duration, and rainfall and dry spells distribution within the growing season, are critical for the success of crop water stress and crop yield modelling. For example, while total annual rainfall was found important for yield simulations of pearl millet cultivars in Niger, a realistic rainfall distribution was also very important for yield estimation of pearl millet in Niger and maize in Benin (Ramarohetra et al. 2013). A range of studies have assessed the relative skill of different SREs to provide information on these parameters through model-based sensitivity analysis (Luetkemeier et al. 2018; Tarnavsky et al. 2018). An example for WRSI over maize growing regions in Tanzania in Fig. 52.4 illustrates the outcomes for a selected location. Using CHIRPS as the rainfall data input, the WRSI calculated for the location shown in Fig. 52.4 falls below 50% (i.e., crop failure) during 2 of the 33 years on record in the 1982–2014 time period and is between 95% and 50% (i.e., crops are water stressed and yield reduction is likely) for 1 of the 33 years on record during the same time period. This information is useful for agro-meteorological drought risk characterisation. Moreover, it shows that the drought spell in 1994/95 growing season is longer and recovery to optimal crop water availability was slower than for the drought events in 1993/94 and 1999/2000.

The skill of six rainfall products, four SREs (i.e., ARC2, CHIRPS 2.0, PERSIANN-CDR, TAMSAT) and two interpolated from rain gauge observations (i.e., GPCC v7, CRU-TS 3.23), was assessed through the APSIM (Agricultural

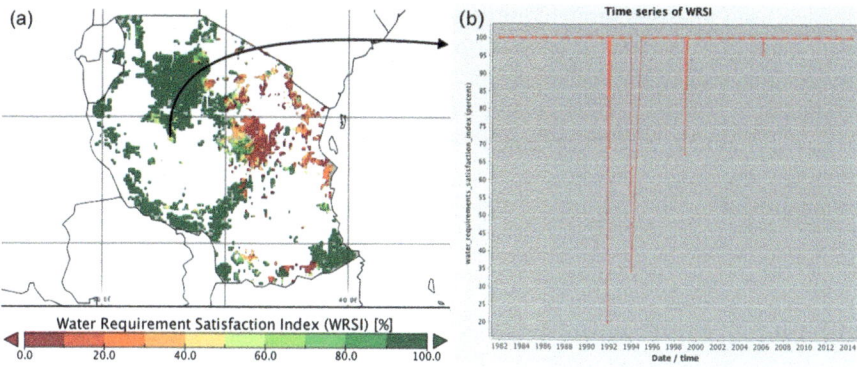

Fig. 52.4 Example of the (**a**) Water Requirement Satisfaction Index (WRSI) for a selected season across Tanzania and (**b**) WRSI time series for the location depicted in (**a**) (after research in Tarnavsky et al. 2018). WRSI was calculated with CHIRPS as the rainfall input dataset

Production System Simulator) crop model (Luetkemeier et al. 2018). Yield estimates from APSIM simulations with each rainfall product were compared to reported millet yield for the 2000–2009 time period. Results showed that the rainfall products calibrated to gauge observations performed better than the non-calibrated in terms of estimating dry spell durations, rainy day counts, and rainfall intensities (R^2 between simulated and observed yield around 0.52). ARC2, CHIRPS, and TAMSAT were also assessed through sensitivity analysis with the WRSI model (Tarnavsky et al. 2018) with results showing that CHIRPS followed by TAMSAT performed best for the assessment of key seasonal parameters such as start, length, and end of season, as well as for estimating crop-yield shortfalls. Such studies highlight the importance of considering the accuracy of SREs, especially in areas of sparse ground observations, while acknowledging that SREs are not independent of ground-based observations.

Although SREs are commonly used in place of station observations for drought monitoring, especially in African countries lacking dense monitoring networks, their use in crop models can lead to large errors in yield estimates (Ramarohetra et al. 2013) as errors in SREs are propagated into biophysical process-based models (Mishra et al. 2013). In a study in West Africa, for example, the TRMM 3B42v6 product performed best showing lowest bias in both rainfall distribution and annual totals (Ramarohetra et al. 2013).

Approaches that tackle crop water stress and yield estimation modelling without direct use of SREs are based on a two-source energy balance model such as the Atmosphere Land Exchange Inverse (ALEXI) model. ALEXI is used to obtain both surface and root zone soil moisture estimates, which in turn are used to force the Decision Support System for Agrotechnology Transfer (DSSAT) crop model (Mishra et al. 2013). Using this approach, ALEXI detected soil moisture signals with sufficient strength to provide adequate inputs for yield simulations with DSSAT over a 10-year period.

The key challenges for crop water stress and yield modelling have been associated with the error and uncertainty in rainfall data inputs and the sensitivity of models used to assess drought impacts on crop production to the choice of rainfall data input from different sources. Efforts to address the latter challenge have led to research and applications that use model data outputs instead of SREs, or SREs in conjunction with satellite-based data on other land surface variables (i.e. temperature/evapotranspiration, soil moisture), to close gaps between atmospheric and land surface processes on the basis of physical process-based relationships.

52.3.4   Cross-Cutting Approaches

52.3.4.1   Drought and Famine Early Warning Systems (EWS)

In the context of drought and famine early warning systems, rainfall data helps to assess the risk of drought during the crop-growing season leading to crop short age/failure and in extreme cases, famine, as well as to assist national governments and international aid organisations in meeting food provision needs. Generally, EWSs combine a prevision (monitoring) and prevention (forecasting) component, respectively aimed at monitoring levels of risk during the growing season and reducing their impacts. Thus, EWSs are inherently based on more than one approach for assessment of the meteorological, hydrological, agricultural, and socio-economic aspects of drought (see Fig. 52.1). SREs, as anomalies/indices or model input datasets (Thornton et al. 1997; Genesio et al. 2011), provide timely and reliable information to support decision-making and intervention planning as an integral part of drought-focused EWS (Genesio et al. 2011).

A review of existing drought monitoring and EWS platforms (Pozzi et al. 2013) outlines the key requirements for effective drought EWSs. These include maximum possible lead time to activate mitigation strategies, underlying information for decision making, both monitoring and forecasting capability, and continental to global coverage, as well as sufficient spatial and temporal detail to identify reliably the variability of current and future drought conditions. SREs and modelling play a key role in drought EWS, especially in regions of sparse in situ observation networks. However, substantial biases between SREs and model outputs relative to observations still exist due to low-density (rainfall) observation networks in most areas prone to drought. An assessment of the achievements and challenges of food security EWSs in West Africa showed that the key factors determining crop productivity that are important for drought and famine monitoring are the onset and end of the rainy season, and the distribution of dry spells during key crop-growth stages, while total seasonal rainfall is less important (Ingram et al. 2002; Genesio et al. 2011).

As part of the Famine Early Warning Systems NETwork (FEWS NET), SREs provide an indication of rainfall variability from year to year, particularly important for regions where rainfall plays a key role in determining yields (Brown 2008). Although other factors such as crop variety, nutrient availability, pest/disease, and management practices affect yield outcomes, their year-to-year variability is less substantial and rainfall remains the main determinant of water availability for crop growth. While gauge data are biased due to wind and other factors, in SREs the bias can be more substantial due to the indirect relationship with rain, i.e. reliance on detecting cold clouds and rain rates through calibration to gauge observations, and thus substantial under–/over-estimation of rainfall can occur (Brown 2008). The CHIRP(S) quasi-global products, specifically developed to minimise biases for the monitoring of rainfall variability and drought in space and time in the context of FEWS NET (Funk et al. 2014), have supported the operational monitoring of the SPI, helping to resolve areas of hydrological extremes relative to a long-term average (climatology) and have been used operationally in the WFP Seasonal Monitor (Fig. 52.2).

The ZAR (Zones A Risque) agro-meteorological model (Di Vecchia et al. 2006) and the SARRA-H (Systéme d'Analyse Régionale des Risques Agroclimatiques, version H) crop model (Genesio et al. 2011) were used together to demonstrate the key components of an operational EWS, i.e. information network, regional-national coordination, and relief operations. The rainfall input for ZAR is provided through Meteosat brightness temperatures correlated with instantaneous precipitation rates from the SSM/I MW passive radiometer; alternatively, FEWS NET dekadal rainfall estimates at 5-km spatial resolution can be used (Di Vecchia et al. 2006). Although present EWSs are focused on monitoring food production through agricultural field campaigns and operational monitoring of agro-climatic events, improvements are required in the areas of identification of risk zones, forecasting of yields, and seasonal forecasting for prediction of impacts, i.e. transition to impact-oriented monitoring, analyses, and prediction. Notably, an effective EWS could increase cereal yields with as much as 20% where the information, particularly on prevision, is disseminated as part of agro-meteorological advisory services (Di Vecchia et al. 2006).

As rainfall in many regions of Sub-Saharan Africa (SSA) is highly variable (i.e. with up to 60% coefficient of variation for annual rainfall) and cereal production fluctuates substantially in response to variable rainfall, monitoring models that can provide timely and accurate crop yield estimates are a critical part of EWS. A prototype famine EWS using weather data from ground observations and SREs, a monthly to daily disaggregation routine SIMMETEO, and historical yield figures at province level for Burkina Faso was demonstrated by Thornton et al. (1997). In the EWS, a range of indicators are monitored over time to provide crop yield estimates as early as up to 2 months before harvest, which are updated in near-real time as harvest time approaches.

Key challenges for EWSs include the appropriate integration of optimally skillful SREs, modelling, and ground observations with reduced errors in satellite-based retrievals, improved consistency in time between products, as well as assimilating remote sensing in models to reduce biases (Pozzi et al. 2013). Additional challenges are to reduce uncertainty and biases in areas of sparse ground observations and to enable sampling and monitoring of multiple stores of water to derive effectively physically-based drought characteristics. A broader challenge for the EWS community is to integrate in situ, satellite, and crop-based information into user-friendly form to assist drought-related response activities and relief efforts globally.

52.3.4.2   Weather Index-Based Insurance (WII)

Weather index-based insurance (WII) has become an attractive financial instrument for coping with weather- and climate-related risks as payouts, based on a weather index such as rainfall over a given period, are automatically processed unlike in traditional agricultural indemnity insurance where crop losses are verified through expensive farm visits. WII can be as simple as a rainfall anomaly over the course of a certain number of days at the start of the rainy season or can use more than one of the approaches described in Fig. 52.1.

A range of single- and multi-variable WII products have been piloted based on deficit and/or excess rainfall, dry spells, low and high temperature, high humidity, and high wind in India; drought based on automatic weather station and SRE data in East Africa; crop yield losses using ARC2 satellite-based indices in Ethiopia and Senegal; insurance against livestock mortality in extreme winter weather in Mongolia; and an NDVI-based insurance for livestock in drought years in Kenya and Ethiopia (Greatrex et al. 2015). Another study (Evkaya 2012) assessed the performance of WII contracts constructed with information from AgroMetShell model outputs, i.e. water deficiency and actual evapotranspiration (AET), in the four growth stages of wheat (initial, vegetative, flowering, and ripening) and the water satisfaction index at harvest, as well as the Normalized Difference Vegetation Index (NDVI) and SPI data inputs. Although SREs were not used (AgroMetShell was forced with station rainfall data), different insurance contracts for wheat were compared in terms of basis risk and a suitable index was selected for each province. In the case of Anatolia in Turkey, the water deficiency index at ripening explained most of the variability in wheat yield and minimised basis risk. Apart from this study, there is a general dearth of literature assessing single- and multi-variable indices over the same region to provide information on optimal region-specific indices.

Most commonly WII products are based on cumulative seasonal rainfall and/or rainfall during the different crop growth stages. A review of WII products based on meteorological indices, and particularly on the lack of rainfall, showed that the main challenge in their wide-scale implementation has been basis risk, i.e. the discrepancy between the index and actual yield outcomes (Leblois et al. 2011). Minimising basis risk requires an index that is a good predictor of yield, and of poor yield in particular; however, most indices insure against the lack of rain and are highly dependent on the data source used to calculate the index. Moreover, the same amount of rainfall deficit does not translate to equal yield shortfalls across regions and over time due to factors influencing growth conditions – such as temperature, humidity, pest and disease outbreaks, nutrient input, etc. – that are not captured in an index based solely on rainfall. As a result, there is an increasing interest in using other satellite-based datasets (e.g., soil moisture) to construct WII based on a more comprehensive analyses of the conditions for crop growth as a function of the interactions between atmospheric and land surface processes.

52.4 Summary and Outlook

State-of-the-art drought risk management research and applications have successfully exploited the key strengths of SREs. In summary, the challenges for SREs and their use in drought risk management are as follows:

- SREs are not independent of ground observations, quite the opposite – the reliability and skill of SREs depends on that of the ground-based observations used in the SRE retrieval algorithms.
- Gaps in ground-based observations affect the quality of SREs as even with the same SRE retrieval algorithm, areas not covered by ground observation networks are likely to have less reliable SREs.
- The uncertainty of SREs when used as model data input, as well as differences between different sources of SREs, propagate through models.
- Gauge data ingested in SREs have declined over time (Nicholson et al. 2018) and information on gauge observations used in the generation of SREs over time and for different locations (countries) is not readily available; for example, while for CHIRPS the information about stations that were used in the algorithm is disclosed, the same is not provided for other SREs.

The above challenges can hinder progress with not only validation and intercomparison studies, but also with applications, as accuracy/skill of the different products varies from region to region. Unless existing SREs have been comprehensively evaluated over a given region/country, a skill assessment would have to be carried out by end-users. Alternatively, coordinated skill assessment for SREs with relevant metrics for different applications could be developed and provided in the future by, for example, the IPWG community involved in the Working Groups on Validation and Applications (see Chap. 19 of volume 1). Additionally, our work highlights the following trends in the development and application of SREs for drought risk management:

- Harmonised and internally consistent long-term datasets on rainfall, temperature, vegetation, and soil moisture have been developed in parallel with a shift away from precipitation-based drought monitoring due to inherent uncertainty. This has increased interest in the use of other Earth observation products in models such as satellite-based soil-moisture (Enenkel et al. 2015) and inputs to models such as ALEXI and DSSAT.
- Although traditionally droughts have been monitored separately from floods and studies focused on both rainfall excess and deficit have been sparse (Zaroug et al. 2014; Toté et al. 2015), efforts have shifted from single- to multi-disaster (or cascading disasters) approaches engaging multi-disciplinary expertise.
- As applications have evolved from simple rainfall monitoring to drought risk management, they have re-focused on impact- and risk-based analyses by sector (often described in climate services terms), which has raised interest in value added data streams where SREs will continue to play an integral role. For example, studies that investigate the sensitivity of model outputs to inputs from

different sources help to inform the evaluation and choice of rainfall products using crop-relevant, impact-oriented metrics such as season onset, length, and cessation (Tarnavsky et al. 2018).

- There has been an increasingly better understanding of when SREs are more skillful than local, gauge-based observations and how their use impacts on the reliability of models (e.g., localised models with gauge observations vs. regional-scale models with SREs) depending on the modelling approach (e.g., deterministic vs probabilistic).

- SREs are one of many data inputs in drought risk management and their relative role and sustained contribution will continue to require better understanding of the uncertainties involved. The choice of suitable SRE source remains within the remits of user-tailored decisions where criteria such as spatial and temporal characteristics, availability/accessibility, reliability, consistency, accuracy, skill, and performance compared to alternative products are considered by stakeholders according to specific applications.

References

AghaKouchak, A., Mehran, A., Norouzi, H., & Behrangi, A. (2012). Systematic and random error components in satellite precipitation data sets. *Geophysical Research Letters, 39*, L09406. https://doi.org/10.1029/2012GL051592.

Amekudzi, L. K., Osei, M. A., Atiah, W. A., Aryee, J. N. A., Ahiataku, M. A., Quansah, E., Preko, K., Danuor, S. K., & Fink, A. H. (2016). Validation of TRMM and FEWS satellite rainfall estimates with rain gauge measurement over Ashanti Region, Ghana. *Atmospheric and Climate Sciences, 6*, 500–518. https://doi.org/10.4236/acs.2016.64040.

Arkin, P. A., & Meisner, B. N. (1987). The relationship between large-scale convective rainfall and cold cloud over the western hemisphere during 1982–84. *Monthly Weather Review, 115*, 51–74. https://doi.org/10.1175/1520-0493(1987)115<0051:TRBLSC>2.0.CO;2.

Awange, J. L., Ferreira, V. G., Forootan, E., Khandu, Andam-Akorful, S. A., Agutu, N. O., & He, X. F. (2015). Uncertainties in remotely sensed precipitation data over Africa. *International Journal of Climatology, 36*, 303–323. https://doi.org/10.1002/joc.4346.

Bergés, J. C., Jobard, I., Chopin, F., & Roca, R. (2010). EPSAT-SG: A satellite method for precipitation estimation; its concepts and implementation for the AMMA experiment. *Annales Geophysicae, 28*, 289–308. https://doi.org/10.5194/angeo-28-289-2010.

Brown, M. E. (2008). *Famine early warning systems and remote sensing data*. Springer, Heidelberg, 309 pp., ISBN: 978-3-540-75369-8.

Di Vecchia, A., Bacci, M., Pini, G., Tarchiani, V., & Vignaroli, P. (2006). Meteorological forecasts and agrometeorological models integration: A new approach concerning early warning for food security in the Sahel. In 6th AARSE international conference on Earth Observation Geoinformation and Science in Support of Africa's Development, 30 Oct–2 Nov, pp. 1–8. Available at http://citeseerx.ist.psu.edu/viewdoc/download?doi=10.1.1.581.9198&rep=rep1& type=pdf, last accessed 25 Nov 2018.

Diem, J. E., Hartter, J., Ryan, S. J., & Palace, M. W. (2014). Validation of satellite rainfall products for Western Uganda. *Journal of Hydrometeorology, 15*, 2030–2038. https://doi.org/10.1175/ JHM-D-13-0193.1.

Dinku, T., Alessandrini, S., Evangelisti, M., & Rojas, O. (2015). A description and evaluation of FAO satellite rainfall estimation algorithm. *Atmospheric Research, 163*, 48–60. https://doi.org/ 10.1016/j.atmosres.2015.01.020.

Dutra, E., Pozzi, W., Wetterhall, F., Di Giuseppe, F., Magnusson, L., Naumann, G., Barbosa, P., Vogt, J., & Pappenberger, F. (2014). Global meteorological drought – Part 2: Seasonal forecasts. *Hydrology and Earth System Sciences, 18,* 2669–2678. https://doi.org/10.5194/hess-18-2669-2014.

Enenkel, M., See, L., Karner, M., Álvarez, M., Rogenhofer, E., Baraldés-Vallverdù, C., Lanusse, C., & Salse, N. (2015). Food security monitoring via mobile data collection and remote sensing: Results from the Central African Republic. *PLoS One, 10,* e0142030. https://doi.org/10.1371/journal.pone.0142030.

Enenkel, M., Steiner, C., Mistelbauer, T., Dorigo, W., Wagner, W., See, L., Atzberger, C., Schneider, S., & Rogenhofer, E. (2016). A combined satellite-derived drought indicator to support humanitarian aid organizations. *Remote Sensing, 8,* 340. https://doi.org/10.3390/rs8040340.

Evkaya, O. O. (2012). *Modelling weather index based drought insurance for provinces in the Central Anatolia Region* (M.Sci. Thesis), 116 pp. Available at http://docplayer.net/21211737-Modelling-weather-index-based-drought-insurance-for-provinces-in-the-central-anatolia-region.html, last accessed 25 Nov 2018.

Funk, C. C., Peterson, P. J., Landsfield, M. F., Pedreros, D. H., Verdin, J. P., Rowland, J. D., Romero, B. E., Husak, G. J., Michaelsen, J. C., & Verdin, A. P. (2014). A quasi-global precipitation time series for drought monitoring. *USGS Data Series, 832,* 4 pp, https://doi.org/10.3133/ds832.

Genesio, L., Bacci, M., Baron, C., Diarra, B., Di Vecchia, A., Alhassane, A., Hassane, I., Ndiaye, M., Philippon, N., Tarchiani, V., & Traoré, S. (2011). Early warning systems for food security in West Africa: Evolution, achievements and challenges. *Atmospheric Science Letters, 12,* 142–148. https://doi.org/10.1002/asl.332.

Greatrex, H., Hansen, J., Garvin, S., Diro, R., Blakeley, S., Le Guen, M., Rao, K., & Osgood, D. (2015). *Scaling up index insurance for smallholder farmers: Recent evidence and insights* (CCAFS Report No. 14 (2015)), pp. 1–32. Available at https://ccafs.cgiar.org/publications/scaling-index-insurance-smallholder-farmers-recent-evidence-and-insights#.W_rQky2h3PA, last accessed 25 Nov 2018.

Grimes, D. I. F., & Diop, M. (2003). Satellite-based rainfall estimation for river flow forecasting in Africa. I: Rainfall estimates and hydrological forecasts. *Hydrological Sciences Journal, 48,* 585–599. https://doi.org/10.1623/hysj.48.4.567.51410.

Grimes, D. I. F., Pardo-Igúzquiza, E., & Bonifacio, R. (1999). Optimal areal rainfall estimation using raingauges and satellite data. *Journal of Hydrology, 222,* 93–108. https://doi.org/10.1016/S0022-1694(99)00092-X.

Heim, R. R., Jr. (2002). Century drought indices used in the United States. *Bulletin of the American Meteorological Society, 83,* 1149–1166. https://doi.org/10.1175/1520-0477-83.8.1149.

Hessels, T. M. (2015). *Comparison and validation of several open access remotely sensed rainfall products for the Nile Basin* (Master Thesis), University of Delft, 110 pp. Available at https://repository.tudelft.nl/islandora/object/uuid:3566f883-16fd-4465-be43-6b2037baa6ff, last accessed 25 Nov 2018.

Hoscilo, A., Balzter, H., Bartholomé, E., Boschetti, M., Brivio, P. A., Brink, A., Clerici, M., & Pekel, J. F. (2014). A conceptual model for assessing rainfall and vegetation trends in sub-Saharan Africa from satellite data. *International Journal of Climatology, 35,* 3582–3592. https://doi.org/10.1002/joc.4231.

Huffman, G. J., Bolvin, D. T., Nelkin, E. J., Wolff, D. B., Adler, R. F., Gu, G., Hong, Y., Bowman, K. P., & Stocker, E. F. (2007). The TRMM Multisatellite Precipitation Analysis (TMPA): Quasi-global, multiyear, combined-sensor precipitation estimates at fine scales. *Journal of Hydrometeorology, 8,* 38–55. https://doi.org/10.1175/JHM560.1.

Husak, G. J., Michaelsen, J., & Funk, C. (2007). Use of the gamma distribution to represent monthly rainfall in Africa for drought monitoring applications. *International Journal of Climatology, 27,* 935–944. https://doi.org/10.1002/joc.1441.

Ingram, K. T., Roncoli, M. C., & Kirshen, P. H. (2002). Opportunities and constraints for farmers of West Africa to use seasonal precipitation forecasts with Burkina Faso as a case study. *Agricultural Systems, 74,* 331–349. https://doi.org/10.1016/S0308-521X(02)00044-6.

Jobard, I., Chopin, F., Bergés, J. C., & Roca, R. (2011). An intercomparison of 10-day satellite precipitation products during West African monsoon. *International Journal of Remote Sensing, 32*, 2353–2376. https://doi.org/10.1080/01431161003698286.

Katiraie-Boroujerdy, P. S., Nasrollahi, N., Hsu, K.-L., & Sorooshian, S. (2013). Evaluation of satellite-based precipitation estimation over Iran. *Journal of Arid Environments, 97*, 205–219. https://doi.org/10.1016/j.jaridenv.2013.05.013.

Kimani, M. W., Hoedjes, J. C. B., & Su, Z. (2017). An assessment of satellite-derived rainfall products relative to ground observations over East Africa. *Remote Sensing, 9*, 430. https://doi.org/10.3390/rs9050430.

Kummerow, C. D., Hong, Y., Olson, W. S., Yang, S., Adler, R. F., McCollum, J., Ferraro, R., Petty, G., Shin, D.-B., & Wilheit, T. T. (2001). The evolution of the Goddard Profiling Algorithm (GPROF) for rainfall estimation from passive microwave sensors. *Journal of Applied Meteorology, 40*, 1801–1820. https://doi.org/10.1175/1520-0450(2001)040<1801:TEOTGP>2.0.CO;2.

Lamptey, B. L. (2008). Comparison of gridded multisatellite rainfall estimates with gridded gauge rainfall over West Africa. *Journal of Applied Meteorology and Climatology, 47*, 185–205. https://doi.org/10.1175/2007JAMC1586.1.

Leblois, A., Quirion, P., Alhassane, A., & Traoré, S. (2011). Weather index drought insurance: An ex ante evaluation for millet growers in Niger. In Intrenational Conference European Association of Agricultural Economists, Zürich, 30 Aug–2 Sept. Available at https://econpapers.repec.org/paper/agseaae11/120378.htm, last accessed 25 Nov 2018.

Lessel, J., Sweeney, A., & Ceccato, P. (2016). An agricultural drought severity index using quasi-climatological anomalies of remotely sensed data. *International Journal of Remote Sensing, 37*, 913–925. https://doi.org/10.1080/01431161.2016.1142689.

Li, J., Lewis, J., Rowland, J., Tappan, G., & Tieszen, L. L. (2004). Evaluation of land performance in Senegal using multi-temporal NDVI and rainfall series. *Journal of Arid Environments, 59*, 463–480. https://doi.org/10.1016/j.jaridenv.2004.03.019.

Lloyd-Hughes, B. (2014). The impracticality of a universal drought definition. *Theoretical and Applied Climatology, 117*, 607–611. https://doi.org/10.1007/s00704-013-1025-7.

Luetkemeier, R., Stein, L., Drees, L., Müller, H., & Liehr, S. (2018). Uncertainty of rainfall products: Impact on modelling household nutrition from rain-fed agriculture in Southern Africa. *Water, 10*, 1–23. https://doi.org/10.3390/w10040499.

Maidment, R. I., Grimes, D. I. F., Allan, R. P., Greatrex, H., Rojas, O., & Leo, O. (2013). Evaluation of satellite-based and model re-analysis rainfall estimates for Uganda. *Meteorological Applications, 20*, 308–317. https://doi.org/10.1002/met.1283.

Maidment, R. I., Grimes, D., Black, E., Tarnavsky, E., Young, M., Greatrex, H., Allan, R. P., Stein, T., Nkonde, E., Senkunda, S., & Alcàntara, E. M. U. (2017). A new, long-term daily satellite-based rainfall dataset for operational monitoring in Africa. *Scientific Data, 4*, 1–17. https://doi.org/10.1038/sdata.2017.63.

Mishra, V., Cruise, J. F., Mecikalski, J. R., Hain, C. R., & Anderson, M. C. (2013). A remote-sensing driven tool for estimating crop stress and yields. *Remote Sensing, 5*, 3331–3356. https://doi.org/10.3390/rs5073331.

Naumann, G., Dutra, E., Barbosa, P., Pappenberger, F., Wetterhall, F., & Vogt, J. V. (2014). Comparison of drought indicators derived from multiple data sets over Africa. *Hydrology and Earth System Sciences, 18*, 1625–1640. https://doi.org/10.5194/hess-18-1625-2014.

Nguyen, P., Ombadi, M., Sorooshian, S., Hsu, K.-L., AghaKouchak, A., Braithwaite, D., Ashouri, H., & Thorstensen, A. R. (2018). The PERSIANN family of global satellite precipitation data: A review and evaluation of products. *Hydrology and Earth System Sciences, 22*, 5801–5816. https://doi.org/10.5194/hess-22-5801-2018.

Nicholson, S. E. (2014). A detailed look at the recent drought situation in the Greater Horn of Africa. *Journal of Arid Environments, 103*, 71–79. https://doi.org/10.1016/j.jaridenv.2013.12.003.

Nicholson, S. E., Fink, A. H., & Funk, C. (2018). Assessing recovery and change in West Africa's rainfall regime from a 161-year record. *International Journal of Climatology, 38*, 3770–3786. https://doi.org/10.1002/joc.5530.

NOAA. (2001). *The NOAA Climate Prediction Center African Rainfall Estimation Algorithm Version 2.0*, 3 pp. Available at http://www.cpc.noaa.gov/products/fews/RFE2.0_tech.pdf, last accessed 25 Nov 2018.

Novella, N. S., & Thiaw, W. M. (2013). African rainfall climatology version 2 for famine early warning systems. *Journal of Applied Meteorology and Climatology, 52*, 588–606. https://doi.org/10.1175/JAMC-D-11-0238.1.

Novella, N. S., & Thiaw, W. M. (2016). A seasonal rainfall performance probability tool for famine early warning systems. *Journal of Applied Meteorology and Climatology, 55*, 2575–2586. https://doi.org/10.1175/JAMC-D-16-0111.1.

Palmer, W. C. (1965). Meteorological drought. *Research paper*, 45, US Department of Commerce, Office of Climatology, Washington, DC, 58 pp. Available at https://www.ncdc.noaa.gov/temp-and-precip/drought/docs/palmer.pdf, last accessed 25 Nov 2018.

Pozzi, W., Sheffield, J., Stefanski, R., Cripe, D., Pulwarty, R., Vogt, J. V., Heim, R. R., Brewer, M. J., Svoboda, M., Westerhoff, R., van Dijk, A. I. J. M., Lloyd-Hughes, B., Pappenberger, F., Werner, M., Dutra, E., Wetterhall, F., Wagner, W., Schubert, S., Mo, K., Nicholson, M., Bettio, L., Nunez, L., van Beek, R., Bierkens, M., Goncalves De Goncalves, L. G., Zell De Mattos, J. G., & Lawford, R. (2013). Toward global drought early warning capability: Expanding international cooperation for the development of a framework for monitoring and forecasting. *Bulletin of the American Meteorological Society, 94*, 776–785. https://doi.org/10.1175/BAMS-D-11-00176.1.

Ramarohetra, J., Sultan, B., Baron, C., Gaiser, T., & Gosset, M. (2013). How satellite rainfall estimate errors may impact rainfed cereal yield simulation in West Africa. *Agricultural and Forest Meteorology, 180*, 118–131. https://doi.org/10.1016/j.agrformet.2013.05.010.

Rhee, J., Im, J., & Carbone, G. J. (2010). Monitoring agricultural drought for arid and humid regions using multi-sensor remote sensing data. *Remote Sensing of Environment, 114*, 2875–2887. https://doi.org/10.1016/j.rse.2010.07.005.

Rienecker, M. M., Suarez, M. J., Gelaro, R., Todling, R., Bacmeister, J., Liu, E., Bosilovich, M. G., Schubert, S. D., Takacs, L., Kim, G. K., Bloom, S., Chen, J., Collins, D., Conaty, A., Da Silva, A., Gu, W., Joiner, J., Koster, R. D., Lucchesi, R., Molod, A., Owens, T., Pawson, S., Pegion, P., Redder, C. R., Reichle, R., Robertson, F. R., Ruddick, A. G., Sienkiewicz, M., & Woollen, J. (2011). MERRA: NASA's modern-era retrospective analysis for research and applications. *Journal of Climate, 24*, 3624–3648. https://doi.org/10.1175/JCLI-D-11-00015.1.

Roca, R., Chambon, P., Jobard, I., Kirstetter, P.-E., Gosset, M., & Bergés, J.-C. (2010). Comparing satellite and surface rainfall products over West Africa at meteorologically relevant scales during the AMMA Campaign using error estimates. *Journal of Applied Meteorology and Climatology, 49*, 715–731. https://doi.org/10.1175/2009JAMC2318.1.

Senay, G. B., & Verdin, J. (2002). Evaluating the performance of a crop water balance model in estimating regional crop production. In Proceedings of PECORA 15 Symposium, Denver, CO, 8 pp. Available at http://www.isprs.org/proceedings/XXXIV/part1/paper/00026.pdf, last accessed 25 Nov 2018.

Senay, G. B., & Verdin, J. (2003). Characterization of yield reduction in Ethiopia using a GIS-based crop water balance model. *Canadian Journal of Remote Sensing, 29*, 687–692. https://doi.org/10.5589/m03-039.

Senay, G. B., Velpuri, N. M., Bohms, S., Budde, M., Young, C., Rowland, J., & Verdin, J. P. (2015). Drought monitoring and assessment. In *Hydro-meteorological hazards, risks and disasters* (pp. 233–263). Elsevier. https://doi.org/10.1016/B978-0-12-394846-5.00009-6.

Serrat-Capdevila, A., Merino, M., Valdes, J. B., & Durcik, M. (2016). Evaluation of the performance of three satellite precipitation products over Africa. *Remote Sensing, 8*, 836. https://doi.org/10.3390/rs8100836.

Smakhtin, V. U., & Hughes, D. A. (2004). *Review, automated estimation and analyses of drought indices in South Asia. Drought Series, Paper 1*, IWMI, Colombo, Sri Lanka, 24 pp. Available at https://www.preventionweb.net/files/1869_VL102136.pdf, last accessed 25 Nov 2018.

Staudinger, M., Weiler, M., & Seibert, J. (2015). Quantifying sensitivity to droughts – An experimental modeling approach. *Hydrology and Earth System Sciences, 19*, 1371–1384. https://doi.org/10.5194/hess-19-1371-2015.

Tadesse, T., Senay, G., Wardlow, B. D., Knutson, C. L., & Haile, M. (2008). The need for integration of drought monitoring tools for proactive food security management in Sub-Saharan Africa. *Natural Resources Forum, 32*, 265–279. https://doi.org/10.1111/j.1477-8947.2008.00211.x.

Tarnavsky, E., Mulligan, M., & Husak, G. (2012). Spatial disaggregation and intensity correction of TRMM-based rainfall time series for hydrological applications in dryland catchments. *Hydrological Sciences Journal, 57*, 248–264. https://doi.org/10.1080/02626667.2011.637498.

Tarnavsky, E., Mulligan, M., Ouessar, M., Faye, A., & Black, E. (2013). Dynamic hydrological modeling in drylands with TRMM based rainfall. *Remote Sensing, 5*, 6691–6716. https://doi.org/10.3390/rs5126691.

Tarnavsky, E., Grimes, D., Maidment, R., Black, E., Allan, R. P., Stringer, M., Chadwick, R., & Kayitakire, F. (2014). Extension of the TAMSAT satellite-based rainfall monitoring over Africa and from 1983 to present. *Journal of Applied Meteorology and Climatology, 53*, 2805–2822. https://doi.org/10.1175/JAMC-D-14-0016.1.

Tarnavsky, E., Chavez, E., & Boogaard, H. (2018). Sensitivity of an adapted Water Requirements Satisfaction Index (WRSI) model to rainfall. *International Journal of Applied Earth Observation and Geoinformation, 73*, 77–87. https://doi.org/10.1016/j.jag.2018.04.008.

Thiemig, V., Rojas, R., Zambrano-Bigiarini, M., Levizzani, V., & De Roo, A. (2012). Validation of satellite-based precipitation products over sparsely gauged African river basins. *Journal of Hydrometeorology, 13*, 1760–1783. https://doi.org/10.1175/JHM-D-12-032.1.

Thornton, P. K., Bowen, W. T., Ravelo, A. C., Wilkens, P. W., Farmer, G., Brock, J., & Brink, J. E. (1997). Estimating millet production for famine early warning: An application of crop simulation modelling using satellite and ground-based data in Burkina Faso. *Agricultural and Forest Meteorology, 83*, 95–112. https://doi.org/10.1016/S0168-1923(96)02348-9.

Toté, C., Patricio, D., Boogaard, H., van der Wijngaart, R., Tarnavsky, E., & Funk, C. (2015). Evaluation of satellite rainfall estimates for drought and flood monitoring in Mozambique. *Remote Sensing, 7*, 1758–1776. https://doi.org/10.3390/rs70201758.

Trambauer, P., Dutra, E., Maskey, S., Werner, M., Pappenberger, F., van Beek, L. P. H., & Uhlenbrook, S. (2014). Comparison of different evaporation estimates over the African continent. *Hydrology and Earth System Sciences, 18*, 193–212. https://doi.org/10.5194/hess-18-193-2014.

Ushio, T., Sasashige, K., Kubota, T., Shige, S., Okamoto, K., Aonashi, K., Inoue, T., Takahashi, N., Iguchi, T., Kachi, M., Oki, R., Morimoto, T., & Kawasaki, Z.-I. (2009). A Kalman filter approach to the Global Satellite Mapping of Precipitation (GSMaP) from combined passive microwave and infrared radiometric data. *Journal of the Meteorological Society of Japan, 87A*, 137–151. https://doi.org/10.2151/jmsj.87A.137.

Verdin, J., & Klaver, R. (2002). Grid-cell-based crop water accounting for the famine early warning system. *Hydrological Processes, 16*, 1617–1630. https://doi.org/10.1002/hyp.1025.

Vernimmen, R. R. E., Hooijer, A., Mamenun, E. A., & van Dijk, A. I. J. M. (2012). Evaluation and bias correction of satellite rainfall data for drought monitoring in Indonesia. *Hydrology and Earth System Sciences, 16*, 133–146. https://doi.org/10.5194/hess-16-133-2012.

Wang, H., Guan, H., Gutiérrez-Jurado, H. A., & Simmons, C. T. (2014). Examination of water budget using satellite products over Australia. *Journal of Hydrology, 511*, 546–554. https://doi.org/10.1016/j.jhydrol.2014.01.076.

Wilhite, D. A., & Glantz, M. H. (1985). Understanding the drought phenomenon: The role of definitions. *Water International, 10*, 111–120. Available at http://digitalcommons.unl.edu/cgi/viewcontent.cgi?article=1019&context=droughtfacpub, last accessed 25 Nov 2018.

WMO. (1992). *METEOTERM: International Meteorological Vocabulary.* https://www.wmo.int/pages/prog/lsp/meteoterm_wmo_en.html, last accessed 25 Nov 2018.

Yatagai, A., Kamiguchi, K., Arakawa, O., Hamada, A., Yasutomi, N., & Kitoh, A. (2012). Aphrodite constructing a long-term daily gridded precipitation dataset for Asia based on a dense network of rain gauges. *Bulletin of the American Meteorological Society, 93*, 1401–1415. https://doi.org/10.1175/BAMS-D-11-00122.1.

Zargar, A., Sadiq, R., Naser, B., & Khan, F. I. (2011). A review of drought indices. *Environmental Review, 19*, 333–349. https://doi.org/10.1139/a11-013.

Zaroug, M. A. H., Eltahir, E. A. B., & Giorgi, F. (2014). Droughts and floods over the upper catchment of the Blue Nile and their connections to the timing of El Niño and La Niña events. *Hydrology and Earth System Sciences, 18*, 1239–1249. https://doi.org/10.5194/hess-18-1239-2014.

Chapter 53
Two Decades of Urban Hydroclimatological Studies Have Yielded Discovery and Societal Benefits

J. Marshall Shepherd, Steven J. Burian, Menglin Jin, Chuntao Liu, and Bradford Johnson

Abstract While only a small fraction of the Earth's total land cover, the majority of the world population resides in cities. Urbanization is critical to the social, political, and economic vibrancy of society, but it also has significant impacts on the environment. NASA-related satellite, modeling, and scientific resources have enabled a systems perspective on the Earth's water cycle, climate, weather, interior and biogeochemical cycles. A particular advancement has been the advancement of knowledge on the complex interactions and pathways by which urbanization interacts with hydrometeorological processes. Herein, a synopsis of the most significant perspectives gained from urban studies funded under the auspices of the Precipitation Measurement Missions program is presented.

Keywords Precipitation · Urbanization · Hydroclimatology · Energy-food-water nexus · Urban precipitation effect · Land use · Urban heat island · TRMM · GPM · IMERG · PMM · MODIS aerosol optical depth · Land hydrological response · Water management · Water stress · Precipitation per person · Urban floods

J. M. Shepherd (✉)
Atmospheric Sciences Program and Department of Geography, University of Georgia, Athens, GA, USA
e-mail: marshgeo@uga.edu

S. J. Burian
Department of Civil and Environmental Engineering, University of Utah, Salt Lake City, UT, USA

M. Jin
Department of Atmospheric and Oceanic Science, University of Maryland, College Park, MD, USA

C. Liu
Department of Physical and Environmental Sciences, Texas A&M University, Corpus Christi, TX, USA

B. Johnson
Department of Geography, University of Georgia, Athens, GA, USA

© Springer Nature Switzerland AG 2020
V. Levizzani et al. (eds.), *Satellite Precipitation Measurement*, Advances in Global Change Research 69, https://doi.org/10.1007/978-3-030-35798-6_29

53.1   Introduction

Urbanization is one of the extreme cases of land use change. Although about 1–3% of the Earth's land is urbanized (Liu et al. 2014), the trends suggest that coverage of impervious surfaces and density of cities will continue to increase in the near future. The majority of humans now live in cities, and it is estimated that by the year 2025, 60% of the world's population will live in them (UN 2014). Urban areas are inherently local to regional scale geographies, the literature has shown that they have impacts on weather, climate, land surface hydrology and biogeochemical cycles (Seto and Shepherd 2009).

Up until the early 2000s, the most commonly studied aspect of urban climatology was the heat island (Zhou and Shepherd 2009). Souch and Grimmond (2006) noted that Shepherd et al. (2002) ushered renewed interest in the research topic of how urbanization affects rainfall. Prior to this decade, there had been a flurry of studies and field campaigns like METROMEX in the 1970s that were trying to answer fundamental questions related to the association of cities with rainfall processes and convective initiation (Shepherd 2013; Changnon et al. 1981). The prevailing conclusion was that cities can modify or initiate precipitating systems. However, Lowry (1998) called into question some of the methodological approaches and casted doubt on many of the findings of that era. He recommended that future studies have the following characteristics:

- Designed experiments—especially legitimate controls and, where appropriate, stratification schemes – in which explicitly stated hypotheses are tested by means of standard statistical methods.
- Replication of the experiments in several urban areas. (Note: The recent emergence of satellite-based precipitation measurement (see Shepherd and Burian 2003) offers a possible resource for achieving this recommendation more efficiently).
- Use of spatially small, and temporally short, experimental units reflecting the discontinuous nature of precipitating systems.
- Disaggregation of standard climatic data to increase sample size and avoid merging effects between dissimilar synoptic weather systems.

Shepherd et al. (2002) leveraged the emergence of satellite-based precipitation radar on the Tropical Rainfall Measuring Mission (TRMM) to counter some of Lowry's criticisms of previous studies that had focused on single cities, seasons, or climatological regimes. This paper ushered in two decades of NASA Precipitation Measurement Mission (PMM)-funded work on the intersections of urbanization and the hydroclimate. This work has led to landmark discoveries and understanding related to the urban hydroclimate, flooding, and landcover-microphysical underpinnings of precipitation. Over 50 peer reviewed publications, grants, and media releases have been generated as a result of the collaborative efforts of the authors. By leveraging the combination of satellite-based precipitation measurement, numerical modeling, hydrological modeling, and analysis, new understanding and potential applications have been advanced in four major areas: (1) the "urban precipitation effect," (2) urban impacts on winter precipitation, (3) land surface hydrological response and water management, and (4) the juxtaposition of urbanization with the energy-water-food nexus (EWFN).

53.2 The "Urban Precipitation Effect"

One of the most pressing questions of the past few decades is whether urban environments can modify or initiate precipitation. Years of research have likely put this question to rest (Shepherd 2013; Haberlie et al. 2015; Ashley et al. 2012; Mitra and Shepherd 2016). The answer is yes. NASA funded research inspired a new generation of studies in this topical area. Shepherd et al. (2002), though flawed in some of its methodology (Shepherd 2004), found evidence of rainfall anomalies downwind of major cities in the southeast United States. That work was inspired by Bornstein and Lin (2000) and their work linked to NASA's Project ATLANTA. They found evidence of warm season precipitation anomalies that were linked to the city.

Further NASA funded work expanded to the coastal regions of Texas. Bouvette et al. (1982) published evidence that the 24-h 100-year storm depth had increased by 15% in suburban areas when compared to the 24-h 100-year storm depth published by the National Weather Service in 1961. They speculated that heavy urbanization was playing a role. Shepherd and Burian (2003) used the TRMM radar and ground-based gauge data to identify anomalies in rainfall around Houston. They hypothesized that the interaction of the urban heat island, mechanical turbulence, and the sea breeze front produced favored regions for convection. Carter et al. (2012) confirmed this possibility using a series of modeling studies on sea breeze- urban heat island interactions. Using a dense rain gage network, Burian and Shepherd (2005) found a statistically significant enhancement to rainfall in Houston. They also revealed that the diurnal cycle had shifted from pre-urban (1940–1958) to post-urban (1984–1999) timeframes. More specifically, the urban and downwind regions had 59% and 30% more rainfall between noon and midnight than an upwind control region. The urban area also had roughly 80% more rainfall occurrences in the same period during the warm season. The aforementioned studies were firmly aligned with other studies that were emerging at the time in other geographic regions. Shepherd et al. (2010) projected Houston's land cover out to 2025 using the UrbanSim growth model and initiated a weather model with that land cover scenario. For the same meteorological forcing, rainfall distribution around the city of Houston changed as a function of whether the landcover was "current" or the "2025" scenario.

Radar-based analysis, hypothetical modeling studies, and field studies all over the world were all converging to similar findings presented by the authors (Ohashi and Kida 2002; Fujibe 2003; Kusaka and Kimura 2004; Mote et al. 2007; Baik et al. 2007; Jin and Shepherd 2008; Ashley et al. 2012; Haberlie et al. 2015). Some combination of Urban Heat Island (UHI)-destabilized boundary layer, urban surface roughness from buildings, and aerosols could initiate convection or modify pre-existing precipitating systems (Niyogi et al. 2011).

A study by Shepherd (2006) explored urban rainfall anomalies in arid regions globally using surface rainfall data and satellite estimates. However, the body of work took an important step in 2009. Hand and Shepherd (2009) applied a new analysis method called a concentration factor. The concentration factor relates spatial variations in rainfall amount and intensity to frequency of occurrence of a certain wind direction. The climatological "downwind" region of Oklahoma City is East-Northeast. It established a prototype methodology for using TRMM and GPM data to conduct analyses of the urban precipitation effect for any city. Figure 53.1 is an example of downwind rainfall anomalies around Oklahoma City represented by this technique. Mitra et al. (2011) expanded PMM-related urban studies to cities in India and identified an urban rainfall effect in the pre-monsoonal rainfall climatology.

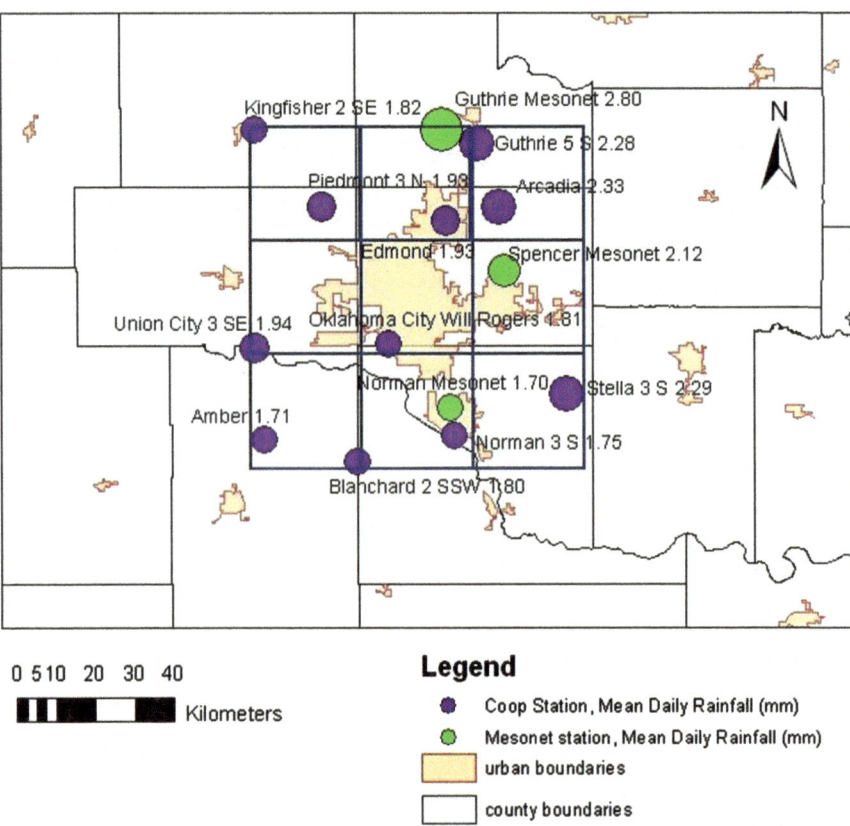

Fig. 53.1 Precipitation anomalies stated in terms of a concentration factor analysis of Hand and Shepherd (2009). The concentration factor relates spatial variations in rainfall amount and intensity to frequency of occurrence of a certain wind direction. The climatological "downwind" region of Oklahoma City is East-Northeast

Studies emerging from the NASA PMM program revealed that not only was the "urban precipitation effect" real, it was evident in heavier rain rates (McLeod et al. 2017). That study also revealed that the rainfall anomaly is a function of the prevailing flow regime. In other words, there may be a predominant downwind anomaly, but the urban effect might be evident at any location for a given flow regime. Debbage and Shepherd (2018b) also found that the urban environment enhanced a pre-existing synoptic scale flooding event in Atlanta (Shepherd et al. 2011). Their results suggested that pre-existing rainfall training was aided by urban-induced convergence zones that were established at the urban – rural interface. This is consistent with finding of Shem and Shepherd (2009) and other scholars (Ashley et al. 2012) that convergence zones on the urban-rural interface and their interactions with prevailing wind flow is critical. Past results, including Shepherd et al. (2002), hypothesized that the storms simply form over the city and translate downwind. While there are likely some aspects of this type of process, it is increasingly clear that the process is more complicated.

Future research will continue to decompose the relative contributions of the urban environment going forward. There is a need to understand what particular atmospheric conditions associated with different types of urban precipitation scenarios, what size and orientation of the city is most associated with the effect, how large-scale precipitation systems interact with the urban environment, representation urban surfaces in models (Jin et al. 2005) and whether aerosols have a suppression or enhancement effect (Jin et al. 2005). For example, the aerosol-cloud-rainfall interaction is a challenging problem that has regional to global implications. Figure 53.2

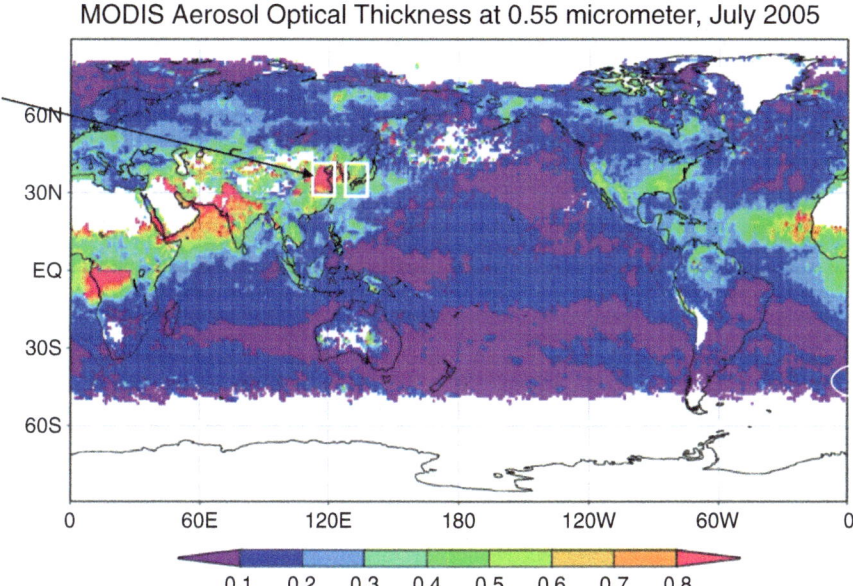

Fig. 53.2 MODIS aerosol optical depth for July 2005. The two boxes represent urban land and oceanic study regions

is an example of how research using TRMM active and passive microwave instrumentation (Jin and Shepherd 2008) can evaluate the precipitation efficiency in high (low) aerosol content regions including highly urbanized regions. The research compared TRMM and MODerate resolution Imaging Spectroradiometer (MODIS) aerosol estimates and found that aerosols affect heavier convective rainfall more so than lighter rainfall. Aerosol processes were also generally more dominant over the ocean region, which suggests that urban land cover dynamics may be the dominant factor there. Such dynamic and thermodynamic mechanisms in an urban system, specifically the urban heat island effect, urban building lifting, and anthropogenic heat fluxes, are proven to be partly responsible for the extreme summer rainfall events in Beijing (Jin et al. 2015).

53.3 Urban Impacts on Winter Precipitation

The launch of the Global Precipitation Measurement (GPM) Mission (Hou et al. 2014) enabled a new dimension in the body of research on urban precipitation. Even with the uncertainties about the "sign" of the change in precipitation due to urbanization, it is widely accepted that urban environments affect rainfall processes (Shepherd et al. 2010; Shem and Shepherd 2009; Lacke et al. 2009; Rosenfeld et al. 2008; Mote et al. 2007). Much of this knowledge has been gained through support from the NASA PMM Project. In fact, a National Academy report on urban meteorology (NAS 2012) heavily acknowledges contributions from NASA funded urban precipitation research. That report also attempted to recommend pathways to bridge the gap between research and application of the knowledge in cities.

However, the impact of the urban environment on snow processes (e.g., snowfall, snowmelt, or snow cover) is not well understood (Shepherd and Mote 2011). The GPM mission enabled a line of inquiry into frozen precipitation analysis because of the additional higher frequency passive microwave channels and the dual-frequency precipitation radar.

Shepherd and Mote (2011) speculated that urban or aerosols effects would also be apparent in winter precipitation events. Changnon (2003) showed that the urban heat island might reduce freezing rain events in urban areas since they are highly dependent on surface temperature. Johnson and Shepherd (2018) found that the influence of the surface and boundary layer urban heat islands, respectively may impact the vertical profile of temperature. This would likely result in "urban signatures" in the distribution of rainfall, mixed precipitation, freezing rain, and snowfall around cities. Their results revealed that 21% of any mixed precipitation observation may be linked to its distance from and dense urban area (Fig. 53.3) along the Washington D.C. to New York City urban climate archipelago (UCA), which is also a new concept (Shepherd et al. 2013) developed during this period of research. UCA's (see Fig. 53.4 for an example) are aggregate systems of urbanized space that

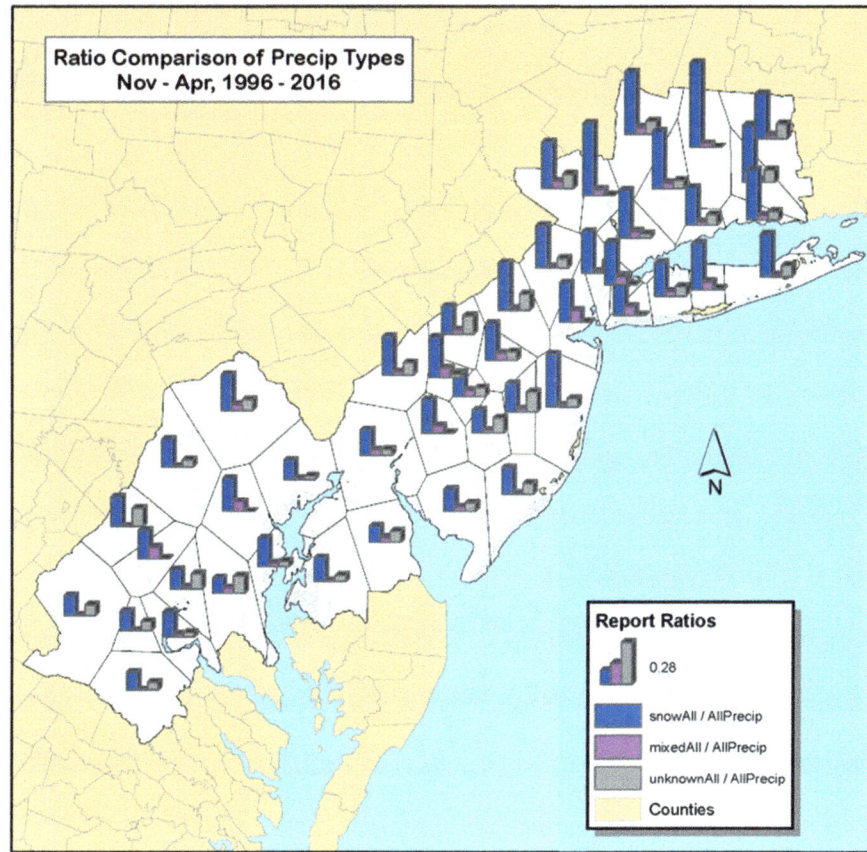

Fig. 53.3 Ratio of precipitation types over the period 1996–2016 (Nov–Apr) in the Washington, Baltimore, Philadelphia, and New York Urban Climate Archipelago. (Adapted from Johnson and Shepherd 2018)

have an impact on some part of the climate system. Johnson and Shepherd (2018) also revealed that reduced wind speeds during mixed phased events allow the boundary layer urban heat island to remain intact. This causes some melting of hydrometeors and more mixed precipitation events. Ongoing studies are using mesoscale models to evaluate the mechanisms behind these observations.

Another thread of work that emerged from these efforts explored the use of PMM and other NASA products to detect snowfall changes and variability. This work has revealed that possible aerosol effects from northern California cities may be modifying, partly, snow cover variability in the downwind mountain regions through various microphysical processes including aerosol-cloud-precipitation interactions, increased snow surface albedo in polluted areas, and snow-albedo feedback.

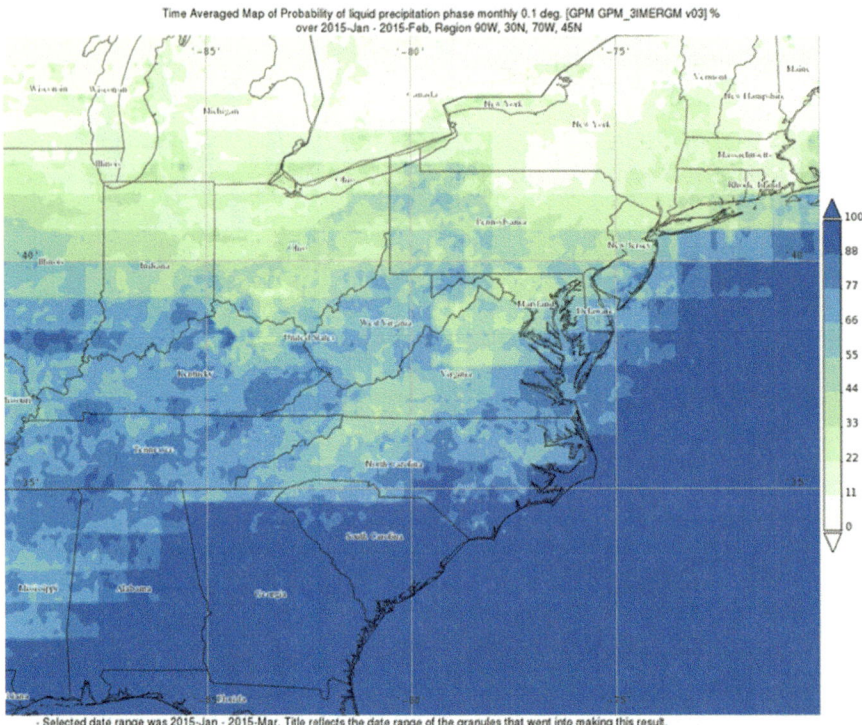

Fig. 53.4 Probability (%) of liquid and frozen precipitation 2015 (January–February) in the DC to NYC Urban Climate Archipelago (UCA) via the GPM IMERG product

53.4 Land Surface Hydrological Response and Water Management

Urban flooding has continued to be a significant hazard to society and a thread of research has focused on the hydrologic response as a function of urbanization. It has been known for some time that impervious surfaces in cities alter the natural water cycle to reduce infiltration and increase surface runoff (Bhaduri et al. 2001). PMM funded research at the University of Utah has explored the use of spaceborne and ground-based rainfall measurements to assess runoff and potential flooding (Reynolds et al. 2008). Using spatial metrics, Debbage and Shepherd (2018a) recently investigated streamflow in the highly urbanized Charlotte, North Carolina to Atlanta, Georgia corridor (Charlanta). They found that increasing urbanization enhanced low and high flow frequency as well as annual peak unit discharge. Contiguous development was the primary driver of the high and low flow enhancement while clustering impervious surfaces in source areas at some distance from the streams amplified the frequency of high flows.

Such findings are important in hydrological and water management design. Reynolds et al. (2005) developed a Geographic Information System (GIS) linkage between mesoscale meteorological model output and an urban watershed model (Fig. 53.5). They tested linkages for simulated urban modified rainfall events and compared them to observations. Burian et al. (2004) performed hydrologic simulations to determine the relative impact of urban enhanced rainfall on urban drainage system performance.

In many parts of the world selected flood scale events are underpredicted or overpredicted (Fig. 53.6) by the global runoff models. Han et al. (2011) found that the common tie among these events was often the location of the event upstream of

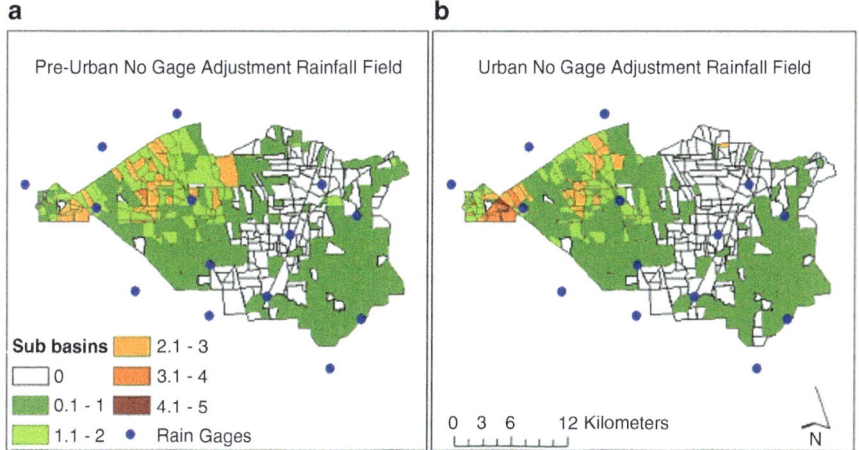

Fig. 53.5 Pre-urban and Post Urban rainfall fields (mm) in the Houston area as represented in a GIS

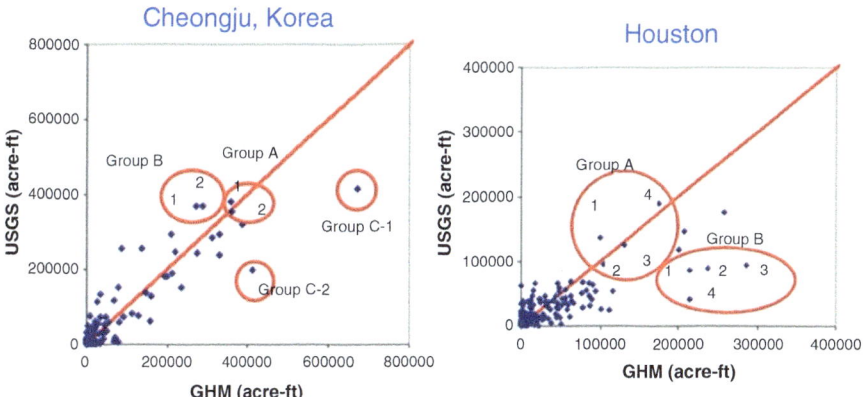

Fig. 53.6 An analysis of runoff values (acre-ft) from a global hydrological model (GHM) and stream gauges. The various groups represent areas that were overestimated or underestimated. Most error was associated with flood control mechanisms upstream. The use of TRMM data products as initialization significantly mitigates these errors

city, a large fraction of which was likely captured by flood control facilities not represented in the model. TRMM data products were tested in a regional flood forecasting and planning system developed for the National Infrastructure Simulation and Analysis Center. Han et al. (2011) investigated the use of TRMM precipitation data products to identify hydrologic changes from urban areas. Using the TRMM rainfall and corresponding runoff records for a set of selected river basins around the world, Han et al. identified urban effects in the rainfall-runoff transformation. They concluded the TRMM precipitation data provides a reasonable means to study global runoff supporting the introduction of a global runoff model. There methods were developed and tested with 17 independent river basins covering a range of slopes. After the slope correction was implemented in the global runoff model, more than 40% of river basins tested showed improved runoff results.

Schroeder et al. (2016) took a different approach to understanding the urbanization and flooding problem. By examining major urban flooding events from 1977 to 2014, this NASA funded research focused on precipitable water. The 99th percentile of the associated climatological precipitable water (PW) dataset was a key threshold for many of the urban flooding cases like the 2013 Colorado floods (Fig. 53.7). They also found that all of the cases exceeded 150% of the climatological mean values. Twenty-nine cases had PW values above two standard deviations from the mean.

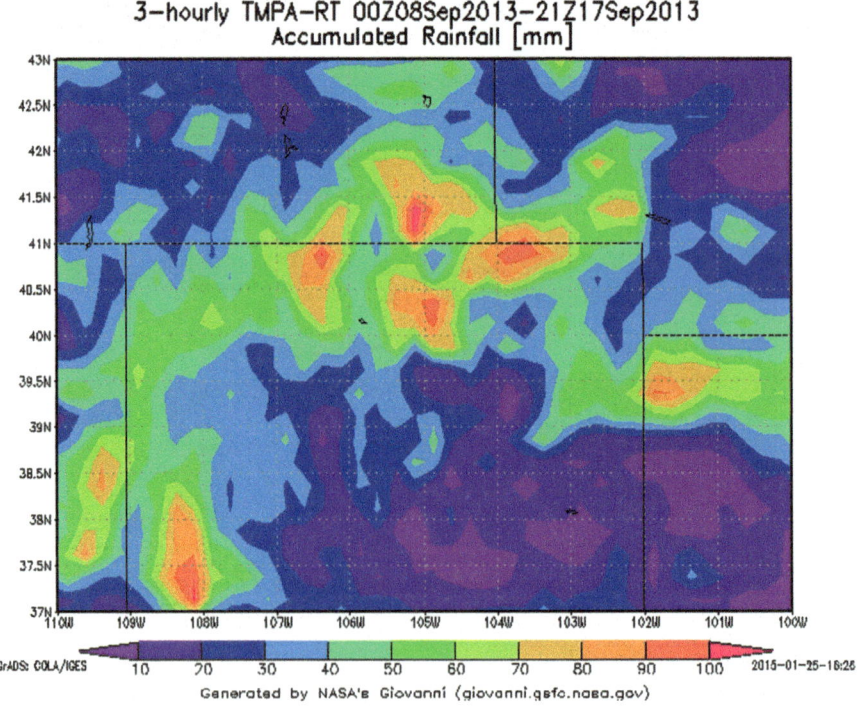

Fig. 53.7 Estimated rainfall rate (mm) for the September 2013 Colorado flood event over the period 8–17 September using the TRMM multi-satellite precipitation analysis

Twenty of which were above the 99th percentile (5 were at/near max value). Eight cases had PW values above 75th percentile. This work was also important because it showed that most of the major urban flooding cases in the United States over the past four decades were associated with a "composite flood sounding" that looked very much like one observed in tropical cyclone environments. This work provides important guidance for weather forecasters trying to assess flooding potential. NOAA's J. J. Gourley described the team's research in this email, *"I just wanted to let you know your research was spot on, and we're now incorporating those findings in a real-time demonstration product - FLASH, http://flash.ou.edu/"—J.J. Gourley/NOAA/NSSL-OU."* (last accessed 3 Apr. 2019).

53.5 Urbanization and the Energy-Food-Water Nexus

Over the past few years, NASA has advanced satellite-based capabilities for collecting precipitation data. Investigators have leveraged this capability for urban hydroclimate studies (Shepherd et al. 2013; Mitra and Shepherd 2016), water availability (Stout et al. 2015), and gross/net primary production (GPP/NPP) for agricultural productivity (Zhao and Running 2010). Published frameworks on the Energy-Water-Food Nexus (EWFN) have not explored hydroclimate topics and have mainly focused on greenhouse gas emissions, resource management, land use, and other aspects of the problem.

A study from the collective body of work of the authors was featured by NASA in a press release (https://www.nasa.gov/feature/goddard/using-nasa-data-to-show-how-raindrops-could-save-rupees, last accessed 26 Nov. 2018). This particular statement from that work (Stout et al. 2015) captures the reasons why this line of important research, "Availability of safe and reliable water supply is an issue in developing nations... Rainwater harvesting (RWH) is a site-specific source control used to satisfy human, agricultural, and safety demands for water..." NASA has repeatedly expressed interest in enabling applications related to freshwater availability and agriculture.

A "first of its kind" set of Precipitation Metrics based on precipitation data, population, and water consumption has been developed through the NASA PMM program (Shepherd and Liu 2018). It is envisioned that such metrics will provide insight on water availability, scarcity, and consumption. Globally, there is limited access to potable water. Rainwater harvesting, capture, and other viable ecosystems services such as water supplementation and ground water recharge can serve as viable alternatives if there is enough available rainfall (Walsh et al. 2014). For example, Stout et al. (2015) found that about 35 gallons of water per person per day is required for cities like Delhi, Kolkata, and Bangalore for basic sustenance. They also noted that rainwater harvesting could provide about 20% of indoor water demand.

Figure 53.8 shows the first prototype global precipitation per person (PPP) metrics. They are derived from the GPM space-borne platform. The Amazon region has a high value because of relatively large precipitation availability in sparsely

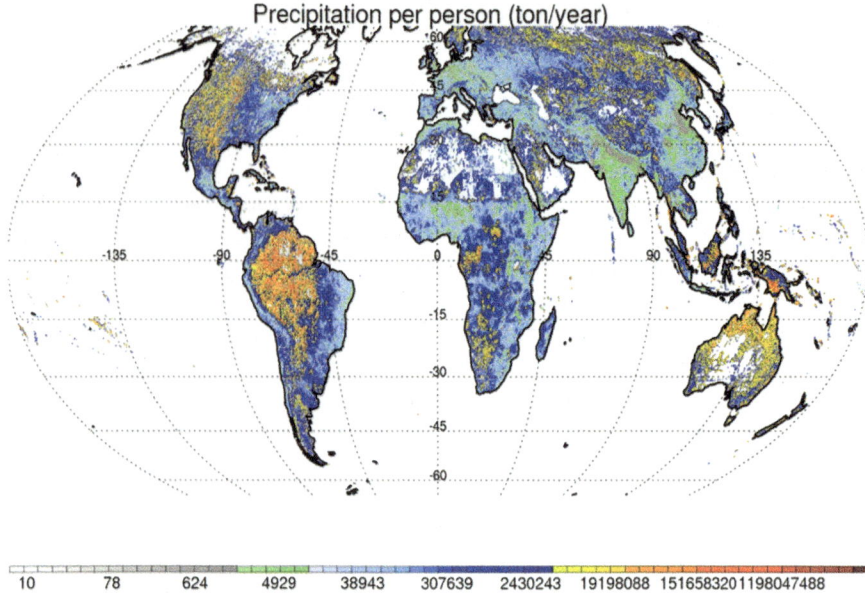

Fig. 53.8 Mean precipitation per person (ton/year) in $0.1° \times 0.1°$ boxes over a 3-year period

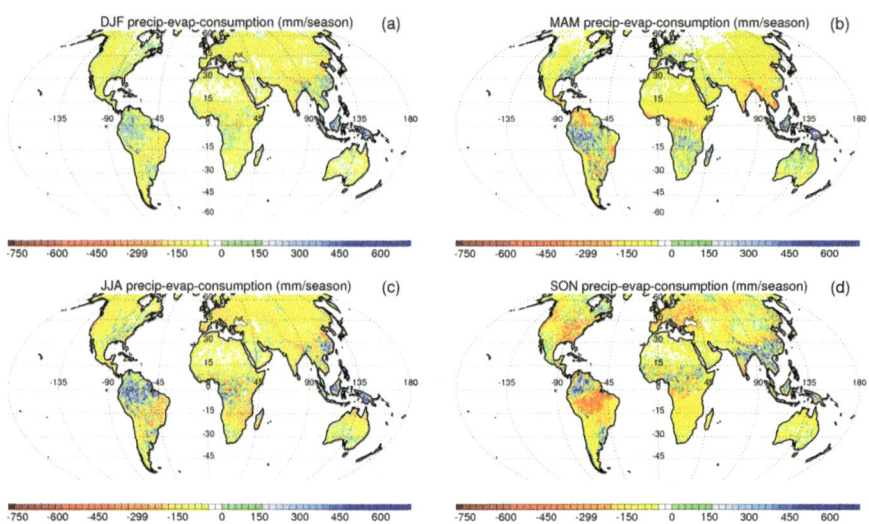

Fig. 53.9 Seasonal estimates of P-E minus consumption over a 4-year period

populated areas. The western United States has relatively high values because of relatively low precipitation availability but a very sparsely populated area. India and China are very densely populated and thought they receive ample precipitation; the PPP is low. This suggests that precipitation alone is not able to support sustenance without ample surficial, groundwater, or managed actions. Shepherd and Liu (2018) have also explored ways to enhance such metrics by including evapotranspiration and consumption (Fig. 53.9).

Figure 53.9 is an example of an analysis of global (Precipitation-Evaporation) minus consumption as a function of season. Colors near 0 indicate likely balance between P-E and consumption. Red colors represent where consumption exceeds (P-E). Though still being tested, this type of value could potentially indicate regions of water stress. For example, much of India experiences a deficit during the pre-monsoonal spring months while the southern part of the country is in a deficit year-round. This research is an innovated way to combine precipitation, urban climate, and EWFN principles. It also supports Mcdonald et al.'s (2011) findings that urban spaces in developing regions will continue to struggle providing water for residents.

Ongoing work continues to analyze the literature. One of the authors, for example, has established relationships with Mehran University of Engineering and Technology faculty and students in Pakistan, and has begun analysis of seasonal monsoon precipitation-agricultural relationships. Research has shown that end of season yields depend on precipitation timing, spatial variability, and consistency of precipitation during the monsoon season (Adnan and Khan 2012). However, the literature only identifies a small quantity of research evaluating precipitation dynamics impact on Barani agricultural production (Mahmood et al. 2012). Collaboration with US-Pakistan Center for Advanced Studies in Water has provided access to Pakistan Bureau of Statistics maize and wheat production data (tons) along with acreage of agricultural area (ha). Due to the spatial and temporal focus of the research, acquisition of NASA TRMM/GPM data products provides average regional precipitation (mm h^{-1}) measurements. Preliminary evaluation of monsoon maize and wheat production (tons ha^{-1}) with weekly and monthly regional precipitation has revealed relationships between quantity and timing of precipitation to end-of-season yields.

The team, led by Menglin Jin, is also exploring another aspect of the EWFN. Water forms a hydraulic force for plant structure and is needed for photosynthesis and metabolic processes to transfer nutrition. Lack of water results in vegetation water stress, or water deficiency. Since water in vegetation strongly absorbs energy in near and middle infrared (NIR and MIR), the reflected solar spectrum can be used in satellite remote sensing technology to monitor wet stress of vegetation. In urban regions, human systems mix with natural and human-induced vegetation systems. Human systems can also be in a water deficiency, which requires responses in resource planning and management to reduce such stress. Green roof technology is a proven practice in urban systems to increase vegetation cover, strengthen soil-atmosphere evapotranspiration, reduce heat and water stress, and introduce other opportunities to supply ecosystem services. A heat-moisture-simulation-enhanced weather forecast model is being used to better represent biophysical and dynamic processes in urban systems (Jin et al. 2014; Zhang et al. 2017).

Water stress, as an important remote sensing-based index, reveals how much water is lacking in a vegetation system. A similar concept can be adapted in an urban system describing lack of water per urban capita or per individual. More importantly, using remote sensing and weather modeling approaches, we examine how much green roof mechanism can reduce urban water stress. The research is attempting to globally assess Urban Water Stress index per Individual (UWSI), using GPM IMERG data and the algorithm developed by PI Shepherd and collaborator Chuntao Liu (Shepherd and Liu 2018). It is also anticipated that urban water stress can be assessed with the intention of understanding the impacts of green roof mechanisms on these indices using regional climate models.

53.6   Concluding Statements

The majority of the world's population is now urban and that will continue to grow. Cities have impacts that scale up to the global climate and back down to the urban space itself. A common thread is the need for water. NASA's innovative satellite-based remote sensing, models, and field studies have enabled an entire generation of studies connecting urbanization and hydroclimate. The initial trajectory of the research in the early part of the twenty-first century was exploratory research to address fundamental questions related to urban precipitation effects. In the past 25 years, the work has evolved to consider some of the most pressing water and EWFN questions of the day. The results are bearing fruit in the scholarly halls, and it is finding its way into decision support activities of nations, agencies, and governments. The most daunting, challenging, and exciting aspects is that new questions continue to emerge.

References

Adnan, S., & Khan, A. H. (2012). Effective rainfall for irrigated agriculture plains of Pakistan. *Pakistan Journal of Meteorolgy, 6,* 61–72. Available at http://www.pmd.gov.pk/rnd/rnd_files/vol6_issue11/6_Effective%20Rainfall%20for%20Irrigated%20Agriculture%20Plains%20of%20Pakistan.pdf, last accessed 26 Nov 2018.

Ashley, W. S., Bentley, M. L., & Stallins, J. A. (2012). Urban-induced thunderstorm modification in the Southeast United States. *Climatic Change, 113,* 481–498. https://doi.org/10.1007/s10584-011-0324-1.

Baik, J.-J., Kim, Y.-H., Kim, J.-J., & Han, J.-Y. (2007). Effects of boundary-layer stability on urban heat island-induced circulation. *Theoretical and Applied Climatology, 89,* 73–81. https://doi.org/10.1007/s00704-006-0254-4.

Bhaduri, B., Minner, M., Tatalovich, S., & Harbor, J. (2001). Long-term hydrologic impact of urbanization: A tale of two models. *Journal of Water Resources Planning and Management, 127,* 13–19. https://doi.org/10.1061/(ASCE)0733-9496(2001)127:1(13).

Bornstein, R., & Lin, Q. (2000). Urban heat islands and summertime convective thunderstorms in Atlanta: Three case studies. *Atmospheric Environment, 34,* 507–516. https://doi.org/10.1016/S1352-2310(99)00374-X.

Bouvette, T. C., Lambert, J. L., & Bedient, P. B. (1982). Revised rainfall frequency analysis for Houston. *Journal of the Hydraulics Division, ASCE, 108,* 515–528. Available at http://cedb.asce.org/CEDBsearch/record.jsp?dockey=0034037, last accessed 26 Nov 2018.

Burian, S. J., & Shepherd, J. M. (2005). Effects of urbanization on the diurnal rainfall pattern in Houston. *Hydrological Processes, 19*, 1089–1103. https://doi.org/10.1002/hyp.5647.

Burian, S. J., Hooshialsadat, P., Reynolds, S., & Shepherd, J. M. (2004). Effect of cities on rainfall and the implications for drainage design. In G. Sehlke, D. F. Hayes, & D. K. Stevens (Eds.), *Critical transitions in water and environmental resources management, proceedings of world water and environmental resources congress.* New York: ASCE.

Carter, W. M., Shepherd, J. M., Burian, S., & Jeyachandran, I. (2012). Integration of lidar data into a coupled mesoscale-land surface model: A theoretical assessment of sensitivity of urban-coastal mesoscale circulations to urban canopy. *Journal of Atmospheric and Oceanic Technology, 29*, 328–346. https://doi.org/10.1175/2011JTECHA1524.1.

Changnon, S. A. (2003). Urban modification of freezing-rain events. *Journal of Applied Meteorology, 42*, 863–870. https://doi.org/10.1175/1520-0450(2003)042<0863:UMOFE>2.0.CO;2.

Changnon, S., Semonin, R., Auer, A., Braham, R., & Hales, J. (1981). *METROMEX: A review and summary* (Meteorological monographs book series) (Vol. 40, p. 181). American Meteor Society. ISBN: 0-933876-52-1.

Debbage, N., & Shepherd, J. M. (2018a). The influence of urban development patterns on streamflow characteristics in the Charlanta Megaregion. *Water Resources Research, 54*, 3728–3747. https://doi.org/10.1029/2017WR021594.

Debbage, N., & Shepherd, J. M. (2018b). Determining the influence of urbanization on the spatiotemporal characteristics of runoff and precipitation during the 2009 Atlanta flood using a coupled land surface-atmospheric model. *Journal of Hydrometeorology, 20*, 3–21. https://doi.org/10.1175/JHM-D-18-0010.1.

Fujibe, F. (2003). Long-term surface wind changes in the Tokyo metropolitan area in the afternoon of sunny days in the warm season. *Journal of the Meteorological Society of Japan, 81*, 141–149. https://doi.org/10.2151/jmsj.81.141.

Haberlie, A. M., Ashley, W. S., & Pingel, T. J. (2015). The effect of urbanisation on the climatology of thunderstorm initiation. *Quarterly Journal of the Royal Meteorological Society, 141*, 663–675. https://doi.org/10.1002/qj.2499.

Han, W. S., Burian, S. J., & Shepherd, J. M. (2011). Assessment of satellite-based rainfall estimates in urban areas in different geographic and climatic regions. *Natural Hazards, 56*, 733–747. https://doi.org/10.1007/s11069-010-9585-7.

Hand, L., & Shepherd, J. M. (2009). An investigation of warm season spatial rainfall variability in Oklahoma City: Possible linkages to urbanization and prevailing wind. *Journal of Applied Meteorology and Climatology, 48*, 251–269. https://doi.org/10.1175/2008JAMC2036.1.

Hou, A. Y., Kakar, R. K., Neeck, S., Azarbarzin, A. A., Kummerow, C. D., Kojima, M., Oki, R., Nakamura, K., & Iguchi, T. (2014). The Global Precipitation Measurement mission. *Bulletin of the American Meteorological Society, 95*, 701–722. https://doi.org/10.1175/BAMS-D-13-00164.1.

Jin, M., & Shepherd, J. M. (2008). Aerosol relationships to warm season clouds and rainfall at monthly scales over East China: Urban land versus ocean. *Journal of Geophysical Research, 113*, D24S90. https://doi.org/10.1029/2008JD010276.

Jin, M., Shepherd, J. M., & King, M. D. (2005). Urban aerosols and their interaction with clouds and rainfall: A case study for New York and Houston. *Journal of Geophysical Research, 110*, D10S20. https://doi.org/10.1029/2004JD005081.

Jin, M., Mullens, T., & Bartholomew, H. (2014). Evaluate CLM skin temperature and soil moisture simulation using ARM ground observation. *Climate, 2*, 279–295. https://doi.org/10.3390/cli2040279.

Jin, M., Li, Y., & Su, D. (2015). Urban-induced mechanisms for an extreme rainfall event in Beijing. *Climate, 3*, 193–209. https://doi.org/10.3390/cli3010193.

Johnson, B., & Shepherd, J. M. (2018). An urban-based climatology of winter precipitation in the Northeast United States. *Urban Climate, 24*, 205–220. https://doi.org/10.1016/j.uclim.2018.03.003.

Kusaka, H., & Kimura, F. (2004). Thermal effects of urban canyon structure on the nocturnal heat island: Numerical experiment using a mesoscale model coupled with an urban canopy model. *Journal of Applied Meteorology, 43*, 1899–1910. https://doi.org/10.1175/JAM2169.1.

Lacke, M., Mote, T. L., & Shepherd, J. M. (2009). Aerosols and associated precipitation patterns in Atlanta. *Atmospheric Environment, 28*, 4359-4373. https://doi.org/10.1016/j.atmosenv.2009.04.022.

Liu, Z., He, C., Zhou, Y., & Wu, J. (2014). How much of the world's land has been urbanized, really? A hierarchical framework for avoiding confusion. *Landscape Ecology, 29*, 763–771. https://doi.org/10.1007/s10980-014-0034-y.

Lowry, W. P. (1998). Urban effects on precipitation amount. *Progress in Physical Geography, 22*, 477–520. https://doi.org/10.1177/030913339802200403.

Mahmood, N., Ahmad, B., Hassan, S., & Bakhsh, K. (2012). Impact of temperature ADN precipitation on rice productivity in rice-wheat cropping system of Punjab province. *Journal of Animal and Plant Sciences, 22*, 993–997. Available at http://thejaps.org.pk/docs/V-22-4/29.pdf, last accessed 26 Nov 2018.

McDonald, R., Green, P., Balk, D., Fekete, B. M., Revenga, C., Todd, M., & Montgomery, M. (2011). Urban growth, climate change, and freshwater availability. *PNAS, 108*, 6312–6317. https://doi.org/10.1073/pnas.1011615108.

McLeod, J., Shepherd, J. M., & Konrad, C. (2017). Spatio-temporal rainfall patterns around Atlanta, Georgia and possible relationships to urban land cover. *Urban Climate, 21*, 27–42. https://doi.org/10.1016/j.uclim.2017.03.004.

Mitra, C., & Shepherd, J. M. (2016). Urban precipitation: A global perspective. In K. C. Seto, W. D. Solecki, & C. A. Griffith (Eds.), *Routledge handbook of urbanization and global environment change* (pp. 152–168). London: Routledge. ISBN: 9781315849256.

Mitra, C., Shepherd, J. M., & Jordan, T. (2011). On the relationship between the pre-monsoonal rainfall climatology and urban land cover dynamics in Kolkata city, India. *International Journal of Climatology, 32*, 1443–1454. https://doi.org/10.1002/joc.2366.

Mote, T. L., Lacke, M. C., & Shepherd, J. M. (2007). Radar signatures of the urban effect on precipitation distribution: A case study for Atlanta, Georgia. *Geophysical Research Letters, 34*, L20710. https://doi.org/10.1029/2007GL031903.

NAS. (2012). *Urban meteorology: Scoping the problem, defining the need*. Washington, DC: National Research Council Committee on Urban Meteorology, The National Academies Press. https://doi.org/10.17226/13328.

Niyogi, D., Pyle, P., Lei, M., Arya, S., Kishtawal, C., Shepherd, M., Chen, F., & Wolfe, B. (2011). Urban modification of thunderstorms: An observational storm climatology and model case study for the Indianapolis urban region. *Journal of Applied Meteorology and Climatology, 50*, 1129–1144. https://doi.org/10.1175/2010JAMC1836.1.

Ohashi, Y., & Kida, H. (2002). Local circulations developed in the vicinity of both coastal and inland urban areas: A numerical study with a mesoscale atmospheric model. *Journal of Applied Meteorology, 41*, 30–45. https://doi.org/10.1175/1520-0450(2002)041<0030:LCDITV>2.0.CO;2.

Reynolds, S., Burian, S. J., Shepherd, J. M., & Manyin, M. (2005). The effect of urbanization-induced rainfall variability on hydrologic response: linking mesoscale meteorological model output to an urban watershed model. In Proceedings of ASCE-EWRI world water and environmental resources conference, 15–19 May, Anchorage, Alaska.

Reynolds, S., Burian, S., Shepherd, J. M., & Manyin, M. (2008). Urban induced rainfall modifications on urban hydrologic response. In W. James et al. (Eds.), *Reliable modeling of urban water systems* (pp. 99–122). Guelph, ON: Computational Hydraulics International.

Rosenfeld, D., Lohmann, U., Raga, G. B., O'Dowd, C. D., Kulmala, M., Fuzzi, S., Reissell, A., & Andreae, M. O. (2008). Flood or drought: How do aerosols affect precipitation? *Science, 321,* 1309–1313. https://doi.org/10.1126/science.1160606.

Schroeder, A., Basara, J., Shepherd, J. M., & Nelson, S. (2016). Insights into atmospheric contributors to urban flash flooding across the United States using an analysis of rawinsonde data and associated calculated parameters. *Journal of Applied Meteorology and Climatology,* 55, 313-323. https://doi.org/10.1175/JAMC-D-14-0232.1.

Seto, K., & Shepherd, J. M. (2009). Global urban land-use trends and climate impacts. *Current Opinion in Environmental Sustainability,* 1, 89–95. https://doi.org/10.1016/j.cosust.2009.07.012.

Shem, W., & Shepherd, J. M. (2009). On the impact of urbanization on summertime thunderstorms in Atlanta: Two numerical model case studies. *Atmospheric Research, 92,* 172–189. https://doi.org/10.1016/j.atmosres.2008.09.013.

Shepherd, J. M. (2004). A reply to Diem et al.'s commentary on "a recent literature contribution focused on urban-induced rainfall in Atlanta". *Journal of Applied Meteorology, 43,* 951–957. https://doi.org/10.1175/1520-0450(2004)043<0951:R>2.0.CO;2.

Shepherd, J. M. (2006). Evidence of urban-induced precipitation variability in arid climate regimes. *Journal of Arid Environments, 67,* 607–628. https://doi.org/10.1016/j.jaridenv.2006.03.022.

Shepherd, J. M. (2013). Impacts of urbanization on precipitation and storms: Physical insights and vulnerabilities. In R. Pielke Sr. (Ed.), *Climate vulnerability.* Elsevier, 1570 pp, ISBN: 9780123847034.

Shepherd, J. M., & Burian, S. J. (2003). Detection of urban-induced rainfall anomalies in a major coastal city. *Earth Interactions, 7,* 1–14. https://doi.org/10.1175/1087-3562(2003)007<0001:DOUIRA>2.0.CO;2.

Shepherd, J. M., & Liu, C. (2018). Global precipitation metrics for assessing Energy-Water-Food Nexus challenges in urban spaces. *Applied Geography,* in review.

Shepherd, J. M., & Mote, T. L. (2011). Can cities create their own snowfall? What observations are required to find out? *IEEE EarthZine.* Available at earthzine.org/can-cities-create-their-own-snowfall-what-observations-are-required-to-find-out/, last accessed 6 March 2020.

Shepherd, J. M., Pierce, H., & Negri, A. J. (2002). On rainfall modification by major urban areas: Observations from space-borne radar on TRMM. *Journal of Applied Meteorology, 41,* 689–701. https://doi.org/10.1175/1520-0450(2002)041<0689:RMBMUA>2.0.CO;2.

Shepherd, J. M., Carter, W. M., Manyin, M., Messen, D., & Burian, S. (2010). The impact of urbanization on current and future coastal convection: A case study for Houston. *Environmental Planning, 37,* 284–304. https://doi.org/10.1068/b34102t.

Shepherd, J. M., Mote, T. L., Nelson, S., McCutcheon, S., Knox, P., Roden, M., & Dowd, J. (2011). An overview of synoptic and mesoscale factors contributing to the disastrous Atlanta flood of 2009. *Bulletin of the American Meteorological Society, 92,* 861–870. https://doi.org/10.1175/2010BAMS3003.1.

Shepherd, J. M., Anderson, T., Bounoua, L., Horst, A., Mitra, C., & Strother, C. (2013). Urban climate archipelagos: A new framework for urban-climate interactions. *IEEE Earthzine.* Available at https://www.fs.usda.gov/treesearch/pubs/46043, last accessed 26 Nov 2018.

Souch, C., & Grimmond, S. (2006). Applied climatology: Urban climate. *Progress in Physical Geography, 30,* 270–279. https://doi.org/10.1191/0309133306pp484pr.

Stout, D. T., Walsh, T. W., & Burian, S. J. (2015). Ecosystem services from rainwater harvesting in India. *Urban Water Journal, 14,* 561–573. https://doi.org/10.1080/1573062X.2015.1049280.

UN. (2014). *World urbanization prospects, the 2014 revision: Highlights.* United Nations Publications. Available at https://esa.un.org/unpd/wup/publications/files/wup2014-highlights.pdf, last accessed 26 Nov 2018.

Walsh, T., Pomeroy, C. A., & Burian, S. J. (2014). Hydrologic analysis of a watershed-scale rainwater harvesting program. *Journal of Hydrology, 508C*, 240–253. https://doi.org/10.1016/j. jhydrol.2013.10.038.

Zhang, H., Jin, M., & Leach, M. (2017). A study of the Oklahoma City urban heat island effect using a WRF/single-layer urban canopy model, a joint urban 2003 field campaign, and MODIS satellite observations. *Climate, 5*, 72. https://doi.org/10.3390/cli5030072.

Zhao, M., & Running, S. W. (2010). Drought-induced reduction in global terrestrial net primary production from 2000 through 2009. *Science, 329*, 940–943. https://doi.org/10.1126/science. 1192666.

Zhou, Y., & Shepherd, J. M. (2009). Atlanta's urban heat island under extreme heat conditions. *Natural Hazards, 52*, 639–668. https://doi.org/10.1007/s11069-009-9406-z.

Chapter 54
Validation of Climate Models

Francisco J. Tapiador

Abstract The validation of climate models using satellite precipitation measurements presents some particularities. By construction, climate models are different from reanalyses and numerical weather prediction models, and several specificities affect the methods of validation and the interpretation of the results. Thus, for instance, an important topic to consider is that climate models are not always intended to provide the temporal succession of the weather but the statistics of the weather over long periods, and that fact affects the validation strategy. In addition, there is a set of quality standards that need to be fulfilled even before analyzing the statistics or the qualitative agreement with observations.

Keywords Precipitation · Rainfall · Climate models · Numerical weather prediction models · Reanalysis · Raingauges · Disdrometers · IPCC · Quality assurance · SST · CRU · GPCC · CPC · CMAP · GPCP · TRMM · TMI · TMPA · Regional climate models · CESM · Global climate models · Annual variability · Seasonal variability

54.1 The Importance of Validation

The scope of climate models has always been testing scientific hypotheses, but now they are increasingly used to inform energy, transportation and urban policies, among other aspects of human life. The basis of model applicability in planning stems from the model's ability to characterize future climates up to the required accuracy and precision (Tapiador et al. 2012, 2017). But before that, it is extremely important to evaluate if the model is appropriate for the application by comparing its results with present-climate measurements (Kidd et al. 2017; Kim et al. 2017; Kucera et al. 2013).

F. J. Tapiador (✉)
University of Castilla-La Mancha, Toledo, Spain
e-mail: francisco.tapiador@uclm.es

© Springer Nature Switzerland AG 2020   1073
V. Levizzani et al. (eds.), *Satellite Precipitation Measurement*, Advances in Global
Change Research 69, https://doi.org/10.1007/978-3-030-35798-6_30

The validation of a climate model is critical to improve its physics and to estimate its ability to provide input to decision makers (Tapiador 2010). The current existence of precipitation measurements with large temporal extent and a global scope has made it possible to build datasets for validating climate models (Adler et al. 2003, 2016; Huffman et al. 1997, 2001, 2007, 2009, 2010; Joyce et al. 2004; Kidd et al. 2003, 2016; Kummerow et al. 2001, 2007; Sorooshian et al. 2000; Xie et al. 2003). In order to fulfill the requirement of global coverage and accuracy, existing products integrate satellite-derived retrievals from many sensors (radars, radiometers) with direct ground observations (gauges, disdrometers, radars), which are used as reference or a calibration source for the satellites (Haddad et al. 1997; Peters-Lidars et al. 2015; Grecu et al. 2009; Retalis et al. 2016; Turk et al. 2000). Unfortunately, there are no perfect data sets that are optimal in both sampling and retrieval error (Katsanos et al. 2016; Kidd 2001; Kidd and Huffman 2011; Kidd and Levizzani 2011; Levizzani et al. 2004; Michaelides et al. 2009; Michaelides 2013, 2014, 2016). The IPCC AR5 (IPCC 2014) acknowledgment of large observational uncertainties in precipitation observations for climate model validation is a reminder of this situation, but also a major motivation for devoting time and resources to improving the methods, possibly through new ingenious techniques (Haddad et al. 2017). On this respect, the advent of the GPM core satellite (Skofronick-Jackson et al. 2017) has greatly improved the ability to precisely measure precipitation.

In spite of the societal impact, there are not many protocols, best practices and quality assurance (QA) methods to validate climate models using observations. However, the public and the decision makers demand that the science behind policies is traceable, transparent and auditable. Drawing on Tapiador et al. (2017), this chapter reviews and summarizes a proposal and provides guidance to validating climate models.

54.2   Precipitation: The Privileged Metric for Validation

It is fair to ask why Tapiador et al. (2017) and this chapter on validation focus on precipitation and neglect other important variables such as temperature, geopotential or humidity. The reason is that precipitation indeed is a privileged metric to validate models (Tapiador et al. 2019). A major reason is that precipitation at high spatial and temporal resolutions is very difficult to model and therefore, if a model compares well with precipitation observations it is a good indicator of good performance (Ebert et al. 1996, 2007).

Apart from the practical applications for validation, there are fundamental (in the sense of 'basic principles') reasons to favor precipitation, which includes the connection between precipitation and global energy budget (L'Ecuyer et al. 2015). Another reason is the fact that precipitation is often considered as a proxy for inferring change statistics in extreme events (Takahashi and Dewitte 2016) and therein can be considered a better metric of ENSO extremes than SST anomalies alone (Tapiador et al. 2019).

Therefore, precipitation is not only a convenient variable to validate models but the ultimate test of model performance. A poor modeling of instantaneous (i.e., model grid step) precipitation is not a major concern, but a good result can certainly make modelers confident in their dynamical core and parameterizations as it is hard to consistently resolve these processes by chance.

Getting precipitation right is challenging for any model, and most modelers do not consider a 200% error in instantaneous, 25 km resolution a major deficiency: the patchy nature of rain and the many geophysical fields that need to be precisely aligned to provide the perfect forecast support such leeway. Thus, it is not uncommon to aggregate the model estimates of precipitation when comparing with observations, either in time or in space, or even to rely on probability distribution functions (PDFs), which are an even more aggregated estimate.

54.3 Basic Comparisons

It is important to be aware of the limitations of precipitation data sets to avoid extracting unsubstantiated conclusions when assessing climate model abilities. The strengths and limitations of the data sets for climate modeling applications need to be considered in order to transcend the first stage of validation: the geographical comparison of the annual means over the model domain. This first comparison is something that has to be done, but a good match between models and observations is a necessary but not sufficient condition to evaluate model performance.

Annual means, such as those shown in Fig. 54.1, can hide compensating errors across the year, especially in areas where the annual cycle is pronounced. Models can underestimate equinoctial precipitation in mid-latitudes and overestimate precipitation near the solstices, the net result agreeing with annual mean observations.

The comparison of latitude or longitude averages is even more crude and can hide radically different performances over land and ocean.

Seasonal comparisons, such as those shown in Fig. 54.2, are a second stage of the validation process. If the model compares well with the observations, then there is increasing confidence in the model's representation of the real climate.

Comparisons considering the uncertainties and limitations of the reference data are rare and cumbersome but nonetheless necessary. The basic set of statistics for validating precipitation in climate models are the annual and seasonal comparisons, the analyses of bias and correlations (through scatterplots or Taylor diagrams), PDFs, latitudinal and longitudinal transects, and time-latitude or longitude diagrams. Spatial indexes such as Geary's or Moran's have also been used to gauge spatial decorrelation (Tapiador 2010, Tapiador et al. 2019). More sophisticated plots, such as the Hovmöller diagram, are useful for model outputs, but not really suited for comparing with small or non-homogenous observational data sets.

A third stage that is far more complex is to delve into direct validation of the physics of the model (Stephens et al. 2002, 2008; Tao et al. 2001, 2006, 2013, 2014, 2016a, b). That implies sufficient knowledge on how the reference datasets are built.

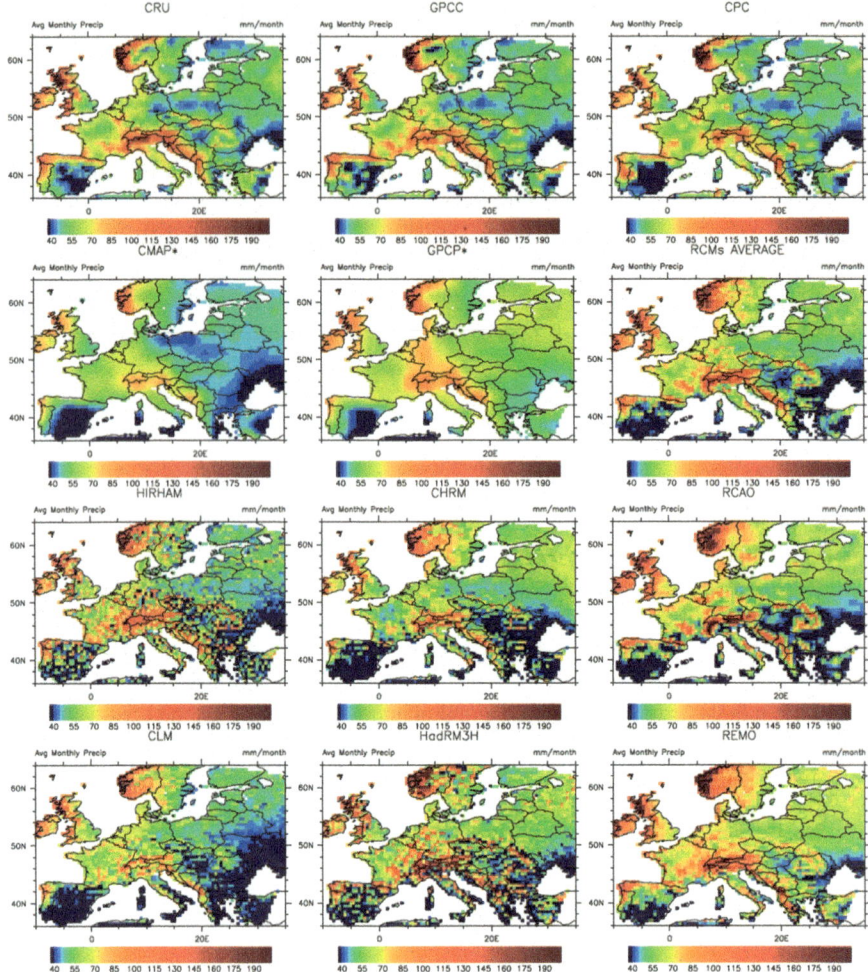

Fig. 54.1 Mean annual precipitation as depicted in five observational databases (CRU, GPCC, CPC, CMAP and GPCP) compared with the ensemble average of several Regional Climate Models (RCMs: first and second rows), and a few individual RCMs (third and fourth rows). The ensemble average compares best with the observations than most of the models but differences are apparent between the observations themselves. [*CRU* Climate Research Unit, *GPCC* Global Precipitation Climatology Centre, *CMAP* CPC Merged Analysis of Precipitation, *GPCP* Global Precipitation Climatology Project.] (Adapted from Tapiador 2010)

Thus, for instance, datasets including infrared data have, by construction, a high dependence of cloud height and a less direct relationship with the ice content in the melting layer, especially at night. This has to be considered when the parameterizations are validated. The details are intricate and specific information about the empirical values and assumptions in the microphysics of precipitation is required.

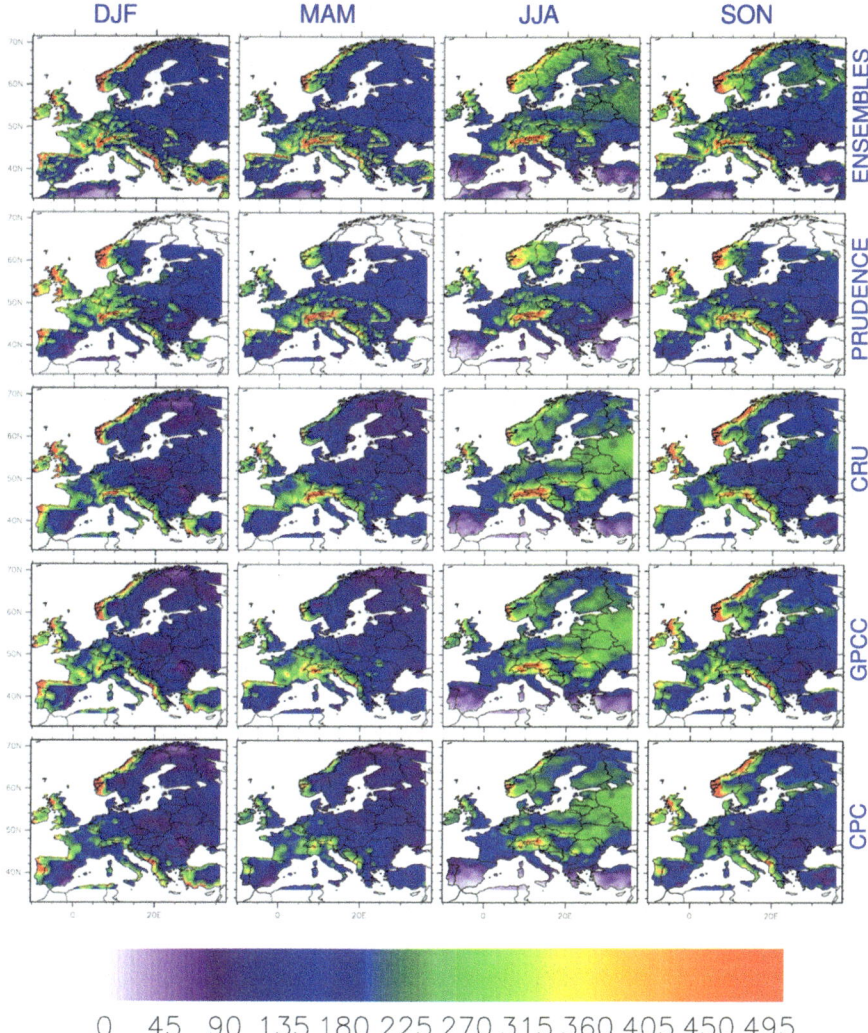

Fig. 54.2 Seasonal precipitation climatologies (1961–2000, columns) derived from ten RCMs nested on reanalysis (ENSEMBLES project data, 25 km grid) compared with ten RCMs forced with GCMs (PRUDENCE project, 50 km grid), and three observational databases at 50 km resolution: CRU rain gauge data, GPCC rain gauge data, and CPC (Climate Prediction Center) rain gauge and satellite-merged data. Units are mm year^{-1}. (Adapted from Tapiador et al. 2011)

54.4   Quality Standards in the Validation of Climate Models

There is growing need of accountability in climate science. Models have long stepped into politics and are now used to drive energy policies and plan large infrastructures such as dams and roads (Tapiador 2009, 2010). Their results are even considered in geopolitical analyses. It is therefore crucial to provide as much confidence as possible not only on their results ('performances'), but also on the way the results are generated ('the quality').

It is important to be mindful of the terminology used, as the meaning of 'quality' differs between communities. In the original realm, which was the industry, 'quality' is precisely defined by the quality assurance (QA) concept. The term does not refer to the intrinsic value of the product but to 'providing confidence that quality requirements will be fulfilled' (e.g., ISO 9000). Thus, QA implies not only that the product is suited for the specific purpose it was built but also that the product has been created following a well-defined set of rules and methods that builds confidence in the whole production process.

QA procedures are designed to minimize errors and mistakes, setting double-blind evaluations, sanity checks and providing a traceable flow of the several stages of the process of generating the product. It is worth clarifying that passing QA does not imply a good performance of the product. All it means is that the product conforms to standards (Tapiador et al. 2017).

Seasonal comparisons, such as those shown in Table 54.1, are a second stage of the validation process. If the model compares well with the observations, then there is increasing confidence in the model's representation of the real climate.

Table 54.1 Basic, extended and full set of requirements of data and climate models involved in validation

Tier	Requirements	
	Validation data	Climate models
Tier 1: Basic confidence	ATBD available	Full technical documentation of the model core is available (incl. Dynamical core, numerical methods, parameterization, and empirical coefficients used for tuning)
	Algorithms or data set creation methodology has been described in peer-reviewed literature	The model has been fully described in a peer-reviewed journal
Plus		
Tier 2: Extended confidence	Products are publicly and freely available	Outputs are publicly and freely available
	There are published comparisons with other data set	Aggregated results compare reasonably well with pre-existing models
	Case studies are available	Benchmarking and case studies are available
Plus		
Tier 3: Full confidence	The code for the algorithms is publicly available/replicable	Source code is publicly available
		The results of the simulations are fully replicable by a third party
	The data set has been successfully used by several independent groups and the results have been published in reputable journals	The model has been used by several independent groups and the results have been published in reputable journals

Adapted from Tapiador et al. 2017

A prior step of QA is about the confidence of the elements involved in the validation. The 'three-tier' scheme of basic, extended, and full-confidence in both data and observations can be useful to gauge the confidence we assign to the validation exercise (Table 54.1). The scheme is intended to provide increasing level of assurance on the quality of the precipitation retrieval algorithm, on the one side, and on the climate model to be validated, on the other side.

Tier 1 is a basic level implying that the Algorithm Theoretical Basis Document (ATBD) of the dataset is accessible (for a canonical example see Huffman et al. 2015), and that the full technical documentation of the model is available (incl. Dynamical core, numerical methods, parameterization, and empirical coefficients used for tuning). It also implies that the algorithms or data set creation methodology have been described in peer-reviewed literature. For models, the requirement is that the model has been fully described in at least one peer-reviewed journal.

The extended, Tier 2 level of confidence is achieved if, in addition to Tier 1 requirements: the products are publicly and freely available and the model outputs are also publicly and freely available; there are published comparisons with other data sets; aggregated models results compare reasonably well with pre-existing models; case studies are available for the observations; and benchmarking and case studies of the model are available.

In Tier 3, full confidence level is a little bit more exigent in that it adds that the code for the retrieval algorithms has to be publicly available/replicable. That means that the source code of the model needs to be publicly ready for scrutiny; the results of the simulations fully replicable by a third party; and the data set successfully used by several independent groups. The results of the retrieval algorithm also should have been published in reputable journals; and the same for the model: having been used by several independent groups and their results published in respectable journals.

It is worth repeating here that once science ventures into providing input for characterizing climate change and setting adaptation and mitigation policies, only models with the higher confidence (Tier 3) should be used. Examples of full confidence level data sets are the Global Precipitation Climatology Project (GPCP) database (Huffman et al. 1997, 2009) on the data side, and the Community Earth System Model (CESM) on the model side. The CESM satisfies all the requirements described above: several versions of the code run in independent research centers, there is a community improving the model, there are many case studies to compare with, and the bugs that might feature in such complex and large codes are swiftly reported and corrected.

54.5 A Checklist for Validation of Climate Models

Tapiador et al. (2017) also proposed a checklist for the validation of climate models. The list (Table 54.2) gathers both common mistakes that sometimes puzzle new-comers to the area, such as taking free-wheeling climate model outputs as the

Table 54.2 A checklist of known issues that must be considered in the interdisciplinary field of validation of precipitation outputs from climate models

1. Rain gauges provide pointwise estimates that may be not fully representative of the area, especially for large areas with a few observations (e.g., the Amazon basin).

2. Rain gauges have known technical limitations and biases and the spatial distribution/length record of the instruments is highly variable.

3. Ground radars are characterized by many sources of uncertainty (i.e., beam blockage, attenuation) that should be taken into account.

4. Precipitation (solid, liquid and mixed phase) has a large spatial and temporal variability making its validation both challenging and important using precipitation datasets in model validation.

5. Satellite estimates are indirect and have limited temporal sampling, and this should be considered in the comparisons.

6. Satellites estimates over land, coast and ocean are derived using different methods and assumptions.

7. Merged precipitation databases are not intended for trend analyses as sensors drift and/or are available over limited time spans.

8. Many of the techniques used in Level-2 products are built upon Bayesian estimates (i.e., they require a prior estimate).

9. The quality of Level-3 precipitation products is driven by microwave observations and therefore is dependent on their availability and quality.

10. The error characteristics resulting from the merging of disparate datasets are not well known.

11. All climate models are tuned to observations, and this must be considered for ensuring a truly independent validation.

12. Global measurements of microphysics are important to avoid overfitting models to empirical parameters.

13. There are known uncertainties in the estimation of diabatic heating fields that affect how models represent some precipitation processes.

14. Model outputs that have been bias-corrected or that are the results of model output statistic techniques cannot be validated.

15. Series derived from GCM-driven RCMs cannot be directly compared with time series of observations.

16. High-resolution global cloud-resolving models (G-CRM) are becoming best suited than RCMs to inform policies and advance our knowledge of the physics of precipitation.

17. Parameterizations can only be validated with data not used in their development and tuning.

18. 'Scope principle': a model cannot claim performances at better resolutions that those at which it has been validated.

19. Blending methods in deriving global precipitation products involves subtleties than must be considered in any validation process.

20. There are significant latitudinal differences in the satellite and ground based estimates in terms of known biases and uncertainties.

21. Parameters and techniques used in the estimation process using satellites and rain gauges may not be universally applicable, both in space and time.

22. Ground validation campaigns are essential for improving the representation of precipitation in models.

23. End-to-end characteristics of the satellite-based retrieval process are not yet fully understood.

24. There is less agreement among satellite products in trends and variability at global scale than in regional variability.

(continued)

Table 54.2 (continued)

25. The precise measurement of shallow and very light precipitation still represents a scientific challenge.
26. While precipitation is a key variable to validate models, there is not agreement in the reference to be compared with. More research and targeted observations are required to fill this gap.
27. Public auditing of model code and precipitation databases algorithms is required if models are used for policy-making and societal applications other than pure research.
28. Every aspect of model and database development should be subject to QC methods and be fully traceable, transparent and auditable.
29. Models must be independently validated by scientists not involved in their development or belonging to the same research network.
30. Users should be made fully aware of the confidence level that can be attributed to model outputs and observational databases.

Adapted from Tapiador et al. 2017

succession of weather instead of as a statistical representation of the climate. It also includes some cross-cutting issues that may not be apparent to scientists working in either modeling or precipitation measurement but that must be considered in the intersection of the fields. That is the case of the empirical and highly assumptive nature of the parameterization of the precipitation microphysics embedded in models (Vivekanandan et al. 1999).

The table includes several important topics sometimes overlooked by modelers. Thus, a tenet of validation is that the spatial and temporal scales at which the comparison is performed represents a limit for judging a model. This is dubbed 'scope principle'. One can hardly claim that a model with a spatial/temporal resolution of 25 km/6 h performs well unless the validation has been performed at such scales and reasonable scores have been found. In other words, the ability of a model to reproduce the precipitation climatology at 2.5° grid spacing does not necessarily imply that the model is suitable at, for instance, daily scale and 25 km spatial resolution. It is worth remembering this principle since some modelers validate their models at aggregated, coarse resolutions and then claim that the outputs can be used at finer resolutions to inform policies.

Another issue worth verbalizing is the topic of 'convergence', i.e. the idea that the agreement of several products of models tells something about their individual worth. It can be argued that convergence is not intrinsically a good estimate of an observational product or model performance since all the developers are equally aware of the range of sensible values for their product and model outputs and tune the free parameters accordingly. Another reason, in the case of models, is that they are not that different: they all have the same set of equations, have similar discretization strategies, share a handful of parameterizations and have the same information about what is a sensible output and what needs to be adjusted in each parameterization. The hypotheses and empirical assumptions are also shared by most models, and that includes spectral and variable-resolution models.

Indeed, the list (Table 54.2) is not exhaustive but a minimum set of items that require attention. Asking those validating models to verify that at least the items in the list have been accounted for would increase the confidence in models of those scientists outside the field and also the public perception of modeling. One way to do that is to make sure that all items have been addressed, explicitly stating that is so. That would help other scientists, decision makers and users to give weight to the validation exercise. Ideally, validation reports on climate models should include an acknowledgment such as 'the validation of the model was performed using the QA procedures described in X'. That would ensure that the model belongs to the type of models that can be used to inform policies. More detailed information such as 'both the observational datasets and the model complained with level-3 quality assurance specifications in X' would also certainly help.

References

Adler, R. F., Huffman, G. J., Chang, A., Ferraro, R. R., Xie, P., Janowiak, J., Rudolf, B., Schneider, U., Curtis, S., Bolvin, D. T., Gruber, A., Susskind, J., Arkin, P. A., & Nelkin, E. (2003). The version-2 global precipitation climatology project (GPCP) monthly precipitation analysis (1979–present). *Journal of Hydrometeorology, 4*, 1147–1167. https://doi.org/10.1175/1525-7541(2003)004<1147:TVGPCP>2.0.CO;2.

Adler, R. F., Sapiano, M., Huffman, G. J., Bolvin, D. T., Wang, J.-J., Gu, G., Nelkin, E., Xie, P., Chiu, L., Ferraro, R. R., Schneider, U., & Becker, A. (2016). New global precipitation climatology project monthly analysis product corrects satellite data drifts. *GEWEX News, 26* (4), 7–9. [Available at https://www.gewex.org/gewex-content/files_mf/1480533350Nov2016_final_opt.pdf, last accessed 27 Nov. 2018].

Ebert, E. E., Janowiak, J. E., & Kidd, C. (2007). Comparison of near-real-time precipitation estimates from satellite observations and numerical models. *Bulletin of the American Meteorological Society, 88*, 47–64. https://doi.org/10.1175/BAMS-88-1-47.

Ebert, E. E., Manton, M. J., Arkin, P. A., Allam, R. J., Holpin, G. E., & Gruber, A. (1996). Results from the GPCP algorithm intercomparison programme. *Bulletin of the American Meteorological Society, 77*, 2875–2887. https://doi.org/10.1175/1520-0477(1996)077<2875:RFTGAI>2.0.CO;2.

Grecu, M., Olson, W. S., Shie, C. L., L'Ecuyer, T. S., & Tao, W.-K. (2009). Combining satellite microwave radiometer and radar observations to estimate atmospheric heating profiles. *Journal of Climate, 22*, 6356–6376. https://doi.org/10.1175/2009JCLI3020.1.

Haddad, Z. S., Sawaya, R. C., Kacimi, S., Sy, O. O., Turk, F. J., & Steward, J. (2017). Interpreting millimeter-wave radiances over tropical convective clouds. *Journal of Geophysical Research, 122*, 1650–1664. https://doi.org/10.1002/2016JD025923.

Haddad, Z. S., Smith, E. A., Kummerow, C. D., Iguchi, T., Farrar, M. R., Durden, S. L., Alves, M., & Olson, W. S. (1997). The TRMM "Day-1" radar/radiometer combined rain-profiling algorithm. *Journal of the Meteorological Society of Japan, 75*, 799–809. https://doi.org/10.2151/jmsj1965.75.4_799.

Huffman, G. J., Adler, R. F., Arkin, P. A., Chang, A., Ferraro, R. R., Gruber, A., Janowiak, J. E., McNab, A., Rudolf, B., & Schneider, U. (1997). The global precipitation climatology project (GPCP) combined precipitation dataset. *Bulletin of the American Meteorological Society, 78*, 5–20. https://doi.org/10.1175/1520-0477(1997)078<0005:TGPCPG>2.0.CO;2.

Huffman, G. J., Adler, R. F., Bolvin, D. T., & Gu, G. (2009). Improving the global precipitation record: GPCP version 2.1. *Geophysical Research Letters, 36*, L17808. https://doi.org/10.1029/2009GL040000.

Huffman, G. J., Adler, R. F., Bolvin, D. T., & Nelkin, E. J. (2010). The TRMM multi-satellite precipitation analysis (TMPA). In F. Hossain & M. Gebremichael (Eds.), *Satellite rainfall applications for surface hydrology* (pp. 3–22). Dordrecht: Springer. https://doi.org/10.1007/978-90-481-2915-7_1.

Huffman, G. J., Bolvin, D. T., Braithwaite, D., Hsu, K.-L., Joyce, R., Kidd, C., Nelkin, E. J., & Xie, P. (2015). *Algorithm theoretical basis document (ATBD) version 4.5: NASA global precipitation measurement (GPM) integrated multi-satellitE retrievals for GPM (IMERG).* Greenbelt: NASA. [Available at https://pmm.nasa.gov/sites/default/files/document_files/IMERG_ATBD_V4.5.pdf, last accessed 27 Nov 2018].

Huffman, G. J., Adler, R. F., Morrissey, M. M., Bolvin, D. T., Curtis, S., Joyce, R., McGavock, B., & Susskind, J. (2001). Global precipitation at one-degree daily resolution from multisatellite observations. *Journal of Hydrometeorology, 2,* 36–50. https://doi.org/10.1175/1525-7541(2001)002<0036:GPAODD>2.0.CO;2.

Huffman, G. J., Bolvin, D. T., Nelkin, E. J., Wolff, D. B., Adler, R. F., Gu, G., Hong, Y., Bowman, K. P., & Stocker, E. F. (2007). The TRMM multisatellite precipitation analysis (TMPA): Quasi-global, multiyear, combined-sensor precipitation estimates at fine scales. *Journal of Hydrometeorology, 8,* 38–55. https://doi.org/10.1175/JHM560.1.

IPCC. (2014). In R. K. Pachauri & L. A. Meyer (Eds.), *Climate change 2014: Synthesis report. Contribution of working groups I, II and III to the fifth assessment report of the intergovernmental panel on climate change, Core writing team.* Geneva: IPCC. 151 pp. [Available at http://ar5-syr.ipcc.ch, last accessed 27 Nov. 2018].

Joyce, R. J., Janowiak, J. E., Arkin, P. A., & Xie, P. (2004). CMORPH: A method that produces global precipitation estimates from passive microwave and infrared data at high spatial and temporal resolution. *Journal of Hydrometeorology, 5,* 487–503. https://doi.org/10.1175/1525-7541(2004)005<0487:CAMTPG>2.0.CO;2.

Katsanos, D., Retalis, A., & Michaelides, S. (2016). Validation of a high-resolution precipitation database (CHIRPS) over Cyprus for a 30-year period. *Atmospheric Research, 169,* 459–464. https://doi.org/10.1016/j.atmosres.2015.05.015.

Kidd, C. (2001). Satellite rainfall climatology: A review. *International Journal of Climatology, 21,* 1041–1066. https://doi.org/10.1002/joc.635.

Kidd, C., & Huffman, G. J. (2011). Global precipitation measurement. *Meteorological Applications, 18,* 334–353. https://doi.org/10.1002/met.284.

Kidd, C., & Levizzani, V. (2011). Status of satellite precipitation retrievals. *Hydrology and Earth System Sciences, 15,* 1109–1116. https://doi.org/10.5194/hess-15-1109-2011.

Kidd, C., Matsui, T., Chern, J., Mohr, K., Kummerow, C., & Randel, D. (2016). Global precipitation estimates from cross-track passive microwave observations using a physically based retrieval scheme. *Journal of Hydrometeorology, 17,* 383–400. https://doi.org/10.1175/JHM-D-15-0051.1.

Kidd, C., Kniveton, D. R., Todd, M. C., Bellerby, T. J., Kidd, C., Kniveton, D. R., Todd, M. C., & Bellerby, T. J. (2003). Satellite rainfall estimation using combined passive microwave and infrared algorithms. *Journal of Hydrometeorology, 4,* 1088–1104. https://doi.org/10.1175/1525-7541(2003)004<1088:SREUCP>2.0.CO;2.

Kidd, C., Becker, A., Huffman, G. J., Muller, C. L., Joe, P., Skofronick-Jackson, G., & Kirschbaum, D. B. (2017). So, how much of the Earth's surface is covered by rain gauges? *Bulletin of the American Meteorological Society, 98,* 69–78. https://doi.org/10.1175/BAMS-D-14-00283.1.

Kim, K., Park, J., Baik, J., & Choi, M. (2017). Evaluation of topographical and seasonal feature using GPM IMERG and TRMM 3B42 over far-East Asia. *Atmospheric Research, 187,* 95–105. https://doi.org/10.1016/j.atmosres.2016.12.007.

Kucera, P. A., Ebert, E. E., Turk, F. J., Levizzani, V., Kirschbaum, D. B., Tapiador, F. J., Loew, A., & Borsche, M. (2013). Precipitation from space: Advancing earth system science. *Bulletin of the American Meteorological Society, 94,* 365–375. https://doi.org/10.1175/BAMS-D-11-00171.1.

Kummerow, C. D., Masunaga, H., & Bauer, P. (2007). A next-generation microwave rainfall retrieval algorithm for use in TRMM and GPM. In V. Levizzani, P. Bauer, & F. J. Turk (Eds.), *Measuring Precipitation from Space.* New York: Springer. 745 pp, ISBN: 978-1-4020-5835-6.

Kummerow, C. D., Hong, Y., Olson, W. S., Yang, S., Adler, R. F., McCollum, J., Ferraro, R. R., Petty, G., Shin, D.-B., & Wilheit, T. T. (2001). The evolution of the Goddard profiling algorithm (GPROF) for rainfall estimation from passive microwave sensors. *Journal of Applied Meteorology, 40*, 1801–1820. https://doi.org/10.1175/1520-0450(2001)040<1801:TEOTGP>2.0. CO;2.

L'Ecuyer, T. S., Beaudoing, H. K., Rodell, M., Olson, W. S., Lin, B., Kato, S., Clayson, C. A., Wood, E., Sheffield, J., Adler, R. F., Huffman, G. J., Bosilovich, M., Gu, G., Robertson, F., Houser, P. R., Chambers, D., Famiglietti, J. S., Fetzer, E., Liu, W. T., Gao, X., Schlosser, C. A., Clark, E., Lettenmaier, D. P., & Hilburn, K. (2015). The observed state of the energy budget in the early twenty-first century. *Journal of Climate, 28*, 8319–8346. https://doi.org/10.1175/JCLI-D-14-00556.1.

Levizzani, V., Adamo, C., Alberoni, P. P., Antonini, A., Battaglia, A., Baptista, J. P. V. P., Bauer, P., Buzzi, A., Capacci, D., Caracciolo, C., Cattani, E., Celano, M., Cimini, D., Costa, M. J., Davolio, S., Dietrich, S., Fantini, M., di Michele, S., Guiliani, G., Hinsman, D. E., Kästner, M., Khain, A., Kidd, C., Kidd, J., Kniveton, D., Lahav, R., Layberry, R., Lensky, I., Malguzzi, P., Mantovani, S., Marzano, F. S., Maurizi, A., Medaglia, C. M., Melani, S., Meneguzzo, F., Messeri, G., Mugnai, A., Natali, S., Orlandi, A., Ortolani, A., Panegrossi, G., Pasqui, M., Pinori, S., Poli, V., Porcù, F., Prodi, F., Purdom, J. F. W., Rosenfeld, D., Sanderson, V., Schmetz, J., Smith, E. A., Solomon, R., Steinwagner, J., Tampieri, F., Tapiador, F. J., Tassa, A., Torricella, F., Tripoli, G. J., Turk, F. J., Vicente, G. A., & Villani, M. G. (2004). Precipitation estimation: From the RAO to Eurainsat and beyond. In H. Lacoste (Ed.), *Proc. 2nd MSG RAO workshop, ESA SP-582, Salzburg, Austria, 9–10 Sept. 2004* (p. 113). Salzburg: ESA Publications Division.

Michaelides, S. (2013). Perspectives of precipitation science: Part I. *Atmospheric Research, 131*, 1–2. https://doi.org/10.1016/j.atmosres.2013.05.017.

Michaelides, S. (2014). Perspectives of precipitation science: Part II. *Atmospheric Research, 144*, 1–3. https://doi.org/10.1016/j.atmosres.2014.03.012.

Michaelides, S. (2016). Perspectives of precipitation science: Part III. *Atmospheric Research, 169*, 401–403. https://doi.org/10.1016/j.atmosres.2015.09.025.

Michaelides, S., Levizzani, V., Anagnostou, E. N., Bauer, P., Kasparis, T., & Lane, J. E. (2009). Precipitation: Measurement, remote sensing, climatology and modeling. *Atmospheric Research, 94*, 512–533. https://doi.org/10.1016/j.atmosres.2009.08.017.

Peters-Lidard, C. D., Kemp, E. M., Matsui, T., Santanello, J. A., Kumar, S. V., Jacob, J. P., Clune, T., Tao, W.-K., Chin, M., Hou, A. Y., Case, J. L., Kim, D., Kim, K. M., Lau, W., Liu, Y., Shi, J., Starr, D., Tan, Q., Tao, Z., Zaitchik, B. F., Zavodsky, B., Zhang, S. Q., & Zupanski, M. (2015). Integrated modeling of aerosol, cloud, precipitation and land processes at satellite-resolved scales. *Environmental Modelling & Software, 67*, 149–159. https://doi.org/10.1016/j.envsoft.2015.01.007.

Retalis, A., Katsanos, D., & Michaelides, S. (2016). Precipitation climatology over the Mediterranean Basin - validation over Cyprus. *Atmospheric Research, 169*, 449–458. https://doi.org/10.1016/j.atmosres.2015.01.012.

Skofronick-Jackson, G., Petersen, W. A., Berg, W., Kidd, C., Stocker, E. F., Kirschbaum, D. B., Kakar, R., Braun, S. A., Huffman, G. J., Iguchi, T., Kirstetter, P.-E., Kummerow, C. D., Meneghini, R., Oki, R., Olson, W. S., Takayabu, Y. N., Furukawa, K., & Wilheit, T. T. (2017). The Global Precipitation Measurement (GPM) mission for science and society. *Bulletin of the American Meteorological Society, 98*, 1679–1695. https://doi.org/10.1175/BAMS-D-15-00306.1.

Sorooshian, S., Hsu, K.-L., Gao, X., Gupta, H. V., Imam, B., & Braithwaite, D. (2000). Evaluation of PERSIANN system satellite–based estimates of tropical rainfall. *Bulletin of the American Meteorological Society, 81*, 2035–2046. https://doi.org/10.1175/1520-0477(2000)081<2035:EOPSSE>2.3.CO;2.

Stephens, G. L., Vane, D. G., Boain, R. J., Mace, G. G., Sassen, K., Wang, Z., Illingworth, A. J., O'Connor, E. J., Rossow, W. B., Durden, S. L., Miller, S. D., Austin, R. T., Benedetti, A.,

Mitrescu, C., & the CloudSat Science Team. (2002). The CloudSat mission and the A-TRAIN: A new dimension to space-based observations of clouds and precipitation. *Bulletin of the American Meteorological Society, 83,* 1771–1790. https://doi.org/10.1175/BAMS-83-12-1771.

Stephens, G. L., Vane, D. G., Tanelli, S., Im, E., Durden, S., Rokey, M., Reike, D., Partain, P., Mace, G. G., Austin, R., S, T., L'Ecuyer, J., Haynes, M., Lebsock, K., Suzuki, D., Waliser, D. W., Kay, J., Gettelman, A., Wang, Z., & Marchand, R. (2008). CloudSat mission: Performance and early science after the first year in orbit. *Journal of Geophysical Research, 113,* D00A18. https://doi.org/10.1029/2008JD009982.

Takahashi, K., & Dewitte, B. (2016). Strong and moderate nonlinear El Niño regimes. *Climate Dynamics, 46,* 1627–1645. https://doi.org/10.1007/s00382-015-2665-3.

Tao, W.-K., Lang, S., Olson, W. S., Meneghini, R., Yang, S., Simpson, J., Kummerow, C. D., Smith, E. A., & Halverson, J. (2001). Retrieved vertical profiles of latent heat release using TRMM rainfall products for February 1998. *Journal of Applied Meteorology, 40,* 957–982. https://doi.org/10.1175/1520-0450(2001)040<0957:RVPOLH>2.0.CO;2.

Tao, W.-K., Santanello, J. A., Chin, M., Zhou, S., Tan, Q., Kemp, E. M., & Peters-Lidard, C. D. (2013). Effect of land cover on atmospheric processes and air quality over the continental United States-a NASA unified WRF (NU-WRF) model study. *Atmospheric Chemistry and Physics, 13,* 6207–6226. https://doi.org/10.5194/acp-13-6207-2013.

Tao, W.-K., Lang, S., Zeng, X., Li, X., Matsui, T., Mohr, K., Posselt, D., Chern, J., Peters-Lidard, C. D., Norris, P. M., Kang, I. S., Choi, I., Hou, A. Y., Lau, K. M., & Yang, Y.-M. (2014). The Goddard cumulus ensemble model (GCE): Improvements and applications for studying precipitation processes. *Atmospheric Research, 143,* 392–424. https://doi.org/10.1016/j.atmosres.2014.03.005.

Tao, W.-K., Smith, E. A., Adler, R. F., Haddad, Z. S., Hou, A. Y., Iguchi, T., Kakar, R., Krishnamurti, T. N., Kummerow, C. D., Lang, S., Meneghini, R., Nakamura, K., Nakazawa, T., Okamoto, K., Olson, W. S., Satoh, S., Shige, S., Simpson, J., Takayabu, Y. N., Tripoli, G. J., & Yang, S. (2006). Retrieval of latent heating from TRMM measurements. *Bulletin of the American Meteorological Society, 87,* 1555–1572. https://doi.org/10.1175/BAMS-87-11-1555.

Tao, W.-K., Wu, D., Lang, S., Chern, J., Peters-Lidard, C. D., Fridlind, A., & Matsui, T. (2016a). High-resolution NU-WRF simulations of a deep convective-precipitation system during MC3E: Further improvements and comparisons between Goddard microphysics schemes and observations. *Journal of Geophysical Research, 121,* 1278–1305. https://doi.org/10.1002/2015JD023986.

Tao, W.-K., Takayabu, Y. N., Lang, S., Olson, W. S., Shige, S., Hou, A. Y., Jackson, G., Jiang, X., Lau, W., Krishnamurti, T., Waliser, D., Zhang, C., Johnson, R., Houze, R., Ciesielski, P., Grecu, M., Hagos, S., Kakar, R., Nakamura, K., Braun, S., & Bhardwaj, A. (2016b). TRMM latent heating retrieval and comparison with field campaigns and large-scale analyses. In *Multi-scale Convection-Coupled Systems in the Tropics. Meteorological Monographs* (pp. 2.1–2.234). AMS. https://doi.org/10.1175/AMSMONOGRAPHS-D-15-0013.1.

Tapiador, F. J. (2009). Assessment of renewable energy potential through satellite data and numerical models. *Energy & Environmental Science, 2,* 1142–1161. https://doi.org/10.1039/B914121A.

Tapiador, F. J. (2010). A joint estimate of the precipitation climate signal in Europe using eight regional models and five observational datasets. *Journal of Climate, 23,* 1719–1738. https://doi.org/10.1175/2009JCLI2956.1.

Tapiador, F. J., Hou, A. Y., De Castro, M., Checa, R., Cuartero, F., & Barros, A. P. (2011). Precipitation estimates for hydroelectricity. *Energy & Environmental Science, 4,* 4435–4448. https://doi.org/10.1039/C1EE01745D.

Tapiador, F. J., Turk, F. J., Petersen, W. A., Hou, A. Y., García-Ortega, E., Machado, L. A. T., Angelis, C. F., Salio, P., Kidd, C., Huffman, G. J., & de Castro, M. (2012). Global precipitation measurement: Methods, datasets and applications. *Atmospheric Research, 104,* 70–97. https://doi.org/10.1016/j.atmosres.2011.10.021.

Tapiador, F. J., Navarro, A., Levizzani, V., García-Ortega, E., Huffman, G. J., Kidd, C., Kucera, P. A., Kummerow, C. D., Masunaga, H., Petersen, W. A., Roca, R., Sánchez, J.-L., Tao, W.-K.,

& Turk, F. J. (2017). Global precipitation measurements for validating climate models. *Atmospheric Research, 197*, 1–20. https://doi.org/10.1016/j.atmosres.2017.06.021.

Tapiador, F. J., Roca, R., Genio, A. D., Dewitte, B., Petersen, W. A., & Zhang, F. (2019). Is precipitation a good metric for model performance? *Bulletin of the American Meteorological Society*, 100, 223-233. https://doi.org/10.1175/BAMS-D-17-0218.1.

Turk, F. J., Rohaly, G., Hawkins, J., Smith, E. A., Marzano, F. S., Mugnai, A., & Levizzani, V. (2000). Meteorological applications of precipitation estimation from combined SSM/I, TRMM and geostationary satellite data. In P. Pampaloni & S. Paloscia (Eds.), *Microwave radiometry and remote sensing of the Earth's surface and atmosphere* (pp. 353–363). Utrecht: VSP International Science Publisher. ISBN: 9789067643184.

Vivekanandan, J., Zrnić, D. S., Ellis, S. M., Oye, R., Ryzhkov, A. V., & Straka, J. (1999). Cloud microphysics retrieval using S-band dual-polarization radar measurements. *Bulletin of the American Meteorological Society, 80*, 381–388. https://doi.org/10.1175/1520-0477(1999)080<0381:CMRUSB>2.0.CO;2.

Xie, P., Janowiak, J. E., Arkin, P. A., Adler, R. F., Gruber, A., Ferraro, R. R., Huffman, G. J., & Curtis, S. (2003). GPCP pentad precipitation analyses: An experimental dataset based on gauge observations and satellite estimates. *Journal of Climate, 16*, 2197–2214. https://doi.org/10.1175/2769.1.

Chapter 55
Extreme Precipitation in the Himalayan Landslide Hotspot

Thomas Stanley, Dalia B. Kirschbaum, Salvatore Pascale, and Sarah Kapnick

Abstract Extreme precipitation from the South-Asian monsoon season combines with significant topographic relief within the Himalayan region to cause landslides that result in hundreds to thousands of fatalities each year. While there are few consistent and publicly available in-situ estimates of rainfall across this region, satellite products and global climate models provide insight into the extreme precipitation patterns that may impact the frequency of landsliding. In this work, we analyzed several extreme precipitation indices using data from a global climate model and the satellite-based Tropical Rainfall Measuring Mission Multi-satellite Precipitation Analysis product to represent extreme precipitation over High Mountain Asia. We then compared the temporal distribution of extreme precipitation to a global database of landslides to better understand the spatiotemporal distribution of potential landslide triggering factors. We found that these indices successfully model the seasonality of landslide activity across the region, but other aspects of spatiotemporal variability require additional information and analysis before they can be applied more broadly.

T. Stanley
Universities Space Research Association, Columbia, MD, USA

Goddard Earth Sciences Technology and Research, Greenbelt, MD, USA

Goddard Space Flight Center, NASA, Greenbelt, MD, USA
e-mail: Thomas.a.stanley@nasa.gov

D. B. Kirschbaum (✉)
Goddard Space Flight Center, NASA, Greenbelt, MD, USA
e-mail: dalia.b.kirschbaum@nasa.gov

S. Pascale
Department of Earth System Science, Stanford University, Stanford, CA, USA

S. Kapnick
Geophysical Fluid Dynamics Laboratory, NOAA, Princeton, NJ, USA

© Springer Nature Switzerland AG 2020
V. Levizzani et al. (eds.), *Satellite Precipitation Measurement*, Advances in Global Change Research 69, https://doi.org/10.1007/978-3-030-35798-6_31

Keywords Precipitation · Rainfall · Asian monsoon · Himalayas · TRMM · TMPA · Landslides · ETCCDI · Global climate models · NASA · GFDL FLOR · Global landslide catalog

55.1 Introduction

The Himalayan region is known as a landslide hotspot, with hundreds to thousands of associated fatalities each year (Nadim et al. 2006; Petley 2012). The pervasive distribution of landslides across this region is due to the combination of the world's greatest relief with heavy monsoon rainfall and occasional major earthquakes. While widespread landsliding can be triggered by a single earthquake or an earthquake with subsequent aftershocks (e.g., the 2014 Gorkha earthquake in Nepal), a large majority of landslides within this region, and around the world, are triggered by rainfall. Landslides within this region, which range from shallow debris flows to larger and more deep-seated failures, are typically triggered by short-duration, high-intensity monsoon rainfall (hours to days). One challenge in characterizing the spatiotemporal patterns in landslide activity within this region is the dearth of both landslide catalogs and in situ precipitation data that can be used to establish the co-occurrence, potential triggers, and even long-term trends in landslide activity. This work considers how satellite and modeled precipitation products can provide additional insight into the region's patterns of landslide activity, which are crucial when considering the global distribution of landslide risk.

Petley et al. (2007) observed an upward trend in the incidence of landslides that cause fatalities throughout Nepal from 1978 to 2005 and attributed this to development within rural development. However, the database from which this conclusion was drawn represents only a small portion of the total number of landslides. It is possible that trends in fatalities are more closely linked to changes in human exposure or vulnerability than trends in the occurrence of landslides in general. Given the difficulty of retrospectively reconstructing a multi-decadal record of landslide activity, gridded precipitation data offers an alternative view of the potential triggering mechanisms of landslide hazard with more consistency across time and space than hazard databases currently allow.

The investigation of extreme precipitation indices, and climate research in general, is justified in part by the belief that it provides information about the future of rainfall-triggered natural hazards, including landslides. The most commonly used extreme precipitation indices were formalized by an Expert Team on Climate Change Detection and Indices (ETCCDI) "to address the need for the objective measurement and characterization of climate variability and change" (World Climate Research Programme 2018). Unfortunately, the linkage between climate indices and landslide activity remains more of an assumption than comprehensively illustrated. Cepeda et al. (2010) used R20mm (which records the number of days with very heavy precipitation) and other indices to corroborate trends from a susceptibility-based rainfall threshold. Chen et al. (2012) noted that 24-hour rainfall greater than 20 mm triggered numerous debris flows in the 5 years after the Chi-Chi earthquake. Statistically significant correlations have been observed between the indices Rx1day and Rx5day (maximum 1- or 5-day rainfall annually) and the annual number of landslides and flash floods reported in Rio de Janeiro, Brazil (Ávila et al. 2016). Similarly, Yang et al. (2016) correlated maximum streamflow with ETCCDI precipitation indices by month in the Huaihe Basin, China. Although the connection between extreme precipitation indices and landslide activity seems obvious, additional research is needed to establish a firm empirical relationship and examine the potential for generalization to other regions.

Warming trends in High Mountain Asia (HMA) climate are more clearly established than observed changes to extreme precipitation. A study of extreme precipitation in Nepal for the years 1970–2012 (overlapping Petley's landslide database, but longer by 15 years) revealed some trends, but the results were mixed (Karki et al. 2017). Likewise, observations of extreme precipitation in eastern and western zones of the Hindu Kush-Himalaya region showed contradictory and statistically insignificant trends from 1960 to 2000 (Panday et al. 2015). However, major increases in extreme precipitation, including 5-day precipitation, have been projected for both the eastern and western Himalayan zones by the end of the twenty-first century in multiple climate scenarios (Panday et al. 2015). The strong probability of increasing precipitation intensity over the next 80 years highlights the need to better understand how changes in extreme precipitation may impact landslide activity.

We have analyzed these indices with data from a global climate model (GCM) that was designed to better represent extreme precipitation in regional climates. For confirmation, we also prepared extreme precipitation indices from satellite observations. Next, we compared the temporal distribution of extreme precipitation to a global database of landslides. The results broadly confirmed the findings of previous research, but a large element of uncertainty regarding future patterns of natural hazards within the region remains.

55.2 Rainfall and Landslide Data

We derived the extreme precipitation indices from both GCM and satellite sources (Table 55.1). Numerical simulations were performed with the Geophysical Fluid Dynamics Laboratory Forecast-oriented Low Ocean Resolution version of CM2.5 (FLOR) (Vecchi et al. 2014). We also generated indices from the Tropical Rainfall Measuring Mission (TRMM) Multi-satellite Precipitation Analysis (TMPA; Huffman et al. 2010). In order to compare the changes in extreme precipitation to

Table 55.1 Data sources

Name	Temporal Coverage	Spatial Coverage	Grid Resolution (Degrees)	References
FLOR	2000–2016	Asia	0.625 × 0.5	Vecchi et al. (2014)
TMPA	1998-2019	50°N-50°S	0.25	Huffman et al. (2010)
GLC	2007–2016	World	Not applicable	Kirschbaum et al. (2015)

a record of landslide activity, we selected a portion of the Global Landslide Catalog (GLC), which provides information on rainfall-triggered landslides from 2007 to 2016 (Kirschbaum et al. 2015). The short time period for which the GLC was available limited a longer-term analysis of all three products (Table 55.1), so we performed other analyses for the joint time period of FLOR and TMPA.

55.2.1   GFDL FLOR

Numerical simulations are performed with the Geophysical Fluid Dynamics Laboratory Forecast-oriented Low Ocean Resolution version of CM2.5 (FLOR; Vecchi et al. 2014). FLOR has been derived from the GFDL Climate Model, version 2.5 (CM2.5; Delworth et al. 2011), which has been successfully used for studies of regional hydroclimatic variability and change (e.g., Delworth and Zeng 2014; Kapnick et al. 2014; Zhang et al. 2015; Delworth et al. 2016; Pascale et al. 2017). FLOR and CM2.5 are identical but differ in horizontal resolution in the ocean-sea ice components (~1 × 1°, with meridional resolution of 1/3° near the equator in FLOR, vs ~0.25° × 0.25°, with gridbox sizes ranging from 28 km at the equator to 8 km in polar regions in CM2.5).

In order to compare the FLOR output with observations over the period 2000–2016, simulated sea surface temperatures (SST) are nudged to the sea surface temperature monthly means observed during 2000–2016 by using the following SST tendency equation:

$$\partial SST(x, y, t)/\partial t = K(x, y, t) + \frac{1}{\tau} \left[SST(x, y, t)_T - SST(x, y, t) \right] \tag{55.1}$$

where $\partial SST/\partial t$ is the time-tendency of sea surface temperature, K is the coupled model's tendency term worked out based on the model governing equations, and the last term on the right-hand side of the equation is the nudging (or restoring) term. This forces the modelled sea surface temperatures to relax, in timescale τ (the nudging timescale, 5 days) toward the target sea surface temperature SST_T, which in this case is the observed temperature. Because the ocean and the atmosphere are not initialized, the model's sea surface temperature would deviate from the observed sea surface temperature without the last term in Eq. 55.1. Therefore, nudging allows the model to run forced by observed sea surface temperatures while still retaining atmosphere-ocean coupling at timescales shorter than 5 days. Similarly, surface pressure and three-dimensional temperature and horizontal winds are nudged every 6 hours towards reanalysis data. Full details are provided in Yang et al. (2018).

55.2.2 TMPA

The TMPA precipitation product provides estimated precipitation from 50°N-S from 1998 to the present. In this work, we used the research version of TMPA Version 7, which is available approximately 3 months after acquisition and is calibrated with the Global Precipitation Climatology Project (GPCP) gridded gauge dataset (Adler et al. 2003). TMPA was developed to utilize an international constellation of microwave and infrared satellite-borne instruments to produce 3-hourly estimates of total precipitation. There are known issues with satellite retrievals over complex terrain due to orographic uplift (Barros et al. 2000, 2004; Bharti and Singh 2015). However, in this study we considered the precipitation retrievals collectively across the entire study domain as well as at the pixel scale to represent potential bias that may occur when evaluating within smaller spatial domains influenced by orographic rainfall processes.

55.2.3 Landslide Data

The Global Landslide Catalog (GLC) provides a global picture of rainfall-triggered landslides based primarily on media reports. The database has over 11,000 entries and includes information on the location (geographic and nominal) and date, impacts, trigger, confidence in the location, qualitative estimate of the size, source of the report, etc. The catalog is known to have biases resulting from reporting sources that are primarily in English, uncertainty about the location and date, as well as issues differentiating landslides from other triggering hazards such as flooding and severe storms or tropical cyclones. The methodology for compiling this inventory and the characterization of the biases is available in Kirschbaum et al. (Kirschbaum et al. 2010, 2015). For this analysis we selected 1076 landslides within an analysis area that includes Nepal, northern India, and small portions of Pakistan and China where the largest proportion of the landslides have been reported (Fig. 55.1).

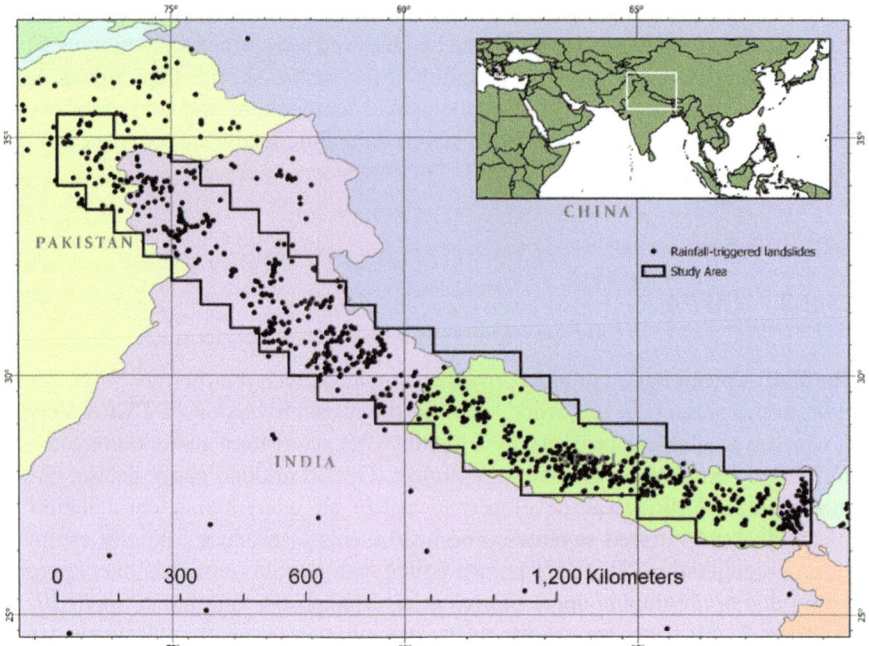

Fig. 55.1 The Global Landslide Catalog reveals a band of terrain with a high number of landslides (black dots indicate landslides from 2007–2016). The Himalayan Mountains and Foothills experience intense monsoon precipitation and occasional seismicity, while comprising some of the Earth's steepest terrain. We focused our analysis on this region by identifying the sites with the greatest elevation difference from neighboring pixels at the half-degree scale (black box). Unless otherwise noted, results are presented for this region

55.3 Methods

Landslides can be triggered by high-intensity, short-duration rainfall, as well as prolonged rain at lower intensities. We selected several of the extreme rainfall precipitation indices from the ETCCDI that are most likely to be relevant to landslide activity. We calculated Rx1day, Rx5day, R10mm, R20mm, CWD, R95pTOT, and R99pTOT (Table 55.2) from both FLOR (https://doi.org/10.5067/W8DR3VBR27PX, last accessed 5 Apr. 2019) and TMPA (https://doi.org/10.5067/5VPZ8AZ9LAKP, last accessed 5 Apr. 2019) products. Rx1day represents the most extreme daily precipitation, a level which is likely to generate landslides. Similarly, Rx5day represents the most extreme precipitation over a 5-day period, which is relevant to the landslides that may be caused by a gradual buildup of groundwater. R10mm and R20mm show the frequency of heavy precipitation, which represents the number of times that landslides are possible, if not guaranteed. Some landslides, especially deep-seated landslides, may be driven more by the duration than the intensity of rainfall. CWD addresses this possibility by showing the length of precipitation events. R95pTOT and R99pTOT represent the annual total amount of extreme precipitation. Because extreme

Table 55.2 Extreme precipitation indices selected for relevance to landslides

Index	Definition
Rx1day	The highest 1-day precipitation.
Rx5day	The highest sum of precipitation from 5 consecutive days.
R10mm	The number of days on which 10 mm or more of precipitation occurred.
R20mm	The number of days on which 20 mm or more of precipitation occurred.
CWD	The longest number of consecutive days with precipitation greater than 1 mm.
R95pTOT	The sum of precipitation on days that exceeded the 95th percentile precipitation, which is typically based on a historical period of 1961–1990. In this case, the base period was set to the 1999–2017 period for the TMPA data, because the earlier period is not available for this dataset.
R99pTOT	The sum of precipitation on days that exceeded the 99th percentile precipitation, which is typically based on a historical period of 1961–1990. In this case, the base period was set to the 1999–2017 period for the TMPA data, because the earlier period is not available for this dataset.

precipitation is defined against local historical conditions, these indices might be more closely associated with landslide activity than single, global thresholds. Although most of these indices can be calculated on a monthly basis, we focused on annual relationships. Monthly means were calculated for Rx1day and Rx5day in order to observe the seasonality of extreme precipitation. The geographic distribution of extreme precipitation was summarized for each index by taking the mean of annual values for the years 2000–2016. All indices were computed in R with the climdex.pcic library (R Core Team 2013; Bronaugh 2015; Hijmans 2015; Revolution Analytics and Weston 2015; Pohlert 2017).

We characterized the trend lines at each location for the precipitation indices based on the computed annual values using the Theil-Sen method for slope estimation (Sen 1968; Theil 1992), with statistical significance from the Mann-Kendall test (Mann 1945; Kendall 1948). These are commonly used in hydrology because nonparametric methods do not require an assumption of Gaussian residuals. The slope is defined as the median of the slopes derived from all possible pairs of data points. Thus, for ($1 \leq i < j \leq n$),

$$slope = median\left(\frac{x_j - x_i}{j - i}\right) \tag{55.2}$$

For this application, i and j represent years in the study period, and x represents the extreme precipitation index at a point. The Mann-Kendall test statistic is defined as:

$$S = \sum_{k}^{n-1} \sum_{j=k+1}^{n} sgn\left(x_j - x_k\right) \tag{55.3}$$

where j and k represent years in this dataset, and x represents the precipitation index at those times. These methods ensured that the volatility over the short time period represented by TMPA would be accounted for when identifying trends.

In order to understand the climate trends most relevant to landsliding, we focused on the region of High Mountain Asia with the highest concentration of landslides: the Himalayan foothills and mountains (Fig. 55.1). The analysis of precipitation extremes in FLOR is therefore restricted to only those grid-points with a slope threshold of 0.6 or larger. While somewhat arbitrary, this choice effectively selects a band of grid points along the zone of rapid elevation increase where most landslides take place. The area was used as a mask, which was then converted with the nearest-neighbor method to match the spatial resolution of TMPA. We applied the mask to all of the precipitation indices before comparing extreme precipitation to landslides. In order to evaluate differences between the eastern and western Himalaya, we divided the study area at 80°E (close to Nepal's westernmost point) into roughly equal parts. The number of landslides reported was also roughly equal (516 in the west and 560 in the east).

55.4 Results

55.4.1 Regional Patterns in Extreme Precipitation

We found very similar geographic distributions of extreme precipitation from FLOR and TMPA and broad similarities between the ETCCDI indices (Fig. 55.2). Indices that represent rainfall frequency more than intensity (CWD and R10) are highest at higher elevations, while indices that solely represent rainfall intensity (Rx1day and Rx5day) are generally higher at lower elevations near the boundary between mountains and plains. R20mm, R95pTOT, and R99pTOT are determined by both the frequency of intense precipitation and while generally related to the elevation have a less clear geographic relationship. In general, FLOR shows the heaviest precipitation at higher elevations than TMPA. The biggest difference between datasets is the high number of consecutive wet days observed in FLOR, more than double the number from TMPA (firm empirical connection 2a). This may be due to the well-known GCM "drizzle-bias", which is often handled using a minimum threshold (e.g., > 1 mm) so that the total number of wet days in the model and observations are equivalent (e.g., Ines and Hansen 2006). There may also be a bias in the opposite direction for TMPA since the passive microwave frequencies onboard TRMM's Microwave Imager (TMI) are not sensitive to light rain. Despite this sharp difference in magnitude, the spatial distribution of CWD is quite similar across datasets. In a few grid cells, mean annual Rx1day exceeds 144 mm, a landslide-triggering rainfall threshold derived empirically from rainfall gauge data across Nepal (Dahal and Hasegawa 2008).

The extreme precipitation indices calculated from TMPA precipitation show decreasing frequency and intensity of extreme precipitation in Nepal, but increases

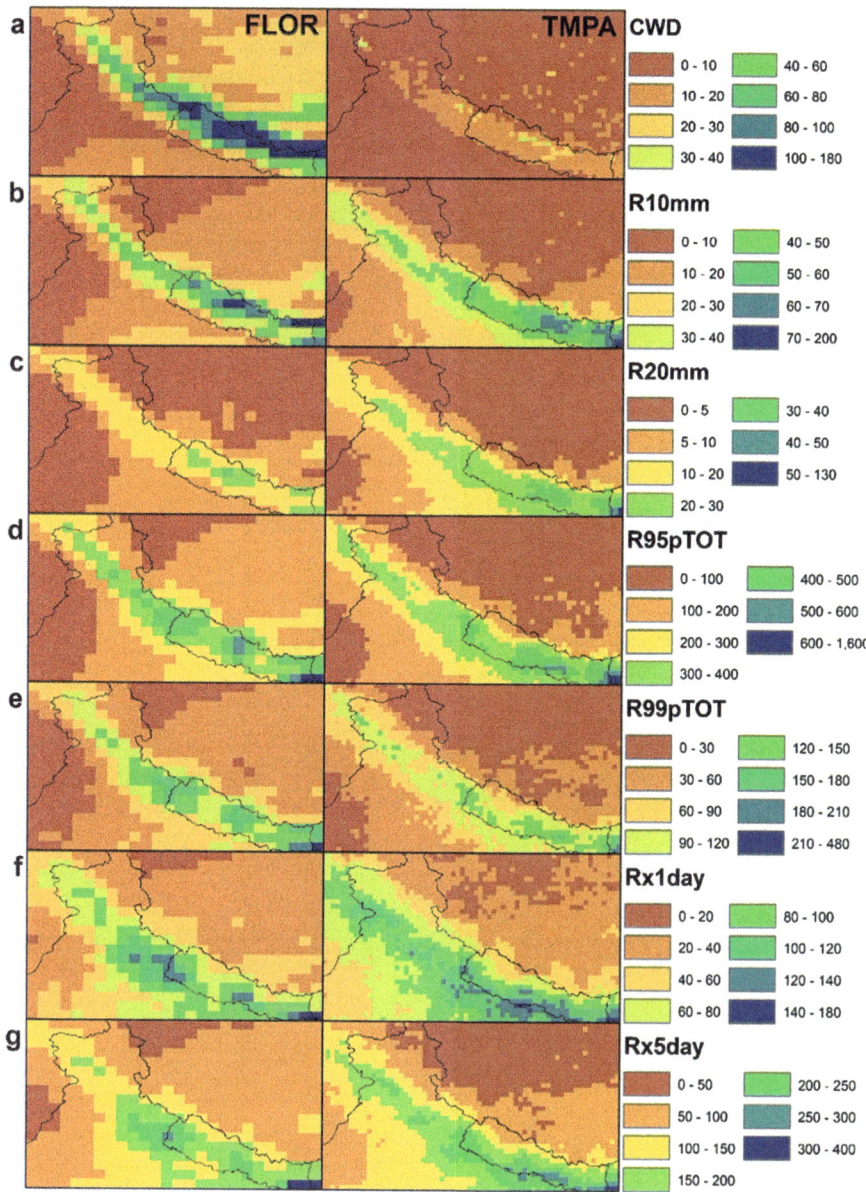

Fig. 55.2 Mean annual values of extreme precipitation indices: (**a**) CWD, (**b**) R10mm, (**c**) R20mm, (**d**) R95pTOT, (**e**) R99pTOT, (**f**) Rx1day, (**g**) Rx5day for GFDL FLOR (left) and TMPA (right). The data products show similar values and geographic distributions for all indices except CWD

in Kashmir, Rajasthan, and a section of Tibet (Fig. 55.3). CWD is the exception to the general trend in some places; it shows some decreases in Kashmir and some increases in southern Nepal. The indices derived from FLOR data show rough

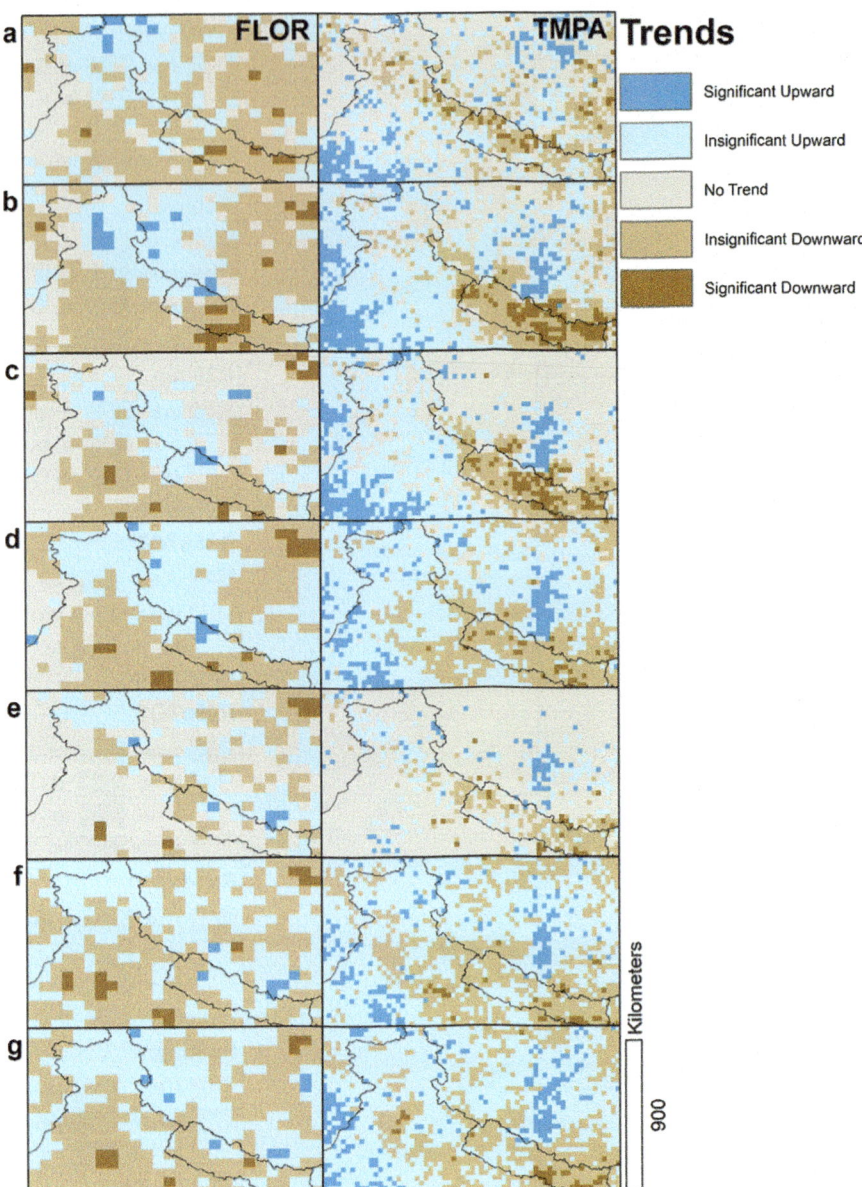

Fig. 55.3 Trends in extreme precipitation indices from 2000–2016: (**a**) CWD, (**b**) R10mm, (**c**) R20mm, (**d**) R95pTOT, (**e**) R99pTOT, (**f**) Rx1day, (**g**) Rx5day for FLOR (left) and TMPA (right)

geographic similarities to those from TMPA, but exhibit strong similarities with each other. In general, FLOR-based indices also show increases in Kashmir, but the trends in Nepal are less consistently negative. For both datasets, trends are not statistically significant in most locations, and the statistically significant trends are not consistently located in the same grid cells across indices and data products.

Although a gauge-based dataset for Nepal covers a much longer time period (Karki et al. 2017), similar trends were observed here. Analysis of R95pTOT, Rx1day, and Rx5day showed mixed but primarily decreasing trends for Central and Eastern Nepal at both meteorological stations and in the gridded products. Similarly, most of Nepal's precipitation gauges showed primarily decreasing trends in R10mm and R20mm, except in the high mountain region. Again, this largely matches the current results from the gridded datasets. CWD presented a mixed record across Nepal, and it is hard to summarize concisely the relationship between the datasets, but Karki et al. (2017) found that CWD was generally decreasing in southern Nepal and increasing in northern Nepal, while Fig. 55.3 shows the opposite pattern. Overall, the trends in extreme precipitation observed at Nepal's meteorological stations mirror the current results, despite the differences in methodology and time period.

55.4.2 Annual Variability

At the annual level, extreme precipitation indices appear to be poorly correlated with landslide activity (Fig. 55.4). In particular, 2007, 2008, 2009, and 2014 have very

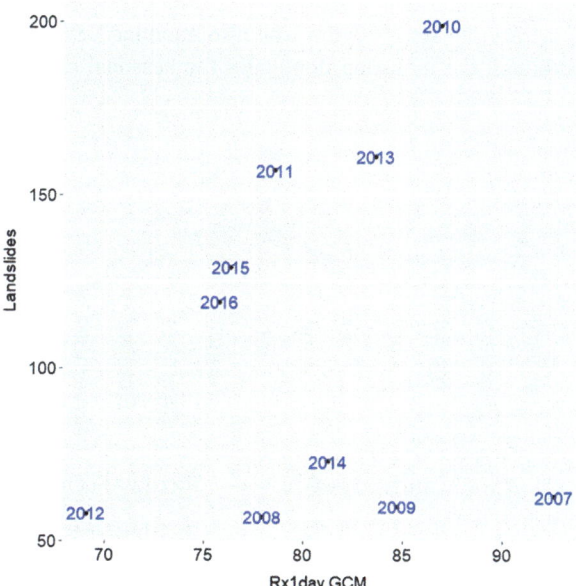

Fig. 55.4 Mean annual Rx1day (x-axis) from FLOR is not highly correlated with landslide activity from the GLC (y-axis). However, the first 3 years of the GLC (2007–2009) may have been underreported. If so, the relationship between extreme precipitation and landslide frequency might be strongly positive

few landslides compared to years with similar Rx1day values. However, the relationship between landslides and extreme precipitation would be fairly strong if evaluated after 2009. The other extreme precipitation indices show a similar pattern.

55.4.3 Seasonal Variability

The seasonal distribution of extreme precipitation in High Mountain Asia is dominated by the monsoon (June–September), although tropical cyclones may occasionally bring heavy rain or snow after the monsoon. As a result, rainfall-triggered landslides are also most abundant in South Asia at this time (Froude and Petley 2018). Landslide activity peaks in July, but is relatively high throughout the monsoon (Fig. 55.5). From October to May, fewer than 5 landslides per month are typically recorded in the GLC. This matches the cycle observed for both Rx1day and Rx5day. FLOR shows a peak of extreme precipitation in February, but this is not matched by a corresponding peak in landslide activity. TMPA also exhibits a peak but it is much less prominent. At the monthly scale, mean Rx5day derived from FLOR roughly doubles Rx1day, with relatively little variation (Fig. 55.6). The eight months with the highest mean value of Rx5day all coincide with the summer monsoon. Thus, both Rx1day and Rx5day encapsulate the seasonality of landslide activity.

Seasonality across the study region was also examined east and west of 80°E (Fig. 55.7). Although the monsoon dominates the seasonal climate cycle of both

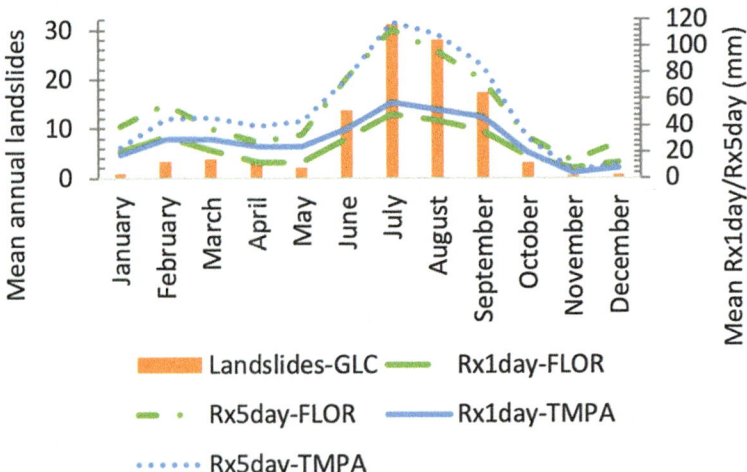

Fig. 55.5 The monthly pattern of landslide activity mirrors the extreme precipitation indices Rx1day (light colors) and Rx5day (dark colors). Landslide activity is represented by the total number of events recorded within the study area for each month from 2007 to 2016 in the GLC (orange). Daily precipitation was obtained from FLOR (green) and TMPA (blue)

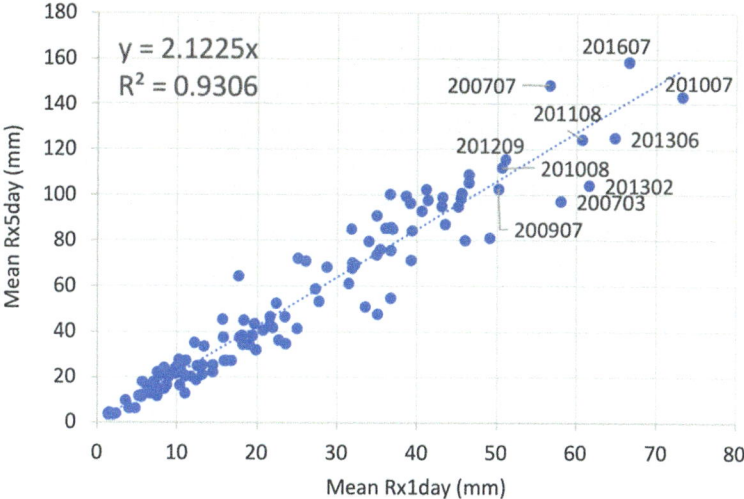

Fig. 55.6 The monthly means for Rx1day and Rx5day derived from FLOR are highly correlated. Precipitation is typically most intense during the monsoon, but February 2013 and March 2007 are exceptions

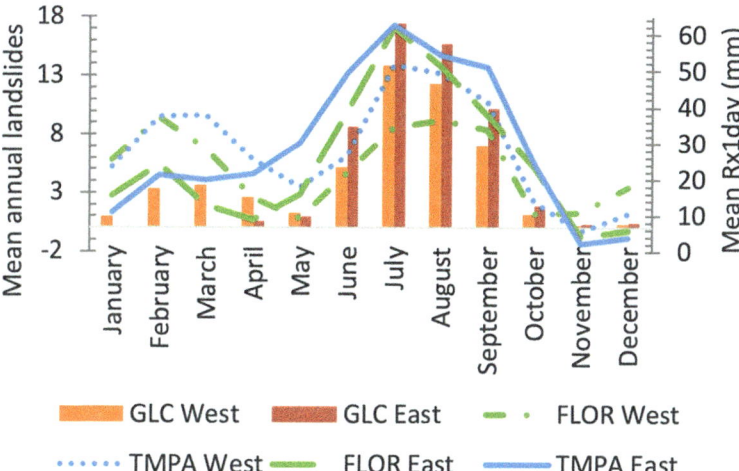

Fig. 55.7 The study area was divided into roughly equal halves at the 80th meridian, which left 516 (560) landslides in the western (eastern) section. The monthly distribution of landslides is similar, but the western portion of the study area experiences a secondary peak of landslide activity in March. Both FLOR and TMPA show a lesser peak in extreme precipitation during February, especially west of 80°

halves of the study area, the western region also experiences a smaller peak in landslide activity in March. Both regions experience high levels of landslide activity from June to September, with a peak in July. In general, this corresponds to the peak

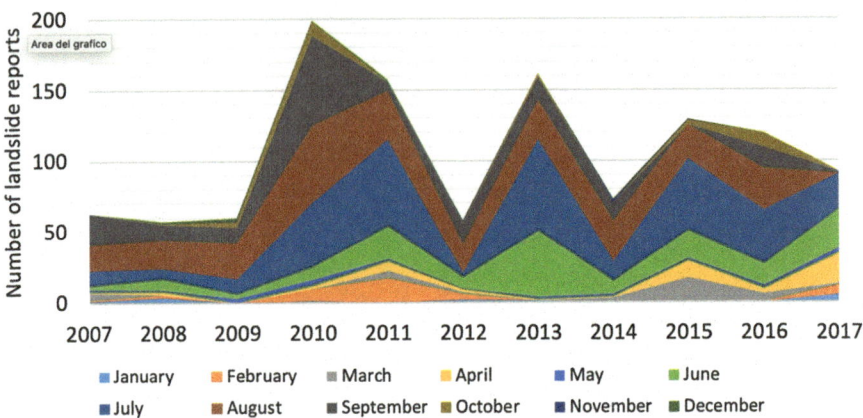

Fig. 55.8 Distribution of the GLC by month and year. June, July, August, and September dominate the record, but some years show little landslide activity in one or more of these months

in extreme precipitation. The exception is the western zone as estimated by FLOR, where the February peak in Rx1day slightly exceeds the monsoonal peak in August. In each region, less than 1 landslide is typically reported for the months of November, December, and January. Both FLOR and TMPA show a low in precipitation during November—nearly zero in the eastern half. This is followed by rising precipitation throughout the winter. This pattern of low landslide activity continues until June in the eastern portion of the study area (Fig. 55.7). On average, a few landslides are reported in the western zone during the months of February, March, and April. However, the mean value reflects a high level of inter-annual variability. 52% of the landslides reported in February occurred in 2011, and 46% of March landslides occurred in 2015 (Fig. 55.8). Rx1day also peaks in February (FLOR) and March (TMPA) over the western zone (Fig. 55.7). Although variable across the decade 2007–2016, the seasonality of extreme precipitation and landslide behavior was well reflected by Rx1day, whether averaged across the entire region or its eastern and western halves.

The total number of landslides reported in the GLC varies substantially, both by year and by month. As of this writing, the GLC has not been completed for 2017, so we have not included it in other analyses. Overall, annual variation is quite strong, with more than 3 times as many landslides in 2010 than in 2012. However, the number of reports is consistently low during the first three years of the GLC. July is the most active month on average, but fewer landslides are reported for July than for August or September 2007–2009, or for most years in the database. The wide range of landslide behavior can be explained, at least in part, by extreme precipitation.

The abnormally high number of landslides recorded in 2010 is explained by the 63 landslides that occurred across the region in September, more than the total number recorded for the years 2008, 2009, and 2012. In contrast, 2015 is notable for the total lack of landslides recorded in September and the record number of landslides in March. Landslides that occurred in 2017 had not been completely catalogued at the time of writing, but the record thus far indicated a relatively active year, especially during the month of April.

55.4.4 Monthly Variability

We analyzed extreme precipitation by individual month to better understand the annual fluctuations in landslide activity. Mean Rx1day appears to have a nonlinear positive relationship with the monthly landslide total (Fig. 55.9). Monthly mean values below 15 mm are associated with very low levels of landslide activity. Above 15 mm, increasing Rx1day appears to generate increasing numbers of landslides, albeit with a high amount of variation. This is consistent with the existence of an intensity threshold below which landslides will rarely be triggered by rainfall (Gabet et al. 2004). More than 30 landslides occurred in nine months, all during the monsoon period. While the precipitation intensity was usually in proportion to number of landslides, the three months with more than 60 landslides (September 2010, July 2011, and July 2013) actually had mean Rx1day less than 40 mm. This is a high number, but not relative to many other months. Another group of outliers can

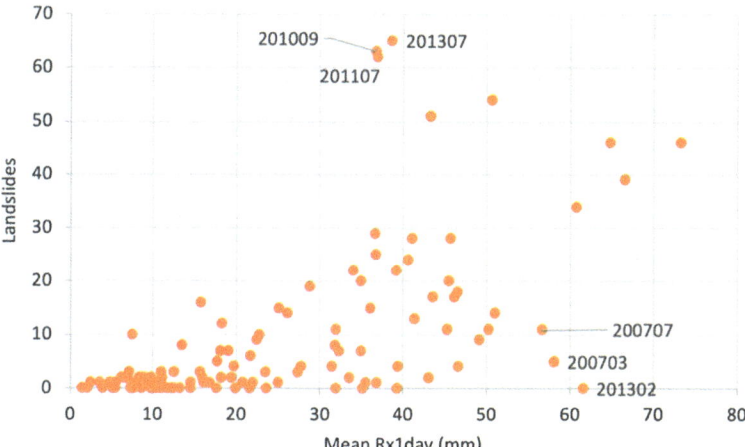

Fig. 55.9 FLOR monthly mean values of Rx1day are weakly correlated to the monthly number of landslides reported in the GLC. Some months that combine extreme precipitation with relatively few landslides (March 2007, July 2007, and February 2013) might be explained by underreporting during the first year of the GLC (March 2007 and July 2007) or the dominance of frozen precipitation (March 2007 and February 2013), which has less of an influence on landslide hazard than rainfall. It is harder to explain the outliers for which more than 60 landslides were reported but mean values of Rx1day were less than 40 mm (September 2010, July 2011, and July 2013). It is possible that pre-conditioning of the soil in previous months that had above average rainfall may have contributed to the clustering of landslides during the following months. The nine months with more than 30 reported landslides are all associated with the summer monsoon, as well as above-average daily precipitation intensity

be seen at the bottom right of Fig. 55.9. These three data points represent months in which intense precipitation failed to trigger landslides; or at least, no landslides were reported for these months. Although these six outliers do not fit the relationship, landslides appear to be associated with increases in Rx1day at the monthly scale, assuming a minimal level of precipitation has occurred.

In general, the mean monthly Rx1day values estimated by FLOR across the study area show a clear linkage between the number of landslides and increasing rainfall estimates. However, there are approximately six months that do not seem to fit this relationship, which we examine in Fig. 55.10. In the three cases where more than 60 landslides were reported, the monthly mean value obscured the existence of local high-intensity precipitation associated with rainfall-triggered landslides. In the three cases where few landslides were reported, we consider how the rainfall distributions may have impacted this result.

Most landslides recorded in September 2010 are clustered near an area of intense precipitation in Himachal Pradesh and Uttarakhand, where Rx1day reaches a maximum of 138 mm. Elsewhere Rx1day only reached 7 mm; the mean value across the

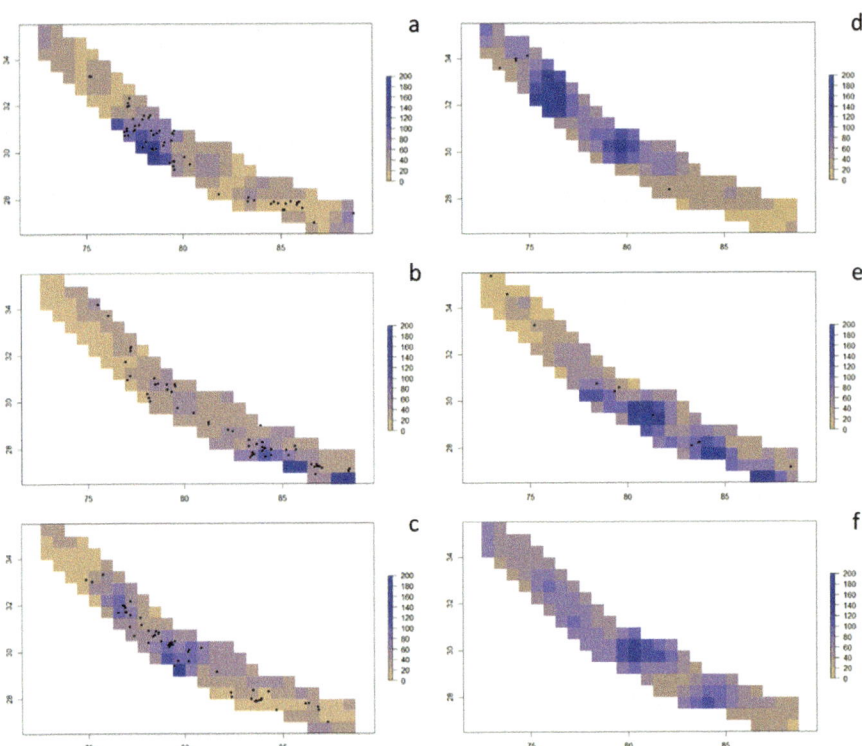

Fig. 55.10 Map of Rx1day monthly estimates (mm) from FLOR and reported landslides across the study region for (**a**) September 2010 (63 landslides), (**b**) July 2011 (62 landslides), (**c**) July 2013 (65 landslides), (**d**) March 2007 (5 landslides), (**e**) July 2007 (11 landslides), and (**f**) February 2013 (0 landslides)

study area was 37 mm. Rx1day for July 2011 ranged from 2 to 166 mm, with a mean of 37. Landslides appear to be clustered in 3 areas: the hills north of Bharatpur and Butwal, Nepal, the hills west of Dhankuta, Nepal, and the mountains of Uttarakhand. A band of extreme precipitation covered the Indian Himalaya and western Nepal in July 2013, with a maximum Rx1day of 167 mm. This appears to be the cause of a loose grouping of reported landslides that occurred from 23–28 July. A second cluster of landslides occurred north of Butwal, Nepal on the 22 July. The timing suggests that both groups of landslides could have been caused by a single meteorological event, but it does not appear to have been modeled by FLOR in the eastern group.

There were also three months that had very few reported landslides but for which Rx1day values were extremely high. Only five landslides were recorded for the month of March 2007, despite the presence of two areas of intense precipitation. Most of the landslides are located along a line between Islamabad, Pakistan, and Srinagar, India. The zones of intense precipitation are located to the southeast, where Rx1day reached a maximum of 187 mm. In this case, the high monthly mean value of Rx1day can be attributed to precipitation in the western half of the study area, but the geographic distribution of landslides leaves the connection between extreme precipitation and slope failure unclear. July 2007 exhibited intense precipitation to much of the study area (peak of 198 mm). In contrast, the western end of the study area experienced no more than 2 mm of precipitation per day, according to FLOR. Despite the clear spatial distribution of extreme precipitation, landslide reporting did not follow this pattern. No landslides were reported for February 2013 despite intense precipitation at multiple locations. Rx1day ranged from 14 to 134 mm. These examples point to the fact that the landslide inventory is highly variable between years and months owing to the availability of reports, time spent by the cataloger, and uncertainty in the timing and location of the GLC reports within this region. The variability may also occur because landslide triggering is a stochastic process with complex failure mechanisms; high rainfall intensity does not guarantee subsequent slope failures.

55.5 Discussion

The extreme precipitation observed within the study area across multiple indices and datasets (Fig. 55.2) suggests that it can be a factor in determining the concentration of landslides at a seasonal scale. Results also show that TMPA and FLOR behave similarly when estimating extreme precipitation indices for both geographic (Fig. 55.2) and seasonal (Figs. 55.5 and 55.7) distributions over the study area. Comparing TMPA and FLOR seasonally highlights several discrepancies, particularly during winter when there is a much more pronounced peak in FLOR relative to TMPA (Fig. 55.5). This may be explained by the presence of frozen precipitation within the winter westerlies that FLOR is able to resolve but is below the sensitivity threshold for TMI's microwave channels onboard the TRMM. Dividing the study area at 80°E reveals that the pre-monsoon landslides were limited to the western side of the study area (Fig. 55.7). This regional variation was mostly reflected by the monthly distribution of Rx1day calculated from both TMPA and FLOR, which showed higher precipitation intensity in spring and lower precipitation intensity in summer for the western zone. However, the distribution of precipitation is much more equally spread between the two zones than the distribution of landslides. We do not assume a linear response to increases in heavy precipitation; rather, landslides might become much more common after local precipitation exceeds historical maximum accumulations. This pattern also appears in the disaggregated monthly means (Fig. 55.9). Overall, extreme precipitation indices derived from both FLOR and TMPA reflect the seasonal cycle of landslides well.

Trends in extreme precipitation indices across the study region are more variable and less statistically significant (Fig. 55.3), largely due to the short time period for which satellite precipitation data is available. The trends for some frequency-based extreme precipitation indices relate to more common events and may reflect medium-term climate cycles rather than the long-term patterns that could be attributed to global climate change. This points to the opportunity to leverage the longer simulations provided by FLOR to provide a more statistically robust representation of extremes. Work is ongoing to evaluate and apply the longer retrospective record provided by FLOR for improved characterizations of regional precipitation extremes.

The co-occurrence of landslide distribution and extreme indices such as Rx1day suggest that some of the ETCCDI indices (specifically Rx1day and Rx5day) are valuable for characterizing seasonal patterns of landslides regionally. However, the presence of similarly high values just south of the study area suggests that rainfall triggering is not the only factor that modulates this relationship (Fig. 55.11). Landslides tend to be broadly distributed and associated with roads or other

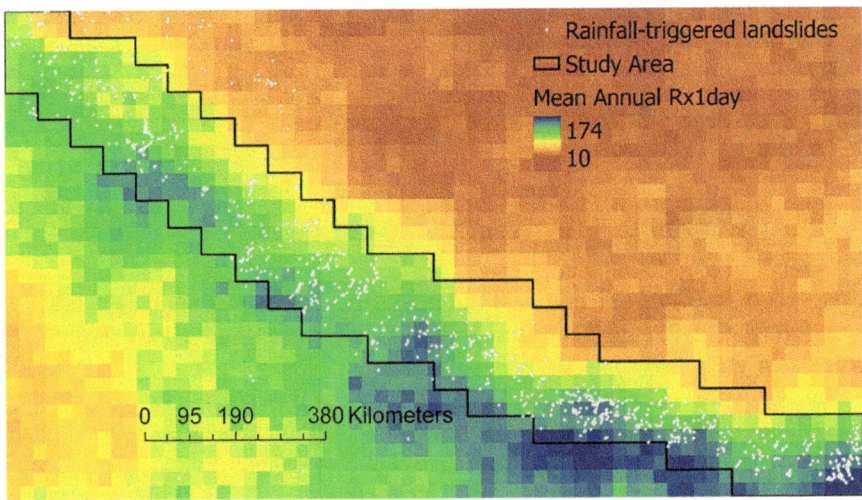

Fig. 55.11 Landslides appear to be slightly more common in grid cells with higher mean values of Rx1day (TMPA shown) and other ETCCDI indices, but the relationship does not appear to be very strong

anthropogenic features, rather than clustered within the rainiest grid cells. Geologic, geometric, and other controls on slope stability must be considered to fully explain the geographic distribution of landslides. Coupling landslide susceptibility maps with the precipitation thresholds or indices can provide both the spatial and temporal characterization of potential landslide activity in near real time (e.g., Kirschbaum and Stanley 2018).

Although the GLC is known to be incomplete (Kirschbaum et al. 2015), no seasonal bias in landslide reporting has been identified. The GLC does not rely on remote sensing, which might be affected by cloud cover or vegetative growth. Although events are entered into the database primarily during the months of June and July, this is unlikely to bias the GLC because the compilation methodology relies on media alerts saved consistently throughout the year. Thus, the monthly distribution is likely to be accurate, even if the annual number of landslides is a gross underestimate. Additionally, we believe the effectiveness of the GLC collection effort has improved over the last decade. In particular, the earliest years are probably underreported relative to later years (Fig. 55.8). If so, the relationship between extreme precipitation indices and annual landslide activity may be much stronger than it appears in Fig. 55.4. Petley (2016) has produced an independent inventory of fatal landslides in Nepal from 1980 to the present with methods similar to the GLC. The year 2008 was the third most active, and a high number of landslides occurred in the years 2007 and 2009 (Petley 2016). The year 2010 has previously been identified as a peak in landslide activity, both in the region (Kirschbaum et al. 2012) and worldwide (Petley 2011; Kirschbaum et al. 2012). The extreme precipitation indices largely corroborate that assessment (Fig. 55.4). Petley (2016) also identified a spike in fatal landslides in 2016, the year after the Gorkha earthquake. However, the GLC does not show a similar increase over the broader study area, and the number of landslides for 2016 appears to be in line with the intensity of extreme precipitation (Fig. 55.4).

The GLC does show a high level of interannual variability, both in annual and monthly totals (Fig. 55.8). Although the first three years of the GLC might not have benefited from advances in Google Alerts and other enhancements of the GLC methodology, we suggest that the variation in subsequent years might be due to a different issue: the nonlinear response of marginally stable slopes to rainfall intensity. Figure 55.9 shows a large number of months with little to no landslide activity. In general, the mean value of Rx1day across the study area is below 15 mm for these months. The remaining months show a positive, but highly variable relationship between increases in extreme precipitation and the number of landslides reported.

We evaluated each of the six months that appeared to defy this relationship. In the three cases where more than 60 landslides were reported, the monthly mean value obscured the existence of regional high-intensity precipitation associated with rainfall-triggered landslides. According to rain-gauge records, heavy precipitation was spread across the state of Uttarakhand, but concentrated on the 18 and 19 September 2010 (Sharma 2012). On the 18th, daily rainfall intensity reached 130, 164, and 458 mm at the Dwarahat, Katarmal, and Nainital meteorological stations, respectively. While the first two are very similar to the modeled precipitation, FLOR did not even approach the latter total. There has been some debate as to whether natural or anthropogenic factors were more important in preparing the ground for a catastrophe (Sati et al. 2011; Haigh and Rawat 2011), but it is almost certain that extreme rainfall over a large area triggered the disaster. Numerous landslides were reported in locations where FLOR precipitation estimates did not show extreme rainfall, which may be a function of the nudging scheme used to represent rainfall for the individual months (Fig. 55.10) or limitations in GCM performance when representing sub-daily variability (Covey et al. 2018). In addition, anthropogenic impacts, landslides preceding peak rainfall intensities, or incorrectly dated events add uncertainty and further complexity to what event(s) accurately triggered the landslides. The relatively low number of landslides in the months of March and July 2007 could be attributed to subsequent improvements in the GLC methodology, but the absence of rainfall-triggered landslides in February 2013 cannot be as clearly explained. In fact, landslides have been reported for this month, but these are associated with falling snow, not rain. Heavy snowfall usually has a less immediate and profound effect on slope stability than rainfall—unless rain is occurring on top of the snowpack, which may increase potential for runoff and infiltration.

Results here suggest that extreme precipitation indices may serve as a useful proxy for future landslide hazard, but the current comparison provides strong evidence only for the long-term seasonality. Given the nearly global availability of satellite-derived precipitation data, the ETCCDI indices can be calculated globally to determine the spatial distribution of extremes at an annual or decadal scale (Fig. 55.12). The potential applicability as well as the limitations of applying these

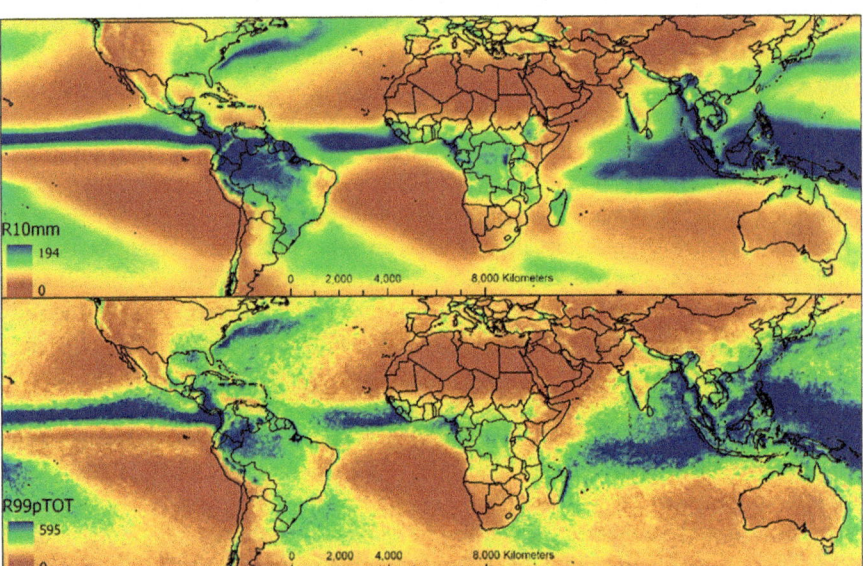

Fig. 55.12 The mean value of R10mm (TMPA) (top) exhibits a smooth spatial distribution due to its emphasis on less extreme precipitation events. In contrast, R99pTOT (TMPA) (bottom) exhibits a noisy spatial pattern, which suggests that the record is too short to contain the most extreme events at every location

indices as a proxy for landslide activity globally are shown in Fig. 55.12, which highlights R10mm and R99pTOT using the full TMPA record. The relative effects of rare and episodic events such as those represented by R99pTOT are clear, while R10mm highlights more consistent patterns globally.

While the focus of this study was on extreme precipitation indices, extreme temperature indices such as number of frost days (FD) and daily minimum temperature (TN_x) might also be linked to landslide activity in HMA. Temperature could affect the phase of the precipitation, which affects landslide triggering: falling snow is less likely to immediately trigger a landslide than falling rain. Melting snow can trigger landslides, especially when combined with simultaneous rainfall. Rising temperatures could affect antecedent conditions even more strongly by melting permafrost that currently enhances rock mass strength, altering the distribution of vegetation and settlement, opening glacial valleys to new mass wasting processes, and forming new lakes in glacier beds (Haeberli et al. 2017). Given that future increases in temperature are more certain than future increases in extreme precipitation (NCVST 2009), modeling the effects of temperature changes to soil moisture, permafrost, snowfields, vegetation, and glaciers on landslide hazard in HMA is a logical direction for future research.

55.6 Conclusions

This work highlights the utility of satellite and GCM precipitation estimates to establish the co-occurrence of extreme precipitation and landslides over the HMA region. While several of the ETCCDI indices demonstrate value in identifying extreme rainfall that could result in landslide activity, incorporation of terrain characteristics is also fundamental to understanding landslide triggering. Furthermore, the distribution of rainfall-triggered landslides within the study area does not appear to be controlled solely by extreme precipitation as measured by mean ETCCDI indices. Antecedent soil moisture and local relief, which may amplify local precipitation maxima through orographic processes, are both critical and were not analyzed in the current work.

Satellite and GCM data agree that the seasonal cycle of rainfall-triggered landslide activity in HMA is captured by the ETCCDI extreme precipitation indices. Even though winter precipitation plays a greater role in the western half of the study area, the landslide cycle is still dominated by the monsoon. This cycle is well known, but it is remarkable that a single number – mean maximum daily precipitation – so thoroughly represents landslide behavior in the study area. The same is unlikely to be true of other regions; in some places, antecedent conditions such as snow cover, vegetative cover, and rainfall may obfuscate the relationship. In cool climates, it will be necessary to treat frozen and liquid precipitation separately, as the effects of snow upon slope stability may be delayed by months. Nevertheless, extreme precipitation indices may be valuable in detecting changes to landslide seasonality in forecasts from global or regional climate models.

The relationship between the annual means of the indices and the annual number of landslides appears to be weak. However, the GLC is not a complete and unbiased record of landslides, nor are alternative datasets, such as remotely sensed inventories or other global landslide inventories (e.g., Froude and Petley 2018). We remain optimistic about the practice of using extreme precipitation indices as a shorthand for landslide hazard under various climate scenarios, but quantification of the relationship will rely on improvements to both the size and comprehensiveness of landslide inventories. Direct modeling of landslides under varying rainfall, soil moisture, and temperature conditions could better explain the historic patterns of landslide activity.

Acknowledgements The authors gratefully acknowledge reviews of this manuscript provided by Robert Emberson (USRA/NASA), Nathaniel Johnson (NOAA), and John Lanzante (NOAA). This work was supported by the NASA High Mountain Asia Project. Specifically, support for D. Kirschbaum and T. Stanley was provided by NASA grant #N5-HMA15-0035. Support for S. Kapnick and S. Pascale was provided by NASA grant #15-HMA15-0016. S. Pascale is also supported by the NOAA CICS grant – NA14OAR4320106.

References

Adler, R. F., Huffman, G. J., Chang, A., Ferraro, R. R., Xie, P., Janowiak, J., Rudolf, B., Schneider, U., Curtis, S., Bolvin, D. T., Gruber, A., Susskind, J., Arkin, P. A., & Nelkin, E. (2003). The version-2 Global Precipitation Climatology Project (GPCP) monthly precipitation analysis (1979–present). *Journal of Hydrometeorology, 4*, 1147–1167. https://doi.org/10.1175/1525-7541(2003)004<1147:TVGPCP>2.0.CO;2.

Ávila, A., Justino, F., Wilson, A., Bromwich, D., & Amorim, M. (2016). Recent precipitation trends, flash floods and landslides in southern Brazil. *Environmental Research Letters, 11*, 114029. https://doi.org/10.1088/1748-9326/11/11/114029.

Barros, A. P., Joshi, M., Putkonen, J., & Burbank, D. W. (2000). A study of the 1999 monsoon rainfall in a mountainous region in Central Nepal using TRMM products and rain gauge observations. *Geophysical Research Letters, 27*, 3683–3686. https://doi.org/10.1029/2000GL011827.

Barros, A. P., Kim, G., Williams, E., & Nesbitt, S. W. (2004). Probing orographic controls in the Himalayas during the monsoon using satellite imagery. *Natural Hazards and Earth System Sciences, 4*, 29–51. https://doi.org/10.5194/nhess-4-29-2004.

Bharti, V., & Singh, C. (2015). Evaluation of error in TRMM 3B42V7 precipitation estimates over the Himalayan region. *Journal of Geophysical Research, 120*, 12,458–12,473. https://doi.org/10.1002/2014JD022121.

Bronaugh, D. (2015). Climdex.pcic: PCIC Implementation of Climdex Routines. Available at https://rdrr.io/cran/climdex.pcic/, last accessed 2 Dec 2018.

Cepeda, J., Hoeg, K., & Nadim, F. (2010). Landslide-triggering rainfall thresholds: A conceptual framework. *Quarterly Journal of Engineering Geology and Hydrogeology, 43*, 69–84. https://doi.org/10.1144/1470-9236/08-066.

Chen, J. C., Huang, W. S., Jan, C. D., & Yang, Y. H. (2012). Recent changes in the number of rainfall events related to debris-flow occurrence in the Chenyulan Stream Watershed, Taiwan. *Natural Hazards and Earth System Sciences, 12*, 1539–1549. https://doi.org/10.5194/nhess-12-1539-2012.

Covey, C., Doutriaux, C., Gleckler, P. J., & Taylor, K. E. (2018). High-frequency intermittency in observed and model-simulated precipitation. *Geophysical Research Letters, 45*, 1–9. https://doi.org/10.1029/2018GL078926.

Dahal, R. K., & Hasegawa, S. (2008). Representative rainfall thresholds for landslides in the Nepal Himalaya. *Geomorphology, 100*, 429–443. https://doi.org/10.1016/j.geomorph.2008.01.014.

Delworth, T. L., & Zeng, F. (2014). Regional rainfall decline in Australia attributed to anthropogenic greenhouse gases and ozone levels. *Nature Geoscience, 7*, 583–587. https://doi.org/10. 1038/ngeo2201.

Delworth, T. L., Rosati, A., Anderson, W., Adcroft, A. J., Balaji, V., Benson, R., Dixon, K., Griffies, S. M., Lee, H.-C., Pacanowski, R. C., Vecchi, G. A., Wittenberg, A. T., Zeng, F., & Zhang, R. (2011). Simulated climate and climate change in the GFDL CM2.5 High-Resolution Coupled Climate Model. *Journal of Climate, 25*, 2755–2781. https://doi.org/10.1175/JCLI-D-11-00316.1.

Delworth, T. L., Zeng, F., Vecchi, G. A., Yang, X., Zhang, L., & Zhang, R. (2016). The North Atlantic Oscillation as a driver of rapid climate change in the Northern Hemisphere. *Nature Geoscience, 9*, 509–512. https://doi.org/10.1038/ngeo2738.

Froude, M. J., & Petley, D. N. (2018). Global fatal landslide occurrence from 2004 to 2016. *Natural Hazards and Earth System Sciences, 18*, 2161–2181. https://doi.org/10.5194/nhess-18-2161-2018.

Gabet, E. J., Burbank, D. W., Putkonen, J. K., Pratt-Sitaula, B. A., & Ojha, T. (2004). Rainfall thresholds for landsliding in the Himalayas of Nepal. *Geomorphology, 63*, 131–143. https://doi.org/10.1016/j.geomorph.2004.03.011.

Haeberli, W., Schaub, Y., & Huggel, C. (2017). Increasing risks related to landslides from degrading permafrost into new lakes in de-glaciating mountain ranges. *Geomorphology, 293*, 405–417. https://doi.org/10.1016/J.GEOMORPH.2016.02.009.

Haigh, M., & Rawat, J. S. (2011). Landslide causes: Human impacts on a Himalayan landslide swarm. *Belgeo, 3*(3–4), 201–220. https://doi.org/10.4000/belgeo.6311.

Hijmans, R. J. (2015). Raster: Geographic Data Analysis and Modeling. R package version 2.4-15. Available at https://cran.r-project.org/web/packages/raster/index.html, last accessed 2 Dec 2018.

Huffman, G. J., Adler, R. F., Bolvin, D. T., & Nelkin, E. J. (2010). The TRMM multi-satellite precipitation analysis (TMPA). In F. Hossain & M. Gebremichael (Eds.), *Satellite rainfall applications for surface hydrology* (pp. 3–22). Dordrecht: Springer. https://doi.org/10.1007/978-90-481-2915-7_1.

Ines, A. V. M., & Hansen, J. W. (2006). Bias correction of daily GCM rainfall for crop simulation studies. *Agricultural and Forest Meteorology, 138*, 44–53. https://doi.org/10.1016/j.agrformet.2006.03.009.

Kapnick, S. B., Delworth, T. L., Ashfaq, M., Malyshev, S., & Milly, P. C. D. (2014). Snowfall less sensitive to warming in Karakoram than in Himalayas due to a unique seasonal cycle. *Nature Geoscience, 7*, 834–840. https://doi.org/10.1038/ngeo2269.

Karki, R., Hasson, S., Schickhoff, U., & Scholten, T. (2017). Rising precipitation extremes across Nepal. *Climate, 5*, 1–26. https://doi.org/10.3390/cli5010004.

Kendall, M. G. (1948). *Rank correlation methods* (p. 272). Oxford: Oxford University Press. ISBN: 0195208374.

Kirschbaum, D. B., & Stanley, T. (2018). Satellite-based assessment of rainfall-triggered landslide hazard for situational awareness. *Earth's Future, 6*, 505–523. https://doi.org/10.1002/2017EF000715.

Kirschbaum, D. B., Adler, R. F., Hong, Y., Hill, S., & Lerner-Lam, A. (2010). A global landslide catalog for hazard applications: Method, results, and limitations. *Natural Hazards, 52*, 561–575. https://doi.org/10.1007/s11069-009-9401-4.

Kirschbaum, D. B., Adler, R. F., Adler, D., Peters-Lidard, C., & Huffman, G. J. (2012). Global distribution of extreme precipitation and high-impact landslides in 2010 relative to previous years. *Journal of Hydrometeorology, 13*, 1536–1551. https://doi.org/10.1175/JHM-D-12-02.1.

Kirschbaum, D. B., Stanley, T., & Zhou, Y. (2015). Spatial and temporal analysis of a Global Landslide Catalog. *Geomorphology, 249*, 4–15. https://doi.org/10.1016/j.geomorph.2015.03.016.

Mann, H. B. (1945). Nonparametric tests against trend. *Econometrica, 13*, 245. https://doi.org/10. 2307/1907187.

Nadim, F., Kjekstad, O., Peduzzi, P., Herold, C., & Jaedicke, C. (2006). Global landslide and avalanche hotspots. *Landslides, 3*, 159–173. https://doi.org/10.1007/s10346-006-0036-1.

NCVST. (2009). Vulnerability through the eyes of vulnerable: Climate change induced uncertainties and Nepal's development predicaments. Boulder, CO. Available at https://www. preventionweb.net/publications/view/12565, last accessed 2 Dec 2018.

Panday, P. K., Thibeault, J., & Frey, K. E. (2015). Changing temperature and precipitation extremes in the Hindu Kush-Himalayan region: An analysis of CMIP3 and CMIP5 simulations and projections. *International Journal of Climatology, 35*, 3058–3077. https://doi.org/10.1002/joc. 4192.

Pascale, S., Boos, W. R., Bordoni, S., Delworth, T. L., Kapnick, S. B., Murakami, H., Vecchi, G. A., & Zhang, W. (2017). Weakening of the North American monsoon with global warming. *Nature Climate Change, 7*, 806–812. https://doi.org/10.1038/nclimate3412.

Petley, D. N. (2011). Global deaths from landslides in 2010 (updated to include a comparison with previous years). *Landslide Blog*. Available at https://blogs.agu.org/landslideblog/2011/02/05/ global-deaths-from-landslides-in-2010/, last accessed 2 Dec 2018.

Petley, D. N. (2012). Global patterns of loss of life from landslides. *Geology, 40*, 927–930. https:// doi.org/10.1130/G33217.1.

Petley, D. N. (2016). Nepal landslides 2016: Losses at the end of the rainy season – The Landslide Blog – AGU Blogosphere. Available at https://blogs.agu.org/landslideblog/2016/08/01/2016-landslide-losses/, last accessed 2 Dec 2018.

Petley, D. N., Hearn, G. J., Hart, A., Rosser, N. J., Dunning, S. A., Oven, K., & Mitchell, W. A. (2007). Trends in landslide occurrence in Nepal. *Natural Hazards, 43*, 23–44. https://doi.org/10. 1007/s11069-006-9100-3.

Pohlert, T. (2017). Tend: Non-Parametric Trend Tests and Change-Point Detection. Available at https://rdrr.io/cran/trend/, last accessed 2 Dec 2018.

R Core Team (2013) R: A language and environment for statistical computing. *R Foundation for Statistical Computing*. Available at http://www.r-project.org/, last accessed 2 Dec 2018.

Revolution Analytics, and S. Weston. (2015) Foreach: Provides Foreach Looping Construct for R. Available at https://rdrr.io/rforge/foreach/, last accessed 2 Dec 2018.

Sati, S. P., Sundriyal, Y. P., Rana, N., & Dangwal, S. (2011). Recent landslides in Uttarakhand: Nature's fury or human folly. *Current Science, 100*, 1617–1620. Available at https://www. researchgate.net/publication/285718177_Recent_landslides_in_Uttarakhand_Nature%27s_ fury_or_human_folly, last accessed 2 Dec 2018.

Sen, P. K. (1968). Estimates of the regression coefficient based on Kendall's Tau. *Journal of the American Statistical Association, 63*, 1379–1389. https://doi.org/10.1080/01621459.1968. 10480934.

Sharma, S. (2012). Catastrophic hydrological event of 18 and 19 September 2010 in Uttarakhand, Indian Central Himalaya – An analysis of rainfall and slope failure. *Current Science, 102*, 327–332. Available at https://www.researchgate.net/publication/224960685_Catastrophic_ hydrological_event_of_18_and_19_September_2010_in_Uttarakhand_Indian_Central_ Himalaya_-_an_analysis_of_rainfall_and_slope_failure, last accessed 2 Dec 2018.

Theil, H. (1992). A rank-invariant method of linear and polynomial regression analysis. In B. Raj & J. Koerts (Eds.), *Henri Theil's Contributions to Economics and Econometrics* (Advanced Studies in Theoretical and Applied Econometrics, 23). Dordrecht: Springer. https://doi.org/10. 1007/978-94-011-2546-8_20.

Vecchi, G. A., Delworth, T., Gudgel, R., Kapnick, S., Rosati, A., Wittenberg, A. T., Zeng, F., Anderson, W., Balaji, V., Dixon, K., Jia, L., Kim, H.-S., Krishnamurthy, L., Msadek, R., Stern, W. F., Underwood, S. D., Villarini, G., Yang, X., & Zhang, S. (2014). On the seasonal forecasting of regional tropical cyclone activity. *Journal of Climate, 27*, 7994–8016. https:// doi.org/10.1175/JCLI-D-14-00158.1.

World Climate Research Programme. (2018). About ETCCDI. Available at https://www.wcrp-climate.org/etccdi, last accessed 2 Dec 2018.

Yang, M., Chen, X., & Cheng, C. S. (2016). Hydrological impacts of precipitation extremes in the Huaihe River Basin, China. *Springerplus, 6,* 1731. https://doi.org/10.1186/s40064-016-3429-1.

Yang, X., Jia, L., Kapnick, S. B., Delworth, T. L., Vecchi, G. A., Gudgel, R., Underwood, S., & Zeng, F. (2018). On the seasonal prediction of the western United States El Niño precipitation pattern during the 2015/16 winter. *Climate Dynamics, 51,* 3765–3783. https://doi.org/10.1007/s00382-018-4109-3.

Zhang, W., Vecchi, G. A., Murakami, H., Delworth, T., Wittenberg, A. T., Rosati, A., Underwood, S., Anderson, W., Harris, L., Gudgel, R., Lin, S.-J., Villarini, G., & Chen, J.-H. (2015). Improved simulation of tropical cyclone responses to ENSO in the Western North Pacific in the High-Resolution GFDL HiFLOR Coupled Climate Model. *Journal of Climate, 29,* 1391–1415. https://doi.org/10.1175/JCLI-D-15-0475.1.

Chapter 56
The Value of Satellite Rainfall Estimates in Agriculture and Food Security

Tufa Dinku

Abstract Agriculture needs to produce more and more food not only to satisfy the current needs, but also to cope with the demand from the increasing population on the planet. However, agriculture today faces multiple challenges. Climate change and variability (extremes) feature at the top of these challenges because agriculture is particularly dependent on the climate. Stability of the food supply system as well as people's livelihoods could be disrupted by climate shocks such as droughts and floods. Stability of food production and people's livelihoods would require making agriculture resilient to climate trends and shocks through adaptation to climate change and finding effective ways to combat uncertainties associated with climate extremes such as droughts and floods. This, in turn, would require effective use of climate information that would enable agricultural practitioners to make better informed decisions at national, institutional and community levels. Provision of weather and climate information will depend on the availability of climate data. Rainfall is one of the major climatic parameters that has a major influencing factor in agriculture. However, rainfall measurements from station observations are inadequate over many parts of the world, and particularly over Africa, due to sparse or non-existent station networks. Satellite-based rainfall products have been used increasingly to complement or replace station observations. The main strength of satellite rainfall estimates is that they provide good spatial coverage, including remote areas, and are freely available. Satellite rainfall estimates are being used to support climate services for agriculture and food security in many different ways. This paper presents some specific examples that include a participatory planning process that is helping farmers in Rwanda to manage climate risks, and weather-based index insurance.

Keywords Precipitation · Rainfall · Climate services · Africa · Droughts · Food security · Agriculture · Global Framework for Climate Services · FAO · National Meteorological and Hydrological Services (NMHS) · Climate Services Partnership ·

T. Dinku (✉)
International Research Institute for Climate and Society (IRI), The Earth Institute at Columbia University, Palisades, NY, USA
e-mail: tufa@iri.columbia.edu

© Springer Nature Switzerland AG 2020
V. Levizzani et al. (eds.), *Satellite Precipitation Measurement*, Advances in Global Change Research 69, https://doi.org/10.1007/978-3-030-35798-6_32

Climate Services for Agriculture · GPCC · TAMSAT · Enhancing National Climate
Services (ENACTS) · Weather index insurance

56.1   Introduction

The United Nations (UN) Department of Economic and Social Affairs has recently
announced that the current world population, which is around 7.6 billion, will reach
8.6 billion in 2030, 9.8 billion in 2050 and 11.2 billion by 2100 (UN 2017). This
increasing trend of 83 million people every year is expected to continue even as
fertility levels will continue to decline. This also means an ever-increasing demand
for food and more stress on food production systems. Agriculture has to produce
more and more food to satisfy the current needs as well as to cope with the demand
from the increasing population. This would require adopting more efficient and
sustainable production systems, as well as adapting to climate change and find
effective ways to manage uncertainties associated with climate extremes.

There are diverse sources of risk to agriculture and food security. Because
agriculture is particularly dependent on the climate, climate change and extremes
are among the most serious challenges to agriculture and the different components of
food security (WMO 2014b; De Leeuw et al. 2014; Hansen 2012; Hansen et al.
2014; Hellmuth et al. 2009; Skees and Collier 2008). Uncertainty associated with
climate variability poses major challenges for agricultural management at all levels.
Stability of food supply as well as people's livelihoods could be disrupted by climate
shocks such as droughts and floods. The frequency of extreme weather is expected to
increase as a result of climate change. This would affect rain-fed smallholder
agricultural systems of the tropics and subtropics, where hunger and rural poverty
are already persistent. And because of the risk-averse nature of small-holder farmers,
climate extremes create disincentives to invest in improved agricultural technology
and market opportunities, and hence suppress opportunities for increased food
production (Hansen et al. 2014).

Wise use of climate information would enable agricultural practitioners to make
better informed decisions at policy, institutional and community levels and help to
make agriculture more resilient in the face of a variable and changing climate.
Effective use of climate information would help to improve the efficient use of
limited resources, and increase food production by reducing impacts of climate
risks and enhancing opportunities (WMO 2014b; Hansen et al. 2014). The provision and effective use of weather and climate information would also be very
useful for anticipating, preparation for and response to agriculture or food security
risks triggered by climate extremes as well as adopting to longer term risks
associated with climate change (WMO 2014b; De Leeuw et al. 2014; Hansen
et al. 2014).

Until recently, there has not been a coherent and comprehensive global approach
to support the management of climate risks to key development sectors, including
agriculture and food security. To bridge this gap, the Third World Climate

Conference held in 2009, and the World Meteorological Congress in 2011 mandated the creation of the Global Framework for Climate Services (GFCS, WMO 2014a). The GFCS has identified agriculture and food security as its top priority (WMO 2014a); disaster risk reduction, water resources, and health are the other priority sectors. The GFCS has now become an overarching strategic priority for the World Meteorological Organization (WMO), guiding all of its work with national meteorological agencies. It is expected to enhance the efforts of WMO, the Food and Agriculture Organization (FAO) and the World Food Programme (WFP), among others, for climate risk monitoring and analysis, and for promoting a more integrated approach to sustainable development and building the resilience of the most vulnerable people (WMO 2014a).

The provision of weather and climate services to agriculture depends on the availability of climate data and climate information products. Climate information may include historical meteorological observations, monitoring of current weather conditions and prediction across a range of lead times. Historical climate data could be used, among many other things, for adapting crops, farming systems and agricultural support to local climatic risks, responding locally to seasonal climate forecasts, improving food security early warning and response, supporting cost-effective index-based agricultural insurance, and understanding drivers of change (Hansen et al. 2014; Hansen 2012).

Rainfall is one of the climatic parameters that is a major influencing factor in agriculture in general, and crop production in particularly. Traditionally, rainfall measurements come from conventional ground weather stations managed mostly by National Meteorological and Hydrological Services (NMHS) across the globe. However, rainfall measurement, historical or current, from station observations are inadequate over many parts of the world, due to sparse or non-existent station networks. This is the case particularly in Africa where the state of the in-situ climate observing system is seriously inadequate, with the number and quality of weather stations in many parts of the continent in decline (Dinku et al. 2016, 2017; Parker et al. 2011; Washington et al. 2006). Thus, satellite-based rainfall products have been used increasingly as complements or in place of station observations. The main strength of satellite rainfall estimates is that they provide good spatial coverage, including remote areas, and are freely available.

Satellite rainfall estimates are being used to support climate services for agriculture and food security in many different ways. The next section will expand on the value of satellite rainfall in agriculture and food security with some specific examples that include combined use of station-satellite data to overcome climate data challenges, a participatory planning process that is helping farmers in Rwanda to manage climate risks, and for expanding weather-based index insurance across Africa.

56.2   The Value of Satellite Rainfall Estimates in Agriculture and Food Security

56.2.1   Climate Services in Agriculture

Weather and climate conditions play crucial roles in agriculture and food security, mainly through extreme events such as droughts, heat waves, frosts, floods and storms. These could lead to crop failure, food insecurity, destruction of key livelihood assets, mass migration of people, and decline in national economic growth. This also means that there are opportunities to improve agricultural productivity and food security by incorporating climate and weather information into agricultural policy, planning and practices. This may be accomplished through integration of climate services into practices and policy for development decisions in agriculture and food security (WMO 2014b).

The need for climate services emanates from society's need to better understand and manage negative impacts as well as benefiting from positive climate outcomes (Hansen et al. 2014). The concept of climate services has been evolving considerably over recent years (Zillman 2009; Hansen et al. 2014). WMO defines climate services as "the provision of one or more climate products or advice in such a way as to assist decision-making by individuals or organizations", (WMO 2014b), while the Climate Services Partnership (CSP) states that climate services involve "the production, translation, transfer, and use of climate knowledge and information in climate-informed decision making and climate-smart policy and planning", (CSP 2018). The common thread between these two definitions is that climate services involves the provision of specific climate information products or services for a specific decision-making process. The interest in climate services at global level has recently resulted in the creation of the GFCS in 2009, with a vision to "enable society to better manage the risks and opportunities arising from climate variability and change, especially for those who are most vulnerable to climate related hazards," (WMO 2014a).

Climate services are critical for strengthening the information and early warning systems on food and agriculture. There are a number of benefits of climate services to agriculture and food security, which include developing sustainable and economically viable agricultural systems, improving production and quality, reducing losses and risks, reducing costs, increasing efficiency in the use of water, labor and energy, conserving natural resources and decreasing pollution by agricultural chemicals or other agents that contribute to the degradation of the environment (WMO 2014b). This may explain why GFCS has made Agriculture and Food Security its priority sector. The GFCS's explanation for this is that "available, accessible, and useful information can help agricultural decision-makers improve their understanding of the mechanisms of climate impact on agricultural development and food systems and estimate populations at risk from food insecurity," (WMO 2014b).

Quality climate data, historical, current, and forecast, are the foundations for any climate services. Climate data are critical inputs in providing climate services to agriculture and for making agriculture more resilient to climate variability and change. If exploited fully, historical climate data could benefit agriculture in many different ways that include the following (Hansen et al. 2014; Hansen 2012):

- Adapting crops, farming systems and agricultural support to local climatic risks including through the analysis of the seasonality, trends, and variability that characterize local climate to provide insights into the risks that farmers and value chain actors face.
- Enabling more local use of seasonal climate forecasts by downscaling seasonal forecasts and packaging them with local historic climate time series, expressed graphically in probabilistic terms, and presented through a structured participatory process.
- Improving food security early warning and response through the use of high-quality local climate records and seasonal forecasts integrated with crop simulation models to increase the lead time and accuracy of production forecasts.
- Supporting cost-effective index-based agricultural insurance through the provision of reliable local near-real-time climate data.
- Understanding the drivers of change through the analysis of local historical climate data records providing a way for farmers and other decision-makers to determine whether perceived changes in agricultural performance are driven by climate, or some other driver of change such as soil degradation.

56.2.2 Challenges to Available Climate Data in Africa

Observations by weather stations managed by the NMHSs are the primary sources of climate data in Africa. However, the weather observation network in Africa is inadequate with the number and quality of weather stations in many parts of the continent has been declining (Dinku et al. 2014, 2017; Parker et al. 2011; Washington et al. 2006; Malhi and Wright 2004). Most of the available stations are unevenly distributed, with most of the stations located along the main roads (Dinku et al. 2014, 2017).

While this problem is prevalent in many parts of the world, it is particularly common in Africa, especially over areas where there are difficult and remote geographies, conflict, and investment in data is a relatively low priority. Figure 56.1 shows that the number of stations per 100 km × 100 km grid box used in the "full data" gridded rainfall product of the Global Precipitation Climatology Centre (GPCC; Becker et al. 2013). There is an uneven distribution of stations with higher densities over South Africa and very few stations over the forested and desert parts of the continent. It should be noted here that stations in Fig. 56.1 are not the only observation stations available in Africa; these are what the NMHS have shared with GPCC. However, Fig. 56.1 does represent the spatial distribution of weather stations across Africa reasonably well.

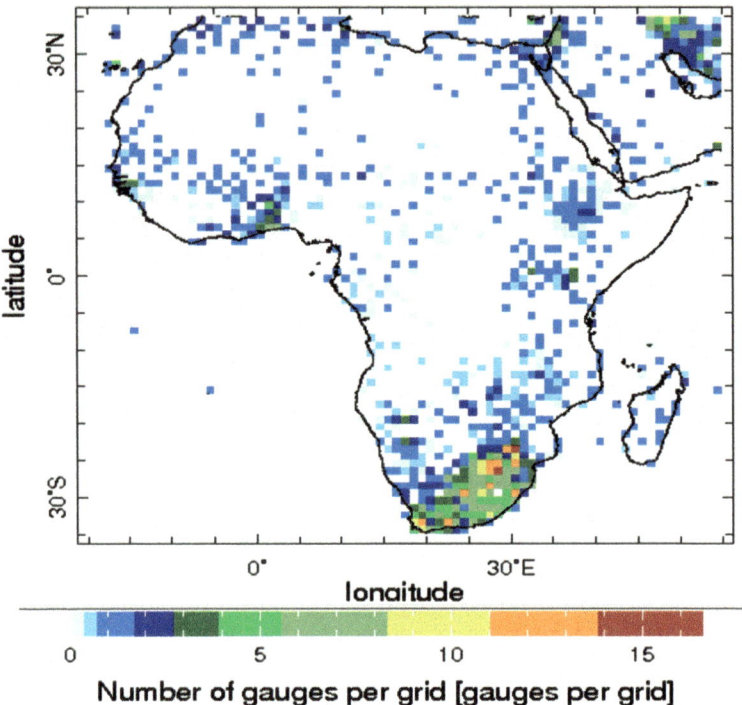

Fig. 56.1 Average (2001–2010) number of stations per 100 km × 100 km grid box used by GPCC gridded rainfall product. (Data source: GPCC)

The number of observation stations have also been declining for several decades now. Figure 56.2 shows time series of average number of stations used in the GPCC full-data product over Africa since 1901. The number of observing stations shows a steady decline since the early 1980's. This decline may be attributed to the decline in observation stations over Africa as well as the decline in the number of stations that the NMHS have been sharing with GPCC. A good example of a decline in weather observation networks at the national level would be that of Uganda (Fig. 56.3), which shows a similar patter to that of the continent in Fig. 56.2.

There are a number of factors contributing to the sparse station network and the decline in the number of weather stations over many parts of Africa. The major ones include, social or political conflict, declining investment in NMHS, and difficult and remote geography. Conflict or political unrest can disrupt weather observation networks resulting in loss of data. An analysis by Auffhammer et al. (2013) has demonstrated the relationship between conflict and the number of missing observations. Rwanda, where the meteorological observation network was devastated during the 1994 genocide, would be a very good example (Fig. 56.4). It took the country nearly 15-years to return to the pre-genocide level of the observation network.

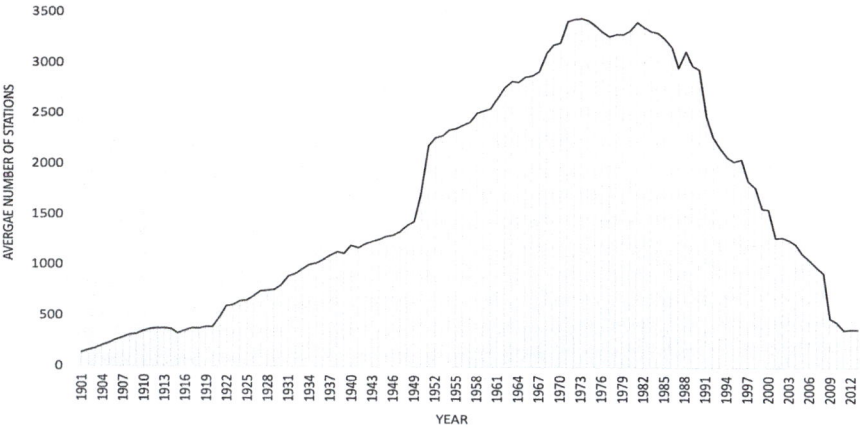

Fig. 56.2 Time series of average number of stations used in the GPCC full-data product over Africa (15°W to 45°E, and 30°S to 30°N). (Data source: GPCC, https://www.dwd.de/EN/ourservices/gpcc/gpcc.html, last accessed 28 Nov. 2018)

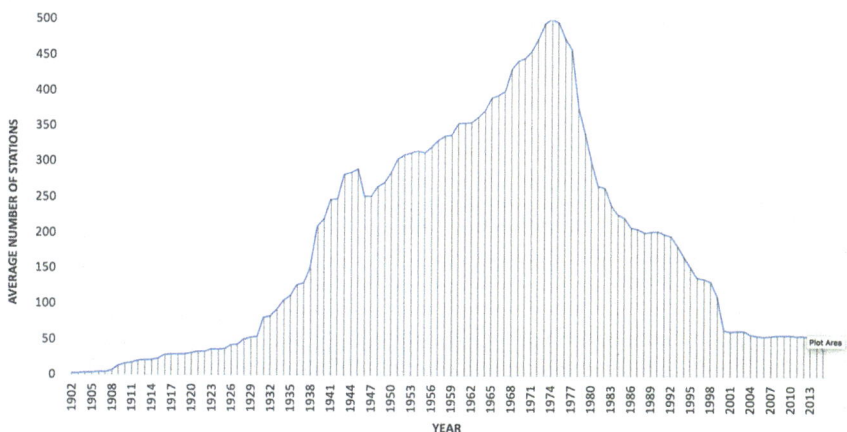

Fig. 56.3 Average number of weather stations (mostly raingauges) reporting each year for Uganda

Difficult geography and terrain contribute to the sparse distribution of the observation network in many parts of the continent. Mountains, forest, and desert areas contribute to the sparse distribution of the observation network by making installation and maintenance of observation network difficult. For instance, there are very few GPCC stations over the Sahara Desert and the Congo basin as shown in Fig. 56.1.

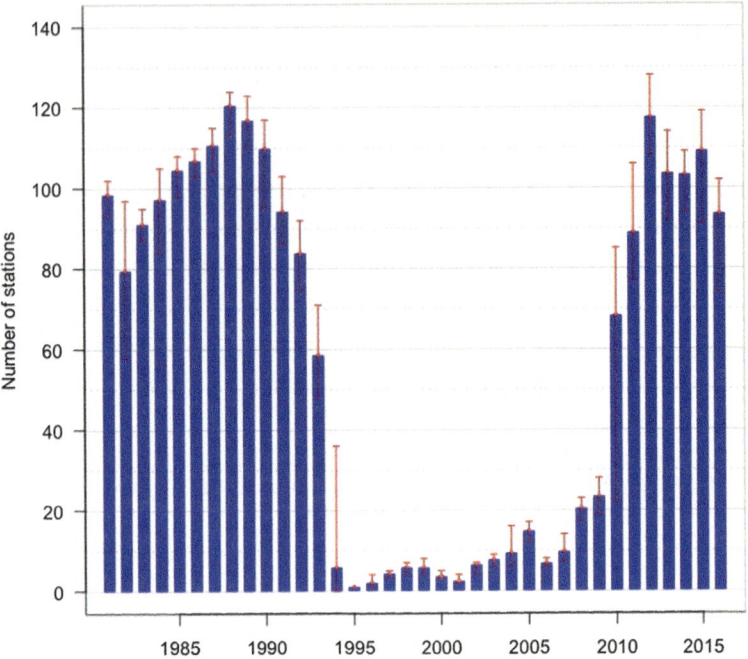

Fig. 56.4 Average number of weather stations reporting each year over Rwanda

56.2.3   The Value of Satellite Data to Overcome Data Availability Challenges

Different approaches have been tried to overcome the challenges of sparse station distribution discussed in the previous section. These include using gridded station rainfall product as well as satellite rainfall estimates. The quality of gridded station data ultimately depends upon the number, spatial distribution, and quality of the station observations. As a result, gridded rainfall products also suffer from the same challenges of sparse station distribution and missing observations. Other weaknesses of gridded rainfall products include the following:

– Temporal inhomogeneity of the historical time series because of the use of different number of stations over time (e.g., Fig. 56.2), which means the quality of the gridded data varies over time;
– They are updated once in a while and may not be suitable for monitoring purposes; and
– Coarse spatial and temporal resolutions are challenging particularly for agricultural applications.

Satellite estimates of rainfall are a widely used climate variable derived from satellite measurements, though they generally incorporate observations to adjust

satellite rainfall estimates for bias corrections. The main strengths of satellite estimates include global coverage, improved temporal and spatial resolution, long
historical time series that now goes back over 30 years, and the fact that a wide
variety of satellite-based rainfall products are now easily available. The large spatial
coverage and high temporal revisit frequency makes satellite rainfall estimates
particularly useful for near real-time monitoring. As a result, satellite-based rainfall
products are being used increasingly in Africa for many different activities, which
include to monitor drought and potential impacts on agriculture and food security
(Sheffield et al. 2014, Verdin et al. 2005; Hutchinson 1991), crop simulation
modelling (Thornton et al. 1997), and weather-based index insurance (Black et al.
2016; Greatrex et al. 2015; De Leeuw et al. 2014; Hellmuth et al. 2009). The next
section presents some examples of specific applications of satellite rainfall estimates.

56.3 Specific Applications of Satellite Rainfall Estimates in Agriculture

56.3.1 Enhancing National Climate Services (ENACTS)

The Enhancing National Climate Services (ENACTS) initiative is an ambitious
effort to simultaneously improve the availability, access and use of climate information (Dinku et al. 2015, 2017). It focuses on the creation of reliable climate data
and information products suitable for national and local decision making. This is
accomplished by working with National Meteorological Services (NMS) to
(i) develop high-resolution spatially and temporally complete gridded historical
meteorological data sets; (ii) generate suites of derived climate information products;
and (iii) disseminate them through online "Maprooms." ENACTS enables the NMS
to provide enhanced services by overcoming the challenges of data quality, availability and access.

ENACTS uses merging techniques to combine satellite rainfall estimates with
station observation to generate gridded data that combines the quality of station
observations with the spatial and temporal coverage of satellite products (Fig. 56.5).
Satellite remote sensing offer spatially complete data with relatively long time series
(> 30 years for satellite rainfall), and which are freely available from many different
centers. Merging the two types of data corrects the errors in the spatially complete
satellite products, which in turn fill spatial and temporal gaps in station records. The
result is a time series of climate data (rainfall and temperature) going back over
30 years at a spatial resolution of about 4 km (Dinku et al. 2014, 2015, 2017). The
main satellite rainfall product used for ENACST is the Tropical Applications of
Meteorology using SATellite and ground-based observations (TAMSAT) rainfall
estimate (Maidment et al. 2014; Tarnavsky et al. 2014). The TAMSAT product starts
from 1983 and is available at daily and dekadal (10-day) time scales and spatial
resolution of about 4 km ($0.0375°$). Climate model reanalysis products are used for

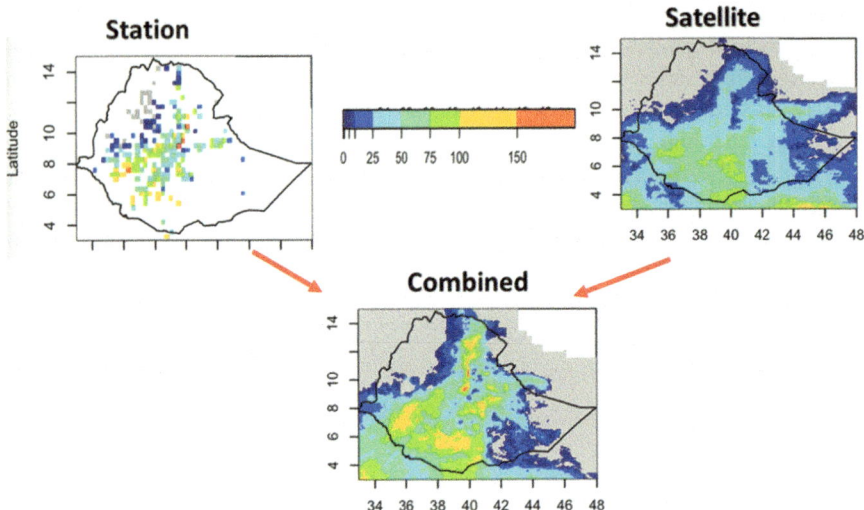

Fig. 56.5 Station observations (top left) of rainfall from the operational network in Ethiopia are combined with the TAMSAT satellite rainfall estimates (top right) to produce a spatially complete and more accurate merged product (bottom)

temperature. Because ENACTS works directly with NMS, which are stewards of much more data than are available to external organizations, the ENACTS data are of higher quality than the best global datasets (Dinku et al. 2018).

ENACTS enables NMS to make climate information products available to the public through an interactive web interface. This is accomplished by installing the IRI Data Library (Blumenthal et al. 2014) at the NMS, which provides a powerful and highly customizable platform for deriving historical, monitored and seasonal forecast tools and products based on the gridded datasets. These products are available online in the form of an interactive "Maproom." The Maprooms, which support analyses, visualization and download of data, open with a map view of statistics (mean, standard deviation or probability-of-exceedance) of the variable and time of year of interest. Users can access additional location-specific products for any selected grid cell or administrative polygon. The Climate Analysis Maproom provides information about year-to-year variability and trends. The Climate Monitoring Maproom supports monitoring of the current season. The Climate Forecasting Maproom provides analysis of relationships between historic variability and predictors such as the El Niño/Southern Oscillation. There are additional "application maproom" for health, agriculture, and water (Fig. 56.6). The Agriculture maproom provides climate information products (wet/dry spells, exceedance/non-exceedance probabilities, and historical rainfall onset and cession dates, etc.), which are relevant for agricultural activities.

ENACTS products and services have already been implemented in 12 countries in Africa and at two Regional Climate Centers in east and west Africa.

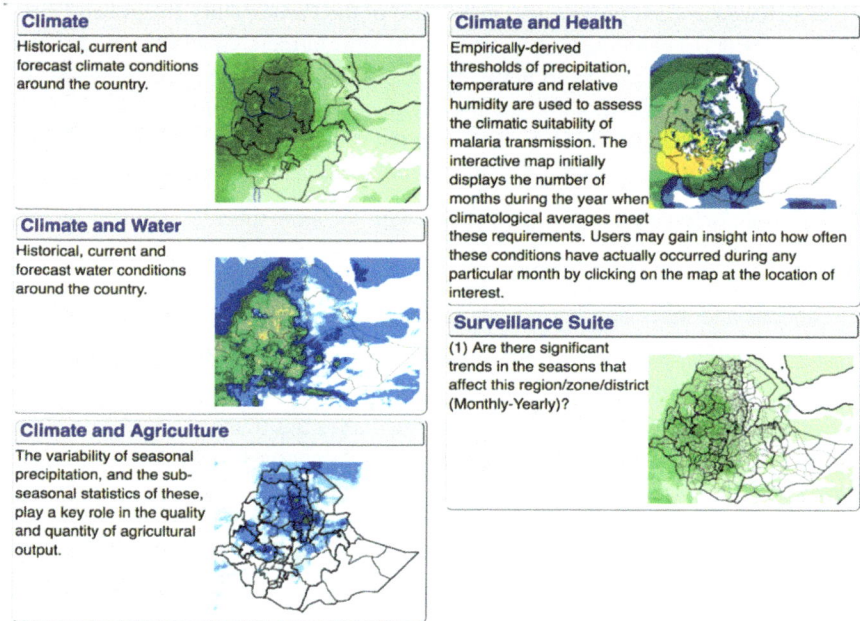

Fig. 56.6 Climate Analysis and Application map room for Ethiopia, consisting of Climate and Application maprooms. (http://www.ethiometmaprooms.gov.et:8082/maproom/)

56.3.2   Rwanda Climate Services for Agriculture Project

The Rwanda Climate Services for Agriculture (CSA) project aims at improving agricultural planning and food security management in the face of a variable and changing climate at both local and government levels. Its primary focus is increasing the resilience of farmers to the changing climate in Rwanda and improve climate risk management skills and hence agricultural productivity for Rwandan farmers. This is accomplished by working directly with technical officers and farmers.

The project builds on the ENACTS approach described above. Rwanda was a big challenge for implementation of ENACTS because of the close to 15 years of interruption in weather observations leading to and following the 1994 genocide (Fig. 56.4). The 1994 civil war and genocide in Rwanda led to a catastrophic disruption of Rwanda's meteorological observation network. Most of the country's weather stations were wrecked, looted or rendered inoperable by the violence. It was not until 2010 that Meteo restored its pre-1994 number of stations (Fig. 56.4). The ENACTS approach helped Meteo Rwanda to use its observational data before and after the genocide to remove biases from satellite rainfall estimates and generate temporally and spatially complete rainfall datasets. The Rwanda Meteorological Agency (Meteo Rwanda) can now provide a range of high-resolution climate data and derived information products.

The Rwanda CSA project also makes use of the Participatory Integrated Climate Services (PICSA) approach to work directly with farmers (Walker Institute 2018). The PICSA approach aims to help "farmers to make informed decisions based on accurate, location specific, climate and weather information; locally relevant crop, livestock and livelihood options; and with the use of participatory tools to aid their decision making," and has three main components (Walker Institute 2018):

- Historical climate data are used along with location-specific crop and livestock information to help farmers assess their risks;
- Planning tools are used by farmers to consider crop, livestock and livelihood options and make appropriate decisions; and
- Farmers update their plans based on seasonal and short-term forecasts.

Through the PICSA approach, agricultural extension staff, non-governmental development agents as well as other intermediaries are trained to integrate climate services into their ongoing work with farming communities across Rwanda's 30 districts. This is possible because the ENACTS approach and the availability spatially complete satellite rainfall products. Satellite rainfall estimate enables implementation of PICSA at any location, while the expanded suite of ENACTS Maproom products fit well with PICSA requirements for providing climate information to farming communities and supporting its use for agricultural decision-making. The ENACTS Maprooms provide an efficient way for the trained intermediaries to access location-specific data and graphs (e.g., Fig. 56.7) as they work with farmers and other local decision makers within the PICSA process. Access to historical rainfall data for any point over Rwanda is made possible only because of the use of satellite rainfall estimate to create a spatially completed dataset.

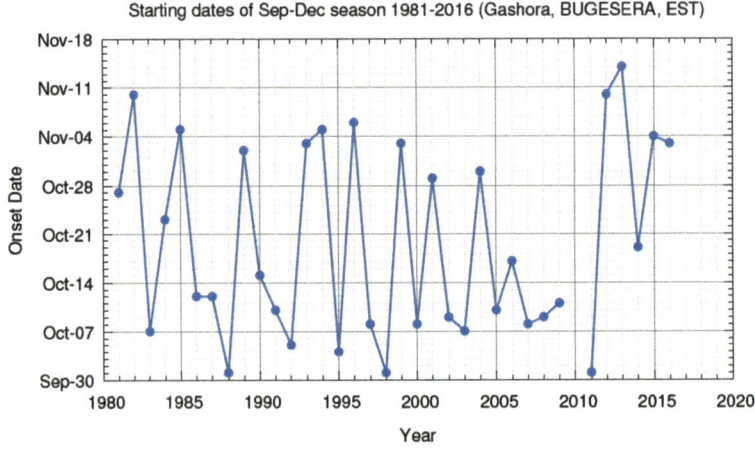

Created: 2018-07-05, 1st 5-d from 1 Oct , 20mm, 3 wet d no 7-d dry in 21d [30.24375E-30.28125E, 2.26875S-2.23125S] in Gashora, BL

Fig. 56.7 An example of PICSA graph showing historical onset dates interactively generated using the online ENACTS Maproom

By the end of the project period, nearly a million farmers are expected to have timely access to useful climate services. This is expected to give farmers better opportunities to transform their livelihoods through improved agricultural productivity. Agricultural planners, policy makers, investors, and food security specialists will be able to respond more effectively to droughts, floods and other climate-related risks.

This project has recently won the Climate Smart Agriculture Project of the Year Award 2018 (https://ccafs.cgiar.org/news/media-centre/press-releases/rwanda-climate-services-agriculture-project-awarded-first-ever#.W_B2EydRdN0, last accessed 28 Nov. 2018), which recognizes outstanding projects that bring together multiple stakeholders in the agriculture ecosystem to form new partnerships that improve productivity, resilience, and efficiency while lowering carbon output.

56.3.3   Weather Index Insurance

Index insurance is an innovative approach for managing risks resulting from weather and catastrophic events, and supports productive opportunities in the agriculture sector (Hellmuth et al. 2009; Hazell et al. 2010; Hansen 2012). Its main objective is to transfer risk so that it is allocated in a more economically and socially desirable fashion, and is linked to an index, such as rainfall, temperature, humidity or crop yields, rather than actual loss (De Leeuw et al. 2014; Brown et al. 2011; Hellmuth et al. 2009). The strength of this approach is that it doesn't require the traditional services of insurance claims, and thus overcomes the challenges of traditional crop insurance in rural parts of developing countries (De Leeuw et al. 2014; Brown et al. 2011; Hellmuth et al. 2009). Some of the advantages of index insurance include (i) low moral hazard and adverse selection; (ii) no expensive loss adjustment for small units; (iii) potentially fewer complex data requirements; and (iv) potentially less complex and more transparent contracts (Skees and Collier 2008).

The most common weather-based index insurance, particularly in developing countries like Africa, is the use of rainfall totals as an index to insure against drought related crop loss (Black et al. 2016; Greatrex et al. 2015; De Leeuw et al. 2014; Hellmuth et al. 2009). In this case, the payouts are based on total rainfall threshold over an agreed period of time. The main difference with the traditional crop insurance is that insurances assessors do not need to visit farmers' fields to assess losses and determine payouts, which would be very difficult particularly in Africa where there are lots of small-holder farmers and the road infrastructure is not well developed.

Availability of a long record of temporally consistent historical climate data is a critical component of designing weather-based index insurance. Climate data is critical for understanding weather-related risks and design and pricing of weather index insurance contracts (Brown et al. 2011; Skees and Collier 2008; De Leeuw et al. 2014). In addition to historical data, weather index insurance also requires current observations for monitoring if crop water requirements are being met.

However, as described in the previous section, lack of availability of quality climate observations is one of the major challenges in Africa. Lack of climate observations is more severe over rural Africa, exactly where index insurance is needed most. This has been the main challenge for expanding weather index insurance over many parts of Africa (Greatrex et al. 2015; Skees and Collier 2008; Hellmuth et al. 2009).

Different remote sensing products are now being used to overcome the challenges of data availability (Greatrex et al. 2015; De Leeuw et al. 2014; Brown et al. 2011; Osgood and Shirley 2012). Satellite rainfall estimates are used to overcome the challenges of sparse rain gauges over most of Africa (Black et al. 2016; Greatrex et al. 2015; De Leeuw et al. 2014). The main strengths of satellite rainfall estimates have already been mentioned above. These advantages apply to rainfall-based index insurance. There are also strengths of satellite rainfall estimates specific to index insurance, which include the following: (i) near real-time availability that can be used to track emerging weather trends as they occur; (ii) more inclusivity compared to weather station data and the potential to lower basis risk for some products; and (iii) are spatially continuous in space and thus can provide actual measurements for all points of interest (Skees and Collier 2008). As a result, now access to index insurance by farmers may not be limited by availability of raingauges station near their fields. This has started making it easier to expand index insurance across Africa (Greatrex et al. 2015; De Leeuw et al. 2014; Brown et al. 2011).

African Rainfall Climatology (ARC, Novella and Thiaw 2013) is one of the widely used satellite rainfall estimates in Africa (Dinku et al. 2015; Greatrex et al. 2015; Osgood and Shirley 2012). The ARC rainfall product has not been chosen because it is the most accurate; it was chosen because it was the only product with consistent long time series at relatively high spatial resolution ($0.1°$) when some of the index insurance projects started in parts of Africa. Currently, there are other similar products such as TAMSAT and CHIRPS (Climate Hazards group Infrared Precipitation with Stations) (Funk et al. 2015), which have higher spatial resolutions and better accuracy (e.g., Dinku et al. 2018), as well as the ENACTS data products for some countries. However, the index insurance community in Africa has been slow to switch to the new satellite-derived products.

56.4   Summary

Climate services involve the production, translation, transfer, and use of climate information for making informed-decisions. Climate services can help in making agriculture more resilient to climate variability and change through better management of negative impacts of climate as well as benefiting from positive outcomes. Quality climate data is the foundation for any climate services. However, in-situ climate observing network is seriously inadequate and has been declining over many parts of the world including Africa. Moreover, in many cases the available stations are unevenly distributed, with most of the stations located along the main roads, limiting availability of data, and hence climate services, to the rural community where these services are needed most.

Different approaches have been used to overcome the challenges of sparse station distribution, and satellite rainfall estimates are the most widely used. Satellite rainfall products are being used for many different activities, which include drought monitoring and potential impacts on agriculture and food security and weather-based index insurance. Three case studies on the use of satellite rainfall estimates in agriculture and food security have been presented in this paper. The first case is the Enhancing National Climate Services (ENACTS) approach, which aims at improving data availability, as well as access to and use of derived climate information products. ENACTS creates temporally and spatially complete rainfall data by blending satellite rainfall estimates with station observation. These data are then used to create an array of climate information products that are made available through an interactive web interface. The second case study, which builds on the first one, aims at improving agricultural planning and food security management for small-holder farmers in Rwanda. This approach combines ENACTS datasets, which are available over any part of the country, and the ENACTS maproom technology with a participatory planning process to help farmers anticipate and prepare for the coming rainfall season. The third case study described the use of satellite rainfall estimates for expanding rainfall-based index insurance across Africa.

The ENACTS approach has been implemented in only 12 countries in Africa, let alone the other parts of the world. There are demands for ENACTS from countries in Africa as well as Asia and South America. The ENACTS approach will also facilitate climate services to agriculture similar to what has been accomplished in Rwanda. Weather index insurance has just started and is expected to expand to different countries across the developing world. All these means that the demand for and use of satellite rainfall products is expected to increase significantly.

References

Auffhammer, M., Hsiang, S. M., & Schelenker, W. (2013). Using weather data and climate model output in economic analyses of climate change. *Review of Environmental Economics and Policy, 7*, 181–198. https://doi.org/10.3386/w19087.

Becker, A., Finger, P., Meyer-Christoffer, A., Rudolf, B., Schamm, K., Schneider, U., & Ziese, M. (2013). A description of the global land-surface precipitation data products of the Global Precipitation Climatology centre with sample applications including centennial (trend) analysis from 1901–present. *Earth System Science Data, 5*, 71–99. https://doi.org/10.5194/essd-5-71-2013.

Black, E., Tarnavsky, E., Maidment, R., Greatrex, H., Mookerjee, A., Quaife, T., & Brown, M. (2016). The use of remotely sensed rainfall for managing drought risk: A case study of weather index insurance in Zambia. *Remote Sensing, 8*, 342. https://doi.org/10.3390/rs8040342.

Blumenthal, M. B., Bell, M., del Corral, J., Cousin, R., & Khomyakov, I. (2014). IRI data library: Enhancing accessibility of climate knowledge. *Earth Perspectives: Transdisciplinarity Enabled, 1*(19), 19. https://doi.org/10.1186/2194-6434-1-19.

Brown, M. E., Osgood, D. E., & Miguel, C. A. (2011). Science-based insurance. *Nature Geoscience, 4*, 213–214. https://doi.org/10.1038/ngeo,1117.

Climate Services Partnership. (2018). *What are climate services?* Available at http://www.climate-services.org/about-us/what-are-climate-services/, last accessed 28 Nov 2018.

De Leeuw, J., Vrieling, A., Shee, A., Atzberger, C., Hadgu, K. M., Biradar, C. M., Keah, H., & Turvey, C. (2014). The potential and uptake of remote sensing in insurance: A review. *Remote Sensing, 6*, 10888–10912. https://doi.org/10.3390/rs61110888.

Dinku, T., Funk, C., Peterson, P., Maidment, R., Tadesse, T., Gadain, H., & Ceccato, P. (2018). Validation of the CHIRPS satellite rainfall estimates over eastern Africa. *Quarterly Journal of the Royal Meteorological Society, 144*(S1), 292–312. https://doi.org/10.1002/qj.3244.

Dinku, T., Thomson, M. C., Cousin, R., del Corral, J., Ceccato, P., Hansen, J., & Connor, S. J. (2017). Enhancing National Climate Services (ENACTS) for development in Africa. *Climate and Development, 10*, 664–672. https://doi.org/10.1080/17565529.2017.1405784.

Dinku, T., Cousin, R., del Corral, J., Ceccato, P., Thomson, M., Faniriantsoa, R., Khomyakov, I., & Vadillo, A. (2016). The ENACTS approach: Transforming climate services in Africa, one country at a time. A World Policy Paper, World Policy Institute. [https://worldpolicy.org/the-enacts-approach, last accessed 28 Nov 2018].

Dinku, T., Block, P., Sharoff, J., & Thmoson, M. C. (2014). Bridging critical gaps in climate services and applications in Africa. *Earth Perspectives, 1*, 15. https://doi.org/10.1186/2194-6434-1-15.

Funk, C., Peterson, P., Landsfeld, M., Pedreros, D., Verdin, J., Shukla, S., Husak, G., Rowland, J., Harrison, L., Hoell, A., & Michaelsen, J. (2015). The climate hazards group infrared precipitation with stations – A new environmental record for monitoring extremes. *Scientific Data, 2*, 150066. https://doi.org/10.1038/sdata.2015.66.

Greatrex, H., Hansen, J. W., Garvin, S., Diro, R., Blakeley, S., Le Guen, M., Rao, K. N., & Osgood, D. E. (2015). Scaling up index insurance for smallholder farmers: Recent evidence and insights. CCAFS Report No. 14 Copenhagen, CGIAR Research Program on Climate Change, Agriculture and Food Security (CCAFS). Available at https://ccafs.cgiar.org/publications/scaling-index-insurance-smallholder-farmers-recent-evidence-and-insights#.W_5ixS2h3UY, last accessed 28 Nov 2018.

Hansen, J. W. (2012). Making climate information work for agricultural development. *World Politics Review*. https://www.worldpoliticsreview.com/articles/11545/making-climate-information-work-for-agricultural-development, last accessed 28 Nov 2018.

Hansen, J. W., Zebiak, S., & Coffey, K. (2014). Shaping global agendas on climate risk management and climate services: An IRI perspective. *Earth Perspectives, 1*, 13. https://doi.org/10.1186/2194-6434-1-13.

Hazell, P., Anderson, J., Balzer, N., Clemmensen, A., Hess, U., & Rispoli, F. (2010). *The Potential for scale and sustainability in weather index insurance for agriculture and rural livelihoods.* IFAD, Rome, Italy. Available at https://documents.wfp.org/stellent/groups/public/documents/newsroom/wfp281391.pdf?_ga=2.75370066.1906432636.1543398487-2016402983.1543398487, last accessed 28 Nov 2018.

Hellmuth, M. E., Osgood, D. E., Hess, U., Moorhead, A., & Bhojwani, H. (2009). Index insurance and climate risk: Prospects for development and disaster management. *Climate and Society, 2*, IRI, Columbia University, New York. Available at https://cgspace.cgiar.org/handle/10568/932, last accessed 28 Nov 2018.

Hutchinson, C. F. (1991). Uses of satellite data for famine early warning in sub-Saharan Africa. *International Journal of Remote Sensing, 12*, 1405–1421. https://doi.org/10.1080/01431169108929733.

Maidment, R. I., Grimes, D. I. F., Allan, R. P., Tarnavsky, E., Stringer, M., Hewison, T., Roebeling, R., & Black, E. (2014). The 30 year TAMSAT African Rainfall Climatology and Time series (TARCAT) dataset. *Journal of Geophysical Research, 119*, 619–10,644. https://doi.org/10.1002/2014JD021927.

Malhi, Y., & Wright, J. (2004). Spatial patterns and recent trends in the climate of tropical rainforest regions. *Philosophical Transactions of the Royal Society, B359*, 311–329. https://doi.org/10.1098/rstb.2003.1433.

Novella, N. S., & Thiaw, W. M. (2013). African rainfall climatology version 2 for famine early warning systems. *Journal of Applied Meteorology and Climatology, 52*(3), 588–606. https://doi.org/10.1175/JAMC-D-11-0238.1.

Osgood, D., & Shirley, K. E. (2012). The value of information in index insurance for farmers in Africa. In R. Laxminarayan & M. K. Macauley (Eds.), *The value of information* (pp. 1–18). Dordrecht: Springer. https://doi.org/10.1007/978-94-007-4839-2_1.

Parker, D., Good, E., & Chadwick, R. (2011). Reviews of observational data available over Africa for monitoring, attribution and forecast evaluation. *Hadley Centre Tech. Note*, 86, 62 pp. Available at https://www.google.com/url?sa=t&rct=j&q=&esrc=s&source=web&cd=1&ved=2ahUKEwjy04_j6vbeAhULGuwKHe9BDVIQFjAAegQIBhAC&url=https%3A%2F%2Fdigital.nmla.metoffice.gov.uk%2Fdownload%2Ffile%2Fsdb%253AdigitalFile%257Ced1eea36-84e4-452c-a1b3-d5b739e6d80c%2F&usg=AOvVaw2zFUU8I-CuO7R_NNEfmfrq, last accessed 28 Nov 2018.

Sheffield, J., Wood, E. F., Chaney, N., Guan, K., Sadri, S., Yuan, X., Olang, L., Amani, A., Ali, A., Demuth, S., & Ogallo, L. (2014). A drought monitoring and forecasting system for sub-Sahara African water resources and food security. *Bulletin of the American Meteorological Society, 95*, 861–882. https://doi.org/10.1175/BAMS-D-12-00124.1.

Skees, J. R., & Collier, B. (2008). The potential of weather index insurance for spurring a green revolution in Africa. CGIAR, Lexington, KY, GlobalAgRisk, Inc. Available at https://cgspace.cgiar.org/handle/10568/934, last accessed 28 Nov 2018.

Tarnavsky, E., Grimes, D. I. F., Maidment, R. I., Stringer, M., Black, E., Allan, R. P., & Stringer, M. (2014). Extension of the TAMSAT satellite-based rainfall monitoring over Africa and from 1983 to present. *Journal of Applied Meteorology and Climatology, 53*, 2805–2822. https://doi.org/10.1175/JAMC-D-14-0016.1.

Thornton, P. E., Running, S. W., & White, M. A. (1997). Generating surfaces of daily meteorology variables over large regions of complex terrain. *Journal of Hydrology, 190*, 214–251. https://doi.org/10.1016/S0022-1694(96)03128-9.

UN. (2017). World population projected to reach 9.8 billion in 2050, and 11.2 billion in 2100. Available at https://www.un.org/development/desa/en/news/population/world-population-prospects-2017.html, last accessed 28 Nov 2018.

Verdin, J., Funk, C., Senay, G., & Choularton, R. (2005). Climate science and famine early warning. *Philosophical Transactions of the Royal Society, B, 360*, 2155–2168. https://doi.org/10.1098/rstb.2005.1754.

Walker Institute. (2018). The Participatory Integrated Climate Services (PICSA). http://www.walker.ac.uk/projects/participatory-integrated-climate-services-for-agriculture-picsa/, last accessed 28 Nov 2018.

Washington, R., Harrison, M., Conway, D., Black, E., Challinor, A., Grimes, D., Morse, J. R., Kay, G., & Todd, M. (2006). Africa climate change: Taking the short route. *Bulletin of the American Meteorological Society, 87*, 1355–1366. https://doi.org/10.1175/BAMS-87-10-1355.

WMO. (2014a). Implementation plan of the global framework for climate services. Geneva, 70 pp. Available at https://www.wmo.int/gfcs/implementation-plan, last accessed 28 Nov 2018.

WMO. (2014b). Agriculture and food security exemplar to the user interface platform of the global framework for climate services. Geneva, 35 pp. Available at https://www.wmo.int/gfcs/sites/default/files/Priority-Areas/Agriculture%20and%20food%20security/GFCS-AGRICULTURE-FOOD-SECURITY-EXEMPLAR-FINAL-14147_en.pdf, last accessed 28 Nov 2018.

Zillman, J. W. (2009). A history of climate activities. *WMO Bulletin, 58*(3), 141. Available at https://library.wmo.int/pmb_ged/bulletin_58-3.pdf, last accessed 28 Nov 2018.

Chapter 57
Using Satellite Estimates of Precipitation for Fire Danger Rating

Robert D. Field

Abstract Fire danger rating systems are cornerstones of forest and land fire management. Different systems are used in different fire prone regions of the world, but all require daily inputs of the weather conditions most important in influencing fuel moisture and potential fire behavior. Because these systems typically have a memory, inaccurate weather data can lead to persistently inaccurate fire danger information. This is particularly the case for precipitation, due to its event-driven nature and complex spatial patterns. This chapter describes different approaches to estimating precipitation for the purposes of fire danger rating, focusing on using satellite precipitation estimates for calculations of the Fire Weather Index System, the most widely used fire danger rating system in the world. Comparisons are made between Fire Weather Index calculations for severe fire events in Canada, Chile, Greece and Indonesia, to illustrate the potential benefits of satellite precipitation for fire management in four very different fire environments.

Keywords Precipitation · Rainfall · Fire Danger Rating System · Fire management · Fire Weather Index · Drought Code · Duff Moisture Code · Global Fire Weather Database · NDVI · FAO · TRMM · MODIS · IMERG

57.1 Introduction

The UN Food and Agriculture Organization (FAO) defines fire danger as "a general term used to express an assessment of both fixed and variable factors of the fire environment that determine the ease of ignition, rate of spread, difficulty of control, and fire impact; often expressed as an index.", and fire danger rating as "a component of a fire management system that integrates the effects of selected fire danger factors into one or more qualitative or numerical indices of current protection needs."

R. D. Field (✉)
Department of Applied Physics and Applied Mathematics, Columbia University, and NASA, Goddard Institute for Space Studies, New York, NY, USA
e-mail: robert.field@columbia.edu

© Springer Nature Switzerland AG 2020
V. Levizzani et al. (eds.), *Satellite Precipitation Measurement*, Advances in Global Change Research 69, https://doi.org/10.1007/978-3-030-35798-6_33

Fig. 57.1 Fire danger rating sign for public display at the Tinanggea Manggala Agni Local Fire Center, South Konawe Regency, South East Sulawesi Province, Indonesia. Fire danger is set daily using the Fine Fuel Moisture Code of the Indonesian Fire Danger Rating System calculated from on-site weather data (Credit: Israr Albar, Indonesian Ministry of Environment and Forestry)

(de Groot et al. 2015; FAO 2007). Fire Danger Rating Systems (FDRS) are cornerstones of fire management around the world (de Groot et al. 2015; Wotton 2009), used for tactical and strategic aspects of fire management, ranging from planning and resource pre-preparedness to prevention and public awareness of the prevailing fire danger (Fig. 57.1).

FDRS complexity will vary, but combining the influence of individual weather components, which are the most dynamic factor of fire danger (Carlson and Burgan 2003), into interpretable indices is always at the core. All FDRS are based on models of fuel moisture for different classes of dead vegetation, and can include measures of fire behavior. The accuracy of an FDRS therefore depends on the underlying models of fuel moisture and fire behavior and on the input weather data from which the indices are calculated. FDRS maps can be calculated using weather data interpolated over a station network (Flannigan and Wotton 1989; Jain and Flannigan 2017; Lee et al. 2002), but in many fire prone regions, the sparseness of surface weather networks limits the accuracy of fire danger information. Because FDRS typically have a 'memory' of past weather conditions, poor characterization of weather inputs can lead to inaccurate fire danger information for weeks afterwards (Lawson and Armitage 2008). The availability of modern reanalyses and numerical weather prediction analysis and forecast fields have provided alternatives regionally

(Bowman et al. 2018; Fang et al. 2015; Venalainen et al. 2014) and globally (Bedia et al. 2015; Jolly et al. 2015), but these too bring with them inevitable regional biases which can translate into biased FDRS information (Bedia et al. 2012; Field et al. 2015; Horel et al. 2014).

There has been a fair amount effort to incorporate satellite information into FDRS calculations to improve estimates in remote regions with sparse weather station coverage. Retrievals of surface temperature and Normalized Vegetation Difference Index (NDVI) have been related with some success to the individual FDRS components over parts of Canada (Han et al. 2003; Jang et al. 2006; Leblon et al. 2001; Merzouki and Leblon 2011; Oldford et al. 2006) and the Mediterranean (Aguado et al. 2003; Fiorucci et al. 2007; Maselli et al. 2003). Backscatter from Synthetic Aperture Radar has been examined as a way of constraining the same components over Alaska (Bourgeau-Chavez et al. 1999; Bourgeau-Chavez et al. 2007, 2013) and the Northwest Territories (Abbott et al. 2007; Leblon et al. 2002) where there is a specific need to capture overwintering effects on soil moisture at the start of the fire season. Similarly, soil moisture retrievals from the Soil Moisture Ocean Salinity mission have been used in calculations of the McArthur Forest Fire Danger Index in Australia (Holgate et al. 2017). Leblon et al. (2016) reviewed the incorporation of satellite remote sensing data in to FDRS, making the important point that whereas surface temperature retrieved from thermal infrared sensors could be used as an input variable to Fire Weather Index (FWI) calculations, the use of NDVI from optical sensors is problematic because it varies with the conditions of overstory vegetation, rather than understory vegetation which is more directly related to fire danger. Chowdhury and Hassan (2015) also reviewed work to incorporate remote sensing data into FDRS, distinguishing between estimating vegetation greenness, weather conditions, surface wetness, and vegetation wetness. The main limitations identified were the inability to capture rapidly changing fire weather conditions, precision of surface temperature estimates, and complications from topography.

Estimating precipitation inputs to FDRS presents a unique challenge in that inaccurate characterization of single precipitation events can have a persistent effect on subsequent fire danger. In trying to understand the effects of precipitation data quality on FDRS accuracy, past work has included the performance of different geostatistical interpolation techniques (Flannigan and Wotton 1989; Hanes et al. 2017; Jain and Flannigan 2017) and the incorporation of surface radar precipitation estimates (Flannigan et al. 1998). Tropical Rainfall Measurement Missions (TRMM) precipitation estimates have been used in examining the US National Fire Danger Rating System performance over Bolivia (Steininger et al. 2013) and blended precipitation estimates (Beck et al. 2017) have been used in a case study of the severe 2017 Chilean fires (Bowman et al. 2018). Beyond those, modern satellite estimates of precipitation have not yet found their way into fire danger rating (de Groot et al. 2015). This chapter describes efforts to expand on limited previous work through enhancements to the Global Fire Weather Database (Field et al. 2015), which is based on the Canadian FWI System. The chapter is organized around an overview of the FWI System, and examples comparing calculations from rain gauge, model analysis and satellite precipitation estimates for recent, representative episodes in different fire environments.

57.2   The Canadian Fire Weather Index System and the Global Fire Weather Database

The Canadian FWI System is used universally in Canada as a component of the Canadian Forest Fire Danger Rating System (Lee et al. 2002; Taylor and Alexander 2006; Wotton 2009), with each province and territory interpreting the different FWI components according to different climate, predominant vegetation type, and fire management needs (Stocks et al. 1989). Because of its modest input requirements and flexibility, the FWI System is the most widely-used FDRS in the world (de Groot and Flannigan 2014; de Groot et al. 2015). It has been included in global fire weather analyses (Jolly et al. 2015; Robinne et al. 2018; Bowman et al. 2017) and forms the basis for emerging global fire danger rating systems (de Groot and Flannigan 2014; de Groot et al. 2015; Field et al. 2015; di Giuseppe et al. 2017, 2018). It is composed of three moisture codes and three fire behavior indices (Fig. 57.2). The moisture codes track the moisture content of litter and forest floor moisture content rather than live fuel moisture, and for all codes, increasing values indicate decreasing moisture content. The Fine Fuel Moisture Code (FFMC) captures changes in the moisture content of fine fuels and leaf litter on the forest floor, where fires can most easily start. The Duff Moisture Code (DMC) captures the moisture content of loosely compacted forest floor organic matter that is a primary source of fuel and also acts as a proxy for the moisture content of dead, medium-size fuels on the forest floor. The

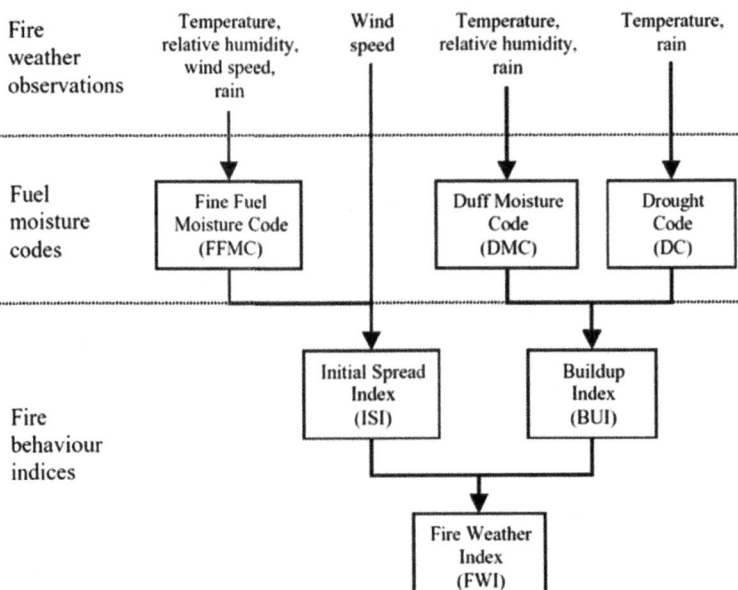

Fig. 57.2 Structure of the Fire Weather Index System. 1200 pm local time surface temperature, relative humidity, wind speed and 24-hour precipitation are used to calculate three moisture codes and three fire behavior indices. (Adapted from de Groot et al. 2007)

Drought Code (DC) captures the moisture content of deep, compacted organic soils and heavy surface fuels. The three moisture codes are calculated on a daily basis using the previous day's moisture codes and the current day's weather. Each has a precipitation threshold below which small amounts of precipitation have no effect on the code, which are 0.5 mm for the FFMC, 1.5 mm for the DMC, and 2.8 for the DC. The moisture codes therefore depend not only on the total amount of precipitation over a period of time, but also on how it is distributed. The three fire behavior indices reflect the behavior of a fire if it were to start. The Initial Spread Index (ISI) is driven by wind speed and FFMC and represents the ability of a fire to spread immediately after ignition. The Buildup Index (BUI) is driven by the DMC and DC and represents the total fuel available to a fire. The FWI combines the ISI and BUI to provide an overall rating of fireline intensity in a reference fuel type, and is intended to provide an overall measure of fire danger. All indices are relative numerical measures, and interpreted differently in local fire environments.

FWI System calculations require measurements of 1200 local time temperature at 2 m, relative humidity at 2 m, sustained wind speed at 10 m, and precipitation totaled over the previous 24 h (van Wagner 1987). Because each day's calculation requires the previous day's moisture codes, weather records must be continuous and any missing data must be estimated (Lawson and Armitage 2008; Taylor and Alexander 2006). Too much missing or erroneous weather data, particularly precipitation, can lead to errors that accumulate over time. Technical details of the FWI System can be found in various technical reports (Dowdy et al. 2009; van Wagner 1987), and the equation source code through publicly available repositories (Cantin 2016).

The NASA GISS Global Fire Weather Database (GFWED) (Field et al. 2015) is a community-oriented FWI database intended for use in understanding the meteorological drivers of fire activity at local, regional and continental scales, and as a starting point for FWI adoption in regions requiring an FDRS. GFWED includes historical versions using NASA's Modern-Era Retrospective Analysis for Research and Applications version 2 (MERRA2) (Gelaro et al. 2017) and near-real time versions using the Goddard Earth Observing System Model, Version 5 (GEOS-5) experimental forecasting system (Molod et al. 2015). For both, there is a small ensemble of different FWI products, each calculated using alternative precipitation estimates. As of 2016, GFWED includes versions using estimates from the Global Precipitation Climatology Project 1-Degree-Daily (GPCP 1DD) product (Huffman et al. 2001) since 1997, the Tropical Rainfall Measuring Mission (TRMM) (Huffman et al. 2007) Multi-satellite Precipitation Analysis (TMPA) 3B42 daily product over 1998–2015, and the Integrated Multi-satellitE Retrievals for GPM (IMERG) (Huffman et al. 2018) estimate available since mid-2014. The different GFWED versions are summarized in Table 57.1.

The addition of FWI versions using satellite-based precipitation estimates was motivated by their application at interannual time scales in understanding controls on fire in the tropics and subtropics (van der Werf et al. 2008a), Indonesia (van der Werf et al. 2008b) and South America (Chen et al. 2013), and the argument that "a comprehensive description of global precipitation can only be achieved from the vantage point of space" (Kidd and Levizzani 2011). Over Indonesia in particular,

Table 57.1 Different versions of the Global Fire Weather Database

Variable	Data source	Period	Latency	Coverage	Resolution
T, RH, wind-speed, snow depth	MERRA-2	1981-present	~2 months	Global	0.5° × 2/3°
	GEOS-5	2014-present (analysis), December 2017-present (forecasts)	~12 h	Global	0.25° × 0.25°
Precipitation	MERRA-2 raw precipitation	1981-present	~2 months	Global	0.5° × 2/3°
	MERRA-2 bias-corrected precipitation	1981-present	~2 months	Global	0.5° × 2/3°
	Sheffield / Princeton precipitation	1981–2010	Variable, 4 + years	Global	0.5° × 0.5°
	NCEP CPC gauge-based analysis of global precipitation	1981-present	1 day	Global	0.5° × 0.5°
	GPCP 1-degree-daily v1.2	1997-present	6+ months	Global	1.0° × 1.0°
	TRMM 3B42	1998–2014	N/A	50S–50N	0.25° × 0.25°
	GPM IMERG – final	May 2014-present	5+ mo.	60S–60N	0.1° × 0.1°
	GPM IMERG – late	May 2014-present	1 day	60S–60N	0.1° × 0.1°

GPCP-based monthly precipitation (Adler et al. 2003) explained considerably more of the variability in fire activity than monthly gauge-based estimates (Field and Shen 2008). The assumption made in incorporating satellite precipitation estimates into GFWED is that improvements using monthly-scale precipitation would translate into similar improvements for integrative fire danger indices, requiring daily precipitation as an input variable. The following examples illustrate the differences in FWI System values for versions using gauge, reanalysis and satellite precipitation estimates for recent fire episodes in North America, South America, Europe, and an analysis of several years of fire activity in Indonesia. Each represents a different fire regime in terms of climate, vegetation, and predominant type of ignition. When possible, regional FWI interpretations are drawn from fire management agencies using individual indices in fire management.

57.3   The 2017 Fire Season in British Columbia, Canada

British Columbia (BC) is characterized by a coastal pacific maritime climate, and an interior montane cordillera region (Amiro et al. 2001), with more arid conditions in the south-central part of the province. Over 1990–2017 there were an average of

1920 fires and 132,419 ha area burned annually reported in the Canadian National Forestry Database (CCFM 2018), but with considerable interannual variability. The FWI forms the basis of fire danger rating in the province (Stocks et al. 1989), which is organized into six fire management regions, for which FWI information is tabulated daily (e.g., https://www2.gov.bc.ca/gov/content/safety/wildfire-status/wild fire-situation/fire-danger, last accessed 3 Apr. 2019).

The 2017 fire season in BC saw 1.2 million ha burned, equivalent to the combined total of the previous 6 years. The severity of the fire season was due to a series of large fires in central and southern BC, rather than to an increase in the number of fires. In mid-August, a cluster of four large fires exhibited extreme, plume-dominated 'pyroCumulonimbus' behavior, injecting smoke directly into the upper troposphere and lower stratosphere (Khaykin et al. 2018), where the smoke persisted in the northern hemisphere for months afterward (Peterson et al. 2018). The aerosol loading of the smoke has been compared to that of a moderate volcanic eruption (Khaykin et al. 2018; Peterson et al. 2018).

Figure 57.3 shows the MODIS active fires and the FWI component computed from rain gauges and IMERG-Final ('F') precipitation estimates over the region in mid-July during the first significant fire outbreak. The FWI categories ranging from Very Low (0–5) to Extreme (>30) are drawn from an early classification from the provincial fire management agency (Stocks et al. 1989), and visual interpretive guides relating FWI values to observed fire behavior in reference jack pine stands (Alexander and de Groot 1988), available at http://cwfis.cfs.nrcan.gc.ca/background/

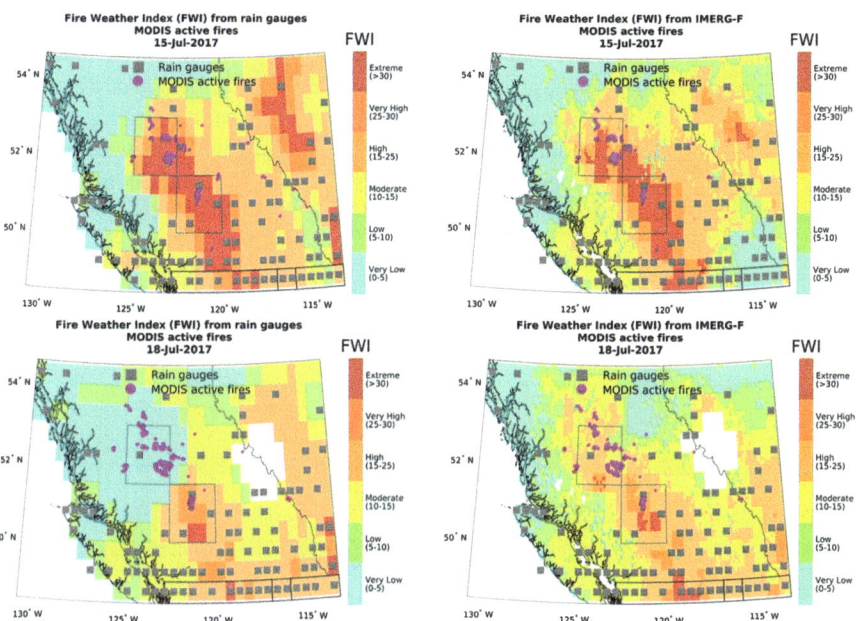

Fig. 57.3 Fire Weather Index calculated from rain gauges (left) and IMERG-F (right) for 15 (top) and 18 July 2017 (bottom). Grey squares indicate grid cells with reporting rain gauges

examples/fwi, last accessed 3 Apr. 2019). For both precipitation estimates on 15 July (top row), there is Low FWI along the coast and to the north, Moderate and High FWI toward the Rocky Mountains in the east, with regions of Extreme FWI in the interior reflecting persistent drying since the spring. The higher resolution of the IMERG-F FWI estimate is more apparent in regions with Very Low to Moderate FWI, compared to regions with higher FWI which degrade toward the gauge-based estimate due to the variability in FWI being driven by MERRA2 temperature and relative humidity fields.

For both FWI estimates, the clusters of fires on 15 July in central (centered on 53.5°N and 124°W) and southern BC (centered on 51°N and 122°W) occurred under predominantly Extreme conditions. Fire activity during this phase continued through 19 July in both regions. Over southern BC, this is consistent with High/Extreme FWI for both precipitation estimates, illustrated for 18 July (middle row). Over central BC, by contrast, there are significant differences between FWI estimates. The gauge-based FWI has dropped to Low due to a single gauge reporting 59 mm of precipitation, which in the absence of nearby gauges, exerts a disproportionate influence on the precipitation field. The absence of precipitation in the IMERG-F estimate results in continued High FWI, more consistent with observed fire activity.

The regional differences in FWI for the two estimates can be seen over the course of the fire season. There is reasonable agreement between the rain gauge and IMERG-F FWI over central BC (Fig. 57.4, left), with Low values at the beginning of May increasing toward Very High values in August, punctuated by periods of precipitation in mid-June and mid-August, and with a return to Low values at the end of the summer. The sharp drop in fire activity on 20 July over both regions corresponds to a sharp drop in IMERG-based FWI, which followed that in the gauge-based FWI over central BC by 2 days. For the next 2 weeks, there were steady increases in FWI consistent with increasing fire activity leading up to the 12 August pyroCumulonimbus' event. This was followed by a moderate precipitation event and lower temperatures associated with the arrival of a cold front (Peterson et al. 2018), and sharp drops in FWI and fire activity. There was a final

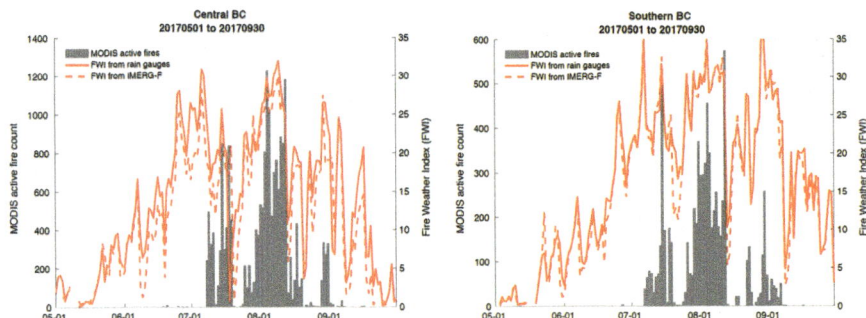

Fig. 57.4 May–September 2017 time series of daily MODIS fire activity and FWI from rain gauges and IMERG-F for central BC (left) and south-central BC (right), corresponding to the regions identified by the grey boxes in Fig. 57.3

increase in FWI and more moderate increase in fire activity toward August, with the fire season ending in both regions by the end of September. Overall, the gauge and IMERG-F FWI were in good agreement over the whole fire season, with a coefficient of determination (r^2) of 0.87 and root-mean squared error of 3.74 for central BC. The r^2 was slightly higher (0.94) and the RMSE slightly lower (2.81) over southern BC, presumably due to there being more rain gauges.

57.4 The 2017 Fire Storm in Central Chile

Most fire in Chile is concentrated in the central and south-central regions spanning 32° to 43°S, with an average of 5100 fire events burning 47,000 ha, with an increase in the number of fires between 1984 and 2014 (Urrutia-Jalabert et al. 2018) attributed to the introduction of flammable pine and eucalyptus plantations (Úbeda and Sarricolea 2016). Most fires in Chile are human caused, with less than 1% thought to be caused by natural sources such as lightning (Úbeda and Sarricolea 2016; Urrutia-Jalabert et al. 2018). The interannual variability in number of fires and burned area is related to higher temperatures and lower precipitation during the summer, and also influenced by conditions during the previous austral winter via increased vegetation growth (Urrutia-Jalabert et al. 2018).

The region was affected by a series of severe fires during the first 6 weeks of 2017 (Urrutia-Jalabert et al. 2018), described by the Chilean President as "The greatest forest disaster in our history" (Anonymous 2017), and since referred to as the 'tormento de fuego' ('the fire storm'). The 5132 km² burned during this period was 5.7 times greater than the next most severe fire season on record (Bowman et al. 2018), due to a combination of ignitions in highly flammable vegetation types and extreme fire weather (Martinez-Harms et al. 2017). Drought captured by high vapor pressure deficits and Palmer Drought Severity Index values was not extreme, but the FWI value of 59 computed from ERA-Interim reanalysis was the highest over central Chile on 25 January 2017 during the 1979–2017 record due to a combination of antecedent drying, record high temperatures, low relative humidity and relatively high wind speeds (Bowman et al. 2018).

Figure 57.5 shows the mean FWI computed from rain gauges and IMERG-F and MODIS active fire activity over the main affected regions of O' Higgins, Maule, and Bío Bío for the weeks of 9, 16 and 23 January. FWI categories are taken from those of the Global Wildfire Information System (GWIS, http://gwis.jrc.ec.europa.eu/about-gwis/technical-background/fire-danger-forecast/, last accessed 28 Nov. 2018) with the addition of a 'Catastrophic' danger class for FWI > 80. During the week of 9 January prior to the most severe fire activity (top row), FWI over the main burning regions ranges from High to Extreme, with good agreement in the FWI computed from rain gauges (left) and IMERG-F (right), despite a very sparse rain gauge network. As in British Columbia, the apparent IMERG-F resolution degrades toward that of the rain gauges in the absence of precipitation over all of central Chile, with the variation in FWI controlled instead by variations in temperature and relative

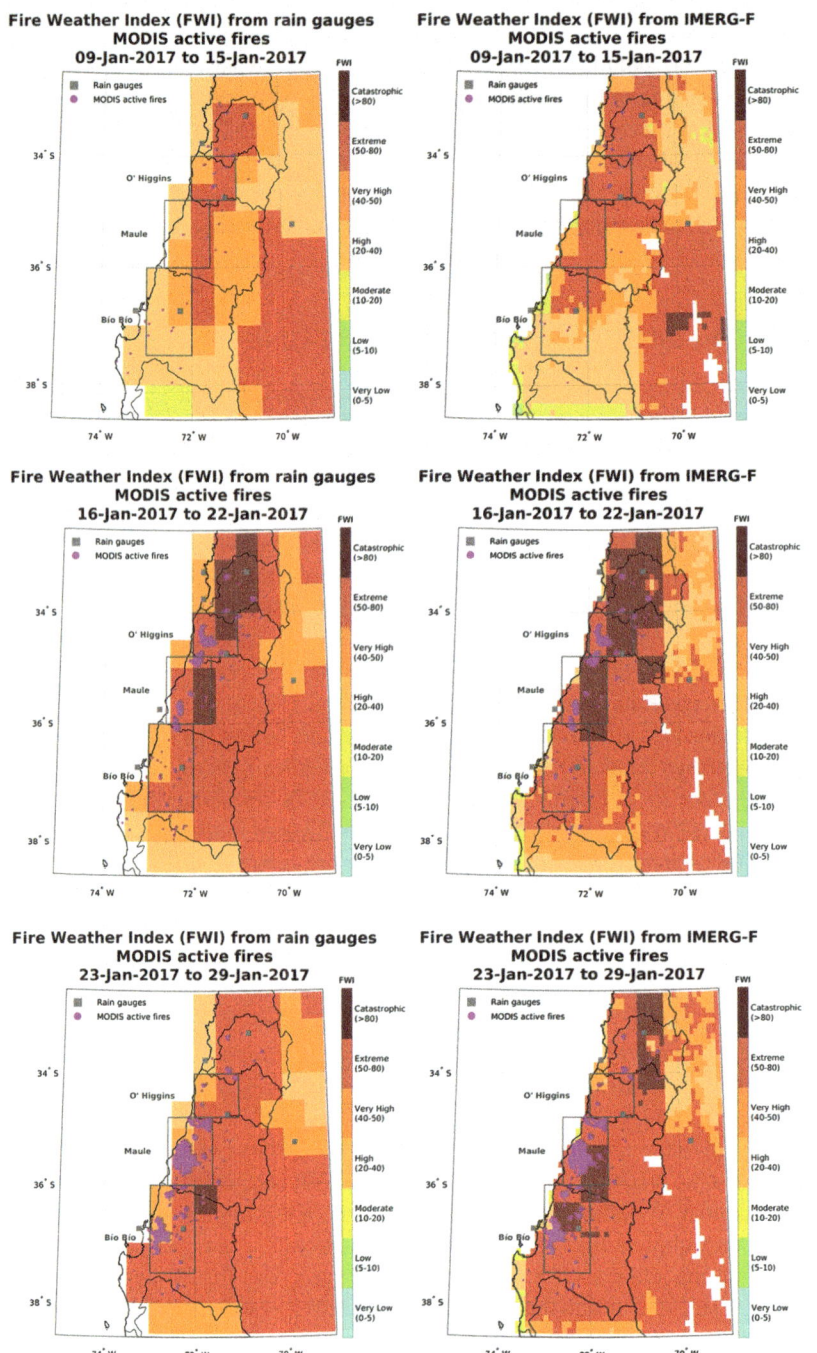

Fig. 57.5 Weekly Fire Weather Index (FWI) and MODIS active fires over areas in Chile affected by catastrophic wildfires during January, 2017. FWI categories are taken from those of the Global Wildfire Information System (GWIS, http://gwis.jrc.ec.europa.eu/about-gwis/technical-background/fire-danger-forecast/, last accessed 28 Nov. 2018) with the addition of a 'Catastrophic' danger class for FWI > 80

humidity which are identical for both estimates. Despite the dry FWI conditions, there is very little fire activity, presumably due to the absence of any ignitions up until that point. The extreme fire activity over O'Higgins and Maule beginning on 16 January (middle row) coincided with an increase in FWI to Very High / Extreme levels, seen in both the gauge and IMERG-F estimates. During the week beginning on 23 January, there was a drop in fire activity in O′ Higgins and a slight decrease in FWI, but an increase in Maule and a major event in Bío Bío, which corresponded to Extreme FWI in the gauge estimate and Catastrophic FWI in the IMERG-F estimate. Over all three regions, FWI timing and magnitude is in reasonable agreement with the reanalysis-based estimate reported previously (Bowman et al. 2018), with the averages in Fig. 57.5 capturing the north-south progression of increasing FWI and fire activity through those 3 weeks in January.

The high FWI during the event reflects the hot, dry and windy conditions during late January, but also antecedent drying, which have been identified as an important driver of fire in central Chile more generally (Urrutia-Jalabert et al. 2018). These are shown for the Maule region (Fig. 57.6), which had the highest fire activity, with similar patterns for O'Higgins and Bío Bío. From August through December 2016, the absence of significant precipitation led to a steady increase in DMC, DC and BUI values (not shown). There was significant precipitation during the third week of December for both gauge and IMERG-F estimates, but which only temporarily lowered the FWI and underlying moisture codes. By the beginning of January 2017, the FWI was 'primed' in the Very High range. The peak in fire activity during the week of 23 January was therefore associated with high temperatures averaging 30 °C and relative humidity averaging 31% and prolonged antecedent drying. Over

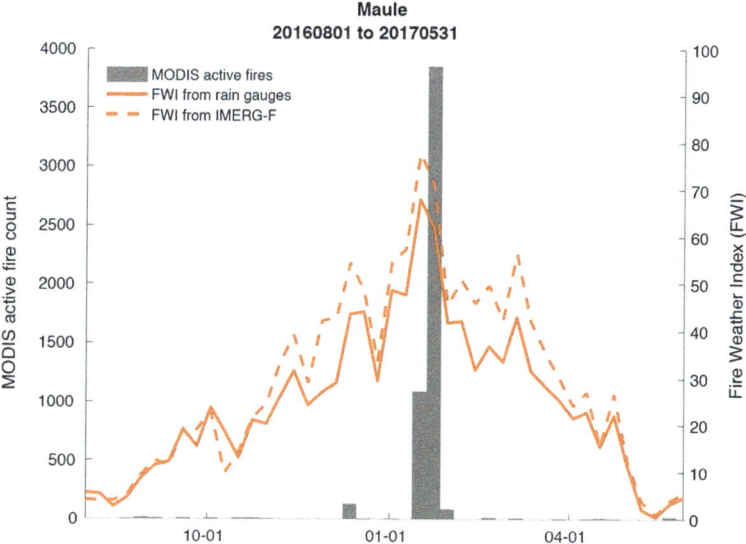

Fig. 57.6 Weekly Fire Weather Index (FWI) and MODIS active fire counts from 1 August 12,016 to 31 May 2017 over the Maule and region of Chile

Maule, there is an apparent regional FWI threshold above 50 for the gauge estimate and 60 for the IMERG-F estimate above which the severe fire activity took place, which was also the same for O′ Higgins and Bío Bío. The slightly higher peak in FWI for the IMERG-F estimate (73) compared to the gauge estimate (66) reflects steadily greater drying since November 2016. This was due not to differences in the total amount of precipitation since August 2016, which was in fact slightly greater for the IMERG-F (14 mm) than from rain gauges (12 mm), but rather to more days with light precipitation and therefore a greater sensitivity to the precipitation thresholds of the underlying moisture codes.

57.5   The 2018 Attica Fires Near Athens, Greece

Based on fire reports from 1983–1997, the fire season in Greece spans May through October, with most burning in July–August (Karali et al. 2014). During 2000–2010, there was an average of 1636 fires burning 45,577 ha annually, but with considerable interannual variation in burned area. Annual burned area during this period was less than 30,000 ha, except for catastrophic fires in 2007, when 225,000 ha burned as the result of two extreme fire events in July and August of 2007 (San-Miguel-Ayanz et al. 2013). The FWI System was adopted for common use over southern Europe in 2006 following an intercomparison of different fire danger rating systems (Viegas et al. 1999), and for southeastern Greece, an FWI of 45 calculated from ERA-Interim data was identified as a critical threshold for the purpose of projecting future fire season length and severity (Karali et al. 2014). Over Crete, extreme FWI was defined for more operational uses as values greater than 60, based on fire occurrence data from fire reports (Dimitrakopoulos et al. 2011). The FWI averaged over Central Greece during the 2007 fires was 50 when calculated from ERA reanalysis, but ranged from 50 to 100 (Athanasopoulou et al. 2014) when calculated from individual weather stations, illustrating that despite weather station density limiting the accuracy of broad fire weather characteristics (Dimitrakopoulos et al. 2011), extreme FWI are not necessarily captured by gridded meteorological data.

In Greece, the worst fires since 2007 occurred near Athens on 23 July 2018, resulting in at least 91 deaths (Lekkas et al. 2018). The extreme fire behavior was the result of prolonged antecedent drying conditions and surface winds of up to 90 km h^{-1}. Figure 57.7 shows the FWI over southern Greece on 23 July for the rain-gauge and IMERG-L based estimates, with all other FWI inputs obtained from GEOS-5 analysis fields. The choice of FWI categories ranging from Very Low to Extreme are also adopted from the Global Wildfire Information System, with the addition of a Catastrophic class for FWI > 80 corresponding to the FWI conditions associated with the severe fires of 2007 (Athanasopoulou et al. 2014). There is reasonable agreement between the two estimates, with FWI increasing toward the southeast, but with a stronger gradient for the IMERG-based estimate. In the

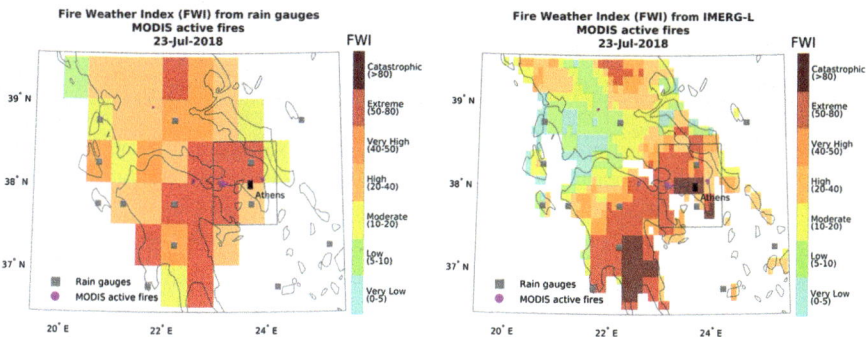

Fig. 57.7 Fire Weather Index (FWI) on 23 July 2018 over southern Greece computed from rain gauges (left) and IMERG precipitation (right)

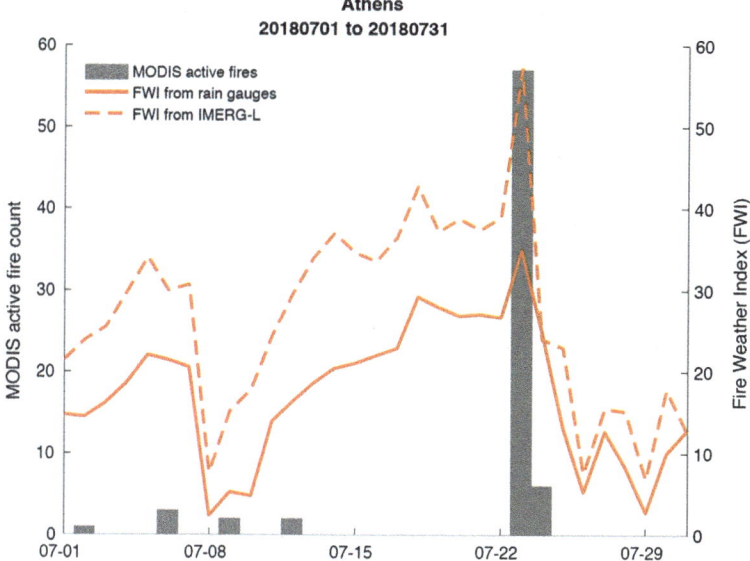

Fig. 57.8 Daily FWI and MODIS active fire totals over the Athens region shown in Fig. 57.7 calculated using IMERG-L and rain gauge precipitation estimates

northwest where rain gauge distribution is particularly sparse, the IMERG FWI ranges from Very Low to Extreme, compared to more uniform values from rain gauges. Near Athens, the FWI ranges from High to Extreme for the rain-gauge based estimate, whereas it peaks in the Catastrophic class for the IMERG-based estimate.

Figure 57.8 shows daily active fire totals and FWI from rain gauges and IMERG for July in the region spanning Athens shown in Fig. 57.7, illustrating the combined effects of antecedent and present weather on fire danger, and its sensitivity to weather inputs. During 8–10 July, the total estimated precipitation was 8.9 mm for IMERG, compared to 25.7 mm for rain gauges. The higher precipitation for rain gauges effectively caused the underlying DMC to 'reset' for the gauge-based estimate, which was not the case for the IMERG-based estimate. For the following 2 weeks, the absence of precipitation and high (>28 °C) temperature caused FWI for both estimates to increase, but more so for the IMERG-based estimate. The peak in FWI on 23 July for both estimates occurred under hot (35 °C) and very dry (RH = 29%) conditions, but was higher (FWI = 59) for IMERG than for the rain gauges (FWI = 42), owing not to differences in precipitation during the preceding days, but mainly to differences 2 weeks before the event.

57.6   Fire Activity Over Western Indonesia, 2015–2018

Severe fire and haze have occurred in Indonesia since the 1960s, as the result of the widespread use of fire in clearing agricultural and logging waste, which can escape and burn underground in peat soils under dry enough conditions. The last severe episode was in 2015, when the region was blanketed in thick smoke for most of September and October (Field et al. 2016), exposing millions of people to dangerously poor air quality (Burki 2017). The FWI System has been adopted as Indonesia's national FDRS following a process of calibration for selected components (de Groot et al. 2007), and its use as fire management tool is central to Indonesia's Program on Forest and Land Fire Control (Nurbaya 2018). Rather than adopt the FWI component as an overall fire danger indicator, three of the sub-indices were selected for particular aspects of the fire problem. The Drought Code was adopted as an indicator for potential burning in disturbed peatland, from which the bulk of smoke originates, and the Initial Spread Index as a difficulty of control metric. The Fine Fuel Moisture Code was adopted as an indicator for the potential for fires to start in light fuels such as tall grass and agricultural residue, where most ignitions occur. FFMC levels were determined from grass ignition tests and satellite fire detections from the late 1990s, with FFMC above 83 considered extreme. This number is the starting point in fire management operations for identifying regions where high fire activity is possible.

To determine the sensitivity of FFMC calculations to input precipitation, the relationship between FFMC and fire activity was examined over May 2015 to March 2018 for three district-level regions where fire activity was most pronounced in the 2015 dry season (Field et al. 2016): southeastern Jambi, eastern South Sumatra, and southeastern Central Kalimantan, shown in Fig. 57.9 with the locations of synoptic weather stations used by the Indonesian Agency for Meteorology, Climatology and Geophysics to calculate FWI maps operationally. For each region, a piece-wise linear regression model was fit between FFMC and MODIS active fire for temporal scales ranging from monthly (common to fire-climate studies such as Bedia et al.

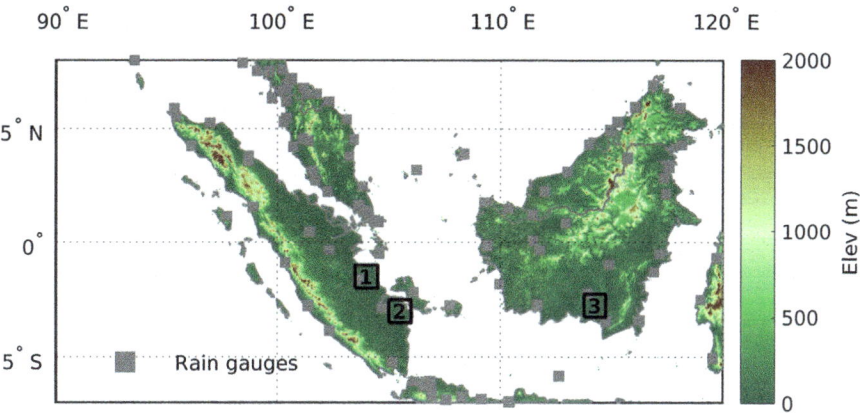

Fig. 57.9 Elevation map of western Indonesia, showing the three low-lying regions in Jambi province (1), South Sumatra (2), and Central Kalimantan (3), with the highest fire activity (Field et al. 2016) and CO emissions (Huijnen et al. 2016) during the 2015 fire season

(2015)), to daily, at which fire management decision making is done. The piecewise linear model was chosen to capture the strongly non-linear relationship between fire weather and fire activity in Indonesia (Field et al. 2016), which is driven by fuels reaching a moisture content threshold below which ignition can occur. These relationships were estimated for FFMC computed from MERRA2 precipitation, MERRA2 precipitation with rain-gauge correction (MERRA2-C), the 'late' IMERG estimate (IMERG-L), and the 'final' IMERG estimate (IMERG-F).

For each region, precipitation estimate, and time-averaging period, Table 57.2 shows the coefficient of determination (r^2) for the fitted FFMC-active fire relationships. At monthly scales, there are very strong relationships between FFMC and active fire totals for all regions and precipitation inputs; the lowest variation in fire activity explained (84%) was for the FFMC computed using IMERG-L in the Jambi region. This is consistent with previous monthly studies showing the uniquely strong climatic controls on fire in Indonesia (Abatzoglou et al. 2018; Bedia et al. 2015; van der Werf et al. 2008a) compared to other regions, attributable to strong interannual variability in climate and the widespread use of fire under sufficiently dry conditions. At shorter time scales more relevant to fire management, these relationships weaken. At weekly timescales, MERRA2-C and IMERG-F were the best performing indices over Jambi, explaining 66% and 63% of weekly fire activity respectively. Over South Sumatra and Central Kalimantan, the IMERG-F based FFMC was the best-performing, explaining 94% and 83% of fire activity respectively. The gains from IMERG-F were reduced at 1- and 2-day time scales, reflecting the hard limits in explaining day-to-day changes in fire activity using weather drivers alone. The variability explained decreased in all regions and precipitation estimates, due mainly to the scatter in fire activity for above-threshold FFMC values. The FFMC remained

Table 57.2 Coefficient of determination (r^2) for fitted linear change-point regression models (Field et al. 2016) between Fine Fuel Moisture Code (FFMC) and MODIS active fire totals over three regions in Indonesia. The FFMC was computed using four different daily precipitation estimates: MERRA2, MERRA2 with rain gauge correction (MERRA2-C), IMERG Late with climatological rain gauge correction (IMERG-L) and IMERG Final with enhanced rain gauge correction (IMERG-F). Estimates were computed for FFMC averaged, and active fires totaled, over 1 month, 7 days, 2 days, and 1 day. For each region, precipitation estimate and averaging period, the highest r^2 is shown in bold

Region	Precipitation estimate	r^2 1-month	7-day	2-day	1-day
1. Jambi	MERRA2	0.91	0.56	0.44	0.32
	MERRA2-C	**0.96**	**0.66**	0.51	**0.37**
	IMERG-L	0.84	0.56	0.49	0.32
	IMERG-F	0.95	0.63	**0.53**	0.36
2. South Sumatra	MERRA2	1.00	0.77	0.74	0.57
	MERRA2-C	1.00	0.82	**0.81**	**0.61**
	IMERG-L	1.00	0.84	0.75	0.58
	IMERG-F	1.00	**0.94**	0.78	0.59
3.Central Kalimantan	MERRA2	0.92	0.76	0.54	0.33
	MERRA2-C	0.96	0.72	0.52	0.32
	IMERG-L	0.96	0.79	0.53	0.35
	IMERG-F	**0.98**	**0.83**	**0.56**	**0.36**

a good separator of low and high fire activity, but presumably due to human-driven ignition patterns, was poor in explaining the exact amount of fire under dry conditions.

At daily time scales, however, there were benefits from IMERG-F not captured in the overall variability explained in Table 57.2. To show this, Fig. 57.10 compares the relationship between daily active fire totals and FFMC calculated from MERRA-2 and IMERG-F. For Jambi (top) there is an increase in fire activity with increasing FFMC calculated from MERRA-2 (left), but with a clearer separation between days with low and high fire activity for IMERG-F (right). In terms of false negatives, there was a total of 692 active fires detected for MERRA-2 when FFMC < 78.9, the estimated change-point threshold. For the FFMC calculated from IMERG-F, this number was reduced to 158 for its threshold of 83.4. Over South Sumatra (middle), the number of active fires occurring under low FFMC conditions was reduced from 461 to 189 when IMERG-F precipitation is used in place of MERRA-2, with a similar reduction of 409 to 134 active fires for Central Kalimantan (bottom). In each region therefore, the improvements yielded by calculating the FFMC from IMERG-F came in reducing the number of false negatives; that is, days with high fire activity occurring under low-FFMC conditions.

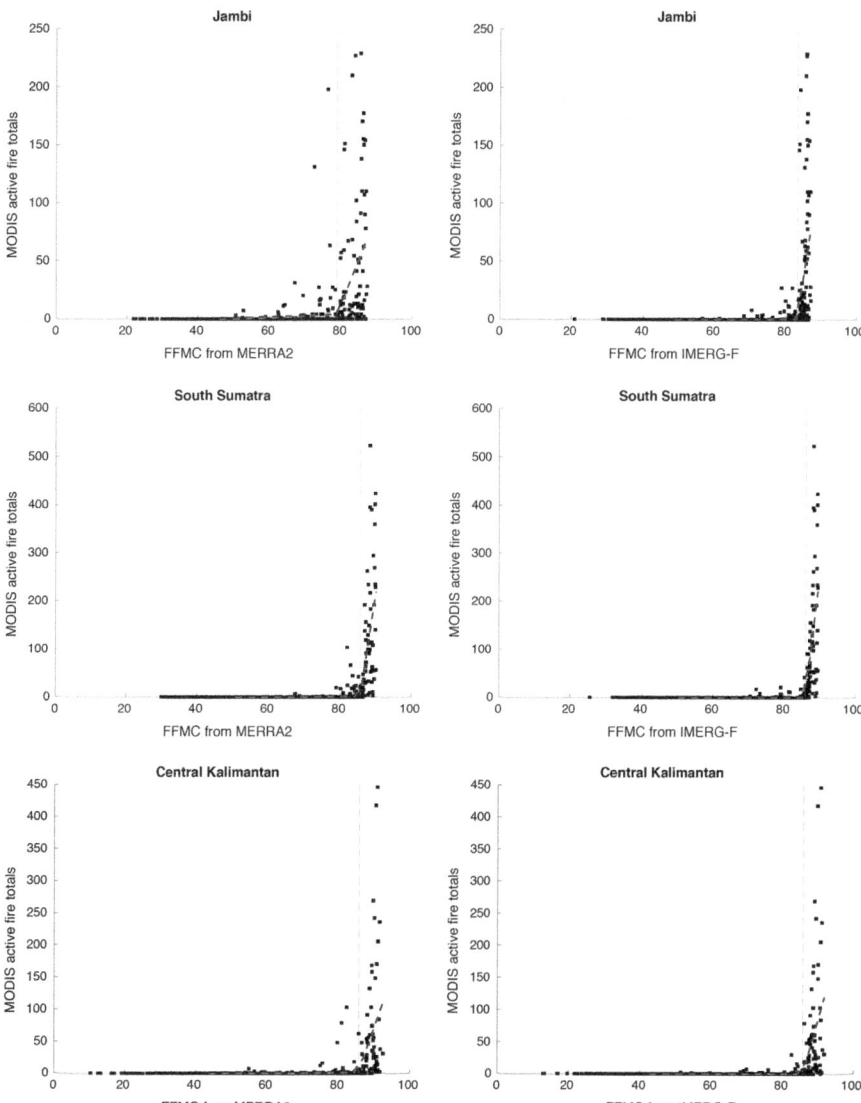

Fig. 57.10 Daily Fine Fuel Moisture Code (FFMC) computed using MERRA2 (left) and IMERG-F (right) precipitation for 1 May 2015 to 31 March 2018 over the regions in Fig. 57.9: (1) Jambi, (2) South Sumatra, and (3) Central Kalimantan. The vertical grey line shows the estimated change point, and the dashed blue line the fitted change-point model

57.7 Conclusions

More accurate and higher-resolution fire danger rating information is always needed by fire managers and researchers (Gould et al. 2013), and satellite data provide a possible means for doing so through improved precipitation estimates (de Groot et al. 2015). For the British Columbia fires of 2017, the IMERG-based estimate led to FWI fields more consistent with observed day-to-day changes in extreme fire activity in the region with fewer rain gauges. For the Chilean fires of January 2017, satellite-based estimates were in good agreement with gauge-based estimates during a period of drought extending back to August 2016, both capturing a north-south progression in extreme fire weather over a 2-week period with high temperatures and low relative humidity, but with slight differences due to different estimates of the timing of small precipitation amounts for several months prior. For the Athens fires of 2018, FWI calculated from IMERG precipitation estimates better characterized explosive fire conditions compared to gauge-based estimates due to an apparent overestimate of precipitation during a single event 2 weeks prior. Over Indonesia, gauge-corrected reanalysis and IMERG precipitation-based FFMC were best able to separate periods of low and fire activity, with the IMERG-based estimate performing best in reducing the number of fires occurring under below-threshold FFMC values. The main point to be taken from these examples is that, while more comprehensive comparisons are required, satellite-based estimates of precipitation have apparent benefits in explaining fire activity in several very different fire environments, and that, at a minimum, careful consideration should be given to the sensitivity of fire danger estimates to precipitation.

The Global Fire Weather Database provides a flexible framework in which the benefits of satellite-based precipitation estimates can be evaluated against more conventional station and reanalysis-based FWI calculations. This approach can be extended to other similar products as they are developed and for other fire danger indices which require precipitation as an input. Three specific areas for further evaluation and development are:

- 'Out-of-sample' evaluation of different gridded fire danger estimates against data from secondary meteorological networks which is not assimilated into reanalysis and satellite estimates of precipitation.
- The incorporation of land cover maps in combination with fuel-specific moisture and fire behavior models such as the Canadian Fire Behavior Prediction System (Wotton 2009) and the US National Fire Danger Rating System (Carlson and Burgan 2003) to capture the influence of different vegetation types on fire danger.
- Evaluation of precipitation estimates from modern geostationary satellites to achieve lower-latency for operational fire danger rating systems, ideally making use of diurnal models of fire danger (de Groot et al. 2015).

Acknowledgements The development of GFWED is supported by the NASA Precipitation Measurement Missions Science Team and the NASA Group on Earth Observations Work Programme. All GFWED data are available at https://data.giss.nasa.gov/impacts/gfwed/ (last accessed 28 Nov. 2018).

References

Abatzoglou, J. T., Williams, A. P., Boschetti, L., Zubkova, M., & Kolden, C. A. (2018). Global patterns of interannual climate-fire relationships. *Global Change Biology, 24*, 5164–5175. https://doi.org/10.1111/gcb.14405.

Abbott, K. N., Leblon, B., Staples, G. C., Maclean, D. A., & Alexander, M. E. (2007). Fire danger monitoring using RADARSAT-1 over northern boreal forests. *International Journal of Remote Sensing, 28*, 1317–1338. https://doi.org/10.1080/01431160600904956.

Adler, R. F., Huffman, G. J., Chang, A., Ferraro, R. R., Xie, P., Janowiak, J., Rudolf, B., Schneider, U., Curtis, S., Bolvin, D. T., Gruber, A., Susskind, J., Arkin, P. A., & Nelkin, E. (2003). The version-2 Global Precipitation Climatology Project (GPCP) monthly precipitation analysis (1979–present). *Journal of Hydrometeorology, 4*, 1147–1167. https://doi.org/10.1175/1525-7541(2003)004<1147:TVGPCP>2.0.CO;2.

Aguado, I., Chuvieco, E., Martin, P., & Salas, J. (2003). Assessment of forest fire danger conditions in southern Spain from NOAA images and meteorological indices. *International Journal of Remote Sensing, 24*, 1653–1668. https://doi.org/10.1080/01431160210144688.

Alexander, M. E., & W. J. de Groot (1988). *Fire behavior in jack pine stands as related to the Canadian Forest Fire Weather Index (FWI) System*. Canadian Forestry Service, Northwest Region. Available at https://cfs.nrcan.gc.ca/publications?id=24310, last accessed 28 Nov. 2018.

Amiro, B. D., Todd, J. B., Wotton, B. M., Logan, K. A., Flannigan, M. D., Stocks, B. J., Mason, J. A., Martell, D. L., & Hirsch, K. G. (2001). Direct carbon emissions from Canadian forest fires, 1959-1999. *Canadian Journal of Forest Research, 31*, 512–525. https://doi.org/10.1139/x00-197.

Anonymous. (2017). Spreading like wildfire. *Nature Climate Change, 7*, 755. https://doi.org/10.1038/nclimate3432.

Athanasopoulou, E., Rieger, D., Walter, C., Vogel, H., Karali, A., Hatzaki, M., Gerasopoulos, E., Vogel, B., Giannakopoulos, C., Gratsea, M., & Roussos, A. (2014). Fire risk, atmospheric chemistry and radiative forcing assessment of wildfires in eastern Mediterranean. *Atmospheric Environment, 95*, 113–125. https://doi.org/10.1016/j.atmosenv.2014.05.077.

Beck, H. E., van Dijk, A. I. J. M., Levizzani, V., Schellekens, J., Miralles, D. G., Martens, B., & de Roo, A. (2017). MSWEP: 3-hourly 0.25° global gridded precipitation (1979–2015) by merging gauge, satellite, and reanalysis data. *Hydrology and Earth System Sciences, 21*, 589–615. https://doi.org/10.5194/hess-21-589-2017.

Bedia, J., Herrera, S., Gutierrez, J. M., Zavala, G., Urbieta, I. R., & Moreno, J. M. (2012). Sensitivity of fire weather index to different reanalysis products in the Iberian Peninsula. *Natural Hazards and Earth System Sciences, 12*, 699–708. https://doi.org/10.5194/nhess-12-699-2012.

Bedia, J., Herrera, S., Gutierrez, J. M., Benali, A., Brands, S., Mota, B., & Moreno, J. M. (2015). Global patterns in the sensitivity of burned area to fire-weather: Implications for climate change. *Agricultural and Forest Meteorology, 214*, 369–379. https://doi.org/10.1016/j.agrformet.2015.09.002.

Bourgeau-Chavez, L. L., Kasischke, E. S., & Rutherford, M. D. (1999). Evaluation of ERS SAR data for prediction of fire danger in a boreal region. *International Journal of Wildland Fire, 9*, 183–194. https://doi.org/10.1071/WF00009.

Bourgeau-Chavez, L. L., Garwood, G., Riordan, K., Cella, B., Alden, S., Kwart, M., & Murphy, K. (2007). Improving the prediction of wildfire potential in boreal Alaska with satellite imaging radar. *Polar Record, 43*, 321–330. https://doi.org/10.1017/S0032247407006535.

Bourgeau-Chavez, L. L., Leblon, B., Charbonneau, F., & Buckley, J. R. (2013). Evaluation of polarimetric Radarsat-2 SAR data for development of soil moisture retrieval algorithms over a chronosequence of black spruce boreal forests. *Remote Sensing of Environment, 132*, 71–85. https://doi.org/10.1016/j.rse.2013.01.006.

Bowman, D., Williamson, G. J., Abatzoglou, J. T., Kolden, C. A., Cochrane, M. A., & Smith, A. M. S. (2017). Human exposure and sensitivity to globally extreme wildfire events. *Nature Ecology & Evolution, 1*. https://doi.org/10.1038/s41559-016-0058.

Bowman, D. M. J. S., Moreira-Muñoz, A., Kolden, C. A., Chávez, R. O., Muñoz, A. A., Salinas, F., González-Reyes, Á., Rocco, R., de la Barrera, F., Williamson, G. J., Borchers, N., Cifuentes, L. A., Abatzoglou, J. T., & Johnston, F. H. (2018). Human-environmental drivers and impacts of the globally extreme 2017 Chilean fires. *Ambio*. https://doi.org/10.1007/s13280-018-1084-1.

Burki, T. K. (2017). The pressing problem of Indonesia's forest fires. *The Lancet Respiratory Medicine, 5*, 685–686. https://doi.org/10.1016/S2213-2600(17)30301-6.

Cantin, A. (2016). Fire weather index system and fire behaviour prediction system calculations. *Comprehensive R Archive Network*. Available from https://cran.r-project.org/web/packages/cffdrs/index.html, last accessed 28 Nov. 2018.

Carlson, J. D., & Burgan, R. E. (2003). Review of users' needs in operational fire danger estimation: The Oklahoma example. *International Journal of Remote Sensing, 24*, 1601–1620. https://doi.org/10.1080/01431160210144651.

CCFM (2018). National Forestry Database. *National forestry database, Canadian Council of Forest Forest Ministers*. Available at http://nfdp.ccfm.org/en/data/fires.php, last accessed 28 Nov. 2018.

Chen, Y., Morton, D. C., Jin, Y., Collatz, G. J., Kasibhatla, P. S., van der Werf, G. R., DeFries, R. S., & Randerson, J. T. (2013). Long-term trends and interannual variability of forest, savanna and agricultural fires in South America. *Carbon Management, 4*, 617–638. https://doi.org/10.4155/cmt.13.61.

Chowdhury, E. H., & Hassan, Q. K. (2015). Operational perspective of remote sensing-based forest fire danger forecasting systems. *ISPRS Journal of Photogrammetry and Remote Sensing, 104*, 224–236. https://doi.org/10.1016/j.isprsjprs.2014.03.011.

de Groot, W. J., & Flannigan, M. D. (2014). Climate change and early warning systems for wildland fire. In Z. Zommers & A. Singh (Eds.), *Reducing disaster: Early warning systems for climate change* (pp. 127–151). Springer. https://doi.org/10.1007/978-94-017-8598-3_7.

de Groot, W. J., Field, R. D., Brady, M. A., Roswintiarti, O., & Mohamad, M. (2007). Development of the Indonesian and Malaysian fire danger rating systems. *Mitigation and Adaptation Strategies for Global Change, 12*, 165–180. https://doi.org/10.1007/s11027-006-9043-8.

de Groot, W. J., Wotton, B. M., & Flannigan, M. D. (2015). Wildland fire danger rating and early warning systems. In J. F. Shroder & D. Paton (Eds.), *Wildfire hazards, risks and disasters* (pp. 207–228). Amsterdam: Elsevier. ISBN: 9780124096011.

Di Giuseppe, F., Remy, S., Pappenberger, F., & Wetterhall, F. (2017). Improving Forecasts of Biomass Burning Emissions with the Fire Weather Index. *Journal of Applied Meteorology and Climatology, 56*, 2789–2799. https://doi.org/10.1175/jamc-d-16-0405.1.

Di Giuseppe, F., Remy, S., Pappenberger, F., & Wetterhall, F. (2018). Using the Fire Weather Index (FWI) to improve the estimation of fire emissions from fire radiative power (FRP) observations. *Atmospheric Chemistry and Physics, 18*, 5359–5370. https://doi.org/10.5194/acp-18-5359-2018.

Dimitrakopoulos, A. P., Bemmerzouk, A. M., & Mitsopoulos, I. D. (2011). Evaluation of the Canadian fire weather index system in an eastern Mediterranean environment. *Meteorological Applications, 18*, 83–93. https://doi.org/10.1002/met.214.

Dowdy, A. J., Mills, G. A., Finkele, K., & de Groot, W. J. (2009). Australian fire weather as represented by the McArthur Forest Fire Danger Index and the Canadian Forest Fire Weather Index. *CAWCR Technical Report, 10*, 84 pp. Available at http://www.cawcr.gov.au/technical-reports/CTR_010.pdf, last accessed 28 Nov. 2018.

Fang, L., Yang, J., Zu, J., Li, G., & Zhang, J. (2015). Quantifying influences and relative importance of fire weather, topography, and vegetation on fire size and fire severity in a Chinese boreal forest landscape. *Forest Ecology and Management, 356*, 2–12. https://doi.org/10.1016/j.foreco.2015.01.011.

FAO. (2007). Fire management – Global Assessment 2006. *FAO Forestry Paper, 151*, 135 pp. Available at http://www.fao.org/docrep/009/a0969e/a0969e00.htm, last accessed 28 Nov. 2018.

Field, R. D., & Shen, S. S. P. (2008). Predictability of carbon emissions from biomass burning in Indonesia from 1997 to 2006. *Journal of Geophysical Research, 113*, 17. https://doi.org/10.1029/2008JG000694.

Field, R. D., Spessa, A. C., Aziz, N. A., Camia, A., Cantin, A., Carr, R., de Groot, W. J., Dowdy, A. J., Flannigan, M. D., Manomaiphiboon, K., Pappenberger, F., Tanpipat, V., & Wang, X. (2015). Development of a global fire weather database. *Natural Hazards and Earth System Sciences, 15*, 1407–1423. https://doi.org/10.5194/nhess-15-1407-2015.

Field, R. D., van der Werf, G. R., Fanin, T., Fetzer, E. J., Fuller, R., Jethva, H., Levy, R., Livesey, N. J., Luo, M., Torres, O., & Worden, H. M. (2016). Indonesian fire activity and smoke pollution in 2015 show persistent nonlinear sensitivity to El Niño-induced drought. *PNAS, 113*, 9204–9209. https://doi.org/10.1073/pnas.1524888113.

Fiorucci, P., Gaetani, F., Lanorte, A., & Lasaponara, R. (2007). Dynamic fire danger mapping from satellite imagery and meteorological forecast data. *Earth Interactions, 11*, 7. https://doi.org/10.1175/EI199.1.

Flannigan, M. D., & Wotton, B. M. (1989). A study of interpolation methods for forest fire danger rating in Canada. *Canadian Journal of Forest Research, 19*, 1059–1066. https://doi.org/10.1139/x89-161.

Flannigan, M. D., Wotton, B. M., & Ziga, S. (1998). A study on the interpolation of fire danger using radar precipitation estimates. *International Journal of Wildland Fire, 8*, 217–225. https://doi.org/10.1071/WF9980217.

Gelaro, R., McCarty, W., Suárez, M. J., Todling, R., Molod, A., Takacs, L., Randles, C. A., Darmenov, A., Bosilovich, M. G., Reichle, R., Wargan, K., Coy, L., Cullather, R., Draper, C., Akella, S., Buchard, V., Conaty, A., da Silva, A. M., Gu, W., Kim, G.-K., Koster, R., Lucchesi, R., Merkova, D., Nielsen, J. E., Partyka, G., Pawson, S., Putman, W., Rienecker, M., Schubert, S. D., Sienkiewicz, M., & Zhao, B. (2017). The modern-era retrospective analysis for research and applications, version 2 (MERRA-2). *Journal of Climate, 30*, 5419–5454. https://doi.org/10.1175/JCLI-D-16-0758.1.

Gould, J. S., Patriquin, M. N., Wang, S., McFarlane, B. L., & Wotton, B. M. (2013). Economic evaluation of research to improve the Canadian forest fire danger rating system. *Forestry, 86*, 317–329. https://doi.org/10.1093/forestry/cps082.

Han, K. S., Viau, A., & Anctil, F. (2003). High-resolution forest fire weather index computations using satellite remote sensing. *Canadian Journal of Forest Research, 33*, 1134–1143. https://doi.org/10.1139/x03-014.

Hanes, C. C., Jain, P., Flannigan, M. D., Fortin, V., & Roy, G. (2017). Evaluation of the Canadian Precipitation Analysis (CaPA) to improve forest fire danger rating. *International Journal of Wildland Fire, 26*, 509–522. https://doi.org/10.1071/WF16170.

Holgate, C. M., van Dijk, A. I. J. M., Cary, G. J., & Yebra, M. (2017). Using alternative soil moisture estimates in the McArthur Forest Fire Danger Index. *International Journal of Wildland Fire, 26*, 806–819. https://doi.org/10.1071/WF16217.

Horel, J. D., Ziel, R., Galli, C., Pechmann, J., & Dong, X. (2014). An evaluation of fire danger and behaviour indices in the Great Lakes region calculated from station and gridded weather information. *International Journal of Wildland Fire, 23*, 202–214. https://doi.org/10.1071/WF12186.

Huffman, G. J., Adler, R. F., Morrissey, M. M., Bolvin, D. T., Curtis, S., Joyce, R., McGavock, B., & Susskind, J. (2001). Global precipitation at one-degree daily resolution from multisatellite observations. *Journal of Hydrometeorology, 2*, 36–50. https://doi.org/10.1175/1525-7541 (2001)0022.0.CO;2.

Huffman, G. J., Bolvin, D. T., Nelkin, E. J., Wolff, D. B., Adler, R. F., Gu, G., Hong, Y., Bowman, K. P., & Stocker, E. F. (2007). The TRMM Multisatellite Precipitation Analysis (TMPA): Quasi-global, multiyear, combined-sensor precipitation estimates at fine scales. *Journal of Hydrometeorology, 8*, 38–55. https://doi.org/10.1175/JHM560.1.

Huffman, G. J., Bolvin, D. T., Braithwaite, D., Hsu, K.-L., Joyce, R., Kidd, C., Nelkin, E. J., Sorooshian, S., Tan, J., & Xie, P. (2018). Algorithm theoretical basis document (ATBD) version 5.2 for NASA Global Precipitation Measurement (GPM) Integrated Multi-satellitE Retrievals for GPM (IMERG). *NASA* 34 pp. Available at https://docserver.gesdisc.eosdis.nasa.gov/public/project/GPM/IMERG_ATBD_V5.pdf, last accessed 28 Nov. 2018.

Huijnen, V., Wooster, M. J., Kaiser, J. W., Gaveau, D. L. A., Flemming, J., Parrington, M., Inness, A., Murdiyarso, D., Main, B., & van Weele, M. (2016). Fire carbon emissions over maritime Southeast Asia in 2015 largest since 1997. *Scientific Reports, 6,* 26886. https://doi.org/10.1038/srep26886.

Jain, P., & Flannigan, M. D. (2017). Comparison of methods for spatial interpolation of fire weather in Alberta, Canada. *Canadian Journal of Forest Research, 47,* 1646–1658. https://doi.org/10.1139/cjfr-2017-0101.

Jang, J. D., Viau, A. A., & Anctil, F. (2006). Thermal-water stress index from satellite images. *International Journal of Remote Sensing, 27,* 1619–1639. https://doi.org/10.1080/01431160500509194.

Jolly, W. M., Cochrane, M. A., Freeborn, P. H., Holden, Z. A., Brown, T. J., Williamson, G. J., & Bowman, D. M. J. S. (2015). Climate-induced variations in global wildfire danger from 1979 to 2013. *Nature Communications, 6,* 7537. https://doi.org/10.1038/ncomms8537.

Karali, A., Hatzaki, M., Giannakopoulos, C., Roussos, A., Xanthopoulos, G., & Tenentes, V. (2014). Sensitivity and evaluation of current fire risk and future projections due to climate change: The case study of Greece. *Natural Hazards and Earth System Sciences, 14,* 143–153. https://doi.org/10.5194/nhess-14-143-2014.

Khaykin, S. M., Godin-Beekmann, S., Hauchecorne, A., Pelon, J., Ravetta, F., & Keckhut, P. (2018). Stratospheric smoke with unprecedentedly high backscatter observed by lidars above southern France. *Geophysical Research Letters, 45,* 1639–1646. https://doi.org/10.1002/2017GL076763.

Kidd, C., & Levizzani, V. (2011). Status of satellite precipitation retrievals. *Hydrology and Earth System Sciences, 15,* 1109–1116. https://doi.org/10.5194/hess-15-1109-2011.

Lawson, B. D., & Armitage, O. B. (2008). *Weather guide for the Canadian Forest Fire Danger Rating System.* Canadian Forest Service, 73 pp. Available at https://cfs.nrcan.gc.ca/publications?id=29152, last accessed 28 Nov. 2018.

Leblon, B., Alexander, M., Chen, J., & White, S. (2001). Monitoring fire danger of northern boreal forests with NOAA-AVHRR NDVI images. *International Journal of Remote Sensing, 22,* 2839–2846. https://doi.org/10.1080/01431160110034663.

Leblon, B., Kasischke, E., Alexander, M., Doyle, M., & Abbott, M. (2002). Fire danger monitoring using ERS-1 SAR images in the case of northern boreal forests. *Natural Hazards, 27,* 231–255. https://doi.org/10.1023/A:1020375721520.

Leblon, B., San-Miguel-Ayanz, J., Bourgeau-Chavez, L., & Kong, M. (2016). Remote sensing of wildfires. In N. Baghdadi & M. Zribi (Eds.), *Land surface remote sensing: Environment and risks* (pp. 55–95). Dordrecht: Elsevier. 9780081012659.

Lee, B. S., Alexander, M. E., Hawkes, B. C., Lynham, T. J., Stocks, B. J., & Englefield, P. (2002). Information systems in support of wildland fire management decision making in Canada. *Computers and Electronics in Agriculture, 37,* 185–198. https://doi.org/10.1016/S0168-1699(02)00120-5.

Lekkas, E., Voulgaris, N., & Lozios, S. (2018). *The July 2018 Attica (Central Greece) Wildfires – Scientific Report (Version 1.3).* Newsletter of Environmental, Disaster and Crisis Management Strategies, National and Kapodistrian University of Athens. Available at https://edcm.edu.gr/images/docs/2018/Newsletter_Attica_Fires_2018_v11.pdf, last accessed 28 Nov. 2018.

Martinez-Harms, M. J., Caceres, H., Biggs, D., & Possingham, H. P. (2017). After Chile's fires, reforest private land. *Science, 356,* 147–148. https://doi.org/10.1126/science.aan0701.

Maselli, F., Romanelli, S., Bottai, L., & Zipoli, G. (2003). Use of NOAA-AVHRR NDVI images for the estimation of dynamic fire risk in Mediterranean areas. *Remote Sensing of Environment, 86,* 187–197. https://doi.org/10.1016/S0034-4257(03)00099-3.

Merzouki, A., & Leblon, B. (2011). Mapping fuel moisture codes using MODIS images and the Getis statistic over western Canada grasslands. *International Journal of Remote Sensing, 32*, 1619–1634. https://doi.org/10.1080/01431160903586773.

Molod, A., Takacs, L., Suarez, M., & Bacmeister, J. (2015). Development of the GEOS-5 atmospheric general circulation model: Evolution from MERRA to MERRA2. *Geoscientific Model Development, 8*, 1339–1356. https://doi.org/10.5194/gmd-8-1339-2015.

Nurbaya, N. (2018). *The State of Indonesia's forests 2018*, 196 pp, ISBN: 978-602-8358-82-8. Available at https://www.google.com/url?sa=t&rct=j&q=&esrc=s&source=web&cd=1&ved=2ahUKEwia5ZKauvfeAhUI-6QKHTh0APwQFjAAegQICRAC&url=http%3A%2F%2Fwww.menlhk.go.id%2Fdownlot.php%3Ffile%3Dthe_state_Indonesia_forests_2018_Book.pdf&usg=AOvVaw3Dg8lu_ci6MFLMx3XRcMSL, last accessed 28 Nov. 2018.

Oldford, S., Leblon, B., Maclean, D., & Flannigan, M. (2006). Predicting slow-drying fire weather index fuel moisture codes with NOAA-AVHRR images in Canada's northern boreal forests. *International Journal of Remote Sensing, 27*, 3881–3902. https://doi.org/10.1080/01431160600784234.

Peterson, D. A., Campbell, J. R., Hyer, E. J., Fromm, M. D., Kablick, G. P., Cossuth, J. H., & DeLand, M. T. (2018). Wildfire-driven thunderstorms cause a volcano-like stratospheric injection of smoke. *npj Climate and Atmospheric Science, 1*, 30. https://doi.org/10.1038/s41612-018-0039-3.

Robinne, F. N., Bladon, K. D., Miller, C., Parisien, M. A., Mathieu, J., & Flannigan, M. D. (2018). A spatial evaluation of global wildfire-water risks to human and natural systems. *Science of the Total Environment, 610*, 1193–1206. https://doi.org/10.1016/j.scitotenv.2017.08.112.

San-Miguel-Ayanz, J., Manuel Moreno, J., & Camia, A. (2013). Analysis of large fires in European Mediterranean landscapes: Lessons learned and perspectives. *Forest Ecology and Management, 294*, 11–22. https://doi.org/10.1016/j.foreco.2012.10.050.

Steininger, M. K., Tabor, K., Small, J., Pinto, C., Soliz, J., & Chavez, E. (2013). A satellite model of forest flammability. *Environmental Management, 52*, 136–150. https://doi.org/10.1007/s00267-013-0073-1.

Stocks, B. J., Lawson, B. D., Alexander, M. E., Vanwagner, C. E., McAlpine, R. S., Lynham, T. J., & Dube, D. E. (1989). The Canadian forest fire danger rating system – An Overview. *Forestry Chronicle, 65*, 450–457. Available from https://cfs.nrcan.gc.ca/publications?id=11347, last accessed 28 Nov. 2018.

Taylor, S. W., & Alexander, M. E. (2006). Science, technology, and human factors in fire danger rating: The Canadian experience. *International Journal of Wildland Fire, 15*, 121–135. https://doi.org/10.1071/WF05021.

Úbeda, X., & Sarricolea, P. (2016). Wildfires in Chile: A review. *Global and Planetary Change, 146*, 152–161. https://doi.org/10.1016/j.gloplacha.2016.10.004.

Urrutia-Jalabert, R., Gonzalez, M. E., Gonzalez-Reyes, A., Lara, A., & Garreaud, R. (2018). Climate variability and forest fires in central and south-central Chile. *Ecosphere, 9*, e02171. https://doi.org/10.1002/ecs2.2171.

van der Werf, G. R., Randerson, J. T., Giglio, L., Gobron, N., & Dolman, A. J. (2008a). Climate controls on the variability of fires in the tropics and subtropics. *Global Biogeochemical Cycles, 22*. https://doi.org/10.1029/2007GB003122.

van der Werf, G. R., Dempewolf, J., Trigg, S. N., Randerson, J. T., Kasibhatla, P. S., Giglio, L., Murdiyarso, D., Peters, W., Morton, D. C., Collatz, G. J., Dolman, A. J., & DeFries, R. S. (2008b). Climate regulation of fire emissions and deforestation in equatorial Asia. *PNAS, 105*, 20350–20355. https://doi.org/10.1073/pnas.0803375105.

van Wagner, C. E. (1987). Development and structure of the Canadian Forest Fire Weather Index System. *Canadian Forestry Service, 37* pp. Available at https://cfs.nrcan.gc.ca/publications?id=19927, last accessed 28 Nov. 2018.

Venalainen, A., Korhonen, N., Hyvarinen, O., Koutsias, N., Xystrakis, F., Urbieta, I. R., & Moreno, J. M. (2014). Temporal variations and change in forest fire danger in Europe for 1960-2012.

Natural Hazards and Earth System Sciences, 14, 1477–1490. https://doi.org/10.5194/nhess-14-1477-2014.

Viegas, D. X., Bovio, G., Ferreira, A., Nosenzo, A., & Sol, B. (1999). Comparative study of various methods of fire danger evaluation in southern Europe. *International Journal of Wildland Fire, 9*, 235–246. https://doi.org/10.1071/WF00015.

Wotton, B. M. (2009). Interpreting and using outputs from the Canadian Forest Fire Danger Rating System in research applications. *Environmental and Ecological Statistics, 16*, 107–131. https://doi.org/10.1007/s10651-007-0084-2.

Chapter 58
Variability of Satellite Sea Surface Salinity Under Rainfall

Alexandre Supply, Jacqueline Boutin, Gilles Reverdin, Jean-Luc Vergely, and Hugo Bellenger

Abstract Two L-Band (1.4 GHz) microwave radiometer missions, the Soil Moisture and Ocean Salinity (SMOS) and the Soil Moisture Active and Passive (SMAP) missions, currently provide salinity measurements in the first centimeter below the sea surface. At this depth, salinity variability at hourly temporal scales is dominated by the impact of precipitation. The dependency of the salinity freshening with the instantaneous rain rate (RR) observed between 50°S and 50°N, with SMOS and SMAP salinities, is very similar. We investigate the influence of rain history on salinity anomalies. By using rain rates retrieved from several microwave satellites measurements including Advanced Microwave Scanning Radiometer 2 (AMSR-2), and Special Sensor Microwave Imager Sounder 17 (SSMIS-17 and SSMIS-16) and by taking advantage of their different crossing times, we estimate the temporal cross-correlation function between salinity freshening and rain rate for different time lags in various tropical and high latitudes regions. Whatever the region, the magnitude of the salinity anomaly associated with precipitation is dominated by the instantaneous RR for each area. The apparent correlation between salinity anomaly and rain history can be explained by RR auto-correlation. The relationship between salinity anomaly (ΔS) and RR is then investigated in six regions, with RR provided using three different algorithms (the Unified Microwave Ocean Retrieval Algorithm (UMORA), the Goddard Profiling algorithm (GPROF) and Integrated MultisatellitE Retrievals for GPM (IMERG)). Differences in RR distribution between the various algorithms lead to

A. Supply (✉) · J. Boutin · G. Reverdin
CNRS, IRD, MNHN, Laboratoire d'Océanographie et du Climat, Expérimentations et Approches Numériques (LOCEAN), Sorbonne Université, Paris, France
e-mail: alexandre.supply@locean.upmc.fr

J.-L. Vergely
ACRI-st, Guyancourt, France

H. Bellenger
Laboratoire de Météorologie Dynamique/IPSL, CNRS, Sorbonne Université, École Normale Supérieure, École Polytechnique, Paris, France

Japan Agency for Marine-Earth Science and Technology (JAMSTEC), Yokosuka, Japan

© Springer Nature Switzerland AG 2020
V. Levizzani et al. (eds.), *Satellite Precipitation Measurement*, Advances in Global Change Research 69, https://doi.org/10.1007/978-3-030-35798-6_34

differences of up to a factor of 2 in ΔS versus RR slopes. For a given RR product, we also observe that part of the variability in ΔS versus RR relationships is related to the variability in wind speed regimes as detected by SMAP wind speed.

Keywords Precipitation · Rain rate · Sea surface salinity · SMOS · SMAP · SSMIS · AMSR-2 · MIRAS · Microwave · UMORA · GPM · GPROF · IMERG

58.1 Introduction

Since 2010, the Soil Moisture and Ocean Salinity (SMOS, Kerr et al. 2010) mission provides the longest record of Sea Surface Salinity (SSS) from space, with a spatial resolution of ~50 km.[1] Since 2015, the Soil Moisture Active and Passive (SMAP, Piepmeier et al. 2017) performs measurements at a similar spatial resolution. Hence, during the last 3 years, the complementarity of the spatio-temporal sampling by the two instruments provides the opportunity to improve the spatial and temporal coverage of salinity measurements from space and to reduce the mean time lag between two salinity estimates in a given pixel.

Although satellite sea surface salinity (SSS) measurements are slightly noisier than in-situ SSS measurements, recent reprocessing of SMOS and SMAP SSS estimates provide very realistic SSS variability with a spatio-temporal resolution not accessible from in situ measurements only (Tang et al. 2017; Boutin et al. 2018). They are of special interest for studying synoptic SSS variability at scales not resolved by Argo measurements, roughly less than 600 km and less than 1 month (e.g. Boutin et al. 2015). In particular, satellite SSS provides new insights into links between river discharge and interannual variability of SSS, e.g. in the Gulf of Mexico (Fournier et al. 2016), and in the western equatorial Atlantic Ocean (Fournier et al. 2017a). At weekly and small spatial scales, strong interaction between fresh river plumes, currents and mesoscale eddies have been evidenced by Fournier et al. (2016) and Fournier et al. (2017b) in the Gulf of Mexico and Bay of Bengal, respectively.

Rainfall influences salinity at various scales. During a year, ITCZ rainfall and Ekman dynamics lead to the displacement of a significant amount of freshwater, greatly reducing salinity in rainy areas but also in non-rainy areas by Ekman transport (Yu 2015; Hasson et al. 2018). Abe et al. (2018) show that accumulation of rain may induce the formation of low salinities eddies with low salinity reaching 70 m depth. These studies, which provide information on various phenomena that impact SSS at temporal scales longer than typically 1 month, are an important background to consider when investigating the effect of rain rate (RR) on SSS at

---

[1]Estimated as the diameter of the equivalent circle, centered on a Grid Point, where SSS is retrieved. The area of the equivalent circle is equal to the mean area of the footprint ellipses of the brightness temperatures (Tb) entering in the SSS retrieval. However, given that the SMOS noisiest Tb are the ones having the lowest resolution, the effective SMOS SSS resolution is likely between 43 km (the mean diameter in the Alias Free Field Of View) and 50 km.

smaller temporal scales. A better understanding of salinity freshening due to rain is needed to improve the interpretation of satellite salinity. Actually, rain events may induce large differences (> 1 pss; dimensionless number of the Practical Salinity Scale) between the first upper centimeter of the ocean and a few meters depth (Henocq et al. 2010; Reverdin et al. 2012). Current ocean general circulation models usually do not include a detailed description of the processes involved in the upper centimeters depth (Bellenger et al. 2017), so that a rain correction should be first applied to the satellite salinity before they can be confronted with ocean model simulations at several tens of centimeters or several meters depth.

After a rain event, salinity is generally quickly restored (on the order of a few hours) to a level close to the one observed before the rain event (Wijesekera et al. 1999; Soloviev et al. 2002; Reverdin et al. 2012; Drushka et al. 2016). However, in some cases, fresh lenses have been observed with a persistence time close to 24 h (Walesby et al. 2015; Dong et al. 2017; Pei et al. 2018). Fresh lens emergence and their spatial and temporal evolution are not well understood (air-sea fluxes of heat and momentum, advection, upper-ocean mixing, etc.). Recent studies succeed to simulate rain-induced fresh lens formation and life cycle by using one-dimensional water column models, the Generalized Ocean Turbulence Model (GOTM, Drushka et al. 2016) or a prognostic model (Bellenger et al. 2017). In these cases, duration of freshwater cool lens is estimated as function of windspeed (WS) and maximum RR. With a maximum RR of 20 mm h^{-1}, a fresh lens may persist less than 2 h in high WS (10 m s^{-1}) conditions and up to 7 h in the low WS (1 m s^{-1}) case (Bellenger et al. 2017). However, these studies are limited to very local scale and have been validated with a limited number of in situ measurements. To predict the influence of rain on salinity, the Rain Impact Model (RIM) applies a one-dimensional turbulent diffusion model from Asher et al. (2014) considering a rain history from the past 24 h that was calibrated using a few rain events observed with the Surface Salinity Profiler (SSP) instrument during the Kilo Moana 2011 cruise (Asher et al. 2014).

Based on satellite observations, a strong correlation between SSS negative anomalies (ΔS) and RR at short temporal (< 1 h) scales have been reported by several studies (see a review in Boutin et al. 2016). Dependency of SSS freshening with RR has a similar order of magnitude to the ones obtained by Schlüssel et al. (1997) with a surface renewal model of the molecular diffusion layer, 0.05 mm for salinity, likely because rain mixing affects salinity in the first centimeters of the surface ocean (Ho et al. 2000; Zappa et al. 2009).

Supply et al. (2017) focuses on the relationship between ΔS and RR in the Inter Tropical Convergence Zone area without considering the potential influence of WS and RR history on the relationship despite the link shown with in-situ measurements. Santos-Garcia et al. (2014), use the RIM model fitted to Aquarius satellite measurements to relate Aquarius ΔS to rain accumulation during the previous 24 h. However, given the temporal under sampling of SSS measured by a single satellite mission (at least 12 h), these studies do not allow one to estimate precisely the hourly evolution of ΔS associated with a rainy event. Thus, to date, a clear freshening signature remaining at the sea surface a few hours after a rain event was not demonstrated in satellite data. We investigate here the role of RR history on SSS anomalies by using several RR satellites combined with SMOS and SMAP data. Models and in-situ measurements indicate that the relationship between RR and ΔS is influenced by WS. We also study this dependency with satellite measurements.

Our study region covers all the oceans between 50°S and 50°N, after removing periods and areas with SSS variability driven by non-rainy processes. A better understanding of the link between RR and ΔS will also benefit satellite RR estimation. Validation of RR measurements over ocean is hindered by the low number of measurements, especially at high latitudes and the difficulty of comparing punctual in-situ and spatially integrated satellite measurements. Compared with RR satellites, SMOS and SMAP SSS measurements have very close spatial resolution and offer an alternative method to complement the rainfall monitoring. We analyze the possibility of using both satellites to assess different RR products.

In Sect. 58.2, we introduce data and methods. In Sect. 58.2.3, results are detailed: (1) intercomparison of RR and SSS products; (2) imprint of RR history on SSS anomaly; (3) variability of ΔS/RR relationship and influence of WS and rain product. In Sect. 58.4, we discuss the results and perspectives of this study.

58.2   Data and Method

SSS, WS and RR data used in this study are summarized in Table 58.1.

58.2.1   Salinity and Wind Speed Data

The SMOS mission carries an L-band Microwave Interferometric Radiometer with Aperture Synthesis (MIRAS) from which SSS measurements are retrieved. It provides SSS from space since 2010 (Reul et al. 2014). We consider only SMOS retrieved SSS at ±400 km from the center of the swath, which are less noisy. SMOS has a revisit time between 3 and 5 days and follows a sun synchronous orbit with a local equator crossing time at 0600 a.m. on the ascending node. SMAP provides SSS from space since April 2015. The SMAP swath is 1000 km with a shorter revisit time (between 2 and 3 days). It crosses the Equator at the same local time as SMOS but in the opposite phase, near 0600 a.m. for descending orbits and near 0600 p.m. for ascending orbits.

In this study, we use the following Level 2 SSS. SMOS SSS are corrected from land-sea contamination and from seasonal latitudinal systematic effects as described in (Boutin et al. 2018) and are available at CATDS (Centre Aval de Traitement des Données SMOS) (SMOS CPDC L2Q products (CATDS 2017)). SMAP SSS are from the L2B version 4 product from JPL (Fore et al. 2017). Satellite SSS are oversampled on an Equal-Area Scalable Earth Grid (EASE) grid of 25 km spacing for SMOS and on a swath grid of 25 km spacing for SMAP. The effective spatial

Table 58.1 Dataset used in this study

Satellite	Instrument	Used variable	Approximative spatial resolution of retrieved variable	Approximative equator crossing time (local solar time) for ascending node	Data source	Algorithm
SMOS	MIRAS	Sea surface salinity	~50 km (averaged resolution; 43 km in the alias free field of view)	6:00 a.m.	CATDS	L2Q V300 (RE05)
SMAP	Radiometer	Sea surface salinity and wind speed	~60 km	6:00 p.m.	Jet Propulsion Laboratory	CAP V4
DSMP-F17	SSMIS	Rain rate	~35 km (based on 37 GHz)	06:30 p.m.	NASA precipitation processing system	GPROF: clim2017v1
					Remote sensing systems	UMORA: v07
DSMP-F16	SSMIS	Rain rate	~35 km (based on 37 GHz)	04:00 p.m.	NASA precipitation processing system	GPROF: clim2017v1
					Remote sensing systems	UMORA: v07
GCOM-W1	AMSR2	Rain rate	~9 km (based on 36.5 GHz)	01:30 p.m.	NASA precipitation processing system	GPROF: clim2017v1
					Remote sensing systems	UMORA: v08
Several satellites	–	Rain rate	0.1°	Each 30 mm	NASA precipitation processing system	IMERG v05B

resolution of these SMOS and SMAP level 2 SSS are quite similar, close to 60 km for SMAP and 50 km for SMOS.

Erroneous SMOS L2Q SSS are removed based on the minimum and maximum acceptable values provided in the files (Vergely and Boutin 2017). We also remove SSS values identified as non-valid during the retrieval process before the application of the bias removal process (Vergely and Boutin 2017) and SSS values very different (difference larger than 3 pss) from a nearby SSS flagged as valid. For SMAP, we use SSS data flagged as valid.

SMOS and SMAP SSS are retrieved using a maximum likelihood approach. For both products, an SSS uncertainty is derived by taking the observed mismatches between observed and modelled brightness temperatures (Tbs) (e.g., like in RFI polluted areas) into account (Boutin et al. 2018; Fore et al. 2017). In our study we only consider SMOS or SMAP SSS with uncertainty lower than 0.8 pss.

We use WS provided in the SMAP files, assuming that the constant $40°$ incidence angle of SMAP and its rotating antenna would provide a WS with a more conservative error than the SMOS estimate that is likely dependent on the geometry in the SMOS field of view and that covers incidence angles from 0 to $\sim 60°$ while horizontal-vertical polarization contrasted signature of roughness is only expected at high incidence angles. The WS retrieved by the SMAP CAP v4 retrieval algorithm uses National Centers for Environmental Prediction Global Forecast System (NCEP GFS) ancillary WS with an error of 1.5 m s^{-1} (Fore et al. 2017).

All SSS and WS data are interpolated for each day (ascending and descending orbits separately) on a regular grid of $0.2° \times 0.2°$ before deriving SSS anomalies from SMOS (ΔS_{SMOS}) and SMAP (ΔS_{SMAP}).

58.2.2 Rain Rate Data

Our study takes advantage of the satellite constellation of the Global Precipitation Measurements (GPM) mission (Hou et al. 2014), dedicated to providing the best possible coverage for monitoring rainfall (Kidd and Huffman 2011), with RR from three different algorithms: Unified Microwave Ocean Retrieval Algorithm (UMORA; Hilburn and Wentz 2008; Wentz et al. 2012, 2014), The Goddard Profiling scheme (GPROF; Kummerow et al. 1996; GPM Science Team 2016, 2017a, b) and the Integrated Multi-satellitE Retrievals for GPM (IMERG, Huffman 2018). GPROF uses a Bayesian approach and cloud resolving models to extract rain information. UMORA simultaneously retrieves sea surface temperature, surface wind speed (in case of no rain), columnar water vapor, columnar cloud water, and surface rain rate from a variety of passive microwave sensors, after a very careful intercalibration of the various sensors (Hilburn and Wentz 2008). It uses an empirical relationship between cloud water liquid path, RR and rain column height. These two different algorithms make different microphysical assumptions, cloud and rain partitioning and rain column height. Previous studies have shown that with temporal and spatial smoothing the two RR estimates are close despite some bias in rainy areas, especially in the East Pacific. In addition, comparisons of individual pixels show significant differences (Hilburn and Wentz 2008). The IMERG interpolated RR combines RR from various satellites of the GPM constellation derived with GPROF algorithm with IR-derived RR to provide RR estimates every 30 min using the morphing technique from the Climate Prediction Center Morphing method (CMORPH, Joyce et al. 2004). IR satellites are used to monitor the displacement of the rain cells. Accuracy of RR derived with morphing method decreases when time shift between considered time and satellite overpass increases (Joyce et al. 2004). Differences between IMERG, GPROF and UMORA algorithm induce differences of RR distribution, as shown in Fig. 58.1.

For UMORA and GPROF, we have considered RR estimated from three different near polar orbit satellites. Their various local equator-crossing times are used to study the effect of temporal shift on the correlation between salinity anomalies and rain rate. DSMP-F17 satellite equator crossing time is close to that of SMOS and SMAP during 2016. DSMP-F16 is more distant in term of time lag from SMOS and SMAP (2 h at the equator) and GCOM-W1 is the most distant in time (more than 4 h from SMOS and SMAP at the equator). These time lags are slightly variable depending on the location of the measurements across the wide satellite swaths.

All RR data are interpolated for each day (ascending and descending orbits separately) on a regular $0.2° \times 0.2°$ grid.

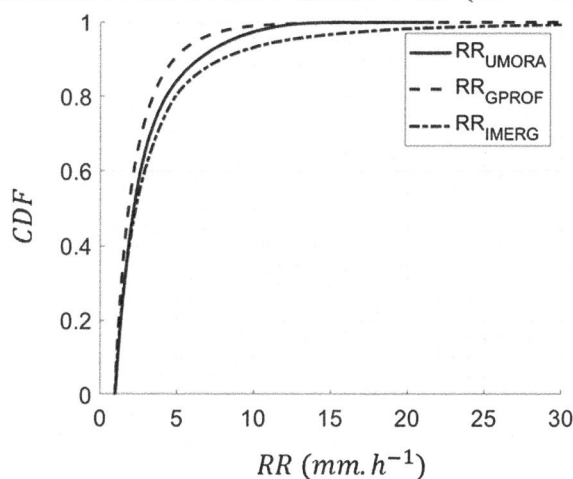

Fig. 58.1 Cumulative Distribution Function (CDF) of rain rates considered during the study (between 50°S and 50°N) obtained with three different algorithms: UMORA (solid line), GPROF (dashed line) and IMERG (dash-dotted line). Only RR higher than 1 mm h^{-1} are considered because RR distribution between UMORA, GPROF and IMERG between 0 and 1 mm h^{-1} strongly differ due to the difficulty of identifying very low rain rates and expected corresponding freshening are within the error of satellite salinities

58.2.3 *Method*

The relationship between RR and SSS anomalies (ΔS) is investigated using the various RR datasets mentioned in the previous section. The influence of time lag between RR and SSS is investigated using RR from different satellites (see Table 58.1).

The study covers the period from January 2016 to December 2016 and the surface ocean between 50°S and 50°N. Complementarily, 6 study areas are defined for testing spatial variability:

1. North Tropical Pacific area (NTPa): between 160° and 120°W and between 0° and 20°N, which includes the Pacific ITCZ,
2. South Tropical Pacific area (STPa): between 160° and 120°W and between 25° and 5°S, which includes the South Pacific Convergence Zone (SPCZ),
3. North Atlantic area (NAa): between 50° and 25°W and between 15° and 40°N;
4. South Atlantic area (SAa): between 40° and 0°W and between 35° and 20°S;
5. South Tropical Indian area (STIa): between 60° and 90°E and between 25° and 0°S;
6. Gulf of Mexico area (GMa): between 100° and 80°W and between 18° and 32°N, in order to illustrate efficiency of river plumes filtering.

Most of these correspond to open ocean areas where rain dominates the salinity variability, whereas GMa offers a more challenging case with high variability due to eddies and Mississippi river plume. These study areas are reported in Fig. 58.2.

58.2.3.1 Salinity Anomalies

The methodology we use to derive ΔS depends only of satellite measurement in order to discard contamination by large scale systematic biases. As shown in Supply et al. (2017), the goal of the ΔS derived spatially with this methodology is to be as close as possible to a temporal ΔS, considering that the pixels surrounded by a rainy area are the most representative of expected salinity in the rainy area before rain

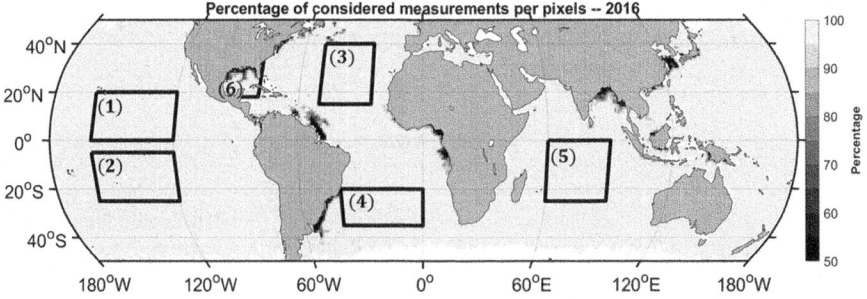

Fig. 58.2 Percentage of measurements retained during 2016 after filtering based on monthly values of $\overline{\Delta S}$ and σ_S. Black lines delimit study areas

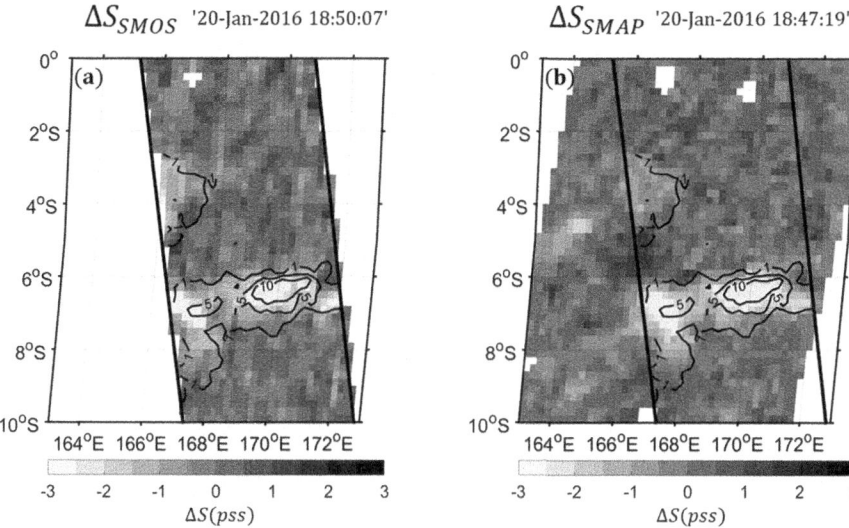

Fig. 58.3 Study case, 20 January 2016, (**a**) SMOS salinity anomaly (**b**) SMAP salinity anomaly (black line are RR isolines in mm h^{-1} from IMERG collocated at less than 15 min with SMOS)

happens (S_{ref}). This methodology makes use of the upper part of the SSS distribution, which is assumed to be affected only by a noise equal to the SSS uncertainty that is provided in the SSS products, to derive a mean SSS used as a reference SSS not affected by rain. In Eq. (58.1), S is the salinity derived from L-Band satellites for the considered pixel. ΔS and S_{ref} are derived from the same swath satellite measurements in a $3° \times 3°$ area by using statistical assumptions:

$$\Delta S = S - S_{ref} \qquad (58.1)$$

Equation (58.1) allows one to detect local SSS anomalies that are related to rain events, but also to other phenomena, creating large local SSS variability such as in river plumes and in regions with large mesoscale variability.

Figure 58.3 shows an example of a freshening event taking place during January 2016 in the SPCZ. During this event, SMOS and SMAP fly over the area of interest at less than 5 min apart. IMERG RR collocated with SMOS SSS in a 15-min radius suggests there was a large RR event in this area, with a spatial pattern very close to the spatial pattern of SMOS and SMAP ΔS.

58.2.3.2   Detection of Rain History

Our goal is to distinguish the impact of the RR history on the observed SSS freshening from the impact of the instantaneous RR given the temporal auto-correlation of the RR ($R_{RR}(\delta t)$). Hence, we compare the observed temporal cross-correlation function between ΔS and RR ($\Gamma_{\Delta S;\ RR}(\delta t)$), with the one inferred from RR temporal auto-correlation function and assuming that only instantaneous RR matters ($\Gamma e_{\Delta S; \ RR}(\delta t)$) following Eq. (58.2) (r(X,Y) correspond to the Pearson correlation between X and Y):

$$R_{RR}(\delta t) = r(RR(t), RR(t + \delta t))$$
$$\Gamma_{\Delta S; RR}(\delta t) = r(\Delta S(t), RR(t + \delta t)) \qquad (58.2)$$
$$\Gamma e_{\Delta S; RR}(\delta t) = \Gamma_{\Delta S; RR}(\delta t = 0) \times R_{RR}(\delta t)$$

Figure 58.4 illustrates two idealized case studies respecting two different hypotheses. In the case of no-influence of RR history on ΔS (first hypothesis (H1)), ΔS depends only on instantaneous RR, ΔS (solid line in Fig. 58.4c) then evolves in phase opposition with RR, the maximum cross-correlation is obtained with a time lag of 0 (Fig. 58.4d, solid line). In the case of the influence of rain history (second hypothesis (H2)), some information on the rain event lasts after the rain abated, in other words, the salinity anomaly survives the rain event (Fig. 58.4d dashed line). Then, the cross-correlation function is flattened towards negative time lags.

RR and ΔS time series are built as follows. For each pixel of the considered study area, we construct a one-year time series of ΔS taken from SMOS and SMAP and a one-year time series of RR_{UMORA} retaining data from SSMIS-F17, SSMIS-F16 and AMSR-2. Then we group all pixel measurements per class of temporal shift (1-h long) and compute auto-correlation and cross-correlation functions. We only retain cases with correlations computed with a minimum number of points of 1000 and a null hypothesis (no relationship between the two considered variables) rejected with a risk of error lower than 1%.

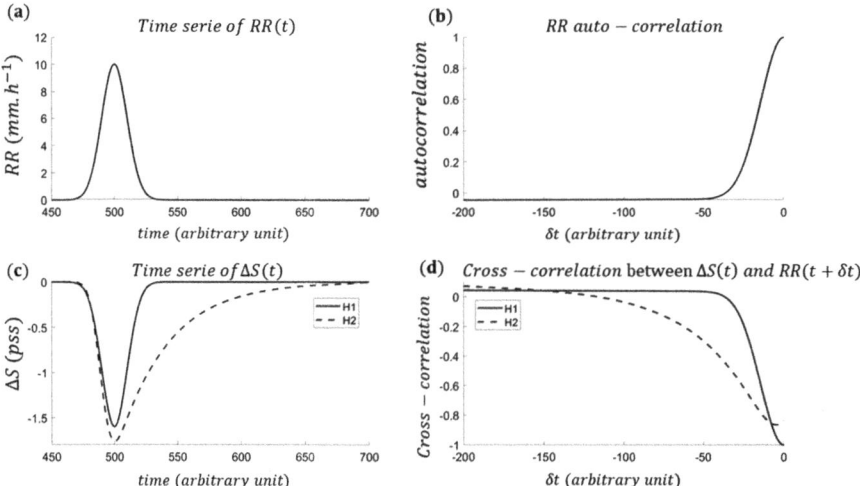

Fig. 58.4 (a) Idealized Rain event (b) Auto-correlation function of RR for idealized rain event (c) Idealized salinity freshening due to the idealized rain event without influence of rain history (line) and with hypothetical influence of rain history (dashed-line). (d) Cross correlation function between RR and salinity freshening without (line) and with (dashed line) influence of time history. In b) and d) negative time lags concern rain before ΔS

58.2.3.3   Filtering of Non-rainy Processes and ΔS Versus Instantaneous RR Relationship

In order to study ΔS versus instantaneous RR relationship, SMOS, SMAP and SSMIS17 measurements are collocated within a 30-mn radius. However, before studying ΔS_{SMOS}, ΔS_{SMAP} and RR coherency, we start by filtering areas where temporal and spatial variability of ΔS are not due to rain. Our filtering methodology considers that rain events are very intermittent, leading to smaller monthly averaged absolute values of ΔS than ΔS associated with other processes that vary more slowly. We compute monthly running mean ($\overline{\Delta S}$) and standard deviation (σ_S) of SSS in each pixel and day. Areas where ΔS variability is not dominated by rain as river plumes or eddies have high absolute values of $\overline{\Delta S}$. We empirically set a threshold value for $\overline{\Delta S}$ and σ_S: each pixel with $\overline{\Delta S}$ lower than −0.65 pss and σ_S higher than 1.75 pss is excluded from the study. Areas impacted by the filtering are shown in Fig. 58.2, which particularly highlights river plumes. Some eddies with low monthly-averaged ΔS magnitude and edges of rivers plumes may be omitted by this filtering methodology, which will be a small source of error in computing the relationship between ΔS and RR. 0.5% of SMOS/SMAP/SSMIS17 collocated grid points are filtered by using this criterium.

To analyze filtering effects, we compare the relationships between ΔS and RR without and with filtering. Observed ΔS_{SMOS} versus RR_{UMORA} and ΔS_{SMAP} versus RR_{UMORA} follow an almost near-linear relationship (Fig. 58.5a, b). Data excluded by the filtering do not show any correlation between ΔS and RR contrary to not-filtered data (Table 58.2). Very low differences are found on correlation values

Table 58.2 Statistics at global scale of ΔS versus RR_{UMORA} relationship for different filtering (Slope is obtained with a linear regression ($Y = \Delta S$, $X = RR_{UMORA}$) and Root Mean Squared Error $RMSD = \sqrt{\overline{(\Delta S - \Delta S_e)^2}} = \sqrt{\overline{(\Delta S - 0.16 \, RR_{UMORA})^2}}$ with ΔS_e estimated considering a relationship between ΔS and RR_{UMORA})

Statistic	Filtering	SMOS <=50 km	SMAP~60 km	Number of considered pixels
Correlation	Without filtering	−0.18	−0.23	10098821
	Data kept after filtering	−0.18	−0.25	10048024
	Data excluded after filtering	−0.04	0.01	50797
Slope	Without filtering	−0.1625 pss. (mm. h^{-1})$^{-1}$ ±0.0002 pss. (mm. h^{-1})$^{-1}$	−0.1645 pss. (mm. h^{-1})$^{-1}$ ±0.0001 pss. (mm. h^{-1})$^{-1}$	10098821
	Data kept after filtering	−0.1607 pss. (mm. h^{-1})$^{-1}$ ±0.0004 pss. (mm. h^{-1})$^{-1}$	−0.1626 pss. (mm. h^{-1})$^{-1}$ ±0.0007 pss. (mm. h^{-1})$^{-1}$	10048024
RMSD	Without filtering	0.61 pss	0.49 pss	10098821
	Data kept after filtering	0.58 pss	0.45 pss	10048024

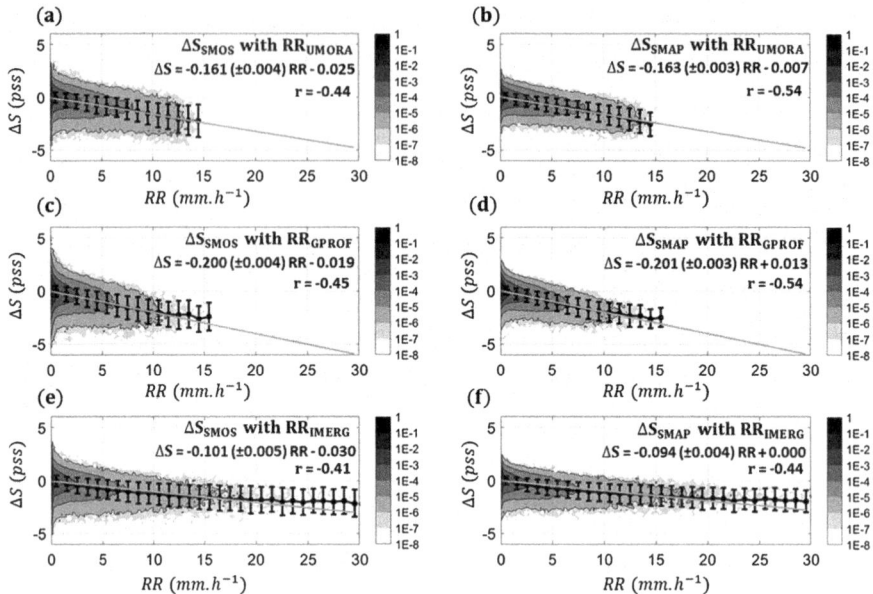

Fig. 58.5 Relationship between ΔS and RR (**a**) for SMOS and RR_{UMORA} (**b**) for SMAP and RR_{UMORA} (**c**) for SMOS and RR_{GPROF} (**d**) for SMAP and RR_{GPROF} (**e**) for SMOS and RR_{IMERG} (**f**) for SMAP and RR_{IMERG} after filtering (black points are average per class of RR and error bar standard deviation per class of RR. Grey lines are linear regression fit. The color scale corresponds to the log10 of the number of occurrences). Pearson correlation coefficient are computed for RR higher than 1 mm h^{-1} because RR distribution between UMORA, GPROF and IMERG between 0 and 1 mm h^{-1} strongly differ due to the difficulty of identifying very low rain rates. The magnitude of the error at the origin derived from linear regression is 10E-4 for all cases

when considering cases with and without filtering, but the RMSD of ΔS versus RR relationship decreases from 0.57 to 0.54 pss for SMOS and from 0.44 to 0.40 pss for SMAP when filtering is applied. The filtering does not significantly influence the slope of the relationship between ΔS and RR. Differences in correlation and RMSD between SMOS and SMAP datasets (Table 58.2) are consistent with the difference in their resolution.

For individual study areas, filtering does not influence the correlation level except in the Gulf of Mexico area. The filtering helps to remove large anomalies related to the river plume of Mississippi River and improves the correlation from -0.15 to -0.19 (-0.37 to -0.43 when taking only rainy cases into account).

58.3 Results

58.3.1 Intercomparison of RR and SSS Products

Considering the three RR products, the ΔS_{SMOS} and ΔS_{SMAP} follow nearly the same dependencies with respect to RR_{UMORA} and RR_{GPROF}. Linear regressions are computed for the three RR products but only UMORA and GPROF follow linear relationship with ΔS contrary to IMERG as shown in Fig. 58.5. Slopes obtained with the linear regression are dependent of the RR product used but the Pearson correlation values (r) are the same for UMORA and GPROF. r values are lower with IMERG due to the non-linearity of the relationship with this RR product. The absolute value of ΔS versus RR slope obtained with GPROF is significantly higher than with UMORA. These differences of slope can be explained by the difference of RR distribution between UMORA and GPROF shown in Fig. 58.1.

Figure 58.5 also illustrates similar values of ΔS standard deviations per class of RR between the different RR products. These values are however lower for ΔS_{SMAP} in comparison with ΔS_{SMOS}. Actually, the noise on ΔS_{SMAP} is less than the noise on ΔS_{SMOS} in almost the same ratio as the spatial integration of both instruments (50 km or less for SMOS, ~60 km for SMAP). The difference in resolution also explains lower values of r with ΔS_{SMOS} in comparison with ΔS_{SMAP}.

58.3.2 Which Is the Imprint of Rain History on Salinity Anomalies?

For this particular study, we do not apply the filtering methodology to remove a part of its possible signature. For this reason, the rain history effect is not studied in the GMa area because of the influence of the Mississippi river plume. IMERG is not used here because of non-linearity between RR and ΔS (see Sect. 58.3.1) and also to avoid the use of interpolated measurements.

Figure 58.6a–e shows that very similar results are obtained in the different study regions. The maximum of ΔS/RR cross-correlation is recorded for the shortest time lag. NTPa area, STPa area and STIa show higher correlation levels compared to other areas because of a higher signal to noise ratio (larger RR in tropical regions and lower noise in satellite SSS in warm waters), and the NAa and SAa areas, which contain the less precise SSS, present the lowest correlation.

Figure 58.6a–e and low error on the observed ΔS versus RR cross-correlation demonstrates that ΔS versus RR cross-correlation derived from RR auto-correlation fits very well with observed ΔS versus RR cross-correlation in all areas. Only cross-correlations considering RR between 1 and 2 h before ΔS measurements seem to show a correlation slightly higher (in absolute value) in comparison with the estimated cross-correlation for NTPa and STPa areas. The differences in RR auto-correlation (Fig. 58.6f) do not influence the correspondence between the observed

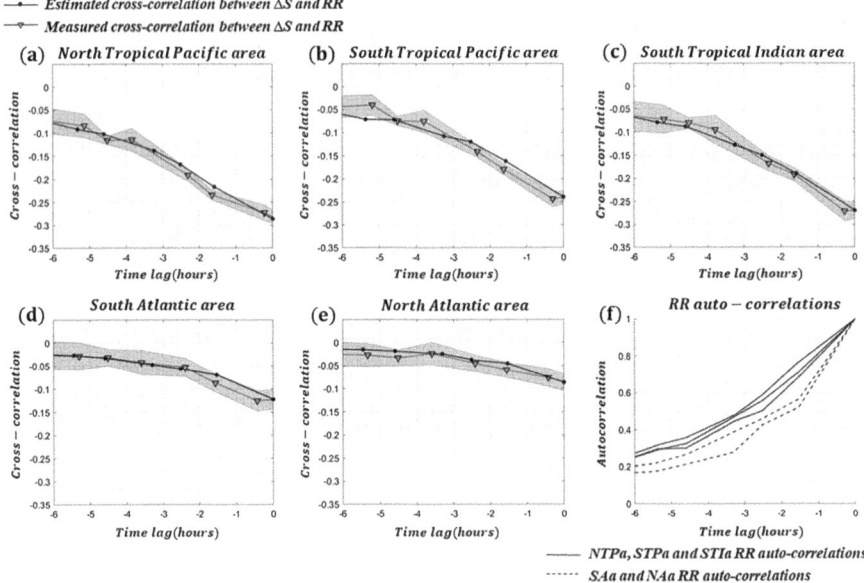

Fig. 58.6 For each study area (**a, b, c, d, e**): (triangle) Temporal cross-correlation between ΔS and RR (individual triangles are linked with a dashed line) (dots) Estimated cross-correlation between ΔS and RR inferred from RR auto-correlation and correlation between ΔS and instantaneous RR (individual dots are linked with a line). Grey colored areas show the 2σ confidence interval for the ΔS versus RR cross-correlation. Negative time lags concern rain before considered freshening. (**f**) RR auto-correlations for study areas

ΔS versus RR cross-correlation and ΔS versus RR cross-correlation derived from RR auto-correlation. This result shows that the correlation between ΔS and past RR is mainly due to the temporal auto-correlation of RR. Similar results are obtained by using RR$_{GPROF}$.

These results do not consider the potential influence of WS. Figure 58.7 shows the computation of ΔS/RR cross-correlation for different classes of SMAP WS in the case of the NTPa area (chosen as one of the areas with the largest WS range (with SIa) and lower WS under rain). The classes of WS are defined as low WS (lower than the 0.3 quantile) and high WS (higher than the 0.7 quantile). If the low WS conditions are encountered over all the NTPa area, high WS conditions are encountered closer to the equator. For salinity, only SMAP measurements are considered and we use SMAP WS. These results clearly show higher correlation between RR and ΔS for lower WS. In addition, despite larger uncertainties, the case with low WS presents higher ΔS/RR cross-correlation than expected considering only RR auto-correlation. This result seems to indicate an influence of rain history for temporal scales less than 5 h in low WS conditions. This part of the study is limited by WS being only known at the time of the SMAP SSS measurement. WS influence on the relationship between RR and SSS freshening is developed in the next section.

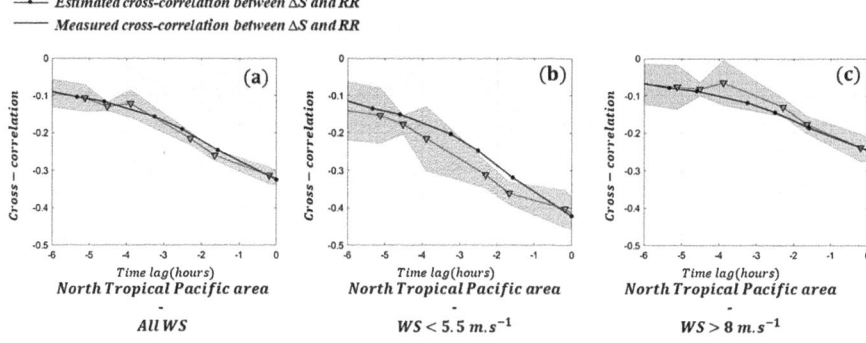

Fig. 58.7 For NTPa study area and SMAP: (**a**) all WS (**b**) WS under 5.5 m s^{-1} (**c**) WS above 8 m s^{-1}. For the three plots: (triangle) Temporal cross-correlation between ΔS and RR (individual triangles are linked with a dashed line) (dots) Estimated cross-correlation between ΔS and RR inferred from RR auto-correlation and correlation between ΔS and instantaneous RR (individual dots are linked with a line). Grey colored areas show the 2σ confidence interval for the ΔS versus RR cross-correlation. Negative time lags concern rain before considered freshening

58.3.3 Variability of the Relationship Between Salinity Freshening and Rain Rate as Function of Wind Speed

We then investigate the WS dependency of ΔS as a function RR over the global ocean, using the filtering methodology (Sect. 58.2.3.3) and SMOS, SMAP as well as the three RR products. The results present a strong dependency of freshening for a given RR according to WS (Fig. 58.8). In no-rain and low rain cases, ΔS_{SMOS} and ΔS_{SMAP} show some biases for high WS: a slight positive bias for ΔS_{SMAP} and a slight negative bias for ΔS_{SMOS}. Despite of these slight biases, SMOS (Fig. 58.8a–c) and SMAP (Fig. 58.8d–f) show very similar patterns, with the three RR products concerning the dependency with WS. Nevertheless, as observed in Sect. 58.3.1, the magnitude of freshening differs strongly between RR products for each RR classes. RR_{GPROF} gives the more different relationship in comparison with RR_{UMORA} and RR_{IMERG}. The latter RR show very similar magnitudes between 0 and 9 mm h^{-1}, but over 9 mm h^{-1} freshening measured for a given RR_{IMERG} are lower than for the same RR_{UMORA} value. Figure 58.8 also shows that significant freshenings are always observed even for high WS (between 10 and 12 m s^{-1}).

Drushka et al. (2016) present a relationship allowing to express maximum ΔS (ΔS_{max}) as a function of maximum RR (RR_{max}) and WS during a freshening event (58.3):

$$\Delta S_{max} = a\,RR_{max}\,WS^{-b} \qquad (58.3)$$

with a $= 0.11$ pss (mm h^{-1})$^{-1}$ and b $= 1.1$. Considering the SMOS and SMAP estimates, we show that the rain history produces a negligible effect on freshening at

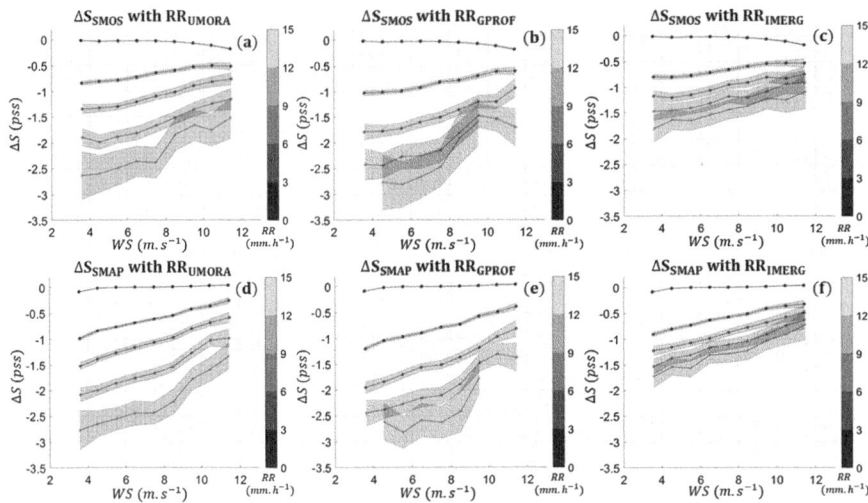

Fig. 58.8 Relationship between ΔS and WS_{SMAP} per class of RR: (**a**) with ΔS_{SMOS} and RR_{UMORA} (**b**) with ΔS_{SMOS} and RR_{GPROF} (**c**) with ΔS_{SMOS} and RR_{IMERG} (**d**) with ΔS_{SMAP} and RR_{UMORA} (**e**) with ΔS_{SMAP} and RR_{GPROF} (**f**) with ΔS_{SMAP} and RR_{IMERG}. Points correspond to ΔS and RR average for a given class of RR. Points are colored with a color corresponding to RR class

Table 58.3 a and b coefficient for Eq. (58.2) considering RR_{UMORA}, RR_{GPROF} and RR_{IMERG}

	a	b
UMORA	-0.37 pss. $(\text{mm. h}^{-1})^{-1}$	0.68
GPROF	-0.90 pss. $(\text{mm. h}^{-1})^{-1}$	0.80
IMERG	-0.35 pss. $(\text{mm. h}^{-1})^{-1}$	0.77

Change of sign compared to Drushka et al. (2016) is due to the fact that ΔS_{max} is defined as the absolute value of the largest-in-magnitude negative value of ΔS in Drushka et al. (2016)

satellite pixel scale. However, the influence of WS is important when considering instantaneous RR. For this reason, we assume that ΔS and RR derived from satellites reflect in-situ ΔS_{max} and RR_{max}. Hence, we replace in this equation ΔS_{max} and RR_{max} with ΔS and RR and derive parameters a and b with RR and WS from satellite data. We found different a and b values according to the RR product used (see Table 58.3). Values of a and b computed with satellites ΔS and RR are different from one RR product to another and very different from values obtained by Drushka et al. (2016).

58.4 Discussion and Conclusion

We have developed a filtering methodology that highlight ΔS_{SMOS} and ΔS_{SMAP} only where freshening is dominated by rain. ΔS_{SMOS} and ΔS_{SMAP} are very consistent. They are combined with RR retrieved from several satellites with different equator crossing times thus in order to investigate rain history imprint on SMOS and SMAP ΔS in various regions. For each of these regions, the apparent correlation between ΔS and past RR intensity is primarily explained by RR auto-correlation while ΔS magnitude is dominated by the instantaneous RR magnitude. Most commonly, the rain history influence seems to be negligible, even though a weak influence of rain history on ΔS_{SMOS} and ΔS_{SMAP} is suggested at low WS (< 5.5 m s^{-1}) for durations up to 5 h despite larger uncertainties. These results are consistent with model results obtained by Bellenger et al. (2017) showing that rain induced fresh lenses have short durations under moderate and high WS. Results obtained in this study show that WS mainly affects the relationship between instantaneous RR and freshening and seems to slightly influence the effect of rain history based on satellite estimations. Nevertheless, this result has to be taken with caution as it has been obtained considering WS$_{SMAP}$. L-band radiometers have been shown to detect high WS under rainfall very well (Reul et al. 2016; Meissner et al. 2017) but the validity of moderate WS retrieved together with SSS under rain events has not been fully assessed. It is due to the difficulty of monitoring the wind speed integrated over ~50 km resolution under rainfall by other means. The WS retrieval also depends on the quality of the prior WS (NCEP GFS in case of SMAP with a 1.5 m s^{-1} error). Hence, further validation would be necessary to fully assess the meaningfulness of our observation. In our analysis, we have confronted SMOS SSS with WS$_{SMAP}$ in order to minimise error correlations between SMAP SSS and WS. Nevertheless, we cannot exclude that rain-induced roughness affects WS$_{SMAP}$ for WS considered in this study (Tang et al. 2015) even if these effects can be moderated by the fact that they are important only at very low wind speeds and are negligible for moderate to high rain rates (we consider only WS higher than 3 m s^{-1} in this study).

We provide a comparison of RR$_{UMORA}$, RR$_{GPROF}$ and RR$_{IMERG}$ products, through ΔS versus RR relationship. We find that despite different slopes in the relationship, UMORA and GPROF record similar correlation levels with ΔS (Table 58.3). This same level of correlation indicates that SMOS and SMAP freshenings are not able to distinguish RR$_{UMORA}$ and RR$_{GPROF}$ pattern globally. The difference of slope is due to the difference of RR distribution between UMORA and GPROF (Fig. 58.1). RR$_{UMORA}$ are higher than RR$_{GPROF}$ from 2 to 15 mm h^{-1} approximatively but higher RR are observed with RR$_{GPROF}$. RR$_{IMERG}$ shows a very close to RR$_{UMORA}$ behavior for the RR versus ΔS relationship (Eq. 58.3) but RR$_{IMERG}$ records higher RR than RR$_{UMORA}$ which saturates at high RR (Fig. 58.1). Additionally, RR$_{IMERG}$ and RR$_{UMORA}$ present a better latitudinal stability of the RR versus ΔS relationship for a given WS (not shown). This proximity between RR$_{IMERG}$ and RR$_{UMORA}$ may be explained by the more thorough calibrations performed in these products. For UMORA, radiometers are intercalibrated at the brightness temperature level. For IMERG, RR are calibrated with respect to the GPM Combined Radar-Radiometer (CORRA; Grecu and Olson 2006) product that uses RR from GPM Microwave Imager (GMI) and Dual-Precipitation Radar (DPR) instruments (Huffman 2018). IMERG RR values are also calibrated using GPCP data thereby modifying RR distribution. Use of the DPR with a finer resolution compared with microwave imagers may explain higher RR observed with IMERG relative to UMORA.

We show that for the range of WS between 3 and 12 m s^{-1}, ΔS versus RR relationship is modulated by WS, leading to different relationships in the different areas. These results are in line with results obtained by Meissner et al. (2014). However, the methodology that we use in this study to compute ΔS allows us to homogenize the ΔS versus RR relationship over all oceans for a given WS. As described in previous studies (e.g., Boutin et al. 2016), the use of the Schlüssel relationship (Schlüssel et al. 1997) describes at first order the ΔS versus RR relationship observed with satellites. However, the Schlüssel relationship takes into account only the molecular diffusion layer for salinity (0.05 mm) but L-band satellite are influenced by rain induced freshening at 1 cm depth. These phenomena are considered by Drushka et al. (2016) and Bellenger et al. (2017) that use RR and WS to explain ΔS spatial and temporal variability. We consider in this study ΔS$_{max}$ and RR$_{max}$ as equal ΔS and RR to retrieve ΔS, RR and WS relationship determined in previous studies using satellite estimations. Derived coefficients differ from the Drushka et al. (2016) estimate. Diagnostic of these differences is a difficult, yet unsolved challenge. Dong et al. (2017) compared in-situ measurements with Drushka et al. (2016) coefficients for Eq. (58.3) and also found different coefficients corresponding to larger freshening for a given RR (using satellite RR) in comparison with freshening values found by Drushka et al. (2016). Differences of ΔS, WS, RR relationships obtained in previous studies reveal the difficulties to compare results obtained at different spatial and temporal scales and for different depths and thus, the difficulties to compare in-situ measurements, model results and satellite estimations.

Scale influence on ΔS, WS, RR relationships raise the question of heterogeneity effects. Rain history influence on freshening observed with in-situ measurements is not observed with satellite measurements. This may induce that part of the rain history effect is contained in an influence of WS considering instantaneous RR. These hypotheses need to be investigated in future studies. Understanding the influence of heterogeneity and its role on ΔS, WS, RR relationship is needed to be able to isolate other phenomenon such as rain splashing.

This study provides a methodology that allows for the removal of the rainfall imprint from each SMOS or SMAP SSS estimation by using instantaneous RR from IMERG provided each 30-min, proposing an alternative solution to RIM model that uses RR history from the past 24 h.

There is a need to conduct detailed process studies to better understand the reasons for the differences between models and in situ measurements. It may lead to estimate distribution of global RR over oceans only considering ΔS measured from SMOS and SMAP and thus to add an independant diagnostic tool of the bias error between different rain products that may reach relative values of 20% over the eastern Pacific Ocean (Adler et al. 2012).

Acknowledgements This work was funded by the ESA STSE SMOS+Rainfall project and by the CNES TOSCA SMOS-Ocean and CATDS projects. Alexandre Supply PhD grant is supported by Sorbonne Université. UMORA rain rates are distributed by Remote Sensing Systems. GPROF and IMERG rain rates are distributed by the Precipitation Processing System at the NASA Goddard Space Flight Center.

References

Abe, H., Ebuchi, N., Ueno, H., Ishiyama, H., & Matsumura, Y. (2018). Aquarius reveals eddy stirring after a heavy precipitation event in the subtropical North Pacific. *Journal of Oceanography, 75*, 37–50. https://doi.org/10.1007/s10872-018-0482-0.

Adler, R. F., Gu, G., & Huffman, G. J. (2012). Estimating climatological bias errors for the global precipitation climatology project (GPCP). *Journal of Applied Meteorology and Climatology, 51*, 84–99. https://doi.org/10.1175/JAMC-D-11-052.1.

Asher, W. E., Jessup, A. T., Branch, R., & Clark, D. (2014). Observations of rain-induced near-surface salinity anomalies. *Journal of Geophysical Research, 119*, 5483–5500. https://doi.org/10.1002/2014JC009954.

Bellenger, H., Drushka, K., Asher, W., Reverdin, G., Katsumata, M., & Watanabe, M. (2017). Extension of the prognostic model of sea surface temperature to rain-induced cool and fresh lenses. *Journal of Geophysical Research, 121*, 484–507. https://doi.org/10.1002/2016JC012429.

Boutin, J., Reul, N., Maes, C., Reverdin, G., Delcroix, T., & Gaillard, F. (2015). *Sea surface salinity from SMOS satellite mission: A synthesis of the main 2010–2015 oceanic results in France.* IUGG 2011–2015 report, 100–103. Available at http://www.iugg.org/members/nationalreports/CNFGG_RapQuad_UGGI_2015.pdf, last accessed 30 Oct 2018.

Boutin, J., Chao, Y., Asher, W. E., Delcroix, T., Drucker, R., Drushka, K., Kolodziejczyk, N., Lee, T., Reul, N., Reverdin, G., Schanze, J., Soloviev, A., Yu, L., Anderson, J., Brucker, L., Dinnat, E., Garcia, A. S., Jones, W. L., Maes, C., Meissner, T., Tang, W., Vinogradova, N., & Ward, B. (2016). Satellite and in situ salinity: Understanding near- surface stratification and sub-footprint variability. *Bulletin of the American Meteorological Society, 97*, 1391–1407. https://doi.org/10.1175/BAMS-D-15-00032.1.

Boutin, J., Vergely, J. L., Marchand, S., D'Amico, F., Hasson, A., Kolodziejczyk, N., Reul, N., Reverdin, G., & Vialard, J. (2018). New SMOS sea surface salinity with reduced systematic errors and improved variability. *Remote Sensing of Environment, 214*, 115–134. https://doi.org/10.1016/j.rse.2018.05.022. [Open Access version: http://archimer.ifremer.fr/doc/00441/55254/].

CATDS. (2017). *CATDS-PDC L3OS 2Q – Debiased daily valid ocean salinity values product from SMOS satellite.* CATDS (CNES, IFREMER, LOCEAN, ACRI). Available at https://doi.org/10.12770/12dba510-cd71-4d4f-9fc1-9cc027d128b0, last accessed 30 Oct 2018.

Dong, S., Volkov, D., Goni, G., Lumpkin, R., & Foltz, G. R. (2017). Near-surface salinity and temperature structure observed with dual-sensor drifters in the subtropical South Pacific. *Journal of Geophysical Research, 122*, 5952–5969. https://doi.org/10.1002/2017JC012894.

Drushka, K., Asher, W. E., Ward, B., & Walesby, K. (2016). Understanding the formation and evolution of rain-formed fresh lenses at the ocean surface. *Journal of Geophysical Research, 121*, 2673–2689. https://doi.org/10.1002/2015JC011527.

Fore, A., Yueh, S., Tang, W., & Hayashi, A. (2017). *SMAP salinity and wind speed data user's guide version 4.0, Jet Propulsion Laboratory, California Institute of Technology.* Available at ftp://podaac-ftp.jpl.nasa.gov/allData/smap/docs/JPL-CAP_V3/JPL_SMAP-SSS-UsersGuide.pdf, last accessed 30 Oct 2018.

Fournier, S., Reager, J. T., Lee, T., Vazquez-Cuervo, J., David, C. H., & Gierach, M. M. (2016). SMAP observes flooding from land to sea: The Texas event of 2015. *Geophysical Research Letters, 43*, 10,338–10,346. https://doi.org/10.1002/2016GL070821.

Fournier, S., Vandemark, D., Gaultier, L., Lee, T., Jonsson, B., & Gierach, M. M. (2017a). Interannual variation in offshore advection of Amazon-Orinoco plume waters: Observations, forcing mechanisms, and impacts. *Journal of Geophysical Research, 122*, 8966–8982. https://doi.org/10.1002/2017JC013103.

Fournier, S., Vialard, J., Lengaigne, M., Lee, T., Gierach, M. M., & Chaitanya, A. V. S. (2017b). Modulation of the Ganges-Brahmaputra river plume by the Indian Ocean dipole and eddies inferred from satellite observations. *Journal of Geophysical Research, 122*, 9591–9604. https://doi.org/10.1002/2017JC013333.

GPM Science Team. (2016). *GPM SSMIS on F16 (GPROF) radiometer precipitation profiling L2 1.5 hours 12 km V05*, Greenbelt, MD, Goddard Earth Sciences Data and Information Services Center (GES DISC), accessed [09/05/2018], https://doi.org/10.5067/GPM/SSMIS/F16/GPROF/2A/05

GPM Science Team. (2017a). *GPM AMSR-2 on GCOM-W1 (GPROF) climate-based radiometer precipitation profiling L2A 1.5 hours 10 km V05*, Greenbelt, MD, Goddard Earth Sciences Data and Information Services Center (GES DISC), accessed [09/05/2018], https://doi.org/10.5067/GPM/AMSR2/GCOMW1/GPROFCLIM/2A/05

GPM Science Team. (2017b). *GPM SSMIS on F17 (GPROF) climate-based radiometer precipitation profiling 1.5 hours 12 km V05*, Greenbelt, MD, Goddard Earth Sciences Data and Information Services Center (GES DISC), accessed [09/05/2018], https://doi.org/10.5067/GPM/SSMIS/F17/GPROFCLIM/2A/05

Grecu, M., & Olson, W. S. (2006). Bayesian estimation of precipitation from satellite passive microwave observations using combined radar–radiometer retrievals. *Journal of Applied Meteorology and Climatology, 45*, 416–433. https://doi.org/10.1175/JAM2360.1.

Hasson, A., Puy, M., Boutin, J., Guilyardi, E., & Morrow, R. (2018). Northward pathway across the tropical North Pacific Ocean revealed by surface salinity: How do El Niño anomalies reach Hawaii? *Journal of Geophysical Research, 123*, 2697–2715. https://doi.org/10.1002/2017JC013423.

Henocq, C., Boutin, J., Reverdin, G., Petitcolin, F., Arnault, S., & Lattes, P. (2010). Vertical variability of near-surface salinity in the tropics: Consequences for L-band radiometer calibration and validation. *Journal of Atmospheric and Oceanic Technology, 27*, 192–209. https://doi.org/10.1175/2009JTECHO670.1.

Hilburn, K. A., & Wentz, F. J. (2008). Intercalibrated passive microwave rain products from the unified microwave ocean retrieval algorithm (UMORA). *Journal of Applied Meteorology and Climatology, 47*, 778–794. https://doi.org/10.1175/2007JAMC1635.1.

Ho, D. T., Asher, W. E., Schlösser, P., Bliven, L., & Gordon, E. (2000). On mechanisms of rain-induced air–water gas transfer. *Journal of Geophysical Research, 105*, 24045–24057. https://doi.org/10.1029/1999JC000280.

Hou, A. Y., Kakar, R. K., Neeck, S., Azarbarzin, A. A., Kummerow, C. D., Kojima, M., Oki, R., Nakamura, K., & Iguchi, T. (2014). The Global Precipitation Measurement mission. *Bulletin of the American Meteorological Society, 95*, 701–722. https://doi.org/10.1175/BAMS-D-13-00164.1.

Huffman, G. J. (2018). *GPM IMERG final precipitation L3 half hourly 0.1 degree × 0.1 degree V05B*, Greenbelt, MD, Goddard Earth Sciences Data and Information Services Center (GES DISC), accessed [08/03/2018], https://doi.org/10.5067/GPM/IMERG/3B-HH/05

Joyce, R. J., Janowiak, J. E., Arkin, P. A., & Xie, P. (2004). CMORPH: A method that produces global precipitation estimates from passive microwave and infrared data at 8 km, hourly resolution. *Journal of Hydrometeorology, 5*, 487–503. https://doi.org/10.1175/1525-7541(2004)005<0487:CAMTPG>2.0.CO;2.

Kerr, Y. H., Waldteufel, P., Wigneron, J. P., Delwart, S., Cabot, F., Boutin, J., Escorihuela, M. J., Font, J., Reul, N., Gruhier, C., Juglea, S. E., Drinkwater, M. R., Hahne, A., Martin-Neira, M., &

Mecklenburg, S. (2010). The SMOS mission: New tool for monitoring key elements of the global water cycle. *Proceedings of the IEEE, 98*, 666–687. https://doi.org/10.1109/jproc.2010. 2043032.

Kidd, C., & Huffman, G. J. (2011). Global precipitation measurement. *Meteorological Applications, 18*, 334–353. https://doi.org/10.1002/met.284.

Kummerow, C. D., Olson, W. S., & Giglio, L. (1996). A simplified scheme for obtaining precipitation and vertical hydrometeor profiles from passive microwave sensors. *IEEE Transactions on Geoscience and Remote Sensing, 34*, 1213–1232. https://doi.org/10.1109/36.536538.

Meissner, T., Wentz, F., Scott, J., & Hilburn, K. (2014). Assessment of rain impact on the Aquarius salinity retrievals. *Proceedings of the Ocean Salinity Science Workshop,* Exeter, United Kingdom, ESA. Available at www.smos-sos.org/presentations-ocean-salinity-science-workshop, last accessed 30 Oct 2018.

Meissner, T., Ricciardulli, L., & Wentz, F. J. (2017). Capability of the SMAP mission to measure ocean surface winds in storms. *Bulletin of the American Meteorological Society, 98*, 1660–1677. https://doi.org/10.1175/BAMS-D-16-0052.1.

Pei, S., Shinoda, T., Soloviev, A., & Lien, R. C. (2018). Upper Ocean response to the atmospheric cold pools associated with the madden-Julian oscillation. *Geophysical Research Letters, 45*, 5020–5029. https://doi.org/10.1029/2018GL077825.

Piepmeier, J. R., Focardi, P., Horgan, K. A., Knuble, J., Ehsan, N., Lucey, J., Brambora, C., Brown, P. R., Hoffman, P. J., French, R. T., Mikhylov, R. L., Kwack, E.-Y., Slimko, E. M., Dawson, D. E., Hudson, D., Peng, J., Mohammed, P. N., De Amici, G., Freedman, A. P., Medeiros, J., Sacks, F., Estep, R., Spencer, M. W., Chen, C. W., Wheeler, K. B., Edelstein, W. N., O'Neill, P. E., & Njoku, E. G. (2017). SMAP L-band microwave radiometer: Instrument design and first year on orbit. *IEEE Transactions on Geoscience and Remote Sensing, 55*, 1954–1966. https://doi.org/10.1109/TGRS.2016.2631978.

Reul, N., Fournier, S., Boutin, J., Hernandez, O., Maes, C., Chapron, B., Alory, G., Quilfen, Y., Tenerelli, J., Morisset, S., Kerr, Y., Mecklenburg, S., & Delwart, S. (2014). Sea surface salinity observations from space with the SMOS satellite: A new means to monitor the marine branch of the water cycle. *Surveys in Geophysics, 35*, 681–722. https://doi.org/10.1007/s10712-013-9244-0.

Reul, N., Chapron, B., Zabolotskikh, E., Donlon, C., Quilfen, Y., Guimbard, S., & Piolle, J. F. (2016). A revised L-band radio-brightness sensitivity to extreme winds under tropical cyclones: The five year SMOS-storm database. *Remote Sensing of Environment, 180*, 274–291. https://doi.org/10.1016/j.rse.2016.03.011.

Reverdin, G., Morisset, S., Boutin, J., & Martin, N. (2012). Rain-induced variability of near sea-surface T and S from drifter data. *Journal of Geophysical Research, 117*, C02032. https://doi.org/10.1029/2011JC007549.

Santos-Garcia, A., Jacob, M. M., Jones, W. L., Asher, W. E., Hejazin, Y., Ebrahimi, H., & Rabolli, M. (2014). Investigation of rain effects on Aquarius Sea surface salinity measurements. *Journal of Geophysical Research, 119*, 7605–7624. https://doi.org/10.1002/2014JC010137.

Schlüssel, P., Soloviev, A. V., & Emery, W. J. (1997). Cool and freshwater skin of the ocean during rainfall. *Boundary-Layer Meteorology, 82*, 437–472. https://doi.org/10.1023/A:1000225700380.

Soloviev, A., Lukas, R., & Matsuura, H. (2002). Sharp frontal interfaces in the near-surface layer of the tropical ocean. *Journal of Marine Systems, 37*, 47–68. https://doi.org/10.1016/S0924-7963 (02)00195-1.

Supply, A., Boutin, J., Vergely, J. L., Martin, N., Hasson, A., Reverdin, G., Mallet, C., & Viltard, N. (2017). Precipitation estimates from SMOS sea surface salinity. *Quarterly Journal of the Royal Meteorological Society, 144*(S1), 103–119. https://doi.org/10.1002/qj.3110.

Tang, W., Yueh, S. H., Hayashi, A., Fore, A. G., Jones, W. L., Santos-Garcia, A., & Jacob, M. M. (2015). Rain-induced near surface salinity stratification and rain roughness correction for Aquarius SSS retrieval. *The Special Issue of the IEEE Journal of Selected Topics in Applied Earth Observations and Remote Sensing (JSTARS) "Aquarius/SACD Mission Calibration/*

Validation Performance and Retrieval Algorithms", *8*(12), 5474–5484. https://doi.org/10.1109/JSTARS.2015.2463768.

Tang, W., Fore, A., Yueh, S., Lee, T., Hayashi, A., Sanchez-Franks, A., King, B., Baranowski, D., & Martinez, J. (2017). Validating SMAP SSS with in-situ measurements. *Remote Sensing of Environment, 200,* 326–340. https://doi.org/10.1016/j.rse.2017.08.021.

Vergely, J. L., & Boutin, J. (2017). SMOS OS level 3: The algorithm theoretical basis document (v300), 05/05/2017. Available at http://www.catds.fr/content/download/78841/1005020/file/ATBD_L3OS_v3.0.pdf?version=3, last accessed 30 Oct 2018.

Walesby, K., Vialard, J., Minnett, P. J., Callaghan, A. H., & Ward, B. (2015). Observations indicative of rain-induced double diffusion in the ocean surface boundary layer. *Geophysical Research Letters, 42,* 3963–3972. https://doi.org/10.1002/2015GL063506.

Wentz, F. J., Hilburn, K. A., & Smith, D. K. (2012). *Remote sensing systems DMSP SSMIS daily environmental suite on 0.25 deg grid, Version 7.* Remote Sensing Systems, Santa Rosa, CA. Available at www.remss.com/missions/ssmi, last accessed 30 Oct 2018.

Wentz, F. J., Meissner, T., Gentemann, C., Hilburn, K. A., & Scott, J. (2014). *Remote sensing systems GCOM-W1 AMSR2 daily environmental suite on 0.25 deg grid, Version V.8, [indicate subset if used].* Remote Sensing Systems, Santa Rosa, CA. Available at www.remss.com/missions/amsr, last accessed 30 Oct 2018.

Wijesekera, H. W., Paulson, C. A., & Huyer, A. (1999). The effect of rain-fall on the surface layer during a westerly wind burst in the western equatorial Pacific. *Journal of Physical Oceanography, 29,* 612–632. https://doi.org/10.1175/1520-0485(1999)029<0612:TEOROT>2.0.CO;2.

Yu, L. (2015). Sea-surface salinity fronts and associated salinity-minimum zones in the tropical ocean. *Journal of Geophysical Research, 120,* 4205–4225. https://doi.org/10.1002/2015JC010790.

Zappa, C. J., Ho, D. T., McGillis, W. R., Banner, M. L., Dacey, J. W. H., Bliven, L. F., Ma, B., & Nystuen, J. (2009). Rain-induced turbulence and air–sea gas transfer. *Journal of Geophysical Research, 114,* C07009. https://doi.org/10.1029/2008JC005008.

Printed by Printforce, the Netherlands